Composites and Advanced Materials Expo (CAMX 2020)

Online
21 – 24 September 2020

ISBN: 978-1-7138-2913-3

Printed from e-media with permission by:

Curran Associates, Inc.
57 Morehouse Lane
Red Hook, NY 12571

Some format issues inherent in the e-media version may also appear in this print version.

SAMPE
21680 Gateway Center Drive, Suite 300
Diamond Bar, California, USA
91765-2454

Phone: +1-626-521-9460

www.sampe.org

Additional copies of this publication are available from:

Curran Associates, Inc.
57 Morehouse Lane
Red Hook, NY 12571 USA
Phone: 845-758-0400
Fax: 845-758-2633
Email: curran@proceedings.com
Web: www.proceedings.com

Track Index

Additive Manufacturing | Advances in Materials | Bonding & Joining | Design, Analysis, and Simulation | Green & Sustainability | Manufacturing & Processing Technologies | Market Applications | Non-Destructive Evaluation & Materials Testing

Additive Manufacturing

Advances in Materials

Bonding & Joining

Design, Analysis, and Simulation

Market Applications

Non-Destructive Evaluation & Materials Testing

THANK YOU CAMX VOLUNTEERS

CAMX Oversight Committee
Dr. Scott Beckwith, Society for the Advancement of Material and Process Engineering (SAMPE®)
John Busel, American Composites Manufacturers Association (ACMA)
Dan Coughlin, American Composites Manufacturers Association (ACMA)
Ted Lynch, Strategic Marketing Innovations (SMI)

CAMX/SAMPE Technical Committee
Dr. Ramy Harik, University of South Carolina
Dr. Ajit Kelkar, Joint School of Nanascience and Nanoengineering
Oleksandr Kravchenko, Det of Mechanical and Aerospace Engineering - Old Dominion University
Alison Lake, GE Aviation
Eric Schmid, Raven Industries - Engineered Films
Dr. Jitendra Tate, Texas State University

CAMX/ACMA Technical Committee
Tony Bennett, Interplastic
Scott Crump, Interplastic
Denny Fink, Resin Research and Recovery, LLC.
Matthew Hayden, Polynt-Reichhold
Richard Higgins, HK Research Corporation
James Jones, Composites One, LLC
Ellen Lacky, University of Mississippi
Ray Liang, West Virginia University
Dana Miloaga, Johns Manville
Hildeberto Nava, Polynt-Reichhold
Rick Pauer, Polynt-Reichhold
Tom Pelt, Retired
Ed Pilpel, Sustainable Composites
Michael Richardson, Crane Composites
JP Schroeder, Scott Bader
Michael Stevens, INEOS Composites
Keith Tellman, ChromaFlo
Amol Vaidya, Owens Corning
Sandeep Vennam, Nippon Electric Glass
Raju Vuppalapati, Razor Composites

THANK YOU CAMX VOLUNTEERS

CAMX Category Chairs
Ehsan Barjasteh, California State University, Long Beach
Timothy Briggs, Sandia National Labs
Lou Dorworth, Abaris
Sana Elyas, Arris Composites
Tanila Gonzales Leal, Embraer
Lessa Grunenfelder, University of Southern California
Alex Hanson, Sandia National Labs
Patrick Jarvis-Shean, Rockwell Collins
Bo Jin, University of Southern California
Caleb Lystrup, Sandia National Laboratories
Sandi Miller, NASA
Sara Moghtadernejad, California State University, Long Beach
James Murray, The University of Edinburgh
Dr. Stacy Nelson, Sandia National Labs
Kedar Murthy, Boston Materials
Eva Pappa, The University of Edinburgh
Alejandro Rodriguez, Solvay
Alyssa Skulborstad, Sandia National Labs
Danning Zhang, Wisk Aero

Manufacturing and Analysis Considerations of Ultem 9085 Resin in Aircraft Structures

Zeaid Hasan*, Bill Skaff, Dusty Dequine

Boom Supersonic, Centennial, CO, USA, 80112

Abstract

The aerospace industry relies on having parts that can be built to very tight tolerances, have high strength to weight ratios, and withstand harsh environments in order to achieve a product that can withstand decades of operation with high reliability and minimal repairs. The materials used in most current civil air transport and military airplanes are metallic materials (e.g aluminum, titanium, steel) and carbon fiber reinforced polymers. It took many years and substantial funding to allow the use of composites on aircraft as primary structures given the stringent regulations imposed by the FAA and other regulatory agencies on the materials used.

Over the past several years a new class of material candidates have emerged that are compatible with additive manufacturing processes, namely thermoplastic resin materials. This can be considered a revolution as the time and cost associated with building parts needed for aircraft structures can be dramatically reduced while allowing for complicated designs.

For the regulatory agencies to allow its use as a mainstream alternative, it will take tremendous effort from the industry to prove its reliability across the different operating environments. This paper will discuss aspects related to the use of Ultem 9085 on flight critical aircraft structure where the primary mode of assembly is using secondary bonding. Aspects related to static, fatigue and vibroacoustic will be discussed and guidance on how to evaluate such parts will be given.

Keywords: composites, bonding, additive manufacturing, analysis, fatigue, vibration, thermoplastics

*Corresponding Author, E-mail: zeaid.hasan@boom.aero

1. Introduction

The use of additive manufacturing (AM) has gained considerable attention recently in many industries including automotive, consumer products, medical and aerospace. The primary attraction comes from its ability to produce net shape 3D components layer by layer directly by automated machines regardless of part design complexity. AM also offers quick fabrication times compared to composite materials, for example, where a typical composite part might need several days to weeks to build, extensive resources, and expertise. By comparison, an AM part can be built within hours from a single computer-aided design (CAD) file [1]. This has a large impact on the cost associated with building an airplane. Boeing is projected to save up to $3 million per plane by 3D printing titanium parts on the 787 Dreamliner [2].

Boom Supersonic began using 3D printing more than two years ago as part of a partnership with Stratasys, a global leader in the market. Since then, the company has manufactured hundreds of 3D printed parts, tools, and prototypes and saved thousands of hours in work time. Example of such parts is shown in Figure 1. Stratasys and Boom recently announced a seven-year partnership extension through 2026 and will continue to implement creative solutions for complex problems [3].

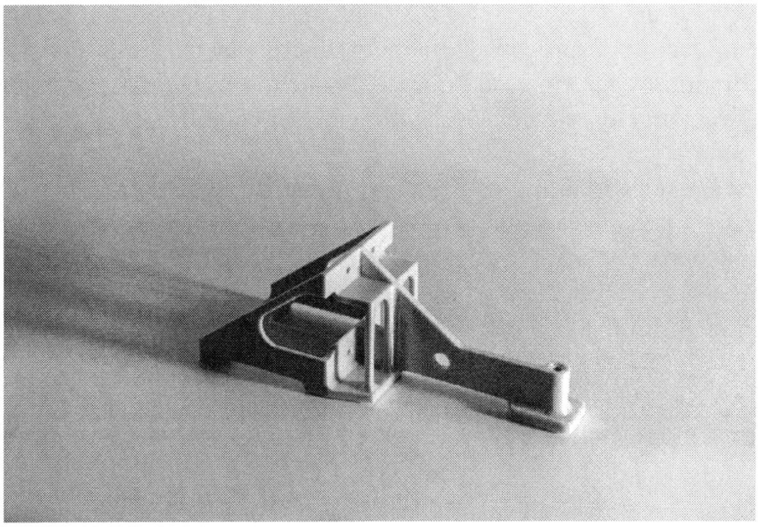

Figure 1. Example pressure regulator bracket built using additive manufacturing

The process also allows for simultaneous production of different parts without the need to have unique setups for each part. Figure 2 shows a 94-hour print job consisting of over 70 parts which was completed on a Stratasys F900 machine with a tray full of flight hardware for various systems on the aircraft. Benefiting the electrical, wing, flight controls, hydraulics, and fuselage system, this single job alone saved thousands of dollars and weeks of lead time when compared to conventional manufacturing methods.

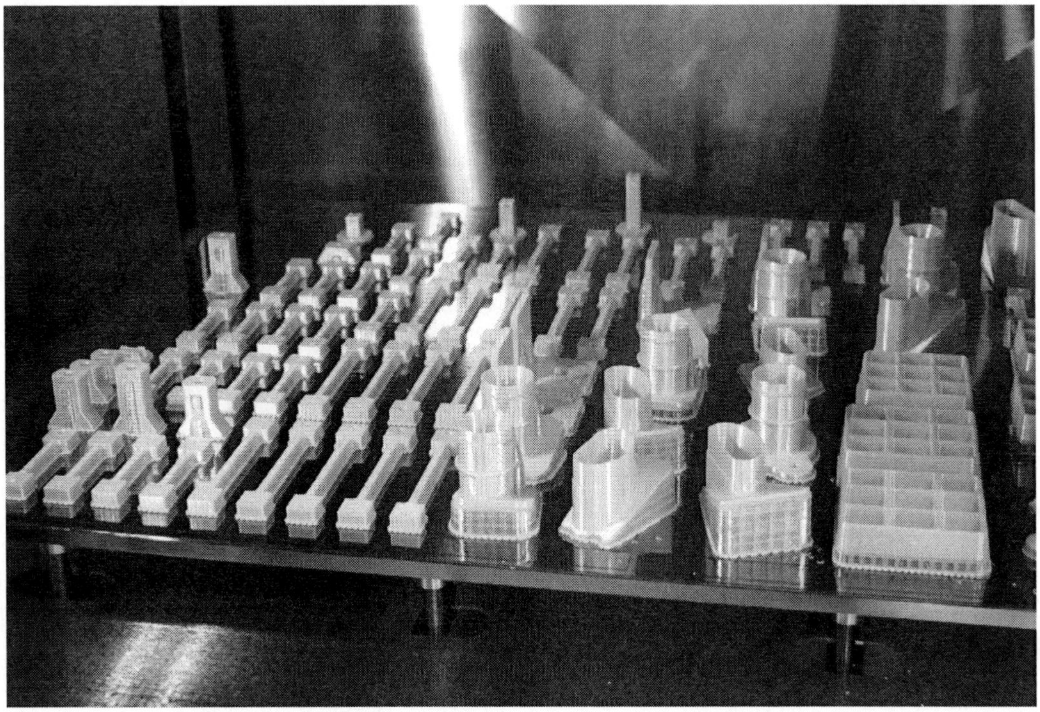

Figure 2. Additive manufactured parts using one single job run

When it comes to the aerospace sector in particular, the stringent requirements set by regulatory agencies such as the Federal Aviation Administration (FAA) and European Aviation Safety Agency (EASA) makes using a new material a challenge that requires extensive analysis and testing to understand the material behavior under its expected operational environments. Singamneni, et al [4] presents a comprehensive review on the use of AM for the aircraft industry and includes the challenges and benefits. The certification of AM aircraft parts is currently a challenging endeavor for many companies that want to use it for their own applications. That stems from the lack of an established certification roadmap and qualified materials in general [5].

As part of the America Makes initiative [6], several companies and institutions worked together over 3 years to generate a database for an AM compatible material. The material quantified was Ultem 9085 and the database features mechanical and physical properties as well as processing parameters. The process used to qualify the material was Fused Deposition Modeling (FDM) in which a polymer filament is melted and then deposited in successive layers to build a 3D component as shown in Figure 3.

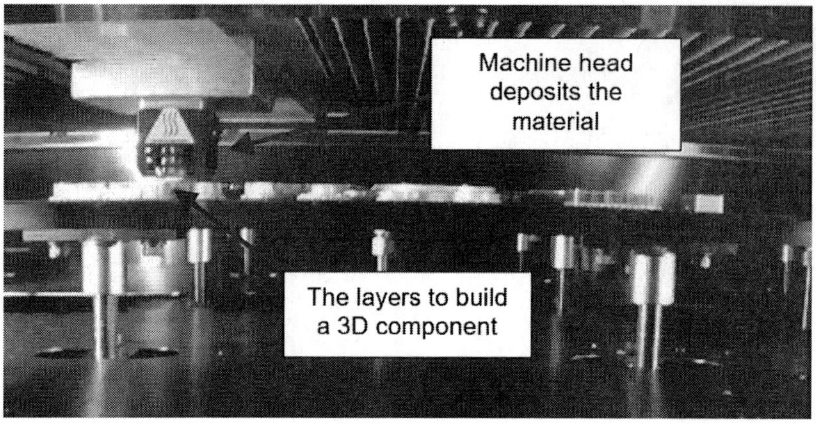

Figure 3. Fused Deposition Modeling process using Ultem 9085 material

Ultem 9085 resin comes in the form of filament and is a PEI (polyetherimide) thermoplastic. It has high strength-to-weight ratio, high thermal and chemical resistance, and meets multiple flame, smoke and toxicity (FST) characteristics per aerospace specifications.

The recent qualification of Ultem 9085 made it a prime candidate for use on the Boom Supersonic XB-1 demonstrator aircraft [7]. Although some static material allowables were available from the qualification testing, there was lack of information on how the material will behave in fatigue and vibroacoustic environments. In addition, the application of interest in our case was the secondary bonding of this material to composite materials, as shown in Figure 4, so the generation of applicable data to substantiate the bondline strength was necessary to size the components.

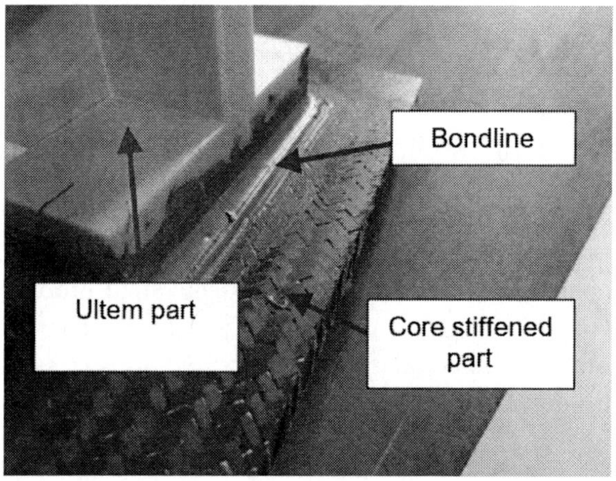

Figure 4. Ultem 3D printed part bonded onto a composite core stiffened part

This paper will shed light on the bonding process of AM Ultem parts onto composite parts including both the process and mechanical data that was generated to be able to compute relevant static margins for the design. The study also offers details about the fatigue and vibroacoustic environments of the material and endeavors to provide suitable data to aid with preliminary design in such environments. It is recommended that this data only be used for initial screening and evaluation; additional testing and evaluation will be needed to match the application of interest.

2. Airframe Design Overview

Boom Supersonic is currently developing the XB-1 prototype aircraft. XB-1 is approximately 1/3 the size of the company's planned passenger airliner, called Overture. Figure 5 shows a prototype of that demonstrator. One of the primary objectives is to evaluate many of the technologies for 21st century commercial supersonic flight such as the materials, engine, systems, and overall design.

Figure 5. XB-1 demonstrator [8]

The airframe is primarily built from composite material parts that are both bonded and bolted depending on the loading scenario and design attributes. Our ongoing build is shown in Figure 6 that highlights the use of composites in our fuselage and wing.

Figure 6. Build progress of the XB-1 demonstrator [9]

One aspect of this demonstrator is our ability to prove the use of materials that have not seen mainstream use on other airframes. We have leveraged the data available and relationships with additive manufacturing suppliers to implement the use of Ultem 9085 on this aircraft where more than 100 parts will be built using this material. Most notably, and a first for the aviation industry, is the use of A.M. Ultem for the leading edges of the inlets, which enabled the performance-critical aero contour to be met at a low weight without the cost and time of fabricating complex multi-piece composite tooling.

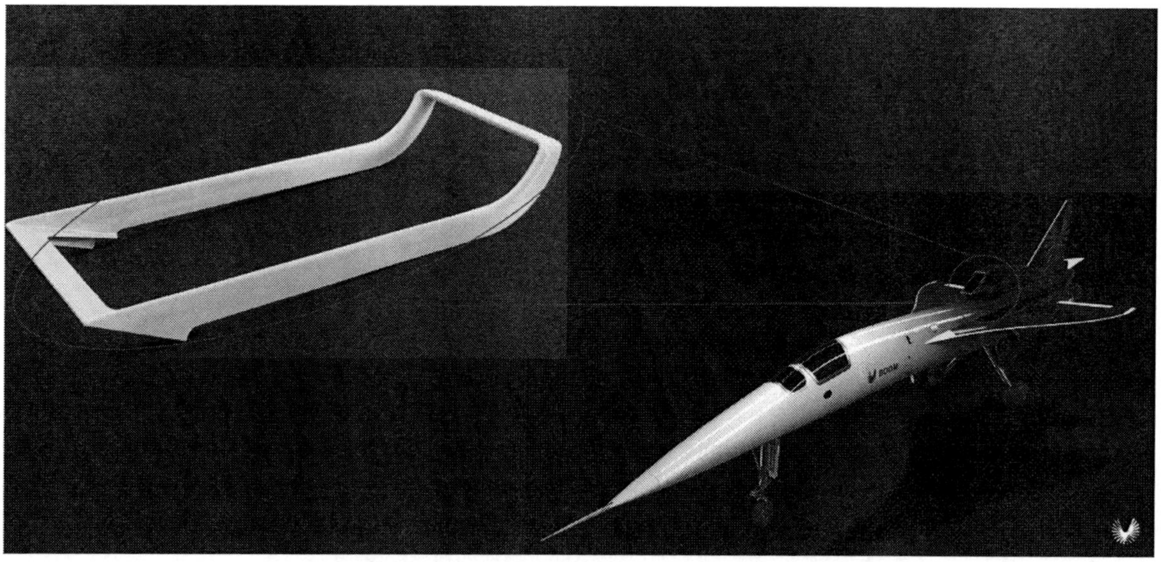

Figure 7. Use of Ultem on XB-1 inlet leading edges

Given our extensive use of this material under different loading and environmental conditions we needed to ensure we understand the different behaviors and how they can impact the overall performance of the structure. This paper is dedicated to discussing those different considerations and report on findings to aid others use this material and highlight some of the limitations observed. Before we dive deeper into the analysis portion, the next section discusses some of the details of secondary bonding and specific details related to bonding of Ultem parts to composite structures.

3. Secondary Bonding Operation

When dealing with composite materials, they are usually joined using several different fabrication processes that can be summarized as follows:

Co-curing: The act of curing a composite laminate and simultaneously bonding it to another uncured material, to a core material such as honeycomb, or to foam core. All resins & adhesives are cured during the same process.

Co-bonding: The curing together of two or more elements, of which at least one is fully cured and at least one is uncured. Requires careful surface preparation of the previously cured substrate. Additional adhesive may be required at the interface.

Secondary Bonding: The joining together, by the process of adhesive bonding, two or more pre-cured composite parts, where the only chemical or thermal reaction is the curing of the adhesive itself. Requires careful preparation of each previously cured substrate at the bonding surfaces. Usually requires well designed fixtures to align & clamp parts during processing.

Figure 8 shows a schematic representation of all three options. Details on each category is documented extensively in the literature [10], but as shown the secondary bonding process does not require having the different parts in sync during the fabrication. Rather, each individual part is built individually and can be assembled using any type of adhesive that is capable of withstanding the requirements set for the design.

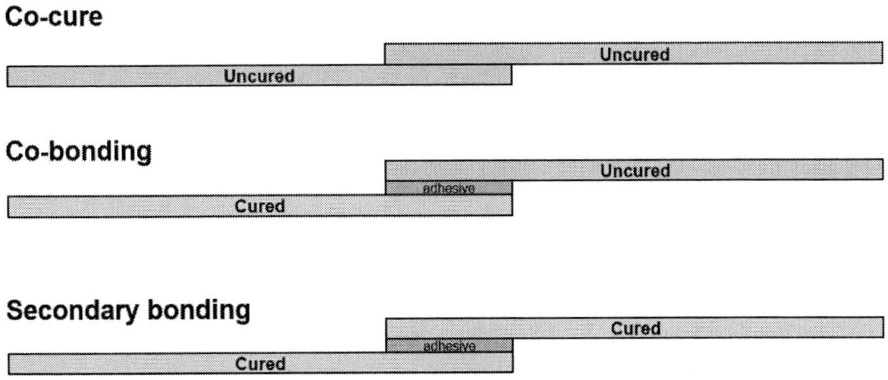

Figure 8. Classification of composite bonding techniques [11]

Secondary bonding is one of the preferred methods when it comes to smaller airframes and demonstrators as the process enables parts to be fabricated independently, followed by a bonding process using structural adhesive as an alternative to mechanical fastening. The long-term durability of adhesively bonded composite joints is critical and relies on several factors that are typically process driven.

The effects of surface preparation for the adherends is critical to initial strength, long-term durability, fracture toughness, and failure modes of bonded joints. Very little information is available on the use of secondary bonds to join Ultem 9085 to primary and secondary composite structure. We thus seek to evaluate the failure of these joints while subject to different surface preparation and loading conditions.

An initial study was first conducted to understand the following aspects:

1. Surfaces preparation of AM Ultem 9085 parts that are subsequently bonded to composite parts using paste adhesive. A comparison between an unprepared and prepared surface is shown in Figure 9.

2. Different loading scenarios of the bonded specimens replicating in-flight loading.

Figure 9. Surface preparation used on the Ultem parts

Loading Type	Schematic	Actual Testing
Tension	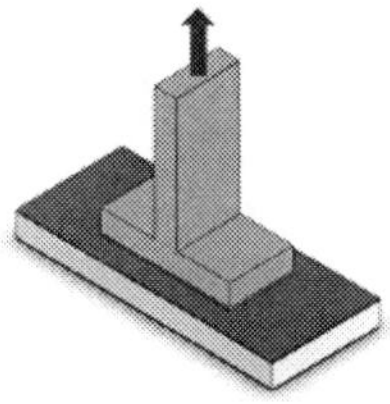	
Bending/Shear	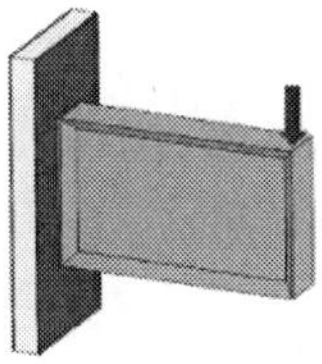	
Bending/Shear	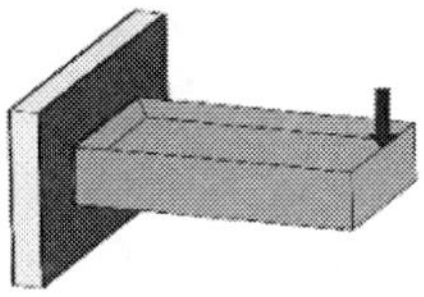	

Table 1. Loading scenarios used to evaluate the failure mode and load for secondary bonded Ultem to composite

The results of the testing for the tension and coupled bending/shear is shown in Figure 10 and 11, respectively, which show the failure loads of those coupons that can be used as preliminary design values for sizing the parts. The results are highly dependent on the geometry and in the case of the coupled bending/shear on the load application point (denoted by L which is the distance from the base to the load application point). The data highlights the linear nature of the failure load for the tension specimens as a function of the specimen base area while the bending/shear has a more non-linear behavior given the added complexity in that loading scenario. Note that several testing replicates were done for each test.

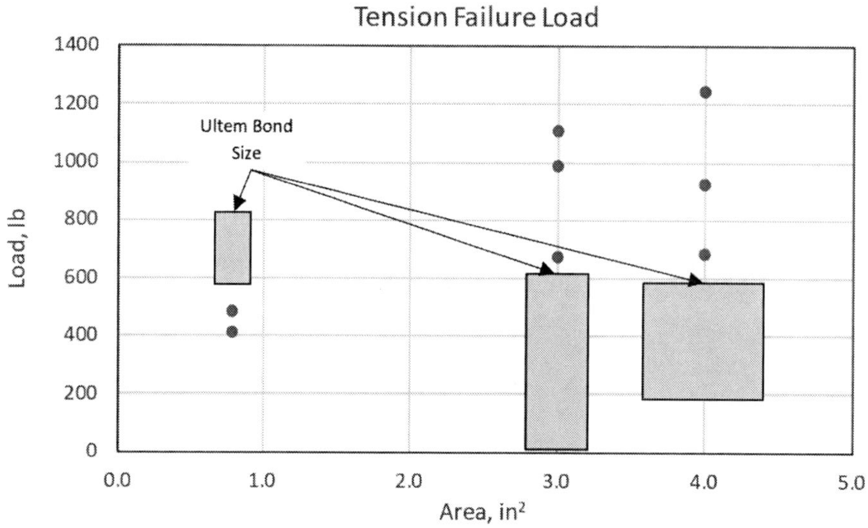

Figure 10. Failure load for the tension specimens as a function of base area

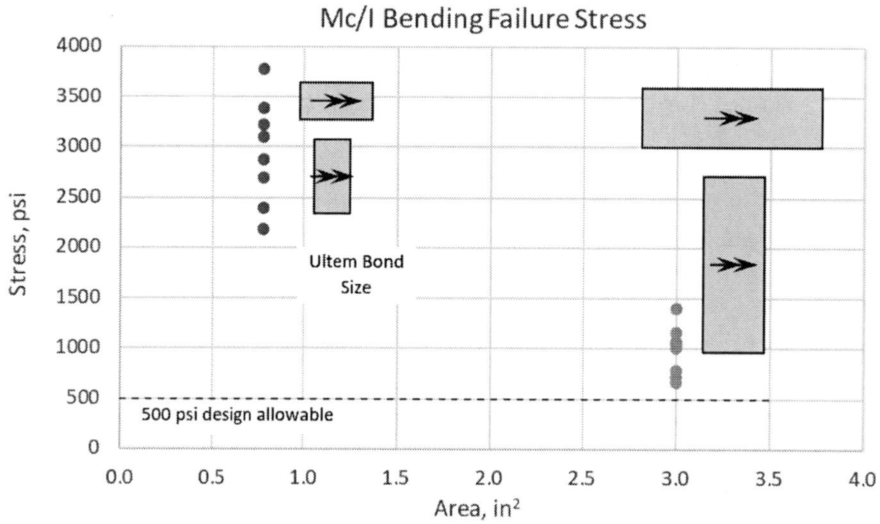

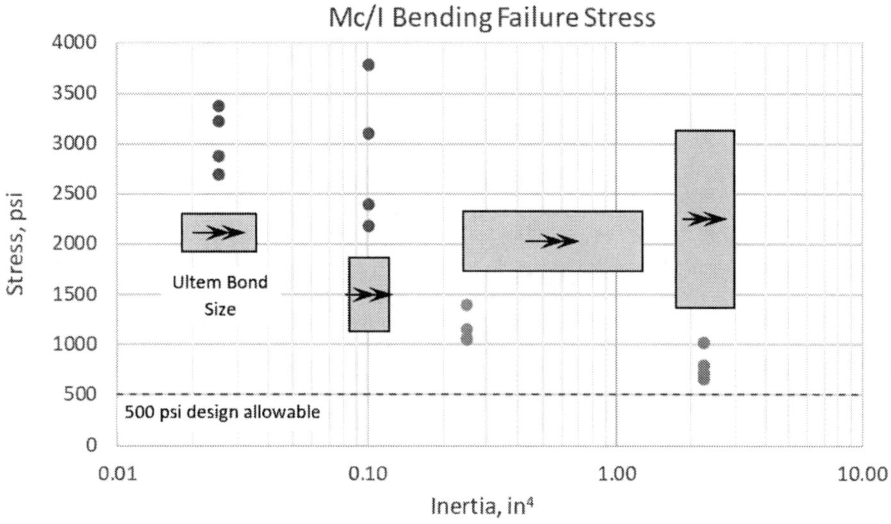

Figure 11. Bending stress as a function of area and specimen's inertia

4. Static Analysis

The first phase of any structural analysis is done by evaluating the static loading against the material allowables. As stated earlier, Ultem 9085 has been tested extensively to achieve a B-basis allowable dataset for the different loading conditions. Table 2 provides a summary of a portion of the material allowable data. As shown in the table, the values are highly dependent on the print direction. Furthermore, minimum allowables do not consistently align with a given direction, and critical directions are not consistent across temperatures. Therefore, it is crucial to perform the analysis with respect to the corresponding print direction and application desired. And since the material is orthotropic in nature, the load application relative to the material direction needs to be considered as well. Von Mises failure criteria, as an example, is not applicable in this case and the use of a maximum or minimum principal strain criteria might be better suited

It is also worth noting that Ultem 9085 strengths are 8%-16% of aluminum but density is 47% so Ultem may not be the best option if a structure is strength critical. Similarly, it may not be the best option for compression (stability) critical structure since the stiffness is approximately 1/30 of aluminum. Ultem 9085 may be the preferable material, even with these disadvantages, for lightly loaded and complex machining scenarios.

Print Direction				Min Values
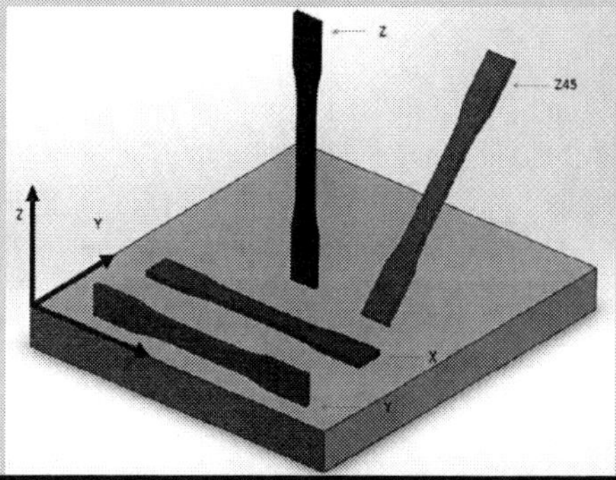				
Room Temperature Dry (70F)				
Property	**X-direction (ksi)**	**Y-direction (ksi)**	**Z-direction (ksi)**	**Z45-direction (ksi)**
Tensile Strength	8.82	9.97	8.21	6.80
Tensile Modulus	337	377	347	341
Compression Strength	6.87	9.09	8.21	7.07
Compression Modulus	385	430	377	364
Flexural Strength	16.20	17.85	8.33	10.45
Flexural Modulus	353	382	333	314
OHT	5.77	8.09	3.20	4.48
OHC	4.90	6.71	7.35	6.97
FHC	5.57	9.65	9.0	8.03
SBS	26.93	25.48	19.25	24.03

Table 2: Material properties Ultem 9085 [12]

5. Fatigue Evaluation

The America Makes initiative generated static allowables for Ultem 9085 but did not include testing pertinent to fatigue. As a result, additional testing was needed to evaluate its effectiveness under cyclic loading. Fisher and Schöppner [13] performed some fatigue testing for different print direction that acted as our initial baseline. Initial Boom testing was performed in the z-direction at two conditions: 1) 1.3 ksi with stress ratio R=-0.3, and 2) 4.0 ksi with stress ratio R=-1. Boom results show good correlation with Fisher and Schöppner, as shown in Figure 12. Based on this data the runout (endurance strength) for Ultem appears to occur at about 400 psi. Comparing that level to aluminum, as shown in Table 3, highlights that Ultem 9085 has significantly and disproportionally lower fatigue capability.

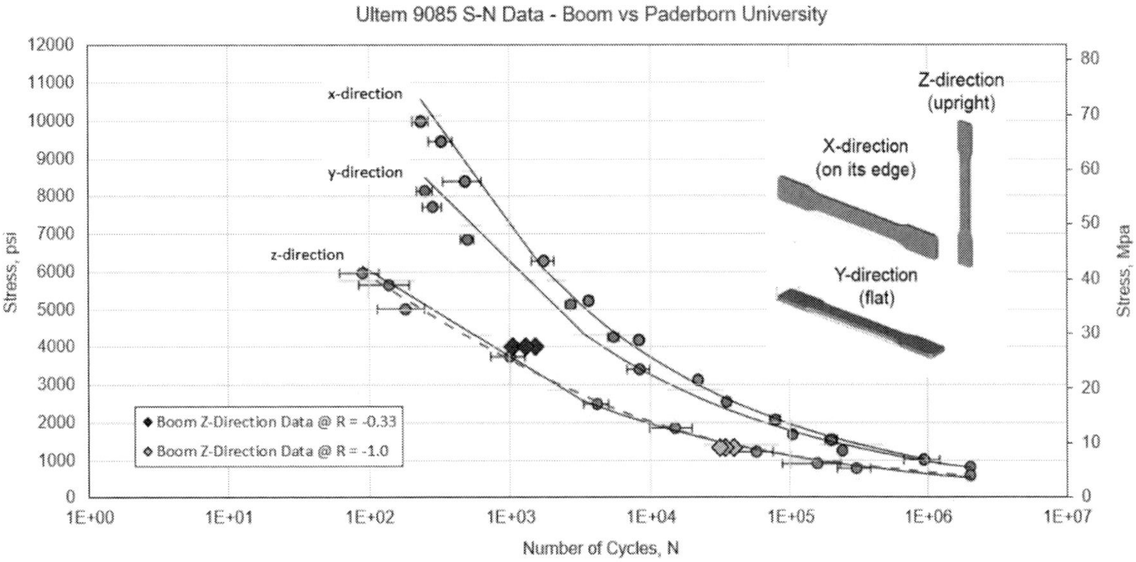

Figure 12. Comparison between published data and data tested internally

Unnotched (Pristine)					
Material	F_{tu}, psi	F_{ty}, psi	$f_{endurance}$, psi	$f_{endurance}/F_{tu}$	$f_{endurance}/F_{ty}$
7050-T7451	68000	59000	14600	21.5%	24.7%
6061-T62	42000	35000	12500	29.8%	35.7%
Ultem 9085	6800	4800	400	5.9%	8.3%

Table 3: Fatigue capability comparison aluminum to Ultem 9085

Additional testing was conducted in order to rigorously fit the fatigue-life equation [14] using a least linear residual solution. This resulted in the parameters shown in Table 4. A summary of the test data and predicted S-N curve is shown in Table 5 and Figure 13, respectively. This preliminary dataset can be utilized for initial sizing when attempting to use on airframe components realizing the deficiencies in fatigue strength compared to other material options.

$$N_f = 10^{(C_1 - C_2 log[S_m(1-R)^{C_3} - C_4])} \quad (1)$$

C_1	C_2	C_3	C_4
9.91503	2.96501	-4.02000	16.94000

Table 4: Fatigue-life equation parameters

Test Data			Predicted Cycles to Failure, N_f	Residual, Cycles	|Residual|, Cycles	% error $\frac{(Predicted - Test)}{Predicted}$
Stress	R-Ratio	Cycles to Failure				
1333	-1.00	39785	34,298	5487	5487	-16.0%
1333	-1.00	34298	34,298	0	0	0.0%
1333	-1.00	31802	34,298	-2496	2496	7.3%
2800	-1.00	3017	2,602	415	415	-16.0%
2800	-1.00	2608	2,602	6	6	-0.2%
2800	-1.00	2457	2,602	-145	145	5.6%
2800	-1.00	2504	2,602	-98	98	3.8%
2800	-1.00	3137	2,602	535	535	-20.6%
4000	-1.00	574	822	-248	248	30.1%
4000	-1.00	869	822	47	47	-5.8%
4000	-1.00	581	822	-241	241	29.3%
4000	-1.00	886	822	64	64	-7.8%
4000	-1.00	759	822	-63	63	7.6%

Table 5: Fatigue test data vs predicted cycles to failure for Ultem 9085

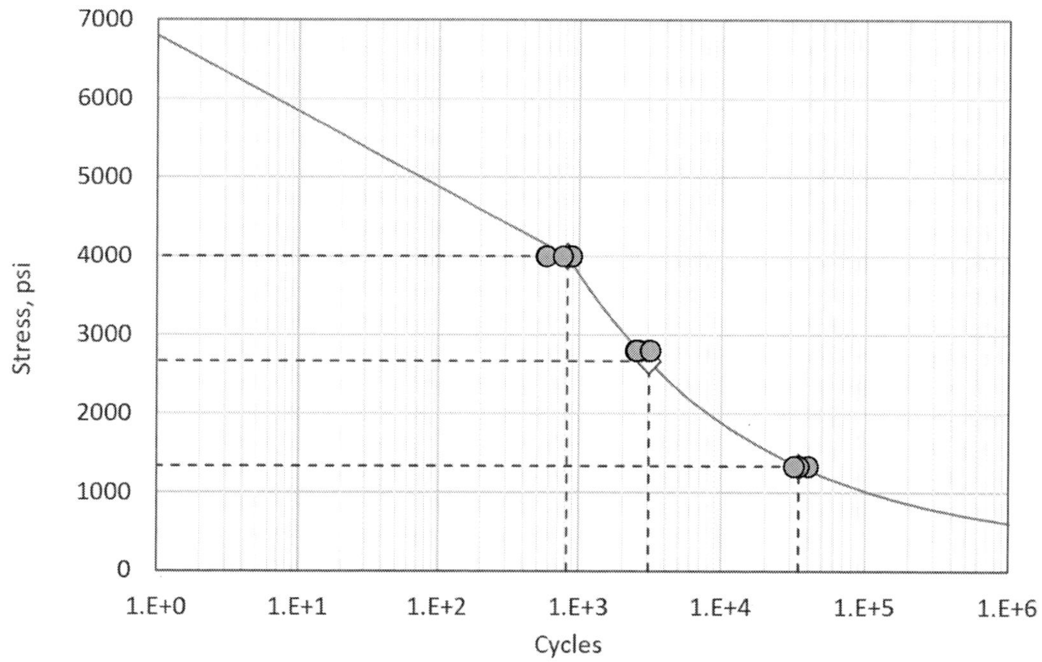

Figure 13. S-N curve for Ultem 9085 for Z-direction printed specimens

6. Vibroacoustic Considerations

Most of the components on the airframe that will be built from Ultem 9085 are system brackets that are generally lighter in terms of weight compared to structural parts. These lighter parts are subject to higher vibratory accelerations than heavy items. Analysts need to consider this mode of failure and size the parts accordingly. One technique to perform this analysis is to consider Miles' equation, as shown below, which can be used to estimate an equivalent static load due to random vibration. We can calculate the acceleration value and multiply it by 3 to get a three-sigma environment which is known to be conservative. This method is commonly used in the design and acceptance testing of hardware mounts in spacecraft.

$$G_{RMS} = \sqrt{\frac{\pi}{2} f_n Q \left[ASD_{input} \right]} \qquad (2)$$

where

G_{RMS} = Root mean square acceleration in G's

f_n = Natural frequency

$Q = 0.5\zeta$

ζ = damping ratio

ASD_{input} = Input acceleration spectral density at f_n. This will be based on reference [15]

Given the limitations of Ultem 9085 to cyclic loading we expect them to be sensitive to acoustic noise that manifests as acoustic excitation and random vibration. Based on the analysis we propose to limit the use of brackets made from Ultem 9085 to a natural frequency $f_n \leq 150$ Hz for all brackets that are 10 lbs or less in weight and can be expanded to any weight depending on the application.

Given the estimated mount stiffness and weight of the hardware one can use the data in Figure 14 to determine the natural frequency of the assembly which is only valid for ratio of bracket weight to supported weight of 0.100. Note that the figure shows part of the table which is based on cantilevered beam frequency calculation and another based on a fuselage like structure per a modified version of DO-160G [16]. The data presented here is an example for a very specific environment, and for a very specific mass ratio. If the structure being analyzed does not satisfy the assumptions the data cannot be used.

① Cantilever Beam Frequency, Hz f > 20 Hz required, f > 30 Hz desired Enter ratio of (bracket weight / supported weight) => 0.100											
Transverse Stiffness, lb/in	0.5 lb	1 lb	2 lb	3 lb	4 lb	5 lb	6 lb	7 lb	8 lb	9 lb	10 lb
100,000	150	150	150	150	150	150	150	150	150	150	150
90,000	150	150	150	150	150	150	150	150	150	150	150
80,000	150	150	150	150	150	150	150	150	150	150	150
70,000	150	150	150	150	150	150	150	150	150	150	150
60,000	150	150	150	150	150	150	150	150	150	150	150
50,000	150	150	150	150	150	150	150	150	150	150	150
40,000	150	150	150	150	150	150	150	150	150	150	150
30,000	150	150	150	150	150	150	150	150	150	150	150
20,000	150	150	150	150	150	150	150	150	150	146	138
10,000	150	150	150	150	150	138	126	117	109	103	98
9,000	150	150	150	150	147	131	120	111	104	98	93
8,000	150	150	150	150	138	124	113	104	98	92	87
7,000	150	150	150	149	129	116	106	98	91	86	82
6,000	150	150	150	138	120	107	98	90	85	80	76
5,000	150	150	150	126	109	98	89	83	77	73	69
4,000	150	150	138	113	98	87	80	74	69	65	62
3,000	150	150	120	98	85	76	69	64	60	56	54
2,000	150	138	98	80	69	62	56	52	49	46	44
1,000	138	98	69	56	49	44	40	37	35	33	31
900	131	93	66	54	46	41	38	35	33	31	29
800	124	87	62	50	44	39	36	33	31	29	28
700	116	82	58	47	41	37	33	31	29	27	26
600	107	76	54	44	38	34	31	29	27	25	24
500	98	69	49	40	35	31	28	26	24	23	22
400	87	62	44	36	31	28	25	23	22	21	20
300	76	54	38	31	27	24	22	20	19	18	17
200	62	44	31	25	22	20	18	17	15	15	14
100	44	31	22	18	15	14	13	12	11	10	10
$f = \frac{1}{2\pi}\sqrt{\frac{K}{m}}$	Supported weight										

② DO-160G Fuselage Environment, G_{rms}, g's 2% Damping, Yellow => f < 30 Hz, Red => f < 20 Hz Ratio of (bracket weight / supported weight) = 0.100											
Transverse Stiffness, lb/in	0.5 lb	1 lb	2 lb	3 lb	4 lb	5 lb	6 lb	7 lb	8 lb	9 lb	10 lb
100,000	23.1	23.1	23.1	23.1	23.1	23.1	23.1	23.1	23.1	23.1	23.1
90,000	23.1	23.1	23.1	23.1	23.1	23.1	23.1	23.1	23.1	23.1	23.1
80,000	23.1	23.1	23.1	23.1	23.1	23.1	23.1	23.1	23.1	23.1	23.1
70,000	23.1	23.1	23.1	23.1	23.1	23.1	23.1	23.1	23.1	23.1	23.1
60,000	23.1	23.1	23.1	23.1	23.1	23.1	23.1	23.1	23.1	23.1	23.1
50,000	23.1	23.1	23.1	23.1	23.1	23.1	23.1	23.1	23.1	23.1	23.1
40,000	23.1	23.1	23.1	23.1	23.1	23.1	23.1	23.1	23.1	23.1	23.1
30,000	23.1	23.1	23.1	23.1	23.1	23.1	23.1	23.1	23.1	23.1	23.1
20,000	23.1	23.1	23.1	23.1	23.1	23.1	23.1	23.1	23.1	22.7	22.1
10,000	23.1	23.1	23.1	23.1	23.1	22.1	21.1	20.3	19.7	19.1	18.6
9,000	23.1	23.1	23.1	23.1	22.8	21.6	20.6	19.8	19.2	18.6	18.1
8,000	23.1	23.1	23.1	23.1	22.1	20.9	20.0	19.2	18.6	18.1	17.6
7,000	23.1	23.1	23.1	23.0	21.4	20.2	19.3	18.6	18.0	17.5	17.0
6,000	23.1	23.1	23.1	22.1	20.6	19.5	18.6	17.9	17.3	16.8	16.4
5,000	23.1	23.1	23.1	21.1	19.7	18.6	17.8	17.1	16.5	16.1	15.6
4,000	23.1	23.1	22.1	20.0	18.6	17.6	16.8	16.2	15.6	15.2	14.8
3,000	23.1	23.1	20.6	18.6	17.3	16.4	15.6	15.1	14.6	14.1	13.8
2,000	23.1	22.1	18.6	16.8	15.8	14.8	14.1	13.6	12.4	11.4	10.5
1,000	22.1	18.6	15.6	14.1	12.4	10.5	9.2	8.9	8.6	8.3	8.1
900	21.6	18.1	15.2	13.8	11.5	9.7	9.0	8.6	8.3	8.1	7.9
800	20.9	17.8	14.8	13.1	10.5	9.1	8.7	8.4	8.1	7.9	7.7
700	20.2	17.0	14.3	11.8	9.5	8.8	8.4	8.1	7.8	7.6	7.4
600	19.5	16.4	13.8	10.5	9.0	8.5	8.1	7.8	7.5	7.3	7.1
500	18.6	15.6	12.4	9.2	8.6	8.1	7.7	7.5	7.2	7.0	6.8
400	17.6	14.8	10.5	8.7	8.1	7.7	7.3	7.0	6.8	6.6	6.4
300	16.4	13.8	9.0	8.1	7.5	7.1	6.8	6.6	6.3	6.2	6.0
200	14.8	10.5	8.1	7.3	6.8	6.4	6.2	5.9	5.7	5.6	5.4
100	10.5	8.1	6.8	6.2	5.7	5.4	5.3	5.0	4.8	4.7	#N/A
$3\sqrt{\frac{\pi}{2} f_n \cdot Q \cdot ASD_{input}}$	Supported weight										

Figure 14. Natural frequency estimates based on estimated mount stiffness and weight of the hardware

A more generic approach is shown in Figure 15. This plot identifies vibroacoustic environment as a function of aircraft region (e.g. fuselage, wing). Maximum g values are also provided that should not be exceeded when conducting analysis during the sizing phase. The analyst should note that environments vary widely by region.

Maximum 3σ Vibroacoustic **LIMIT** Environment per DO-160G w/XB-1 Adjustments $f_n \leq 150$ Hz, All Component Weights					
Fuselage	Wheel Well Adjacent Bays	Wheel Wells	Wing & Empennage	Instrument Panel	Nacelle & Pylon
23.1g	39.9g	56.4g	39.9g	14.8g	46.0g

Full 3σ Vibroacoustic **LIMIT** Environments per DO-160G w/XB-1 Adjustments
Systems and Structures Bracket Environments V1.2

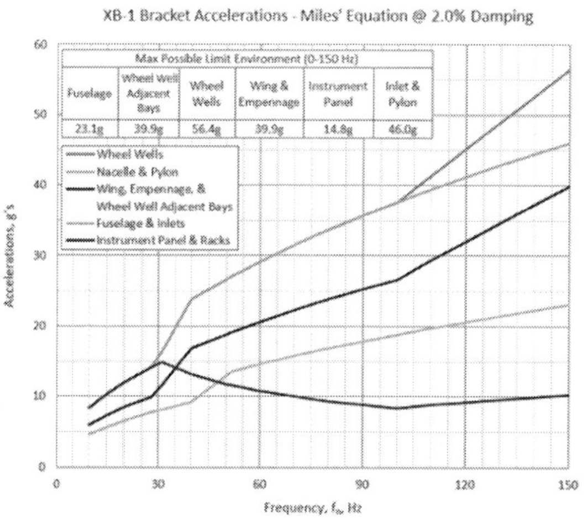

Figure 15. Acceleration as a function of natural frequency for different airframe structures

7. Conclusion

This paper presented details regarding the use of Ultem 9085 for aerospace structural components and ways to analyze these new and emerging material as it pertains to structural analysis. As stated earlier, the use of additive manufacturing in the aerospace industry is gaining widespread acceptance but the methods of how to evaluate such parts has not been fully developed and additional guidance is required.

This study primarily focused on the static, fatigue and vibroacoustic aspects for Ultem parts fabricated using FDM. The assembly process using bonding onto composite parts was discussed and showed promising results. The surface preparation of the substrates was shown to have a large impact on the final failure load and mode. Moreover, loading scenario and part design has a large impact on the bondline allowables and must be considered during the design phase.

It was shown that the material exhibits orthotropic characteristics and failure modes. As a result, max stress or strain need to be considered for the analysis. It was also shown that Ultem 9085 has disproportionally lower fatigue capability compared to aluminum, and a fatigue life equation was developed based on a limited set of testing that can be utilized for preliminary analysis. Finally, vibroacoustic environments pertaining to smaller parts such as those made of Ultem 9085 were evaluated. Guidance on how to achieve an appropriate acceleration value was given and shown to rely on the structure being evaluated and stiffness of the part.

8. References

1. https://www2.deloitte.com/content/dam/insights/us/articles/additive-manufacturing-3d-opportunity-in-aerospace/DUP_706-3D-Opportunity-Aerospace-Defense_MASTER2.pdf

2. https://venturebeat.com/2017/04/11/printed-titanium-parts-expected-to-save-millions-in-boeing-dreamliner-costs/

3. https://investors.stratasys.com/news-releases/news-release-details/boom-supersonic-signs-seven-year-partnership-extension-stratasys

4. Singamneni S, Yifan LV, Hewitt A, Chalk R, Thomas W, et al. (2019) Additive Manufacturing for the Aircraft Industry: A Review. J Aeronaut Aerospace Eng 8: 214. doi:10.4172/2329-6542.1000214

5. Solution Guide "Certified Additive Manufacturing for Aircraft Interiors"

6. http://additivemanufacturing.com/2019/02/07/america-makes-announces-complete-qualified-database-of-material-properties-for-fused-deposition-modeling-fdm-additive-manufacturing-of-ultem-9085-resin/

7. https://blog.boomsupersonic.com/an-inside-look-at-the-build-of-booms-supersonic-demonstrator-e540bbcb3a90

8. https://newatlas.com/boom-xb-1-prototype/46452/

9. https://blog.boomsupersonic.com/an-inside-look-at-the-build-of-booms-supersonic-demonstrator-e540bbcb3a90

10. Budhe, S. et al, 2016, "An updated review of adhesively bonded joints in composite materials" International Journal of Adhesion & Adhesives 72 (2017) 30–42

11. Design, analysis and fabrication of thick co-cured wing structures

12. https://www.wichita.edu/research/NIAR/Documents/NCP-RP-2018-007-Rev-NC-ULTEM-9085-Stat-Analysis-Report-April-16-2019.pdf

13. Fischer, M., Schöppner, V. Fatigue Behavior of FDM Parts Manufactured with Ultem 9085. JOM 69, 563–568 (2017). https://doi.org/10.1007/s11837-016-2197-2

14. Metallic Materials Properties Development and Standardization (MMPDS) Part 5, DOT/FAA/AR, Office of Aviation Research, Washington, D.C. 20591

15. MIL-STD-781C, Reliability Design Qualification and Production Acceptance Tests: Exponential Distribution. Department of Defense, Washington D.C, 1977.

16. DO-160G

Abbreviations

FAA: Federal Aviation Administration

AM: Additive Manufacturing

FDM: Fused Deposition Molding

CAD: Computer Aided Design

EASA: European Aviation Safety Agency

3D PRINTED DUAL-CURE EPOXY: ENABLING HIGH-PERFORMANCE HYBRID COMPOSITE STRUCTURES

Alec J. Redmann[1], Paul V. Oehlmann[2], Rui Zhang[3], Sue J. Mecham[3], Tim A. Osswald[1]

1- Polymer Engineering Center, Department of Mechanical Engineering, University of Wisconsin-Madison, Madison, WI

2- Chair of Carbon Composites (LCC), Department of Mechanical Engineering, Technical University of Munich, Garching, Germany

3- Department of Chemistry, University of North Carolina at Chapel Hill, Chapel Hill, NC

Abstract

A new process is described in this paper which provides an improved alternative to traditional joining and manufacturing methods for pre-impregnated fiber reinforced plastic (FRP). Utilizing additive manufacturing and a dual-cure epoxy resin, joining members are first 3D printed using a UV process; resulting in a semi-rigid, but only partially cured part. This part still has chemical potential and bonding availability when it is integrated with pre-impregnated fiber reinforcement. The assembly is then heated to activate the second curing reaction and co-cure the two materials, forming a permanent, void-free cohesive bond. The bonding quality of pre-impregnated fiber-epoxy composites joined with this co-curing method are compared to an industrial adhesive by single lap-shear and block shear tests, and show competitive results. A current aerospace application is also examined.

Introduction

Fiber reinforced plastics (FRP) are used for the design of high-performance applications in aerospace, automotive and many more industries due to their excellent material characteristics. Low density fibers embedded in a matrix form a highly anisotropic composite on a lamina basis which results in stiffness-to-weight and strength-to-weight ratios unachievable with other manufacturing methods. With the use of FRP composite materials in high performance structures, the benefits of saving weight are closely related to efficiency, operating- and life cycle costs. This justifies a wide use of FRP materials in applications with high operating costs, such as aerospace. Current estimates for carbon composite utilization are up to 50% by weight in commercial aerospace and up to 85% by weight in military aerospace applications [1, 2].

While there are many advantages to FRP composites, there are challenges with experience and understanding of how to best manufacture these components. Many designers and engineers revert to manufacturing composite structures with the same approaches as metal structures in so-called "black metal design". This leads to a suboptimal design, especially during assembly where drilled holes for mechanical fasteners reduce the laminate strength due to damaging of fibers and inducing stress concentrations [3, 4]. Adhesive bonding is generally preferred in composite structures because, unlike mechanical fasteners, adhesive bonding does reduce the laminate strength or decrease the strength-to-weight ratio by adding mass of the metal joints to the structure. Galvanic corrosion is also minimized, which can be responsible for more than 20% of the maintenance costs in aerospace and automotive applications [5]. Despite this, adhesives are often avoided due to quality control concerns. Current adhesive technology can introduce air in the bond line, decreasing the bond strength and causing unpredictable failure. Due to a very labor-intensive process, increased costs for preparation, assembly and quality control must be taken into account.

In order to solve these problems, a new process has been developed for joining composite members utilizing additive manufacturing and a dual-cure epoxy resin. The resin is first 3D printed using a UV process; resulting in a semi-rigid, but only partially cured part. This part still has chemical potential and bonding availability when it is integrated with pre-impregnated fiber reinforcement. The assembly is then heated to activate the second curing reaction and co-cure the two materials, forming a permanent, void-free cohesive bond.

Previous work has shown this process can be used to create hybrid composite lattice structures, increasing the specific strength of panels of composite beams by creating a co-cured sandwich structure using a 3D printed epoxy lattice core [6]. Another application has been shown for T-joint structures common in aerospace structures, where the deltoid region formed between the perpendicular flange and web section was 3D printed and co-cured in the same process, resulting in increased geometric accuracy and fewer defects compared to traditional joining methods [7]. This paper expands on the prior work, demonstrating the bonding quality of composite adherends co-cured with the 3D printed epoxy adhesive. An application for an experimental aircraft is also explored, where small quantities of customized composite parts are required.

Experimental Methods

Materials

This study investigates the material characteristics of hybrid composite structures with the use of a dual-curing epoxy resin, EPX 82. The material, produced by Carbon Inc., is manufactured with an M1 printer using the Digital Light Synthesis (DLS) process. Carbon's EPX 82 epoxy resin is a rigid material with improved temperature resistance for various applications. A partially cured part is manufactured with the DLS process in the first stage, using UV-light to form an acrylate network. This intermediate part is dimensionally stable, but shows weak mechanical properties. A second stage consists of a thermal curing cycle to fully crosslink the material.

The material is stored in cartridges containing part A and B in separate chambers to inhibit a reaction. A static mixer with a 1:1 ratio is used to fill the resin tank for printing. Samples for this study are printed with default settings and a layer height of 100 μm. Partially cured parts have to be removed from the build platform and cleaned to remove excess resin. The recommend cleaning processes involves using 99% dipropylene glycol methyl ether (DPM) as a solvent for removing the excess resin from the printed parts, limiting the solvent exposure to a maximum of four minutes. The parts are then rinsed with isopropyl alcohol (IPA) and placed on a tray in a Yamato DKN 602C programmable convection oven to complete the second, thermal curing stage (Figure 1). Table 1 shows the material properties of fully-cured EPX 82.

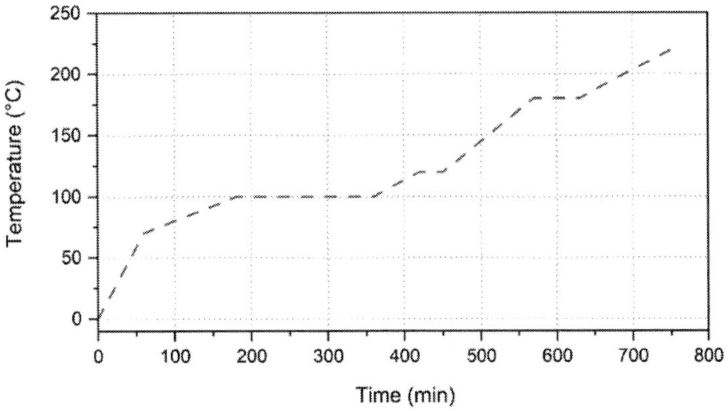

Figure 1: Thermal curing cycle for EPX 82

Table 1: Material Properties for EPX 82 [8]

Material Properties	Value	Unit
Elastic Modulus	2.8	GPa
Ultimate Tensile Strength	82	MPa
Elongation at break	5.9	%
Glass Transition Temperature	150 – 155	°C

Two pre-impregnated FRP materials with different curing properties are used in this study. The first is Hexcel's HexPly F263 carbon fiber prepreg, which contains a structural and heat resistant epoxy resin for aerospace applications. The manufacturer recommends a curing temperature at 179 ± 6 °C. The second material is a 7781 E-glass fiber reinforced pre-impregnated material, NB-EP4030 D, produced by Mitsubishi. This is a low-temperature curing epoxy resin which is processed at 135 ± 15 °C. The recommended curing cycles for these two materials are shown in Figure 2. The low- and high temperature curing resins provide a wider processing window to characterize their bonding capabilities with the 3D printed adhesive EPX 82. Table 2 shows the properties of the neat resin matrix materials and Table 3 shows the final properties with the fiber reinforcements.

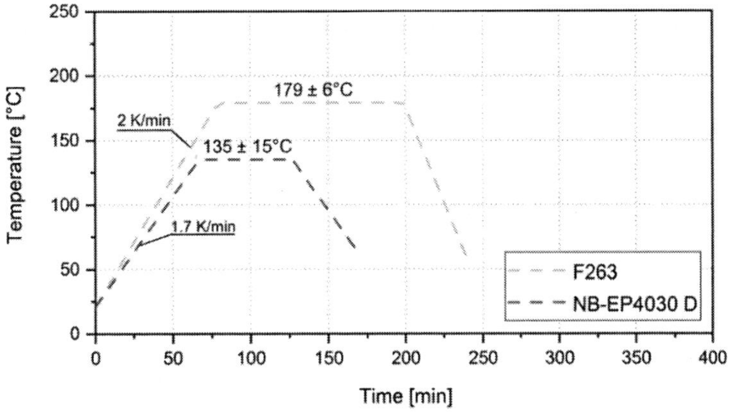

Figure 2: Curing cycle for F263 and NB-EP4030 D materials

Table 2: Material Properties for Neat Resin Matrix

Material Properties	Value	Unit
Hexcel HexPly F263 [9]		
Tensile Modulus	4.27	GPa
Tensile Strength	75	MPa
Elongation at break	2.4	%
Glass Transition Temperature	188	°C
Mitsubishi NB-EP4030 [10]		
Tensile Modulus	-	GPa
Tensile Strength	-	MPa
Elongation at break	-	%
Glass Transition Temperature	130	°C

Table 3: Material Properties for Cured Composites

Material Properties	Value	Unit
Hexcel HexPly F263 (W3-282) [9]		
Tensile Modulus	60.7	GPa
Tensile Strength	670	MPa
Mitsubishi NB-EP4030 (7781 E-glass) [10]		
Tensile Modulus	32.8	GPa
Tensile Strength	448	MPa

Shear Testing

Two different types of shear testing were conducted in this study: single lap-shear and block shear. Test specimen for both are manufactured using the same process with different dimensions. In the first step, the pre-impregnated material is cut to the required dimensions and stacked to achieve the specified sample height. An aluminum plate is used as the lower mold. The first component of the upper mold is aligned with the stacked materials. This creates a functional surface that is needed later during shear testing. A second mold is then placed on top of the materials to form the gap for placement of the adhesive material. The partially-cured EPX 82 adhesive is placed into the gap in the next step. In the fourth step, pre-impregnated plies are stacked on top of the EPX 82 adhesive layer. A diagram of the manufacturing steps for block shear is shown in Figure 3.

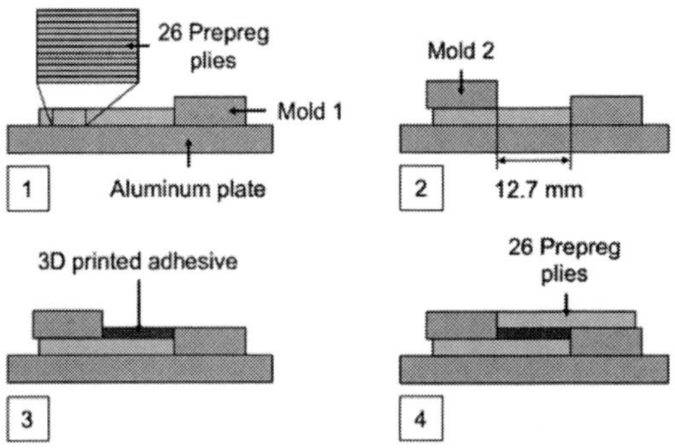

Figure 3: Manufacturing process for block shear test specimen

A vacuum bag is then placed over the stack as shown in Figure 4 to create a sufficient pressure for consolidation during the thermal curing. This ensures that no air is remaining in the sample which could lead premature failure of the sample. A single layer of peel ply and a breather fabric is placed between the vacuum bag and the mold to simplify the removal of the component after curing and to ensure that air can escape from the structure.

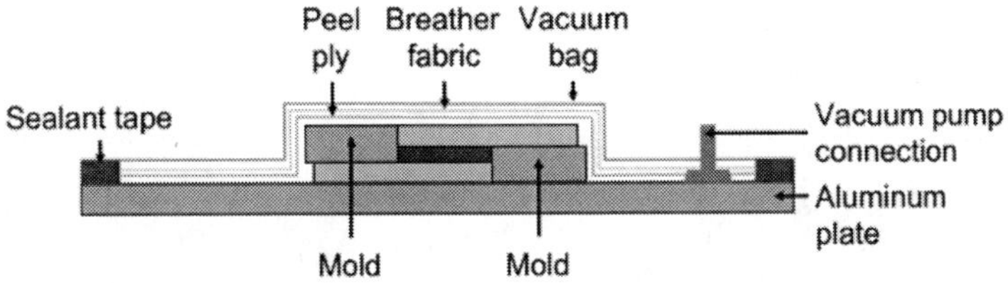

Figure 4: Vacuum bag set-up for consolidation during thermal curing

For single lap-shear testing, the specimen geometry is used as specified in ASTM D5868 [11] and depicted in Figure 5. NEMA Grade G-10 glass epoxy laminate alignment tabs are glued onto the specimens to ensure the centering of the specimens between the tensile testing grips. An Instron 59767 tensile testing machine with a 30 kN load cell was used to apply a loading rate of 13 mm/min.

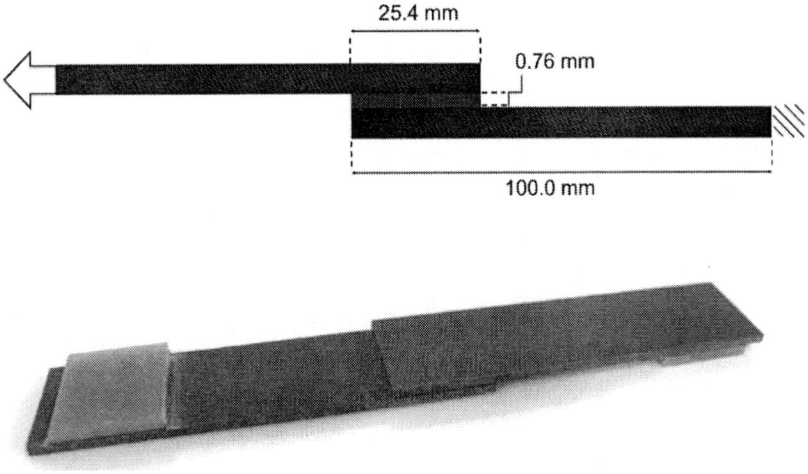

Figure 5: Sample dimensions for single lap-shear testing (top) and a finished sample (bottom)

The block shear test specimens are cut to dimensions as specified by ASTM D4501 [12] and shown in Figure 6. An Instron 5967 tensile testing machine is used to characterize the shear strength of the samples. Block shear testing requires a special fixture to clamp the sample in place, reducing the peeling effects which occur during single lap-shear tests. The tool is manufactured out of 6 mm thick steel plates and is attached with pins on both ends to form a ridged connection. The block shear sample is placed in the right half of the fixture, which is referred to as the adapter plate. A toggle clamp is then attached to the sample to prevent slipping during testing. Tension is applied to one side of the test fixture (shearing tool), while the other side remains static. The sample is sheared apart until failure occurs. A testing speed of 1.26 mm/min is used.

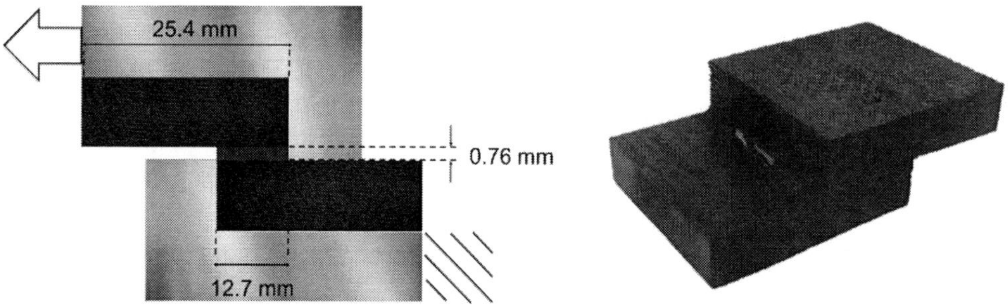

Figure 6: Sample dimensions for block shear testing (Left) and a finished sample (Right)

Results

Single Lap-Shear

The results of the single lap-shear tests are shown in Figure 7 with the sample designation (Adherend-Adhesive). For these tests, only the NB-EP 4030 material from Mitsubishi was used for the adherend. The 3M DP190 adhesive had the highest average single lap-shear strength, but also the highest standard deviation. This indicates issues with consistent application of adhesives in secondary bonding applications. EPX 82, the 3D printed adhesive, demonstrated a failure strength within one standard deviation of the commercial adhesive comparison. Both adhesive materials outperform the standard single lap-shear joint with the NB-EP 4030 bonded directly to itself. This is due to the nature of the single lap-shear testing and the peeling stresses inherent in the test geometry. While combined loading conditions with normal stresses are useful and common in real applications, these results do not clearly indicate shear strength. Block shear testing was subsequently conducted to evaluate a pure-shear loading situation.

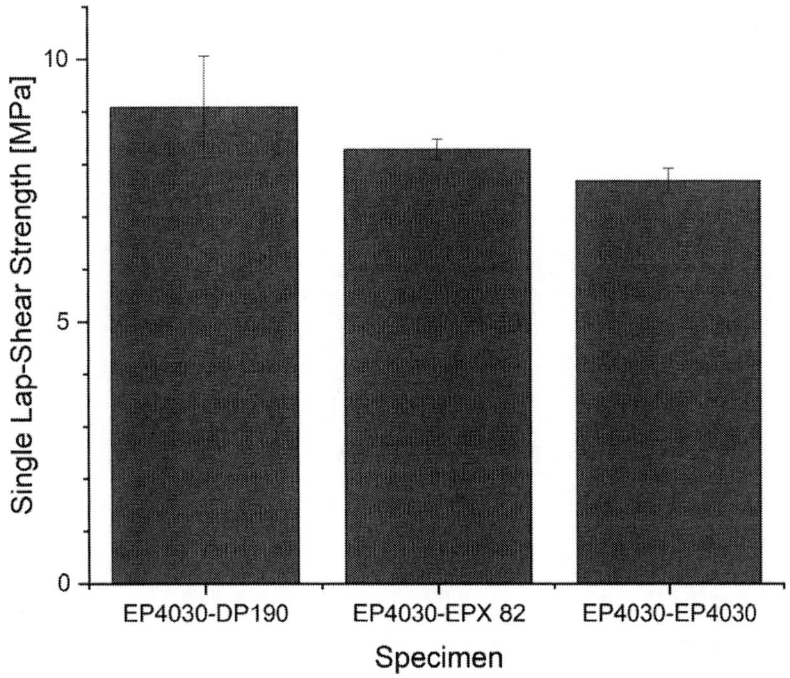

Figure 7: Results of single lap-shear tests for various materials [Adherend-Adhesive]

Block Shear

Results of the block shear testing are shown in Figure 8 with the sample designation [Adherend-Adhesive]. The same adhesive materials were used for comparison in this test, however, the Hexply F263 material was included for the composite adherend as a comparison with the NB-EP4030 due to the significantly different curing cycles and maximum cure temperatures of the materials. The block shear tests show that the EPX 82 performs well in comparison with the commercial DP190 adhesive and is within one standard deviation of the bonding strength of the adherends themselves. Results for all combinations are at least a factor of two higher than the measured single lap-shear strengths, indicating that a pure shear loading condition results in a significantly stronger bond compared to the combined loading that occurs in single lap-shear.

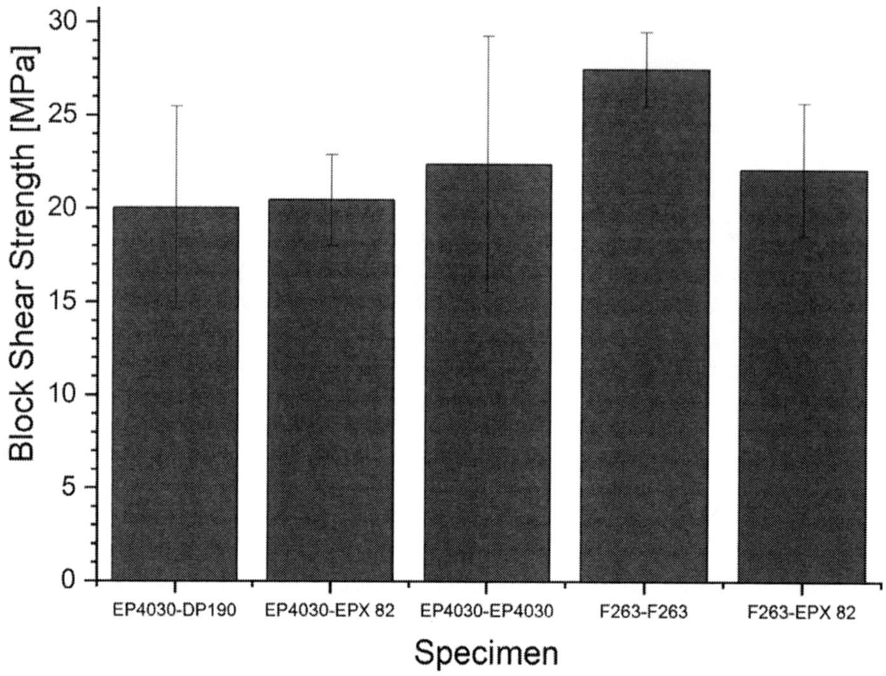

Figure 8: Results of block shear tests for various materials [Adherend-Adhesive]

Application

A first aerospace application has been developed using the characterized co-curing process in collaboration with DarkAero from Madison, Wisconsin. The DarkAero 1 is the first two-passenger experimental aircraft with a full carbon fiber fuselage. Elevator bellcranks for the control system were printed using the DLS process and co-cured with pre-impregnated F263 carbon fiber sheets on both sides. The final components, the assembled control system, and the position in the airplane is shown in Figure 9. These two components were previously milled from a solid billet of material with a high scrap rate, tool wear, and material waste. The DLS printing process allows four partially UV-cured parts to be printed simultaneously under 30 minutes using a small M1 machine. Dozens of these hybrid composite structures can be manufactured within hours and the efficiency can be increased further using a larger build volume or different orientation approaches.

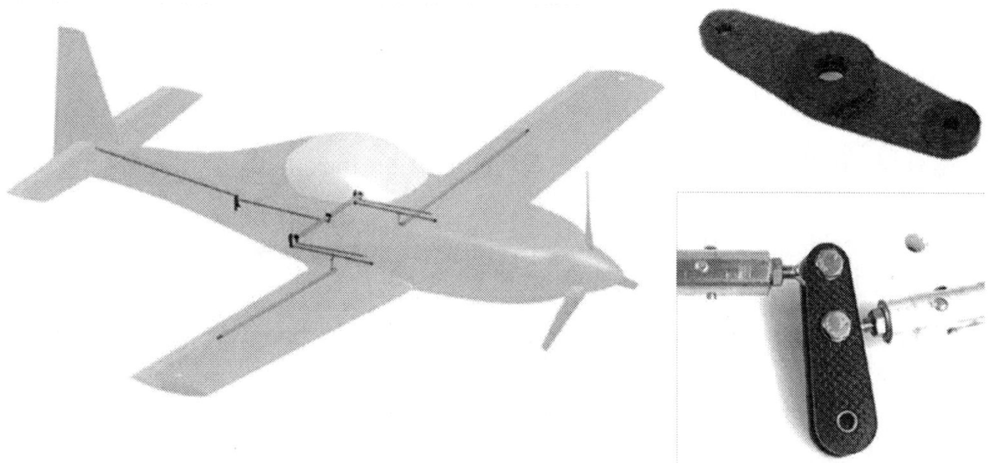

Figure 9: Application of the hybrid-composite manufacturing process for a vertical tail rudder bellcrank used on the DarkAero 1

Conclusions

The aim of this study is to develop new methods for manufacturing composite structures without the use of mechanical fasteners or industrial adhesives in order to minimize the stress concentration induced by drilling or improve a difficult to control process in current bonding technology. The approach described uses an epoxy-based 3D printing dual cure resin which is UV-cured using Carbon's DLS process in the first step to form a partially cured part in the printing process. Being in a semi-ridged gelled state, the part has to be thermally cured in a second step to develop the full mechanical properties of the material. By using pre-impregnated fiber reinforced composite sheets on epoxy base, the UV-cured part and the composite material can be joined and co-cured under pressure with the formation of chemical bonds in an optimized thermal curing cycle. This results in a lightweight hybrid composite structure with improved material characteristics and design freedom due to the 3D printing process. Mechanical tests showed bonding strength comparable to commercial structural adhesives for certain load conditions. This combination of 3D printed joining elements and FRP enables a new family of hybrid, high-performance composite structures.

Acknowledgements

This material is based upon work supported by the National Science Foundation Graduate Research Fellowship Program under Grant No. DGE-1747503 and the Draper Technology Innovation Fund (TIF) through the University of Wisconsin-Madison and Wisconsin Alumni Research Foundation. Any opinions, findings, and conclusions or recommendations expressed in this material are those of the authors and do not necessarily reflect the views of these organizations. The authors would like to thank Daniel Miller for the material and design advice. Support for the DLS printing at the University of North Carolina was provided through a sponsored research program with Carbon, Inc. of Redwood City, CA.

References

[1] Mathes, V. "The composites industry: plenty of opportunities in heterogeneous market." Reinforced Plastics 62 (1), pp 44-51, 2018.

[2] Jones, R. M. "Design of Composite Structures: An Introduction to the Many Aspects of Composite Structures Design." Bull Ridge Publishing, 2015.

[3] Vinson, J.R. "Mechanical fastening of polymer composites" Polymer Engineering and Science, 29(19), pp. 1332–1339, 1989.

[4] Schürmann, H. "Konstruieren mit Faser-Kunststoff-Verbunden." 2nd Ed. Verlag Berlin Heidelberg, 2007.

[5] "Estimated Impact of Corrosion on Cost and Availability of DoD Weapon Systems FY18 Update." LMI, 2018.

[6] Austermann, J., Redmann, A., Dahmen, V., Quintanilla, A., Mecham, S., and Osswald, T.A. "Fiber-Reinforced Composite Sandwich Structures by Co-Curing with Additive Manufactured Epoxy Lattices." J. Compos. Sci., 3, 53, 2019.

[7] Redmann, A., Dahmen, V., Austermann, J., Quintanilla, A., Mecham, S., and Osswald, T.A. "Fabrication of hybrid composite T-joints by co-curing with 3D printed dual cure epoxy." Composites: Part B, 183, 107728, 2019.

[8] "EPX 82." Technical Data Sheet 107172, Rev A, Carbon, 2018.

[9] "F263." Technical Data Sheet, Hexcel. 2013.

[10] "4030: 250-300°F (121-149°C) Cure Epoxy Resin System". Technical datasheet. Mitsubishi Chemical, 2017.

[11] ASTM D5868-01: Standard Test Method for Lap Shear Adhesion for Fiber Reinforced Plastic (FRP) Bonding. ASTM International, 2014.

[12] ASTM D4501-01: Standard Test Method for Shear Strength of Adhesive Bonds Between Rigid Substrates by the Block-Shear Method. ASTM International, 2014.

MONOLITHIC FRAME FABRICATION

Warren R. Ronk and Blaise F. Bergmann
Spirit AeroSystems, Inc.
Wichita, Kansas USA

ABSTRACT

Spirit AeroSystems, Inc., (Spirit) is the largest Tier-1 supplier of composite fuselage structures. For commercial programs, Spirit fabricates and assembles co-cured composite skin stringer body panels with composite body frames. Body frames are structural members providing reinforcement to aircraft fuselages. Frames are fabricated from polymer matrix composites are typically mechanically fastened to intermediary components known as shear ties, which are then fastened to the fuselage. Monolithic frames reduce part and fastener counts, weight, and installation time by integrating the shear tie into the frame geometry such that the frame can be directly fastened to the fuselage. However, the use of monolithic composite frames is currently limited due to the need for technology advancement and development of a robust manufacturing process. This paper will present a manufacturing approach that reduces the cost and increases the quality of monolithic composite frames. This approach consists of 1) preparation of a flat prepreg charge using automated fiber placement (AFP), 2) using a hot drape forming (HDF) technique to form the flat prepreg charge into the frame geometry, and 3) autoclave curing of the frame. Experimental application of this technology to multiple frame geometries, including Z frames and C channel frames with a variety of bend radii, will be presented. Results of inspection will also be discussed, including non-destructive inspection using pulse echo ultrasonic (PE) to determine part quality.

Keywords: AFP, HDF, Wrinkles
Corresponding author: Warren Ronk

1. INTRODUCTION

Frames are structural components that reinforce the aircraft fuselage. Current composite frames are a two part design with the frame attached to the fuselage through a shear tie. Monolithic frames, also known as one part frames, incorporate the shear tie into the frame geometry and directly attach the frame to the fuselage (Figure 1).

Figure 1. Monolithic Frames.

Monolithic frames reduce part count, weight, and installation time. However, the use of monolithic composite fames is currently limited due to the need for technology advancement and development of a robust manufacturing process. Current composite fabrication is accomplished by several methods including labor intensive hand lay-up. This process can be automated by the use of automated fiber placement (Figure 2) which lays down slit tape in controlled fiber geometries. AFP utilizes slit tows up to 13mm slit width.

Figure 2. AFP of a Flat Charge.

The flat charge generated by the AFP laying down fibers can be performed in a hot drape forming process. HDF is the application of heat and vacuum or pressure to form a three dimensional preform [1, 2]. In this study, two types of HDF were used, vacuum and double bladder techniques. The two processes were used for two types of frame geometry. They are Z frames and C channel frames. These frames are named after the profile they make (Figure 3 and Figure 4).

Figure 3. C channel frame.

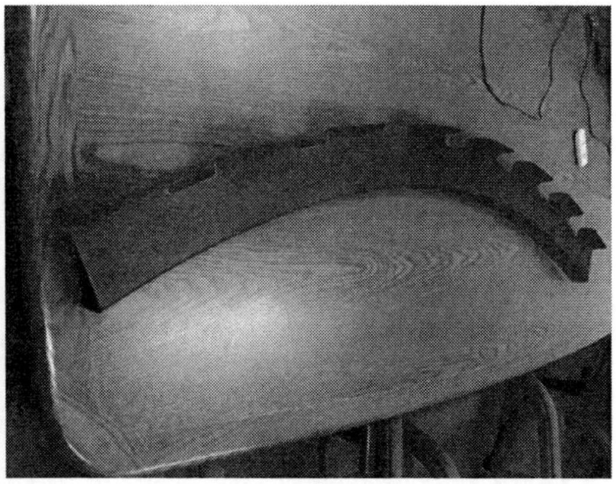

Figure 4. Z frame.

The double bladder HDF is used to for Z-frames and vacuum forming HDF is used for forming C channels.

2. EXPERIMENTATION

An experimental study of the fabrication of frames was performed. Z frames and C channel frames with radius of curvature varying from 0.36 m to 2.5 m were fabricated (Figure 4). Fabrication was all done by maximizing automation combining AFP and HDF.

2.1. Automated Fiber Placement

Carbon/epoxy slit tape was laid up using an Electroimpact gantry machine. Tapes with widths from 3mm wide to tapes 13mm wide were used, varying with the radius of curvature of the frame being laid up.

2.2. Hot Drape Forming

2.2.1 Vacuum Forming

A vacuum table with a mosite bladder was placed under a table of IR heaters and the temperature raised and vacuum was applied. The apparatus used is shown in Figure 5.

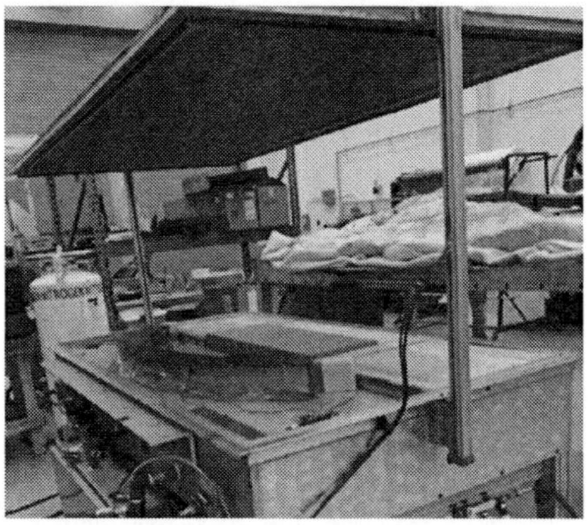

Figure 5. Vacuum Forming HDF.

2.3. Double Bladder Forming

7A custom forming device was built which utilized two air bladders. Figure 6 shows the external image of the former. Figure 7 shows the loading of the device with an AFP charge. The device's two mosite air bladders work in opposition on each side of the forming die. Mosite is a durable silicone rubber with good high performance temperature capabilities. The result is the Z shape frame Figure 3. The forming was accomplished with moderate heat in the chamber and pressure applied in each bladder.

Figure 6. Full external view of double bladder machine

Figure 7. Loading flat charge into the double blader apparatus.

2.4. PE Ultrasound

The ultrasonic images were created using a Maus a 3.5 MHz 3 inch phased array pulse echo probe. It was used to manually scan the frames. Afterwards, on a separate machine, a manual a-scan was performed using a 5 MHz 0.25in probe in pulse echo to verify indications found in the original scan results.

2.5. Dimensional Inspection

The equipment used to scan the parts was a FARO Quantum-s PCMM Arm with 3mm probe, and a blue-line laser scanner. The data was collected using Polyworks software.

3. RESULTS

3.1 Laminate Quality

Internal laminate quality was assessed. Frame taper was sanded to expose and measure fiber angle of internal plies, Figure 8. Ply Orientation demonstrates that all plies fall within +/-5°. .This is a measure of success for fiber lay-up and cure.

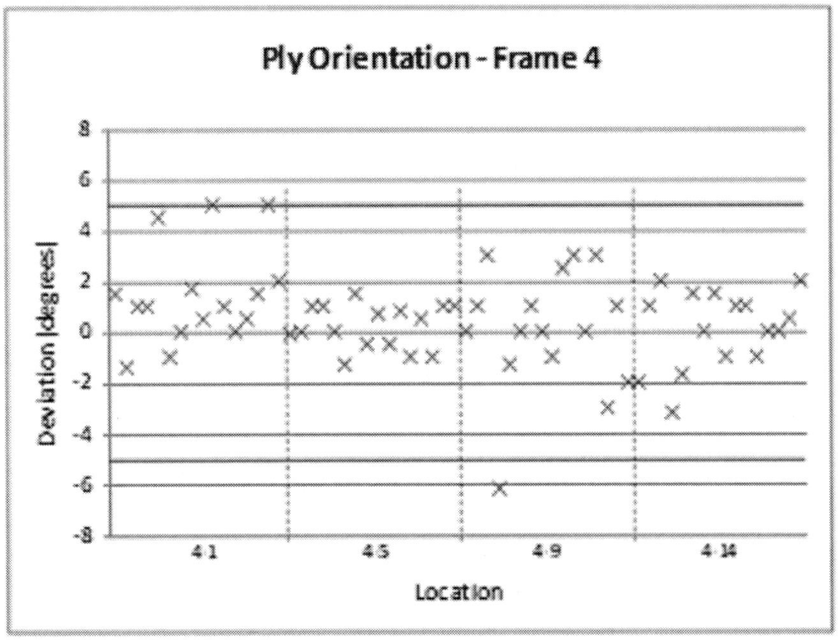

Figure 8. Internal Ply Orientation.

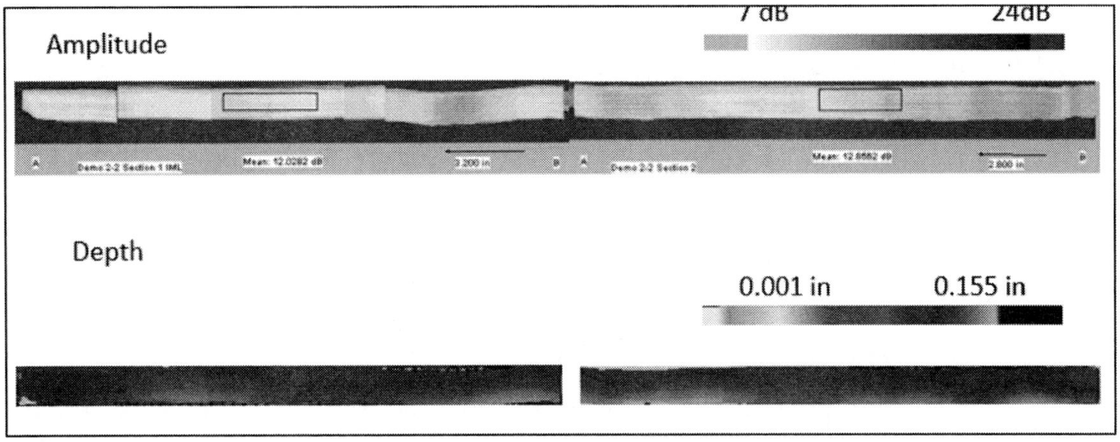

Figure 9. PE Ultrasound Image, No Indications.

3.2. NDI (PE Ultrasound)

Twenty two parts were inspected by PE Ultrasound. Figure 9 is an example of a clean scan with no indications greater than 6dB/12dB. This indicates that the process can yield good parts with little or no voids or porosity.

3.3. Hot Drape Forming Results

The results of the HDF process were judged by the number of wrinkles. Surface wrinkles were not visible in the frames fabricated according to the HDF process listed in the experimental section. Thus in the process of forming the 3D preform no wrinkles were created.

4. CONCLUSIONS

The results of this work indicate that AFP is capable of automating the lay-up process while maintaining the required fiber geometries. It is also shown that HDF is a robust process which can be used to perform complex shapes, such as Z's, which curve around various radii of curvature, or C channels. The two techniques AFP and HDF were combined here to make an automated process of laying up three dimensional preforms with controlled fiber geometries. Monolithic frames were produced using these automated methods. The resulting one part frames reduced part count and reduced labor over two part frames.

5. REFERENCES

1. H Alshahrani and M Hojjati, Optimum Processing Parameters for Hot Drape Forming of Out-of-Autoclave Prepreg Over Complex Shape Using a Double Diaphragm Technique. 20th International Conference on Composite Materials. Copenhagen, 2015.
2. Sorrentino and C Bellini, Potentiality of Hot Drape Forming to Produce Complex Shape Parts in Composite Material, The International Journal of Advanced Manufacturing Technology, 85, 945-954 (2016).

ADDITIVE MANUFACTURING OF A SCINTILLATOR FOR RADIATION DETECTION

Peter J. Joyce, Brad Baker, Marshall Millett
U.S. Naval Academy
Annapolis, MD USA

ABSTRACT

The objective of this work was to create an unobtrusive, discreet radiation sensor. In this paper we describe the additive manufacturing process used to fabricate an organic scintillation detector for use in detecting ionizing radiation. An off the shelf stereolithography printer was used with standard transparent resin which was doped with varying scintillant powders in varying amounts. The concentration of the dopant, the method of mixing, and the overall printing process was investigated to produce an additively manufactured scintillator which was then mated with a standard photomultiplier tube to detect and measure ionizing radiation. The overall detector was tested with gamma and neutron sources over exposure periods. The detector was able to clearly distinguish the sources from background with an expected continuum energy response relative to the strength and energy of the source. Future work will include customization of the geometry, coupling to other light sensors, and expansion to other printing methods, and evaluation of the ability to further discriminate non-ionizing radiation sources to create an effective and discreet detector.

Keywords: organic scintillator, plastic scintillator, additive manufacturing, vat polymerization, stereolithography, radiation detection

Corresponding author: Peter J. Joyce (pjoyce@usna.edu)

1. INTRODUCTION

Our goal was to use additive manufacturing (AM) techniques to develop a scintillation detector consistent in terms of cost and effectiveness with commercially available options while adding inherent benefits of AM. The primary benefit of using additive manufacturing for this application is the flexibility and speed with which it can be used to fabricate a geometry for the scintillator that is potentially covert or incorporated into another object specific to a customer's needs. One common means of radiation detection involves the sensing of fluorescence from a substance that has been excited by interactions with ionizing radiation [1]. Such sensors are called scintillation detectors and may be made with inorganic or organic materials. Inorganic scintillants such as sodium iodide are effective at detecting ionizing radiation over a wide range of energies due to the higher density of these materials [1], and there has been some research on the use of additive manufacturing of inorganic scintillants [2] . Organic scintillation detectors made of aromatic polymers have lower intrinsic efficiency due to lower density [3]. The incorporation of special scintillating materials such as anthracene, POPOP, and P-terphenyl enhances the response of the detector [1, 4]. The production of organic scintillating radiation sensors by additive manufacturing has also been shown. [5, 6, 7, 8].

Nuclear and radiological materials emit radiation, including gamma rays and neutrons (the focus of this research) as well as other particles such as betas and alphas. As nuclear terrorism has become a larger threat to U.S national security, the ability to detect nuclear and radiological materials that are out of regulatory control has been a concern to the Department of Homeland Security. Terrorist groups may weaponize nuclear or radiological material into improvised nuclear devices or radiological dispersal devices that may cause great destruction. Radiological materials may be obtained from common devices that use radiological isotopes such as blood irradiators, therapy tools in hospitals, well loggers, or thermoelectric generators [9]. Nuclear and radiological materials may be hidden and shielded, so it is vital that the use of radiation detectors expands and improves, including use in covert applications.

The proposed detector in this research responds to ionizing radiation incident on an AM component fabricated with a scintillating material additive. As radiation interacts with the material, scintillation produces visible wavelength photons, a light sensor is used to detect the energy deposited. Light sensors include photomultiplier tubes that convert photons into an amplified electrical signal corresponding to the energy deposited. The electrical signal is provided to a pulse counter or multichannel analyzer to provide a useful output. A typical commercial detection system is shown in Fig. 1.

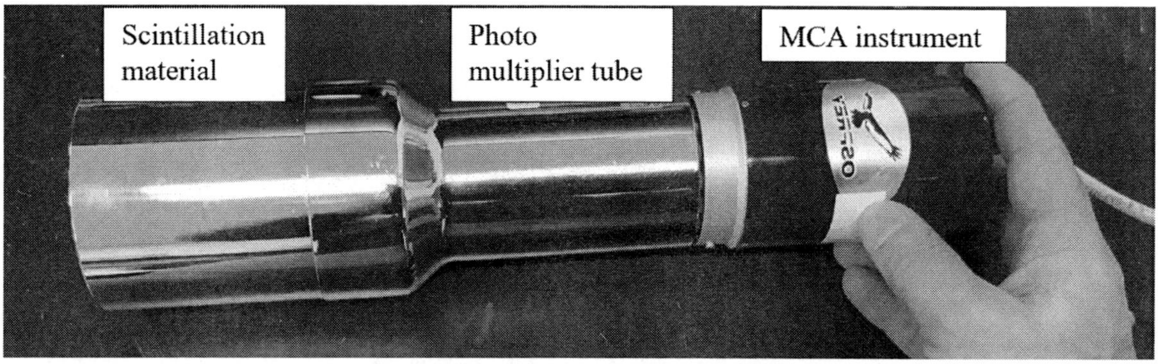

Figure 1. Commercial scintillating radiation detector with photomultiplier tube and multi-channel analyzer (MCA) instrument.

In developing a scintillator that is intended for a specific application, additive manufacturing techniques are beneficial in that they allow the device to be built on demand and to any size or configuration. Intended customers include federal, state, and local law enforcement agencies involved in protective measures related to homeland security. Because the detection of illicit nuclear or radiological materials likely represents an added mission to the customer, it may be desirable to incorporate radiation sensors in equipment that is already in use for other purposes. Additive manufacturing of radiation sensors allows objects to be created that serve a dual purpose and provide the ability to conduct radiation surveys covertly. Examples of possible dual purpose devices that incorporate an additively manufactured radiation detection aspect include plastic linings, coverings, or inserts that could be incorporated into personal protective equipment such as helmets, vests, or aprons. Additionally, it is possible to incorporate additively manufactured sensors into extremity monitoring systems such as ring detectors or integration into other personal devices such as gloves, booties, or other worn devices. Additively manufactured sensors can be also be used to mimic common devices such as lapel pins, nametags, belt buckles, a coffee cup or other common devices. Use of additive manufacturing in radiation sensors is not limited to small personal devices and can be applied to larger applications such as large-scale screening devices including portal monitoring devices at airports, ports of entry, or other large screening locations. This same technology can also be applied to portable detectors including handheld screening devices for personnel or equipment or vehicle mounted detectors on cars, trucks, or other vehicles, and also unmanned vehicles such as unmanned aerial vehicles, drones, or robotic devices. The low density, adequate mechanical properties, and ability to directly integrate with structural materials make organic scintillators by AM well suited for applications where weight savings is relevant.

The overarching goal of this project was to design and fabricate a scintillation detector that is unobtrusive or covert.

2. MATERIALS AND PROCESSING

2.1 Doping the resin

The mixing process used in this research evolved from small scale tests to evaluate the miscibility and solubility of the dopant in the resin to mixing on a larger scale that was better suited to the designed function of the 3D printers. To begin with small amounts of scintillating dopant were mixed into small quantities of the vendor supplied polymer resin as shown in Fig. 2. The mixture was heated and stirred to promote thorough mixing but the quantities were small. Once the practical limits of solubility were investigated the mixing process was scaled up to mix batches of resin equivalent to an entire cartridge of resin for 3D printing (Fig. 3) so that the printer could operate as intended with no modifications due to using the doped resin. The exact details of the mixing process are available in US patent application 16-567-80216/567,802 [7].

(a) (b)

Figure 2. Mixing anthracene into the proprietary Formlabs clear resin. A small amount is measured into a scintillation vial (a) and then the solution is heated (b) and stirred until the anthracene is suspended in the resin.

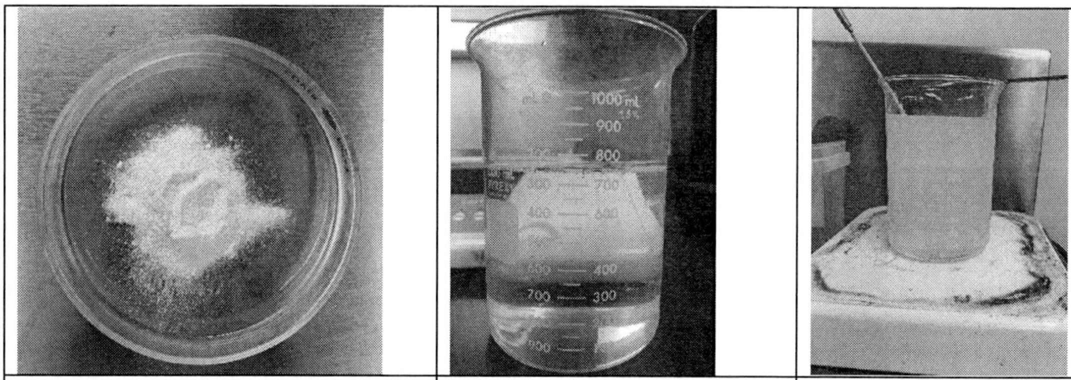

(a) (b) (c)

Figure. 3. (a) Anthracene powder before mixing, (b) Formlabs clear resin before mixing, (c) 1 L of resin and anthracene being heated and mixed.

2.2 Additive manufacturing

Stereolithography (SLA) a variant of the vat polymerization printing process [10], is an AM technique in which a photopolymer is cured using a UV light [11,12]. The Form 2 3D printer from Formlabs is an inverted stereolithography printer [13] that prints from the bottom up so the print bed is lowered into the resin and then cured with a focused laser beam. An illustration of the inverted SLA process can be seen below in Fig. 4.

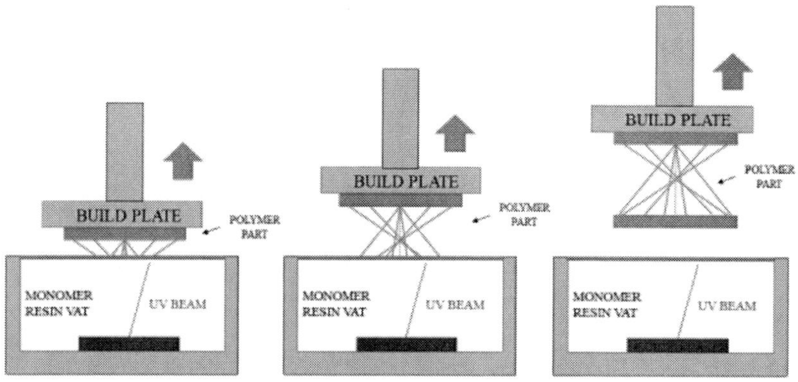

Figure 4.: Inverted SLA schematic [14]

(a) 0.3 wt. % anthracene	(b) 0.3 wt. % anthracene + 0.1 wt. % boron powder	(c) 0.3 wt. % anthracene + 0.1 wt. % boron (from boric acid)

Figure 5. AM scintillation sensors as produced (a & b) and following support trimming and surface preparation with reflective paint (c).

3. TESTING

Initial proof of principle testing showed that doping (or contaminating) Formlabs clear resin with a small amount of anthracene powder did not affect the cure of the resin during the SLA printing process and the 3D printed parts did scintillate when exposed to gamma radiation. The goal of this project was to explore the design space both in terms of materials and geometry. First, samples were created to replicate the test results from the initial testing and then to verify that anthracene was the best scintillant additive. Samples were also created to investigate the effects of scintillant concentration. Sample configurations were altered to explore the effects of geometry on system performance. Finally, samples were created to explore sensitivity to neutrons.

To test the detector, radiation sources were placed 33 mm away from the face of the detector and gamma radiation interactions incident to the detector were counted and sorted into different energy bins using a multichannel analyzer (MCA), providing a spectrum of energy deposition events over the course of the test.

The design test plan is marked by several test parameters varied to determine the effect on the number of distinguishable ionizing radiation events for each configuration. The first test parameter is the type of resin that is used. The second parameter tested was the scintillating dopant. Resin type and dopant are considered to be the most important design factors because if the resin is not transparent then the visible light released from the interaction of the radiation with the scintillating material may not make it to the light sensor. Initial testing of Formlabs resins ruled out every resin with the exception of standard clear resin and engineering high temperature resin because of their optical transparency. A review of the literature provided a list of candidate scintillant additives including anthracene, POPOP {1,4-bis[2-(5-phenyloxazolyl)]-benzene}, and P-Terphenyl [6] . Boron dopant was also added to the test matrix to investigate neutron sensitivity.]. Boron is a strong neutron absorber and the energetic particles produced in the absorption reaction provide the same detectable energy deposition as with gamma interactions, so boron doping is also combined with one of the other three scintillants. In this work we investigate both the type of scintillant as well as the scintillant concentration. The test matrix for some of the variables is shown in Table 1 below.

Table 1: Test matrix to compare scintillants, scintillant concentration, and thicknesses

Scintillant	Concentration (wt. %)	Thickness (in.)
Gamma detection		
Anthracene	0.1	1.0
Anthracene	0.3	0.5
Anthracene	0.3	1.0
Anthracene	0.3	1.5
Anthracene	1	1.0
POPOP	0.3	1.0
P-terphenyl	0.3	1.0
Gamma + Neutron detection		
Anthracene + Boron	0.3 Anthracene + 0.1 B	1.0
Anthracene + Boric Acid	0.3 A + 0.1 H_3BO_3	1.0
(all samples 2" diameter)		

The final test parameter is the geometry of the test puck. By varying the thickness of the puck, the transmission of photons through the medium will be affected. With a perfectly transparent material, the performance will continue to increase with thickness, but with opacity, some optimum thickness is expected. The test plan was to vary thickness while holding a constant diameter of 2 inches, thus keeping incident surface area constant.

3.1 Blacklight testing

A simple qualitative test of the samples was conducted by placing the samples under a black light and observing the response. The ultraviolet radiation emitted by a black light is adequately energetic to cause the same excitation and fluorescence within a scintillation detector as seen with ionizing radiation. The photons emitting from the sample, visible as a blue glow in Fig. 6, are similar to those that would be seen upon exposure to ionizing radiation, though at a much lower level (not visible to the human eye).

Figure 6. 0.3 wt. % anthracene sample under black light exhibiting fluorescence.

3.2 Radiation testing

All samples were prepared for testing by painting the entire surface of the cylindrical sample (less one face as shown in Fig. 7) with a reflective white coating[15]. The white paint provides diffuse reflection of emitted photons, increasing the likelihood of a photon crossing the mating surface and entering the photo-multiplier tube (PMT).

Figure 7. AM sample painted with reflective white paint to optimize photon capture by the PMT.

The unpainted end of the sample or PMT mating surface was then coated in optical grease (Fig 8(a)) to promote good optical coupling between the sample and the PMT. The scintillating sample was then secured to the PMT with electrical tape and the entire assembly was wrapped in aluminum foil to shield the PMT from visible light. All tests were conducted using a Canberra

Osprey® multichannel analyzer instrument attached to the PMT. The digital output signal was then provided by the Osprey instrument to a desktop computer through an ethernet connection, and the energy dependent response was visually displayed as in Fig. 9. Tests were conducted with identical instrument settings and test configurations, for example, all tests used the same shielding configuration, a 33 mm distance from the radiation source to detector face, and all exposures were conducted for five minutes. The test setup, shown in Fig. 8(c), was shielded with lead bricks in order to minimize the effect of background radiation.

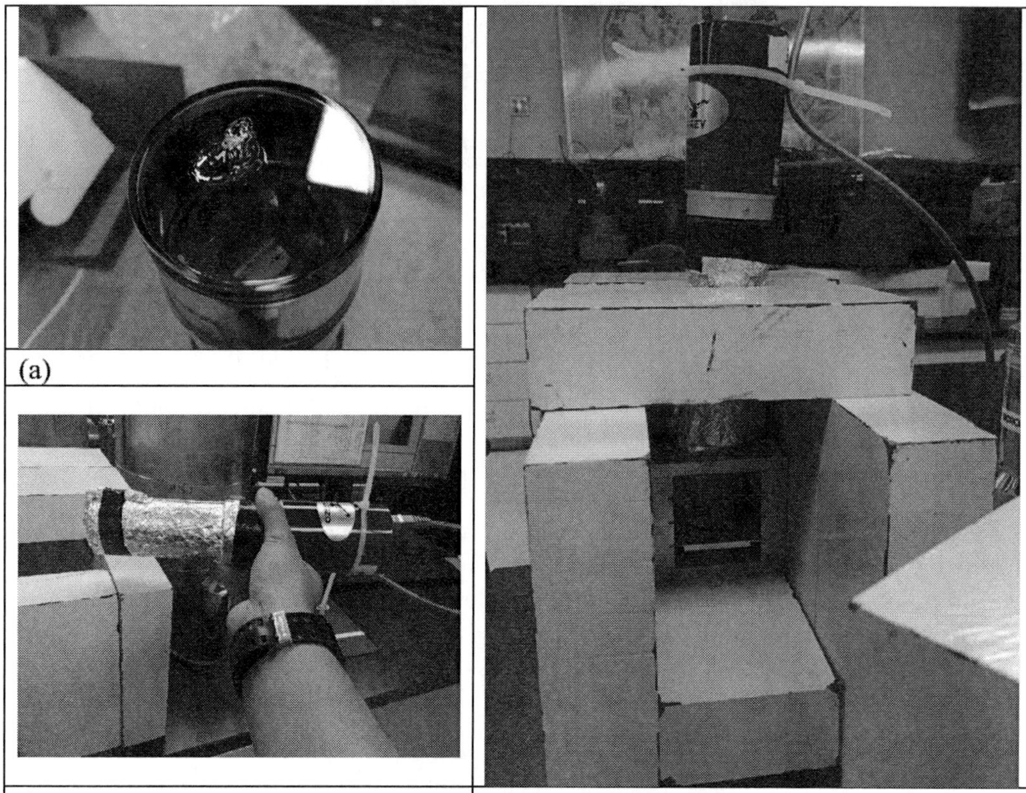

(a) (b) (c)

Figure 8. Testing radiation sensitivity of the scintillating detector. 8(a) shows the AM scintillator prepped with optical grease for optimal coupling with the PMT, 8(b) shows the scintillator taped to the PMT with aluminum foil surrounding to shield from outside light, 8(c) shows the test setup with lead bricks to shield the radiation source.

Two different gamma radiation sources were used: [1] a moderate energy source, cesium 137 (Cs-137) and [2] high energy source, cobalt 60 (Co-60). The Cs-137 source provides a single gamma ray at an energy of 662 keV and the Co-60 source provides two gamma rays at energies of 1173 and 1332 keV. (For comparison, a typical visible photon has an energy of around 2 eV.) Because the amount of energy deposited in the detector is related to the energy of the incident gamma ray, it was expected that exposure to the Co-60 source would yield a higher energy response. This is clearly seen in the Results section to follow. The testing process for each sample involved a total of three five minute runs consisting of one run with Cs-137, one with Co-60, and one background test with no sources present.

4. RESULTS

The overall results of this research are promising. Raw data clearly confirms that incident radiation is detectable above background and is repeatable. Improvements in mixing processes over the course of this research have yielded an improved response. Several candidate scintillating dopants were evaluated with anthracene exhibiting the greatest response. The intrinsic efficiency of the AM sensor compares favorably to commercially manufactured organic scintillators. Initial analysis of geometry has also been conducted, and finally preliminary results of neutron detection using boron doping is also promising.

4.1 Scintillant results

The dominant mode of gamma interaction in the sensor is by Compton scattering, where a gamma ray interacts with an electron in a scattering collision. The gamma ray imparts energy to the electron and then exits the sensor and the high energy ionized electron causes excitation and fluorescence as it slows down within the scintillating material. Fig. 9 shows the energy dependent spectrum results of final recommended concentration of 0.3 wt. % anthracene for both cobalt and cesium, and compared to background. For all test results, the response seen below the relative energy value of 50 is attributed to system noise. Therefore, results below this threshold were ignored. Fig. 9 shows several important results: (1) while there is some self-scintillation effect due to aromatics in the neat resin, the response was notably enhanced by the addition of anthracene, (2) both gamma sources can clearly be discriminated from background, (3) the response from both sources is from interaction with gammas in the Compton continuum range of energies thereby producing a continuum response rather than a photopeak response similar to all inorganic scintillators, and (4) the response from both detectors has a minimum range of sensitivity above background and inspection of these results suggests integration across a range of energies to simplify discrimination. The self-scintillating response of the neat resin to ionizing radiation may be of interest to the resin manufacturer. Finally while the Cs-137 and Co-60 sources do not have the same energy spectrum, two important features are indicated in Fig. 9 Both the Cs-137 and Co-60 sources produced a response that can be discriminated from background and the response to the higher energy Co-60 gammas yielded a continuum with a much higher energy endpoint to the Cs-137 response. This was expected, and indicates that some level of energy spectrometry may be possible, supporting the identification of radiation sources.

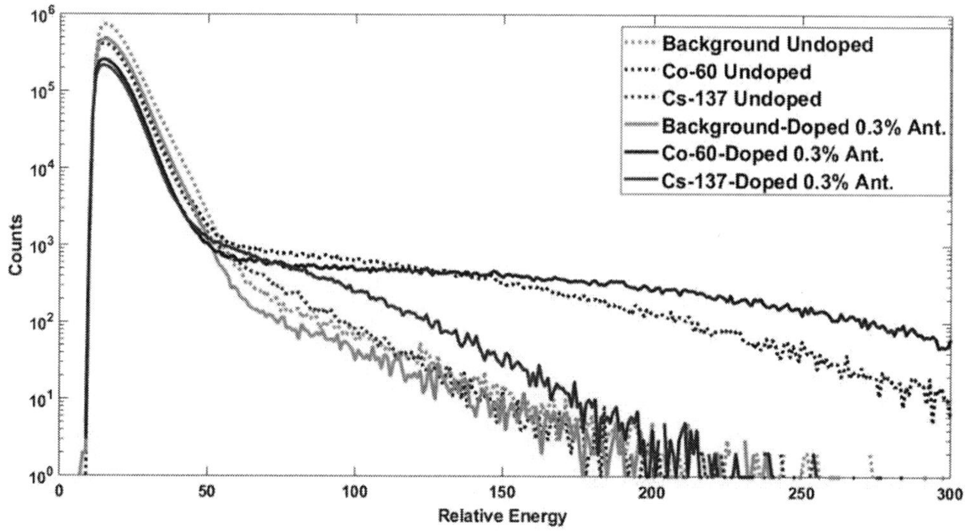

Figure 9. Spectrum response for several configurations, with the effect of incident radiation on the sensor indicated

Since the detector's response is determined by single Compton scattering (non-capture) gamma interactions with all scattering angles possible, incident gammas at a discrete energy show a broad continuum of deposition events. In order to discriminate from the background as well as to eliminate the low energy region corresponding to noise, the spectral response curves from each test, e.g. Fig. 9, were integrated from a lower cutoff energy range through the Compton continuum to give a single integrated value for counts above background. This approach also allows for the subtraction of background counts to give a clear representation of the total detector source response. Integrated results from this approach are presented in bar graph format representing counts above background. Figure 10 shows both initial results with 0.3 wt. % anthracene and results obtained from a second batch that involved an improved and scaled up mixing process better suited to the normal function of the Form 2 printer resulting in improved dissolution of the scintillant and improved sensor response.

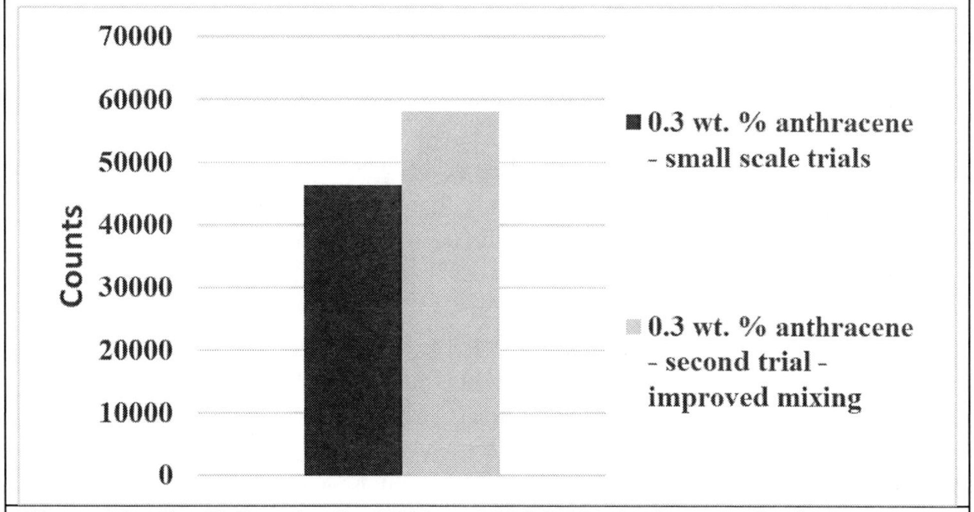

Figure 10. Co-60 Results for 2019 vs 2020 0.3% Anthracene Pucks

Using an empirical approach based on nominal values of scintillating mediums in conventionally manufactured organic scintillators [5], a target range of dopants and dopant concentrations was identified for testing as shown in Table 1. Varying concentrations of the anthracene scintillant were tested to see if they would mix with the resin, if the doped resin would print in the Form 2, and if the resulting printed part would scintillate and to what extent. Anthracene powder was added to the clear resin at 0.1, 0.3, and 1.0 wt. %. The 1.0 wt. % anthracene mixture was difficult to keep in solution and was difficult to print with. The 1.0 wt. % anthracene mixture was also cloudy before and after printing, with opacity detracting from the light transmission through the clear resin. At lower concentrations, mixing was easier and the dopant had much less effect on the transparency of the mixture.

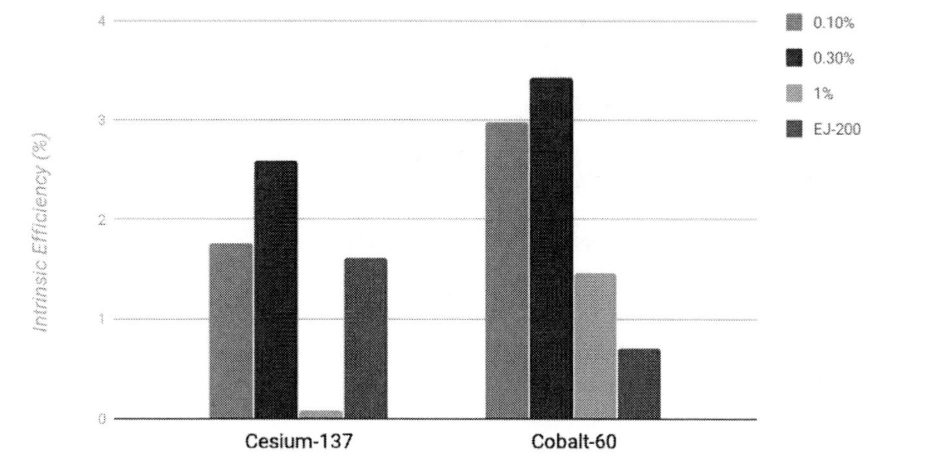

Figure 11. Intrinsic Efficiency of AM scintillators (0.10 wt. %, 0.30 wt. % and 1 wt. % anthracene) versus commercially manufactured scintillator EJ-200 for different radiation sources.

Radiation testing with both Cs-137 and Co-60 showed the maximum efficiency for the 0.3 wt. % anthracene mixture (Fig. 11), this result matched similar results in the literature for conventionally manufactured scintillators [1]. The intrinsic efficiency refers to the number of source gamma rays detected divided by the number of gamma rays incident on the scintillator [1]. The intrinsic efficiency of each configuration was determined and a comparison with a commercial plastic scintillator of similar size was conducted, showing the samples produced by additive manufacturing to have a higher efficiency than an off the shelf plastic scintillator (EJ-200) [16]. It is also obvious from Fig. 11 why the 0.3 wt. % concentration was chosen.

Fig. 12 illustrates subsequent test results using a Co-60 source with other scintillating dopants: POPOP and P-Terphenyl. Both Anthracene and POPOP dissolved at 0.3 wt. % added but P-Terphenyl was not fully soluble at 0.3 wt. % as indicated by small amounts of powder visible in solution. The bar chart shows the integrated gamma response for channels 100 through 600. Each channel corresponds to a distinct energy level (however the energy levels could not be calibrated due to the lack of photopeaks.) The results suggest what was already thought to be true, that anthracene yields the greatest response when using the same weight percent of scintillant; 0.3 wt. % anthracene performed 17.9% better than the next best scintillant, 0.3 wt. % POPOP doped.

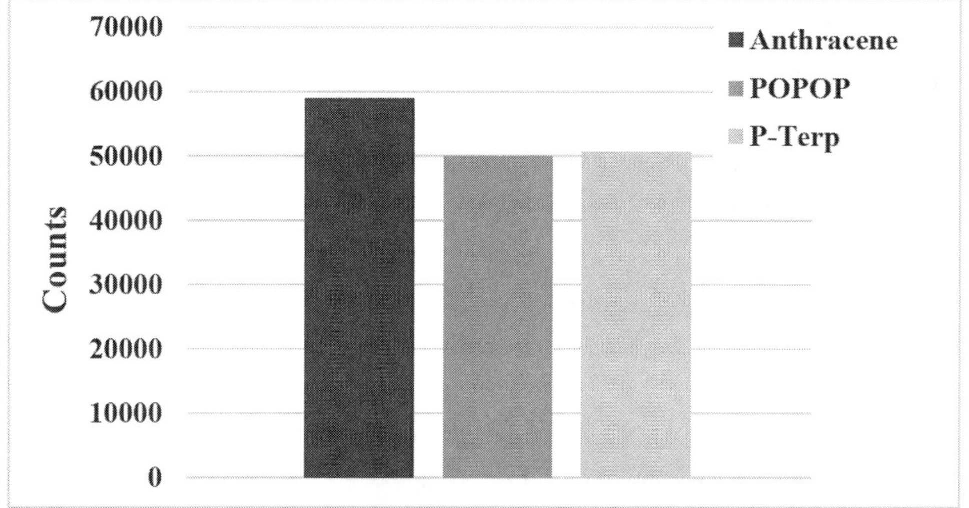

Figure 12. Comparison of different scintillants all at 0.3 wt. %

Tests were conducted as per Table 1 to evaluate the response with varying sample thickness with the mating surface area held constant as well as tests where the mating surface area was varied while the volume of the sample was varied. Results were inconclusive but suggest that the surface area of the sensor with a yet to be determined minimum thickness is the dominant geometric design parameter.

4.2 Neutron sensitivity

Figs. 13 and 14 show the initial test results for trying to implement a boron dopant into the mixing and printing process. Boron was added by dissolving boron powder at 0.1 wt. % in addition to 0.3 wt. % anthracene. While the powder was soluble and printable, the printed part came out opaque and was dark brown in color. This new boron and anthracene doped puck was tested against the anthracene doped puck from previous tests. As a note, a new coat of reflective paint was applied to any puck that was tested multiple times. This attempt at neutron sensitivity was unsuccessful for two reasons: the opacity of the sensor and the use of an AmBe neutron source. The opacity resulting from the boron significantly reduces the light transmission and the AmBe source is not a pure neutron source and produces a high-energy gamma ray that is difficult to shield and interferes with the detector response. Results shown in Figs. 13 and 14 may indicate some neutron sensitivity, but are ambiguous for the reasons mentioned. Further analysis of the pulse shape might distinguish neutron interactions from gamma interactions.

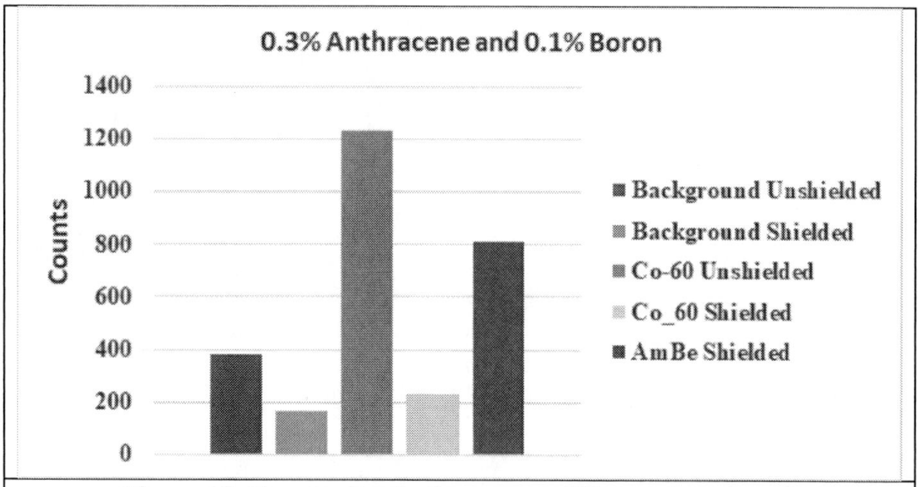

Figure 13. Results for Anthracene and Boron doped puck

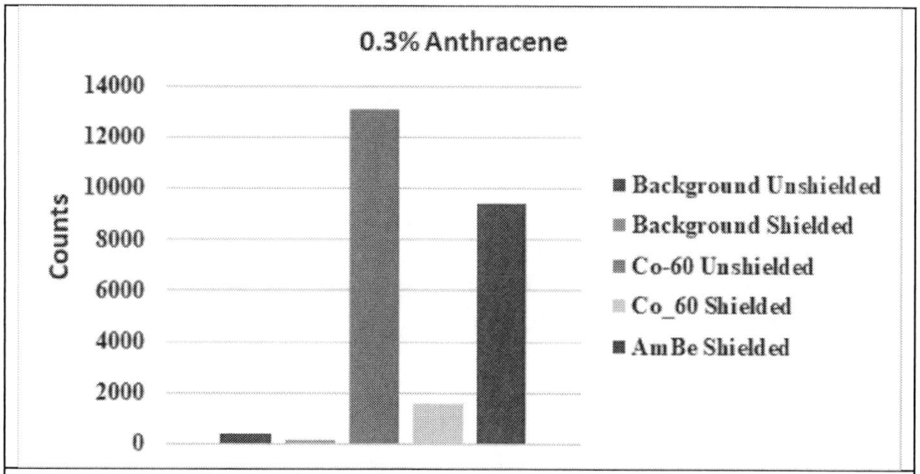

Figure 14. Results for Anthracene doped puck

5. CONCLUSIONS

- Scintillant powders can be added to commercial SLA resins with little modification to the normal function of the SLA printer to produce an AM sensor that will scintillate under ionizing radiation.
- AM organic scintillators can discriminate radiation sources from background radiation.

- As hypothesized, of the scintillant dopants tested, anthracene provides the highest response and there appears to be optimal concentration in the vicinity of 0.3 wt. %.
- AM can be used to print scintillation sensors in complex shapes such as the coffee cup shown in Fig. 15, opening the way to small, discreet, unobtrusive, even covert radiation detectors.
- Boron doping was demonstrated and although results thus far are inconclusive they suggest feasibility of neutron detection.

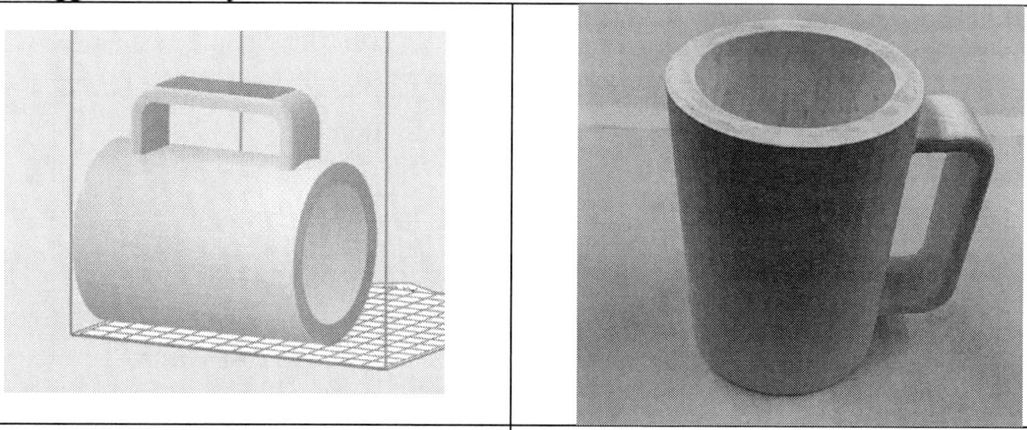

(a) (b)

Fig. 15. Computer rendering of a coffee cup (a) and a 3D printed coffee cup (b)

6. FUTURE RESEARCH

This research is ongoing and there are currently more tests planned to achieve a series of prototype devices by the Spring of 2021. One of the primary goals moving forward is to continue varying geometries in an attempt to explore light transmission through complex shapes. Because one of the main objectives is to produce a device that is covert by design, we plan to test smaller volume 3D printed samples together with smaller PMTs. PMTs do not have the design flexibility of the additively manufactured sensors, but are available as "micro" devices, though these have small light sensing windows and require a high voltage power supply. We plan on looking into the use of SiPM arrays which will give us the ability to test micro-scale puck volumes and geometries. Lastly, we plan to print and to test real world objects that can be converted to fit on either a micro PMT or a SiPM-based based light sensor. Before testing was interrupted, the coffee cup shown in Figure 15 was printed with 0.3 wt. % anthracene but not tested.

Future tests of neutron sensitivity include incorporating other organic compounds of boron compounds to reduce opacity. The use of boric acid powder will be explored, but there are concerns about the solubility and the effect of the oxygen content on light transmission. Additional testing with a different neutron source is also planned.

Our evaluation of improved mixing methods appropriate to the volume of our 3D printers is ongoing. We have yet to establish rigorous solubility limits for the scintillant dopants in the Formlabs resin. From our initial experiments with P-Terphenyl we know that the miscibility of different dopants varies. We hope to migrate this program of work to the advanced Formlabs Form 3 printers in order to print with greater optical clarity. The Form 3 takes advantage of a

Low Force Stereolithography printing technology together with a flexible resin tank and linear illumination; we hope this yields better resolution and improved transparency in complex geometries.

7. ACKNOWLEDGMENTS

This work was conducted with help from Midshipmen: James Agan, Sydney Barber, Zeke Durham, James Reneau, and Ed Valdez during the 2019/2020 academic year and Midshipmen Jared Antonio, Ben McHale and Tyler Kiyota during the 2018/2019 academic year. Funding was provided by the Defense Threat Reduction Agency's Nuclear Science and Engineering Research Center (DTRA/NSERC). Special thanks also to USNA Technical Support Division staff for all their help in the USNA MakerSpace.

8. REFERENCES

1. G.F. Knoll, *Detection and Measurement.* 4th ed. Wiley, 2010.[
2. G.A. Dosovitskiy, et al., 'First 3D-printed complex inorganic polycrystalline scintillator'. *CrystEngComm*, 19, pp-4260-4264, 2017.
3. F.D. Brooks, 'Development of Organic Scintillators'. *Nucl.Instrum.Meth.* 162 pp477-505, 1979.
4. N. Tsoulfanidis and S. Landsberger, Measurement and Detection of Radiation, 4th Ed., CRC Press, 2015.
5. Y. Mishnayot et al., '3D Printing of Scintillating Materials'. *arXiv*: 1406.4817, 2014.
6. A. Perrin, 3D Printing Scintillating Detectors for Field Emission Detection in Niobium SRF Cavities. Thesis, 2015.
7. B. W. Baker et al, *Additively Manufactured Plastic Scintillation Detector*, Non Provisional Patent Application No. 16/567,802, 2019.
8. B.W. Baker et al., Design and Evaluation of an Additively Manufactured Plastic Scintillation Detector, presented at TMS 2020, San Diego, CA, February 23-26, 2020.
9. R. L. Murray and K. E. Holbert, Nuclear Energy: An Introduction to the Concepts, Systems, and Applications of Nuclear Processes, Eighth Edition, Elsevier, 2020.
10. ASTM Standard 52900:2015, '*Additive manufacturing – General principles – Terminology*'. ASTM International, West Conshohocken, PA, 2015.
11. C. W. Hull, Apparatus for production of three-dimensional objects by stereolithography, U.S. Patent No. 4575330A, 1986.
12. Formlabs, *The Ultimate Guide to Stereolithography (SLA) 3D Printing*..Formlabs, Inc. Accessed 10 May 10, 2020. https://formlabs.com/blog/ultimate-guide-to-stereolithography-sla-3d-printing/
13. Formlabs, *Form 2 Desktop Stereolithography 3D Printer Installation and Usage Manual* Formlabs Inc. Accessed May 10, 2020. https://media.formlabs.com/m/7253926716b40054/original/-ENUS-Form-2-Manual.pdf

14. E. Z. Delannoy et al., 'Additive Manufacturing of Optical Components Using Commercial Off The Shelf 3D Printers'. To be presented at CAMX 2020, Orlando, FL, September 21-24, 2020.

15. Eljen Technology, *Reflective Paint EJ-510, EJ-520*. Eljen Technology. Accessed May 10, 2020. https://eljentechnology.com/images/products/data_sheets/EJ-510_EJ-520.pdf

16. Eljen Technology, *General Purpose Plastic Scintillator EJ-200, EJ-204, EJ-208, EJ212*. Eljen Technology. Accessed May 10, 2020 https://eljentechnology.com/images/products/data_sheets/EJ-200_EJ-204_EJ-208_EJ-212.pdf

ADAPTIVE MANUFACTURING WITH LASER MATERIAL DEPOSITION USING INTERMITTENT MACHINE-INTEGRATED LINE SCANNING

Jan Bremer[1], Sven Linden[1], Tim Fleischhauer[1], Andres Gasser[1], Johannes Henrich Schleifenbaum[1,2]

1: Fraunhofer Institute for Laser Technology
2: RWTH Aachen University – Digital Additive Production
Aachen, Germany

ABSTRACT

Laser Material Deposition (LMD), also known as Directed Energy Deposition (DED), has the potential to decrease production costs and lead time in repair and manufacturing applications of high-value and high-performance components. The ability to produce near-net-shape parts yields low material loss and is especially valuable when working with high-cost materials such as titanium alloys or nickel-based superalloys. LMD is commonly used for repair applications in industries such as aerospace, tooling and turbomachinery. As for most direct material deposition processes, geometrical process stability and robustness of the LMD process can be a limiting factor for applications in additive manufacturing (AM).

In this work, the principle of intermittent process control based on component geometry scans using a machine-integrated line scanner is validated and tested for tolerance to geometric deviations of the substrate. The developed prototype software is validated by hybrid-additive manufacturing of a boroscope flange geometry in 2.5 D on an industrial robotic system.

To evaluate transferability into full 3D geometry applications, algorithms are presented to fit the tool center point (TCP) path to freeform surfaces, where surfaces are rasterized as a depth maps (Z-Map), clipping non-build geometry from the measurements and filtering for stability improvements of the algorithm. The layer height control strategy algorithm based on intermittent scanning and using Z-maps is tested for continuous adaption of layer height from substrate to target geometry to enable stable near-net shape processing. Based on this Z-Map algorithm as input, a free-form slicer is further developed to compute TCP paths with surface normal data, which is simulated on 3 D curved components.

Keywords: Laser Material Deposition, Directed Energy Deposition, laser line scanner, adaptive processing, layer height control, point cloud manipulation
Corresponding author: Jan Bremer

1. INTRODUCTION

Laser Material Deposition (LMD), also known as Directed Energy Deposition (DED) is a technology commonly applied for coating and AM using metals. In AM using LMD, a three-dimensional part is deposited layer-by-layer on a substrate. A high power laser beam is used to create a local melt pool into which filler material in the form of powder or wire is injected. Tracks are formed through relative movement of the processing head to the substrate. After solidification,

the filler material forms a track on the substrate. Layers of material are formed by deposition of multiple tracks. Three-dimensional parts are manufactured through stacking of layers. [1, 2]
In Figure 1 a schematic view of a powder LMD process is depicted. The tool center point (TCP) is defined in Z at a constant standoff distance (SOD) between the powder nozzle and the substrate and in the center of the laser beam. The focus point of the powder gas jet is set with a specific offset from the TCP. Negative values for the powder focus offset are defined as the powder focus being shifted into the substrate.

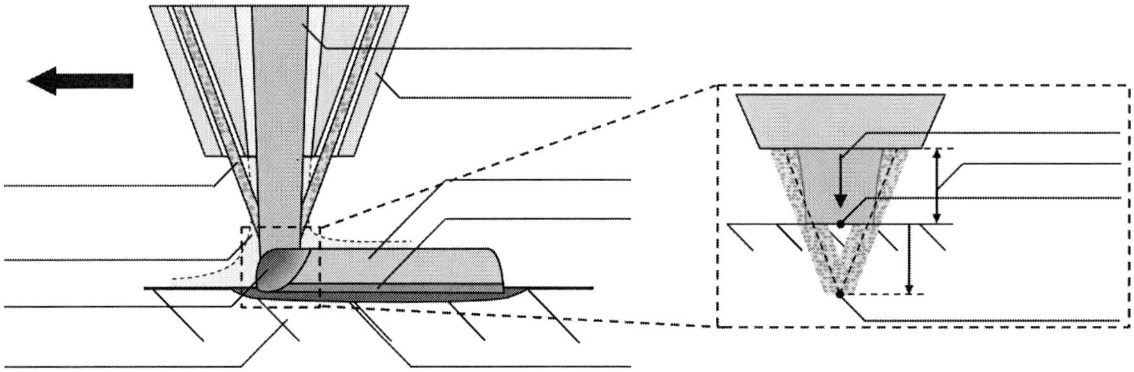

Figure 1. Schematic of the LMD process

Compared to other metal-AM technologies, LMD has advantages such as high deposition rates commonly higher than 1 kg/h, high material efficiencies of more than 90 %, a wide range of processable materials and a large work envelopes in the meter-range. It is commonly used in industries such as aerospace, turbomachinery, or tooling for manufacturing of functional parts and repair of worn production parts. [3]
The integration of LMD in a manufacturing process chain can reduce lead time and production costs. Monolithic parts, deposited by LMD, can substitute component assemblies consisting of multiple parts, reducing assembly complexity and time as well as saving raw material. [4]
In the tooling industry, dies and molds often have limited in-service lifetimes resulting from exposure to high loads under elevated temperatures, thermal cycling, cyclic loading conditions and contact with corrosive materials. LMD based remanufacturing of worn or damaged tools can yield considerable cost savings during the tool's life cycle. Furthermore, investigations indicate that the application of LMD in tooling production can reduce lead times by up to 35 % compared to conventional machining processes. [5]
Increasing process stability and geometric accuracy of deposited parts are key objectives of LMD research. By increasing geometric accuracy, less post-processing with conventional manufacturing technologies is needed, resulting in reduced production costs and lead time. Increasing process stability further reduces scrap rates and can increase the economic potential of the process.
Process stability is influenced by correct estimation of layer and build height during path planning. After deposition of a layer, the distance between TCP and substrate is increased by the estimated deposition height. If the estimated height does not match the deposited height, the SOD is altered, resulting in a disturbed interaction between laser beam and powder gas jet. This can lead to reduced deposition quality, geometrical inaccuracies and to instability of the LMD process. [6]
Online measurement and control of the deposited height is difficult due to high emittance at the melt pool and a wide range of influencing factors such as thermal conditions and geometrical

variations. One approach to height control is to predict the deposited height by using analytical or semi-empirical models as well as neuronal networks. These models are provided with information about the melt pool temperature distribution, which is measured by using various arrangements of charge coupled device (CCD) cameras, and process parameters such as feed rate or powder feed rate [7–13]. These approaches can be limited by the amount of computation time needed to process the CCD images and the amount of training necessary to tune the model parameters.
Another approach towards height control in LMD is to interrupt the deposition process for intermittent measurements of the deposited height and then adjust the process parameters in the subsequent deposition process based on empirical models. Laser line scanners (LLS) and structured light sensors are used to measure the deposited height [14–18]. Powder feed rate, feed rate and laser power are found to influence deposited height the most [19–21]. The powder feed rate is rarely considered in deposition height control due to its comparatively long response time [19]. The feed rate can be adjusted locally along the TCP path based on the measured differences between deposited height and estimated height [15]. Deposited height can also be controlled by adjusting the number of deposited layers in the subsequent deposition phase [14] or by adjusting the estimated layer height to fit the deposited layer height [16].
This work further contributes to the evaluation of the layer height control strategy developed in previous work [22]. It is based on the contributions of *Garmendia et al.* [15]. The control strategy is briefly presented in this work. Experiments are conducted to evaluate the capabilities of the control strategy to compensate local defects and produce demonstrator parts.

2. EXPERIMENTATION

2.1 System setup

Experiments are conducted on a LMD system consisting of a six axis industrial robot system and a cladding head. The cladding head consists of beam shaping optics unit, a continuous coaxial powder delivery nozzle and a Laser Line Scanner (LLS). Laser source is a diode laser which provides a maximum laser beam output of 5500 W and a wavelength of 940 – 1060 ± 10 nm at a beam parameter product of 30 mm*mrad. The system setup is depicted in Figure 1.

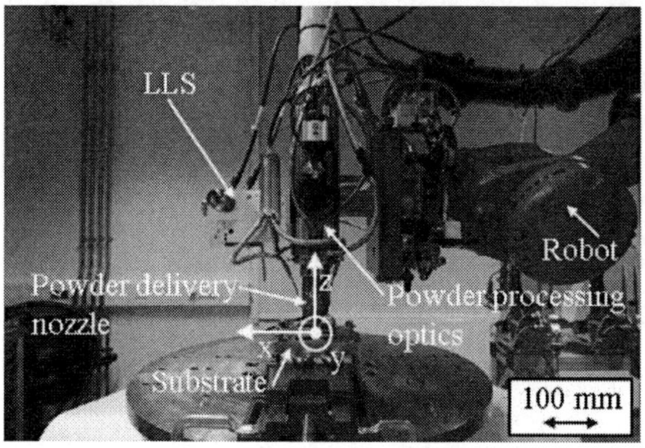

Figure 1. System setup

Plasma atomized Inconel 718 powder with a grain size distribution of 15 – 45 µm is used in the experiments. The chemical composition is depicted in Table 1.

Table 1. Chemical Composition of Inconel 718

Element	Cr	Co	Fe	Mo	Ni	Nb	Ta	Ti
Mass fraction [%]	17 – 21	0 – 1	13 – 25	2.8 – 3.3	50 – 55	4.75 – 5.5	0 – 0.05	0.65 – 1.15

2.2 Process parameters

All experiments are conducted using the process parameters depicted in Table 2. Nominal feed rate is 33.3 mm/s. Feed rate variations in experiments and within the height control strategy are denoted as deviation percentages from the nominal feed rate.

Table 2. Process parameters

Parameter	Value
Feed rate	33.3 mm/s
Laser power	1000 W
Beam diameter	1.6 mm
Powder feed rate	9.5 g/min
Powder delivery gas flow rate	3.5 NL/min

Parameter	Value
Powder focus offset	–1 mm
Shielding gas flow rate	10.7 NL/min
Track Overlap	50 %
Estimated layer height	0.65 mm

2.3 Measurement technologies

LLS measure two-dimensional height profiles by utilizing the light section principle. Three-dimensional scanning of an object is achieved by synchronizing measurement profiles with relative movement data of workpiece and scanner. In this work, height profiles are measured at a frequency of 83.3 Hz. Resolution of the scan line is 1280 points, which yields a distance of 78 μm between adjacent points on a nominal scan line of 100 mm. Reference resolution in the z-axis is 12 μm for the given configuration. A commercial software module is used to automatically acquire and synchronize line scanning and machine data, resulting in output of a point cloud of the scanned geometry in the working coordinate system.

2.4 Height control strategy

The fundamentals of the investigated layer height control strategy have been developed and implemented in previous work [22]. The approach is based on constant material deposition rates and with increase of feed rate to decrease layer height and vice versa. All other process parameters remain constant throughout the investigation.

Support points are inserted along the tool center point (TCP) path to allow for local adaptions of the feed rate. After deposition of four layers, the deposited geometry is scanned with a LLS. Scanned data is obtained as a point cloud. Compensation algorithms are applied to the point cloud to minimize impacts of disturbances such as mounting tolerances, sensor latencies or unprecise teaching of the LLS as a tool in the robot control. Local component build height is calculated from the point cloud for each support point within an influence area surrounding it. The influence area is a rectangle with edge lengths being equal to the support point distance Δs_{eff} and the beam diameter d. For each support point, a local feed rate is calculated with respect to the difference between local height and target height. The next four layers are deposited with locally adjusted feed rate at each support point. Control parameters and functional relations are determined experimentally. [22] A schematic view of the height control strategy is depicted in Figure 2.

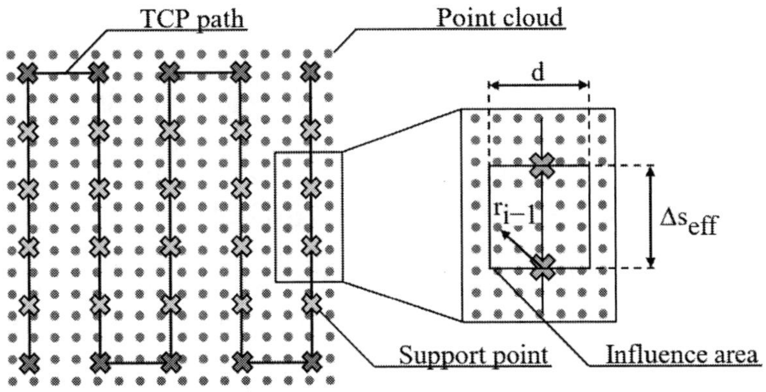

Figure 2. Schematic view of the height control strategy

2.5 Experimental investigations

Two experiments are conducted to further evaluate the capabilities of layer height controlled LMD processes. Cuboid specimens as well as a demonstrator geometry are built using LMD with and without applied layer height control strategy for comparison purposes.

Experiment 1
Six grooves with semicircular cross sections are machined into two substrates using a 10 mm spherical cutter. Depth is increased from 1 mm to 6 mm in increments of 1 mm. Three cuboids with an edge length of 25 mm are deposited on top of each groove. Two cuboids are deposited on each groove with an applied layer height control strategy. The substrate is measured with the LLS. Target geometry is a planar surface. The grooves represent local deviations. A third cuboid is deposited on each groove without applied layer height control strategy for comparison purposes. After deposition of four layers, the next control cycle starts with a LLS measurement of the deposited geometry. Deposition is aborted, if surface defects become visible.
Samples of each successfully deposited groove depth are prepared metallographically. Cross sections are recorded and evaluated.

Experiment 2
A boroscope flange is deposited with and without applied layer height control strategy on a planar substrate in ten blocks consisting of four layers each. A Computer Aided Manufacturing (CAD) model of the boroscope flange is depicted in Figure 3.

Figure 3. CAD model of the deposited boroscope flange

Experiment 3

A TCP path for the deposition of a cuboid with an edge length of 10 mm on the outer face of a cylinder with a diameter of 100 mm is generated with and without applied layer height control strategy. The cuboid is positioned at an angle of 20° and inserted half-way into the cylinder. Therefore, an algorithm is implemented to compute a slicing between the substrate surface and the protruding part of the cuboid and subsequent transformation into TCP path. Applied layer thickness is 1 mm, for the layer height control strategy a compensating layer thickness tolerance of 0.5 mm is applied. For both scenarios, the build-up vector is defined as $\vec{v} = (0,0,1)$.

Figure 4. Cross-section of cuboid intersecting a horizontal cylinder at an angle of 20°

3. RESULTS

In Experiment 1, cuboids were deposited successfully on top of substrates prepared with grooves up to a maximum groove depth of 3 mm. Surface defects are observed during the deposition on top of the 4 mm groove as depicted in Figure 4. Defects occur in the layer height controlled as well as in the uncontrolled process.

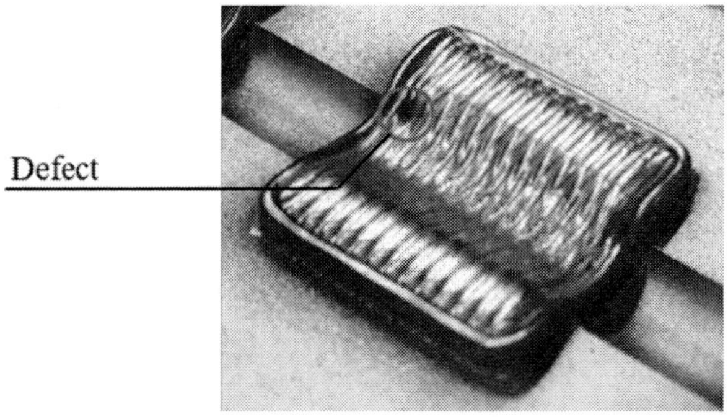

Figure 4. Observed surface defects during deposition on top of a 4 mm groove

For the deposition on top of the 3 mm deep groove, a single LLS measurement profile of each deposited block is depicted in Figure 5. The deviation from the target height decreases with each deposited block in the layer height controlled process.

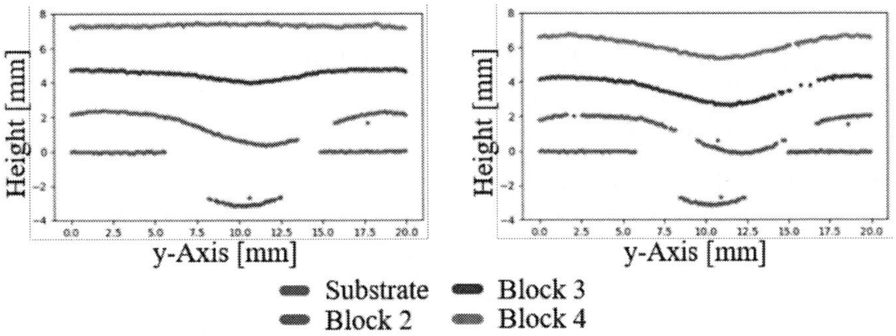

a) Controlled process b) Uncontrolled process

Figure 5. LLS profile measurements of layer height controlled and uncontrolled deposition

After deposition of three blocks, the average deviation from the target height in the layer height controlled process (Figure 5a) is 0.12 mm, whereas in the uncontrolled process (Figure 5b) the average deviation is 1.22 mm. The difference between maximum and minimum measured height value in the uncontrolled process is 1.35 mm and in the layer height controlled process 0.38 mm. The deviation is by a factor 3.6 greater in the uncontrolled process than in the layer height controlled process. Similar results with less magnitudes are obtained for the deposition on top of the 1 mm and 2 mm deep grooves. On average, a local deviation of 1 mm can be compensated within a single layer height control cycle and the deposition of one block consisting of four layers. The height values after deposition of three blocks are depicted in table 3. Target height is 7.45 mm.

Table 3. Comparison of resulting height values after deposition of three blocks

	Controlled	**Uncontrolled**
Minimum height [mm]	7.12	5.43
Maximum height [mm]	7.5	6.78
Average height [mm]	7.33	6.25
Deviation from target height [%]	1.6	16.1

Local deviations from the target geometry are compensated when applying the layer height control strategy to the process. In the uncontrolled reference process, local deviation decreases slightly with each consecutively deposited block due to self-stabilization of the process. Specimen deposited on top of the 3 mm deep groove in the uncontrolled and in the layer height controlled process are depicted in Figure 6.

a) Controlled process

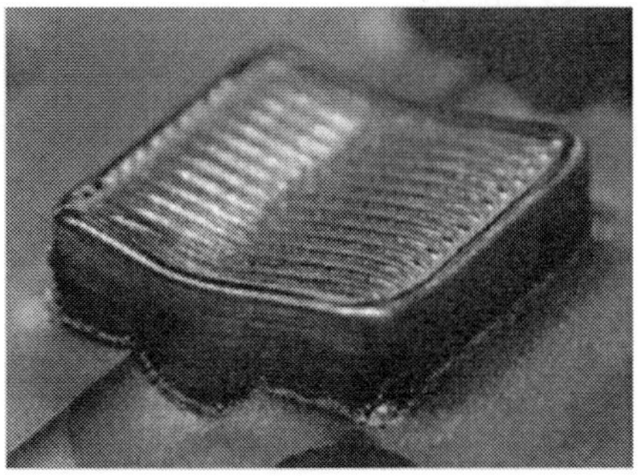

b) Uncontrolled process

Figure 6. Specimen deposited on top of the 3 mm deep groove

For each groove depth in the range of 1 mm to 3 mm one cuboid is inspected metallographically regarding the formation of defects in the groove area. In all investigated samples, occasional bonding defects as well as pores within the material deposited on top of the grooves are observable. The observed defects have a maximum diameter of 290 µm for a groove depth of 1 mm, 217 µm for a groove depth of 2 mm and 340 µm for a groove depth of 3 mm. Light microscopy images of the investigated cross sections are depicted in Figure 7.

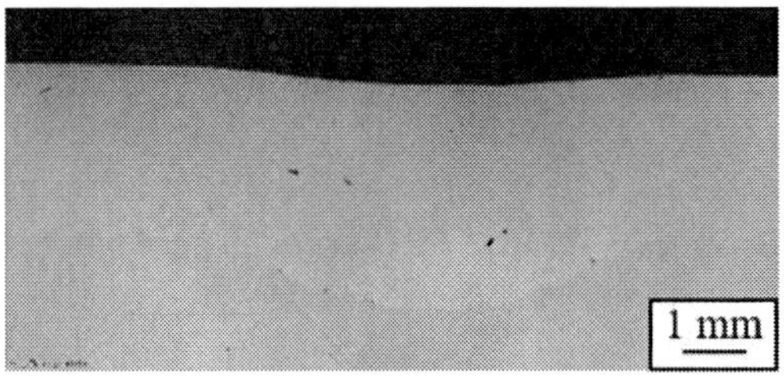

a. Slot depth 1 mm

a. Slot depth 2 mm

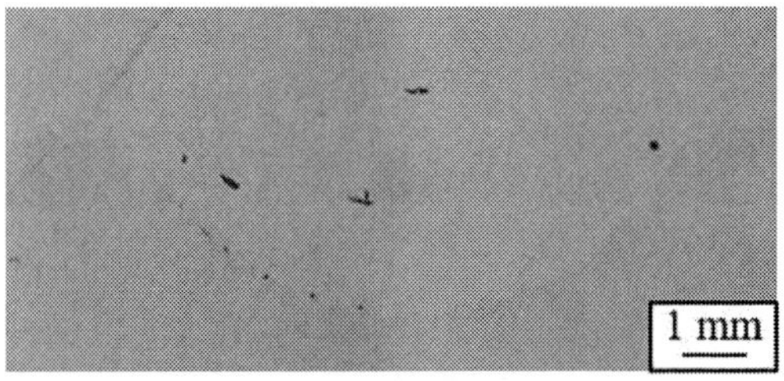

a. Slot depth 3 mm

Figure 7. Metallographic inspection of sample cuboids

The layer height control strategy is applied to the deposition of a demonstrator component in Experiment 2. The difference between target height and deposited height is depicted in Figure 7 for each deposited block consisting of four layers. After deposition of ten blocks, deviation from target height is 85 % higher in the uncontrolled process than in the layer height controlled process. In the layer height controlled process, an average deviation from target height of 0.44 mm is observed, which fluctuates slightly during the deposition process, whilst deviation is increasing steadily in the uncontrolled process. The layer height control strategy is not capable of reducing the difference between deposited height and target height to zero.

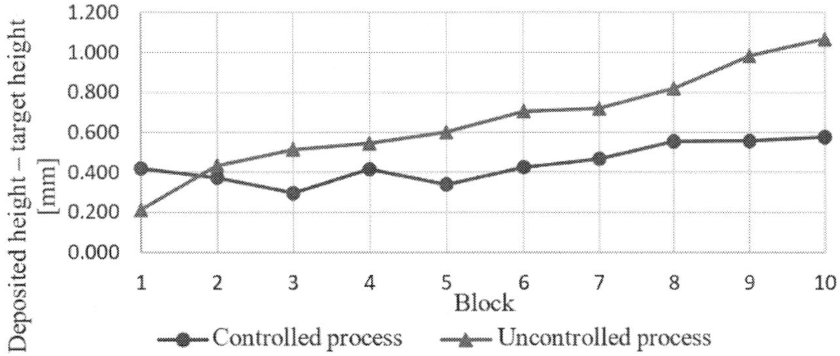

Figure 8. Progression of the difference between deposited height and target height

Deposited height is calculated in Figure 8 as the average height value of all point cloud points measured with the LLS that belong to the deposited geometry. The standard deviation of the point heights for each LLS measurement is depicted in Figure 9. It increases with each block deposited in the uncontrolled as well as in the layer height controlled process. No notable improvement is observed in the layer height controlled process.

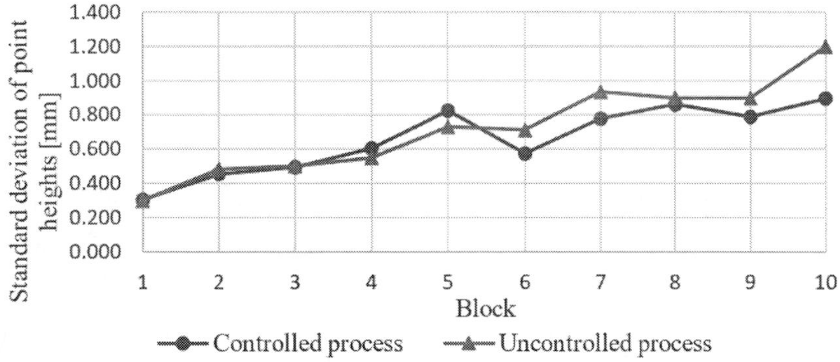

Figure 9. Standard deviation of the point heights for each deposited block

Increasing standard deviation of the point heights within a LLS measurement results from decreasing flatness of the deposited geometry. Especially in the areas surrounding the small holes, the flatness of the deposited geometry is disturbed by material accumulation. The deposited boroscope flange is depicted in Figure 10.

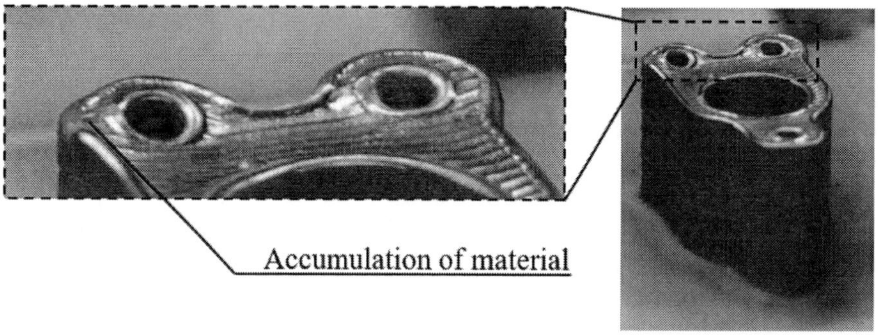

Figure 10. Deposited boroscope flange

With increasing height of the deposited geometry, the amount of disturbances obtained within a LLS measurement increases. The LLS falsely detects points belonging to the side walls of the previously deposited geometry or powder adhering to the side walls as points belonging to the deposited surface. These falsely detected points have comparably low height values and they drastically decrease calculated local feed rates. The observed effect depends on the scanning direction of the LLS. In Figure 11a a LLS measurement of the deposited boroscope flange is depicted. False detection of side wall points occurs only on walls whose normal vector $\bar{n}$ points against the direction of the LLS movement (black arrow). For walls with normal vector points in the direction of the LLS movement, side walls are not detected because of shadowing effects. The target height of the depicted control cycle is 26 mm. Points with a height value less than 20 mm are displayed in red color. They accumulate around the rear holes and that part of the inner hole, whose normal vector points against the direction of the LLS movement. White areas in Figure 11a indicate areas, where no points are obtained due to shadowing effects. An expanded view of the corresponding TCP path in the area of a rear hole is depicted in Figure 11b. Falsely detected side wall points lie within the influence areas of support points belonging to the outer contour path. Their height values are used in the feed rate calculation resulting in low overall local height values corresponding to low local feed rates. Low feed rates yield further material accumulation creating instability, because in the subsequent step side wall points with low height values are again falsely detected.

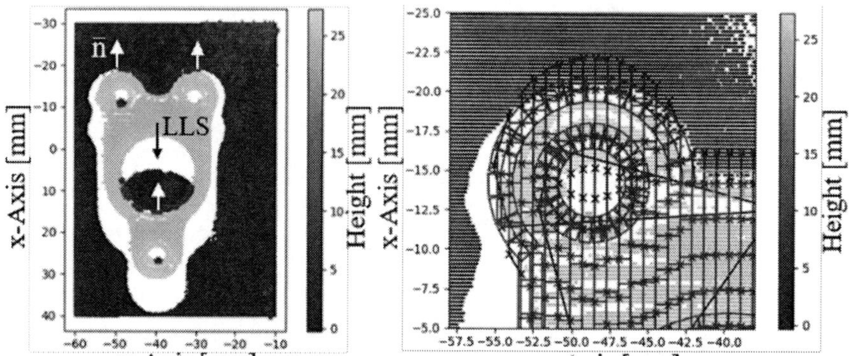

Figure 11. Falsely detected points in an LLS measurement and corresponding TCP path

In Experiment 3, a software is used to generate TCP paths for the deposition of a cuboid inserted at the side onto a horizontal cylinder at an angle of 20°, see Figure 14. To obtain a more robust and automatized computation of the height layer control strategy, a slicing algorithm has been implemented to compute TCP paths with surface normal data on 3D curved components. The software pipeline is shown in Figure 12. Through user input a base geometry is selected and a target geometry is placed into said base geometry. A machine-integrated LLS is used to digitalize the corresponding substrate area of the workpiece. The scanned area is loaded into the slicing algorithm together with the CAD model of the target part and serves as the base layer for the slicing algorithm. Slicing is conducted by sequentially shifting the base layer through the build-up volume and, if needed, adapting its shape towards a desired strategy, which is depicted in Figure 13. In the following, the slicing algorithm with and without applied layer height control strategy and subsequent TCP path generation are explained and evaluated.

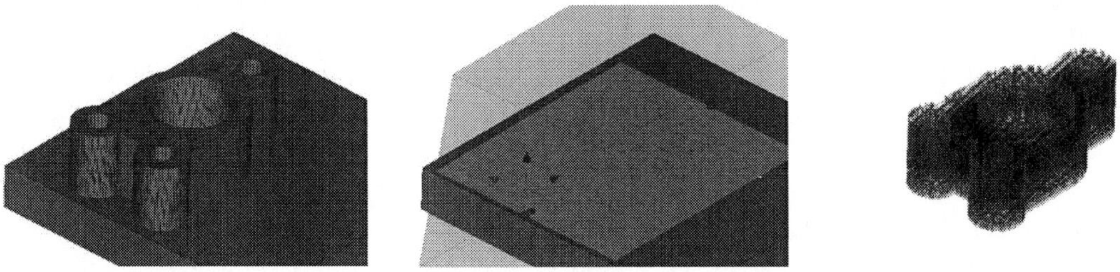

Figure 12. Software pipeline of the slicing algorithm: First, a target geometry is placed onto workpiece (left). Then a line scanner digitalizes the surface (middle). Finally, build-up is computed between surface scan and target part (right) using a newly developed slicing algorithm.

Slicing requires a build-up volume. It is computed from a 3D curved volume by rasterizing its vertical surfaces into Z-maps, where each pixel within the map can be defined as a finite element with height and normal vector. To achieve connection with the workpiece surface, the substrate surface forms the bottom shape of the 3D curved volume. The top shape of the 3D curved volume features the protruding part of the build-up volume.
Using the CAD model as a bounding box, the base layer can be filtered to contain the build geometry only.

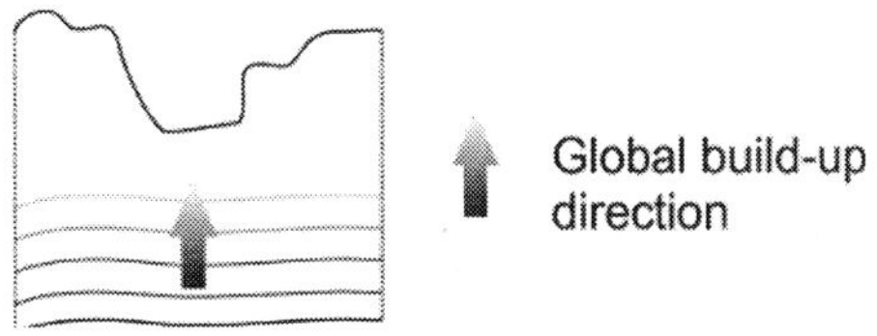

Figure 13. Build-up of a volume with offset strategy. Shifting and evolving the lowest Z-map results in slicing the build-up volume with Z-maps as slices. A global build-up direction helps to identify the number of residual slices.

Shifting the base layer towards the top of the volume with a build-up vector produces an ordered set of slices. A singular shift requires each pixel to be shifted individually. To compute the build-up vector for a pixel, several approaches are valid: Build-up vector linear to the build-up direction, linear to the normal of the pixel, or a mixture of both. *Bold et al.* [23] indicate that a combination of both strategies leads to the most realistic build-up.

Once every pixel's height is above its comparison point of the top layer, the slicing process is defined as finished.

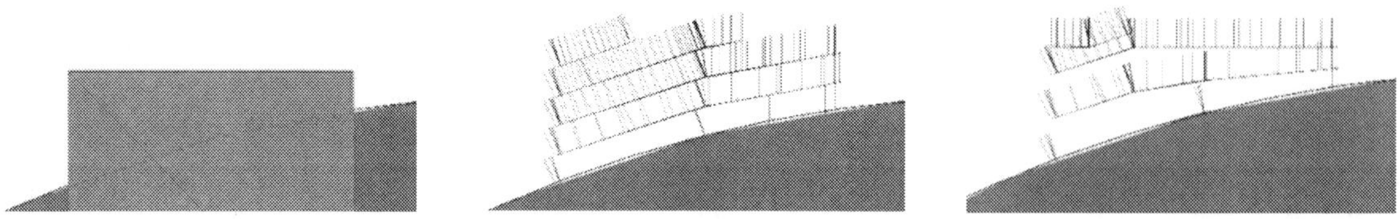

Figure 14. Build-up of cube on cylinder (left) with intersection line in purple. Offset strategy (middle) produces five layers [layer height = 1 mm]. Layer height control strategy (right) produces four layers [layer height = 1 mm, compensation range= [0.5, 1.5] mm].

When executed without layer height control strategy, the magnitude of the build-up vector is always constant and equal to the chosen layer thickness. This strategy is depicted in Figure 14.b. Build-up layers are displaced towards build-up direction using a build-up along the global build-up vector $\vec{v} = (0,0,1)$. A total of five layers are computed, which are visualized as black outlines above the cylinder. The first three layers spread over the whole area of the build part, while the upper two layers only cover a part of the build-up area. Once blocks of the base layer exceed the top of the build-up volume, the algorithm deactivates those to prevent unnecessary build-up.
The red lines located above a layer in the image visualize the normal vectors of its underlying support point. In most cases, the normal vectors are computed correctly with regard to the layer, however for some support points, it points in the direction of the build-up vector, see Figure 15. In these cases, the algorithm is not able to retrieve enough information from the Z-map to compute a functional normal vector and falls back to the build-up vector.
For the build-up with height layer control strategy, layer height is derived globally and adapted locally. The goal of this strategy is for the base layer to evolve into the shape of the top of the volume. Therefore the average height of base layer and top layer of the volume are compared and produce a global constraint, which is locally balanced with the height difference of each pixel of the base layer compared to its comparison pixel at the top layer. This ultimately produces an individual height with which the base layer is able to evolve individually and detached from all other pixels step-by-step into the top layer.
This strategy is depicted in Figure 14.c. Here, a layer thickness of 1 mm is used and a height control tolerance of 0.5 mm is applied at each layer, which results in an applicable height range of [0.5, 1.5] mm. In the first three layers, the build-up layers are fanned out at the lower end using the height control strategy to compensate for the widening height gap between surface and build-up part. The topmost build-up layer is calculated using the lower end of the height range to match the upper shape as closely as possible.

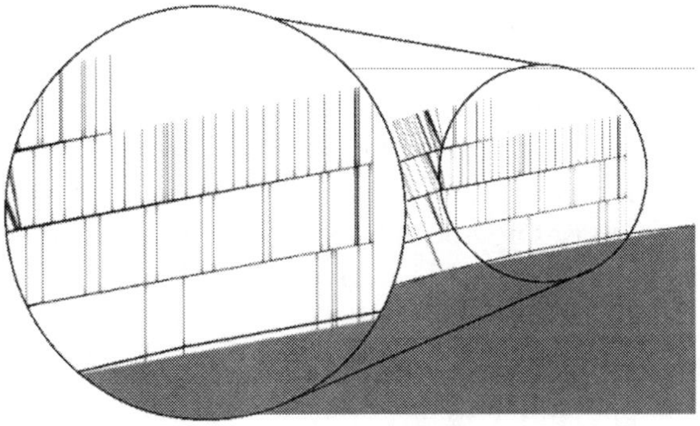

Figure 15. View from the side: Layers with normal vectors (red lines) falsely pointing in build-up direction.

After build-up, each layer is transformed from a Z-map into toolpaths. Contour toolpaths are computed using the outermost pixels on a Z-map including inner outermost pixels of implied holes. Hatching lines are computed using identified contours and the surface terrain of the Z-map. Similar to the approach described in 2.4, support points are inserted using the neighborhood of the Z-map at appropriate points. For each support point a normal vector is derived from the surface region of the Z-map.

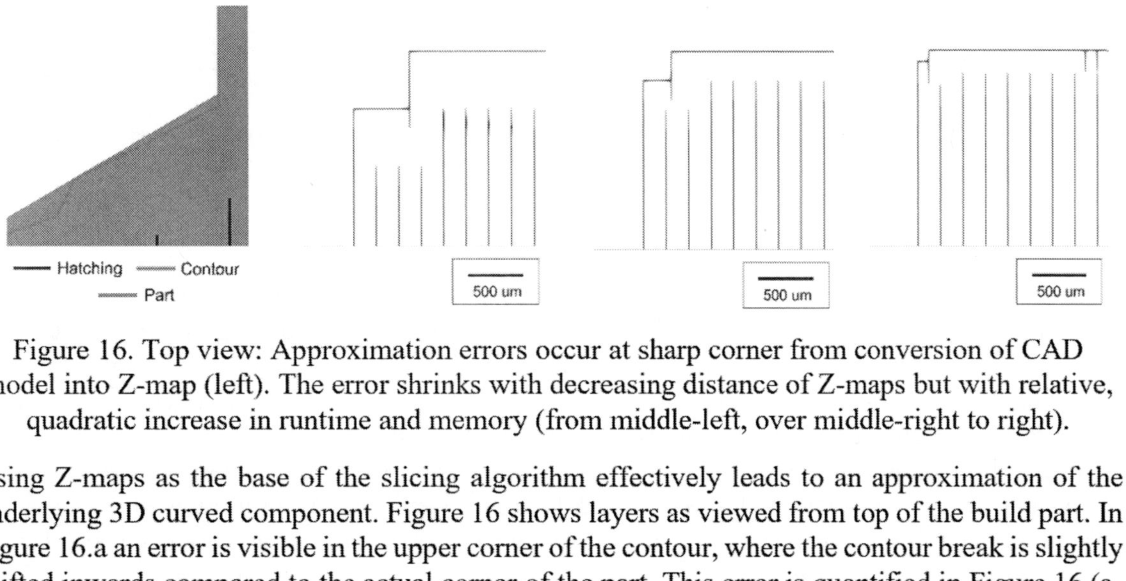

Figure 16. Top view: Approximation errors occur at sharp corner from conversion of CAD model into Z-map (left). The error shrinks with decreasing distance of Z-maps but with relative, quadratic increase in runtime and memory (from middle-left, over middle-right to right).

Using Z-maps as the base of the slicing algorithm effectively leads to an approximation of the underlying 3D curved component. Figure 16 shows layers as viewed from top of the build part. In Figure 16.a an error is visible in the upper corner of the contour, where the contour break is slightly shifted inwards compared to the actual corner of the part. This error is quantified in Figure 16.(a-c) with pixel resolutions of 0.5, 0.25 and 0.1 mm. With increasing Z-map resolution, the approximation error shrinks, leading to sharper corners and contours. This however results in a squared relative increase in runtime and memory consumption.

4. CONCLUSIONS

The capabilities of a geometry-based layer height control strategy for LMD processes based on machine-integrated laser line scanning are evaluated in this work. In the first experiment, cuboids are deposited on substrates prepared with grooves of varying depth to evaluate the capability of the control strategy for compensating local height deviations. With application of the control strategy, local defects can be compensated up to a depth of 3 mm without indications of surface defects.

Deposition of a demonstrator part reveals current limitations of the developed layer height control strategy. False detection of points belonging to side walls as points belonging to the deposited surface increases with increasing absolute height. As a consequence, low feed rate values are calculated in areas of the deposited geometry, where higher feed rates are required, resulting in excessive accumulation of material. The observed effect depends on the direction of the LLS movement. Deviation from the target height of the component is reduced by 85 % with application of the layer height control strategy compared to an uncontrolled reference process.

A slicing algorithm is implemented to produce automatized TCP paths with or without layer height control strategy. The layer height control strategy addressed within the slicing algorithm is currently implemented in a naïve way, where a global number of build-up layers is predicted for the whole build-up and enforced locally to compensate the necessary local layer height deviations. First results produced for a cuboid deposited onto a cylinder are evaluated. The clipping of build to non-build geometry from the measurements is addressed using an additional CAD model of the build part as a cut template. Using a Z-map as the basis for layers, TCP paths can be computed on free-form surfaces and adapted to the shape of the given substrate. The computation of normal vectors for each support point enables a DED process in 3D. The choice of Z-maps comes at the cost of approximated part shape and quadratic runtime and memory costs.

5. OUTLOOK

Further investigation of LLS scan strategies should be conducted to increase the stability of the developed layer height control strategy. Through variation of the LLS orientation during the scan process as well as through involving several LLS movement directions, the influence of shadowing might be reduced, and the yield of data points might be increased. This will, together with further development of the control algorithm, reduce errors in the control strategy.

Functional relations between layer height and feed rate have so far only been investigated in layers with track lengths by magnitudes larger than the laser beam diameter. Further investigation of functional relations between layer height and feed rate should be conducted in terms of short tracks and small layers to increase geometric accuracy when depositing filigree structures.

Comprehensive metallographic analysis should be conducted to evaluate the metallurgical effects of this process control strategy to define stable parameter windows.

Further development is planned for the slicing algorithm to increase geometric flexibility and applicability to a wider range of geometric features. Equidistant track spacing on steep component angles and strategies for stable implementation of the layer height strategy on arbitrary component features are current subjects for research.

6. REFERENCES

[1] F. Klocke, *Fertigungsverfahren 5.* Berlin, Heidelberg: Springer Berlin Heidelberg, 2015.

[2] E. Toyserkani, A. Khajepour, and S. Corbin, *Laser cladding.* Boca Raton, Fl: CRC Press, 2005.

[3] F. Klocke, K. Arntz, M. Teli, K. Winands, M. Wegener, and S. Oliari, "State-of-the-art Laser Additive Manufacturing for Hot-work Tool Steels," *Procedia CIRP*, vol. 63, pp. 58–63, 2017, doi: 10.1016/j.procir.2017.03.073.

[4] D. Konitzer, S. Duclos, and T. Rockstroh, "Materials for sustainable turbine engine development," *MRS Bull.*, vol. 37, no. 4, pp. 383–387, 2012.

[5] W. R. Morrow, H. Qi, I. Kim, J. Mazumder, and S. J. Skerlos, "Environmental aspects of laser-based and conventional tool and die manufacturing," *Journal of Cleaner Production*, vol. 15, no. 10, pp. 932–943, 2007.

[6] S. Donadello, M. Motta, A. G. Demir, and B. Previtali, "Monitoring of laser metal deposition height by means of coaxial laser triangulation," *Optics and Lasers in Engineering*, vol. 112, pp. 136–144, 2019, doi: 10.1016/j.optlaseng.2018.09.012.

[7] L. Song, V. Bagavath-Singh, B. Dutta, and J. Mazumder, "Control of melt pool temperature and deposition height during direct metal deposition process," *Int J Adv Manuf Technol*, vol. 58, 1-4, pp. 247–256, 2012, doi: 10.1007/s00170-011-3395-2.

[8] A. Heralić, A.-K. Christiansson, M. Ottosson, and B. Lennartson, "Increased stability in laser metal wire deposition through feedback from optical measurements," *Optics and Lasers in Engineering*, vol. 48, no. 4, pp. 478–485, 2010, doi: 10.1016/j.optlaseng.2009.08.012.

[9] A. Das, S. Mukherjee, S. K. Giri, and A. K. Lohar, "An Improved Learning Based Multilayer Height Control Strategy in LMD Process," in *2017 14th IEEE India Council International Conference (INDICON)*, Roorkee, Dec. 2017 - Dec. 2017, pp. 1–6.

[10] A. Fathi, A. Khajepour, E. Toyserkani, and M. Durali, "Clad height control in laser solid freeform fabrication using a feedforward PID controller," *Int J Adv Manuf Technol*, vol. 35, 3-4, pp. 280–292, 2007, doi: 10.1007/s00170-006-0721-1.

[11] D. Hu and R. Kovacevic, "Sensing, modeling and control for laser-based additive manufacturing," *International Journal of Machine Tools and Manufacture*, vol. 43, no. 1, pp. 51–60, 2003, doi: 10.1016/S0890-6955(02)00163-3.

[12] M. Iravani-Tabrizipour, M. Asselin, and E. Toyserkani, "DEVELOPMENT OF AN IMAGE-BASED FEATURE TRACKING ALGORITHM FOR REAL-TIME CLAD HEIGHT DETECTION," *IFAC Proceedings Volumes*, vol. 39, no. 16, pp. 914–920, 2006, doi: 10.3182/20060912-3-DE-2911.00157.

[13] M. Iravani-Tabrizipour and E. Toyserkani, "An image-based feature tracking algorithm for real-time measurement of clad height," *Machine Vision and Applications*, vol. 18, no. 6, pp. 343–354, 2007, doi: 10.1007/s00138-006-0066-7.

[14] I. Garmendia, J. Leunda, J. Pujana, and A. Lamikiz, "In-process height control during laser metal deposition based on structured light 3D scanning," *Procedia CIRP*, vol. 68, pp. 375–380, 2018, doi: 10.1016/j.procir.2017.12.098.

[15] I. Garmendia, J. Pujana, A. Lamikiz, J. Flores, and M. Madarieta, "Development of an Intra-Layer Adaptive Toolpath Generation Control Procedure in the Laser Metal Wire Deposition Process," *Materials (Basel, Switzerland)*, vol. 12, no. 3, 2019, doi: 10.3390/ma12030352.
[16] I. Garmendia, J. Pujana, A. Lamikiz, M. Madarieta, and J. Leunda, "Structured light-based height control for laser metal deposition," *Journal of Manufacturing Processes*, vol. 42, pp. 20–27, 2019, doi: 10.1016/j.jmapro.2019.04.018.
[17] M. Gegel, A. Nisbett, D. Bristow, and R. G. Landers, "Laser Line Scan Characterization Of Geometric Profiles In Laser Metal Deposition," *Solid Freeform Fabrication 2016: Proceedings of the 26th Annual International*, vol. 27, pp. 1103–1111, 2016.
[18] A. Heralić, A.-K. Christiansson, and B. Lennartson, "Height control of laser metal-wire deposition based on iterative learning control and 3D scanning," *Optics and Lasers in Engineering*, vol. 50, no. 9, pp. 1230–1241, 2012, doi: 10.1016/j.optlaseng.2012.03.016.
[19] H. Qi, M. Azer, and P. Singh, "Adaptive toolpath deposition method for laser net shape manufacturing and repair of turbine compressor airfoils," *Int J Adv Manuf Technol*, vol. 48, 1-4, pp. 121–131, 2010, doi: 10.1007/s00170-009-2265-7.
[20] L. Dubourg and L. St-Georges, "Optimization of Laser Cladding Process Using Taguchi and EM Methods for MMC Coating Production," *Journal of Thermal Spray Technology*, vol. 15, no. 4, pp. 790–795, 2006, doi: 10.1361/105996306X146785.
[21] L. Tang, J. Ruan, T. E. Sparks, R. G. Landers, and F. Liou, "Layer-to-layer height control of Laser Metal Deposition processes," in *2009 American Control Conference*, St. Louis, MO, USA, Jun. 2009 - Jun. 2009, pp. 5582–5587.
[22] J. Bremer, T. Fleischhauer, A. Gasser, and H. Schleifenbaum, "Development of an Intermittent Height Control using Line Scanning for Successive Toolpath Adaption in Laser Material Deposition with Industrial Robots," 2020.
[23] M. Bold, N. Pirch, S. Ziegler, and H. Schleifenbaum, "Evaluation of Offline Path Planning for Laser Metal Deposition on Freeform Surfaces," *Lasers in Manufacturing Conference*, 2019.

DEVELOPMENT OF A TEST BENCH FOR FUSED FILAMENT FABRICATION

Alexander Matschinski, Alberto Juan Alburquerque, Benedikt Kriegl, Kalle Kind, Klaus Drechsler
Technical University of Munich, Chair of Carbon Composites
Munich, Germany

Corresponding author: Alexander Matschinski

ABSTRACT

Current investigations in Fused Filament Fabrication (FFF) only use the typical process parameters to optimize the printed products. Therefore, FFF is not reaching beyond technology readiness level (TRL) 3 for most industrial applications. In order to increase the TRL, detailed process understanding, especially the interactions of process parameters, has to be gained. This paper presents a novel test bench for the detailed analysis of the extrusion process. The setup includes three load cells. This enables the investigation of the interrelationships between process parameters and the material behavior (e.g. pressure) in the melt zone. The investigations include parameter sets at extrusion temperatures between 195 and 230 °C and extrusion velocities between 100 and 600 mm/min. The decrease in feed pressure due to friction in the bowden tube is significant. As expected, with increasing feed speed and/or decreasing extrusion temperature, the counter pressure in the nozzle increases. Measured counter pressures with common PLA as feedstock material are between 1 and 27 N. Next steps are to extend the investigations to high performance materials such as polyether-aryl-ketone (PEAK) and to investigate feed slippage.

1. INTRODUCTION

The production of additively manufactured polymer components has significantly increased in several industries, especially in aeronautical and medical applications [1]. One of the commonly used processes is Fused Filament Fabrication (FFF). Main benefits of the process are the relatively easy dimensional scalability and processability of most thermoplastic polymers as well as the possibility to add (endless) fiber reinforcements [2,3]. However, the certification of printed products is quite challenging due to instabilities in processing and comparatively poor properties of the produced parts.

1.1 State of the art

System manufacturers like Stratasys have introduced a closed system concept to solve these problems. Material, its storage as well as the process parameters and conditions are specified without the possibility of adaptation. This has made it possible to generate certifications for blends of polyetherimide (PEI) and polycarbonate (PC) in the automotive, medical and aerospace sectors [4]. For this material class (also called ULTEM 9085), the manufacturer specifies 33 to 47 MPa for the tensile strength, depending on the print orientation [5]. Depending on the orientation, these mechanical properties are only 46.5 and 85.8 % of those of samples from an injection molding process [6]. These results show the unsolved issue of anisotropy even in a certified system. In addition to this issue, other research work has identified the following categories for the challenges in Additive Manufacturing: Anisotropy [6,7], porosity [8,9], geometric fidelity [10,11], surface quality [12,13], repeatability [14] and residual stresses [15].

Previous research work uses the typical process parameters to overcome these limitations. Extensions to other high-performance polymers and optimization of the printing results are carried out using parameters such as extrusion temperature, printing speed and material throughput [16,17]. The lack of understanding of the interactions between process parameters and the material behavior of the polymers limits evolution. Melt properties such as viscosity [18] or crystallization on cooling [19] influence the performance of the printed products. Further correlations and influences by the extrusion process itself are only partially known. These are effects such as nozzle swelling of the polymer at the exit point of the extrusion process [20]. This phenomenon is caused by conveying the material through a die geometry. In addition, the nozzle geometry influences, among other things, the fiber orientation when printing composite materials [21].

The use of high-performance materials requires a more detailed understanding of the polymer melt beyond the typical process parameters. Integrating fiber reinforcements further increases this need. Duty et al. [22] have applied a viscoelastic model for this purpose. In this way, requirements for pressure-driven flow, bead formation, bead functionality, and component-level functionality are set in order to be able to make a prediction about the printability of a material. Yan et al. [23] already use a combination of simulated temperature distribution and the viscosity-temperature relationship for optimized process planning for handling high-performance materials in the sintering process. For extrusion-based processes, this approach can be successful with the modelling according to Osswald [24]. Osswald presents a model for the liquifying process in the hot-end. For the modeling, the main parameters are the die angle and length and the pressure through the filament feeding. A good basis for an experimental investigation is created. Nevertheless, for a closer examination the internal pressure of the nozzle has to be measurable.

The feeding force of the filament determines the internal nozzle pressure. The power of the feeder motor set this. Initial investigations are only considering the motor and its feed force [25]. Further work uses a separate measurement by using a universal testing machine to determine the feeding behavior of the feeder [26]. In this way, no direct measurement or online monitoring of the process is possible. Only by using vibration sensors could Tlegenov et al. [27] and Li et al. [28] make this possible. The sensor used is located on the print head and can be calibrated to the feed force by coupling it to the print head. The measurements are not direct quantitative values of the feeding force. Only with test setups such as those of Go et al. [29] and Nienhaus et al. [30] can direct measurement data be determined during extrusion. Both use a load cell directly at the print head. For processing high-performance materials of the PEAK class, printers with heated build chambers are used. To protect the electronics of the feeder from the temperature, a bowden tube is inserted between the print head and the feeder. Therefore, the research work from state of the art does not offer a holistic view of the extrusion process. Furthermore, the presented test setups for online measurement cannot handle high temperature or abrasive material. These setups are not suitable for investigations of high performance composite materials.

1.2 Concept

In order to be able to investigate the interactions between the process parameters, the state variables of the extrusion such as the melt pressure and the extrudate, a test bench is presented in this thesis. The focus is on the properties of the polymer melt in combination with the properties inner shape of the nozzle. With the test bench, the interrelationships can be determined under real conditions. Based on this knowledge, threshold values for selected parameters can be defined in the future. These allow the derivation of guidelines for ensuring part quality.

2. EXPERIMENTATION

The concept of the test bench arranges the elements of the extrusion process vertically. This includes the process from the spool holder to the extrusion die. For better flexibility the construction is modular.

2.1 Extrusion system

The extrusion system consists of four elements. These are the feedstock holder, feeding mechanism, bowden tube connection and print head. The elements are controlled by an Arudino microcontroller including a RAMPS 1.4 as printer controller and an IC MAX6675 board for temperature measurement.

2.1.1 Filament feedstock and feeding

The filament holder is made of four bearings. This means that spools of different sizes can be used. The friction losses are minimal due to this arrangement. An additional guiding element guarantees a controlled unwinding process. The following feeding mechanism is a Bondtech. This system has been improved by substituting the idler bearing with a second hobbed roller. Both rollers incorporate a gear and are counter rotating. This enables the hobbed rollers to grip the filament from both sides, which increases the extrusion force. The Bontech includes a simple stepper. According to the manufacturer, this design can generate up to 14 kg of feed force without slippage. For this reason, this feed mechanism is used in a wide range of professional 3D printers.

2.1.2 Bowden tube and print head

A bowden tube with 2 mm diameter is used to simulate real operating conditions. The tube is clamped between two Festo-adapters and connects the feeding unit with the print head. In industrial 3D printers (e.g. from GEWO GmbH) this tube can have a length of up to 2 meters. Therefore, the length and the course can be varied for the test bench. The print head system is an interchangeable system. A water cooling system ensures the ideal operating temperature. In the current arrangement, the print head operates with the E3D standard nozzle system. Two heating cartridges with 40 W each allow a hot-end temperature of more than 500 °C. The heating length is 22 mm.

2.2 Sensor system

The measuring system consists of three measuring points in total. The measuring points are 5 Kg load cells. The measurement data can be continuously recorded by using a microcontroller in combination with a HX711.

As in the designs by Go et al. [29] and Nienhaus et al. [30], the load cell is connected to the print head via a lever arm. The assembly cuts the print head completely free. The measurable force is thus the pushing force transmitted through the filament. A schematic representation of this arrangement is shown in Figure 1.

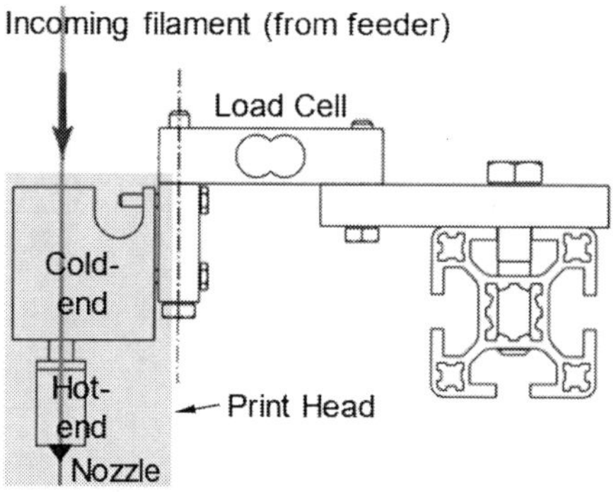

Figure 1. Construction of print head (measuring point 3).

For processing PEAK, most printer systems use a tempered built chamber. Mounting the feeding mechanism outside the chamber protects it from the temperature. This favors the use of a bowden tube. Friction effects occur when the filament is feeded through such a tube made of polytetrafluoroethylene (PTFE). For this reason, such a bowden tube is added to the test setup for the investigation. This bowden tube connects the feeder with the print head. Designing both connections as measuring points enable the measuring of the friction effects. Two additional load cells are used. These are each equipped with an adapter to which the ends of the tube are attached. The first load cell is mounted directly below the Bontech and the second directly above the cold-end of the print head. Figure 2 shows a schematic representation.

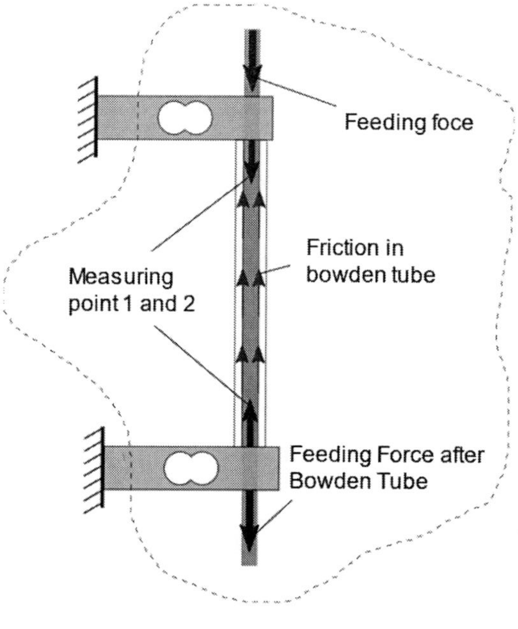

Figure 2. Construction for bowden tube (measuring point 1 and 2).

The combination of these individual concepts results in an overall concept. Figure 3 shows on the left a schematic and on the right an isometric representation of the conversion into a test design. The three installed load cells are numbered according to the time of the filament passing through. In addition, the schematic illustration shows the occurring forces and their effect by arrows.

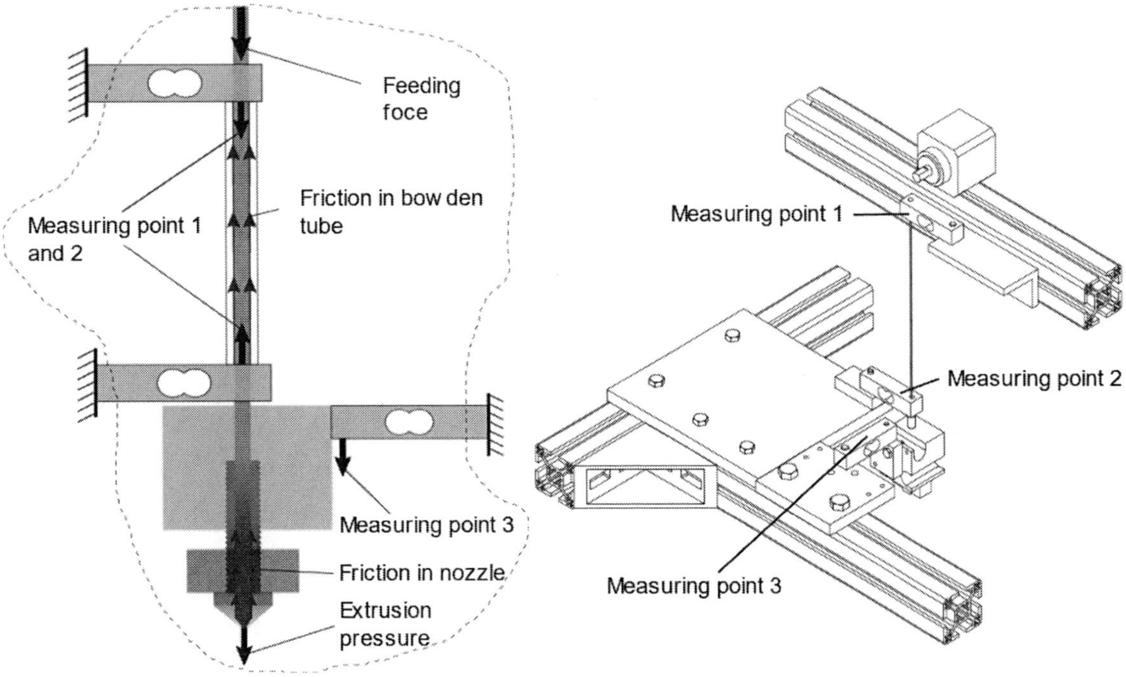

Figure 3. Overall concept of the monitoring system.

3. RESULTS AND DISCUSSION

A routine over 60 seconds performs the extrusion process. The temperature of the hot-end and the speed of the extrusion are specified by a simple movement code. The experimental matrix comprise hot-end temperature and feeding speed as parameters. For a substantiated analysis 195, 215 and 230 °C for the temperature and 100, 300 and 600 mm/min for the speed are examined. Extruded Material is common poly-lactic-acid (PLA).

3.1 Force progression of measuring point 1 to 3

Figure 4 shows the force curve for the three measuring points. The fluctuation during the recording is approximately 1.5 N. This corresponds to more than 50 % of the average value. The loss due to friction caused by the bowden tube can be determined by a comparison between measuring points 1 and 3. This is consistently at one Newton over the duration of the recording. The quantitative deflection of the force measurement is not large enough to derive more detailed cause-and-effect relationships. In addition, a tensile effect is noticeable at the second measuring point.

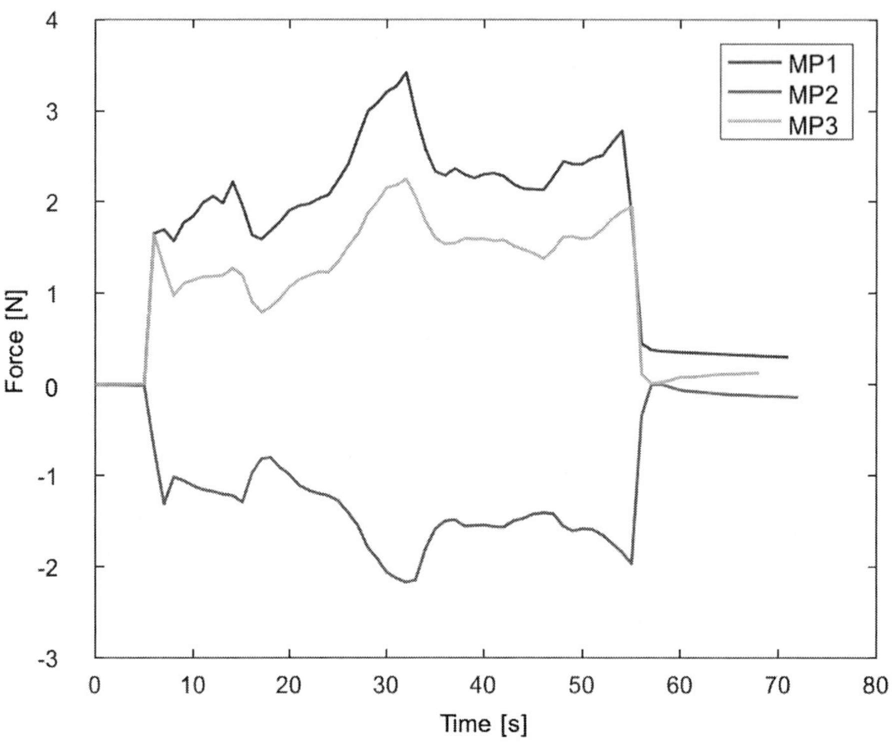

Figure 4. Results for the three measuring point 1-3 (T=195 °C; 100 mm/min).

3.2 Influences of the bowden tube arrangement on the feeding

In order to examine the tensile force at the second measuring point more precisely, the bowden tube is shortened to the minimum length. In Figure 5 this arrangement is shown in the left section of the picture. The right section shows the same arrangement in operation. The second measuring point is loaded in tension. This causes the pipe to break out sideways during operation. For a further consideration of the influence, the pipe was lengthened so far that it forms a helix. This arrangement was then used for another series of measurements. The results are compared in Figure 6. In contrast to the first measurements, the extrusion speed is 600 mm/min. The deflection at the measuring points is therefore many times greater. The fluctuations of the results during recording are smaller. Thus, a process-related fluctuation becomes visible. In addition, the force deflection at the measuring points is 40 % smaller in a helical pipe. This can be explained by the alignment-related increase in friction. The dependence of the transfer capacity of the feeding force on the arrangement of the pipe can thus be confirmed as high.

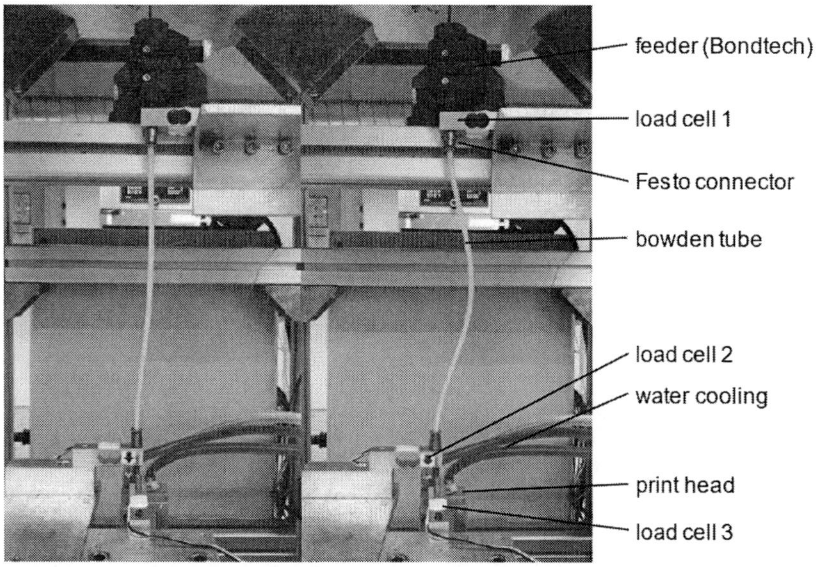

Figure 5. Setting of bowden tube without force (left) and with force (right).

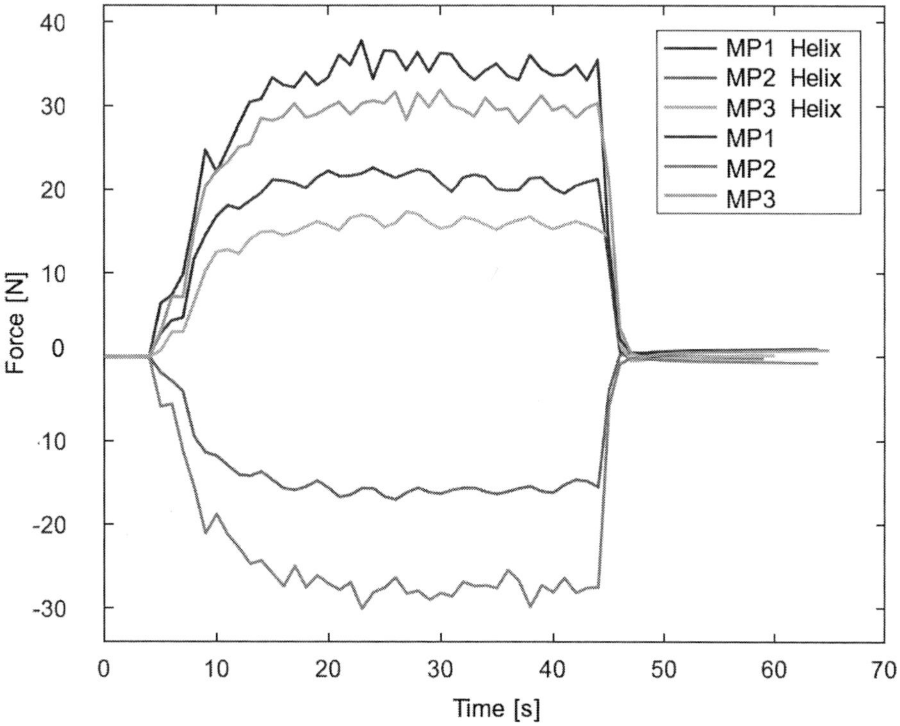

Figure 6. Comparison of different arrangements of the bowden tube (T=215 °C; 600 mm/min).

3.3 Influences of temperature and speed on the feeding

For these tests the tube has been shortened to a medium length. This configuration most closely resembles the condition during operation of a printer system. Figure 7 shows the results of measuring point 3 of all nine test series from three parameters for temperature (195, 215 and 235 °C) and feeding speed (100, 300 and 600 mm/min). The dependencies are as expected. A decreasing temperature and/or increasing feeding speed increases the force acting on the print head. The maximum load occurs with the parameter set with 195 °C extrusion temperature and 600 mm/min feeding speed. At this bowden length the measured force varies around the average value of 25 N. Due to this extreme pairing of parameters, the fluctuation in the feeding force is significant. The mean value of the fluctuation is ±1.5 N. If the extrusion temperature is increased by 20 °C, the average measured force drops to 17 N. The fluctuation in this graph is not significantly different. Only an increase by further 20 °C to 235 °C lowers the average force to 11 N and also reduces the fluctuation to ±1 N. In comparison, halving the feeding speed to 300 mm/min reduces the measurable force from the initial value (at 195 °C) to 7 N. Now, the fluctuation in the signal is not significant in any way. The other parameter sets cause a maximum fluctuation of even less than 5 N. This relatively small deflection and the short recording time make a more detailed interpretation difficult. Due to the small deflection, measurement errors or signal noise lead to relatively large shifts in the results. For further investigations, the signal for these parameters has to first be verified by a significance analysis.

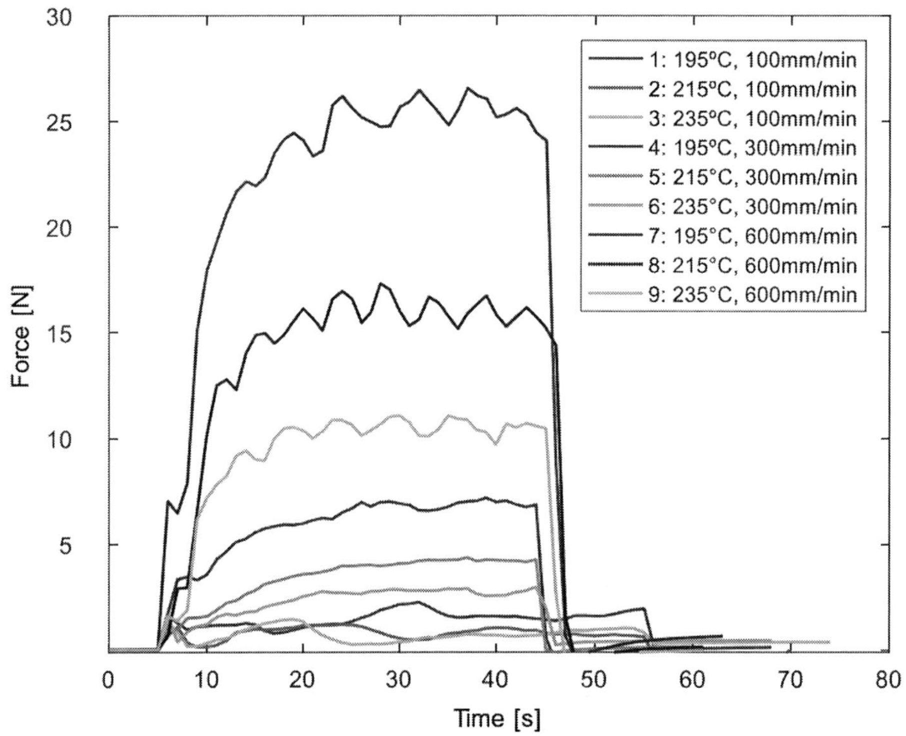

Figure 7. Comparison of different extrusion sets.

4. CONCLUSIONS AND OUTLOOK

The developed test set-up allows a detailed investigation of the extrusion behavior in the FFF. The measured data obtained are within the expected measuring range and confirm the assumptions derived from the state of the art. The measurable data characterize the polymer melt as well as the influence of the feeding section itself. The frictional effects of the feeding section can be quantified for the first time. For example, friction causes a 30 % drop in force when the tube is helical in shape. For the determination of the inner nozzle pressure, the feeding speed is an essential parameter, as is the extrusion temperature. With increasing temperature and decreasing velocity, the inner nozzle pressure drops. These interactions can be quantitatively investigated with this design. The parameter combination of 195 °C and 600 mm/min causes the highest counter pressure in the nozzle with a peak of 27 N.

Extending the investigation to PEAK-class material including short fiber reinforcement is the next step. In the test setup presented, the feedstock material (filament) is not further investigated, instead it is assumed to be ideal. The industrial variation of the filament diameter is ± 0.05 mm. A filament diameter of 1.75 mm results in a minimum cross-sectional area of 2.27 mm² and a maximum of 2.54 mm². The mass flow is thus 12 % greater at the upper limit than at the lower limit. The analysis of the incoming filament with laser cut sensors is the next development step. This enables the detection of diameter variations and their influence on the process. Furthermore, the system presented does not detect the effective feed rate of the filament. Current systems (e.g. the one from Nienhaus et al. [30]) use an additional wheel with encoder to determine the true feed velocity. Slip detection is also not possible with this design. Therefore, a new detection approach is integrated according to [31].

5. REFERENCES

[1] Nee AYC. Handbook of Manufacturing Engineering and Technology. London: Springer London; 2015.

[2] Brenken B, Barocio E, Favaloro A, Kunc V, Pipes RB. Fused filament fabrication of fiber-reinforced polymers: A review. Additive Manufacturing 2018;21:1–16. https://doi.org/10.1016/j.addma.2018.01.002.

[3] van Der Klift F, Koga Y, Todoroki A, Ueda M, Hirano Y, Matsuzaki R. 3D Printing of Continuous Carbon Fibre Reinforced Thermo-Plastic (CFRTP) Tensile Test Specimens. OJCM 2016;06(01):18–27. https://doi.org/10.4236/ojcm.2016.61003.

[4] Cicala G, Ognibene G, Portuesi S, Blanco I, Rapisarda M, Pergolizzi E et al. Comparison of Ultem 9085 Used in Fused Deposition Modelling (FDM) with Polytherimide Blends. Materials (Basel) 2018;11(2). https://doi.org/10.3390/ma11020285.

[5] Stratasys. Ultem 9085 Data sheet.

[6] Zaldivar RJ, Witkin DB, McLouth T, Patel DN, Schmitt K, Nokes JP. Influence of processing and orientation print effects on the mechanical and thermal behavior of 3D-Printed ULTEM® 9085 Material. Additive Manufacturing 2017;13:71–80. https://doi.org/10.1016/j.addma.2016.11.007.

[7] Raut S, Jatti VS, Khedkar NK, Singh TP. Investigation of the Effect of Built Orientation on Mechanical Properties and Total Cost of FDM Parts. Procedia Materials Science 2014;6:1625–30. https://doi.org/10.1016/j.mspro.2014.07.146.

[8] Wang X, Zhao L, Fuh JYH, Lee HP. Effect of Porosity on Mechanical Properties of 3D Printed Polymers: Experiments and Micromechanical Modeling Based on X-ray Computed Tomography Analysis. Polymers (Basel) 2019;11(7). https://doi.org/10.3390/polym11071154.
[9] Papon EA, Haque A, Mulani SB. Process optimization and stochastic modeling of void contents and mechanical properties in additively manufactured composites. Composites Part B: Engineering 2019;177:107325. https://doi.org/10.1016/j.compositesb.2019.107325.
[10] Gurrala PK, Regalla SP. DOE Based Parametric Study of Volumetric Change of FDM Parts. Procedia Materials Science 2014;6:354–60. https://doi.org/10.1016/j.mspro.2014.07.045.
[11] Chohan JS, Singh R, Boparai KS, Penna R, Fraternali F. Dimensional accuracy analysis of coupled fused deposition modeling and vapour smoothing operations for biomedical applications. Composites Part B: Engineering 2017;117:138–49. https://doi.org/10.1016/j.compositesb.2017.02.045.
[12] Chen L, Zhang X, Gan S. Effects of laser polishing on surface quality and mechanical properties of PLA parts built by fused deposition modeling. J. Appl. Polym. Sci. 2020;137(3):48288. https://doi.org/10.1002/app.48288.
[13] Singh R, Singh S, Singh IP, Fabbrocino F, Fraternali F. Investigation for surface finish improvement of FDM parts by vapor smoothing process. Composites Part B: Engineering 2017;111:228–34. https://doi.org/10.1016/j.compositesb.2016.11.062.
[14] Kun K. Reconstruction and Development of a 3D Printer Using FDM Technology. Procedia Engineering 2016;149:203–11. https://doi.org/10.1016/j.proeng.2016.06.657.
[15] Kantaros A, Karalekas D. Fiber Bragg grating based investigation of residual strains in ABS parts fabricated by fused deposition modeling process. Materials & Design 2013;50:44–50. https://doi.org/10.1016/j.matdes.2013.02.067.
[16] Deng X, Zeng Z, Peng B, Yan S, Ke W. Mechanical Properties Optimization of Poly-Ether-Ether-Ketone via Fused Deposition Modeling. Materials (Basel) 2018;11(2). https://doi.org/10.3390/ma11020216.
[17] Dey A, Yodo N. A Systematic Survey of FDM Process Parameter Optimization and Their Influence on Part Characteristics. JMMP 2019;3(3):64. https://doi.org/10.3390/jmmp3030064.
[18] Phan DD, Swain ZR, Mackay ME. Rheological and heat transfer effects in fused filament fabrication. Journal of Rheology 2018;62(5):1097–107. https://doi.org/10.1122/1.5022982.
[19] Zhang Y-H, Liu Q-H, Niu Y-M, Zhang S-L, Wang D, Jiang Z-H. Properties and crystallization kinetics of poly(ether ether ketone)-co-poly(ether ether ketone ketone) block copolymers. J. Appl. Polym. Sci. 2005;97(4):1652–8. https://doi.org/10.1002/app.21395.
[20] Wang K. Die Swell of Complex Polymeric Systems. In: Vicente J de, editor. Viscoelasticity - From Theory to Biological Applications. InTech; 2012.
[21] Blanco I. The Use of Composite Materials in 3D Printing. J. Compos. Sci. 2020;4(2):42. https://doi.org/10.3390/jcs4020042.
[22] Duty C, Ajinjeru, Christine, Vidya, Kishore, Brett C, Hmeidat N, Chen, Xun, Liu, Peng, Arabi Hassan, Ahmed, Lindahl J et al. A viscoelastic Model for Evaluating Extrusion-based print conditions.
[23] Yan M, Tian X, Peng G, Li D, Zhang X. High temperature rheological behavior and sintering kinetics of CF/PEEK composites during selective laser sintering. Composites Science and Technology 2018;165:140–7. https://doi.org/10.1016/j.compscitech.2018.06.023.

[24] Osswald TA, Puentes J, Kattinger J. Fused filament fabrication melting model. Additive Manufacturing 2018;22:51–9. https://doi.org/10.1016/j.addma.2018.04.030.
[25] Boman B. Extruder Drive Force Measurements; Available from: http://www.vinland.com/Extruder_Drive_Force.html.
[26] Geng P, Zhao J, Wu W, Ye W, Wang Y, Wang S et al. Effects of extrusion speed and printing speed on the 3D printing stability of extruded PEEK filament. Journal of Manufacturing Processes 2019;37:266–73. https://doi.org/10.1016/j.jmapro.2018.11.023.
[27] Tlegenov Y, Hong GS, Lu WF. Nozzle condition monitoring in 3D printing. Robotics and Computer-Integrated Manufacturing 2018;54:45–55. https://doi.org/10.1016/j.rcim.2018.05.010.
[28] Li Y, Zhao W, Li Q, Wang T, Wang G. In-Situ Monitoring and Diagnosing for Fused Filament Fabrication Process Based on Vibration Sensors. Sensors (Basel) 2019;19(11). https://doi.org/10.3390/s19112589.
[29] Go J, Schiffres SN, Stevens AG, Hart AJ. Rate limits of additive manufacturing by fused filament fabrication and guidelines for high-throughput system design. Additive Manufacturing 2017;16:1–11. https://doi.org/10.1016/j.addma.2017.03.007.
[30] Nienhaus V, Smith K, Spiehl D, Dörsam E. Investigations on nozzle geometry in fused filament fabrication. Additive Manufacturing 2019;28:711–8. https://doi.org/10.1016/j.addma.2019.06.019.
[31] Matschinski A. Additive Manufacturing: European Patent(EP3587077).

THERMOPLASTIC ADDITIVE MANUFACTURING – ULTEM 9085 FOR STRUCTURAL APPLICATIONS

Daniel R. Klenosky, Brian R. Kitt
Spirit AeroSystems, Inc.
Wichita, Kansas, USA

ABSTRACT

ULTEM 9085 is a thermoplastic Fused Deposition Modeling (FDM) material with structural potential; high thermal and chemical resistance; adherence to industry flame, smoke, and toxicity (FST) standards; and high strength-to-weight ratio (about 25% of the Ti-6-4 ratio). It not only meets the traceability and process verification requirements necessary for certification and use in the aerospace industry but there are published, statistically significant material property values. Finally, the ability to produce complex, net-shape parts via FDM with no need for post-processing opens the door for a number of design features not currently practical for other materials, especially hard-to-machine materials like Ti-6-4. With these many advantages, ULTEM 9085 has become a common material for a variety of non-structural aerospace applications.

The natural next step for this material is to investigate its potential for use in secondary structural applications. In this work, a number of generic Ti-6-4 bracket-like parts were thickened based on published ULTEM 9085 and Ti-6-4 properties to have equivalent stiffness to their titanium counterparts. ULTEM 9085 parts were mechanically tested at room temperature via proof testing followed by loading to failure under service-like loads to verify strength predictions, understand failure modes, and learn about trade-offs between the FDM process and the conventional Ti-6-4 forming and machining process. Test results shed light on the relative advantages and disadvantages of the material and the FDM process for use in secondary aerospace structural applications.

Keywords: ULTEM, Additive Manufacturing, FDM
Corresponding author: Daniel R. Klenosky

1. INTRODUCTION

1.1 Background

Spirit AeroSystems, Inc. has identified a research and development opportunity in the manufacturing of brackets from ULTEM 9085 material, additively manufactured (AM) via fused deposition modeling (FDM) (Figure 1).

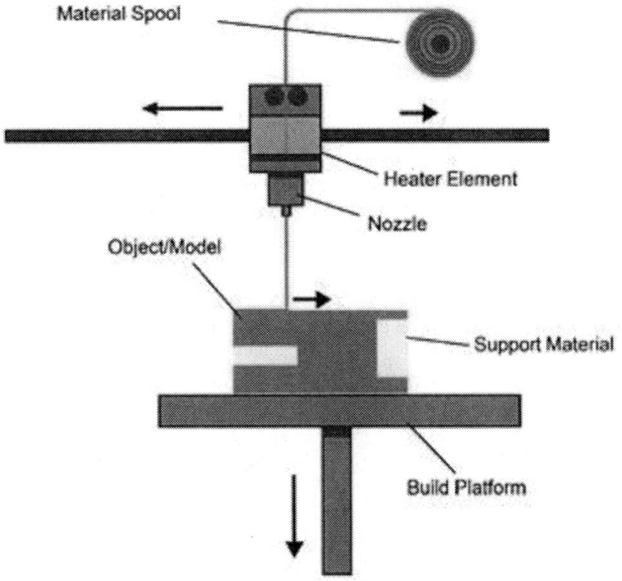

Figure 1. Schematic Description of Fused Deposition Modeling

The major advantage of this material is the high strength-to-weight ratio (competitive to Ti-6-4) and the existence of B-basis allowables [1], shown in Table 1. These allowables are a major cost barrier associated with polymer AM materials and other new materials and processes. As these allowables exist, it is necessary to verify the mechanical behavior of prototype parts against the predicted behavior. This work describes strength, stiffness, and fatigue testing of generic systems bracket-type aerospace parts that was performed in order to evaluate and advance the technology.

Table 1. ULTEM 9085 B-Basis Allowables Properties [1]

	Orientation	**$F_{t,u}$ (ksi)**	**$F_{t,y}$ (ksi)**	**Elongation [2]**	**Density (lbs/in^3) [2]**
ULTEM 9085	X	8.82	4.87	5.4%	0.048

1.2 Prototype Bracket Parts

The generic bracket parts used in this work were based off of generic Ti-6-4 aerospace systems brackets. Brackets such as these are often sized based on stiffness requirements, rather than other common sizing requirements such as tensile strength, fatigue, or damage tolerance. For this initial attempt at comparing ULTEM 9085 brackets to those made from Ti-6-4, the thickness of the ULTEM brackets was selected as the thickness required so that the longest arm of the bracket is at least as stiff as its Ti-6-4 counterpart. The brackets are shown below from Figure 2-Figure 4, along with the thickness of both the original generic Ti-6-4 part and the proposed ULTEM 9085 replacement part.

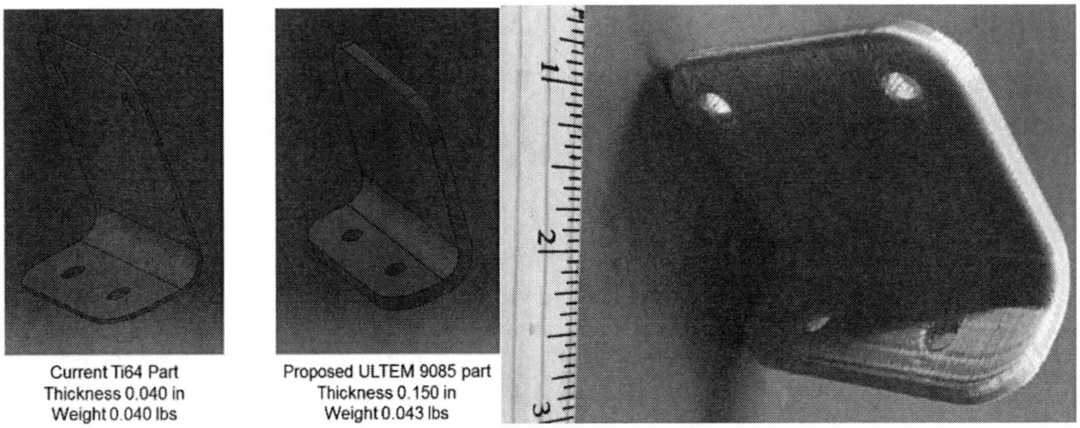

Figure 2. Prototype Bracket -1

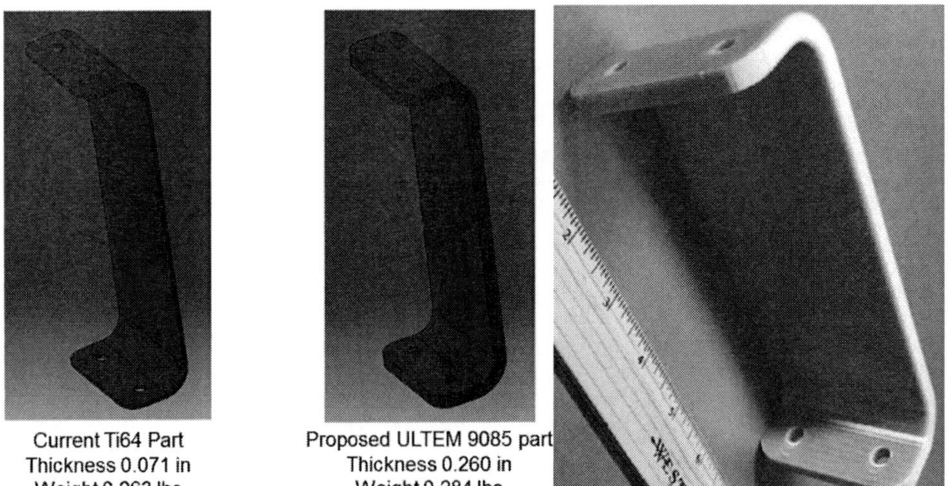

Figure 3. Prototype Bracket -2

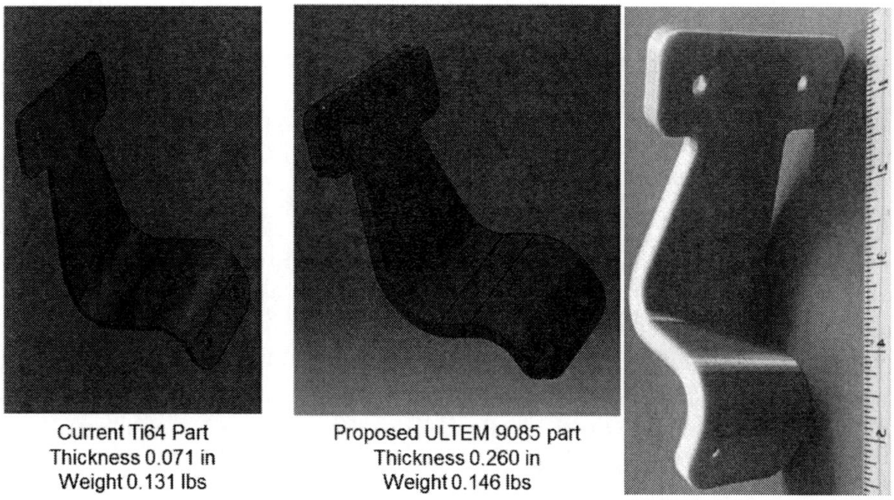

Figure 4. Prototype Bracket -3

These prototype parts were formed on a Stratasys Fortus 900mc (Figure 5), located at the National Institute for Aerospace Research (NIAR) on the campus of Wichita State University. All ULTEM 9085 material used in this test was procured from Stratasys, Inc. and adhered to NCAMP Material Specification NMS 085/1 Rev IR and was printed according to NCAMP Process Specification NPS 89085 Rev C (Reference 2,3).

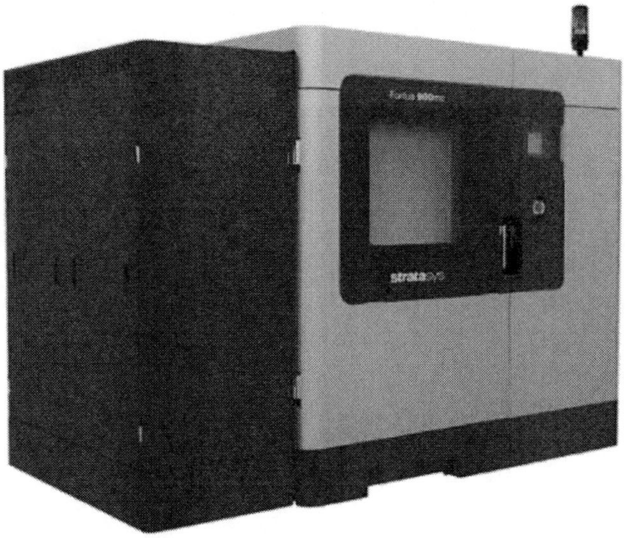

Figure 5. Stratasys Fortus 900mc FDM Machine

1.3 Purpose of Testing

The generic brackets in this work were subjected to both static and dynamic testing. The static testing was intended to verify that the observed stiffness and strength properties correspond to what is expected based on the available material property data. The dynamic testing was intended to understand if the failure mechanism and location would match the observed failure in static testing, or if there are other mechanisms at play that will limit fatigue performance.

With that in mind, the tests in this work were intended to serve two larger purposes: 1) To increase understanding about the behavior of ULTEM 9085 material printed via FDM and 2) to understand how this behavior may deviate from conventional (metallic) behavior models and impact design considerations. This is important not only to understand how this material will serve as a drop-in replacement for existing structural parts, but it also has implications for the potential to re-design and/or consolidate multiple parts into one – two key concepts that make additive manufacturing such an attractive alternative for aerospace structural parts in the first place.

2. EXPERIMENTATION

2.1 Static Testing

The intent of the static tests was to verify that ULTEM 9085 FDM structural parts can be designed using traditional Spirit design methods, given the published allowables property data. The results from this test can be used to verify that this material behaves predictably in deflection at low loads and in failure at extreme loads. This test was guided by the specifications in ASTM D3039/D3039M [5] although there were deviations due to the geometry of these parts. There were 3 tests each of the 3 different part shapes, for a total of 9 tests.

The intended free body diagrams are shown in Figure 6, along with the parts loaded onto the test fixtures in Figure 7, Figure 8, and Figure 9.

The test procedure was as follows:

1. Load the part to 20 lbf (abuse load) at 0.02 in/min (displacement control per ASTM D3039/D3039M) according to the appropriate free body diagram described in Figure 6.
2. Pause for 3 seconds, record deflection (via deflectometer), and inspect the part.
3. Unload and inspect the part for cracking, especially the area where it is mounted to the load frame.
4. Re-load at 0.2 in/min until failure.
5. Record load at failure (Fu) and max deflection.

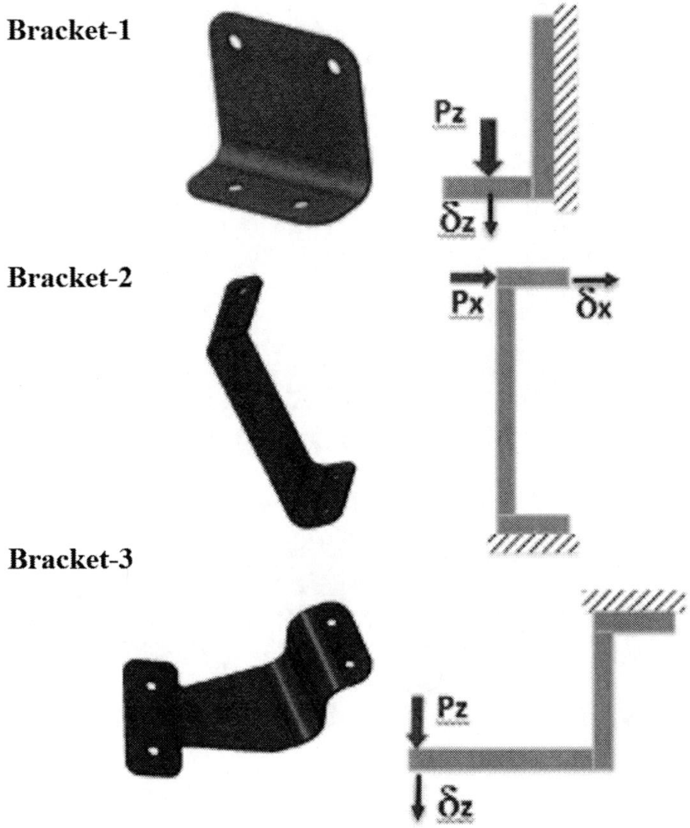

Figure 6. Free Body Diagrams for Each of the Three Static Generic Bracket Tests

Figure 7. Bracket-1 Test Setup

Figure 8. Bracket-2 Test Setup

Figure 9. Bracket-3 Test Setup

2.2 Dynamic Testing

The purpose of this test was to identify potential problems associated with crazing and/or poor fatigue behavior associated with ULTEM 9085 FDM material, especially located around the holes used to mount this material to the test fixture (or to an aerostructure). These properties are not well-understood for this material. This test was guided by the specifications in ASTM D3479/D3479M (Reference 6), although there were deviations due to the geometry of this part.

In this initial screening test, only four bracket -1 parts were tested. Brackets -2 and -3 were not tested at this time. The free body diagram and test fixture for this test are identical to those used in the static test, shown in Figure 6 and Figure 7.

This repeated load test was intended to resemble a fatigue test (per ASTM D3479/D3479M) with a stress ratio R=0.05 and cyclic frequency of 4 Hz. The max load was to be approximately 0.80* F_u (obtained from the static test). These test parameters were designed for the test specimens to fail between 1,000-10,000 cycles.

3. RESULTS

3.1 Static Test Results

All nine test brackets were successfully subjected to their 20 lbf abuse loads for at least 3 seconds. Inspections after abuse loading did not reveal any indication of cracking, permanent deformation, or other damage. An example of the typical response to each part loaded to 20 lbf is shown in Figure 10, along with the corresponding average deflection as measured by the deflectometer.

Bracket 1	Bracket 2	Bracket 3
Avg. Deflection 0.018 in	Avg. Deflection 0.382 in	Avg. Deflection 0.061 in

Figure 10. Each Bracket Loaded to its 20 lbf Abuse Load

The results of the subsequent load-to-failure test are summarized in Table 2. For each part, the location and mode of failure was consistent between all three tests. All statically failed specimens are shown from Figure 11 to Figure 13.

Table 2. Static Test Results

Bracket Number	Average Deflection at Abuse Load (in)	Average Max Deflection at Failure (in)	Average Load to Failure (lbf)
-1	0.018	0.236	163
-2	0.382	2.261	245
-3	0.061	0.676	242

Figure 11. Statically Failed Bracket-1 Parts

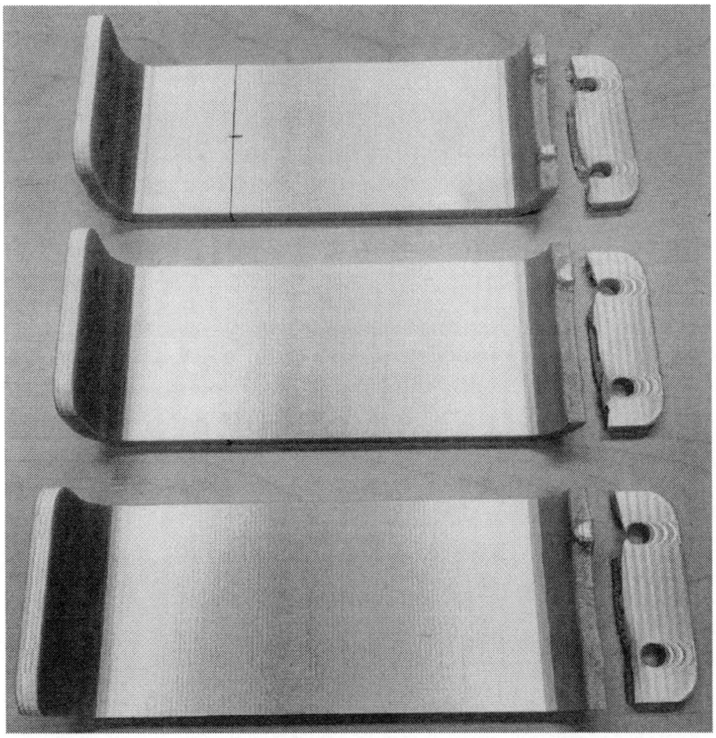

Figure 12. Statically Failed Bracket-2 Parts

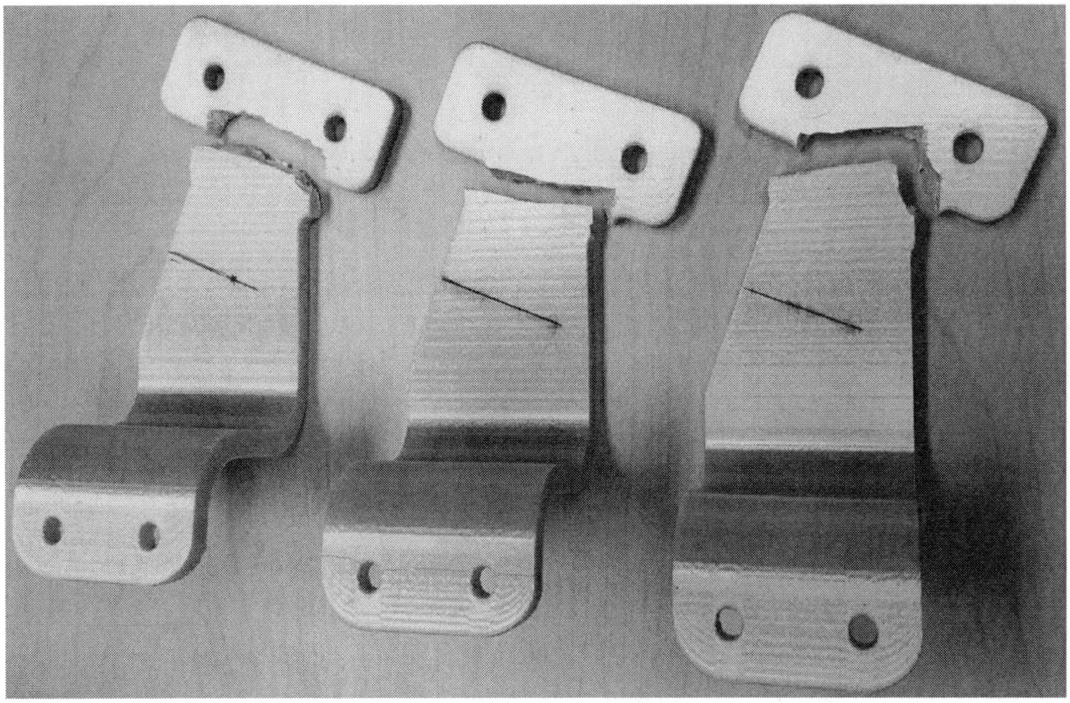

Figure 13. Statically Failed Bracket-3 Parts

3.2 Dynamic Test Results

The calculated maximum load to failure (F_u) from the static testing of bracket -1 was 163 lbf, so $0.8*F_u$=131 lbf was chosen as the maximum load for the dynamic repeated load test. With this max load and the other test conditions specified in Section 2.2, the four tested parts survived 239, 305, 1389, and 2802 cycles to failure, for an average of 1184 cycles to failure. It should be noted that the scatter for this testing was quite large, with a standard deviation of 1200 cycles. This large deviation is likely due to amount of part movement made possible by the test setup. This setup should be corrected in future testing.

All four test brackets that were subjected to this test failed at the same place as the static test brackets - at the radius in the part – indicating that the test geometry and load is leading to these failures rather than flaws that may be inherent to the FDM process. All four dynamically failed brackets are shown in Figure 14.

Figure 14. Dynamically Failed Bracket-1 Test Specimens

4. CONCLUSIONS

4.1 Implications of Results

The major finding from this initial screening test is that ULTEM 9085 manufactured via FDM may be a suitable material for systems bracket-type structural aerospace parts. No permanent damage was observed during or after the 20 lbf abuse test for any parts. All three brackets demonstrated the capacity to support a load at least 8x the abuse load without signs of failure. All observed failure occurred in the expected locations, indicating there are not inherent flaws associated with the FDM process that are causing unexpected failure.

Additional failure characterization, a re-design of the fatigue test fixture, and larger sample sizes are required to make larger conclusions about the materials and parts in this work. These tests were only ever intended to serve as an initial screening.

Future work should focus on characterizing the differences between the observed behavior and the expected behavior (perhaps obtained via FEM analysis or via comparison to existing parts) of the ULTEM 9085 test brackets and/or a bracket made from a conventionally produced material such as titanium. The long-term goal of this work should aim to identify the relationship between certain AM design shapes and deviations from expected mechanical behavior, in order to create a set of more general guidelines to inform the future part design that can take full advantage of additive manufacturing processes.

5. REFERENCES

1. *Stratasys Certified ULTEM 9085 Fortus 900mc Additively Manufactured Polymer Material Qualification Statistical Analysis Report*. Report No. NCP-RP-2018-007 Rev N/C. NCAMP, April 16, 2019.
2. *ULTEM 9085 Production-Grade Thermoplastic for Fortus 3D Production Systems*. Stratasys, Inc. 2016.
3. *NCAMP Material Specification – Stratasys Inc. ULTEM 9085 Resin*. Document No. NMS 085 Rev B. NCAMP. April 12, 2019.
4. *NCAMP Process Specification – Polymer Additive Manufacturing Materials, Machine, Processing, and Quality Requirements Specification for ULTEM 9085 and Stratasys Inc. Fortus 900mc Machine*, Document No. NPS 89085 Rev C. NCAMP. April 12, 2019.
5. *ASTM D3039/D3039M, Standard Test Method for Tensile Properties of Polymer Matrix Composite Materials*. ASTM International. 2017.
6. *ASTM D3479/D3479M, Standard Test Method for Tension-Tension Fatigue of Polymer Matrix Composite Materials*. ASTM International. 2019.

VIRTUAL INVESTIGATION OF RESIDUAL PART DEFORMATION DUE TO BUILD PLATE SUPPORT CHARACTERISTICS IN MATERIAL EXTRUSION ADDITIVE MANUFACTURING

Eduardo Barocio, Akshay Jacob Thomas, R. Byron Pipes
Composites Manufacturing and Simulation Center
West Lafayette, IN

ABSTRACT

The Material Extrusion Additive Manufacturing (MEAM) process with fiber-reinforced thermoplastics has enabled scaling the dimensions of geometries that can be printed to the scale of multiple meters. Multiple applications including tooling for processing composites and prototype vehicle components have been successfully demonstrated with large scale MEAM systems such as BAAM and LSAM. The highly non-isothermal processing conditions developed during the printing process give rise to residual stresses and part deformations. To prevent the part from deforming excessively during the printing process multiple methods have been adopted in MEAM to adhere parts to the build plate. These include adhesives, thin sheets of polymeric material adhered with vacuum, and textured surfaces. In addition, the build plate is in some cases actively heated. Depending on the part geometry and the stresses developed during the printing process, debonding of the printed part from the build plate can occur during printing. To prevent the exhaustive process of empirically calibrating processing conditions for different configurations of part adherence, this work investigates virtually the effect of three different methods for adhering the part to the build plate; namely an adhesive layer, a thin sheet of polymer (build sheet) held with vacuum to the build plate, and a rigid constraint at the bottom provided by some mechanisms that provide mechanical interlocking between the part and the build plate. Two geometries which include a plate and a double stringer tool are investigated under the different conditions. The investigation is carried out with ADDITIVE3D© and with a material card for which temperature and deformation predictions have been validated experimentally. It was observed that, for both geometries, the adhesive model showed maximum pull-off from the build plate whereas the model with the rigid constraint at the bottom showed the least pull-off. This preliminary study indicates that a stiff interaction between the part and the build plate reduces the part deviation from the designed geometry.

Keywords: Material Extrusion Additive Manufacturing, AM Process Simulation, Build plate debonding, AM Residual Stress and Deformation, Fiber Reinforced Thermoplastics

Corresponding author: Eduardo Barocio (ebarocio@purdue.edu)

1. INTRODUCTION

Material Extrusion Additive Manufacturing (MEAM) is a process by which fiber-filled polymeric material is melted and mixed in an extruder followed by the deposition on to a build platform prescribed by a machine code. The process includes pelletized feedstock being melted in an extruder and exiting a deposition nozzle. This enables the layer-by-layer generation of three-dimensional geometries at the scale of multiple meters in length and with high geometrical complexity. In some cases, the build plate moves relative to the nozzle whereas in some machines the nozzle is mobile. Examples of commercial large-scale MEAM systems include the BAAM system by Cincinnati Incorporated [1] and the LSAM system by Thermwood Corporation©, which are capable of depositing beads of molten material at rates up to 45 kg/hr [2] and 227 kg/hr, respectively. The Composites Additive Manufacturing Research Instrument (CAMRI), a mid-scale MEAM system, was developed at Purdue University for investigating the phenomena involved in the MEAM process [3]. Love et. al [4] discussed the effect of the addition of discontinuous carbon fibers (CF) to acrylonitrile butadiene styrene (ABS) and shows the enhancement of mechanical properties of ABS parallel to the printing direction. However, the enhancement is not pronounced to be used in structural applications. Therefore, the prominent application of MEAM parts have been in the production of composite tooling and molds, printed at net near shape and subsequently machined to the final size [4], [5].

The highly non-isothermal printing process coupled with material anisotropy leads to residual stresses and deformations. The extent of the deformation varies with varying printing strategies, material system, temperature of the build platform (in the case of a heated bed), etc. Therefore, it remains critical to adhere the part to the build platform during the printing process. Multiple methods exist in MEAM to adhere the part to the build platform and these include adhesives, thin sheets of polymeric material adhered with vacuum, textured surfaces, etc. Depending on the part geometry and the printing conditions (material and machine parameters) the part may debond from the build platform. To circumvent the cumbersome task of empirical calibration of the various methods to attach the part to the build platform, this work looks at virtually investigating the effect of different boundary conditions on the deformation of the part. To that end, two methods of adhering the part to the build plate will be studied here – using an adhesive and using a build sheet held down by vacuum. To illustrate the methods, a schematic has been shown in Figure 1.

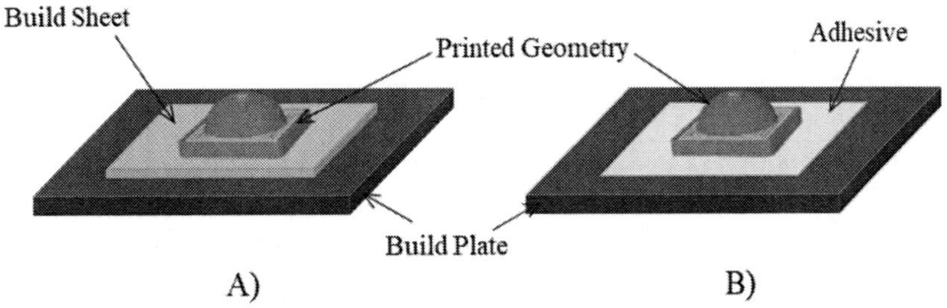

Figure 1. Schematic representation of methods used to attach geometries to the build plate during printing process. A) Geometry printed on top of build sheet held with vacuum to the build plate. B) Geometry printed on top of a thin layer of adhesive applied on the build plate.

Multiple thermo-mechanical analysis have been reported for predicting the residual stresses and deformation resulting from the material extrusion additive manufacturing process or the related Fused Filament Fabrication (FFF) process [6]–[13]. However, these previous studies have considered rigid bonding of the printed geometries to the build plate. The integrated simulation workflow Additive3D© [7], [14] was used in this work to investigate the effects on the residual deformation of the different methods used to attach the material deposited during the printing process to the build plate of the machine. Additive3D© is a physics-based simulation workflow that allows replicating the process of depositing beads of molten material by activating initially dormant elements in a finite element mesh based on the trajectory of the print head. The sequentially coupled thermomechanical analysis carried out in Additive3D© captures phenomena such as orthotropic heat transfer, local convection conditions, heat losses during material compaction, anisotropic shrinkage, crystallization shrinkage, polymer crystallization and polymer crystal melting, and thermoviscoelasticity [7], [14]–[16].

Different methods have been adopted for adhering the first layer of printed material to the build plate of the additive manufacturing system including adhesives [15], [16], thin sheets (i.e., between 1 and 4 mm) of polymer held with vacuum to the build plate which are referred to as build sheets, the bead board patented by Thermwood® which consist of plywood with pellets of acrylonitrile butadiene styrene (ABS) glued to the surface [17], and systems that provide mechanical interlocking of the printed material with the build plate like the system developed by CEAD [18]. The thermal residual stresses developed during the printing process of a geometry can cause debonding of the part from the bead board or failure of an adhesive layer. By the same token, the build sheet can be deformed, the printed geometry can debond from the build sheet, or the build sheet can be deformed to the extent of compromising the vacuum attaching the build sheet to the build plate. Hence, the next section focuses on investigating through simulations the part deformation at the end of the printing process resulting from utilizing a build sheet, an adhesive, and a rigid constraint at the bottom of the printed geometry. The modeling techniques and assumptions made for each method for adhering the part to the build plate are presented first and are followed by a simulation study performed for two geometries susceptible to warp printed with the three different part-build plate attachment methods. The material used in this study was polyphenylene sulfide (PPS) reinforced with 50% by weight of carbon fiber (CF-PPS) produced by Celanese© by pultrusion and chopped into pellets. The CF-PPS has been characterized for simulating the printing process with Additive3D©, and simulations results have been validated against experimental measurements [7], [15], [16].

2. METHODS

To investigate the effects of using a build sheet, a layer of adhesive, and a rigid constraint on the residual deformation of a printed part, two geometries that are susceptible to deform during the printing process were chosen. The first geometry consists of a plate printed with four layers through the thickness and with two layers oriented at zero degrees and the subsequent two layers oriented at ninety degrees [0/0/90/90]. Such a printing configuration leads to internal stresses due to the anisotropy of the printed fiber-reinforced material in addition to the thermal residual stresses introduced with the additive process. These stresses can cause the plate to debond from the build plate during the printing process and makes the plate susceptible to attaining an anticlastic bending shape upon removal from the build plate. Figure 2.-A shows the dimensions of the plate considered in this work.

The second geometry is referred to as double stringer tool and consists of a single-sided tool used for autoclave manufacturing with fiber reinforced thermosets of a representative aircraft stringer component. Figure 2-B shows the geometry and dimensions of the double stringer tool. The stringer tool is printed with layers stacked along the height shown in Figure 2-B and with two beads across the thickness of the wall. Beads with dimension of 6 mm and 2 mm in width and height, respectively, were used for both geometries. The double stringer tool was chosen since it is susceptible to warp up at the ends due to the thermal residual stresses developed during the printing process.

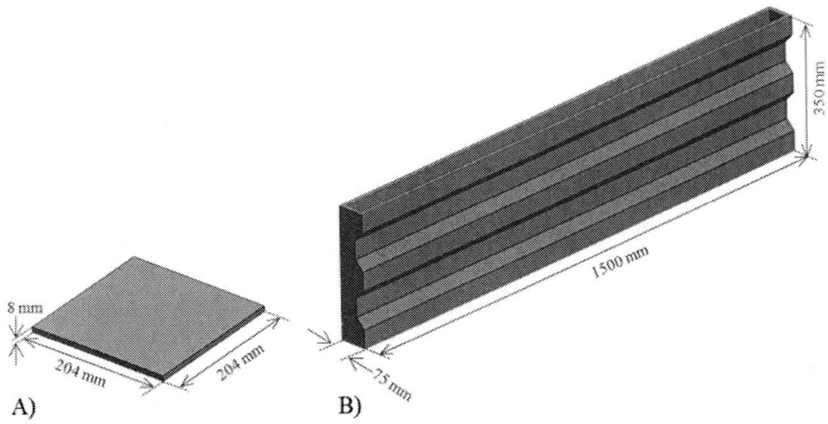

Figure 2. Geometries used in investigation. A) Horizontal plate printed with layer orientation 0/0/90/90. B) Geometry of double stringer tool.

The approaches for modeling the build sheet, the layer of adhesive, the rigid constraint, and the assumptions made for each method were as follows. The build sheet was modeled explicitly in the finite element mesh and was assumed perfectly bonded to the printed part (element connectivity enforced between the build sheet and printed part). Though, it can occur that the part debonds from the build sheet during the printing process, which was not considered in the current model. Further, the build sheet material was modeled as a thermoelastic solid and with constant heat capacity, thermal conductivity, and coefficient of thermal expansion. The material properties for the polycarbonate (PC) considered for the build sheet were obtained from a materials database [19]. The pressure load resulting from the vacuum applied underneath the build sheet was modeled through connector elements (CONN3D2) [20] that imposed a constant force at the bottom nodes of the finite elements in the build sheet. The second end of the connector elements was connected to ground. Each nodal force was computed by integrating the vacuum pressure over one quarter of the area of each element sharing the node at the bottom of the build sheet. Thus, a constant pressure is approximated despite variations in mesh density. To model the tape applied around the edges of the build sheet to secure it on the vacuum table, connector elements with a relatively low stiffness (e.g., 1 N/mm) in the three directions were used. Figure 4.-C shows the finite element models of the plate and the stringer tool printed on a build sheet.

To represent the layer of adhesive, the interaction between the part and the build plate was modeled using a suitable cohesive interaction. A contact pair between the bottom surface of the printed part and an analytical rigid plane connected to ground was used for implementing the cohesive interaction. The cohesive interaction serves to model the failure of the adhesive allowing the part to pull off from the build plate during the printing process. A bi-linear traction separation law as shown in Figure *2* describes the constitutive behavior of the cohesive interaction. The initiation of damage (debonding) is triggered based on the maximum stress criterion and evolves according to a displacement-based damage evolution law. Currently, only the normal traction is considered to model the pulling off the part from the build plate. Referring to Figure *3*, the following values are used- $K_n = 0.125\ (N/mm^3),\ t_n = 0.0125\ (N/mm^2), and\ \delta_n^f = 5\ mm$. Figure *4*-B shows the finite element models of the plate and the stringer tool with a rigid plane at the bottom of the parts.

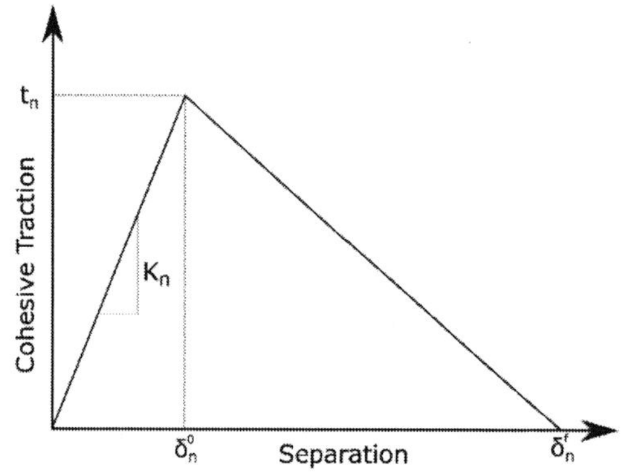

Figure 3. Illustration of the bi-linear traction separation law. δ_n^0 represents the displacement at the onset of failure and δ_n^f represents the displacement at failure.

To model the rigid constraint, a kinematic boundary condition was applied at the bottom nodes of the printed part to constraint all the degrees of freedom. Figure *4*-A shows the finite element models of the plate and the stringer tool used for the condition with the rigid constraint.

For the heat transfer analysis both the geometries were discretized using DC3D8 (8 node linear brick element for diffusive heat transfer) elements and for the mechanical analysis the geometries were discretized using C3D8 (8-node linear brick) elements available in ABAQUS© [21]. Each bead in the plate geometry is represented using 2 elements through the thickness (stacking direction) and 2 elements through the width (transverse to printing direction). In the case of the stringer tool geometry each bead was discretized using 1 element through the thickness and 1 element though the width of the bead. The build sheet was also discretized using the same elements for the case of the build sheet model (Figure *4*-C). The build sheet was 3 mm in thickness and was discretized using 1 element through the thickness. The mesh density in the in-plane dimensions of the build sheet was made 4 times finer in the region adjacent to the part (in comparison to the rest of the build sheet). Multiple steps were modeled in the process simulation to closely represent the actual process of printing a geometry. These steps are listed in Table 1 with the corresponding time (s) for each process.

Table 1. Steps in the process simulations

Step	Time (s) for the Plate Geometry	Time (s) for the Stringer Tool Geometry
Material Deposition	312	10762
Cooling on build plate	300	900
Spring back	NA	NA
Cooling off bed	900	900
Removal of part from build sheet (applicable only to build sheet model)	NA	NA

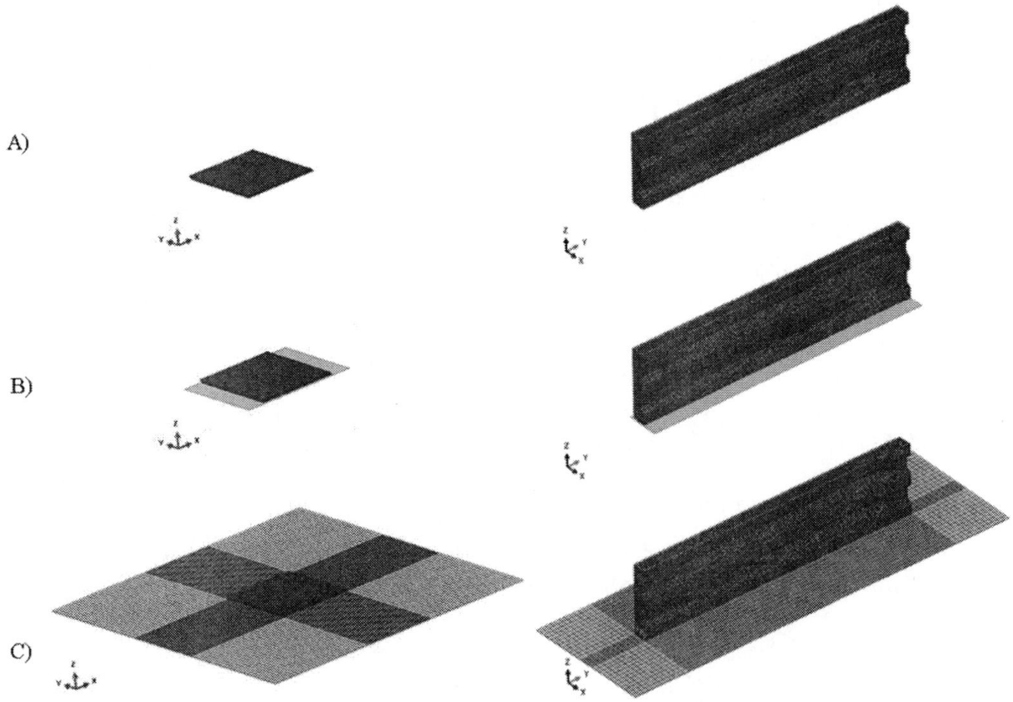

Figure 4. Finite element meshes for the different models for attaching printed parts to the build plate. A) Kinematic boundary condition used for rigid constraint. B) Cohesive contact used for adhesive layer. C) Build sheet with connector elements.

The plate was printed at speed of 5.3 m/min and the stringer tool was printed at a rate of 6 m/min. A table listing the printing parameters is shown in Table 2.

Table 2. Process and material parameters

Parameter	Value
Ambient Temperature Convection (°C)	**30**
Ambient Temperature Radiation (°C)	**165**
Material Deposition Temperature (°C)	**300**
Emissivity	**0.92**

3. RESULTS

3.1 Plate Geometry

First, the deformation results for the plate geometry for the three modeling techniques are presented. Figure *5* shows the deformation of the plate at different times during the process simulation for the three different modeling techniques. Each column in Figure *5* represents one model at different times during the process simulation. Figure *5* (A-I), which corresponds to the adhesive model, shows one end of the plate pulling off from the build sheet as indicated by the small displacement in the positive 3-direction. It should be noted that the 3-direction corresponds to the layer stacking direction or the Z-direction indicated in Figure *5*. However, for the other two models (Figure *5* (A-II) and Figure *5* (A-III)), a dominant negative displacement developed due to the thermal and crystallization shrinkage of the printed material. At the end of the deposition process, the adhesive model showed a greater displacement in the 3-direction (Figure *5* (B-I)) than the other two models. The build sheet model and the rigid constraint model showed a negligible displacement in the 3- direction at this instant of time (Figure *5* (B-I) and (B-II)). Further, considerable deformation was developed in the plate at the end of the cooling on bed step for the adhesive model. This can be seen in Figure *5*(C-I) and corresponds to the failure of the cohesive interaction caused by the stresses developed during the printing process. Finally, once the plate was removed from the build plate (and the build sheet for the case of the build sheet model) the plate attained the anticlastic bending shape for all three models. This is shown in Figure *5*(D-I) – Figure *5*(D-III). However, it is worthwhile to compare the magnitude of the deformation the plate developed once it was cooled to the room temperature at the end of the cooling off bed step and removed from the build sheet. While the adhesive model developed the largest maximum positive displacement (5.6 mm), the rigid constraint model developed the smallest maximum positive displacement (~4.6 mm). This results from the fact that the adhesive model allows deformation in the 3-direction as debonding occurs during the printing process. Further, the build sheet model developed a maximum positive displacement greater than the rigid constraint model but less than the adhesive model. However, while comparing the magnitude of negative displacement a reverse trend was observed. The rigid constraint model attains the maximum negative displacement followed by the build sheet model and the adhesive model (Figure *5*). This can be attributed to an increase in bending stiffness in the direction transverse to the direction where the deformation of the plate originated during the printing process [21]. Consequently, the adhesive model which developed the largest positive displacement in the 3-direction (before being removed from the build plate) developed the largest transverse bending stiffness which resulted in the smallest negative curvature. Conversely, the rigid constraint model which developed the smallest positive curvature in the longitudinal direction led to the lowest transverse stiffness among the three models and therefore produced the highest negative curvature.

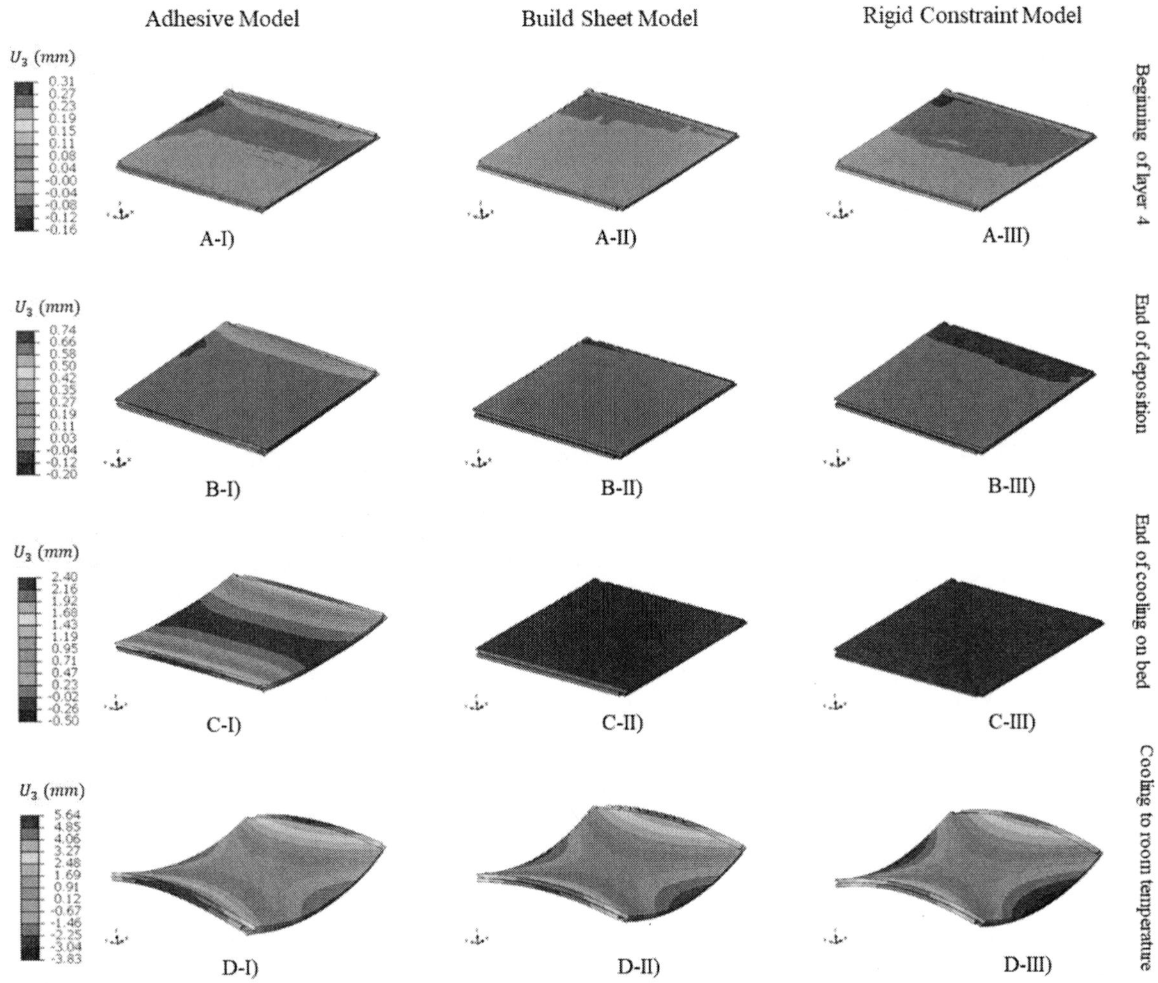

Figure 5. Deformation in the 3-direction of the plate at different times during the process simulation for the three different modeling techniques. A) Deformation during the printing process after 3 layers. B) Deformation after the deposition process C) Deformation after cooling of the part on the build plate. D) Deformation after detaching part from the build plate and cooling to the room temperature.

To compare the anticlastic profile of the plate attained from the different models after being cooled to room temperature, the 3-direction displacements along the two in-plane directions of the plate are plotted in Figure *6*. The paths along which the displacement was extracted is also indicated in Figure *6*. As mentioned earlier, the adhesive model attains the maximum positive curvature whereas the rigid constraint model attains the least positive curvature. In the case of the negative curvature, the rigid constraint model shows the maximum curvature with the adhesive model showing the smallest negative curvature.

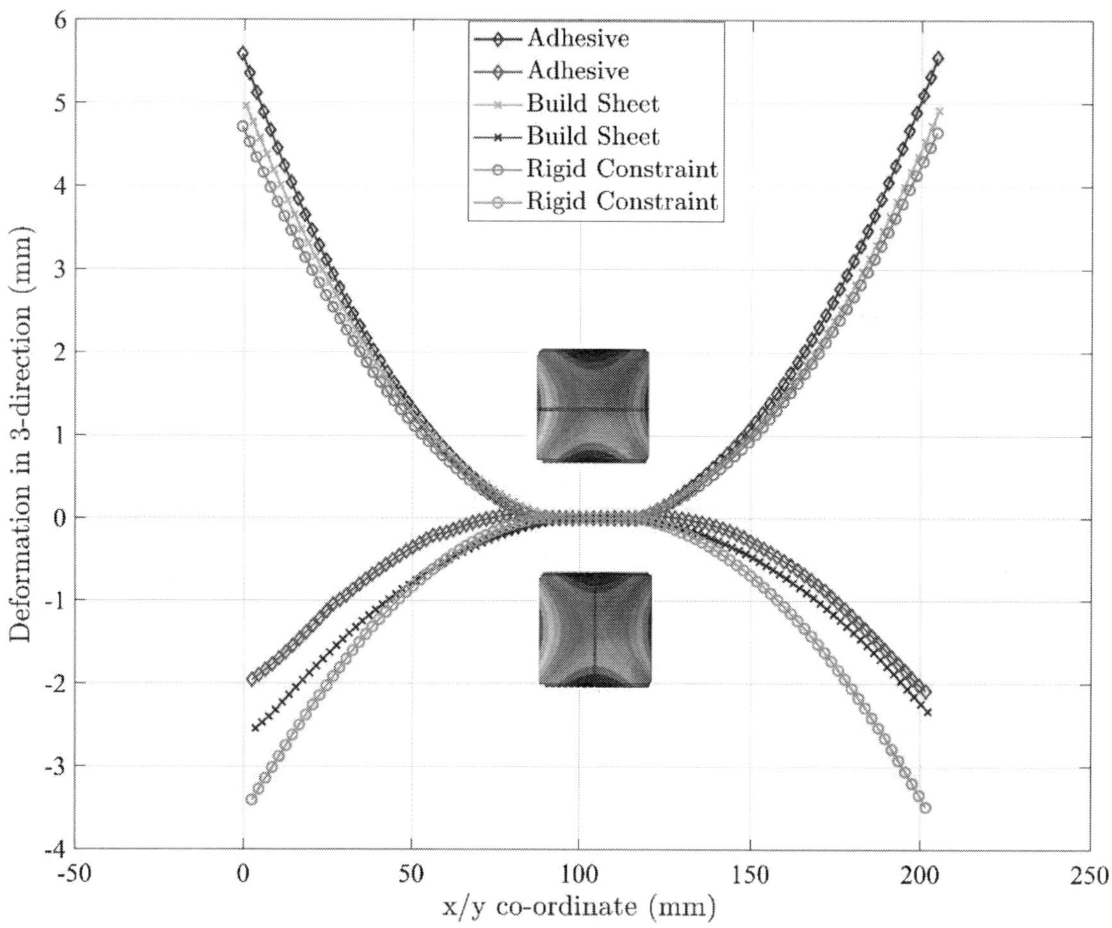

Figure 6. Comparison of the positive and negative curvatures of the plate for the different models.

3.2 Stringer Tool Geometry

Figure 7 shows the evolution of the deformation in the 3-direction at different steps of the printing process and for the three build plate modeling approaches considered in this study. The first row corresponds to the deformation developed after depositing 106 layers. At this point in the printing process, the adhesive started to fail in the regions near the ends of the stringer tool, allowing the ends of the part to warp up due to the thermal residual stresses. The asymmetric debonding process can be related to the printing strategy where the two beads across the walls are printed from the inside to the outside and starting at the lower right corner of the part (opposite to the end where debonding is more pronounced). However, further investigations considering the in-plane shear behavior for the cohesive model should be carried out to confirm the observed deformation behavior. For the same time in the process, small deformations developed at the bottom of the part printed on the build sheet since the build sheet is being pulled off as a result of the thermal residual stresses, and no deformation was observed in the part printed with the rigid constraint since it remains attached to the build plate. The second row shows the deformation developed after cooling the printed part on the heated build plate for 15 minutes. For the part printed on the adhesive, the damage of the adhesive progressed more allowing the part to warp up more. For the build sheet, slightly more deformation developed at the bottom of the stringer tool. Finally, the third row shows the deformation developed after the printed parts have been removed from the build plate and cooled to the room temperature and with the build sheet removed for the geometry printed on the build sheet. The part printed on the layer of adhesive developed the largest residual deformation followed by the part printed on the build sheet, and the part printed with the rigid constraint developed the smallest deformation.

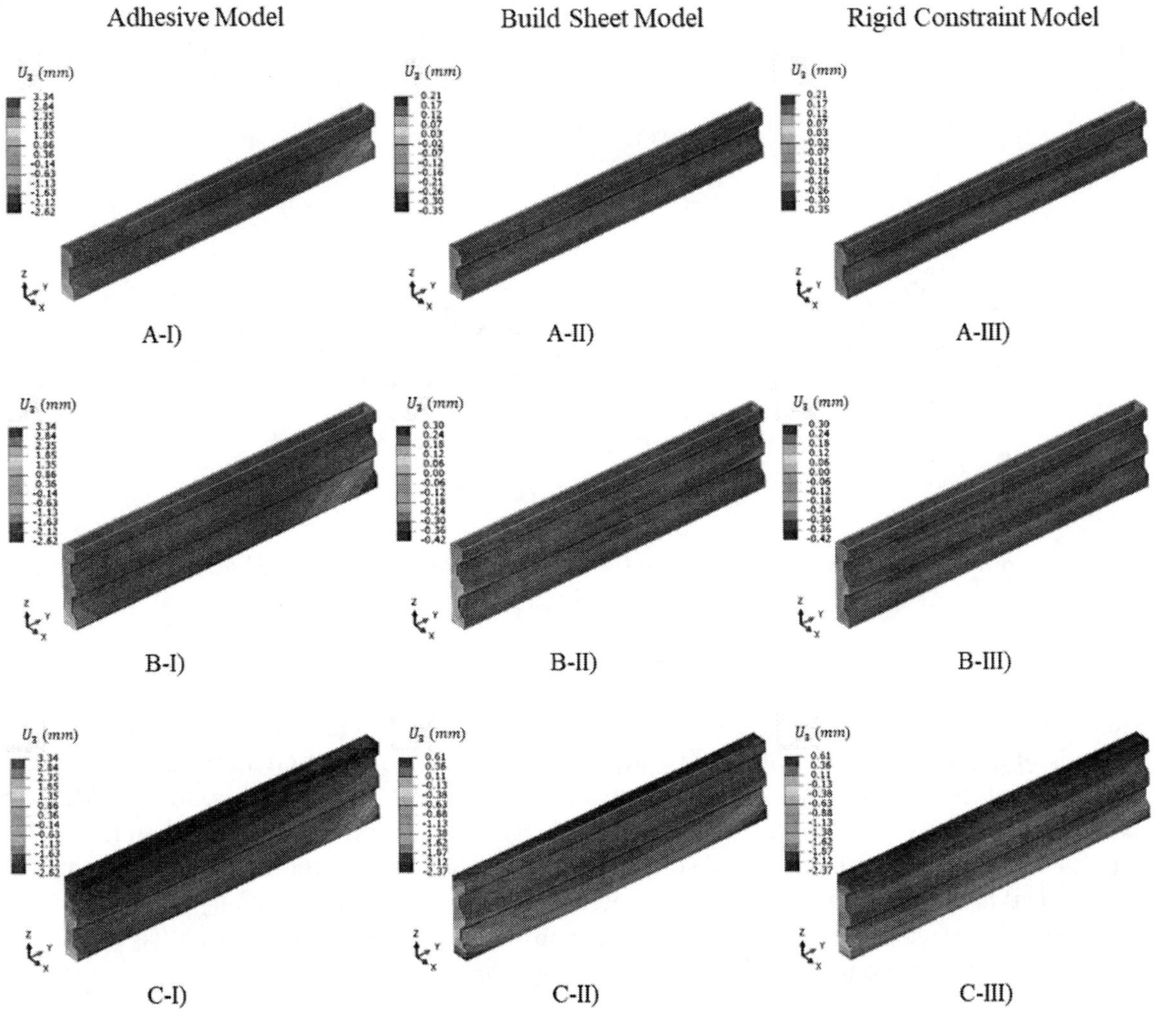

Figure 7. Deformation in the 3-direction of the stringer for the three methods for adhering the part to the build plate. A) Deformation during the printing process after 106 layers. B) Deformation after cooling of the part on the build plate. C) Deformation after detaching part from the build plate and cooling to the room temperature.

To provide a quantitative comparison of the deformation developed at the bottom of the stringer tool, the deformation in the 3-direction was extracted at the bottom nodes located at the front face of the part. Figure 8 compares the deformation in the 3-direction developed at the bottom of the stringer tool for the three methods considered for adhering the part to the build plate.

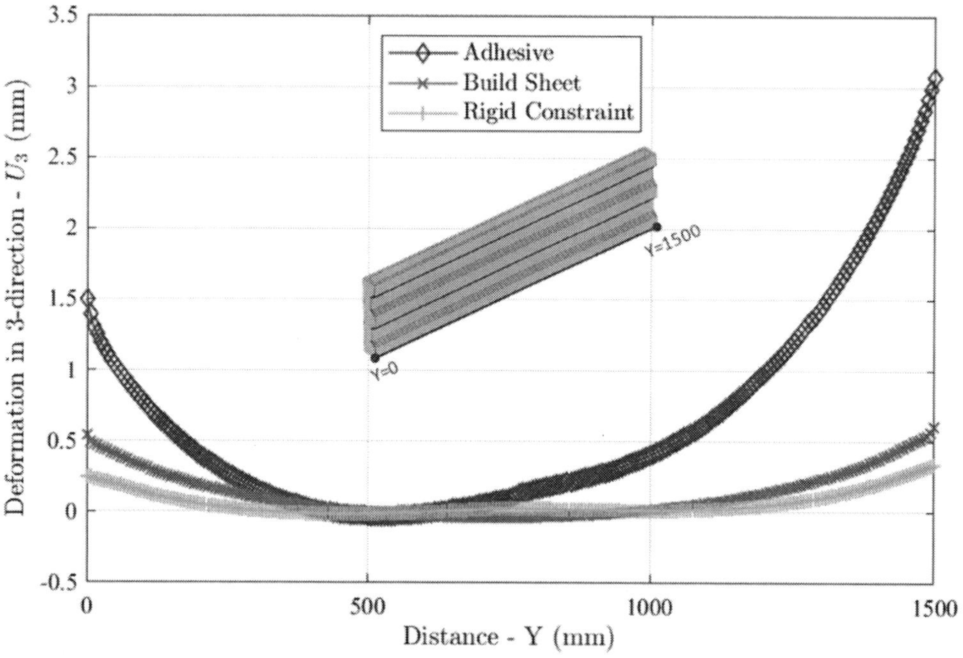

Figure 8. Comparison of the deformation in the 3-direction developed at the bottom of the stringer tool for the three methods investigated for adhering the part to the build plate.

4. SUMMARY AND CONCLUSION

This work was motivated by the need to have a simulation tool to predict the effect of different methods of adhering a part to the build plate. To that end, the simulation suite in Additive3D©, developed at Purdue University, was extended to model the material extrusion additive manufacturing process under the influence of different configurations of part adherence to the build plate. The adhering methods considered included an adhesive layer, a thin sheet of polymer (build sheet) held with vacuum to the build plate, and a rigid constraint potentially attained through mechanical interlocking of the printed material and the build plate. The adhesive layer allowed debonding of the part from the build plate and was modeled through a cohesive interaction between the bottom surface of the printed part and a rigid surface connected to ground. The constitutive behavior of the cohesive interaction was provided by a traction separation law. Currently, only the behavior in the direction normal to the build plate was considered yet the contributions of the in-plane shear behavior will be investigated in the future. For the build sheet, a thin sheet of polymer was modeled explicitly in the finite element model. The bottom of the build sheet was restrained by a constant pressure implemented through connector elements that applied a constant nodal force. The edges of the build sheet were also attached to connector elements with a relatively low stiffness to simulate the tape used in the installation of the build sheet on the vacuum table. Furthermore, the modeling methods implemented in this work are applicable to other techniques for adhering parts to the build plate including the bead board or preforms printed with similar materials.

Two geometries that are susceptible to deform due to residual stresses developed during the printing process were used to investigate the effect of the three different methods of adhering a part to the build plate on the residual deformation. The first geometry investigated was a plate with four layers through the thickness (0/0/90/90). The second geometry investigated was a single-side tool used in autoclave manufacturing with fiber reinforced thermosets of a representative aircraft stringer component. Both studies showed that the longer the geometry is retained attached to the build plate and the higher the stiffness of the attachment, the lower the out of plane deformation that develops. While this is true for the stringer tool and for deformation of the plate in the positive direction, the inverse behavior was observed for the side of the plate that deformed in the negative 3-direction. This was attributed to the increase in the bending stiffness of the plate in one direction resulting from the warpage of the plate in the opposite direction which originated during the printing process.

Ongoing work focuses on characterizing the behavior of an adhesive used in the CAMRI as well as the bead board used in the LSAM system. Experimental validation of the simulation predictions will be performed in the future to provide more confidence on the simulation predictions and the modeling methods implemented for the build sheet and adhesive layer. Finally, this work exposed some of the implications of the different methods for adhering parts to the build plate with the aim of contributing to reduce the experimental trials required to successfully print a part.

5. REFERENCES

1. A. Roschli et al., "Designing for Big Area Additive Manufacturing," Addit. Manuf., vol. 25, pp. 275–285, Jan. 2019, doi: 10.1016/j.addma.2018.11.006.
2. C. E. Duty *et al.*, "Structure and mechanical behavior of Big Area Additive Manufacturing (BAAM) materials," *Rapid Prototyp. J.*, vol. 23, no. 1, pp. 181–189, 2017, doi: 10.1108/RPJ-12-2015-0183.
3. E. Barocio, B. Brenken, A. Favaloro, J. Ramirez, V. Kunc, and R. B. Pipes, "Fusion bonding simulations of semi-crystalline polymer composites in the extrusion deposition additive manufacturing process," in *32nd Technical Conference of the American Society for Composites 2017*, 2017, vol. 4.
4. E. Barocio, B. Brenken, A. Favaloro, and R. B. Pipes, "Extrusion deposition additive manufacturing of composite molds for high-temperature applications," in *International SAMPE Technical Conference*, 2017.
5. A. A. Hassen *et al.*, "The durability of large-scale additive manufacturing composite molds," in *CAMX 2016 - Composites and Advanced Materials Expo*, 2016.
6. P. Kim *et al.*, "Analysis on Part Distortion and Residual Stress in Big Area Additive Manufacturing with Carbon Fiber-Reinforced Thermoplastic using Dehomogenization Technique," 2019.
7. B. Brenken, E. Barocio, A. Favaloro, V. Kunc, and R. B. Pipes, "Development and validation of extrusion deposition additive manufacturing process simulations," *Addit. Manuf.*, vol. 25, pp. 218–226, 2019, doi: 10.1016/j.addma.2018.10.041.

8. Y. Zhang and Y. K. Chou, "3D FEA simulations of fused deposition modeling process," in ASME 2006 International Manufacturing Science and Engineering Conference, 2006, pp. 1121–1128.

9. Y. Zhang and K. Chou, "A parametric study of part distortions in fused deposition modelling using three-dimensional finite element analysis," Proc. Inst. Mech. Eng. Part B J. Eng. Manuf., vol. 222, no. 8, pp. 959–968, 2008, doi: 10.1243/09544054JEM990.

10. M. R. Talagani et al., "Numerical simulation of big area additive manufacturing (3D printing) of a full size car," SAMPE J., vol. 51, no. 4, pp. 27–36, 2015.

11. A. Cattenone, S. Morganti, G. Alaimo, and F. Auricchio, "Finite element analysis of Additive Manufacturing based on Fused Deposition Modeling (FDM): distortion prediction and comparison with experimental data.," J. Manuf. Sci. Eng., 2018.

12. B. Courter, V. Savane, J. Bi, S. Dev, and C. J. Hansen, "Finite Element Simulation of the Fused Deposition Modelling Process," in Proceedings of the NAFEMS World Congress, 2017, pp. 11–14.

13. A. El Moumen, M. Tarfaoui, and K. Lafdi, "Modelling of the temperature and residual stress fields during 3D printing of polymer composites," Int. J. Adv. Manuf. Technol., Jun. 2019, doi: 10.1007/s00170-019-03965-y.

14. E. Barocio, B. Brenken, A. Favaloro, M. Bogdanor, and R. B. Pipes, "Extrusion Deposition Additive Manufacturing with Fiber-Reinforced Thermoplastic Polymers," in Structure and Properties of Additive Manufactured Polymer Components, 1st ed., K. Friedrich and R. Walter, Eds. Woodhead Publishing, 2020, p. 450.

15. B. Brenken, "Extrusion Deposition Additive Manufacturing of Fiber Reinforced Semi-crystalline Polymers," Ph.D. Dissertation, Purdue University, 2017.

16. E. Barocio, "Fusion Bonding of Fiber Reinforced Semi-Crystalline Polymers in Extrusion Deposition Additive Manufacturing," Ph.D. Dissertation - Purdue University, 2018.

17. D. Marrett, "The Secret to LSAM Print Quality… A Different Process," Thermwood Blog, 2018. http://blog.thermwood.com/the-secret-to-lsam-print-quality-a-different-process (accessed May 24, 2020).

18. "CEAD-AM," *Solutions: Print bed.* https://cead-am.com/home/solutions/ (accessed May 24, 2020).

19. MatWeb CE, "Overview of materials for Polycarbonate, Extruded." [Online]. Available: http://www.matweb.com/search/datasheet.aspx?matguid=501acbb63cbc4f748faa7490884cdbca&ckck=1. (Accessed: 05-Mar-2020).

20. "Simulia User Assistance 2019," Dassault Systemes, 2019.

21. V. Pini, J. J. Ruz, P. M. Kosaka, O. Malvar, M. Calleja, and J. Tamayo, "How two-dimensional bending can extraordinarily stiffen thin sheets OPEN," Nat. Publ. Gr., 2016, doi: 10.1038/srep29627.

INVESTIGATING INTER-WELD BONDS UNDER TENSION IN MECHATRONIC AM PROCESSING

Sean Psulkowski, Jolie Frketic, Helen Parker, Raquel Werner, Tarik Dickens
Department of Industrial and Manufacturing Engineering, FAMU-FSU College of Engineering
Tallahassee, FL,32310, USA

ABSTRACT

Additive manufacturing (AM) for rapid product development, has seen continued growth in adoption to facilitate complex low volume part production. However, the current process lacks true simultaneous deposition of multi-materials in a single operating space and is limited by lengthy build times and end part functionality. Attempts to correct this through unique nozzles or filament slicing still retain significant idle time between distinct material deposition. In tandem with industrial automation acquisition, a novel application of collaborative manipulators has been developed to fulfill the need for truly simultaneous AM operations in a shared space. Robotic manipulators occupy an established foothold in current industrial manufacturing settings, establishing the infrastructure for future wide scale adoption of collaborative printing. This paper aims to investigate and quantify the optimal operational scheme of a collaborative AM system sharing a workspace of non-uniform resolution. In particular, the effects of printing speed, nozzle diameter, time elapsed between layer extrusion and material overlap in dual-printed parts. The study showed larger overlap between layers and slower printing speed to have contributed significantly to higher mechanical strength in parts. Finally, this study benchmarks operational variance of multiple printing agents as it compares to modern manufacturing systems, acting as a cornerstone to facilitate a fluid workflow of collaborative AM.

Keywords: Additive Manufacturing; Multi-Material; SCARA; Collaborative Robotics; Material Extrusion;
Corresponding author: Tarik Dickens

1. INTRODUCTION

With the recent paradigm shift to rapidly manufacturing unconventional structures for customized function, Additive Manufacturing (AM) continues its trajectory towards a $35.6 billion industry by 2024 [1]. Recent work highlights the inclusion of robotics to augment the AM workflow with a particular interest towards multi-agent systems that collaboratively print larger scale structures [2]. In principle, the inclusion of multiple agents enables multi-material printing, on-the-fly operational control, and higher throughput of printed structures [3]. The detriment to this is the added difficulties of complex collision avoidance schemes as well as a joining weld-line where the separated part meets; Both having the potential to compromise the manufacturing process and functionality of the printed structure.

From the many different modes of AM, fused deposition modeling (FDM), also called material extrusion (ME), demonstrates a high potential for fabricating multi-functional 3D-printed parts competitive to conventional manufacturing techniques. Moreover, the potentiality for ME to grow in frequency of application is massive, driven largely by the comprehensive material library available with the extrusion process [4]. Material extrusion is enabled through filament material fed via a motor through a heated channel, like an automated hot glue gun. Overall, there exist numerous benefits inherent to the ME process: such as reduced material waste, ease in application deployment, and growing popularity amongst industrial manufacturers and hobbyists alike. However, extrusion printed structures are mechanically limited in engineering applications due to their orthotropic properties induced by the consecutive heating and cooling of molten filament between discontinuous lines and layers; contributing weaker tensile properties orthogonal to tool path. The linchpin of the ME process are these interfacial bonds formed between the adjacent strands of deposited filament; The bonding quality among polymer beads in the ME process determines the integrity and mechanical properties of resulting prototypes[5]. The "bonding potential", or adhesion, between two deposited thermoplastics is defined by equation 1:

$$\varphi = \int_0^t (T - T_c)\, d\tau \tag{1}$$

Here, T is the interface temperature, Tc is the critical bonding temperature, t is time, and is the time between the deposition of rows. In general, the higher the average part temperature during the printing process, the better the bonding quality [6]. As a metric for printed part strength, bond quality formed between individual roads depends on the growth of the neck formed between adjacent beads and the molecular diffusion and randomization at the weld-line, shown in Figure 1.

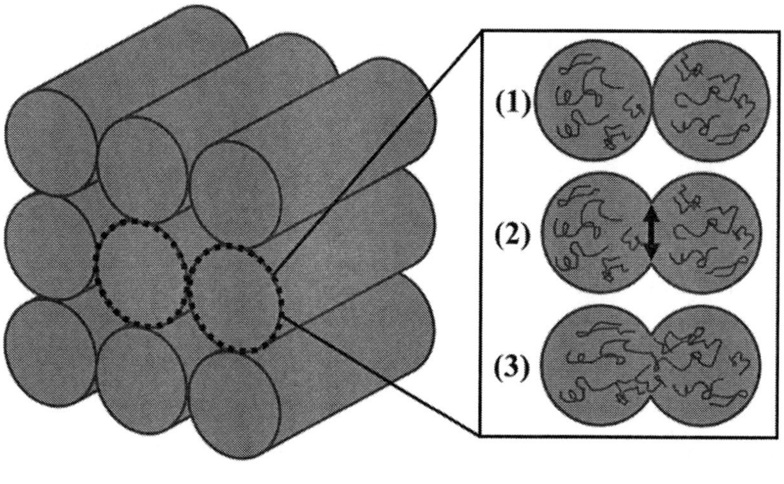

Figure 1. Bond formation between two thermoplastic strands (1) surface contact (2) neck growth along weld-line (3) molecular diffusion at interface and randomization

Previous research utilized such neck growth to develop a processing procedure that experimentally assess bond quality as a function of this neck growth and sample failure under flexural loading [7]. Similar investigations claim disentanglement occurring in the heated channel combined with delay in entanglement recovery as the key mechanisms responsible for reduced mechanical strength at the weld-line [8]. Owing to the property-process relationship in AM, the operating parameters of the system strongly correlate to the mechanical performance of the manufactured parts. Therefore, a collective of research has centered around identifying and optimizing manufacturing parameters to maximize part strength and morphology in regard to interfiber bonding at the weld-line. The following analyze printing parameters which influence thermo-temporal properties corresponding to bond strength: heating elements and printing speed [9], slicing parameters and interlocking geometry [10], [11], and material overlap at the bond interface [12], [13].

Overall, 3D printing is an effective tool to realize designs of higher complexity via a plethora of materials currently available. However, the need still exists for operators to achieve an adequate level of understanding how the physical system influences the mechanical strength of parts manufacturing through the emerging robotic AM process. The following study utilizes two collaboratively printing selective compliance articulated robot arms (SCARA) [14] to identify the extent to which the robotic agents influence the polymer-polymer bonding mechanism in extrusion printing and whether a singular parameter surmounts the printing process in this new setup.

2. EXPERIMENTATION

Operating parameters were selected based on the significance towards the weld-line strength of a multi-material print under tension. Each parameter has an upper, middle, and lower bound, with a category specifying the driving force of the parameter in this experiment - the robotic agent used in this study (notated as "Process"), or the dual material extrudate ("Material"). The operating parameter driven by both material and process are shown in Table 1.

Table 1. Printing Parameters

Tested Parameter	Min	Mid	Max	Category
Nozzle Diameter (mm)	1	1.5	2	Process
Printing Speed (mm/s)	15	27.5	40	Process
Time Elapsed (s)	15	97.5	180	Material
Overlap (%)	0-10	10-25	25-50	Material

The Nozzle Diameter is the exit channel where thermoplastic flows out of the robot's mounted hot end. Printing Speed is the traversal speed of the robot's end effector during the extrusion process. Time Elapsed is the time interval between the start of either strands of extrudate which compose a sample and overlap. The Overlap refers to the surface area shared between two side-by-side strands of extrudate if a cutaway is made orthogonal to the traversal axis of the strand. For brevity, overlap between the two strands is coded as 0, 0.5, 1 for the lower, middle, and upper bounds respectively.

To better image the deposited roads, and for ease of determination of the interface material selection consisted of Iron-Composite PLA and Stainless Steel-Composite PLA. The base polymer lends itself to more flexible operating conditions such as a low temperature hot end, faster extrusion speeds and lack of a heated bed. Moreover, the visual differences between the two materials provided a visual aid in determining deposition accuracy (in overlap and diameter). Printed samples were then fixed to tabs with an epoxy adhesive placed opposite to the weld-line, and the printed sample was measured for length, width, and height via optical microscopy. Operating procedure consisted of adhering to ASTM standards for "Determining the Short Term Tensile Weld Strength of Chemical-Resistant Thermoplastics"[15] and "Cohesive Strength of Sheet Plastics"[16]. Mechanical failure was tested via Dynamic Mechanical Analysis (DMA), with a Force ramp of 3N/min until the samples reached failure under tension. Modulus and ultimate tensile strength were collected, as these properties are largely of interest when comparing printed part performance to bulk material in functional applications [17]–[20].

3. RESULTS AND DISCUSSIONS

Facilitating this study in multi-material bond strength were two SCARA simultaneously extruding material within a single shared space without the constraints of conventional cartesian platforms. Orchestrating this collaborative process invites higher complexity manufacturing towards multifunctional part production. A diagram of this system can be seen in Figure 2.

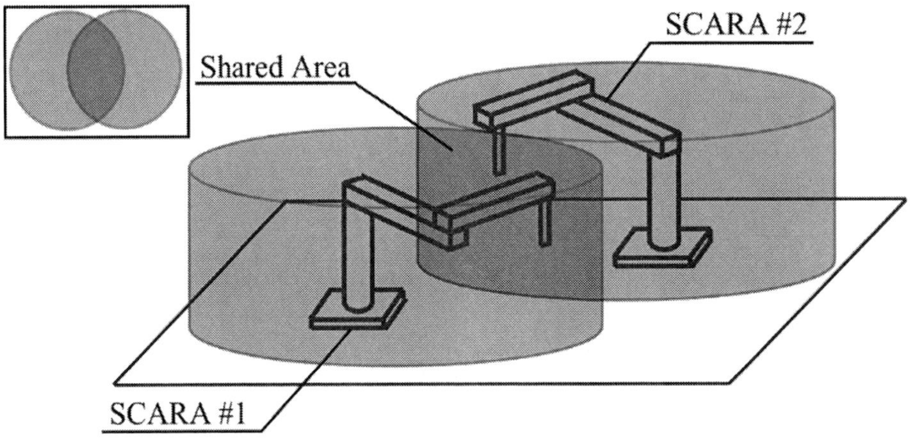

Figure 2. Dual-SCARA AM platform diagram highlighting shared area.

The dynamics of the system are largely influenced by the angular velocity and acceleration of the joints traversing through the build area. Equation 2 translates the inertial effects into cartesian space for a direct comparison to conventional systems.

$$F \cdot \delta x = \tau \cdot \delta \vartheta \tag{2}$$

Where F is the cartesian force-moment vector acting on the manipulators end effector, δx is infinitesimal cartesian displacement of the end effector , τ is vector of torques at the joints , and δϑ is vector of infinitesimal joint displacements [21]. The result, the operating resolution of the SCARA's tooling shown in Figure 3.

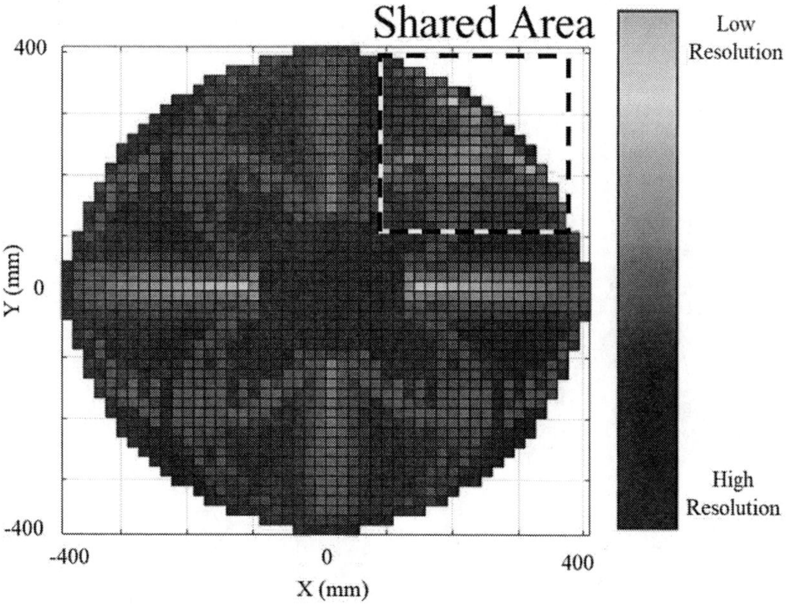

Figure 3. Calculated resolution of a single SCARA's operating window, highlighting the potential error propagation in the shared space

A single SCARA's center axis of rotation is at the plot's origin, and the circle encompasses full operating range for printing. With darker shade representing low inertial effects and therefore higher resolution, the variation of inertial effects across position indicate non-uniform resolution for printing. Factoring in the cohabitation of collaborating manipulators, in what is labeled "Shared Area", propagating inertial effects appear more readily, and therefore have larger influence to the end part's structure. Tool path resolution and printed structure resolution are treated as analogous in this study and are subject to the same variation across the shared build platform.

The afore-mentioned dual robotic agents then collaboratively printed multi-material samples in regard to the array of parameters listed in Table 1. These samples were affixed for mechanical testing as shown in Figure 4. Fiber orientation in between the two tabs placed emphasis on the interfacial strength along the bonded weld line; the point of highest failure due to reduced entanglement from latency/variability in a collaborative print. Measuring mechanical response emphasizes which parameters contribute the most to reduced functionality in collaborative prints. Overall, the samples experienced brittle failure in tension, all fracturing along the weld-line where the two strands met. Some samples experienced both adhesive and cohesive bonding at this weld-line, which is discussed later in terms of polymer entanglement.

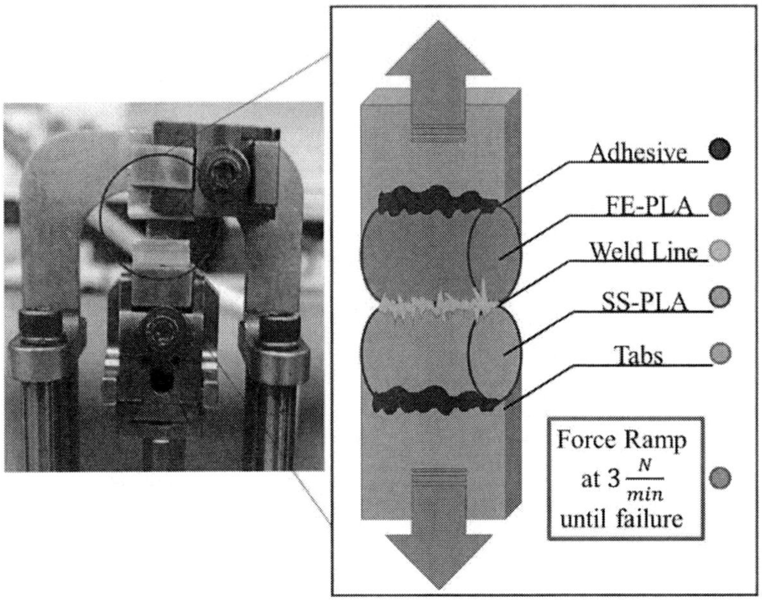

Figure 4. Dual-SCARA printed multi material sample diagram under tension

The following are the mechanical responses of tensile samples representative of all operating parameters via a full two factor design of experiment with midpoints and two replicates. Figure 5 shows the ultimate tensile strength of the samples.

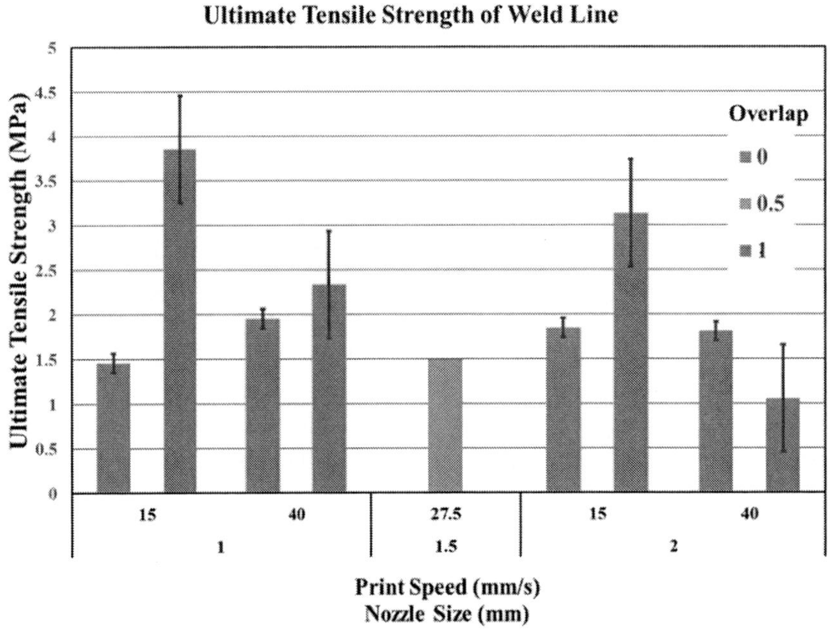

Figure 5. Ultimate Tensile Strength of samples across parameter ranges

Among all criteria, larger overlap corresponds to the highest UTS, but only when coupled with slower printing speeds. This highlights the importance of extrudate morphology in relation to strength. With more overlap in the sample, there is a larger surface area along the weld-line. This successfully achieves greater bonding between the two roads and results in a larger neck. However, the print speed parameter is proportional to extrudate mass flow through the heated channel of the hot end. At the upper bound, this results in the extrudate possessing a lower thermal gradient and therefore a lower chance for polymer entanglement to occur between the strands. Moreover, higher mass flow of thermoplastic invites more instances of undesirable porosity upon extrusion, which lowers the contact area at the weld-line of the sample and negatively affects the resultant properties. The samples presenting the highest UTS (3.65 MPa) were those printed at 15mm/s, with 0 overlap and a nozzle diameter of 1mm, seconded by its counterpart of larger nozzle diameter. This underlines the trend that increased print speed results in a smaller UTS difference across samples. The lowest UTS (1.46 MPa) was from the sample with the fastest printing speed, greatest overlap, and largest nozzle diameter, compounding the importance of a high thermal gradient for cohesive bonding. Moreover, preliminary factorial analysis declared the interaction of velocity and overlap to be significant, split between process and material parameters, where maximum UTS could be achieved by increasing overlap and decreasing velocity. This is likely due to the elongation of polymer chains through shear in the heated nozzle. However, the model possesses higher variability than desired, and would benefit from increased replicates. Figure 6 shows the main effect(A) and interaction(B) between these parameters.

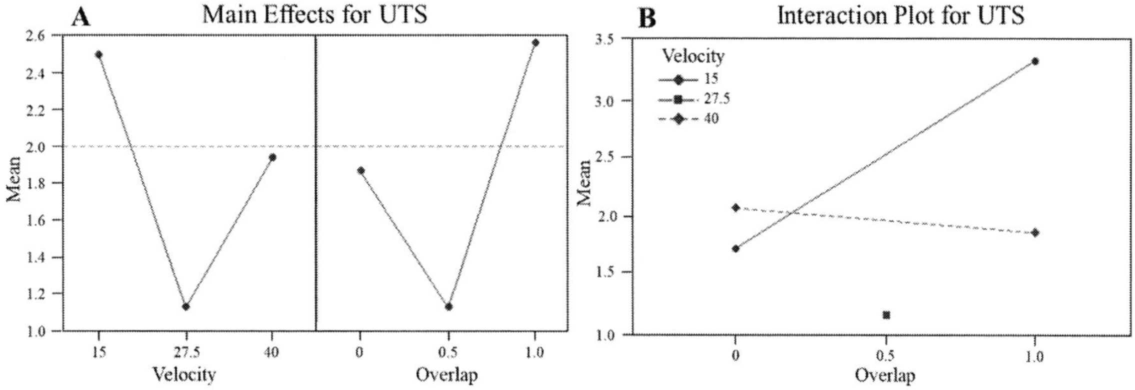

Figure 6. Factorial response of ultimate tensile strength including significant main effects (A) and interaction plot (B)

Another important metric to monitor is stiffness measured in the dual-print samples. Figure 7 shows the results for Young's modulus found during the same tensile test.

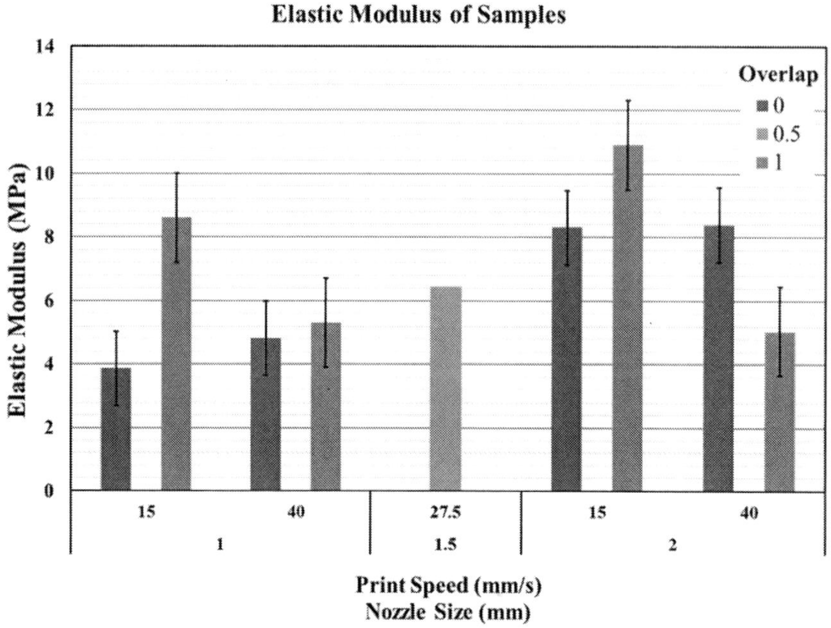

Figure 7. Elastic modulus of samples across parameter ranges

Regarding elastic modulus of the samples, higher overlap coupled with larger nozzle diameter and lower speeds produced the stiffest sample at 10.91 MPa. Overall, slower print speed resulted in a greater modulus along with larger nozzle diameter, which had a higher average than its smaller counterparts. The exception in each case was the upper bound for all parameters at 5.04 MPa, again signaling the role operating conditions have on cohesive bonding. Larger nozzle size also appears to increase elasticity but can also lead to lesser weld strength, as an increase in sizes gives higher strength at the weld-line but also introduces more porosity during printing. Additionally, a larger extrudate width caused by a larger nozzle may increase the contact area, and thus neck formation at the weld-line which would yield higher polymer entanglement and ultimately higher strength. Factorial Analysis showed the interaction of nozzle diameter, time and overlap to be significant, a majority of process parameters, with respectively increased values resulting in maximizing the modulus response These factorial effects(A) and interactions(B) are visualized in Figure 8.

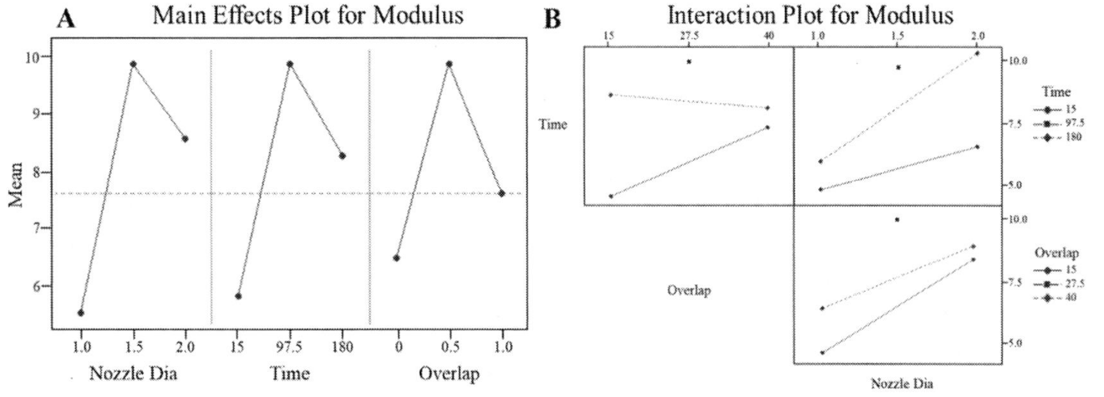

Figure 8. Factorial response of elastic modulus including significant main effects (A) and interaction plot (B)

A scanning electron microscope (SEM) provided imaging of the polymer fracture surfaces to yield greater insight into the amount of polymer-polymer welding that occurred along the weld line of the samples. The extremes of such occurrences are shown in Figure 9.

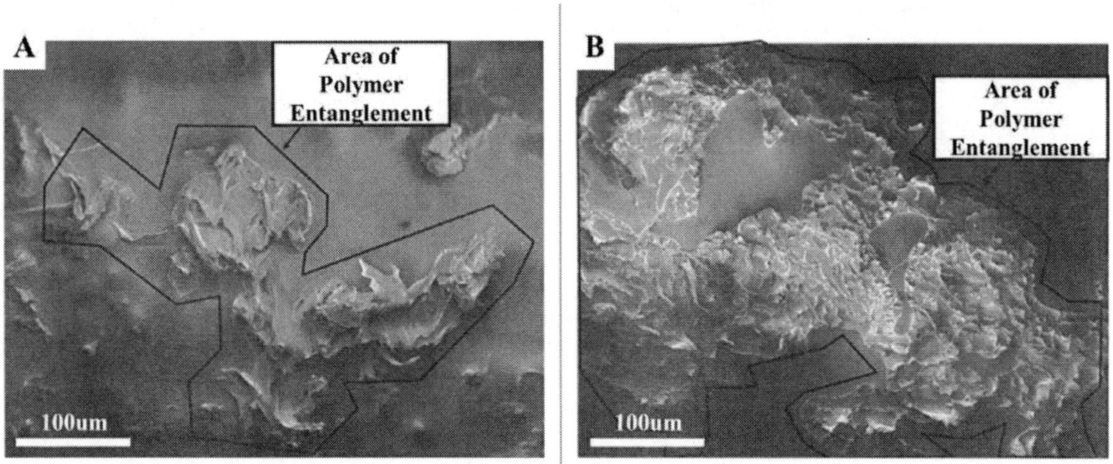

Figure 9. Micrographs with difference in weld line as seen in two different fracture surfaces, outlined in red. A) The area of fracture on a sample with low tensile strength (0.87MPa) , B) Fracture area of the sample with higher fracture strength (1.68 MPa).

Here, the sample in (A) was formed at the upper bound of the printing speed parameter and lower bound of overlap, while (B) was printed under the inverse. As expected, the SEM images showed that samples made with greater overlap had larger areas of polymer welding at the fracture surface of the weld-line. Furthermore, the sample in (A) possessed a smaller area of plastic deformation with a greater portion of the surrounding polymer being smooth, denoting no deformation occurring and therefore a smaller occurrence of strand adhesion in these areas along the weld-line. This is further qualified by the strength found during testing of these samples, as the sample in (A) had a strength of 0.87 MPa while the sample in (B) performed nearly twice as well at 1.68 MPa. It can be correlated that the area of polymer entanglement is directly responsible for an increase in UTS, as the SEM images show almost double the amount of deformation shown in the second micrograph as in the first.

4. CONCLUSIONS

True simultaneous deposition of multi-materials in a single operating space enables heightened complexity in end part functionality, at the cost of introducing weakness along the joining wield of both materials. An investigation between crucial operations conditions has shown interaction between both process and material parameters towards mechanical performance. Preliminary results have demonstrated a decrease in print speed and increase of overlap to result in a larger area of polymer entanglement and therefore increased strength. Additionally, larger nozzle diameter, strand overlap and increased time elapsed between prints yields lower elasticity. Despite this, preliminary factorial models developed from this experiment require additional replicates, indicating 1.) Ultimate tensile strength and modulus might not be adequate metrics to predict the complexities of polymer entanglement 2.) The variability inherent to collaborative printing might be too large for the preliminary sample size. Future work should aim to quantify and mitigate the effects variable resolution has towards the bond strength of joining collaborative parts. Similar studies can be developed to investigate the influence of other process parameters towards the thermo-mechanical operations which drive robotic AM. Continued research in multi-material printing furthers a future of scaled collaborative AM.

5. REFERENCES

[1] "3D Printing | Wohlers Associates." https://wohlersassociates.com/ (accessed Nov. 29, 2019).

[2] X. Zhang *et al.*, "Large-scale 3D printing by a team of mobile robots," *Autom. Constr.*, vol. 95, pp. 98–106, Nov. 2018, doi: 10.1016/j.autcon.2018.08.004.

[3] H. Han, H. Bae, H. Kang, J. Son, and H. Kim, "Multi agent 3D printer and robot system for mass personalization faas platform," in *2016 International Conference on Information and Communication Technology Convergence (ICTC)*, Oct. 2016, pp. 1129–1131, doi: 10.1109/ICTC.2016.7763386.

[4] T. D. Ngo, A. Kashani, G. Imbalzano, K. T. Q. Nguyen, and D. Hui, "Additive manufacturing (3D printing): A review of materials, methods, applications and challenges," *Compos. Part B Eng.*, vol. 143, pp. 172–196, Jun. 2018, doi: 10.1016/j.compositesb.2018.02.012.

[5] P. G. De Gennes, "The Formation of Polymer/Polymer Junctions," in *Tribology Series*, vol. 7, J. M. Georges, Ed. Elsevier, 1981, pp. 355–367.

[6] M. A. Yardimci, T. Hattori, S. I. Guceri, and S. C. Danforth, "Thermal Analysis of Fused Deposition," p. 10.

[7] Q. Sun, G. M. Rizvi, C. T. Bellehumeur, and P. Gu, "Effect of processing conditions on the bonding quality of FDM polymer filaments," *Rapid Prototyp. J.*, vol. 14, no. 2, pp. 72–80, Mar. 2008, doi: 10.1108/13552540810862028.

[8] C. McIlroy and P. Olmsted, "Disentanglement Effects on the Welding Behaviour of Polymer Melts during the Fused-Filament-Fabrication Method for Additive Manufacturing," *ArXiv170309295 Cond-Mat*, Mar. 2017, Accessed: Jan. 14, 2019. [Online]. Available: http://arxiv.org/abs/1703.09295.

[9] J. Yin, C. Lu, J. Fu, Y. Huang, and Y. Zheng, "Interfacial bonding during multi-material fused deposition modeling (FDM) process due to inter-molecular diffusion," *Mater. Des.*, vol. 150, pp. 104–112, Jul. 2018, doi: 10.1016/j.matdes.2018.04.029.

[10] A. C. Abbott, G. P. Tandon, R. L. Bradford, H. Koerner, and J. W. Baur, "Process-structure-property effects on ABS bond strength in fused filament fabrication," *Addit. Manuf.*, vol. 19, pp. 29–38, Jan. 2018, doi: 10.1016/j.addma.2017.11.002.

[11] F. Tamburrino, S. Graziosi, and M. Bordegoni, "The influence of slicing parameters on the multi-material adhesion mechanisms of FDM printed parts: an exploratory study," *Virtual Phys. Prototyp.*, vol. 14, no. 4, pp. 316–332, Oct. 2019, doi: 10.1080/17452759.2019.1607758.

[12] T. J. Coogan and D. O. Kazmer, "Healing simulation for bond strength prediction of FDM," *Rapid Prototyp. J.*, vol. 23, no. 3, pp. 551–561, Jan. 2017, doi: 10.1108/RPJ-03-2016-0051.

[13] H. R. Brown, "Relation between the Width of an Interface between Two Polymers and Its Toughness," *Macromolecules*, vol. 34, no. 11, pp. 3720–3724, May 2001, doi: 10.1021/ma991821v.

[14] J. B. Frketic, S. Psulkowski, A. Sharp, and T. Dickens, "Dexterous Printing and Fabrication of Multi-Functional Parts: Design for Science and Engineering Education," *Procedia Manuf.*, vol. 10, pp. 1087–1096, Jan. 2017, doi: 10.1016/j.promfg.2017.07.099.

[15] D20 Committee, "Practice for Determining the Short Term Tensile Weld Strength of Chemical-Resistant Thermoplastics," ASTM International. doi: 10.1520/C1147-14.

[16] D20 Committee, "Test Method for Bond or Cohesive Strength of Sheet Plastics and Electrical Insulating Materials," ASTM International. doi: 10.1520/D0952-15.

[17] S. Garzon-Hernandez, D. Garcia-Gonzalez, A. Jérusalem, and A. Arias, "Design of FDM 3D printed polymers: An experimental-modelling methodology for the prediction of mechanical properties," *Mater. Des.*, vol. 188, p. 108414, Mar. 2020, doi: 10.1016/j.matdes.2019.108414.

[18] S. Verma and V. Chaturvedi, "Multiresponse Optimization of Fused Deposition Modeling Process Using Utility-Based Taguchi Approach," *J. Adv. Manuf. Syst. Singap.*, vol. 17, no. 4, p. 551, Dec. 2018, doi: http://dx.doi.org.proxy.lib.fsu.edu/10.1142/S0219686718500312.

[19] J. D. Berrio Bernal, this link will open in a new window Link to external site, E. C. N. Silva, and W. M. Rubio, "Characterization of effective Young's modulus for Fused Deposition Modeling manufactured topology optimization designs," *Int. J. Adv. Manuf. Technol. Heidelb.*, vol. 103, no. 5–8, pp. 2879–2892, Aug. 2019, doi: http://dx.doi.org.proxy.lib.fsu.edu/10.1007/s00170-019-03747-6.

[20] A. Peng, X. Xiao, and R. Yue, "Process parameter optimization for fused deposition modeling using response surface methodology combined with fuzzy inference system," *Int. J. Adv. Manuf. Technol. Heidelb.*, vol. 73, no. 1–4, pp. 87–100, Jul. 2014, doi: http://dx.doi.org.proxy.lib.fsu.edu/10.1007/s00170-014-5796-5.

[21] J. J. Craig, *Mechanics and Control Third Edition*. .

ADDITIVE MANUFACTURING OF OPTICAL COMPONENTS USING COMMERCIAL OFF-THE-SHELF 3D PRINTERS

Ethan Z Delannoy, Peter J. Joyce, R. Brian Jenkins,
Charles Nelson, Deborah M. Mechtel, David P. Durkin
United States Naval Academy
Annapolis, Maryland

ABSTRACT

The goal of this research is to use 3D printing with polymers to fabricate rigid or flexible optical components (such as optical waveguides or lenses) to control light propagation. Vat polymerization and fused deposition modeling printing techniques were used to fabricate optical components. Scanning electron microscopy, infrared spectroscopy, thermogravimetric analysis and differential scanning calorimetry were used to evaluate the structure/composition and thermal properties of the printed materials. These data identified defects in the original printed structures that were responsible for light scattering, and guided the development of novel post-processing techniques to optimize the optical properties of the additively manufactured components. Post-process annealing delivered components with significant improvements in transparency and reduced light scattering. As a result, lenses have been printed that successfully focus light, and optical refraction and reflection of laser light was observed in 3D-printed waveguide structures. Optical losses were measured to determine the extinction coefficient for the Beer-Lambert Law and a double slit experiment was used to demonstrate if the 3D printed parts maintained spatial coherence.

1. INTRODUCTION

The use of commercial off-the-shelf 3D printers in the private sector and military to rapidly prototype and construct equipment is increasingly common. For example, private companies are using 3D printing to supplement the supply of personal protective equipment like face shields during the COVID-19 crisis [1]. With respect to the military, 3D printers are being utilized on forward deployed Marine Corps bases and Navy ships in order to make small repairs to equipment and spaces [2]. 3D printing gives a distinct advantage in circumventing the usual lengthy processes needed to construct complex designs.

The ability to control or guide the propagation of optical waves is important for myriad applications. Components like optical waveguides and lenses often must be produced with complex machinery and techniques, requiring time for construction and delivery to their customers [3]. However, for the military or private sector this time delay associated with receiving these optical components may be detrimental to their mission or safety. The goal of this research is to determine the viability of using off-the-shelf 3D printers that are currently being used in industry and the military to create functional optical components.

To determine the feasibility of using additive manufacturing to print optical components, commercial off the shelf printers and transparent materials were used to print lenses and optical waveguides. Techniques to improve the optical transparency of the 3D printed components were tested to verify their effectiveness. The materials used to construct the components were characterized using scanning electron microscopy, thermogravimetric analysis, and differential scanning calorimetry. Propagation of light through 3D printed lenses and optical waveguide structures was observed to investigate their ability to guide light coherently. The beam coherence and attenuation of light transmitted through the constructed optical components was determined though a double slit experiment and optical loss measurements, respectively.

2. EXPERIMENTATION & RESULTS

2.1 Printing Techniques

Two printing techniques that use commercial off the shelf printers were investigated as part of this study: fused deposition modeling (FDM) and stereo lithography (SLA). The FDM printing technique is comprised of a heated extruder that is guided by a gantry system that layers heated thermoplastic filament on the build plate, creating a solid plastic structure when cooled [4]. The FDM process is shown in Figure 1. FDM prints created from thermoplastic filament cannot be post-processed after fabrication to improve transparency and do not require supports in order to print for this application.

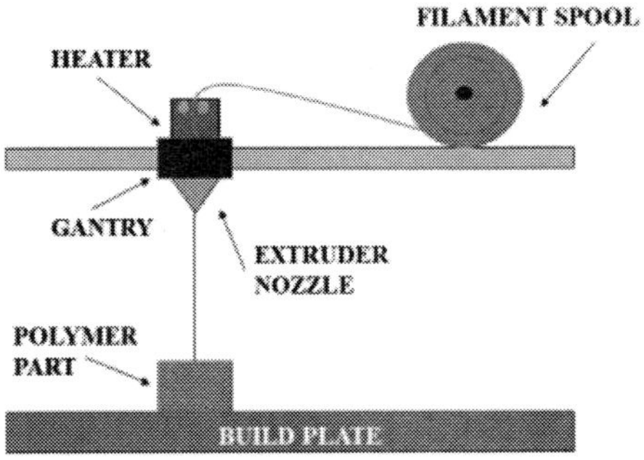

Figure 1. Fused deposition modeling with thermoplastic filaments.

Vat-Polymerization, also referred to as a Stereolithographic (SLA) printing technique, uses a UV laser to polymerize monomer resin on a build plate [5]. The UV laser forms polymer layers one upon another in the monomer resin tank to create a polymer structure. Once the structure is created from the polymer layering, it is lifted from the resin tank to be washed. The SLA process is shown in Figure 2. Parts printed using the SLA technique must be washed in an alcohol bath for 20 minutes after printing in order to remove excess monomer resin. Heating and UV post-curing can be used to increase the rigidity of the parts after washing.

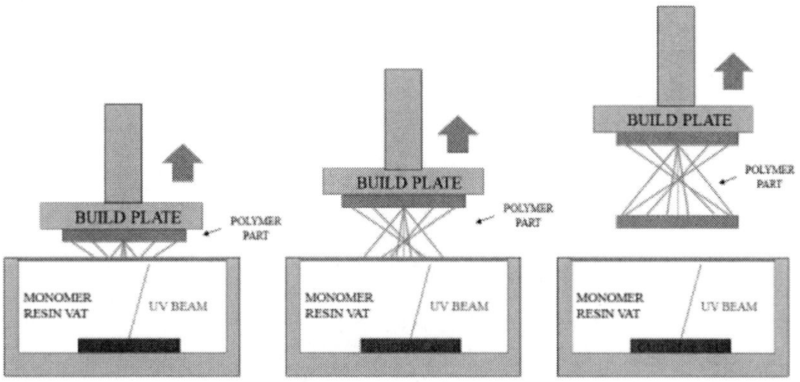

Figure 2. Vat polymerization of a polymer structure from monomer resin.

2.2 Initial Printing and Materials Testing

To analyze the feasibility of FDM and SLA printing techniques and materials, sample slabs were initially printed to test for transparency and to observe reflection and refraction. Preliminary slab waveguides were fabricated with the dimensions of 0.508 x 1.27 x 2.54 cm.

The filaments and resins that were used were selected as prospective materials to construct the optical components due to their initial optical transparency. The transparency of each slab was determined by viewing a black dot through the material; if the dot was distinct and clearly visible, the material was at least partly transparent. The selected printers, materials, printing specifications, and 3D structures that were printed are listed below in Table 1.

Table 1. Sample slab printing test matrix [6-10]

Structure	Printer Used	Filament (FDM) Resin (SLA)	Print Time (hrs)	Layering (with respect to direction of light propagation)	Resolution	Post Processing
Small Slab	Ultimaker S5 ™ (FDM)	Nylon and PLA filament	0.25-1	Not Parallel	0.05 mm	No
	Form2™ (SLA)	Elastic, High Temp, Clear V.4 resin			100 μm	
Large Slab	Form2 ™ Form 3 ™ (SLA)	Elastic, High Temp, and Clear V.4 resin	10-13	Parallel Not Parallel	25-100 μm	No
Cylinder	Form2 ™ (SLA)	Elastic, High Temp, and Clear V.4 resin	3-5	Parallel Not Parallel	100 μm	No
Lens	Form2 ™ Form 3 ™ (SLA)	High Temp, and Clear V.4 resin	0.25-2	Parallel Not Parallel	25-100 μm	Yes

Both slabs printed using FDM exhibited unacceptable warping. The PLA slab was moderately transparent, but the nylon slab was not at all transparent even at the settings for transparent prints. The amount of warping the sample underwent during printing makes FDM a poor candidate for printing optical waveguides.

Figure 3. Representative PLA "small slab" sample printed by the FDM method, showing warped shape and moderate transparency.

Slabs printed using SLA resin were dimensionally accurate and transparent. An example of a successful Clear V.4 resin print is shown in Figure 4. Elastic and High Temp resins printed with approximately the same amount of dimensional accuracy but with slightly less transparency than the Clear V.4 print.

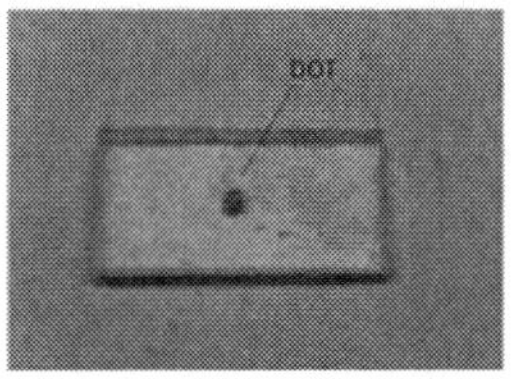

Figure 4. Representative Clear V.4 resin "small slab" sample printed by the SLA method, showing uniform shape and transparency.

2.3 Post Processing

After a part is 3D printed using one of the techniques described above, components often had a rough surface and features internal to the part that contribute to scattering or distortion of the laser beam propagating through the component [11]. An example is shown in Figure 5.

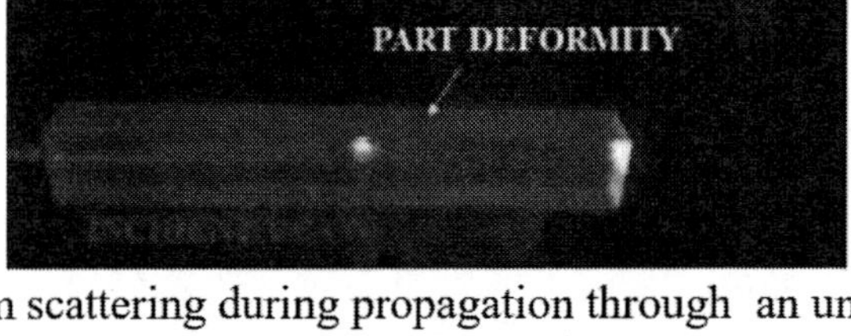

Figure 5. Laser beam scattering during propagation through an unprocessed SLA part.

In order to reduce scattering and distortion, the surface of the polymer component was post-processed after being printed. The post-processing technique consisted of sanding the surface of the part and then post-curing in the Formcure^{TM}. The SLA component was manually sanded in ascending order of grit as follows: 220, 320, 400, 600, and 1000 grit. The flakes from sanding the polymer part were washed off in the Formwash^{TM} for 30 seconds.

To refinish the rough surface the sanded part was then dipped in a pool of the same monomer resin used to print the 3D part. During the post-curing process, the monomer polymerizes and becomes part of the component. Therefore, anything touching the resin during the polymerization will be connected to the part. In order to avoid this, a stand was created using aluminum foil tape to support the part during the curing, with minimal surface contact as shown in Figure 6.

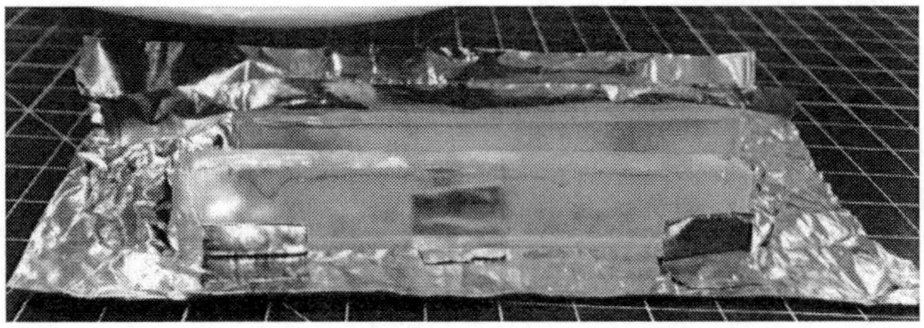

Figure 6. Aluminum tape stand for supporting waveguide structure during UV curing.

The heating of the polymer anneals the part and relieves its internal stresses and microstructures. The finished part is placed in an alcohol wash in order to remove any uncured resin. Figure 7 illustrates the basic steps used during post-processing.

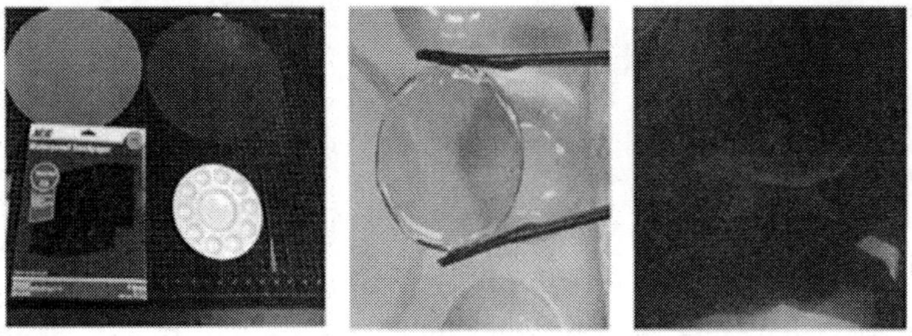

a) b) c)

Figure 7. Post-processing techniques for 3D printed optical components include (a) sanding materials for surface finishing part, (b) dipping lens in Clear V.4 resin for curing, and (c) post-curing of lens.

2.4 Evaluating the Physiochemical Properties of the Optical Components

2.4.1 Evaluating Surface Morphology of 3-D printed Optical Components via Scanning Electron Microscope

The optical components that were printed have surface features and roughness on the order of the HeNe laser wavelength (632.8 nm). Such features result in unwanted distortion of the light that propagates through the device. To verify that the post processing removed the surface roughness of an SLA print, a scanning electron microscope (SEM) was used. The polymer samples were carbon coated under high vacuum prior to SEM analysis to prevent charging from the electron beam.

The SEM imaging revealed that the post-processing techniques were effective at reducing the surface roughness of the polymer structures. Figure 8 shows the comparison between unprocessed and processed component surfaces for both the lens and the waveguide.

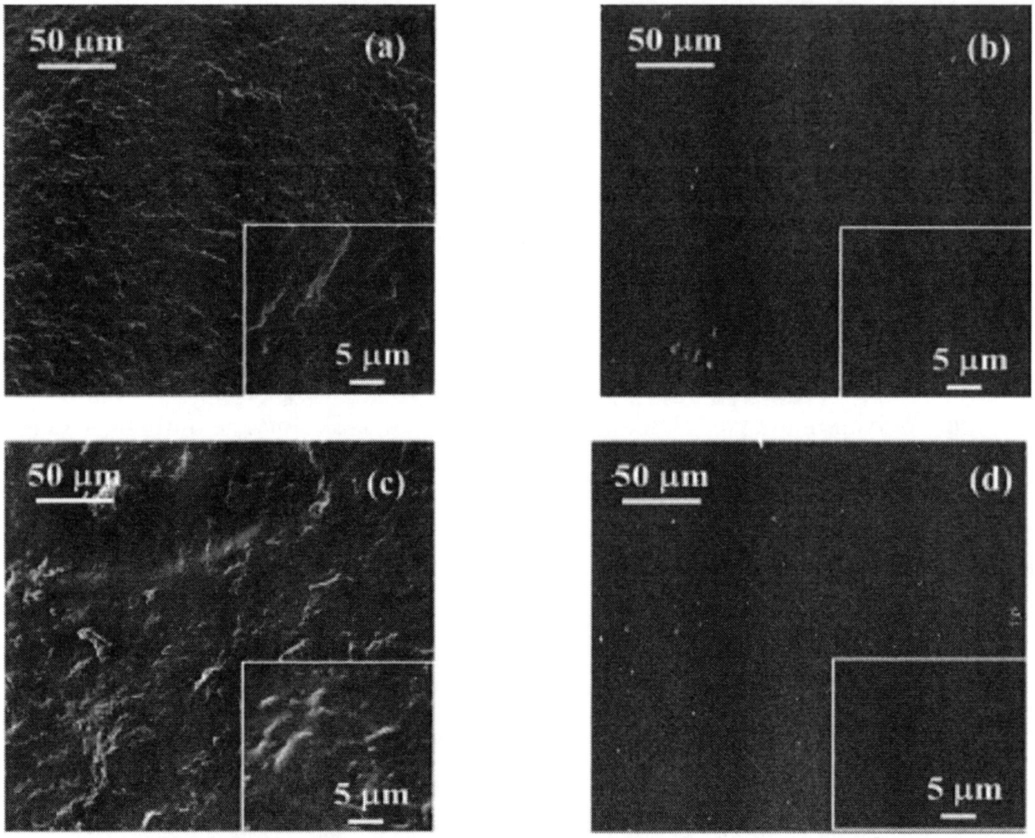

Figure 8. Scanning Electron Microscopy (SEM) of polymerized SLA (a) waveguide before thermal annealing, (b) waveguide after thermal annealing, (c) lens before thermal annealing and (d) lens after thermal annealing. In each case, polymerized SLA was annealed at 80 °C. SEM images were collected at 10 kV. Low resolution images are presented in the main frames, and higher resolution images of the surface are presented in each insert.

The SEM images show features on an unprocessed part with dimensions on the order of approximately 1 µm, similar in scale to the HeNe laser wavelength 632.8 nm. Thus, the wavefront distortion of the transmitted light is significantly reduced when the roughness of the polymer component was decreased.

2.4.2 Evaluating Thermal Properties of 3-D Printed Optical Components via Thermogravimetric Analysis

Thermogravimetric (TGA) analysis was used to evaluate the material mass loss as a function of temperature during pyrolysis. The data helped to inform the optimal post-processing temperature that could be used to anneal the 3D printed optical components without degrading the polymer. Samples (5 mg ±0.5mg) were evaluated in triplicate at a continuous heating rate of 10 °C/min from room temperature to 900 °C in a N_2 atmosphere. A representative thermogram is presented in Figure 9.

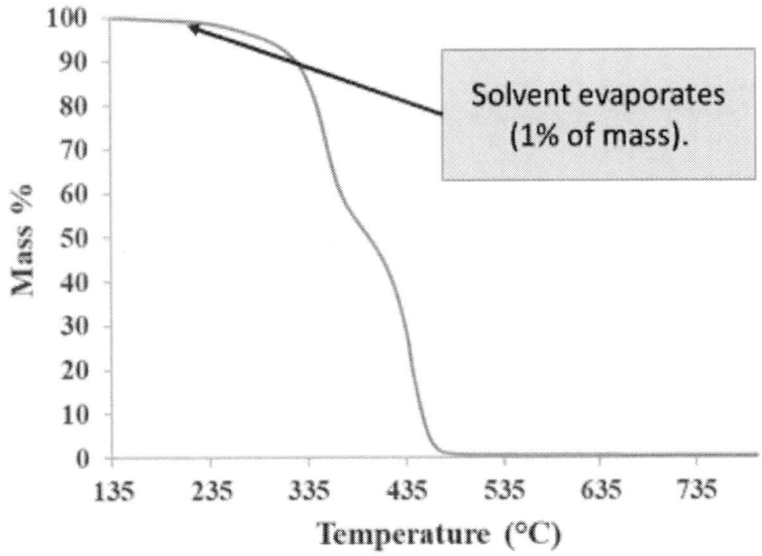

Figure 9. Representative thermogram of polymerized SLA polymer evaluated under nitrogen.

The thermogram in Fig. 9 shows some solvent loss at low (<100 °C) temperatures, and clearly indicates that the polymer degradation begins at approximately 210 °C. The polymer degradation profile occurs in at least three stages, as has been observed in previous experiments with PMMA [12].

2.4.3 Differential Scanning Calorimetry

Differential scanning calorimetry (DSC) measures the amount of heat absorbed (endothermic) or released (exothermic) when a sample is heated in a controlled atmosphere [13]. For polymers, some characteristics of the heat flow observed on a DSC graph are indicative of smoothing of structures and relief of internal stress. The glass-transition (T_g) of a polymer occurs at a specific temperature and results in the polymer changing from rigid (glassy) to softer material as result of internal stresses being relieved [13]. It is suspected that the internal microstructures that are annealed inside the polymer part contribute to additional distortion in the wavefront of the transmitted light.

For this experiment, a SLA printed sample was heated from 22 °C to 190 °C and cooled back to 22 °C (two complete cycles in N_2) and the endothermic and exothermic heat flows were recorded. The DSC data are presented in Figure 10.

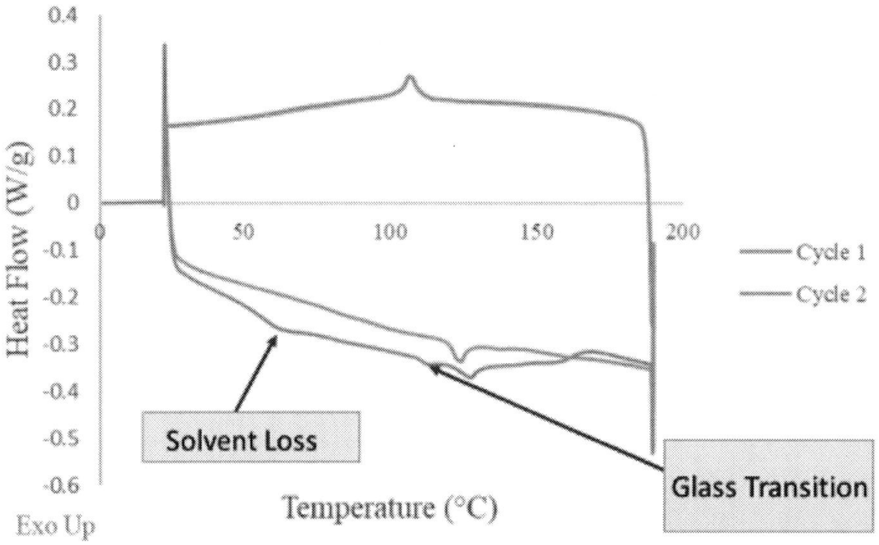

Figure 10. Differential Scanning Calorimetry (DSC) of a polymerized sample under nitrogen. Data represents two repeat cycles from 20 to 190 °C at 10 °C/min.

In the first cycle, solvent is released from the glassy polymer beginning at approximately 55 °C. In the second heating and cooling cycle, that endotherm is not observed, however the glass transition is observed at 100-110 °C, indicative of PMMA [14]. Based on the combined data from the TGA and the DSC, 80 °C was selected for post-processing the 3D printed optical components. This temperature allowed solvent to be released while maintaining and smoothing internal structures of the polymer structure that would contribute to additional scattering.

2.5 Optical Components

The optical components selected in this study for construction using 3D printing were lenses and optical waveguides.

2.5.1 Lenses

Rays of light that are incident on a lens either focus or diverge, depending on the type of lens that is used. The direction of light rays propagating through the lens change their direction due to refraction at the curved surface. The 2.5 cm diameter SLA lenses were printed on the Form2™ and Form3™ printers. Clear V.4 Resin was used to construct the lenses since it was the most transparent of all the resins observed during the initial tests. Both bi-convex and planar-convex lenses were fabricated and manually sanded to fix surface imperfections. The cross-sections of bi-convex and planar-convex lenses are shown in Figure 11.

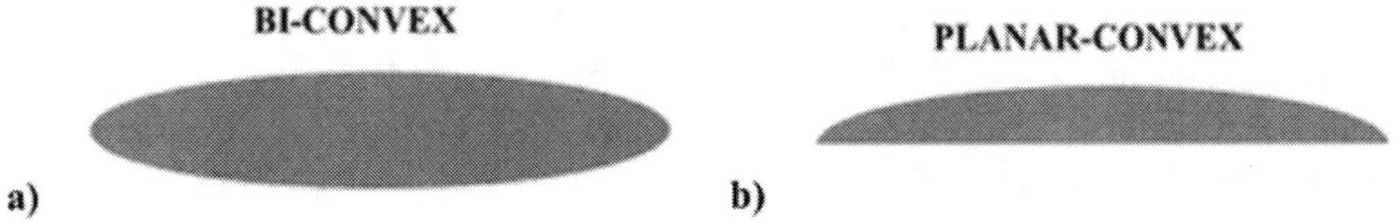

Figure 11. Lens designs for 3D printing, (a) bi-convex, and (b) planar-convex.

During post-processing both lenses were post-cured for 70 minutes. However, the planar-convex lens was post-cured at an oven temperature of 80 °C while the bi-convex was not heated during post curing. The variation in heating during post-curing was done in order to determine if heat was necessary to adequately refinish the surface of the lenses. The result of the post processing for each lens is shown in Figure 12.

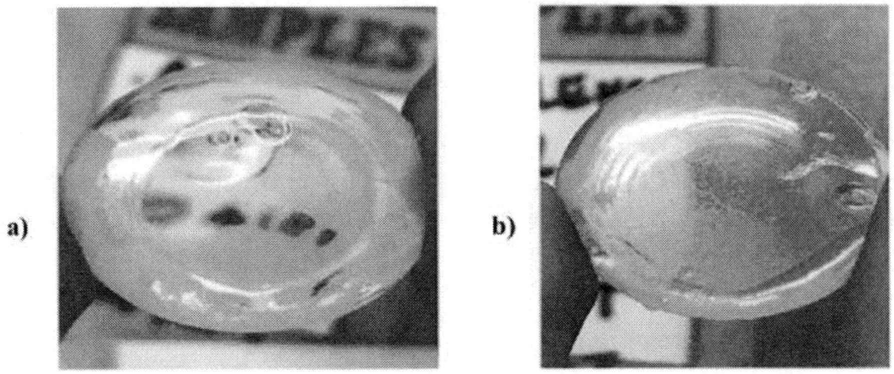

Figure 12. Post-processed (a) bi-convex lens and
(b) planar lens constructed from Clear V.4 resin.

The resin polymerized over the entire surface area of the planar-convex lens while the bi-convex lens had regions that did not exhibit full polymerization, because the bi-convex lens was post-cured without heat. Light that propagated through each lens experienced significantly less scattering at the post-cured surfaces. A comparison between the focusing properties of a lens without post-processing and the post-processed bi-convex lens is shown in Figure 13.

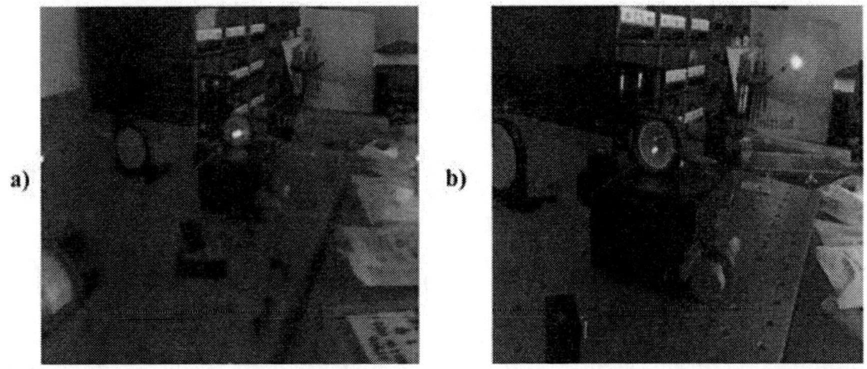

Figure 13. Laser propagation through (a) "as printed" lens and (b) post processed bi-convex lens.

2.5.2 Optical Waveguide

An optical waveguide is a spatial channel in which light propagates as modes - the superposition of plane waves propagating at different angles down the guide [15]. A slab waveguide consists of cladding and core layers. The cladding material has a lower index of refraction and typically surrounds the core. An example of the structure of a slab waveguide is illustrated in Figure 14.

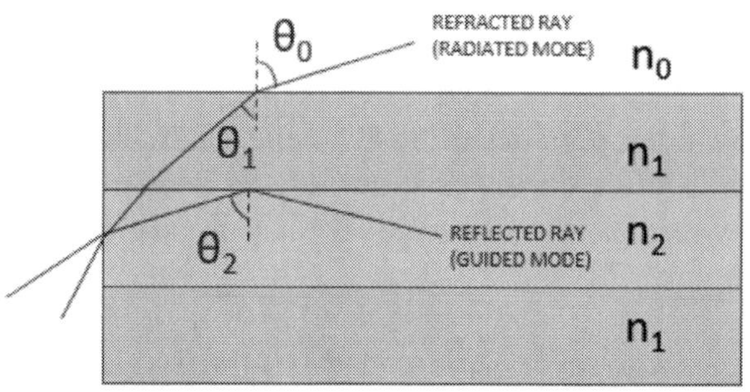

Figure 14. Slab waveguide showing refraction and internal reflection at layer boundaries.

The difference in the refractive index (n_i) of the core and the cladding results in refraction of the wave when it crosses each material interface. Refraction at a material interface is governed by Snell's Law, given in Equation (1) for the top interface in Figure 14.

$$n_1 \sin \theta_1 = n_0 \sin \theta_0 \tag{1}$$

For light that is incident at angles greater than the critical angle, total internal reflection (TIR) occurs and the light completely reflects back into the incident material. The angle (θ_c) at which this occurs is defined as

$$\theta_c = \tan^{-1}\left(\frac{n_1}{n_2}\right), \tag{2}$$

as illustrated by the reflected ray at the interface between materials n_2 and n_1 in Figure 14, where $\theta_2 \geq \theta_c$.

Clear V.4 resin was selected as the core for the optical waveguide and the material has an index of refraction of 1.5403 [16]. High Temp resin was selected for the cladding of the optical waveguide and the index of refraction is unknown. Preliminary slab waveguides were fabricated with the dimensions of 10.16 x 1.27 x 0.635 cm.

The surfaces of the waveguide structures were post-processed at 80 °C for 15 minutes on each face. The cladding slabs were secured to the core slab by putting a layer of Clear V.4 resin on each face that was in contact with the core and then post-curing the parts. The constructed waveguide is shown in Figure 15a) and an example test result that demonstrates optical propagation through the part is shown in Figure 15b). As shown in Figure 15b), the laser beam propagates in the same direction at each interface between layers of the slab, with little to no refraction (other than in the incident beam at the entrance) This suggests that the index of refraction of the High Temp resin and Clear V.4 resin is similar.

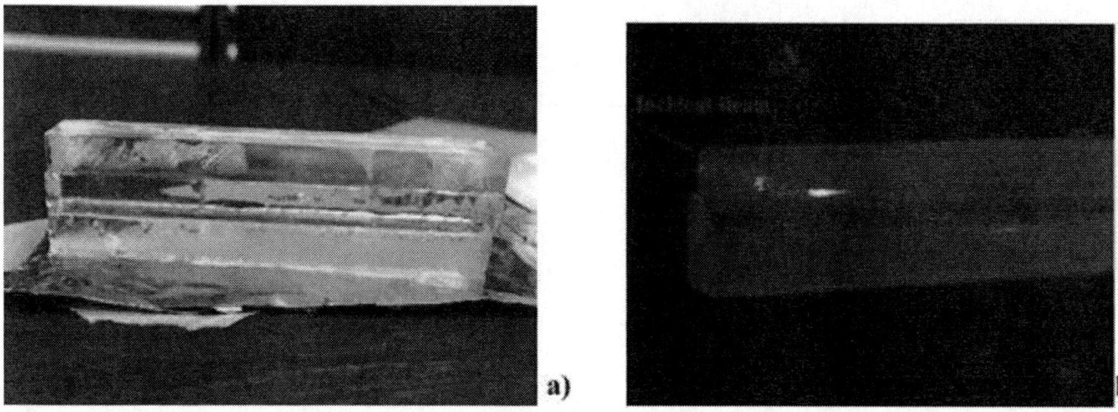

a) b)

Figure 15. (a) Constructed 3D printed waveguide with High Temp resin and Clear V.4 resin as the cladding and core, respectively (b) Laser testing with optical waveguide to observe refraction at layer boundaries.

2.6 Double Slit Experiment

Wavefronts of laser light are spatially coherent over a certain distance, referred to as the coherence length. Ideally, a functional optical waveguide will maintain the mode of the laser light propagating through it along with its degree of coherence. In order to verify that the light transmitted through the 3D printed waveguide maintained its spatial coherence, a double slit experiment was conducted. The planar wavefront of the light incident to the double slit with a slit width on the order of the optical wavelength will result in two point sources of spatially coherent light [17]. The two point sources generate waves that constructively and destructively interfere, resulting in interference patterns, as shown in Figure 16.

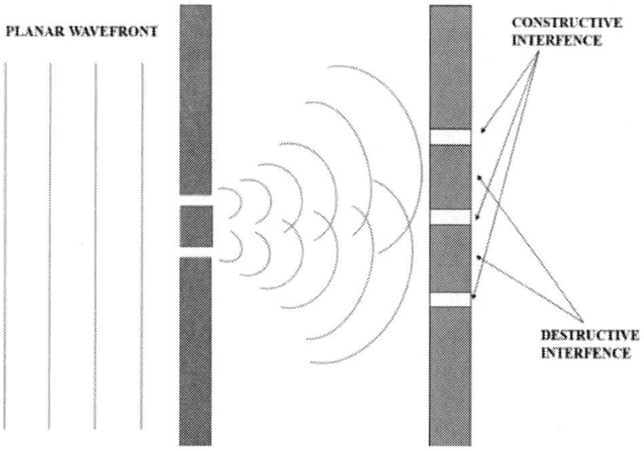

Figure 16. Planar wavefront creating interference pattern from double slit.

For the post-processed lens, light transmitted through the double slit resulted in constructive and destructive interference patterns. The existence of these interference patterns indicates the beam wavefronts maintained a degree of spatial coherence through the waveguide. The setup and resulting interference patterns are shown in Figure 17.

a) 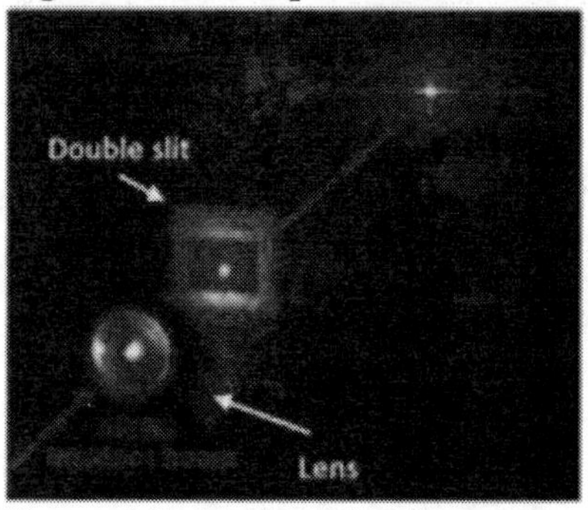

b) 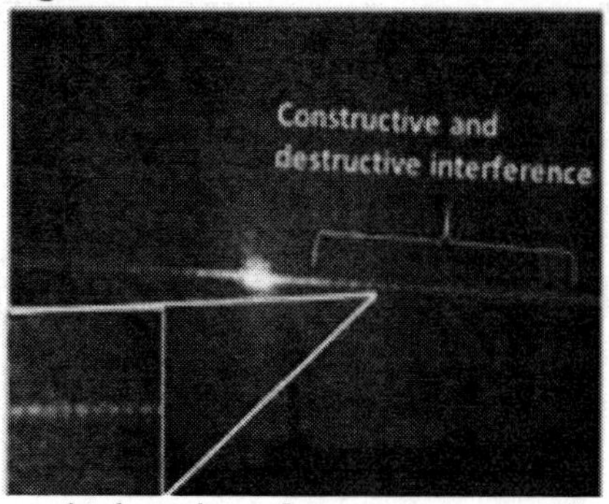

Figure 17. (a) Double slit setup with lens (b) lens interference patterns from double slit (see expanded inset image).

The spatial coherence of light transmitted through post-processed waveguide structures was also tested. Figure 18a) shows the far-field pattern produced by the double slit for light after it propagated through the waveguide structure. Figure 18) shows the far-field pattern without a double slit.

a) 

b) 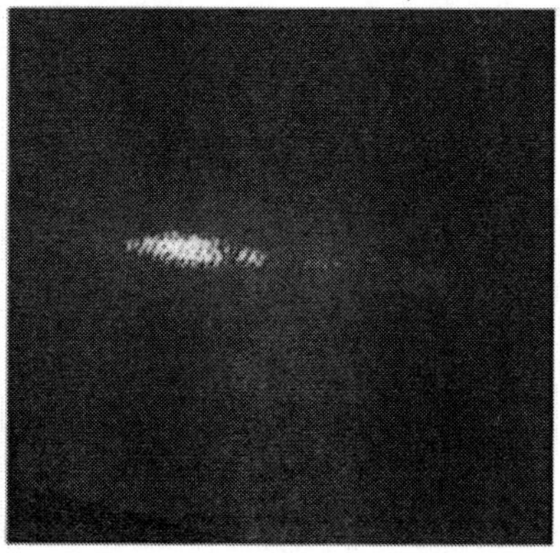

Figure 18. Far-field pattern created from light transmitted through waveguide with the double slit and (b) without the double slit.

The transmitted light through the slab structure and then through the double slit in Figure 18a) suggests that there may be partial spatial coherence in the transmission, though the far-field pattern is not particularly well defined. From Figure 18b), the interference pattern that occurs without the double slit is likely caused by the layering of the 3D printed structure that is parallel with (planar to) the direction of propagation. Each layer of the polymer part may cause scattering and distortion of the laser beam (Figure 19) since the beam diameter is significantly greater than layer thickness of the AM process.

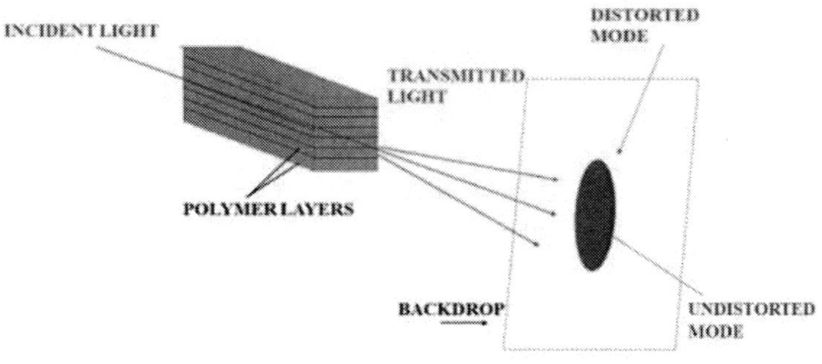

Figure 19. Layering of polymer waveguide structure with respect to the direction of laser propagation.

2.7 Optical Loss Testing

As light propagates through a medium, it can experience spreading (diffraction), scattering, and absorption [18]. Light propagating through the medium of a 3D printed polymer experiences each of these effects. The optical loss of the polymer media can be estimated using Beer-Lambert Law: [18]

$$P = P_0 e^{-\beta d} \quad (3)$$

Given that the initial optical power (P_o), transmitted power (P), and distance of propagation (d) are known, Equation (3) can be rearranged to find the absorption coefficient (β).

$$\beta = -\frac{\ln\left(\frac{P}{P_0}\right)}{d} \quad (4)$$

The HeNe laser light was transmitted through post-processed Clear V.4 parts of various lengths and the transmitted light through the part was recorded on an optical power meter, Thorlabs^{TM} PM100D Optical Power Meter and Thorlabs^{TM} S120C Sensor, shown schematically in Figure 20.

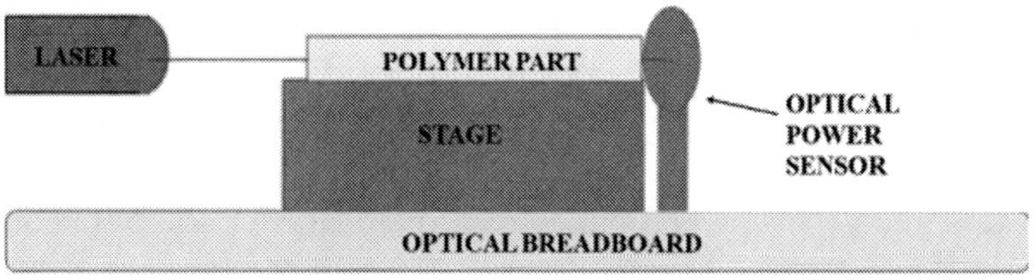

Figure 20. Optical loss testing setup with HeNe laser serving as the coherent light source.

The printed parts varied in length from 2.54 cm to 10.16 cm and the resulting transmitted power was recorded over three trials (Table 2). This data was used to estimate the average absorption coefficient.

Table 2. Optical power received after transmission over varying lengths of propagation. Data represents average values of tests performed in triplicate

Length of Part (cm)	Received Optical Power (mW)		
	Trial 1	Trial 2	Trial 3
Free-space	1.51	1.56	1.54
2.54	1.1	1.25	1.23
5.08	1.08	1.19	1.11
7.62	0.97	1.03	1.0
10.16	0.94	0.95	0.94

Using the data in Table 2 and Equation (4), the absorption coefficient of the Clear V.4 resin was estimated to be 0.0629 cm^{-1}. Using this estimated coefficient for the Trial 1 data, the measured transmitted power was graphed against the theoretical transmitted power in Figure 21.

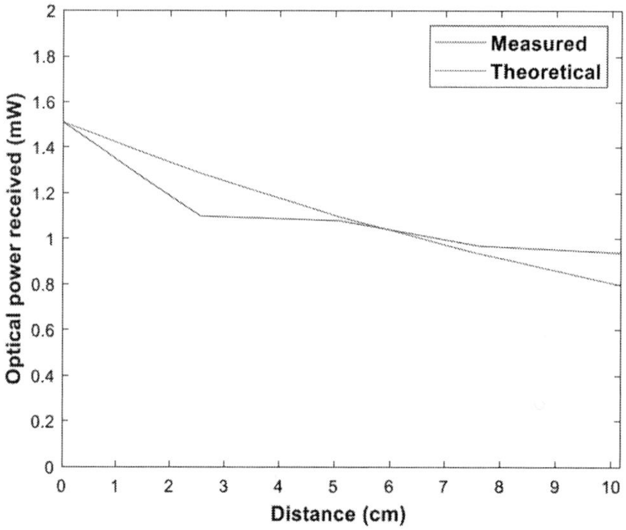

Figure 21: Measured versus theoretical optical loss characteristic based on Trial 1 data.

3. DISCUSSION

Using FDM printing is not feasible for this application. Warping of the thermoplastic filament prevented precise alignment with other 3D printed parts. Difficulty manufacturing a well-defined optical interface with a separate core and cladding made FDM impractical for waveguide components. Moreover, the nylon filament prints failed, resulting in the nylon burning as well as becoming attached to the extruder nozzle. The SLA printed materials printed with better precision as well as being more transparent. However, the light that propagated through the "as printed" SLA materials scattered.

While beam coherence was demonstrated in the lenses there was still light scattering on the backdrop of double slit experiment. This scattered light observed in the backdrop was separate from the interference pattern shown in Figure 17b) and is most like caused by contaminants and imperfections in the lens during the printing process. These obstructions present in the components could also contribute to further attenuation of the laser light.

Despite the use of post-processing on the waveguide structures, the flat surfaces of the slabs structures exhibited slight curvature after polymerizing due to a meniscus that formed due to the surface tension of the resin. This curvature could further contribute to distortion in the wavefronts when light enters or exits the curved faces. Additionally, the SLA specimen with layers parallel to the direction of propagation had more scattering than the parts with layers perpendicular to the direction of propagation. Given that the beam diameter is considerably wider than the layering of the 3D print, this suggests that impurities such as dust or containments may cause a slight variation in the index of the refraction or scattering sites within each layer, or microstructures may exist at the interface between parallel layers causing the additional wavefront distortion in parallel layered parts.

4. CONCLUSIONS

The PLA filament slabs that were printed using FDM warped during their construction. The uneven cooling of the thermoplastic during the print caused the warping. Therefore, the PLA and nylon filaments are not desirable materials to construct the waveguide.

The SLA materials printed with better precision and with greater transparency. However, the light that propagated through the unprocessed SLA materials was scattered. Post-processing SLA components greatly decreased distortion of the laser propagation. Double slit testing revealed that it is possible for laser light to maintain a degree of spatial coherence while propagating through a polymer component. Optical loss testing showed that attenuation of light propagating through the components remains significant.

SEM revealed rough surface microstructures in the layers of SLA printed components on the order of the HeNe wavelength that caused the light to scatter. Additionally, DSC and TGA revealed that the SLA specimens would begin to release solvent at approximately 55°C and will begin to degrade at 210℃. Therefore, 80 °C was selected as the temperature for annealing the 3D printed optical components.

5. REFERENCES

[1] Ackerman, Dan. "Making 3D Printed Face Shields, Mask Accessories and Other Coronavirus Tools." *CNET*, 21 Apr. 2020. <www.cnet.com/news/making-3d-printed-face-mask-accessories-and-coronavirus-tools/>.

[2] Fuentes, Gidget. "Marines At Sea Leaning on 3D Print-to-Order Parts to Stay in the Fight." *USNI News*, 25 Nov. 2019. <news.usni.org/2019/11/25/marines-at-sea-leaning-on-3d-print-to-order-parts-to-stay-in-the-fight>.

[3] Yariv, Amnon, and Pochi Yeh. *Photonics: Optical Electronics in Modern Communications.* Oxford Univ. Press, 2009.

[4] Palermo, Elizabeth. "Fused Deposition Modeling: Most Common 3D Printing Method." *LiveScience*, Purch, 19 Sept. 2013. <www.livescience.com/39810-fused-deposition-modeling.html>.

[5] "The Ultimate Guide to Stereolithography (SLA) 3D Printing (Updated for 2019)." *Formlabs.* <formlabs.com/blog/ultimate-guide-to-stereolithography-sla-3d-printing/>.

[6] "Technical Data Sheet- Ultimaker Nylon". *Ultimaker*, 18 July 2017, <ultimaker.com/download/68517/TDS Nylon v3.012.pdf>.

[7] "Technical Data Sheet- Ultimaker PLA Ultimaker". *Ultimaker*, 18 July 2017, <ultimaker.com/download/74599/UM180821 TDS PLA RB V10.pdf>.

[8] "Ultimaker S5 Installation and User Manual." Ultimaker.com, *Ultimaker*, 2018, <ultimaker.com/download/74047/UM180509 UMS5 Manual V18.1.pdf>.

[9] "Compare Formlabs SLA 3D Printers Tech Specs." *Formlabs*, <formlabs.com/3d-printers/form-3/tech-specs/>.

[10] "Materials Data Sheet - Photopolymer Resin for Form 1 and Form2." Formlabs.com, *Formlabs*, 18 Sept. 2018, archive-media.formlabs.com/upload/XL-DataSheet.pdf.

[11] Rank, Manuel, and Andreas Heinrich. "Three-Dimensional Printing of Optics." *Society ofPhoto-Optical Instrumentation Engineers*, 2018, DOI:10.1117/3.2324763.

[12] David P. Durkin, Miranda J. Gallagher, Benjamin P. Frank, Elizabeth D. Knowlton, Paul C. Trulove, D. Howard Fairbrother, Douglas M. Fox. "Phosphorus-Functionalized Multi Wall Carbon Nanotubes as Flame Retardant Additives for Polystyrene and Poly (methyl methacrylate)". *Journal of Thermal Analysis and Calorimetry*, DOI: 10.1007/s10973-017-6432-z (2017).

[13] Hatakeyama, T., and F. X. Quinn. *Thermal Analysis: Fundamentals and Applications to Polymer Science*. John Wiley y& Sons, 1994.

[14] Teng, Hongxiang, et al. "High Glass Transition Temperatures of Poly(Methyl Methacrylate) Prepared by Free Radical Initiators." *Journal of Polymer Science Part A: Polymer Chemistry*, vol. 47, no. 1, Jan. 2009, pp. 315–317., doi:10.1002/pola.23154.

[15] Yariv, Amnon, and Albert Pochi Yeh. Optical Waves in Crystals Propagation and Control of Laser Radiation. J. Wiley & Sons, 2003.

[16] "Creating Camera Lenses with Stereolithography." *Formlabs*. <formlabs.com/blog/creating-camera-lenses-with-stereolithography/>.

[17] Jackson, David P, et al. "Subtleties with Young's Double-Slit Experiment: Investigation of Spatial Coherence and Fringe Visibility." *American Journal of Physics*, vol. 86, no. 683, 23 Aug. 2018. DOI:10.1119/1.5047438.

[18] Payne, Craig M. *Principles of Naval Weapon Systems*. Naval Institute Press, 2010.

ADDITIVE MANUFACTURING OF POLYMER DERIVED SILICON OXYCARBIDE CERAMIC STRUCTURES

Shengheng Gu, Jayanta Kapat, Jihua Gou*
Composite Materials and Structures Laboratory
Department of Mechanical and Aerospace Engineering
University of Central Florida
4000 Central Florida Blvd, Orlando, FL 32816

ABSTRACT

This study presents the synthesis of a photo-polymeric resin towards 3D printing of polymer derived ceramic to create amorphous SiOC ceramic structures with complex geometries. The SiOC polymer derived ceramic resin was synthesized by a photosensitive methyl-silsesquioxane as pre-ceramic polymer and 3-(trimethoxysilyl)propyl methacrylate as polymer chain graft, namely MK-TMSPM. Through the hydrolysis sol-gel reaction, the MK-TMSPM system condensed into a cross-linked network with multiple photosensitive function groups, which are key to the photo-curing process while free radical photo-initiator excited by the UV source. During the Digital Light Processing (DLP), the liquid resin was locally cured to the desired shape on each layer based on the input CAD model. Once the first layer cured, the building plate moves upwards to pull out resin layer by layer with a cured thickness of 0.05 mm. After DLP, the green body was sintered in a tube furnace in the nitrogen atmosphere at 1,000°C. This method allows to fabricate a component with specific 3D geometries that can be converted into ceramic structures, offering a number of potential applications, such as medical devices, nano-electromechanical systems, high temperature materials, and anti-corrosive materials.

Keywords: Additive Manufacturing; Digital Light Processing (DLP); Polymer Derived Ceramics; UV-Curable Resins; Pyrolysis

*Corresponding Author: Email: jihua.gou@ucf.edu, Phone: +1(407) 823-2155

1. INTRODUCTION

Additive manufacturing has attracted significant attention recently because it offers more advantages against traditional manufacturing methods, such as mold-free fabrication, customizability, and rapid prototyping[1, 2]. It is based on computer aid design model, which is sliced into g-code to construct an object via layer-by-layer structure using the principle of material accumulation[3]. Compared to additive manufacturing of polymers and metals, the development of 3D printing of ceramics is relatively slow [4]. Ceramic materials have good advantages in high temperature and corrosive environments [3, 5]. Additive manufacturing of ceramics has attracted more and more research interests in the development of new 3D printing process and feedstock materials.

Direct ink writing (DIW) process uses ceramic slurries to fabricate the green body of the ceramic structures [3, 6]. Selective laser sintering (SLS) and fused deposition modeling (FDM) processes are based on ceramic particles and powders[7]. PDCs such as ZrO_2, SiO_2, Al_2O_3, SiC, and SiOC

[10] have been used for the pre-ceramic resin. Zanchetta et al. [2] reported 3D printing of SiOC ceramic micro-components with strut size down to 200 um. Eckel et al. [3] reported that preceramic monomers could be cured with the UV light through stereolithography by a patterned mask, forming a complex shaped preceramic part[6, 11]. In this study, it was reported the fabrication of pre-ceramic SiOC resin with a cubic structure by using photosensitive methyl-silsesquioxane polymer via sol-gel approach[12] and stereolithography technique.

2. EXPERIMENTATION

Material synthesis

A commercially available methyl-silsesquioxane resin powder (SILRES MK, Waker-Chemie GmbH, Germany) was dissolved into the mixture of two organic solvents of THF (Tetrahydrofuran) and TPM (Tripropylene Glycol Monomethyl Ether) with the volume ratio 1:1. After stirring and sonication for 1h, 3-(trimethoxysilyl) propyl methacarlate (TMSPM) was added to the solution with a weight ratio of 4:1 to MK powder, to obtain photosensitive function group network preceramic polymer. After stirring and sonication for another hour, the MK-TMSPM system was hydrolyzed in the acidic condition with $2.5 < pH < 4.5$ at room temperature for 12h.

The main chemical structure of photocurable siloxane is shown in Figure 1. During the hydrolysis process, the methyl group was replaced by the hydroxyl. TMSPM acted as chain grafts to the Si-O cross-linked network. After the hydrolysis and condensation with TMSPM, the MK siloxane structure became photosensitive SiOC network with high densification.

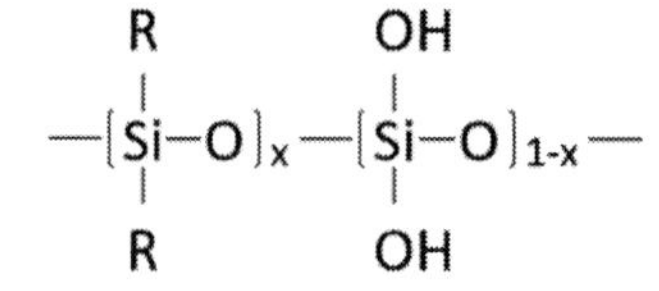

Figure 1 General chemical formula of polysiloxane, in this study, x=0.75

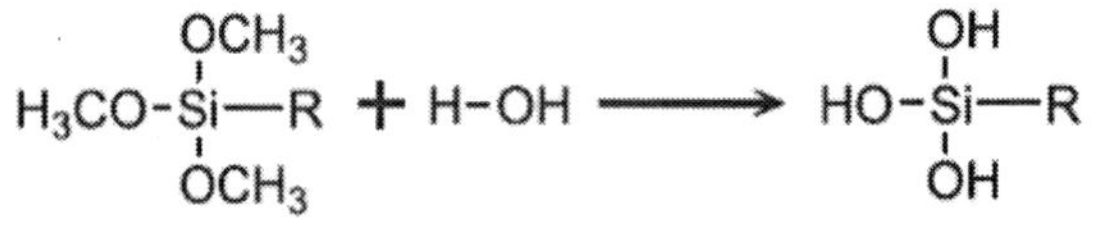

Figure 2 Siloxane hydrolysis

Before printing, the organic solvent THF was removed to form high solid loading and high viscosity resin system. The THF solvent was evaporated in a vacuum oven and 45℃ for 60 min. The solvent removal was verified by calculating the weight loss. A photo-initiator, phenylbis (2,4,6-trimethlbenzoyl) phosphine oxide was added into the photosensitive resin. The photo-initiator was excited by the UV light and formed radicals which could further initiate the radical photo-polymerization along with the chain growth. The SiOC network was cross-linked, forming a pre-ceramic green body. All printing parameters were adjusted to achieve an optimal UV curing dosage and exposure time.

3D printing

3D printing process was conducted using digital light processing (DLP) process. The CAD model was imported to 3D printer as STL files. Complex geometric 3D model was sliced into multiple layers with 2D image features and sliced thickness. The CAD model and printing parameters are shown in Figure 3. The scanning speed is key to the UV dosage of the resin system.

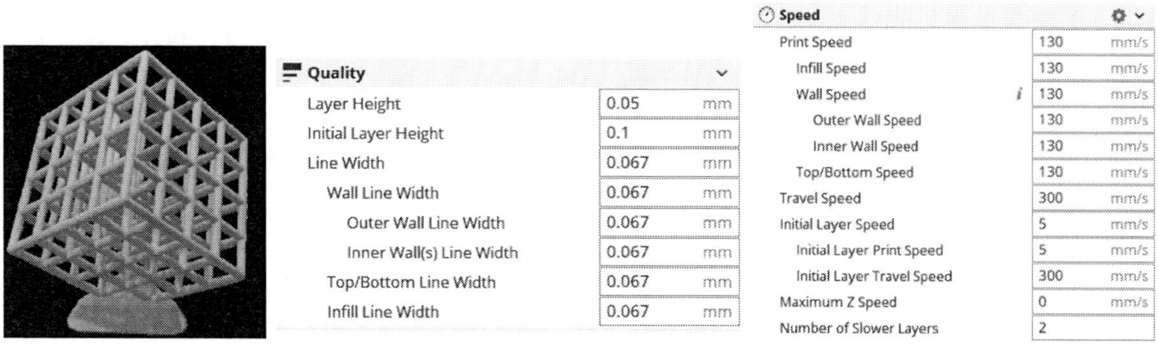

Figure 3 (a) CAD model for printing; (b) layer thickness parameter setting; (c) scanning speed and exposure time setting

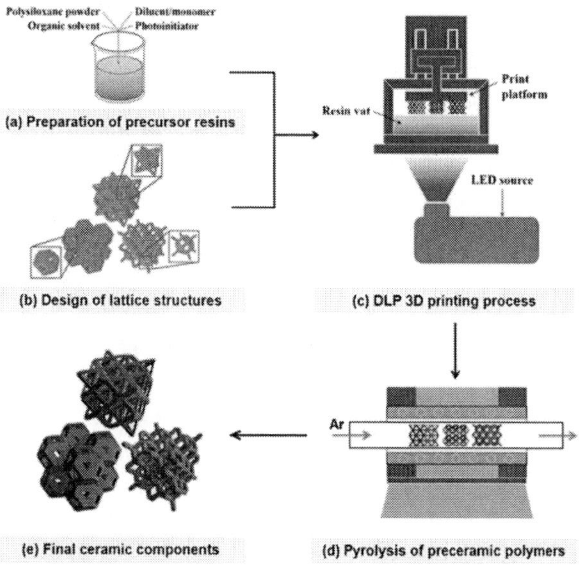

Figure 4 schematic process of synthesis and 3D printing, (a-b) synthesis of preceramic resin; (c) 3D printing; (d) sintering; (e) pyrolyzed SiOC ceramic component.

After the first layer cured, the build plate moved up and the vat tilted for a fixed angle. The resin refilled the gap between the build plate and the resin vat, and exposed to the UV light to form a new layer on the build plate at the same time. The gap between the build plate and the resin vat corresponding to the slicing thickness, and the projection of the subsequent images continued the cured part until the printing of entire model was accomplished. In this study, the layer resolution and lateral thickness were set as 50 um to maintain a high surface quality. After printing, the green body was peeled off from the build plate, and then washed with ethyl alcohol to removed residual photoinitiator and solvent for 3 minutes. The post-curing process was conducted in the UV box to complete the crosslinking of the resin by exposing in soft UV light for 60 seconds. Finally, the green body was sintered in a tube furnace at 1,000°C for 2h in the argon atmosphere. The heating up rate and cooling rate were both 2°C/min. The process schematic is shown in Figure 4. After printing and sintering, the SiOC ceramic part was made.

3. RESULTS AND DISCUSSION

The X-ray diffraction (XRD) pattern of sintered SiOC ceramic parts was collected using an X-ray diffractometer (XRD Basic diffraction PANalytical Empyrean). The microstructure of green body and sintered preceramic parts were studied through scanning electron microscope (SEM, Zeiss ULTRA-55 FEG SEM). A Fourier transform infrared spectroscope (FTIR, Perkin Elmer Spectrum 100 series) in the range of 400-4500 cm^{-1} was used to investigate the chemical reaction process such as hydrolysis and photopolymerization, which provided bonding and functional groups at different stages.

XRD of SiOC Ceramic

The XRD pattern of SiOC ceramic shown in Figure 5 indicating that the resulting SiOC is amorphous, compared to manufacturing of typical SiOC at temperature around 1,000°C, the pattern of X-ray diffraction is well established that the sintered ceramic parts should be amorphous[5]. The scattered pattern obtained by XRD has no atomic periodicity, as well as no sharp or narrow high intensity peaks appeared, and thus it is supposed to be composed mainly of amorphous SiOC, which agree with literatures about SiOC fabrication XRD patterns[5]. At the sintering temperature of 1,000°C, the preceramic green body transformed into amorphous SiOC ceramics and only a broad peak at about 24-25° was identified. Brinckmmann et al.[5] reports a fabrication of SiOC via 3D printing, the XRD pattern of SiOC ceramic in Figure 6, after pyrolysis at 1,000°C in argon confirming it has an amorphous structure without obvious porosity.

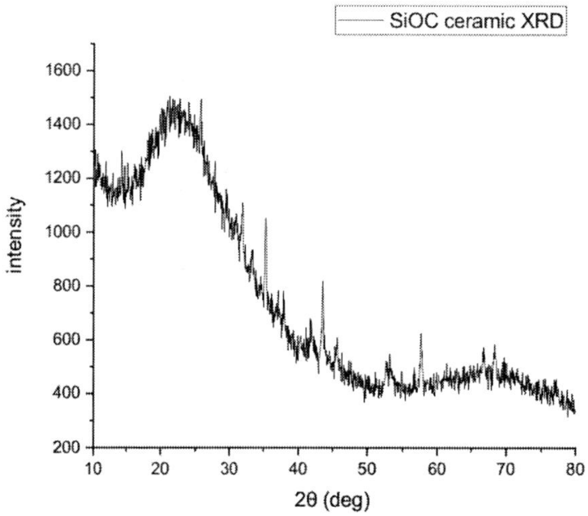

Figure 5 XRD pattern of sintered SiOC ceramic

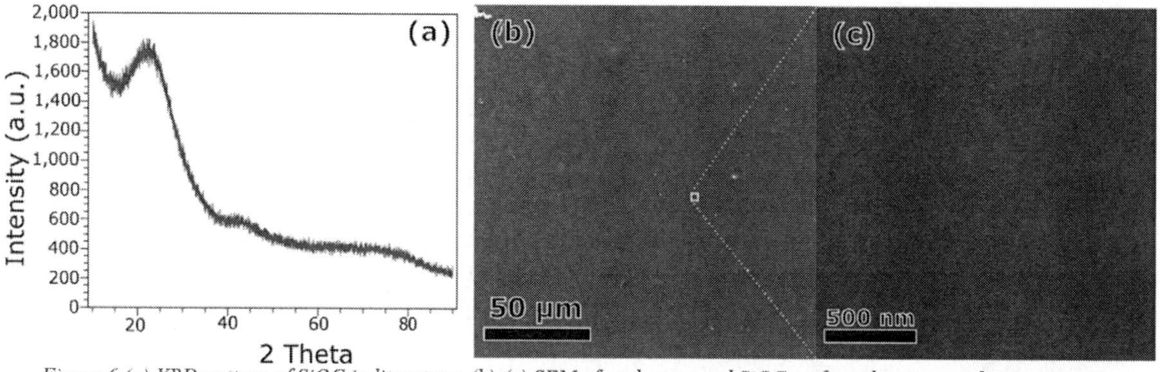

Figure 6 (a) XRD pattern of SiOC in literature; (b)-(c) SEM of carbon coated SiOC surface showing no obvious porosity.

Microstructure of SiOC Ceramics

The cubic lattice structure of SiOC ceramic after the sintering at 1,000°C has a good surface quality without obvious defects on the surface as shown in Figure 7. The pyrolyzed SiOC ceramic shrank due to the thermal stress, resulting in visible cracks and fractured layers as shown in Figure 8.

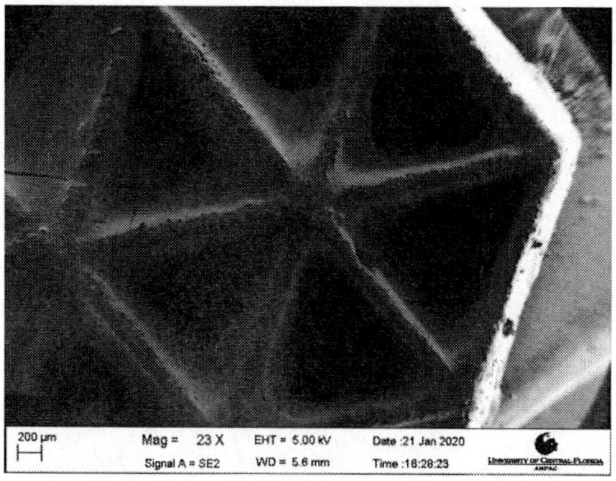

Figure 7 SEM image of SiOC sintered ceramic parts in 23x magnification

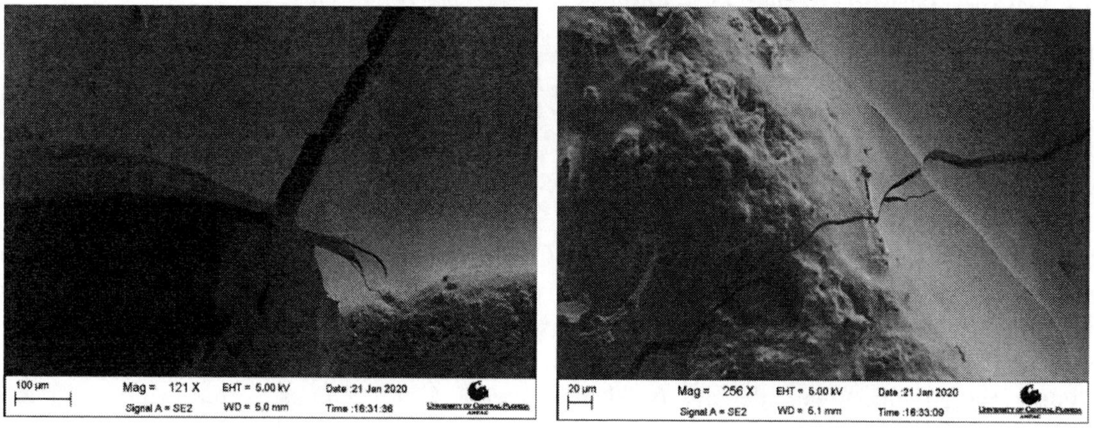

Figure 8 SEM image of partial parts of SiOC component showing some cracks and fracture along in edges.

The surface of layered ceramics components is smooth and without porosity while some of edges have short cracks due to volume shrinkage resulting from thermal stress. The thickness of each layer is about 50um, which is almost same with slicing parameter of printing layer thickness, therefore, the shrinkage occurs in thickness scale has not been observed obviously. The obtained SiOC ceramic components possess high density with rare porosity, which can be observed from the fracture layer surface of SEM image. In 256x magnification, cracks can be seen, though the fracture and crack are very clear and indicates a cleavage fracture defect, the SiOC ceramic component would not be significantly affected in perspective of mechanical properties. The smooth surface with few pores can be identified at 256x magnification microscale since the gases escaping during the pyrolysis, which has non-significant effect on ceramic component's mechanical properties. The surface details verify the XRD results about formation of amorphous SiOC.

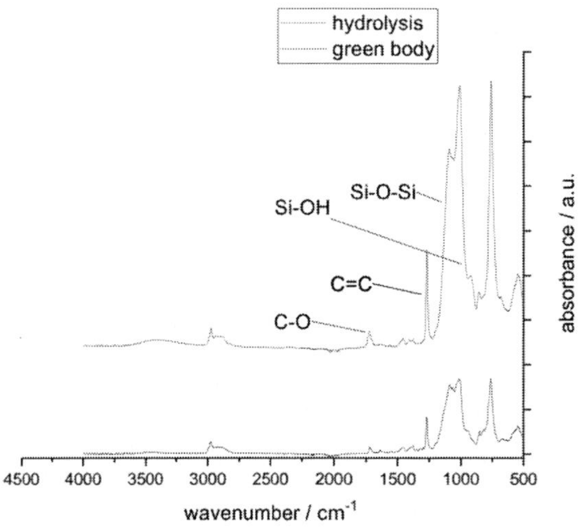

Figure 7 FTIR spectrum of synthesized resin system before and after photopolymerization

3.3 FTIR Spectroscopy of SiOC Ceramic

Through FTIR of the resins before and after printing (figure 9), the observed FTIR absorbance peaks corresponding to characteristic function group, including C-H band at around 2950 cm^{-1}; C=O unsaturated band at around 1700 cm^{-1}; C=C appears at around 1300 cm^{-1}; and Si-O-Si peak at 1100 cm^{-1}.[7] All these bonds appear on the FTIR spectrum before and after photopolymerization, in MK-TMSPM resin system, the TMSPM provides Si-O-C and C=C function group, which are backbone of preceramic cross-linking network and photosensitive unsaturated bond respectively. Therefore, the C=C intensity indicates the crosslinking degree during polymerization process. After the UV polymerization, the FTIR spectrum with fewer C=C absorbance reflects C=C function groups decrease so that photopolymerization occurs. And C=O bonds at 1700 cm^{-1} are also unsaturated bonds which can be polymerized by radicals to form crosslinking network. The intensity or absorbance of C=O decrease after printing indicates siloxane crosslinked. In other words, the lower the peak of the C=C and C=O bonds from specific group of resin system, the more crosslinking can be found in the preceramic polymer. The decreasing of Si-O-Si network, demonstrates the densification process of the resin system, in process of photopolymerization, Si-O-Si bonds were induced by TMSPM, then further play a role as backbone of crosslinking network to condense the resin and polymerize the monomers into full network structure. From the FTIR spectrum, process of polymerization and chemical reaction can be determined, concluding that sol-gel MK-TMSPM preceramic resin is UV curable and photo sensitive. The green body chemical structure with specific function groups should be also obtained.

4. CONCLUSIONS

In the study, a siloxane preceramic polymer was created via stereolithography 3D printing and the resulting pyrolyzed SiOC was obtained by the sintering process. High solid loading mixture of siloxane and silicone with high stability and good curing property was developed during the synthesis. Photopolymerization process between monomers and photoinitiator was conducted by UV stereolithography, the silicone-based precursor was crosslinked. The pyrolysis process generated highly dense, complicated cubic lattice structure with a theoretical density of 96.5%. The ongoing work will focus on the improvement in mechanical properties and thermal properties of SiOC ceramic structures by optimizing resin formation, 3D printing parameters and pyrolysis conditions.

5. ACKNOWLEDGMENTS

This work was sponsored by the Federal Aviation Administration (FAA) Center of Excellence for Commercial Space Transportation (FAA COE CST) under Task 253: Ultra-high Temperature Composite Thermal Protection Systems. The authors acknowledge the administrative and technical support of Ken Davidian and Nickolas Demidovich.

6. REFERENCES

1. Li, Z.Y., et al., *Additive manufacturing of lightweight and high-strength polymer-derived SiOC ceramics.* Virtual and Physical Prototyping, 2020.
2. Zanchetta, E., et al., *Stereolithography of SiOC Ceramic Microcomponents.* Advanced Materials, 2016. **28**(2): p. 370-376.
3. Fu, Y., et al., *Multiple metals doped polymer-derived SiOC ceramics for 3D printing.* Ceramics International, 2018. **44**(10): p. 11030-11038.
4. Wei, L., et al., *Fabrication of SiOC ceramic with cellular structure via UV-Assisted direct ink writing.* Ceramics International, 2020. **46**(3): p. 3637-3643.
5. Brinckmann, S.A., et al., *Stereolithography of SiOC Polymer-Derived Ceramics Filled with SiC Micronwhiskers.* Advanced Engineering Materials, 2018. **20**(11).
6. Pierin, G., et al., *Direct Ink Writing of micrometric SiOC ceramic structures using a preceramic polymer.* Journal of the European Ceramic Society, 2016. **36**(7): p. 1589-1594.
7. Santoliquido, O., P. Colombo, and A. Ortona, *Additive Manufacturing of ceramic components by Digital Light Processing: A comparison between the "bottom-up" and the "top-down" approaches.* Journal of the European Ceramic Society, 2019. **39**(6): p. 2140-2148.
8. Brigo, L., et al., *3D Nanofabrication of SiOC Ceramic Structures.* Adv Sci (Weinh), 2018. **5**(12): p. 1800937.
9. de Hazan, Y. and D. Penner, *SiC and SiOC ceramic articles produced by stereolithography of acrylate modified polycarbosilane systems.* Journal of the European Ceramic Society, 2017. **37**(16): p. 5205-5212.

10. Zocca, A., et al., *Additive Manufacturing of Ceramics: Issues, Potentialities, and Opportunities.* Journal of the American Ceramic Society, 2015. **98**(7): p. 1983-2001.
11. Mei, H., et al., *Strengthening three - dimensional printed ultra - light ceramic lattices.* Journal of the American Ceramic Society, 2019. **102**(9): p. 5082-5089.
12. Costakis, W.J., et al., *Additive manufacturing of boron carbide via continuous filament direct ink writing of aqueous ceramic suspensions.* Journal of the European Ceramic Society, 2016. **36**(14): p. 3249-3256.

COMPRESSION MOLDING TOOL PRODUCTION VIA ADDITIVE MANUFACTRURING - DESIGN AND PART OPTIMIZATION

John F. Unser[1], Peter Wang[2], Andrzej Nycz[2]

[1] IACMI The Composites Institute, Knoxville, TN

[2] Oak Ridge National Laboratory (ORNL) Manufacturing Demonstration Facility (MDF), Knoxville, TN

ABSTRACT

Advanced Composites Manufacturing Innovation (IACMI), Oak Ridge National Lab Manufacturing Demonstration Facility (ORNL MDF), Century Tool, and Lyndoll Bassell have teamed together to focus on metal printing of tooling for compression molding composites. The goal of this technical collaboration was to evaluate the use of large-scale metal additive manufacturing (AM) to print metal molds for the mainstream (>100k units/year) production of large composite components by compression molding. Today, metal tools are typically made by subtractive machining of large blanks of forged tool steel, or sometimes by metal casting in conjunction with post-cast machining. Lead time for procuring a large forged blank is often many months. This drives program schedules and generates significant metal scrap. Metal AM has the potential to reduce lead time and waste, which would increase the viability of large-scale composites. Additionally, metal AM can provide unique advantages such as increasing geometric complexity and enabling design changes mid-program. For this project, baseline metrics for additively manufactured metal tooling were established, and small-scale metal tool was additively manufactured and evaluated. The results indicate that it is feasible to produce metal tooling for composites using metal AM.

Keywords: additive manufacturing, compression molding, metal AM, robotics
Corresponding author: John F. Unser

1. INTRODUCTION

Today, metal tools for the production of large composite parts are typically made by subtractive machining of large blanks of forged tool steel, or sometimes by metal casting in conjunction with post-cast machining. These traditional manufacturing methods limit geometric complexity and generate significant metal scrap that must be collected for recycle. Additionally, the lead time for procuring a large forged blank is often many months, which significantly lengthens program schedules. The cost and lead time for procuring large metal molds inhibits commercial deployment of composites. If cost and lead time of metal molds can be significantly reduced without sacrificing performance, it will favorably impact the business case and reduce the risk of implementing composites in high volume platforms, such as mainstream automobiles. Lead time reduction also allows more time for material development and validation before the platform design must be frozen, thus increasing the probability that composites will be implemented on a platform.

Institute for Advanced Composites Manufacturing Innovation (IACMI) run by Collaborative Composite Solutions Corporation (CSS) is growing the US composites market by advancing materials and manufacturing technologies from basic research and development to deployment-ready products. IACMI's roadmap specifically includes additively manufactured molds for composites manufacturing. In contrast to traditional methods of metal tool production, metal AM can eliminate the lead time for blank procurement, which would result in faster build times. Additionally, AM can dramatically reduce subtractive machining and enable geometrical complexity with enhanced functionality, such as conformal heating and cooling.

However, there are many challenges to print a compression molding tool that need to be overcome. One is just understanding the requirements. Preliminary requirements and targets for mold performance, cost, and lead time have been established, but need to be confirmed and/or refined as appropriate for metal AM. Molds produced by metal AM need to deliver the same performance of today's tool steel molds, with sufficient size and complexity to address ≥ 80% of mainstream automotive applications in (semi) structural (excludes Class A) compression molding. Beyond current performance, the metal AM should enable design changes at any time in the program (i.e., mold dimensions can be changed by adding material to the mold face). Additionally, the AM could offer further extensions into tooling for other processes, such as autoclave curing, thermoforming and injection molding.

The Manufacturing Demonstration Facility (MDF) at Oak Ridge National Laboratory (ORNL) in collaboration with Lincoln Electric developed unique capabilities in manufacturing large scale metal parts [1, 2]. It is based on Gas Metal Arc Welding (GMAW) and advanced robotics. The research spans the entire workflow of complex part manufacturing. It starts with part design and optimization as described in [3], followed up by toolpath planning [4] and simulation [5]. A critical component is process development [6] and material characterization [7]. This made the MDF an ideal partner to collaboratively address these challenges with CCS. The goal of this this work was to evaluate the feasibility of AM to meet the established requirements and provide further improvements. This involved two main tasks. The first one was to establish baseline metrics for tooling, and the second task was to develop and analyze a small-scale tool design. The results of this project indicate that metal AM can be used to fabricate a metal tool that meets the established criteria.

2. EXPERIMENTATION

2.1 Establish baseline metrics for tooling

The baseline metrics were obtained through discussions with industrial partners with experience in making metal molds, like Ford and Core Molding Technologies, Tier 1 suppliers, and tool manufacturers. The required metrics are presented below:

Parameter	Goal	Stretch Goal
Modelled cost reduction	25%	50%
Modelled lead time reduction	50%	80%
Durability	1 M mold cycles	3 M mold cycles
Molding pressure	1,300 psi	2,500 psi
Molding temperature	300°F	--
Surface temperature uniformity	± 10°F	± 5°F
Surface dimensional stability	≤ 0.12 mm after 10 thermal cycles	≤ 0.05 mm after 10 thermal cycles
Surface finish	SPI B2 after polishing	--
Maintenance and cleaning	Equal to baseline	--

In addition to these metrics, the following goals were also identified:

- Produce integrated heating channels using hot oil as heating media
- The system capability for allowing design changes during program life (materials can be added or subtracted as needed)
- Materials properties that are equal to or better than P20 steel
- Provide hydraulic ejection on bottom surface (cavity) of part in 4 places
- Mount all ejectors/plugs/sensors within ± 0.05mm of flush to the core/cavity surface
- Tool stiffness adequately maintains deflection below 0.05mm in any direction, assuming molding pressures of 1,000 psi. Design should accommodate molding pressures in the tool to 1300 psi without failure.
- A profile tolerance for the core and cavity tool surfaces within 0.05mm (will require CMM report). Press interface reference surfaces shall be parallel within 0.05mm over area.

2.2 Complete small-scale tool design and structural analysis

2.2.1 Selection of the tool

Based on discussions with partners, two existing tools were considered. The tool selected from those initially considered is shown in Figure 1. This design contained the details of the machined surfaces, alignment hardware, mounting and sensor details. The model below shows the upper and lower molds as well as all of the associated hardware. There are four guideposts (10) control the molds movement. There are contact plates (11) that act as hard stops to ensure the mold stops at the correct height. There are two holes (12) for thermocouples and one port (13) for pneumatic ejection. The sets of six bolts (14) on the top and bottom are to fasten the molds to the platens on the hydraulic press. The tool in its original form is a very complex assembly.

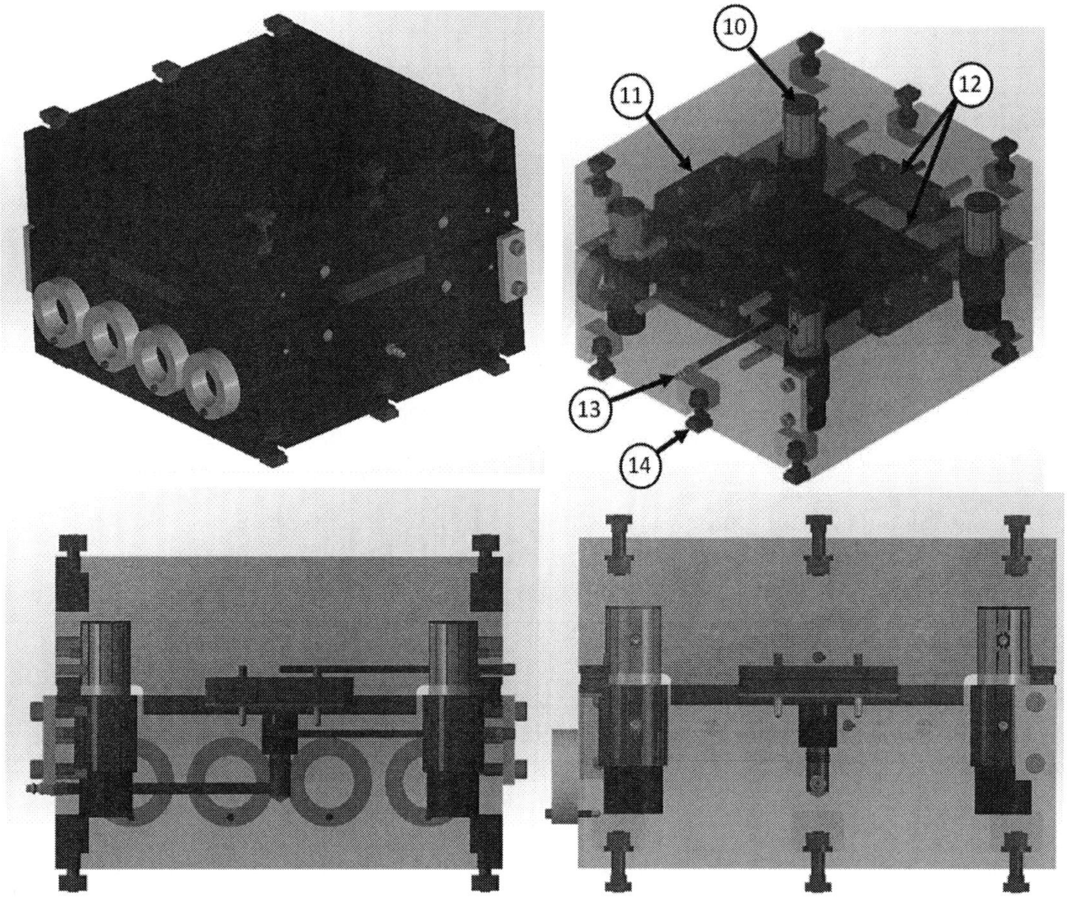

Figure 1. Baseline tool without ejection pin system

A design for a mechanical ejection system was also developed. Figure 2 provides the details of the ejection system designed for the selected tool. The model shows the nine plates required for the ejection system. The pins and associated bushings are all flanged and retained between thelayers of the bolted plates. The plates 1 and 2 are bolted together. The plates 3,4, 5, 6 and 9 are bolted together. The plates 7 and 8 are bolted together. The plates 2 and 3 are the upper and lower molds, respectively. The plates 7 and 8 drive ejector pins to push the part out of the mold and the position of the plates is reset using the push back pins. There are currently 30 screws to

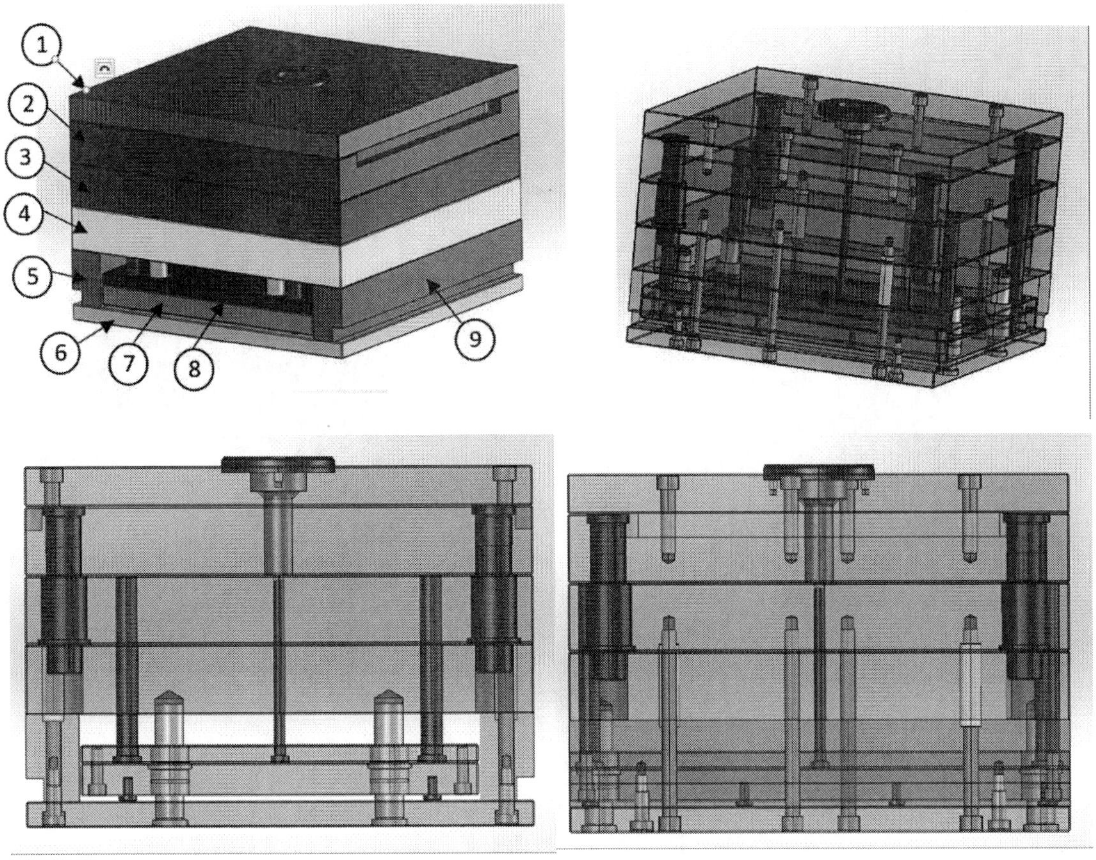

Figure 2. Candidate tool with mechanical ejection system

fasten 9 metal parts together. The indents on the side of the assembly are to use clamps to attach the mold to the platens.

2.2.2 Tool subcomponent merging

The first goal before redesigning the tool for printability was to minimize the number of parts. The first iteration was focused on combining the design of the molds with the design of the ejector system. The number of machined parts was able to be reduced by taking advantage of the additive manufacturing process. Four of bottom mold pieces (3, 4, 5 and 9 in the ejector system illustrated in Figure 2) were combined to form the printed bottom part. Two mold pieces (1 and 2) were combined to make the printed top part. These combinations and the resulting parts are illustrated below in Figures 3 and 4.

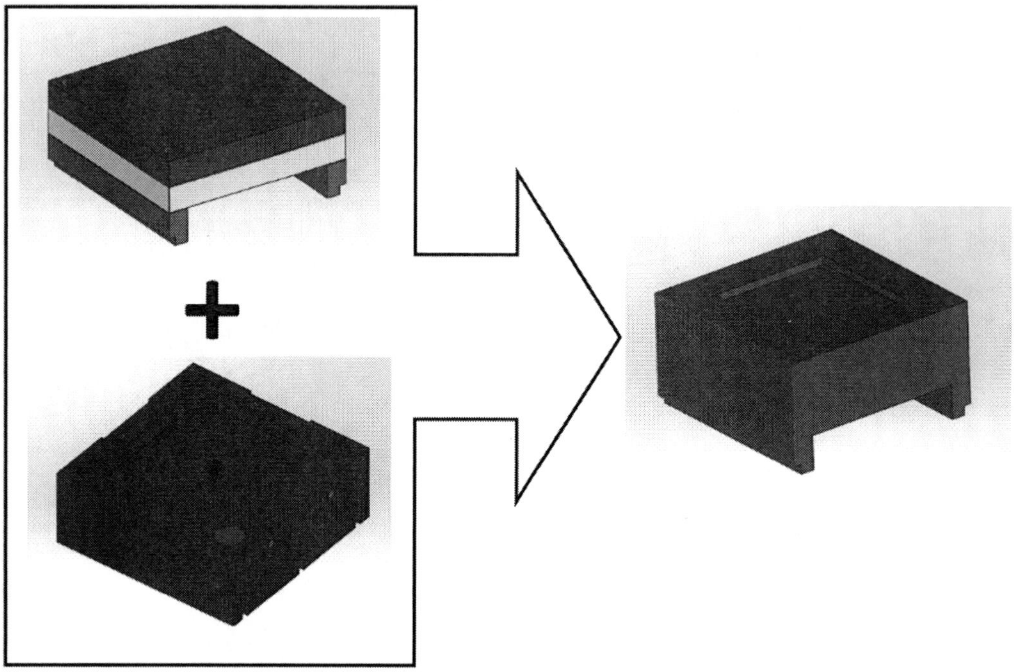

Figure 3. Merging the top (female) part of the tool into a single part

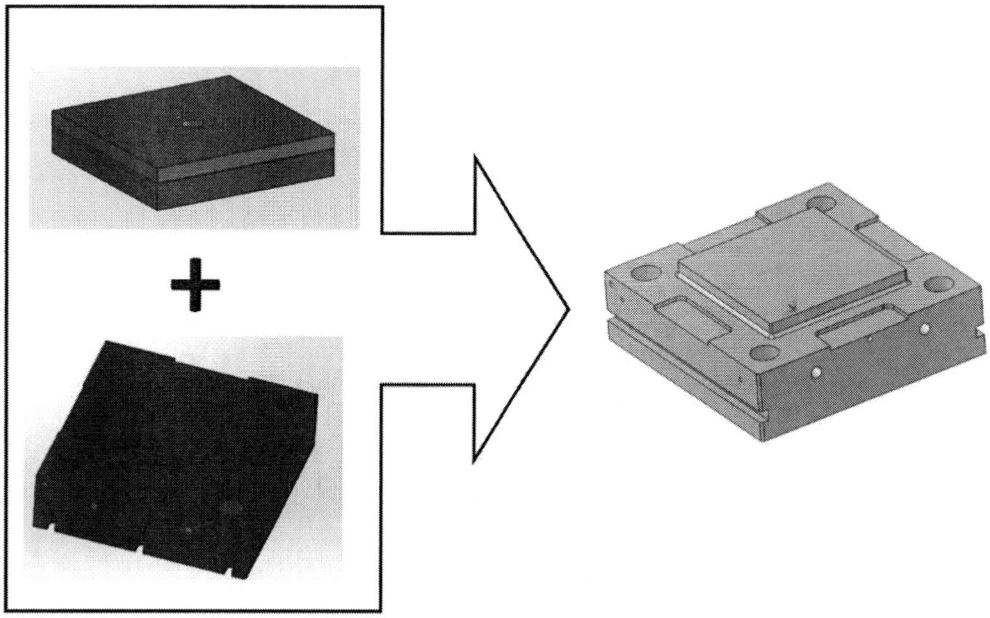

Figure 4. Merging process for the bottom (male) part of the tool.

2.2.3 Design of the heating channels

The next critical part of the design was to embed the heating channels in the tool. In certain tools, adding heating channels adds unnecessary complexity to the tool and complicates its maintenance, as illustrated in Figure 5.

Figure 5. Complexity of heating channels

Heating of the blocks is required by the molding process and can be accomplished by heating the hydraulic press platens or the mold blocks themselves. The mold blocks are heated using electric cartridge heaters or tubing with heated fluid. The tubing can provide more controlled and uniform heating to the surface of the mold. However, tubing is difficult and time consuming to install. Therefore, integrated heating channels were added to the design. Figure 6 shows the top and bottom of molds with the addition of the integrated heating channels.

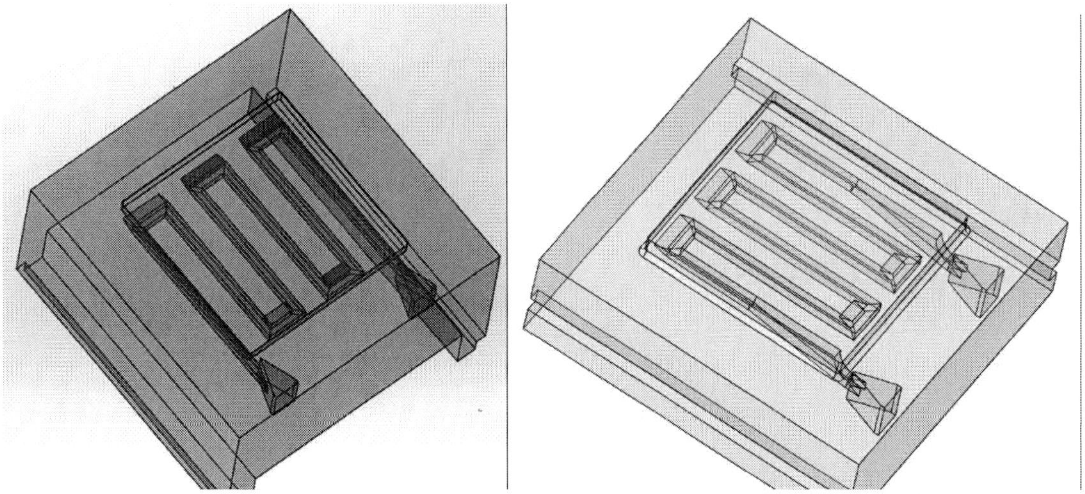

Figure 6. Tool design with heating channels

The heating channel in the printed design follows a serpentine path, which allows for the heating source to be closer to the mold surface without the need for complex machining and tubing installation. The heating channels have been added to both the top and bottom mold sections. Cavities provided at either end of the heating channel allow for a threaded connection to be machined directly into the block, which eliminates the extra step of welding on a fitting and removes a potential leak point.

3. RESULTS

3.1 Initial design analysis

The preliminary design with the integrated heat pipes and merged parts mitigated the need for machining and fasteners. The top (male) part of the mold weighs 463 lb and the bottom (female) part weighs 620 lb. The designs were analyzed using finite element analysis (FEA). The results indicated that male part had a predicted deflection of .003 mm under 600 psi of molding pressure and a maximum stress of 3,300 psi under the 1300 psi maximum load, as shown in Figure 7. The female part had a deflection of .020 mm under 600 psi of molding pressure and has a maximum stress of 900 psi under the 1300 psi maximum load, illustrated in Figure 8.

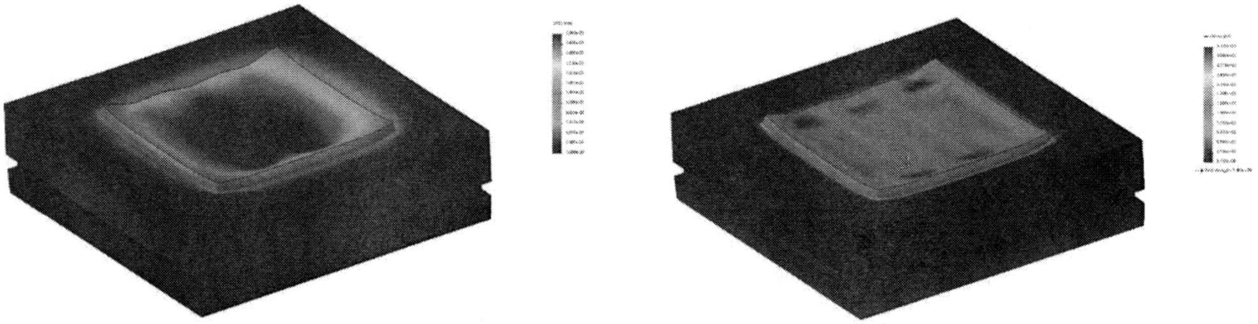

Figure 7. Male tool deflection (left; 0.03mm at 600 psi) and maximum stress (right; 3.3 ksi with 1300 psi and uniform loading)

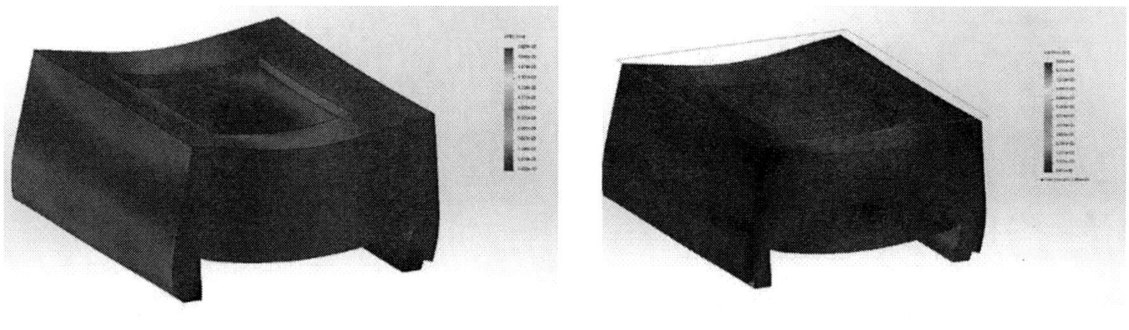

Figure 8. Female tool deflection (left; 0.02mm at 600 psi) and maximum stress (right; 900 psi with 1300 psi and uniform loading)

The top mold has 223 layers and is estimated to take 126 hours to print. The bottom mold has 223 layers and is estimated to take 151 hours to print. The toolpaths for printing one layer with the wire-arc process are shown in Figure 9.

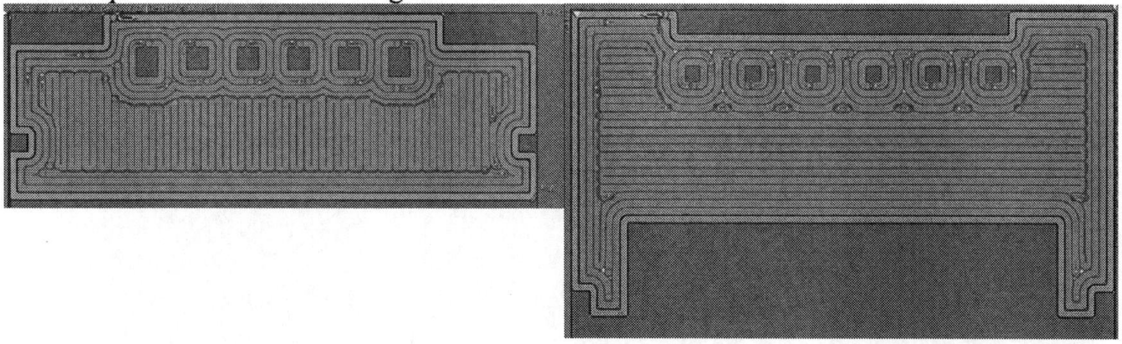

Figure 9. Initial toolpaths for the male (left) and female (right) parts, using the wire-arc process

The molding surfaces, hard stop mounting surface, mounting groove surfaces and bottom mating surface must be machined. These areas are highlighted in blue in Figure 10.

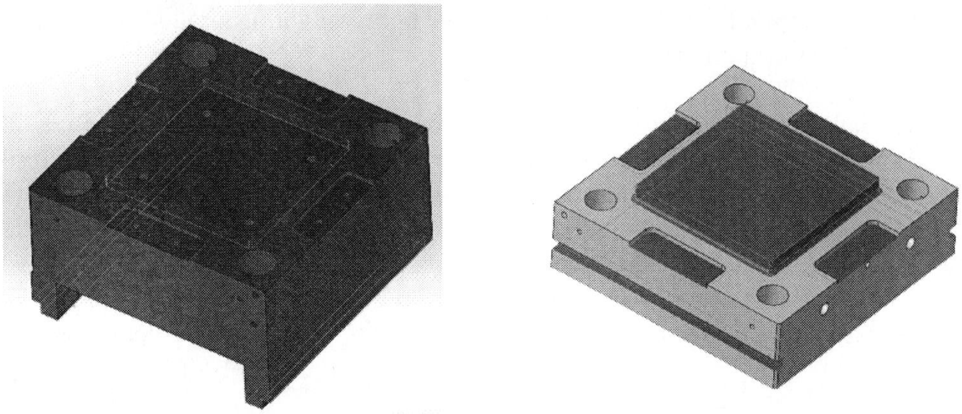

Figure 10. New design showing areas requiring machining (in blue)

3.2 Weight reduction

In order to reduce both the weight and the printing time, several strategies to lightweight the molds were investigated. Initially, a variety of patterns to hollow out the interior structure of the molds were developed, as shown in Figure 11.

Figure 11. Different approaches to part lightweight

The patterns were analyzed using FEA and passed the deflection and stress criteria for the given loads. However, it was found that the majority of these light-weighting strategies would make the printing significantly slower and more difficult.

Effective light-weighting was achieved by creating through holes that penetrated through the entire mold. The locations and sizes of the holes, shown in Figure 12, were selected to avoid hitting any internal features while maintaining a minimum offset of a ½". Additionally, on the bottom (female) mold, two support posts were placed to stiffen the mold structure. The modified top (male) mold weighs was 300 lb and the bottom (female) mold weighs 484 lb, saving 163 and 136 lb for the top and bottom molds, respectively. The estimated build times for the top and bottom molds are 96 hours and 128 hours, saving 30 hours and 23 hours, respectively, with the toolpath plan shown in Figure 13.

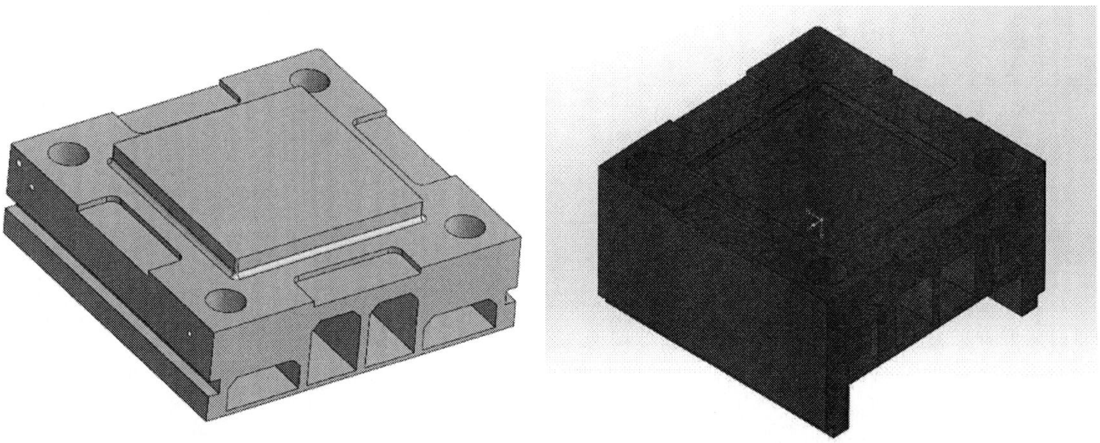

Figure 12. Weight reduction via internal ribbing of the male part (left) and female part (right)

Figure 13. Toolpaths with weight reduction cavities for the male part (left) and female part (right)

The use of multiple materials is also being explored for this build. The outer forming surfaces will be constructed of 410 Stainless Steel while the infill will be a mild steel. The faster deposition rate of the mild steel as well as the significantly reduced cost will save both money and build time of the part. The deposition system will be set up such that it can quickly switch between the two different materials. See Figure 14 for a slice showing the two different materials, with the 410 stainless steel having a narrow bead path than the mild steel.

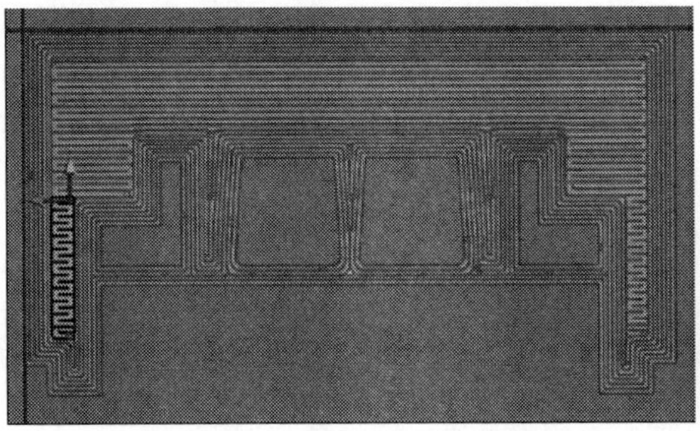

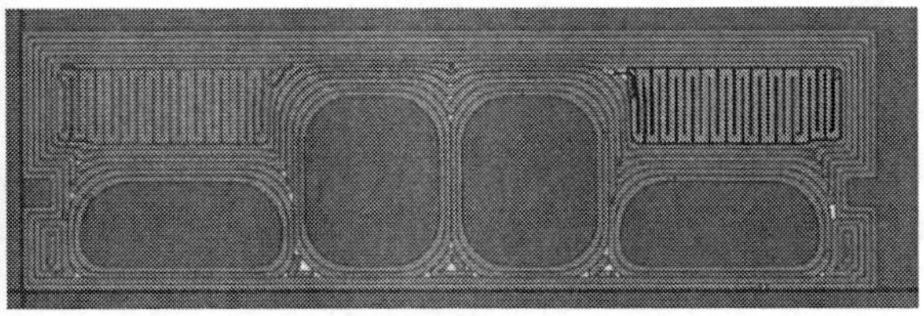

Figure 14 Multi material tool paths for the male (right) and female parts (left). The mild steel can be seen to have a much wider bead than the stainless steel

The deflections and maximum stresses of the top (male) and bottom (female) parts, as predicted by FEA, are illustrated in Figures 14 and 15, respectively. The top mold had a predicted deflection of .012 mm under a 600 psi load and a maximum stress of 9ksi under a 1300 psi load. The bottom mold had a deflection of .039 mm under 600 psi load and a maximum stress of 19 ksi under a 1300 psi load. Maximum stress was observed at the contact between the support posts and the bottom mold.

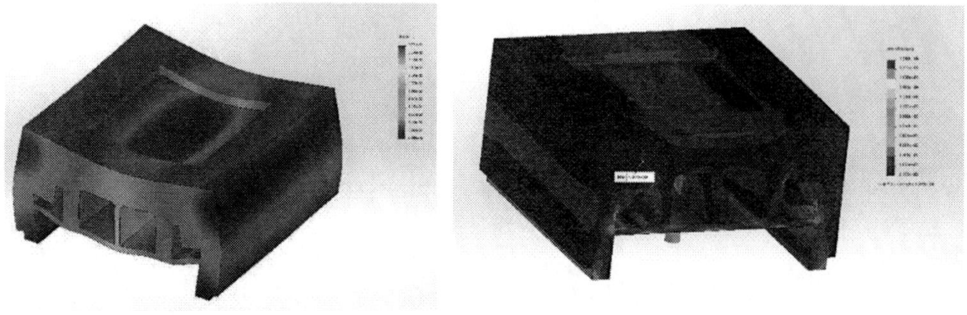

Figure 16. Top mold deflection map (left) and stress map (right)

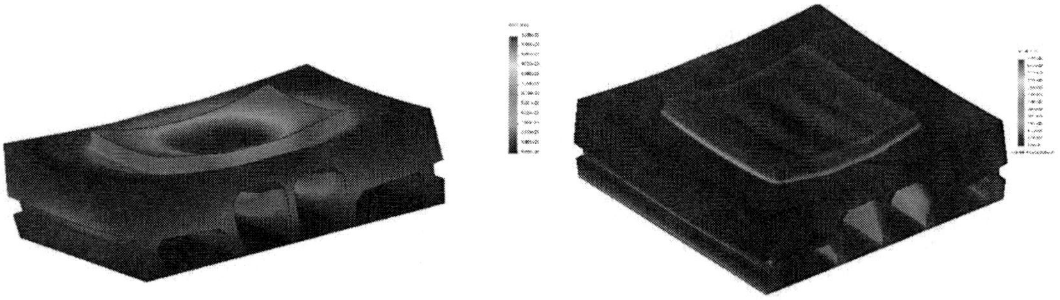

Figure 15. Bottom mold deflection (left) and stress map (right)

This new design reduces the weight, number of parts, fasteners, and machining while maintaining deflection and max stress requirements.

A CFD thermal analysis was also performed to theoretically verify the performance of the heating channels. The oil temperature is 350F, the pressure is 30 PSI and the flow rate is 20 GPM. The thermal conductivity of the mold is 40 W/m K and the ambient temperature is 68 F. The analysis was run using SolidWorks Flow Simulation and found that the average surface temperature of the molding surface was maintained at 305F which was within the specified range, shown in **Error! Reference source not found.**.

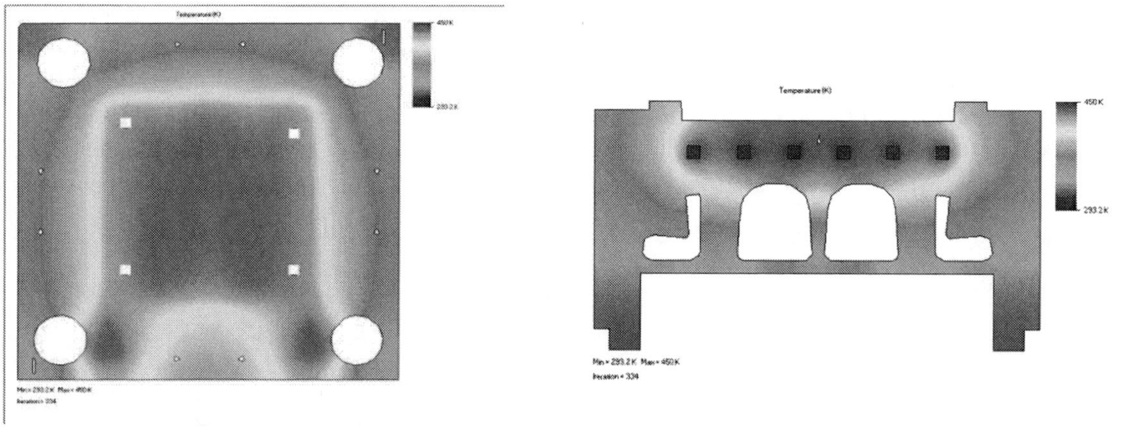

Figure 17 Thermal analysis of heating channel performance

4. CONCLUSIONS

This project advances the ability to produce molds for manufacturing mainstream automotive structural composite components with ≥ 50% reductions in cost and/or lead time of mold procurement. It is a vital first step toward an audacious goal: printed metal molds used in mass production of large, complex automotive parts such as roofs, floors, and body sides at rates ≥ 100k units annually, and > 1M components produced per mold. The proposed tool was redesigned for the AM wire-arc process based on prior experience in tool manufacturing. The optimization results showed a decrease in number of components and total weight compared to original part. The AM design allowed for introduction of internal heating channels and manifolds. The further optimization decreased the estimated print time. The proposed changes will also decrease the machining time while meeting the mechanical criteria. The next step is the manufacturing of the tool using the wire-arc system.

5. ACKNOLEGMENTS

This material is based upon work supported by the U.S. Department of Energy, Office of Science, Office of Advanced Manufacturing Office under contract number DE-AC05-00OR22725.

6. REFERENCES

1. Andrzej Nycz, Mark Noakes, Bradley Richardson, Andrew Messing, Brian Post, Johnathan Paul, Jason Flamm, Lonnie Love, *Challenges in Making Metal Large-Scale Complex Parts for Additive Manufacturing: A Case Study Based on the Additive Manufacturing Excavator*, in *Solid Freeform Fabrication*. 2017: Austin, TX.
2. Hassen, A.A., et al., *Scaling Up metal additive manufacturing process to fabricate molds for composite manufacturing.* Additive Manufacturing, 2020. **32**: p. 101093.
3. Greer, C., et al., *Introduction to the design rules for Metal Big Area Additive Manufacturing.* Additive Manufacturing, 2019. **27**: p. 159-166.
4. Patrick, S., A. Nycz, and M. Noakes, *MULTI-MATERIAL PROCESS PLANNING FOR ADDITIVE MANUFACTURING.*
5. Simunovic, S., et al. *Metal big area additive manufacturing: Process modeling and validation*. in *NAFEMS World Congress*. 2017.
6. Vaughan, D., et al., *SS410 Process Development and Characterization*. 2019, Oak Ridge National Lab.(ORNL), Oak Ridge, TN (United States).
7. Shassere, B., et al., *Correlation of Microstructure and Mechanical Properties of Metal Big Area Additive Manufacturing.* Applied Sciences, 2019. **9**(4): p. 787.

Thermal Cycling of Dahltram I-350CF

Zachary I. Skelton, Aubrey G. Jackson, Philip A. Lunn, Oliver P. Bottler
Airtech International Inc.
Huntington Beach, CA U.S.A

ABSTRACT

Using Large Format Additive Manufacturing, a layup mold was fabricated using Airtech's Dahltram I-350CF for development testing. Thermal cycle testing was then completed to validate the performance of high temperature polymers for autoclave use. All testing was completed at Airtech Huntington Beach using their in house testing capabilities. During the course of 100 thermal cycles dimensional stability and vacuum integrity were deemed critical. In addition to thermal cycle testing, a thermal survey was also performed on the mold to confirm that the printed tool could meet processing conditions for the greater aerospace industry.

Keywords: Large Format Additive Manufacturing, Thermal Cycle Testing, Dahltram I-350CF, Autoclave Tooling, Thermoplastics, Thermal Survey

Corresponding author: Zachary I. Skelton

1. INTRODUCTION

The acceptance of Large Format Additive Manufactured tools has slowly been gaining momentum over the last several years. Several development programs have been started for limited use or prototype tooling with the goal of fabricating one to several production parts. While the technology has been shown to be achievable with resin infusion [1] as well as in autoclave processing [2], very little research has been completed showing how this tooling will respond with continued use. Several development programs have been initiated because of the success in fabricating autoclave tooling, but their intent was for limited prototype production [3]. The purpose of this testing is to validate Dahltram I-350CF for continued 176 °C (350 °F) autoclave service and establish tool life from autoclave thermal cycles.

Key metrics for this testing were dimensional stability and vacuum integrity. While thermosets have a proven track record when used for tooling, thermoplastics have not typically been used. Thermoplastics are known to have creep and slowly move over time when exposed to heat. Another possible issue with thermoplastics is the exposure to oxygen at elevated temperature. While the amount of oxygen present in the autoclave is minimal during processing, this is the primary application for Dahltram I-350CF when used for the fabrication of 176 °C (350 °F) cure thermoset composites. Autoclave processing was completed with nitrogen at 7 bar (100 psi), but oxygen is still present from the initial one atmosphere. Several composite laminates were made on the demonstrator tool during this testing, but pulling a laminate off for every cure was not included in order to expedite testing. Following the conclusion of 100 thermal cycles, a thermal survey was performed on the tool to validate heat up rates and thermal gradient to ensure that it was capable of meeting industry standards.

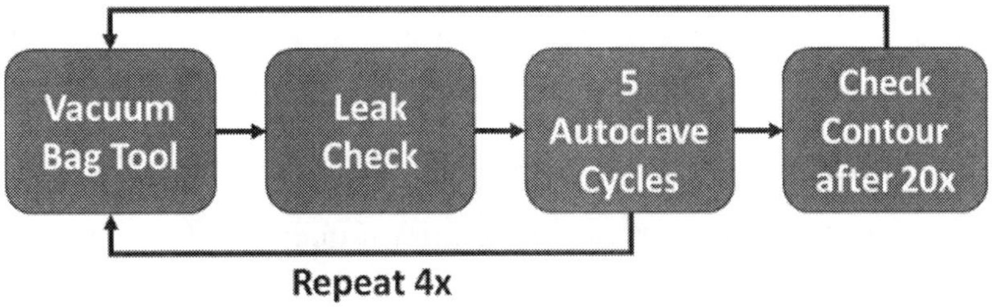

Figure 1: Process Overview

2. EXPERIMENTATION

2.1 Multi-run Cure Cycle Used

The cure cycle used for the testing consisted of the following:

1. Pull full vacuum and ramp at 5 °C/min (10 °F/min) to 176±5 °C (350±10 °F) while pressuring to 7 bar (100 psi) with nitrogen.
2. Once lagging thermocouple reaches 171 °C (340 °F), begin 2 hour hold.
3. Cool at 5 °C/min (10 °F/min) to below 49 °C (120 °F) while maintaining pressure.
4. Once high TC reaches under 49 °C (120 °F), hold for 45 minutes.
5. Repeat cure cycle 4 times.
6. On completion of 5th cure, dump pressure and release vacuum.

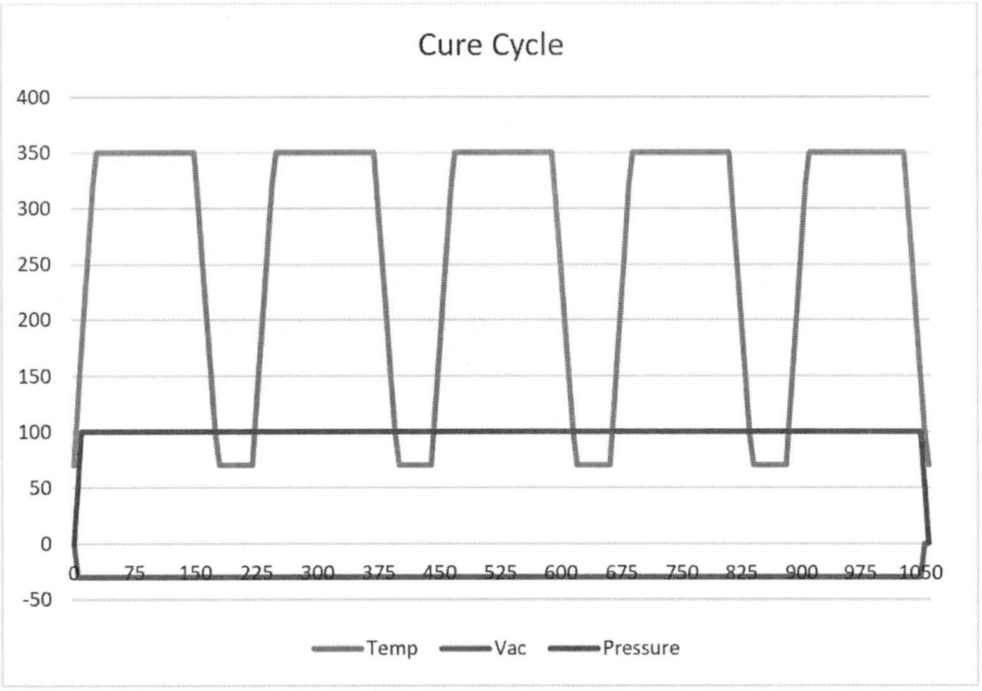

Figure 2: Multi-run Cure Cycle (All Units are Imperial)

2.2 Failure Criteria

For failure criteria in the testing, the following was selected. First, the tool must be able to hold vacuum throughout the cure cycle. Second, the tool must maintain its geometry with ±0.127 mm (±0.005 in). While not a critical part of the testing, a visual inspection was also included to look for any discoloration, cracking, or degradation of the tool.

Vacuum integrity would be validated with room temp drop tests over ten minutes (< 25.4 mmHg in 10 min, < 1 inHg in 10 min). During the cure, vacuum is monitored, and several 176 °C (350 °F) leak checks were ran during the testing. Leak checks during every thermal cycle were not ran since the majority of the testing was ran over night. Geometry would be validated with a Creaform Laser Scanner after every 20 cycles.

2.3 Tool and Equipment Used

A U-shaped tool with two flat flanges and complex contour was used for the testing. The tool is minimally reinforced and built to Airtech design standards for a tool of this size. Finished face sheet is two printed beads and is 35.6 mm (1.4 in) thick, nominally. Reinforcement headers are 20.3 mm (0.8 in) thick with a contact area of approximately 63.5 mm (2.5 in). Overall tool dimensions were selected due to autoclave sizing. The tools and its dimensions can be seen in the following Figure 3.

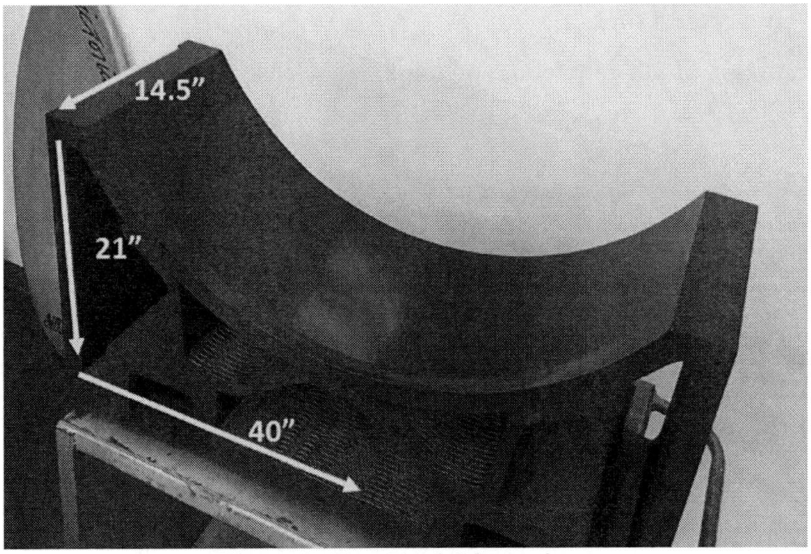

Figure 3: Test Tool Dimensions
(1,106mm x 368mm x 533mm)

Post NC machining, the tool was hand finished to better than 0.8 Ra (32 Ra) finish. This surface finish was achieved prior to the application of sealer. Sealer selected was Chemlease MPP 2180 and the tool was coated with four coats to manufacturer recommendation. Sealer selected for the testing was Marbacoat TRE 45 ECO with 2 coats being applied. Materials for seal and release were selected because of availability. Standard recommended process by Airtech for sealing printed mold tools is 4 coats of MPP2180, 3 coats of Zyvax Sealer GP, and 2 coats of Zyvax Release Ease 2.0. With all of them being applied per the manufacturer recommendations.

Airtech's Huntington Beach, CA autoclave was used for the thermal cycling and elevated leak checks. Due to tool weight, a cart is needed. Available space and autoclave dimensions are shown in the following Figure 4.

Figure 4: Tool Loaded in Autoclave for Testing

Dimensional analysis was completed with a Creaform Laser Scanner shown in Figure 5. Using the scanner an optimized mesh was created and then best fit to tool. The critical surfaces for the analysis was the U-shaped portion of the tool, with the other surfaces considered excess. The accuracy or repeatability for the scanner is ±0.0381 mm (±0.0015 in).

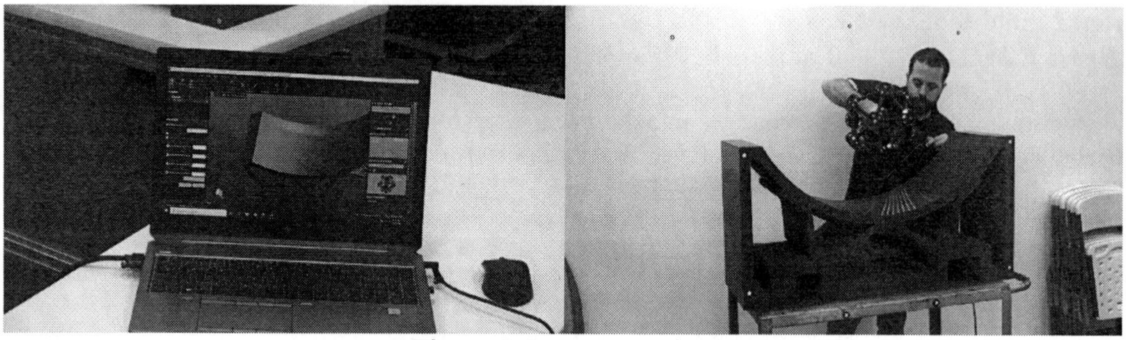

Figure 5: Laser Scanning

3. RESULTS

3.1 Visual Inspection of Tool

Upon completion of 100 cycles, the tool and surface was visually inspected for any signs of degradation. Apart from the buildup of release coating, no obvious changes are noted. This includes no mark off of the fiber weave or TC wire in the tool surface. As typical with mold release, it periodically needs to be removed. A light sanding, re-sealing, and re-releasing could be applied to correct this. Figure 6 shows the tool surface before thermal cycles, and after 100 cycles and no removal of the applied release.

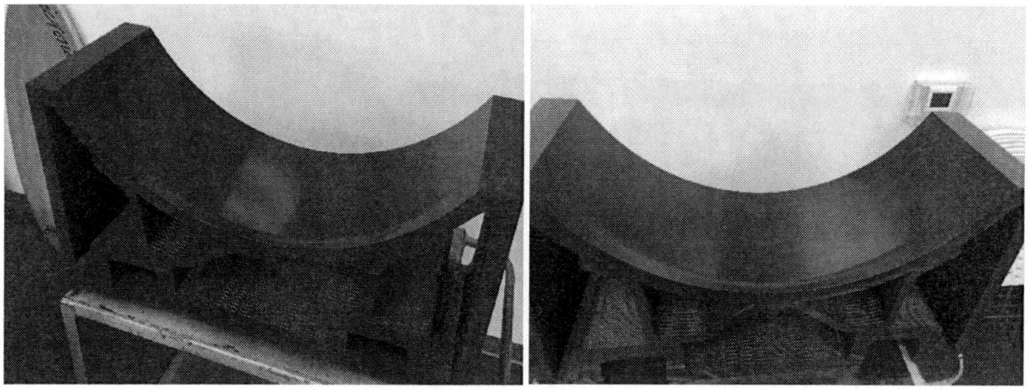

Figure 6: Surface Finish Comparison

In thermoset tooling, micro-cracking of the resin will eventually lead to tool failure. As micro-cracks propagate they create failure propagation points and eventually lead to leak paths in the tool surface. This tool degradation is further complicated by the fact that part resin can infiltrate that cracks potentially causing pull out of fibers and resin or disbonds in the tool if release does not properly penetrate the cracks. Figure 7 shows close up micrographs of the surface, no problems were noted during microscopic inspection at 1000x magnification. The lack of mico-cracking in the tool surface could indicate that tool life for printed tools is significantly longer than that of thermoset composite tools.

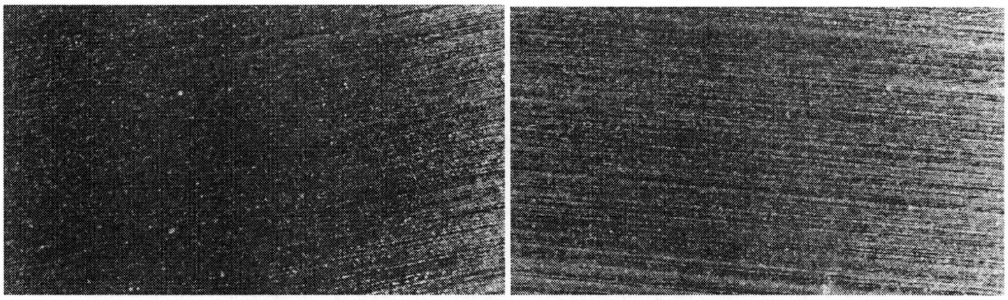

Figure 7: Micrographs of Tooling Surface.

A slight discoloration from the bag sealant tape was noted after repeated thermal cycling. This discoloration is common with all tooling approaches with no known detrimental effects, even after 500+ cycles with a bag being applied each time. No adverse effects were noted on the printed surface, and sealant tape easily stuck to the surface. Vacuum integrity was unaffected. The noted discoloration can be seen in Figure 8.

Figure 8: Sealant Tape Close up

3.2 Sample Laminates

Sample laminates were pulled from the tool surface at several points during the thermal cycling. Figure 9 is an image of a Toolmaster TMGP 4200, epoxy laminate that was pulled from the mold. Additional Epoxies, Benzoxazine, and Bismaleimids (BMIs) were also pulled from the tool surface without issue but are not shown. All laminates exhibited similar finish and gloss with only minor differences in color due to resin chemistry.

Figure 9: Cured Epoxy Laminate

3.3 Vacuum Integrity

During the start of every 5x cure cycle a leak check was performed at ambient temperature to check vacuum integrity. Vacuum was also monitored continuously during the cure without any issues. As an additional measure, the vacuum integrity of the tool was checked at 182 °C (360 °F) after 100 cycles with 7 bar (100 psi) applied. Table 1 shows the leak check data gathered through the Autoclave in Huntington Beach, CA. Total leak was 0 kPa and 0.6 kPa (0.2 inHg) in 10 minutes while at temperature and pressure. All units have been converted to metric from imperial for clarity as monitoring equipment is not programmed for metric display or data collection.

Time (H:M:S)	Press. (bar)	Temp. (°C)	Vac. P1 (kPa)	Vac. P2 (kPa)
2:24:20	6.89	182.8	-95.5	-94.1
2:26:20	6.91	183.1	-95.5	-94.1
2:28:20	6.92	183.4	-95.5	-93.8
2:30:20	6.94	183.6	-95.5	-93.5
2:32:20	6.95	183.4	-95.5	-93.5
2:34:20	6.88	183.1	-95.5	-93.5

Table 1: Autoclave Leak Check

3.4 Dimensional Stability

Following every twenty thermal cycles, the contour was checked with the laser scanner. Critical surface was the U-shaped portion of the tool for best fitting the scanned mesh. The post 100 cycle scanned data was best fit to the original scan for the tooling surface post hand finishing and sealing. To clarify, the tool surface was compared at 100 cycles to the original scan data to monitor movement. Figure 9 shows the comparative data with a color map set to the failure criteria. Where yellow areas are shown ±0.114 mm (±0.0045 in) and red areas are areas of no data, or areas that exceed ±0.127 mm (±0.005 in).

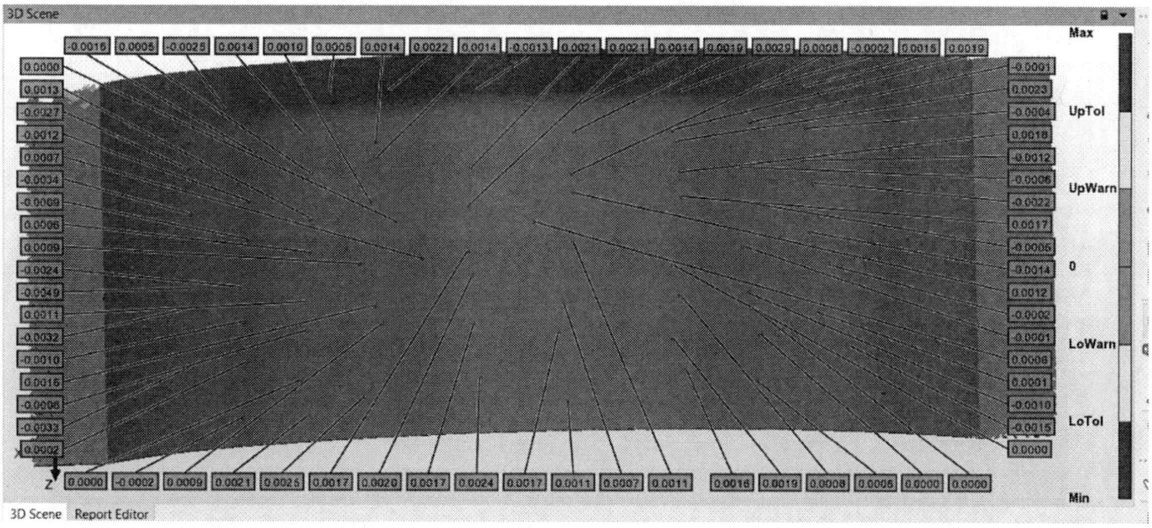

Figure 9: Color Map of Tooling Surface (All Units are Imperial)

3.5 Thermal Profile Analysis

During the course of the thermal profile testing, a thermal survey was done on the tool in an effort to help qualify the heat up and cooling rates of the mold. Overall the thermal conductivity and density of the printed material is comparable to that of a traditional composite laminate in all degrees of freedom. The only exception is that for a 3D printed tool the Z-direction is no longer through the face sheet thickness, but is in plane and oriented in the stacking direction of the print beads. While X-direction conductivity is fiber dominated and relatively high, the Z-direction conductivity is lower. While conductivity is a key variable in tool heat up rates, mass concentrations and airflow are also critical and typically play a more significant role.

Density values are similar to that of a traditional thermoset composite tool laminate, but the printed tool have a much thicker face sheet. Current practice in additive manufacturing is to print the face sheet two beads thick for autoclave vacuum integrity. This yields a face sheet that is approximately 35.6 mm (1.4 in) thick post machining. Traditional thermoset composite face sheets vary from 6.35 mm (0.25 in) to 12.7 mm (0.5 in) thick. This added thermal mass on the printed tool is further complicated by the backup structure contact points are typically 20.32 mm (0.8 in) thick. This yields locations in the critical face sheet area that vary from 35.6 mm (1.4 in) thick, up to 55.88 mm (2.2 in) thick.

For the thermal survey, 10 thermocouples were placed on the tool surface in both low and high mass concentration areas. Figure 10 shows the TC locations, approximate header locations (shaded blue), and the direction of the airflow during the cure.

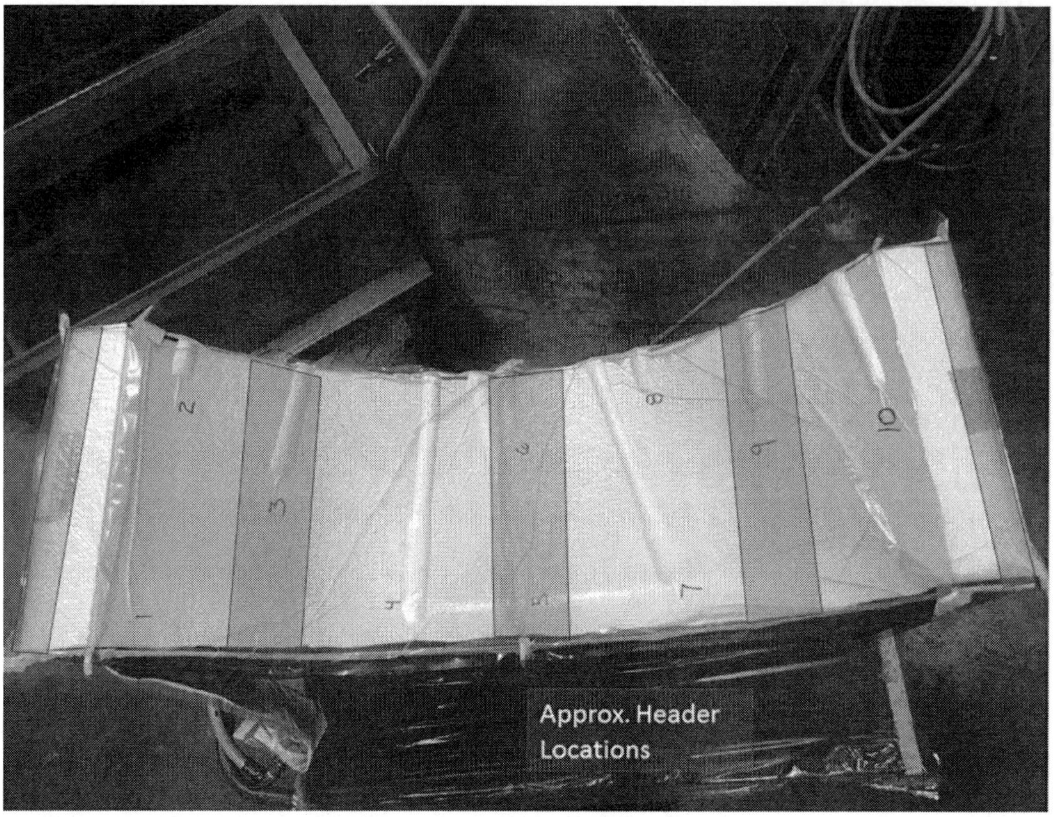

Figure 10: Thermal Profile Setup

Cure cycle for the testing is out lined below and was selected to simulate an actual part cure. Note that the cure is a vented cure, but vacuum integrity was maintained with no evidence of positive pressure. Vacuum readings for the duration of the cure cycle would alternate between 0 and 3 mbar (0 to +0.1 inHg) which is the resolution of the monitoring equipment and well within requirements.

1. Apply 0.7 bar (22 inHg) minimum vacuum to bag.

2. Pressurize autoclave to 5.9 bar (85 psig) with nitrogen. Vent vacuum bag to atmosphere when pressure reaches 1.4 +0.7/-0 bar (20 +10/-0 psig). Keep vented until end of cure. Pressure under vacuum bag shall be 0+0.3/-0 bar (0 +5/-0 psig) throughout remainder of cure cycle.

3. Apply heat after full pressurization of autoclave.

4. Heat-up rate shall be 0.5 to 2.7 °C/min (1 to 5 °F/min).

5. Hold for 120 +60/-0 min at 180 ± 5 °C (355 ± 10 °F) and 5.9 +1/-0 bar (85 +15/-0 psig). Temperature is based on lagging thermocouple. Cure begins when last thermocouple reaches the minimum cure temperature. All thermocouples shall remain in cure temperature range.

6. Cool under pressure until the part temperature reaches 60°C (140°F) or below. Natural pressure drop (1.3 bar, 19 psig max) in autoclave due to cool down is allowed. Cool-down rate shall be 2.7 °C (5 °F/min) maximum.

7. When part temperature reaches below 60 °C (140 °F), release pressure and remove part.

Cure trend chart is shown in Figure 11. Overall the cure was well within spec despite poor airflow. Airflow was not directed through the tool and was restricted by a header on either side. No ventilation holes are present in the side headers of the tool, or in the central headers. The lagging thermocouple ramped at 1.6 °C/min (2.8 °F/minute) overall, with the max air temp reaching 183 °C (361 °F) during the ramp up. Maximum thermal gradient peaked at 11.6 °C (20.9 °F) for approximately 15 minutes, with the average gradient being in the 6.6 - 8.9 °C (12 – 16 °F) range. No cascade was programmed into the cure cycle and thermocouples were allowed to deviate to a maximum. Overall run time for the tool was 268 minutes, with a maximum pressure under the bag of 7 mbar (0.1 psi). Note that 7 mbar (0.1 psi) is the resolution of the monitoring equipment.

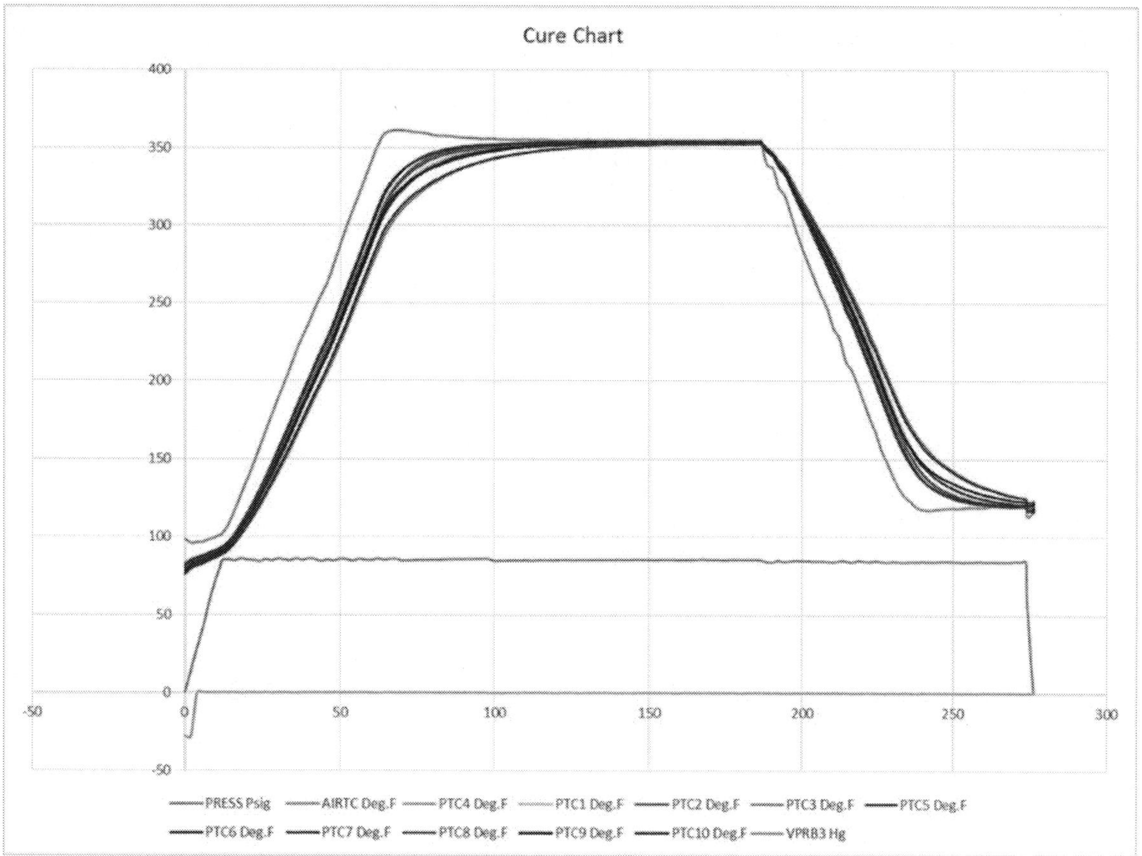

Figure 11: Thermal Profile Trend Chart (All Units are Imperial)

4. CONCLUSIONS

Overall Dahltram I-350CF exhibits excellent vacuum integrity and dimensional stability in an autoclave environment. Successfully completing 100 thermal cycles is major milestone for Large

Format Additive Manufacturing (LFAM), and it was completed without failure. Previous to this testing the major target for LFAM was low volume development work, but the results of this testing help open the door for LFAM to a larger market. Low volume production in addition to prototype production are easily achievable, helping to grow the value proposition of LFAM in the industry.

Future testing of Dahltram I-350CF includes the continued thermal cycling of the resin system, with the goal of taking the tool to failure. Laminates will continue to be pulled from the tool during this time. Vacuum integrity and dimensional stability will continue to be monitored.

5. REFERENCES

[1] Vlastimil K., Hassen A., Lindahl J., Seokpum K., Post B., Love L. "Large Scale Additive Manufactured Tooling for Composites." *15th Japan International SAMPE Symposium and Exhibition.* Japan 2016.

[2] Vlastimil K., Lindahl J., Dinwiddie D., Post B., Love L., Duty C., Matlack M., Fahey Jr. R., Hassen A. "Investigation of In-Autoclave Additive Manufacturing Composite Tooling." *The Composites and Advanced Materials Expo.* Anahiem, CA 2016.

[3] "Thermwood and Bell Collaborate on 3D-Printed Helicopter Blade Mold." *Composites World.* Francis S. February 15th, 2019. May 8th, 2020. <https://www.compositesworld.com/blog/post/thermwood-and-bell-collaborate-on-3d-printed-helicopter-blade-mold>.

HYBRID MANUFACTURING TECHNIQUE USING LARGE-SCALE ADDITIVE MANUFACTURING AND COMPRESSION MOLDING FOR HIGH PERFORMANCE COMPOSITES

Vipin Kumar[1], Seokpum Kim[2], Vidya Kishore[2], Kaustubh V. Mungale[3], Avery Nowlin[3], Uday Vaidya[3], Craig Blue[2], Vlastimil Kunc[2], and Ahmed Arabi Hassen[1]

[1] Materials Science and Technology Division (MSTD), Oak Ridge National Laboratory (ORNL), Oak Ridge, TN 37830
[2] Energy & Transportation Science Division (ETSD), Oak Ridge National Laboratory (ORNL), Oak Ridge, TN 37830
[3]Department of Mechanical, Aerospace, and Biomedical Engineering, University of Tennessee, Estabrook Rd, Knoxville, TN 37996, USA

ABSTRACT

Additive Manufacturing (AM) or 3-D printing has moved from small-scale prototypes to large-scale functional structures. With the introduction of large-scale 3D printers into the market a new industry has been born. For the first time, we are presenting the idea of preparing large-scale 3-D printed β-stage preforms to be used in traditional compression molding (CM) manufacturing technique to prepare final functional structures. Short carbon fiber (CF) reinforced material acrylonitrile butadiene Styrene i.e. ABS with 20% CF is used in present work. It is shown that highly aligned 3D printed short carbon fiber composite maintains its orientation after compression molding. A sandwich panel is prepared to show the capability to prepare complicated parts by integrating AM and CM. Interphase of the prepared dissimilar material is also studied using SEM. Samples prepared using AM-CM integration showed 11% and 79 % improvement in tensile strength and tensile modulus respectively compared to the traditional extrusion compression molding (ECM) process.

Keywords: Additive Manufacturing, Integrated Composite Processing, Compression Molding
Corresponding author: kumarvi@ornl.gov

1. INTRODUCTION

3D printed boat, 'built' by the world's largest 3D printer (Ingersoll Inc.) [1], largest 3D printed trim tool for Boeing 777x winglet (Oak Ridge National Laboratory, ORNL) and first 3D printed Car (ORNL and Cincinnati Inc) are some of the examples of large-scale polymer additive manufacturing (AM) using extrusion deposition process. Extrusion deposition process involves thermoplastic pellets melted through a heated extruder and deposited layer by layer via a nozzle. By producing these life-size functional structural components, large-scale polymer AM has changed the way 3D printing industry was used to be seen few years back. ORNL and Cincinnati Inc. introduced Big Area Additive Manufacturing (BAAM) system in 2014 [2]. Since then, there is huge advancement in this area [3]. The additional breakthrough in this series is the addition of dual material feeding mechanism to the BAAM system to print multi-materials in a single structure. Two dryer unit located next to the BAAM system keeps two types of different materials ready for deposition through extrusion nozzle. Nevertheless, the choice of multi-material used is not only limited to two material and can be extended to many different materials. Feedstock materials in the form of pellets provides the flexibility of using a larger variety of polymer systems in BAAM.

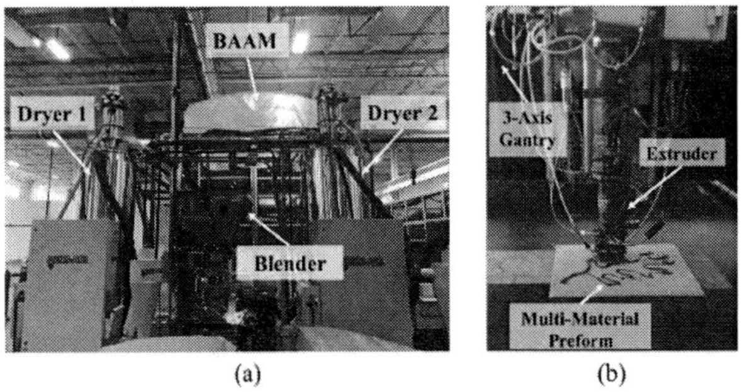

Figure 1. (a) Multi-material printing setup on BAAM system, and (b) Extruder setup showing multi material deposition

With the additional ability to print multi-materials using BAAM technique, AM can offer new opportunities to reduces cost, weight and can improve functionalities of the printed parts [4,5]. A good aesthetic is a bonus of multi-material printing as shown in **Figure 1**. The compatibility of different materials with each other, function-driven design, desired printing cost and weight decides the combinations of the materials to be used in BAAM. For present study, the most common material for large scale printing acrylonitrile butadiene Styrene (ABS) and carbon fiber (CF) have been used. An idea of integrating AM technology with Compression Molding (CM) manufacturing technique is laid within this work. Large-scale polymer AM has advantages such as reduced cost, reduced waste and ability to make complicated structures, but still possess serious drawbacks such as high porosity levels and relatively low mechanical properties across the deposition direction (Z-direction). Present work focuses on the capability of BAAM system to produce multi-material parts and use them as β-stage preforms for the CM technique. For demonstration, a complicated structure like a sandwich panel, ABS/CF on top and bottom and neat ABS in middle, was printed and compressed to obtain a final structure whose mechanical performance was compared with the traditional Extrusion Compression Molding (ECM) process.

2. EXPERIMENTATION

BAAM system used in this study consists of a vertical extruder attached to 3 axis gantry system. Extruder is attached to two feeding hoppers. The extruder on a 3-axis gantry crane allows material deposition at very high rate and can control the direction of the depositions. In this technique the part is first modelled using computer added design (CAD) software or other methods and further sliced using another software namely "slicer" to convert the CAD model into layer-wise representation of the model. This slicer file is fed into the BAAM system which produce the final part layer-by-layer. It is already established that in this extrusion process the fiber orientation within the deposited bead tends to have a higher alignment at the deposition direction [6].

Neat ABS and 20% by weight CF filled ABS pellets were provided by Techmer PM. The pellets were dried for 8 hours at 70°C. The AM preforms were manufactured with a designed deposition path in order to align the orientation of the CF in a desired direction. Three panels were manufactured; (1) Neat ABS, (2) ABS/CF, and (3) Sandwich ABS/CF -ABS- ABS/CF (See **Figure 2**). Each panel had three layers printed in 0° orientation (i.e. direction of the bead deposition). Hot-press (Carver Press-30 Ton) was utilized for the CM process of the AM preform. The compression parameters for the AM-preform were carefully optimized to obtain desired thickness of the panels. Over heating or excess pressure could be detrimental for the AM-preform, see **Figure 3**. The platens of the hot-press were set to 135°C and the pressure was increased incrementally from 1 to 10 Ton. The molded plates were hold under pressure for 10 minutes before removing the pressure. The AM preforms panels mechanical properties were compared against the properties of panels made by conventional ECM process. In this process the same feedstock, neat ABS and ABS/CF, material was fed through a single screw extruder to produce a viscous charge. The charge was then transferred to a closed mold press in a fast-acting press to produce the composite panels. 30 Ton pressure was applied, and dwell time of 3 minutes was maintained. Printed direction in AM preforms and the charge direction in CM molded parts will be referred as "X-direction" in this manuscript. Size of all panels was 279.4 mm × 279.4 mm (11" × 11"). Samples for tensile test were cut from the prepared panels.

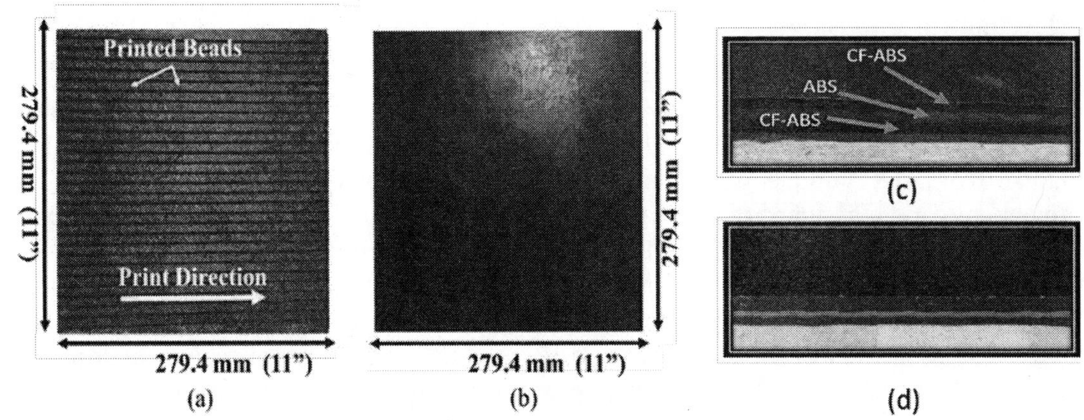

Figure 2. Example of single material AM-preform (ABS/CF); (a) Before compression molding, and (b) After compression molding. Example of multi-material AM-preform; (c) before compression molding, and (d) after compression molding. Sandwich panel is an example that cannot be prepared using conventional ECM or injection molding processes easily.

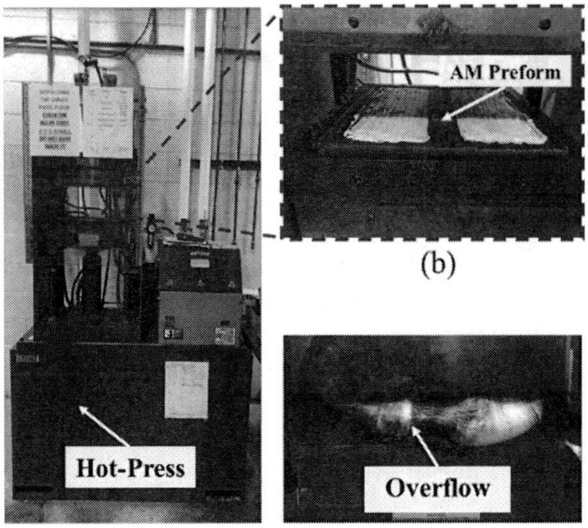

Figure 3. Optimization of compression molding parameters of the AM-preforms; (a) Compression molding setup, (b) Close up for the mold and the AM preform, and (c) Material Overflow resulted from un-optimized parameters.

3. RESULTS

3.1 Mechanical Properties

Tensile test results are shown in **Figure 4**. Tensile strength and modulus of AM-CM samples in X-direction are 11% and 79 % higher, respectively, than that of ECM samples. Tensile strength and modulus of neat ABS was same in both manufacturing techniques, but tensile strength and modulus of ABS/CF samples were significantly high in AM-CM compared to ECM only. The improvement can be assigned to the highly orientated CF in the AM-CM samples compared to ECM samples. High porosity in the AM produced parts and poor adhesion of the adjacent printed layers in Z-direction are the main reasons for poor mechanical properties of 3D printed structures [7]. As the extrusion process involves the melting and mixing of the feedstock material, this thermal process generates gases which may causes heterogeneous micro/meso pores in the structures [8]. Air trapping during feeding process, moisture in feedstock material or volatile additives increase chance of porosity. These heterogeneous pores create concentrated stress during loading on the polymer structure making them inherently weak. Although extrusion of melted polymer through a small orifice help to align the filler fibers into the deposition direction due to the applied shear force in the nozzle during extrusion (i.e. shear flow phenomena). By controlling the deposition direction, preferred fiber alignment can be achieved.

On the other hand, traditional ECM technique is composed of two method, extrusion of the charge and compression in a mold. Compression under pressure help to reduce porosity in CM produced parts[9]. However, due to the absence of the control over the material flow in the mold, fiber orientation is highly heterogeneous in the CM produced structures. Using the AM preform in place of the extruded charge for compression molding all the advantages of AM are combined with all the advantages of the CM processes. CF orientation in AM samples makes it a good candidate to prepare samples with enhanced mechanical properties compared to ECM process. Tensile properties shown in **Figure 5** of AM-CM samples confirm that the low porosity and high fiber alignment assisted in the improvement of the mechanical properties.

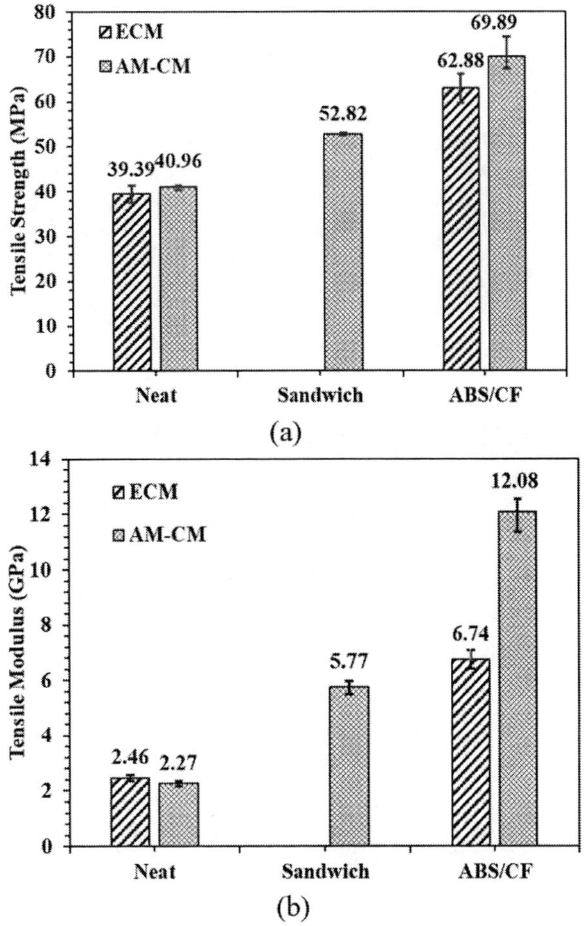

Figure 4. (a) Tensile strength (b) Tensile modulus of various samples prepared by CM and AM-CM integrated system.

Flexibility to prepare complex parts, which are not possible with convocational ECM process is another benefit of using AM preform. With AM-CM complex geometries, with specific properties in predefined directions can be obtained. This property can be an advantage over the over-molding processes or in cases where weight saving is needed. Combination of AM with conventional CM technique can produce complicated multi-material structures that cannot be produced by any other conventional processes easily, hence the lack of the data points in **Figure 4**, tensile strength and stiffness, for sandwich structures made by ECM process.

3.2 Morphology

Optical microscopy and Scanning Electron Microscopy (SEM) images were utilized to study the porosity, interphase and CF orientation. **Figure 5a** shows AM preforms before compression.

Large void content is clearly visible in the figure. It can be noticed in **Figure 5b** that the ECM resulted in random fiber orientation and porosity levels that are less than the ones observed in the AM preforms. It is found that the AM-CM process is highly effective in reducing porosity from the AM produced parts while maintain its high fiber orientation as shown in **Figure 5c**. To show the capability to modify the internal microstructure which eventually affect the mechanical properties, an ORNL word was printed using the fiber orientation by designing the deposition direction of the CF based PLA material through a 0.25 mm diameter nozzle as shown in **Figure 5d**.

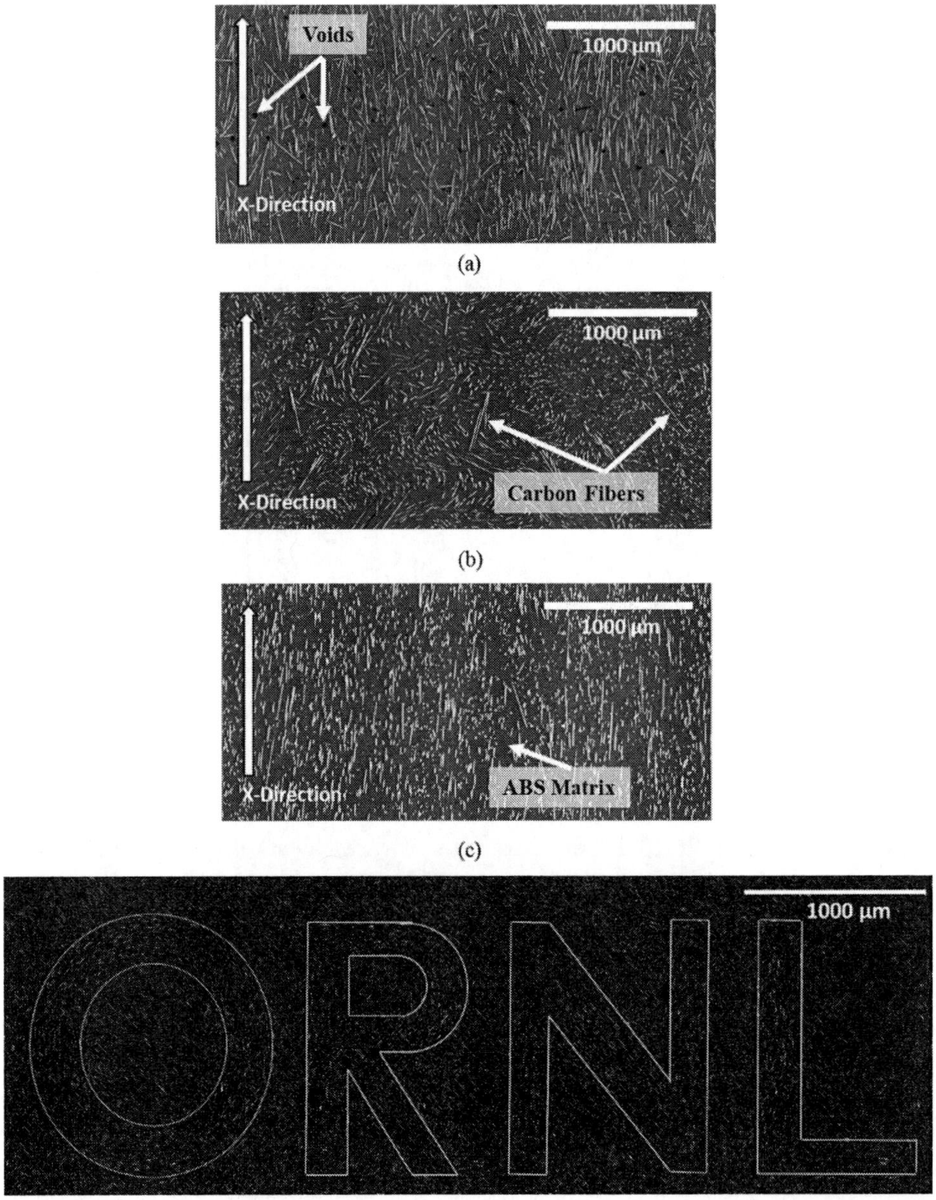

(d)

Figure 5. Optical microscopy showing the microstructure for; (a) AM preform, (b) ECM molded panel, and (c) AM-CM molded panel, and (d) Writing ORNL with fiber orientation in PLA Matrix.

A good interphase between ABS and ABS/CF was observed from SEM the images shown in **Figure 7**. Good interphase confirms the effective over molding process between neat ABS and ABS/CF materials. Over molding techniques are used to selectively reinforce part in targeted areas. Conventional over molding process uses a continuous unidirectional (UD) reinforced tapes, where they are strategically placed in a mold cavity then the mold is filled with (i.e. over molded) with short/long fiber-reinforced thermoplastics. In such techniques, slipping of the UD tapes during the mold filling, is a common risk. With our newly introduced concept of integrating AM and CM process, highly aligned short/long fibers can be integrated into the molded part and can be printed in a way (i.e. highly aligned fibers) to get the maximum strength and stiffness in the desired direction. In our proposed technique of over-molding parts, it was noticed that there is no slipping between the two dissimilar materials. However, direct comparison of mechanical properties between traditional over-molding technique and proposed technique has not been done yet and will be considered in the future work.

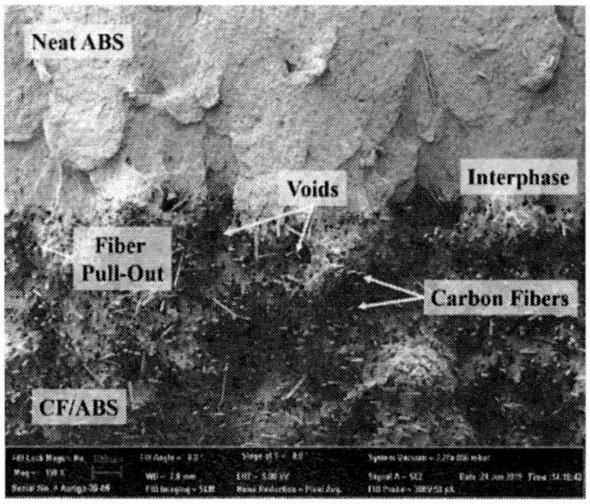

Figure 7. SEM image for the fracture surface of multi-material (neat ABS/CF-ABS) produced by AM-CM process showing the interphase.

4. CONCLUSIONS

In the present paper, we have demonstrated the ability to print multi-materials using Big Area Additive Manufacturing (BAAM) technique. We showed that the combination of additive manufacturing (AM) with conventional compression molding (CM) technique can produce complicated multi-material structures that cannot be produced easily by any other conventional processes. Integrating AM technology with an established manufacturing technique, CM, was used to demonstrate the manufacturing of sandwich panel. AM multi-material preforms were compressed to obtain the final molded components. Morphology of the composite were studied using optical microscopy and scanning electron microscope to understand the fiber orientation, porosity and interphase properties between multi-materials. The enhanced mechanical performance based on design are discussed with correlation to the fiber orientation in the AM-CM produced coupons. Tensile testing confirmed the effectiveness of AM-CM integrated technique over conventional CM technique.

5. ACKNOWLEDGMENTS

Research sponsored by the U.S. Department of Energy, Office of Energy Efficiency and Renewable Energy, Advanced Manufacturing Office, under contract DE-AC05-00OR22725 with UT-Battelle, LLC. For large format additive manufacturing, the printing equipment was provided by Cincinnati Incorporated, a manufacturer of metal and additive manufacturing equipment, headquartered in Harrison, Ohio (www.e-ci.com). The printing material for large format additive manufacturing was provided by Techmer PM, a material design and manufacture company headquartered in Clinton, TN. The design and execution of the research was done at the Manufacturing Demonstration Facility of Oak Ridge National Laboratory.

6. REFERENCES

1. D. Sharp, *World's largest 3D-printed object – a 25-foot boat – unveiled in Maine*, in *The Enterprise News*. 2019.
2. L. Love, C. Duty, B. Post, R. Lind, P. Lloyd, V. Kunc, and C. Blue, *Breaking Barriers in Polymer Additive Manufacturing*. Oak Ridge National Laboratory (ORNL); Manufacturing Demonstration Facility (MDF). 2015.
3. A.A. Hassen, J.o. Lindahl, X. Chen, B. Post, L. Love, and V. Kunc, *Additive Manufacturing of Composite Tooling Using High Temperature Thermoplastic Materials*, in *SAMPE Conference*. 2016: Long Beach, CA,.
4. V. Kumar, R. Spencer et al., *3D printed layer of polyaniline-based conductive polymer for lightning strike protection of CFRP*. International Conference on Lightning and Static Electricity, ICOLSE 2019, Wichita, Kansas, USA. 2019
5. V. Kumar, T. Smith et al., *Replacing metal-based lightning strike protection layer of CFRPs by 3d printed electrically conductive polymer layer*. CAMX – The Composites and Advanced Materials Expo, Anaheim, CA, USA
6. H. Tekinalp, V. Kunc, G. Velez-Garcia, C. Duty, L. Love, A. Naskar, and S. Ozcan, *Highly oriented carbon fiber–polymer composites via additive manufacturing*. Composites Science and Technology, 2014. **105**: p. 144-150.
7. C. Duty, J. Failla, S. Kim, T. Smith, J. Lindahl, and V. Kunc, *Z-Pinning approach for 3D printing mechanically isotropic materials*. Additive Manufacturing, 2019. **27**: p. 175-184.
8. X. Zhao, H. Tekinalp, X. Meng, D. Ker, B. Benson, Y. Pu, and S. Ozcan, *Poplar as Biofiber Reinforcement in Composites for Large-Scale 3D Printing*. ACS Applied Bio Materials, 2019. **2**(10): p. 4557-4570.
9. T.S. Lundström and A. Holmgren, *Dissolution of Voids during Compression Molding of SMC*. Journal of Reinforced Plastics and Composites, 2010. **29**(12): p. 1826-1837.

TRANSPORT PROPERTIES OF CARBON FIBERS DERIVED FROM PETROLEUM- BASED MESOPHASE PITCH WITH MODIFIED TRANSVERSE MICROSTRUCTURES FOR ENHANCED TENSILE STRENGTH

Sagar Kanhere[1], Dr. Victor Bermudez[1], Prof. Amod Ogale[1]
[1]Center for Advanced Engineering Fibers and Films (CAEFF), Department of Chemical Engineering, Clemson University, Clemson, SC USA

ABSTRACT

Carbon fibers are extensively used in aerospace, defense, and automobile industry due to their lightweight and superior strength. Some of these applications demand high electrical and thermal conductivity, which glass fibers cannot provide because glass is an insulating material (in contrast to conducting carbon). For instance, in spacecraft electronic housing, it is important that carbon fiber reinforced composite be able to dissipate heat. Mesophase pitch-based carbon fibers possess excellent electrical/thermal conductivity due to their high graphitic crystallinity. Also, they can be produced from petroleum-based mesophase pitch, potentially a low-cost precursor. In this study, transport properties of such carbon fibers are reported.

Precursor fibers possessing a wide range of micro-texture (radial to circular alignment) of discotic liquid crystalline molecules were produced using novel spinneret design. The design is well-suited for high-volume production of pitch fibers using continuous extruders. After thermo-oxidative stabilization in an air environment at 220°C and atmospheric pressure, fibers were graphitized at 2100°C. The tensile strength of carbon fibers with circular (non-radial) micro-texture was measured at 2.9 ± 0.4 GPa whereas that for radial texture was 2.4 ± 0.2 GPa. The higher strength for non-radial texture is a consequence of deflection of crack in the hoop direction along circular layers.

Often, such an improvement in strength is accompanied by a reduction in other properties. However, the current results indicate that the longitudinal electrical resistivity of carbon fibers was measured to be about 5 μΩ·m for both types of micro-textures. As estimated from the Issi-Lavin correlation, this corresponds to a thermal conductivity of 300 W/m·K, which is about 1500% greater than that of the best PAN-derived carbon fibers. Thus, micro-textural modification that led to enhanced tensile properties did not deteriorate transport properties. Further, X-ray diffraction revealed that the graphitic d_{002} layer spacing was 0.34 nm for both micro-textures, indicating about 20% graphitic crystallinity. Interestingly, the circular (non-radial) micro-texture that helped to enhance tensile strength did not result in reduced graphitic content, so conductivity did not deteriorate. It is hypothesized that the longitudinal texture of the graphitic crystallite remains unchanged even as the transverse texture changes from radial to circular, a phenomenon that will be investigated in future studies.

1. INTRODUCTION

Carbon fibers are finding increased use in structural and semi-structural composite applications due to their high specific strength and stiffness. Carbon fibers also offer a unique advantage of being electrically and thermally conductive, in contrast to other reinforcing fibers like glass or aramid. This is especially beneficial where heat dissipation or electromagnetic interference (EMI) shielding is required. High conductivity of carbon fiber makes the composite material effective for EMI shielding [1].

Carbon fibers must be produced from organic precursor fibers since carbon itself does not melt or dissolve in any solvent. Polyacrylonitrile (PAN), pitch, and rayon are typical commercially used precursors. Polyacrylonitrile (PAN) based carbon fibers dominate the market, but the cost of the carbon fiber is still a major drawback to use them in cost-sensitive applications like mass transit or automotive manufacturing. 25% of the cost of carbon fiber comes from the cost of precursors [2]. Mesophase pitch is potentially a low-cost precursor and it offers high carbon yield, melt-spinnability and fibers do not need stretching during heat treatment. Mesophase pitch-based carbon fibers (MPCF) have better transport properties than PAN-based due to their graphitic crystallinity. Mesophase pitch is a discotic liquid crystalline polynuclear aromatic polymer produced from the bottom product of the petroleum or coal tar processing. Viscosity of the mesophase pitch is extremely sensitive to temperature, which makes it harder to control the cross-sectional microstructure of the resultant fibers [3]. Microstructural orientation achieved during spinning is retained during oxidation and carbonization steps [4]. Enhanced perfection and orientation of graphitic crystallite during graphitization improve the transport properties of mesophase pitch-based carbon fibers [5].

Radial transverse microstructure for fibers with circular cross-section allows more graphitic crystallinity than other microstructures, which leads to higher thermal conductivity [5]. Edie et al. [6] produced ribbon-shaped carbon fibers with line origin transverse texture with transport properties comparable to that of commercial P-100 and P-120 carbon fibers.

Due to the sheet-like structure, MPCF allows easy formation and propagation of transverse cracks [7]. Radial alignment of graphitic layer planes leads to radial cracks forming Pac-man splits whereas fibers with non-radial microstructure deflect crack propagation in the hoop direction along circular layers. So, carbon fibers with non-radial microstructure show higher strength than fibers with radial microstructure [7,8]. So, the primary goal of this study was to generate petroleum mesophase pitch-based carbon fibers with non-radial microstructure using scalable fiber production processes.

2. EXPERIMENTAL

2.1 Carbon fiber processing

Petroleum-based mesophase pitch containing about 75% mesophase content was used to spin fibers. A conventional spinneret with circular capillaries was used to obtain fibers with radial microstructure, whereas a spinneret with a modified, proprietary design was used to make fibers with non-radial microstructure. Pitch fibers were spun at about 350°C and stabilized in a forced-air convection oven at temperatures about 220°C. Stabilized fibers were then carbonized and graphitized at temperatures up to 2100°C in Astro 1000 furnace (Thermal Technology LLC).

2.2 Analysis

2.2.1 Electrical resistivity measurement

Electrical resistance (R) of single fibers was measured by first mounting single filaments on paper tabs that were then placed between two copper wires separated by 10 mm [9]. To reduce contact resistance, flash dry silver paint (SPI Supplies, West Chester, PA) was applied at contact points of carbon fibers and copper wire. For all measurements, the contact resistance was only about 1 Ω whereas the resistance of an individual carbon fiber was close to 1,000 Ω. Resistivity (ρ) of the fiber was calculated using equation 1, where A is the cross-sectional area of the fiber and L is the length of fiber between the copper wires. Cross-sectional area of the fiber was determined by measuring the diameter of individual filaments using Olympus BX60 optical microscope. Thermal conductivity (κ) of the carbon fibers was estimated from Issi-Lavin correlation (Equation 2) in W/m·K using electrical resistivity values in μΩ·cm.

$$\rho=R\frac{A}{L} \qquad (1)$$

$$\kappa=440000/((\rho+258)-295) \qquad (2)$$

2.2.2 Tensile testing

Tensile analysis of carbon fibers was based on the ASTM D3379-75 standard test procedure. A single fiber was mounted on a paper tab with a 10 mm gauge length and bonded to the tab using epoxy resin. After curing of epoxy resin, the paper tab was aligned carefully within the pneumatic grips of a MTI Phoenix testing machine, and the sides of the tab were burned away. Fiber was then stressed to failure at a constant strain rate of 0.5 mm/min.

2.2.3 SEM and X-ray analysis

To analyze the cross-sectional microstructure and longitudinal surface of the carbon fibers, scanning electron microscopy was conducted on samples using high-resolution FESEM scopes (S4800/SU5000, Hitachi). Wide-angle X-ray diffraction was conducted on carbon fiber bundles using Bruker D8 Venture diffractometer with a copper Incoatec Microfocus Source IμS, with a wavelength of 0.15406 nm and operated at 50 kV and 1 mA.

3. RESULTS

3.1 Microstructure

SEM micrographs, displayed in Figure 1 (a) and (b), show different micro-textures that were generated in resulting carbon fibers using standard circular capillary (left) and specially designed spinneret (right). With a standard capillary, radial microstructure was obtained. In contrast, specially designed capillary design led to non-radial microstructure with circular core with radial outer rim, often referred to as "onion-skin". Longitudinal and cross section in Figure 1 (c) and (d) show uniform fiber diameter and consistency in cross sectional fiber microstructure.

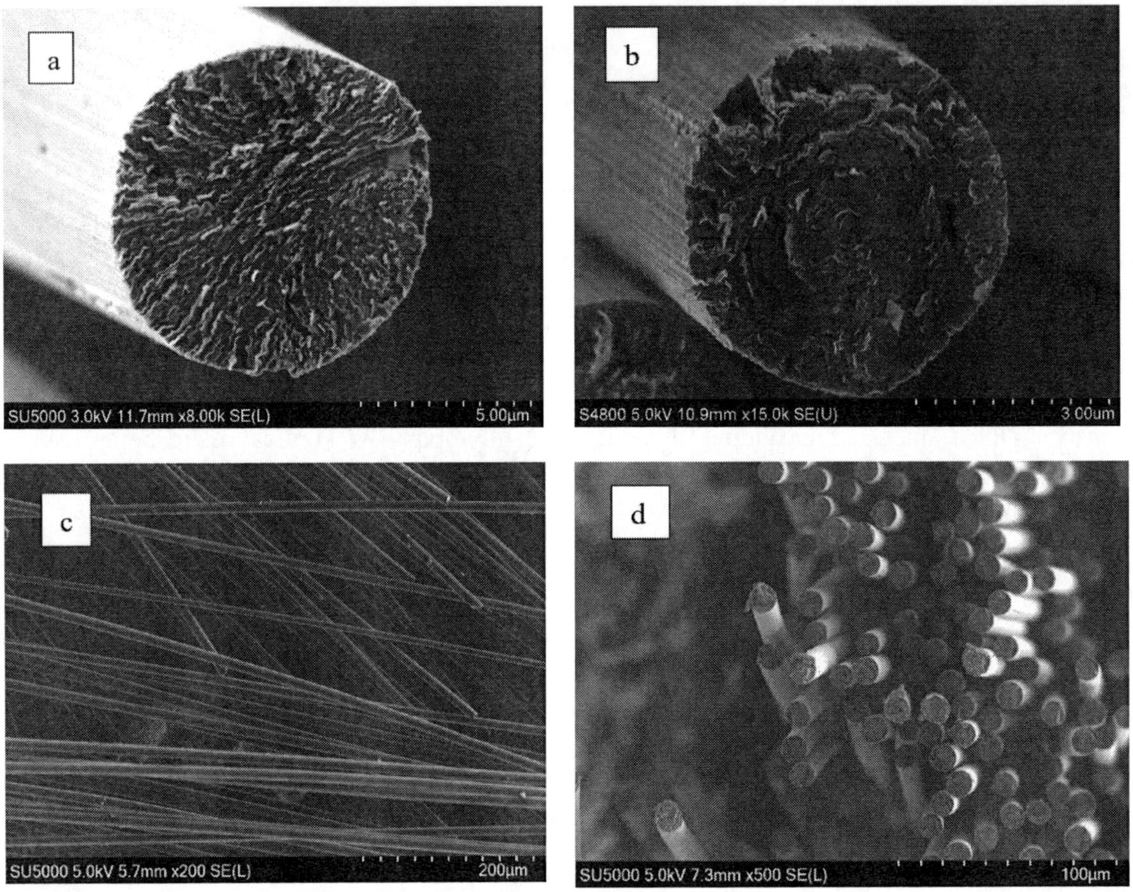

Figure 1. SEM micrographs of carbon fibers with (a) radial microstructure (b) onion-skin microstructure (c) longitudinal section of the fiber and (d) cross section of carbon fibers

Wide-angle x-ray diffractograms are displayed in Figure 2. Position of d_{002} peaks at approximately 26° indicates that there is no measurable difference in degree of crystallinity between the fibers with radial versus onion-skin micro-textures.

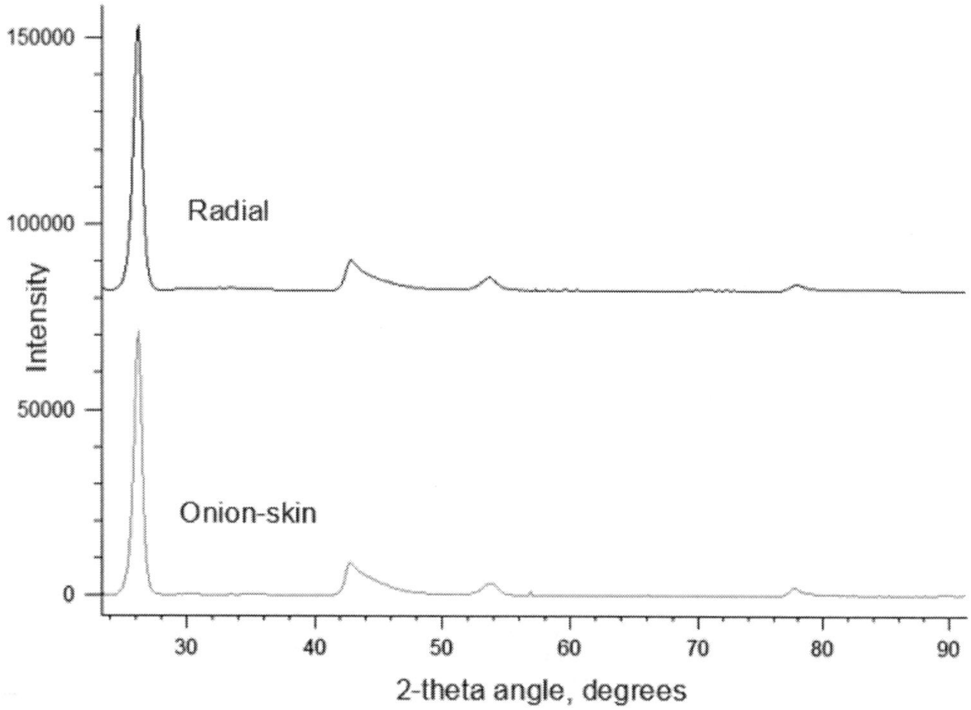

Figure 2. Wide-angle x-ray diffraction (WAXD) spectra for carbon fibers possessing radial (top) and onion-skin (bottom) microstructure

3.2 Tensile properties

Carbon fibers possessing circular microstructure display a decreasing trend in tensile strength with increasing diameter, with a range of 15 down to 7 μm; average strength was close to 2.9 ± 0.4 GPa. CF with radial microstructure have average tensile strength of about 2.4 ± 0.2 GPa. Difference between means is 0.5 with 95% confidence interval from 0.12 to 0.87. Compliance corrected modulus value was about 550 GPa whereas apparent modulus values ranged from 300-400 GPa for both types of fibers.

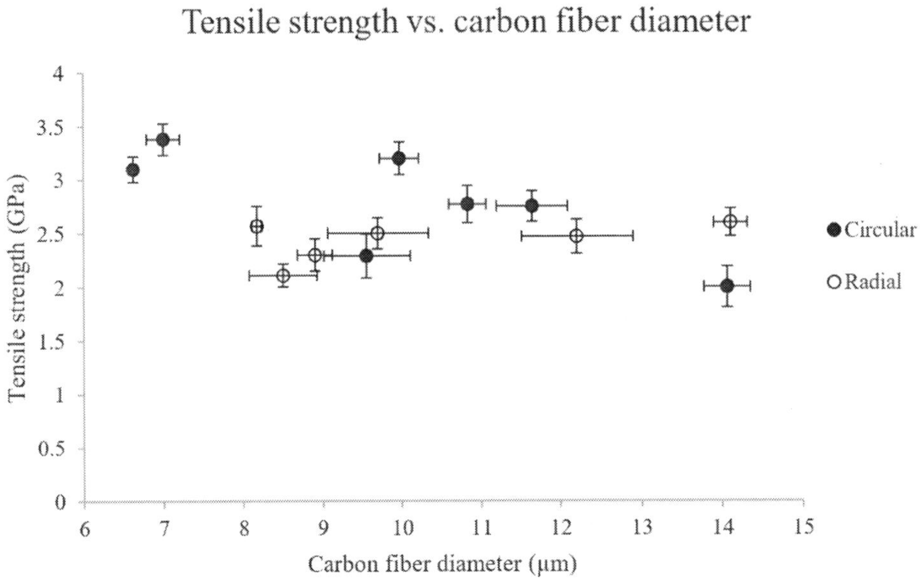

Figure 3. Tensile strength of carbon fibers as a function of fiber diameter for radial (open symbols) and circular (filled symbols) micro-textures.

3.3 Electrical resistivity and thermal conductivity

Figure 4 displays the measured electrical resistivity and estimated thermal conductivity of the fibers of different microstructure. Electrical resistivity of the carbon fibers did not change significantly with diameter. Also, average electrical resistivity of fibers with radial microstructure was 4.8±0.7 μΩ·m, whereas that for circular microstructure was 4.9±0.3 μΩ·m. Thermal conductivity of carbon fibers estimated using Issi-Lavin correlation ranges from 200 to 600 W/m·K. Just like for electrical resistivity, there is no significant difference between thermal conductivity of radial or onion-skin microstructure. Average thermal conductivity for radial microstructure is 316 ±65 W/m·K and ones with onion-skin microstructure is 306±21 W/m·K, indicating no deterioration of electrical/thermal properties even as strength improved.

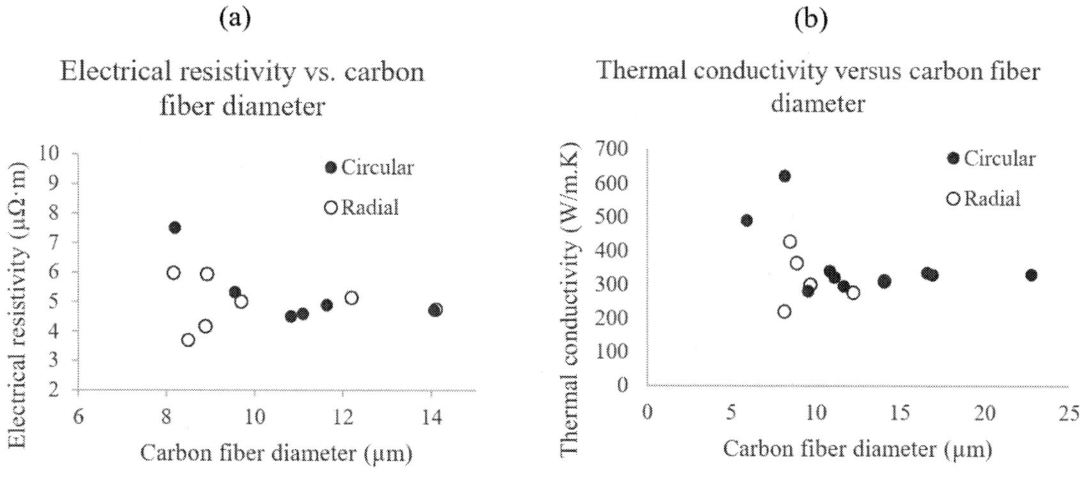

Figure 4. Transport properties of carbon fibers with radial (open symbol) and circular (filled symbol) micro-textures: (a) measured electrical resistivity and (b) estimated thermal conductivity using Issi-Lavin correlation.

4. CONCLUSIONS

Carbon fibers with circular micro-texture, often referred to as "onion-skin" microstructure, were successfully produced using a novel, scalable spinneret design. These fibers with circular microstructure displayed tensile strength of 2.9 ± 0.4 GPa, as compared with 2.4±0.17 GPa for radial microstructure. The combination of tensile strength and modulus values are among the highest reported for carbon fibers derived from petroleum pitch with less than 100% mesophase content. Even as tensile strength improved with micro-texture modification, transport properties did not deteriorate. Rather, the absolute value of electrical conductivity remained high at about 2000 S/cm (corresponding to a resistivity of 5 μΩ·m) and thermal conductivity was estimated at 300 W/m·K, which is comparable to commercial P-75 carbon fibers.

5. ACKNOWLEDGEMENT

The effort depicted was sponsored by the Defense Advanced Research Projects Agency and was accomplished under Cooperative Agreement Number HR0011-16-2-0014. The views and conclusions contained in this document are those of the authors and do not necessarily reflect the position or the policy of the Government. No official endorsement should be inferred.

The authors are very thankful to Caroline Christopher and Elijah Taylor for their assistance in tensile testing and electrical resistivity measurement, and George Wetzel and Dr. Collin McMillen for their help with SEM and x-ray testing, respectively. The project used CAEFF facilities originally supported by NSF ERC grant EEC 9731680.

Distribution Statement "A" (Approved for Public Release, Distribution Unlimited)

6. REFERENCES

[1] Q. Zhao, K. Zhang, S. Zhu, H. Xu, D. Cao, L. Zhao, R. Zhang, W. Yin, Review on the Electrical Resistance/Conductivity of Carbon Fiber Reinforced Polymer, Appl. Sci. 9 (2019) 2390. https://doi.org/10.3390/app9112390.

[2] C.D. Warren, Carbon fiber precurosors and conversion, Dep. Energy. (n.d.). https://www.energy.gov/sites/prod/files/2016/09/f33/fcto_h2_storage_700bar_workshop_3_warren.pdf.

[3] D. Edie, THE EFFECT OF PROCESSING ON THE STRUCTURE AND PROPERTIES OF CARBON FIBERS, Carbon N. Y. 36 (1998) 345–362. https://www.sciencedirect.com/science/article/pii/S0008622397001851.

[4] V. Bermudez, S. Lukubira, A.A. Ogale, 1.3 Pitch Precursor-Based Carbon Fibers, in: P.W.R. Beaumont, C.H. Zweben (Eds.), Elsevier, Oxford, 2018: pp. 41–65. https://doi.org/https://doi.org/10.1016/B978-0-12-803581-8.10312-1.

[5] M.G. Huson, High-performance pitch-based carbon fibers, in: Struct. Prop. High-Performance Fibers, Woodhead Publishing, Oxford, 2017: pp. 31–78. https://doi.org/https://doi.org/10.1016/B978-0-08-100550-7.00003-6.

[6] D.D. Edie, K.E. Robinson, O. Fleurot, S.P. Jones, C.C. Fain, High thermal conductivity ribbon fibers from naphthalene-based mesophase, Carbon N. Y. 32 (1994) 1045–1054. https://doi.org/10.1016/0008-6223(94)90213-5.

[7] Y. Huang, R.J. Young, Microstructure and mechanical properties of pitch-based carbon fibres, J. Mater. Sci. 29 (1994) 4027–4036. https://doi.org/10.1007/BF00355965.

[8] S. Kanhere, S. Lukubira, M.C. Tang, O. Ozdemir, A.A. Ogale, Petroleum Pitch-Based Carbon Fibers With Modified Transverse Microstructure, in: Carbon Conf., Lexington, 2019.

[9] V. Bermudez, Anomalous Effect of Spinning Conditions on the Mechanical and Transport Properties of Mesophase Pitch-Based Carbon Fibers, Clemson University, 2019. https://tigerprints.clemson.edu/all_dissertations/2370.

SYNTHESIS OF AN ARYL ETHER KETONE-CONTAINING POLYBENZOXAZINE

J. Scott Murphy and Jeffrey S. Wiggins*
The School of Polymer Science and Engineering, The University of Southern Mississippi
Hattiesburg, Mississippi

ABSTRACT

Thermoset networks, such as epoxide-amines and polybenzoxazines, typically display brittle behavior as a result of their rigid backbones limiting sub-glass transition temperature (T_g) relaxations of the polymer network. Methods to decrease the brittleness and increase the toughness of aerospace-grade thermoset networks traditionally include the incorporation of aliphatic curatives, rubbers, or thermoplastics. However, these methods often reduce the ultimate T_g of the system and increase network heterogeneity. Alternatively, integrating thermoplastic moieties, specifically polyaryletherketone moieties, into the backbone of a polymer network is an effective approach to increase the toughness of a polymer network while simultaneously minimizing the reduction of ultimate T_g and eliminating network heterogeneity. This research incorporates aryl ether ketone (AEK) backbones into polybenzoxazine networks to increase network strain-absorbing and energy-dissipating mechanisms. Herein, an AEK containing benzoxazine monomer was synthesized, and monomer synthesis is confirmed via proton nuclear magnetic spectroscopy. Differential scanning calorimetry is used to further prove monomer purity, endothermic melting, and determine the onset and peak polymerization temperatures. Dynamic mechanical analysis is employed to quantify the T_g and glassy storage modulus of the AEK containing polybenzoxazine network. Thermogravimetric analysis is utilized to determine the thermal stability and char yield of the AEK monomer and the cured polybenzoxazine network. This work demonstrates the successful synthesis and thermomechanical characterization of an AEK containing polybenzoxazine.

Keywords: Benzoxazines, Polybenzoxazines, Thermosets, Monomer Synthesis, Polymer Synthesis
Corresponding author: Email Jeffrey.Wiggins@usm.edu

1. INTRODUCTION

Polybenzoxazine networks have been researched heavily for their unique thermoset properties such as near-zero shrinkage, hydrogen bonding capabilities, and high glass transition temperature (T_g).[1][2] Benzoxazine monomers are prepared by reacting a phenolic derivative, primary amine, and paraformaldehyde. Significant work has gone into synthesizing a library of benzoxazine monomers from phenolic and diphenolic derivatives with primary amines, and paraformaldehyde. What is lacking in the literature is the investigation of diamines, specifically aromatic diamines, for the synthesis of benzoxazine monomers. Work on aromatic diamine-based benzoxazines is limited due to the formation of an insoluble triazine network in aromatic diamine-based benzoxazine synthesis. [3][4] Diamine-based benzoxazines show promising improvements in the

T_gs and mechanical properties of polybenzoxazine network which is important for aerospace matrices, but the reported increases in mechanical properties come from the aliphatic diamine which lack the thermal performance of aromatic networks.[5]

Recent work by Varley *et al.* has shown that tri-aryl ether and tri-aryl ketone substituted anilines show great potential in epoxide-amine systems.[6][7] Both the tri-aryl ether and tri-aryl ketone systems showed increases in flexural modulus, strength, and distortional behavior as compared to a commercial available epoxide-amine standard. Similar work by Misasi and Wiggins investigated aryl ether ketone (AEK) diamines for use in epoxide-amine networks.[8] Misasi studied the effect on mechanical properties by varying the aryl groups from 4 to 6 in AEK diamines. In his work, Misasi found that by incorporating just 5 mol% of a 5 aryl AEK diamine into an epoxide-amine system showed a 20% increase in Young's modulus, 8% increase in yield stress, and 32% increase in yield strain compared to the neat system. Both the degree and nature of these mechanical property enhancements is not only unprecedented in the literature, and it demonstrates how distortional modes of deformation can be designed into glassy polymer network architectures between the crosslinks to increase toughness of the network. For this reason, this work aimed to synthesize a benzoxazine monomer from an AEK diamine, which has been shown to increase mechanical properties in an epoxide-amine networks.

2. EXPERIMENTATION

2.1 Materials

4-phenoxyaniline (97%, Alfa Aesar), Methanesulfonic acid (99%, Acros Organics), Isophthalic acid (99%, Aldrich), Eaton's reagent (Acros Organic), Terephthalic acid (99+%, Acros Organics), Phenol (99.5%, Acros Organics), Paraformaldehyde (95%, Sigma-Aldrich).

2.2 Monomer Synthesis

2.2.1 Synthesis of 4-phenoxybenzenaminium methanesulfonate (1)

4-phenoxyanline (50 g, 270 mmol) was added to an Erlenmeyer flask with 600 mL of ethyl acetate and stirred at room temperature. After dissolution of 4-phenoxyanline in ethyl acetate, methanesulfonic acid (19.3 mL, 297 mmol) was added dropwise. During the addition of methanesulfonic acid the reaction mixture became increasingly more viscous, so additional ethyl acetate (1200 mL) was added to maintain proper stirring. The reaction mixture was stirred for 30 minutes, vacuum filtered, and washed well with ethyl acetate until the filtrate did not show a spot *via* TLC. 4-phenoxybenzenaminium methanesulfonate (**1**) was obtained as a light lilac powder after drying for 24 hours under vacuum at 60 °C (69.9 g, 92%).

2.2.2 Synthesis of 1,3-phenylenebis((4-(4-aminophenoxy)phenyl)methanone) (2)

To a flame-dried, one neck 500 mL round bottom flask was added a dried stir bar, **1** (128.4 g, 456 mmol), isophthalic acid (36.1 g, 217 mmol), and Eaton's reagent (638 mL). The mixture was blanketed with dry nitrogen and then heated to 60 °C with rapid stirring. After 24 h, the reaction mixture was first allowed to cool and then slowly poured into 2 L of ice water while stirring. The resultant suspension was then alkalized to pH 10 with 30 wt% aqueous sodium hydroxide and then vacuum filtered. The precipitate was washed with deionized water until the filtrate obtained a pH of 7. The precipitate was then baked at 50 °C under nitrogen for 24 h and then dried under vacuum at 50 °C. 1,3-phenylenebis((4-(4-aminophenoxy)phenyl)methanone (**2**) was obtained as a light-yellow powder (104 g, 96%).

2.2.3 Synthesis of 1,3-phenylenebis((4-(4-(2H-benzo[e][1,3]oxazin-3(4H)-yl)phenoxy)phenyl)methanone) (3)

Synthetic procedure adapted from Baqar *et at.*[9], to a one neck 100 mL round bottom flask was added a stir bar, **2** (2 g, 4.00 mmol), phenol (0.79 g, 8.4 mmol), and paraformaldehyde (0.48 g, 16 mmol). Xylenes (10 mL) was added into the flask, and the mixture was heated to reflux at 140 °C. After 3 h, the reaction mixture was first allowed to cool to room temperature and then the reaction mixture was precipitated in hexanes (225 mL) and vacuumed filtered to afford a light-yellow powder. The powder was then dissolved in chloroform and washed three times with 1 N sodium hydroxide and then deionized water till the aqueous layer ran clear. The organic layer was then dried over magnesium sulfate anhydrous and filtered, and the chloroform was removed by rotary evaporation. Benzoxazine monomer 1,3-phenylenebis((4-(4-(2H-benzo[e][1,3]oxazin-3(4H)-yl)phenoxy)phenyl)methanone) (**3**) was obtained as a light-yellow powder (1.6 g, 56%).

2.3 Preparation of Polybenzoxazine Network

To a silicon mold, was added the benzoxazine monomer **3**. The mold with **3** was then placed in a vacuum oven to melt and degas the monomer before curing. The silicon mold with degassed **3** was then placed in an oven, and the monomer was cured at 180 °C, 200 °C, and 220 °C for 2 hours at each temperature to form polybenzoxazine network, **Poly(3)**. The cured network was allowed to cool gradually overnight in the oven before being removed for further experiments.

Scheme 1: *General synthesis for the AEK containing benzoxazine monomer (3).*

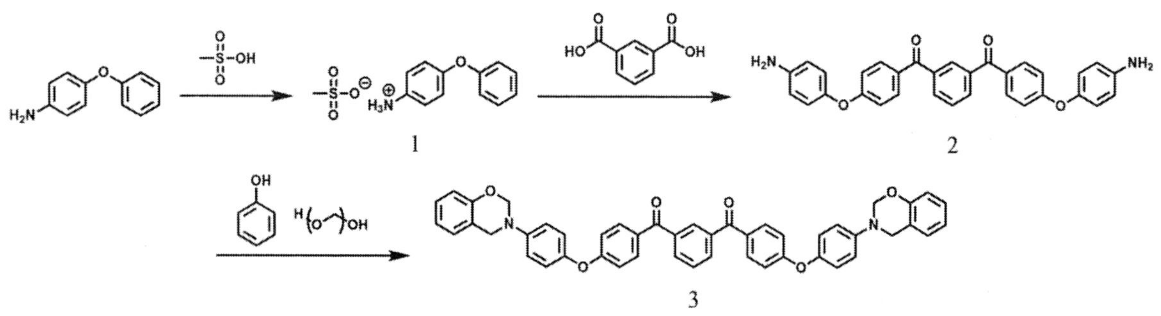

2.4 Characterization

2.4.1 Spectroscopy Analysis

2.4.1.1 Nuclear Magnetic Resonance

Solution state Proton Nuclear Magnetic Resonance (^{1}H NMR) spectra were acquired in deuterated dimethyl sulfoxide (DMSO-d_6) with tetramethylsilane as an internal standard using a Bruker Ascend™ 600 MHz spectrometer. All ^{1}H NMR spectra were collected with 32 scans.

2.4.1.2 Fourier-transform Infrared Spectroscopy

Attenuated total reflectance Fourier-transform infrared (ATR-FTIR) spectroscopy was performed using a PerkinElmer Frontier™ MIR spectrometer equipped with a deuterated triglycine sulfate detector and a diamond coated zinc selenide ATR crystal. All ATR-FTIR spectra were collected with 16 scans at a resolution of 2 cm^{-1}.

2.4.2 Thermomechanical Analysis

2.4.2.1 Thermogravimetric Analysis

A TA Instruments™ Q50 Thermogravimetic Analyzer (TGA) with a platinum pan was utilized to investigate the degradation of samples upon heating. Approximately 10 mg of each sample was heated from 30 °C to 600 °C at a heating rate of 10 °C/min under nitrogen at a flow rate of 60 mL/min.

2.4.2.2 Differential Scanning Calorimetry

A TA Instruments™ Q2000 Differential Scanning Calorimeter (DSC) with Tzero pans and Tzero hermetic lids was utilized to investigate the endothermic, melting, and exothermic, reactive, behavior the benzoxazine monomer. Approximately 1.5 mg of monomer was heated from 30 °C to 300 °C at a heating rate of 20 °C/min under nitrogen at a flow rate of 50 mL/min.

2.4.2.3 Dynamic Mechanical Analysis

Dynamic mechanical analysis (DMA) was performed utilizing a TA Instruments™ Q800 DMA using a single cantilever fixture holding a specimen of dimensions 20 x 5 x 0.5 mm. Storage and loss moduli and tan δ were measured under a temperature ramp of 2 °C/min, oscillatory strain of 0.05%, and frequency of 1 Hz.

3. RESULTS

3.1 Monomer and Polymer Network Synthesis

To the knowledge of the authors, this work demonstrates the first synthesis of an aryl ether ketone (AEK) containing benzoxazine. The synthetic procedure used to afford the AEK containing benzoxazine monomer (**3**) is shown in **Scheme 1**. The AEK diamine precursor (**2**) was synthesized in a two-step process developed by Patterson and Wiggins.[10] The first step in this synthesis involves protecting the free amine on 4-phenolxyaniline by forming a salt with methanesulfonic acid to yield precursor **1**. Friedel-Crafts acylation is then performed with **1** and isophthalic acid in Eaton's reagent to yield **2**. The method for synthesizing AEK diamines in this work has been shown to be more saleable and higher yielding (104 g, 96%) than methods used by Lassila *et al.* (6.22 g, 93%)[11] and Smith *et al.* (62 mg, 80%)[11], and this work avoids the use of high-pressure Parr autoclave reactors, catalyst, and halogenated starting materials.

As mentioned in the synthesis section of the experimental, the conditions of the AEK benzoxazine synthesis were adapted from work by Baqar *et al.*[9] Much like this work, they were making a benzoxazine monomer from an ether derivative of aniline. The ^{1}H NMR spectra for **1**, **2**, and **3** are shown in **Figure 1**, **2**, and **3**, respectively.

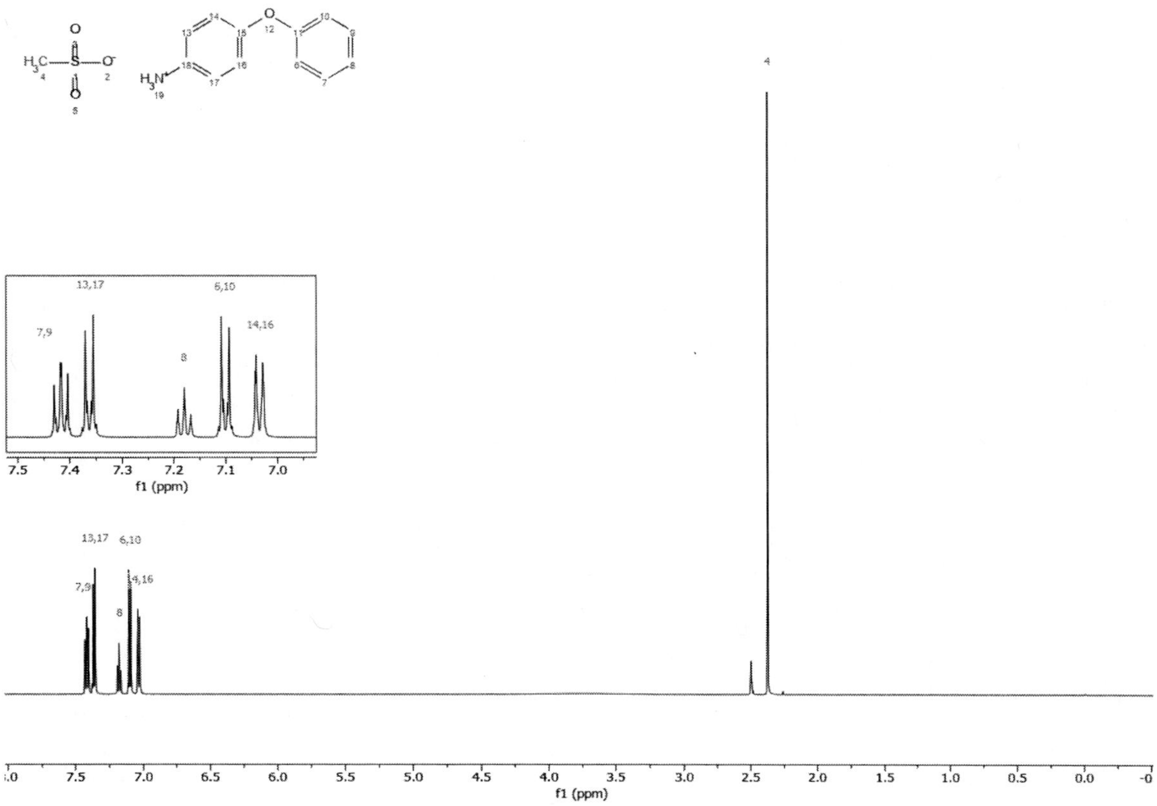

Figure 1: *^{1}H NMR spectrum of **1** in DMSO-d₆.*

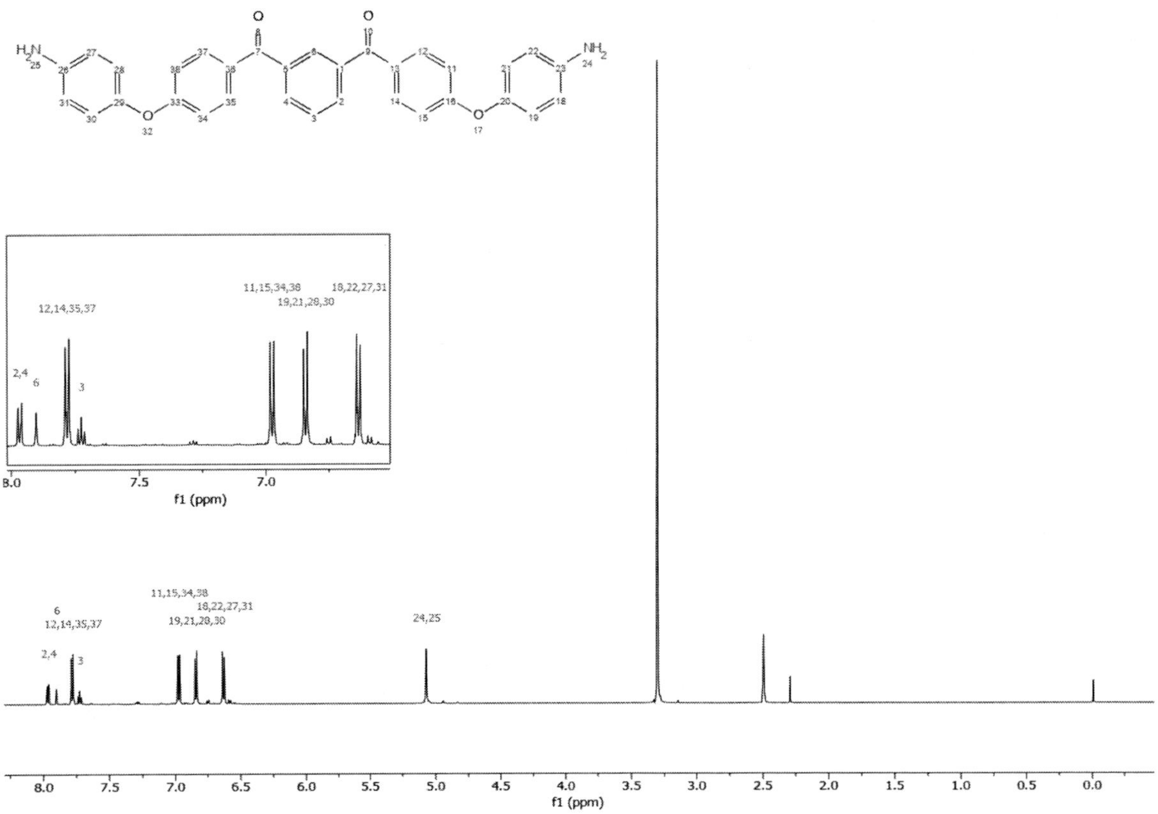

***Figure 2:** ^{1}H NMR spectrum of **2** in DMSO-d_6.*

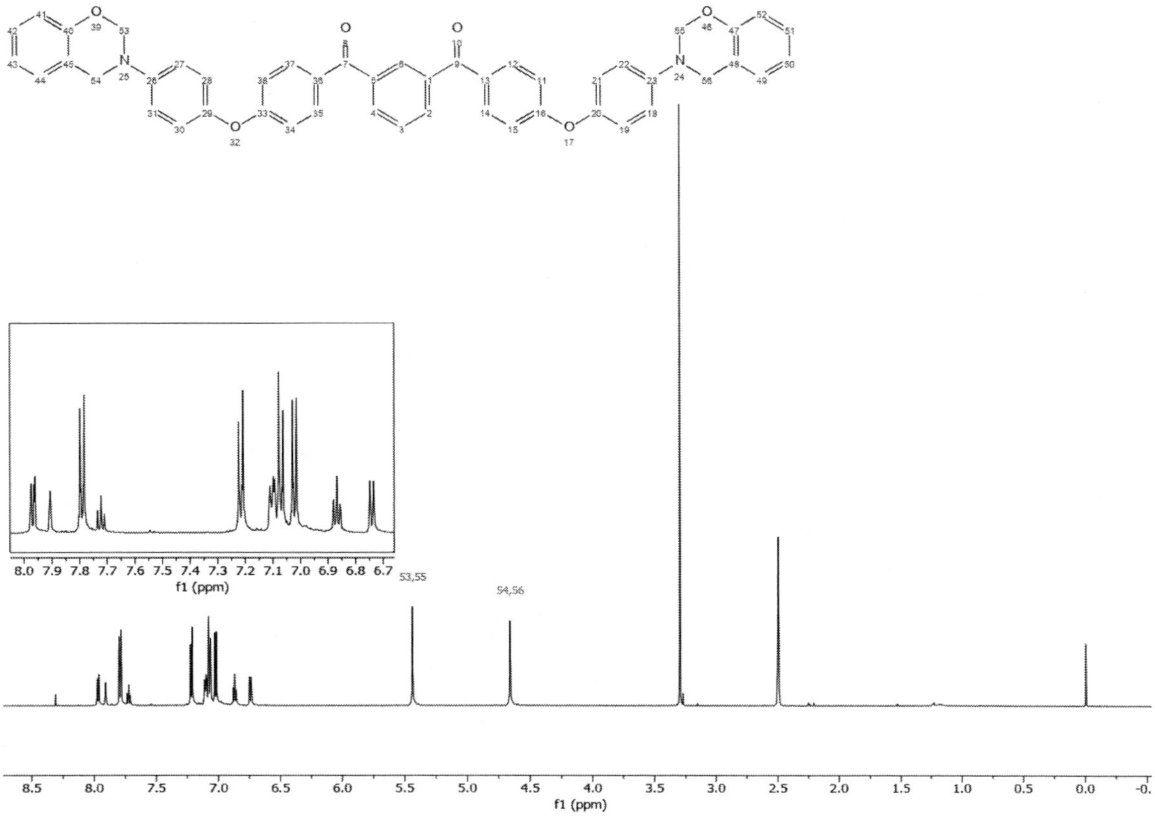

*Figure 3: ^{1}H NMR spectra of **3** in DMSO-d_6.*

Once **3** was obtained, it was thermally polymerized to form **Poly(3)**. To examine if the network had reached a fully cured state, ATR-FTIR was used to compare the 940 cm^{-1} out-of-plane stretch mode of the benzene which the oxazine ring is attached in **3** to **Poly(3)**. The ATR-FTIR spectra for **3** and **Poly(3)** are shown in **Figure 4**.

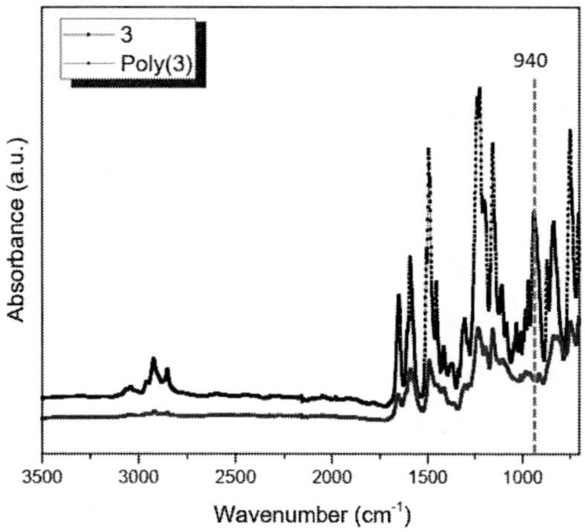

*Figure 4: ATR-FTIR spectra of **3** and **Poly(3)**.*

3.2 Thermomechanical Properties of Monomer and Polymer Network

3.2.1 Thermogravimetric Analysis

TGA was utilized to characterize the thermal stability of **3** and **Poly(3)** under nitrogen atmospheric conditions. **Figure 5** shows the thermal degradation profiles for **3** and **Poly(3)** up to 600 °C. The 5% ($T_{d,5\%}$) and 10% ($T_{d,10\%}$) mass loss events for **3** were found to be 242 °C and 335 °C and for **Poly(3)** 424 °C and 508 °C. There is an early mass loss event in **3** starting approximately at 69 °C that is assumed to be from residual chloroform from the synthesis of **3**. **Poly(3)** does not appear to have a single mass loss events but rather a complex slow degradation process. The char yields for **Poly(3)** was also found to be exceptionally high at 82% at 600 °C. The relevant TGA data can be found summarized in **Table 1**.

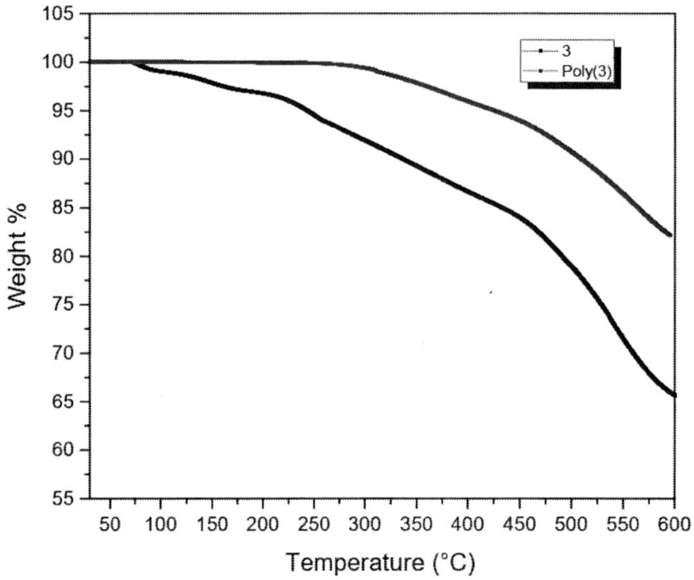

Figure 5: *TGA thermogram of **3** and **Poly(3)** at a heating rate of 10 °C/min.*

3.2.2 Differential Scanning Calorimetry

DSC was utilized to characterize the thermal transitions of **3**. **Figure 6** shows endothermic, down, melting temperature (T_m) as well as the exothermic, up, onset (T_{onset}) and peak reaction (T_p) temperatures of **3**. The T_m for **3** was found to be relatively high at 166 °C, but the T_{onset} and T_p were approximately 225 °C and 268 °C which gave a large enough processing window to degas and cast samples of **3**. The heat of reaction (ΔH) was observed to be 183.8 J/g. The relevant DSC data can be found summarized in **Table 1**.

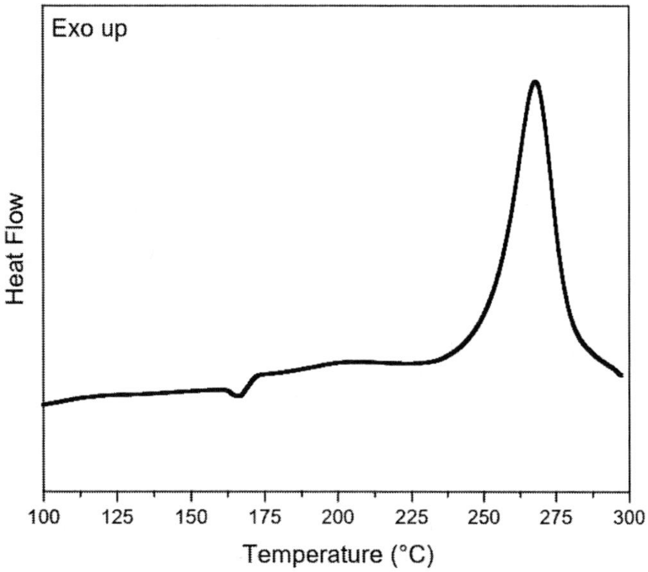

Figure 6: *DSC thermogram of 3 at a heating rate of 20 °C/min.*

3.2.3 Dynamic Mechanical Analysis

DMA was utilized to understand the thermomechanical properties of **Poly(3)**. **Figure 7** shows the storage modulus (E') and tan δ vs temperature of **Poly(3)**. The glass transition temperature (T_g) is taken as the peak of the tan δ. The tan δ cure for **Poly(3)** has a peak at 232 °C and a lower temperature shoulder which is indicative of an under cured network. At 30 °C, the glassy E' is 3619 MPa. The relevant DMA data can be found summarized in **Table 1**.

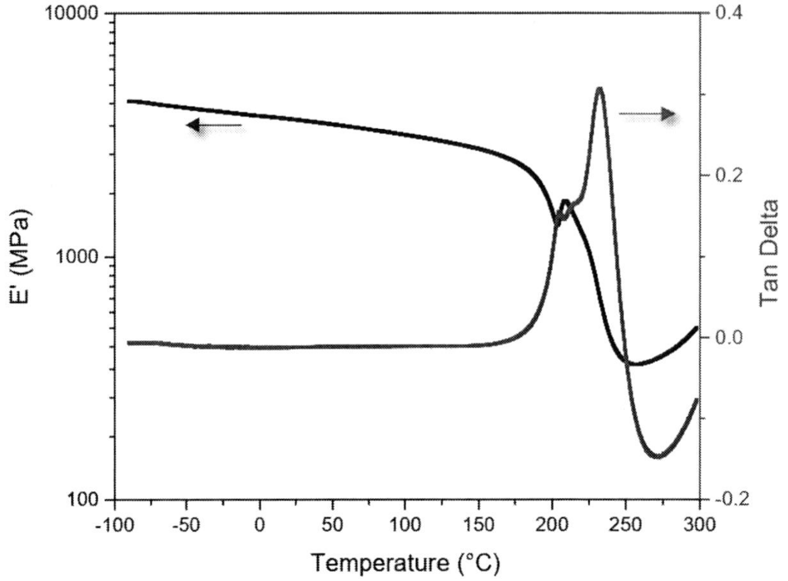

Figure 7: *DMA thermogram of Poly(3) at a heating rate of 2 °C/min.*

Table 1: *Summary of thermomechanical data of **3** and **Poly(3)**.*

Property	3	Poly(3)
$T_{d,5\%}$ (°C)	242	424
$T_{d,10\%}$ (°C)	335	508
Char Yield (%)	66	82
T_m (°C)	166	
T_{onset} (°C)	225	
T_p (°C)	268	
ΔH (J/g)	184	
E' at 30 °C (MPa)		3619
T_g (°C)		232

4. CONCLUSIONS

An AEK benzoxazine monomer and its polymer network were synthesized. The AEK diamine synthesis was accomplished using a method developed by Patterson and Wiggins that is both scalable and high yielding.[10] It was determined by DSC that **3** has a T_m of 166 °C and a T_{onset} and T_p of 225 °C and 268 °C, respectively. TGA shows **Poly(3)** displays a high char yield at 82%. DMA results show **Poly(3)** to display a high T_g of 232 °C, and it is probable that the ultimate T_g of the network is even higher due to the slight under cure of the network shown by the lower temperature shoulder. So, the cure schedule for **Poly(3)** will need further investigation. The novel benzoxazine **3** shows promise as a high performance next-generation aerospace thermoset matrix due to its superior thermal performance, and if it were combined with a low T_m reactive diluent it could be used in application such as resin transfer molding. Additional studies must be conducted to investigate the mechanical performance of **Poly(3)**, and various isomers of **3** are being synthesized to compare network properties.

5. REFERENCES

[1] H. Ishida and X. Ning, "Phenolic Materials via Ring-Opening Polymerization: Synthesis and Characterization," *J. Polym. Sci. Part A Polym. Chem.*, vol. 32, pp. 1121–1129, 1994.

[2] H. Ishida and D. J. Allen, "Physical and mechanical characterization of near-zero shrinkage polybenzoxazines," *J. Polym. Sci. Part B Polym. Phys.*, vol. 34, no. 6, pp. 1019–1030, 1996.

[3] S. Shamim Rishwana, A. Mahendran, and C. T. Vijayakumar, "Studies on structurally different benzoxazines based on diphenols and diamines: Kinetics of thermal degradation and TG-FTIR studies," *Thermochim. Acta*, vol. 618, pp. 74–87, 2015.

[4] Z. Brunovska, J. P. Liu, and H. Ishida, "1,3,5-Triphenylhexahydro-1,3,5-triazine - active intermediate adn precursor in the Novel Synthesis of Benzoxazine Monomers and Oligomers," *Macromol. Chem. Phys.*, vol. 200, no. 7, pp. 1745–1752, 1999.

[5] D. J. Allen and H. Ishida, "Effect of phenol substitution on the network structure and properties of linear aliphatic diamine-based benzoxazines," *Polymer (Guildf).*, vol. 50, no. 2, pp. 613–626, 2009.

[6] R. J. Varley *et al.*, "Effect of aromatic substitution on the kinetics and properties of epoxy cured tri-phenylether amines," *J. Appl. Polym. Sci.*, vol. 136, no. 18, pp. 1–11, 2019.

[7] L. Q. Reyes *et al.*, "Synthesis of tri-aryl ketone amine isomers and their cure with epoxy resins," *Polym. Adv. Technol.*, vol. 31, no. 4, pp. 827–837, 2020.

[8] J. Misasi, "Hybrid Aryl-Ether-Ketone and Hyperbranched Epoxy Networks," *Dissertation*. 2015.

[9] M. Baqar, T. Agag, H. Ishida, and S. Qutubuddin, "Methylol-functional benzoxazines as precursors for high-performance thermoset polymers: Unique simultaneous addition and condensation polymerization behavior," *J. Polym. Sci. Part A Polym. Chem.*, vol. 50, no. 11, pp. 2275–2285, 2012.

[10] M. A. Patterson, "Distortional Aryl Ketone Ether Hybrid Epoxy Matrices," *Dissertation*. 2019.

[11] K. R. Lassila and M. E. Ford, "An Improved Route to 1,3-bis-[4-(4-Aminophenoxy)-benzoyl]benzene and Related Compounds," *Synth. Commun.*, vol. 21, no. 2, pp. 229–234, 1991.

PHENYLPHOSPHINE OXIDE CATALYSIS OF EPOXY POLYMERIZATION

Witold K. Fuchs, Catherine, A. Sarantes, Jeffrey S. Wiggins
University of Southern Mississippi
Hattiesburg, MS/United States of America

ABSTRACT

Atomic oxygen (AO) attack to carbon-fiber reinforcing polymers (CFRP's) on spacecraft surfaces in low earth orbit (LEO) threatens spacecraft safe operation and service life. Incorporating phenylphosphine oxide (PPO) groups into epoxies has been shown to offer a self-regenerating method of protection from AO, but has catalytic effects on the polymerization reaction of epoxies. Herein, epoxides and diamines containing PPO groups were synthesized and the reactivity of the prepared monomers was evaluated for shelf-life and workability. The catalytic effect of PPO on the epoxy polymerization was investigated via differential scanning calorimetry (DSC), rheokinetic evaluation, real-time Fourier-transfer infrared spectroscopy (FT-IR), and real-time heated ^{31}P nuclear magnetic spectroscopy (NMR). The catalytic mechanism was further investigated through electron rich and electron deficient PPOs to elucidate the effect of aromatic substituents on the catalytic effect. It was found that the incorporation of electron rich phosphines had significant catalytic effects on the epoxy polymerization, shifting the onset of polymerization earlier by 50 °C. Moreover, the material remains processable as evaluated by rheology and a mechanism for the catalytic reaction is proposed. This work builds upon previously reported research to fully evaluate PPO epoxies and furthermore suggests that PPO epoxies are excellent candidates for AO resistant CFRPs.

Corresponding author: Witold K. Fuchs witold.fuchs@usm.edu

1. INTRODUCTION

Advanced composite materials are promising candidates in astronautical applications, particularly as deployable structures and spacecraft structural components.[1] Of the many available chemistries for CFRPs, epoxy polymers are the most commonly used composite matrices in satellites and spacecraft due to their versatility and cost-effectiveness.[1], [2] However, LEO conditions including high vacuum (approx. 10^{-4} – 10^{-5} Pa), extreme thermal cycling (-150 °C – 150 °C), VUV radiation (< 200 nm), and AO exposure render LEO particularly aggressive towards organic polymers.[3] Among all these factors, AO, by far, plays the most important role in the erosion processes of organic materials.[4]

Recently, AO resistant polymer development has focused on incorporating phenylphosphine oxide (PPO) groups into the backbone of aromatic polymers.[5]–[12] It has been reported that the AO resistance of phosphorus (P) containing polymers is due to the formation of a passivating phosphate surface layer that prevents underlying material from further erosion.[13], [14]

Herein, epoxides and diamines containing PPO groups were synthesized and the reactivity of the prepared monomers was evaluated for shelf-life and workability. It is commonly understood that Lewis bases catalyze epoxy polymerizations. However, to the best of the author's knowledge, no report on the catalytic activity of phenylphosphine oxide in epoxies has been reported. The substantial potential to alter the electron density of an aromatic phosphine oxide offers significant tuneability of potential catalytic effects.

2. Experimental

2.1 Chemicals and Materials

All reagents were used as received.

2.2 Phosphine Oxide Epoxy Synthesis

2.2.1 Characterization

Nuclear magnetic resonance (NMR) spectroscopy was conducted on a Bruker Ascend 600 MHz spectrometer for ^{1}H NMR and ^{31}P NMR with tetramethylsilane as an internal standard. Fourier transform infrared spectroscopy (FT-IR) was carried out on a PerkinElmer Frontier spectrometer with the scanning wavenumber in range of 400-4000 cm^{-1} at 32 scans. Thermal analyses were performed using a TA instruments Q50 thermogravimetric analysis (TGA) instrument at a heating rate of 10 °C·min^{-1} with the temperature ranging from 25 °C to 800 °C under N_2. Differential scanning calorimetry (DSC) was performed using a TA instruments Q200 DSC at a heating rate of 10 °C·min^{-1} from ambient temperature to 300 °C. Rheological analyses were performed using an ARES-G2 rheometer equipped with 25 mm disposable aluminum parallel plates.

2.2.2 Phosphine Oxide Epoxide (PO-E) Synthesis

Phosphine oxide functional hydroxyl compound (1 mol) was added to a 1000 mL three-neck round-bottom flask equipped with a mechanical stirrer, a refluxing condenser, an addition funnel, and a N_2 inlet/outlet. Epichlorohydrin (20 mols) and 30 % benzyltrimethylammonium chloride (BMAC) in methanol were added and heated until an intermediate was formed. The intermediate was concentrated using reduced pressure.

The intermediate was dissolved in a suitable solvent, and 20 % aqueous NaOH and saturated Na_2CO_3 were added to the reaction. The reaction was heated and stirred vigorously until product formation was confirmed by ^{1}H NMR. The product was then extracted and washed with deionized water (HCC = 0.064 wt %). Product purity was confirmed by ^{1}H NMR, ^{31}P NMR, and FT-IR.

2.2.3 Synthesis of bis-3-methoxyphenylphosphine oxide (BMPPO)

Bis-3-methoxyphenylphosphine oxide (BMPPO) was synthesized according to previous literature reports with some modification. The details are as follows:[15]

To a 1000 mL three-neck round-bottom flask equipped with a mechanical stirrer, addition funnel, condenser, and N_2 inlet/outlet, magnesium turnings (2 eq, 0.86 mol) were dissolved in dry THF (700 mL) as much as possible overnight. The following day, 4-bromoanisole (2 eq, 0.86 mol,

162.72 g) was transferred to an addition funnel and added over three hours, maintaining solution temperature at -5 °C. The solution was then stirred for three hours. To this solution mixture, phenylphosphonic dichloride (1 eq, 0.43 mol, 83.56 g) was added dropwise through an addition funnel. Diethyl ether was added in order to separate the solution into organic and aqueous phases. The organic layer was collected by extraction and washed 3x with 1 M NaOH and 3x with DI H_2O. The collected organic layer was dried over $MgSO_4$ and rotary evaporated to give a gel-like product (yield approx. 95 %). (SI 1).

2.2.4 Synthesis of bis-3-nitrophenylphosphine oxide (BNPPO)

Bis-3-nitrophenylphosphine oxide (BNPPO) was synthesized according to previous literature reports with some modification. The details are as follows:[15], [16]

To a 1000 mL three-neck round-bottom flask equipped with a magnetic stirrer, reflux condenser, addition funnel, and an N_2 inlet/outlet triphenylphopshine oxide (1 eq, 0.162 mol, 45.1 g) and 200 mL of 96 % sulfuric acid were added. The reagent was dissolved with vigorous stirring, and the reaction system was cooled to -5 °C. A solution of fuming nitric acid (2 eq, 0.323 mol, 20.35 g) in sulfuric acid (200 mL) was added dropwise, with care being taken to not allow the reaction to exceed 10 °C during the addition. The reaction was stirred and allowed to react for eight hours. The reaction mixture was then hydrolyzed over ice, extracted with chloroform, and washed with saturated aqueous sodium hydroxide solution until neutral pH. The solution was then concentrated using reduced pressure and recrystallized from absolute ethanol to give a yield of 67 % (^{1}H NMR 600 MHz δ 7.6 2H ddd; δ 7.75 2H dddd, 1H tt; δ 7.8 2H dddd; δ 8.2 2H ddd; δ 8.45 2H dt, δ 8.65 2H ddd). (SI 2).

2.2.5 Synthesis of bis-3-aminophenylphosphine oxide (BAPPO)

Bis-3-aminotriphenylphosphine oxide (BAPPO) was synthesized according to previous literature reports with some modification. The details are as follows:[15], [16]

A 500 mL three-neck round-bottom flask equipped with a magnetic stirrer and an N_2 inlet/outlet was charged with BNPPO (1 eq, 0.03 mol, 10 g) and anhydrous tin(II)chloride (15.83 eq, 0.475 mol, 90 g). A solution of 100 mL fuming hydrochloric acid in 200 mL of absolute ethanol was introduced into the flask and stirred at ambient temperature for five hours. The solution was then concentrated under reduced pressure, neutralized with 25 % aqueous NaOH solution, re-dissolved in ethanol and filtered through celite to remove tin(II)chloride. The filtered product was extracted with chloroform, washed three times with deionized water, and concentrated to give a pure product with 95 % yield (^{1}H NMR 600 MHz, 4H δ 5.45 4H dd; δ 6.6 2H ddd; δ 6.7 2H ddd, 1H tt; δ 6.8 2H ddd; δ 7.2 2H ddd; δ 7.5 2H dddd, δ 7.55 3H m). (SI 3).

3. Results & Discussion

The increasing effort for re-usable commercial spacecraft has motivated the development of high-performance carbon-fiber reinforcing polymers with longer space environment service-life. Our strategy has been to incorporate the PPO group into the versatile epoxy-amine polymer.

Surprisingly, the addition of phosphine oxide has significant catalytic effects on epoxy-amine polymerization, which we report below.

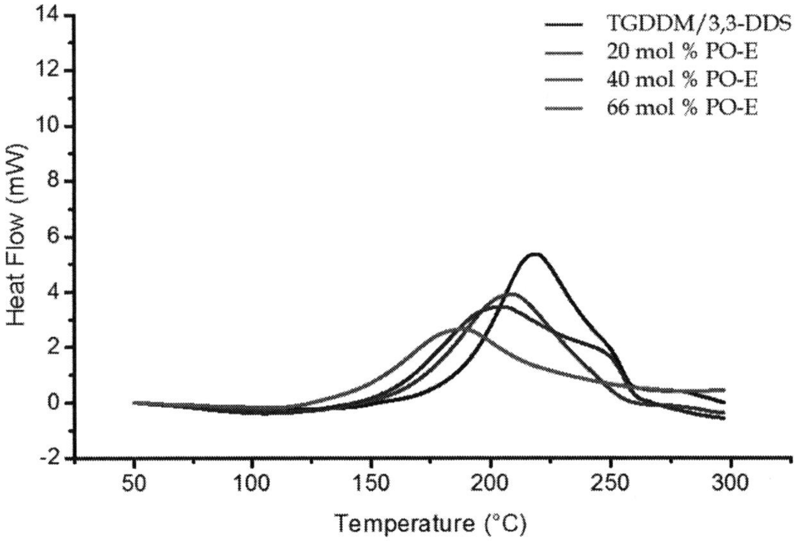

Figure 1. Polymerization behavior of PO-E epoxy-amines

Differential scanning calorimetry (DSC) results of the PO-E monomer demonstrate the effect of phosphine oxide addition on the epoxy-amine polymerization reaction (Figure 1). The substantial affect on the reaction was initially unexpected, as no previous report suggests that phosphine oxide monomers behave this way. One report regarding PPO diamines does suggest that a PPO diamine has a slightly higher reactivity than that of 3,3-DDS but lower than DDM.[16] Our initial results demonstrated as much as a 50 °C earlier onset temperature of polymerization and a 30 °C earlier peak polymerization temperature (Table 1). However, due to the presence of the reactive epoxide on this monomer, the apparent catalysis could not be solely attributed to the presence of the phosphine oxide, so an unreactive monomer with similar electron density (BMPPO) was analyzed.

Bis-4-methoxyphenylphosphine oxide (BMPPO) was synthesized and analyzed *via* DSC for potential catalytic behavior in a TGDDM/3,3-DDS epoxy-amine (Figure 2). The addition of BMPPO definitively shifted the onset of polymerization towards lower temperatures. In addition, the trend is in good agreement with results obtained for the PO-E molecule. This supports the hypothesis that phosphine oxide plays a catalytic role in the polymerization reaction.

Table 1. Polymerization behavior of PO-E epoxy-amines

	TGDDM/3,3-DDS	20 mol % PO-E	40 mol % PO-E	66 mol % PO-E
Onset Temperature (°C)	173.92	150.42	145.23	128.02
Peak Polymerization Temp (°C)	218.68	209.10	204.17	188.31

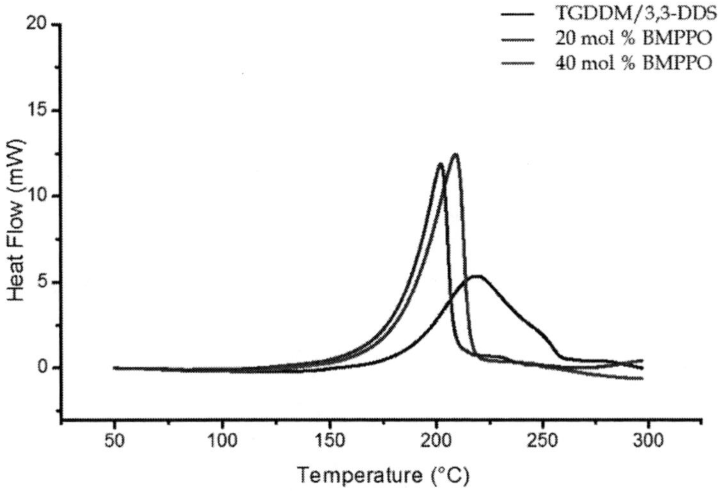

Figure 2. Catalytic effect of electron-dense, unreactive bis-3-methoxyphenylphosphine oxide (BMPPO).

Based on the results in Figures 1 and 2 it was apparent that phosphine oxide has an effect on the reaction between an amine and an epoxide. However, it remained to be seen whether the apparent catalysis impacted etherification behavior. To investigate this, BMPPO at a loading level of 40 mol % was added to neat TGDDM and the reaction behavior was monitored by DSC (Figure 3). It was observed that TGDDM with BMPPO at 40 mol % reacted earlier than neat TGDDM. Whereas the etherification of TGDDM can be observed beginning at approximately 270 °C, the onset of etherification in the presence of BMPPO occurs at approximately 180 °C. The magnitude of this apparent catalytic effect therefore appears to be greater for the etherification reaction (approx. 90 °C) compared to the primary amine reaction (approx. 30 °C).

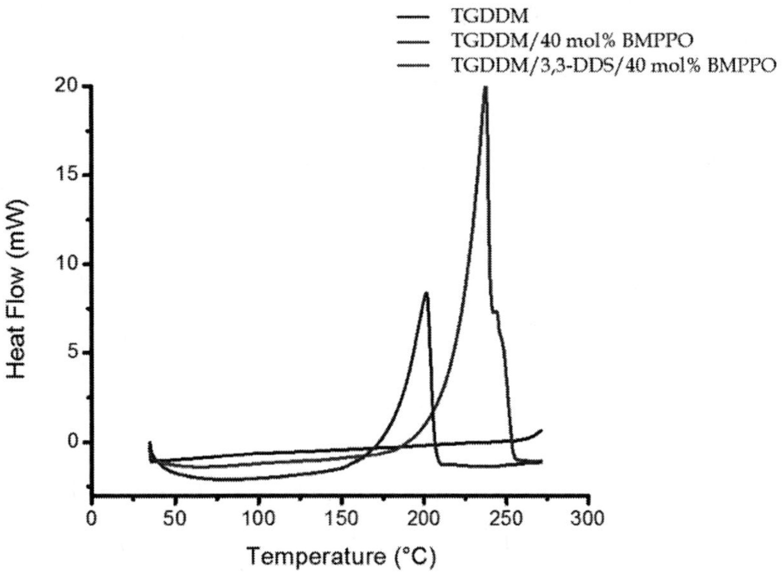

Figure 3. Catalytic effect of electron-dense, unreactive BMPPO on epoxy-amine polymerization (blue), and epoxy etherification (red).

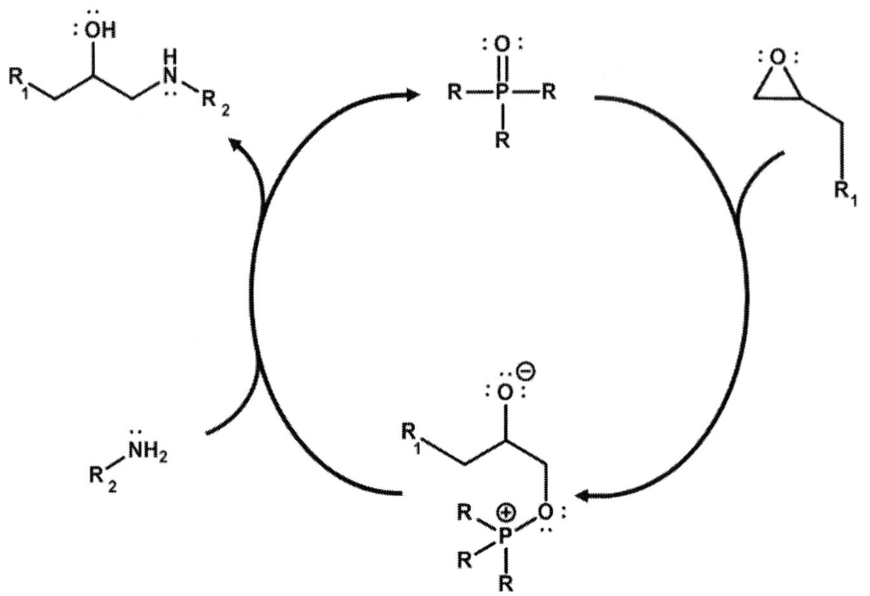

Scheme 1. Proposed catalytic mechanism of phosphine oxide on epoxy-amine polymerization.

Based on the observed catalytic effect, we propose a mechanism for the catalysis of the epoxy-amine reaction by phosphine oxide (Scheme 1). Previous literature on the isomerization of epoxides also suggests that phosphine oxide interacts with an epoxide ring *via* interactions with the oxygen atom, giving further confidence in the proposed mechanism.[17] If the proposed mechanism were true, increasing electron density on the phosphorus atom would increase the magnitude of the catalytic effect. To test our hypothesis, we prepared triphenylphosphine oxide molecules with varied electron density relative to the phosphine (Figure 4). A clear trend of increasing catalytic behavior with increasing electron density was observed, in agreement with literature reports on phosphine oxide catalysis in other systems (Figure 5). Unfortunately, samples containing BNPPO thermally degraded before a full heat flow peak could be observed. The peak at approximately 267 °C was attributed to the peak polymerization and used for comparison (Table 2). It further became clear that in a TGDDM/3,3-DDS system the catalytic effect was only observable with a high electron density on the phosphorus.

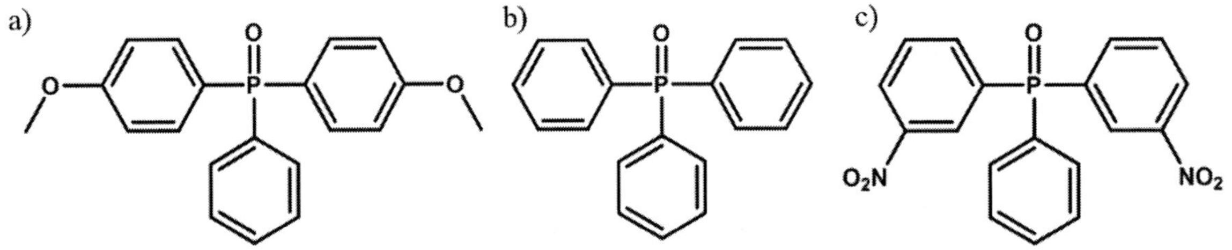

Figure 4. a) structure of bis-4-methoxy phenylphosphine oxide (BMPPO), b) triphenylphosphine oxide (tPPO), c) bis-3-nitrophenylphosphine oxide (BNPPO)

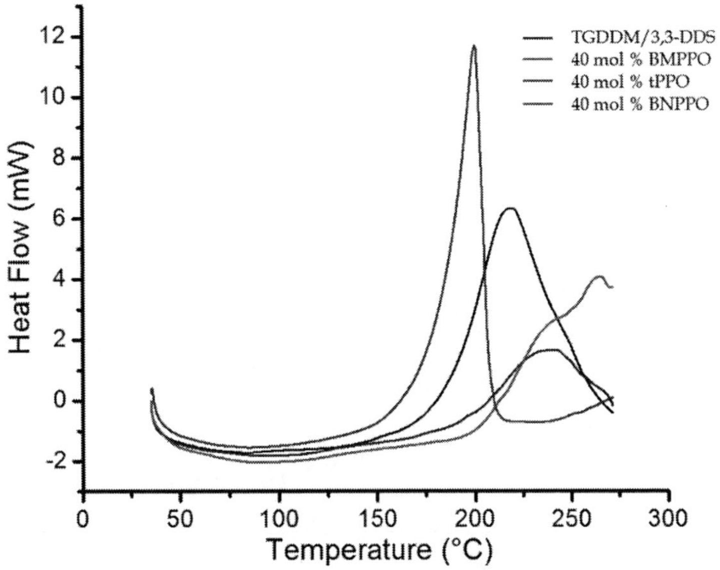

Figure 5. Catalytic effect of phosphine oxide on epoxy-amines polymerization

The observed catalysis appears to be most significant in reactions with high activation energies (ie. etherification), whereas low activation energy reactions (primary amine with epoxide) appear to experience a less significant catalysis. It is likely that as amine reactivity increases, the catalytic effect of phosphine oxide would be overshadowed by the reactivity of the amine. Further research into this catalysis is warranted, as slight catalytic effects may have pronounced impact on material processability.

The effect of the catalysis on rheological behavior during polymerization is quite pronounced. At high concentrations of phosphine oxide (which are desirable for atomic oxygen resistance) the PPO containing epoxy-amine slurry begins increasing viscosity at approximately 90 °C (Figure 6). The viscosity well shifts to approximately 105 °C with 66 mol % PO-E whereas a TGDDM/3,3-DDS viscosity well occurs at approximately 140 °C. Furthermore, the viscosity of PO-E slurries is somewhat higher than that of TGDDM/3,3-DDS at the same temperature

Table 2. Catalytic effect of phosphine oxide on epoxy-amine polymerization

Sample Composition	Peak Polymerization Temp (°C)
TGDDM/3,3-DDS	219.44 ± 0.09
40 mol% BMPPO	200.89 ± 0.76
40 mol% tPPO	241.83 ± 1.07
40 mol% BNPPO	267.11 ± 2.23

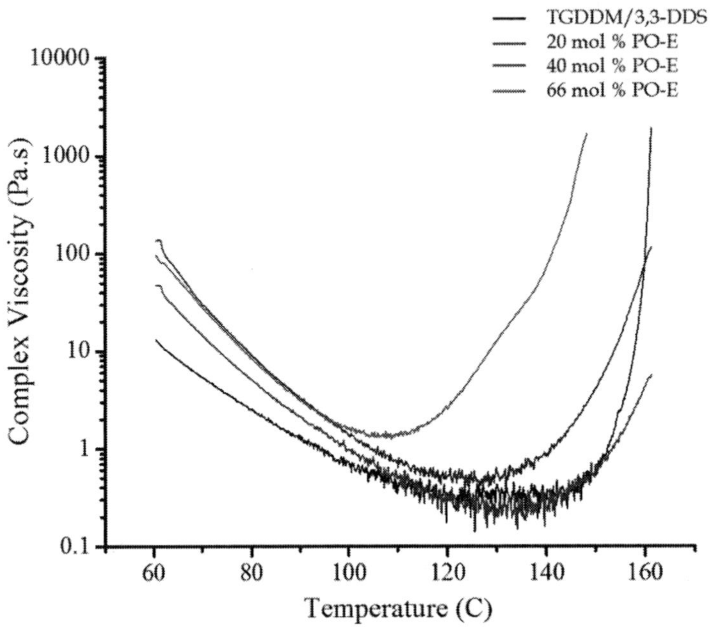

Figure 6. Rheological analysis of the polymerization behavior of PO-E epoxy-amines.

Notwithstanding the effect on reactivity and viscosity of high concentrations of PO-E, the material remains processable at 90 °C (Figure 7). Furthermore, the identified trend of reactivity with electron density on the phosphorus provides a tunable handle to alter the catalytic effect, paving a pathway for improving the processing behaviors of phosphine oxide epoxy-amines.

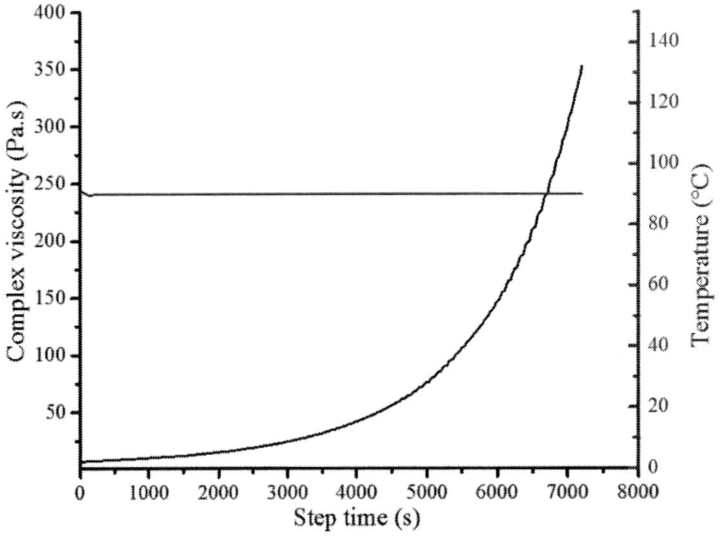

Figure 7. Polymerization behavior of 66 mol % PO-E epoxy amines. Isothermal polymerization rheokinetics at 90 °C.

4. Conclusions

In this work we have demonstrated the catalytic effect of PPO on epoxy-amine polymerization. Furthermore, it was found that increasing electron density on the phosphorus increased the magnitude of the catalytic event. DSC and rheometry were used to investigate the catalysis on TGDDM/3,3-DDS systems. Further investigation will be performed using real-time ^{31}P NMR. In addition, activation energy studies will be performed to quantify the magnitude of the catalysis. Finally, real-time FT-IR monitoring of the polymerization will be performed to elucidate whether etherification is significant in epoxy-amine polymerizations catalyzed by PPO.

Phosphine oxide containing epoxy-amines provide an attractive platform for the preparation of LEO CFRPs. Their high Tg and relative ease of preparation makes them suitable for applications in rigorous LEO conditions. Although their catalytic behavior was unexpected, methods to tune its magnitude have been identified should its effect on the processability of the material be necessary.

5. References

[1] Y. He, A. Suliga, A. Brinkmeyer, M. Schenk, and I. Hamerton, "Atomic oxygen degradation mechanisms of epoxy composites for space applications," *Polym. Degrad. Stab.*, vol. 166, pp. 108–120, 2019, doi: 10.1016/j.polymdegradstab.2019.05.026.

[2] A. Suliga, E. M. Jakubczyk, I. Hamerton, and A. Viquerat, "Analysis of atomic oxygen and ultraviolet exposure effects on cycloaliphatic epoxy resins reinforced with octa-functional POSS," *Acta Astronaut.*, vol. 142, pp. 103–111, 2018, doi: 10.1016/j.actaastro.2017.10.018.

[3] "Atomic Oxygen Protections in Polymeric Systems: A Review," *The Journal of Space Technology and Science*, vol. 18, no. 1. pp. 1_34-1_48, 2002, doi: 10.11230/jsts.18.1_34.

[4] J. Kleiman, Z. Iskanderova, Y. Gudimenko, and S. Horodetsky, "Atomic oxygen beam sources: A critical overview," *Eur. Sp. Agency, (Special Publ. ESA SP*, no. 540, pp. 313–324, 2003.

[5] J. W. Connell, P. M. Hergenrother, and J. G. Smith, "The Effect of Oxygen Plasma on Some Experimental Polymers Containing Silicon and Phosphorus," *J. Fire Sci.*, vol. 11, no. 2, pp. 137–146, 1993, doi: 10.1177/073490419301100203.

[6] J. G. Smith, J. W. Connell, and P. M. Hergenrother, "Oxygen plasma resistant phenylphosphine oxide-containing poly(arylene ether)s," *Polymer (Guildf).*, vol. 35, no. 13, pp. 2834–2839, 1994, doi: 10.1016/0032-3861(94)90314-X.

[7] A. J. Gavrin, S. I. U. W. A. Au-yeung, and B. O. B. Mojazza, "Atomic oxygen resistant, low alpha, anti-static polyimides for potential space applications arthur j. gavrin, siu wa au-yeung, bob mojazza," pp. 435–441, 1995.

[8] J. W. Connell, J. G. Smith, and P. M. Hergenrother, "Oxygen plasma-resistant phenylphosphine oxide-containing polyimides and poly { arylene ether heterocycle) s : 1," vol. 36, no. 1, pp. 5–11, 1995.

[9] J. W. Connell, J. G. Smith, C. G. Kalil, and E. J. Siochi, "The effect of low earth orbit atomic oxygen exposure on phenylphosphine oxide-containing poly(arylene ether)s," *Polym. Adv. Technol.*, vol. 9, no. 1, pp. 11–19, 1998, doi: 10.1002/(SICI)1099-1581(199801)9:1<11::AID-PAT728>3.0.CO;2-T.

[10] J. W. Connell, "Effect of low Earth orbit atomic oxygen exposure on phenylphosphine oxide-containing polymers," *High Perform. Polym.*, vol. 12, no. 1, pp. 43–52, 2000, doi: 10.1088/0954-0083/12/1/304.

[11] J. W. Connell and K. A. Watson, "Space environmentally stable polyimides and copolyimides derived from bis(3-aminophenyl)-3,5-di(trifluoromethyl)phenylphosphine oxide," *High Perform. Polym.*, vol. 13, no. 1, pp. 23–34, 2001, doi: 10.1088/0954-0083/13/1/303.

[12] K. A. Watson, F. L. Palmieri, and J. W. Connell, "Space environmentally stable polyimides and copolyimides derived from [2,4-bis(3-aminophenoxy)phenyl] diphenylphosphine oxide," *Macromolecules*, vol. 35, no. 13, pp. 4968–4974, 2002, doi: 10.1021/ma0201779.

[13] J. H. Wei *et al.*, "Atomic oxygen resistant phosphorus-containing copolyimides derived from bis[4-(3-aminophenoxy)phenyl] phenylphosphine oxide," *Polym. Sci. Ser. B*, vol. 56, no. 6, pp. 788–798, 2014, doi: 10.1134/S1560090414060086.

[14] L. L. Fewell, "Atomic-oxygen induced structural changes in polyphosphazene films and coatings," *J. Appl. Polym. Sci.*, vol. 41, no. 1–2, pp. 391–406, 1990, doi: 10.1002/app.1990.070410131.

[15] K. Demchuk, O; Stankevic, M; Pietrusiewicz, *Arylphosphine Oxides and Heteroatom Derivatives (Update 2015)*. 2015.

[16] Y. L. Liu, G. H. Hsiue, R. H. Lee, and Y. S. Chiu, "Phosphorus-containing epoxy for flame retardant. III: Using phosphorylated diamines as curing agents," *J. Appl. Polym. Sci.*, vol. 63, no. 7, pp. 895–901, 1997, doi: 10.1002/(SICI)1097-4628(19970214)63:7<895::AID-APP9>3.0.CO;2-L.

[17] D. E. Bissing and A. J. Speziale, "The Phosphine Oxide Catalyzed Isomerization of Epoxides," *J. Am. Chem. Soc.*, vol. 87, no. 6, pp. 1405–1406, 1965, doi: 10.1021/ja01084a061.

6. Supporting Information

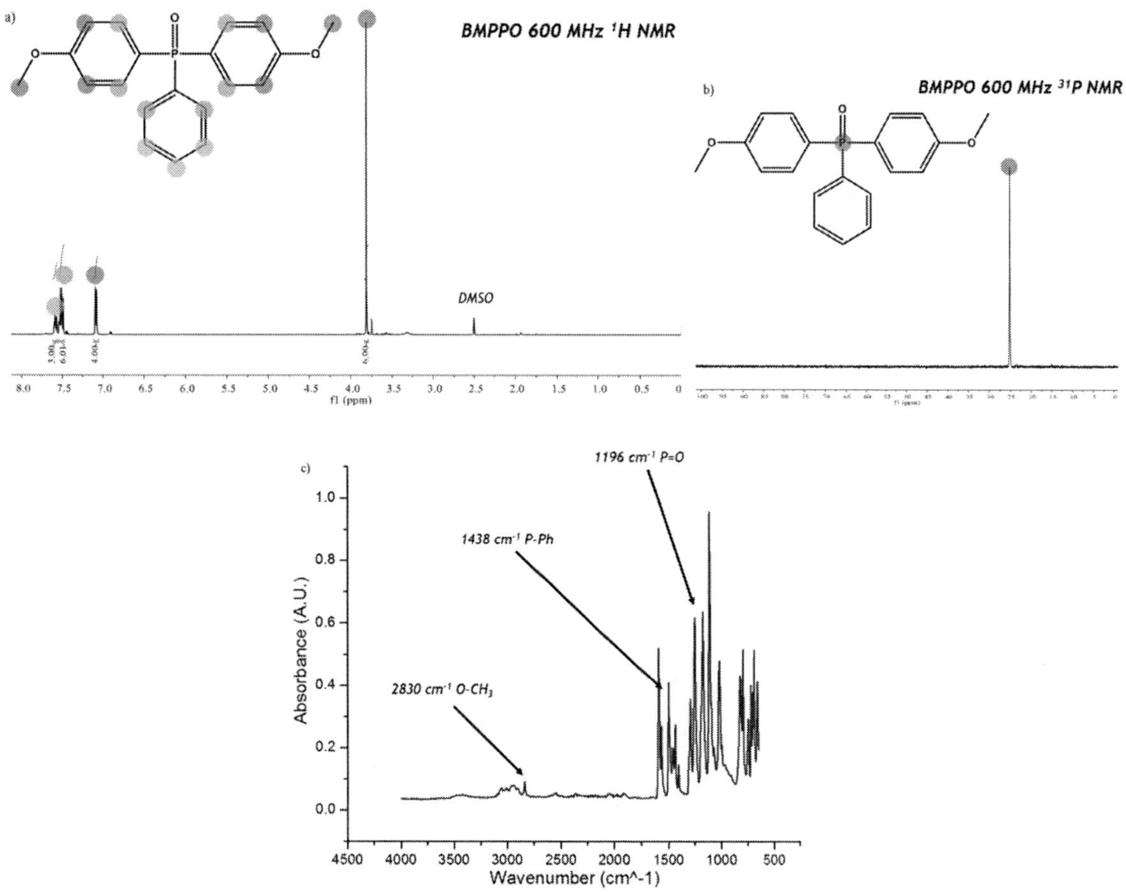

SI.1 (a) ^{1}H NMR of synthesized BMPPO (b) ^{31}P NMR of synthesized BMPPO (c) FT-IR of synthesized BMPPO

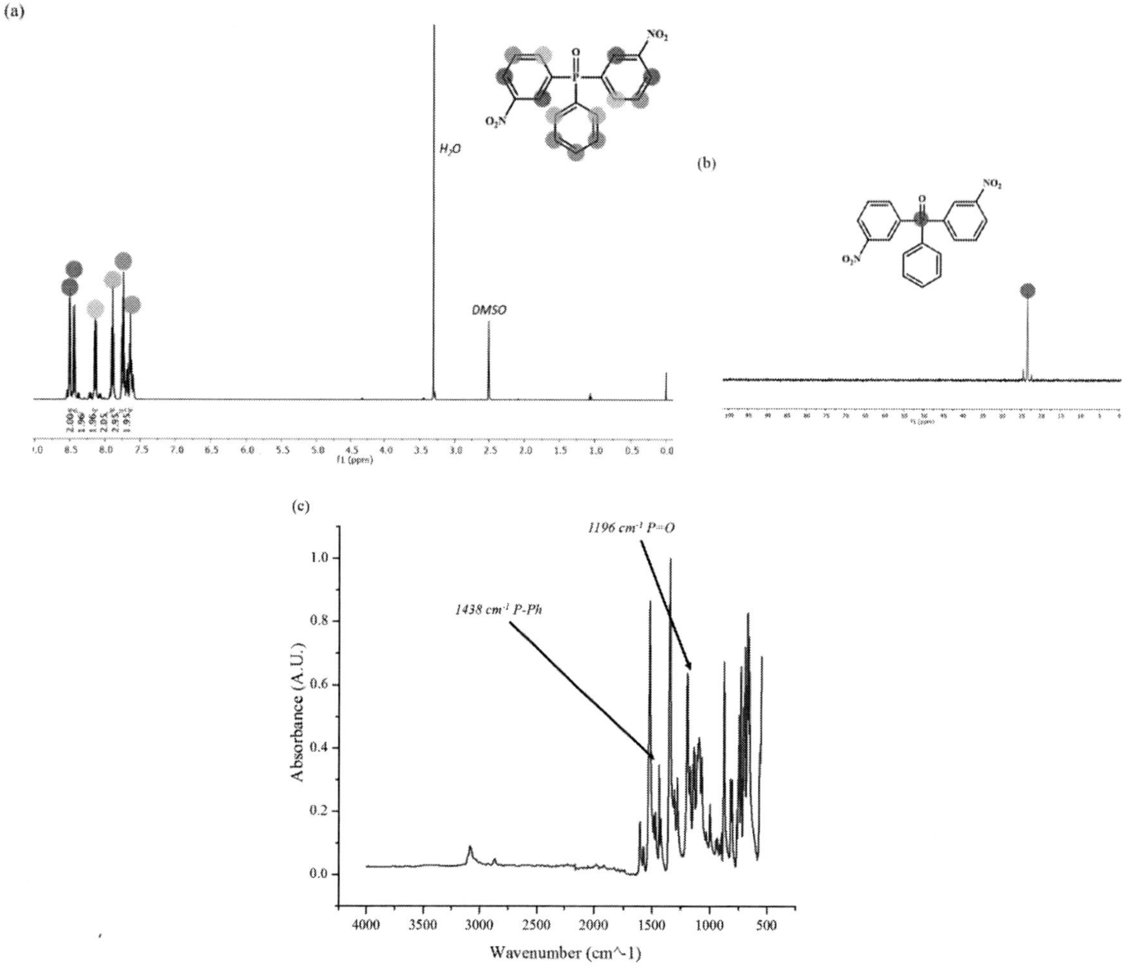

SI.2 (a) ^{1}H NMR of synthesized BNPPO (b) ^{31}P NMR of synthesized BNPPO (c) FT-IR of synthesized BNPPO

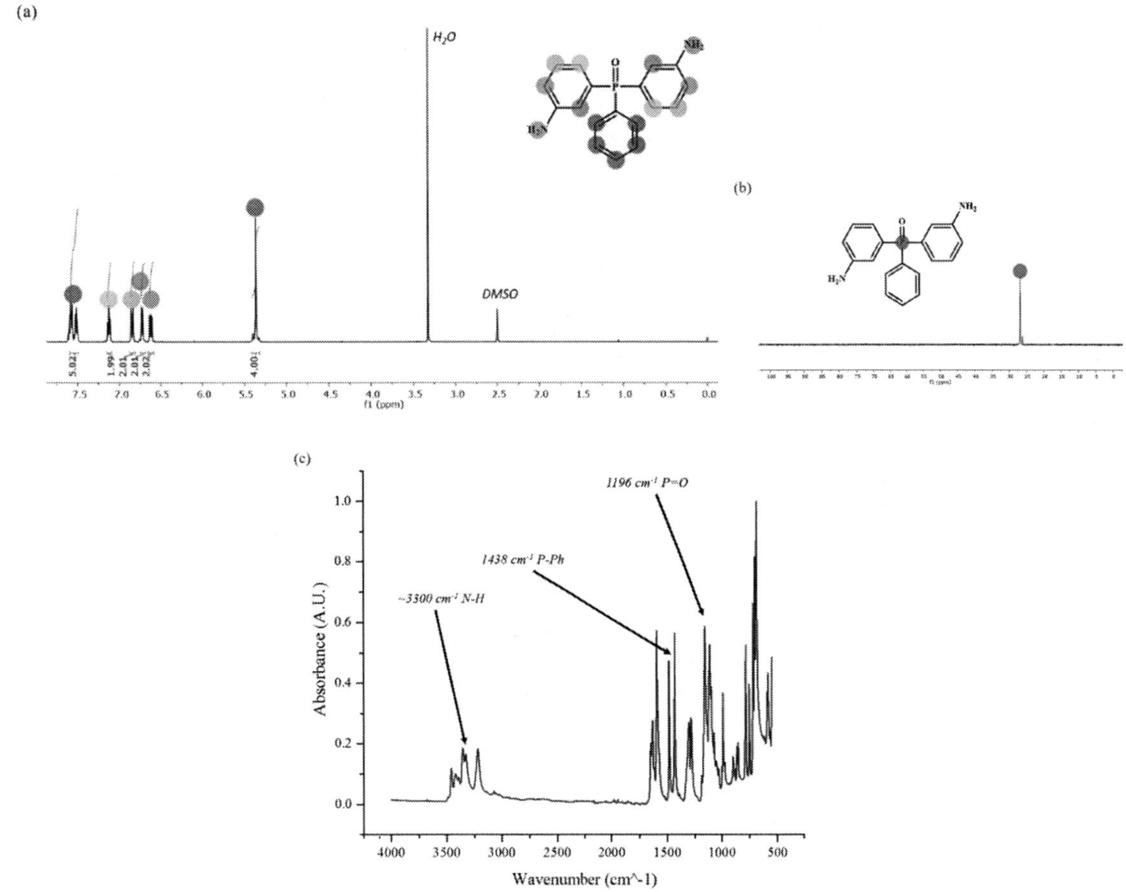

SI.3 (a) ^{1}H NMR of synthesized BAPPO (b) ^{31}P NMR of synthesized BAPPO (c) FT-IR of synthesized BAPPO

MANUFACTURING OF POLYMER COATED GLASS FIBRES FOR INCREASED FIBRE VOLUME CONTENT

Sandesh. Shirke[1], Richard. Haas[1], Alexander. Lüking[1], Robert. Brüll[1], Thomas. Gries[1], Thomas. Exlager[2],
[1]RWTH Aachen University, Aachen, North Rhine-Westphalia/Germany, [2]Coatema® Coating Machinery GmbH, Dormagen, North Rhine-Westphalia/Germany

ABSTRACT

In order to overcome the shortcomings of poor resin infiltration and increasing void content in products made using Glass Fibre Reinforced Plastic (GFRP), it is important to address the issue at the micro-level. The proposed solution is to coat the glass fibres at the production stage. This is achieved by coating the individual glass fibres with the polymer matrix during the spinning process. A process chain for coating individual glass filaments during production is developed at the Institut für Textiltechnik (ITA), RWTH Aachen University, Aachen, Germany. The glass filaments are spun at 1000 m/min speed this helps in ensuring high efficiency and productivity. A coating technology with a selection of an appropriate coating resin as per the end-use application is chosen. All fibres are completely coated with the matrix which ensures uniform distribution of forces and prevents the formation of voids. Coating of glass fibres also eliminates the intermediary process thereby saving cost, time and labour efforts.

Keywords: Fibre reinforced plastics, Glass fibre reinforced plastics, Glass filaments, Reinforcement fibres, Matrix, Fibre volume fraction, Coating module

Corresponding author: Alexander Lüking, Sandesh Shirke

1. INTRODUCTION

Fibre-reinforced plastics (FRP) is a combination of two different materials. These materials are mostly reinforcing textile fibres along with matrix. The matrix protects the fibre from mechanical damage and permits the transfer of applied forces to the reinforcing fibres. The combination of fibres along with the matrix absorbs the structural and applied forces. The FRP components are light in weight with improved mechanical properties. Due to such a unique offering of the properties by the fibre reinforced composite, they are being used in different mobility sectors like automotive, aerospace and energy sectors.

In particular, endless glass fibre reinforced plastics (GFRP), are important because of their lightweight properties achieved with low price. They also offer high strength and stiffness values, combined with low density. Therefore, GFRP's offer high potential in the sectors where high strength with low weight is required. The current state of the art for GFRP manufacturing have improved over a period of time, however, they have disadvantages which restricts optimum use of the fibre properties.

Current process and techniques used to manufacture the GFRP such as VARTM, RTM, Heat press and Hybrid yarn (Glass fibre + thermoplastic yarn) do not offer products with optimum use of desired properties. The current state of the art has disadvantages such as low fibre volume fraction (FVF) and high void content. Also, these techniques are time consuming and cost intensive because of multiple manufacturing steps. Due to the inhomogeneous distribution of fibre-matrix, as well as the low fibre volume fraction, the field of glass fibre reinforced plastics offer a high research potential. The schematic view of current manufacturing process compared with new coating process can be seen in the Figure. 1.

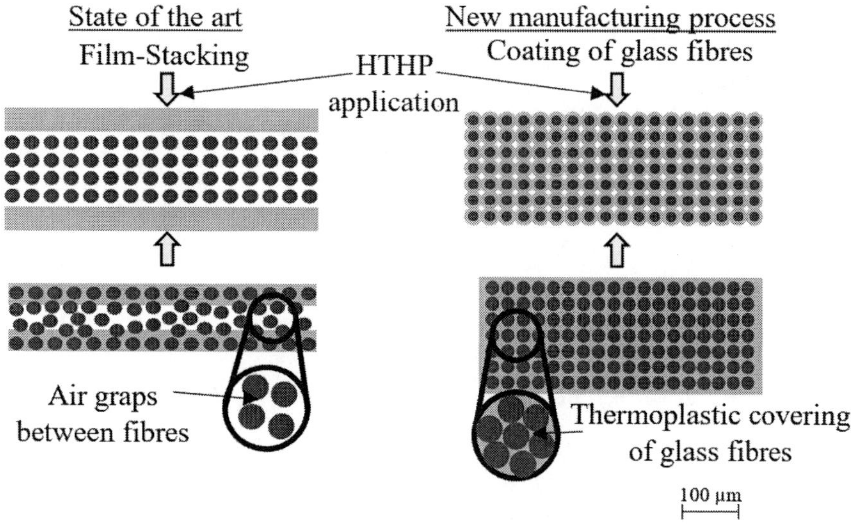

Figure 1. Schematic representation of defects in composite manufacturing.

In today's manufacturing process of GFRP, prepregs are produced. These prepregs offer certain disadvantages, due to which their applications are limited. When these prepregs are consolidated, ideally all the glass filaments should get covered by resin in order to obtain the optimum mechanical properties. However, this is not achieved as the glass fibres are not entirely impregnated with resin. An inhomogeneous distribution of fibre and resin, results in low fibre volume content of 40 – 60 vol.% in the composite material. The maximum potential of 91 % fibre volume content is achievable. Therefore, the production of thermoplastic, glass fibre reinforced plastics (GFRP) is considered in detail in this work.

1.1 State of the art

Fibre composites provide high strength to weight ratio. This distinguishes this material class from other materials such as steel or aluminium. Therefore, fibre composites are preferred for lightweight construction. Fields of application include energy, aviation and automotive sectors. To give a comprehensive overview of the characteristics of this material group, specific features in the structure and the basic manufacturing principle are explained below.

Fibre composites consist of at least two components. The first component is fibre material which is completely enclosed by the second material which is matrix material. The fibre component absorbs tensile stress in fibre direction. The matrix component, on the other hand, provides the cohesion between the fibres and protects them from environmental influences. In most cases, a sizing is applied in the boundary layer between fibre and matrix. On the one hand, the sizing serves to create a chemically compatible bond between the fibre and the surrounding matrix. On the other hand, it assumes a protective function of the fibres in the manufacturing process. Thus, it reduces the friction between the fibres, resulting in less abrasion stress. This reduces the fibre damage during production. As fibre material glass filaments with unidirectional structures of GFRP are present which is shown in Figure. 2. In addition to the thermoplastic matrix and the glass fibre, the boundary layer between the two components can be recognised.

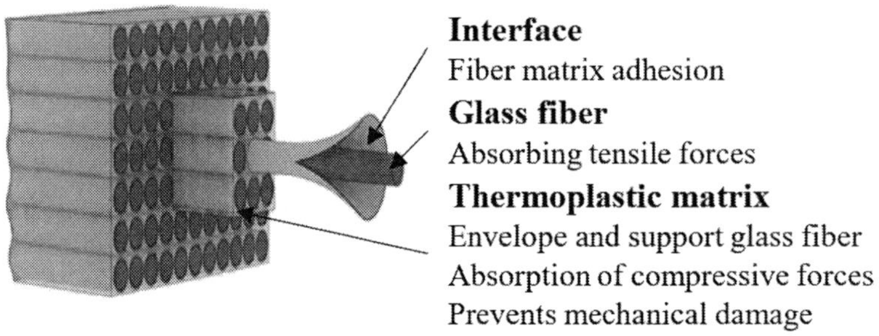

Figure 2. Pictorial representation of construction of GFRP. [4]

1.1.1 Production of a glass fibre reinforced plastics

The manufacture of a GFRP takes place in several steps, starting first with preparation of fibre components. For a load-oriented design, it is necessary to bring the glass fibres into the structure of a textile semi-finished product. Subsequently, impregnated structures, so called prepregs are manufactured. There are a number of different methods, the most common ones include winding process, vacuum process and pressing process [1]. In the pressing process shown in Figure 3, a semi-finished textile product is first placed in a tempered tool. This is followed by pressing under constant pressure and constant temperature. While maintaining the pressure, the component is cooled and demoulded. The implementation of each step is crucial for the quality of the final product. Thus, both the total cost and the mechanical properties of the product depend crucially on the quality of the intermediates. [2] [3]

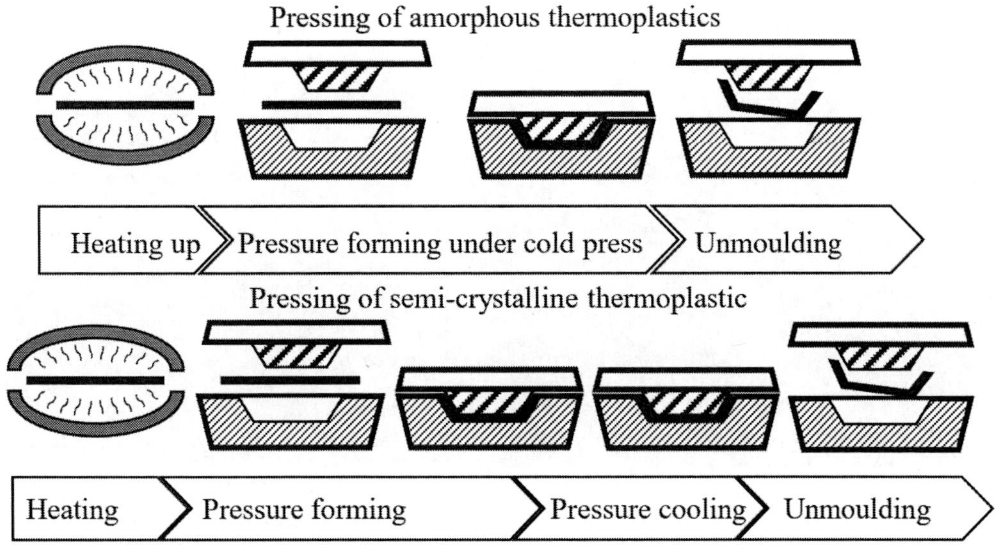

Figure 3. Schematic representation of manufacturing of composite using pressing process.

During production, the flow behaviour of the matrix component is a decisive parameter. This depends on the temperature, pressure, time and shear rate. For a uniform fibre matrix distribution, it is necessary due to the process that the matrix travels along certain flow paths in the production process. If the melt solidifies too early in the flow process, an unequal distribution occurs. Therefore, long flow paths lead to an inhomogeneous distribution of the plastic in the GFRP component. This is a manufacturing difficulty especially for thermoplastic materials with a high melt viscosity. Often it is not possible in the production of GFRP components directly to produce the desired geometry. This is followed by one or more forming processes in which the fibre matrix composite is brought into the final form of the GFRP component. [3]

1.1.2 Deficit in current manufacturing process

The challenges that arise with all state-of-the-art processes is that, only the exterior of each roving is coated with the matrix material. This means that are fibres inside the rovings that are not immersed to the matrix resulting in air pockets present in the rovings final composite part. Since the mechanical forces in the fibre reinforced composite are absorbed and transmitted from the matrix to the fibres, these non-connected fibres cannot contribute to the load-bearing capacity of the component. However, they add on to the weight of fibre composites. In Figure 4, we can clearly observe that there are air pockets present in the components produced using the commingling or Film-Stacking process. Apart from voids, there is an occurrence of fibre to fibre contact that is not desirable in the composite manufacturing.

Commingling Film- Stacking

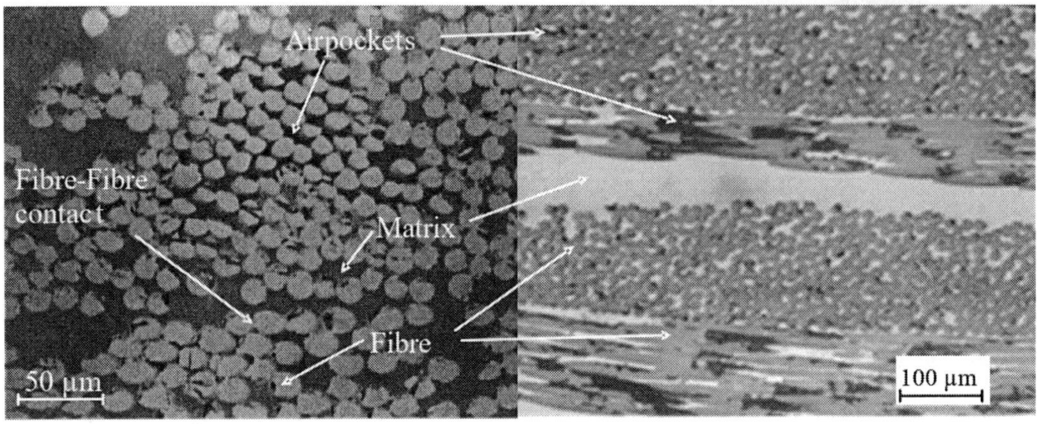

Figure 4. Microscopic view of defects investigated in composites made from commingling process (left) and Film-stacking process (Right). [Source ITA]

2. EXPERIMENTATION

As part of the research work, the intent is to develop a coating process that overcomes the challenges mentioned in all above process. The coating of glass filaments with a thermoplastic is integrated directly into the spinning process. These coated glass filaments can then be used to manufacture the fibre reinforced composites. The manufacturing of prepregs can thus be significantly simplified and the costs can be further reduced, since the processing step for subsequent coating can be omitted.

2.1 Construction of coating module

The schematic view of the coating on the glass spinning line is shown below in Figure. 5. In the figure it can be seen the position of the coating module in line with the glass fibre production. The important components of the coating module are shown in the figure with a schematic drawing of the coating module.

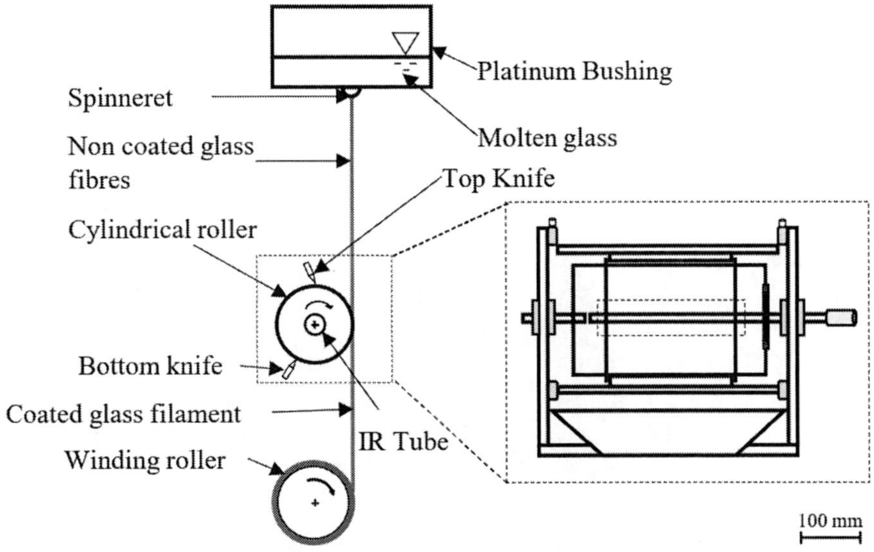

Figure 5. Schematic representation of coating module in line with glass fibre spinning.

The coating module consist of a cylindrical roller that rotates at same surface speed as the glass filaments are spinning as shown in Figure. 6. The cylindrical roller is rotated on a shaft which is divided into two sections. One section consists of a left side shaft which has the main drive connected to the motor. This drives the main cylinder at the required speed which matches the winding roller. The other section is a right side shaft or stationery cylindrical shaft which houses the electrical cable which is connected to the infrared tubes. The infrared tubes are also mounted on the right side cylindrical shaft. The infrared tubes heat up the cylindrical roller from inside which ensures the polymer viscosity and uniform coating of the glass filaments is maintained throughout the coating process.

The coating module is built on left and right vertical side plates. The vertical plates house both the cylindrical shafts in it with the help of bearings. The right side vertical plate has wide space with wedge plate connected to the stationary cylindrical shaft. The purpose of having wide plate is to facilitate maintenance work inside the cylindrical roller when the need arises.

There are two knives used in the coating module namely as top and bottom knives. Both the knives play an important role in the coating module. The top knife and bottom knife are fixed in a set of holders or clamping devices, knife connector and knife connector cap. The top knife is use to provide a uniform coating layer around the cylindrical roller. The amount of polymer that could pass through the knife and the cylindrical roller surface can be adjusted with the help of the linear guide device.

The bottom knife is used to clean the cylindrical roller from the polymer melt that is not used in the coating process. This would also ensure that a new polymer coat in applied on the cylindrical roller surface. Bottom knife is pressed against the cylindrical roller with the help of a spring mounted on the bottom knife connector cap. The pressed condition of bottom knife helps to clean the polymer completely from the cylindrical roller. The tip of bottom knife is made up of rubber which does not erode or damages the cylinder surface.

At the base of the coating module a bottom plate is placed on which the two vertical side plates are erected. Base plate serves the purpose of giving a platform to the entire coating module. There is also a funnel sheet placed at the bottom of the cylindrical roller. The purpose of the funnel sheet is to collect the polymer which is removed with the help of the bottom knife. The removed polymer can be collected and later can be reused. This would help to reduce waste and also ensure no damage to the parts which are placed at bottom of the coating module.

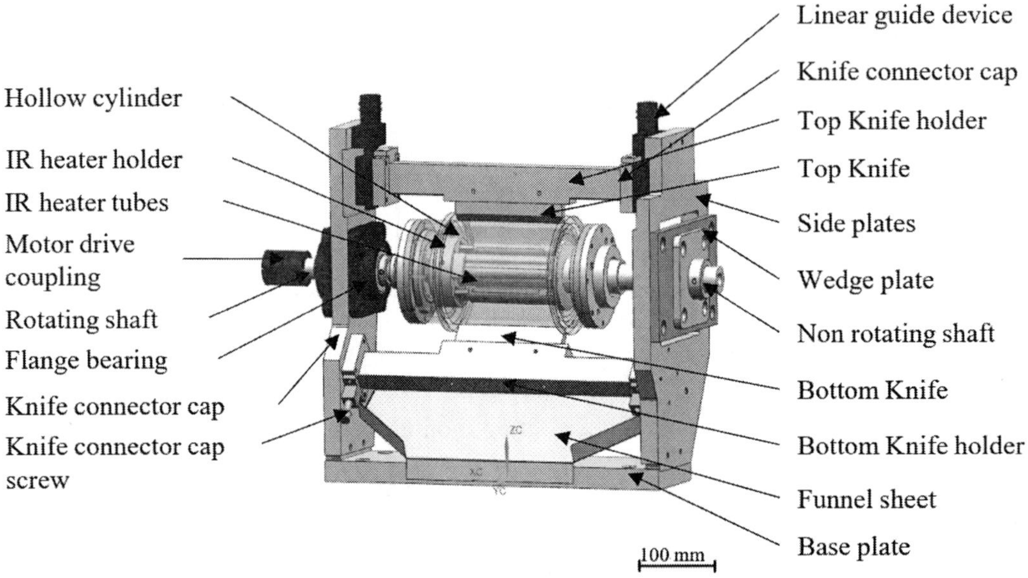

Figure 6. Pictorial view of coating module with individual parts.

3. RESULTS

The aim of this research is to develop a process chain for coating individual glass filaments during the glass fibre production. The aim is also to overcome the drawbacks offered by the current manufacturing techniques and processes. A novel coating application module is developed and a suitable coating material is selected according to the end application. Subsequently, the process is integrated into the institute's own glass fibre spinning system at ITA, RWTH Aachen University, to produce individual impregnated glass fibres.

From the performed tests and experiments, it can be seen that impregnation of individual glass filaments with the help of suitable polymer is possible. There are several tests performed and samples were analysed under the SEM and microscope. In the images Figure 7, the coating of glass filament with polymer is shown. The prepregs produced are examined by means of a scanning electron microscope (SEM). The images are examined for its pattern and the uniformity of the coating.

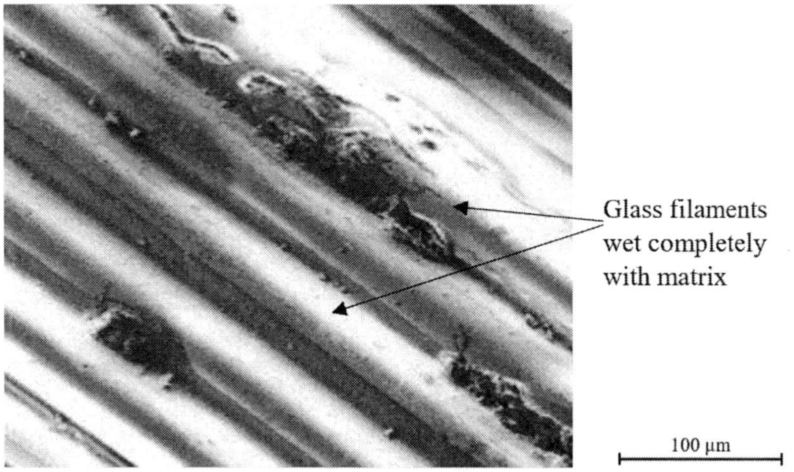

Figure 7. SEM image of coated glass filaments in composite material using the ITA developed coating process.

In Figure 8 below SEM image of single polymer coated glass filament is shown. It was carefully pulled out of the composite as shown in Figure 7. A core diameter of glass filament 10 μm with polymer coating of 0.9 μm is achieved.

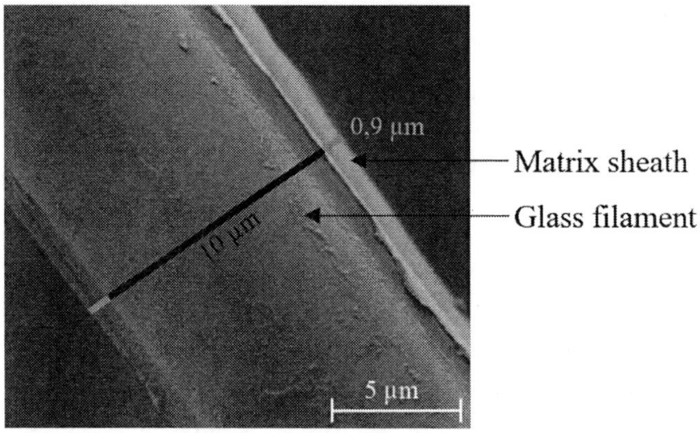

Figure 8. SEM image of single glass filament coated with polymer matrix.

4. CONCLUSIONS

The aim of research is to coat glass filament in line during the production of glass filament. With current research and experiments, it is feasible to realise coating of individual glass filaments in the production stage. The coating process of glass filaments at the production stage helps in shortening of fibre composite production by eliminating the intermediately process step of prepreg production. Coating the glass filaments at the production stage, helps in realizing the potential of every single glass filament which contributes to manufacturing the composites of GFRP. With coating of glass filament, the fibre volume content of composite increases.

Fibre coating is a breakthrough process to fulfil the requirements of automotive and the aerospace industry. It helps in achieving objectives of having lightweight composites material and at the same time has better mechanical properties as compared to the conventional composite components. Innovative products like fibre coated glass filaments with low price is essential in a highly competitive market. The fibre that we obtained is already coated with a polymer (pre-impregnated), thereby omitting intimidatory process and helps in shortening the entire process flow chain and saving the manufacturing cost.

5. REFERENCES

[1] Henninger, H Frank., Beitrag zur Entwicklung neuartiger Fertigungsverfahren zur Herstellung von Bauteilen aus kontinuierlich faserverstärkten Thermoplasten Kaiserslautern: Technische Univ. Dissertationsschrift: Kaiserslautern, 2005.

[2] Flemming, M., Ziegmann, G., Roth S., Faserverbundbauweisen: Halbzeug du Bauweisen 1, Heidelberg: Springer-Verl.,1996.

[3] AVK - Industrievereinigung Verstärkte Kunststoffe e.V. (Hrsg.): Handbuch Faserverbundkunststoffe / Composites - Grundlagen, Verarbeitung, Application, 4. Wiesbaden: Springer-Verlag, 2014.

[4] Cherif, C.: Textile Werkstoffe für den Leichtbau, Berlin, Heidelberg: Springer-Verlag, 2011

THE MISSING LINK: HOW COUPLING AGENTS ENHANCE MECHANICAL PERFORMANCE

Lee R. Gunning, Brian Kleinheinz, René Nagelsdiek
BYK USA Inc.
Wallingford, CT

ABSTRACT

Thermoset based composites have become an essential part of our daily life. Their benefits are not limited to current technologies but also in pioneering future trends. The increasing importance of electric vehicles, wind energy, and aviation results in a higher demand of lightweight but mechanically extremely robust and reliable reinforced materials. Though it is possible to improve single components such as resins, fillers, or fibers, there is a certain limit due to the influence of each individual component. In an alternative approach, the link between the resin matrix and a dispersed phase providing reinforcement (fillers, fibers) can be optimized. An easy way to achieve the desired coupling are additives which have the ability to cross-link with the resin and to form a bond to the filler or fiber. These additives are known as Coupling Agents. The structure of Coupling Agents needs to be adjusted according to the chemistry of the matrix and dispersed phase. The Coupling Agents of the BYK-C 8000 family solve this task by creating a strong bridge between resin and reinforcement, giving access to more durable and long-lasting materials and allowing higher freedom of design. In this article, we provide an overview about the general mode of action of coupling agents and illustrate their versatile potential by means of various application examples.

Corresponding author: Brian Kleinheinz (brian.kleinheinz@altana.com)

1. INTRODUCTION

1.1 Performance-limiting Interfaces

Over the last decades, composites based on thermoset materials have become an essential part of our daily life. They find application in established goods such as for example sanitary ware or sporting equipment. Moreover, they are also pioneering for future trend applications.

In many cases, composite materials are comprised of a combination of a cross-linked organic polymer (matrix) and inorganic materials, such as particulate fillers or fibers. While the inorganic material can also have the function of a cost-effective filler, it often improves surface quality, flame retardancy, or abrasion resistance of the final composite. Fibrous materials (like glass fibers or carbon fibers) help to significantly enhance the mechanical properties of the composite, such as tensile, flexural, and shear strength while contributing to a lower weight of the final part. Without this inorganic reinforcement, high-tech applications such as aviation, automotive industry, or wind energy production would not be possible using the desired light-weight materials.

Despite the beneficial nature of the inorganic components in the composite, the boundary between the matrix (resin) and the inorganic part may turn out to be the weakest link of the overall system, limiting its maximum performance. With respect to this boundary effect, though it may be possible to improve single components, such as resins, fillers, or fibers, there is a certain limit to the possible optimization of each individual component. Taking into consideration the relevance of the organic/inorganic boundary, an improved link between the resin matrix and the dispersed phase (fillers, fibers) can result in improved mechanical properties. An easy way to achieve the desired coupling are additives which have the ability to cross-link with the resin and to form a bond to the filler or fiber. These additives are known as Coupling Agents.

1.2 General Structure and Working Mechanism of Coupling Agents

Coupling Agents provide a connecting function between the organic resin matrix and particulate filler or fiber materials by establishing strong bonds to the surface of the filler/fiber and cross-linking into the matrix during the thermoset curing process. This strong connection between the organic and the inorganic components of the formulation results in increased mechanical properties of the composite material prepared. To become efficient, the Coupling Agent needs to be employed in a system where the weakest boundary of the composite is the interface between filler / fiber and resin (cf. Figure 1). This means a Coupling Agent cannot prove value if the material failure is caused by cohesive fracture of the resin or the inorganic component.

Figure 1. General working principle of a coupling agent in a composite material. The adhesion between resin and inorganic dispersed phase is strengthened.

As a rule of thumb, a Coupling Agent will contain at least two kinds of functional groups: (1) a reactive group which is able to copolymerize with the matrix during the curing process, and (2) a highly affinic group which is able to adhere strongly to the surface of the particle or the fiber (cf. Fig. 2). For instance, in a radically curing system (like an unsaturated polyester or acrylate system), the Coupling Agent likely comprises polymerizable double bonds. If the thermoset system is cured via a two-component mechanism (2-pack system, like most epoxy systems or polyurethane systems), the Coupling Agent may be reactive towards the resin or the hardener.

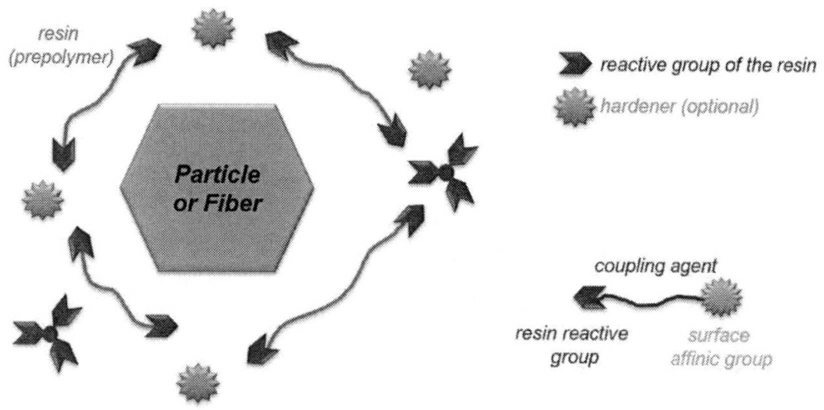

Figure 2. Thermoset system, inorganic reinforcement, and Coupling Agent before curing of the composite.

The general mode of action is shown in Figure 3 in which the composite system is shown after curing is completed. Surface affinic groups of the Coupling Agent adhere to the surface of the inorganic component(s) used. At the same time, the additive has cross-linked into the matrix.

Since the Coupling Agent works as a bridge builder between matrix and dispersed phase, it is apparent that this additive has to be designed for a certain combination of matrix and inorganic phase: There is no universal Coupling Agent that will provide benefit for any combination.

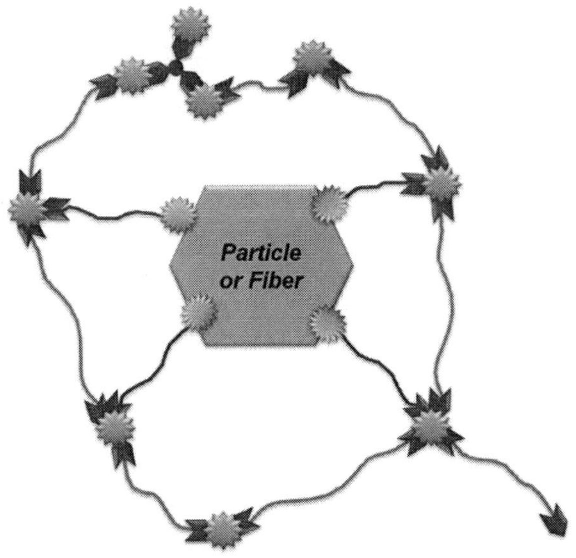

Figure 3. Thermoset system, inorganic reinforcement, and Coupling Agent after curing of the composite.

2. EXPERIMENTATION

2.1 Sample preparation

The processing of reinforced plastics depends primarily on the structural parameters of the reinforcement, for example length and shape. Short fibers or fillers are simply added to the polymer mass. Long fibers must be placed thoroughly to guarantee an optimum of mechanical properties.

In this review, the following three processing methods are used:

1. Casting process for fillers as reinforcement
2. SMC process for short fibers (a) and unidirectional fabrics (b) as an example for closed mold applications
3. Infusion molding process for unidirectional fabrics, e.g., for cold cure applications.

1. Casting process: All liquid components are prepared as masterbatch and the fillers are filled into a separate container. The resin system is weighed into a new cup and the additive is added and mixed in by hand. The curing agent is also added and homogenized. The filler is mixed in by using a dissolver for 90 s. The formulation is then filled into an appropriate die (10 cm x 3.5 cm x 20.5 cm) and distributed homogenously to guarantee consistent thickness of the samples. After curing of 1 hour at room temperature the formulation is cured at 60 °C overnight. The following day specimens can be cut according to DIN EN ISO 178.

2. (a) SMC production with chopped fibers: All liquids (including the coupling agent) and an anti-shrink component are weighed into a metal container and dispersed starting with a rotation speed of 700 rpm (circumferential speed: 4.58 m/s) and then ramped up to 2000 rpm (circumferential speed: 13.09 m/s), finished when reached 35 °C. Last step is to add the thickener to the paste. The paste is dispersed again with 1500 rpm (circumferential speed: 11.78 m/s) for 1 minute to reach a homogeneous distribution of the thickener. The SMC line is started, and the paste is split into the two doctor boxes. The cutter is started, and SMC is produced. The SMC stored for at least 1 day at 35 °C.

2. (b) SMC production of unidirectional fabrics: Fabrics are cut into a 60 x 60 cm^2 layer and every SMC is produced with four layers for each panel. Two different SMCs (control and additive sample) are produced as in the following: All liquids and an anti-shrink component are weighed into a metal container and dispersed starting with a rotation speed of 700 rpm (circumferential speed: 4.58 m/s) and then ramped up to 2000 rpm (circumferential speed: 13.09 m/s), finished when reached 35 °C. Last step is to add the thickener to the paste. The paste is dispersed again with 1500 rpm (circumferential speed: 11.78 m/s) for one minute to reach a good distribution of the thickener. The SMC line is started, and the paste is split into the two doctor boxes. Before the first package (with four layers of fabric) is inserted, the glass cutter is started and some glass is drizzled onto the paste (as a barrier, so that no extra paste is squeezed back on to the fabric when it runs through the impregnation area of the line). Directly afterwards, the package is inserted. At the end of the SMC line, each package is cut off separately and placed flat onto a table. After all the packages have passed through, it is packed as one and stored for one day at 35 °C in a heated storage room.

The coupling agent can be added to the composite in several ways:

I. To the fiber sizing or the resin before application

II. To the fiber before application

III. During SMC production

Method II is a new approach (Second Sizing) of adding a coupling agent to the system. In this case the additive is sprayed onto the dry fiber prior to application.

After the production of the SMC and a storage time of at least 24 hours, the panels are molded. The mold coverage depends on the SMC, here 94% mold coverage of the tool surface when using unidirectional fibers and 80% when molding chopped fiber SMC. Parameters for the molding process are: pressure: 1600 KN (133 bar); closing speed: 2.0 mm/s; molding time: 120 s, temperature: 150 °C.

After molding the panels, the following test specimens are cut:

- 250 mm x 25 mm Transverse Tensile Strength according to DIN EN ISO 527-5
- 80 mm x 15 mm Flexural Strength (transverse) according to DIN EN ISO 14125

3. The infusion mold is prepared with sealing in the notches and polished with release agent. The fabrics (four layers) are cut into 380 x 380 mm^2 sheets and placed symmetrically in the mold (orientation of the fabrics: 90°/0°, 0°/90 // 90°/0°, 0°/90). A distance frame (2.5 mm) is used to set the fiber volume/weight content via the thickness of the panel. Four hoses on the bottom of the mold are used for the infusion, vacuum is applied from the top. Vacuum must be below 10 mbar and should not rise over 25 mbar after closing the valve. The formulation is filled into a bucket and mixed in a vacuum dissolver for 1 min at 500 rpm and then for 9 min at 250 rpm. If an additive is added, this is done to the resin or hardener before mixing the complete formulation. After mixing, the bucket is placed beneath the mold. The resin is slowly infused into the mold. After infusion and storing 1 h at RT, the whole mold with hoses is placed in a convection oven and cured for 8 h at 40 °C to guarantee a demolding. After pre-curing, the panel is demolded and placed in the oven to post cure for 8 h at 80 °C. The panel is cut into 10 specimens (250 x 25 x 2.5 mm^3) for transverse tensile testing according to DIN EN ISO 527-5B.

2.2 Measurement of mechanical properties

The adhesion of the reinforcement to the matrix can be tested in several different ways. The most common test method is mechanical testing, such as tensile, flexural or shear strength measurements.

Tensile testing is a destructive test procedure that gives information on the tensile strength, tensile Young's modulus and strain. This testing method measures the force that is needed to break the composite specimen and the elongation to that breaking point. Tensile properties for fiber reinforced composites are tested according to DIN EN ISO 527-4, DIN EN ISO 527-5, ASTM D 638 and ASTM D 3039. The alignment of the sample is critical, as composites are anisotropic and generally brittle. This means the properties of the material depend on the applied force. Tensile strength is very high when tested in fiber direction but does not give an indication on the bonding of the reinforcement and matrix.

For measurement of the flexural properties the tests are carried out according to DIN EN ISO 14125 or DIN EN ISO 178. The specimen is placed down horizontally, and a force is applied to the top of the material through a one-point contact until failure. For composites with unidirectional reinforcements this is done parallel to the fiber to test the adhesion between fiber and matrix.

Fatigue testing can also be done to determine the bonding of the reinforcement to the matrix, especially in fiber reinforced composites. This test method is very complex and time intensive.

Prior to measurement of mechanical properties, the specimen is stored at 23 °C and a relative humidity of 50 % for a period of 24 h.

Visual tests help to identify the residue of resin on the filler or fiber. This gives an identification of the bonding of the matrix to the reinforcement. Electron microscopic pictures show leftovers of the cured system on the surface of the filler or fiber, which underlines a failure of the composite in the matrix and not on the interface. If reinforcements removed from a composite do not expose any resin leftovers at their surface, it can be assumed that the adhesion between the resin matrix and the fiber leaves room for improvement.

The analysis of the fracture was used to determine the cause of failure in the composite material.

In this paper tensile tests according to DIN EN ISO 527 and flexural testing in accordance to the standard DIN EN ISO 178 are mainly carried out.

3. RESULTS

3.1 Quartz-filled unsaturated polyester system as an example for a casting application

Typical applications for quartz-filled unsaturated polyester systems are for example polymer concrete and artificial marble. In these systems, it is desirable to increase the mechanical performance for a given part or to reduce the thickness of the construction part while having the same level of mechanical properties (save material, cost reduction). Both can be achieved by using a coupling agent. As can be seen in Diagram 1, the flexural strength can significantly be increased using a dosage as low as 0.2 wt.% (based on filler). Once the ideal amount of connecting spots has been promoted by the coupling agent, a further increase of its dosage cannot result in higher mechanical properties.

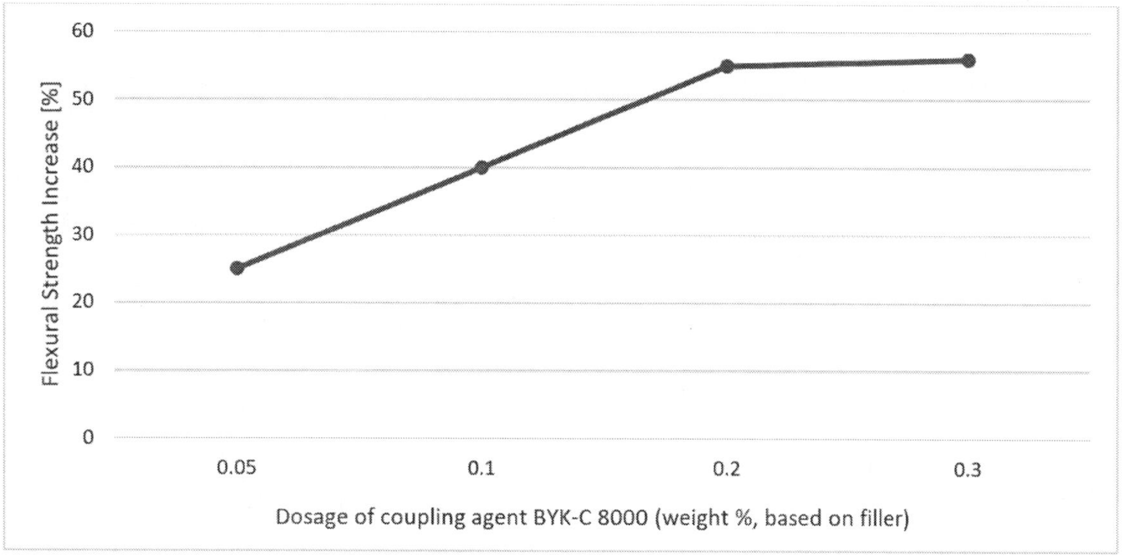

Diagram 1: Relative improvement of the flexural strength in a quartz filled unsaturated polyester system as a function of the dosage level of the coupling agent.

This effect can also be visualized via microscopic imaging (cf. Fig. 4): In the presence of the coupling agents, cavities are minimized, resin and filler are connected to each other in a robust manner.

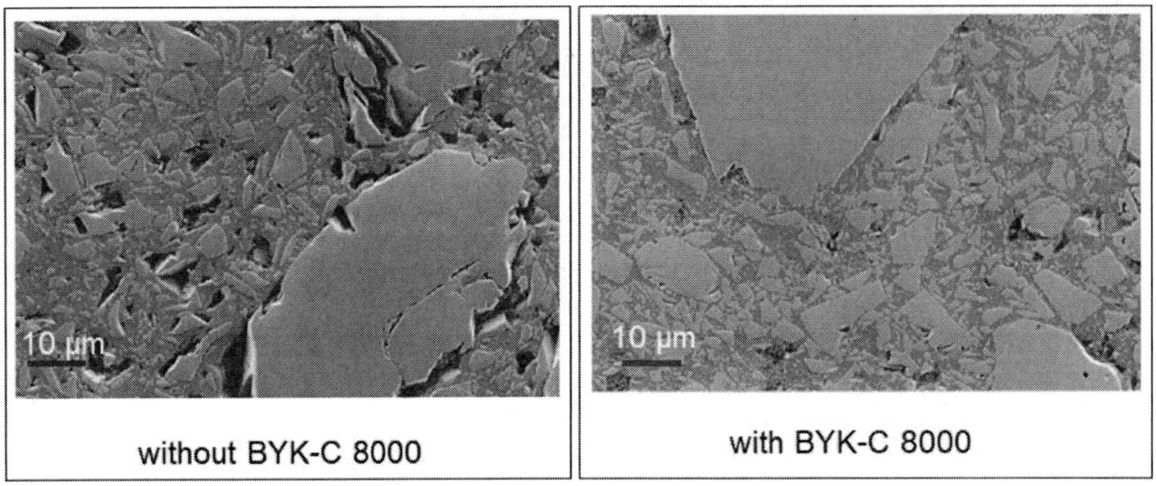

Figure 4: Electron microscopic image of a quartz filled unsaturated polyester resin after exposure to hot water in the absence (left) and presence (right) of a coupling agent.

In addition to mechanical improvement, the systems equipped with the coupling agent can provide increased hot water and chemical resistance. This prolongs the lifetime of a construction part and minimizes the formation of blisters when exposed to humidity and temperature.

3.2 Carbon-fiber reinforced vinyl ester system as an example for SMC

Carbon fiber reinforced plastics are used when the need for high material strength meets the demand for lightweight properties of the composite parts. Examples for these applications comprise aerospace and general automotive industry, electro mobility, and wind energy applications.[1]

Appropriate coupling agents can provide benefits for flexural and tensile strength both for chopped fibers and fabrics (cf. Diagram 2).

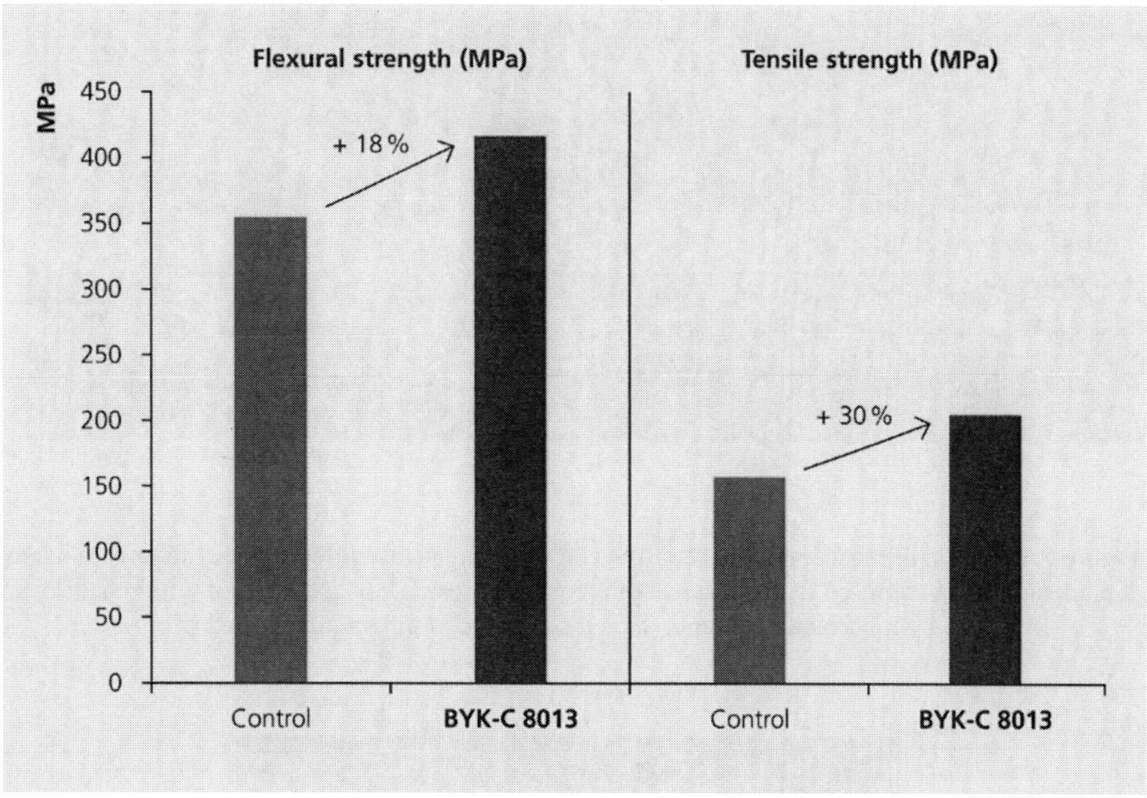

Diagram 2: Improvement of mechanical properties in a carbon fiber reinforced vinyl ester system using a coupling agent.

Again, the mechanical improvement was visualized by electron microscopy (cf. Figure 5): In the absence of a coupling agent, only few resin leftovers can be detected at the fiber surface after material failure; this clearly underlines a poor adhesion between fiber and matrix. In contrast, in the presence of a suitable coupling agent, significant amounts of resin residues can be seen at the fiber surface. In some cases, when using the coupling agent, the adhesion between carbon fiber and matrix is so strong that a fiber fracture occurs (cf. Figure 6).

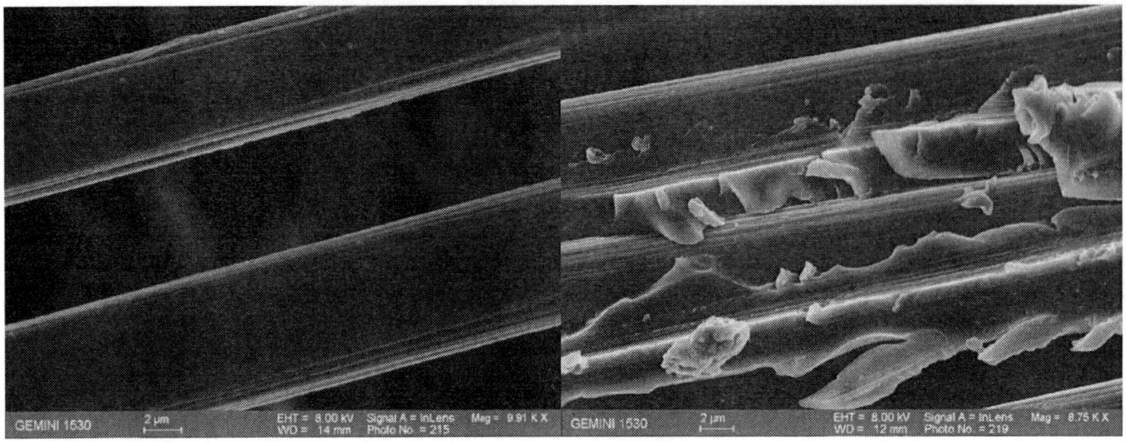

Figure 5: Electron microscopic images of the carbon fiber surface after material failure. Whereas a plain surface is observed in the absence of a coupling agent (left), significant leftovers of resin on the fiber surface are found when a suitable coupling agent is used (right).

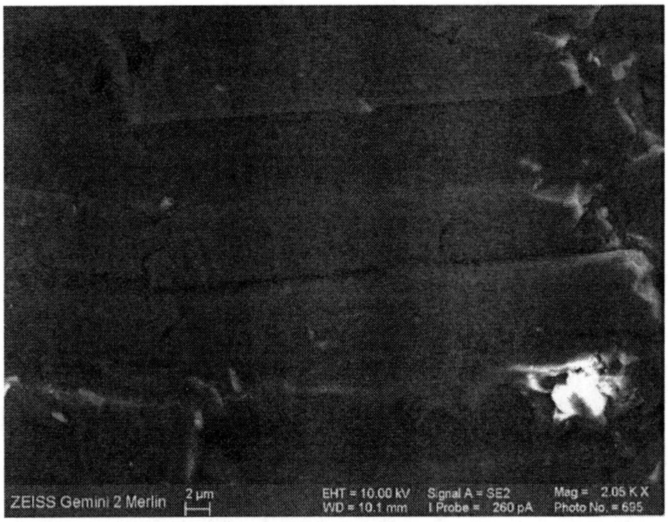

Figure 6: An extremely strong adhesion between carbon fiber and vinyl ester matrix has been promoted using a coupling agent. Material failure resulted in a fiber fracture instead of an adhesive fracture of the specimen.

3.3 Glass fiber reinforced epoxy system as an example for infusion molding

The most commonly used fibrous material to reinforce composites are glass fibers. As shown for carbon fiber reinforced materials, the transverse tensile strength can be significantly improved by addition of an appropriate coupling agent. Mechanical improvements found in different resin systems are shown in Diagram 3.

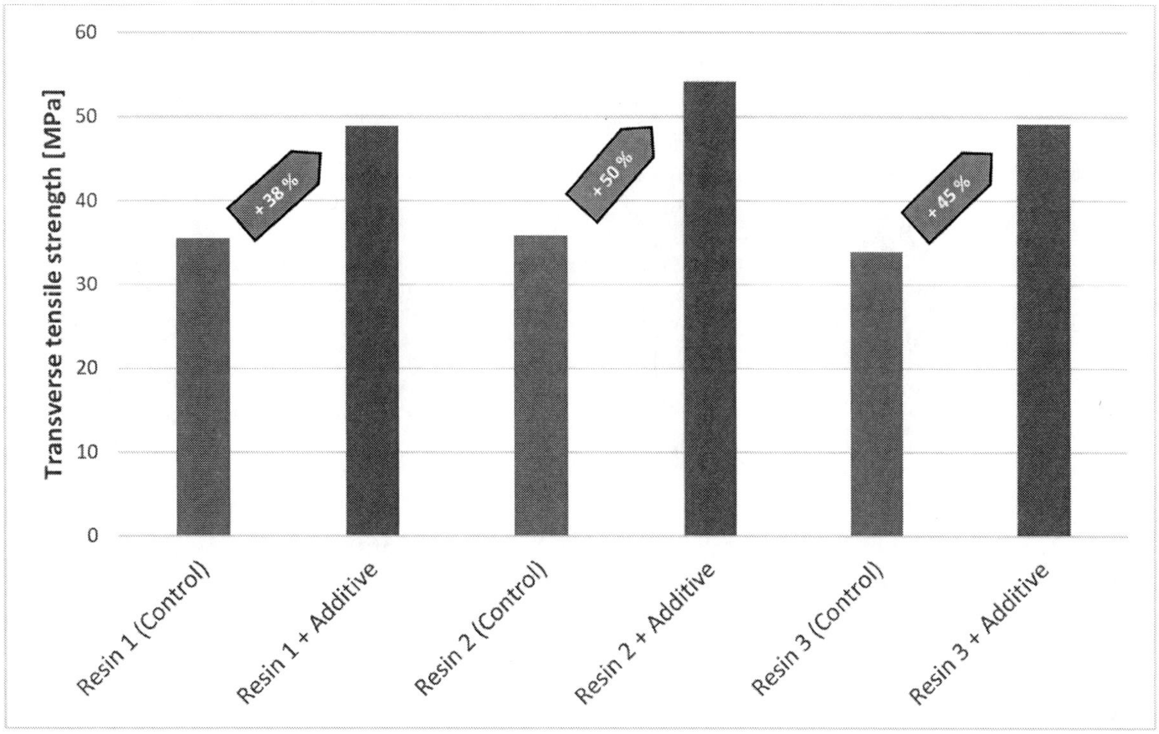

Diagram 3: Increase of transverse tensile strength in various glass fiber reinforced epoxy composites.

As an additional effect to the general improvement, glass fiber sizings deteriorate with age, leading to inferior performance of the final system. Again, the coupling agent can assist to compensate this aging process, yielding composites with similar properties like those based on virgin fiber material. This effect helps to improve process robustness for the composite manufacturer due to higher reliability and lower batch variations of the mechanical properties of the finished part as well as the final construction.

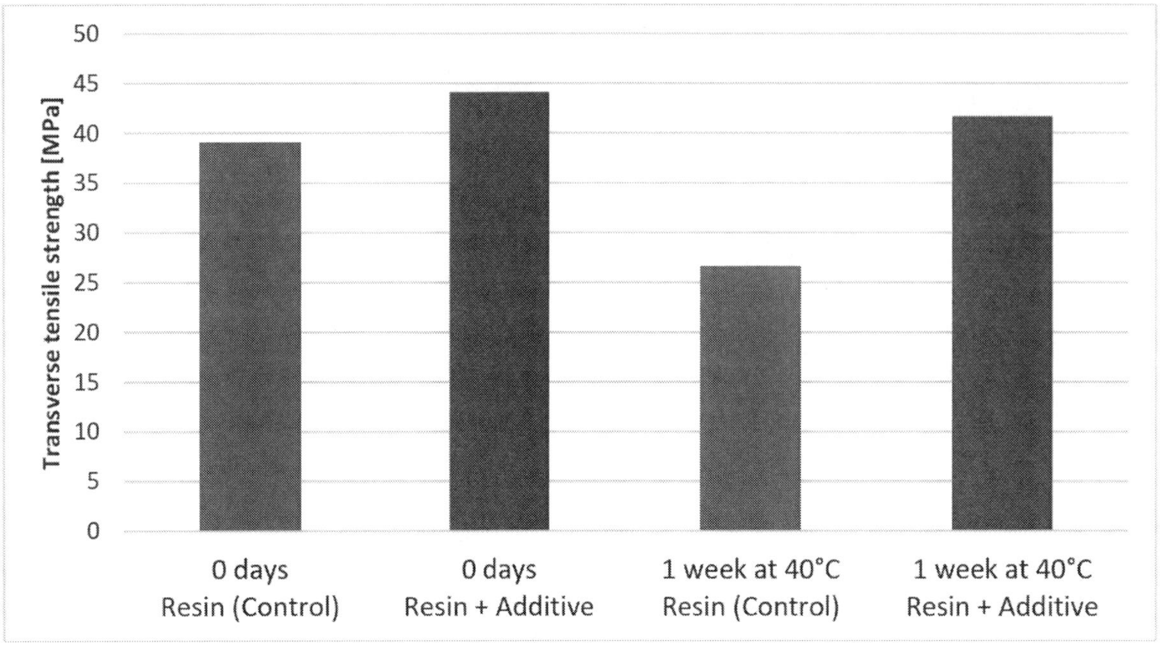

Diagram 4: Transverse tensile strength of a glass fiber reinforced epoxy resin before and after aging at elevated temperature (light columns: no coupling agent; dark columns: coupling agent employed)

3.4 Overview of Suitable Product Combinations

There is a multitude of resins systems with different curing mechanisms and characteristics. Most of these can be combined with various fillers and/or fibers. This results in a vast number of material compositions and therefore in numerous possible boundary combinations. To ensure a good bonding link for all these reinforcement/matrix combinations, a variety of coupling agents represents the toolbox. The following table summarizes suitable coupling agents which can be employed in different systems to provide improved mechanical properties and reduced quality fluctuation.

Table 1: Recommended coupling agents for different combinations of resins and reinforcements.

Mechanism	*Resin Systems*	*SiO_2*	*ATH, $CaCO_3$*	*Glass Fiber*	*Carbon Fiber*
Radical Curing	UP, VE	**BYK-C 8000**	BYK-C 8002	**BYK-C 8003**	**BYK-C 8013** **BYK-C 8014**
	Acrylate	BYK-C 8000	**BYK-C 8002**	BYK-C 8003	BYK-C 8013
Polyaddition	Epoxy	**BYK-C 8001**		**BYK-C 8001**	

4. CONCLUSIONS

The mechanical properties of composites are often improved by enhancements of their components. Though it is possible to improve single components, such as resins, fillers, or fibers, there is a certain limit due to the influence of each individual component. In an alternative approach, the link between the resin matrix and a dispersed phase providing reinforcement (fillers, fibers) can be optimized. An easy way to achieve the desired coupling are additives which have the ability to cross-link with the resin and to form a bond to the filler or fiber. These additives are known as Coupling Agents; their structure needs to be adjusted according to the chemistry of the matrix and dispersed phase. The Coupling Agents of the BYK-C 8000 family solve this task by creating a strong bridge between resin and reinforcement, giving access to more durable and long-lasting materials and allowing higher freedom of design.

Coupling Agents have to be tailored for a certain matrix/reinforcement combination by adjustment of functional groups suitable for the very application system. Their improving properties can be of advantage to boost cost/performance levels of certain application systems. Therefore, the use of Coupling Agents provides access to both new technical solutions and cost advantages as well.

Coupling Agents can be used at different stages in process, such as (i) during the production of reinforcements (surface treatment of fillers and fibers), (ii) a part of the resin formulation (resin additive use), and (iii) upon composite production (e.g., as second sizing or directly before or during application). They can therefore be employed by all members of the value chain to differentiate from the state of the art, including manufacturers of fillers and fibers, resin houses, and composite producers.

5. REFERENCES

1. R. Nagelsdiek and L. R. Gunning, *Reinforced Plastics* **2019**, *63(2)*, 97. [DOI: 10.1016/j.repl.2018.09.001]

CUSTOMER FRIENDLY NOVOLAC EPOXY VINYL ESTER RESINS

Kevin R. Lambrych, Lisa M. Adkins, and Achille Bivigou Koumba
INEOS Composites – INEOS Corrosion Science Center
5220 Blazer Parkway
Dublin, OH 43017

ABSTRACT

Novolac epoxy-based vinyl ester resins (NEVERs) were commercially introduced in the early 1970's to deliver improved thermal performance and improved chemical resistance properties as compared to bisphenol-A epoxy vinyl ester resins (EVERs). They provide high resistance to solvents, acids and oxidizing substances such as chlorine. They also offer high retention of strength and toughness at elevated temperatures.

A key reason NEVERs have improved thermal and chemical resistance is due to their higher crosslinking density. Every molecule in a NEVER has more reactive sites than a molecule in an EVER. As a byproduct of this increased functionality, NEVERs are more reactive. In thicker laminate constructions or at elevated shop temperatures, this increased reactivity makes NEVERs more difficult to process due to higher heat development during the cure. Fabrication efficiency is compromised as operators are limited in the number of layers of reinforcement that can be laid up during one application.

This paper will touch on the history of NEVERs and review a new novolac epoxy vinyl ester resin technology called Derakane™ Signia™ 470 resin. This new NEVER technology offers improved cure characteristics that allow fabricators to lay-up thicker parts in a single step, thus improving fabrication efficiency. This resin also has reduced styrene emissions, better surface cure characteristics, and greatly reduced grinding requirements. Together, these benefits provide a greatly improved customer experience. The new resin is built upon the same polymer backbone as previous generations of INEOS Derakane™ 470 type NEVERs so the mechanical, thermal and corrosion performance is unchanged.

General mechanical properties, corrosion performance (ASTM C581 testing) and secondary bonding data will be presented to establish that this new range of resins maintain the same level of performance as previous generations of novolac epoxy vinyl ester resins. Performance data from the lab and the field demonstrating improved processing and fabrication efficiency will also be presented.

Keywords: Corrosion, thermoset, epoxy vinyl ester resin, novolac, EVER, NEVER, composite, fiber reinforced polymer, FRP, glass reinforced polymer, GRP, fabrication, styrene emission

Corresponding author: Kevin R. Lambrych (krlambrych@ashland.com)

1. INTRODUCTION

Epoxy vinyl ester resins have been used in anti-corrosion applications since the early 1960's. The first commercially available resins were based on epichlorohydrin bisphenol-A epoxy co-reacted with methacrylic acid and thinned in styrene. The challenge in producing these early vinyl ester resins was preventing resin gelation in the reactor during production and establishing an acceptable shelf-life. Advances in resin manufacturing knowhow eventually led to improved EVER shelf-life, usability and performance.

After introduction of the bis-A EVERs, it became apparent that a better resin was needed for solvents and highly oxidative chemical environments. This led to the invention of novolac epoxy-based vinyl ester resins in the late 1960's. The first commercially available NEVER was introduced under the Derakane™ 470 resins (D470) name in 1972. With improved crosslinking, this new family of vinyl ester resins allowed fiber-reinforced polymer (FRP) to be used in higher temperature applications. However, this improved performance comes at a cost. NEVERs are well known for their high reactivity when compared to EVERs thus making good resin manufacturing knowhow and stability technology even more critical. Since the introduction of NEVERs in 1972, there have been very few leaps of new resin production and stability technology. The most recent new technology, Derakane™ Signia™ (DS), was introduced in 2018. Prior to that, the last significant new vinyl ester technology was Derakane™ Momentum™ (DM) in 2000.

In our 2018 CAMX paper, "New Epoxy Vinyl Ester Resins for Use in Corrosion Resistant Applications," we introduced the new DS resin technology.[1] DS combines improved shelf-life with styrene suppression technology, reduced need for grinding and cure characteristics that allow for more efficient FRP fabrication. The ability for DS technology to improve a shop's environment and drive production efficiency is discussed in the 2019 CAMX paper, "Leveraging Modern Resin Technology for Shop Efficiency Improvements."[2] In these two papers, we focused on bis-A type EVER, Derakane™ Signia 411 resin (DS411). In this paper, we will demonstrate how the DS benefits have been incorporated into the D470 series novolac epoxy vinyl ester resins. D470 resins combine corrosion resistance with superior high temperature performance. They deliver outstanding resistance to oxidation and mixtures of chemicals, including solvents. In the following sections, we touch on how the new Derakane™ Signia 470 resin (DS470) is user friendly and provides the legacy performance design engineers and owners of corrosion resistant FRP equipment require.

1.1 For FRP Fabricators

Today fabricators are challenged with finding qualified and experienced operators to meet current order demands. DS resins were designed to deliver improved processing characteristics to drive faster laminate consolidation, lower foaming, and greatly reduced sanding for application of secondary laminations. These features, combined with reduced styrene emissions, faster Barcol development upon cure and reduced odor from finished parts, leads to greatly improved shop efficiency and cleanliness resulting in a much more desirable workplace.

The DS resins also offer significantly improved resin stability, allowing for an extension of the commercial warranty period. This is particularly attractive to fabricators seeking to comply with warranty requirements in specifications.

1.2 For Designers and Engineers

Performance is critical to process design engineers. DS470 resin is built on an unchanged polymer backbone so engineers can be assured that all previous corrosion studies, historical data and field case histories remain relevant to demonstrate performance for their design requirements. Combined with improvements for the fabricator, this development is expected to contribute to the availability of better, more efficiently produced and environmentally friendly FRP equipment.

1.3 For Asset Owners and Equipment End Users

DS resins also incorporate a unique proprietary detection system that can conclusively verify a DS resin was used to fabricate a specific piece of FRP equipment. This provides greater assurance that the asset owners equipment design specification has been followed for desired performance and service life expectations.

2. EXPERIMENTATION

2.1 Fourier-Transform Infrared Spectroscopy

Fourier-transform infrared spectroscopy (FTIR) was used to compare the polymer backbone of DS470 and DM470 resins. Neat resin castings of each resin were prepared, cross sectioned, polished, and analyzed. FTIR spectra were acquired with an Agilent 670 spectrometer and an Agilent 620 IR microscope following ISO-9001 SOP SAM-3461 requirements.

2.2 Test Laminate Preparation

Test laminates were made using DS470 and DM470 to compare mechanical properties. Laminate construction was V/M/M/Wr/M/Wr/M where V is C-glass veil, M is chopped strand mat (450 g/m^2 (1½ oz/ft^2)), and Wr is woven roving (800 g/m^2 (24 oz/yd^2)) for a final thickness of 6 mm (¼ in). Laminates were cured using cumene hydroperoxide (CHP) and cobalt naphthenate (CoNap). The cure schedule was 24 hours at room temperature and 6 hours at 80°C (175°F). Flex and tensile mechanical properties were determined using ASTM methods and Instron mechanical testing equipment[3,4,5].

2.3 ASTM C581 Screening[6]

ASTM C581 screening was performed according to the standard and accepted methods. Four chemical environments were selected – deionized water (DI) at 80°C (180°F), 15% hydrochloric acid (HCl) at 80°C (180°F), 15% sodium hydroxide (NaOH) at 50°C (122°F), and toluene at 50°C (122°F). In this testing DS470 was compared to DM470. Corrosion coupons were constructed using one ply of C-glass veil followed by three plies of 450 g/m^2 (1.5 oz/ft^2) chopped strand mat, followed by another ply of C-glass veil. The coupons were edge coated with a mixture of the resin and fumed silica. The retention of properties such as Barcol hardness, flex strength, flex modulus, and general appearance were evaluated after one month of exposure. Coupons will also be evaluated after 3, 6, 9, and 12 months.

2.4 Secondary Bonding Test Laminate Preparation

Fiberglass-reinforced laminates were prepared for fiber tear evaluation using standard hand lay-up preparation methods. Laminates were prepared in the INEOS Corrosion Science Center in Dublin, Ohio, and also at various customer locations. Several resin gel times based on methyl ethyl ketone peroxide (MEKP), CHP, CoNap, and dimethylaniline were used. To understand the effect of reactive mass on cure and secondary bonding fiber tear, thick and thin laminates were prepared. Primary laminates were fabricated as a representative FRP substrate, aged, subjected to surface preparation or no preparation, and finally a secondary lamination was applied to the primary laminate. The final laminate constructions were split at the interface between the primary and secondary layers using an INEOS established delamination wedge test using an Instron Model 5582 in compression setup. The degree of fiber tear due to the failure of the secondary bond was evaluated according to an established scale.

2.4.1 Thick Laminate Preparation

Primary substrate laminates were constructed using 7 plies of 450 g/m^2 (1½ oz/ft^2) chopped strand mat and promoted and initiated epoxy vinyl ester resin. The primary laminates were cured at ambient conditions for 24 hours. In some cases, primary laminates were post cured for 4 hours at 82°C (180°F) to simulate extended aging. A secondary layup of 7 plies of 450 g/m^2 (1½ oz/ft^2) chopped strand mat and promoted and initiated epoxy vinyl ester resin was applied on top of the primary laminate. For ease of driving the wedge between the primary and secondary layers, a one-inch border of tape was applied to the primary laminate before application of the secondary layer (Figure 1). The laminates were cured at ambient conditions (17°C to 26°C) for a minimum of 72 hours, followed by separation of the primary and secondary layers to evaluate bond peel resistance and fiber tear.

2.4.2 Thin Laminate Preparation

Thin laminates were prepared as described above in the Thick Laminate Preparation section, with three plies of 450 g/m^2 (1½ oz/ft^2) chopped strand mat being used in place of 7 plies.

2.5 Dimensional Stability of Thick Laminate Construction

To evaluate the cure characteristics of DS470 and their effect on warping, pullback or burning due to excessive exotherm temperatures, thick laminates were produced at customer trials and in lab conditions with several applications of reinforcement applied in a single step. Gel times between 30 and 50 minutes were used for these studies with CHP or CHP/MEKP/CoNap cures. The laminates were made with alternating layers of 12 plies of chopped strand mat (450 g/m^2 (1.5 oz/ft^2)) and 11 plies of woven roving (800 g/m^2 (24oz/yd^2)) for a final thickness of approximately 18 mm (0.7 in). Initial test laminates were 46 cm x 46 cm (18 in x 18 in) where later test laminates are 30 cm x 24 cm (12 in x 24 in) to promote the chance of warp or twist due to asymmetric shrinkage.

2.6 Styrene Emission Suppression Testing

Low styrene emission testing was performed, and the results were calculated according to rules promulgated by U.S. EPA regulations establishing the RPC MACT (Subpart WWWW). Test methods used are defined on page 46, Appendix A to Subpart WWWW – Test Methods Vapor Suppressant Effectiveness Test Protocol.[7]

2.7 Styrene Emission Testing by Headspace Gas Chromatography

The amount of styrene emitted by neat castings of cured resin was measured using dynamic headspace gas chromatography (HSGC) with a mass spectrometry detector (MSD).

2.7.1 Dynamic HS/GC/MS Analysis

Samples were prepared as described below. Each vial was heated at 70°C (160°F) for 10 minutes. After equilibration, dual needles were inserted into the headspace and the entire headspace volume was swept onto a Tenax trap maintained at 28°C (83°F). The headspace volume was continuously swept for an additional two minutes. After dynamic headspace sampling, the Tenax trap was thermally desorbed onto a 25-meter fused silica HP-624 capillary column that was attached to a mass selective detector. Resin samples were analyzed 24 hours and 120 hours after cure.

2.7.2 Sample Preparation

Test specimens were prepared by adding approximately 20 mg of promoted and initiated resin with a 15-minute cure to a 22 mL headspace vial. Unsealed vials were allowed to sit for 24 or 120 hours at 25 °C (75 °F). Prior to HSGS analysis, the vials were flushed with clean air for 10 seconds then capped, trapping all unreacted styrene. A typical sample within a 22 mL headspace test specimen is represented in Figure 1.

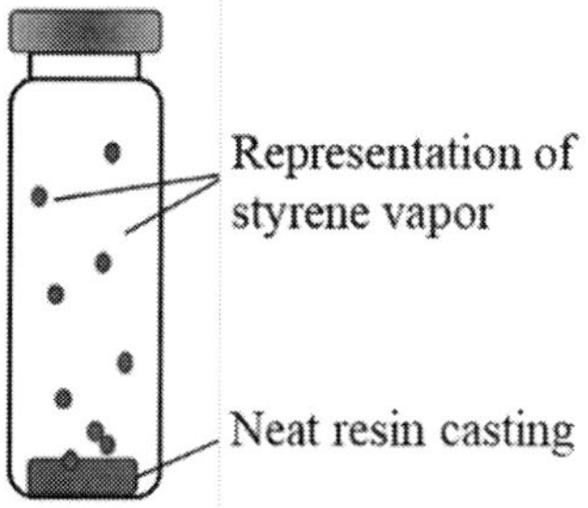

Figure 1. Sealed headspace gas chromatography vial with cured resin casting.

3. RESULTS

3.1 FTIR Spectra Comparison

The FTIR spectra of DS470 and DM470 resins casting were compared to confirm the polymer backbone was unchanged, Figure 2. We also did comparison by ^{1}H and ^{13}C nuclear magnetic resonance (NMR) to confirm the unchanged polymer backbone. However, due to the proprietary nature of this information, these spectra cannot be shared. In Figure 2, the overlaid IR spectra are identical indicating the polymer backbones of the new DS470 and the previous generation DM470 are the same.

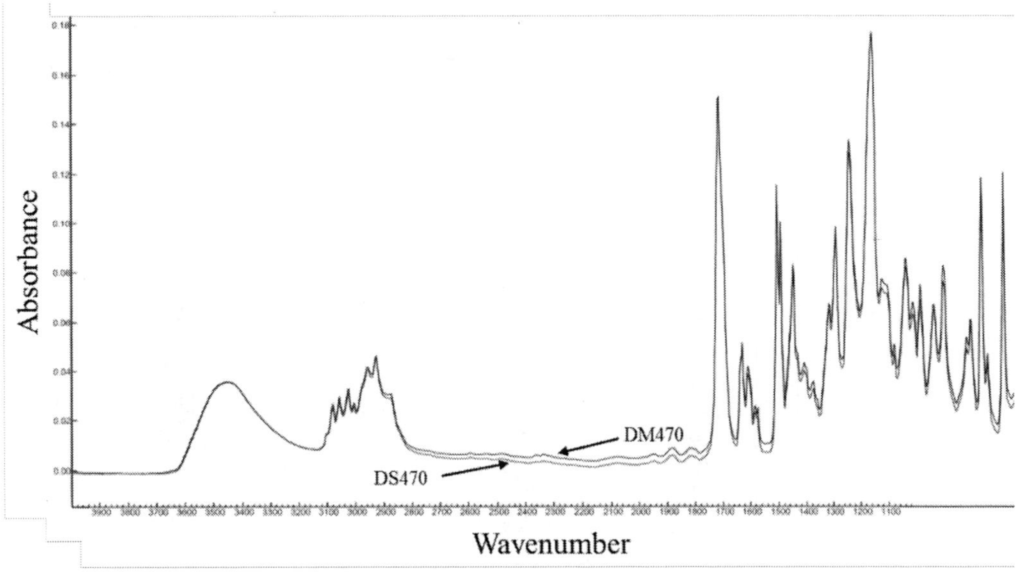

Figure 2: FTIR Spectra of DS470 compared to DM470

3.2 Laminate Mechanical Properties Comparison

Laminate mechanical properties were measured to confirm performance is unchanged for DS470 versus the previous generation of NEVER technology, DM470. The values are reported in Table 1. Taking standard deviation into account for each of the test methods, the resins perform essentially the same.

Table 1. Mechanical properties of DS470 and DM470 laminates. Laminates were cured with CHP and CoNap for 24 hours at ambient conditions and post-cured.

Laminate Property	DS470	DM470	Unit (SI)	Method
Tensile Strength	179	172	MPa	ASTM D638
Tensile Modulus	12.9	12.8	GPa	ASTM D638
Flexural Strength	209	190	MPa	ASTM D790
Flexural Modulus	9.8	9.1	GPa	ASTM D790
Neat Resin Casting Heat Distortion Temperature	150	150	°C	ASTM D648

3.3 ASTM C-581 Test Results

Four basic chemical environments were chosen for ASTM C-581 corrosion testing to compare DS470 and DM470 corrosion resistant properties, Figure 3. The chemical environments evaluated were deionized water (DI) at 80°C (180°F), 15% hydrochloric acid (HCl) at 80°C (180°F), 15% sodium hydroxide (NaOH) at 50°C (122°F), and toluene at 50°C (122°F). At the time publication, only 30-day results were completed due to the constraints of the CoVid-19 crisis. Corrosion testing is continuing and will be reported in future updates of this paper. Looking at Figure 3, DM470 and

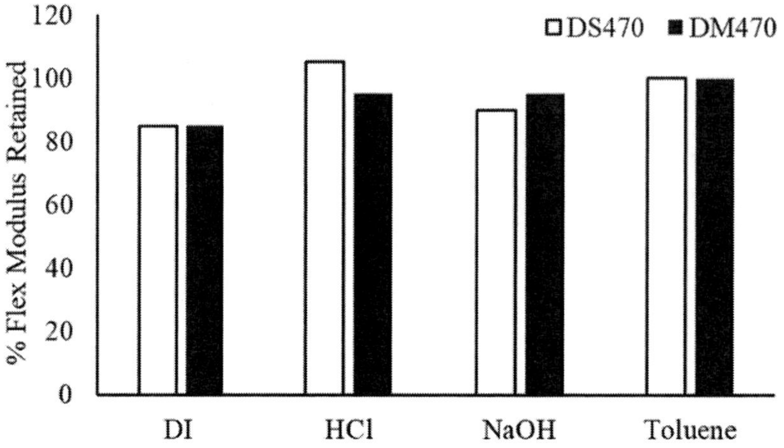

DS470 perform similarly which is not unexpected given they share the same polymer backbone.

Figure 3. ASTM C581 corrosion study comparing the performance of DS470 and DM470.

3.4 Secondary Bonding Test Results

Secondary bonding by wedge or peel test is difficult to evaluate as it is largely based on subjective observations of the observer/operator. A wedge test standard was developed by INEOS Composites Technical Service for previous secondary bonding studies. Figure 4 shows the relative level of fiber tear at 20, 60 and 100%. For safe separation of the primary and secondary layer, an Instron Model 5582 outfitted with a wedge fixture in compression configuration was used.

Figure 4. Secondary Bonding Wedge Test example scale.

The tested specimens shown in Figure 5 exhibited excellent fiber tear and are representative of the fiber tear observed with specimens made in fabrication shops during DS470 trials and in the INEOS Corrosion Science Center.

Figure 5. Typical secondary bonding fiber tear results for DS470 laminates.

Similar to results in our 2018 CAMX paper, in laminates where the corrosion barrier or a resin rich surface is the interface with the secondary layer, we typically found poor secondary bonding. This is common with most resins. It is thought that dis-bonding occurs due to a lack of reinforcement at the interface. These observations suggest the surface should be profiled with light grinding to prevent dis-bonding if a continuous resin rich layer is present.

When constructing equipment to a manufacturing standard such as RTP-1 or an end user specific specification where grinding is mandated between primary and secondary overlays, these protocols should be followed, and surface preparation should be performed as directed by the standard or specification.

3.5 Thick Laminate Dimensional Stability Results

One of the challenges in working with reactive resins, like NEVERs, is the amount of resin and reinforcement that can be applied in one application. Like DS411, DS470 has technology that provides cure kinetics which allow for layup of thicker laminations in one step. The novel styrene suppression technology present in DS470 dramatically reduces styrene evaporation during cure making shrinkage on the air side of the laminate like that on the mold side. This serves to minimize warpage in the finished part, Figure 6. In these tests, when the laminate cooled to room temperature, the Barcol values on the air side were close in value to those on the mold side.

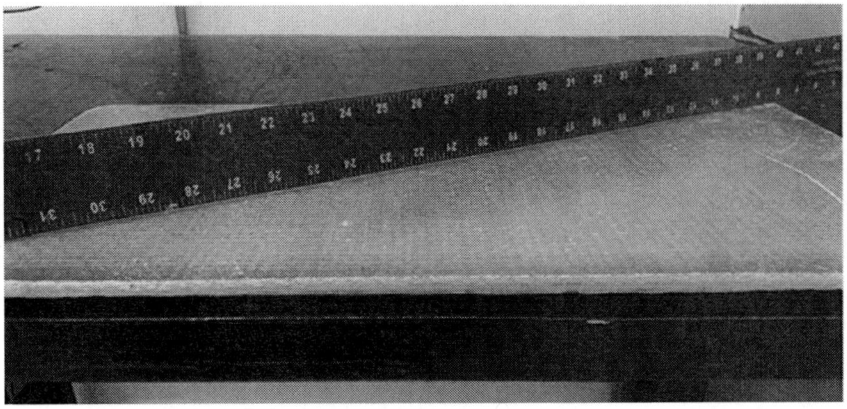

Figure 6. Common thick laminate test specimen. This 30 cm x 60 cm x 1.8 cm (12 in x 24 in x 0.7 in) laminate showed no mold pre-release, warp, twist, or burning due to excessive exotherm.

3.6 Styrene Emission Test Results and Observations

With the introduction of the US National Emission Standards for Hazardous Air Pollutants and the associated MACT and regulations regarding human exposure to regulated chemicals, styrene emission has become important to fabricators over the past several years. DS resins contain a novel styrene suppression technology that reduces styrene emissions depending on the fabrication method. Each DS resin has a styrene suppression system that is specific to that resin. This allows for improved stability over a range of temperatures. The calculated Vapor Suppression Effectiveness (VSE) factor for DS 470 and Equivalent Non-Vapor Suppressed (also known as non-LSE) Styrene Content are reported in Table 2.

Table 2. DS 470 Resin Emissions Calculations

Resin Application Method	Manual	Non-Atomized Mechanical	Filament Winding
DS 470 Resin Emissions (lbs/ton)	75	64	79
Equivalent Non-Vapor Suppressed Styrene Content (%)	31.5	31.0	25.2
Table 1 40 C.F.R. Part 63 Subpart WWWW Emission Calculation Factor (VSE Factor)			0.2

The actual styrene content in DS 470 is 33%, but the styrene suppression technology in DS resins allows them to perform the same as resins with less styrene content. This is seen in the second row of Table 2 – the percent of styrene shown here is what a typical resin would have to contain to achieve the same emissions as the EPA calculates for manual lay-up, non-atomized spray-up or by filament winding. Styrene suppression technology has a larger effect for processes like filament winding were faster suppression film formation is promoted by air moving over the composite surface due to mandrel rotation.

As with DS411 trials, shop workers noted less styrene odor during fabrication. Finished laminates had considerably less styrene odor compared to non-DS resins. Laminates made with DS470 that were transported in closed containers had notably less styrene odor upon opening the container as compared to traditional EVER and UPR based composites. To quantify the difference in odor, we compared styrene emissions from DS470 and DM470 neat resin castings by HSGC. The data in Figure 7 presents residual styrene emitted from neat resin castings 24 hours and 120 hours after cure. The data shows DS470 emits roughly 3.5 times less styrene than DM470.

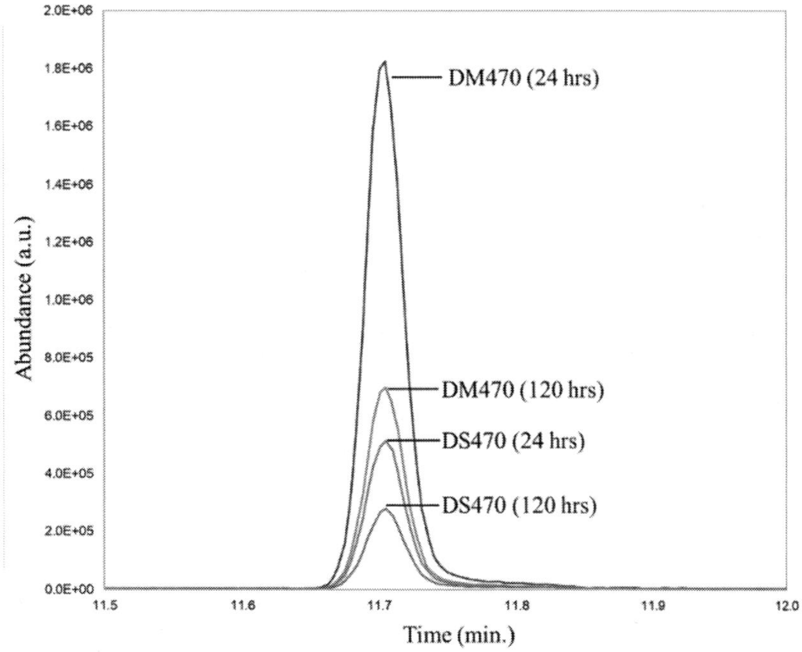

Figure 7. Residual styrene emission by headspace gas chromatography for DS470 and DM470 24 and 120 hours after cure.

4. CONCLUSIONS

Much of the data presented in this paper is based on experiments conducted at DS470 trials with corrosion resistant equipment fabricators globally.

Thirty-day ASTM C581 results indicate that DS470 has the same corrosion-resistant performance as the previous generation of technology, DM470. Results for 3, 6, 9, and 12 months will be reported in future updates.

As with the introduction of DS411, feedback from fabricators has been that the DS470 resin technology package significantly improves manufacturing efficiency through lower foaming with MEKP initiators, faster rollout and consolidation of reinforcement layers, reduced styrene emissions, and less grinding. Additional studies are underway to further quantify these features for future publications.

Improved resin storage stability has been identified by fabricators as one of the most significant benefits of the Signia resin technology package. Extending the commercial warranty period represents a leap forward in EVER technology that will significantly impact how they manage inventory and streamline shop production.

5. ACKNOWLEDGEMENTS

The authors are deeply grateful to colleagues throughout the industry who helped with content and constructive input for this paper – particularly those within INEOS Composites and the teams at Kenway Composites, Plas-Tanks Industries, and Tri-Clor (alphabetical order) for participating in DS 470 beta testing.

6. REFERENCES

[1] K. Lambrych, L. Adkins, A. Bivigou Koumba, *New Epoxy Vinyl Ester Resins for Corrosion Resistant Composites*; CAMX 2018 Proceedings, ACMA, Arlington, VA 22201, www.acmaeducationhub.org.

[2] K. Lambrych, L. Adkins, A. Bivigou Koumba Leveraging Modern Resin Technology For Shop Efficiency Improvements; CAMX 2019 Proceedings, ACMA, Arlington, VA 22201, www.acmaeducationhub.org.

[3] ASTM Standard D638-14, 2014, *Standard Test Method for Tensile Properties of Plastics*, ASTM International, West Conshohocken, PA, 2014, www.astm.org.

[4] ASTM Standard D790-17, 2017, Standard Test Methods for Flexural Properties of Unreinforced and Reinforced Plastics and Electrical Insulating Materials, ASTM International, West Conshohocken, PA, 2017, www.astm.org.

[5] ASTM Standard D648-18, 2018, Standard Test Method for Deflection Temperature of Plastics Under Flexural Load in the Edgewise Position, ASTM International, West Conshohocken, PA, 2018, www.astm.org.

[6] ASTM Standard C581-15, 2015, Standard Practice for Determining Chemical Resistance of Thermosetting Resins Used in Glass-Fiber-Reinforced Structures Intended for Liquid Service, ASTM International, West Conshohocken, PA, 2015, www.astm.org.

[7] 40 C.F.R. Part 63, Subpart WWWW, Table 1.

ADDITIVE MANUFACTURING OF STRAIN GAUGE SENSORS USING CONDUCTIVE POLYMER BY FUSED DEPOSITION MODELING AND ITS APPLICATIONS ON LINERLESS COMPOSITE PRESSURE VESSELS

Shamim Mondal, Ryan D. Smith, Jeremy J. Eavey, Robert M. Villarreal
Infinite Composites Technologies
10738 E. 55th Pl.
Tulsa OK 74146

ABSTRACT

This investigation reports on the use of additively manufactured biaxial strain gauge sensors for assessing the loads and deformations in a linerless (type V) composite pressure vessel (CPV). Fused deposition modeling (FDM) is a widely used rapid prototyping/additive manufacturing technique due to its relatively fast processing time and low cost. However, additive manufacturing of strain gauge sensors using FDM is uncommon because there are very few suitable conductive filaments commercially available. The filament selected for the strain gauge sensors in this study is a conductive polymer with low resistivity and high melting point which are crucial for the strain gauge to maintain its shape and integrity to obtain accurate, low power readings after being subjected to high temperatures during the post-curing process of CPVs. A small footprint and the lack of a permanently attached substrate allows the placement of strain gauges integrated into the composite layer to have minimal impact on the structural integrity of the pressure vessel. The applications for this study include but are not limited to pressurant and propellant tanks for launchers and spacecraft, self-contained breathing apparatus, space habitats and alternative fuel tanks.

Corresponding author: Shamim Mondal

1. INTRODUCTION

1.1 Studies on strain gauges

Strain gauges are electrical devices that provide real-time information about the strain on the material that the gauge is attached to. One such material that would greatly benefit from the usage of strain gauges for health monitoring purposes is a type V pressure vessel. Furthermore, it would be beneficial if a CPV were to have strain gauges embedded within the vessel for subsurface monitoring. While there are many types of strain gauges, this paper will be focusing on the implementation of bi-axial strain gauges contained within a CPV.

Commercially available strain gauges are sealed within two layers of various types of substrates. Due to this substrate it is inadvisable to utilize such a gauge because the composite layers of a CPV can delaminate from one another and cause mechanical failure. Furthermore, commercially manufactured strain gauges are expensive. In contrast, additively manufactured strain gauges can be produced at a very low cost which makes them attractive for applications that require multiple strain gauges. With localized strain monitoring being of utmost importance for a CPV, multiple strain gauges would be utilized to capture localized strain measurements to allow for global CPV monitoring.

1.2 Studies on conductive filament

Truly conductive additively manufactured filament is a very new concept. In 2016 the first conductive filament capable of reliably operating as an electrical circuit was created by Multi3D LLC [1]. In 2017, Flowers et. al published a paper detailing the construction of radio frequency (RF) electronics showcasing a successful application of their conductive filament [3]. There are multiple filaments that are generally conductive, but this was the first product to make it to market that had a sufficiently high conductivity to reliably function as an electrical circuit. Soon after this filament was created, additively manufactured electrical sensors were being produced including strain gauges.

1.3 Objective

Many within the pressure vessel market know that type V tanks have the benefit of increased storage volume with a decrease in weight. One point of contention is their viability. With the ability to monitor strain, users can monitor the health of type V pressure vessels ensuring their safety and reliability.
From an additive manufacturing viewpoint, utilizing FDM in the creation of electrically conductive circuits is attractive because it allows for rapid prototyping. Finally, additively manufactured strain gauges can be produced at a significantly lower cost than current commercially available strain gauges. This paper will be covering the implementation of additively manufactured strain gauges that do not utilize a substrate but are instead affixed within the structure of CPVs.

2. EXPERIMENTATION

2.1 Strain Gauge Design

Strain gauges are typically evaluated with a Wheatstone bridge circuit. Since the strain gauges change in resistance, ΔR (Ω), under load typically increases by just a few ohms at most (depending on the nominal resistance value), the Wheatstone bridge can measure the small differences in voltage between the two sides of the bridge. We will be using the Wheatstone type I quarter-bridge setup [5] with one strain gauge acting as a variable resistor under load, and three external resistors at the same value near the gauge's nominal value. These three additional resistors complete the quarter bridge circuit.

In the design phase, we took a critical look at how the resistivity of available electrically conductive filament shapes the resistance of a typical grid strain gauge. Utilizing the relationship between resistivity ρ (Ω-cm), length L (m), and cross-sectional area A (m^2) of the conductive filament in Equation 1 below, we designed both strain gauges (vertical and horizontal) to have a nominal resistance of ~350 Ω. We chose this value to compare our additively manufactured gauge against other industry standard ones close to the same resistance.

$$R = \frac{\rho L}{A} \qquad [1]$$

Since our target resistance was approximately 350 Ω, we needed to use a conductive filament with good conductivity and the ability to withstand temperatures up to 125 °C which is the temperature we post-cure our CPVs. We decided to use Multi3D's Electrifi high-temp filament for its 0.01 Ω-cm resistivity and ability to withstand temperatures up to 140 °C [7]. Table 1 on the following page shows the different products compared for our design, and Figure 1 shows the design of the strain gauge. Originally, the gauge on the right was flipped vertically as shown in Figure 5 but was changed to facilitate electrical connection inside of the CPV.

Table 1. Conductive filament selection criteria.

Conductive Filament	Thermal Limit (°C)	Resistivity (Ω-cm)
Proto-Pasta	195	15
Multi3D Original	60	0.006
Multi3D High Temperature	140	0.01

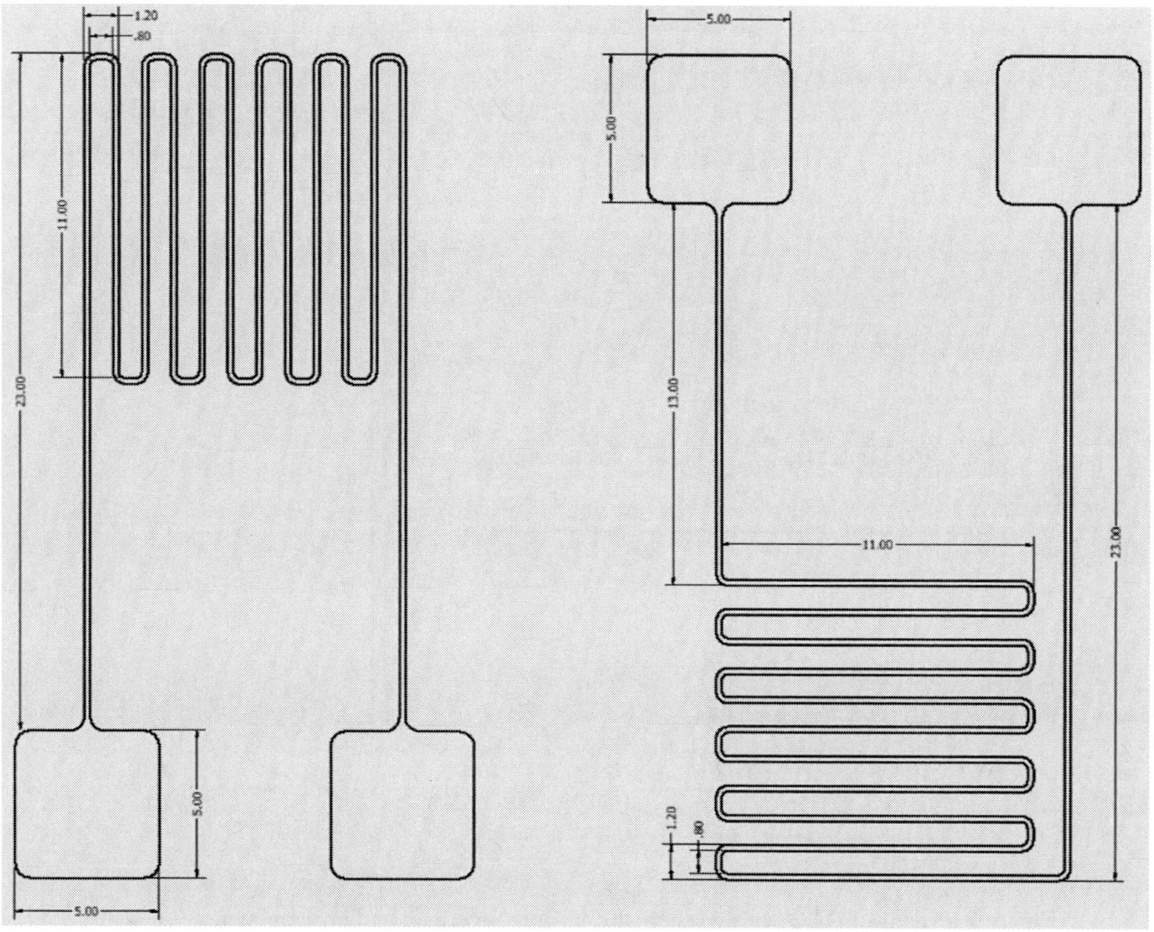

Figure 1. Biaxial strain gauge design (mm)

2.2 Electronics Configuration

A planned future measurement setup includes employing a Raspberry Pi as a data acquisition device with analog/digital converters (ADCs) and multiplexers. This lowers the sensor setup manufacturing cost with little theoretical effect on the data fidelity and quality. The ADCs send sensor data to the Raspberry Pi via SparkFun's Qwiic connect system with an adjustable programmable amplifier, yielding a full-scale range (FSR) resolution from 125 V to 3 mV step sizes [6]. Since this system uses I2C protocol, each ADC has a physical address of 0x48 with up to three other addresses available. Our plan is to connect multiple ADCs that represent zones of interest on a pressure vessel to one Pi; this requires utilizing a Qwiic multiplexer breakout board. This mux allows eight devices with the same physical I2C address to be connected on the same I2C bus. The proposed wiring schematic is shown below in Figure 2.

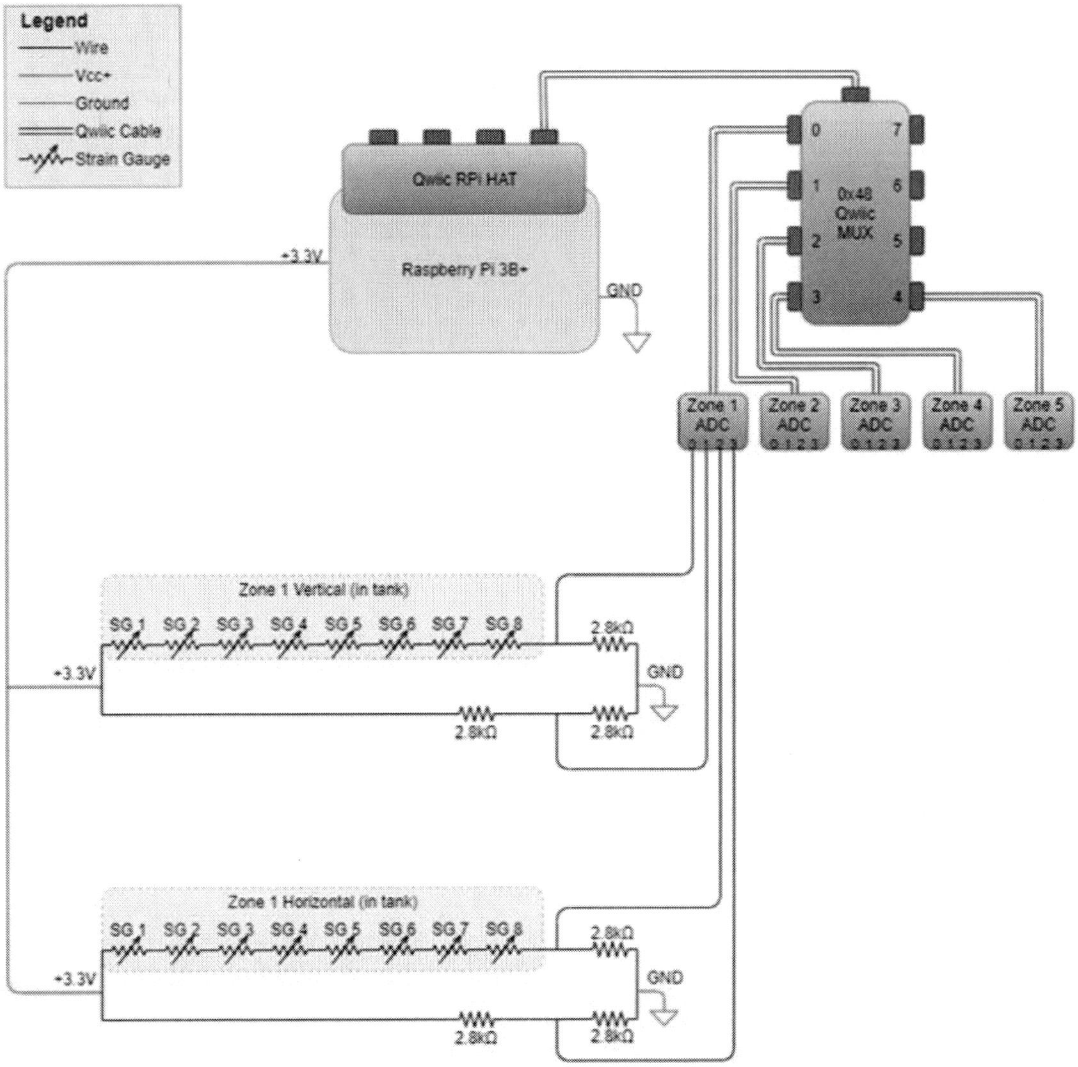

Figure 2. Proposed DAQ wiring schematic for $N = 8$ vertical and horizontal strain gauges.

The above diagram illustrates the proposed system model. Each ADC is connected to one zone of interest on the tank, each zone consisting of two Wheatstone quarter-bridge setups (one vertical strain gauge and one horizontal). This setup allows us to place multiple vertical/horizontal strain gauges in a zone of interest and measure their equivalent resistance by connecting them in series as shown below in Equation 2.

$$R_{eq} = R_1 + R_2 + \cdots + R_N \quad [2]$$

As an example, we chose $N = 8$ strain gauges for the vertical and horizontal parts of the zone each for a total of 16 strain gauges per zone. The size of the individual CPV will determine the N value required. Using Equation 2 yielded an equivalent resistance of 2.8 kΩ which determines the nominal resistance of the other three resistors in the quarter bridge. The ADC measures the difference in voltage across the bridge arms and amplifies this difference through the programmable amplifier. Since our setup requires multiple zones of interest, we decided to employ the use of a I2C multiplexer to allow up to eight ADCs to be connected to the Pi on the same bus with the default physical address of 0x48. Figure 3 below shows the five zones of a CPV to be measured.

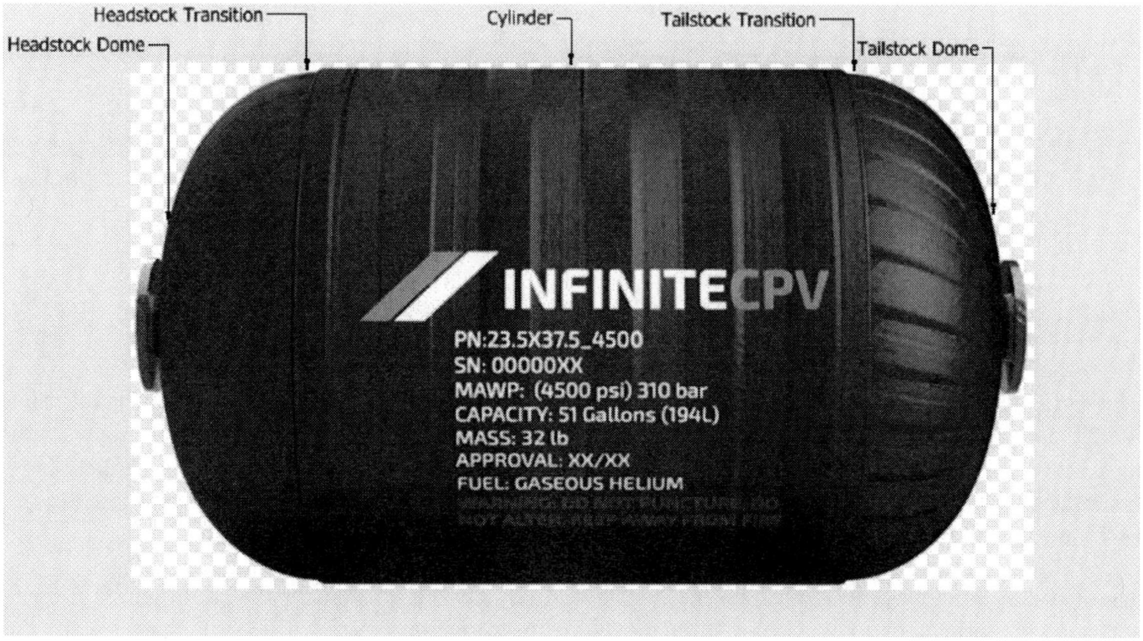

Figure 3. Composite pressure vessel zones.

Each zone will be a 3.3 V circuit powered by the 3.3 V rail of the Raspberry Pi. A Raspberry Pi 3B+ can output up to 50 mA on the 3.3 V rail conservatively [4]. To ensure we have the proper current available we utilize Equation 3 below to determine that our 5 zones collectively will require approximately 11.8 mA from the power source.

$$I = \frac{V}{R} \quad [3]$$

Regarding message transmission, we will be using MQTT protocol to transmit strain gauge data from the Raspberry Pi to a computer through an online broker such as HiveMQ. Alternatively, one could set up an additional Raspberry Pi with additional layers of security to act as a broker to transmit data. This system functions on a publish/subscribe basis; multiple Pi's can publish their data, and any computer with access can subscribe to published topics such as strain and resistance. The Python script running this system can be programmed to set off alarms in the case of strain events and massive pressure changes. If a strain event occurs (i.e. localized strain reaches above 1.00 %, or any other configurable threshold) an alarm would be triggered that would warn personnel about the pressure vessel's health.

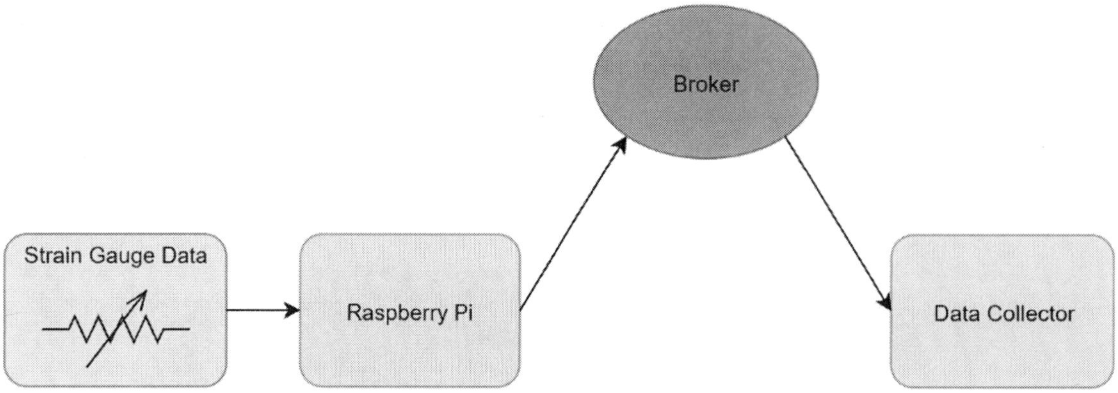

Figure 4. Proposed MQTT operation flow diagram.

2.3 Manufacturing

The width of each individual trace is 0.2 mm which necessarily requires very tight manufacturing tolerances which can be difficult to achieve with consumer level printers. This width was chosen as one of the variables to achieve the target resistance value of 350 Ω. The depth of the design proved to be challenging to manufacture, and ultimately required an alteration in the design to make the gauges 0.2 mm thick instead of 0.1 mm thick.

The material to be used as a substrate that will eventually be able to transfer the strain gauge from the surface it was printed on to the CPV is still undecided. For testing purposes, we are using both Kapton tape as well as blue painter's tape as the substrate. These will be used as substrates for testing strain gauges on the exterior of a tank before embedding them into a CPV. Following the additive manufacturing process, another layer of tape is placed on top to secure the strain gauge between two layers. This allows for exterior testing of the strain gauges and increases the structural integrity.

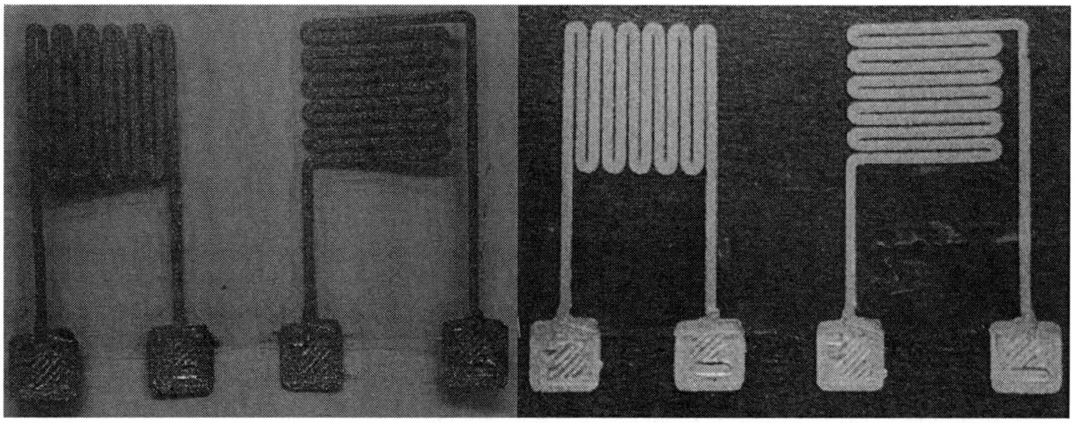

Figure 5. Additively manufactured strain gauges.

2.4 Testing

To measure the prototypes properly, wiring needed to be attached to the additively manufactured gauges. Due to the relatively low melting point of the filament, it is imperative to use a solderless connection. To accomplish this, we used silver epoxy on the contact pads and wiring. To take accurate measurements good bonding must be made not only on the strain gauge pad but also on the wire. Once sufficiently bonded, a measurement is obtained by connecting to the other end of the wires.

3. RESULTS

3.1 Initial Results

Initial testing was done with a multimeter to get resistance measurements just to ensure the strain gauge functioned on a basic circuitry level. Following the resistance measurements obtained with a multimeter, a Data Acquisition unit (DAQ) was utilized to record resistance readings shown in Table 2 below. An initial test has also been done with an additively manufactured strain gauge on a CPV during a burst test shown in Figure 6.

Table 2. Initial DAQ resistance measurements.

Strain Gauge Prototype Number	Horizontal Gauge Initial Resistance (Ω)	Horizontal Gauge Strained Resistance (Ω)	Vertical Gauge Initial Resistance (Ω)	Vertical Gauge Strained Resistance (Ω)
1	346	680	338	700
2	338	668	357	717

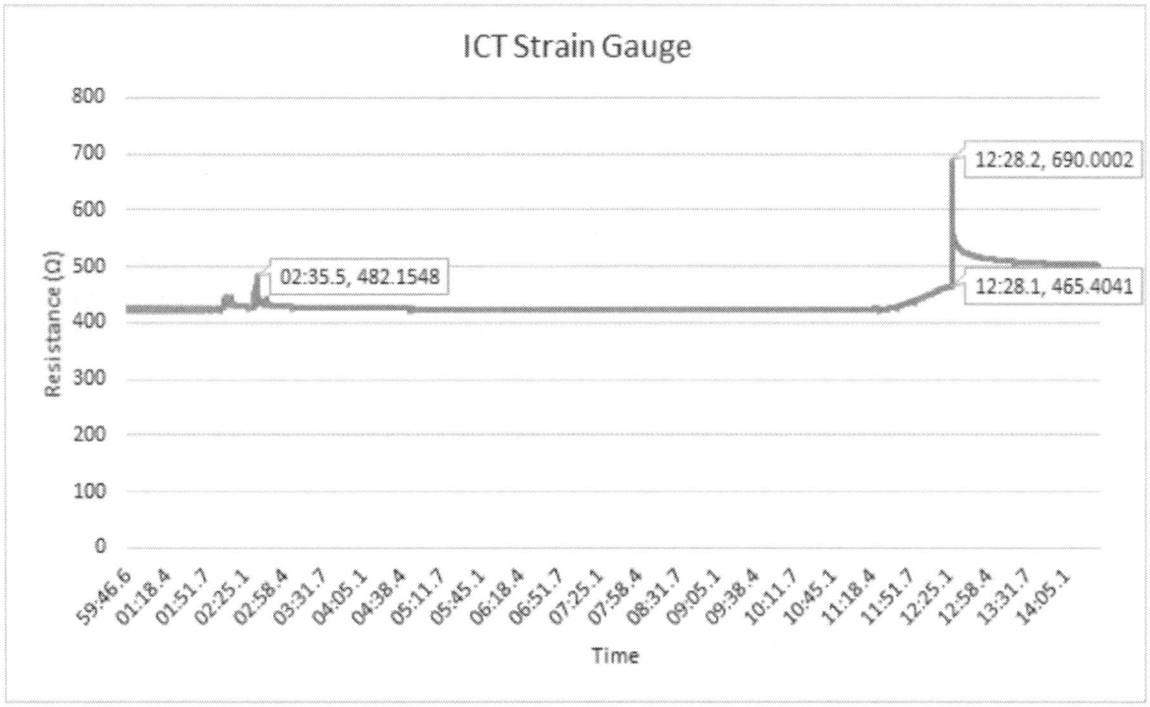

Figure 6: Additively manufactured strain gauge on CPV test.

The test shown in Figure 6 measured raw resistance rather than utilizing a quarter-bridge setup. Due to this, the sensitivity of this test is vastly reduced. The spike to 482 Ω at the beginning of the test was the result of a technician manually stimulating the strain gauge to ensure proper operation and data acquisition. As the pressure inside of the CPV began to rise around 11:18.4 we saw the resistance rise correlating to that pressure increase. The next major event shown at 12:28.1 and 465 Ω was the resistance measured the moment before the CPV burst. As the CPV burst, we saw a spike to 690 Ω as denoted in the data flag at 12:28.2. The peak value is of little concern to us as this is the point in which a catastrophic failure has already occurred. The most important data point is the 465 Ω moment. Knowing the pressure and correlating strain gauge behavior will allow us to predict failure points and warn users before a catastrophic failure occurs. This shows our proof of concept will function as intended and will only become more sensitive as the quarter-bridge setup is implemented.

3.2 Planned Future Testing

The circuit will be made into zones, all connected to one Raspberry Pi as shown earlier in Figure 2. This allows one Raspberry Pi to handle all desired sensors up to 32 zones and can give alerts as to which zone is alarming. Planned future testing includes more strain gauges placed on the outside of a CPV and then testing gauges embedded within all utilizing the quarter-bridge setup for increased sensitivity. Typically the carbon fiber's conductivity would make the entire CPV conductive, but due to the non-conductive nature of the resin applied to each layer that is an insulator no part of the CPV is conductive and thus the strain gauge circuits are electrically isolated from the CPV. These tests will allow fine tuning of the added bridge resistors to obtain optimal sensitivity. To make electrical connections within the CPV, a very thin electrical conductor is necessary. This depth requirement necessitates that traditional wire not be used, but rather some form of paint or epoxy. Generally, conductive paints cannot withstand the high temperatures CPVs are subjected to during the curing process. For this reason, high temperature silver infused conductive epoxy will be used for all electrical connections made within the tank and traces will be ran to the end-fitting. The silver epoxy chosen can withstand a continuous operating temperature of 150 °C because we post-cure our CPVs at 125 °C. The metallic end-fitting would electrically connect all individual silver epoxy traces together so a very thin layer of polyurethane will be applied to the end-fitting to maintain electrical isolation of all strain gauge circuits.

4. CONCLUSION

Creation of strain gauges without a permanently attached substrate presents unique opportunities to embed strain gauges within materials allowing for the detection of strain on individual layers of materials. Relevant to composite based applications, this opportunity allows assurance that the composite structure remains uncompromised on a layer by layer basis. Providing real-time assurance about the functionality and safety of CPVs via strain measurements is vital in promoting widespread adoption of this emerging technology.

5. REFERENCES

[1] "About Us." *Multi3D LLC.* Accessed 17 January 2020 https://www.multi3dllc.com/about-us/.

[2] "CDP1xxxx Safety Data Sheet." *Proto-pasta.* 5 October 2018. Accessed 17 January 2020 https://cdn.shopify.com/s/files/1/0717/9095/files/CDP1xxxx_SDS.pdf?1992606272897634343.

[3] Flowers, P., Reyes, C., Ye, S., Kim, M., Wiley, B. "3D printing electronic components and circuits with conductive thermoplastic filament." *Additive Manufacturing* 18 (2017) 156-163.

[4] "Introducing the Raspberry Pi Model B+." *Adafruit.* Accessed 17 January 2020 https://learn.adafruit.com/introducing-the-raspberry-pi-model-b-plus-plus-differences-vs-model-b/power-supply.

[5] "Quarter Bridge Type I." *National Instruments.* January 2019. Accessed 17 January 2020 http://zone.ni.com/reference/en-XX/help/370466AH-01/measfunds/quarter1/#GUID-F34636DC-B311-4257-BDD0-88B684CF662B.

[6] "SparkFun Qwiic 12 Bit ADC – 4 Channel (ADS1015)." *SparkFun.* Accessed 9 January 2020 https://www.sparkfun.com/products/15334.

[7] "What is the melting point for Electrifi Filament?" *Multi3D LLC.* Accessed 17 January 2020 https://www.multi3dllc.com/faqs/.

[8] "Wheatstone Bridge Circuit." *HBM.* HBM, Inc., USA. Accessed 27 December 2019 https://www.hbm.com/en/7163/wheatstone-bridge-circuit/.

MECHANICALLY IMPROVED AND MULTIFUNCTIONAL CFRP ENABLED BY RESINS WITH HIGH CONCENTRATIONS EPOXY-FUNCTIONALIZED FLUOROGRAPHENE FILLERS

Junhua Wei, Eugene Beh, Rahul Pandey, Gabriel Iftime, Jessy Rivest, Sean Garner

PARC, A Xerox Company

3333 Coyote Hill Rd, Palo Alto, CA, USA

ABSTRACT

To meet the maximum potential of the mechanical properties of carbon fiber reinforced plastics (CFRP), stress transfer between the carbon fibers through the polymer matrix must be improved. A recent promising approach reportedly used reinforcing particles as fillers dispersed in the resin. Carbon based fillers are an excellent candidate for such reinforcing particles due to their intrinsically high mechanical properties, structure and chemical nature similar to carbon fiber and high aspect ratio. They have shown great potential in increasing the strength, elastic modulus and other mechanical properties of interest of CFRPs. However, a percolation threshold of ~1 % of the carbon-based particle concentration in the base resin has generally been reported, beyond which the mechanical properties deteriorate due to particle agglomeration. As a result, the potential for further increase of the mechanical properties of CFRPs with carbon-based fillers is limited. We report a significant increase in the strength and elastic modulus of CFRPs, achieved with a novel reinforced thermoset resin that contains high loadings of epoxy-reacted fluorographene (ERFG) fillers. We found that the improvement in mechanical performance of CFRPs was correlated with increase in ERFG loading in the resin. Using a novel thermoset resin containing 10wt% ERFG filler, CFRPs fabricated by wet layup technique with twill weaves showed a 19.6% and 17.7% increase in the elastic modulus and tensile strength respectively. In addition, because of graphene's high thermal conductivity and high aspect ratio, the novel resin enhanced CFRPs possessed 59.3% higher through-plane thermal conductivity and an 81-fold reduction in the hydrogen permeability. The results of this study demonstrate that high loadings of functionalized particles dispersed in the resin is a viable path towards fabrication of improved, high-performance CFRP parts and systems.

Keywords: CFRP, functionalized graphene, tensile properties, hydrogen permeability, through-plane thermal conductivity
Corresponding author: Junhua Wei, jawei@parc.com

1. INTRODUCTION

Carbon Fiber Reinforced Plastics (CFRP) have been widely used as lightweight materials replacing metals because of their superior specific strength, stiffness, and corrosion resistance [1]. In recent years, their application has been extended from high-end industries, like aerospace, aircraft, sports, and military to cost-sensitive industries, like automotive and energy [2]. This new trend requires reducing the cost of parts made from CFRP. This price reduction can be achieved through two approaches: 1) enhancing mechanical properties of CFRPs resulting in lower usage for same function and 2) eliminating functional parts by enabling CFRPs with these functions.

Improving the stress transfer between the carbon fibers (CF) and improvement of stiffness of polymer matrix are most popular research topics. Besides development of highly efficient CFRP fabrication techniques, low price CF manufacturing and material-saving structure designs, these studies also aim at cost reduction via decrease in CFRP usage. Due to the weak interfacial interaction between the load bearing CFs and the polymer matrix that integrate the structure, delamination is one of the major failure modes for CFRP. One of the key reasons for this failure is that CF has relatively low density of surface functional groups, which result in lesser surface interactions. To address this problem, increasing the interface interaction through sizing, a polymer coating that improves stress transfer through chemical bond and physical interlock, is required as part of the CF manufacturing process [3]. Inspired by the sizing mechanism, rigid particles have been added either into polymer sizing or by spraying on the CFs, to encourage strong physical interlock.
New approach to lower the cost and weight of systems containing CFRPs is to make CFRP a functional part in the system. For example, Type V storage vessels used for hydrogen storage are designed with only a single layer of CFRP which endures the compressed gas pressure and simultaneously acts as an efficient hydrogen barrier. Compared with other tank designs (e.g. Type IV), Type V tank is more cost competitive via process, material, and weight saving through limiting the hydrogen barrier liner. However, to realize this design, a large amount of fillers with good gas barrier properties need to be present into the CFRP to form an effective gas barrier layer.

Carbon-based materials use in polymer sizing have drawn a lot of attention to this approach because of their superior mechanical properties, chemical structure like that of CFs, and most importantly due to their high aspect ratio. Methods including electrophoretic deposition [4, 5], spray [6, 7], and dip coating [8-12] have been employed to deposit graphene and carbon nanotubes onto CFs surfaces. CFRPs made with these coated CFs show ~20-70% improvement on interlaminar shear strength. Up to 10 wt% of particles laden sizing coated onto CF by such approaches have been demonstrated [8]. However, this approach generally does not improve the tensile or flexural modulus [5, 8, 13]. This is because the polymer matrix is much weaker compared to CF, and as a result, the polymer matrix deforms under stress. To reinforce both the modulus and strength of CFRP, researchers have tried to add carbon-based particulate materials directly into the resin matrix as well, [14-19] but this approach achieved only limited improvement in mechanical properties. The primary limitation is particle aggregation, which prevents high particle loading resins [16, 20]. Typically, CFRP with ~1 wt% of carbon-based fillers in the polymer matrix have shown the best properties. A polymer matrix containing a high loading of dispersed carbon-based materials would therefore be likely to radically increase both, strength and modulus, of CFRPs. Meanwhile, carbon-based materials have been appraised for their excellent properties as electrical and thermal conductors, gas barrier, fire retardant et al.[21, 22] The advantages of their multifunction can introduce CFRP with special properties to replace other functional part and reduce the device price, when the particle concentrations are high enough.

Fluorographene (FG) is a unique carbon-based material which maintains the excellent mechanical properties of graphene but with excellent chemical, electronic, tribological, and hydrophobic properties, due to the presence of fluorine atoms [23]. One distinctive advantage of FG is that it is easily forms dispersion and exfoliates because fluorination disrupts the van der Waals forces between FG sheets [24]. Other researchers have studied the electrical, thermal and mechanical properties of FG-reinforced polyimide [25-27]. We have previously reported on dispersing epoxy-monomer–functionalized FG in epoxy with 30 wt% loading, which achieved ~90%, ~60%, and ~170% improvement in the tensile strength, modulus, and toughness respectively [28]. In this work, resin containing ERFG was used to make CFRP by wet lay-up method. The high particle loading in the polymer matrix improves stress transfer between the CF, both by physically interlocking with the CFs to improve strength, and by chemical linkage within the polymer matrix to improve modulus (Figure 1). Additionally, the high particle loading also enables the CFRP to exhibit properties of FG present within the CFRP, such as low gas permeability, high thermal conductivity, et al. This mechanically improved and multifunctional CFRP is a very competitive composite for cost sensitive CFRP applications.

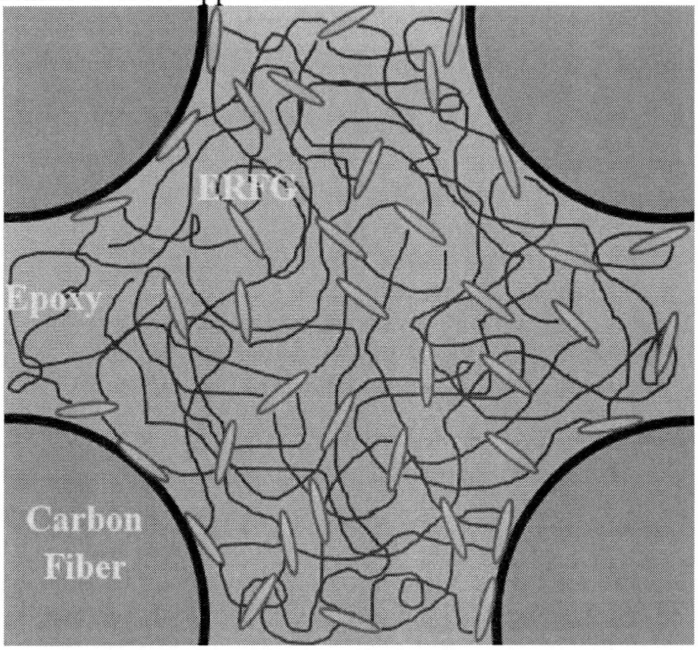

Figure 1. Illustration of CFRP containing a high loading of epoxy-reacted fluorographene (ERFG, orange ovals) in the polymer matrix.

2. EXPERIMENTATION

2.1 Materials

Graphite fluoride (FG, also called fluorographite), triethylenetetramine (TETA), and acetone were bought from Sigma Aldrich. Bisphenol A based epoxy resins (EPON 826) with 178 g/mol EEQ (Epoxy Equivalent Weight[29]) was purchased from Miller-Stephenson Chemical Co., Inc. Fibre Glast 2000 (Epoxy), Fibre Glast 2060 (Hardener), 2K 2×2 twill weave carbon fiber fabrics and all the materials for wet layup were from Fibre Glast. Toray T700 12K 2×2 plain weave fabrics and T300 3K 2×2 plain weave fabrics were purchased from Composite Envisions. T800S 12K 2×2 twill weave fabrics and M55JB 6K 2×2 plain weave fabrics were supplied by Rock West

Composites. MGS L285 (epoxy) and H287 (hardener) were purchased from Aircraft Spruce & Specialty Co.

2.2 ERFG particle synthesis

ERFG particles were synthesized using the same experimental procedure as in our earlier published work [28, 30], and is presented in Scheme 1. Briefly, fluorographite was dispersed and solvent exfoliated in TETA, then the C-F bonds on the fluorographite were substituted with amine groups from TETA at high temperature under argon gas protection, thereby producing amine-reacted fluorographene (ARFG). The ARFG particles were recovered from the TETA via filtration, washed, and then dispersed into epoxy resin (EPON 826), to synthesize epoxy-monomer-reacted fluorographene (ERFG).

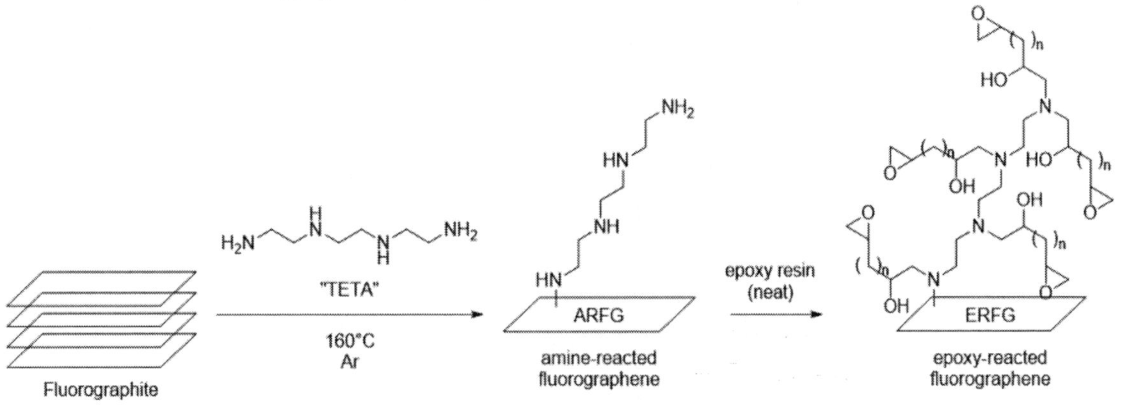

Scheme 1. Synthesis of ARFG and ERFG from fluorographite.

2.3 CFRP fabrication

The formulation was prepared by a two-step procedure. Firstly, ERFG particles were added into an epoxy resin by centrifugal mixing. Immediately before use for CFRP fabrication, the epoxy resin containing ERFG particles was mixed with a hardener. The optimum weight ratio between the epoxy/ERFG mixture to the hardener was determined by a dynamic DSC screening for the maximum curing enthalpy. The CFRP panels were fabricated according to literature [31]. Next, the formulation consisting of epoxy with ERFG particles and hardener was spread onto each layer of CF fabric. After piling and spreading, accessories like the vacuum bagging were added. The impregnated CF fabrics were cured at room temperature for 5 hr under vacuum, and then post-cured at 100℃ for another 5 hr. The cured CFRP was removed from vacuum and bagging and was then cut into specific shapes according to ASTM D3039 by a water-cooled diamond saw for tensile testing.

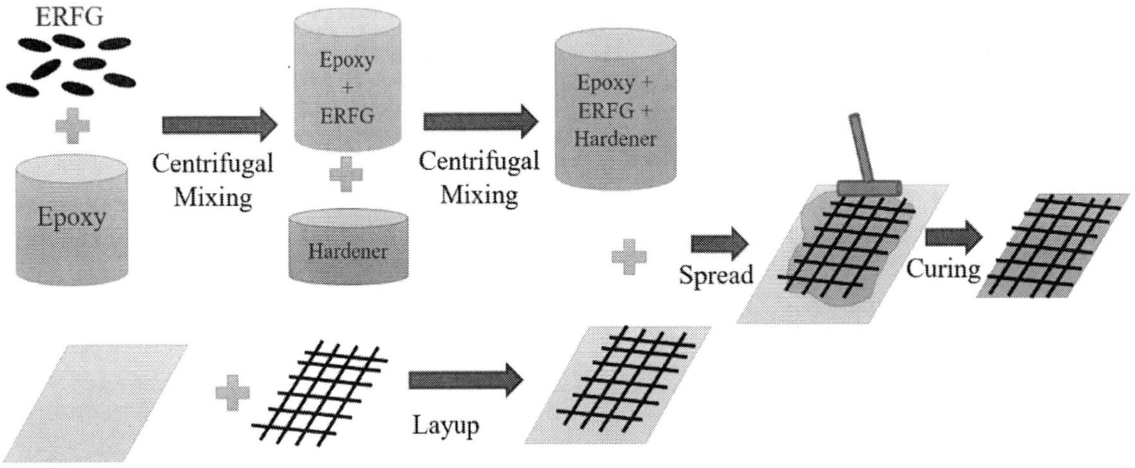

Scheme 2. Fabrication of CFRP using wet layup method.

2.4 Characterization

The chemical functionalization of the particles was characterized by an ATR FT-IR (Perkin Elmer). The morphology of the ERFG particles was imaged by a TEM (Jeol 1200).
The mechanical properties of CFRP were measured according to ASTM 3039 using a universal tester (AGS-X, Shimadzu). The through-plane thermal conductivity was measured as per ASTM D7984 using a thermal conductivity analyzer (TCi, C-Therm). Hydrogen permeability tests were performed by a professional testing lab (Versaperm Ltd.) in accordance with ASTM G148.

3. RESULTS

3.1 Particle Characterization

The mechanism of attaching molecular amine groups to fluorographene through substitution of surface fluorine atoms has been previously reported[32]. In this work, TETA served both as solvent to exfoliate fluorographite and as a reagent for introduction of amino functional groups onto the exfoliated fluorographene to produce Amine Reacted Florographene(ARFG) particles. The color of fluorographite changed from white to black during this substitution reaction. In the second step, the amine groups on ARFG were capped with EPON 826, a low molecular weight diglycidyl epoxy resin derived from bisphenol A. FT-IR spectroscopy was used to identify the chemical groups present on the particles during each step of synthesis. In Figure 2a, a prominent feature at 1260 cm^{-1} characteristic of covalent C-F stretching of fluorographite was detected[33]. There are three identifying absorptions peaks for TETA: 1050-1200 cm^{-1}, 1450 cm^{-1}, 1600 cm^{-1} representing C-N stretching, C-H bending, and N-H bending, respectively[34, 35]. In the ARFG FTIR spectrum, a broad absorption peak appears from 1000 cm^{-1} to 1200 cm^{-1} as a result of C-N stretching appeared in place of C-F stretching. The C-F stretching absorption peak at 1260 cm^{-1} is greatly diminished in comparison to the starting material, fluorographite. The 1580 cm^{-1} absorption on the ARFG is inherited from TETA representing N-H bending. In the FTIR spectrum ofERFG, the epoxy functionality display a large number of additional absorptions peaks: 1034 cm^{-1}, 1184 cm^{-1} and 1241 cm^{-1}, 1300 cm^{-1}, 1509 cm^{-1} representing C-O-C ether stretching, C-C-O-C stretching, stretching of cyclic ether, and C-C stretching of aromatic ring, respectively[36, 37]. In

ERFG, the C-O-C absorptions extended the broad feature seen in ARFG (1000 cm^{-1} to 1200 cm^{-1}) to a broader range (1000 cm^{-1} to 1300 cm^{-1}), which indicates the presence of epoxy groups onto the ERFG particles. The high aspect ratio, single-digit microsized, and few-layer sheet[38] ERFG was shown in the TEM micrographs of ERFG particles (Figure 2b).

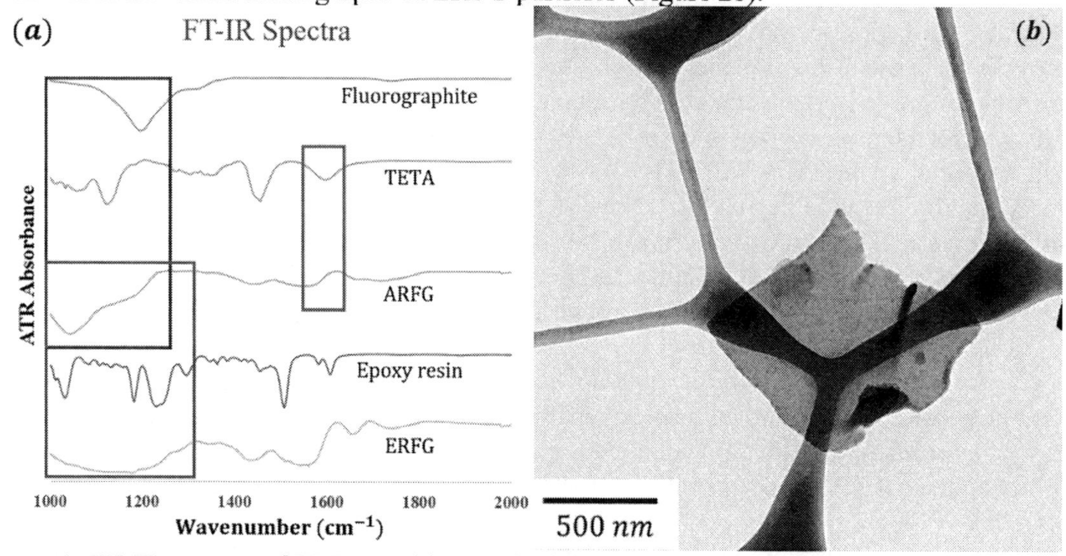

Figure 2. FT-IR spectra of ERFG and its synthetic precursors (a), and TEM image of an ERFG particle (b).

3.2 CFRP characterization

CFRPs were fabricated with Fibre Glast CF fabrics and Fibre Glast 2000/2060 resin base with different weight loadings of ERFG according to the procedure shown in Scheme 2. The mechanical, thermal, and gas barrier properties of these samples were investigated and are summarized in Table 1.

Unlike the carbon fiber surface particle coating from previous reports, adding particles to the polymer matrix increased both the elastic modulus and the tensile strength. The increase of the elastic modulus of the CFRP made by ERFG resin correlated well with the increase of the ERFG loading below 15wt%. However, the tensile strength decreased for CFRP with 15wt% ERFG. There are two possible explanation for this phenomenon. First, the viscosity of resin with ERFG particles increased when adding more particles. For highly viscous resin, higher vacuum pressure and elevated temperature are required for uniform wetting. Second, 15wt% of particles in the resin can form aggregates or liquid crystal domains. Both these structures prevent the uniform distribution of the particles through the CF fabric. Since the enhanced interfacial interaction between CFs and polymer matrix is based on the premise of attachment of the ERFG present in the polymer matrix to the surface of CF to form a physical interlock. The formation of aggregates or liquid crystal domains decreases the density of ERFG coated ont the CF surface and causes less effective stress transfer and results in decrease of CFRP's overall tensile strength. Further investigations are underway to reveal which mechanism dominates at higher particle loading CFRPs.

The in-plane thermal conductivity of C FRP can reach ~1000 W/(m×K) due to the continuous nature of the carbon fibers [39]. In contrast, the through-plane thermal conductivity of CFRP is in the 0.2-0.5 W/(m×K) range [40] because the polymer matrix separates the carbon fiber tows.

Historically reported approaches added highly thermally conductive particles like graphene [40] or diamond [41] into the polymer matrix to increase the through-plane conductivity, but this compromises the other properties of CFRP, mainly mechanical properties. In this work, a significant improvement in the through-plane conductivity is achieved. The ERFG particles serve as bridges to facilitate the conduction of heat between individual carbon fibers. A 60% improvement in the through-plane thermal conductivity was realized in this way (Table 1). This significant improvement on the through-plane thermal conductivity can improve the CFRP's stability under thermal cycling and will enable the replacement of metal built electronic box in weight essential applications [42], The 2D structure and high aspect ratio of the ERFG particles dispersed within the cured epoxy matrix were efficient in blocking the through-plane permeation of hydrogen gas. The hydrogen permeability of CFRP made with 10wt% ERFG was 81 times lower than that of CFRP made with pure, unfilled epoxy resin (Table 2). These results make ERFG-enhanced CFRPs as state-of-the-art hydrogen barrier materials, far exceeding the properties of conventional polymer liner used in COPV (Composite Overwrapped Pressure Vessels) tanks [43-45]. The decreased hydrogen permeation is a huge benefit for building COPV tanks for hydrogen or natural gas storage. Additionally, use of ERFG-enhanced CFRPs eliminates one manufacturing step – polymer liner pre-molding – required to manufacture the current Type IV COPV tanks. This multifunctional CFRP is very promising for direct manufacturing of liner-less Type V COPV tanks.

Table 1. Mechanical properties, through-plane thermal conductivity, and hydrogen permeability of CFRP made by Fibre Glast CF fabrics and Fibre. Glast 2000/2060 resin with different wt% of ERFG (standard deviation)

Sample	Pure (0wt% ERFG)	5 wt% ERFG	10 wt% ERFG	15 wt% ERFG
Elastic Modulus (GPa)	51.5 (1.3)	55.1 (7.6)	61.6 (4.4)	63.9 (3.1)
Tensile Strength (MPa)	500.7 (15.6)	536.3 (14.9)	608.9 (22.8)	583.9 (22.0)
Thermal conductivity (W/(m×k))	0.425 (0.005)	0.578 (0.001)	0.677 (0.006)	NA
Hydrogen permeability (mol/(m×s×Pa))	1.79E-14 (0.01)	NA	2.20E-16 (0.02)	NA

Note: the carbon fibers take 37 vol% to 42 vol% in the tested CFRPs.

Table 2. Hydrogen permeability of conventional polymer liner for COPV tanks [43]

Material	**Hydrogen permeability** 10E-16 mol/(m×s×Pa)	Ref.
ERFG-enhanced CFRPs	2.20	This work
HDPE	11.20-8.09	[46, 47]
LDPE	24.50	[47]
PA	6.69-2.23	[47]
PTFE	36.20	[48]

3.3 Discussion

In this work, the idea of adding high loadings of graphene derivative into the polymer matrix to increase both the elastic modulus and tensile strength of CFRP has been achieved with Fibre Glast fabrics and 2000/2060 resin. To further demonstrate the universality of this idea, we tested the tensile properties of CFRP made with another widely used resin system, MGS 285/287, and with different classes of CFs. To maintain consistency, all CF fabrics used were made with Toray CF. In these carbon fibers, T300 and T700 represent the standard modulus CF. T300 is primarily used for automotive parts and sports products, and T700 is used in gas tanks. The T800 represents the intermediate modulus carbon fiber used in aircraft applications. The M55 is a high modulus carbon fiber mainly used in spacecraft. MGS resin systems have been widely used in CFRP wet layup fabrication with applications in aircraft. By testing CFRP made using widely popular commercial resin with different grades of CFs, we demonstrated the universality of our idea.

Table 3 summarizes the elastic modulus and tensile strength of CFRPs made with different CF using MGS 285/287 resin with 10wt% ERFG. the elastic modulus of all the CFRP samples made with ERFG show a 15-20% increase in these mechanical properties, due to formation of highly uniform dispersion of ERFG in the polymer matrix and from stronger interfacial interaction between polymer and ERFG through epoxy monomer functionalization. However, the trend observed in tensile strength is more complex. With T700, T800, and M55 CFRPs, the tensile strength increased by ~15-20% when 10 wt% ERFG was added in the resin. When compared, the strength of CFRPs with T300 CF increased by only ~10%. The variation with use of different carbon fibers could be the result of different sizing used on the carbon fibers and structure of fabrics. The thickness, stiffness, and chemical structure of the sizing radically affects the interface between carbon fiber and the polymer matrix. The addition of ERFG amplifies the effect of sizing on interfacial interactions. Different CF weaving structure and fabric density (gsm) affect the resin penetration through the fabric, which together with the increased resin viscosity due to dispersed ERFG particles, resulted in a non-uniform distribution of the particles across CFRP thickness.

Table 3. Mechanical properties of CFRP made by different types of CF fabrics and 10wt% ERFG MGS resin

	Elastic Modulus (GPa)		**Tensile Strength (MPa)**	
CF	**Pure (0 wt% ERFG)**	**10 wt% ERFG**	**Pure (0 wt% ERFG)**	**10 wt% ERFG**
T300	45.9 (3.4)	51.5 (0.9)	402.9 (14.2)	432.0 (25.1)
T700	51.9 (3.7)	59.1 (7.1)	586.9 (26.7)	701.8 (48.0)
T800	56.8 (3.1)	65.2 (4.7)	683.7 (45.9)	762.5 (37.3)
M55	94.5 (2.0)	111.8 (2.5)	376.2 (28.7)	427.1 (38.0)

Note: the carbon fibers take 38 vol% to 45 vol% in tested CFRPs.

4. CONCLUSIONS

In this work, significant improvements to both the elastic modulus and the tensile strength of CFRP were achieved through high ERFG particle loadings in the polymer matrix of CFRP. The low surface tension of C-F groups in the fluorographite precursor material, the epoxy monomer functionalization, and the small size, collectively result in high loading of well-dispersed exfoliated ERFG in epoxy resin. Furthermore, the loading of ERFG is high enough to enhance other properties of CFRP composite. The through-plane thermal conductivity of CFRP was increased by ~60% and the hydrogen permeability of CFRP was decreased by a factor of ~81. This work provides a universal method to reduce the cost of CFRPs by 1) decreasing the weight of CFRP parts through improved mechanical properties, and by 2) reducing system complexity by having CFRP parts multiple desirable properties which would ordinarily require separate parts made from different materials. The approach reported in our work represents a viable strategy for low cost parts made with CFRP for applications in automotive, sporting goods and energy sectors.

5. REFERENCESS

[1] C. Soutis, *Fibre reinforced composites in aircraft construction*. Progress in Aerospace Sciences, vol. 41, no. 2, pp. 143-151, 2005/02/01/ 2005.

[2] *Carbon Fiber Market by Raw Material (PAN, Pitch, Rayon), Fiber Type (Virgin, Recycled), Product Type, Modulus, Application (Composite, Non-composite), End-use Industry (A & D, Automotive, Wind Energy), and Region - Global Forecast to 2029.* MARKETSANDMARKETS2019.

[3] Y. Luo, Y. Zhao, Y. Duan, and S. Du, *Surface and wettability property analysis of CCF300 carbon fibers with different sizing or without sizing.* Materials & Design, vol. 32, no. 2, pp. 941-946, 2011/02/01/ 2011.

[4] T. Sun, M. Li, S. Zhou, M. Liang, Y. Chen, and H. Zou, *Multi-scale structure construction of carbon fiber surface by electrophoretic deposition and electropolymerization to enhance the interfacial strength of epoxy resin composite.s* Applied Surface Science, vol. 499, p. 143929, 2020/01/01/ 2020.

[5] C. Wang, J. Li, S. Sun, X. Li, F. Zhao, B. Jiang, and Y. Huang, *Electrophoretic deposition of graphene oxide on continuous carbon fibers for reinforcement of both tensile and interfacial strength.* Composites Science and Technology, vol. 135, pp. 46-53, 2016/10/27/ 2016.

[6] A. K. Srivastava, V. Gupta, C. S. Yerramalli, and A. Singh, *Flexural strength enhancement in carbon-fiber epoxy composites through graphene nano-platelets coating on fibers.* Composites Part B: Engineering, vol. 179, p. 107539, 2019/12/15/ 2019.

[7] F. Awan and T. Subhani, *Preparation and Characterization of Carbon Nanotube Deposited Carbon Fiber Reinforced Epoxy Matrix Multiscale Composites.* Advanced Nano Research, vol. 1, no. 1, p. 14, 2017.

[8] X. Zhang, X. Fan, C. Yan, H. Li, Y. Zhu, X. Li, L. and Yu, *Interfacial Microstructure and Properties of Carbon Fiber Composites Modified with Graphene Oxide.* ACS Applied Materials & Interfaces, vol. 4, no. 3, pp. 1543-1552, 2012/03/28 2012.

[9] W. Qin, C. Chen, J. Zhou, and J. Meng, *Synergistic Effects of Graphene/Carbon Nanotubes Hybrid Coating on the Interfacial and Mechanical Properties of Fiber Composites.* Materials (Basel, Switzerland), vol. 13, no. 6, p. 1457, 2020.

[10] L. Liu, F. Yan, M. Li, M. Zhang, L. Xiao, L. Shang, and Y. Ao, *Improving interfacial properties of hierarchical reinforcement carbon fibers modified by graphene oxide with different bonding types.* Composites Part A: Applied Science and Manufacturing, vol. 107, pp. 616-625, 2018/04/01/ 2018.

[11] J. Fu, M. Zhang, L. Jin, L. Liu, N. Li, L. Shang, M. Li, L. Xiao and Y. Ao, *Enhancing interfacial properties of carbon fibers reinforced epoxy composites via Layer-by-Layer self assembly GO/SiO2 multilayers films on carbon fibers surface.* Applied Surface Science, vol. 470, pp. 543-554, 2019/03/15/ 2019.

[12] C. Xiao, Y. Tan, X. Wang, L. Gao, L. Wang, and Z. Qi, *Study on interfacial and mechanical improvement of carbon fiber/epoxy composites by depositing multi-walled carbon nanotubes on fibers.* Chemical Physics Letters, vol. 703, pp. 8-16, 2018/07/01/ 2018.

[13] A. Ashori, H. Rahmani, and R. Bahrami, *Preparation and characterization of functionalized graphene oxide/carbon fiber/epoxy nanocomposites.* Polymer Testing, vol. 48, pp. 82-88, 2015/12/01/ 2015.

[14] Y. Li, H. Zhang, Z. Huang, E. Bilotti, and T. Peijs, *Graphite Nanoplatelet Modified Epoxy Resin for Carbon Fibre Reinforced Plastics with Enhanced Properties.* Journal of Nanomaterials, vol. 2017, p. 5194872, 2017.

[15] S. S. Reddy, C. Yuvraj, and K. P. Rao, *Design, Analysis, Fabrication and Testing of CFRP with CNF Composite Cylinder for Space Applications.* International Journal of Composite Materials, vol. 5, no. 5, p. 2166, 2015.

[16] E. Bekyarova, E. Thosternson, A. Yu, M. Itkis, D. Fakhrutdinov, T. Chou, and R. Haddon, *Functionalized Single-Walled Carbon Nanotubes for Carbon Fiber−Epoxy Composites.* The Journal of Physical Chemistry C, vol. 111, no. 48, pp. 17865-17871, 2007/12/01 2007.

[17] S. Aldajah and Y. Haik, *Transverse strength enhancement of carbon fiber reinforced polymer composites by means of magnetically aligned carbon nanotubes.* Materials & Design, vol. 34, pp. 379-383, 2012/02/01/ 2012.

[18] T. Glaskova-Kuzmina, A. Aniskevich, M. Zarrelli, A. Martone, and M. Giordano, *Effect of filler on the creep characteristics of epoxy and epoxy-based CFRPs containing multi-walled carbon nanotubes.* Composites Science and Technology, vol. 100, pp. 198-203, 2014/08/21/ 2014.

[19] A. K. Pathak, M. Borah, A. Gupta, T. Yokozeki, and S. R. Dhakate, *Improved mechanical properties of carbon fiber/graphene oxide-epoxy hybrid composites.* Composites Science and Technology, vol. 135, pp. 28-38, 2016/10/27/ 2016.

[20] J. Kim, J. Cha, B. Chung, S. Ryu, and S. H. Hong. Composites Science and Technology, vol. 192, p. 108101, 2020/05/26/ 2020.

[21] J. Wei and J. Qiu, *Allyl-Functionalization enhanced thermally stable graphene/fluoroelastomer nanocomposites.* Polymer, vol. 55, no. 16, pp. 3818-3824, 2014/08/05/ 2014.

[22] Y. Cui, S. I. Kundalwal, and S. Kumar, "Gas barrier performance of graphene/polymer nanocomposites," *Carbon,* vol. 98, pp. 313-333, 2016/03/01/ 2016.

[23] J. Robinson, J. Burgess, C. Junkermeier, S. Badescu, T. Reinecke, F. Perkins, M. Zalalutdniov, J. Baldwin, J. Culbertson, P. Sheehan, and E. Snow, *Properties of Fluorinated Graphene Films.* Nano Letters, vol. 10, no. 8, p. 3001, 2010.

[24] X. Ye, P. Gong, J. Wang, H. Wanga, S. Ren, and S. Yang, *Fluorinated graphene reinforced polyimide films with the improved thermal and mechanical properties.* Composites: Part A, vol. 75, p. 96, 2015.

[25] Z. Chen, H. Huang, S. Yan, Z. Zheng, S. Liu, Y. Yuan, J. Zhao, and Y. Fu, *New Synthetic Approach of Fluorine-Containing Graphene Oxide for Improving Dielectric and Mechanical Properties of Polyimide Composites.* Ind. Eng. Chem. Res., vol. 26, p. 9926, 2017.

[26] W. Li, W. Zhao, L. Mao, S. Zhou, C. Liu, Z. Fang, and X. Gao, *Investigating the fluorination degree of FG nanosheets on the tribological properties of FG/PI composite coatings.* Progress in Organic Coatings, vol. 139, p. 105481, 2020.

[27] X. Yin, Y. Feng, Q. Zhao, Y. Liu, S. Li, H. Dong, W. Hu, and W. Feng, *Highly transparent, strong, and flexible fluorographene/fluorinated polyimide nanocomposite films with low dielectric constant.* Journal of Materials Chemistry C, vol. 6, p. 6378, 2018.

[28] E. Beh, G. Iftime, J. L. B. Rivest, J. Wei, and R. Pandey, *Composite Materials Comprising Chemically Linked Fluorographite-Derived Nanoparticles.* US, 2019, Application number: 16/225125.

[29] E. P. Douglas, H. R. Hamilton, J. C. Nino, A. Stewart, and J. Tatar, *Highly accelerated lifetime for externally applied bond critical fiber-reinforced polymer (FRP) infrastructure materials.* Department of Materials, Science & Engineering University of Florida, 2014.

[30] J. Wei, G. Iftime, J. L. B. Rivest, and S. Garner, *Fiber Reinforced Plastic Enhanced by Functionalized Graphene.* US, 2018, Application number 16/228456.

[31] J. S. Tomblin, Y. C. Ng, and K. S. Raju, *Material Qualification and Equivalency for Polymer Matrix Composite Material Systems: Updated Procedure.* DOT/FAA/AR-03/19, 2003.

[32] H. Barès, A. Bakandritsos, M. Medved, J. Ugolotti, P. Jakubec, O. Tomanec, S. Kalytchuk, R. Zboril, and M. Otyepka, *Bimodal role of fluorine atoms in fluorographene chemistry opens a simple way toward double functionalization of graphene.* Carbon, vol. 145, pp. 251-258, 2019/04/01/ 2019.

[33] K.-J. Jeon, Z. Lee, E. Pallak, L. Moreschini, A. Bostwick, C. Park, R. Mendelsberg, V. Radmilovic, R. Kostecki, T. Richardson, and E. Rotenberg, *Fluorographene: A Wide Bandgap Semiconductor with Ultraviolet Luminescence.* ACS Nano, vol. 5, no. 2, pp. 1042-1046, 2011/02/22 2011.

[34] Z. ALOthmana, E, Yilmazb, M. Habilla, I. Alsohaimia, A. Aldawsaria, N. Al-Harbic, and M. Soylak, *Triethylenetetramine modified multiwalled carbon nanotubes for the efficient preconcentration of Pb(II), Cu(II), Ni(II) and Cd(II) before FAAS detection.* RSC Advances, vol. 5, pp. 106905-106911, 2015.

[35] N. Ngamkitpinyo and A. Imyim, *Selective extraction of Pb(II) using triethylenetetramine-modified polystyrene-divinylbenzene resin.* J. Mater. Environ. Sci., vol. 6, no. 5, pp. 1562-1569, 2015.

[36] S. Jahanshahi, A. Pizzi, A. Abdulkhani, and A. Shakeri, *Analysis and Testing of Bisphenol A—Free Bio-Based Tannin Epoxy-Acrylic Adhesives.* Polymers, vol. 8, p. 143, 2016.

[37] R. A. Chowdhury, M. Hosur, M. Huruddin, A. Tcherbi-Narteh, A. Kumar, V. Boddu, and S. Jeelani, *Self-healing epoxy composites: preparation, characterization and healing*

performance. Journal of Materials Research and Technology, vol. 4, no. 1, pp. 33-43, 2015/01/01/ 2015.
[38] N. Liaros, A. B. Bourlinos, R. Zboril, and S. Couris, *Fluoro-graphene: nonlinear optical properties.* Opt. Express, vol. 21, pp. 21027-21038, 2013.
[39] C. A. Silva, E. Marotta, M. Schuller, L. Peel, and M. O'Neill, *In-Plane Thermal Conductivity in Thin Carbon Fiber Composites.* Journal of Thermophysics and Heat Transfer, vol. 21, no. 3, pp. 460-467, 2007.
[40] F. Wang and X. Cai, *Improvement of mechanical properties and thermal conductivity of carbon fiber laminated composites through depositing graphene nanoplatelets on fibers.* Journal of Materials Science, vol. 54, no. 5, pp. 3847-3862, 2019/03/01 2019.
[41] M. Srinivasan, P. Maettig, K. Glitza, B. Sanny, A. Schumacher, M. Duhovic, and J. Schuster, *Out of Plane Thermal Conductivity of Carbon Fiber Reinforced Composite Filled with Diamond Powder.* Open Journal of Composite Materials, vol. 6, pp. 41-57, 2016.
[42] M. Martins, R. Gomes, L. Pina, C. Pereira, O. Reichmann, D. Teti, N. Correia, and N. Rocha, *Highly Conductive Carbon Fiber-Reinforced Polymer Composite Electronic Box: Out-Of-Autoclave Manufacturing for Space Applications.* Fibers and Polymers, vol. 6, p. 92, 2018.
[43] N. L. Newhouse, *Development of Improved Composite Pressure Vessels for Hydrogen Storage. Hexagon Lincoln* Department of Energy Efficiency and Renewable Energy Fuel Cell Technologies Office 2016.
[44] J. Humpenöder, *Gas permeation of fibre reinforced plastics.* Cryogenics, vol. 38, no. 1, pp. 143-147, 1998/01/01/ 1998.
[45] D. Schultheiß, *Permeation Barrier for Lightweight Liquid Hydrogen Tanks*. PhD, Mathematisch-Naturwissenschaftlichen, Universit¨at Augsburg, Publikationsserver der Universit¨at Augsburg, 2007.
[46] P. D. L. Staff, W. Woishnis, Ed. *Permeability and Other Film Properties of Plastics and Elastomers*. Plastics Design Library, 1995.
[47] H. Saechtling and F. Pabst, *Kunststoff-Taschenbuch* (no. v. 24). Hanser, 1989.
[48] D. O. Stodilka, N. P. Kherani, W. T. Shmayda, and S. J. Thorpe, *A tritium tracer technique for the measurement of hydrogen permeation in polymeric materials.* International Journal of Hydrogen Energy, vol. 25, no. 11, pp. 1129-1136, 2000/11/01/ 2000.

CURE PATH DEPENDENCE OF REACTION-INDUCED PHASE SEPARATION IN GLASSY AMORPHOUS POLYMER NETWORKS

Matthew C. Hartline, Jared Bates, Nathaniel L. Prine, David C. Walker, David E. Garcia, Xiaodan Gu, Jeffrey S. Wiggins
Department of Polymer Science and Engineering, University of Southern Mississippi
Hattiesburg, Mississippi, USA

ABSTRACT

Reaction induced phase separation of high T_g thermoplastics from epoxide/amine networks during cure is one common method for the toughening of networks. However, the toughness realized in the final network entirely depends on the morphology, which can range from droplet dispersed, co-continuous, to phase inverted. An additional complication the dependence of mechanism of phase separation on the cure path on the. Spinodal decomposition is the preferred phase separation mechanism, but the competing process is nucleation growth phase separation. To realize maximum toughening effect, co-continuous morphology is desired, and the most consistent morphologies are developed through the spinodal decomposition phase separation mechanism. For this work, tetraglycidyl -4,4'-diaminodiphenylmethane (TGDDM), 4, 4'-diaminodiphenylsulphone (44DDS), and non-reactive polyetherimide (PEI) are dissolved in solvent until homogenous, then the solvent is removed and the 44DDS is clarified at 125 °C until only the b-staged, transparent yellow film remains. Rheology from clarified sample through gelation resolves the phase separation mechanism. Samples are cured from 80 °C to 180 °C at ramp rates of 1 and 5 °C/min and remain at the 180 °C isotherm for two hours. Samples are further post cured at 220 °C for an additional hour to ensure full cure. Morphology of cured networks is examined with atomic force microscopy-based infrared spectroscopy (AFM-IR), which additionally establishes the chemical identity of each phase. Morphology is confirmed with optical microscopy, and dynamic mechanical analysis (DMA), where thermomechanical response correlates with morphology type. Final morphology depends heavily on the thermoplastic loading level, while the mechanism of phase separation can depend upon the cure rate. At all cure rates, co-continuous networks are realized when PEI loading $\geq$ 15 wt.%.

Keywords: Co-continuous, Thermoplastic, Thermoset, Phase Separation
Corresponding author: Jeffrey S. Wiggins

1. INTRODUCTION

The brittle nature of highly crosslinked thermosets often limits their intended uses. In lower moduli networks with reduced crosslink density, the typical strategy to extend the utility and toughen the matrix is to include a secondary, ductile phase. The secondary phase is typically a rubber, which allows for energy dissipation, but reduces T_g and limits upper use temperatures.[1] In high T_g thermoset strategies, an alternative strategy is to include a thermoplastic with a similar T_g.[2] After solubilization of the thermoplastic into the thermoset, the epoxide:amine reaction proceeds to the moment where the thermoplastic loses solubility, and reaction induced phase separation begins. The most beneficial morphology from a toughening standpoint is when a co-continuous morphology is attained, which only occurs when the phase separation proceeds through a nucleation growth mechanism.[3] Monitoring phase separation in real time is possible through monitoring rheology through gelation, where an increase in viscosity is associated with the more desirable spinodal decomposition, while a decrease in viscosity is due the nucleation growth mechanism.[4,5] This work determines the effects of loading level and cure rate on morphology development during reaction induced phase separation of high T_g thermoplastics from highly crosslinked thermoset systems.

2. EXPERIMENTATION

2.1 Materials

The epoxide is low viscosity tetraglycidyl -4,4'-diaminodiphenylmethane (TGDDM) (Gabriel RAR921). The amine curative used is 4, 4'-diaminodiphenylsulphone (44DDS) (Gabriel). Polyetherimide (PEI) (Sabic, Ultem 1000P), with a molecular weight (Mn) of 39,000 g/mol and particle size 350 μm. HPLC grade dichloromethane (99.9%, DCM) is utilized for solubilization of 44DDS and PEI into TGDDM. All materials are used as received.

2.2 Preparation of Networks

Solvent casting is used to prepare homogenous films. The TGDDM and 44DDS are weighed and combined stoichiometrically in scintillation vials, with the desired loading level of PEI. DCM is added up to the neck, and the vial is vortexed until a milky, yellow homogenous mixture results. The homogenous mixture is poured into a shallow aluminum dish, and the DCM is removed in a well-ventilated convection oven at 100 °C. After the DCM is removed, the opaque, milky film is heated to 125 °C to clarify the 44DDS. After clarification, the films are cooled and cryo-pulverized and stored in the freezer until cast into sample geometry. Samples are degassed and consolidated, then cured at the indicated cure rates (1, 5, or 20 °C) from 80 to 180 °C, followed by a 2-hour isothermal hold at 180 °C. Small specimens are post-cured at 220 °C for 1 hour.

2.3 Rheology through Vitrification

Small angle oscillatory shear rheology is performed on consolidated specimens using 25 mm disposable aluminum parallel plates, with a 0.5 mm gap and 5% strain. The test is terminated shortly after vitrification, denoted by the crossover of G'/G". Samples are heated from 80 to 180 °C at the indicated cure rate and held at 180 °C until vitrification when required.

2.4 Optical Microscopy

Cured specimens are cryo-fractured in liquid nitrogen and examined using the Laxco SEBA 2 digital microscope using 4x through 100x magnification for observation of phase separated morphology.

2.5 Dynamic Mechanical Analysis

Dynamic mechanical analysis of post cured specimens is performed in tension mode with a ramp rate of 3 °C/min from room temperature to 300 °C. The appearance of two T_gs is examined, and the ratio of the height of the thermoplastic to thermoset tanδ is correlated with observed morphology.

2.6 Atomic Force Microscopy- Based Infrared Spectroscopy

To more closely examine the submicron morphology and to determine the partitioning of thermoplastic and thermoset material into phase separated domains, AFMIR is used. Samples are microtomed to and the imide peak at 1728 cm^{-1} is monitored to determine which domains are thermoplastic filled, as the imide moiety is limited to the polyetherimide-containing domains.

3. RESULTS

3.1 Rheology through Vitrification

The viscosity well through vitrification for all neat specimens at all cure rates displays a featureless curve, with no increase that would indicate phase separation, as expected, Figure 1. In the 15 and 20 wt% PEI specimens, the increase in viscosity with all cures denotes the desired spinodal decomposition phase separation. The 10 wt% PEI displays the most interesting rheological behavior, with a drop in viscosity indicationg nucleation growth phase separation at the 1 °C/min cure rate, but an increase in viscosity with the 5 and 20 °C/min cure rates, denoting spinodal decomposition.

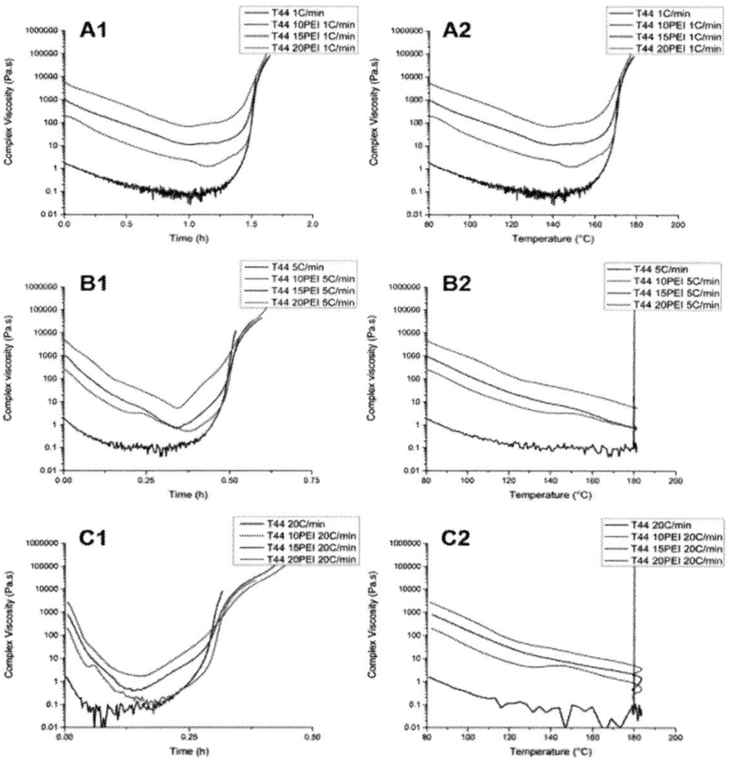

Figure 1. Rheology through cure of PEI filled T44. Plots are sorted by ramp rate: A, 1 °C/min; B, 5 °C/min; C, 20 °C/min, each with respect to time (1) and temperature (2).

3.2 Optical Microscopy

The cryo-fractured samples are examined using optical microscopy. At both cure rates, the neat T44, 15 and 20wt% PEI samples display different states of co-continuous morphology, Figure 2, and Figure 3. While the 15 wt% PEI specimens display a globally co-continuous morphology, large thermoplastic-rich and thermoset-rich domains of several hundred microns. The 20 wt% PEI samples display a much finer, nearly uniform nodular morphology with nodules only a few microns in size. The 10 wt% specimens display different morphologies depending on the cure rate, with the 1 °C/min cure path appearing to have a droplet dispersed morphology, while the 5 °C/min posseses a much more heterogenous morphology, with large, phase separated features.

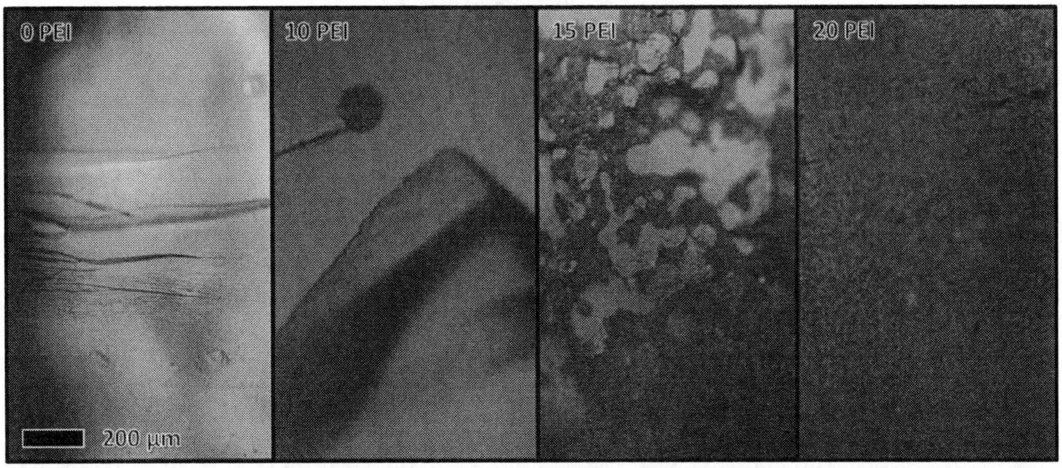

Figure 2. Low magnification optical micrographs of T44/PEI networks cured at 1 °C/min.

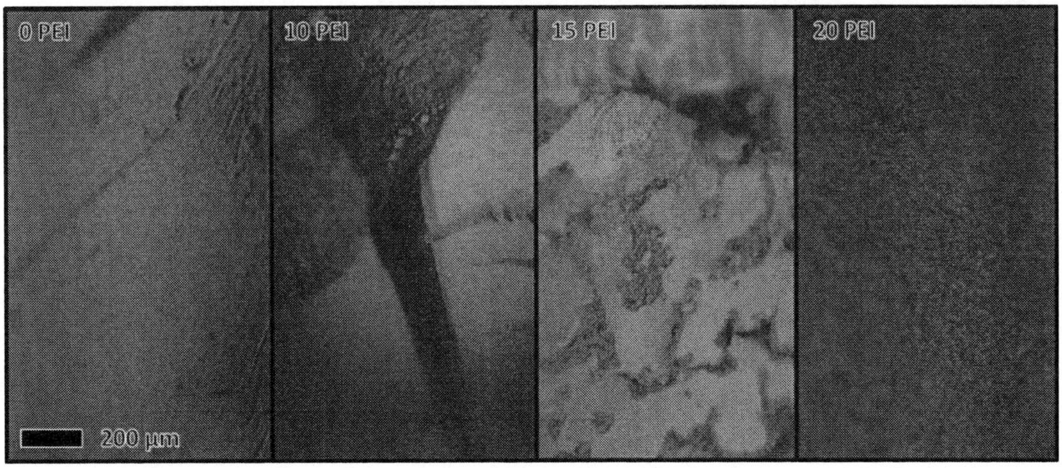

Figure 3. Low magnification optical micrographs of T44/PEI networks cured at 5 °C/min.

The dramatic difference in the morphology of the 10 wt% PEI specimens is likely a manifestation of the different phase separation mechanisms. The nucleation growth mechanism at the slow 1 °C/min cure rate would result in the formation of small droplets that slowly grow, while the spinodal decomposition mechanism would coarsen and continue to demix until gelation fixed the morphology. This demixing would likely result in fewer, large droplets, but due to the tetrafunctional epoxide and amine used, the early gelation process will likely prematurely lock in large domains of unconnected thermoplastic-rich domains.

3.3 Dynamic Mechanical Analysis

To probe the mechanical response of the network in relation to the developed morphology, dynamic mechanical analysis is performed, Figure 4 and Figure 5. The increasing response with increased loading level is consistent with the results previously observed.[6]

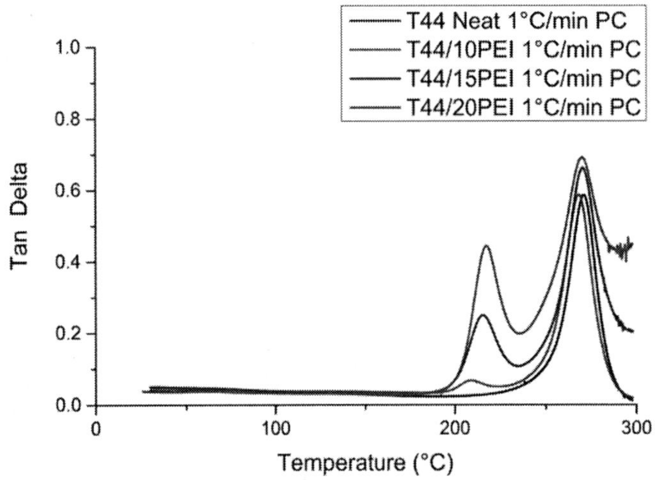

Figure 4. Tension DMA of T44/PEI post cured networks cured at 1 °C/min.

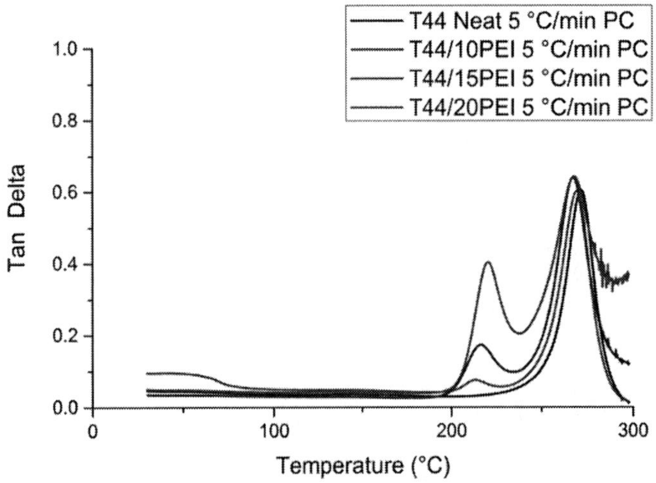

Figure 5. Tension DMA of T44/PEI post cured networks cured at 5 °C/min.

However, the 15 wt% PEI response was not noted by Girard-Reydet et al, however the intensity is consistent with the morphology.[6] While the global network is co-continuous and connected, the smaller domains are thermoplastic- or thermoset-rich with the inverse phase droplet dispersed, the intermediate response of the tan ratio is consistent with the intermediate response.

Table 1. Tabulated DMA results of phase separated

	Tα	Hα	Tβ	Hβ	β/α
T44_1C	271.89	0.586	-	-	0
T44_10PEI_1C	268.14	0.586	208.45	0.067	0.11
T44_15PEI_1C	270.48	0.661	215.39	0.249	0.38
T44_20PEI_1C	270.15	0.691	217.54	0.440	0.64
T44_5C	271.39	0.604	-	-	0
T44_10PEI_5C	268.92	0.601	212.54	0.075	0.12
T44_15PEI_5C	266.78	0.639	215.59	0.173	0.27
T44_20PEI_5C	268.98	0.641	220.05	0.434	0.68

3.4 Atomic Force Microscopy- Based Infrared Spectroscopy

The atomic force microscopy allows for the specific chemical identity of different regions to be identified in a heterogenous sample. The top image is the AFM-height image, and the bottom is the AFM-IR is the lower image. In the case of the T44/PEI reaction induced phase sepration, the regions containing PEI are highlighted in orange, while the thermoset regions appear blue, Figure 6.

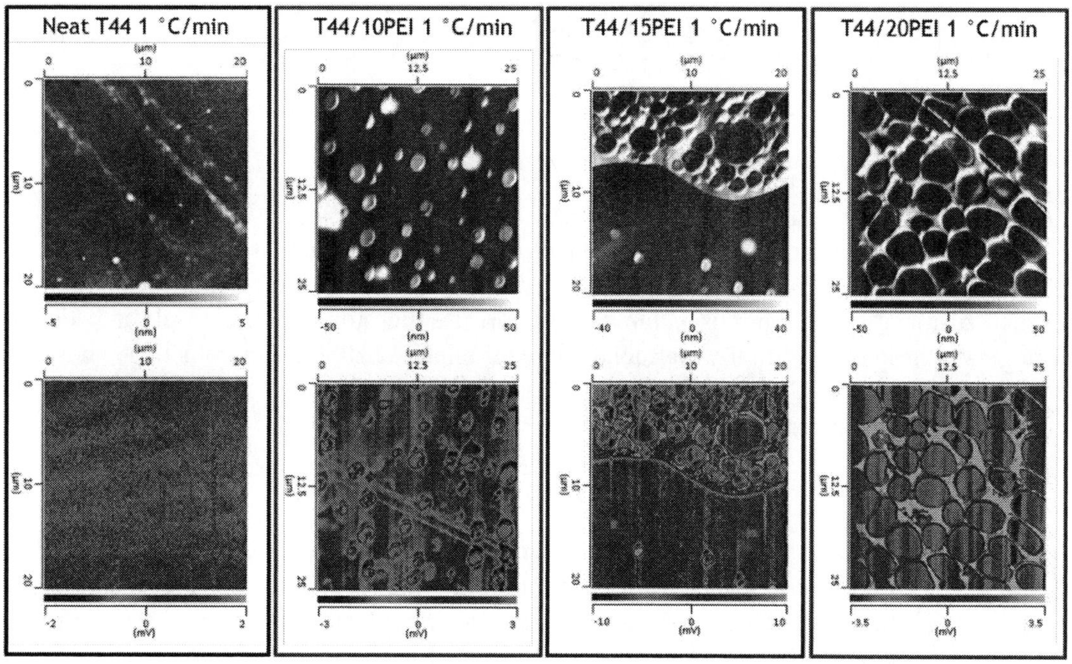

Figure 6. Morphology of T44/PEI cured at 1 °C/min. AFM (top) and AFM-IR (bottom).

With the 1 °C/min cure, the morphology of PEI loaded samples ranges from droplet dispersed to co-continuous domains. The 10 wt% PEI contains fairly large droplets of thermoplastic in a thermoset-rich continuous phase. The size of these thermoplastic-droplets is consistent with the droplets seen in the thermoset-rich domain of the 15 wt% PEI micrograph, and the top of the same image displays a thermoplastic-rich domain where thermoset droplets coalesced and could not pierce through the phase boundary after the thermoplastic vitrified. The 20 wt% PEI displays a thermoplastic web with thermoset droplets, however the volume fraction of the droplets is significantly higher than that of the thermoplastic. This is consistent with the co-continuous mechanical response of the network in the DMA.

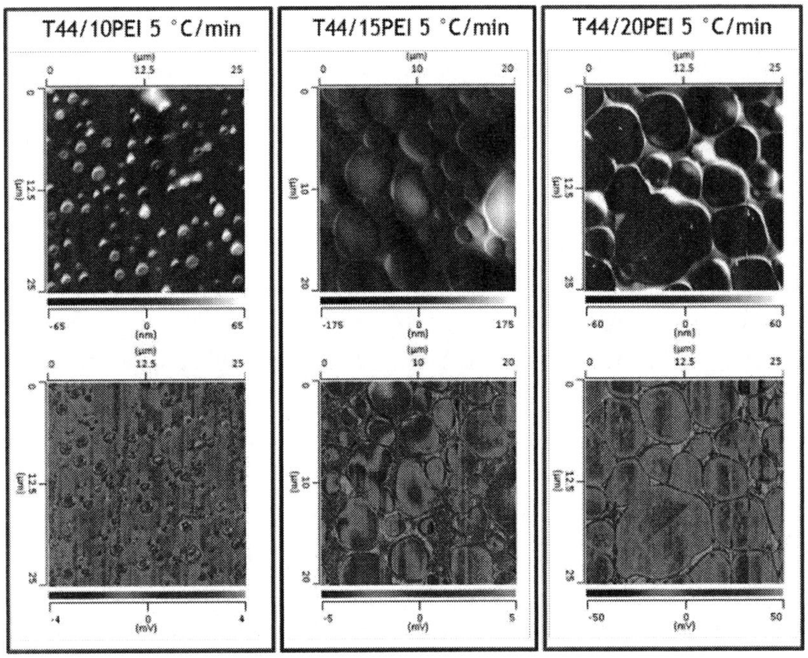

Figure 7. Morphology of T44/PEI cured at 5 °C/min. AFM (top) and AFM-IR (bottom).

The behavior is similar when the networks cured at 5 °C, Figure 7. The area observed from the 10 wt.% PEI has an increased number of thermoplastic domains, that on average have a smaller size. This is likely due to the spinodal decomposition of the faster cure path. The 15 wt.% PEI specimen displays small thermoset domains within the thermoplastic-rich webbing, but no real thermoplastic droplets within the thermoset-rich domains. At 20 wt.% there is little phase mixing, but the larger thermoset domains are likely a result of the lower viscosity during the early stages of the 180 °C cure isotherm.

4. CONCLUSIONS

Reaction induced phase separation is a complex series of events controlled by both kinetic and thermodynamic phenomena, with added complexities due to the viscoelastic nature of polymers. With the evidence of these studies, the morphology development is more consistent along the entirety of the specimen with more rapid cure protocol, which saves process time and improves uniformity of morphology development. However, the cure rate is limited by heating method used. This is confounded by the differences in rheology through cure, where the lower viscosity allows for more facile diffusion which allows for more regular features. Future studies will elucidate the effect on temperature gradients during cure on reaction.

5. REFERENCES

[1] Bucknall, C. B.; Yoshii, T. Relationship Between Structure and Mechanical Properties in Rubber-Toughened Epoxy Resins. *Br Polym J* **1978**, *10* (1), 53–59. https://doi.org/10.1002/pi.4980100110.

[2] Bucknall, C. B.; Gilbert, A. H. Toughening Tetrafunctional Epoxy Resins Using Polyetherimide. *Polymer (Guildf).* **1989**, *30* (2), 213–217. https://doi.org/10.1016/0032-3861(89)90107-9.

[3] Yu, G.; Wu, P. Effect of Chemically Modified Graphene Oxide on the Phase Separation Behaviour and Properties of an Epoxy/Polyetherimide Binary System. *Polym. Chem.* **2014**, *5* (1), 96–104. https://doi.org/10.1039/c3py00878a.

[4] Poncet, S.; Boiteux, G.; Pascault, J. P.; Sautereau, H.; Seytre, G.; Rogozinski, J.; Kranbuehl, D. Monitoring Phase Separation and Reaction Advancement in Situ in Thermoplastic/Epoxy Blends. *Polymer (Guildf).* **1999**, *40* (24), 6811–6820. https://doi.org/10.1016/S0032-3861(99)00104-4.

[5] Park, S. J.; Bae, K. M.; Seo, M. K. A Study on Rheological Behavior of MWCNTs/Epoxy Composites. *J. Ind. Eng. Chem.* **2010,** *16* (3), 337–339. https://doi.org/10.1016/j.jiec.2010.01.051.

[6] Girard-Reydet, E.; Vicard, V.; Pascault, J. P.; Sautereau, H. Polyetherimide-Modified Epoxy Networks : Influence of Cure Conditions on Morphology and Mechanical Properties. *J. Appl. Polym. Sci.* **1996**, *65* (12), 2433–2445. https://doi.org/10.1002/(SICI)1097-4628(19970919)65:12<2433::AID-APP15>3.0.CO;2-1.

New high-temperature polyimides to push high-performance polymer boundaries

Christian Maurer, Günter Gasparin, Dieter Danzer
Evonik Fibres GmbH, Lenzing, AUSTRIA

Ankur Kant*, Nathan Schindler
Evonik Corporation
Parsippany, NJ/USA

ABSTRACT

In this paper the authors review the main properties of polyimides and explain the influence of certain chemical structures on the obtained mechanical and thermal properties. Based on two examples new developments are presented, with which limitation of existing products have been overcome and the operating temperature could be expanded. A new type of polyimide molding powder shows extraordinarily high thermo-oxidative stability with constant strength over more than 1000 hours, when storing at 340°C in air. A new polyimide fiber with higher T_g demonstrates less shrinkage, enabling use in filtration processes with surge temperatures up to 280°C.

Keywords: Polyimides, composites, fibers
*Corresponding author: Ankur Kant (ankur.kant@evonik.com)

1. INTRODUCTION

Polyimides are a very interesting class of high-performance materials due to their high mechanical strength and thermal stability. They are widely used as parts, films and fibers. In the following paper the basic properties of this material are reviewed, and new product strategies are presented, showing how temperature-based boundaries can be overcome.

2. REVIEW ON POLYIMIDES

Polyimides are thermoplastic material with a fully aromatic backbone. The polymers are synthesized in a polycondensation reaction from dianhydrides and diamines/diisocyanates. Figure 1 gives a brief overview of typical monomers used for polyimides.

The reaction of PMDA and ODA was first developed by DuPont around 1960 [1]. It requires a two-step synthesis and forms a soluble intermediate (polyamic acid, PAA), which is transferred in a polyimide by cleavage of H_2O by heat or by chemical substances. The process is applicable for all diamines and dianhydride in order to form films, powders or coatings.

Upjohn developed a one-step reaction of diisocyanates and dianhydrides and the formation of CO_2. [2] The one-step synthesis is limited to soluble polyimides (e.g. P84®) and a small selection of aromatic diisocyanates.

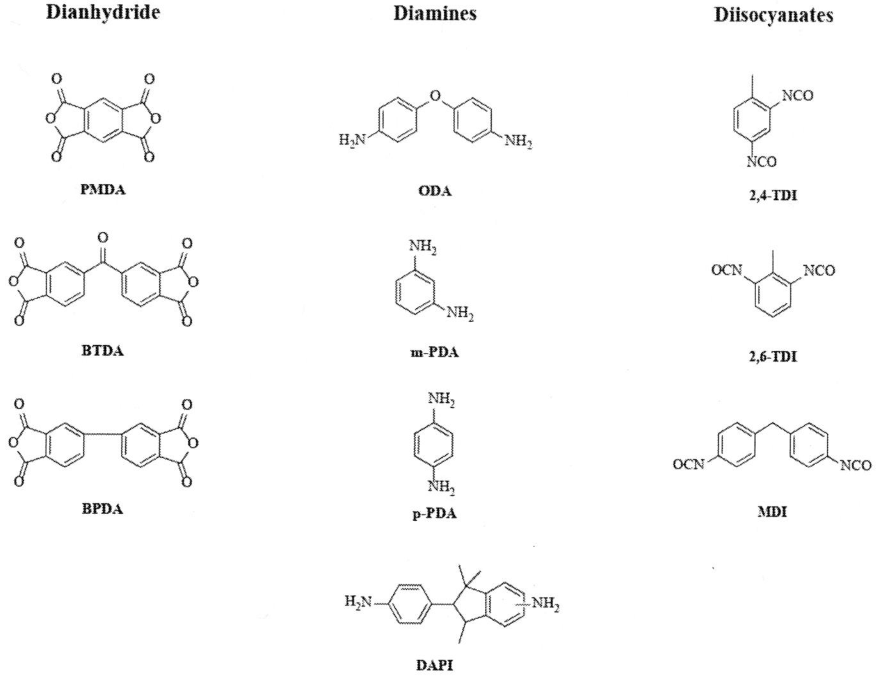

Figure 1: Typical monomers used in polyimide products

The fully aromatic nature of the monomers ensures an outperforming thermal stability and high glass transition temperature (T_g). Polyimides are also inherently resistant to flame combustion and can therefore be used in high fire safety standard applications such as in aerospace.

Table 1: Glass transition temperature and limited oxygen index of commercially available polyimides [3]

	PMDA-ODA	BPDA-ODA	BPDA-*p*-PDA	BTDA-TDI/MDI
Tg [°C]	> 500	270	> 500	330
LOI	53	49	66	38

Table 1 gives an overview of the glass transition temperature of typical polyimides. All polyimides show a T_g above 300°C. The high glass transition temperature enables use of the material at continuous operating temperatures of up to 250°C and ability to withstand peak operating temperatures far above 300°C for short durations. The LOI of all polyimide materials is significantly above 21% and all polyimide materials are also rated V-0 in UL94 test.

The data in Table 1 also highlight the fact that alkyl side groups of the monomers lowers the glass transition temperature and LOI. The main reason is the lower oxidative stability of alkyl versus aryl groups.

Due to its fully aromatic nature polyimides are insoluble in common solvents such as alcohols, esters, amines and hydrocarbons. Rare exceptions of this rule are polyimides where diamines have alkyl side group next to the amino group (e.g. P84® or Matrimid®). These polymers show reasonable solubility in aprotic dipolar solvents such as DMF or NMP. While the imide group is stable under acidic conditions, the polymers are sensitive to hydrolysis under basic conditions. Thus, polyimides are preferably used under dry conditions. In general polyimides have a limited tendency to absorb moisture. Especially PMDA and BPDA based polyimides demonstrate moisture absorption far below 0.5% after 24hrs at 23°C.

Molded polyimide parts show high mechanical strength and extremely high flexural modulus without any reinforcing fillers. Table 2 compares the tensile test data of three different polyimides. Polyimides based on BPDA shows the highest tensile modulus. Typical for all polyimides the strain at break is rather low compared to PEEK or TPI.

Table 2: Tensile and flexural test data of molded stock shapes (prepared by HCM or isostatic pressing) [4-6]

	PMDA-ODA	BPDA-ODA	BTDA-TDI/MDI
Density [g/cm³]	1,43	1,39	1,34
Tensile strength [MPa]	86	120	116
Tensile strain at break [%]	7,5	4,5	3,8
Flexural Strength [MPa]	110	160	170
Flexural Modulus [MPa]	3102	4200	3450

3. APPLICATION OF POLYIMIDES

3.1 High Performance Polymers and Polyimides

Polyimides are used in many high-performance demanding applications such as composites, fuel cells, displays, and to replace metal and ceramic parts. With their characteristic amber color, polyimides are known for excellent thermal stability, very good radiation resistance, and good chemical resistance. Polyimide's superior performance under the combination of high temperature and load sets them apart from other high temperature polymers.

High temperatures or frictional wear at high speeds and loads often circumscribes the use of ordinary engineering plastics; hence, advanced high-performance polymers have taken their place in demanding applications. Plastics processors can use polyimides where ordinary plastics would soon melt or decompose. For example, polyimides exhibit remarkable heat stability and creep resistance even at an elevated temperature of 250°C or higher — (Figure 2)

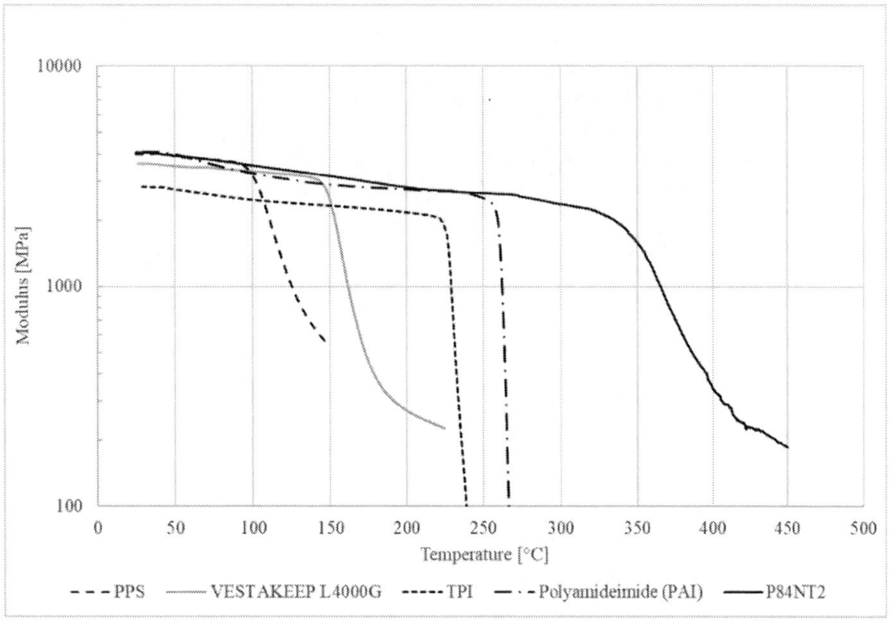

Figure 2: DMA analyses of various high-performance polymers (1 Hz, 2K/min)

3.2 Polyimide Market and Applications

Polyimides found a home in aerospace and electrical insulation markets soon after their introduction. This was primarily driven by space and military applications. Later, metal replacement in commercial aviation and automotive sector played a significant role in further development of polyimides. Polyimide powders are traditionally used to make parts for metal and ceramic replacement in applications ranging from Aerospace to glass-handling components. In addition, polyimides have found a niche in many tribological applications.

Here is a short summary of key applications of Polyimides in Automotive and Aerospace markets:

Automotive:

Superior tribological properties of polyimides combined with ease of adding fillers such as PTFE helped polyimides gain strong foothold in brakes, fuel systems and transmission applications in automotive industry. Some applications are as follows:

Brakes: friction pad (see Figure 3), valve seat, wear pad.

Fuel system: bushing, seal ring, valve seat bearings, charging valve and bearing retainer in turbo chargers.

Transmission: seal ring, thrust washer, valve seat, ball check valve.

Other areas: thrust plugs and washer in electrical motors, vane pump, hinge bushing in doors.

Figure 3: Friction pads used in brakes.

Aerospace:

With 45 percent lower specific gravity than Aluminum, Polyimides offer great weight saving opportunities to replace metal parts in many aerospace applications. Some examples are as follows:

Flight dynamics: Primary and secondary control actuators

Jet engine components: Stator vane bushes, thrust washer

Landing gear: locknut inserts, bearings for shock strut

In addition to applications in Automotive and Aerospace markets, Polyimides have found home in many industrial applications such as wafer processing and plasma torch tips. Superior

Polyimide fibers, such as P84®, have been used in high temperature filtration applications for thirty years. Polyimide fibers can withstand peak operating temperatures up to 260°C, which places it above most other polymeric fibers (Figure 4). Polyimide fibers exhibit excellent textile characteristics, such as tensile strength and elongation at break, making them suitable for non-woven and woven applications.

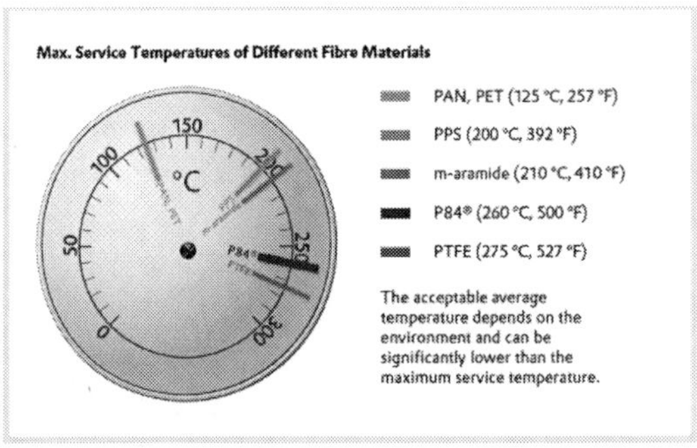

Figure 4: Maximum peak operation temperature of different polymeric fibres.

Finally, P84 polyimide fibers bring an additional benefit in filtration performance. The unique, irregular multilobal shape of P84® fiber increases the specific surface area by 65% compared to traditional round fibers. Since P84® fibers do not melt when heat treated or singed, the unique fiber shape is retained (Figure 5). Therefore, P84® fibers provide filtration efficiencies approaching the same level as membranes, without the pressure drop caused by the tight pore structure of a membrane.

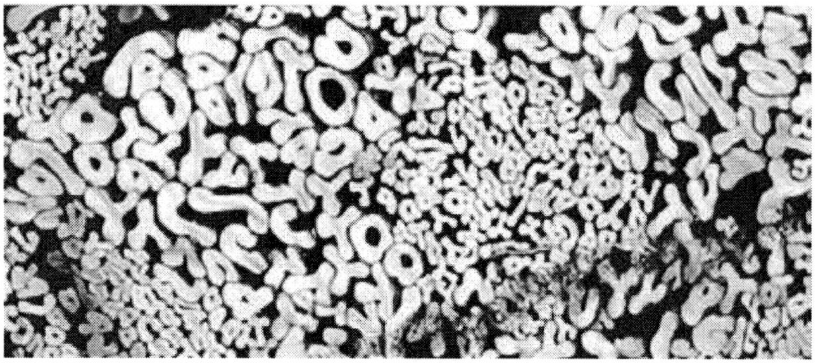

Figure 5: Typical cross section of P84® fibers.

Polyimide fibers have also been adopted in high tech protective clothing and insulative products.

4. NEW POLYIMIDES TO PUSH HIGH PERFORMANCE POLYMER BOUNDARIES

Evonik Fibres GmbH recently developed and launched two new polyimide products to overcome limitations of existing polyimide materials in terms of operating temperature.

4.1 Polyimide molding powders with high thermal-oxidative stability

Polyimides are well known for high thermal stability (up to 250°C) concerning degradation and deformation as well as low outgassing. However, above 250°C the available polyimide materials have some drawbacks compared to aluminum or other inorganic materials that limits the use of polyimides. Thus, there was a clear need to develop a new material with improved thermal stability and mechanical strength compared to the existing polymers. We are pushing the boundaries for the application of polyimide materials to temperatures up to 300°C, providing light weight solutions under extreme conditions.

Based on the above described requirements a new fully aromatic polyimide material (P84®UHT) for higher continuous operating temperature was developed. The new polymer can be processed via standard sintering processes such as hot compression molding or direct forming.

The mechanical properties of the molded material (via HCM, density 1.44 g/cm³) have been tested in a tensile test (ISO 527).

Table 3: mechanical properties of new polyimide material.

	Tensile strength [MPa]	Elongation at yield [%]	Tensile modulus [MPa]
23°C	142	4,4	5588
200°C	71	7,9	3307

The material shows high strength and stiffness without the need for reinforcing fillers. Even at elevated temperature (200°C) the strength remains at a reasonable level.

It was demonstrated in a simple thermo-oxidative stability test, that the new P84®UHT polyimide material can withstand high temperatures with significantly lower mass loss (40 to 70% lower). The polyimide exhibits the lowest mass loss with around 0,1 % after 100 hours at 315°C (Figure 6)

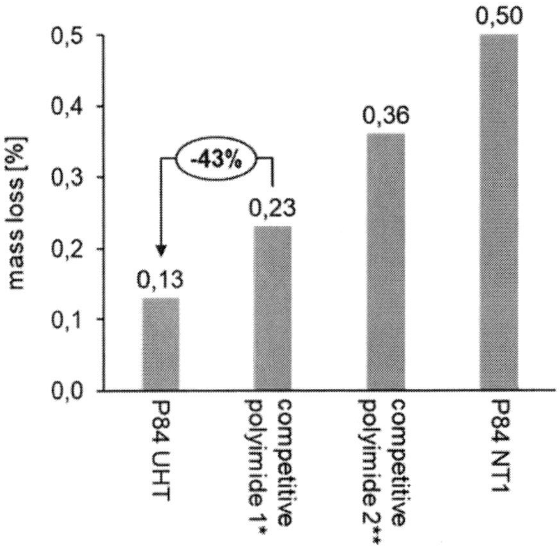

*Figure 6: Mass loss of polyimide parts after 100 hrs at 315°C (*competitive polyimide 1 is a BPDA based polyimide, **competitive polyimide 2 is a standard PMDA-ODA based polyimide).*

This polyimide material has also been investigated in long term thermal stability trials, in which the test bars (80x10x4 mm) were stored at 340°C under air in a convection furnace. Test bars have been analyzed after 100 to 3000 hours concerning residual strength (flexural test according to ISO 178 at ambient temperature). 30 samples have been placed at the same time into the furnace. Every 300 hours three samples were taken out of the furnace, cooled to ambient temperature and analyzed in a three-point bending strength test.

In the flexural strength test (ISO 178) it was demonstrated that even after 1000 hours the new polyimide material P84®UHT shows no change in mechanical strength (Figure 7). The flexural strength remains above 200 MPa, while the benchmark material (competitive polyimide 1) shows a drop of 20% from its initial strength within the first 1000 hours. The new P84®UHT shows extraordinarily high stability in the first 1500 hours and an equivalent performance to the benchmark in the long-term.

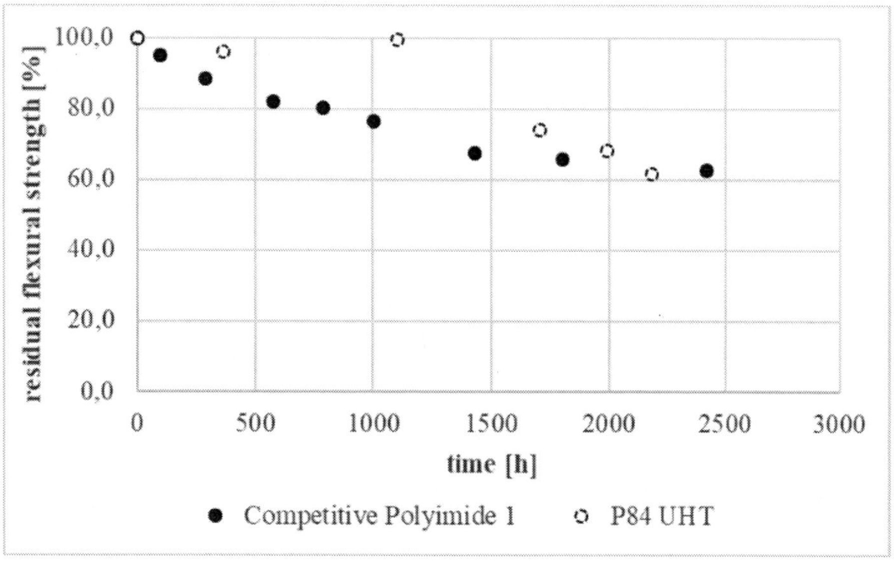

Figure 7: Residual flexural strength of polyimide materials after storage at 340°C.

The new polyimide material was also designed to withstand high loads in tribological contacts. The improved thermo-oxidative stability helps to improve the wear resistance of this new polyimide material. The explanation is that friction is causing high temperatures in the wear contact, which results in degradation of the polymeric material (= wear). P84®UHT can withstand those high temperatures resulting in less wear.

For a first basic comparison the material was tested in an unlubricated pin-on-disc (PoD) test setup (load: 8 MPa, speed: 1 m/s, metal counterpart (disc) is 1.3505 (100Cr6) [23°C] or 1.4108 (Cronidur) [250°C]). For the test an unfilled material as well as a compound with 15wt% graphite have been used. Both P84®UHT materials have been processed via direct forming (density: 1,36 g/cm³ [unfilled], 1,46 g/cm³ [filled]), a stock shape of competitive polyimide 1 was used as received.

Figure 8 shows data for specific wear rate and coefficient of friction (CoF) at 23°C and 250°C. For both - unfilled and 15% graphite filled grades - the material demonstrates an extraordinarily low wear rate which increases by raising the temperature. Compared to other high-performance polyimide, unfilled P84®UHT exhibits low wear at ambient temperature.

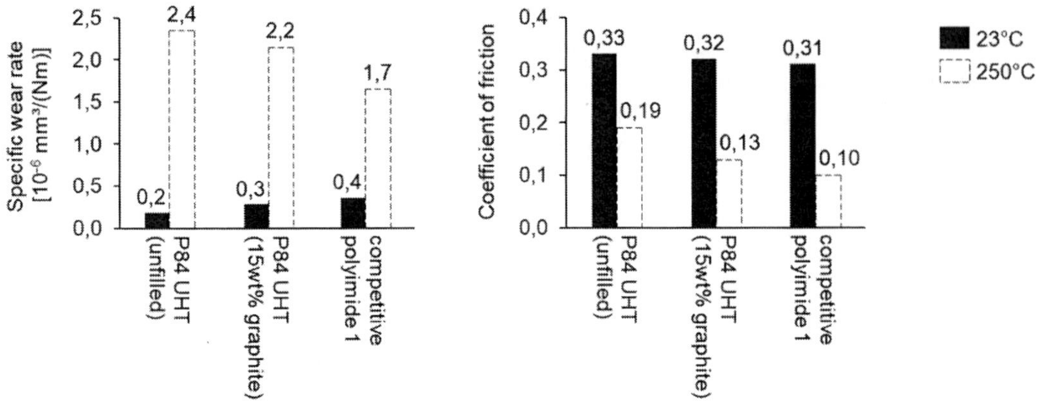

Figure 8: Results of PoD test with P84®UHT.

In parallel, the CoF at 23°C is not affected when the new polyimide material is compounded with solid lubricants (graphite). The neat polyimide P84®UHT shows similar performance compared to unfilled competitive polyimide 1. Enhancing the temperature to 250°C significantly decreases the CoF of all materials.

Further application tests with P84®UHT highlighted excellent wear resistance results in industrial applications such as bearings or thrust washers. Figure 9 shows results of sliding bearing tests in which P84®UHT with graphite has been compared to other high-performance polymers. The test has been performed with plastic bearings at a rotational speed of 1 m/s and a normal force of 100 and 250N at ambient temperature. The bearings (outer diameter: 34 mm, inner diameter: 30 mm, length 20 mm) were prepared by direct forming (P84®UHT) or injection molding (TPI, PEEK). The results demonstrate the high wear resistance of P84®UHT compared to PEEK and TPI. Depending on the Fv level the wear resistance of P84®UHT is decreasing slightly while keeping a good coefficient of friction.

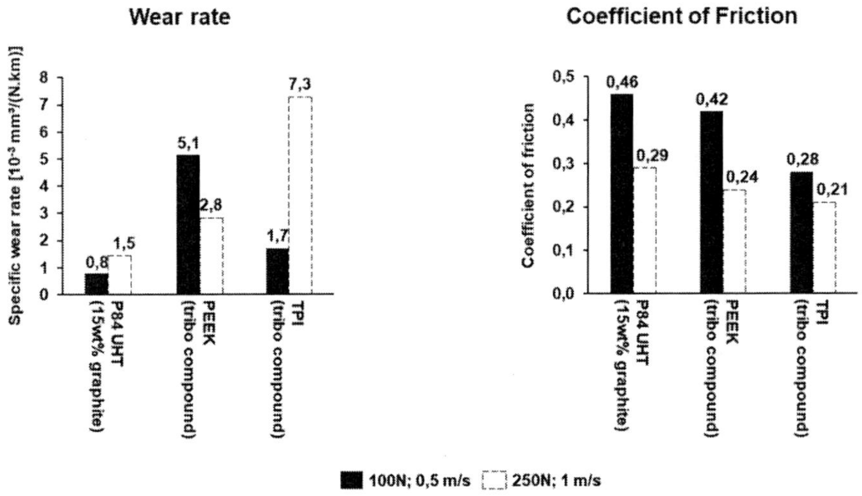

Figure 9: Results of sliding bearing tests with P84 UHT (15wt% graphite) and other high-performance polymers

4.2 Polyimide fibers with reduced shrinkage

Existing polyimide fibers (e.g. P84®fibres) are beneficial in terms of thermal stability compared to aramides, polyphenylensulfide and polyamideimide. Polyimide textile fibers can be used up to 260°C peak temperature, while keeping their textile character. They are established in the hot gas filtration market for filtration non-wovens to remove dust from hot waste gases of cement kilns or coal-fired power plants. In addition to the thermal properties, the unique multilobed cross-section of P84® ensures high surface area which is beneficial for high efficiency in filtration processes.

Although the polymer enables high thermal stability, the fiber starts shrinking above 240°C (P84® shrinks about 3% at 240°C (15 min)). [7] The shrinkage limits the use of polyimide fibers at operating temperatures > 240°C. For specific applications, e.g incineration processes higher operating temperatures would be beneficial, so glass/ceramic fibers can be replaced by polyimide fibers, which textile character enables higher ductility. Shrinkage is a function of the glass transition temperature. An increase in T_g compared to standard P84® fibers can overcome the existing limitations.

Thus, a new base material for polyimide fibers, which exhibits an enhanced T_g, was developed. Fibers made from the new polyimide benefit from lower shrinkage and push the boundaries for the application of polyimide fibers up to 280°C.

The new polyimide fiber P84®HT is based on a fully aromatic backbone. Table 4 shows the textile properties of the new polyimide fiber compared to standard P84® fibers.

Table 4: basic textile properties of P84 HT

	P84®	**P84® HT**
Min. Tenacity [cN/tex]	35	35
Elongation [%] min./max.	25/35	25/35
Shrinkage at 240°C/15 Minutes [%]/max.	3	1
Shrinkage at 270°C/15 Minutes [%]/max.	> 10	3
Glass Transition Temperature – T_g [°C]	333	381

The P84®HT fibers show a significant increase of about 48°C in T_g, which results in a significant reduction of shrinkage at 240 and 270°C. Due to stretching of the fibers during the production process, the fibers are oriented and start to relax already significantly below the T_g.

It is important to check if a similar shrinkage behavior can be observed in the real textile fabric too. Thus, additional work with PI non-woven (550 g/m²) have been performed to understand the textile characteristics at elevated temperature. Non-woven prepared from P84® and P84®HT fibers have been placed in a forced-air furnace at 240 and 290°C. The dimensional changes have been measured after a defined time period. Figure 10 summarizes the results.

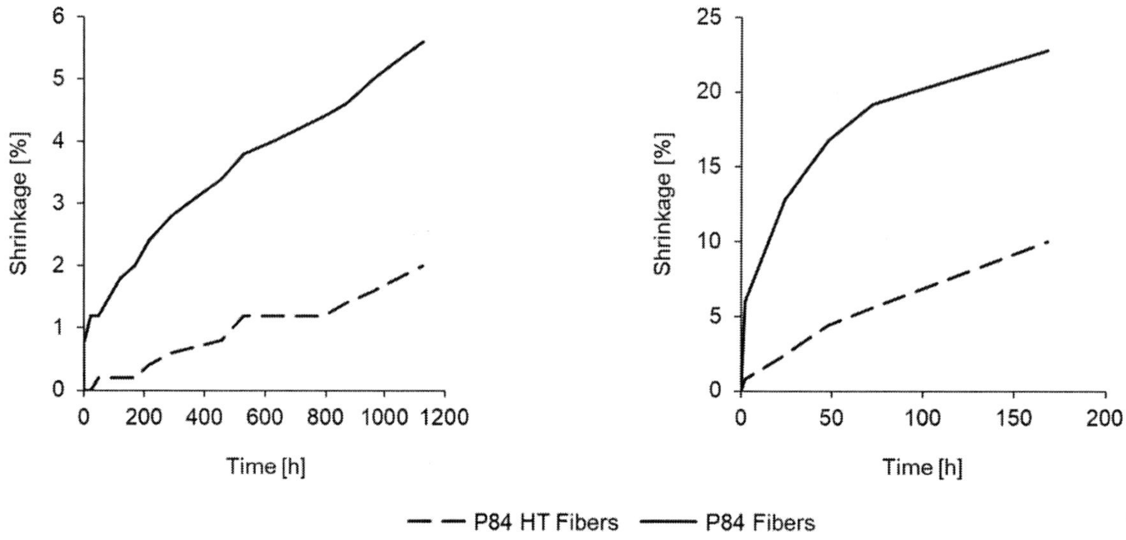

Figure 90: Comparative shrinkage of P84® and P84®HT non-woven (orthogonal to process direction) at 240°C(left) and 290°C (right)

.

Both tests demonstrate significantly lower shrinkage of P84®HT. At 240°C textile fabrics prepared from this new polyimide fiber exhibit a reduction in shrinkage by 64% compared to standard P84® products. After 1100 hours the shrinkage has roughly reached 2,2%. The fabric is still flexible and shows a textile characteristic.
At 290°C the conditions are more extreme, and the textile shrinkage accelerates. After 168 hours the new polyimide fiber shrinks only by 10%. When compared to standard P84® products, the shrinkage has been reduced by 56%. While the standard fiber indicates brittleness, the P84®HT remains flexible.

5. CONCLUSIONS

In this paper the basic properties and benefits of polyimide were discussed. Based on two examples it could be shown how recent developments expand the operating temperature range of these high-performance materials. It has been demonstrated that P84®UHT offers exceptionally high thermal-oxidative stability and provides high mechanical strength even after >1000 hours at 340°C in air. This new material has the potential to further expand the replacement of metals in constructive parts. With P84®HT fibers it has been shown that the higher T_g of the polyimide material has overcome the high shrinkage of existing polyimide products. The new polyimide fibers enhance the operating temperature for textile fibers in filtration processes. The remaining flexibility of the non-woven ensures a proper performance.

6. REFERENCES

[1] Ghosh, Malay K., and K. L. Mittal. *Polyimides: Fundamentals and Applications.* USA: Marcel Dekker, Inc., 1996

[2] Kirby, A.J.. *Polyimides: Materials, Processing and Applications*. USA, iSmithers Rapra Publishing, 1992

[3] Ohya, H; Kudryavsev, V V; Semenova, S I. Polyimide *Membranes: Applications, Fabrications and Properties*, Taylor and Francis, 1997

[4] DuPont, technical datasheet Vespel® SP-1 Polyimide isostatic shapes; downloaded April 28th 2020; https://www.dupont.com/content/dam/dupont/amer/us/en/transportation-industrial/public/documents/en/VPE-A10861-00-B0614.pdf

[5] UBE Industries, LTD. UIP polyimide powder product information sheet; downloaded April 28th 2020, https://www.ube.com/upilex/catalog/pdf/uip_e.pdf?20190220

[6] Ensinger, Technical Data TECASINT 1011, downloaded April 28th 2020 https://www.ensingerplastics.com/de-de/halbzeuge/produkte/polyimid-tecasint-1011-natural#/product-technical-detail-collapse-item-0-lvl-1

[7] Evonik Fibres GmbH, P84 filtration brochure, downloaded May 14th 2020, https://www.p84.com/sites/lists/re/documentshp/p84-filtration-brochure.pdf

CLOSE THE GAP – A DURABILITY STUDY OF EPOXY-BASALT FIBER COMPOSITE REBAR FOR CONCRETE REINFORCEMENT

Yi-Ling (Ivan) Liang[1], Huifeng Qian[1], Francisco De Caso[2], Prof. Antonio Nanni[2]
1. Olin Epoxy, 2301 N. Brazosport Blvd., Building B-1603, Freeport, TX 77541.
2. Univ. of Miami, Dept. of Civil, Arch., and Env. Engineering,
1251 Memorial Dr., Coral Gables, FL 33146.

ABSTRACT

Nonconventional rebar made by fiber-reinforced polymer (FRP) composite has been available to the industry for many years, mainly due to its unique value propositions, including: (a) high mechanical performance to weight ratio, and (b) corrosion free nature in those constructions exposed to harsh environments. Unfortunately, even nowadays, engineers have to adopt relative excessive safety factors when using FRP reinforcement because of the lack of durability-performance information. To address this need, this study presents laboratory results conducted at a Florida Department of Transportation approved testing facility. The tested bars were pultruded using epoxy-anhydride thermoset and basalt fibers. The durability performance was evaluated after an accelerated environmental treatment that involves a 90-day immersion in alkaline solution at 60 °C. Details, including testing method, specification, and promising outcomes are discussed in this paper.

Keywords: Basalt fibers, concrete, composites, durability, rebar

Corresponding author: Ivan Liang, Olin Epoxy, YLiang2@Olin.com

1. INTRODUCTION

Fiber-reinforced polymer (FRP) rebar has a unique value proposition for niche civil engineering industries, particularly for those projects requiring corrosion-resistant, lightweight/easy handling, non-conductive in electricity, and transparent in electrical field/radio frequency materials (Figure 1). For critical projects like water treatment plant, marine structure, toll booth, high voltage substation, bridge, and tunnel, infrastructure architects have already started to adopt FRP rebar. In addition to the aforementioned advantages, builders and project owners can typically be financially compensated by the increase of construction productivity, reduction of labor, lifting machine time, and rehabilitation cost when using lightweight FRP rebar, despite higher raw material cost than steel counterparts. Therefore, it is not surprising to see a high average growth rate (12.9%) of the global sales in composite rebar from 2012 to 2017. [1]

Figure 1. Image of FRP rebars

FRP rebars are typically produced by pultrusion, a continuous manufacturing process, to assemble glass fiber and thermoset plastic into glass fiber reinforced polymer (GFRP) rebar. Among various options of thermoset, unsaturated polyester and vinyl ester are probably the most popular ones for rebar pultruders. While vinyl ester provides rapid production speed and adequate protection to the load-bearing fiber, pultruders have not stopped seeking alternative polymer matrix solutions, due to the needs in volatile organic compound (VOC)-free demand and superior performance. In fact, epoxy thermoset has drawn rebar pultruders' attention because of its flexible processability from the versatile chemistry, higher thermal and chemical resistance, and the steady supply to FRP composite industry.

Regardless of those technical merits of using epoxy thermoset, concerns such as production speed and lack of durability information have been identified as the main barriers for pultruders to adopt the new material. In response to the unmet need, Olin Epoxy decided to team up with industrial and academic partners, including Dixie Chemical, No Rust Rebar Inc., and the Department of Civil, Architectural, and Environmental Engineering at the University of Miami, to conduct comprehensive studies in this field. It is hoped, by providing a thorough durability investigation, the industry may apply pultruded FRP rebar to solve the engineering challenges in a more efficient manner.

For the epoxy resin, the most commonly seen liquid epoxy resin is bisphenol A diglycidyl ether-based epoxy (BADGE, Figure 2) [2]. The aromatic rings on the molecular backbone from the bisphenol A moiety provide rigidity, hydrophobicity, and elevated temperature performance. The ether linkage from the diglycidyl ether moiety shows superior chemical resistance compared to other polymers containing ester bonds. Other epoxy resin families such as bisphenol F epoxy resin, cycloaliphatic epoxy resins, epoxy phenol novolac, and other multifunctional and modified epoxy resins have also been used for FRP composite applications.

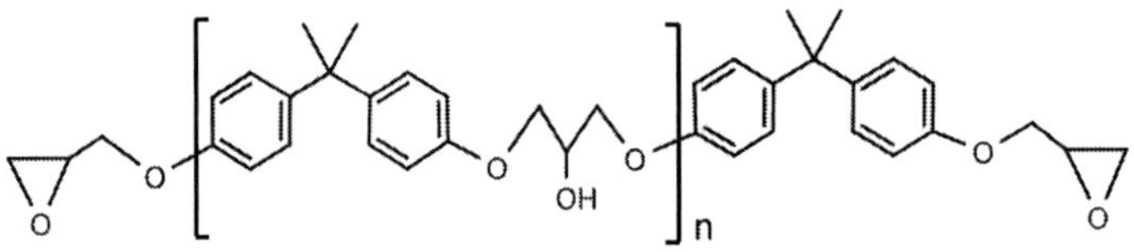

Figure 2. Schematic of bisphenol A diglycidyl ether-based epoxy (BADGE)

Numerous reports and scientific studies can be found in the public domain, particularly for the interest in the durability of cured epoxy thermosets exposed to water, seawater, and concrete pore solution [3-8]. Generally speaking, the water uptake of epoxy thermoset is slightly higher than that of vinyl esters and unsaturated polyesters, probably due to the degree of hydrophobicity and crosslink density of the polymer matrix [3]. The interaction between water and crosslinked epoxy thermoset does not disrupt the network structure [4-5]. Many researchers observed the water enter the free volume between the polymer chains, then temperately form weak hydrogen bonding with the polar portion of the network, and also severs as a plasticizer to enhance the local polymer chain movement on a molecular level [4-6]. Such impact is reversible once the epoxy thermoset is fully dried (recovered to 0% weight gain from water adsorption) [6].

In addition, epoxy thermosets exhibit a similar level of durability to that of vinyl esters [6-7]. In those side-by-side studies, after extended water/moisture treatment (1 to 18 months) at elevated temperature (20-50 °C), the mechanical properties of tested FRP composites made with epoxy systems usually maintain at a similar level to the counterparts made by vinyl esters [7-8]. In contrast, the durability of unsaturated polyesters to the concrete pore/alkaline solution and seawater is inferior [2]. Evidence suggests merely those ester linkages in polyester matrix experienced hydrolysis, and polymer chain scission and weight loss occurred after aggressive environmental treatment (immersed 60 °C concrete pore solution for 2 weeks), while there were no such signs seen in either the amine-epoxy system or vinyl ester matrix. Vinyl esters are generally considered to be more stable to hydrolysis than polyester, due to the fact that ester linkages in vinyl esters are terminal and shielded by methyl functional groups.

In this study, basalt fiber reinforced polymer (BFRP) rebars were pultruded with basalt fiber and epoxy-anhydride based formulation. The physio-mechanical, and more importantly, durability properties of BFRP rebar were evaluated at a Florida Department of Transportation qualified testing facility.

2. EXPERIMENTS

2.1 Durability tests (alkaline resistance) of epoxy BFRP rebar

2.1.1 Tensile Load Retention with Load (ASTM D7705-B/D7205)

This durability exposure applies a dual-mode of aging; on the one hand, the high temperature combined with the high alkalinity solution: the solution reproduces the concrete alkalinity, while the high temperature aims at accelerating any potential degradation mechanisms that may occur in the composite rebars. On the other, the sustained load, which is equivalent to a serviceability level of strain, aims at opening any potential micro-cracks in the composite rebar, thus allowing the solution to penetrate further in the rebar and further degrade the composite rebar. This exposure represents an aggressive environment and thus provides a measure of durability of the rebar qualitatively. Typical protection of the rebar to aggressive environments is attributed to both the manufacturing quality process and the cured resin properties. It should be noted that this test exposure is not a requirement of the ASTM specification however, it is a requirement by the FDOT Specification 932-3. The BFRP results are significantly high, with a relatively high load retention values, which reflects durable composite rebar. Moreover, the tensile mode of failure for all rebars samples was adequate, engaging the full cross-section resulting in a 'broom' or explosive failure.

2.1.2 Transverse Shear Strength (ASTM D7705-A/D7617)

Although the strength retention of transverse shear is not a composite rebar specification, it was conducted to evaluate the possible degradation of the basalt fibers within the BFRP rebars post immersion in the high-temperature alkaline solution. Therefore, there is not an applicable specification requirement. Nevertheless, the specification percentage used in the tensile load retention (without load) may be used as a suggested specification, which is equivalent to 80%. To this end, the BFRP rebar results in adequate performance, where #4 and #5 rebars do not show any degradation and is a reflection of the durability of basalt fibers when used in composite rebars for concrete reinforcement.

2.1.3 Total Enthalpy of Polymerization and Glass Transition Temperature (ASTM D7705-A/E2160/E1356)

Similar to the strength retention of transverse shear, the degree of cure and glass transition temperature post immersion to alkaline solution at high temperature is not a composite rebar specification. It is used as a reference to understand possible changes in the degree of cure and glass transition temperature. Therefore, the are no specification values, however, it is assumed that the same levels as the control (benchmark) tests should be maintained; therefore, degree of cure should be more than 95%, and the glass transition temperature should also be more than 100°C based on DSC method. The BFRP rebar results show that there are no significant changes in the degree of cure or glass transition temperate with retention values close to 100% (i.e., no change).

2.1.4 Scanning Electron Microscopy (SEM) Imaging

SEM investigation to the BFRP rebars post exposure to the alkaline environment was conducted as a qualitative examination to evaluate if there was any visible degradation at the core of the rebar specimen or the surface or between interfaces. Based on the results, no signs of inadequate compatibility between fiber/sizing/epoxy were visible, nor changes to the surface of the overall core of the rebar specimen. In other available studies, hairline cracks have been detected in GFRP rebars at the fiber/sizing/epoxy interfaces and attributed to the degradation of the overall composite rebar. The SEM study post exposure provides additional evidence on the durability provided by the BFRP rebars under evaluation.

3. RESULTS AND DISCUSSIONS

3.1 Epoxy-Anhydride Pultrusion System

Tables 1-4 show the typical properties of epoxy pultrusion system components, which are based on epoxy resin and pre-catalyzed anhydride hardener. The viscosity of epoxy resin is about 11,000 to 14,000 mPa·s at 25 °C, while the pre-catalyzed anhydride hardener is 75 to 150 mPa·s at 25 °C. The mixing ratio of epoxy resin to pre-catalyzed anhydride hardener is 100 to 92. The mix viscosity of epoxy pultrusion system is about 500 mPa·s, which is good for the fiber impregnation. The gel time at 23 °C is more than 24 hours, and gel time at 140 °C is only 90 seconds, indicating the reactivity of epoxy pultrusion resin system fits well with the requirement of pultrusion process. The thermal and mechanical properties of the cured epoxy pultrusion system (clear casting) are listed in Table 4. Glass transition temperature (Tg) measured by DSC method is 130-140 °C. It is much higher than typical cured unsaturated polyester and vinyl ester resin (100-120 °C). Tensile strength is about 90 MPa, and tensile modulus is 3.2 GPa. The mechanical properties are typically better than other thermosetting resins like unsaturated polyesters and vinyl esters.

Table 1. Typical properties of epoxy pultrusion system (component level)

Property	Epoxy Resin	Pre-catalyzed Anhydride Hardener
Viscosity @ 25 °C [mPa.s] ASTM D445	11000 to 14000	75-150
Density @ 25 °C [g/cm^3] ASTM D4052	1.1-1.2	1.1-1.2
Shelf Life (months)	24	12

Table 2. Mixing ratio of epoxy pultrusion system components

	Epoxy Resin	**Pre-catalyzed Anhydride Hardener**
Parts by weight	100	92

Table 3. The properties of mixing epoxy pultrusion system

Property	**Value**
Viscosity @ 25 °C [mPa·s] ASTM D2983	500
Gel Time @ 23 °C [hrs] Gardner Standard Model Gel Timer	>24
Geltime @ 140 °C (secs) Gelnorm Gel Timer	98

Table 4. The properties of cured epoxy pultrusion system (clear casting)

Property[(1)]	**Value**
Tg [°C] DSC, midpoint, 10 K/min	130-140
Tensile Strength [MPa] ASTM D638	90
Tensile Modulus [GPa] ASTM D638	3.2
Elongation at peak [%] ASTM D638	5.2
Flexural Strength [MPa] ASTM D790	114
Flexural Modulus [GPa] ASTM D790	2.2
(1) Properties measured after cure cycle of 1hr @90°C + 2hr @140°C	

3.2 A durability study of Epoxy-based BFRP rebar

This report provides experimental results to evaluate fundamental physio-mechanical and more importantly durability properties of nominal #3, #4 and #5 BFRP rebars made with an epoxy-anhydride based resin formulation and a sand coated surface with a helically wrapped cord around it. Based on the results presented herein, the following preliminary conclusions can be inferred based on the data set presented.

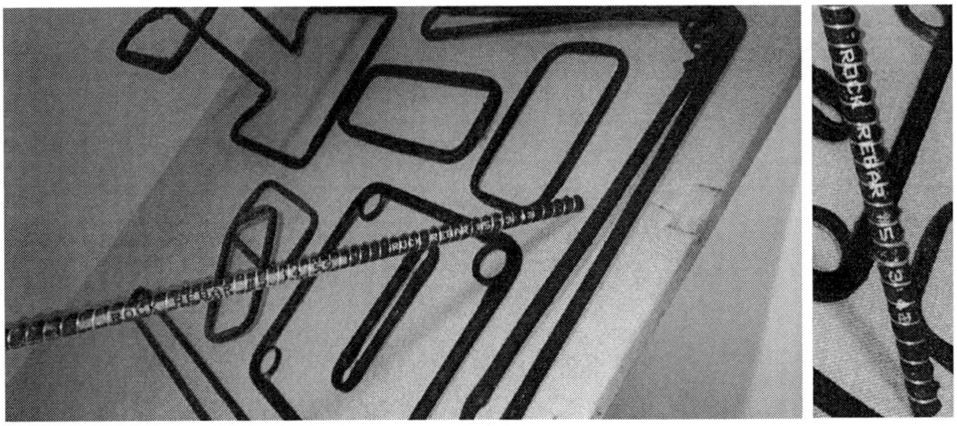

Figure 3. BFRP rebar made with Epoxy/anhydride resin

The results are highlighted as follows:

I. Overall qualitatively, based on SEM imaging post exposure no visible degradation within the cross-section of the rebar or the surface are present; moreover, no signs of inadequate compatibility between fiber/sizing/epoxy is visible, providing additional evidence for the durability of the BFRP rebars under evaluation.

II. Overall, the evaluated mechanical properties of aged BFRP rebars result in equivalent or higher magnitudes than existing minimum ASTM/FDOT specifications or available glass fiber reinforced polymer (GFRP) rebars as provided by GFRP manufactures.

III. More specifically, the tensile properties under the aggressive exposure post immersion in alkaline solution with a sustained load exceed the required retention strength FDOT 932 specification requirements.

3.2.1 Thermal/mechanical properties of specimens prior to durability immersion tests

This section shows the thermal/mechanical properties of specimens prior to durability immersion tests. All experimental tests are conducted in accordance with ASTM and/or ISO standard test methods (refer to Experimental section), and thus results are applicable towards

BFRP rebar certification and material approval per applicable requirements. The testing results are summarized in Table 5.

Table 5. Summary of thermal/mechanical properties prior to durability tests

Standard Test Method	Test Description	Spec. ASTM D7957	Test Value*	Test Result
	SAMPLE #3			
ASTM D7617	Guar. Transverse Shear Strength	>19 ksi	19.7 ksi	Pass
ASTM D2584	Fiber Content (by weight)	>70%	79.9 %	Pass
ASTM D7205	Guar. Tensile Force	13.2 kip	15.5 kip	Pass
	Tensile Modulus of Elasticity	≥ 6.5 Msi	7.0 Msi	Pass
	Strain	≥ 1.1%	2.0 %	Pass
ASTM D792	Measured Cross Sectional Area	0.104 in^2 to 0.161 in^2	0.109 in^2	Pass
ASTM D570	Moisture Absorption Short Term	≤ 0.25 %	0.20 %	Pass
	Moisture Absorption Long Term	≤ 1.00 %	0.73 %	Pass
ASTM D7913	Guar. Bond Strength	>1100 psi	1359 psi	Pass
ASTM E2160	Degree of Cure	>95 %	99.6 %	Pass
	Glass Transition Temperature (DSC)	>100 °C	138 °C	Pass
	SAMPLE #4			
ASTM D7617	Guar. Transverse Shear Strength	>19 ksi	21.4 ksi	Pass
ASTM D2584	Fiber Content (by weight)	>70%	83.3 %	Pass
ASTM D7205	Guar. Tensile Force	21.6 kip	26.3 kip	Pass
	Tensile Modulus of Elasticity	≥ 6.5 Msi	7.2 Msi	Pass
	Strain	≥ 1.1 %	1.9 %	Pass
ASTM D792	Measured Cross Sectional Area	0.185 in^2 to 0.263 in^2	0.207 in^2	Pass
ASTM D570	Moisture Absorption Short Term	≤ 0.25 %	0.21 %	Pass
	Moisture Absorption Long Term	≤ 1.00 %	0.86 %	Pass
ASTM D7913	Guar. Bond Strength	>1100 psi	1711psi	Pass
ASTM E2160	Degree of Cure	>95 %	99.2 %	Pass
	Glass Transition Temperature (DSC)	>100 °C	133 °C	Pass
ASTM D7914	Strength of Bend	> 60 %	165 %	Pass
	SAMPLE #5			
ASTM D7617	Guar. Transverse Shear Strength	>19 ksi	21.0 ksi	Pass
ASTM D2584	Fiber Content (by weight)	>70 %	79.8 %	Pass
ASTM D7205	Guar. Tensile Force	29.1 kip	33.1 kip	Pass
	Tensile Modulus of Elasticity	≥ 6.5 Msi	7.0 Msi	Pass
	Strain	≥ 1.1%	1.6 %	Pass
ASTM D792	Measured Cross Sectional Area	0.288 in^2 to 0.388 in^2	0.292 in^2	Pass
ASTM D570	Moisture Absorption Short Term	≤ 0.25 %	0.17 %	Pass
	Moisture Absorption Long Term	≤ 1.00 %	0.67 %	Pass
ASTM D7913	Guar. Bond Strength	>1100 psi	1227 psi	Pass
ASTM E2160	Degree of Cure	>95%	99.1 %	Pass
	Glass Transition Temperature (DSC)	>100 °C	139 °C	Pass

n/a - not applicable

* Guaranteed values provided for reference purposes only based on a smaller sample group.

Overall, as indicated quantitatively herein, the evaluated mechanical properties of un-aged BFRP rebars result in equivalent or higher magnitudes than existing minimum ASTM/FDOT specifications or available GFRP rebars as provided by GFRP manufactures. Overall qualitatively, based on SEM imaging the BFRP bars result in a homogeneous cross-section were not anomalies or large voids are found, comparable within all bar sizes. The relative increasing of the cross-sectional area of the tested rebars to be in the mid-upper- bound of the specified ASTM allowable range is recommended.

3.2.2 The durability tests post immersion in a high alkaline solution for 90 days at 60 °C

Durability exposures immersed in alkaline solution (without sustained load as applicable) are conducted in accordance with ASTM and/or ISO standard test methods (refer to Table 6 for the summary results), and thus results are applicable towards BFRP rebar certification and material approval per applicable requirements.

Table 6. Summary of thermal/mechanical properties after durability tests

Standard Test Method	Test Description	Spec. ASTM D7957	Test Value*	Test Result
SAMPLE #3				
Test Properties Post Alkaline Resistance per ASTM D7705, 90 days at 60 °C				
ASTM D7205	Tensile Load Retention (with load)[1]	>70 %	78.3 %	Pass
ASTM D7617	Trans. Shear Strength Retention	n/a	84.2 %	n/a
ASTM E2160	Degree of Cure	>95 %	99.06 %	Pass
	Glass Transition Temperature (DSC)	>100 °C	139 °C	Pass
SAMPLE #4				
Test Properties Post Alkaline Resistance per ASTM D7705, 90 days at 60 °C				
ASTM D7705	Tensile Load Retention (with load)[1]	>70%	92.3 %	Pass
ASTM D7617	Trans. Shear Strength Retention	n/a	107.6 %	n/a
ASTM E2160	Degree of Cure	>95 %	99.4 %	Pass
	Glass Transition Temperature (DSC)	>100 °C	128 °C	Pass
SAMPLE #5				
Test Properties Post Alkaline Resistance per ASTM D7705, 90 days at 60 °C				
ASTM D7705	Tensile Load Retention (with load)[1]	>70 %	89.8 %	Pass
ASTM D7617	Trans. Shear Strength Retention	n/a	106.2 %	n/a
ASTM E2160	Degree of Cure	>95 %	99.1 %	Pass
	Glass Transition Temperature (DSC)	>100 °C	139 °C	Pass

n/a - not applicable

* Guaranteed values provided for reference purposes only based on a smaller sample group.
1 Note: this is not an ASTM specification requirement; the criteria is provided by FDOT Specification 932.

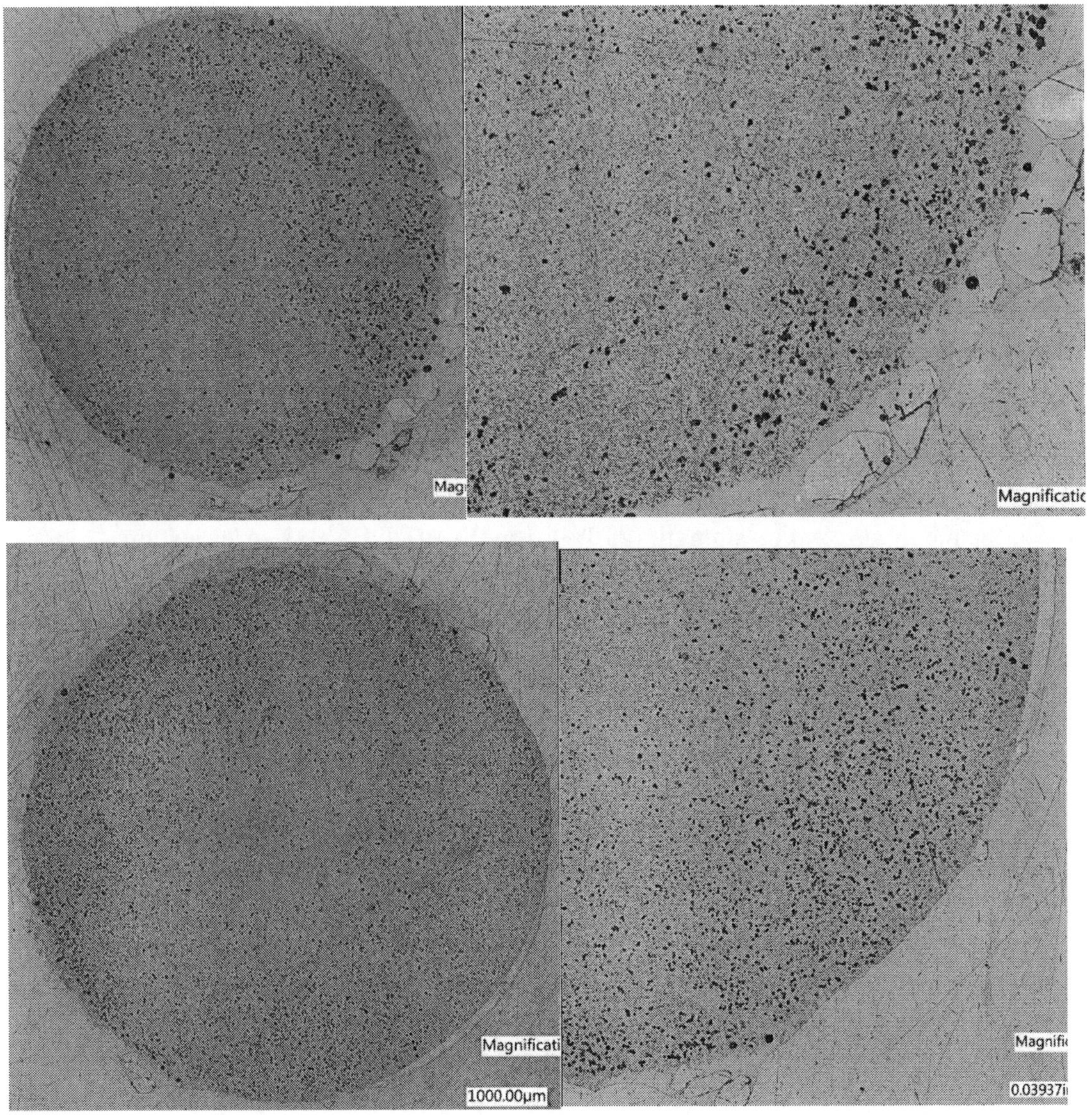

Figure 4. SEM images from Sample #3 (top) and Sample #4 (bottom)

Overall as indicated quantitatively herein, the evaluated mechanical properties of aged BFRP rebars result in equivalent or higher magnitudes than existing minimum ASTM/FDOT specifications or available GFRP rebars as provided by GFRP manufactures. Overall qualitatively, based on SEM imaging post exposure no visible degradation within the cross-section of the rebar or the surface are present; moreover, no signs of inadequate compatibility between fiber/sizing/epoxy is visible, providing additional evidence on the durability of the BFRP rebars under evaluation. More specifically, the tensile properties under the aggressive exposure post immersion in alkaline solution with a sustained load exceed the required retention strength FDOT 932 specification requirements.

4. SUMMARY

We reported an epoxy pultrusion system which can be used as polymer matrix for Fiber Reinforced Polymer (FRP) rebar manufacture. Nominal #3, #4, and #5 BFRP rebars were made with the epoxy pultrusion system. The bar had a sand-coated surface with a helically wrapped cord around it.

A thorough laboratory study of BFRP rebar was conducted at a Florida Department of Transportation qualified testing facility. Overall, the evaluated physio-mechanical properties of aged BFRP rebars meet or exceed existing minimum requirements from ASTM/FDOT specifications or from available glass fiber reinforced polymer (GFRP) rebars as provided by GFRP manufactures. The durability performance was evaluated after an accelerated environmental treatment that involves a 90-day immersion in alkaline solution at 60 °C. The tensile properties of BFRP rebar under the aggressive exposure post immersion in alkaline solution with a sustained load exceed the required retention strength (>70%) from FDOT DEV 932 specification.

5. REFERENCES

1. QYR Construction Research Center, *Global FRP Rebar Sales Market Report,* 2017.
2. Pham, H. Q. and Marks, J. M., *Epoxy Resins. In Ullmann's Encyclopedia of Industrial Chemistry,* 7th edition. Wiley, 2010.
3. Chin, J. W., Nguyen, T., Aouadi, K. "Sorption and diffusion of water, salt water, and concrete pore solution in composite atmrices", J. *Appl. Polym. Sci.*, 71 (1999): 483-492.
4. Jelinski, L. W., Dumais J. J., Cholli, A. L., Ellis, T. S. and Karasz, F. E. "Nature of the water-epoxy interaction" *Macromolecules*, 18 (1985): 1091-1095.
5. Antoon, M. K., Koenig, J. L. "Fourier-transform infrared study of the reversible interaction of water and a crosslinked epoxy matrix" *J. Polym. Sci., Polym. Phys.*, 19 (1981): 1567-1575.
6. Zinck, P., Gerard, J. F. "Polyepoxide–water interactions: Influence of the chemical structure of the network" *Polym. Degard. Stab.,* 93 (2008): 1231-1237.
7. Miller, D., Mandell, J. F., Samborsky, D. D., Hernandez-Sanchez, B. A., Griffith, D. T., *AIAA SDM,* Wind Energy Session, (2012).
8. Marouani, S., Hamelin, L. C. "Ageing of carbon/epoxy and carbon/vinyl ester composites used in the reinforcement and/or the repair of civil engineering structures" *Composites,* 43 (2012): 2020-2030.

A BREAK-THROUGH TECHNOLOGY REDUCING RESIDUAL FORMALDEHYDE IN PHENOL-FORMALDEHYDE RESINS

Carlos Maldonado, Frank Ludvik, Ramji Srinivasan
Georgia-Pacific Chemicals

Atlanta, GA/United States

ABSTRACT

Reducing residual formaldehyde content in phenol-formaldehyde resins (phenolic resins) is a frequent request in composites and other industries. Phenolic resins used for industrial applications, such as abrasives, filtration, honeycomb, foaming, etc., historically contain free formaldehyde levels ranging from 0.3 % to greater than 1.5 %. A new break-through technology reduces the free formaldehyde in phenolic resins to below 1000ppm (or less than 0.1 %). This new ultra-low free formaldehyde resin was compared to a commercial phenolic resin used for industrial applications. The resin characterization (gel permeation chromatography and high-performance liquid chromatography) and thermal analysis (differential scanning calorimetry, dynamic mechanical analysis) as compared to the standard commercial phenolic resins will be discussed.

Keywords: Low free formaldehyde, phenol-formaldehyde resins, composites
Corresponding author: Carlos Maldonado

1. INTRODUCTION

Phenol-formaldehyde resole resins are often used in industrial applications such as abrasives, filtration, honeycomb, foaming, etc [1]. It is well known that these resins have formaldehyde levels (historically ranging from 0.3 % to greater than 1.5 %) [2]. Some degree of free formaldehyde escape can be expected during processing. Resin producers typically work to reduce the level of free formaldehyde in the resins and its release during processing of such resins.

Chemists and resin producers have used numerous techniques to reduce free formaldehyde from formaldehyde-based resins. A variety of formaldehyde scavengers (i.e., chemicals added to the resin during or after its preparation) have been used to reduce formaldehyde emissions [3].

Another technique for lowering formaldehyde emission levels is to increase the relative molecular weight of the phenolic resin. But, most of the industrial applications require low molecular weight resins to improve properties of the finished products [4].

2. EXPERIMENTATION

2.1 Phenol-formaldehyde resins

A regular base-catalyzed phenol-formaldehyde resole resin currently used in the coated abrasives market was chosen as a control resin (S1), and a new resole resin made with the break-through technology (S2) was chosen as the experimental resin. Both resins are waterborne and have the same formaldehyde/phenol molar ratio. A regular solvent-based resole currently used in the honeycomb composite industry, designated as S3, was used to compare the performance of S2 as a honeycomb reinforcing resin. The typical properties of the resins are shown in Table 1 below.

Table 1. Typical properties

	S1 - Control	S2 – Experimental	S3 - Control
Non-volatiles (%)	75	75	43
Viscosity (cP)	2900	3520	70
pH	8.7	9.3	8.5
Free Formaldehyde (%)	0.8	0.034	0.5
Free Phenol (%)	5.6	4.6	16.5
Solvent	Water	Water	Isopropanol

2.2 Thermal analysis

Differential Scanning Calorimetry (DSC) was performed utilizing equipment from TA Instruments. The samples were weighed into a hermetic pan with a laser-drilled pinhole in the lid to release off-gasses and analyzed under 800 psi N_2 from 40 °C to 250 °C at 10 °C/min.

Dynamic Mechanical Analysis (DMA) was performed utilizing equipment from TA Instruments. A 4-ply fiberglass braid was impregnated with resin and clamped to a dual cantilever mount. Samples were analyzed from 50 °C to 300 °C at 10 °C/min with a 20 mm amplitude at 20 Hz.

2.3 Chromatography

High-performance liquid chromatography (HPLC) was performed utilizing a Waters system. For formaldehyde, hydrophilic interaction liquid chromatography (HILIC) separation and post column derivatization with Nash reagent was used with 3-point external calibration and UV detection at 410 nm. For phenol, reverse phase separation with 3-point external calibration and UV detection at 274 nm was utilized.

Gas phase chromatography (GPC) analysis was performed utilizing a Waters system. Samples were analyzed using UV detection at 254 nm on an instrument calibrated with polystyrene.

2.4 Honeycomb sample preparation

Assessments were performed on S2 and S3 resins using the following procedures.
Uncoated Nomex® honeycomb sheet with 0.5 inch thickness was cut into 7 by 7 inches sheets and the mass recorded for each. The dimensions and mass were utilized to calculate the density after

each dip. After initial weighing, the uncoated sample was placed in a large, aluminum tray and resin solution was subsequently poured over it, submerging the honeycomb completely for about 30 seconds. Using large tweezers, the honeycomb was flipped and allowed to soak for another 30 seconds. It was then removed and placed on a horizontal metal drying rack and allowed to drip into a second aluminum tray. Once the dripping ceased after 30 seconds, the honeycomb was placed on a slow, miniature conveyor belt and passed under an 80 psi air knife located 4 inches above the sample. The honeycomb was flipped after each complete pass through the air knife for a total of four passes. Once drying was completed, the sample was placed in a vertical drying rack. Initially the samples were dried at 82 °C between dips to ensure target density was achieved. After obtaining the correct density, the samples were held for 7 minutes each progressively at 82 °C, 121 °C, 132 °C, 177 °C and 193 °C to achieve a complete and even cure.

The cured honeycombs were then cut into 2 by 2 inches samples and subjected to compressive strength testing using an Instron® Universal Testing System. The instrument is a model 5984 fit with compression fixtures and has a maximum capability of 150 kN. The end of test criteria was set for the peak load to drop by 10 %, with a compression rate of 0.05000 in/min.

3. RESULTS

3.1 Molecular weight

The molecular weight of the two resins was determined by GPC. Sample S1 obtained Mn of 374, Mw of 599 and Mz* of 928. Sample S2 obtained Mn of 381, Mw of 606 and Mz of 902. The results are very similar for both resins with respect to molecular weight distribution, as shown on Figure 1.

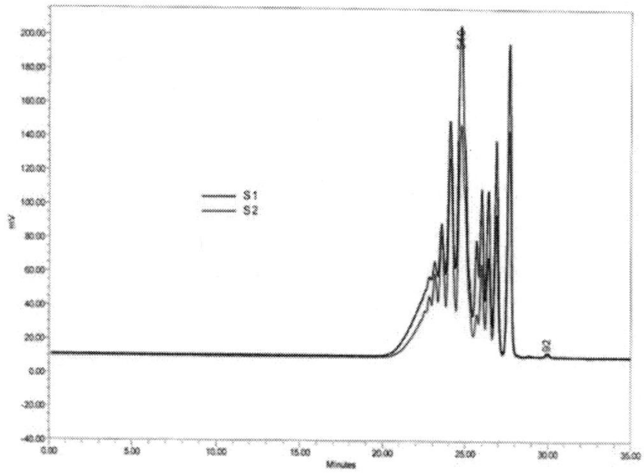

Figure 1. Distribution of Molecular Weights by GPC

3.2 Free monomers

The free monomers content were obtained by HPLC, see Figure 2. Sample S1 had a free phenol content of 5.6 % and free formaldehyde content of 0.8 %. Sample S2 had a free phenol content of 4.6 % and a free formaldehyde content of 0.038 %. The S2 free formaldehyde content is significantly lower than 0.1%. This result is reproduceable. The lower free formaldehyde is attributed to the break-through technology used in the resin S2. As a result of lower free formaldehyde, the free phenol was also reduced in the sample S2 by about 18 % compared to the resin S1.

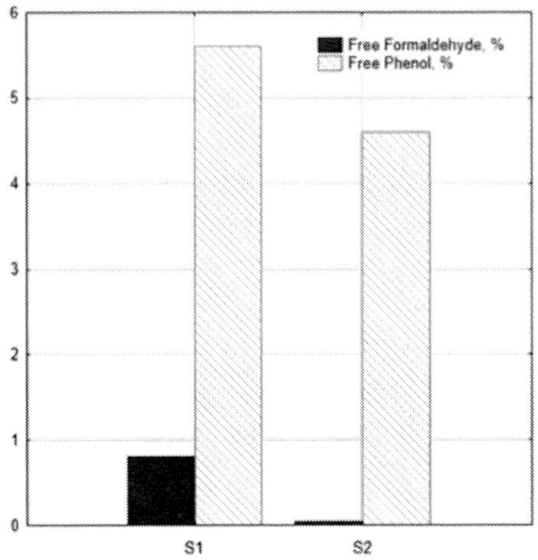

Figure 2. Free Monomers Content

3.3 DMA results

The storage modulus is qualitative only since the addition of the modulus from the interaction of the sample to the braid alone cannot be subtracted. Time and temperatures are separate from the braid and are quantitative. Figures 3a and 3b show the cure profiles of both resins. The S1 resin had an onset of cure of 157 °C, end of cure of 180 °C, loss modulus maximum of 170 °C and tan delta maximum of 155 °C. The S2 resin had an onset of cure of 151 °C, end of cure 171 °C, loss modulus maximum of 162 °C and tan delta maximum of 149 °C. All the results show that the S2 resin is a faster curing resin when compared to S1.

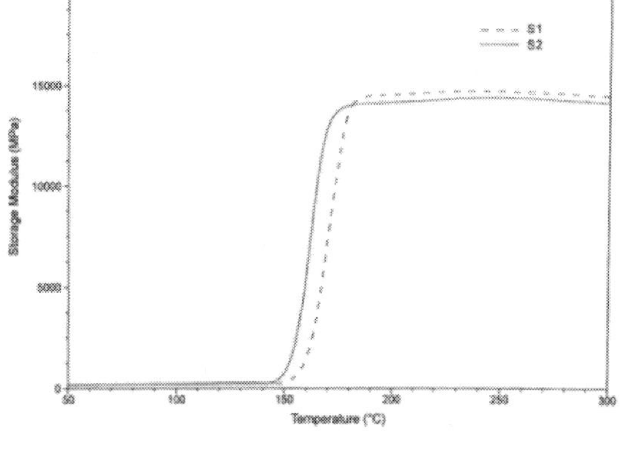

Figure 3a. Storage Modulus Performance

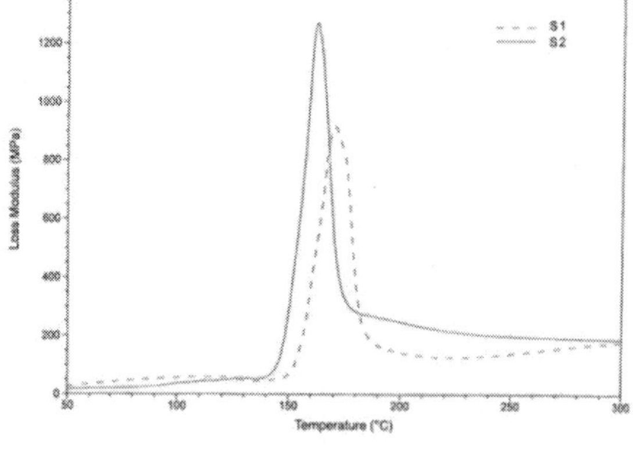

Figure 3b. Loss Modulus

3.4 DSC results

Like the DMA results, DSC shows that the pre-cure starts at a lower temperature for S2 resin (79 °C), than S1 resin (84 °C). See Figure 4. The onset of cure was similar for both resins, 117 °C for S1 and 115 °C for S2. The cure maximum is lower for S2 (147 °C) than S1 (158 °C). The heat of reaction was much lower for S2 (274 J/g) than S1 (373 J/g).

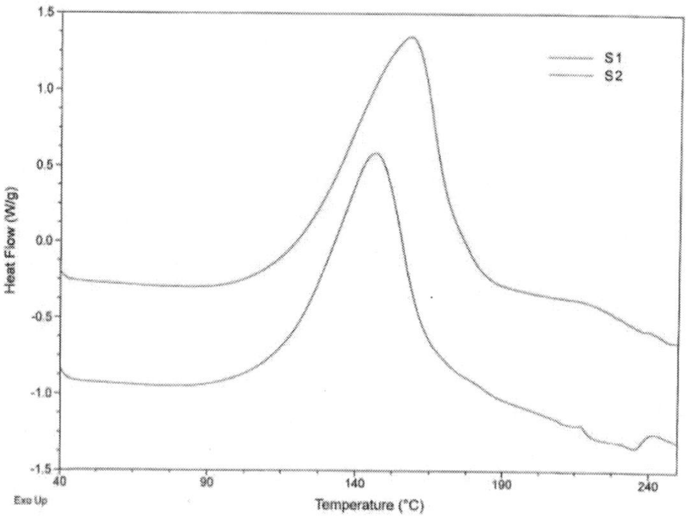

Figure 4. Heat Flow Performance

3.5 Honeycomb assessment

Sample S2 required 2 and 3 dips to obtain the honeycomb density for 30 % and 40 % resin content, respectively. Sample S3 required 1 dip and 2 dips to obtain the honeycomb density for 30 % and 40 % resin content, respectively.

Figures 5a and 5b show the densities and compressive strengths obtained for each resin with different non-volatiles content. Figure 5a shows the densities of the sets indicating they are statistically equivalent. Figure 5b shows the compressive strengths of the resins. The S2 waterborne resin made with the new technology showed improved compressive strength compared to that of the commercially-used resin S3 which is solvent-based. Eliminating a solvent that constitutes volatile organic content (VOC) in the resin can be an added advantage.

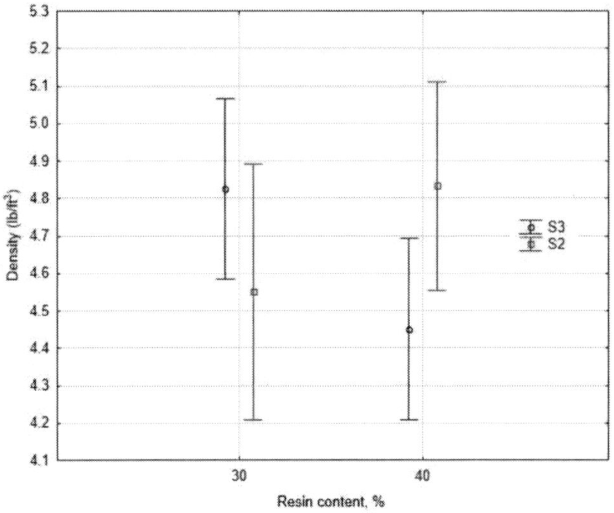

Figure 5a. Densities of Honeycomb made with S2 and S3 Resins with Different Resin Content

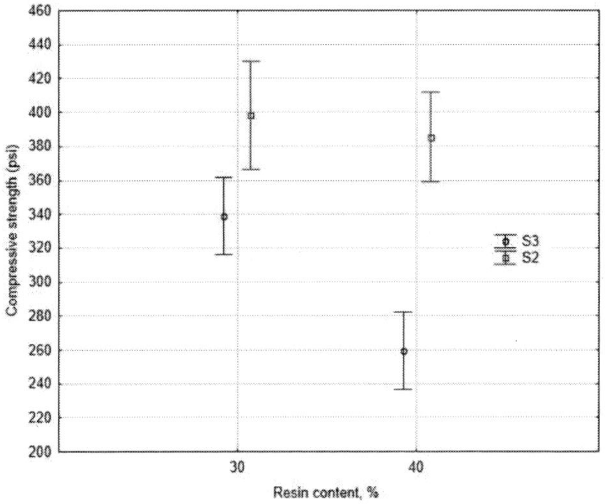

Figure 5b. Compressive Strengths of S2 and S3 Resins with Different Resin Content

4. CONCLUSIONS

GPC results indicate similar molecular weights distribution for both resins. HPLC confirmed that the free formaldehyde obtained for S2 is lower than 0.1 %. The free phenol also is lower with the new technology indicating that the free formaldehyde was condensed with phenol. The curing profiles indicate that S2 has a faster reaction profile than S1. In addition, the honeycomb composite compressive strength study indicated new resin S2 provided improved performance. Based on the analytical results and the initial compressive strength performance, the new S2 breakthrough technology shows promise for various industrial applications by meeting or possibly exceeding standard phenolic resin performance at lower residual monomer levels for both water and solvent soluble resins. Unexpectedly, the S2 resin also provides an opportunity to eliminate VOCs generated by solvent-based resins such as the S3 control resin.

5. REFERENCES

[1] Johannes Fink, "Phenol/Formaldehyde Resins." *Reactive Polymers: Fundamentals and Applications.* 3rd ed. Cambridge, MA: Elsevier, 2018.

[2] L.B. Manfredi, O. de la Osa, N. Galego Fernández, and A. Vásquez "Structure-properties relationship for resols with different formaldehyde/phenol molar ratio." *Polymer.* June, 1999: 3867-3875.

[3] A. Pizzi and C.C. Ibeh, "Phenol-Formaldehydes." *Handbook of Thermoset Plastics.* 3rd ed. San Diego, CA: Elsevier, 2014.

[4] Beata Strzemiecka, Adam Voelkel, Danuta Chmielewska, and Tomasz Sterzynski, "Influence of different fillers on phenolic resin abrasive composites. Comparison of inverse gas chromatographic and dynamic mechanical-thermal analysis characteristics" *International Journal of Adhesion and Adhesives.* Volume 51 (2014) 81-86.

Nomex is a trademarked owned by DuPont Safety & Construction, Inc.

Instron is a trademark owned by Illinois Tool Works Inc.

***Mw**: weight-average molecular weight. **Mn**: number-average molecular weight. **Mz**: z-average molecular weight.

MULTI-WALLED CARBON NANOTUBE DISPERSION FOR AEROSPACE ADHESIVES VIA CONTINUOUS HIGH SHEAR REACTOR

Matthew C. Hartline[1], Christopher D. Croshaw[1], Hayden A. Hanna[1], David E. Garcia[1], Kyzysztof K.K. Koziol[2], Sameer S. Rahatekar[2], Jeffrey S. Wiggins[1]

[1]Department of Polymer Science and Engineering, University of Southern Mississippi Hattiesburg, Mississippi, USA

[2]Enhanced Composites & Structures Center, Cranfield University

Cranfield, Bedford, UK

ABSTRACT

Often, carbon nanotubes are utilized to increase the conductivity and/or mechanical performance of a polymer matrix. The improvements in mechanical performance are directly related to the high modulus and high aspect ratio of the carbon nanotubes. The conductivity of the nanocomposite is related to the purity, quality of the dispersion, and dispersion state of the nanotubes. The high aspect ratio multi-walled carbon nanotubes (MWCNTs) used in this work are grown in aligned "carpets", which is in contrast to the more common "tangled balls" of MWCNTs. This reduces the required shear to untangle and disperse the nanotubes, which are bound together through strong Van der Waals interactions. Once a dispersion is created, it can be disrupted during the cure of the material, as viscosity drops and diffusion increases. This work prepares MWCNT dispersions in the continuous high shear reactor, which greatly reduces the time required to adequately disperse the nanotubes, while simultaneously b-staging the epoxide-amine network. The dispersion states of the uncured and cured nanocomposites are examined using optical microscopy and transmission electron microscopy. Dispersion throughout cure is monitored using real-time optical microscopy digital imaging and oscillatory shear rheological experiments to determine the critical viscosity, above which reagglomeration is minimized. Mechanical, thermomechanical, and electrical characterization of nanocomposite networks is presented.

Keywords: Nanocomposite, Adhesive, Thermoset, Conductive
Corresponding author: Jeffrey S. Wiggins

1. INTRODUCTION

The use of multiwalled carbon nanotubes (MWCNTs) to increase the mechanical strength is hardly a novel concept in contemporary literature.[1–4] While the inherent conductivity of the MWCNTs provide the framework for conductivity, the resulting conductivity of the network depends on the dispersion quality of the nanotubes in the cured network, for which the dispersion throughout the entirety of the cure must be maintained.[5] In an aerospace epoxy, the B-staged matrix must be heated to high temperatures to drive conversion of the epoxide-amine reaction to cure the network. This increase in temperature causes a significant drop in viscosity leading to significant reagglomeration of dispersed particles during cure, which ultimately controls the final conductivity of the nanocomposite.[5] The dispersion state of MWCNTs also greatly alters the viscosity of the curing nanocomposite.[6]

Traditionally, carbon nanotubes are produced as tangled bundles in bulk processes, and often include metallic contaminants. The nanotubes provided by Cranfield University are grown in aligned mats, or carpets, for facile dispersion through shear mixing. These aligned mats of carbon nanotubes are dispersed through simple high-speed mixing, but the mixing process degrades the tube length and limits ultimate conductivity of these networks.

Herein, this work probes the use of the continuous high-shear reactor to prepare B-staged dispersions of MWCNTs and focuses on the final processing of the nanocomposite matrices to control dispersion through the cure, to maintain dispersion and maximize conductivity of the cured networks.

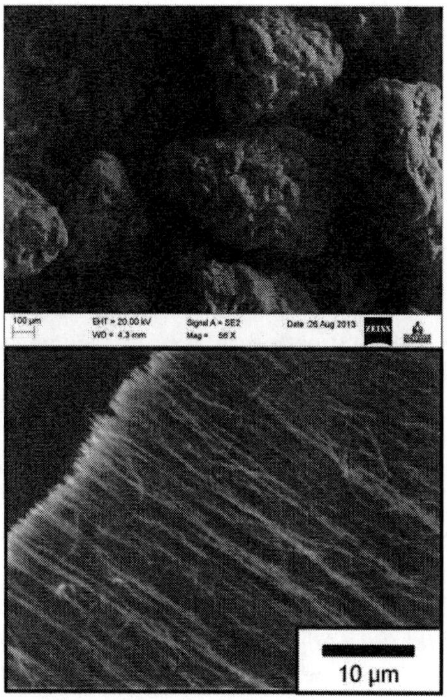

Figure 1. Traditional bundles of MWCNTs (Baytubes C150P, top) [5] *and aligned MWCNTs (Cranfield University, bottom)* [3].

2. EXPERIMENTATION

2.1 Materials

All materials are used as received unless otherwise noted. Low viscosity tetraglicidyl 4,4'-diaminodiphenylmethane (TGDDM, RAR921) and the amine 4,4'-diaminodiphenylsulfone (44DDS) are sourced by Gabriel. Aligned mats of multi-walled carbon nanotubes (MWCNTs) are provided by Cranfield University, and are grown through the root growth process in a quartz tube furnace using acetylene as the source hydrocarbons.

2.2 Network Preparation

Nanocomposite networks are prepared in the continuous high shear reactor, the Coperion 26mm TSE. Preheated TGDDM is fed through a preheated line and metered through a pump for gravimetric monitoring. The 44DDS is fed by a loss-in-mass feeder to maintain stoichiometry with the TGDDM. The MWCNTs are fed through a prototype volumetric nanofeeder, which is found to provide accurate feeding down to 0.02 g/min and facilitates feed rates of up to 1.0 g/min with variance less than 0.05 g/min at maximum flow rates. Variance at low feed rates is within 10% of target feed rates. Residence time in the continuous reactor is 1 minute, 5 seconds for the MWCNT nanocomposite production, and due to lower viscosity, and the neat T44 has a residence time of 55 seconds. The material is quenched in the barrel to minimize reagglomeration due to low temperatures. Upon discharge, material is collected in aliquots, labeled, and stored in the freezer until use.

2.3 Thermogravimetric Analysis

Thermogravimetric analysis (TGA) is performed to verify the relative loading levels of MWCNT included in the nanocomposite networks. A heating rate of 10 °C/min is used and samples are ramped from room temperature to 800 °C to examine residual char, and the onset of degradation. Additionally, the residual mass is measured.

2.4 Differential Scanning Calorimetry

Differential scanning calorimetry (DSC) is performed to determine the effects, if any, on the reaction kinetics of MWCNT nanocomposites when compared with that of the neat network. The samples are run from room temperature to the onset of degradation, determined via TGA, at ramp rates of 1, 5, 10 and 20 °C/min.

2.5 Rheology through Vitrification during Cure

Small angle oscillatory shear rheology is performed to determine rheological well from 40 to 180 °C at 1 °C/min ramp rates, and isothermally held at 180 °C until vitrification, denoted by the G'/G" crossover. Experiments are performed using ARES G2 rheometer, and strain is set to 5%, and auto-adjust is disabled. The 25mm parallel disposable plate geometry is selected due to the network curing to the plates.

2.6 Optical Microscopy

Both dispersion and the reagglomeration of MWCNTs during network cure are examined using optical microscopy and videos of the evolution of the dispersion through cure. The same temperature profiles from the rheological experiments are used on the programmable Mettler FP82HT hot state, with specimens sandwiched between glass slides and sealed with a high temperature silicone adhesive. The samples are recorded using the Nikon OPTIPHOTO2-POL microscope, and the analysis is carried out using Olympus Stream Essentials.

3. RESULTS

3.1 MWCNT Feeding

The twin screw nanofeeder allows metering of the MWCNTs at constant rates over extended periods of time. For the purposes of these experiments, the steady and accurate dosing of <1 g/min is needed to create the low loading levels to examine the dispersion. Above 1 wt.%, MWCNT nanocomposites become dark and opaque, and the dispersion quality becomes increasingly difficult to examine, and improvements in mechanical and electrical properties either plateau or decrease. Typically, the MWCNT would be prepared in a slurry and pumped into the continuous reactor, but due to the viscosity increase and the ease of exfoliation of the Cranfield MWCNTs the slurry method is not appropriate for these experiments. Extended mechanical stirring of the slurry would lead to increased viscosity in the slurry, which would slow the feed rate of the slurry through a peristaltic pump, up to a point where the peristaltic pump could no longer pump the slurry. This inconsistency over time limits the scalability of nanocomposite production, however using the volumetric nanofeeder allows for the consistent feeding over extended periods of time. The feeder was found to have a minimum feed rate of 0.02g/min with MWCNTs, and possessed a linear calibration up to 1 g/min. Rates beyond 4000 pulses/sec are not explored, as the target flow rates for this work are limited to 0.2 and 0.5 g/min.

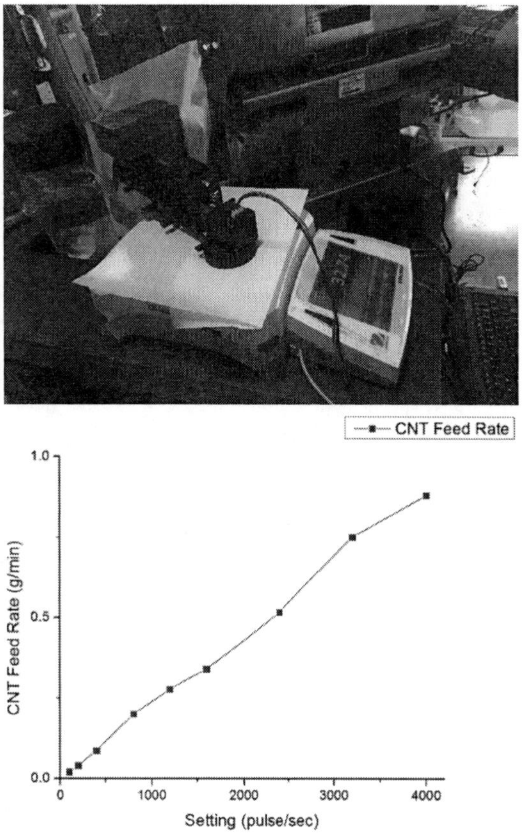

Figure 2. Prototype nanofeeder for MWCNTs (top), and feed rate calibration of feeder with MWCNTs (bottom).

3.2 Thermogravimetric Analysis

Thermogravimetric analysis indicates that the presence of low loading levels of MWCNTs does not significantly alter the degradation profile of the T44 network in air. Residual masses are in general agreement with the desired loading levels, although the T44/5%MWCNT specimen is slightly high, Figure 3. The 5% mass loss occurs 22 degrees earlier in the T44 neat network than the networks that include MWCNTs, Table 1.

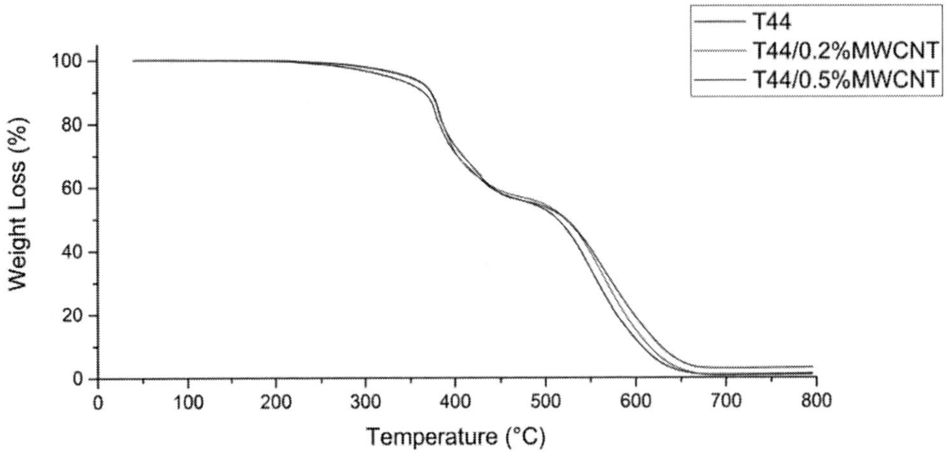

Figure 3. TGA in air of various T44/MWCNT networks.

Table 1. TGA

Network	$T_{5\%}$ (°C)	Residual Mass at 800 °C (%)
T44 Neat	322.8	1.18
T44/0.2%MWCNT	344.2	0.73
T44/0.5%MWCNT	346.5	2.91

3.3 Rheology through Vitrification during Cure

The rheology through vitrification displays similar profiles for the neat T44 and the T44/0.2%MWCNT networks, Figure 4. The T44/0.5%MWCNT network maintains a higher minimum complex viscosity throughout the entirety of the rheological well, due to a different regime of dispersion interactions. As the loading is increased, the elastic nanoparticle to nanoparticle interactions lead to an increased viscosity.

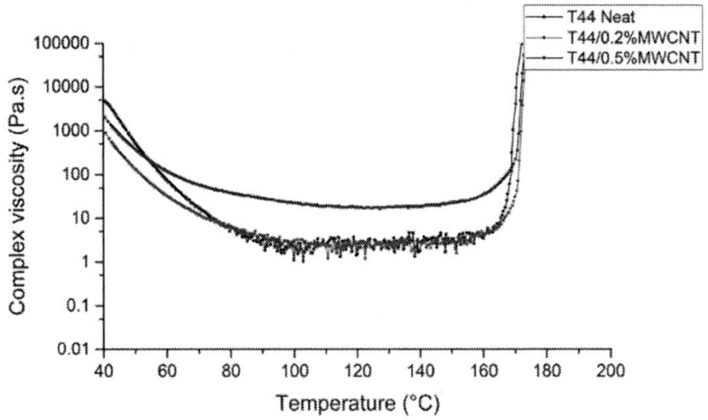

Figure 4. Viscosity through cure

Monitoring the storage and loss moduli through gelation, the T44/0.5%MWCNT network appears to vitrify at ~75 °C and remain vitrified during cure until shortly before gelation at 171 °C prior to the 180 °C isotherm during cure, Figure 5. This first G'/G" crossover is not a true vitrification of the network, but the average mechanical response of the filled network with high aspect ratio fillers exhibiting a "solid-like" response. At the 0.5 wt.% loading level, the exfoliated nanotubes begin to tangle, and this physical entanglement leads to the "solid-like" response of the network. Optical videos reveal the diffusion occurring through the nanotube movement during this "vitrified period". Also notable is the earlier gelation of the neat network at 169 °C when compared to that of the MWCNT filled networks where gelation occurs at 171 °C.

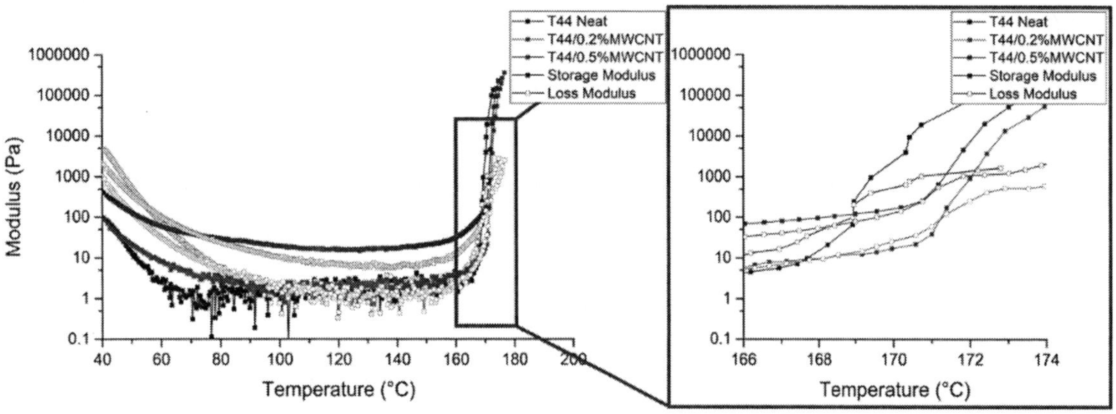

Figure 5. Storage and loss modulus through cure for T44/MWCNT networks.

3.4 Optical Microscopy

Examination of the MWCNT nanocomposites reveals the beginning of exfoliation of the bundles of nanotubes, and regions of dispersed and exfoliated nanotubes between the remaining bundles, Figure 7. This level of exfoliation is typically achieved in 1-2 hours in a batch, and often results in the degradation of nanotube length during processing (REF).

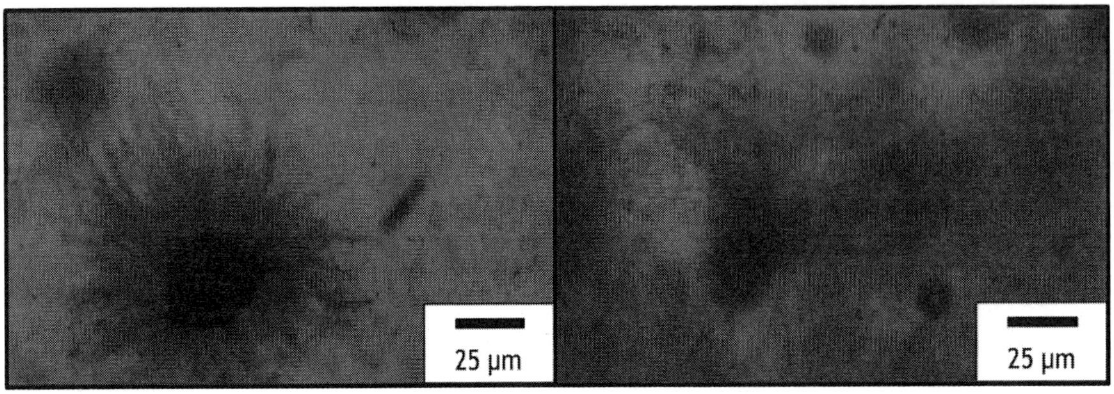

Figure 6. TGDDM and 0.3wt%MWCNT bundles after processing in continuous high-shear reactor.

Using the same TGDDM and MWCNT flow rates, increased screw speed results in fewer large agglomerates remain after processing, Figure 7. The increase in screw speed did not greatly alter the residence time, which remained between 1:05 and 1:15 for all examined screw speeds.

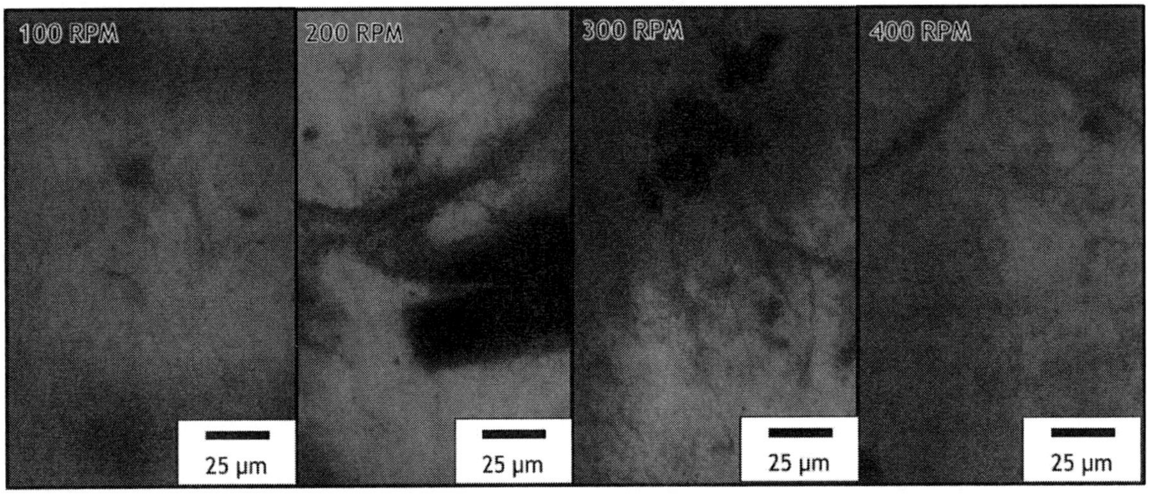

Figure 7. Effects of screw speed on 0.3wt%MWCNT bundles in TGDDM after processing in continuous high-shear reactor.

With the specimens to examine the effects of screw speed on dispersion and exfoliation, the 44DDS curative is added to produce a B-staged nanocomposite. The dispersion of the uncured, B-staged nanocomposite at both the 0.2 and 0.5 wt.% loading levels consists of regions of well dispersed MWCNT, as well as regions where the nanotubes remain in bundles, Figure 9 and Figure 10. The heterogenous dispersion results in drastically differing micro-viscosities, which manifest as significant reagglomeration during typical networks cures observed during optical videos. The critical temperature is difficult to determine due to the heterogenous nature of the specimens, but significant diffusion begins to occur at 75-80 °C, which correlates well with previous research (Greenhoe REF). This correlates with a complex viscosity of ~ 1 Pa.s.

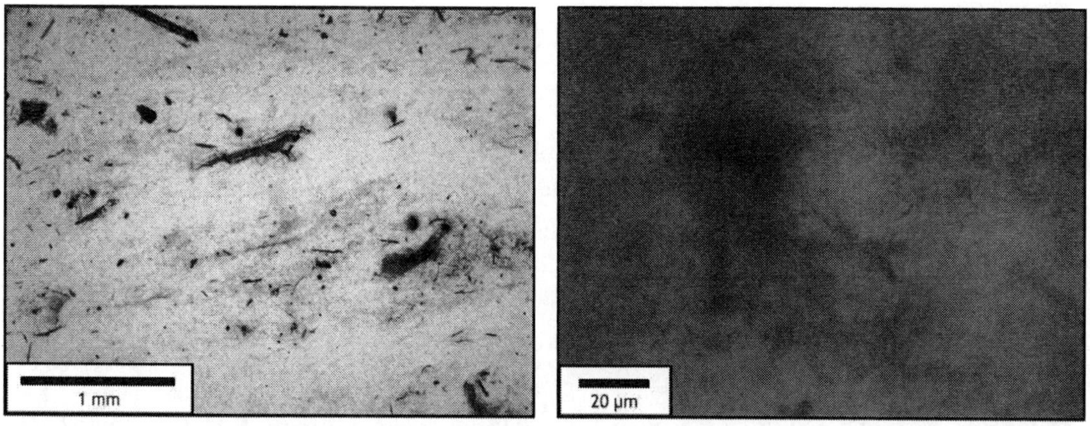

Figure 8. T44/0.2MWCNT nanocomposite prepared in the continuous high-shear reactor.

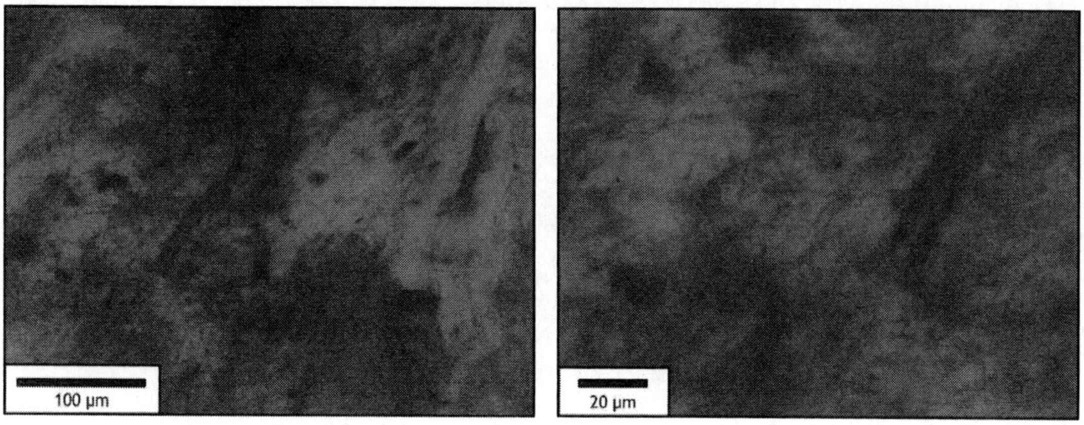

Figure 9. T44/0.5MWCNT nanocomposite prepared in the continuous high-shear reactor.

The 0.5 wt% MWCNT nanocomposite possesses a more even nanoparticle distribution, likely owed to the higher viscosity during processing in the continuous reactor, Figure 10. Most importantly through this work, most visible MWCNTs remain >25 µm. Traditional batch processing of the aligned MWCNTs with mechanical mixing degrade tube length during the dispersion process, and at the highest loading levels tube length is usually degraded to ~5 µm. This short tube length limits conductivity and mechanical reinforcement due to a reduction in the aspect ratio.

4. CONCLUSIONS

This work presents an efficient, scalable approach for the dispersion of nanomaterials. The aligned nanotubes are readily dispersed using the continuous reactor with short residence times and moderate shear. The short residence time with the current screw configuration led to incomplete unbundling of the aligned MWCNTs but maintained the length of the individual dispersed nanotubes. With the heterogenous dispersion and incomplete exfoliation of the MWCNT bundles, future experiments will alter the screw design in the continuous reactor to increase the residence time to increase the potential to untangle the MWCNT bundles.

5. REFERENCES

(1) Sandler, J.; Shaffer, M. S. P.; Prasse, T.; Bauhofer, W.; Schulte, K.; Windle, A. H. Development of a Dispersion Process for Carbon Nanotubes in an Epoxy Matrix and the Resulting Electrical Properties. *Polymer (Guildf).* **1999**, *40* (21), 5967–5971. https://doi.org/10.1016/S0032-3861(99)00166-4.

(2) Rahatekar, S. S.; Hamm, M.; Shaffer, M. S. P.; Elliott, J. A. Achieving Electrical Percolation in Polymer-Carbon Nanotube Composites. *Mater. Res. Soc. Symp. - Proc.* **2004**, *788*, 271–276. https://doi.org/10.1557/proc-788-l5.8.

(3) Rahatekar, S. S.; Koziol, K. K.; Kline, S. R.; Hobbie, E. K.; Gilman, J. W.; Windle, A. H. Length-Dependent Mechanics of Carbon-Nanotube Networks. *Adv. Mater.* **2009**, *21* (8), 874–878. https://doi.org/10.1002/adma.200802670.

(4) Russ, M.; Rahatekar, S. S.; Koziol, K.; Farmer, B.; Peng, H. X. Length-Dependent Electrical and Thermal Properties of Carbon Nanotube-Loaded Epoxy Nanocomposites. *Compos. Sci. Technol.* **2013**, *81*, 42–47. https://doi.org/10.1016/j.compscitech.2013.03.011.

(5) Greenhoe, B. M. Nanostructured Morphologies in Glassy Polymer Networks, 2016.

(6) Rahatekar, S. S.; Koziol, K. K. K.; Butler, S. A.; Elliott, J. A.; Shaffer, M. S. P.; Mackley, M. R.; Windle, A. H. Optical Microstructure and Viscosity Enhancement for an Epoxy Resin Matrix Containing Multiwall Carbon Nanotubes. *J. Rheol. (N. Y. N. Y).* **2006**, *50* (5), 599–610. https://doi.org/10.1122/1.2221699.

Role of Surface Functionality and Polyamic Acid in Carbon Fiber/PEI Interface

Munetaka Kubota[1, 2], Joseph M. Deitzel[1], John W. Gillespie Jr.[1, 2, 3, 4, 5]
[1] University of Delaware, Center for Composite Materials, Newark DE, 19716
[2] University of Delaware, Department of Material Science and Engineering, DE, USA
[3] University of Delaware, Department of Mechanical Engineering, DE, USA
[4] University of Delaware, Department of Civil and Environmental Engineering, DE, USA
[5] University of Delaware, Department of Electrical and Computer Engineering, DE, USA

Keywords: Carbon Fiber, Interface, PEI, PAA, Thermal Oxidation
Corresponding author: Munetaka Kubota

Abstract

The promise of chemical resistance, high toughness, recyclability, and fast cycle times have reignited the interest in fiber-reinforced thermoplastics. To fully realize the synergistic effects of fiber-reinforced composites, the fiber/matrix interface becomes critical. Fiber manufacturers apply a sizing package to protect the fibers from damage during handling and weaving, but also include adhesion-promoting coupling agents onto fibers designed for thermosetting resins such as epoxy. Analogous coupling agent options become scarce since many of the organic oligomers used for adhesion promotion degrade at the high processing temperatures of the engineering thermoplastics such as polyetherimide (PEI) or polyether ether ketone (PEEK). Furthermore, unlike thermosetting resins, which have an opportunity to covalently bond to the coupling agent, the mechanism to improve the fiber/thermoplastic interface is unclear. This work investigates the role of the surface functionality and polyamic acid in improving carbon fiber/PEI interface employing the single fiber fragmentation test (SFFT). Surface functionalization was done using a thermal oxidation technique, and polyamic acid was selected for its ability to imidizes at the process temperature of PEI. The interfacial shear strength was increased by about 20% with a combination of surface functionalization and a coating of polyamic acid.

1. INTRODUCTION

As virgin carbon fiber becomes more affordable, and recycled carbon fiber becomes more accessible, manufacturers evaluate the opportunities to replace traditional materials with composites[1]. Fiber-reinforced thermoplastics have reemerged into the industry consciousness for their promise of chemical resistance, high toughness, recyclability, and fast cycle times. The fiber and matrix must be mated by their interface properties to realize the full potential of fiber-reinforced plastics. An interface that is too weak will result in unstable growth of debonding at the fiber resin interface and not efficiently share load among fibers ,resulting in poor composite tensile properties, while one that is too strong can cause a localized fracture clustering in the same plane[2], again inducing premature failure on the macro scale. Therefore, to optimize the fiber/matrix system, one must consider the interface, which will maximize the composites' performance[3–6].

Methods to improve the fiber/matrix interfaces for thermosetting systems have been studied for decades[2,7]. Traditionally, fiber manufacturers may decide to functionalize the surface

CAMX Virtual Conference Proceedings. September 21-24, 2020.

of the fibers and apply a sizing package composed of adhesion promoting coupling agents[8–10]. The fiber manufacturers have different fiber/sizing systems recommended for different resin systems. Glass manufacturers frequently use amine and epoxide functionalized silanes. Carbon fiber manufacturers, however, are more secretive about the composition of their sizing package, but others have theorized the use of epoxy and polyurethane-based adhesion promotors. These conventional sizing packages cannot survive the processing temperatures for some high-performance engineering thermoplastics such as PEI, PEKK, and PEEK.

In this study, a multistep batch process was utilized to improve the interface properties between a commercial PAN-based carbon fiber and PEI. This method includes a thermal oxidation step and a subsequent polyamic acid (PAA) coating step, which can easily be scaled up to treat and coat large batches of chopped fibers, including recycled carbon fiber. The effectiveness of the surface treatment process was monitored using contact angle measurements and XPS. The fiber integrity through this oxidation process was monitored using single fiber tensile tests, and the changes to the interface properties were evaluated through single fiber fragmentation tests (SFFT).

2. EXPERIMENTATION

2.1. Materials

2.1.1. Carbon Fiber

All experiments were conducted on unsized Hexcel IM7 carbon fiber. These fibers came treated with Hexcel's standard surface treatment.

2.1.2. Polyether Imide

SABIC Ultem 1000 polyether imide (PEI) film was used for all of the interface test. 0.127mm thick PEI film purchased from Professional Plastics was used for the SFFT.

2.1.3. Polyamic Acid Salt

The commercially available low molecular weight polyamic acid (PAA) salt system Michelman Hydrosize® HP-1632 was used to coat the carbon fiber in order to promote adhesion between carbon fiber and PEI.

2.2. Surface Functionalization

Modifying the surface chemistry is a common practice to manipulate the adhesion behavior and, thus, its interfacial properties. Carbon fiber manufacturers use a proprietary surface treatment process to clean and functionalize the surface to encourage wetting and bonding to the sizing package and resin[11]. However, the degree and type of functionalization may not be optimal for all matrices. Various approaches have been taken to functionalize the surface of carbon fibers, including electrochemical and plasma techniques[12–15]. A few recent studies have shown successful oxidation of carbon fiber using an ozone-based technique[16–21]. This type of surface treatment process is more accommodating for discontinuous fibers compared to the electrochemistry-based treatments. In this study, an ozone base thermal oxidation technique was used to change the carbon fiber surface's oxygen content. A simple equipment description for the thermal oxidation reactor designed used for this study can be seen in Figure 1. All thermal oxidation was conducted in this custom thermal ozone reactor for 180 minutes and approximately 4800 – 5000 ppm of ozone. The ozone was produced using the Enaly 5000BF Industrial Ozone Generator using dry air as the feed gas. To achieve this concentration, the reactor was set at full power with a continuous flow rate of 2 lpm of dry air. The ozone concentration was measured using the 2B Technology UV-106-MH. Both the ozone reactor and ozone detector were allowed to warm up for at least 20 minutes on bypass before the reaction was started to reduce output variability from batch to batch. Once the ozone generator and detector were warmed up, the reactive gas mixture was released into the glass reactor from the bottom of the reactor, and the ozone was allowed to percolate through the carbon fiber and exit at the top of the reactor. After the reaction was complete, the reactor was purged thoroughly using dry air till the reactor ozone concentration falls below 1 ppm.

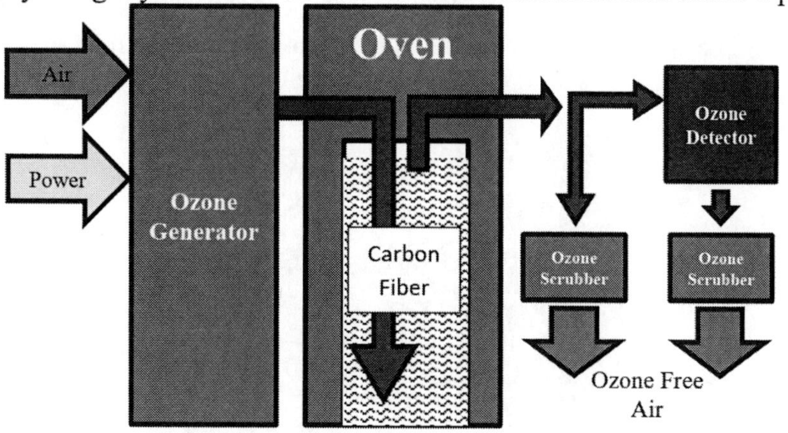

Figure 1: Thermal oxidation reactor setup

2.3. Surface characterization

The effectiveness of the ozone treatment was monitored by the change in contact angle and the oxygen content. To measure the contact angle changes, a modified Wilhelmy plate technique for a cylinder shape was employed using the Cahn dynamic contact angle (DCA) analyzer seen in Figure 2[22]. The contact angle between a single carbon fiber and distilled water was measured. This DCA measures the load as a function of displacement as the fiber advances and retracts from the liquid. The surface topography was measured using an AFM (the Bruker Dimension Icon). All scans were captured in tapping mode using a ScanAsyst-Air AFM tip, which has a silicon nitride tip with a spring constant of 0.4 N/m. The oxygen content on the surface of the carbon fiber was measured using an X-ray Photoelectron Spectroscopy (XPS) and monitored the oxygen to carbon ratio.

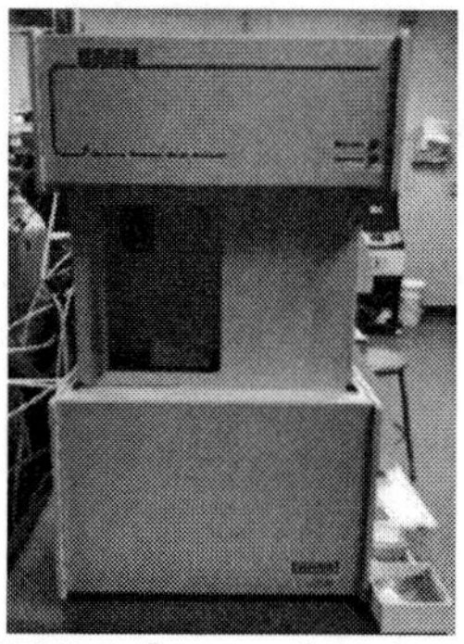

Figure 2: Cahn DCA analyzer used to measure contact angle

2.4. Filament Tensile Testing

The goal of this study was to tailor the fiber/matrix interface to meet the design needs of the end composite component. The surface chemistry modification process must not damage the fiber integrity. Single filament tensile tests were conducted to determine if the thermal oxidation process negatively impacts the fiber strength. The tests were conducted on the automated tensile tester developed by Dia-Stron shown in Figure 3. All experiments were conducted with 12 mm gauge length and an extension rate of 0.01 mm/sec. Single fibers were separated and placed into Dia-Stron's proprietary sample holder and glued onto the plastic tabs using UV curable epoxy Loctite AA3201. The epoxy was cured using a UV spotlight and finished in the UV flood lamp chamber. Individual fiber diameters were measured using the Keyence VK-X200 laser confocal microscope.

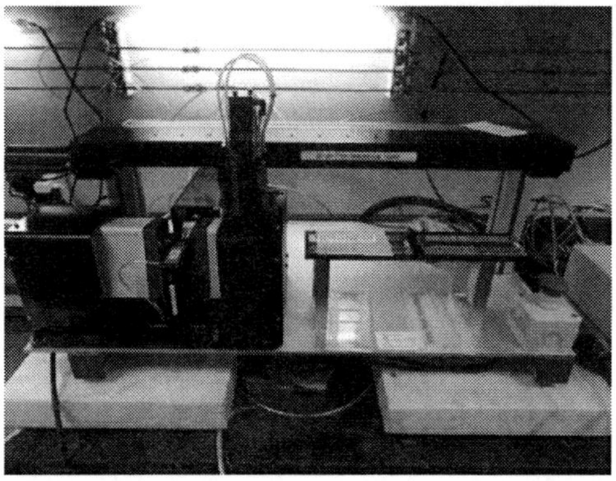

Figure 3: Dia-Stron automated single filament tensile tester

2.5. Adhesion Promotor Application

PAA adhesion promotor was applied to the carbon fiber. To prevent these fibers from sticking together and causing damage during the separation process, the thermally oxidized fibers were separated first. Individual fibers were dipped at room temperature into the PAA solution for 30 seconds, and a pipet was used to ensure the fiber was thoroughly coated. The coated fibers were hung and dried overnight in ambient conditions.

2.6. Interface Testing Equipment

Quantification of interface properties was necessary to understand the impact of the surface functionalization and added adhesion promotor. There are a few commonly used micromechanical tests, such as the microdroplet and the fiber pushout tests. For thermoplastic resins and carbon fiber, these two tests are difficult to manufacture the samples. To characterize the interfacial shear strength (IFSS) between the carbon fiber and the PEI matrix and understand the effects of the surface functionalization and sizing, the single fiber fragmentation test (SFFT) was chosen[2,23–25]. SFFT was conducted using the MTI Instruments SEMTester 1000 and the Nikon compound microscope with cross-polarized light in transmission mode. The SFFT setup used can be seen in Figure 4.

Figure 4: MTI Instruments SEMTester 100 tester in the Nikon compound microscope

3. RESULTS

3.1. Surface Characterization

The effectiveness of the thermal oxidation was evaluated using the changes in the wettability of the carbon fiber as a function of treatment time. The wettability of the carbon fiber was measured using a modified Wilhelmy plate technique to measure cylindrical geometry[22]. A single filament of the carbon fiber was adhered to the end of a thin metal rod using the UV activated epoxy. The rod was hung on the balance in the Cahn DCA analyzer, and the motor slowly lowered the fiber into the liquid. The DCA analyzer recorded force and displacement. The fiber was first analyzed using toluene, which would perfectly wet the fiber. This experiment was used to determine the fiber diameter using the modified Wilhelmy equation can be seen in Equation 1. By measuring the weight change due to the fiber making contact with the liquid and knowing the surface tension of the liquid, one can calculate the fiber's effective circumference and, therefore, the diameter.

Equation 1: Modified Wilhelmy equation for a rod geometry

$$\theta = cos^{-1}(\frac{m \cdot g}{\pi d_f \gamma})$$

d_f: fiber diameter

m: mass

g: gravitational constant

θ: contact angle

γ: surface tension

This sample was tested using distilled water using the same approach as before to determine the contact angle of the fiber with water. Equation 1 can be used by using the fiber diameter determined by the toluene experiment. Figure 6 shows the contact angle of the as-received IM7 carbon fiber and that of IM7 that has gone through the thermal oxidation process. It can be seen that the contact angle was successfully reduced with the oxidation process.

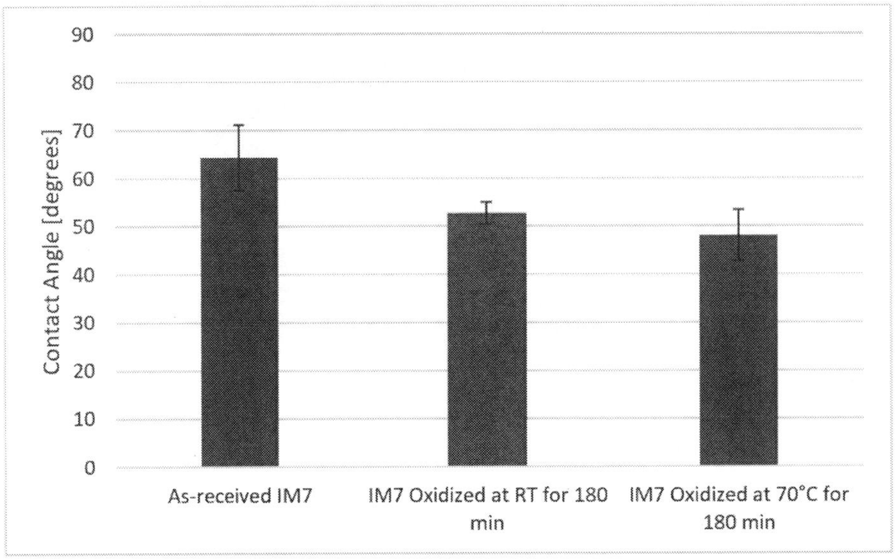

Figure 5: Contact Angle of IM7 with Water

AFM of the treated surface was conducted to identify any changes in the surface morphology due to the ozone functionalization. Plasma and UV/Ozone surface treatment studies conducted by other groups showed evident changes in the surface topology or roughness[26,27]. These two treatments were more aggressive than the thermal ozone process used in this study. Figure 6 shows representative AFM scans of the as-received IM7 and that of a surface of the IM7 the oxidation process at 70°C 180 minutes treatment. IM7 retained the ridge-like features, which runs along the fiber surface. The surface roughness was measured after removing the cylindrical background to eliminate the effect of the curvature. The surface roughness measured was not statistically different, seen in Table 1. However, qualitatively, the ridge features were less distinct, which was unsurprising since the oxidation process was removing material from the fiber's surface.

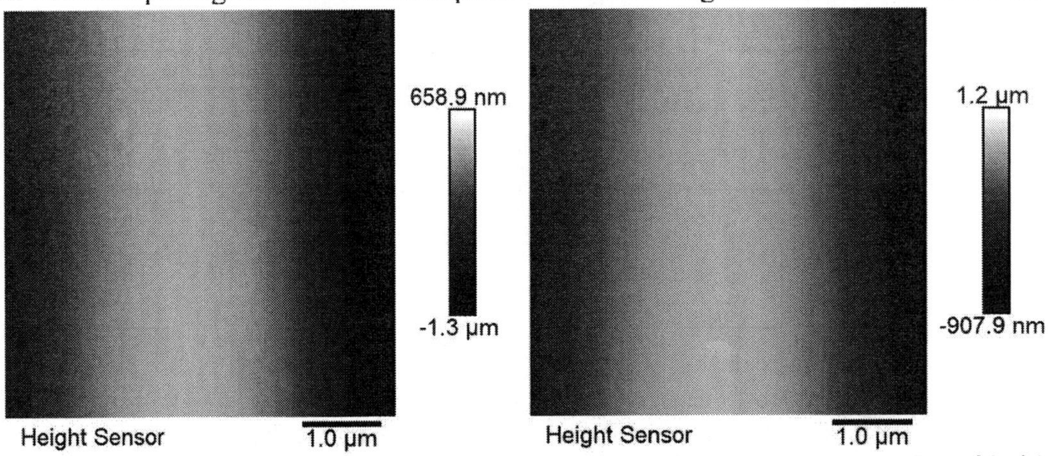

Figure 6 AFM Scans of the IM7 Surfaces, As Received IM7 (left), IM7 Oxidized at 70°C 180 Minutes Treated (right)

Table 1 Surface roughness measurements from AFM scans

	R_{rms} [nm]	STDEV
As-received IM7	16.3	4.5

IM7 Oxidized at 70°C for 180 min	17.5	1.4

The surface chemistry evaluation by XPS was conducted to monitor the O:C ratio. Table 2 shows the normalized O:C ratio. Consistent with the contact angle measurements, the oxidation process at room temperature and at 70°C shows a definite change of the O:C ratio, suggesting a surface chemistry change.

Table 2: Normalized Surface O:C Ratio for the Ozone Treated IM7

Conditions	Normalized Surface O:C Ratio
As-received IM7	1.00
IM7 Oxidized at RT for 180 min	1.59
IM7 Oxidized at 70°C for 180 min	1.75

3.2. Filament Tensile Testing

Single fiber tensile tests were conducted on the as-received fibers as well as the oxidized fibers to determine the effect of the thermal oxidation process had on the fiber integrity. The single filament tensile test was conducted on the Dia-Stron automated tensile tester as described in the experiment section. At least 20 replicates were tested for each condition. Summary, shown in Figure 7, shows that the oxidation process did not negatively affect the fiber strength. T-tests were performed to determine significance of the tensile strength variabilities with an α=0.05. This analysis suggested that the 90-minute thermal oxidation resulted in no change in strength with a P-Value of 0.776. For the 180-minute thermal oxidation, both room temperature and 70°C treatment resulted in a statistically significant increase in strength with P-values of 0.023 and 0.001 appropriately. This result was similar to what was reported by others[28–30]. In Askeland's work, their UV/Ozone treatment was believed to remove material, which includes surface defects that weaken the fibers[30]. Since no clear fiber strength reduction was observed for any of the thermal oxidation processes, thermal oxidation treatment at 70°C was selected for interface testing because of the lowest contact angle and highest O:C ratio.

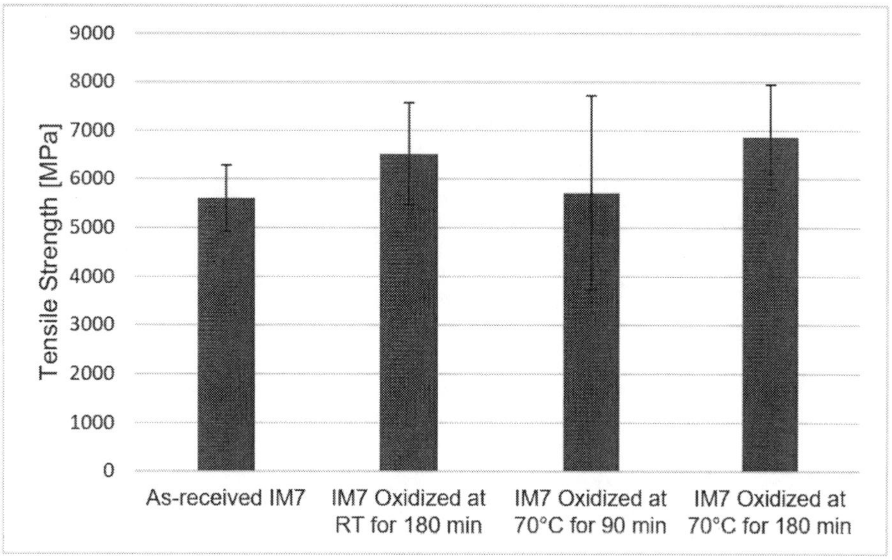

Figure 7: Tensile Strength of IM7 with Ozone Treatment

3.3. Single Fiber Fragmentation Test

SFFT was conducted on the as-received IM7 carbon fiber, the thermally oxidized carbon fiber and a PAA coated thermally oxidized carbon fiber to monitor the effect of surface chemistry on the interfacial properties. To make the SFFT samples, a high-temperature mold release was applied on two aluminum plates as well as two sheets of Kapton film. The Kapton film allowed for a smooth defect-free surface on the SFFT samples. While the mold release was allowed to dry, the 0.127 mm thick PEI film was cleaned using acetone to remove any residue from the manufacturing process. The Kapton film was placed on the aluminum plate, and a cleaned PEI film was placed on the mold release coated Kapton film. An array of carbon fibers was carefully placed onto the PEI film with ~20mm spacing from each other, and ~25mm of fiber was exposed past the Kapton film on each side. The fiber ends were taped to the aluminum plate using high-temperature aluminum tape to prevent the fibers from moving during the rest of the sample preparation process while maintaining some tension in the fiber. The second piece of PEI film was placed on top of the fibers, followed by another Kapton film and an aluminum plate. This entire setup was then vacuum bagged using Kapton film and high-temperature vacuum seal tape. This package was leak checked and then placed into a press. This package was heated to 335°C from room temperature over 45 minutes under contact pressure, and then pressed under 0.7MPa pressure for 45 minutes at temperature. The package was cooled under pressure to room temperature over 30-45 minutes. The samples were carefully removed and disassembled. The resulting film was then cut into 6.6 mm x 40.6 mm (0.26"x1.6") strips using a template, making sure that the sample strips were centered on each fiber. These samples were tabbed using card stock and superglue, maintaining a 20.3 mm (0.8") gauge length. The tabs were necessary to protect the samples from premature failure and tearing due to the serrated grips of the SEMTester.

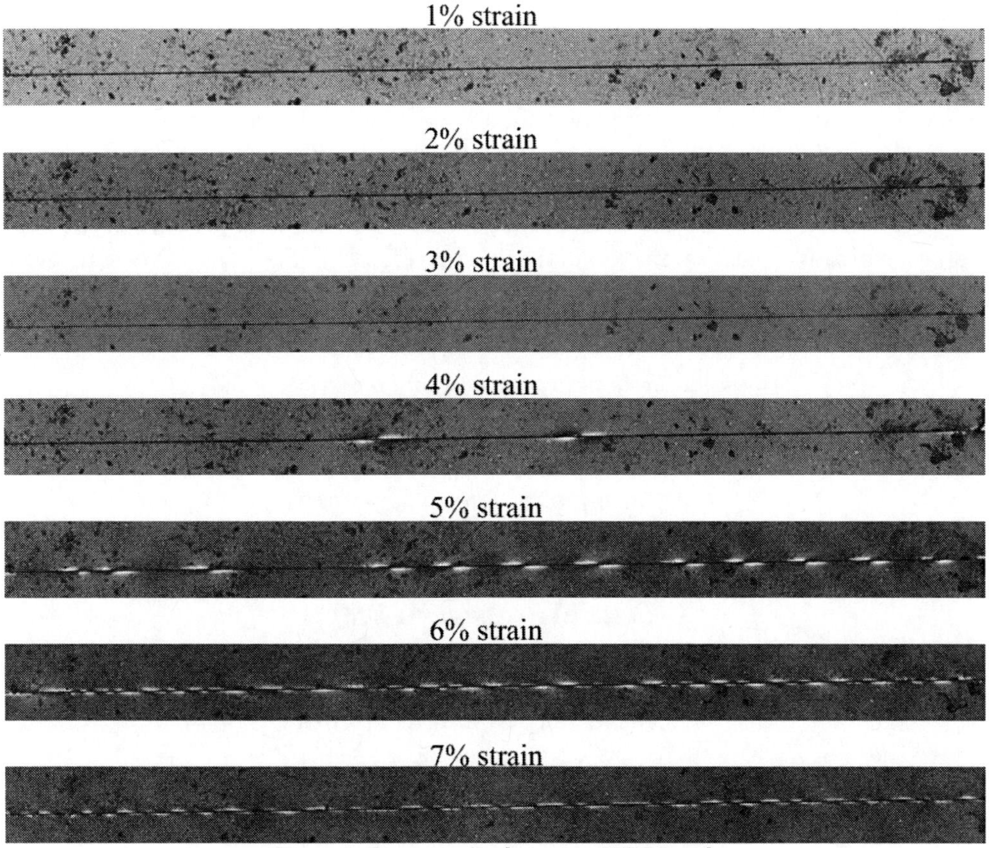

Figure 8: Fragments Evolution in a SFFT Sample

These micro-composite samples were tested using the SEMTester in the Nikon compound microscope in transmission mode under cross-polarized light. The samples were strained at 1% strain/minute and imaged at 0.25% increments for analysis. The number of fragments grew as the sample was strained. The cross-polarized light highlights the fiber breaks as fringes regions seen in Figure 8. The sample was strained until the number of fragments stagnates. At this point, the fragment lengths are too short to transfer enough load to overcome the fiber strength. The number of fragments has saturated, and the average and critical fragment lengths could be determined for analysis.

For this study, the simple Kelly-Tyson model was used to determine the average effective fiber/matrix interfacial shear strength for the baseline, oxidized, and PAA coated systems[2,31,32]. Equation 2 shows the Kelly-Tyson equation used for data reduction to compare the IFSS of the various treatments. Figure 9 shows the IFSS for the different treatment conditions. From these results, only the thermal oxidation at 70°C for 180min with the PAA coating produces a definite statistical increase in the IFSS. Even though the oxidation process showed significant changes in the contact angle and O:C ratio, it did not result in a statistically significant improvement in the IFSS according to a T-test analysis with an $\alpha=0.05$. This result suggests that surface oxides alone do not directly improve the interface between carbon fiber and PEI and require a coupling species. The P-value between the as-received IM7 and the IM7 with thermal oxidation was 0.194, thus there is not enough evidence to conclude that the interface is statistically different. Same could be said about the IM7 with the 90-minute thermal oxidation and PAA coating, with a P-value of 0.089.

The 180-minute thermal oxidation, in conjunction with the PAA salt, did show an increase of ~20% improvement in the IFSS and decreased the variability in the measured results from a coefficient of variability of 14% for the as-received IM7 to 6% for the thermally oxidized IM7 with PAA. This is statistically significant with a P-value of 0.005 and can conclude that there is sufficient evidence that these two interfaces are statistically different. The improvement in the interface properties is likely due to an increase in uniformity of the surface adhesion at the interface. Reactions between the oxides on the carbon fiber surface reacting with the carboxylic acids on the PAA in an esterification reaction may explain this increase in IFSS. The PAA salt also imidized during the sample processing coupling the interface between the carbon fiber and PEI matrix likely through molecular entanglement.

Equation 2: The Kelly-Tyson equation used to calculate the IFSS

$$\tau = \frac{\sigma_f d_f}{2l_c}$$

$$\tau: Interfacial\ Shear\ Strength$$

$$\sigma_f: Fiber\ Strength$$

$$d_f: Fiber\ Diameter$$

$$l_c: Critical\ Fragement\ Length$$

$$l_c = \frac{4}{3}\bar{l}$$

$$\bar{l}: Average\ Fragement\ Length$$

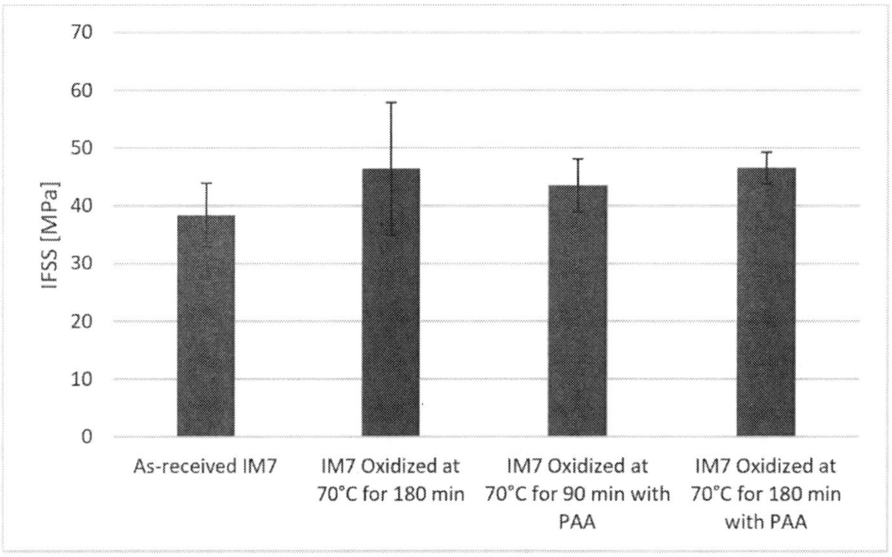

Figure 9: IFSS Determined by SFFT

4. CONCLUSION

This study investigated the role of surface chemistry and PAA on the interfacial properties between commercial PAN-based carbon fiber and PEI. A scalable ozone-based batch process was successfully utilized to increase the surface oxygen content by 75% while successfully reduce the contact angle for IM7 with water by 15°. Filament tensile test was used to confirm that this oxidation process did not adversely affect fiber integrity. Slight improvements in the tensile strength were observed for the long dwell process conditions, consistent with previous observations in the literature. The impact of the surface oxidation and a PAA-based adhesion promoter was tested using SFFT, and the results were evaluated using the Kelly-Tyson approach. The results were analyzed using a T-test and showed for α=0.05, only 180-minute thermal oxidation with PAA coating condition created an interface that was statistically different from the as-received IM7. This condition and coating increased the IFSS by 20% and a reduction in interface variability. Further investigation is required to determine the interface improvement mechanism. This process may be adapted to treat recycled carbon fiber, which may lose surface oxygen due to the pyrolysis or solvolysis process.

5. ACKNOWLEDGEMENTS

The effort depicted was sponsored by the Defense Advanced Research Projects Agency and was accomplished under Cooperative Agreement Number **HR0011-16-2-0014**. The views and conclusions contained in this document are those of the authors and do not necessarily reflect the position or the policy of the Government. No official endorsement should be inferred. The authors would like to acknowledge and thank Michelman Inc. for providing the HP-1632 as well as Dr. Daniel Knorr and Dr. Ngon Tran of the U.S. Army Research Laboratory for providing the XPS data for the oxidized carbon fiber surface.

6. REFERENCES

1. Kraus, T. & Kuhnel, M. Global carbon fibre market remains on upward trend. *Reinf. Plast.* **58**, 38–45 (2014).
2. Swolfs, Y., Verpoest, I. & Gorbatikh, L. A review of input data and modelling assumptions in longitudinal strength models for unidirectional fibre-reinforced composites. *Compos. Struct.* **150**, 153–172 (2016).
3. Zhao, F. . & Takeda, N. Effect of interfacial adhesion and statistical fiber strength on tensile strength of unidirectional glass fiber/epoxy composites. Part I: experiment results. *Compos. Part A Appl. Sci. Manuf.* **31**, 1203–1214 (2002).
4. Madhukar, M. S. & Drzal, L. T. Fiber-Matrix Adhesion and Its Effect on Composite Mechanical Properties. III. Longitudinal (0°) Compressive Properties of Graphite/Epoxy Composites. *J. Compos. Mater.* **26**, 310–333 (1992).
5. Shih, G. C. & Ebert, L. J. Theoretical Modelling of the Effect of the Interfacial Shear Strength on the Longitudinal Tensile Strength of Unidirectional Composites. *J. Compos. Mater.* **21**, 207–224 (1987).
6. Ganesh, R., Sockalingam, S., Haque, B. Z. & Gillespie, J. W. Dynamic effects of single fiber break in unidirectional glass fiber-reinforced composites. *J. Compos. Mater.* **51**, 1307–1320 (2017).
7. Johannson, O. K., Stark, F. O., Vogel, G. E. & Fleischmann, R. M. Evidence for Chemical Bond Formation at Silane Coupling Agent Interfaces. *J. Compos. Mater.* **1**, 278–292 (1967).
8. Viswanathan, H., Rooke, M. A. & Sherwood, P. M. A. X-ray Photoelectron Spectroscopic Studies of Carbon-FIber Surfaces. 21. Comparison of Carbon Fibers Electrochemically Oxidized in Acid using Achromatic and Monochromatic XPS. *Surf. Interface Anal.* **25**, 409–417 (1997).

9. Zhang, J., He, D. L., Wagner, H. D., Wiesel, E. & Bai, J. B. Interfacial studies of carbon fiber/epoxy composites using single fiber fragmentation test. *Compos. Interfaces* **20**, 421–429 (2013).
10. Yang, Y., Zhao, Y., Li, Y., Dong, Q. & Chen, D. Effect of sizing on the interfacial shear strength of carbon fiber/epoxy resin monofilament composite. *Journal Wuhan University of Technology, Materials Science Edition* **29**, 483–487 (2014).
11. Viswanathan, H., Wang, Y. Q., Audi, A. A., Allen, P. J. & Sherwood, P. M. A. X-ray photoelectron spectroscopic studies of carbon fiber surfaces. 24. Interfacial interactions between polyimide resin and electrochemically oxidized PAN-based carbon fibers. *Chemistry of Materials* **13**, 1647–1655 (2001).
12. Wang, Y. Q., Zhang, F. Q. & Sherwood, P. M. A. X-ray photoelectron spectroscopic study of carbon fiber surfaces. 23. Interfacial interactions between polyvinyl alcohol and carbon fibers electrochemically oxidized in nitric acid solution. *Chemistry of Materials* **11**, 2573–2583 (1999).
13. Wang, T. & Sherwood, P. M. A. X-ray Photoelectron Spectroscopic Studies of Carbon Fiber Surfaces. 18. Interfacial Interactions between Phenolic Resin and Carbon Fiber Electrochemically Oxidized in Ammonium Carbonate Solution and Their Effect on Oxidation Behavior. *Chemistry of Materials* **7**, 1020–1030 (1995).
14. Sun, H. *et al.* Recycling of carbon fibers from carbon fiber reinforced polymer using electrochemical method. *Composites Part A: Applied Science and Manufacturing* **78**, 10–17 (2015).
15. Tiwari, S. & Bijwe, J. Surface Treatment of Carbon Fibers - A Review. *Procedia Technol.* **14**, 505–512 (2014).
16. Fu, X., Lu, W. & Chung, D. D. L. Ozone Treatment of Carbon Fiber for Reinforcing Cement. *Carbon N. Y.* **36**, 1337–1345 (1998).
17. Zielke, U., Hüttinger, K. J. & Hoffman, W. P. Surface-Oxidized Carbon Fibers : I . Surface Structure and Chemistry. *Carbon N. Y.* **34**, 983–998 (1996).
18. Park, S. J. & Kim, B. J. Roles of acidic functional groups of carbon fiber surfaces in enhancing interfacial adhesion behavior. *Mater. Sci. Eng. A* **408**, 269–273 (2005).
19. Jin, Z., Zhang, Z. & Meng, L. Effects of ozone method treating carbon fibers on mechanical properties of carbon/carbon composites. *Mater. Chem. Phys.* **97**, 167–172 (2006).
20. Osbeck, S., Bradley, R. H., Liu, C., Idriss, H. & Ward, S. Effect of an ultraviolet/ozone treatment on the surface texture and functional groups on polyacrylonitrile carbon fibres. *Carbon N. Y.* **49**, 4322–4330 (2011).
21. Deitzel, J. M. *et al.* Surface treatment of TuFF pitch-based carbon fiber for adhesion promotion in high Tg thermoplastic composites. in *International SAMPE Technical Conference* **2019-May**, (2019).
22. Tretinnikov, O. N. & Ikada, Y. Dynamic Wetting and Contact Angle Hysteresis of Polymer Surfaces Studied with the Modified Wilhelmy Balance Method. *Langmuir* **10**, 1606–1614 (1994).
23. Lacroix, T., Keunings, R., Desaeger, M. & Verpoest, I. A new data reduction scheme for the fragmentation testing of polymer composites. *J. Mater. Sci.* **30**, 683–692 (1995).
24. Wu, W., Verpoest, I. & Varna, J. An improved analysis of the stresses in a single-fibre fragmentation test - II. 3-phase model. *Compos. Sci. Technol.* **58**, 41–50 (1998).
25. Shioya, M. & Takaku, A. Estimation of Fibre and Interfacial Shear Strength by Using a

Single-Fibre Composite. *Compos. Sci. Technol.* **55**, 33–39 (1995).
26. Xie, J. *et al.* Improving carbon fiber adhesion to polyimide with atmospheric pressure plasma treatment. *Surf. Coatings Technol.* **206**, 191–201 (2011).
27. Rich, M. J., Drown, E. K., Askeland, P. & Drzal, L. T. Surface Treatment of Carbon Fibers By Ultraviolet Light + Ozone : Its Effect on Fiber Surface Area and Topography. *the 19Th International Conference on Composite Materials* (2013).
28. Rich, M. J., Drzal, L. T., Rook, B. P., Askeland, P. & Drown, E. K. Novel carbon fiber surface treatment with ultravilet light in ozone to promote composite mechanical properties. *ICCM Int. Conf. Compos. Mater.* (2009).
29. Rich, M. J. *et al.* Novel Carbon Fiber Surface Treatment With Ultravilet Light in Ozone To Promote Composite Mechanical Properties.
30. Askeland, P. A., Fukushima, H., Rich, M. & Drzal, L. T. UV-Ozone Surface Modification of Carbon Based Reinforcements for Composite Materials.
31. Kelly, A. & Tyson, W. R. Tensile Fibre-Reinforced Metals : Copper/Tungsten and Copper/Molybsenum. *J. Mech. Phys. Solids* **13**, 329–350 (1965).
32. Berg, J. & Jones, F. R. The role of sizing resins, coupling agents and their blends on the formation of the interphase in glass fibre composites. *Compos. Part A Appl. Sci. Manuf.* **29**, 1261–1272 (1998).

COHESIVE ZONE MODELING SIMULATION OF THERMAL STRESSES IN DISSIMILAR MATERIAL JOINTS

Grant Percy and Seyed Soltani
Florida Polytechnic University
Lakeland, FL/USA

ABSTRACT

This paper describes the finite element simulation of an adhesively bonded composite/metal single lap joint undergoing cyclic thermal loading with cohesive zone modeling (CZM) in Ansys Workbench. Single-lap joints are widely used in structural applications to join two parts via an overlapping bond. The Ansys model was comprised of a carbon fiber composite laminate bonded with an epoxy adhesive to an aluminum plate. The model was used to study the aggregate structural effects of cyclic thermal loadings applied through periodic temperature boundary conditions simulating the service temperatures typical for the aircraft. The simulation results revealed that the heat flux was highest around the adhesive. This was attributed to the extremely low thermal conductivity of the adhesive. Additionally, since there was such a large difference in the thermal conductivity of the aluminum and the carbon fiber composite, there ended up being almost no heat transfer at the carbon fiber composite and adhesive interface. The modeling approach described in this paper is a promising technique not only for modeling dissimilar material joints, but also for modeling the composites individually. This is mainly because the model can be easily modified to incorporate similar and dissimilar materials per application requirements.

Keywords: Adhesively Bonded Composites, Finite Element Analysis, Thermal Stresses
Corresponding author: seyed.soltani@gmail.com

1. INTRODUCTION

The past decade of structural materials research has been marked by a wave of new applications for materials like structural adhesives and composites. In industries like aerospace engineering, composites and structural adhesives have been replacing more traditional materials and joining methods because they offer higher strength to mass ratios, vibration resistance, and more flexibility in designs. But due to their inherent complexity, effective modeling and analysis can be difficult to create [1-11]. Composites are a multilayered structure composed of thinner layers of differing materials adhered together. This allows for combination of the properties of each of the individual materials that comprise the composite, resulting in a material that is fully customizable to its application. These composites can be lighter, stronger and more flexible than any traditional material. But this customizability comes at the cost of increased complexity in manufacture, application and analysis. Manufacturing composites is much more difficult than manufacturing metals because defects from the manufacturing process, like voids, can easily cause premature material failure. Additionally, the mixing of materials means that the differences in their intrinsic properties, like coefficient of thermal expansion (CTE), can generate substantial stresses that were negligible in previous materials. Finally, performing analyses on composites is much more complex because unlike metals, which can usually be assumed to be isotropic, composites are always anisotropic as their properties are only uniform within each layer. This makes analyzing composites particularly difficult because, instead of modeling one material, each layer and the interlayer interactions should be modeled. Despite these complexities, composite's promises of lighter, stronger and more flexible materials are the reasons why currently there is a considerable amount of both money and effort going into the understanding and modeling of composites and their applications. Adhesives have been around since the dawn of civilization, but recently there has been an influx of easier-to-use, stronger and weather-resistant structural adhesives [12]. Combine that with an increase in mixed material joints, which often cannot be bonded through more traditional means like welding or riveting, and the result has been a surge in structural adhesive use. Structural adhesives are lighter and stronger, and they can be bonded to almost any material and provide resistances to vibration, electricity and thermal loadings [13]. This makes structural adhesives very attractive to modern manufacturers, especially in the case of a mixed material joints [14]. If mixed material joints welding doesn't work, neither does soldering or brazing, but adhesives can easily be applied to almost any material and still outperform traditional bonding methods [15-27].

2. METHODOLOGY

The Finite Element Analysis (FEA) was employed for this study. The FEA takes a geometry, subdivides it into smaller domains called elements, defines the loadings and properties through a series of boundary equation, and then uses reactions in each domain to describe the reaction of the whole geometry. This allows for problems too complex to solve to be simulated with relative accuracy. In modern engineering, advances in computers allow for relatively easy simulations of structures and assemblies through software packages. Ansys Workbench was the FEA software package used for this study.

2.1 Adhesive Model

The single lap joint model was comprised of three components: the top solid, the bottom solid and the middle adhesive. Modeling adhesives is notoriously complex due its highly ambiguous nature, the unusual nature of its crack propagation and the high variability of the material's manufacture. Therefore, to simplify this simulation while maintaining accuracy, the adhesive was separated three regions. The center part of the adhesive was modeled as a bed uncoupled linear shear and normal springs using the Winkler foundation model [22]. This meant that with the center part could be assumed as isotropic and easily modeled. The interfaces between the top and bottom solids and the adhesive were modeled using cohesive zone modeling (CZM) [28]. CZM doesn't represent a physical section of the model, but the forces in between the two solids. Specifically, as the surfaces between the solid and the adhesive separate, they are resisted by cohesive tractions until the maximum stress is reached and failure occurs.

2.2 Material Selection

In the developed model, there were five regions of importance: the top aluminum solid, the bottom carbon fiber & epoxy composite, the middle adhesive and the interfacial layers between the middle adhesive and each of the top and bottom layers. Because of the method chosen to model the adhesive, a material profile had to be built for each one of these regions. When modeling these complex systems, it is important to note which material properties are available, because in Ansys each type of analysis requires a specific set of properties to be known. This was primary consideration in choosing materials because the availability of the necessary material properties was limited in some cases. Tables 1 & 2 contain the material properties used for this study.

Table 1: Material properties used for the thermal and structural analysis [29-31].

Region	Material	Young's Modulus (Pa)	Poisson's Ratio	CTE (1/C)	Tensile Yield Strength (Pa)	Compressive Yield Strength (Pa)	Shear Yield Strength (Pa)	Thermal Conductivity (W/m.K)
Top Solid	Aluminum Alloy	7.1×10^{10}	0.33	2.3×10^{-5}	2.8×10^{8}	2.8×10^{8}	NA	144-165
Middle Adhesive	3M Cohesive Zone	4.5×10^{9}	0.32	6×10^{-5}	1.28×10^{7}	5.79×10^{7}	1.03×10^{7}	0.185
Bottom Solid	Epoxy Carbon Prepreg	1.21×10^{11} (x) 8.6×10^{9} (y) 8.6×10^{9} (z)	0.27 (xy) 0.4 (yz) 0.27 (xz)	-4.7×10^{-7} (x) 3×10^{-5} (y) 3×10^{-5} (z)	8.3×10^{8} (x) 8.3×10^{8} (y) 5×10^{7}(z)	4.4×10^{8} (x) 4.4×10^{8} (y) 1.4×10^{8} (z)	1.2×10^{8} (x) 5×10^{7} (y) 5×10^{7} (z)	10.46
Top Interface	3M Bonded with Aluminum	4.5×10^{9}	0.32	6×10^{-5}	1.28×10^{7}	5.8×10^{7}	1.03×10^{7}	0.185
Bottom Interface	3M Bonded with Acrylic	4.5×10^{9}	0.32	6×10^{-5}	1.28×10^{7}	5.8×10^{7}	6.9×10^{5}	0.185

Table 2: Material properties used for the CZM analysis [29-31].

Region	Debonding Interface Mode	Tangential Slip under Normal Compression	Maximum Normal Contact Stress (Pa)	Contact Gap at the Completion of Debonding (m)	Maximum Tangential Slip at the Completion of Debonding (m)	Maximum Equivalent Tangential Contact Stress (Pa)	Dampening Coefficient (s)
Top Interface	Mixed	Yes	7×10^{6}	5×10^{-4}	5×10^{-4}	7×10^{6}	1×10^{-3}
Bottom Interface	Mixed	Yes	2×10^{7}	5×10^{-4}	5×10^{-4}	6×10^{6}	1×10^{-3}

2.3 Geometry

The primary goal of this analysis was to build an adhesively bonded composite and metal joint, then simulate it under a cyclic thermal loading and return the thermal stresses that result in the mixed material joint. The single lap joint was chosen for this analysis (Figure 1), because of its simplified geometry . This geometry is comprised of three extruded rectangles. The top and bottom solids are each 100mm x 30 mm × 5 mm and have a 30mm overlapping portion. The middle section is 30 mm × 30 mm × 5 mm and aligns with the edges of the top and bottom solids. It represents the adhesive. Additionally, the faces of both top and bottom solids that come into contact with the adhesive section was split at the edge of the adhesive. This was done to create two adhesive interfacial regions for the purpose of analyzing them using CZM.

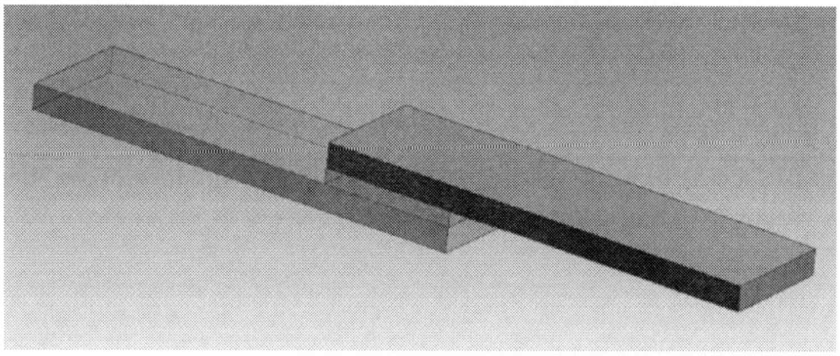

Figure 1: Geometry of the single lap joint used for this study.

2.4 Meshing

Figure 2 shows the mesh used for this study. The mesh was built with the default size (8.36 mm) hexahedral elements. A more refined mesh was used at the contact surfaces between the adhesive and the top and bottom solids. The element size for the quadrilateral elements in the refined mesh was 2.5 mm.

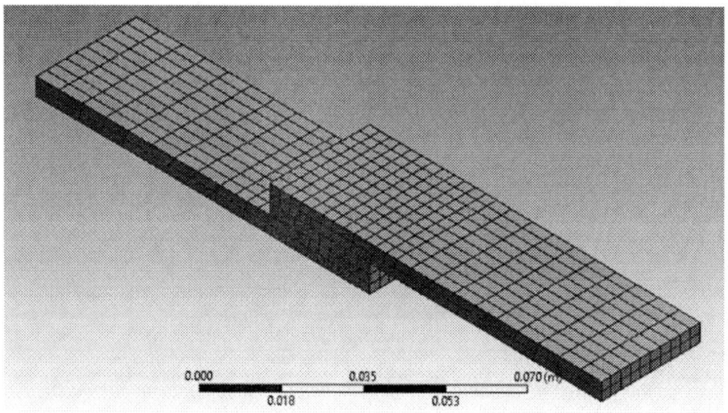

Figure 2: The meshed model used for this study.

2.5 Fracture Mechanics

The modeling of the adhesive/solid interfaces was done through fracture tool in the model section of the static structural analysis system in Ansys. As previously stated, the interfacial layers were modeled using the cohesive zone model applied using the contact debonding tool. When a loading is applied, Ansys simulates the stress in the cohesive zone using the linear traction-separation law until the maximum traction is reached at which point debonding starts and the adhesive begins to fail.

2.6 Loadings and Boundary Conditions

The purpose of this analysis is to simulate the effects of a cyclic thermal loading on the adhesively-bonded mixed material joint. The initial temperature boundary condition was set to 20°C and temperature loading was applied to the outer surfaces of the model. The temperature loading was meant to simulate the varying temperature an aircraft might experience moving from the hot tarmac (55°C) to ambient temperature (20°C) and to the upper atmosphere (-45°C). The temperature loading was applied as a ramped loading at 30 second intervals. This loading was then imported into static structural, a fixed support was placed at either end of the model and the resulting thermal deformation, thermal stresses and the contact debonding were simulated.

3. RESULTS

3.1 Thermal Results

The transient thermal analysis showed that the as the model goes through 3 full cycles of the -45°C to 55°C external temperature loading, there is an excessive amount of heat flux at the edges of the aluminum and adhesive interface (Figures 3 & 5). This result is not mirrored at the carbon fiber composite and adhesive interface (Figure 4).

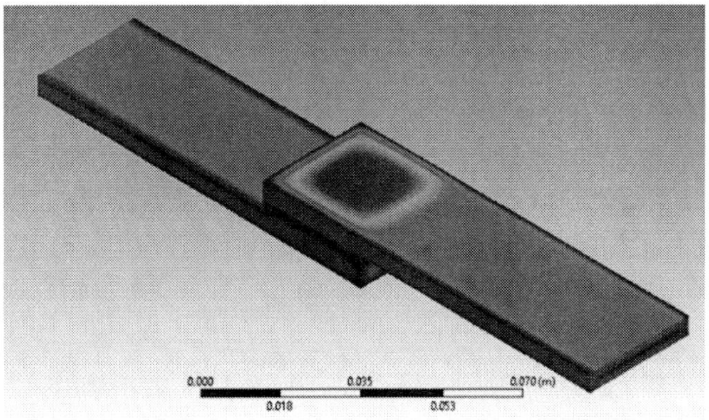

Figure 3: Heat flux contour plot of the model.

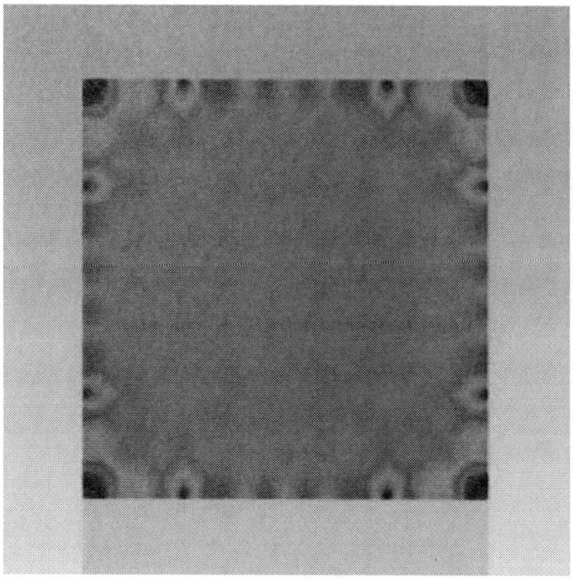

Figure 4: Heat flux contour plot of the aluminum/adhesive interface.

Figure 5: Heat flux contour plot of the adhesives & carbon fiber interface.

3.2 Structural Results

The structural analysis of the thermal loading resulted in a slight expansion in every part of the model, with the greatest expansion happening near the carbon fiber composite and adhesive interface (Figure 6). Additionally, the expansion of the model caused a transverse deformation of the carbon fiber composite and slight bowing in the aluminum. The greatest deformations for each happened at the end of the adhesive facing the carbon fiber composite. The thermal expansion of each of the parts of the model also caused internal stresses to build up within the model (Figure 7). The highest thermal stresses occur at the edge of the aluminum and adhesive interface on the underside of the aluminum solid. Additionally, the carbon fiber composite resulted in a much lower and more uniform stress distribution than the aluminum. This is likely due to the higher Young's modulus of the carbon fiber composite. Finally, the interfacial layers had both a noticeable increase in both frictional stress and pressure (Figures 7-11) primarily at the edges of the interfacial layers.

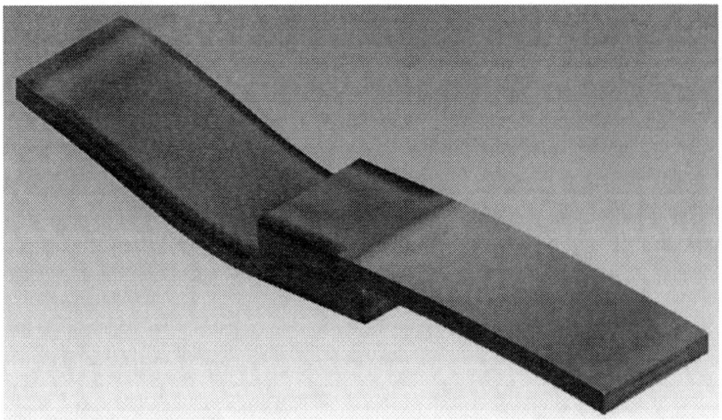

Figure 6: Equivalent stress contour plot for the model under thermal loading.

Figure 7: Total deformation contour plot for the model under the applied temperature loading.

Figure 8: Frictional stress contour plot of the aluminum and adhesive interfacial layer.

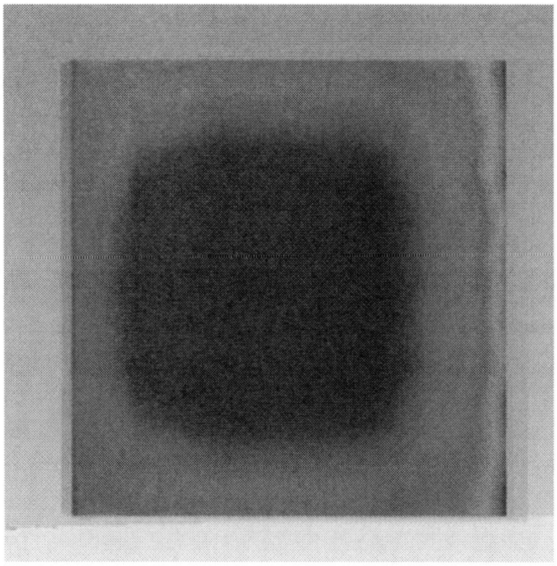

Figure 9: Frictional stress contour plot of the carbon fiber and adhesive interfacial layer.

Figure 10: Equivalent stress contour plot of the aluminum and adhesive interfacial layer.

Figure 11: Equivalent stress contour plot of the carbon fiber and adhesive interfacial layer.

3.3 Further Discussion

When analyzing the thermal data (Figure 3) one of the most notable results is how the heat flux is highest around the adhesive. This is likely due to the extremely low thermal conductivity of the adhesive. As the temperature varies on the external surfaces of the model, the heat does not get conducted throughout the adhesive layer, but instead just builds up on the outer surface and the edges, which leads to a significantly higher heat flux on the edges of the adhesive. Additionally, since there is such a large difference in the thermal conductivity of the aluminum and the carbon fiber composite, there ends up being almost no heat transfer at the carbon fiber composite and adhesive interface (Figure 5) in contrast to the aluminum and adhesive interface (Figure 4) where there is variation up to a few orders of magnitude. This will likely be the point of failure for sustained loadings, because even though this simulation was for only 300s there was a 10°C temperature difference between the two interfaces and a sustained loading will increase the likelihood of thermal stresses exceeding maximum traction and initiate debonding at the aluminum and adhesive interface. In the structural analysis data, those thermal stresses buildup in the model. This is especially seen in the equivalent stresses plot (Figure 7) where the highest stresses are at the edges of the adhesive, particularly on the underside of the aluminum. Similarly, both the frictional stress and pressure contour plots show highest concentrations at the edges. Notably, the frictional stress at the aluminum and adhesive interface reached the maximum contact stresses for the interface, indicating that under this temperature loading contact debonding would start to occur at the edge of the aluminum and adhesive interface.

4. CONCLUSIONS

The purpose of this simulation was to analyze an adhesively bonded mixed material joint under cyclic thermal loading. The developed model demonstrates most of the tools necessary to accomplish this task to real world standards, but there are some missing pieces to this simulation. The Ansys model was comprised of a carbon fiber composite laminate bonded with an epoxy adhesive to an aluminum plate. The model was used to study the aggregate structural effects of cyclic thermal loadings applied through periodic temperature boundary conditions, simulating the service temperatures typical for the aircraft. The simulation results revealed that the heat flux was highest around the adhesive. This was attributed to the extremely low thermal conductivity of the adhesive. Additionally, since there was such a large difference in the thermal conductivity of the aluminum and the carbon fiber composite, there was almost no heat transfer at the carbon fiber composite and adhesive interface, in contrast to the aluminum and adhesive interface where there was variation up to a few orders of magnitude. It would be relatively simple to improve with modest experimentation or a deeper dive into experimental data. The presented finite element analysis of the materials joints is a promising technique and should be usable not just for modeling mixed material joints, but also modeling the composites individually. This is because composites are often joined in a similar manner with similar material and this simulation could be easily modified to accommodate them.

5. REFERENCES

[1] A. Abdul-Baqi, P. J. G. Schreurs, and M. G. D. Geers, "Fatigue damage modeling in solder interconnects using a cohesive zone approach," *International Journal of Solids and Structures,* vol. 42, pp. 927-942, 2005 https://doi.org/10.1016/j.ijsolstr.2004.07.026.

[2] K. S. Al-Athel, A. F. M. Arif, and S. Pashah, "Behavior and failure of adhesive bonds in pin fin heat sinks using cohesive zone model," *International Journal of Adhesion and Adhesives,* vol. 68, pp. 397-406, 2016 https://doi.org/10.1016/j.ijadhadh.2015.12.016.

[3] R. P. Alessio, N. M. Andre, S. M. Goushegir, J. F. dos Santos, J. A. E. Mazzaferro, and S. T. Amancio-Filho, "Prediction of the mechanical and failure behavior of metal-composite hybrid joints using cohesive surfaces," *Materials Today Communications,* p. 101205, 2020 https://doi.org/10.1016/j.mtcomm.2020.101205.

[4] D. L. Alves, R. D. S. G. Campilho, R. D. F. Moreira, F. J. G. Silva, and L. F. M. da Silva, "Experimental and numerical analysis of hybrid adhesively-bonded scarf joints," *International Journal of Adhesion and Adhesives,* vol. 83, pp. 87-95, 2018 https://doi.org/10.1016/j.ijadhadh.2018.05.011.

[5] L. Benabou, Z. Sun, and P. R. Dahoo, "A thermo-mechanical cohesive zone model for solder joint lifetime prediction," *International Journal of Fatigue,* vol. 49, pp. 18-30, 2013 https://doi.org/10.1016/j.ijfatigue.2012.12.008.

[6] R. D. S. G. Campilho, M. D. Banea, J. A. B. P. Neto, and L. F. M. da Silva, "Modelling adhesive joints with cohesive zone models: effect of the cohesive law shape of the adhesive layer," *International Journal of Adhesion and Adhesives,* vol. 44, pp. 48-56, 2013 https://doi.org/10.1016/j.ijadhadh.2013.02.006.

[7] M. C. Corbett, P. A. Sharos, M. Hardiman, and C. T. McCarthy, "Numerical design and multi-objective optimisation of novel adhesively bonded joints employing interlocking surface morphology," *International Journal of Adhesion and Adhesives,* vol. 78, pp. 111-120, 2017 https://doi.org/10.1016/j.ijadhadh.2017.06.002.

[8] M. El Zaroug, F. Kadioglu, M. Demiral, and D. Saad, "Experimental and numerical investigation into strength of bolted, bonded and hybrid single lap joints: Effects of adherend material type and thickness," *International Journal of Adhesion and Adhesives,* vol. 87, pp. 130-141, 2018 https://doi.org/10.1016/j.ijadhadh.2018.10.006.

[9] R. L. Fernandes and R. D. S. G. Campilho, "Numerical evaluation of dissimilar cohesive models to predict the behavior of Double-Cantilever Beam specimens," *Procedia Structural Integrity,* vol. 1, pp. 42-49, 2016 https://doi.org/10.1016/j.prostr.2016.02.007.

[10] M. Gaiotti, A. Zamarin, B. L. Josefson, and T. Matulja, "A numerical sensitivity analysis on the cohesive parameters of a carbon-steel single lap joint," *Ocean Engineering,* vol. 198, p. 106010, 2020 https://doi.org/10.1016/j.oceaneng.2019.04.091.

[11] M. R. Gheibi, M. H. Shojaeefard, and H. Saeidi Googarchin, "Experimental and numerical analysis on the cohesive behavior of an automotive adhesive improved by MWCNT subjected to mode I and II loadings," *International Journal of Mechanical Sciences,* vol. 153-154, pp. 271-286, 2019 https://doi.org/10.1016/j.ijmecsci.2019.02.009.

[12] I. Skeist, *Handbook of Adhesives*: Springer US, 2012.

[13] E. M. Petrie, *Handbook of Adhesives and Sealants*: McGraw-Hill Education, 2007.

[14] N. R. C. C. S. A. A. use, *Aerospace Structural Adhesives*: National Academy of Sciences-National Academy of Engineering, 1974.

[15] N. R. Hamilton, J. Wood, A. Galloway, M. B. Olsson Robbie, and Y. Zhang, "The metallurgy, mechanics, modelling and assessment of dissimilar material brazed joints," *Journal of Nuclear Materials,* vol. 432, pp. 42-51, 2013 https://doi.org/10.1016/j.jnucmat.2012.07.022.

[16] F. Hirsch, S. Müller, M. Machens, R. Staschko, N. Fuchs, and M. Kästner, "Simulation of self-piercing rivetting processes in fibre reinforced polymers: Material modelling and parameter identification," *Journal of Materials Processing Technology,* vol. 241, pp. 164-177, 2017 https://doi.org/10.1016/j.jmatprotec.2016.10.010.

[17] J.-n. Hu, T. Fukahori, T. Igari, Y. Chuman, and A. C. F. Cocks, "An evaluation of creep rupture strength of ferritic/austenitic dissimilar weld interfaces using cohesive zone modelling," *Procedia Structural Integrity,* vol. 2, pp. 934-941, 2016 https://doi.org/10.1016/j.prostr.2016.06.120.

[18] P. Hu, X. Han, L. F. M. da Silva, and W. D. Li, "Strength prediction of adhesively bonded joints under cyclic thermal loading using a cohesive zone model," *International Journal of Adhesion and Adhesives,* vol. 41, pp. 6-15, 2013 https://doi.org/10.1016/j.ijadhadh.2012.10.009.

[19] R. A. Hunter-Alarcon, J. Leyrer, E. Leal, A. Vizan, J. Perez, and L. F. M. da Silva, "Influence of dissimilar composite adherends on the mechanical adhesion of bonded joints for small blade wind turbine applications," *International Journal of Adhesion and Adhesives,* vol. 83, pp. 178-183, 2018 https://doi.org/10.1016/j.ijadhadh.2018.02.018.

[20] I. Katsivalis, O. T. Thomsen, S. Feih, and M. Achintha, "Development of cohesive zone models for the prediction of damage and failure of glass/steel adhesive joints," *International Journal of Adhesion and Adhesives,* vol. 97, p. 102479, 2020 https://doi.org/10.1016/j.ijadhadh.2019.102479.

[21] M. Lißner, B. Erice, E. Alabort, D. Thomson, H. Cui, C. Kaboglu, *et al.*, "Multi-material adhesively bonded structures: Characterisation and modelling of their rate-dependent performance," *Composites Part B: Engineering,* vol. 195, p. 108077, 2020 https://doi.org/10.1016/j.compositesb.2020.108077.

[22] S. Stapleton, "The Analysis of Adhesively Bonded Advance Composite Joints using Joint Finite Elements," Ph.D. Dissertation, Aerospace Engineering, The University of Michigan, 2012.

[23] F. A. Stuparu, D. A. Apostol, D. M. Constantinescu, C. R. Picu, M. Sandu, and S. Sorohan, "Cohesive and XFEM evaluation of adhesive failure for dissimilar single-lap joints," *Procedia Structural Integrity,* vol. 2, pp. 316-325, 2016 https://doi.org/10.1016/j.prostr.2016.06.041.

[24] P. Tsokanas, T. Loutas, G. Kotsinis, V. Kostopoulos, W. M. van den Brink, and F. Martin de la Escalera, "On the fracture toughness of metal-composite adhesive joints with bending-extension coupling and residual thermal stresses effect," *Composites Part B: Engineering,* vol. 185, p. 107694, 2020 https://doi.org/10.1016/j.compositesb.2019.107694.

[25] H. Wan, J. Min, J. Zhang, J. Lin, and C. Sun, "Effect of adherend deflection on lap-shear tensile strength of laser-treated adhesive-bonded joints," *International Journal of Adhesion and Adhesives,* vol. 97, p. 102481, 2020 https://doi.org/10.1016/j.ijadhadh.2019.102481.

[26] F. Yan, X. Fang, L. Chen, C. Wang, J. Zhao, F. Chai, *et al.*, "Microstructure evolution and phase transition at the interface of steel/Al dissimilar alloys during Nd: YAG laser welding," *Optics & Laser Technology,* vol. 108, pp. 193-201, 2018 https://doi.org/10.1016/j.optlastec.2018.06.039.

[27] B. Zhang, X. Chen, K. Pan, and C. Yang, "J-integral based correlation evaluation between microstructure and mechanical strength for FSSW joints made of automotive aluminum alloys," *Journal of Manufacturing Processes,* vol. 44, pp. 62-71, 2019 https://doi.org/10.1016/j.jmapro.2019.05.039.

[28] M. Elices, G. V. Guinea, J. Gómez, and J. Planas, "The cohesive zone model: advantages, limitations and challenges," *Engineering Fracture Mechanics,* vol. 69, pp. 137-163, 2002 https://doi.org/10.1016/S0013-7944(01)00083-2.

[29] H. Dodiuk and S. H. Goodman, *Handbook of Thermoset Plastics*: Elsevier Science, 2013.

[30] S. Zhang and D. Zhao, *Aerospace Materials Handbook*: CRC Press, 2016.

[31] 3M. (2019). *3M™ Scotch-Weld™ Epoxy Adhesive DP100 Clear*. Available: https://3m.citrination.com/pif/000163?locale=en-US

SURFACE ACTIVATION METHODS FOR BONDING TO THERMOPLASTICS: PLASMA AND A NEW TAILORED UV BASED METHOD

Sina Chaeichian, Kaspar Schaerer, Ruairi O'Kane, Michael D. Halbasch
Henkel Adhesive Technologies, Aerospace
Bay Point, CA 94565

ABSTRACT

The constant trend to optimize the costs of raw materials and related processing costs has accelerated the use of thermoplastic composites as alternatives to traditional thermoset composites in the Aerospace Industry. While thermoplastics can be easily fused together above their melting temperatures, it is difficult to bond them to dissimilar substrates, including thermosets and aluminum. The high melting temperatures of engineered thermoplastics like Polyarylether Ketone family (350 – 430 °C) limit their process. Therefore, lower temperature bonding systems would be desirable.

This paper discusses the development of a new and rapid UV pretreatment method for thermoplastic composites allowing improved bonding to dissimilar substrates. In addition, plasma pretreatment is also investigated as a common method since it can have good proficiency in some cases depending upon the chemistry of adhesives. Contact angle measurement was done for surface analysis of treated substrates where regarding to the new tailored UV method, complementary analysis such as XPS and TOF-SIMS was conducted. To evaluate the bond strength, mechanical tests including fracture toughness (G_{IC}) and tensile lap shear tests were conducted.

Contact angle measurements, XPS, and TOF-SIMS analysis revealed an increase in polarity and wettability of the thermoplastic surface after pretreatments due to the formation of new oxygen-containing functional groups. This rapid UV surface treatment method led to a significant improvement in bonding of PAEK thermoplastic composites with both low and elevated temperature-cure adhesives verified by mechanical test results. This method offers new opportunities for fast and safe bonding to the thermoplastic materials resulting in excellent bond strength comparable with plasma pretreatment.

1. INTRODUCTION

In aerospace industry, the need to develop high-performance structural materials has been continuously growing where the unrelenting passion of the industry demands more replacement of metals with light weight composite materials in primary structures like fuselages and wings [1]. Composite materials typically consist of strong fibers in a tough resin matrix which have been traditionally thermosetting resins. However, the constant trend to optimize the costs of raw materials and related processing expenses has accelerated the application of thermoplastic composites as alternatives to thermoset composites in the aerospace industry. Unlike thermosets, thermoplastics do not need chemical crosslinking for solidification and can be easily formed under sufficient heat and simply solidified by cooling to maintain their shapes at speeds much faster than curing of thermosets. Therefore, high-performance engineering thermoplastics have attracted a lot of attention in composite industries for aerospace applications where polyaryletherketones (PAEKs) represent the most promising class of materials for this purpose.

Polyaryletherketones (PAEKs) are semi-crystalline thermoplastic resins with outstanding thermal and mechanical properties where the ratio of keto- and ether- groups defines different families within this class. Polyetheretherketone (PEEK) with two ether groups in its repeat unit (Figure 1) has a growing interest in manufacturing of carbon fiber reinforced plastics (CRFP) due to its outstanding properties such as good dimensional stability at high temperature due to its high melting point (about 335 °C), high yield strength (about 90 MPa) with an elongation at break of 120-170% [2]. However, high-performance engineered thermoplastic polymers especially PEEK show poor wettability and adhesive bonding strength due to the low surface energy. These characteristics do not allow satisfactory adhesion of surface finishes such as paints or effective adhesion in structural bonding of thermoplastic composite structures.

As a solution, surface treatment of the thermoplastic substrate is required to modify their surface chemistry and improve their wetting characteristic [3]. For this purpose, various chemical and physical surface treatments have been developed including solvent cleaning, etchants such as chromic acid, corona discharges, flame and plasma pretreatment [4,5]. However, physical pretreatment methods are preferred in the industry. Plasma surface treatment is often one of the preferred methods as it can provide stronger and more stable surface energy enhancement. Low pressure plasma (LPP) can result in a higher quality of surface treatment, but it has a limiting factor for the size of parts due to the restricted size of the vacuum chambers [5]. To eliminate the expensive limiting vacuum systems, many research studies have been done on atmospheric-pressure plasma (APP) leading to significant improvement in adhesion strength of the thermoplastic composite parts, but still it calls for more investigations [2,5-8].

Another physical method for surface treatment of the thermoplastics is ultraviolet (UV) irradiation technique. UV excimer lamps allow the surface treatment at low surface temperature. In addition, the relative simplicity in using and building a continuous treatment system makes it appropriate for industrial applications [2].

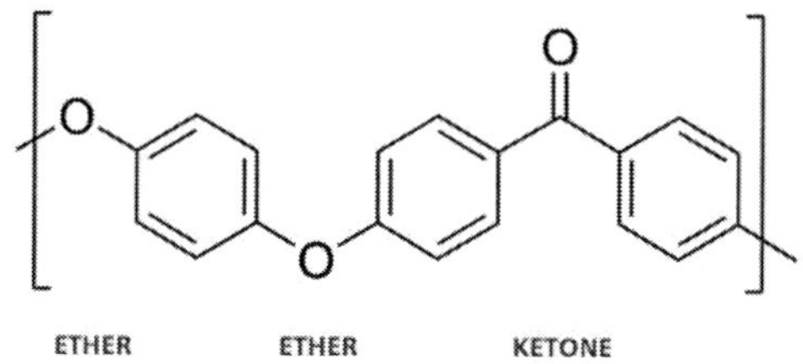

Figure 1. Chemical structure of PEEK.

A recent UV surface treatment technique developed at Henkel Aerospace offers a significant improvement in the adhesion strength to the PAEK family represented by PEEK with the advantage that a very short processing time is sufficient. The emitted photons by UV irradiation can activate or break chemical bonds at the surface of the polymeric substrates. Additionally, the UV irradiation can generate highly reactive ozone formed from oxygen through photolysis. Consequently, the treatment results in the formation of new functional groups at the surface of the thermoplastic substrates increasing the polarity and wettability where many of the new formed functional groups are able to form covalent bonds with the epoxy adhesive leading to very strong bonds.

2. EXPERIMENTATION

2.1 Materials

Adhesives with different curative classes designed for distinct service temperatures were used in this study. Three films adhesives with high temperature cure system which are commercially available including LOCTITE EA 9696, EA 9695, and EA 9658 and two paste adhesives including a room temperature cure adhesive, LOCTITE EA 9394, and an elevated temperature cure paste adhesive, LOCTITE EA 9394/C-2 were used as bonding adhesives for preparing mechanical test specimens including fracture toughness and tensile lap shear specimens. All film adhesives are unsupported (no scrim) with a specific mass of 0.29 kg/m^2. Proprietary customer prepregs for preparing thermoset composite panels with unidirectional fiber orientation as well as PEEK composite panels (carbon reinforced with 5HS fiber pattern) were used to prepare the test specimens for mechanical testing. All thermoset and thermoplastic composite panels had a thickness of 0.16 cm (0.063 in). For secondary bonding trials, LOCTITE EA 9895 WPP was used as a wet peel ply.

A surfacing Film LOCTITE EA 9845 LC with cupper mesh was used to evaluate the effect of the pretreatments on bonding strength of PEEK substrates and epoxy-based surfacing films.

2.2 Treatment Methods

Plasma treatment was done with an Open-air Plasma system with a rotation jet at atmospheric pressure. The treatment was done with a 14-degree nozzle with the following conditions: a travel speed of 6.5 mm/s, a pressure of 10 psi, and a specimen distance of 10 mm.

UV surface treatment was done by using a specific UV excimer lamp and variation in UV intensity was done by adjusting the distance to the lamp and exposure time. Equation 1 shows the relation between UV intensity, lamp-specimen distance and exposure time. UV intensity has a linear correlation with exposure time where it exponentially reduces by increasing the distance to the excimer lamp mainly due to absorption of UV radiation in the ambient atmosphere [2].

$$I_{treatment} = I\ (d).t\ \left[\frac{mJ}{cm^2}\right] \qquad [1]$$

Where $I_{treatment}$ is indicating intensity, d is representing the distance between the lamp and specimen and t is time in seconds.

2.3 Test Methods

Double cantilever testing (DCB) was done with rectangular laminated specimens consisting of one carbon-reinforced PEEK panel and one carbon-reinforced epoxy panel bonded by the adhesives. First, a large thermoplastic panel (15.24 cm x 30.48 cm) was treated by UV or plasma pretreatment method. The specimens layup consists of thermoplastic panel, adhesive film, and thermosetting prepregs with UD alignment where bonding was done by co-curing of the adhesive and prepregs mainly in an autoclave at 177 °C (350 °F) under a pressure of 0.6 MPa (90 psi) for 2 h. The large bonded panels were cut into 1.27 cm (0.5 in) wide strips with a length of 30.48 cm (12 in). The test was conducted according to a customer specification to measure mode I fracture toughness energy (G_{IC}). The testing was done with a crosshead speed of 2.54 cm/min where loading was applied to reach a crack with a length of about 10 cm. During the test, each crack arrest position was marked on the edge of the specimen. The area in the load deflection curve between two known crack length positions were measured to calculate the fracture energy.

Tensile Lap Shear tests were done according to ASTM D1002 [9]. First the PEEK panels were abraded in overlap area by a sandpaper with a grit size of 400 and then treated by UV or plasma pretreatment method. To make the specimens, 2 large panels of PEEK with a size about 23 cm x 10 cm where bonded with adhesives along their length with an overlap of 1.27 cm and bond-line of about 0.127 mm. Curing was done in an oven at a specific curing temperature recommended for each adhesive under a pressure of 0.3 MPa provided by clamps. Then the large bonded panels were cut into 2.54 cm wide strips. In the case of notched lap shear specimens, two large pretreated panels with dimensions of 20.3 cm x 20.3 cm where bonded by the film adhesives where curing was done in an autoclave at 177 °C (350 °F) under a pressure of 0.6 MPa (90 psi) for 2 h. The bonded panels were cut into 2.54 cm wide strips notched to give a 1.27 cm overlap at the end.

3. RESULTS

3.1 Surface Analysis

Many studies have been done on surface analysis of PAEK thermoplastic family treated by atmospheric plasma method. Techniques such as contact angle measurement and XPS showed that plasma treatment increases wettability and polarity of the surface of PEEK thermoplastics and composites by forming some oxygen-containing polar functional groups [5-8]. In this study, several surface analysis techniques including XPS, TOF-SIMS, and contact angle measurements have been used to investigate the effect of this new tailored UV method on the surface characteristics.

3.1.1 X-ray photoelectron spectroscopy (XPS)

X-ray Photoelectron Spectroscopy (XPS) was conducted to obtain a quantitative and chemical composition analysis of the surface of the thermoplastic composite substrates before and after UV treatment. The information provided by XPS analysis is displayed in Table 1.

Table 1. Elemental analysis of PEEK panels surface with different UV irradiation time.

Treatment method	Treatment time (s)	Average Carbon (%)	Average Nitrogen (%)	Average Oxygen (%)
No treatment	0	78.5	0.9	17.5
UV method	10	66.8	1.6	26.7
UV method	90	64.5	1.0	28.1

Table 1 shows content changes of three elements on the substrate surface due to UV treatment. A very short UV exposure time (10 sec) resulted in a significant increase of oxygen content by about 52% where the longer exposure time resulted in an increase by about 60%. Similar observations have been reported regarding plasma treatment of PEEK substrates in the literature whereas the oxygen content increase by plasma activation is greater (about 75%) [7] suggesting higher proficiency of plasma treatment in increasing surface polarity. These pretreatments lead to the formation of new oxygen-containing functional groups such as carbonyl, carboxyl and hydroxyl groups which significantly increase the polarity and wettability of the surface. In addition, many of these new functional groups are able to form covalent bonds with reactants in adhesives such as epoxy adhesive leading to robust bond lines.

Contact angle measurement was conducted to evaluate the change of wettability after both UV and plasma treatment (Figure 2). As expected, the test showed both activation methods resulted in a significant reduction in water contact angle indicting an increase in polarity and consequently wettability of the surface. Interestingly, plasma activation resulted in lower contact angle values confirming XPS analysis observation where correlation between the level of surface polarity and adhesion strength will be evaluated in the next section.

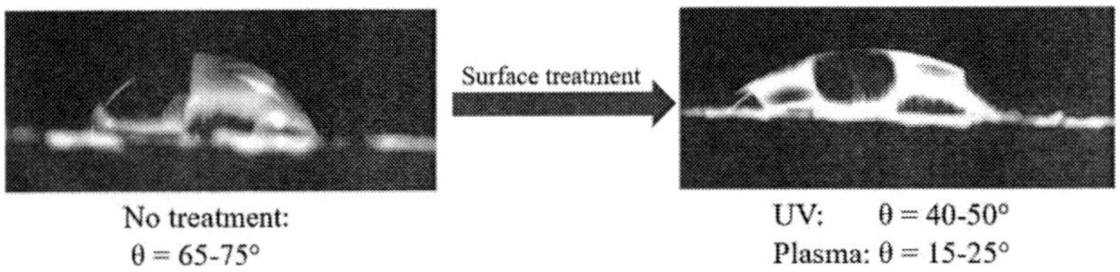

Figure 2. Contact angle (θ) of a water droplet on PEEK panel surface before and after surface pretreatment.

Time-of-Flight Secondary Ion Mass Spectrometry (TOF-SIMS) is an extremely sensitive analytical technique providing elemental, chemical state, and molecular information from the surface of solid materials. In this work, TOF-SIMS was conducted to evaluate and identify the effect of UV irradiation on the chemical structure and composition of the surface of PEEK composite panels. This analytical test confirmed the formation of various new functional groups such as alcohols, aldehydes, ketones and carboxyl groups.

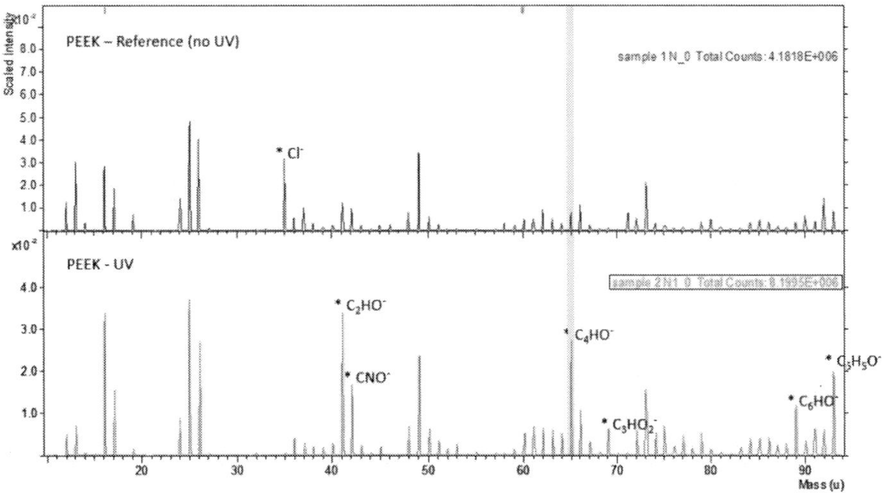

Figure 3. TOF-SIMS spectrum diagrams of PEEK panels before and after UV pretreatment.

The multiple surface analytical techniques demonstrated the efficiency of the UV and plasma pretreatment methods in improving the polarity and wettability of the composite surface by the formation of new functional groups which allow to form covalent bonds with the epoxy adhesives.

3.2 Mechanical properties

3.2.1 Fracture toughness

Fracture toughness was chosen as the key test method to evaluate the bonding capability of the treated thermoplastic PAEK substrates, represented by PEEK, with high temperature-cure film adhesives. This test method measures the toughness of the adhesive captured in G_{IC} value and the adhesive bond failure mode allows for an excellent evaluation of the adhesion performance between adhesive and substrate. For this purpose, LOCTITE EA 9696, EA 9695 and EA 9658 were selected to prepare double cantilever beam (DCB) test specimens for G_{IC} fracture toughness measurements.

Figure 4 shows the effect of surface treatment methods on fracture toughness energy (G_{IC}) of DCB specimens bonded with different epoxy-based film adhesives. The figure illustrates an improvement in adhesive bonding of CF-PEEK to CF-Epoxy substrates by conducting the surface pretreatments; however, the efficiency of each activation method is different. This discrepancy in the performance of UV and plasma activation was confirmed by surface analysis. Plasma treatment results in a greater increase in polarity and wettability of PEEK substrate, but it does not necessarily mean a better performance in improving adhesive bonding of the thermoplastic. In fact, the chemistry of the adhesive also plays an important role on the proficiency of each treatment method. For instance, in the case of PEEK bonding with LOCTITE EA 9658, UV irradiation has a better performance whereas plasma activation is more efficient in PEEK bonding with LOCTITE EA 9695.

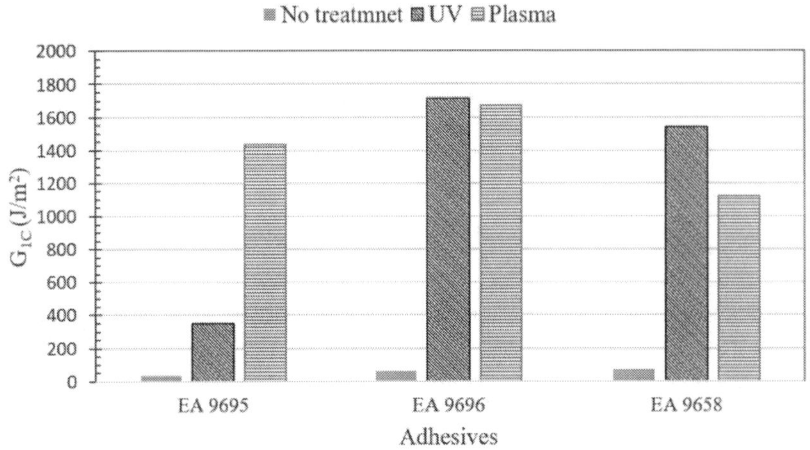

Figure 4. Fracture toughness (G_{IC}) measurements of PEEK samples bonded with EA 9696, EA 9695, and EA 9658 with and without UV and plasma pretreatment.

Adhesive bond failure mode as illustrated in Figure 5 clearly indicate the effect of the both treatments on improving adhesion strength. The failure mode for all untreated substrates was 100% adhesion failure to the PEEK substrate. In contrast, UV and plasma activation directed the fracture failure mode to mainly substrate failure except in case of bonding UV treated substrate with EA 9695 where the main failure is adhesion failure with some level of cohesion failure of the adhesive.

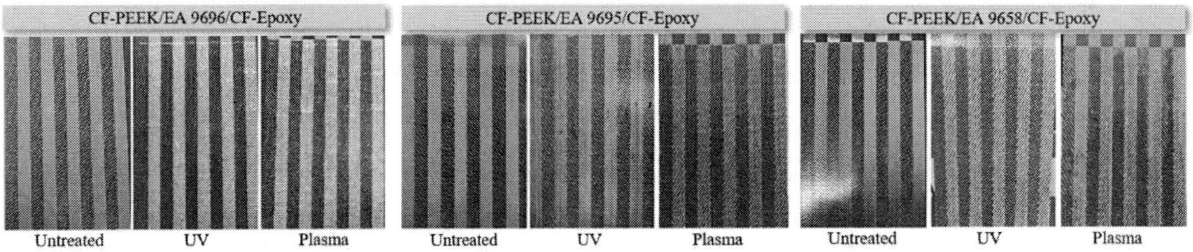

Figure 5. Adhesive bond failure mode of DCB specimens with and without UV and plasma pretreatment. The thermoplastic and thermoset strip from each DCB specimen are paired where the black 5HS woven pattern strip is PEEK.

A deeper study in the chemistry and formulation of LOCTITE EA 9695 was done for better understanding the cause of the substantial discrepancy in the performance of UV and plasma activation in PEEK bonding. Theoretically, the adhesion strength of bonding with a specific adhesive depends upon rheological behavior, polarity and surface tension, as well as chemistry and reactive functional groups of the adhesive. In this study, we found out the curing mechanism and rate of the adhesive plays a very important role in bonding of UV treated PEEK substrates. In this regard, slight modification in curative formulation was done to change the reactivity of EA 9695. The results suggested that an optimization in the adhesive curing rate is favorable to achieve strong adhesion strength likely by controlling chemical bonding between adhesive and new functional groups formed on the activated substrate. Figure 6 shows how the curing rate modification strongly enhanced bonding of UV treated PEEK and improved failure mode from adhesion failure to mainly substrate failure.

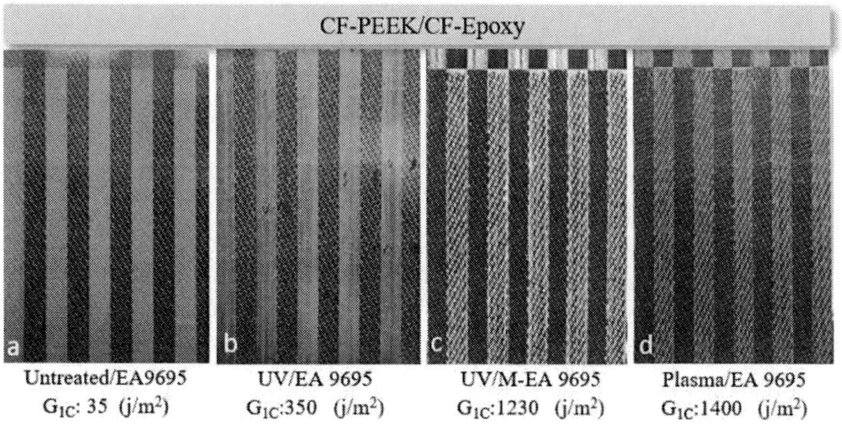

Figure 6. Adhesive bond failure mode of DCB specimen a) untreated/EA 9695, b) UV/EA9695, C) UV/Modified (M)-EA 9695, d) Plasma/EA9695. The thermoplastic and thermoset strip from each DCB specimen are paired where the black 5HS woven pattern strip is PEEK.

Although UV and plasma treatment have almost similar performance in adhesive bonding of PEEK substrates in most cases, it is noteworthy that the areal activation rate (m^2/s) of UV method is about 10 times faster than that of plasma activation which favors the UV treatment with regard to high throughput applications especially for large parts and substrates. For instance, required treatment time for a DCB size panel (20 cm x 35cm) by a UV bulb with a size of 10 cm is about 20 - 40 sec, but by one plasma jet is about 5-7 min.

3.2.2 Bonding Types: Co-Bonding and Secondary bonding

To determine the scope of application opportunities that the newly developed UV based method provides, the fracture toughness samples prepared with multiple bonding methods including co-bonding inside autoclave, co-bonding Out of Autoclave (OoA), and secondary bonding have been evaluated. LOCTITE EA 9696 adhesive was selected for this side by side comparison. In Figure 7, for both co-bond inside and out of autoclave, G_{IC} values ($\approx$1600 J/m^2) are nearly identical and mainly substrate failure was observed. The lower G_{IC} values (1170 J/m^2) for the secondary bonding can be explained by the observed peel ply failures on the thermoset composite panel and no adhesion failure to the thermoplastic PEEK substrate was noticed.

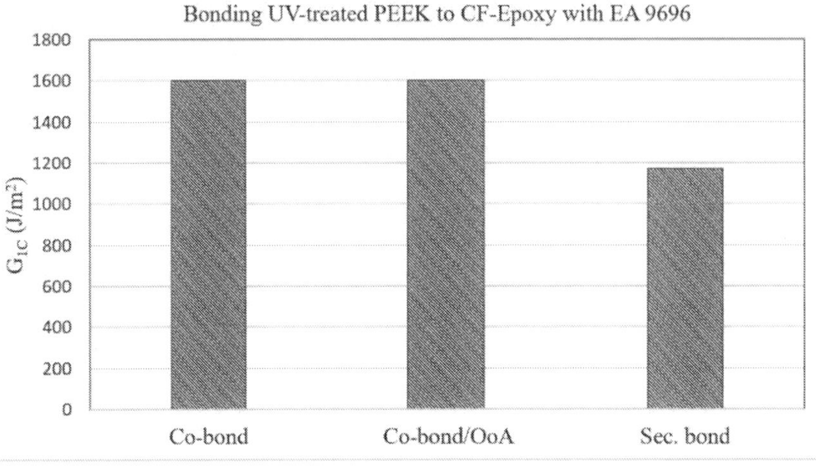

Figure 7. Comparing different bonding of UV treated PEEK to CF-Epoxy

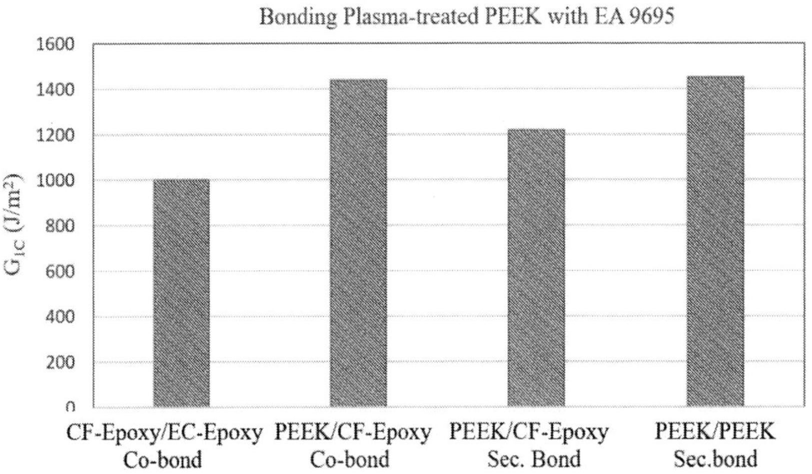

Figure 8. Comparing different bonding of Plasma treated PEEK to CF-Epoxy

The performance of plasma treatment for secondary bonding of PEEK substrate with EA 9695 has been evaluated. Figure 8 shows DCB results for co-bonding and secondary bonding of PEEK which demonstrates the high efficiency of the activation for different bonding types. In other words, the treatment leads to almost identical strength for co-bonding and secondary bonding. It is noteworthy that the flexibility of the substrate affects G_{IC} values via affecting stress distribution during running the test. G_{IC} values of co-bonded PEEK/CF epoxy is higher than that of co-bonded thermoset CF-epoxy since PEEK composite (5HS) has less rigidity compared to CF-Epoxy (UD).

3.2.3 Tensile lap shear

Tensile lap shear is a common test method to evaluate the adhesion strength of adhesive bonding of various substrates in Aerospace industry. In this section, single tensile lap shear and notched tensile lap shear were used to evaluate the effect of UV and plasma treatment on improving PEEK substrates bonding with room and elevated temperature-cure paste adhesives and high temperature-cure film adhesives respectively.

Figure 9 shows tensile lap shear strength of PEEK specimens treated by UV or plasma and bonded with LOCTITE EA 9394 and EA 9394/C-2 which are room-temperature and elevated temperature cure (>93 °C) adhesives respectively. The results proved a significant improvement in PEEK bonding with UV and plasma activation almost with the same performance. However, the failure mode was mainly adhesion failure with some level of substrate failure.

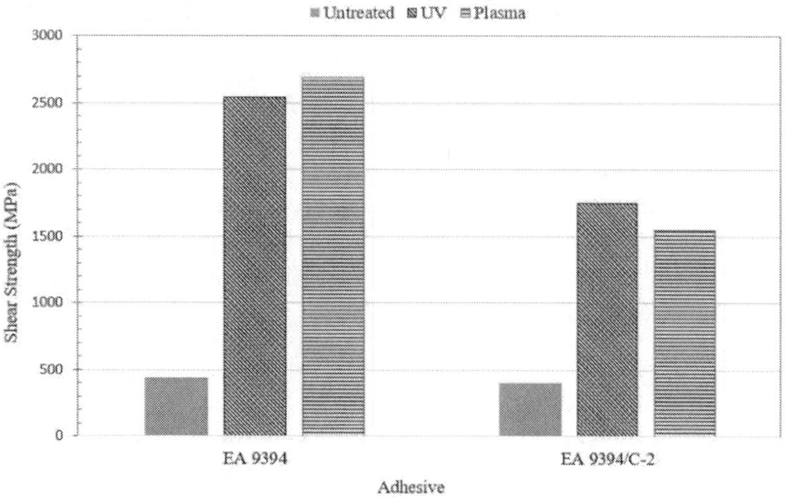

Figure 9. Tensile lap shear of PEEK substrates bonded with a room-temperature cure (EA 9394) and an elevated-temperature cure (EA 9394/C-2) paste adhesive with and without pretreatments

To further study this new tailored UV treatment effect on PEEK bonding with film adhesives, notched tensile lap shear test has been used to validate DCB test results. Table 2 and Figure 10 show shear strength values and failure mode of PEEK specimens bonded with LOCTITE EA9696 and EA9658. High shear strength and substrate failure mode confirmed that UV treatment significantly enhances the bonding strength.

Table 2. Shear streangth and failure mode of UV treated PEEK specimens bonded with film adhesives.

Specimen	Surface treatment	Shear strength (MPa)	*CV (%)	Failure mode
CF-PEEK/EA 9658	Untreated	3.5	18	Adhesion failure
CF-PEEK/EA 9658	UV treated	25.3	5	Mainly substrate failure
CF-PEEK/EA 9696	Untreated	3.2	12	Adhesion failure
CF-PEEK/EA 9696	UV treated	26.7	2	Mainly substrate failure

*CV: Coefficient of variation

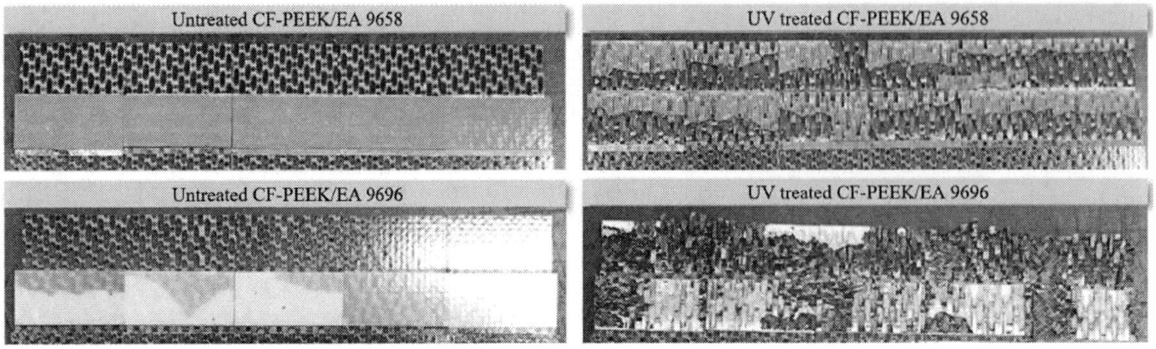

Figure 10. Adhesive bond failure mode of notched tensile lap shear specimens with and without UV treatment.

Overall the mechanical tests including fracture toughness and tensile lap shear tests conclude that the new tailored UV pretreatment method allows for forming very strong bonds to the thermoplastic PEEK substrate comparable with plasma activation. However, the efficiency of each pretreatment on improving adhesive bonding of PEEK substrates strongly depends upon the chemistry and characteristics of the adhesive as well.

3.4 Surfacing film application

Surface pretreatment of thermoplastic composites is not only essential for adhesive bonding, but it is also required for cosmetic applications and painting. Thermoplastic and thermoset composite structures that are exposed to environmental conditions or need to meet certain aesthetic requirements typically include a surfacing film with or without lightning strike protection to provide the required surface properties to the composite structures prior painting.

To demonstrate the efficiency of UV and plasma pretreatment on adhesion strength between a surfacing film and a PEEK composite panel, the epoxy-based LOCTITE EA 9845 LC (with cupper mesh as lightning strike protector) was used as a surfacing film. The adhesion strength evaluation of the surfacing film cured at 177 °C inside autoclave (0.62 MPa pressure) and out of autoclave (vacuum pressure) was done by Cross Hatch test method (Figure 11). Untreated substrate has very poor adhesion to the surfacing film whereas pretreated samples (Figure 11b and c) showed excellent crosshatch rating of 10/10 indicating strong adhesion. It is noteworthy that the test showed similar efficiency of the both pretreatment methods for improving bonding of the surfacing film to PEEK substrate. In addition, similar results were observed for specimens cured at a lower temperature (121 °C).

This example indicates that the UV and plasma pretreatment methods can be used for bonding a wide range of epoxy-based thermosetting materials and are not exclusively limited to epoxy adhesives.

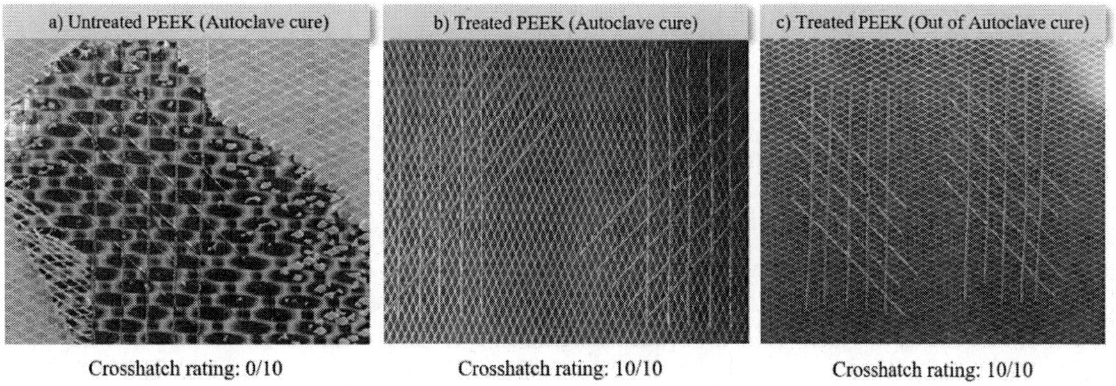

Figure 11. Effect of pretreatment on bonding of surfacing film to PEEK composite panels: Inside and out of autoclave cured at 177 °C.

4. CONCLUSIONS

In a side by side comparison, the effect of a new tailored UV surface pretreatment versus plasma activation on adhesive bonding strength of PEEK thermoplastic composite substrate was evaluated. In this regard, analytical techniques were used to learn the effect of the treatments on surface characteristics of the substrate and mechanical tests including double cantilever beam and tensile lap shear test were conducted to examine bonding strength.

After only 10 seconds of UV irradiation, surface analysis by XPS and TOF-SIMS revealed a significant increase in oxygen content (17% to 26.7%) due to the formation of new polar groups such as carboxylate, carbonyl, hydroxyl, aldehyde, and ketone groups at the surface. Similar observation related to plasma treatment of PEEK have been reported in the literature except with a higher oxygen elemental increase. Water contact angle measurements showed a substantial reduction in the values after treatment where plasma activation resulted in much lower contact angle values indicating higher efficiency of plasma activation in increasing wettability and polarity of PEEK surface. These chemical modifications at the surface overcome the low surface energy of the thermoplastic and lead to a much better wettability of the surface. In addition, the new formed chemical groups allow for covalent bonding to the epoxy-based adhesives resulting in dramatically improved bonding strength (G_{IC}: 1600 J/m^2) with mainly substrate failure. Plasma treatment results in a greater increase in polarity and wettability of PEEK substrate, but it does not necessarily mean a better performance in improving adhesive bonding of the thermoplastic, as the mechanical results showed that the performance of each treatment is not only defined by surface characteristics of the pretreated substrates, but also depends upon the chemistry, curing rate and mechanism of the adhesive.

The capability of UV and plasma activation was evaluated for different bonding types. The results showed high efficiency of UV treatment for co-bonding inside and outside of autoclave and secondary bonding in autoclave. Similar results were observed regarding the proficiency of plasma activation in improving different bonding types such as co-bonding and secondary bonding of PEEK substrates. Tensile lap shear tests showed high efficiency of the both treatment techniques for improving bonding of the thermoplastic composite with epoxy-based adhesives containing different curatives including high temperature cure (121 and 177 °C), elevated temperature cure (93 °C), and room temperature cure (25 °C) formulation.

UV and plasma pretreatment methods can be used for bonding a wide range of epoxy-based thermosetting materials and are not exclusively limited to epoxy adhesives. In regard to a cosmetic application, the adhesion strength of an epoxy-based surfacing film to a thermoplastic PEEK substrate by using the both treatment methods was evaluated where the surfacing film was cured as different conditions including inside and out of autoclave at 121 and 177 °C. In a side by side comparison, the crosshatch testing showed nearly perfect adhesion of the treated samples with both activation methods whereas the untreated specimens displayed adhesion failure to the PEEK surface.

In conclusion, the new UV treatment method offers new opportunities for fast and safe bonding to thermoplastic materials resulting in excellent bond strength comparable with plasma activation with the advantage of a much shorter activation time. However, the efficiency of each pretreatment method strongly depends on the chemistry of the bonding adhesive as well.

5. REFERENCES

1. Cognard, Philippe. *Handbook of adhesives and sealants.* Oxford, UK: Elsevier ltd., 2005.

2. Arikan, E., Holtmannspotter, J., Hofmann, T. & Gudladt, J., "Vacuum-UV of polyetheretherketone (PEEK) as a surface pre-treatment for structural adhesive bonding." *The Journal of Adhesion* 95(10) (2020): 917-944.

3. Mathieson, I., & Bradley, R., H., "Improved adhesion to polymers by UV/ozone surface oxidation." *International Journal of Adhesion and Adhesives* 16(1) (1996): 29-31.

4. Brewis, D., M., Dahm, R., H., & Mathieson, I., "A new general method of pretreating polymers" *Journal of Materials Science Letters* 116(2) (1997): 93-95.

5. Iqbal, H., M., S., Bhowmik, S., & Benditcus, R., "surface modification of high performance polymers by atmospheric pressure plasma and failure mechanism of adhesive bonded joints." *International Journal of Adhesion and Adhesives* 30(6) (2010): 418-424.

6. Dupuis, A., Ho, T., H., Fahs, A., Lafabrier, A., Louarn, G., Bacharouche, J., Airoudj, A., Aragon, E. & Chailan, J., F., "Improving adhesion of powder coating on PEEK composite: Influence of atmospheric plasma parameters." *Applied Surface Science* 357 (2015): 1196-1204.

7. Li, w., Sang, L., Jian, X. & Wang, J., "Influence of sanding and plasma treatment on shear strength of 3D-printed PEI, PEEK, and PEEK/CF." *International Journal of Adhesion and Adhesives* 100 (2020): 102614.

8. Gravis, D., Poncin, E., F. & Coulon, J., F., "Correlation between the surface chemistry, the surface free energy and the adhesion of metallic coatings onto plasma-treated poly(ether etherketone)." *Applied Surface Science* 501 (2020): 144242.

9. ASTM Standard D1002-10, 2019, "Standard test method for apparent shear strength of single lap-joint adhesively bonded metal specimens by tension loading (Metal-to-Metal)" ASTM International, West Conshohocken, PA, 2019, DOI:10.1520/D1002-10, www.astm.org.

Nanomechanical Property Characterization of Composite Adhesive Bonding Systems

Rita J. Olander*, Brian D. Flinn*, Ashley C. Tracey†

*Department of Materials Science and Engineering
302 Roberts Hall, MS 352120
University of Washington, Seattle, WA 98195

†The Boeing Company
P.O. Box 3707, MC 0R-103
Seattle, WA 98124

ABSTRACT

This research is directed toward further understanding how manufacturing methods affect initial bondline formation between matrix resins and adhesives. Nanoindentation techniques were used to characterize various regions of adhesively bonded carbon fiber epoxy samples including the matrix resin, adhesive, and bondline mixing zone (interface/interphase). Bondlines from two co-bonded systems (Toray 3900/3MAF 555 and Toray 3900/Solvay FM309-1) and one secondarily bonded (Toray 3900/Cytec Metlbond® 1515-4) were characterized to understand the differences in interface and/or interphase development. Co-bonds produce a mixing of the matrix resin and the adhesive, resulting in a mixed interface, or interphase region, as compared to a sharper interface produced by secondarily bonded samples. For this preliminary study, three specific bonding systems were characterized using nanoindentation extreme property mapping (XPM™). By measuring mechanical responses on the micron scale, the data identifies various structures and regions, such as where the adhesive/adherend interphase begins, and if it is heterogeneous or homogeneous in nature. This work is unique in that other mechanical property characterization methods have not been able to isolate the adhesive/adherend interphase due to the micron scale of this region. These methods not only measure the adhesive/adherend interphase thickness but also allow for the evaluation of properties that may be related to bond quality and performance.

Corresponding author: Rita J. Olander*, Brian D. Flinn*, Ashley C. Tracey†

1. INTRODUCTION

Previous work has shown that nanoindentation techniques are able to evaluate structures and properties of adhesively bonded carbon fiber epoxy structures on a micron length scale. This research characterizes three composite bonding systems commonly used by the aerospace industry. [1]

1.1 Motivation

A bonding system is defined as components of the bonding process and design space. This includes the substrate material, adhesive, cure cycle, and surface preparation technique. Each parameter has a significant impact on the bond quality of an adhesive joint and it is important to understand how each contributes to the bonding system performance. In industry, application of adhesively bonded parts are typically dictated by the bulk properties of the material bonding systems. For simplicity, adhesives and prepreg resins are often considered homogeneous materials during product design. Bondlines, however, are heterogeneous in nature and are composed of several micron-scale regions that have not been well characterized due to the difficulty in testing such small features. Characterizing these regions may provide more understanding of process dependent variables as well as potentially expand the engineering application design space.

1.2 Background

Adhesives are used to bond composite substrates in the aerospace industry as they provide several benefits over mechanical fastening, including more evenly distributed loads and weight reduction. However, bonded part performance is process dependent and difficult to inspect post-cure. [2]

There are three main adhesive bonding techniques used to join thermoset resin composites: co-curing, secondary bonding, and co-bonding (**Figure 1**). Co-curing involves joining two uncured composite adherends with an adhesive and curing all three components simultaneously, creating adherend-adhesive mixing zones. Secondary bonding joins two fully cured composite adherends, with treated surfaces, using an adhesive. This results in limited mixing zones between the surface-treated adherend and adhesive. Co-bonding involves joining an uncured composite adherend to a fully cured, surface treated composite adherend using an adhesive. The sections are then cured together, allowing a limited mixing zone between the surface treated adherend and a larger mixing zone between the uncured adherend and adhesive. The properties and characteristics of these mixing zones are theorized to be dependent on the joining mechanisms: adhesion between cured adherends and adhesive; and chemical cross-linking between adhesive and uncured adherends. [3]

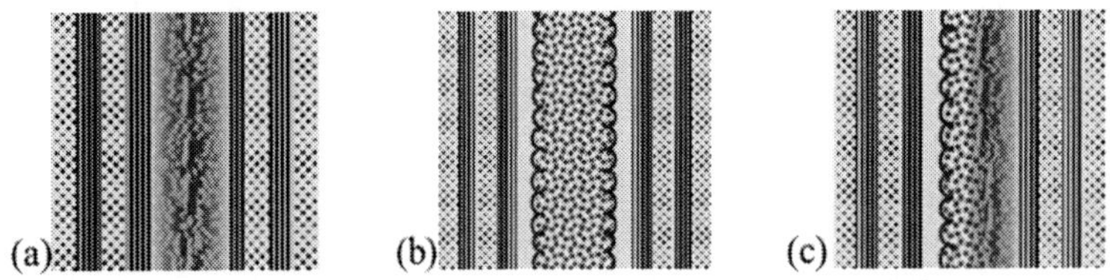

Figure 1. Simple schematics of the three bonding types: co-curing (a), secondary bonding (b), and co-bonding (c).

An interface can be described as the boundary between two zones of different chemistry and/or microstructure. During the cure cycle, it is common for these boundaries to have limited chemical interactions. The term interphase is used to describe these three-dimensional mixing zones where interactions and mixing between two dissimilar materials occur. Within these interphase regions, properties are unique from the bulk materials and gradated, whereas the properties of an interface are more characteristic of the bulk materials on either side of a specific grain boundary. [4]

A detailed schematic of a co-bond is shown in **Figure 2**. Region A is the cured carbon fiber reinforced polymer (CFRP) substrate comprised of fibers traveling in either the 0° or 90° orientation. Region B includes excess matrix resin that develops on the surface of the previously cured adherend. Region C is the peel ply prepared surface with the characteristic scalloped edges, thought to result in a *peel ply/matrix resin interphase*. The previously cured adherend is then bonded to an uncured adherend with the use of a film adhesive. On the left-side of the image, this creates an interface region between the *peel ply/matrix resin interphase* and adhesive. On the center and right-sides of **Figure 2**, the film adhesive contains the bulk adhesive region and a suggested *adhesive/matrix resin interphase*, or *co-bond interphase* region. This interphase region is not visibly evident but occurs near the previously uncured adherend.

Using nanoindentation techniques, mechanical characterization of these regions at the micron-scale are possible which could be used to model adhesive systems as well as provide information on the bond quality and performance.

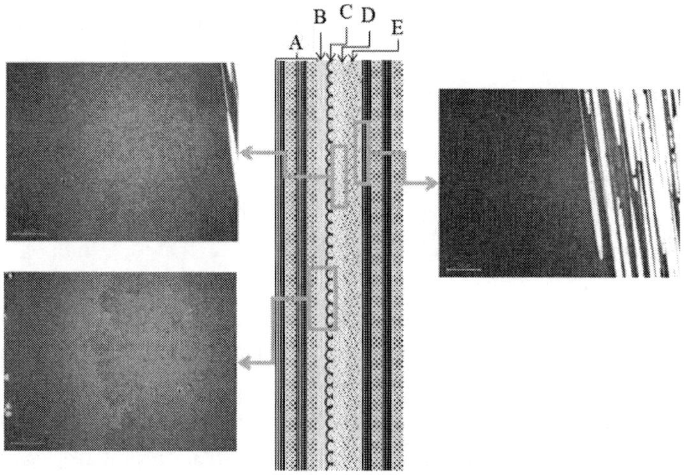

Figure 2: Schematic of bondline regions including CFRP [0/90] (A), matrix resin rich region (B), peel ply preparation imprint, possible *peel ply/matrix resin interphase*/adhesive interface region (C), adhesive (D), potential *adhesive/matrix resin interphase,* or *co-bond interphase* region (E).

1.3 Characterizing Bondlines through Nanoindentation

Nanoindentation has proven to be a powerful tool to study mechanical behavior on a micron scale. The nanoindenter is able to perform highly localized, semi-destructive tests with high resolution and accurate positioning, providing an advantage over traditional techniques [5] [6]. Previous nanoindentation studies have been performed on several adhesive systems including bonding to spruce wood, dental materials, and carbon fiber composite structures. [5] [6] [7] Zheng and Ashcroft's 2004 study on film adhesives investigated the distribution of mechanical properties across a bondline and compared them to the mechanical properties of bulk adhesive samples. They discovered that nanoindentation could effectively characterize the mechanical properties across the bondline, though an interphase could not be detected within the spatial resolution of the tests performed. The lowest spatial resolution, or the distance between indents, they were able to achieve was 10 μm using a 10,000 μN force. [6] New nanoindentation technology allows for more precise load and displacement measurements, which can reduce the size of the nanoindentation and improves spatial resolution to approximately 1 μm, depending on the method parameters. Additionally, newer nanoindenters can perform many functions other than nanoindentation including scanning probe microscopy, extreme property mapping (XPM™), scratch and wear tests, nano-Dynamic Mechanical Analysis (nanoDMA), and modulus mapping.

1.3.1 Indentation Methods

For simple nanoindentation tests, the nanoindenter can be operated in a load-controlled mode where the indenter applies a specified load on the sample for a given time before releasing, or a displacement-controlled mode where the indenter goes down a specified depth for a given time before releasing. Previous research showed that load control provided more accurate hardness and reduced modulus values due to less sensitivity to surface tilt and was therefore used for the reported testing. As the indenter penetrates the surface, load and displacement are measured and plotted. From the load-displacement curve, hardness (*H*) is calculated using **equations (1)** and **(2)**, and reduced modulus (*Er*) can be found by taking the tangent of the unloading curve at the instant point of unloading.

$$A_c = k_1 h_c^{\ 2} + k_2 h_c \tag{1}$$

$$H = \frac{P}{A_c} A_c \tag{2}$$

A_c = contact area of the indenter tip, k_1, k_2 = fitted constants, P = maximum load [6]

1.3.2 Extreme Property Mapping (XPM™)

Extreme property mapping (XPM™) mode is a nanoindentation technique that involves performing quick nanoindentations within a specified array using the TriboScanner piezo scanner, developed by Hysitron, to move the indenter tip. The hardness and reduced modulus values are measured at every location, mapped on an X-Y graph, and given a color dependent on the value. Using a traditional single indentation mode, the nanoindenter head moves using motors, which requires a two-minute dwell time between indents as the system waits for vibrations of the motor to dissipate. However, in XPM™ mode, the indenter tip moves using the piezo scanner, which does not produce significant vibrations and requires no dwell time but limits the array size as the piezo scanner can only move the indenter tip within a 94 by 94 μm area. Although the high loading rates and shallow indentations associated with standard XPM™ settings can influence the measured hardness, the high collection rate allows for comparison analysis across several regions [7]. This technique, overall, can be used to quickly acquire mechanical properties and map on various regions throughout the bondline.

2. EXPERIMENTATION

2.1 Experimental Materials

Three adhesive bonding systems, using common aerospace materials, were investigated in this research, as shown in **Table 1**. The samples were mounted in Bakelite and polished to 0.1 µm surface finish. Nanoindentation was performed on a Hysitron Triboindenter 980.

Table 1. Bonding systems characterized by nanoindentation.

	Bond Type	Adherend	Surface Preparation (cured adherend only)	Adhesive
Bonding System 1	Co-bond	Toray T800S/3900 resin [F1]	Diatex 1500EV6 woven polyester peel ply [F2]	3M AF 555 polyester supported [F3]
Bonding System 2	Co-bond	Toray T800S/3900 resin [F1]	Diatex 1500EV6 woven polyester peel ply [F2]	Solvay FM® 309-1 Knit or mat carrier [F3]
Bonding System 3	Secondary Bond	Toray T800S/3900 resin [F1]	Diatex 1500EV6 woven polyester peel ply [F2]	Cytec Metlbond® 1515-4 modified epoxy supported [F3]

[F1] 350 °F cured carbon fiber reinforced polymer matrix
[F2] Peel ply removed just prior to bonding
[F3] 350 °F cured film adhesive

2.2 Characterization Techniques

2.2.1 Nanoindentation – Extreme Property Mapping (XPM™)

In this study, the same indentation parameters traditionally used for simple, single nanoindents were used for XPM™ indents. These XPM™ indents were much deeper and took longer to conduct, as compared to typical XPM™ parameter indents, but provided both reliable relative values and absolute values. XPM™ mode was used to gather many data points within a desired region quickly and to map the distribution of mechanical properties across the bondline to analyze the size of each region. All XPM™ measurements were performed at 2000 µN of force with a 5 second load time, 10 second hold time, 5 second unload time, and indentations spaced 4 µm from each other. First, XPM™ maps parallel to the bondline were collected to identify existing regions within the bonding system through the mechanical property gradients. Perpendicular to the bondline XPM™ maps were then performed for each region within the given bonding system to understand region specific mechanical properties.

3. RESULTS & DISCUSSION

3.1 XPM™ of Bonding System 1: Co-bond with Toray 3900 and 3M AF 555

XPM™ arrays were performed parallel and normal to the bondline to identify the various regions of the adhesive. Reduced modulus maps were created to show the change in mechanical properties. An example is shown in **Figure 3** with parallel and perpendicular XPM™ maps offset to the nanoidentation arrays. The dark red and blue regions of the XPM™ map represent areas of higher and lower reduced modulus, respectively. The mappings show that the matrix resin has the highest values, while most of the adhesive has much lower values. The *co-bond interphase* is shown on the right side of the perpendicular XPM™ image. This interphase thickness is approximately 40-50 μm in this region but was shown to vary in thickness throughout the bondline. A well-defined interface between the matrix resin and adhesive is shown in the parallel XPM™ image parallel to the blondline (**Figure 3B**). Both maps support the existence of the resin/adhesive interface and *co-bond interphase* regions suggesting they may be identified on a micron-scale through the assessment of mechanical properties.

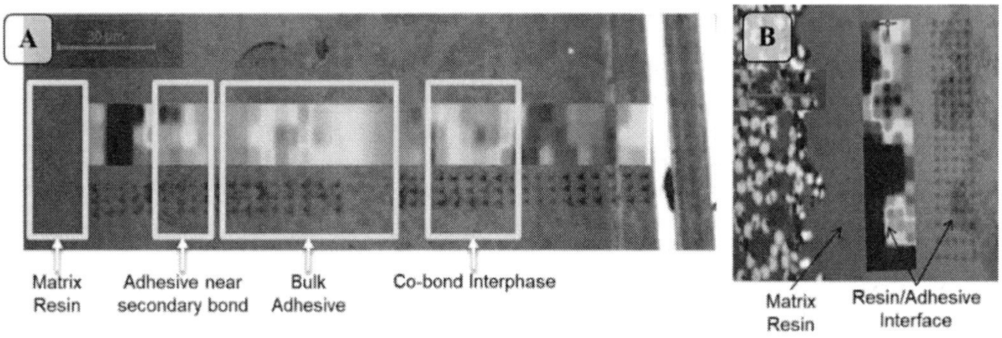

Figure 3. Images of perpendicular XPM™ maps across bondline (A), parallel XPM™ map of *matrix resin/adhesive interface* (secondarily bonded region)(B).

Parallel XPMs™ were performed to determine the mechanical properties of specific regions in the bondline (**Figure 4**). Reduced modulus and hardness were measured within four regions: resin near the interface, adhesive near the interface, the center/bulk of the adhesive, and the *co-bond interphase*. 135 indents were performed in the three adhesive regions and 30 in the matrix resin region. The variation in hardness and reduced modulus values throughout the adhesive could be due to the heterogeneous nature of the adhesive (i.e. particles and scrim that can affect the measured mechanical properties), especially on the micron scale.

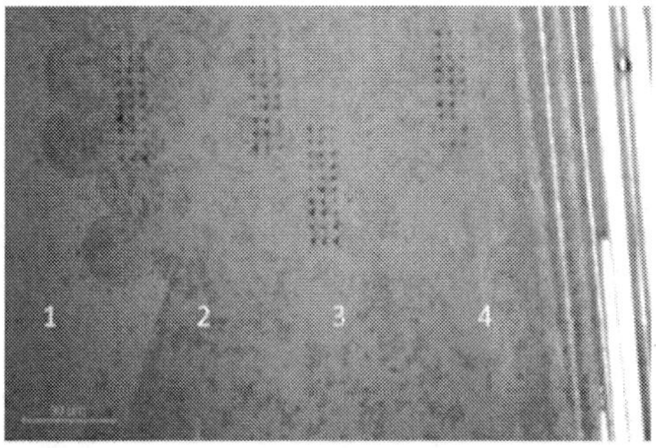

Figure 4. Example of parallel XPM™ locations of matrix resin (1), adhesive near interface (2), bulk adhesive (3), *co-bond interphase* (4) regions within bondline.

As shown in **Figure 5**, the matrix resin has the highest average reduced modulus and hardness. The adhesive regions, on the other hand, showed varying average values of reduced modulus and hardness. The adhesive nearest to the interface had the lowest average mechanical properties while the center of the adhesive, referred to as the bulk adhesive, showed intermediate values. Most interestingly, the *co-bond interphase* region had average mechanical property values between the resin and adhesive. Additionally, a trendline of increasing average values can be seen across the adhesive from the interface to the *co-bond interphase*, as shown in **Figure 5** and **Figure 6**. This observed phenomena supports the theory that mixing zones between the bulk materials develops as the adhesive and adherend are cured. This is especially shown in the highest average adhesive property values *co-bond interphase*, suggesting the uncured adherend and adhesive can mix more readily. Lower average values at the interface suggest that cured adherends somewhat inhibit mixing with the adhesive at the secondary bond region.

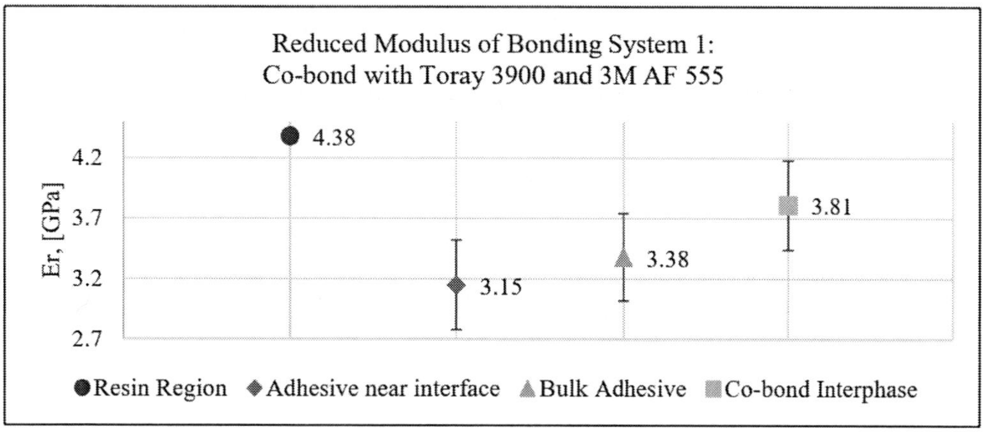

Figure 5. Reduced modulus measurements of AF 555 adhesive regions with increasing trendline from interface to *co-bond interphase* regions.

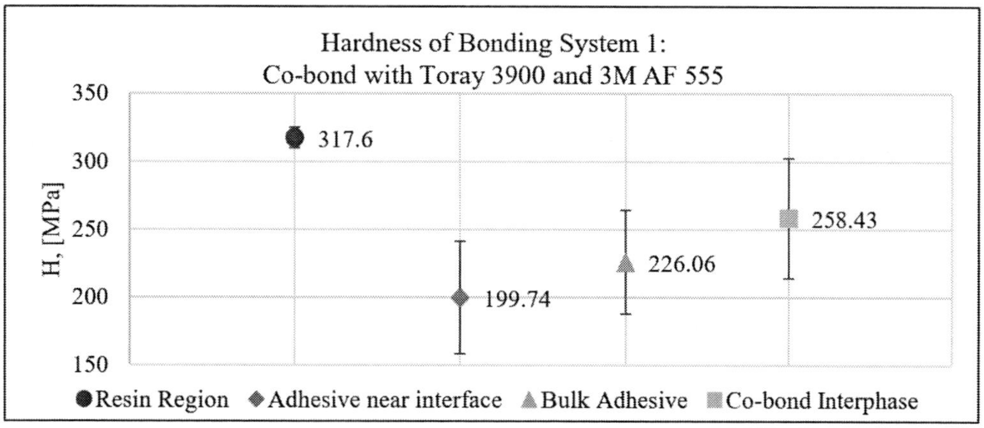

Figure 6. Hardness measurements of AF 555 adhesive regions with increasing trendline from interface to *co-bond interphase* regions.

3.2 XPM™ of Bonding System 2: Co-bond with Toray 3900 and Solvay FM® 309-1

Figure 7 shows the reduced modulus and hardness gradients across the bondline using XPM™ maps perpendicular to the bondline. Due to the uneven surface of the sample, a few of the edge indentations overlapped as indicated by the red edges. This measured plastic deformation effects from the previous indents and therefore are displaying artificially high mechanical properties. Classifying these artificially high measurements as outliers and omitting them from the data set resulted in mechanical properties following similar trends to bonding system 1 (co-bond) with the lowest values near the interface, intermediate values in the bulk and *co-bond interphase*, and highest values in the resin.

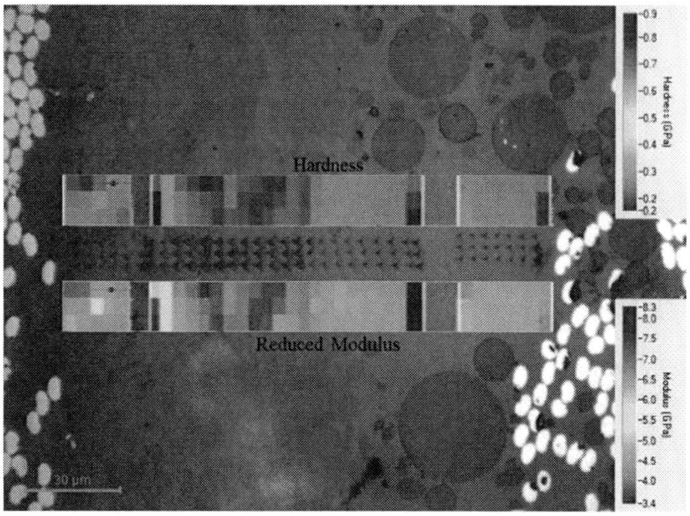

Figure 7. Perpendicular to the bondline XPM™ map of bonding system 2 with resin area (outermost left), *co-bond interphase* (inner left), bulk adhesive (left-middle), interface (middle), opposite resin rich area region (right).

The thickness of the co-bonded interphase was more challenging to characterize in bonding system 2. This may be attributed to the viscosity differences between AF 555 and FM 309-1. As-received AF 555 adhesive minimum viscosity is 290 Poise (P) at 120°C with a 1°C per minute temperature rate from room temperature. [8] FM 309-1 (M), on the other hand, had a measured as-received minimum of 4,200 P between 75°C and 135°C with a 1°C per minute temperature rate from room temperature, which is approximately one order of magnitude more viscous than AF 555. [8] Because FM 309-1 in bonding system 2 has a higher viscosity, mixing between adherend resin and adhesive is likely to be significantly less during cure, as compared to the AF 555 bonding system. This is supported by multiple perpendicular maps indicating that the *co-bond interphase* region thickness is approximately 20 μm.

Parallel XPM™s across the entire bondline were performed, as shown in **Figure 8**. 100 indentations were averaged to determine the reduced modulus and hardness values (**Figure 9** and **Figure 10**). Again, similar to bonding system 1, a distinctive trend can be seen across the bondline with higher mechanical properties at the co-bond and lower mechanical properties at the adhesive near the interface. To understand if these properties were statistically significant and unique to each region, T-tests were performed (**Table 2**).

Table 2. Two tail T-test evaluations of regions across bondline for bonding system 2.

T-Test #	Property	Regions	T stat.	P-value	T critical (two-tail)	Evaluation
1	Er	Adhesive near interface Bulk Adhesive	-3.63	3.57E-04	1.972	Statistically Unique
2	Er	Bulk Adhesive Co-bond Interphase	-10.10	1.63E-19	1.972	Statistically Unique
3	Er	Co-bond Interphase Resin[F1]	-4.72	6.08E-06	1.978	Statistically Unique
4	Er	Co-bond Interphase Resin Region	-9.42	4.00E-17	1.975	Statistically Unique
5	Er	Resin[F1] Resin Region	-8.10	1.11E-13	1.974	Statistically Unique
6	H	Adhesive near interface Bulk Adhesive	-3.42	7.59E-04	1.972	Statistically Unique
7	H	Bulk Adhesive Co-bond Interphase	-3.13	2.49E-03	1.993	Statistically Unique
8	H	Co-bond Interphase Resin[F1]	-8.83	2.50E-15	1.976	Statistically Unique
9	H	Co-bond Interphase Resin Region	-5.81	2.91E-08	1.974	Statistically Unique
10	H	Resin[F1] Resin Region	-0.20	8.44E-01	1.979	Statistically **NOT** Unique

[F1] resin between two adherend plies.
Significance level (α): 0.05

The T-test shows that all of the regions across the adhesive have a unique reduced modulus. This suggests *that adhesive/resin interfaces* and *co-bond interphases* can be specifically defined within this bonding system. With this phenomenon apparent in this co-bonding system as well, it can be theorized that a larger mixing zone, and true three-dimensional interphase, develops during a co-bond due to the uncured adhesive's and uncured adherend's ability to mix more readily. Conversely, mixing is limited between the cured, surface treated adherend and adhesive, resulting in a bonded interface. Parallel to the bondline XPM™ maps of the resin between prepreg adherend plies and the resin rich region nearest the interface were compared as well. In **Table 2**, the resin between the adherend plies and resin rich regions are statistically unique for reduced modulus but not statistically unique for hardness values. The significance value (α) was varied between 0.05 and 0.005 to determine if this discrepancy was a result of the T-test error margin, however, all values resulted in the same disposition. There are many potential explanations for this difference, including the heterogenous nature of the resin system, XPM™ measuring particles within the resin, or the interphase or interface development between the resin and adjacent regions.

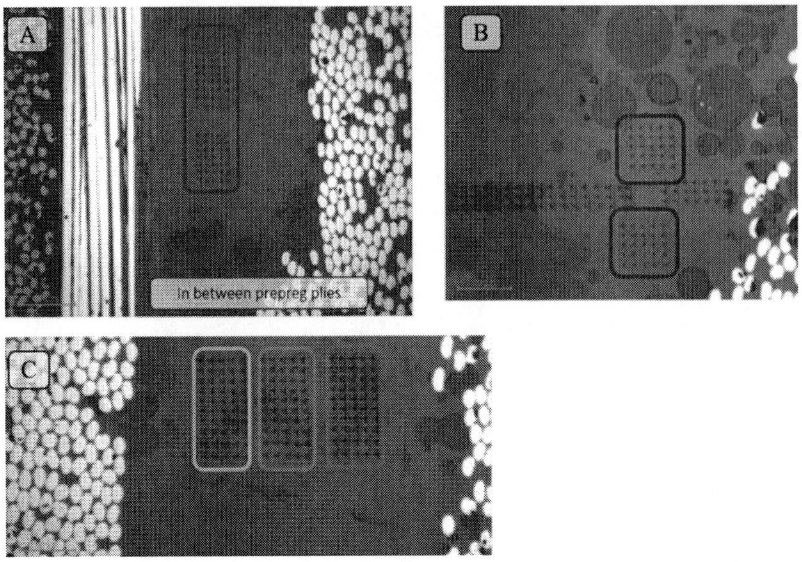

Figure 8. Example of parallel XPM™ locations for resin (A), resin regions (near interface, B), and adhesive *co-bond interphase* (yellow, C), bulk adhesive (green, C), and adhesive near the interface (blue, C) parallel maps for bonding system 2.

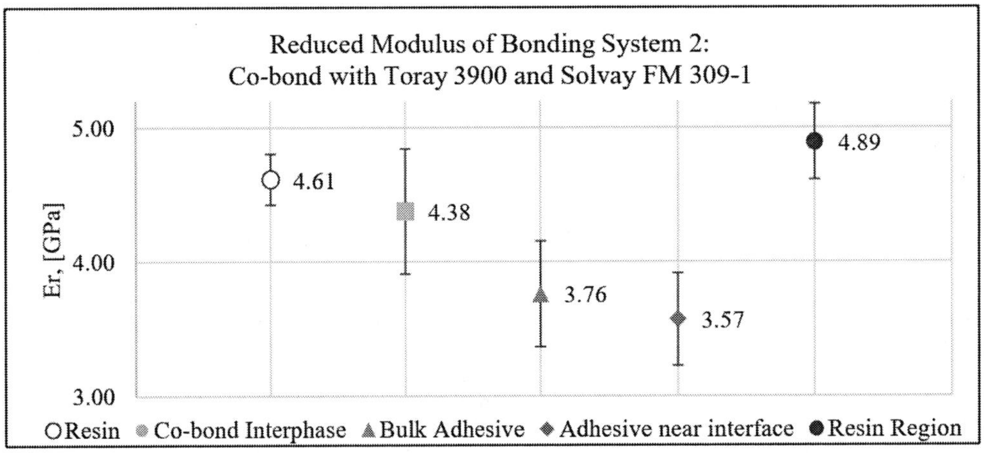

Figure 9. Reduced modulus measurements of bonding system 2 with increasing trendline from interface to *co-bond interphase* adhesive regions.

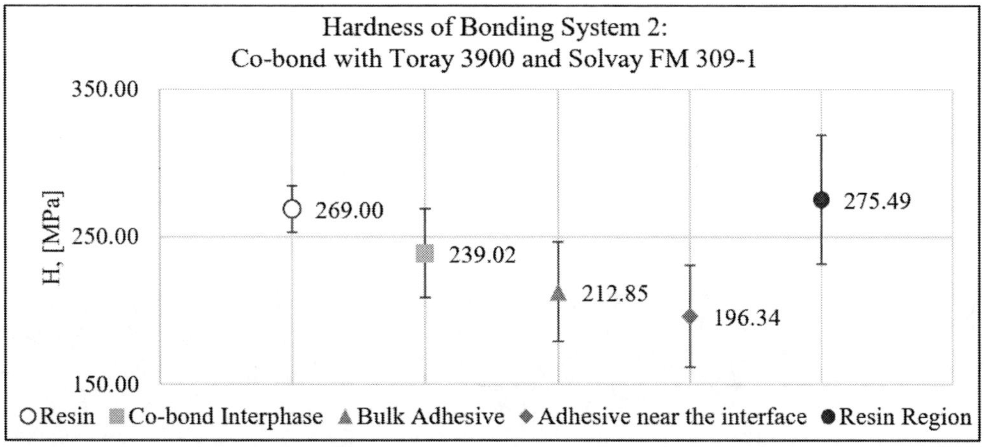

Figure 10. Hardness measurements of bonding system 2 with increasing trendline from interface to *co-bond interphase* adhesive regions.

3.3 XPM™ of Bonding System 3: Secondary Bond with Toray 3900 and Cytec Metlbond® 1515-4

Figure 11 shows the reduced modulus and hardness gradients across the bondline using perpendicular to the bondline XPM™ maps. In contrast to the previous co-bonding systems, the adhesive within bonding system 3 has only one distinctive region, the "bulk adhesive". Mixing between the adhesive and the resin may be restricted due to both adherends in this bonding system being fully cured prior to the bonding cure. While there is only one significant region in the adhesive, a subtle gradient was seen across the bondline. The small variation in the mechanical properties may be attributed to the heterogeneous nature of this region (e.g. surface preparation (at the secondary bond), adhesive scrim, particles).

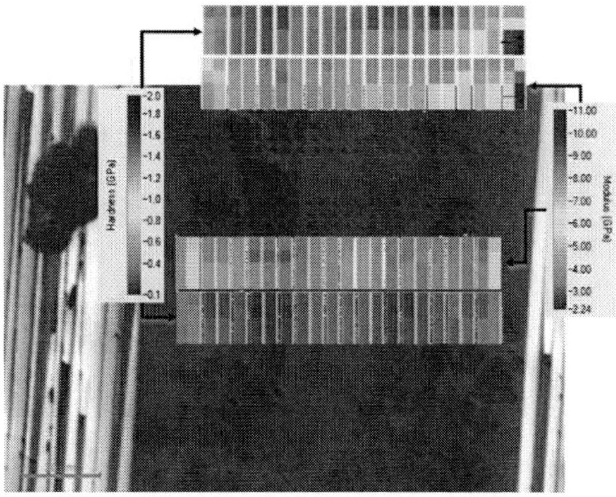

Figure 11. Perpendicular to the bondline XPM™ maps of bonding system 3 with resin area (right and left edges) showing the highest values and the adhesive regions (middle sections) showing similar mechanical properties across the bondline.

Parallel XPM™ maps across the entire bondline were performed, as shown in **Figure 12**. With bonding system 3, 100 indentations were averaged to determine the reduced modulus and hardness values (**Figure 13** and **Figure 14**). The averaged resin region reduced modulus and hardness values were 4.82 GPa and 301 MPa, respectively. As expected with a secondary bonding system, the mechanical properties measured within the adhesive regions were all within their respective standard deviations and did not exhibit any increasing reduced modulus or hardness trendlines as measured on the co-bonded samples.

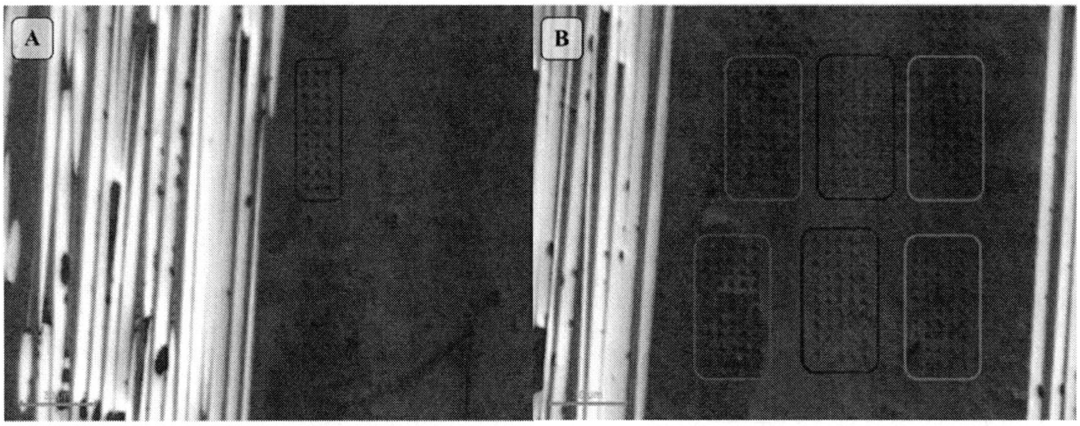

Figure 12. Example of parallel XPM™ locations for resin (A), and adhesive regions resin regions (near interface, B), and adhesive near the left interface (B, left), bulk adhesive (B, middle), and adhesive near the right interface (B, right) region parallel maps for bonding system 3.

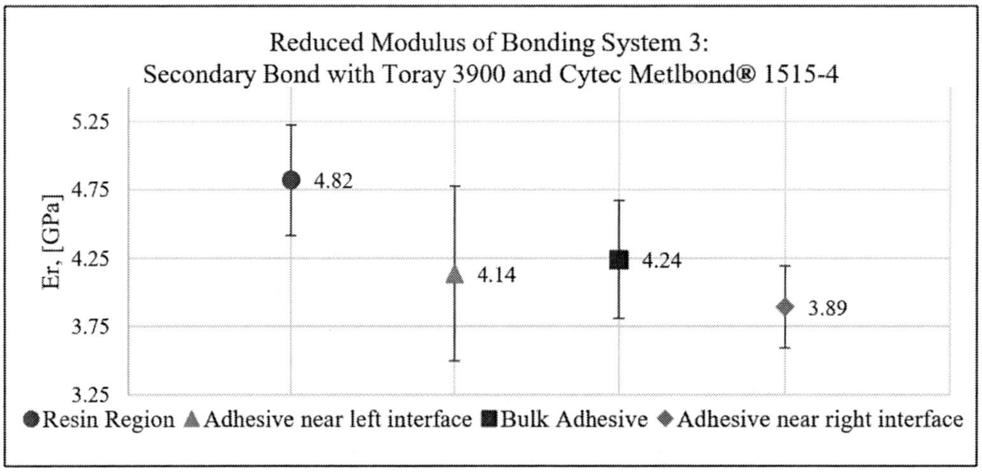

Figure 13. Reduced modulus measurements of bonding system 3 regions.

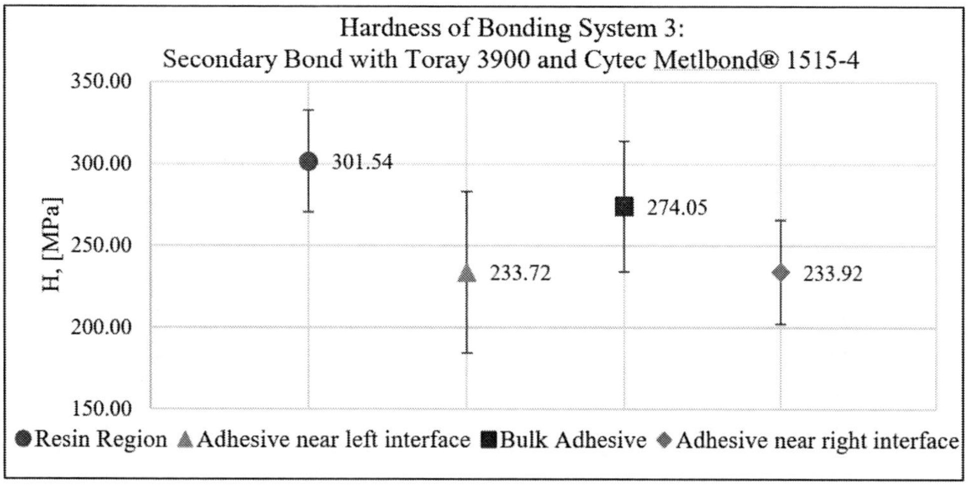

Figure 14. Hardness measurements of bonding system 3 regions.

To further understand if these adhesive data sets are unique, statistical analysis was performed on between each of the adhesive regions (**Table 3**). The T-tests revealed that only the adhesive near the right interface had a unique reduced modulus and hardness properties compared to the bulk. The adhesive near the left interface T-test values support that this region has characteristics of the bulk adhesive, but this may be skewed due to the location of the XPM™ maps or the thickness of the regions. More investigation into the surface topography, region thickness, and interface characterization is needed to further understand this trend.

Table 3. Two tail T-test evaluations of regions across adhesive regions for bonding system 3.

T-Test #	Property	Regions	T stat.	P-value	T critical (two-tail)	Evaluation
1	Er	Adhesive near right interface Bulk Adhesive	-6.60	4.66E-10	1.973	Statistically Unique
2	Er	Bulk Adhesive Adhesive near left interface	-1.34	1.83E-01	1.974	Statistically **NOT** Unique
3	Er	Adhesive near right interface Adhesive near left interface	-3.45	7.43E-04	1.977	Statistically Unique
4	H	Adhesive near right interface Bulk Adhesive	-7.25	9.39E-12	1.972	Statistically Unique
5	H	Bulk Adhesive Adhesive near left interface	-6.34	1.67E-09	1.973	Statistically Unique
6	H	Adhesive near right interface Adhesive near left interface	0.03	9.74E-01	1.974	Statistically **NOT** Unique

Significance level (α): 0.05

4. CONCLUSIONS

4.1 Summary

4.1.1 Reduced Modulus and Hardness

The reduced modulus and hardness measurements of the adhesive regions in each of the co-bonded systems showed unique regions. Each region's averaged mechanical property values ranked the adhesive at the interface (lowest), bulk adhesive, *co-bond interphase*, and resin regions (highest) for both co-bonded systems. The co-bonded interphase regions showed intermediate values between the bulk properties of the adhesive and resin which suggests there is significant mixing during cure. The thickness of this region may be dictated by the adhesive material properties (e.g. viscosity), as shown by comparing bonding systems 1 and 2. Evaluation of both bondlines showed that the co-bonded interphase of the AF 555 system was approximately twice as large as the FM 309-1 system's *co-bond interphase* region. This may be attributed to the AF 555 adhesive used in bonding system 1 having a significantly lower viscosity than FM 309-1 used in bonding system 2.

In bonding system 3, only subtle differences between the various regions can be observed. All regions within the resin and the adhesive have overlapping standard deviations. These high standard deviations suggest that the bondline is very heterogeneous in nature. Statistical analysis determined that there are unique properties of adhesive near the interface and the bulk adhesive. At this time, it is unclear why this phenomenon is observed.

4.2 Limitations

4.2.1 Characterizing Interfaces

Although there have been many advances to measure bondline properties at the micron scale, the interface is still challenging to characterize. Due to the scale and complex surface characteristics of this region, nanoindentation techniques are limited to measuring properties of the materials near the interface. Evaluation of the interface is currently best done prior to bonding and cure, yet this does not allow for investigation into the adhesive and adherend interaction. Other nanoindentation techniques, such as modulus mapping, will be explored in future work to further attempt to characterize interfaces.

4.2.2 Equipment Limitations

Additionally, there are several limitations with evaluating composite bonding systems with nanoindentation. Nanoindentation locations are chosen on the surface of a sample but when the indenter tip penetrates the surface there is uncertainty of what lies beneath. This is most significant in highly heterogeneous systems (i.e. when scrim or particles are present in the system). Further, plastic zones are created within the sample surface when indented. These plastic zones can affect other indents that are performed too close. Large indentation spacing can prevent any interactions between plastic zones, though the spatial resolution decreases as indentation spacing increases.

4.3 Future Work

Additional methods to evaluate bondline properties are nano-dynamic mechanical analysis (nanoDMA) and modulus mapping. NanoDMA involves applying an oscillating force via the indenter tip to measure the viscoelastic response of the material as the temperature of the sample is varied. From this technique, tan delta measurements of the viscous response over the elastic response can be plotted versus temperature to identify a glass transition temperature (Tg). Tg measurements may enable more investigation into how a bondline performs over its lifecycle. This is significant because industry determines the design and application space by a system's Tg.

Modulus mapping is a feature on the Hysitron Triboindenter 980 that combines scanning probe microscopy with nanoDMA capabilities. The indenter tip scans across a surface while oscillating, measuring the viscoelastic properties of the region at a given temperature. A number of properties are able to measured, including the loss modulus and storage modulus, across the desired region. This technique may be able to further evaluate small scale regions, such as the interface. Additionally, by varying the temperature during these mapping trials, larger scans with a smaller spatial resolution of the bondline region may be performed.

Future work will include a variety of bonding systems, both with ideal and suboptimal bonding surfaces to characterize process variables effects on the bondline. This future work will focus on the interface and interphase development and its relation to bonding performance. Additionally, there is a growing interest in bondline specific properties, such as hot-wet exposure, aging, heat damage, and characterization of additives. Although bulk properties of adhesives and composite substrates have been thoroughly characterized, there are still many unknowns regarding interface and interphase behavior in various exposure scenarios and these region's overall effects on bonding performance. Future work will use nanoindentation technology and other macro mechanical property evaluation methodologies to investigate exposure and service effects on bonding systems.

5. ACKNOWLEDGMENTS

This work was conducted at the Molecular Analysis Facility, a National Nanotechnology Coordinated Infrastructure site at the University of Washington, which is supported in part by the National Science Foundation (grant NNCI-1542101), the University of Washington, the Molecular Engineering & Sciences Institute, and the Clean Energy Institute.

This work was partially supported by the Federal Aviation Administration (FAA) Research Grant, Improving Adhesive Bonding of Composites through Surface Characterization. Ahmet Oztekin was the FAA grant monitor. The authors wish to thank the FAA Center of Excellence at the University of Washington (Advanced Materials in Transport Aircraft Structures) and The Boeing Company for sponsoring this on-going research.

6. REFERENCES

[1] A. C. Zukaitis, R. J. Olander and B. D. Flinn, "Nanomechanical Property Characterization of Adhesive Bondlines," in *CAMX2019*, Anahiem, 2019.

[2] M. Song, J. Kweon, J. Choi and J. Byun, "Effect of manufacturing methods on shear strength of composite single-lap bonded joints," *Composite Structures,* vol. 92, no. 9, pp. 2194-2202, 2010.

[3] T. Kruse, "Bonding of CFRP Primary Areospace Structures: Overview on the Technology Status in Contect of the Certification Boundary Conditions Adressing Needs for Development," in *ICCM19*, Montreal, 2013.

[4] D. A. Jesson and J. F. Watts, "The Interface and Interphase in Polymer Matrix Composites: Effect on Mechanical Properties and Methods for Identification," *Polymer Reviews,* vol. 32, no. 3, pp. 321-354, 2012.

[5] A. Sadr, Y. Shimada, H. Lu and J. Tagami, "The viscoelastic vehavior of dental adhesive: A nanoindentation study," *Dental Materials,* vol. 25, no. 1, pp. 13-19, 2009.

[6] S. Zheng and I. A. Ashcroft, "A depth sensing indentation study of the hardness and modulus of adhesives," *International Journal of Adhesion and Adhesives,* vol. 25, no. 1, pp. 67-76, 2005.

[7] E. D. Hintsala, U. D. E. Hangen and D. D. Stauffer, "High-Throughput Nanoindentation for Statistical and Spatial Property Determination," *The Journal of Minerals, Metals, & Materials Soceity* , vol. 70, no. 4, pp. 494-503, 2018.

[8] T.-H. Hou, J. M. Baughman, T. J. Zimmerman, J. K. Sutter and J. M. Gardner, "Evaluation of Sandwich Structure Bonding In Out-Of-Autoclave Processing," SAMPE Technical Conference, Salt Lake City, UT, 2010.

Improved Adhesion in Pultruded Composite Materials And Test Methods for Evaluation

Leigh Nolen, Yves Cordeau, Christiaan Mauldin, Dr. Mac Puckett
PolyOne Advanced Composites
3850 Pinson Valley Pkwy
Birmingham, AL 35217

Abstract:

Many composite materials are subcomponents of a larger assembled structure. Adhesives are an excellent way to attach these smaller components into a larger structure, distributing load across a large surface area without damaging the composite, as compared to mechanical fasteners which can concentrate the load at what is often a site of damage in the composite. Typically pultruded composites are completely cured coming from the die and will have a mold release on the surface. Both of these factors can reduce the ability to achieve maximum adhesion to a pultruded surface. Depending on part geometry, a peel ply material may be pultruded onto the part, which can be removed to prepare the surface before adhesive is applied. Alternatively, sanding or abrading the part followed by thoroughly cleaning the surface is a typical – and labor-intensive – method for preparing a part to maximize adhesion.

This paper will discuss the addition of a modifier to the resin formulation that acts as a "primer" on the pultruded surface of the part. This eliminates the need to prepare the surface beyond the removal of dust and other particulates that accumulate on most surfaces. Cleanliness of the surface to be bonded is still essential to creating a bond of maximum strength. This unique process of surface preparation eliminates the time consuming and expensive methods typically required to prepare an adhesively bonded structure. This novel "priming" process does not significantly affect the thermal or mechanical properties of the underlying composite. Comparison of composite properties and adhesive bond strength of parts made and bonded in various ways will be discussed.

1. Introduction

Adhesives are an excellent way to aggregate smaller composite components into a larger structure, distributing load across a large surface area without damaging the composite. The value of adhesive use as an assembly technique is balanced by the necessity of adhering to the rules that are generally understood [1]:

1. Prepare and clean the surface to be bonded
2. Application of a thin uniform layer of adhesive (forming a thin uniform bondline)
3. Fixturing the components to be bonded with application of pressure to prevent movement
4. Hold in place until adhesive is completely cured.

As composite surfaces to be bonded get larger and as topography becomes more complex the preparation of the surface and thoroughly cleaning those surfaces before application of adhesive is increasingly more difficult and time consuming. Nevertheless, the preparation and cleaning of surfaces before bonding is absolutely essential to creating optimum bonding. Typical methods of preparing pultruded composites for adhesive bonding include running the pultrusion with peel ply and then removing the peel ply before use, or abrading with sand paper or grit blasting to remove mold release and create a rougher surface that provides additional mechanical interlock during adhesion. Whether using peel ply or abrasion to prepare the surface, additional inspection and cleaning must occur to insure that foreign matter is removed and the surface is clean before bonding.

This paper is divided into two distinct portions. First, an evaluation of bondline thickness parameters and their effect on adhesion is measured using a standard test method, ASTM D3165-07 STM for Strength of Adhesives in Shear by Tension Loading of Single-Lap-Joint Laminated Assemblies. [2] See Figure 1. This test method allowed us to test parameters that were meaningful to this work, but at other times we have used and sometimes prefer alternative test methods such as ASTM D 3528-96, ASTM D1002-99, ISO 4587. [3]. Production of a thin uniform bondline, is a well-established principle of good adhesion. But as relatively complex parts are assembled with adhesives, in the real world of manufacturing, a certain amount of variation is anticipated in the thickness of the bondline. The initial part of this work evaluates the effect of bondline thickness on the measured shear strength of the bond.

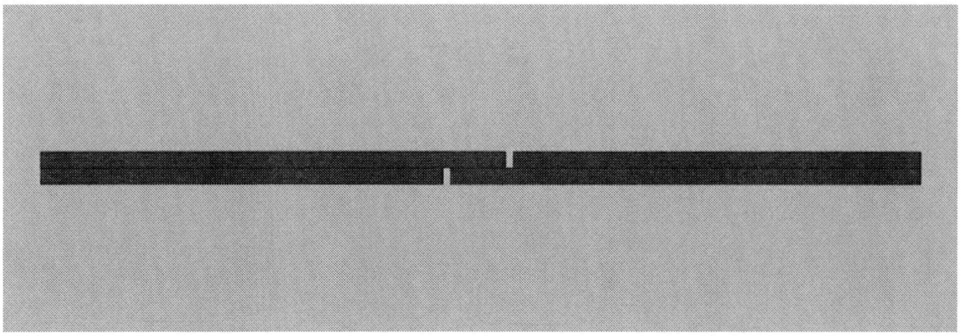

Figure 1: Test specimen for ASTM D3165 standard. (1).

Typically pultruded composites are completely cured coming from the die and will have a mold release on the surface. Both of these factors can reduce the ability to achieve maximum adhesion to a pultruded surface. In the second portion of this paper, an evaluation of the strength of adhesively bonded joints produced with pultruded laminates is performed. The pultruded laminates were materials that were both abraded and then cleaned with a solvent compared to laminates that were only wiped with a solvent to remove dirt and debris.

2. Experimentation:

2.1 Description of the samples.

Specimens were prepared from two pultruded bars measuring 2 inches wide (51 mm), 0.125 inches (3.2 mm) thick and 8 inches (203 mm) long were stacked. A 1-ounce or 2-ounce nylon veil was used between the bars to control the bondline thickness of adhesive. Table 1 describes the samples that were adhesively bonded in this study.

Table 1: Description of samples.

Sample Description	Surface Preparation	Resin Matrix Type	Bondline Thickness mm / in
1 oz. Nylon veil, 1 layer	Acetone Wipe	Modified Isophtalic Polyester	0.254 / 0.010
2 oz. Nylon veil, 1 layer	Acetone Wipe	Modified Isophtalic Polyester	0.406 / 0.016
2 oz. Nylon veil, 2 layers	Acetone Wipe	Modified Isophthalic Polyester	0.584 / 0.023
Shim, 0.018" thick	Acetone Wipe	Modified Isophtalic Polyester	0.457 / 0.018
Shim, 0.019" thick	Wet-Sand, Acetone Wipe	Unmodified Isophtalic Polyester	0.483 / 0.019

Figure 2 provides an image of a typical sample.

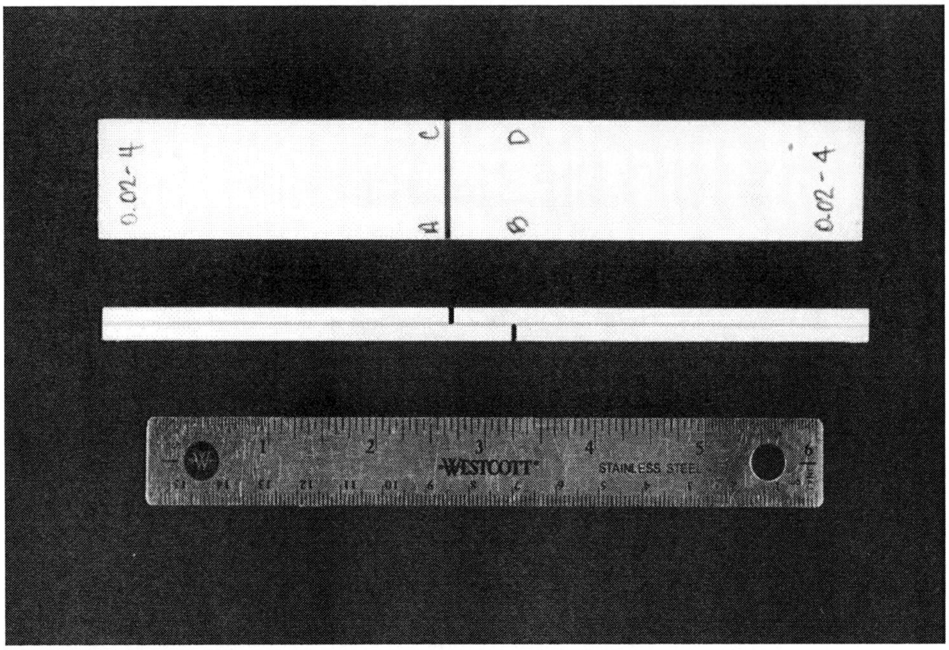

Figure 2: Typical Sample based on ASTM D3165-07.

2.2 Preparation of the samples.

Each pultruded bar was marked in the notch areas and the part thickness measured in four locations. The surface of the composite was cleaned using a cloth and acetone. The veil was impregnated with a two part epoxy adhesive (SikaPower 1200) [4] using a roller, and then placed onto the composite. The second composite bar was placed on top and the stack was placed against an alignment rail as shown in Figure 3. The parts were secured using c-clamps and a steel bar to provide even pressure distribution.

Two additional sets of specimens were fabricated using shims to control the bondline thickness. Shims were placed on each side of the notched area at a distance of one inch (25.4 mm). The first set of specimens were fabricated with pultruded bar containing the modified polyester. With the polyester "primed" for adhesion this specimen did not require the surface to be sanded.

The second set of specimens were fabricated with a pultruded bar containing standard polyester with no modifier. These specimens were prepared in a more typical manner, with the specimen prepared by wet sanding the surface with 120 grit sandpaper. The surface was then cleaned with acetone and shims used to control the bondline thickness.

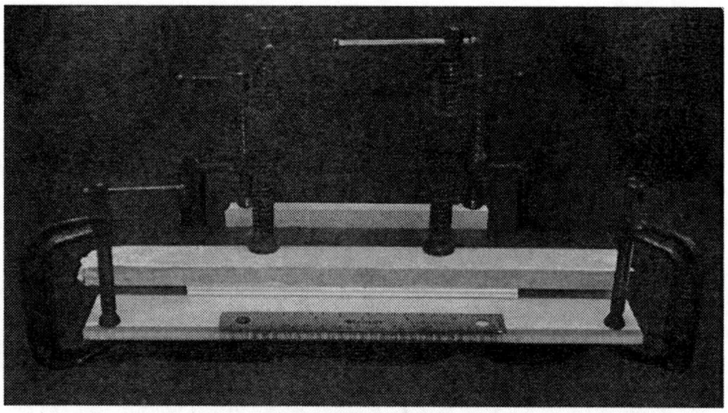

Figure 3: Alignment fixture.

All specimens were cured at 70°C for two hours. Excess resin was removed from the specimen sides as needed. The fixture was allowed to cool to room temperature prior to disassembly. Thicknesses were measured in the four previously measured locations and the adhesive thickness calculated.

The specimens were cut to a length of 7 inches (178 mm) by cutting 0.5 inches (13 mm) from each end to remove excess adhesive. Notching was performed using a surface grinder equipped with a diamond-cutting wheel 0.053 inches (1.35 mm) thick. The specimens were machined to a one inch (26 mm) width by cutting 0.5 inches (13 mm) from each side.

3. Results

3.1 Results of Bondline Thickness Study

This bondline thickness study was performed to develop a better understanding of the role bondline thickness plays in determining the strength of the part. It is well-established that a thin uniform adhesive bondline is essential to optimizing strength in the finished part. The results suggest that the bondline thickness over the range of 0.007 inches (0.18 mm) to 0.017 inches (0.44 mm) provides essentially the same strength, with apparent shear strength tapering off and coefficient of variation in the test values increasing. This result is very advantageous for manufacturing where maintaining tight tolerances across large and or complex parts can be very difficult. A specimen installed into the test fixture is shown in Figure 4. Test results for the samples prepared with nylon veil are shown in Table 2.

Figure 4: ASTM D3165-07 Test Setup.

Table 2: Tensile shear results of specimens fabricated with nylon veil.

Sample Description	Bondline Thickness (mm / in)	Tensile Shear Stress (psi)	Standard Deviation (psi)	Coefficient of Variation (%)
1 oz. Nylon veil, 1 layer	0.254 / 0.010	1,960	70	4
2 oz. Nylon veil, 1 layer	0.432 / 0.017	1,940	40	2
2 oz. Nylon veil, 2 layers	0.584 / 0.023	1,650	180	11

3.2 Results of Abraded vs Non-Abraded Adhesion Testing

Even though it is widely believed that it is essential to spend significant time and labor to modify the surface of pultruded composites followed by a thorough cleaning to enhance adhesion, the results of this testing suggest that there is a better way. By using a pultrusion resin with a "primer" built into the resin, it is possible to get excellent adhesion to a pultruded part by simply wiping off the surface to remove dirt and debris. Results are shown in Table 3.

Table 3: Tensile shear results of specimens fabricated with shims.

Sample Description	Bondline Thickness (mm / in)	Tensile Shear Stress (psi)	Standard Deviation (psi)	Coefficient of Variation (%)
Shim, 0.018" thick	0.457 / 0.018	1,860	150	8
Shim, 0.019" thick, wet-sanded	0.482 / 0.019	1,790	300	17

4. Conclusions

Thin uniform bondline thickness is considered essential to maximizing the strength in adhesive bonds. Test results from this study show that strength remains relatively constant across bondlines as thin as 0.007 inches (0.18 mm) up to 0.017 inches (0.44 mm), while shear properties decreased with bondlines thicker than 0.017 inches (0.44 mm). Coefficient of variation in the properties also increased with the thicker bondlines. This result provides an unanticipated benefit to those assembling parts with adhesive, with the relatively wide variation in thickness of applied adhesive providing very good shear properties. There are many benefits associated with using adhesive versus mechanical fastening, including typically lighter weight, less damage to composite structures, and improved fatigue performance. The relatively high consistency of bond strength up to bondline thicknesses of 0.017 inches (0.44 mm) provides additional reason to evaluate adhesion as a preferred manufacturing technique.

The ability to formulate a pultrusion resin to provide a priming effect on the surface of the part is an exceptionally valuable addition to the pultrusion process. There is a significant amount of time, labor, and ultimately cost put into modifying pultruded composite surfaces to allow them to be bonded into larger structures. The removal of the waste associated with these secondary processes will be a real benefit to those that manufacture bonded composite structures.

5. References

1. Plastics Engineering Handbook of the Society of the Plastics Industry, Inc. 5th Edition, Michael L. Berins, Editor, pp. 693- 750 "Joining and Assembling Plastics", Van Nostrand Reinhold, NY, 1991.

2. ASTM D3165-07 *Standard Test Method for Strength of Adhesives in Shear by Tension Loading of Single-Lap-Joint Laminated Assemblies*

3. a. ASTM D 3528-96 *Standard Test Method for Strength Properties of Double Lap Shear Adhesive Joints by Tension Loading*

   b. ASTM D 1002-99 *Standard Test Method for Apparent Shear Strength of Single-Lap-Joint Adhesively Bonded Metal Specimens by Tension Loading*

   c. ISO 4587, 3rd Edition 2003-03-15, *Adhesives – Determination of Tensile Lap-Shear Strength of Rigid-to-Rigid Bonded Assemblies*

4. SikaPower 1200, Fast curing, tough, high strength assembly adhesive, Product Data Sheet, **https://usa.sika.com/dms/getdocument.get/bb8bae71-e83b-395a-a205-2e02a8e4b0ef/pds_SikaPower-1200_e%203.2016.pdf**

PLASMA TREATMENT OF COMPOSITES AND CARBON FIBERS TO INCREASE BOND STRENGTH

Daphne Pappas, Nathaniel Eternal, Raul Gonzalez, Tim Smith
Plasmatreat
Hayward, CA

ABSTRACT

Atmospheric pressure plasma (APP) is a surface engineering method that has been widely used in the manufacturing of lightweight composite materials. APPs have proven effective in overcoming the poor interfacial strength between the fibers and matrix through the plasma-induced surface functionalization. Due to the propensity of composite structures to delaminate under loading, it is necessary to develop composite structures with inherent resistance to such damage.

Traditional methods aiming to improve performance include thickness reinforcements and toughened matrices. However, these methods do not enhance the fiber and matrix interface strength. Non-thermal atmospheric–pressure plasmas have gained popularity due to their ability to easily produce non-hazardous waste processes, low operating costs, fast materials processing speed and industrial scalability. APPs have the ability to modify polymeric fibers via mild plasma etching and cross-linking and the grafting of surface functional groups that can be tailored to enhance adhesive strength in polymer-based composites without affecting the bulk mechanical properties of the material.

In this paper, results from the air plasma modification of carbon fiber materials and carbon fiber reinforced polymer (CFRP) composites will be presented. Lap shear specimens made with carbon fiber reinforced polymers were bonded with an epoxy-based adhesive were tested before and after short exposure to air plasmas. A 40 % increase of lap shear strength was observed for the specimen that were plasma treated prior to the application of the adhesive. Characterization of treated CFRPs and carbon fiber fabric was also performed by Water Contact Angle (WCA) goniometry measurements and X-ray Photoelectron Spectroscopy (XPS) analysis. While the as-received materials were hydrophobic and had low surface energy, few seconds of air plasma exposure led to complete surface wetting. This can be attributed to the mild oxidation of the surface and the attachment of polar groups as confirmed by XPS analysis.

While the study involved the bonding of identical composites, bonding to other composites or dissimilar materials can benefit from plasma treatment as well. The mechanical properties of composites such as tensile strength is increased when the reinforcing material is treated with plasma before being molded to the matrix material. Analyzing the strength by performing other fracture tests like a four point flexular test can be performed to further demonstrate the plasma effect. Inspection of the fracture surface will also determine the mode of failure and the contribution of plasma treatment to increasing strength.

Keywords: plasma, surface modification, surface activation, carbon fibers, composites, adhesion
Corresponding author: Daphne Pappas, daphne.pappas@plasmatreat.com

1. INTRODUCTION AND BACKGROUND

The need for finding alternative to metals high strength material candidates has led the aerospace and automotive industry to promote research and development of composite materials. These materials enable structural weight reduction while maintaining mechanical strength. Carbon fiber composites have been widely used to replace metals and polymers. One of the main challenges with composites are failures due to their poor interfacial properties when they are incorporated in epoxy or resin matrix systems for the development of carbon fiber reinforced composites. The mechanism of failure can be attributed to the delamination of the fiber from the matrix material. By improving the bond of the fiber to the matrix, the strength of the material can be improved significantly.

Adhesives play a key role in composite manufacturing as adhesive bonding is one of the powerful joining methods because of its high fatigue resistance and high strength-to-weight ratio [1]. Adhesively bonded joints in polymeric composite structures are capable of bonding thick material components to thin layers without distortion. The levels of localized stress forces that are present in mechanical joints are reduced enabling the bonding of dissimilar types of materials [2].

The ability to form a bond between two materials is mainly dependent on: a) the chemical composition of the two components and b) their surface topography. The chemical compatibility between two materials (polymer-polymer or polymer-adhesive, for example) will lead to the formation of a chemical bond. In the second case, enhanced surface roughness will contribute to higher levels of mechanical interlocking which is a rather physical interaction and not chemical. The surface energy of the bonded surfaces is a critical property that can provide some insight on its capability to bond. This property is often expressed as improved "wetting" of the surface and is often tested with dyne inks. Most polymer-based materials exhibit low levels of surface energy or poor wettability. As an example, the surface energy of polytetrafluoroethylene (PTFE) is typically around 20 mN/m and is non-wettable, while polyamide (nylon) that exhibits better wettability has a surface energy closer to 40 mN/m [3].

Traditionally, surface treatments involving the use of chemical agents, like solvent wipe and acid etching are employed to improve bonding. Other techniques employ more energetic treatments such as corona discharge, plasma and flame. The latter exhibit better performance -without requiring the use of toxic chemicals- that can be attributed to the cleaning and combined physical and chemical modification of the treated surfaces. Overall, the impact of the various surface modification methods is dependent on the type of materials being treated and the formulation of the adhesive.

Plasma surface treatment is of particular interest for the composites industry as it utilizes a dry process approach that does not involve chemical baths, harmful chemicals, waste removal and open flames sources. Vacuum-based plasmas have been used for the cleaning and activation of polymeric materials since the 1960s [4]. Recently, plasmas operating under atmospheric pressure conditions (APPs) have been gaining a lot of traction due to their capability to upscale and generate high throughput. Several academic groups and equipment manufacturers have developed APPs for the surface activation of polymers. The process is based on the application of an electric field on a gas flowing through it at high rates. Nitrogen, argon, oxygen, helium and air are typically the gases being used. Air appears to be the most attractive gas due to the low cost of operation in manufacturing.

The Openair® system used in this study uses clean dry air (CDA) to generate plasma inside a plasma jet. The air becomes partially ionized leading to electron-induced excitation of the nitrogen and oxygen molecules that compose the air. Depending on the type of jet and nozzle-head used, the plasma glow can have a diameter ranging from a few millimeters to several centimeters enabling the system to be used in either very localized, narrow treatment areas or larger areas. The jet is typically connected to 3 or 6-axis robotic systems that allow the movement of the plasma over large areas and the continuous or selective treatment. The effect of these APP jets can be primarily controlled via the control of the jet speed and the gap between the plasma source and the treated substrate.

The mechanism of surface modification of carbon-based materials under atmospheric pressure plasmas can be summarized as a surface functionalization effect through the grafting of new functional groups onto the treated surfaces that originate in the plasma phase. In air-based plasmas, the presence of nitrogen ions (N^+, N_2^+), NO, N_2O, NO_2, NO_3, atomic oxygen and nitrogen, OH radicals and excited atomic and molecular species produce a very reactive phase where these radicals are allowed to react with the activated polymer surfaces. A mild oxidation of the surface mainly due to the presence of the above-mentioned oxygen-based free radicals has been confirmed by several literature reports. It is important to note that the chemical activation (oxidation) of the surface does not cause any degradation of the bulk material properties of the treated materials since it the treatment is surface specific and affects only the top 10 nm of the material [5].

Several literature references have reported carbon fiber strength increase after the surface modification of APPs for composite applications. The mild oxidation of the fiber surface leads to an enhancement of the hydrophilicity of the fiber leading to improved adhesion to the matrix material. Erden et al. studied the effect of Openair® plasma treatment on carbon fibers used in the fabrication of carbon fiber/polyamide-12 composites [6]. Results from this work showed an oxygen uptake of 8.9 % due to the plasma modification, while the O/C surface atomic concentration increased by 12 %. Tensile tests were also performed on the single fiber composite samples to measure the interfacial shear strength (IFSS). The IFSS increased from 40.2 to 82.7 MPa after Openair® plasma treatment. The single fiber composite samples are a good representation of the interfacial bond strength between the fiber and matrix.

In this paper, 2 types of carbon-based materials were exposed to an APPs aiming to improve their surface properties. Characterization of the treated surface was conducted by using water contact angle goniometry and X-ray photoelectron spectroscopy, methods commonly used to assess surface energy and chemistry, respectively. Analysis of the mechanical properties such as strength was performed with lap shear testing.

2. EQUIPMENT AND MATERIALS

The APP treatment was performed using a Plasmatreat Openair® system equipped with a 5001 generator and RD1004 jet. The plasma generation was based on the arc discharge effect, using clean dry air as the process gas. The generator provided power at frequencies between 19 kHz and 24 kHz. A plasma control unit (PCU) enabled the measurement of power delivered at the site of plasma generation. The flow rate of the ionization gas was also controlled by the PCU, allowing gas flow delivery in the (40-70) L/min range. The RD1004 plasma jet rotated at a speed of 2800 rpm while it was rastered over the studied materials by a cartesian robot, creating a treatment path of about 19 mm. The distance between the nozzle tip and the various materials that were treated ranged between 6 mm and 12 mm. The effective treatment width is dependent on the sample-nozzle distance and the velocity of the jet moving over the treated surfaces.

Two types of carbon fiber substrates were used: a) uni-directional carbon fiber mats (Hexcel, IM7 - intermediate modulus) and b) carbon fiber bars (2.5 mm thick, 25.4 mm x 100 mm, McMaster Carr).

The surface wettability was assessed using contact angle goniometry. Water droplets of 5 μl were dispensed on the samples before and after the plasma exposure A BTG Surface Analyst 3001 system was used to measure the water contact angles formed. A total of 5 drops per sample were measured and the average values were calculated.

Near-surface analysis of the surface chemistry was performed with an Omicron XPS/UPS system equipped with an Argus detector using Omicron's DAR 400 dual Mg/Al X-ray with 0.8 eV resolution. The X-ray source used in this study was a Mg Kα (hv: 1256.6eV) X-ray source that was operated at 15 kV, 300 W and had a take off angle of 40 °. The analysis was conducted under ultra high vacuum conditions as the pressure was maintained at $7x10^{-7}$ Torr. Survey and high resolution scans were performed on the as-received and plasma treated samples to identify the changes in surface elemental composition due to the plasma treatment.

Lap shear strength was measured using the Instron 3366 mechanical tester. Specimen were prepared according to the ASTM D 1002, ASTM D3163, and DIN ISO 1465 standards. Throughout all testing, the pull rate was maintained at 0.03 mm/s. The lap shear bar specimens were 2.5 mm thick, 25 mm wide and 100 mm long. A controlled volume of Henkel EAM-121 HP Hysol adhesive was applied on the end of each strip, covering an area of 12.5 mm^2.

3. RESULTS AND DISCUSSION

3.1 Plasma Treatment of Hexcel carbon fiber mats

The effect of Openair® plasma treatment on small size coupons of the carbon fiber mat was studied. The samples had a size of approximately 15 cm^2 and were placed on the aluminum stage of the robotic system.

Air plasmas were generated at a frequency of 21 kHz, applied power of 450 W, while the air flow was maintained at 45 L/min for all processes. The velocity of the scans varied from 50 mm/s to 100 mm/s while the distance between the jet exit and substrate was maintained at 8 mm – see figure 1.

Figure 1. Treatment of Hexcel carbon fiber fabric with Openair® plasma

Prior to the plasma activation, the carbon fiber fabric surfaces were found to be hydrophobic; the average water contact angle was 66 °. Following plasma treatments at 50 mm/s and 100 mm/s, the surfaces became extremely wettable. The applied water drops post plasma exposure were absorbed instantly and could not be measured. This proves that the surface physicochemical properties were changed due to the short plasma exposure leading to significant improvement of the surface wettability and enhancement of the surface energy.

The hypothesis that the surface composition was altered due to the plasma treatment was also confirmed by the results of the XPS analysis. Survey and high resolution spectra of carbon fiber mat samples that were treated at a speed of 100 mm/s were obtained.

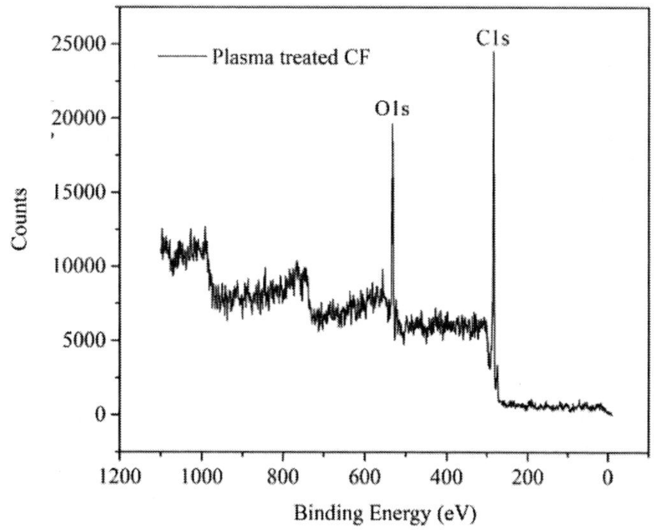

Figure 2. XPS survey scan of plasma treated carbon fabric

As seen in figure 2, besides the presence of carbon at 285 eV, a new peak appeared at the binding energy of 532 eV that can be attributed to the formation of oxygen functional groups on the fiber surface. The analysis showed that the surface composition post plasma was: C1s: 66.75 % and O1s: 33.25 %. The increase of the oxygen atomic concentration can be attributed to the grafting of oxygen functionalities that are formed in the plasma gas phase due to the strong presence of reactive atomic oxygen.

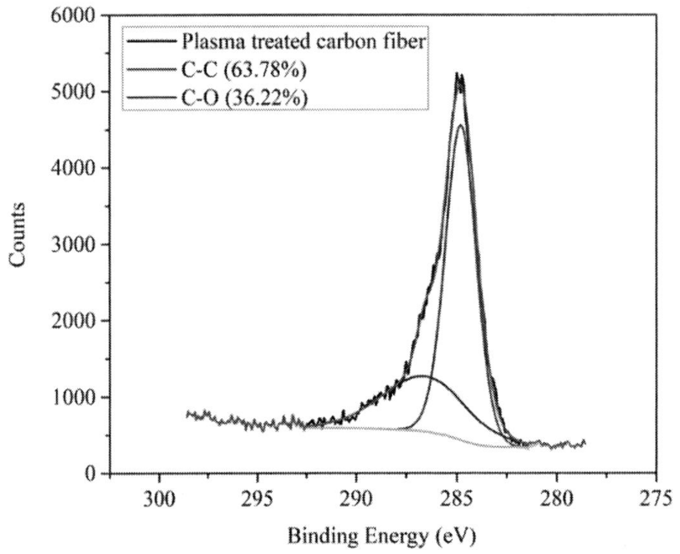

Figure 3. C1s high resolution spectrum of plasma treated carbon fabric

The deconvolution of the C1s peak revealed the presence of two components: C-C and C-O at atomic concentrations of 63.78 % and 36.22 %,respectively- see figure 3.

3.2 Plasma Treatment of carbon fiber reinforced composite bars

The Openair® system described earlier was also used for the plasma treatment of carbon fiber composite surfaces as shown in figure 4. Air plasmas were generated at a frequency of 21 kHz, applied power of 443 W, while the air flow was maintained at 45 L/min for all processes. The samples were cut in bars that had a width of 2.54 cm and length of 10 cm. The distance between the plasma and the samples was kept constant at 6 mm while the robotic system enabled the movement of the jet at a constant speed of 100 mm/s.

The surface wettability after the plasma exposure was found to be improved. The average WCA post plasma treatment was 35 °, about 50 % lower than the values recorded from the testing of the untreated CFRC bars (average WCA: 70 °). The enhancement of the surface hydrophilicity can be attributed in this case to the attachment of oxygen-based functionalities to the surface, as described in section 3.1.

Figure 4. Treatment of CFRC bars with Openair® plasma

A total of twelve specimen (6 untreated and 6 plasma-treated) were prepared for mechanical testing in order to investigate the effect of increased surface energy to the adhesive strength of bonded CFRC bars. The pull-test specimen had a bond area of 25.4 mm x 12.5 mm. The adhesive was applied immediately after the plasma treatment and the samples were allowed to cure for 24 hrs at room temperature conditions. Then each set of bars was mounted on the mechanical tester using non-slip grips. The testing was conducted with a pull rate of 0.03 mm/s. All specimen studied were fractured at the interface between the 2 bonded bars; no material failure was observed. The recorded maximum force at break and the average lap shear strength is shown in Table I:

Table 1. Pull test results

Treatment type	Force at Break (kN)	Lap Shear Strength (MPa)	Type of failure
Untreated	5.5 ± 0.8	17.0	Adhesive
Plasma	7.8 ± 0.6	24.1	Cohesive

Results show an increase of about 24 % in the force required to separate the bonded materials after plasma treatment and therefore higher lap shear strength is achieved: 24.1 MPa versus 17.0 MPa. Failure analysis was conducted after the completion of the pull tests that revealed the adhesive mode of failure for the as-received specimen. Traces of the adhesive still remained present on the bonded areas of the plasma treated samples indicating cohesive failure mode.

4. CONCLUSIONS

Air based plasmas were applied as a surface engineering method to improve the adhesion properties of carbon fiber materials. The treatments were fast and resulted in significant increase of lap shear strength of samples that were plasma treated and bonded with a two-part epoxy.

Surface characterization showed an increase of the surface oxygen content that led to improved wettability. As the method requires only air and electricity, industrial upscaling is cost effective and can be beneficial to the automotive, aerospace and defense industry.

5. ACKNOWLEDGEMENTS

The authors would like to thank Professor Amir Asadi and his team at Texas A&M University for providing the XPS data from the analysis of the carbon fiber materials.

6. REFERENCES

1. Wahab, Magd Abdel, *Joining Composites with Adhesives: Theory and Applications* Lancaster, PA: DEStech Publications, 2016.
2. Baldan, A., "Adhesively-bonded Joints and Repairs in Metallic Alloys, Polymers and Composite Materials: Adhesives, Adhesion Theories and Surface Pretreatment", *Journal of Materials Science*, 39(1) (2004): 1-49.
3. Murthy, Vidyarani Sangnal Matt Durandhara, *Improving the Adhesion of Glass/Polypropylene (Glass/PP) and High-Density Polyethylene (HDPE) Surfaces by Open Air Plasma Treatment*, Masters Thesis, University of Tennessee, 2017.
4. Liston, Edward, "Plasma Treatment for Improved Bonding: A Review", *The Journal of Adhesion*, 30(1-4) (1989): 199-218.
5. Pappas, Daphne, "Status and Potential of Atmospheric Plasma Processing of Materials", *Journal of Vacuum Science & Technology A*, 29 (2011): 020801.
6. Erden, Seckin, Ho, Kingsley, Lamorineire, Steven, Lee, Adam, Yildiz, Hasan, Bismarck, Alexander, "Continuous Atmospheric Plasma Oxidation of Carbon Fibres: Influence on the Fibre Surface Bulk Properties and Adhesion to Polyamide 12", *Plasma Chemistry and Plasma Processing*, 30 (2010): 471-487.

VIRTUAL TESTING OF CFRP COUPON USING DIGIMAT VA

P. MARTINY, A. CHERUET*

e-Xstream Engineering, part of Hexagon's Manufacturing Intelligence

(corresponding author: Anthony.cheruet@hexagon.com)

ABSTRACT

In the research of light weighting solutions, the use of CFRP has dramatically increased during the last two decades both in aerospace and automotive industries. However, designers are still facing the challenge to accelerate the insertion of new materials for applications. Traditionally, screening, characterization and even design of new materials is done by physical testing. However, composites materials offer an extraordinary choice of material combinations so that such traditional approaches become inefficient at best.

Simulation accelerates these test campaigns, providing insights and answers well before physical coupons can be ordered, created, tested and reported. It will help to address some of the inefficiencies in the process and reduce the total time from initial screening to final, fully characterized materials.

Tools, like DigimatTM propose a complete framework to conduct simulations. This framework includes a continuum damage model to accurately capture the damage initiation and propagation that take place during the loading. Additional features are added to this model, such cohesive element to model the interface, in-situ strengths, mesh sensitivity control and effect of manufacturing stresses. The framework is applied on several coupon configurations with several aerospace composite grades.

1 INTRODUCTION

Continuous fiber reinforced polymers offer a unique combination of mechanical properties and density, allowing substantial weight saving in structural applications compared to metal based solutions. However, compared to metals, composites also offer a much increased complexity and significantly higher characterization cost - which hinder their adoption.

At the first step of the building block approach, a composite material must be characterized through numerous laminate tests, covering several coupon type (unnotched tension/compression, open-hole tension/compression…), several layups and several conditions. Due to the inherent variability of composite mechanical behavior, each test configuration must be repeated many times in order to obtain proper statistical evaluations of the laminate strength.

This process is heavily time and money consuming. For this reason, simulation-based solutions using micromechanics and multi-scale models involving progressive failure and non-linear finite element analysis is developed.

2 MOTIVATIONS FOR VIRTUAL TESTING

The motivation for using virtual testing in the implementation of new composite materials is simple: to make it possible to develop and implement new materials within a product development cycle; in other words, for the material itself to become a design variable. The vision is for market studies and preliminary design to identify needs for specific enhanced materials performance, for material suppliers to develop new materials to meet those requirements, and for those materials to be used in new product design. Unfortunately, with the traditional test-based approach to materials development and characterization, the design cycle for new materials is far longer than the design cycle for new products, requiring thousands of expensive, time consuming tests. The time to develop and introduce a new material system in aerospace is often over 10 years from proposal to production. Also, the cost for such a program is high, given the amount of testing required. These factors have severely hindered development and implementation of new materials. Virtual testing is one part of the solution for breaking the time and cost curve, and enabling new materials in design.

The other motivation for using virtual testing is to expand the design space of existing material systems to enable more efficient designs. With composite materials, the ability to optimize part performance through the intelligent alignment of fibers is virtually unlimited. However, traditional composite designs are often quasi-isotropic which gives rise to the term "black aluminum." They use a limited number of permitted ply angles, generally restricted to 0°, ±45°, and 90° plies. Further design rules restrict the minimum number of plies for any given ply angle in the layup, the maximum number of consecutive plies of the same orientation that can be used, etc. These rules serve to limit the design space to those layups that can be characterized with a test program, and stay away from designs that experience has shown, exhibit undesirable characteristics. However, restricting the design space in this way throws away much of the potential efficiency that composites have to offer.

Virtual testing can open up the design space by using simulation to characterize the performance of non-traditional layups using an expanded number of possible layup angles. As you add to the number of possible ply angles, the number of possible layup combinations grows geometrically and the number of tests required to characterize that non-traditional space becomes staggering, and not practical using physical testing alone. Virtual testing provides the means to intelligently explore the design space and discover the most efficient designs quickly.

3 MODELING STRATEGY OF THE COMPOSITES

The modeling strategy implemented in the framework is based on several academic publications [1], [2], [3].

In this approach, a continuum damage model is used to model the ply and a cohesive zone model is used to model the interface between the plies.

A brief description of these models is given below but the reader can refer to the articles previously mentioned for further details.

3.1 Continuum damage model of the ply

3.1.1 *Constitutive model*

The model relies on a continuum damage model used to predict the initiation and propagation of intralaminar damage mechanisms.

In this case, the free energy density is expressed as a function of damage variables associated to the different failure modes:

$$\Psi = \frac{\langle\sigma_{11}\rangle^2}{2(1-d_1)E_1} + \frac{\langle-\sigma_{11}\rangle^2}{2(1-d_1)E_{1c}} - \upsilon_{12}\sigma_{22}\left(\frac{\langle\sigma_{11}\rangle}{E_1} + \frac{\langle-\sigma_{11}\rangle}{E_{1c}}\right) + \frac{{\sigma_{22}}^2}{2(1-d_2)E_2} + \frac{{\sigma_{12}}^2}{2(1-d_6)G_{12}}$$

$$+(\alpha_{11}\sigma_{11} + \alpha_{22}\sigma_{22})\Delta T$$

Where σ_{ij} (i,j=1,2) are the stress components, d_i (i=1,2,6) are the scalar damage variables, α_i (i=1,2) are the coefficients of thermal expansion.

With such an expression, it is assumed that damage can be initiated only by the in-plane stresses with no effect from the out-of-plane stresses.

3.1.2 *Damage initiation criteria*

The damage failure initiation is based on a LARC03 failure criteria and used to predict four intralaminar failure mechanisms. The meaning of each parameter is given Table 1.

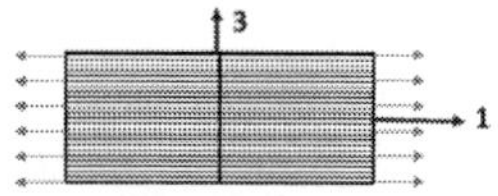	$\phi_{1+} = \frac{E_1}{X_T}\varepsilon_{11} = \frac{\tilde{\sigma}_{11} - \nu_{12}\tilde{\sigma}_{22}}{X_T}$ if $\widetilde{\sigma_{11}} \geq 0$
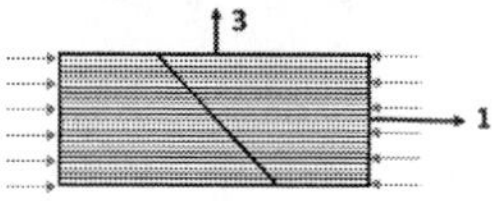	$\phi_{1-} = \frac{\langle \mid \tilde{\sigma}_{12}^m \mid + \eta^L \tilde{\sigma}_{22}^m \rangle}{S_L}$ if $\widetilde{\sigma_{11}} \leq 0$
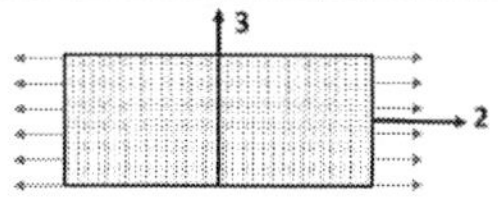	$\phi_{2+} = \begin{cases} \sqrt{(1-g)\frac{\tilde{\sigma}_{22}}{Y_T} + g\left(\frac{\tilde{\sigma}_{22}}{Y_T}\right)^2 + \left(\frac{\tilde{\sigma}_{12}}{S_L}\right)^2} & \text{if } \tilde{\sigma}_{22} \geqslant 0 \\ \frac{1}{S_L}\langle \mid \tilde{\sigma}_{12} \mid + \eta^L \tilde{\sigma}_{22} \rangle & \text{if } \tilde{\sigma}_{22} < 0 \end{cases}$
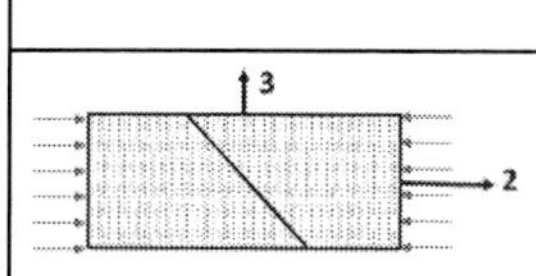	$\phi_{2-} = \sqrt{\left(\frac{\tilde{\tau}_{eff}^T}{S_T}\right)^2 + \left(\frac{\tilde{\tau}_{eff}^L}{S_L}\right)^2}$ if $\tilde{\sigma}_{22} < 0$

Figure 1: Four failure mechanisms

The effective (undamaged) stress tensor ($\tilde{\sigma}$) is a function of the undamaged stiffness and the elastic strain, $\tilde{\sigma} = C_0 \varepsilon_e$.

The components of the effective stress tensor in the coordinate system associated with the rotation of the fibres are calculated as:

$$\sigma_1^m = \sigma_1 cos^2(\varphi) + \sigma_1 sin^2(\varphi) + 2\tau_{12} sin(\varphi)cos(\varphi)$$

$$\sigma_2^m = \sigma_1 sin^2(\varphi) + \sigma_2 cos^2(\varphi) + 2\tau_{12} sin(\varphi)cos(\varphi)$$

$$\tau_{12}^m = -\sigma_1 sin(\varphi)cos(\varphi) + \sigma_2 sin(\varphi)cos(\varphi) + \tau_{12}(cos^2(\varphi) - sin^2(\varphi))$$

With φ is the misalignment angle at failure of the unidirectional subjected to uniaxial longitudinal compression. The computation of φ is given by the expression:

$$\varphi = tan^{-1}\left(\frac{1 - \sqrt{1 - 4\left(\frac{S_L}{X_c} + \eta_L\right)\left(\frac{S_L}{X_c}\right)}}{2\left(\frac{S_L}{X_c} + \eta_L\right)}\right)$$

The effective shear stress for the matrix compression are computed as:

$$\tau_{eff}^T = -\sigma_2 \cos(\alpha)(\sin(\alpha - \eta_T \cos(\alpha))$$

$$\tau_{eff}^L = \cos(\alpha)\,|\tau_{12}| + \eta_L \sigma_2 \cos(\alpha)$$

with longitudinal friction coefficient η_L given by:

$$\eta_L = \frac{S_L * \cos(2\alpha)}{Y_C * \cos^2(\alpha)}$$

And the transverse shear stress S^T given by:

$$S^T = Y_c \cos(\alpha)\left(\sin(\alpha) + \frac{\cos(\alpha)}{\tan(2\alpha)}\right)$$

3.1.3 Damage propagation laws

The damage initiation thresholds, ri (i=1+, 1-, 2+, 2-) are related to the loading functions ϕ_i (i=1+, 1-, 2+, 2-) damage variables by the damage evolution laws.

In the longitudinal and transverse directions, the following evolution are used:

$$r_{1+} = \max\left\{1, \max_{s=0,t}\{\phi_{1+}^s\}, \max_{s=0,t}\{\phi_{1-}^s\}\right\}$$

$$r_{1-} = \max\left\{1, \max_{s=0,t}\{\phi_{1-}^s\}\right\}$$

$$r_{2+} = \max\left\{1, \max_{s=0,t}\{\phi_{2-}^{s}\}, \max_{s=0,t}\{\phi_{2+}^{s}\}\right\}$$

$$r_{2-} = \max\left\{1, \max_{s=0,t}\{\phi_{2-}^{s}\}\right\}$$

It can be seen in the above expressions that r1+ depends upon both ϕ_{1+} and ϕ_{1-} while r1- depends only upon ϕ_{1-}. This represents the crack closure phenomenon upon load reversal. A crack created under tensile loads will close up under compressive loads and the material will recover part of its load carrying capability (r1- depends only upon ϕ_{1-}) while the cracks created under compressive loads will remain open under tensile loads (r1- depends not only upon ϕ_{1+} but also upon ϕ_{1-}).

To properly reproduce the mechanical response of the material, different damage propagation laws used for each direction: di = di(ri) (i=1+, 1-, 2+, 2-). These propagation laws are also function of the intralaminar fracture toughness associated at each failure modes. This allows a correct energy dissipation during the damage propagation.

In addition, these propagation laws are also function of the characteristic length of the finite element. This allows to limit the mesh-size dependency by ensuring the same energy dissipation per unit area of crack propagation regardless of the size of the elements.

Longitudinal Tension	Longitudinal compression	Transverse Tension and compression	In-plane shear

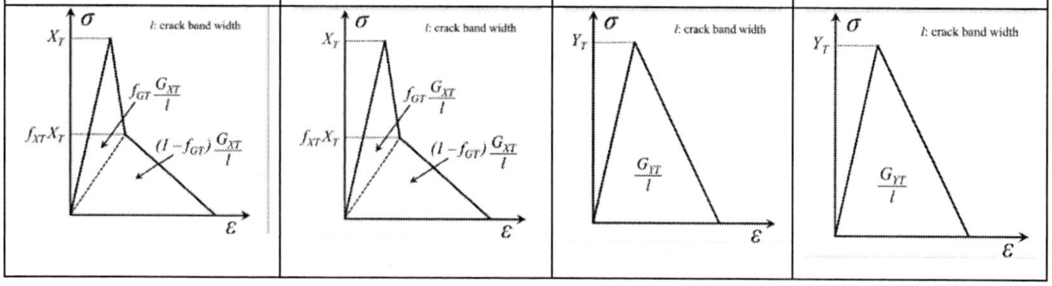

Figure 2: Damage propagation laws

3.1.4 In-situ strengths

The model implemented considers also that the ply strengths which are controlled by matrix failure phenomena (YT, YC and SL) are in situ properties, i.e. they depend upon the thickness of the ply and upon its location in the laminate thickness. The motivation behind is that a ply within a laminate is subject to some constraints coming from the adjacent plies which delays the onset of matrix failure phenomena and increases the corresponding strength values. This constraining effect is larger with thinner plies, since the adjacent plies are closer, and larger for embedded plies than for surface plies which have only one adjacent plies.

The formulation implemented was proposed by [3] and allows to derive the transverse tension and in-plane shear in-situ strengths from their lamina values, from the ply thickness and from the position of the ply within the laminate.

3.2 Cohesive zone model for the interface

Delamination phenomena are predicted using the cohesive zone model suggested by [8]. To the contrary of the cohesive zone models typically available in commercial finite element software, it offers the advantage of being thermodynamically consistent. Non-physical energy dissipation or restauration phenomena are therefore avoided. It further relies on the energy-based criterion proposed by [9] to predict delamination under mixed-mode loading conditions:

$$G_{TC} = G_{IC} + (G_{IIC} - G_{IC})\left(\frac{G_{II}}{G_T}\right)^m$$

where GTC is the critical energy release rate required for delamination at a mode-mixity equal to (GII/GT), GIC and GIIc are the critical energy release rates in mode I and in mode II, and the exponent m is a parameter of the model.

4 SIMULATION OF THE COUPON BEHAVIOR

4.1 Modeling of the coupon using FEM

All coupon finite-element models use 8-node linear hexahedral element, with one element per ply thickness. Height-node cohesive elements with zero initial thickness are inserted between the plies with dissimilar orientations, sharing the same nodes than the adjacent plies.

4.2 Material data

The table below summarizes the material data required to define properly the material model.

Parameters	Comments
X_T	Longitudinal Ply Strength in Tension
E_1	Longitudinal Ply Stiffness in Tension
$nu_{12\,t}$	In-plane Poisson ratio in Tension
Y_T	Transverse Ply Strength in Tension
E_2	Transverse Ply Stiffness in Tension
X_C	Longitudinal Ply Strength in Compression
$E_1\,C$	Longitudinal Ply Stiffness in Compression
$nu_{12\,c}$	In-plane Poisson ratio in Compression
$E_{2\,c}$	Transverse Ply Stiffness in Compression

Y_C	Transverse Ply Strength in Compression
G_{12}	In-plane shear modulus
In-plane stress-strain curve	To determine non-linearity and S_L
α	Fracture angle in transverse compression
GXT (kJ/m2)	Fracture toughness axial tension
GYT (kJ/m2)	Fracture toughness axial compression tension
GXC (kJ/m2)	Fracture toughness transverse tension
GYC (kJ/m2)	Fracture toughness transverse compression
GXY (kJ/m2)	Fracture toughness in-plane shear
Alpha_L (/deg.C)	Axial CTE ply
Alpha_T (/deg.C)	Transverse CTE ply

Table 1: Ply material input data

The material data required for the interface are summarized below.

Parameters	Comments
G1c (kJ/m2)	Fracture toughness Mode I
G2c (kJ/m2)	Fracture toughness Mode II
Normal Stress (Mode I)	Damage initiation stress in Mode I
Shear Stress (Mode II)	Damage initiation stress in Mode II
m	Exponent for fracture toughness in mixed-mode

Table 2: Interface input data

4.3 Manufacturing stresses

It has been proven that the residual stresses coming from the manufacturing process have a significant effect on the final performance of the composite. As a consequence, a thermo-mechanical load step is performed prior the mechanical loadings. During this thermo-mechanical load step, a temperature drop is applied on the coupon, from the curing temperature down to the test temperature, under simply supported mechanical boundary conditions.

The effect of these manufacturing stresses is shown in the table 3, applied on the IM7/977-3 material system.

Test	Layup	Experimental [MPa]	Digimat-VA with manufacturing stresses: Prediction [MPa]	Digimat-VA with manufacturing stresses: Error [%]	Digimat-VA without manufacturing stresses: Prediction [MPa]	Digimat-VA without manufacturing stresses: Error [%]	Effect of manu-facturing stresses on strength [%]
UNT	[30/60/90/-60/-30]_2s	473	474	0%	542	15%	-12%
UNT	[0/45/90/-45]_2s	866	868	0%	917	6%	-5%
UNT	[60/0/-60]_3s	1005	984	-2%	1037	3%	-5%
UNC	[30/60/90/-60/-30]_2s	382	358	-6%	440	[illegible]	-18%
UNC	[0/45/90/-45]_2s	603	598	-1%	666	10%	-10%
UNC	[60/0/-60]_3s	765	671	-12%	778	2%	-14%
OHT	[30/60/90/-60/-30]_2s	409	402	-2%	444	8%	-9%
OHT	[0/45/90/-45]_2s	554	493	-11%	504	-9%	-2%
OHT	[60/0/-60]_3s	543	503	-7%	507	-7%	-1%
OHC	[30/60/90/-60/-30]_2s	295	266	-10%	292	-1%	-9%
OHC	[0/45/90/-45]_2s	341	404	[illegible]	404	[illegible]	0%
OHC	[60/0/-60]_3s	358	394	10%	387	8%	2%

Table 3: Effect of manufacturing stresses on coupon strength [10]

5 APPLICATION

The complete workflow is shown on the figure 3. It embeds FEA technology, composites expertise and material failure models. The objective here is to avoid repetitive work of finite-element models definition, material parameters that account for the variability.

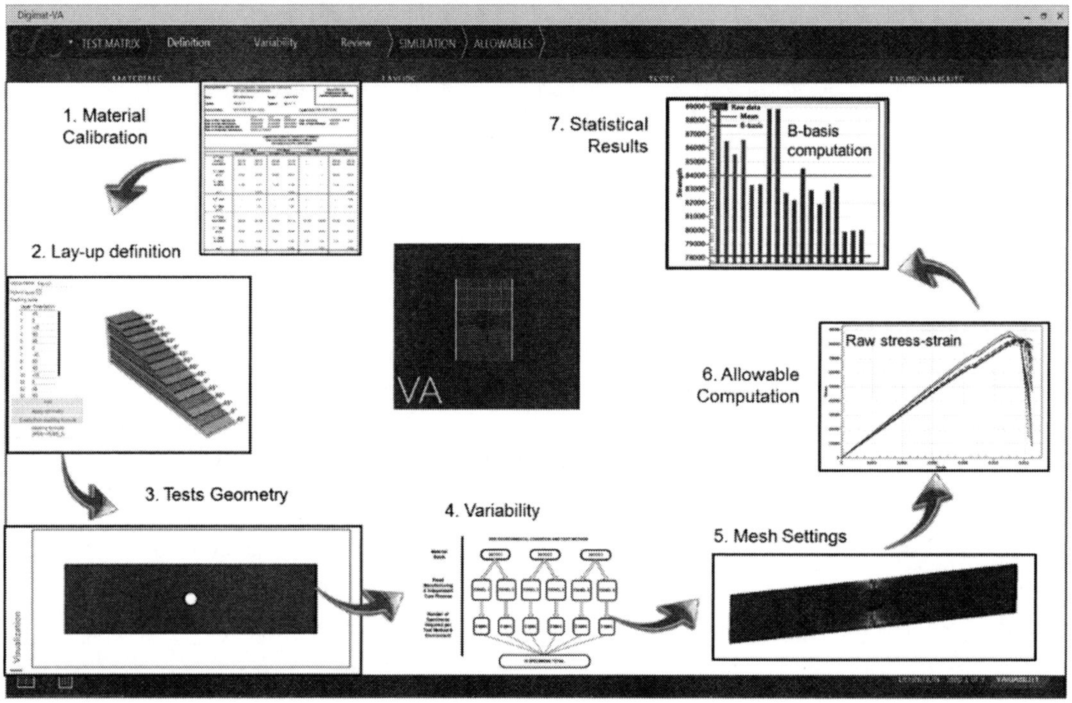

Figure 3: Test matrix definition in Digimat VA

The finite element model of standard tests, such as the first tests performed during the screening phase or the allowable generation phase of a material development cycle are available in the software. These tests library includes the unnotched specimen in tension or compression, the open-hole specimen in tension/compression, the filled-hole specimen with different types of bolt in tension and compression. In addition, the bearing tests, in single and double shear configuration are also available.

At this point, only the unnotched and open-hole tests are considered for the validation of the framework as the modeling doesn't involve the out-of-plane stresses in the damage initiation computation.

The validation of the framework is conducted for several material systems used in different layups for several coupon configurations.

- The material system tested are all aerospace grades using intermediate or high modulus and high strength carbon fibers and epoxy resin.
- Layups varies from classical quasi-isotropic (25/50/25), hard (50/40/10) and soft layup (10/80/10) but include some non-traditional layups. Open-hole test configuration includes several hole diameters as well to capture the hole-size effect.

The objective is to demonstrate that the framework is suitable for any configurations, without refining the analysis settings for a specific material or a specific test. In this context, it is important to mention that all the analyses concerning the unnotched tests are run with exactly the same mesh settings and analyses settings no matter in tension or compression. The same is done for the open-hole configurations.

All the results are summarized in the graph below:

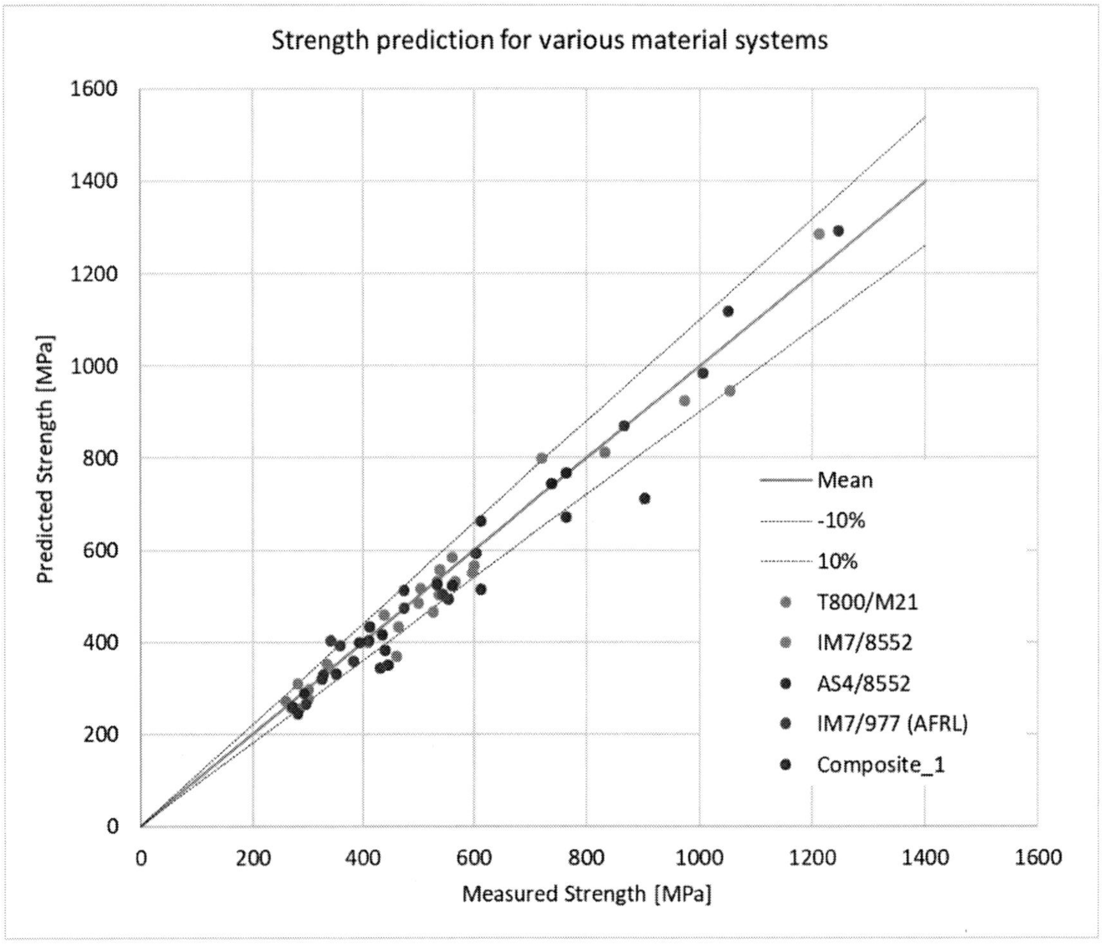

Figure 4: Strength prediction for various coupon configurations

Detailed results are also given for the IM7/8552 and T800/M21 material systems in table 4 and table 5.

6 CONCLUSIONS

To speed-up the integration of new composite materials, a solution based on simulation is proposed in order to generate coupon strength. The process is based on advanced modeling of the composites behavior that describes realistically the progressive failure of such materials. The automatic generation of the finite-element model of the coupon coupled with this material model is embedded in a vertical environment called Digimat-VA. The process has been successfully demonstrated for several aerospace material systems in unnotched and open-hole coupon configurations. Further developments include modeling of the out-of-plane stresses is already considered.

REFERENCES

[1] P. Maimi, PP. Camanho, JA. Mayugo, CG Davila. A continuum damage model for composites laminates: Part I – Constitutive model. Mechanics of Materials 39 (2007).

[2] P. Maimi, PP. Camanho, JA. Mayugo, CG Davila. A continuum damage model for composites laminates: Part II – Computational implementation and validation. Mechanics of Materials 39 (2007).

[3] C. Furtado, G. Catalanotti, A. Arteiro, PJ. Gray, BL. Wardle, PP. Camanho. Simulation of failure in laminated polymer composites: Building block validation. Composites Structures 226 (2019).

[4] M. Bogdanor, C. Oskay. Prediction of progressive damage and strength of IM7/977-3 composites using eigne-deformation based homogenization approach. Material Science. 2017.

[5] Simulation of delamination in composites under quasi-static and fatigue loading using cohesive zone models. A. Turon. PHd dissertation, University of Girona (2007).

[6] Hexcel 8552 AS4 Unidirectional prepreg at 190 gsm. Qualification Material Property Data report NCAMP report number CAM-RP-2010-002 Rev A.

[7] Hexcel 8552 IM7 Unidirectional prepreg at 190 gsm. Qualification Material Property Data report NCAMP report number CAM-RP-2009-015 Rev A

[8] A. Turon, PP. Camanho, J. Costa, CG. Dávila. A damage model for the simulation of delamination in advanced composites under variable-mode loading. Mechanics of materials, 38 (2006).

[9] ML. Benzeggagh, MJCS. Kenane. Measurement of mixed-mode delamination fracture toughness of unidirectional glass/epoxy composites with mixed-mode bending apparatus. Composites science and technology 56 (1996).

[10] S. Clay, P. Knoth. Experimental results of quasi-static testing for calibration and validation of composite progressive damage analysis methods. Journal of Composite Materials, Vol 51, Issue 10 (2017).

A TOOL ANALYSING THE DRAPEABILITY OF UNIDIRECTIONAL CARBON FIBER TEXTILES ON ANY GIVEN GEOMETRY

Santino Wist
Institut für Textiltechnik der RWTH Aachen University
Aachen, Germany

ABSTRACT

The combination of lightweight design and high mechanical properties of fiber composite materials plays a crucial role in meeting the current demands for higher energy and resource efficiency. Unidirectional carbon fiber textiles can be produced based on preforms. During the forming of the preform the textile undergoes drape-mechanics to alter its initial flat shape to three-dimensional geometry. Undulations and folds are adverse effects of these necessary drape mechanics. It is of great importance to design the final part's geometrical shape according to the draping abilities of the carbon fiber textile to achieve the desired physical properties.
This research project designed a software tool for geometrical analysis of any given surface geometry. It analyses the geometry for problematic zones for unidirectional draping processes and visually highlights them.
Local vector lengths are representing the unidirectional fibers. By comparing local lengths, the effect of relative fiber motion can be modeled. Through experiments, critical relative fiber motion has been quantified and implemented into the calculation. Resulting is a tool that requires geometry and basic information about the textile as input. The output, however, is the geometry with local scaled colors representing the *drapeability* in that area

Keywords: Composite, Preform, Draping, Geometry, Unidirectional
Corresponding authors: Santino Wist, David Malik, Stefan Hesseler

1. INTRODUCTION

Over the last decades, a scientific trend has been identified: the combination of lightweight design while maintaining high mechanical properties through the use of fiber composite material. Such trend plays a key role in meeting the current demands for higher energy and resource efficiency. Therefore, in a time of rising product complexity, production processes should be optimized to remain cost-efficient. Unidirectional carbon fiber textiles have the potential to take lightweight design one step further by reducing the weight with equal mechanical properties. Each textile layer can be orientated exactly in the direction of the carried load exploiting the anisotropic properties of carbon fiber to the maximum. [1] [2]
Moreover, fiber composites can be produced based on preforms. To summarize, preforms are, as the name implies, geometrically preformed textiles that are created through a multi-step process (Figure 1). The molding step ultimately defines the decisive characteristics of the product by establishing the load-carrying fiber orientation. During this process, the textile undergoes certain

drape-mechanics to alter its flat shape to a three-dimensional geometry. Nonetheless, undulations and folds are negative effects of these necessary drape mechanics. Following these principles, it is of great importance to design the final part's geometrical shape according to the draping abilities of the carbon fiber textile to achieve the desired physical properties without decline. [3] [4]

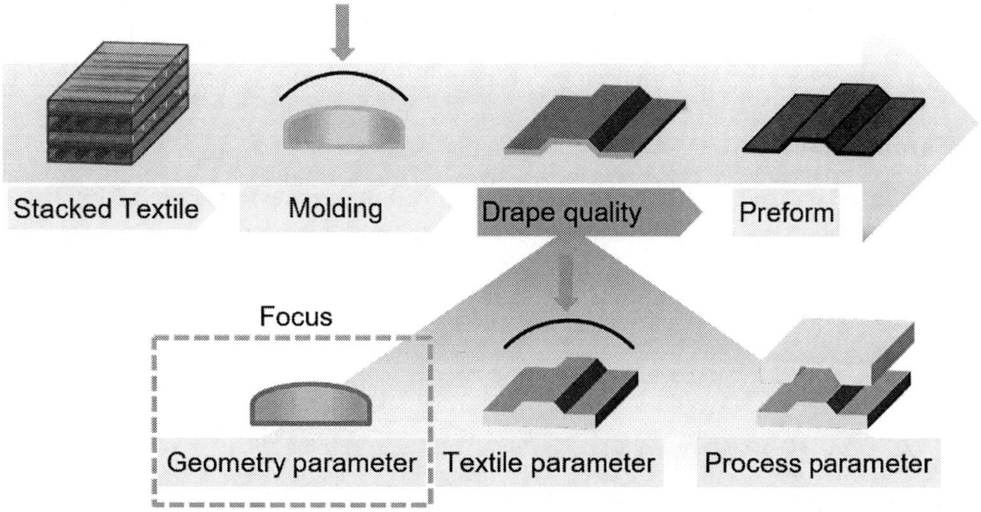

Figure 1: The multistep preforming process and the parameters for the quality of the preform

Furthermore, quality of the highest order in preforms is achieved if the geometry can be draped accurately without adverse drape effects. In other words, high quality preforming is dependent on three parameters: geometry, textile, and the molding process. Unfortunately, the missing standardization and investigation of these parameters prevent a wide-ranged simulation for unidirectional fiber textiles. This results in a great number of experiments and a sequential design process. A valid definition of the drape mechanics and the effects for any geometry carries the potential of a substantial optimization loop. As a central deficit, today stands the incomplete description of the quality affecting parameters, especially the parameter geometry. There are no current tools that reliably connect the parameters of geometry to the quality of the preform for unidirectional carbon fibers. Trial and error is economically inefficient and has to be replaced by a pre-production tool. Therefore, it is expedient to reach an understanding of quality endangering zones in the planned part directly from the proposed geometrical shape. [5] [6] This research project aims to design a software tool for geometrical analysis of any given geometry. The tool analyses the geometry, identifying problematic zones for unidirectional draping processes and visually highlights them.

1.1 Audience

To understand this technical paper, basic experience in composite manufacturing, especially in the handling of textiles, is advised. For people working in the development and design of new geometries that are to be produced with unidirectional composites, this research could provide new insights into the mechanisms used during the draping. Furthermore, developers of fiber simulation programs may find new ways to implement these mechanisms into their code.

1.2 Overview of contents

This technical paper introduces the state-of-the-art of preforming with unidirectional textiles and the significant draping mechanisms that are property to these textiles. Then, a short insight is given to why new simulation approaches for unidirectional textiles are necessary.
Furthermore, the development of the geometry analysis tool is described the following three chapters: the physical process, the mathematical model and the numerical implementation. Finally, the results of a given geometry are compared to an actual preforming experiment.

1.3 State of the Art

The state-of-the-art is structured according to the preforming process (Figure 2). Firstly, stands an overview of the properties of fibers and their interaction with the matrix where they are embedded. Secondly, the description of the molding process and the simultaneous draping mechanisms. Thirdly, the presentation of already existing geometry analysis tools.

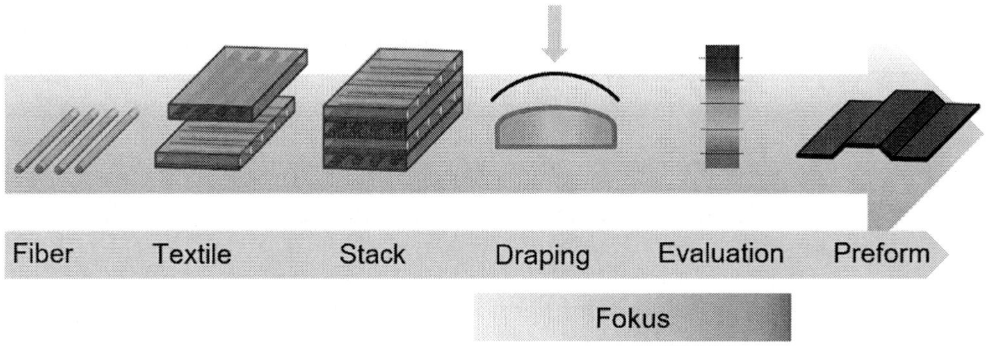

Figure 2: Preforming process and focus of the research

1.3.1 Composite Materials

The combination of high strength and E-modulus with low density predestine composite materials for lightweight design [7]. Additional properties are outstanding energy absorbability and dampening ability.
Structurally composites consist of three components: the fiber, the matrix, and the layer connecting both. The fiber carries the load while the matrix protects the fiber from external influences and distributes the load to all the fibers through the connecting layer (Figure 3). [1][8]

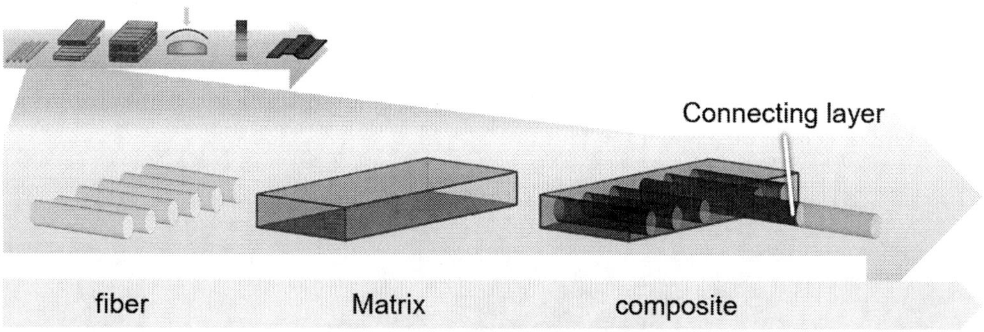

Figure 3: Built of a composite

Notably, the strength and stiffness of the fiber are very directional (Figure 4). This behavior is called anisotropic and contrasts composite materials sharply from metal counterparts. These properties emphasize the necessity of designing composite parts in a way that fiber orientation and direction of the load coincide. [1]

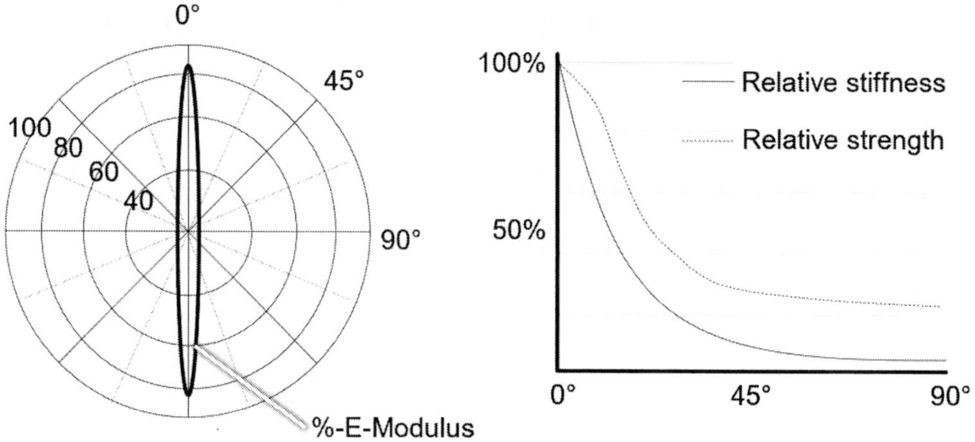

Figure 4: The directional properties of carbon fiber [9]

1.3.2 Preform molding process

The molding process is the main work step during the preforming. The textile layers are covered with a binder and stacked. Embedded into the different molding forms, they are heated and pressed into their final form. The heat melts the binder and hardens it. As a result, the hardened binder keeps the preforms in shape after the forms are opened. Table 1 shows different variations of the stamp molding forms.

During the forming process, multiple draping mechanisms take place. Draping describes the process of a two-dimensional textile adopting its form to a three-dimensional geometry. The molding process puts great strain on the dry textile. The more complex the geometry, the greater the local forces are. Especially simultaneous convexity in multiple directions has been notoriously challenging to high-quality draping. Good drapability aims for a formed result without any folds, gaps, or undulations within the textile. [1][9]

Draping mechanisms for woven textiles are characterized by sheer mechanisms. Unfortunately, sheer mechanisms do not apply for the non-woven unidirectional textiles. Instead, the unidirectional textiles limit their draping mechanisms to changes in fiber orientation, fiber relative movements, and fiber widening (Figure 5). [5]

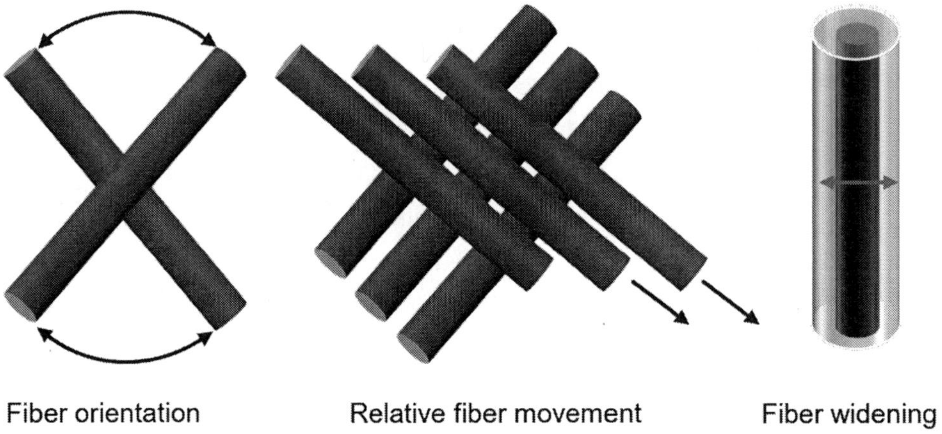

Figure 5: Draping mechanisms of unidirectional textiles [5]

The draping mechanisms both enable the adaptation of textiles to three-dimensional shapes and cause folds, gaps, or undulations. On the one hand, for the adaptation, the textile needs to distort locally. Subsequently, a distortion is desired to the extent that no negative draping effects take form.

On the other hand, the fibers follow the local forces and change their orientation within the textile. Through their movement relative to each other, they approach neighboring fibers until a maximum accumulation of fibers. Here a fold occurs. Besides folds, other adverse draping effects are gaps and undulations, as depicted in Table 2.

Table 1: Draping effects

Fold	Gap	Undulation
	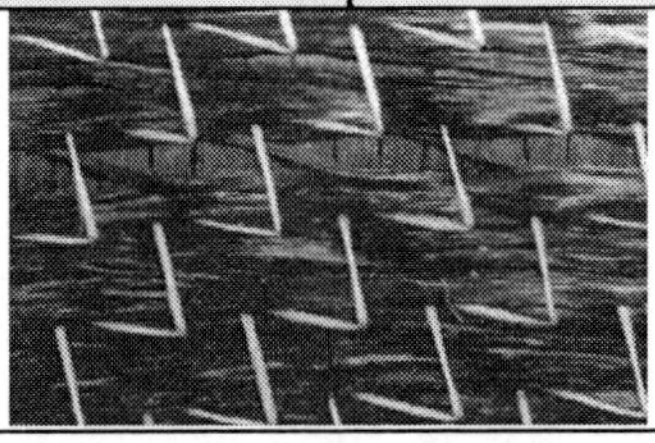	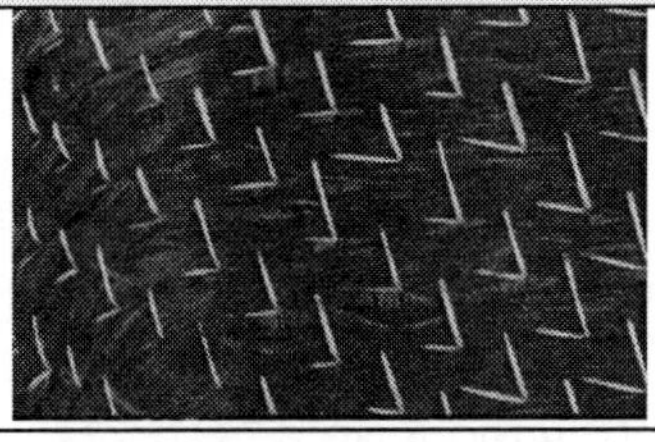

1.3.3 Existing approaches to geometry analysis

Kinematic molding analysis is an established solution for weaved textiles. The kinematic draping simulation uses mapping algorithms to predict the behavior of the two-dimensional textiles when draping onto three-dimensional geometries. Fiber deformation and orientation simulations are based on its incompressible and inextensible character— resulting in pure shear deformation for fabrics. The simplifications lead to reduced processing time, but also inaccuracies, especially with complex geometries. [12]

Moreover, the textile is represented by a network of knots and connecting lines. Connected lines are defined as constant in length. Following this approach, the deformation of the textile can be simulated solely by shifting the knots and the associated shear of the length-constant connecting lines [13]. In practice, a starting point is defined, and two constant lengths at a 90° angle define two more points. Now spheres, with a radius equal to the constant length, are drawn around the two new points. Where the two spheres intersect with each other, and the surface of the geometry, new points are generated. This process is visualized in Figure 6. Unfortunately, the shear mechanism is not the dominating draping mechanism for unidirectional textiles and therefore proves not useful for simulation. [9]

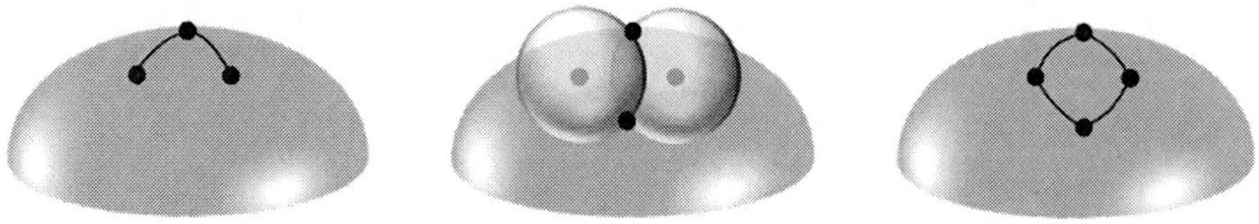

Figure 6: Concept of the kinematic analysis

A more convenient approach that suits the unidirectional fibers draping mechanisms is the geodesic analysis. A geodesic length defines the shortest connection between two points over a given surface. For the geodesic analysis, multiple lengths in fiber direction over a given geometry are being calculated and compared to each other. Different lengths are to result in a relative fiber movement, according to Härtel [5]. Figure 7 visualizes two geodesic lengths over a cut of pyramid shape. This research paper uses geodesic analysis as a basis for calculations.

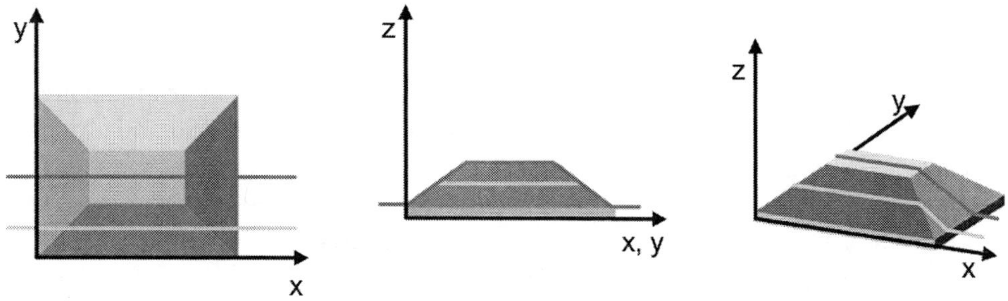

Figure 7: Two geodesic lines crossing a cut-off pyramid

2. DEVELOPMENT OF THE GEOMETRICAL ANALYSIS TOOL

There are several steps in the development of this research's tool to analyze geometric complexity. First, the actual physical process is presented, and mathematical models are applied to it for its quantification. Second, the mathematical models are translated into a numerical model that can be used to simulate the original physical process.

2.1 Physical process

The textile of Figure 8 presents one layer of unidirectional filaments. Tapes are placed at equal distances to hold the filaments in place during handling. Additionally, points are added to the filaments to enable better references for optical measurements.

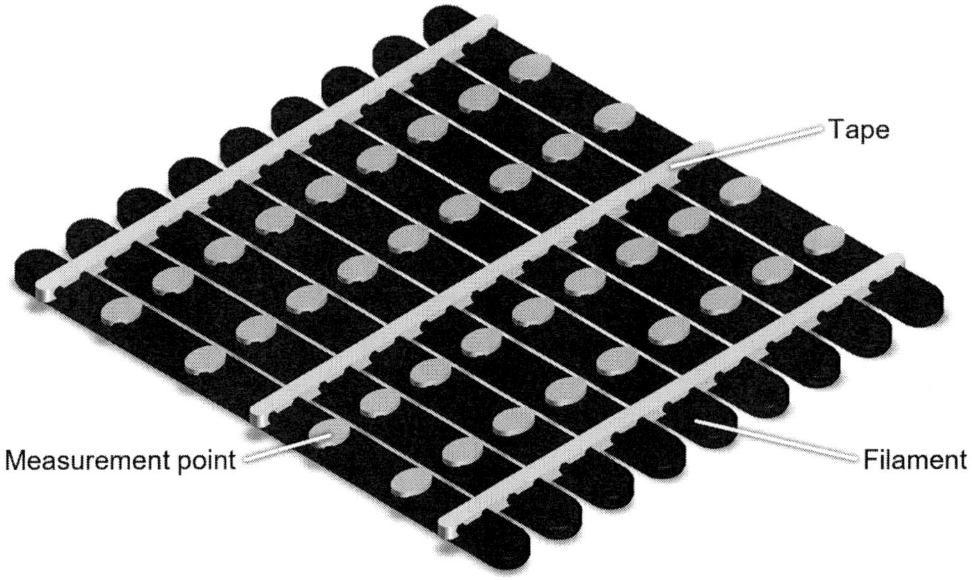

Figure 8: Schematic representation of the unidirectional textile with tapes and measurement points

Figure 9 shows the molding process, consisting of a negative and a positive molding form that represents the geometry. The textile is preheated, placed between the molding forms, and pressed into the desired geometry. The chosen geometry is a combination of many of the commonly used geometry shapes of the automotive and aerospace industry. A spherical-end and a pyramid-end are applied. Corners but also convex and concave shapes are found. This combination allows for a consistent evaluation of the tool's findings without needing a great number of experimental shapes to cover all possibilities.

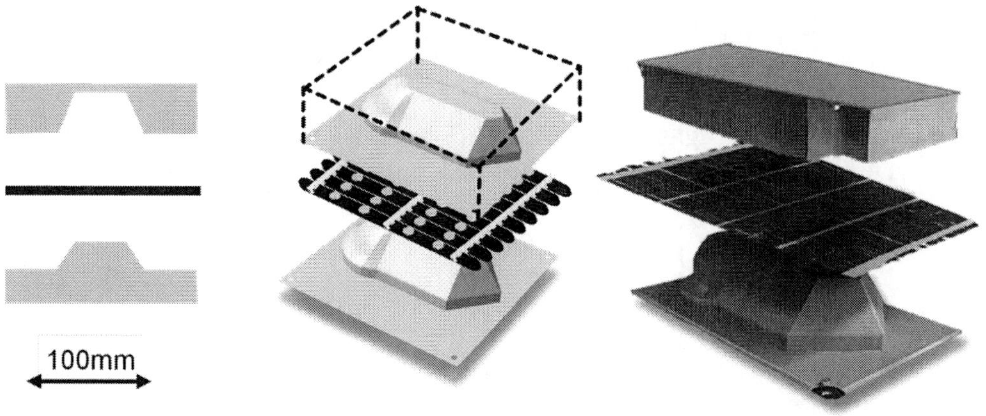

Figure 9: The preform molding process schematically (left), as a model (middle) and in reality (right)

2.2 Mathematical model

Figure 11 shows the projection of two filaments in a geometric shape. Following the geodesic approach, the lengths of the two filaments over the geometry are differing in length. The mathematical model calculates the local lengths l1,3 and l2,3 between the measurement points and compares them to the neighboring lengths.
Plus, carbon fiber filaments are not able to stretch in lengths. A mathematically occurring difference would, therefore, result in relative fiber motion in reality, as the only possible compensation for the difference in lengths is a buckling of the red filament.

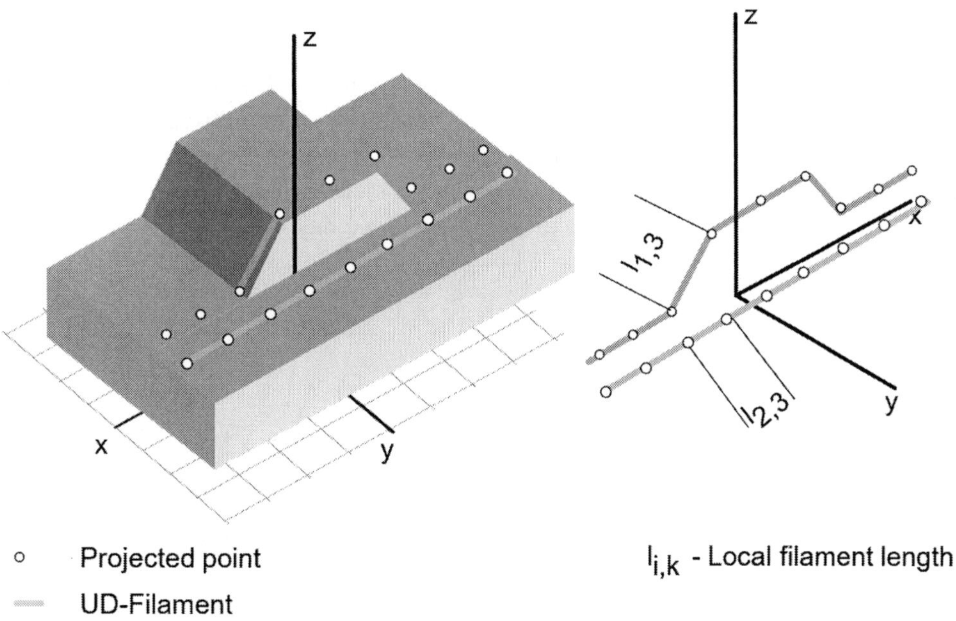

Figure 10: Projection of two unidirectional filaments with measurement points

The same approach is used in the orthogonal direction. Instead of lengths, this approach compares the widths $b_{1,3}$ and $b_{1,4}$ between two filaments. A significant increase in width signals the creation of a fiber gap.

Other local comparisons of lengths or widths, for instance, comparing $l_{1,3}$ with $l_{1,4}$ or $b_{1,3}$ with $b_{2,3}$, have also been tested. However, they were ruled insignificant to the drape mechanisms by test results.

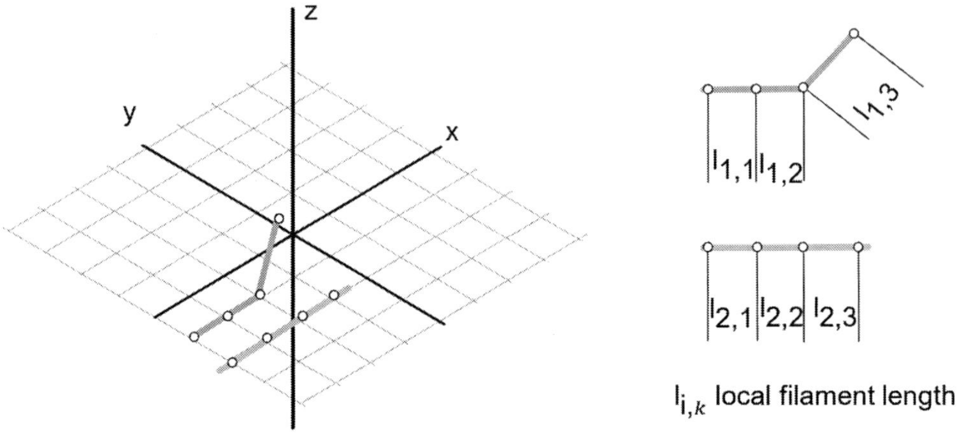

Figure 11: The approach to local width comparison

2.3 Numerical model

The mathematical approach is going to be translated into a numerical model. For this to happen, the textile's measurement points need to be projected automatically onto a geometric shape in a CAD program. This technique has been used in Virtual Basic for Applications (VBA) in Catia, Dassault Systèmes, France. A custom net of points is created above the geometry and projected on it. The point net can be adjusted in density and width between the filaments to represent the textile as good as possible. A denser net should analyze smaller geometrical shapes within the designed part, while for simpler and bigger traits, a broader point net is sufficient. Also, a denser net amplifies the processing time. Figure 13 exposes the steps to simplify the geometry into a point cloud. Here the measurement points are projected on a half-sphere.

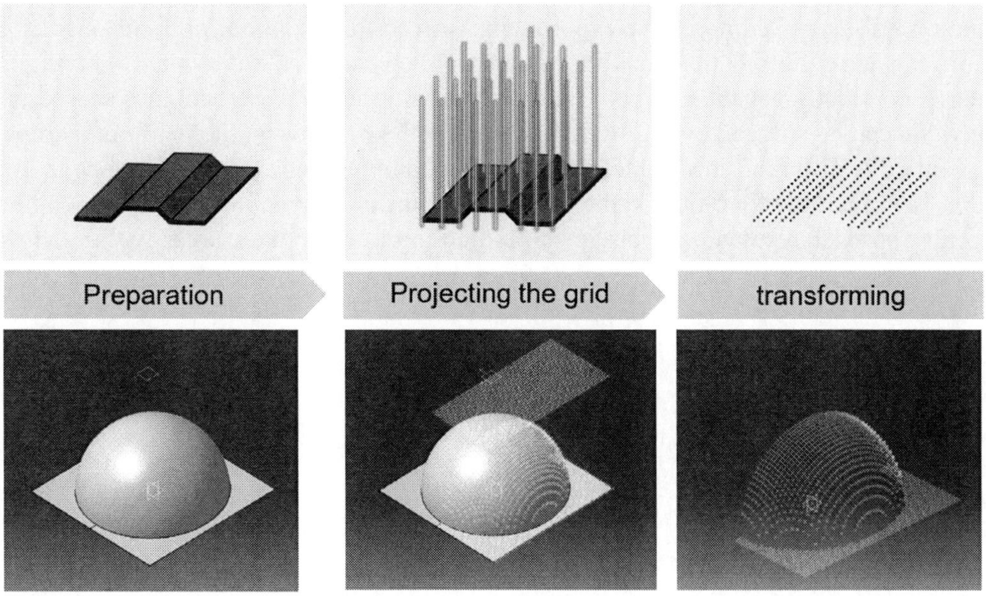

Figure 12: Generation and projection of a point cloud with the example of a half sphere

Following these steps, the length and width differences can be calculated, as established by chapter 2.2 on the mathematical model. Figure 14 presents how length and width differences can appear at any four neighboring points over the geometry. The two results of the calculation can be added up and converted into a value specific to the area limited by the four points. The higher the field value, the more difficult it is to drape that specific area of the geometry. Previous experiments have compared field values to occurring draping effects resulting in an applicable scale of different colors.

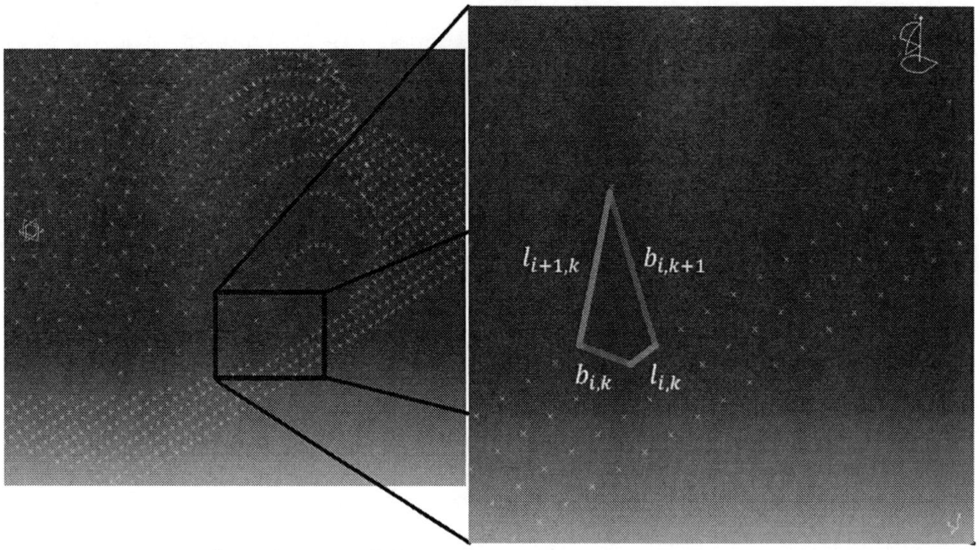

Figure 13: Length and width differences calculated with the example of a half sphere

Investigation efforts and comparative experiments proved that neighboring fields also affect each other. For instance, a field with a positive value would be canceled out by the following field with a correspondingly negative value. This statement implies that the buckling caused by the positive value can be reduced by the stretch caused by the negative value, without a drape defect. Consequently, the different field values were calculated by adding up the difference in length and width. However, the final field value is built from the average of all neighboring 8 field values. This approach resulted in a far closer prediction of problematic areas within the analyzed geometry.

3. RESULTS

Figure 14 visualized the color-differentiated, three-dimensional representation of the tool's calculations. The findings depict the geometry's influence on the unidirectional fabric and highlight problematic areas. The tool remains variable by querying parameters and can, therefore, be used on various geometric models. The results of the calculation are validated by comparison with the test results.

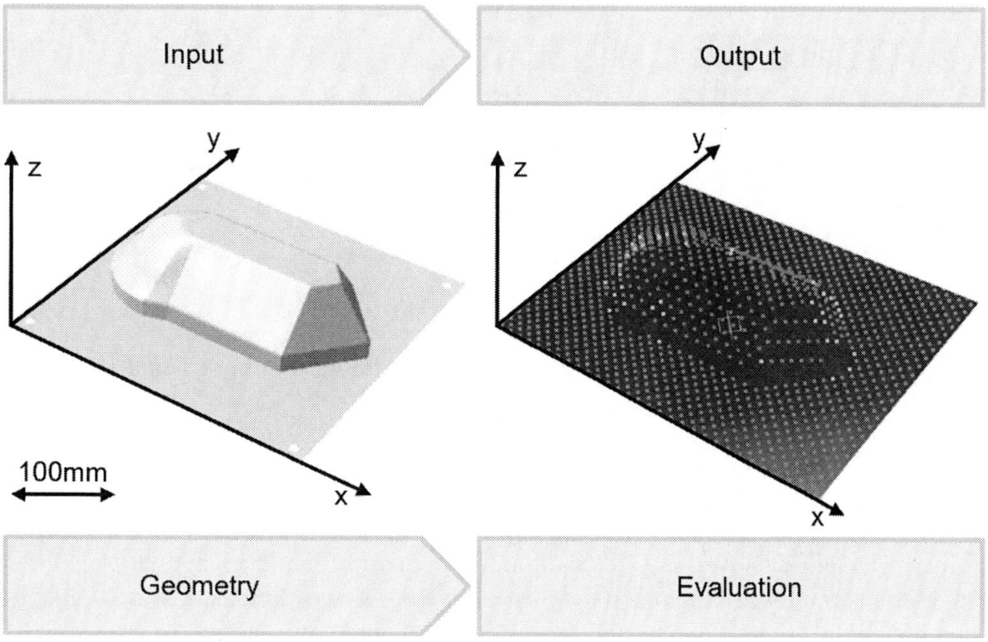

Figure 14: Input and output of the tool

Figure 15's intention is threefold. First, it shows the draping result on the left. Second, the result of the calculation on the right. Third, an overlaid view in the middle. It is crucial that the marked places do not necessarily represent the location of the draping error but instead determine the location of the error's origin. Depending on the starting point of the draping, the error can spread from the place of origin in different directions. At a greater distance from the source of the error, the fulfillment of the quality requirements can no longer be guaranteed.

Figure 15 shows the draping and calculation results at 0 ° fiber orientation. It can be seen both in the actual experiment and in the calculation that the critical zones lie in the lower area of the pyramid-end and the spherical-end.

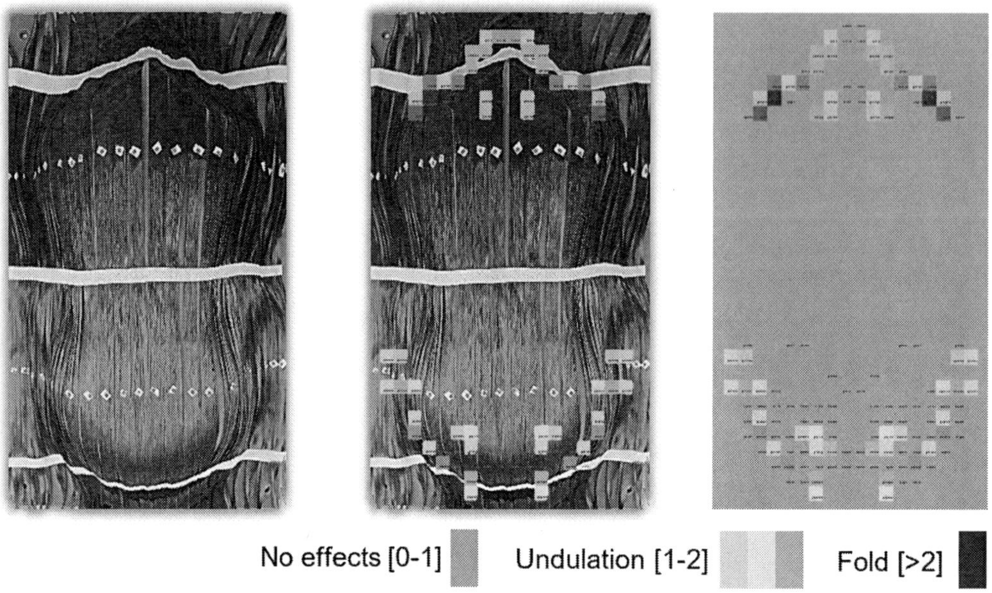

Figure 15: Draping (left) and calculated (right) result under 0° fiber orientation with an overlaid view (middle)

4. CONCLUSIONS

Preforms must follow the designed geometry as precisely as possible. Deviations from the desired fiber orientation as well as folds and undulations represent critical quality deficits. The requirement to form a three-dimensional component from a two-dimensional textile is a significant challenge [11]. The expected local expansion and compression of the surface disturb the vital orientation of the direction-dependent material. These disturbances must be minimized and controlled in a targeted manner [1]. For this purpose, a geometry analysis tool was developed. The geometry analysis tool examines the geometry of areas that are difficult to drape and makes them visually recognizable.

Adverse draping effects, such as undulations and folds, are attributed to deformation mechanisms. The state-of-the-art shows that deformation mechanisms of the unidirectional semi-finished products are based only to a limited extent on shear mechanisms. Therefore, existing analysis tools for unidirectional semi-finished products are not realistic [5]. The tool developed in this work bases its calculations on the deformation mechanisms of fiber relative movement and filament widening typical for unidirectional materials. For this, the causes of the mechanisms in the geometries to be draped are highlighted. The finite difference method reduces the sometimes highly complex component geometries to simple local differences in length and width. The tool's approach is based on functions that are integrated into Catia, Excel, and Windows. The geometry to be examined is discretized in points in Catia, and local length and width differences are calculated in Excel. Here, the individual filament bundles within the unidirectional fabric and their course are compared with one another via the component geometry. Scripts that can be run from Windows establish the connection between Excel and Catia and allow the geometry evaluation to be made variable according to the requirements of the user.

The tool is used to create a three-dimensional, color-coded evaluation from a geometric model of the preform in Catia. Red and yellow areas represent locally intricate geometrical lines, while

green areas can be draped without defects. It should be noted that this is not a draping simulation. The points are created in the original fiber direction but do not insist on the validity after draping to reflect the position of the filaments. It is only a question of being able to calculate and evaluate the local displacements that are caused by the filaments. The results of the calculation are exported as a two-dimensional top view in Excel, but also as a three-dimensional model in Catia.
To validate aforementioned calculations, tests were carried out with a unidirectional textile and compared with the evaluation of the analysis tool. For this, the preform process, as shown in Table 2, is run for a set of geometries and fiber orientations. Both the analysis tool and real experiments are run simultaneously, and the results are compared.

Table 2: Calculation of the tool compared to the original process

	Textile	Molding	Preform
schematically	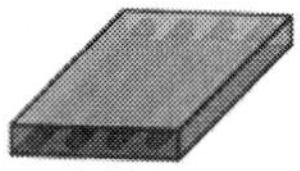		
Analysis tool			
Real process			

In this work, it is shown that the early evaluation of a designed component geometry regarding its drapability with unidirectional fabrics is possible with the geometry analysis tool. Areas that endanger quality can be identified via the visual output.

5. REFERENCES

[1] N.N.: Handbuch Faserverbundkunststoffe/Composites, Grundlagen Verarbeitung Anwendung, 3. Aufl. Wiesbaden: Vieweg+Teubner/GWV Fachverlage, 2010

[2] R. Lässig, M. Eisenhut, A. Mathias, R. Schulte, F. Peters, T. Kühmann, W. Waldemann, W. Begemann, Serienproduktion von hochfesten Faserverbundbauteilen Studie, Roland Berger Strategy Consultants, 2012

[3] FOREL: Projektstart für automatisierte Herstellung von komplexen FVK-Bauteilen für Elektrofahrzeuge, Lightweight-design, 2015

[4] F. Henning, B. Thoma, L. Kärger, T. Fürst, F. Schrimaier: Cost-efficient preforming as leading process step to achieve a holistic and profitable RTM product, development, Stuttgart, Germany 2015

[5] F. Härtel: Entwicklung eines Vorentwurfansatzes für die automatisierte Drapierung von Faserhalbzeugen, Universität Stuttgart, Dissertation, 2015

[6] W. Siebenpfeiffer: Leichtbau-Technologien im Automobilbau, ATZ/MTZ-Fachbuch in Technologieplanung zur automatisierten Fertigung von Preforms für CFK-Halbzeuge, Wiesbaden, Springer Fachmedien, 2014

[7] W. Weißbach: Werkstoffkunde: Strukturen, Eigenschaften, Prüfung, Viewegs Fachbücher der Technik, 17. Auflage, Springer-Verlag, Berlin 2010

[8] C. Cherif, O. Diestel, T. Engler, E. Hufnagl, S. Weiland: Weiterverarbeitungsaspekte und Anwendungsbeispiele, Textile Werkstoffe für den Leichtbau, Springer-Verlag Berlin, Heidelberg, 2011

[9] N. Schur: Drapieren und Handhaben von unidirektionalen Faserhablzeugen in einer automatisierten Prozesskette zur Herstellung komplexer Prefroms, Stuttgart Universität, Dissertation, 2016

[11] J. Klingele: Produktorientierte Auswahl von Verfahren zur Vorfixierung textiler Preforms, RWTH Aachen, Dissertation, 2014

[12] S. Hancock, K. Potter: The use of kinematic drape modelling to inform the hand lay-up of complex composite components using woven reinforcements, Composites Part A: Applied Science and Manufacturing, 2006

[13] F. Van der Weeen: Algorithms for draping fabrics on doubly-curved surfaces, International Journal for numerical methods in engineering, 1991

VALIDATION OF THE ITRAC MATERIAL RESPONSE CODE FOR GLASS/PHENOLIC COMPOSITE

Colin Yee[1], Jon Langston[1], Hao Wu[2], William Fahy[2], and Joseph H. Koo[1,2*]

[1]The University of Texas at Austin, Walker Dept. of Mechanical Engineering, Austin, TX 78712

[2]KAI, LLC, Austin, TX 78739

*jkoo@mail.utexas.edu

ABSTRACT

A simulation of the glass/phenolic ablative MXB-360, produced by Sioux Manufacturing, is validated in an oxidative hyperthermal environment using the Northrop Grumman Insulation Thermal Response and Ablation Code (ITRAC). Temperature profile and surface behavior data are generated by testing MXB-360 samples at heat fluxes of 500, 1,000, and 1,500 W/cm^2 using an Oxyacetylene Test Bed (OTB). K-type thermocouples embedded at staggered depths within the test models (30 mm diameter and 15 mm thick) provide in-depth temperature profiles. These data are used to demonstrate the accuracy of the 1D material response model using ITRAC. Complete Kinetic Modeling Parameter (KMP) factors for MXB-360 are developed through the use of Thermogravimetric Analysis (TGA), while both pyrolysis gas enthalpy tables and B-prime tables are developed through the use of Electron Dispersion X-ray (EDX) mass spectrometry and the ACE surface thermochemistry code, respectively.

1. INTRODUCTION

1.1 Background

Rocket and missile technology have become increasingly available throughout the world due to the rapid commercialization of space and Thermal Protection Systems (TPS) are necessary to protect launch structures from the extreme hyperthermal environment generated by rocket exhaust plumes. While TPS are typically thought of as a means of protecting an airframe during hypersonic flight or atmospheric reentry [1], ensuring the reusability of fueling and launch structures is important in reducing the cost of space flight. Full scale testing is traditionally utilized to test TPS as they are the most reliable way to subject material samples to the combination of heat, vibration, and thermal shock associated with launch conditions [2, 3]. However, full scale testing for incremental design improvements is cost prohibitive, and ablation and heat transfer models must be reliable enough to reduce design cycle times [4].

Acceptably accurate ablation models must be developed and validated through aerothermal ablation testing and material characterization. This testing typically involves the use of low cost, small scale systems utilizing heated plasma [5, 6] or oxyacetylene flames [7, 8] to generate a high-temperature oxidative environment. Temperature data obtained from ablation testing can then be used to validate Kinetic Modeling Parameters (KMPs) and material composition changes can be utilized to generate net gaseous outflow (*B'*) tables. KMPs are determined via Thermogravimetric Analysis (TGA) in an oxidative environment and then validated against experimental data. Depending on the complexity of the material, KMPs may be determined for each reaction step that occurs during decomposition, which is also dependent on the availability of reactants in the environment. Accurate KMPs and heats of formation are critical in modeling the ablation of TPS materials as they are the foundation of an accurate decay state function and determine the rate of energy consumption while the material is decomposing [9].

Figure 1 shows the dominant effects associated with the ablation of TPS materials. Heat input to the material comes primarily from the exhaust jet of the rocket nozzle and the combustion of fuel. The TPS material pyrolyzes as it is heated, releasing decomposition gases which mostly diffuse back through the lower density material near the ablation front. The gases may then flow through the char layer, providing further heat input to the material as they either ignite or radiate. The diffusion of pyrolysis gases additionally results in internal stresses throughout the TPS material.

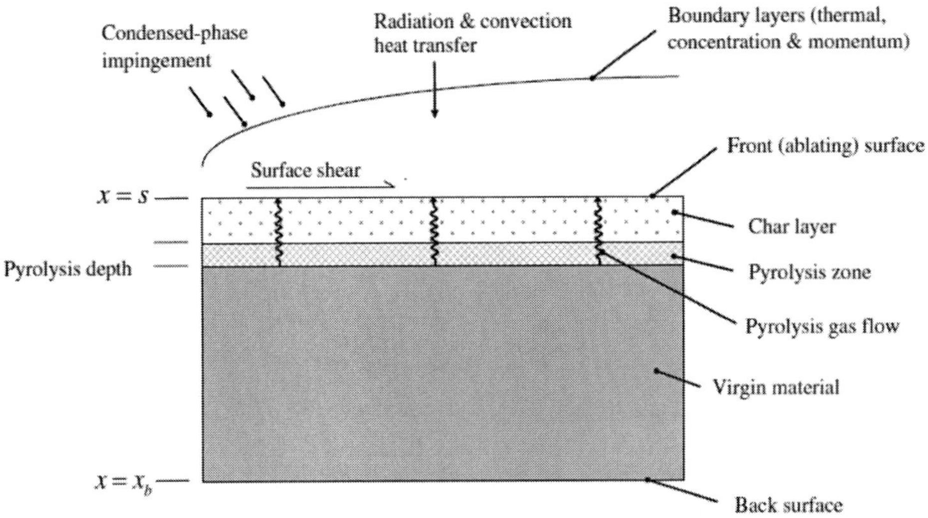

Figure 1. Ablation phenomena [10].

As the material continues to heat, the fully pyrolyzed char is removed by shear forces and the pyrolysis zone penetrates further into the virgin material. The virgin material and the back surface maintain a lower temperature than the ablating surface due to heat being removed by the heat of transformation, mass loss through the removal of heated char, and mass loss through the diffusion of pyrolysis gases. Structural components are maintained at a safe temperature while the ablating front is subjected to the extreme hyperthermal environment. Predicting the expected mass loss of the shield, internal temperature gradient, and depth of ablation are critical for design engineers in order to ensure adequately protection.

MXB-360 is a low-cost glass/phenolic ablative composite manufactured by Sioux Manufacturing Corporation and protects structural components from solid rocket motor exhaust. It is in use by the US Navy on the bottom plenum of the MK41 Vertical Launch System (VLS) on surface ships in order to prevent damage to the ship's frame [11]. Its low cost and availability make it ideal for protecting commercial launch structures, but it does not have a well validated ablation model in open literature due to difficulties in simulating the dynamics of the glass mat reinforcement in an extreme hyperthermal environment. The Insulation Thermal Response and Ablation Code (ITRAC) developed by Northrop Grumman Innovation Systems was utilized to produce a material response model for MXB-360 in high heat flux oxidative environments.

2. EXPERIMENTATION

2.1 Thermogravimetric Analysis (TGA)

Thermogravimetric Analysis (TGA) utilizes a scale within a furnace to measure the mass of a material sample while it degrades at high temperature. It is more expeditious than multiple material degradation tests at constant temperature and compensates for degradation that occurs during the heat up to a constant temperature test [12].

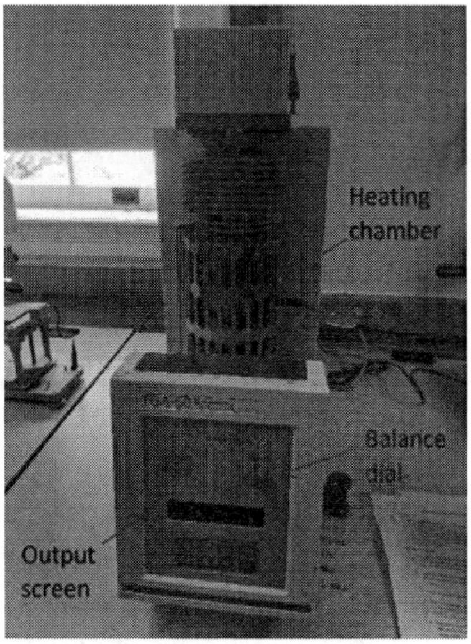

Figure 2. Shimadzu Scientific TGA-50 [13].

The low temperature TGA data was generated using a TGA-50 from Shimadzu Scientific Instruments pictured in Figure 2, while high temperature TGA data was referenced from the manufacturer material data sheet. 10 mg of pulverized MXB-360 samples were heated in a dry air environment from room temperature to 100°C and then maintained at that temperature for 30 minutes in order to remove any moisture. At the end of the drying period, the TGA-50 directly heated the sample at a constant heating rate up to 1,000°C. The purge rate of the air was maintained at 50 ml/min in order to adequately remove pyrolysis gases while minimizing risk of disturbing the material sample in the furnace. TGA data for the carbon-silica reactions from 1,000 to 1,600°C were referenced from the MXB-360 data sheet.

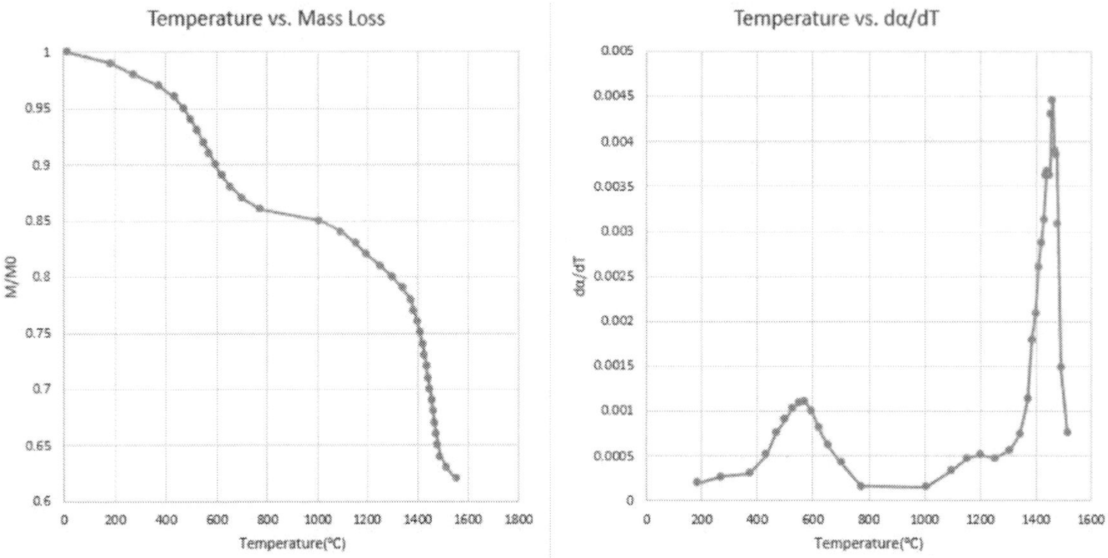

Figure 3. TGA at a heating rate of 20°C/min in air for the glass/phenolic composite MXB-360. (Left) Mass fraction versus temperature. (Right) dTGA versus temperature. α is the extent of reaction used to normalize the mass loss.

Heating rates of 10, 20, and 50°C/min were utilized for all tests in order to provide adequate curve separation during Kinetic Modeling Parameter (KMP) development. Three separate reactions are present based on the TGA and dTGA pictured in Figure 3, the first two are due to the decomposition of the phenol formaldehyde resin, while the last is the decomposition of the char due to interactions with the silica glass.

2.2 Oxyacetylene Test Bed (OTB)

An Oxyacetylene Test Bed (OTB) utilizes an oxygen-acetylene flame to simulate the hyperthermal oxidizing conditions of solid rocket motor exhaust [14]. The University of Texas OTB shown in Figure 4 uses a moveable lathe chuck to fix the material test model. The distance from the test model to the torch is used to control heat flux. Heat flux is calibrated using a Vatell Corporation TG 1000-1 water cooled Gardon gage connected to a National Instruments Data Acquisition Unit (NI DAQ).

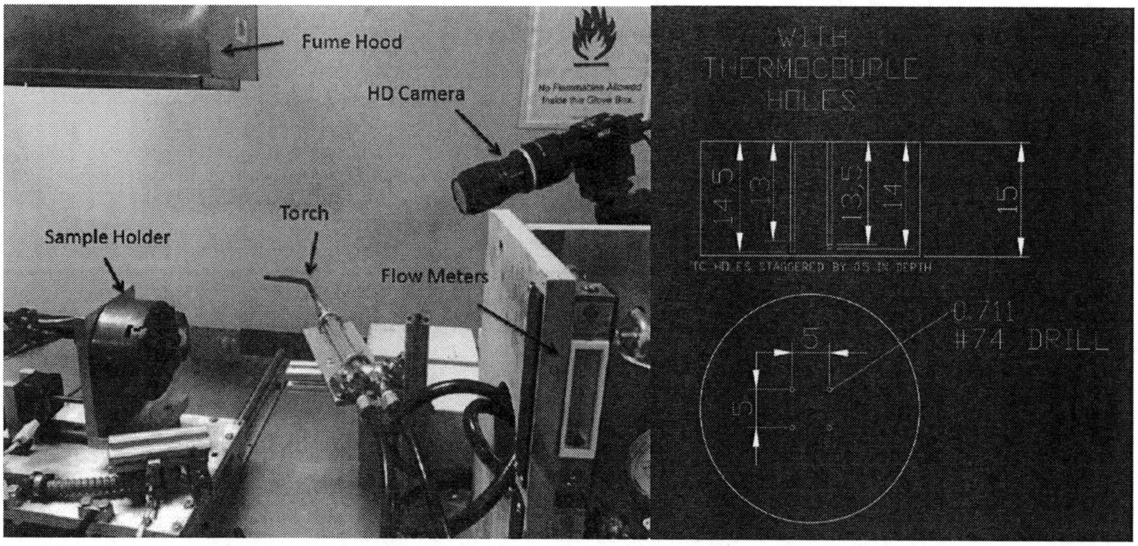

Figure 4. (Left) The University of Texas OTB [15]. (Right) OTB test model standard with all dimensions in mm and four embedded 0.5 mm diameter K-type thermocouples. The cylinder is 15 mm thick, 30 mm in diameter, and cut by water jet.

An HD camera records real time morphological and topological changes, while a two-color IR pyrometer and an IR camera provide noncontact surface temperature measurement. Four K-type 0.5 mm thermocouples produced by Omega Engineering are fixed in the test model using high-temperature graphite adhesive at the back face. The thermocouples are connected to a NI DAQ along with a timing switch. The timing switch it positioned at the back of the sample holder rail assembly so that the exposure time of the test model is measured precisely.

Figure 5. (Left) MXB-360 test model subjected to a 1,000 W/cm^2 heat flux. (Right) The virgin MXB-360 test model.

The in-situ ablation sensing data used for validation was generated using heat fluxes of 500, 1,000, and 1,500 W/cm^2 for 30 seconds as shown in Figure 5. As the ablation front approaches the thermocouples, they extend through the char layer and are destroyed, permitting real time tracking of ablation rate.

2.3 Energy Dispersive X-Ray (EDX) Spectroscopy

EDX utilizes the electron beam from a scanning electron microscope (SEM) to excite the electrons of the atoms of the observed material. As the electrons move back into a lower energy state shell, they emit X-rays, which are then counted by a sensor and sorted by energy. Different elements are then correlated to their signature X-ray energy and counts per electron volt, providing a quantitative elemental composition of the material.

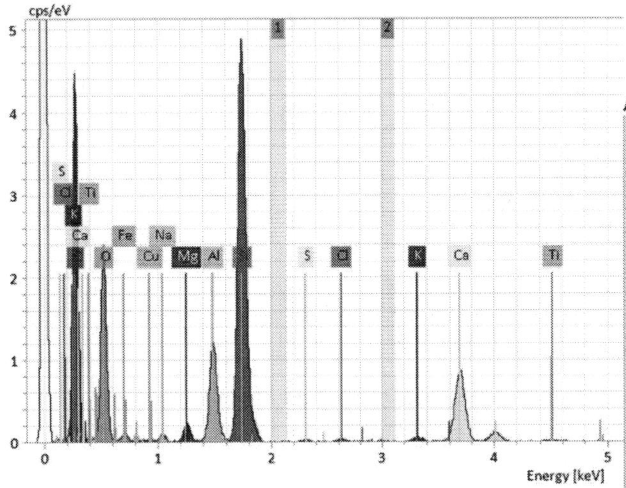

Acquisition

Element	At. No.	Mass [%]	Mass Norm. [%]	Atom [%]	abs. error [%] (3 sigma)	rel. error [%] (3 sigma)
C	6	64.48	48.87	61.01	22.43	34.78
O	8	41.24	31.26	29.29	14.86	36.03
Na	11	0.57	0.43	0.28	0.19	34.20
Mg	12	0.74	0.56	0.35	0.20	27.28
Al	13	3.47	2.63	1.46	0.57	16.28
Si	14	14.14	10.72	5.72	1.84	13.03
S	16	0.07	0.05	0.03	0.09	121.25
Cl	17	0.11	0.09	0.04	0.09	81.65
K	19	0.20	0.15	0.06	0.10	50.63
Ca	20	4.53	3.43	1.28	0.49	10.78
Ti	22	0.07	0.05	0.02	0.09	131.30
Fe	26	2.17	1.64	0.44	0.29	13.26
Cu	29	0.14	0.10	0.02	0.11	80.24
		131.92	100.00	100.00		

Figure 6. EDX spectroscopy of virgin MXB-360 on a Hitachi S5500 STEM.

The EDX data used for surface thermochemistry analysis was taken from a Hitachi S5500 STEM. Several samples of virgin MXB-360 were examined to establish an elemental composition baseline and identify contaminants present from the production process as shown in Figure 6. Because the silica in MXB-360 separates from the phenolic during ablation, EDX spectroscopy was performed on the glass char and the carbonaceous char separately using area scan mode. Only the carbonaceous char was compared against the virgin material for the purposes of determining the pyrolysis gas composition because the silica-oxide glass either evaporates or advects vice reacting with decomposition products to produce gas components.

3. MODELING

3.1 Kinetic Modeling Parameter (KMP) Development

KMPs refer to the reaction model function curve fit parameter (m), the Arrhenius pre-exponential term, and the activation energy of a particular material in a specific environment, typically either inert or oxidizing. KMP analysis starts with TGA data and the extent of reaction. The extent of reaction α can be described as a ratio of the density difference of the material between the virgin and char states compared to the current density:

$$\alpha = \frac{\rho_v - \rho_s}{\rho_v - \rho_c} \tag{1}$$

With ρ_v being the virgin density, ρ_s being the current density, and ρ_c being the char density. At a virgin state, ρ_s is equal to ρ_v and α is equal to zero; at a fully charred state, α is one. The changes in all material properties from the virgin to the charred material state are approximated by α, which serves as a linear interpolation term.

While heat and mass conservation equations can be utilized for fully virgin or char state material ablation, the density of a given section of material and its state of thermal decay are determined by the decay form of the Arrhenius reaction equation. The Arrhenius decay equation for change in extent of reaction with respect to temperature is modeled as follows:

$$\frac{d\alpha}{dT} = Ae^{-\frac{E_A}{RT}}f(\alpha) \tag{2}$$

Where α is the extent of reaction between virgin and char states, A is the Arrhenius pre-exponential, E_A is the activation energy of the material, R is the universal gas constant, and f(α) is the reaction model function [9].

Mass versus temperature data from constant heating rate TGA testing in air was normalized to provide both percent mass loss and extent of reaction versus time. By rearranging the Arrhenius decay equation and the reaction model function, the KMPs for each decay reaction were determined. The Kissinger form of the decay function f(α) is:

$$f(\alpha) = (1 - \alpha)^m \tag{3}$$

The Kissinger single reaction order parameter form of the decay function was used for compatibility with ITRAC. Utilizing the Kissinger equation will result in singular activation energy (E_A) and reaction order (m) values while the pre-exponential (A) will change as a function of heating rate. While the Kissinger decomposition equation may not provide adequate simulation data above α values of 0.7, it is simple optimize the KMPs and is less computationally intensive than other methods [16].

As the pre-exponential changes with heating rate for the Kissinger method, equation (2) is altered in order to arrive at another form of the Arrhenius equation:

$$\frac{d\alpha}{dT} = \frac{A}{\beta}e^{-\frac{E_A}{RT}}f(\alpha) \tag{4}$$

With β representing the heating rate dT/dt. Equation (4) is further rearranged such that:

$$\ln\left[\beta\frac{\frac{d\alpha}{dt}}{f(\alpha)}\right] = -\frac{E_A}{RT} + \ln A \tag{5}$$

E_A and A are extracted from a linear curve fit of the left side of equation (5) against 1/T based on TGA data. While the 10, 20, and 50°C/min heating rates are not representative of the approximately 200°C/sec peak heating rate experienced by the MXB-360 test models during OTB testing, the use of three or more TGA data sets from differing heating rates can be used to predict realistic KMPs for all heating rates as demonstrated by Sihn et al. [16] The temperature versus normalized mass loss graph in an oxidizing environment is shown below in Figure 7.

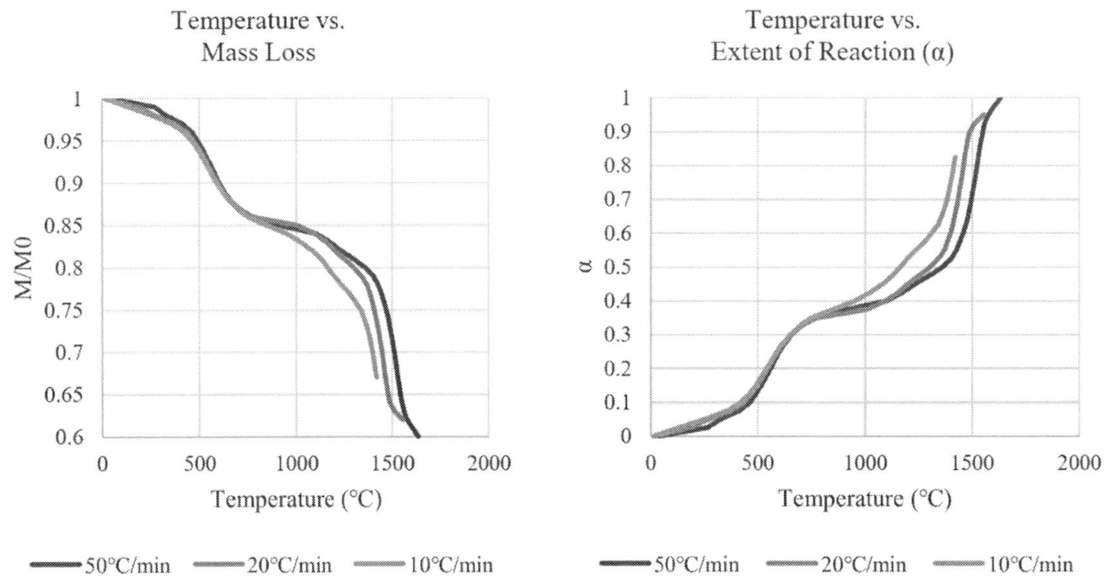

Figure 7. (Left) TGA temperature versus normalized mass loss. (Right) TGA temperature versus extent of reaction.

By normalizing the data based on the initial and final mass fractions using equation (6), the temperature versus overall extent of reaction (α) graph from Figure 7 is developed.

$$\alpha = \frac{M_0 - M}{M_0 - M_f} \tag{6}$$

Where M_0 is the initial mass fraction, M_f is the final mass fraction, and M is the mass fraction as a function of temperature. Equation (6) is also used to describe the extent of reaction for individual decomposition reactions ongoing in the material based roughly on the mass fractions at which they initiate and terminate.

Three reactions take place based on Figure 7, the first of which peaks at about 550°C, the second at 1,200°C, and the third at 1,450°C. As the heating rate increases, the second and third decomposition reactions begin to occur simultaneously. Figure 8 shows the combined and separated dα/dT data sets.

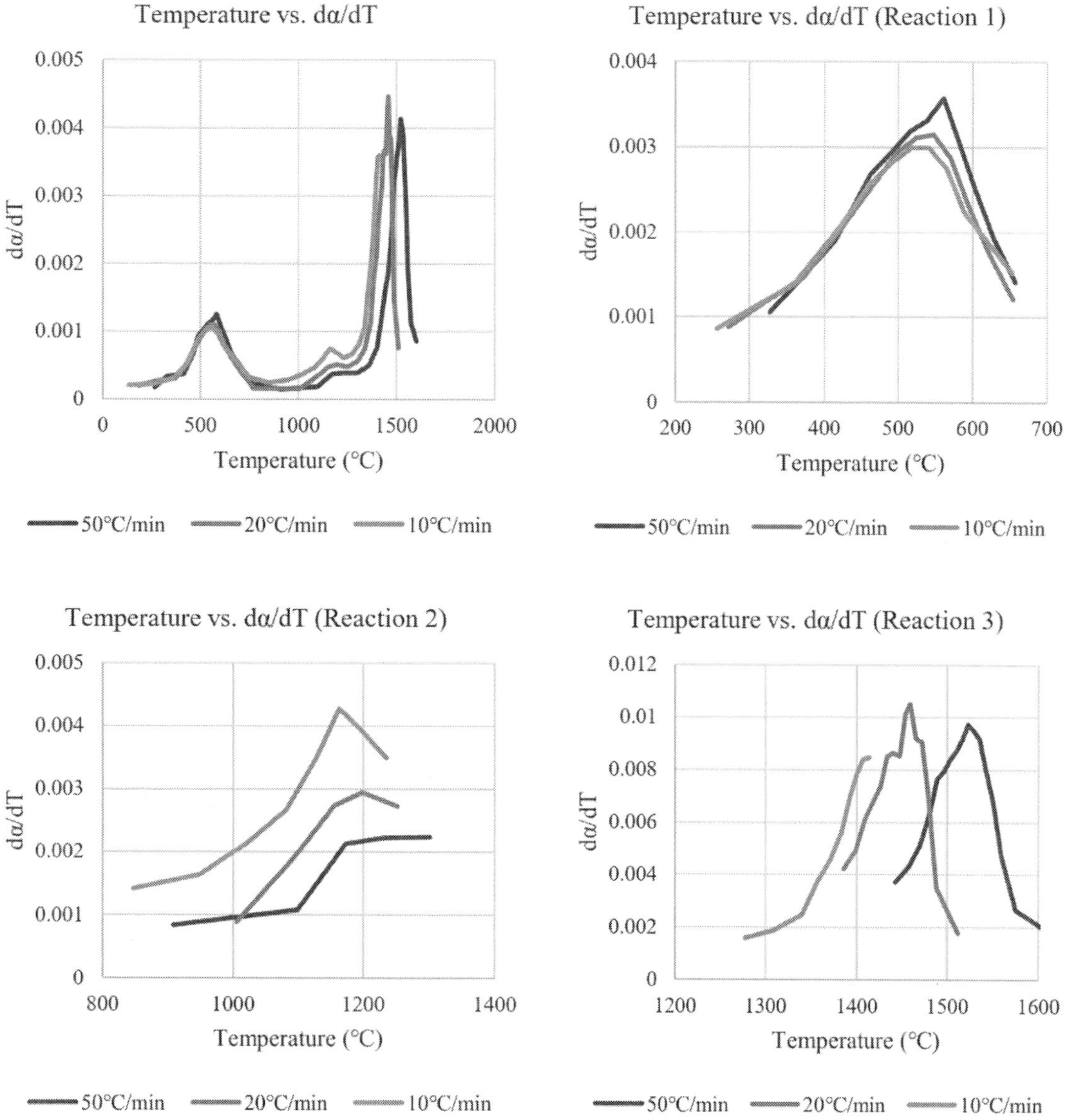

Figure 8. The combined temperature versus dα/dT curve for the MXB-360 TGA data and the individual temperature versus dα/dT curves for reactions one, two, and three.

Using numerical methods, f(α) was chosen to linearize the curves as much as possible as shown in Figure 9.

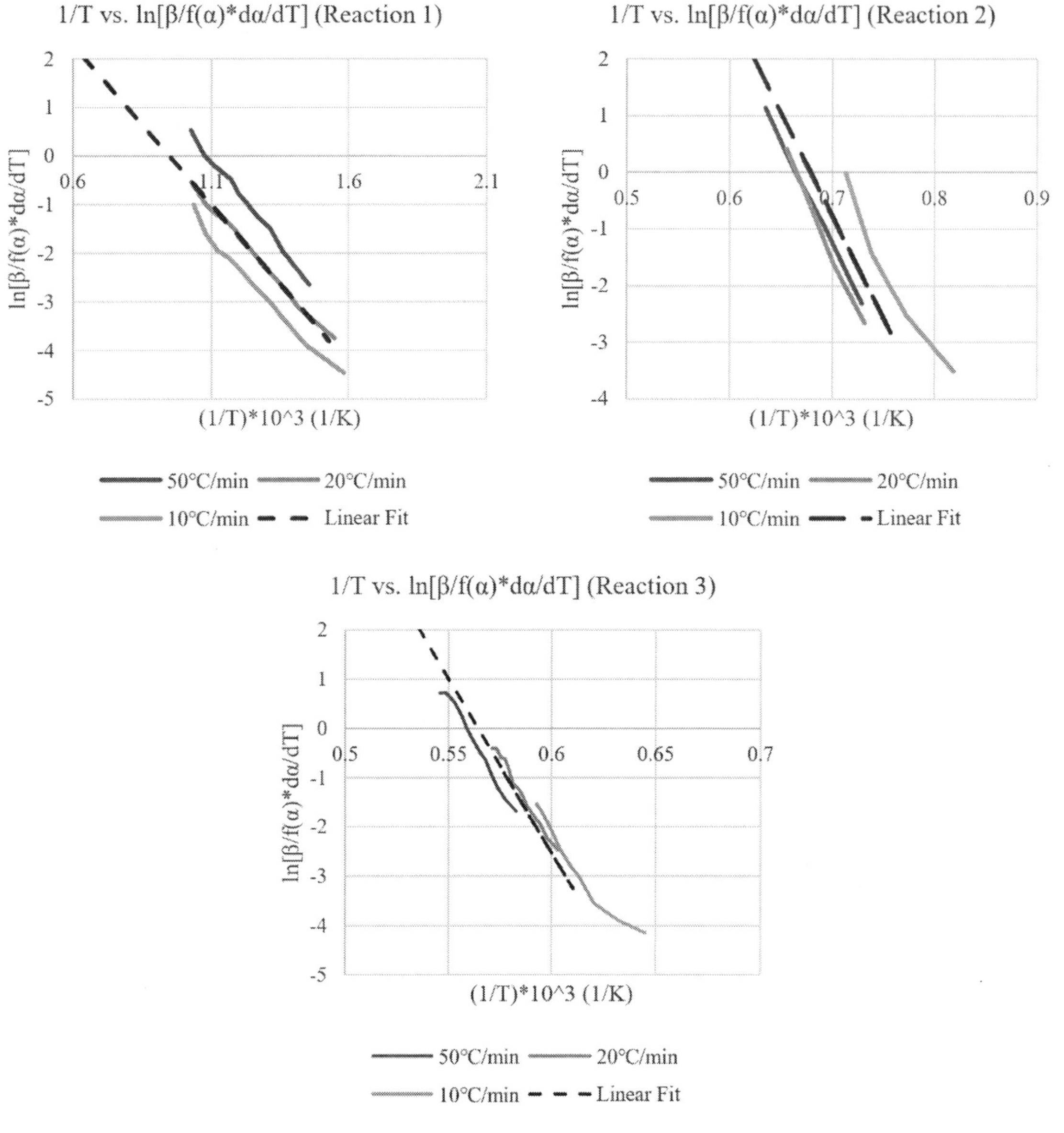

Figure 9. 1/T versus ln[β/f(α)*dα/dT] for reactions one, two, and three. A linear curve fit is performed on the average of the three curves for each reaction, generating the red dotted line, whose intercept is logarithmically proportional to the pre-exponential term and whose slope is proportional to the activation energy.

A least squares linear curve is separately fit to the data for each of the reactions. These fits are averaged together to generate the activation energy and pre-exponential term. In order to linearize the data as much as possible, the reaction order term *m* from equation (3) is varied in order to minimize the sum of the squares of Y-axis error from points on the fitted curves to points derived from data. The final KMPs for the three degradation reactions are presented in Table 1.

Table 1. KMPs for the three decomposition reaction chains based on the TGA data in an oxidative environment and the Kissinger method

Reaction	**1**	**2**	**3**
m (unitless)	1.205	3.927	1.244
E_A (kJ/mol)	54.6	301.9	587.2
A (1/sec)	1.009×10^{1}	1.005×10^{9}	4.034×10^{15}
Mass Fraction (M/M0)	0.17	0.10	0.23

3.2 Surface Thermochemistry Analysis

In order to adequately model the heat and mass transfer conditions at the boundary, the composition of the pyrolysis gas resulting from decomposition of the material must be well modeled. Net gaseous outflow (B') must be tabulated as a function of pyrolysis gas (B_g') and char consumption (B_c') for a given material through thermochemical *B'* tables and T-chem tables. These tables were generated using the Aerothermal Chemical Equilibrium (ACE) code and composition changes from EDX analysis of the virgin and charred MXB-360 [10].

The elemental composition differences between the virgin and charred state of the material are taken and the compared to mass fraction data from TGA. Because EDX cannot detect lighter elements such as hydrogen, the assumption was made that all hydrogen was contained in phenol formaldehyde and all hydrogen was liberated during decomposition. Enthalpy tables and *B*' tables for the pyrolysis gases were then utilized along with the elemental composition to determine thermodynamic properties and the equilibrium conditions at the surface at each step in the ITRAC model [17].

4. RESULTS

The MXB-360 ITRAC model utilized the standard temperature, standard pyrolysis, and thermomechanical erosion solution options. The simplified advection with no pore pressure option was chosen because the wicking and advection of the glass mat results in a stiff, porous carbonaceous char layer which is not prone to failing due to excessive pore pressure. The in-depth thermocouple temperatures are compared against the ablation code output in Figure 10.

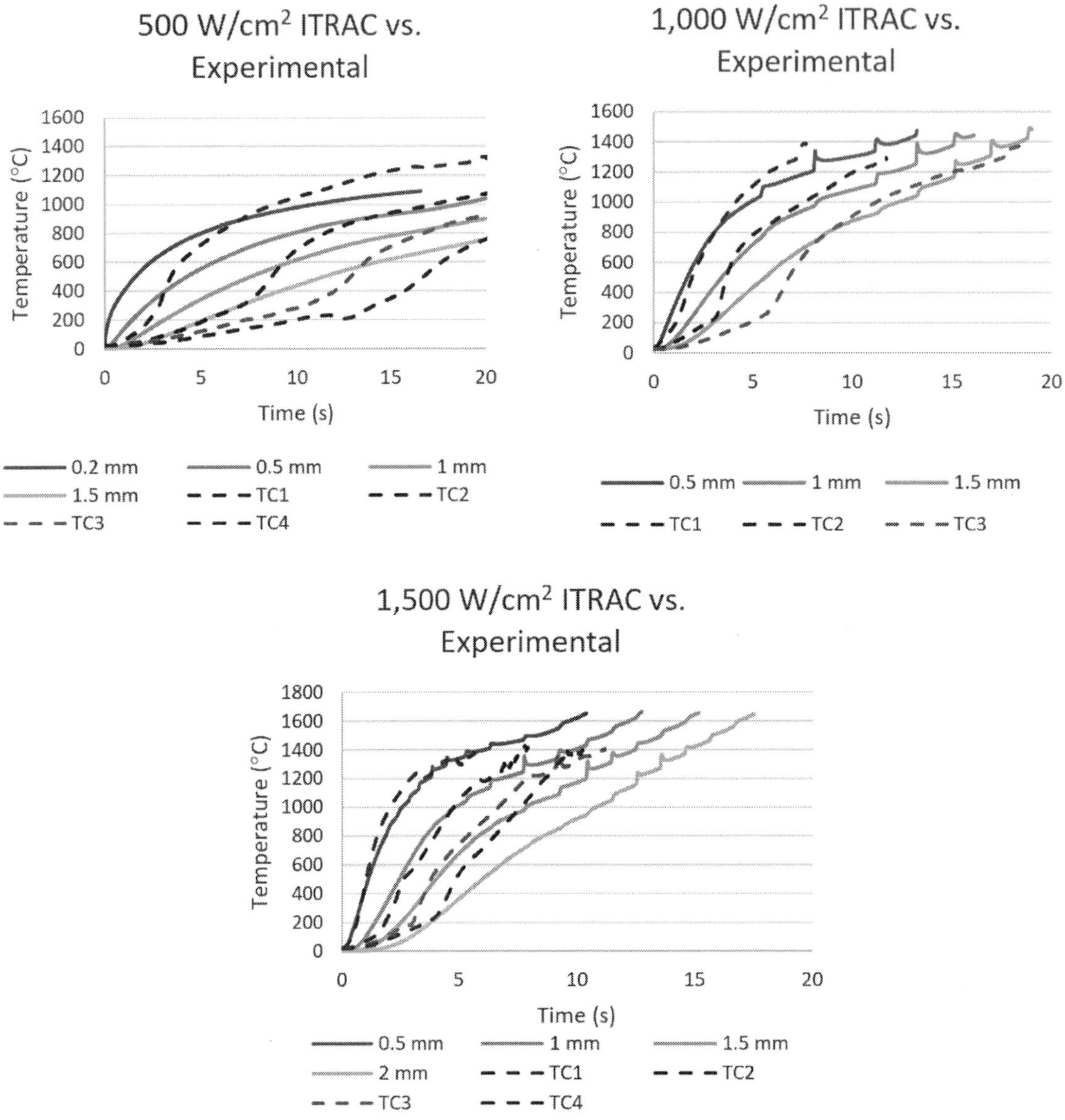

Figure 10. Simulated versus experimental in-depth temperature profiles over time for the first 20 seconds of the 500, 1,000, and 1,500 W/cm^2 OTB tests. Simulated temperature profiles are measure from the ablating edge. Experimental results are highlighted in red. TC4 for the 1,000 W/cm^2 test was omitted due to loss of connection.

The simulated temperature profiles tracked closely with the experimental profiles and exhibited similar slopes. Significant error on thermocouple channel 1 (TC1) on the 500 W/cm^2 test was due to the low heat flux and temperatures below 1,400°C permitting the thermocouple to survive outside of the char layer, as well as inaccuracy in machining placing the thermocouple extremely close to the virgin surface. The similarity of TC3 and TC4 on the 1,500 W/cm^2 test was due to the thermocouple hole depth spacing being 0.25 mm vice the nominal 0.5 mm due to errors in machining. Experimental and simulated ablation characteristics are shown in Table 2. Simulated mass loss and linear ablation rate were greater than the experimental values.

Table 2. Comparison of experimental and simulated ablation metrics

Heat Flux (W/cm²)	Experimental Ablation Rate (by TC method) (mm/s)	Simulated Ablation Rate (mm/s)	Experimental Mass Loss	Simulated Mass Loss
500	0.040	0.118	6.2%	10.6%
1,000	0.079	0.175	13.7%	22.9%
1,500	0.205	0.213	14.7%	31.1%

The low experimental initial heating rate experienced by the thermocouples is most likely due to the rapid wicking and advection of the glass mat reinforcement from the composite. The large thermal conductivity mismatch between the reinforcement and the matrix results in a large amount of heat being channeled to the glass mat, which melts and is drawn out of the matrix by pressure effects as pictured in Figure 11. The high density and thermal capacity of the silica glass mat compared to the phenolic results in a significant amount of heat removal ahead of the nominal pyrolysis zone, resulting in a temporarily depressed temperature in the matrix.

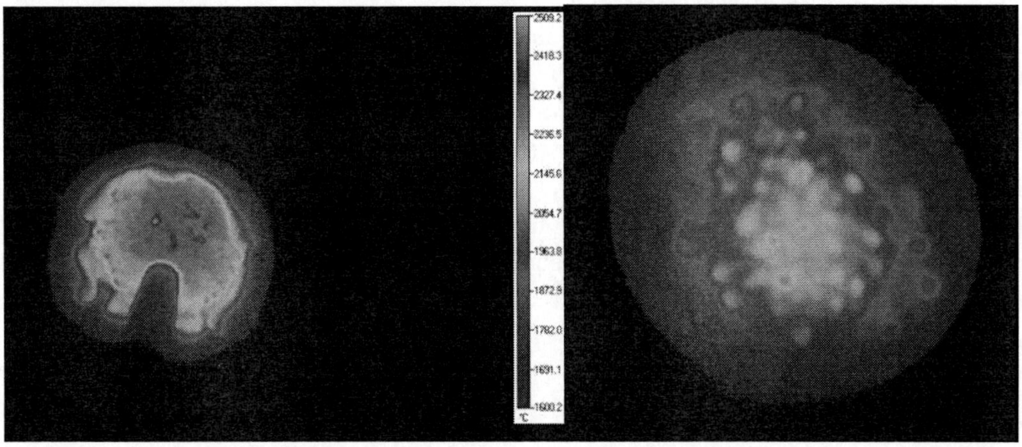

Figure 11. IR camera showing the ablation craters of the 1,000 W/cm^2 (Left) and 500 W/cm^2 (Right) tests.

The characterization of MXB-360 as a single bulk component and limitations in ITRAC make the glass mat wicking difficult to solve. Because TGA samples are held in a crucible below the melting point of the silica glass mat, there is no instrument data demonstrating at what temperature the glass mat begins to separate from the matrix. The *B'* tables in ITRAC only account for pyrolysis gases and there is no intended method for simulating the advection of individual components of a single composite material.

Melted glass mat solidifying around the edges of the test model resulted in a high level of mass retention in the test models, as shown in Figure 12. Because the melt was not fully advected from the test model, the mass loss was significantly lower than the simulated mass loss for all heat flux cases. Additionally, because heat flux is controlled by standoff distance from the torch, the flow field velocity was not high enough to fully remove melted silica away from the central ablation crater during the 500 W/cm^2 test as shown in Figure 12. This static glass melt layer acts as an oxidation barrier and further reduce the overall ablation rate.

Figure 12. (Left) MXB-360 subjected to 500 W/cm^2 heat flux. (Center and Right) Test model after exposure to 1,000 W/cm^2 heat flux for 30 seconds. Note the re-solidified silica glass around the ablation crater.

5. CONCLUSION

KMPs, *B'* tables, and pyrolysis gas enthalpy tables were determined for the glass/phenolic MXB-360 ablative composite using TGA and EDX spectroscopy. These properties were used to develop an ablation material response model using Northrop Grumman's ITRAC to simulate the exposure of MXB-360 to a hyperthermal oxidative environment. The code output was validated against in-depth temperature profile data obtained from OTB tests with heat fluxes of 500, 1,000, and 1,500 W/cm^2. The in-depth temperature profile from the code output matched closely with the experimental results, while other ablation metrics, such as linear ablation rate and mass loss were not as accurate due to the separation of the glass mat reinforcement from the phenolic matrix during testing.

6. FUTURE WORKS

The MXB-360 material response model will be further refined in order to provide more accurate outputs. This will likely require extensive thermal and mechanical characterization of the phenolic char and a method for simulating the early advection of glass mat in ITRAC. Similar testing and model development will be performed for MXBE-350, a rubber modified version of MXB-360 designed to survive high speed particle ejection from solid rocket motor exhaust plumes. SEM microstructural analysis of the post-test MXB-360 ablative will provide us with more insights of the different affected zones of the material.

7. REFERENCES

1. Starkey, R.P. and M.J. Lewis, *Critical Design Issues for Airbreathing Hypersonic Waverider Missiles*. Journal of Spacecraft and Rockets, 2001. **38**(4): 510-519.

2. Szczepanowska, H. and T.G. Mathia, *Space Heritage: The Apollo Heat Shield; Atmospheric Reentry Imprint on Materials' Surface.* MRS Proceedings, 2011. **1319**.
3. Suzuki, T., et al., *Postflight Thermal Protection System Analysis of Hayabusa Reentry Capsule.* Journal of Spacecraft and Rockets, 2014. **51**(1): 96-105.
4. Knight, D., et al., *Assessment of predictive capabilities for aerodynamic heating in hypersonic flow.* Progress in Aerospace Sciences, 2017. **90**: 39-53.
5. Paglia, L., et al., *Carbon-phenolic ablative materials for re-entry space vehicles: plasma wind tunnel test and finite element modeling.* Materials & Design, 2016. **90**: 1170-1180.
6. MacDonald, M.E., C.M. Jacobs, and C.O. Laux, *Interaction of Air Plasma With Ablating Heat Shield Material.* IEEE Transactions on Plasma Science, 2014. **42**(10): 2658-2659.
7. Natali, M., et al., *An in-situ ablation recession sensor for carbon/carbon ablatives based on commercial ultra-miniature thermocouples.* Sensors and Actuators B: Chemical, 2014. **196**: 46-56.
8. Koo, J.H., et al., *In Situ Ablation Recession and Thermal Sensor for Thermal Protection Systems.* Journal of Spacecraft and Rockets, 2018. **55**(4): 783-796.
9. Ewing, M.E. and B. Pincock, *Heat Transfer Modeling of a Charring Material Using Isoconversional Kinetics.* Heat Transfer Engineering, 2016. **38**(13): 1189-1197.
10. Ewing, M.E., T.S. Laker, and D.T. Walker, *Numerical Modeling of Ablation Heat Transfer.* Journal of Thermophysics and Heat Transfer, 2013. **27**(4): 615-632.
11. Koo, J.H., et al., *Silicone Polymer Composites for Thermal Protection of Naval Launching System.* Journal of Spacecraft and Rockets, 2011. **48**(6): 904-919.
12. Ozawa, T., *A New Method of Analysing Thermogravimetric Data.* Bulletin of the Chemical Society of Japan, 1965. **38**.
13. Kim, S.D., et. al., *Development of Polyamide 6/Polyamide 66 Copolymer Nanocomposite and Its Additive Manufacturability.* The Composites and Advanced Materials Expo. CAMX Conference Proceedings, 2019. Anaheim, CA.
14. Yee, C., et al., *In Situ Ablation Recession Sensor for Ablative Materials Based on Ultraminiature Thermocouples.* Journal of Spacecraft and Rockets, 2014. **51**(6): 1789-1796.
15. Menz, R., et al., *Characterization of 3-D Woven Carbon/Phenolic using In-Situ Ablation Sensing, Video Imaging, and Numerical Simulation*, in *2018 AIAA/ASCE/AHS/ASC Structures, Structural Dynamics, and Materials Conference*. 2018.
16. Sihn, S., et al., *Identifying unified kinetic model parameters for thermal decomposition of polymer matrix composites.* Journal of Composite Materials, 2018. **53**(20): 2875-2890.
17. Ewing, M.E. and D.A. Isaac, *Thermodynamic Property Calculations for Equilibrium Mixtures.* Journal of Thermophysics and Heat Transfer, 2018. **32**(1): 118-128.

Characterization of GFRP Bars and Couplers for Prestressed Concrete

Nafiseh Kiani[a], Marco Rossini[a], and Antonio Nanni[a]

[a] Department of Civil, Arch. and Env. Engineering, University of Miami, Coral Gables, FL, USA

Abstract

Glass fiber-reinforced polymer (GFRP) bars have significant potential in prestressed concrete (PC) applications. Recent technical literature indicates that the creep rupture threshold for state-of-the-practice GFRP bars can be higher than 40% of their guaranteed tensile strength, thus allowing a sufficient level of tensioning at the time of construction. To develop field-deployable systems for PC elements with particular focus on partially-prestressed applications (i.e., combination of tensioned and non-tensioned reinforcement), an experimental investigation was conducted. This paper reports on the mechanical properties and behavior of two different GFRP bar diameters, an anchorage system that uses polymeric wedges, and a bar splicing solution that may be necessary to meet the length of conventional prestressing beds. Experimental results obtained in this study support the use of GFRP bars in PC elements. The composites investigated in this study were deployed in the construction of two partially-prestressed piles serving as a real-case demonstrator for the viability of GFRP prestressing in substructure applications.

Keywords: GFRP bars; GFRP splicing; prestressed concrete; mechanical couplers; wedge anchors.

1. Introduction

Prestressing approaches are widely accepted in concrete constructions to improve concrete performance under serviceability limit states. The concept of applying a prestressed load and eliminating tension stresses in the concrete has been developed to prevent the crack and control deflection at service loads. However, limited cracks may be permitted in the concrete [1]. Therefore, partial prestressing in terms of combination of both prestressed and non-prestressed reinforcement would result in effective control of crack width, reduction of camber, control of deflection, and efficient constructability.

Fiber-reinforced polymer (FRP) reinforcing bars have been developed as alternatives to steel bars in corrosive environments. The mechanical properties of FRP composites differ from those of steel with the advantages of corrosion resistance, higher tensile strength, low modulus of elasticity, higher durability, and lighter weight, electrical and thermal non-conductivity and ease application. In coastal and bridge structures, corrosion of steel reinforcement can result in high repair costs and may affect the failure mechanism of concrete members as well. Using FRP reinforcement will eliminate the corrosion problem and thus increase the service life of concrete structures.

The advantages of FRP reinforcement in terms of corrosion resistance and high durability in reinforced concrete (RC) structures have been also extended to prestressed concrete (PC) structures. Among different types of fibers, Carbon FRP (CFRP) is commonly preferred in PC elements because of high strength, high creep rupture strength, and low relaxation. However, CFRP drawbacks include low strain to failure, high material cost and difficulties in the efficient use of wedge anchors for tensioning [2,3].

The first application of GFRP tendons was in a post-tensioned two-span highway bridge in Germany in 1986 by using an epoxy-socket anchor with a post-tensioning force of 135 kips [4]. GFRP bars are more economical alternatives in substructure applications where low fatigue loading is present, and limited level of prestressing is required. At low levels of prestress, GFRP may also be pulled using conventional steel anchors [5]. Furthermore, the estimated cost of GFRP bars is a portion of the cost of CFRP reinforcement [5]. Basalt FRP (BFRP) tendons may also be an alternative. In fact, previous studies have shown that the creep rupture stress can be as high as 52% of ultimate tensile strength [7].

The design tensile strength f_{fu} is defined as the guaranteed tensile strength (i.e., the mean tensile strength minus three standard deviations) multiplied by an environmental reduction factor C_E (0.7 for concrete exposed to earth and weather, or 0.8 for not exposed to earth and weather). Glass FRP (GFRP) bars have been historically known as the most vulnerable to creep rupture. The ratio of GFRP creep rupture stress to the guaranteed tensile strength was estimated as 0.29 according to ACI.440.1R [8] and conservatively limited to $0.20f_{fu}$ for design. If a higher creep rupture limit were to be demonstrated, GFRP could be possibly used in PC substructure applications. According to more recent experiments, various creep rupture limits have been suggested for GFRP bars to date ranging from 30% to more than 40% of the design strength f_{fu} [9,10].

The type of anchorage system plays a significant role in transferring the prestress load during pre-tensioning [11]. The stress concentration could be reduced by using proper anchors during the tendon tensioning. The slippage of anchors is another important factor in premature tendon failure. The type of anchor and coating of GFRP bars affect the bond between tendon and anchor as well. Different anchorage systems are available including bonded, mechanical, spike, wedge and clamping anchors. The investigation of the wedge anchorage system with CFRP tendons has been conducted before and the results showed that the wedge type anchors have sufficient strength during the tensioning with low probability of premature failure in the anchorage zone [12].

In this study, to develop field-deployable systems for GFRP-PC elements with particular focus on partially prestressed applications, an investigation on mechanical properties of bars and performance (i.e., the ultimate capacity and behavior) of wedge anchors as well as mechanical couplers was conducted. Mechanical couplers are used to connect two reinforcing bars when lap splices are not practical. Since the length of GFRP bars is limited by logistic considerations, steel swaged couplers can be used at precast plants to meet the length of conventional prestressing beds splicing GFRP bars to traditional steel prestressing strands.

2. GFRP Partially Prestressed Pile

The data obtained in this study were used to demonstrate the potential of GFRP partial prestressing in substructure applications. Accordingly, two partially prestressed piles were constructed using twelve No.4 GFRP bars (0.5 in. diameter) pre-tensioned at 30% of their guaranteed tensile strength after seating losses, twelve non-prestressed No.8 GFRP bars (1 in. diameter), and No.4 GFRP stirrups (0.5 in. diameter and 1.5 in. bent radius) at a maximum spacing of 6 in. The piles were designed adapting to the case of GFRP partial prestressing the provisions of: a) the second edition of AASHTO LRFD Bridge Design Guide Specification for GFRP-Reinforced Concrete; and b) the first edition of the AASHTO Guide Specifications for the Design of Concrete Bridge Beams Prestressed with CFRP Systems. The cross section of the pile and reinforcement layout are shown in Figure 1. The steel swaged couplers are shown in Figure 2 and used for splicing the No. 4 GFRP bars with the 0.5 in. steel strands in the prestressing bed.

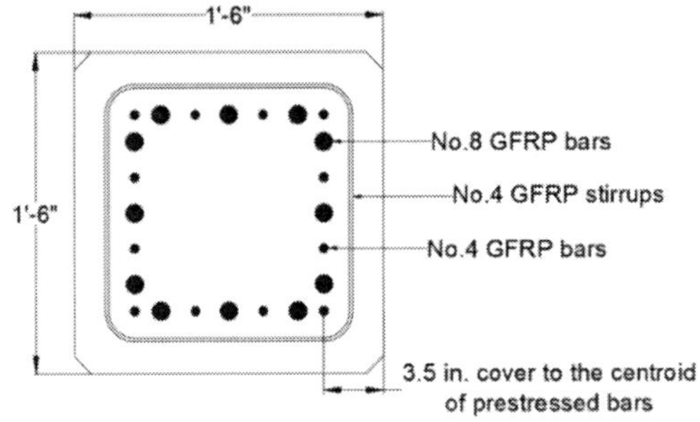

Figure 1. Cross section of GFRP pile

(a) (b)

Figure 2. (a) GFRP partially prestressed pile (b) Steel swaged couplers

3. Mechanical Characterization

The mechanical properties of GFRP bars and stirrups were evaluated according to ASTM D7957, ACI 440.3R, and FDOT specifications [13–15]. Two different GFRP bar sizes with diameter of 0.5 in. and 1 in. (No.4 and No.8 bars), and a pre-bent GFRP stirrup of 0.5 in. diameter (No.4 stirrups) were tested. The stirrups are sand-coated and are bent at a 90-degree angle with a 1.5 in. bent radius. GFRP bars cannot be bent at the construction site, but only at the time of manufacturing prior to the polymerization of the thermoset resin.

Fiber content, moisture absorption, measured cross sectional area, guaranteed ultimate tensile load, and tensile modulus of elasticity were determined through laboratory testing. For each test, a minimum of three specimens were used per related test conditioning specifications. For stirrups, the specimens were prepared at the bent location for fiber content and moisture absorption tests, while straight part of the stirrups (i.e., legs) are used for the remaining properties according to ASTM D7957 and FDOT requirements. The measurement results are summarized in Table 1.

The results showed that all requirements were met according to ASTM D7957 and FDOT specifications. The moisture absorption of pre-bent stirrups was observed to be higher than that of straight GFRP bars. This can be explained because of: a) the dense sand-coating present on the surface of the bends, and b) fiber separation resulting from the bending process. Regarding tensile properties, the guaranteed ultimate tensile strengths were determined as 21.70 kip for No.4 GFRP bars and 91.13 kip for No.8 GFRP bars. Tensile modulus of elasticity were determined as 6,720 ksi for No.4 GFRP bars and 7,230 ksi for No.8 GFRP bars.

Table 1. Mechanical properties summary

Test	Test Method	Bar Size				ASTM D7957 Requirement
			No.8	No.4	Stirrup No.4	
Fiber Content (%)	ASTM D2584	Avg.	77.8	73.3	73.2	≥ 70
		SD	2.55	0.13	1.31	
Moisture Absorption in 24 h (%)	ASTM D570	Avg.	0.02	0.18	0.25	≤ 0.25
		SD	0.00	0.05	0.10	
Measured Cross Sectional Area (in^2)	ASTM D792	Avg.	0.780	0.201	0.195	No.4: 0.185-0.263, No.8: 0.738-0.913
		SD	0.005	0.000	0.000	
Tensile Modulus of Elasticity (ksi)	ASTM D7205	Avg.	7230	6720	NA	≥ 6500
		SD	0.41	0.22		
Guaranteed Ultimate Tensile Load (kip)		Avg. -3SD	91.13	21.70		No.4:21.6, No.8: 66.8

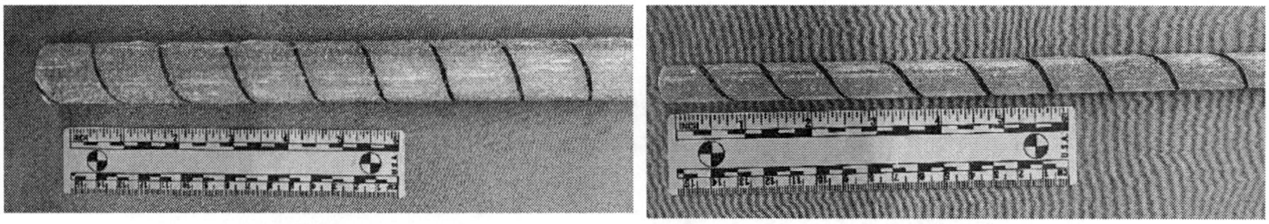

(a) (b)
Figure 3. (a) No.8 GFRP bar and (b) No.4 GFRP bar

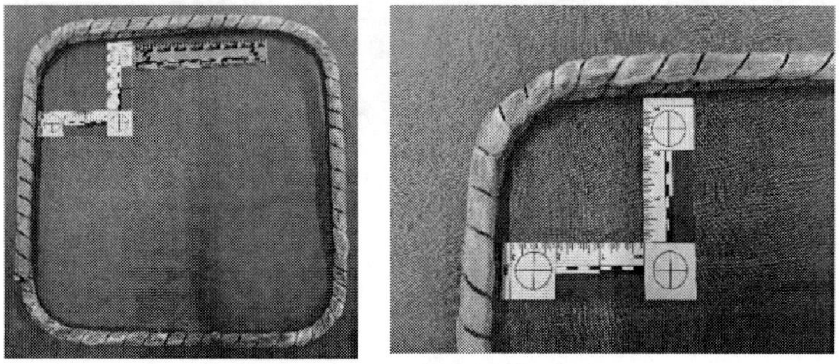

Figure 4. No.4 GFRP stirrup

4. Pull Test with Coupler

A pull test was carried out on spliced GFRP bars with a swaged mechanical coupler adapting the provisions of ACI 440.3R and ASTM D7205 specifications. The spliced GFRP bar fitted with wedge anchors, is placed in a 200-kip screw-driven loading frame and uniformly loaded in displacement control at a rate of 0.08 in/min while recording force and cross-head displacement. From a pull test, the ultimate strength of the coupler, and the load-displacement curve of the mechanical system can be obtained.

The wedge anchors used in the experiment consist of a steel barrel with two angled plastic wedges fitted to the inside of the barrel and the No.4 GFRP bar. The angled wedges may be changed to accommodate different GFRP bar sizes. The wedge anchors are shown in Figure 5. In the test setup, the anchors were secured on each end of the bar, while a steel swaged coupler was placed at mid-length of the specimen to connect two GFRP bars. The specimen was positioned in the cross-head of the loading machine as shown in Figure 6.

A typical load-displacement curve resulting from the test is shown in Figure 7. The results showed that failure at the coupler's location occurred at 13 kip corresponding to about 60% of the guaranteed tensile strength of the GFRP bar. In the load-displacement curve, there are two sudden reductions of load which may be due to anchors or coupler slippage. Failure occurs at the coupler's location because of shear lag within the GFRP bar as shown in Figure 8a. The capacity of the coupled system is deemed to be adequate to achieve effective prestressing of GFRP bars. The slip of wedge anchors was measured to be less than 1.0 in. after the testing. No failure occurred in the anchorage zone.

Figure 5. Wedge anchorage system test setup

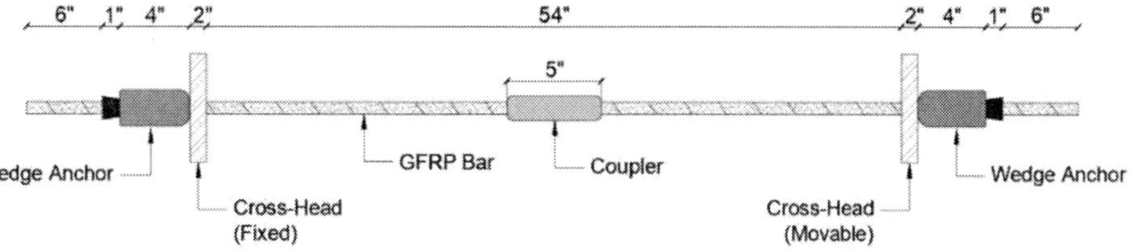

Figure 6. Pull test on spliced GFRP bars with a swaged coupler

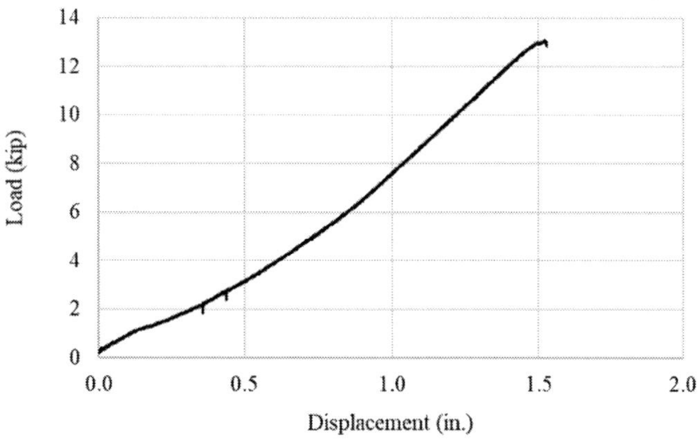

Figure 7. Load-displacement curve of GFRP with coupler

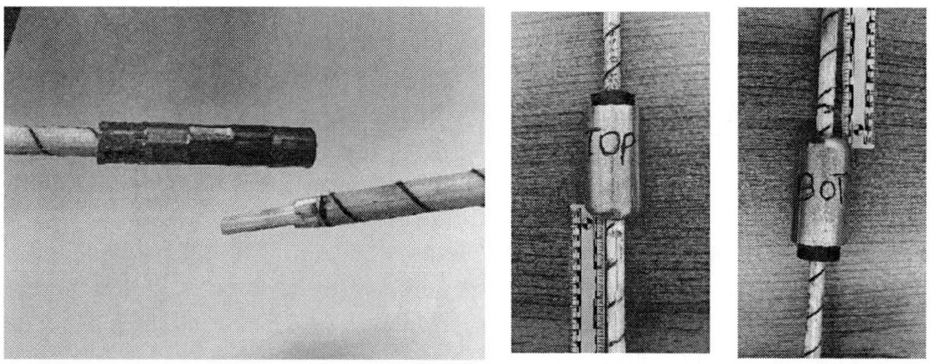

(a) (b) (c)

Figure 8. (a) Failure mode of coupler; (b) top wedge anchor; and, (c) bottom wedge anchor

5. Conclusions

Test results obtained in this study support the viability of prestressed GFRP bars in substructure applications. As part of a study investigating partially prestressed piles reinforced with prestressed and non-prestressed GFRP bars, the mechanical properties of GFRP bars and pre-bent stirrups were evaluated based on required specifications. The guaranteed tensile strength and modulus of elasticity of GFRP bars were determined. Moreover, the ultimate capacity of coupled GFRP bars was evaluated. The results showed that failure occurred at the coupler's location at 60% of the guaranteed tensile strength of the GFRP bar which is twice the prestressing value of 30% of the guaranteed tensile strength adopted in the field. The measurements of the wedge anchors during pull test with coupler revealed that the slip of wedges was as expected and no premature failure in the anchorage zone was observed. The mechanical properties and slip of wedge anchors and couplers need a comprehensive investigation before commercial deployment.

6. Acknowledgements

The authors gratefully acknowledge the financial support from: (1) ''MILDGLASS: GFRP Strand for Resilient Mild Prestress Concrete" National Academy of Sciences, NCHRP-IDEA/207; (2) ''I/UCRC Center for the Integration of Composites into Infrastructure (CICI)" National Science Foundation, United States 1439543; (3) SIREG for providing material and technical support to conduct the experimental portion of this study; and, (4) Gate precast for technical support and for donating the partially prestressed demonstrative piles mentioned in this study.

7. References

[1] P. Ghoddousi, A.M. Abbasi, E. Shahrokhinasab, M. Abedin, Prediction of Plastic Shrinkage Cracking of Self-Compacting Concrete, Adv. Civ. Eng. 2019 (2019). https://doi.org/10.1155/2019/1296248.

[2] N. GRACE, Bridge Street Bridge: the first CFRP prestressed concrete bridge in the United States, Concr. Eng. 7 (2003) 33–37.

[3] O. Alsheraida, S. El-gamal, A. Al-saidy, Wedge Anchorage System for Pre-stressed CFRP Reinforcement – A Literature Review, (2015) 227–234.

[4] J.W. Schmidt, A. Bennitz, B. Täljsten, P. Goltermann, H. Pedersen, Mechanical anchorage of FRP tendons - A literature review, Constr. Build. Mater. 32 (2012) 110–121. https://doi.org/10.1016/j.conbuildmat.2011.11.049.

[5] M. Rossini, A. Nanni, Composite strands for prestressed concrete: State-of-the-practice and experimental investigation into mild prestressing with GFRP, Constr. Build. Mater. 205 (2019) 486–498. https://doi.org/10.1016/j.conbuildmat.2019.02.045.

[6] T. Cadenazzi, G. Dotelli, M. Rossini, S. Nolan, A. Nanni, Life-Cycle Cost and Life-Cycle Assessment Analysis at the Design Stage of a Fiber-Reinforced Polymer-Reinforced Concrete Bridge in Florida, Adv. Civ. Eng. Mater. 8 (2019) 20180113. https://doi.org/10.1520/acem20180113.

[7] X. Wang, J. Shi, J. Liu, L. Yang, Z. Wu, Creep behavior of basalt fiber reinforced polymer tendons for prestressing application, Mater. Des. 59 (2014) 558–564.

[8] ACI 440.1R-03, Guide for the design and construction of concrete reinforced with FRP bars (ACI 440.1R-03), 2005. https://doi.org/10.1061/40753(171)158.

[9] B. Benmokrane, V.L. Brown, K. Mohamed, A. Nanni, M. Rossini, C. Shield, Creep-Rupture Limit for GFRP Bars Subjected to Sustained Loads, J. Compos. Constr. 23 (2019) 1–7. https://doi.org/10.1061/(ASCE)CC.1943-5614.0000971.

[10] M. Rossini, E. Saqan, A. Nanni, Prediction of the creep rupture strength of GFRP bars, Constr. Build. Mater. 227 (2019) 116620. https://doi.org/10.1016/j.conbuildmat.2019.08.001.

[11] A. Nanni, C.E. Bakis, E.F. O'Neil, T.O. Dixon, Performance of FRP tendon-anchor

systems for prestressed concrete structures, PCI J. 41 (1996) 34–44.
[12] A. Al-Mayah, K. Soudki, A. Plumtree, Development and assessment of a new CFRP rod-anchor system for prestressed concrete, Appl. Compos. Mater. 13 (2006) 321–334. https://doi.org/10.1007/s10443-006-9019-6.
[13] ASTM, Standard Specification for Solid Round Glass Fiber Reinforced Polymer Bars for, i (2019) 1–5. https://doi.org/10.1520/D7957.
[14] ACI440 3R, Guide Test Methods for Fiber-Reinforced Polymers (FRPs) for Reinforcing or Strengthening Concrete Structures, Am. Concr. Inst. (2004) 1–40.
[15] FDOT, Standard Specifications for Road and Bridge Construction, (2013).

MODELING SPRING-IN VARIABILITY DUE TO RESIDUAL STRESS IN CONTOURED AUTOCLAVE COMPOSITE PARTS

Matthew L. Kirby[a], Geoffrey J. Frank[b], David S. Riha[a], Andy Ko[c], Robert W. Koon[d], and Michelle U. Serna[e]

[a]Southwest Research Institute
San Antonio, Texas

[b]University of Dayton Research Institute
Dayton, Ohio

[c]Phoenix Integration
Blacksburg, Virginia

[d]Lockheed Martin Aeronautics Company
Marietta, Georgia

[e]Northrop Grumman Corporation
Clearfield, Utah

ABSTRACT

Spring-in variability due to variations in the residual stress field of contoured autoclave composite parts is a primary source of nonconformances that result in costly part-rework. The main driver of spring-in and warpage is the difference between through-thickness and in-plane cure shrinkage and thermal expansion. State-of-the-art finite element process modeling frameworks are capable of capturing this phenomenon, and thus could be used to predict variability in spring-in for contoured parts. However, simulations of full three-dimensional representations of these parts remain computationally expensive and are generally not tractable for manufacturing environments. The objective of this work was to develop a tractable process model for predicting spring-in variability in a representative contoured part. A formal model verification and validation framework was used to develop models to predict the final part state due to variations in the materials, lay-up, and processing. The models developed in this effort included a reduced-order finite element process model using the COMPRO® plug-in for ABAQUS and a general analytical model was implemented in ModelCenter® for use in a manufacturing environment. Although there remain variations in the materials and processes that the models were not able to capture, both the reduced-order model and analytical model were relatively successful at capturing trends in spring-in variability.

1. INTRODUCTION

The aerospace industry is increasingly using carbon fiber-reinforced autoclave composites for critical applications due to substantial weight savings that can be realized without a considerable effect on structural integrity, as compared to metallic components. However, fabrication of autoclave composites is challenging due to variability in the incoming material and the intricate and interrelated manufacturing process necessary to create finished composite parts. These challenges ultimately result in variability in the final parts, which often leads to difficulties in assembly, part rework, or even part scrappage. Thus, part variability is a key contributor to increased manufacturing costs and schedule delays associated with autoclave composites. Reducing and, if possible, preventing composite part variability is of significant importance to the aerospace industry in order to reduce cost/schedule impacts and ensure part quality and performance. The focus of this study was to understand the primary sources of residual stress-induced spring-in variability, which is one of the most common process-induced deformations in contoured autoclave composites [1], and develop a tractable model for predicting spring-in variability in a representative contoured part that could be implemented in a manufacturing setting.

Spring-in, or springforward, is the distortional reduction in composite flange angle that manifests itself after the composite has been cured and subsequently removed from the tool or mold [1]. In a typical manufacturing operation, composite part designs are empirically tested for spring-in, and compensation approaches for spring-in, such as tool modifications [2] or shimming procedures, are developed heuristically [3]. This trial and error process for developing a compensation approach can result in extensive testing, which is a significant cost burden. If shimming is selected as the compensation approach, this can also lead to nontrivial cost and schedule impacts because current shimming procedures are performed manually and can be time-consuming [3].

Residual stress, which causes shape distortions such as spring-in, develops in a contoured composite part during manufacturing as a result of several intrinsic (resin system, fiber architecture, lay-up stacking sequence, part geometry, etc.) and extrinsic (processing techniques and parameters, tool and bag design, etc.) factors [4]. According to the literature, the primary mechanisms of residual stress-induced spring-in are generally categorized as thermoelastic or non-thermoelastic [5, 6]. The thermoelastic component of spring-in is reversible and is related to the mismatch between the through-thickness and in-plane coefficient of thermal expansion (CTE), whereas the non-thermoelastic component is non-reversible and is related to cure shrinkage, part-tool interaction, consolidation, and other cure-evolution phenomena [5]. Although the thermoelastic component is reversible in the sense that the strains are elastic and are developed and recovered as a function of temperature during processing, thermoelastic spring-in is complicated by the fact that the CTE of the material is evolving during the cure process as a function of degree of cure. The primary spring-in mechanisms of through-thickness and in-plane thermal strain and cure shrinkage strain and their relationship to spring-in angle have been described by the following analytical equation [6, 7]:

$$\Delta\theta = \theta\left(\frac{\varepsilon_l - \varepsilon_t}{1+\varepsilon_t}\right) = \Delta\theta_{CTE} + \Delta\theta_{CS} = \theta\left(\frac{(\alpha_l - \alpha_t)\Delta T}{1+\alpha_t \Delta T}\right) + \theta\left(\frac{\Phi_l - \Phi_t}{1+\Phi_t}\right) \qquad (1)$$

Where θ is the original contour angle, ε_l is the total (cure shrinkage plus thermal strain) composite effective longitudinal strain, ε_t is the total (cure shrinkage plus thermal strain) composite effective transverse strain, $\Delta\theta_{CTE}$ is the change in contour angle due to thermal strain, $\Delta\theta_{CS}$ is the change in contour angle due to shrinkage strain, α_l is the composite effective longitudinal CTE, α_t is the composite effective transverse CTE, Φ_l is the composite effective longitudinal cure shrinkage strain, and Φ_t is the composite effective transverse cure shrinkage strain.

Notable secondary phenomena related to residual stress-induced spring-in include part shape (C-shaped or L-shaped) [8], tool shape (male or female) [9, 10], tool material (specifically the CTE of the tool material relative to the part CTE) [8, 11], and changes in resin volume during cure (possibly due to resin bleed or volatile off-gassing) [3, 4]. Additional disputed secondary phenomena include post-cure method (free-standing post-cure or post-cure in the tool) [4], autoclave cure parameters (ramp rate, hold temperature, and autoclave pressure) [4, 12], part thickness [6, 11, 13, 14], radius of curvature [11, 13, 14], lay-up stacking sequence [6, 8, 11, 14], and tool surface (with or without release agent) [4, 8, 11]. Other phenomena may exist, but they were not found in the open literature.

The aerospace industry has been a major catalyst behind recent efforts to model these primary and secondary phenomena and ultimately to predict residual stress-induced spring-in. Since residual stress development during cure, as a result of these phenomena, is described by complex and interrelated differential equations, finite element analysis (FEA) has been the typical approach for performing detailed spring-in simulations [9, 13, 15, 16]. However, general purpose commercial off-the-shelf FEA software does not account for degree of cure and thus cannot capture viscoelastic effects, which results in inaccurate residual stress and absolute deformation prediction [15]. Therefore, user-defined subroutines have been employed with commercial off-the-shelf FEA software to define a fully viscoelastic model that accounts for degree of cure, such as a user material subroutine (UMAT) in ABAQUS [15]. A software package for computing thermo-viscoelastic resin properties as a function of time and temperature within the framework of an FEA simulation of an autoclave cure cycle, COMPRO, has recently been developed as a plug-in for ABAQUS, ANSYS, and MSC Marc and is the most-referenced commercial software tool in the open literature for modeling residual stress in composite structures and components [17-20]. COMPRO employs sequentially coupled thermo-chemical, flow-compaction, and stress-deformation FEA to model residual stress development during processing. The thermo-chemical simulation is governed by heat transfer, and resin properties are updated according to cure kinetics models to predict the temperature field as a function of time during autoclave cure [18]. The flow-compaction simulation is governed by Darcy's law, in which the resin and fiberbed are modeled as a viscous fluid and poroelastic medium, respectively, to predict resin flow and fiberbed compaction during the pre-gelation regime of resin cure advancement [18]. The stress-deformation simulation is governed by solid mechanics, and resin is modeled as a cure-hardening instantaneously linear elastic (CHILE) material, as a function of degree of cure, which is combined with the linear-elastic fiber properties via micromechanics approaches to predict ply-level mechanical properties and stress-strain development in the laminate during cure [18]. To reduce the computational cost of these simulations, multiple plies can be represented within a single finite element by "smearing" the ply-level properties through the element [18]. Despite the representation of multiple plies within a single element along with other techniques to improve computational efficiency, these process simulations remain expensive, even for today's computer systems, and the adoption of these simulations in manufacturing environments remains limited. Therefore, along with understanding and modeling residual stress-induced spring-in variability, another objective of this study was to develop models and software tools that could capture trends in spring-in variability at a reduced computational cost to more detailed finite element models in the hope of making spring-in prediction capability more accessible in a manufacturing setting.

2. MATERIALS AND METHODS

To accomplish the objectives of understanding and modeling residual stress-induced spring-in variability, a contoured aircraft edge component was selected as the demonstration case for spring-in evaluation. A three-dimensional half-part finite element model of the component was developed using the COMPRO plug-in for ABAQUS, and eight components were fabricated and tested to support uncertainty quantification (UQ) and model calibration and validation. Additionally, to reduce computational costs, a three-dimensional finite element model of a small region of the component (reduced-order model) was developed, referred to as the "strip" or "feature" model, and the analytical model described by Equation 1 was implemented in ModelCenter. The models were developed within a model verification and validation (V&V) framework and both NESSUS® [21] and ModelCenter software tools were used to propagate uncertainties and perform probabilistic studies, such as global sensitivity analysis.

2.1 Verification and Validation Framework for Model Development

Model V&V is a methodology for developing computational models that is concerned with building credibility in the models and quantifying confidence in model predictions. V&V is used to systematically identify required phenomena, approximations and uncertainties in models and data. UQ plays a key role in model V&V. The goal of UQ activities is to quantify the uncertainties in both experimental and simulation results such that the model fidelity requirements can be assessed (validation) and the predictive accuracy of the model quantified. The model development approach using a V&V framework proposed in the Integrated Computational Materials Engineering (ICME) V&V Guidelines [22, 23] provides tools for consistent documentation of model predictive capability for ICME models. These guidelines provide a communication mechanism to understand the current state of models using the tool maturity level (TML), flow of information in the models using model hierarchies, and the current uncertainties and limitations in the models using phenomena importance and ranking tables (PIRT). The TML and PIRT provide a mechanism for programmatic decisions, such as what experiments or model development activities are needed to achieve a required TML. In addition, the ICME V&V guideline checklists provide documentation that the required steps were addressed.

2.2 Demonstration Case Fabrication

Since the intended use of the model is to predict residual stress-induced spring-in variability in contoured composite parts, it was important to choose a representative aerospace component for model validation activities. Therefore, a contoured aircraft edge component was selected as the demonstration case. The edge component had a U-shaped cross-section with closed ends to form a "canoe-shaped" part. This particular component offered a good combination of variability challenges, with a history of nonconformances requiring corrective action, and the ability to model the most likely variables leading to the nonconformances. The contoured aircraft edge component is formed by a hand lay-up process on a male steel tool using four harness satin woven fabric-bismaleimide (BMI) prepreg. Cytec Solvay Group's CYCOM® 5250-4 curing BMI [24] is the prepreg resin system, and both E-glass and carbon fabric, made of Hexcel's HexTow® IM7 carbon fiber [25], are used in the lay-up of the component. During lay-up, prepreg plies are wrapped around the ends of the tool to close the ends of U-shaped component. This wrapping procedure introduces additional complexity in the hand lay-up process and results in variability in the location and angle of the wrapped edges of the ply. Vacuum debulks are performed at regular intervals throughout the lay-up process. After the component lay-up process, the lay-up is fitted with a caul and then cured on the tool in an autoclave, based on the resin manufacturer's recommended cure cycle provided in the resin technical data sheet [24]. An expert elicitation was conducted to identify possible sources of variability and uncertainty in the manufacturing process of the contoured aircraft edge component. Table 1 provides a list of the candidate process variables that were considered in this study, along with the primary effect(s) of the process variable based on the present understanding of the underlying physics in the component manufacturing process, and the capability of modeling the primary effect in an ABAQUS-COMPRO process simulation.

Table 1. Summary of expert elicitation and potential process variables for investigation

Candidate Process Variable	Primary Effect(s)	Model in ABAQUS-COMPRO Simulation?
1. Wrapped end of ply location	1a. Ply slip relieves stress 1b. Altered stacking sequence	1a. No 1b. Complex setup
2. Wrapped end of ply angle	2a. Altered stacking sequence	2a. Complex setup
3. Prepreg out-time	3a. Changes in initial degree of cure 3b. Changes in prepreg tack	3a. Yes 3b. No
4. Pressure at local features	4a. Resin pressure & flow around features	4a. Yes, but only for simple models
5. Min/max time-temperature cure conditions	5a. Changes in time history of property evolution	5a. Yes
6. Min/max resin content and fiber areal weight	6a. Changes in initial properties 6b. Changes in property evolution	6a. Yes 6b. Yes
7. Autoclave time-temperature variations	7a. Changes in time history of property evolution	7a. Yes
8. Ply thickness	8a. Part thickness 8b. Resin pressure & flow around features	8a. Yes 8b. Yes, but only for simple models
9. Caul variations	9a. Resin pressure & flow around features	9a. Yes, but only for simple models

To validate the finite element process models of the component and the analytical spring-in model described by Equation 1, eight demonstration case aerospace edge components were fabricated with minor changes being made to specific process variables. No process variables were altered in the fabrication of three baseline components. A single process variable was altered for each of the other five components fabricated in this study, in an attempt to isolate phenomena and understand variability in the manufacturing process. Table 2 provides a list and brief description of the eight demonstration case edge components. After the completion of the first baseline (component 1) the lay-up technician was changed. Since a number of lay-up and bagging features in the fabrication of the edge components are known to be "lay-up technician dependent", a new baseline (component 2) was fabricated by the new lay-up technician and then replicated (component 3) to assess the reproducibility of the data when all control variables are held constant. Components 2 through 8 were fabricated by the same lay-up technician. The process variables that were altered in the fabrication of components 4 through 8 were selected based on both the perceived importance of variations in the process variable and the capability of capturing the effect of variations in an ABAQUS-COMPRO process simulation, in accordance with the findings of the expert elicitation (Table 1).

Table 2. Summary of demonstration case aerospace edge components fabricated in this study

Component #	Description
1	Baseline 1 (fabricated by lay-up technician A)
2	Baseline 2 (fabricated by lay-up technician B)
3	Baseline 3 (fabricated by lay-up technician B)
4	Increased prepreg out-time (ambient-temperature aging of prepreg close to the specification limit for the resin)
5	Reduced number of debulks
6	Reduced resin bleed (fortified resin dam for additional resin containment)
7	Minimized time-temperature history during autoclave cure, referred to as "cold-fast" cure cycle (used lower specification limit for hold temperature and hold time and upper specification limit for ramp rate)
8	Maximized time-temperature history during autoclave cure, referred to as "hot-slow" cure cycle (used upper specification limit for hold temperature and hold time and lower specification limit for ramp rate)

2.3 Experimental Measurements

Several measurements of the eight edge components were taken during and after fabrication. Many of these measurements are not commonly recorded in a production environment but were specifically taken to support the model development effort in this study. The weight, thickness, and wrapped end angle of each ply was recorded during lay-up. Ply weight was determined by measuring the total weight of the prepreg material cut for lay-up and then subtracting the weight of all scrapped material after lay-up. Ply thickness was measured with a flat head micrometer. Wrapped end of ply angle was determined using a protractor with the 0° orientation parallel to the tool surface. In addition, the start and end time of each debulk in the process was recorded.

After cure, the weight and thickness of each component was measured, and a dimensional inspection by Coordinate Measuring Machine (CMM) was performed to measure the geometric variation of each component from the as-designed state. The dimensional inspections included over 738 points on each component. Spring-in was estimated from these inspections by fitting surfaces to the CMM data of the web and sides of the component mid-way along the long axis of the component. The resulting surfaces were used to compute the deviation from the nominal angle between the web and sides 12.7 mm from the web. Following the dimensional inspections, the components were sectioned for photomicroscopy and acid digestion. Photomicroscopy was used to estimate local thickness and acid digestion was used to measure local resin content and fiber volume fraction in the different regions of the component (i.e. web, corners, and sides). Preliminary model validation involved the qualitative comparison of trends identified in the experimental data and trends predicted by the models.

3. SPRING-IN MODEL DEVELOPMENT

3.1 Phenomena Importance and Ranking Table

Another way to synthesize the information from the expert elicitation related to modeling spring-in variability is in the form of a PIRT. The PIRT shown in Table 3 is a notional ranking of the key phenomena, ranked first according to relative importance and secondarily according to confidence in current models to capture the phenomena.

Table 3. Phenomena Importance and Ranking Table for Modeling Spring-in Variability

	Source of Spring-in Variability	Importance	Confidence	Representation in Model	Comments
Incoming Material and Preparation					
A.1	Resin content/ fiber volume fraction (V_f)	**High**	Medium	Location-specific cure kinetics	Avg. resin content/V_f from material certification package (material certs.), but spatial distribution not known
A.2	Thermo-chemical properties	Medium	Medium	Location-specific cure kinetics	Confidence in cure kinetics models, but uncertainty/variability in model parameters not investigated
A.3	Uncured ply thickness	Medium	Medium	Input geometry	Could potentially be used to predict ply initial V_f, measured with micrometer
A.4	Prepreg out-time	Low	Medium	Initial degree of cure (DOC)	Estimated initial DOC based on out-time with COMPRO, but near lower bound of COMPRO's predictive range
Part Lay-up and Vacuum Compaction (Debulk)					
B.1	Debulk schedule	**High**	Low	Location-specific geometry and V_f	Impact of debulk schedule is not fully understood nor captured by models
B.2	Resin bleed & migration → heated debulk	**High**	Medium	Flow-compaction boundary conditions	Has significant impact on local residual stresses and modeled with some success using COMPRO
B.3	Wrapped end of ply location	Low	Low	Input geometry (no ply sliding)	Complex model setup and models do not capture ply sliding
B.4	Bulk factor	Low	Low	Thickness (no gas phase)	No representation of gas phase in models, but assumed to be insignificant
B.5	Stack sequence & wrapped end of ply angle	Low	Medium	Input geometry	Influences homogenized laminate properties, especially around wrapped corners, but assumed to be insignificant
Consolidation, Cure, and Post-cure					
C.1	Pressure at local features (caul interactions)	**High**	Low	Location-specific geometry, cure kinetics, and contact with caul	Significant impact on local residual stresses, but difficult to measure and relationship to V_f not known (assumed perfect caul-part fit → no interferences)
C.2	Resin bleed & migration → consolidation	**High**	Medium	Flow-compaction boundary conditions	Significant impact on local residual stresses, modeled with some success using COMPRO
C.3	Cure cycle	**High**	High	Thermal history	Significant impact, drives cure/material properties, captured with COMPRO
C.4	Local thermal gradients	Medium	High	Location-specific heat transfer and cure kinetics	Yields spatial variations in cure/material properties captured by COMPRO thermo-chemical simulation

The PIRT highlights the most significant sources of uncertainties and variations and opportunities for further model development. For example, debulk schedule and pressure at local features (caul interactions) are both listed as some of the most important phenomena but relatively low confidence that these phenomena are being captured in the current models. Thus, future resource allocation should focus on model development efforts that improve understanding of and capability for predicting the impact of debulk schedule and pressure at local features and their relationship to local resin content (RC) and fiber volume fraction (V_f). Additional phenomena that are considered to be of highest importance are initial RC and V_f (especially the spatial distribution), resin migration and bleed related to both heated debulks and autoclave consolidation, and cure cycle variations. Since the difference between through-thickness and in-plane thermo-chemical properties (i.e., cure shrinkage and CTE) has been reported to be the main driver of spring-in and is closely related to RC and V_f, phenomena associated with local variations in RC and V_f, such as resin migration/bleed and pressure at local features, are therefore some of the most important.

3.2 ABAQUS-COMPRO Models

Two ABAQUS-COMPRO models were developed for predicting spring-in and spring-in variability in this effort. A half-part model was developed for high fidelity spring-in simulations; however, this model is cost prohibitive and is not conducive for use in manufacturing environments. Therefore, a reduced-order "strip" model (Figure 1) of the part, which is a thin slice through the center of the part along the short axis, was developed to capture trends in spring-in variability at a reduced computational cost. The half-part model was used to confirm the spring-in trends predicted by the strip model and to account for features that were not represented in the strip model, such as wrapped end of ply locations.

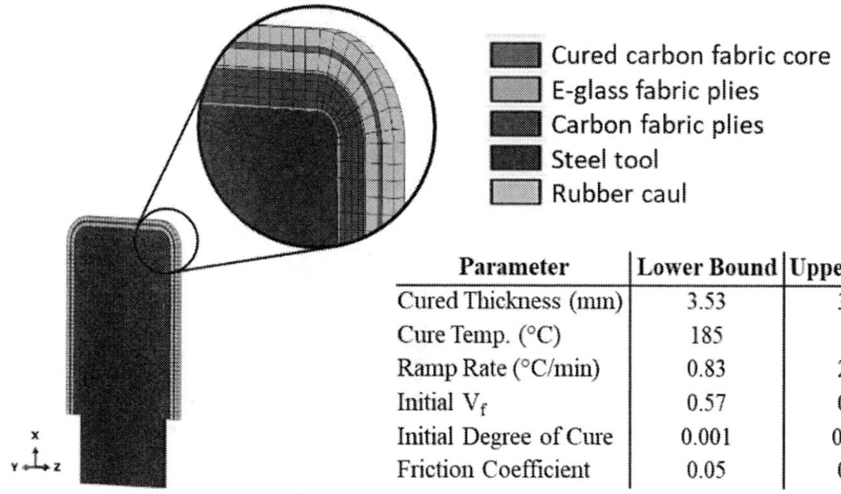

Parameter	Lower Bound	Upper Bound
Cured Thickness (mm)	3.53	3.68
Cure Temp. (°C)	185	196
Ramp Rate (°C/min)	0.83	2.78
Initial V_f	0.57	0.63
Initial Degree of Cure	0.001	0.140
Friction Coefficient	0.05	0.30

Figure 1. Detail of ABAQUS-COMPRO strip model and parameters varied in trade study

The half-part model had 77,919 total elements and the part was comprised of linear hexahedral elements (C3D8), with six elements through the thickness. Additionally, the wrapped end of ply angles were modeled in the half-part model, and the simulations captured variations in the end of ply angles. The symmetry boundary condition was imposed through the middle of the part along the short axis, which is not exactly correct because the ends of the aerospace edge component differ slightly. It took approximately one day of computation (wall clock) time[1] to complete a simulation of the half-part model. Due to the complexity of the half-part model, only the thermo-chemical and stress-deformation simulations were performed, flow-compaction was not performed. Five half-part (Table 4) cases were simulated in this effort and compared to the strip model to understand the capabilities of the reduced-order strip model to predict spring-in trends observed in the high-fidelity simulations. In contrast to the high-fidelity simulations, the strip model simulations shown in Table 4 used quadratic hexahedral elements (C3D20) and took only two to four hours of wall clock time[1] to complete, which included flow-compaction analysis.

Table 4. Comparison of spring-in results from half-part model and comparable strip model

Simulation Definitions						**At Mid-side from Half-part Model**	**From Comparable Strip Model**
Ply Angle Condition*	Cured Thickness (mm)	Cure Cycle	Initial V_f	Initial DOC	Friction Coefficient	Avg Spring-in Angle at 20°C after Post-cure (deg)	Avg Spring-in Angle at 20°C after Post-cure (deg)
Uniform	3.530	Cold-Fast	0.57	0.001	0.05	0.95	1.07
Baseline	3.530	Cold-Fast	0.57	0.001	0.05	0.95	1.07
Baseline	3.530	Cold-Fast	0.63	0.001	0.05	0.84	0.97
Baseline	3.530	Hot-Slow	0.57	0.001	0.05	0.95	1.14
Baseline	3.530	Hot-Slow	0.63	0.001	0.05	0.85	1.05

*wrapped end of ply angles modeled in half-part simulations only

The results show that the half-part (Table 4) simulations predict slightly lower mid-side spring-in than the comparable strip model simulations, likely due to the added constraint of the end of the part which is not represented in the strip models. The half-part results also indicate that different wrapped end of ply angles, within the bounds of the simulations in this study, have a negligible effect on spring-in. The only difference between the first and second half-part cases in Table 4 is the ply angle condition used in the simulation – the "uniform" ply angle condition represents a significantly different condition than the "baseline", yet the average spring-in angle results were identical. In general, the key trends from the half-part and strip models agree: increased fiber volume fraction results in decreased spring-in and the cold-fast cure cycle slightly reduces spring-in compared to the hot-slow cycle.

A trade study using the strip model was conducted to ascertain significant factors effecting spring-in variability and to assess modeling practices and element types. The trade study involved forty-six unique strip model simulations that were generated by varying the parameters listed in Figure 1 within the specified bounds. The bounds were determined based on experimental observations (cured thickness), the resin technical datasheet (cure temperature and ramp rate), the prepreg material certification package (initial V_f), preliminary study (initial degree of cure), and "best practice" assumptions (friction coefficient). A preliminary study estimated the upper bound of the initial degree of cure (DOC) by exercising the resin cure kinetics models to predict degree of cure after an isothermal hold at 22.2°C (room temperature) for the maximum allowable out-time. According to this simulation, the upper bound for the initial DOC was predicted to be 0.14. This DOC value was greater than expected and it should be noted that room temperature predictions

[1] Reported wall clock times are based on running 4 cores on a 2016-era computer and should be considered relative.

are near the lower bound of COMPRO's predictive range. Figure 2 displays a graphical summary of the findings from the strip model trade study.

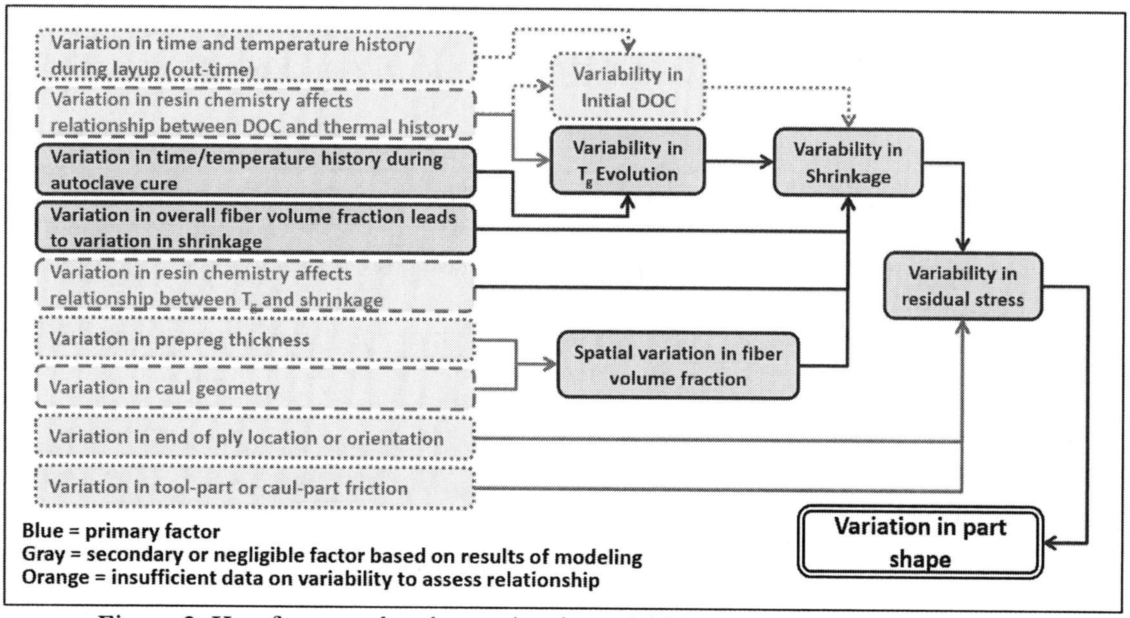

Figure 2. Key factors related to spring-in variability as predicted by the models

The results of the strip model trade study indicated that the factors with the greatest impact on spring-in variability are variations in V_f and time-temperature history during autoclave cure. Of the two, V_f variation is the most significant. In all trade study cases the strip model simulations sequentially ran all three COMPRO analyses (thermo-chemical → flow-compaction → stress-deformation). The development of spatial variation in V_f due to resin migration/bleed during consolidation in the pre-gelation regime, predicted by the flow-compaction analysis, was used in the stress-deformation analysis to predict residual stress development and subsequently spring-in. It should be noted that the flow-compaction analysis of the strip model does not capture the full three-dimensional effects of resin flow along the length of the part. Based on these analyses, spring-in is predicted to decrease with increasing V_f and increase with increasing time-temperature during autoclave cure. Other factors that were not sufficiently addressed in this study but may also influence spring-in variability include variations in resin chemistry/cure kinetics model parameters and caul fit/geometry. The caul is assumed to fit perfectly in the strip model, and the effect of caul preload, which moves resin and results in local changes in V_f, is not accounted for in the strip model simulations. Secondary or negligible factors that did not contribute significantly to spring-in variability within the parameter ranges investigated in this study include variations in initial DOC as a result of prepreg out-time, prepreg thickness, wrapped end of ply location and angle, and friction coefficient.

Results of a portion of the trade study related to modeling practices showed that the mesh refinement in the strip model was sufficient to achieve a converged solution and that element type has a minimal effect on the spring-in predictions over the range of element types assessed (C3D8, C3D8I, and C3D20). In fact, the mesh resolution in the strip models could be relaxed without a significant impact on the spring-in predictions. This result is meaningful because the original strip model simulations used C3D20 elements and had greater mesh refinement than the half-part model that was compared in Table 4, which was primarily comprised of first-order C3D8 elements. This result provides confidence in the model predictions and the findings from the comparison between the high-fidelity and reduced-order simulations.

4. IMPLEMENTATION OF ANALYTICAL MODEL IN MODELCENTER

An objective of this effort was to develop models and software tools for predicting spring-in variability at a reduced computational cost compared to more detailed finite element models and thus provide an accessible process modeling tool for manufacturing environments. The reduced-order strip model showed capabilities for capturing trends in spring-in variability at a significant cost savings compared to the high-fidelity simulations, completing in only two to four hours of wall clock time using four cores on a 2016-era computer. However, the time required for even these simple models is not ideal for probabilistic study of spring-in variability, which requires the execution of the model multiple times. A decrease in analysis time can be achieved by using the analytical spring-in model described in Equation 1. In fact, the two most important factors related to spring-in variability as predicted by the ABAQUS-COMPRO models, variations in V_f and time-temperature history during autoclave cure, are captured by the analytical model. Thus, based on the analytical model, an integrated software framework for fast prediction of spring-in variability was developed using ModelCenter.

4.1 Integrated Software Framework for Fast Prediction of Spring-in Variability

ModelCenter is an application developed by Phoenix Integration that enables the creation of simulation workflows by integrating and automating individual engineering models and provides an integrated environment for performing trade studies and probabilistic analysis. ModelCenter was selected as the software platform for implementing the spring-in model due to its accessibility and use in manufacturing environments, built-in analysis and visualization capabilities, and open-architecture that allows for integration of additional user-developed or third-party analysis tools.

To make the spring-in model more accessible to users in manufacturing settings, an Excel-based input file/database was created. This Excel-based input file is used to generate and visualize the time-temperature profile for autoclave cure and define random variable distributions for probabilistic analyses, and the composite material system to use in the analysis is selected from the included database. ModelCenter is used to integrate the input file with the cure kinetics models for the composite material system. ModelCenter exercises the cure kinetics models and then calculates the composite effective longitudinal (α_l) and transverse (α_t) CTE and composite effective longitudinal (Φ_l) and transverse (Φ_t) cure shrinkage strain according to the following micromechanics equations [26]:

$$\alpha_l = \frac{CTE_{11f}E_{11f}V_f + CTE_rE_r(1 - V_f)}{E_{11f}V_f + E_r(1 - V_f)} \quad (2)$$

$$\alpha_t = \left(CTE_{22f} + \nu_{13f}CTE_{11f}\right)V_f + \left(CTE_r + \nu_rCTE_r\right)\left(1 - V_f\right) - \left[\nu_{13f}V_f + \nu_r(1 - V_f)\right]\alpha_l \quad (3)$$

Where CTE_{11f} is the fiber longitudinal CTE, E_{11f} is the fiber longitudinal elastic modulus, CTE_r is the resin CTE, E_r is the resin longitudinal elastic modulus, CTE_{22f} is the fiber transverse CTE, and ν_{13f} is the Poisson's ratio describing contraction in the through-thickness direction given an extension in the longitudinal direction. Equations 2 and 3 are also used to calculate the effective cure shrinkage factor (CS) in the longitudinal and transverse directions by replacing the CTE terms in the equation with CS terms. Note that the fiber CS is zero. The effective CS is then integrated over DOC to obtain the effective cure shrinkage strain. After the composite effective properties are calculated, ModelCenter passes these values to the analytical spring-in model described by Equation 1. Spring-in can be calculated for a deterministic case or a full cumulative distribution or sensitivity analysis can be performed based on the specified input variation using embedded probabilistic components in the ModelCenter workflow.

4.2 Global Sensitivity Analysis

In this effort, a Global Sensitivity Plugin was developed for ModelCenter that performs variance based sensitivity analysis [27]. This global sensitivity analysis component is embedded in the ModelCenter workflow and was used to perform a sensitivity analysis of spring-in variability to understand the relative importance of variations in the input parameters on spring-in variability. The input parameters investigated in this sensitivity study (initial V_f, cure cycle ramp rate, cure temperature, and initial degree of cure) were the same as those investigated in the strip model trade study, with the exception of cured thickness and friction coefficient since the analytical spring-in model does not account for thickness effects or tool-part/caul-part interactions. All input parameter variations were described by uniform distributions based on the upper and lower bound values used in the strip model trade study (Figure 1). Figure 3 shows the results of the spring-in sensitivity study for a nominal initial component angle of 90 degrees, with the input parameter distributions annotated on the plot.

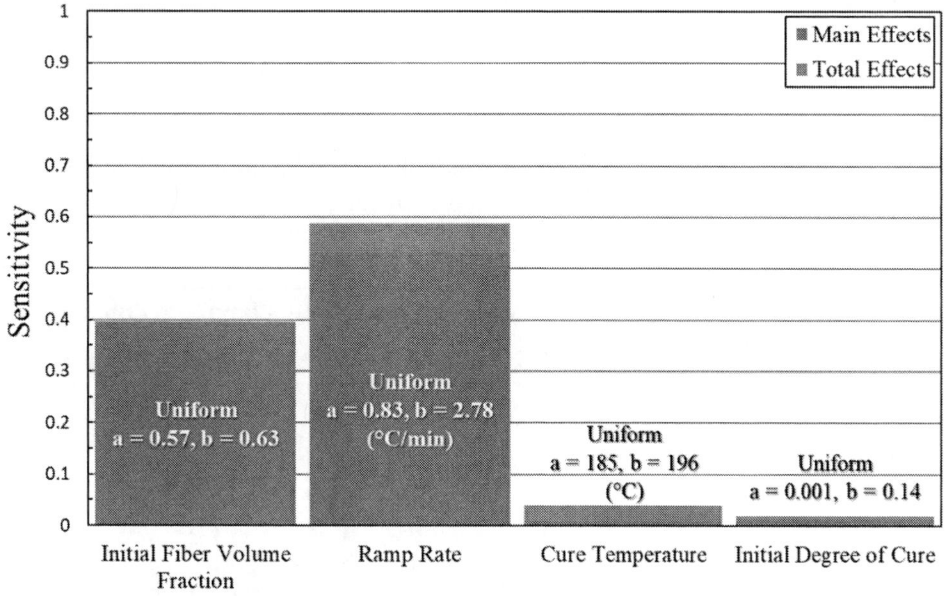

Figure 3. Spring-in sensitivity study using analytical model implementation in ModelCenter

In total, 6,000 instances of the analytical spring-in model were exercised in the sensitivity study, which took roughly four and a half hours of wall clock time. In comparison to the ABAQUS-COMPRO models, it took roughly the same amount of time to perform the entire sensitivity study as it would take to run one to two strip model simulations. Based on the sensitivity study, variation in the initial V_f and cure cycle variation have the greatest impact on spring-in variability, and variation in the initial degree of cure as a result of prepreg out-time has a negligible effect on spring-in variability, which is similar to the findings from the ABAQUS-COMPRO simulations. The analytical spring-in model also captures the V_f and cure temperature trends observed from the ABAQUS-COMPRO simulations: spring-in is predicted to decrease with increasing V_f and increase with increasing cure temperature during autoclave cure. However, the analytical model predicts spring-in to decrease with slower ramp rates, which was not expected based on the ABAQUS-COMPRO simulations. In addition, the spring-in sensitivity study results indicate that cure cycle variation is a primary factor related to spring-in variability, whereas the ABAQUS-COMPRO models show variation in the initial V_f to be a primary factor and cure cycle variation to be a secondary factor. A possible reason for the differences between the ABAQUS-COMPRO and analytical spring-in model results is that the analytical spring-in model does not capture changes in the V_f during processing as the ABAQUS-COMPRO strip model does. It should also be noted that the analytical spring-in model is not able to account for spatial variations in properties as the ABAQUS-COMPRO models can. Overall, the analytical spring-in model captured similar trends to the ABAQUS-COMPRO models, with some minor differences, at a reduced computational cost, making it feasible to perform probabilistic studies.

5. EXPERIMENTAL RESULTS AND DISCUSSION

Measurements of the eight aerospace edge components fabricated in this study were used to identify trends and perform preliminary model validation activities. Each component was sectioned at three locations along the long axis of the part, and for each location thickness and V_f were investigated at four regions: one side, both corners, and the web. Using photomicroscopy, three thickness measurements were estimated in each of the four regions, for a total of thirty-six photomicroscopy thickness measurements for each component. Acid digestion was then used to measure local V_f at each of these same four regions, for twelve V_f measurements per component.

The strongest correlation observed in the data was between local cured thickness estimated from photomicroscopy and local V_f measured via acid digestion. Figure 4 shows the relationship between local thickness and local V_f in the components after cure.

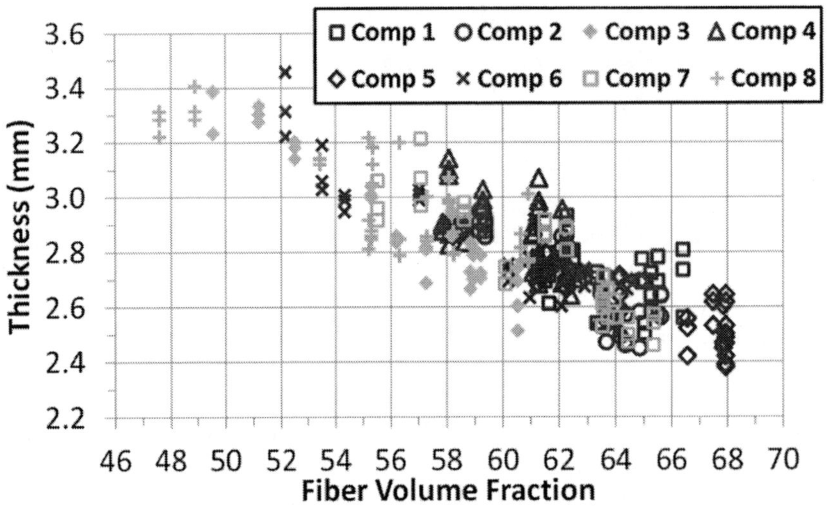

Figure 4. Relationship between local thickness and V_f in aerospace edge components after cure

The data in Figure 4 show that there is variation of up to 10 percent in V_f within each component. In general, the V_f in the corners was higher than the V_f in the web and sides. Comparing across all components the V_f in the corners ranged from roughly 55 to 68 percent, a 12 percent variation. There is less variation within a component when comparing one region (i.e. web, corner, or side) across the three locations along the long axis of the part, but some regions exhibit as much as 6 percent variation across the three locations. Note that for each V_f measurement in Figure 4 there are three corresponding thickness measurements, since three thickness measurements were made at each region compared to just one V_f measurement. The experimental results show that there is a strong inverse relationship between local thickness and local V_f. Due to the fact that the fiber is denser than the resin, increased V_f results in decreased thickness. This relationship is not surprising and indicates that local thickness could be used to predict local V_f. The relationship between local thickness and local V_f could be meaningful, given the strong relationship between initial V_f and spring-in predicted by the models.

To determine whether the experimental results confirm the trend between V_f and spring-in that the models predict, the average V_f for each component measured from acid digestion was compared to the spring-in angle estimated from the CMM inspection data (Figure 5). The V_f shown in Figure 5 is an average of the twelve V_f measurements (from four regions at three locations) made for each component. The values of deviation from the nominal angle between the sides, estimated from the CMM inspections, can be considered to represent the variability in the overall shape of the parts, with greater values corresponding to more spring-in. The correlation shown in Figure 5 exhibits the same trend as predicted by the model: spring-in variability is sensitive to variation in V_f and increased V_f results in decreased spring-in. This serves as a qualitative validation of the spring-in model predictions as the models show that they are capable of predicting key trends that exist in the experimental data. Attempts to show correlations between spring-in variability and the other variables controlled during aerospace edge component fabrication were not successful. The fact that only one aerospace edge component was fabricated for each of the controlled variable conditions limits the trend assessment. Furthermore, the large amount of scatter in Figure 5, and the apparent lack of correlation between the controlled variables in part fabrication, imply that some of the variables that were not controlled are important.

Comp #	Deviation from Nominal Angle Between Sides (deg)	Average Fiber Volume Fraction (%)
1	-0.270	63.6
2	0.131	62.2
3	0.525	56.5
4	0.000	60.0
5	-0.025	65.1
6	0.360	59.7
7	0.285	61.7
8	0.395	55.6

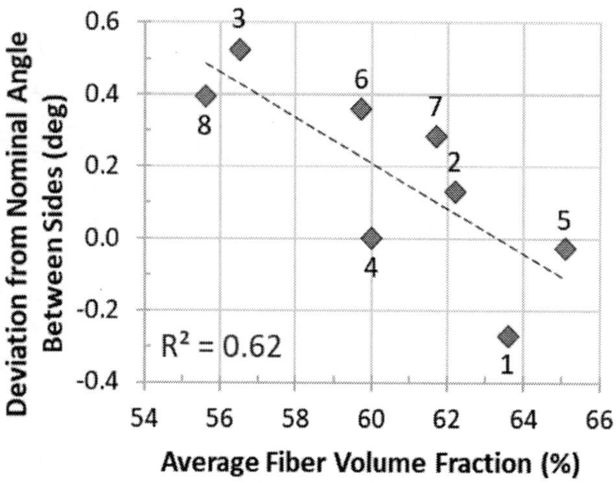

Figure 5. Relationship between spring-in and V_f in aerospace edge components after cure

6. CONCLUSIONS

A V&V framework for model development was employed to develop models for predicting spring-in variability. Two ABAQUS-COMPRO finite element models were developed in this effort: one high-fidelity and one reduced-order model. The reduced-order "strip" model simulations showed good agreement with the high-fidelity simulations at a significantly reduced computational cost. Tool maturity level (TML) was assessed using the ICME V&V guidelines and the model achieved TML 3 (on a scale from 1 to 5). TML 3 means that the model can predict key effects of incoming material and process variations on spring-in with reduced testing. In addition, an integrated software framework for fast prediction of spring-in variability was developed by implementing an analytical spring-in model in ModelCenter, which predicted similar trends to the ABAQUS-COMPRO models with some minor differences. The ModelCenter workflow for predicting spring-in has automated probabilistic capabilities, such as global sensitivity and full cumulative distribution analysis, and was specifically designed for use in a manufacturing setting.

All of the models indicated that, variation in V_f is a primary factor related to spring-in variability and increased V_f results in decreased spring-in. The model results also suggest that controlling initial V_f and cure cycle variation could help control spring-in and reduce spring-in variability. The ABAQUS-COMPRO models predicted cure cycle variation to be a secondary factor related to spring-in variability, however the analytical model appeared to predict a different relationship between spring-in and ramp rate and predicted cure cycle variation to be a primary factor. The same relationship between V_f and spring-in was observed in the experimental measurements of the eight aerospace edge components that were fabricated in this study. This agreement between the models and experimental V_f results served as a qualitative validation of the model predictions. However, the experimental results did not show a significant trend between cure cycle variation and spring-in variability as the models predicted, which is a somewhat inconclusive result due to limited experimental data (only one component was fabricated for each of the controlled variable conditions). The literature seems to indicate that there should be a significant trend between cure cycle variation and spring-in variability, but additional experiments are needed to confirm this. Attempts to show correlations between spring-in variability and the other variables controlled during aerospace edge component fabrication were not successful. The large amount of scatter in the experimental data, and the apparent lack of correlation between the controlled variables in part fabrication, imply that some of the variables that were not controlled in this study, such as caul fit and caul material integrity which influence pressure at local features, are important.

In conclusion, a reduced-order finite element process model and a fast-running integrated software framework, capable of capturing trends in spring-in variability, were successfully developed. These models provide significant computational savings relative to high-fidelity process models. Use of these tools could make spring-in prediction capability more accessible and reduce cost and schedule impacts of spring-in variability on composite manufacturing operations.

7. ACKNOWLEDGEMENTS

The authors wish to acknowledge the funding support for this work from the Air Force Research Laboratory, Materials and Manufacturing Directorate, Manufacturing and Industrial Technologies Division under Contract FA8650-15-C-5707. The authors also wish to acknowledge Chris Madsen of Northrop Grumman Corporation for his help with the experimental analysis and Convergent Manufacturing Technologies for their support throughout the model development process.

8. REFERENCES

[1] R. Sweeting, X. Liu, and R. Paton, "Prediction of processing-induced distortion of curved flanged composite laminates," *Composite structures,* vol. 57, no. 1-4, pp. 79-84, 2002. https://doi.org/10.1016/S0263-8223(02)00070-3

[2] K. Kosugi, S. Maekawa, Y. Hirose, T. Sana, T. Hatakeyama, and H. Tamura, "Low cost manufacturing approach for composite outer wing of SST," *Advanced Composite Materials,* vol. 10, no. 2-3, pp. 229-236, 2001. https://doi.org/10.1163/156855101753396708

[3] E. Kappel, "Forced-interaction and spring-in–relevant initiators of process-induced distortions in composite manufacturing," *Composite Structures,* vol. 140, pp. 217-229, 2016. https://doi.org/10.1016/j.compstruct.2016.01.016

[4] G. Fernlund, A. Poursartip, G. Twigg, and C. Albert, "Residual stress, spring-in and warpage in autoclaved composite parts," *TECHNICAL PAPERS-SOCIETY OF MANUFACTURING ENGINEERS-ALL SERIES-,* 2003.

[5] N. Ersoy, K. Potter, M. R. Wisnom, and M. J. Clegg, "Development of spring-in angle during cure of a thermosetting composite," *Composites Part A: Applied Science and Manufacturing,* vol. 36, no. 12, pp. 1700-1706, 2005. https://doi.org/10.1016/j.compositesa.2005.02.013

[6] D. Radford and T. Rennick, "Separating sources of manufacturing distortion in laminated composites," *Journal of Reinforced Plastics and Composites,* vol. 19, no. 8, pp. 621-641, 2000. https://doi.org/10.1177/073168440001900802

[7] D. Radford and R. Diefendorf, "Shape instabilities in composites resulting from laminate anisotropy," *Journal of Reinforced Plastics and Composites,* vol. 12, no. 1, pp. 58-75, 1993. https://doi.org/10.1177/073168449301200104

[8] G. Fernlund *et al.*, "Experimental and numerical study of the effect of cure cycle, tool surface, geometry, and lay-up on the dimensional fidelity of autoclave-processed composite parts," *Composites part A: applied science and manufacturing,* vol. 33, no. 3, pp. 341-351, 2002. https://doi.org/10.1016/S1359-835X(01)00123-3

[9] Q. Zhu, P. H. Geubelle, M. Li, and C. L. Tucker III, "Dimensional accuracy of thermoset composites: simulation of process-induced residual stresses," *Journal of composite materials,* vol. 35, no. 24, pp. 2171-2205, 2001. https://doi.org/10.1177/002199801772662000

[10] C. Huang and S. Yang, "Study on accuracy of angled advanced composite tools," *Material and Manufacturing Process,* vol. 12, no. 3, pp. 473-486, 1997. https://doi.org/10.1080/10426919708935158

[11] C. Albert and G. Fernlund, "Spring-in and warpage of angled composite laminates," *Composites Science and Technology,* vol. 62, no. 14, pp. 1895-1912, 2002. https://doi.org/10.1016/S0266-3538(02)00105-7

[12] S. R. White and H. Hahn, "Cure cycle optimization for the reduction of processing-induced residual stresses in composite materials," *Journal of Composite Materials,* vol. 27, no. 14, pp. 1352-1378, 1993. https://doi.org/10.1177/002199839302701402

[13] H. Wiersma, L. Peeters, and R. Akkerman, "Prediction of springforward in continuous-fibre/polymer L-shaped parts," *Composites Part A: Applied Science and Manufacturing,* vol. 29, no. 11, pp. 1333-1342, 1998. https://doi.org/10.1016/S1359-835X(98)00062-1

[14] J. Patterson, G. Springer, and L. Kollar, "Experimental observations of the spring-in phenomenon," in *Proceedings of the 8th international conference on composite materials (ICCM8)*, 1991.

[15] S. Clifford, N. Jansson, W. Yu, V. Michaud, and J.-A. Månson, "Thermoviscoelastic anisotropic analysis of process induced residual stresses and dimensional stability in real polymer matrix composite components," *Composites Part A: Applied Science and Manufacturing,* vol. 37, no. 4, pp. 538-545, 2006. https://doi.org/10.1016/j.compositesa.2005.05.008

[16] B.-S. Kim, N. Bernet, P. Sunderland, and J.-A. Manson, "Numerical analysis of the dimensional stability of thermoplastic composites using a thermoviscoelastic approach," *Journal of Composite Materials,* vol. 36, no. 20, pp. 2389-2403, 2002. https://doi.org/10.1177/0021998302036020882

[17] "COMPRO Model Documentation, release 1.5.13," Convergent Manufacturing Technologies, Nov. 2017.

[18] A. Johnston, R. Vaziri, and A. Poursartip, "A plane strain model for process-induced deformation of laminated composite structures," *Journal of composite materials,* vol. 35, no. 16, pp. 1435-1469, 2001. https://doi.org/10.1106/YXEA-5MH9-76J5-BACK

[19] T. Sreekantamurthy, T. B. Hudson, T.-H. Hou, and B. W. Grimsley, "Composite Cure Process Modeling and Simulations using COMPRO® and Validation of Residual Strains using Fiber Optics Sensors," in *Proceedings of the American Society for Composites: Thirty-First Technical Conference*, 2016.

[20] A. A. Hanson, S. M. Nelson, T. Briggs, B. T. Werner, B. L. Volk, and T. Storage, "Experimental Measurement and Finite Element Modeling of Residual Stresses in Simple Composite Structures," Sandia National Lab.(SNL-CA), Livermore, CA (United States), 2016.

[21] "NESSUS User Manual," Southwest Research Institute, San Antonio, TX, 2015.

[22] B. Cowles, D. Backman, and R. Dutton, "Verification and validation of ICME methods and models for aerospace applications," *Integrating Materials and Manufacturing Innovation,* vol. 1, no. 1, p. 2, 2012. http://dx.doi.org/10.1186/2193-9772-1-2

[23] B. A. Cowles, D. G. Backman, and R. E. Dutton, "Update to recommended best practice for verification and validation of ICME methods and models for aerospace applications," *Integrating Materials and Manufacturing Innovation,* vol. 4, no. 1, p. 2, 2015. http://dx.doi.org/10.1186/s40192-014-0030-8

[24] Cytec Solvay Group, "CYCOM 5250-4 Prepreg System," Technical Data Sheet, Mar. 2012 [Rev 01].

[25] Hexcel Corporation, "HexTow IM7 Carbon Fiber," Product Data Sheet, 2018.

[26] "CCA Material Constitutive Model Documentation, release 1.5.18," Convergent Manufacturing Technologies, Dec. 2018.

[27] A. Saltelli, P. Annoni, I. Azzini, F. Campolongo, M. Ratto, and S. Tarantola, "Variance based sensitivity analysis of model output. Design and estimator for the total sensitivity index," *Computer Physics Communications,* vol. 181, no. 2, pp. 259-270, 2010. http://dx.doi.org/10.1016/j.cpc.2009.09.018

UNDERSTANDING SIZE EFFECTS ON FLEXURAL PROPERTIES IN DISCONTINUOUS FIBER REINFORCED NYLON COMPOSITES

Siavash Sattar[1], Benjamin Beltran Laredo[1], Diego Pedrazzoli[2], Mingfu Zhang[2], Oleksandr G. Kravchenko[1]
[1]Old Dominion University, Norfolk, Virginia/USA
[2]Johns-Manville Corp., Denver, Colorado/USA

ABSTRACT

Discontinuous fiber reinforced composites are increasingly becoming more common among the aerospace and automotive industry. Their manufacturing cost and material characteristics make them quite versatile in their uses, especially for secondary structures. Due to discontinuous fiber morphology of the reinforcement these materials typically exhibit a size effect in the mechanical behavior, including both strength and stiffness. The flexural properties of long discontinuous fiber composites are subject to the span effect typical for high-performance continuous fiber composites. However, the meso-structure of long fiber- discontinuous composite is significantly different from continuous fiber laminates. Therefore, it becomes important to understand how the meso-structure in composite is affecting the mechanical response under flexural loading, especially taking into consideration the size effects due to proximity in scales between the fiber reinforcement and macroscopic composite. The present work considered organosheets produced using 1" (25.4mm) long glass fiber, which is embedded within a in situ polymerized polyamide 6 (nylon) matrix. The meso-scale morphology of the glass fiber bundles is preserved during the impregnation process, resulting in the collimated fibers. The flexural properties of the material were tested with different span ratios in 3-point. The size of the tested coupon had thickness of about 2.5mm, while the loading span was varied. Digital image correlation was used to understand the strain field deformation under different loading configurations. The material system revealed strong dependence of the flexural response with the loading span size. The initial increase of the flexural modulus with span below 1" is explained by the large contribution of shear deformation in case of short beam, while further increase in span lead to more gradual increase of the flexural modulus reaching the plateau behavior at spans exceeding 2" (50.8mm). The flexural strength on the other hand showed gradual decrease and plateau-like behavior with span increase from 1" to 2". Increasing the spans beyond that point was not considered due to large deformation and hard to reach complete failure. The behavior was explained by the reduced fiber damage mechanism as the span increases beyond the nominal fiber length, causing more substantial matrix plasticity and local disbanding between the collimated fiber bundles. To support the understanding of the present results, a computational modeling was established to describe the observed mechanical variability for this material system, including the dependence of the flexural properties.

Keywords: compression molding, discontinuous composites, size effect
Corresponding author: Oleksandr Kravchenko (okravche@odu.edu), Diego Pedrazzoli (diego.pedrazzoli@jm.com)

1. INTRODUCTION

Discontinuous fiber reinforced polymer composites are becoming increasingly more common among the aerospace and automotive industry, particularly for complex geometries. High production rates with low cost are significant advantages of short and long discontinuous fiber reinforced composites compared with continuous fiber composites. The mechanical properties of discontinuous fiber composites strongly depend on the fiber length and fiber orientation [1]. The fiber orientation in injection molded composites is controlled by the manufacturing parameters such as position and number of injection gates, pressure and temperature of molds, and filling rate [2]. Short discontinuous fiber composites typically do not provide a sufficient reinforcement for structural applications. Therefore, long-fiber thermoplastic composites can be used to provide the required mechanical properties for structural applications, while allowing for increased formability. Furthermore, due to the discontinuous nature of the reinforcement, the resulting mechanical properties strongly depend on the local orientation states.

Microscopic methods such as X-ray CT imaging, optical microscopy, and scanning electron microscopy (SEM) can be used to analyze the orientation distribution of fibers by computing the average local fiber distribution through the specimen thickness and cross-section [3]. Thi et al. [4] used X-ray computed tomography to measure fiber orientations of the injection-molded short-glass fiber reinforced PA 6 . The standard tests for determining the mechanical properties in short fiber injection molding composites typically consider the role of fiber orientation state produced by the flow. Specifically, conducting tensile and flexural tests on specimens processed in different mold flow direction is a method to analyze the effect of the fiber orientation on the material's properties. Wang et al. [5] conducted tensile tests for short glass fiber reinforced PA 6 specimens parallel and perpendicular to the mold flow direction to see the effect of loading direction on the material properties. Kravchenko et al. [6], [7] presented a parametric study to show the effect of the mesostructure, orientation state and the aspect ratio of the prepreg platelet molded composite on the effective tensile stiffness and strength, as well as the failure mode. Abrishambaf et al. [8] applied four-point flexural tests for steel fiber reinforced self-compacting concrete manufactured with different flowing directions to investigate the effect of the fiber orientation and dispersion on the creep behavior. Since the size of the individual fiber reinforcement in short fiber systems is on the order of 100-200 μm, which does not result in very high degree of mechanical anisotropy, the ASTM standard for measuring flexural properties typically recommends thickness-to-span ratio in the range of 16-20 following ISO 14125 [9] (described as Class II material). It is important to point out that measuring flexural properties of the continuous fiber composites considers the importance of shear-induced deformation due to high degree of mechanical anisotropy in composites even with relatively high span-to-thickness ratios (above 20). These size effects are known to cause apparent flexural modulus and strength variation with span-to-thickness ratio. In the case of continuous fiber material systems, the apparent flexural modulus is reported as a function of the span-to-thickness ratio and typically the modulus increases with increase in span until it reaches plateau value, which is characteristic of the tensile modulus of the composite [10]. Since long fiber thermoplastic composites present a discontinuous fiber material system with a very high aspect ratio (above 1000), the span ratio effect, characteristic for high-anisotropy composites, will be affected by the fiber orientation state. Due to the discontinuous fiber reinforcement having the same length magnitude as the span of the beam, additional size effects are expected to develop in long fiber thermoplastic composites. Therefore, it is important to understand how orientation state is affecting the apparent flexural properties in long fiber thermoplastic composites.

The present paper considers the analysis of the size effect in long fiber thermoplastic composites made of glass fiber reinforced Nylon 6 produced via in-situ anionic polymerization of caprolactam with non-uniform orientation of long fibers with uniform fiber length. This study presents the statistical microscopy characterization of fiber and bundle orientation distribution within the composite, which was used for establishing computational model. The flexural properties of long fiber thermoplastic composite specimens were tested with different span-to-thickness ratios in 3-point bending to analyze the size effect on the flexural mechanical properties for the given global fiber bundle orientation state. The future work will use the established computational model to predicting different failure modes in the composites, which can provide the explanation to experimentally observed behavior.

2. EXPERIMENTATION

2.1 Materials & Methods

The present work considered organosheets produced using 1" long glass fiber impregnated with a polyamide 6 matrix produced by Johns Manville Co. The mesoscale morphology of the glass fiber is preserved during the impregnation process, resulting in collimated fiber regions, called bundles. The matrix rich pockets are formed between the individual bundles. Therefore, the organosheet composites can be viewed as a two-phase system, consisting in bundles, which are represented by impregnated fibers with high fiber volume fraction, and PA6 matrix pockets (Fig. 1a). The volume fraction of fibers and the orientation distribution affects the mechanical and physical properties of the material. Furthermore, the mechanical properties of organosheets containing random bundles of fibers are governed by the bundle aspect ratio [6], defined as bundle length to thickness ratio, where increase in aspect ratio provides higher stiffness and strength. In this study, fiber orientation state, bundle orientation state and associated volume fractions were measured high-resolution microscopy analysis (Fig.1-a). The global fiber volume fraction (FVF) in the organosheet was computed based on the image binarization. The micrograph is 8-bit grayscale images with 256 different intensity levels. Using ImageJ software micrograph was binarized with an adaptive threshold and 256 levels reduced to 2 levels of black and white color scheme referred to the matrix and fiber region, respectively, (Fig. 1-b). In adaptive binarization method, the brightness of the background and the saturation of the black areas along the line is measured in order to find optimal binarization parameters for each separate line fragment. The high-resolution micrographs with visible and recognizable fiber boundaries were used in this study for adaptive binarization and the threshold was selected to ensure accurate representation of the fiber boundaries while choosing the threshold intensity during FOD measurements. The fiber volume fraction within a single bundle is an important variable governing the effective (homogenized) properties of the composite. The fiber volume fraction in bundles was measured by identifying the individual bundles and computing the corresponding FVF in each bundle (Fig1-c-d). As evident from equation (1), bundle volume fraction can be calculated from the below relationship.

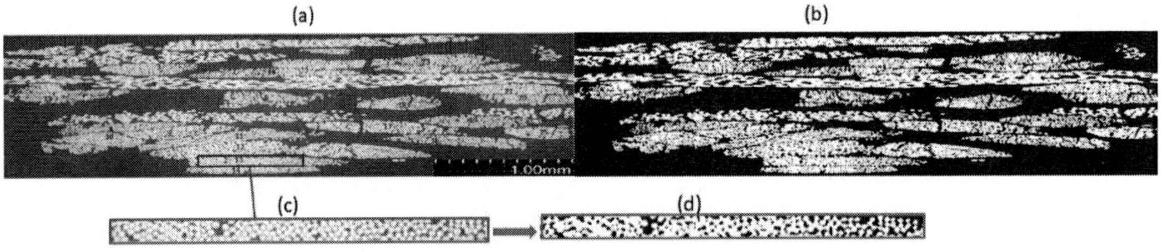

Fig.1 (a) Cross-section micrographs of an organosheet with labeling distinguished bundles by ID numbers; (b) binary micrograph; (c) cropped labeled fiber bundle; (d) binary fiber bundle

$$FVF_{glob.} = FVF_{bundle} \times BVF \quad (1)$$

where FVF_{global} is the FVF in the entire micrographs, FVF_{bundle} is the FVF in the bundles and BVF is the bundle volume fraction.

Since the planar fiber orientation was assumed, the spatial orientation of fibers in each bundle was defined as the 2nd order symmetric 2 × 2 tensor (Eq. 2) calculated as the dyadic product of its orientation vector p_i (Eq. 3-4) based on angle ψ_i between the plane's normal and the cylinder's longitudinal axis of the fiber shown in Fig 2-a. The angle of ψ_i is calculated by Eq. 5, where l_i and h_i, shown in Fig 2-b, are the major and minor diameter of the ellipse shape fiber, respectively. Fig 2-c shows the distribution of a cylindrical shape fiber diameters within the selected bundle in the range of 10-16μm, indicating that over two-thirds of fiber diameters were distributed in the range of 12 to 14 μm. Eq. 6 was used to calculate the individual selected bundle orientation by taking the average over the calculated fibers orientation in the bundle, where n is the number of the fibers within the bundle.

$$A_{ij} = \begin{bmatrix} a_{11} & a_{12} \\ a_{21} & a_{22} \end{bmatrix} \quad (2)$$

$$a_{ij} = p_i p_j \quad (3)$$

$$p_i = \begin{bmatrix} cos\psi_i \\ sin\psi_i \end{bmatrix} \quad (4)$$

$$\psi_i = \text{acos}\,(h_i/l_i) \quad (5)$$

$$A'_{ij} = \frac{\sum_n A_{ij}}{n} \quad (6)$$

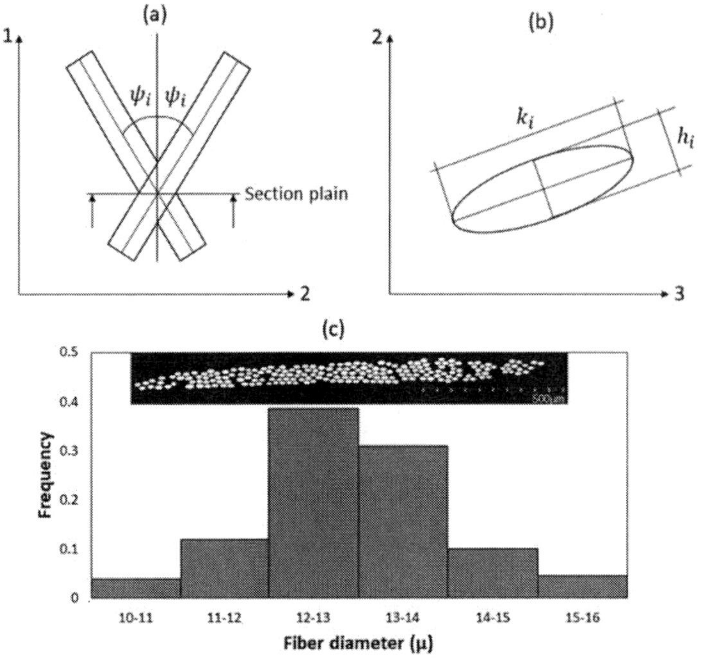

Fig.2 (a) Schematic of orientations for a fiber, (b) elliptical cross-section of fiber and (c) fiber diameter distribution within a selected bundle

In order to calculate the global bundle orientation tensor in the organosheet, equation 7 was applied using the area weighting factor, where k is the number of the recognized bundles within the micrograph and w_k is the area weighing factor of the selected bundle, calculated by Equation 8, where A_i is the elliptical cross-sectional bundle area in the plane of the cut, and ψ_i' is the average of fibers angle in the bundle. Ellipse fitting on the recognized labeled bundle was performed by Ellipse Fit software to measure the area of the bundles, and the major and minor diagonals, shown in Fig 3 for a single bundle, where k'_i and h'_i is the measured major and minor diagonals, respectively. Measured major and minor diagonals values in the plane of cut were corrected using the average of fibers angle in the bundle.

$$A''_{ij} = \frac{\sum_k w_k A'^k_{ij}}{\sum_k w_k} \tag{7}$$

$$w_i = A_i \times cos\,(\psi_i') \tag{8}$$

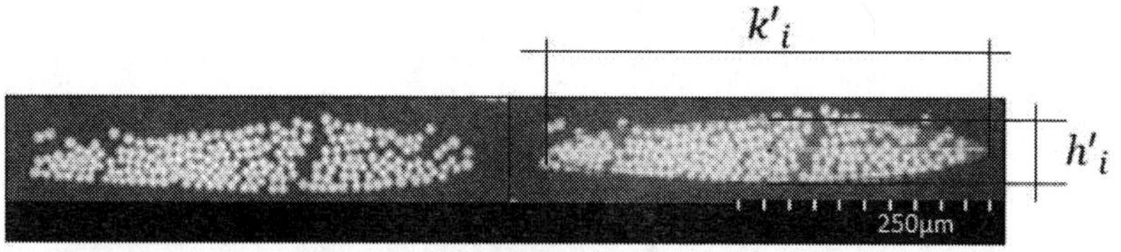

Fig.3 An example of ellipse fitting on the identified bundle

The experimental approach was designed to evaluate the flexural behavior of fiber reinforced composite under 3-point bending with different span ratios. In order to produce specimens with a thickness of 2.5 mm, compression molding was done using two layers of organosheets in a stainless-steel mold. Hot press was used for melting and consolidation process. The temperature of the press was increased to 230°C and 5-ton load was applied on the 10×10 inch panel for 15min. Specimens with dimensions of 127×25 mm were cut by a water jet machine.

2.2 Experimental procedure

The specimens were tested under 20, 30, 50, 70, and 100 mm loading span using the MTS test machine with a 10kN load cell. Digital image correlation (DIC) was used to evaluate the strain distribution pattern on the surface of the specimen for calculating strain field by the induced displacement. DIC method is based on capturing images of the surface pattern on the sample during the test. White scattered dots were spray-painted on the black surface of the samples to create the pattern for computing and visualizing the strain deformation. ARAMIS GOM Correlate software was used to analyze strain components. Fig.4 shows the testing set up for 70mm span with the DIC recognized surface pattern. Flexural stress and modulus were computed by equations 8 and 9, respectively.

$$\sigma = \frac{3FL}{2bt^2} \tag{8}$$

$$E = \frac{L^3F}{4bt^3d} \tag{9}$$

where, F is the applied load, L is the span length, b is the width of the specimen, t is the thickness and d is the deflection.

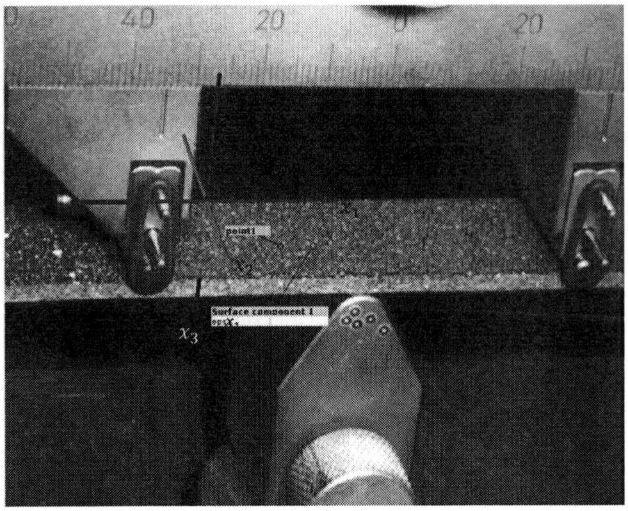

Fig.4 DIC image of three-point bending with 70 mm span

2.3 Modeling procedure

The microstructures of the model were generated using Digimat-FE containing bundles with transversely isotropic elastic behavior and matrix pockets. The effective properties of the bundle were calculated by analytical micromechanics using self-consistent field model based on the average fiber volume fraction in the bundles. The global bundle orientation was specified through an orientation tensor.

Based on the constructed organosheet, predicted effective properties of the unidirectional bundles, and the isotropic effective properties of the PA 6 matrix, a three-point bending test was implemented in ABAQUS with the different span lengths of 30, 50 and 70mm. Figure 5-a and b shows the top and the cross-sectional view of the generated organosheet, where the black and white region show the bundle and matrix pocket distribution, respectively. A voxel based finite element mesh was used for generation the structure of the organosheets. The voxelized representation of the bundles is shown in Fig. 5-c. Figure 5-d shows the configuration of the three-point bending model for the span of 70mm. Three different sample configurations were generated and analyzed for these span providing means to assess the mechanical properties and the effect of span ratio on this material system. The details about the models including, bundle volume fraction, bundle orientation tensor, bundle dimensions, and the homogenized material properties of the bundles is discussed in section 3-1, computed based on the microstructural analysis of the micrographs.

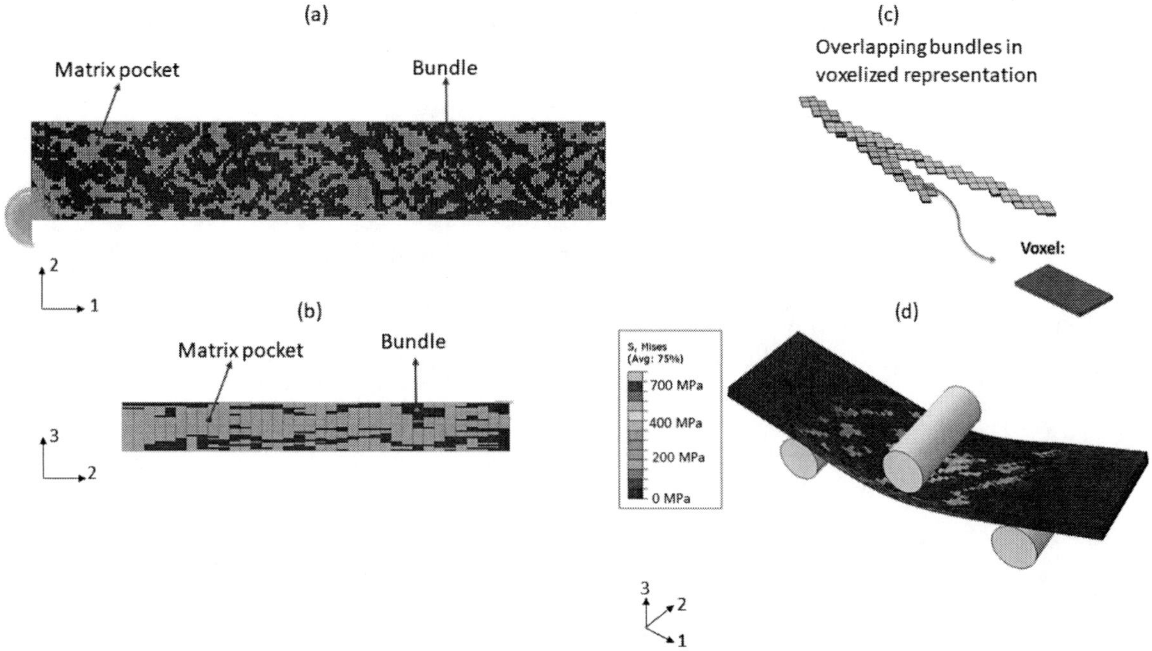

Fig.5 (a) Top view and (b) Cross-sectional view of the generated organosheet morphology. (c) Overlapping bundles in voxelized representation (d) Deformed state for 3-point bending analysis

3. RESULTS

3.1 Microstructural analysis

Table 1 presents the calculated FVF, average bundle volume fraction, and FVF in bundles by microstructural image processing on the 2-D cross-section micrograph, while bundle volume fraction, BVF, was calculated from Eq. 1. Table 2 shows the calculated effective material properties of the bundles with 68% FVF using self-consistent field micromechanical model. The corresponding bundle orientation tensor components are provided in Table 3, used for the bundle orientation in the constructed organosheets for the simulations. Figure 6 shows the bundle orientation distribution in range of 20°-80°, indicating that over half of all the bundles were oriented in the range of 30° to 40°. Figure 7 indicates the fiber orientation distribution for a selected bundle containing 191 individual fibers with orientation in the range of 50° to 70°, with the peak and average value being in the range of 60°-65. Figures 8 shows the distribution of the measured major and minor diagonal of the bundles, respectively, scattered in the range of 250-1650 μm and 50-120μm. The mean and standard deviation for major and minor diagonals was 786±228μm (coefficient of variance: 29.08) and 76±12μm, respectively. Since variations of the bundle major and minor diagonals were not significant, the mean value is considered as the bundle dimensions in the computational analysis, with the length of 1 inch.

Table 1. Characterized microstructural properties

Property	Value
Global FVF	0.34
Average FVF in bundles	0.68±0.03
Bundle volume fraction	0.55

Table 2. Effective material properties of the bundles

Property	Value
FVF in bundles, V_f	0.68
Glass fiber Young's modulus, E_f	85 GPa
Glass fiber Poisson's ratio, υ_{12f}	0.21
PA6 Young's modulus, E_m	1.8 GPa
PA6 Poisson's ratio, υ_m	0.43
Longitudinal modulus, E_{11}	58.4 GPa
Transverse modulus, E_{22}	9.1 GPa
In-plane shear modulus, G_{12}	3 GPa
Out-plane shear modulus, G_{23}	2.8 GPa
Longitudinal Poisson's ratio,υ_{12}	0.27
Transverse Poisson's ratio, υ_{23}	0.45

Table 3. Average bundle orientation tensor

Property	Value
a_{11}	0.53
a_{22}	0.47
a_{12}	0.39

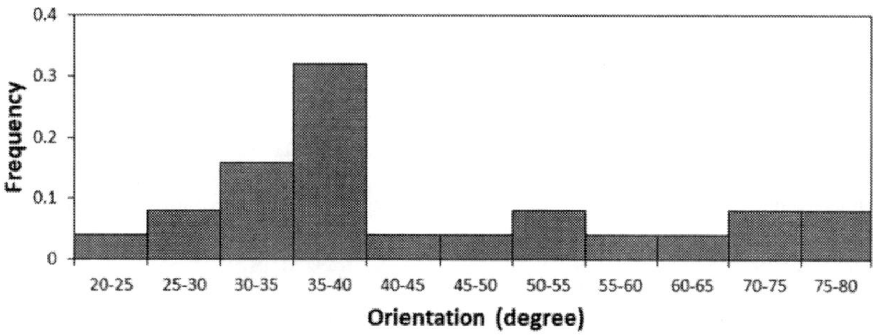

Fig.6 Bundle orientation distribution

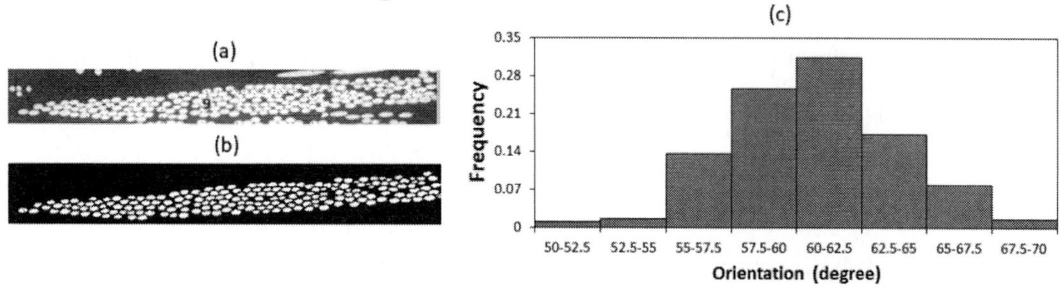

Fig.7 (a) Cropped image of the identified bundle; (b) Binary image of bundle (c) Distribution of fiber orientations in the bundle

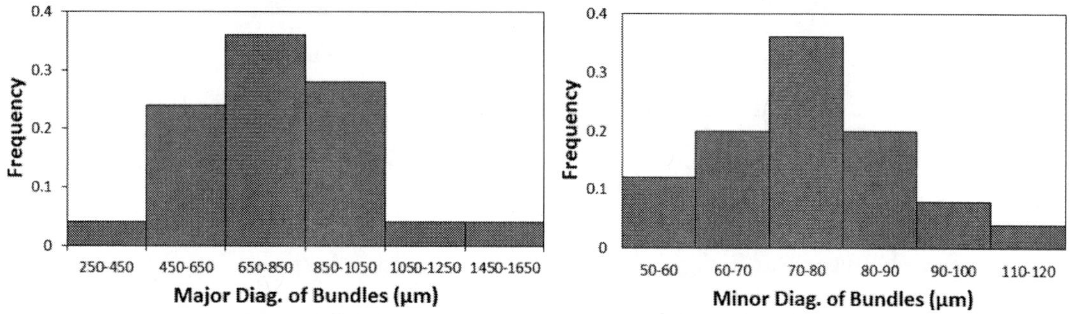

Fig.8 Distribution of the bundle major and minor diagonals

3.2 Flexural modulus and strength properties from experimental and modeling results

Figure 9 shows the DIC strain pattern in x_1 direction at the failure under 3-point bending for 30mm, 50mm and 70 spans. The strain fields show in Fig 9 were calculated based on the induced displacement on the surface of the sample, where the maximum strain was concentrated on a strip at the bending line, and the 30mm span testing has a higher strain value when compared with 50mm and 70mm span. Figure 10 shows the flexural stress-strain behavior in three-point bending for the span length of 20mm, 30mm, 50mm, 70mm and 100mm. The flexural strain value, calculated by the applied cross-head displacement of the MTS test machine, indicated a decline trend from 6.8% to 2.7% by increasing the span from 20mm to 100mm. Figure 11 shows the variation of flexural modulus and strength with respect to span-to-thickness ratios from experimental results. The initial increase of the average flexural modulus from 7.9 GPa to 9.6 GPa by increasing the span ratio from 8 to 11 can be related to effect of shear deformation in highly anisotropic long fiber composites in case of the shorter beam, while further increase in span leads to a more increase of modulus, following the gradual decrease to reach a plateau-like behavior at the span ratios exceeding 25. The initial increase of the modulus with span is similar to the behavior observed in continuous fiber composites, which is governed by the diminishing role of the intralaminar shear deformation in the laminates. A similar effect develops in the composites as span increased up to the double nominal fiber length (about 2" span for 1" nominal fiber length). However, as the span increases further, the drop in the flexural modulus is explained by the discontinuous fiber morphology governed by the orientation tensor.

The average flexural strength showed a gradual decrease and plateau like behavior from 408MPa to 389MPa by increasing the span ratio from 8 to 11, following the decline to 202 MPa by increasing the span ratio to 33. The decline of flexural strength is explained by a reduction in fiber damage when the span is exceeding the fiber length, causing higher matrix damage and bundle debonding. The dashed lines in figure 11-a-b show the average measured tensile modulus (12.5 GPa) and strength (174 MPa), respectively, from the tensile testing following the ISO 527-4 [11]. Tensile strength compared very closely with the flexural strength for the higher span ratio, while tensile modulus was higher than the flexural modus in different loading span. Fig 12 shows a comparison between the flexural modulus calculated from the modeling and experiment for the three-point bending test with 30 mm, 50 mm, and 70 mm span. The predicted flexural modulus agreed well with the experiments and was the most favorable at the 50 mm span test. As mentioned in section 2-3, three different sample configurations were generated and analyzed for these spans, and the error bars in Fig 12 shows the variation of the flexural strength for three analysis conducted for each span ratio. It is evident that the global flexural response can be adequately predicted by the proposed bundle-matrix based morphology. Future work will use the developed computational model to predict the damage propagation and the failure modes in the composites.

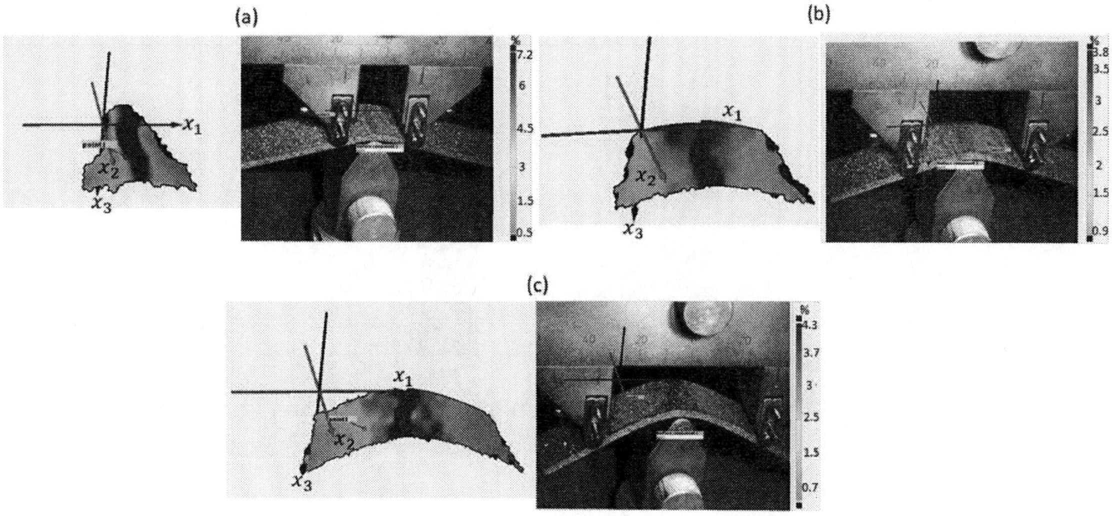

Fig. 9 DIC image of the strain field distribution at failure for three-point bending test with (a)30mm (b)50mm and (c)70mm span

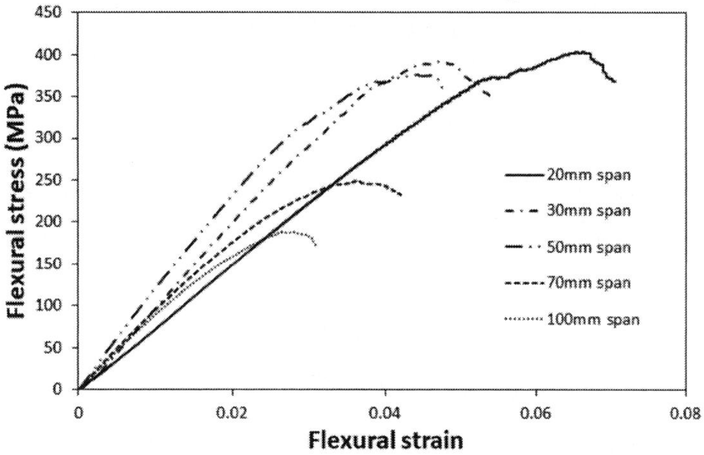

Fig. 10 Stress-strain behavior for 20mm, 30mm, 50mm and 100mm span length

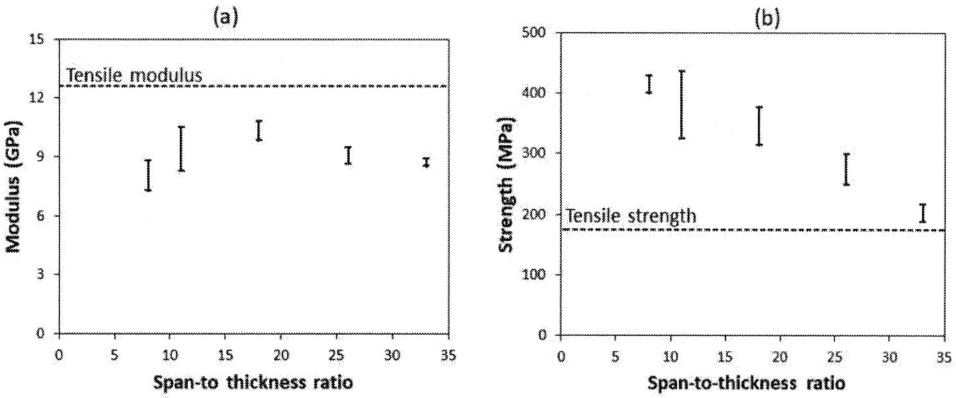

Fig. 11 (a) Flexural modulus and (b) Flexural strength with respect to span-to-thickness ratio

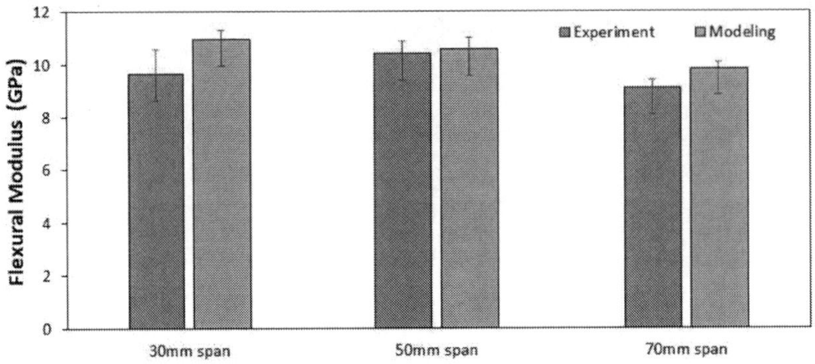

Fig. 12 Flexural modulus computed from three-point bending analysis and experiments

4. CONCLUSIONS

In this study, the effect of the loading span size on flexural modulus and strength of glass fiber reinforced PA6 composite was analyzed by performing three-point bending tests with different span-to-thickness ratios. The results revealed high dependence of flexural response with the loading span size in the long fiber thermoplastic composites. The flexural modulus exhibited a gradual incline by increasing the span to 30 mm (beam length to thickness ratio of 11), which is close to the nominal fiber length. The high contribution of shear deformation during beam bending for spans lower than the fiber length explains lower flexural modulus. However, the strength at spans below 30 mm was significantly higher, which is likely to be the cause of the increased fiber-governed damage. As span increased from 30 mm to 50 mm (beam length to thickness ratio of 18), a moderate increase in the flexural modulus was observed, while the flexural strength remained the same. Exceeding the span over 50 mm showed a gradual decrease of flexural modulus by reaching a plateau behavior. On the other hand, flexural strength exhibited a decline, which is expected due to the considerable matrix plasticity in the case of the higher loading span size.

The statistical microscopy analysis was performed to characterize the fiber and bundle orientation distribution and volume fraction, used to develop a computational model for analyzing the variability in the mechanical properties of the long fiber thermoplastic composite. The flexural modulus calculated from the computational modeling showed a close agreement with the experimental results. In future work, the model will be used to predict the failure mode and the damage propagation in the material.

5. REFERENCES

[1] S. Mortazavian and A. Fatemi, "Effects of fiber orientation and anisotropy on tensile strength and elastic modulus of short fiber reinforced polymer composites," *Composites Part B: Engineering*, vol. 72, pp. 116–129, Apr. 2015, doi: 10.1016/j.compositesb.2014.11.041.

[2] S. Mortazavian and A. Fatemi, "Fatigue behavior and modeling of short fiber reinforced polymer composites: A literature review," *International Journal of Fatigue*, vol. 70, pp. 297–321, Jan. 2015, doi: 10.1016/j.ijfatigue.2014.10.005.

[3] M. De Monte, E. Moosbrugger, and M. Quaresimin, "Influence of temperature and thickness on the off-axis behaviour of short glass fibre reinforced polyamide 6.6 – cyclic loading," *Composites Part A: Applied Science and Manufacturing*, vol. 41, no. 10, pp. 1368–1379, Oct. 2010, doi: 10.1016/j.compositesa.2010.02.004.

[4] T. B. Nguyen Thi, M. Morioka, A. Yokoyama, S. Hamanaka, K. Yamashita, and C. Nonomura, "Measurement of fiber orientation distribution in injection-molded short-glass-fiber composites using X-ray computed tomography," *Journal of Materials Processing Technology*, vol. 219, pp. 1–9, May 2015, doi: 10.1016/j.jmatprotec.2014.11.048.

[5] Z. Wang, Y. Zhou, and P. K. Mallick, "Effects of temperature and strain rate on the tensile behavior of short fiber reinforced polyamide-6," *Polymer Composites*, vol. 23, no. 5, pp. 858–871, 2002, doi: 10.1002/pc.10484.

[6] S. G. Kravchenko, D. E. Sommer, B. R. Denos, W. B. Avery, and R. B. Pipes, "Structure-property relationship for a prepreg platelet molded composite with engineered meso-morphology," *Composite Structures*, vol. 210, pp. 430–445, Feb. 2019, doi: 10.1016/j.compstruct.2018.11.058.

[7] S. G. Kravchenko *et al.*, "Tensile properties of a stochastic prepreg platelet molded composite," *Composites Part A: Applied Science and Manufacturing*, vol. 124, p. 105507, Sep. 2019, doi: 10.1016/j.compositesa.2019.105507.

[8] A. Abrishambaf, J. A. O. Barros, and V. M. C. F. Cunha, "Time-dependent flexural behaviour of cracked steel fibre reinforced self-compacting concrete panels," *Cement and Concrete Research*, vol. 72, pp. 21–36, Jun. 2015, doi: 10.1016/j.cemconres.2015.02.010.

[9] "ISO14125:1998," *ISO*. https://www.iso.org/cms/render/live/en/sites/isoorg/contents/data/standard/02/36/23637.html.

[10] L. A. Carlsson, D. F. Adams, and R. B. Pipes, *Experimental Characterization of Advanced Composite Materials, Fourth Edition*. CRC Press, 2014.

[11] "ISO527-4:1997," *ISO*. https://www.iso.org/cms/render/live/en/sites/isoorg/contents/data/standard/00/45/4595.html.

Improvements in the Sustainability of SMC and BMC Composites

Mike Gruskiewicz and Tim Pfister
LyondellBasell Premix LLC
Conneaut, Ohio, USA

ABSTRACT

Continued environmental and market pressures underscore the need to improve the sustainability aspects of all materials, including composites. Products which effectively achieve the most favorable life cycle performance will tend to gain market share in the long run. The requirement for cost and performance parity relative to traditional materials is a given; in fact, those factors largely play into the overall sustainability equation.

Uses of composites in many current applications are driven by the performance benefits during the use phase of the product. Composites are used in many cases *because* they improve sustainability in the application. For example, composites offer weight savings that save fuel in vehicles ranging from cars, to trucks, trains, and aircraft. In other examples, design flexibility allowing complex shapes achieves greater energy output in wind turbines, and lower energy consumption in HVAC blower housings. The durability and corrosion resistance of composites drives sustainability as well—a material that resists corrosion through multiple replacement cycles of alternative material inherently has a more favorable environmental footprint than shorter-life alternatives.

The challenge for composites is to build upon the great sustainability achieved in the use phase of the product by improving sustainability in the manufacturing phase of the product, and at the end-of -life phase. This paper examines some of the options to improve sustainability during the initial manufacturing phase. Specifically, SMC's (Sheet Molding Compounds) and BMC's (Bulk Molding Compound) have been compounded using raw materials that include renewable and recycled sources. These sustainably-sourced raw materials comprise components of the resin matrix, and can include functional fillers and reinforcing fibers as well. Through well-designed formulation approaches, molding characteristics and end-use performance attributes are maintained with little or no compromise in performance relative to traditional composites. Key examples of sustainable composites in high performance applications will be presented.

Keywords: Sustainability, recycling, SMC, BMC, composite,

Corresponding author: Mike Gruskiewicz

1. INTRODUCTION

The use of composites in many current applications is driven by the performance benefits during the use phase of the product. Composites are used in many cases *because* they improve sustainability in the application. For example, composites offer weight savings that save fuel in vehicles ranging from cars, to trucks, trains, and aircraft. In other examples, design flexibility allowing complex shapes achieves greater energy output in wind turbines, and lower energy consumption in HVAC blower housings.

Such environmentally-driven applications come under scrutiny for their overall environmental impact. One must legitimately consider the additional environmental impact during raw material extraction, during the manufacturing phases, and at the end-of-life. The concept of life cycle impacts has been around at least since the 1960's, and is now embodied in the ISO Standard

14040 (1). The overall concept for life cycle assessments is captured in Figure 1. The environmental impact of a product may take into account all activities and impacts from cradle to grave.

Figure 1: The Life Cycle

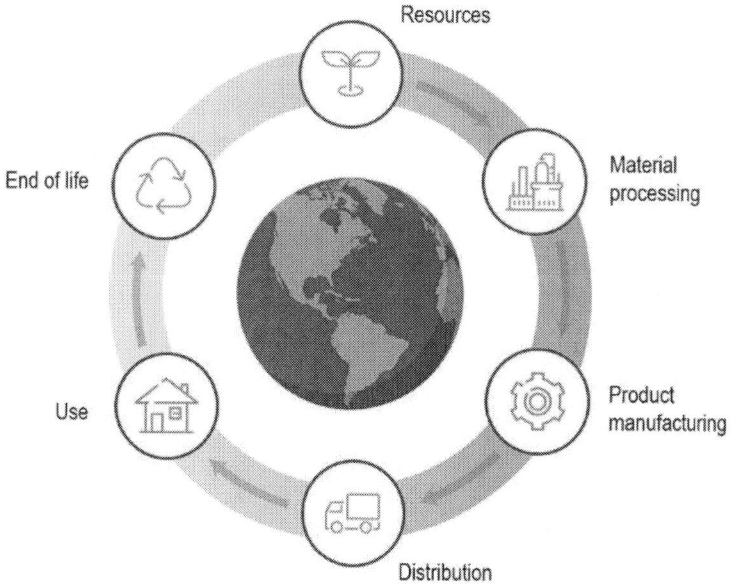

Our message, and the value creation for our customers related to performance benefits in the use phase, can potentially be diminished in other phases of the life cycle. Composites face much scrutiny since they are not easily or commonly recycled, particularly the dominant thermoset composites. While the industry is working aggressively to develop end-of-life solutions, this paper examines cases where creative approaches are being used to incorporate renewable and recycled raw materials as feedstocks to the composites—IE, reducing the footprint during extraction of raw materials and during manufacture of the composite.

Admittedly, this paper makes some assumptions, we believe reasonable assumptions, that incorporating recycled and renewable components into the products improves sustainability. That assumption must be proven quantitatively as the industry develops life cycle assessment tools and databases.

2. DISCUSSION

2.1 Semi-structural FR SMC Optimized for Sustainability

Our first example examines the development of a Sheet Molding Compound (SMC) designed for semi-structural panels or enclosures, where flammability resistance is desired. This work is a continuation and extension of work reported at CAMX2018 by Tim Pfister (2). The new development relates to a semi-structural SMC suitable for a variety of applications, including, industrial and recreational covers and enclosures, and utility vaults.

Pfister's work involved the development of an SMC which utilized both renewable and recycled content reacted into the polyester resin. The renewable component is achieved by replacing petroleum-derived propylene glycol with the same chemical from a renewable source derived from corn or soy. A recycled component is achieved via chemical digestion of PET thermoplastic. Since both of those feedstocks result in direct analogs, or very close analogs, of their petroleum counterparts, performance of the resin in the application is assured to be essentially identical, and this has been verified in the development.

An SMC formulation typically includes a multitude of components. Replacing just the resin portion with a more sustainable solution is a good step toward sustainability, but the impact is diluted unless the other components are addressed. A semi-structural SMC usually includes the components shown in Table 1:

Table #1: Composition of a typical semi-structural SMC	
Raw Material component	**Weight fraction**
Resin, with monomer	10-15%
Shrinkage control additive, with monomer	5-10%
Initiators and inhibitors	0.2-1%
Pigment	0.5-2%
Mineral filler	40-50%
Chemical thickening agent	0.5 to 2%
Fiber reinforcement	25-35%

It is apparent that, in a semi-structural SMC, filler comprises a substantial portion of the product weight. (That is not the case in structural SMC grades—as the fiber level is raised, the mineral filler must be reduced.) The target application which is the subject of this study requires flammability resistance, leading to the use of ATH to achieve flame retardance. ATH imparts flame resistance by releasing water vapor from its crystalline structure at about 220 deg C, drawing energy away from combustion. (This is another example where the baseline product offers environmental advantages in the "use phase", compared to alternate materials that may require halogenated additives to achieve FR performance.).

Although the ATH has a very critical role as a functional filler, our screening assessments suggested that we could displace 5% of the total weight of the ATH and still achieve the required flame resistance (targeted to achieve performance equivalent to UL 94 5V).

To that end, we incorporated a filler derived from soy hulls as a replacement for the ATH used in the baseline SMC. Note that the soy hulls are a low value byproduct from soy bean utilization. The hull is the outer skin of the bean, not the fibrous pod that is left in the field at harvest. The value of the soybean is driven most by the soy oil, which is used as a feedstock for printing inks, as feedstocks for other chemicals, or in the production of bio-diesel fuels. The oil is extracted from the bean, leaving the soy meal, and the soy hull, as depicted in Figure 2. The soy meal is highly valued as a human and animal feed, containing a high percentage of digestible protein. The soy hull, on the other hand, is of limited value and is used mostly as an extender in animal feeds.

Figure 2: Composition of the Soy Bean

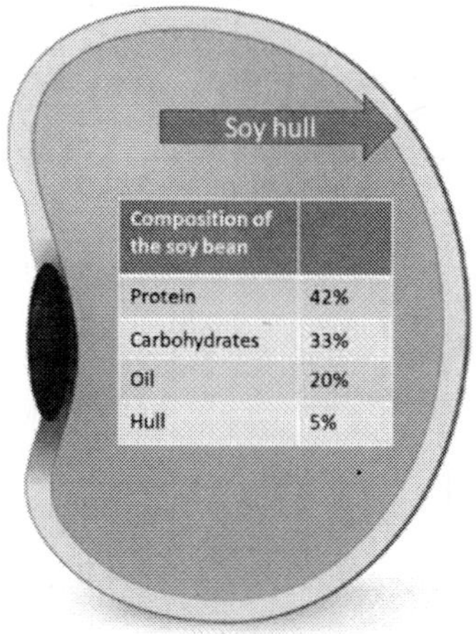

Like other plant based fillers, soy hulls can have a deleterious impact on the performance of the composite due to moisture uptake during the use phase of the composite. This downside has been mitigated in previous work, including work presented by Dr Paula Watt at CAMX2014 (3). By subjecting the soy hulls to torrefaction, they can be rendered to become an effective filler for composites.

In the current work, we utilized a resin that includes the bio-PG and rPET, and then further replaced 5% of the filler weight with the soy filler. The resulting properties are detailed in Table #2, where the control includes the sustainable resin with all ATH as the filler, and the experimental formula replaces 5% of the ATH with the torrefacted soy hull filler.

Table #2 details the outcome of the flammability testing, conducted as an unofficial screening evaluation, for a 500W vertical burn test. It is concluded that the SMC with the 5% addition of the torrefacted soy filler clearly passed the flammability requirement, although there is an inconsequential compromise in burn time of the individual specimens.

Table 2: Flammability test per 500W vertical burn

Sample/conditioning	Specimen #	Flame (sec)	Afterglow (s)	Total (s)	Max allowed (s)
Control, 48 hr	1	0	0	0	60
	2	4	0	4	60
	3	1	0	1	60
	4	2	0	2	60
	5	1	0	1	60
Outcome				**Pass**	
Experimental, 48 hr	1	5	0	5	60
	2	8	0	8	60
	3	6	0	6	60
	4	8	0	8	60
	5	9	0	9	60
Outcome				**Pass**	

In continuation of the work, Table #3 details a comparison of the mechanical properties of these same materials.

Property (units)	Method	Control avg	Control SD	Experimental Avg	Experimental SD
Tensile Strength (MPa)	ASTM D638	84.5	10.7	58.7	8.82
Flexural Strength (MPa)	ASTM D790	170	31.7	144	12.3
Flex Mod (MPa)	ASTM D790	9080	927	7480	792
Izod (J/m)	ASTM D256	527.7	1.42	528.6	0.74
Notched Izod (J/m)	ASTM D4812	527.4	1.45	527.4	1.11

It is noted that the tensile and flexural properties are somewhat compromised relative to the all-ATH control. This attests to the challenges encountered to incorporate bio-based fillers. Surprisingly, the impact properties were not compromised at all. Overall, while the experimental SMC product meets some of the design objectives, and may be a viable material in some applications, further work is required to achieve a product fully on par with the control.

2.2 A Novel BMC Solution Toward Improved Sustainability

Our next case study showcases a unique solution where a recycled polymer comprising another component of the formulation, the shrinkage control agent, has been implemented on a large-scale production basis within an application space. The application space is automobile lighting, where BMC has a critical role as the material of choice for many forward lighting reflectors. The application is very demanding, with critical parameters including thermal stability, dimensional stability, precise replication of molded angles and geometry, and suitability for basecoating and subsequent metallization. Approximately 30 million pounds of BMC is used in such applications annually in North America alone.

The tight dimensional requirements dictate that the BMC must be designed for low shrinkage, leading to a requirement for a low profile additive in the polymer matrix. Low profile additives for BMC include thermoplastic polymers, which are dissolved in monomer, usually styrene, to impart a phase structure that counteracts shrinkage that would otherwise occur upon polymerization of the polyester resin.

The selection of that thermoplastic polymer was recognized as an opportunity to achieve a more sustainable solution. Taillight and turn signal lenses are molded using acrylic polymers, such polymers are known to have utility as shrinkage control agents in BMC (4). Although one would think of these thermoplastics as being readily recyclable, in fact due to multi-color injection molding, it is not cost effective to separate,

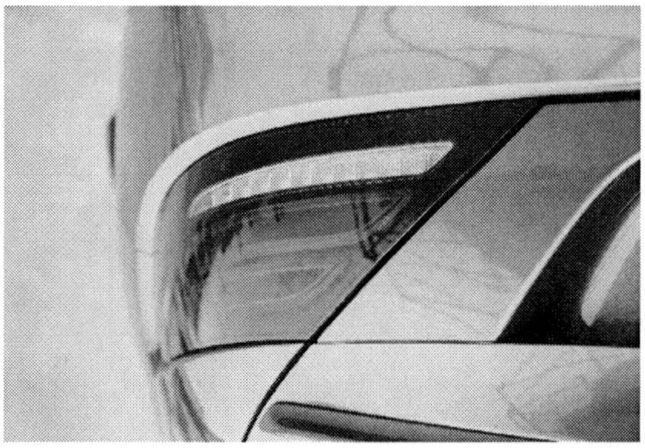

regrind, and re-process molded lenses in many situations, even on post-industrial basis. We have found that the scrap lenses can be ground to a powder, dissolved into the reactive monomer for our thermoset BMC, and achieve the desired function as shrinkage control agents. The mixed colors are not a problem since the BMC application is not color critical, and the fillers in the BMC obscure any effect of the colorant. Obviously, the process does require careful monitoring and quality assurance at each step. Controls have been developed which enable this circularity to work on a full production basis, now proven by years of reliable performance.

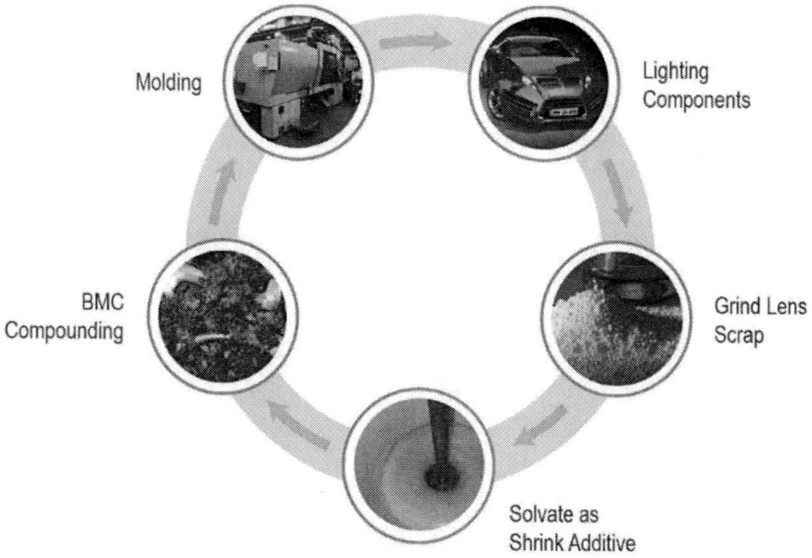

The fact that this can be achieved in a demanding, critical application is a testament to what can be achieved. This approach has already led to the utilization of more than two million pounds of lens material that otherwise would have been landfilled as scrap.

3. CONCLUSIONS

As we continue to work aggressively toward end-of-life solutions for our thermoset composites, we can continue to move forward toward greater sustainability by incorporating renewable and recycled components into the virgin compounds. The work discussed in this paper reveals some of the challenges that must be overcome to maintain performance, but also notes the achievement that can be had, with more than two million pounds of recovered product re-introduced to a high performance application. Advancing the overall sustainability of our products is key since the selling point for so many of our applications relates to sustainability in the use-phase.

4. REFERENCES

1. International Organization for Standardization, *ISO 14040*, Geneva, Switzerland

2. Pfister, *Sheet Molding Compounds Containing Renewable and Recycled Content for Improved Sustainability*, CAMX 2018, Orlando FL, 2018

3. Watt, Paula; *"Study of Cure Inhibition in Low Density High Bio-Based on Carbon (BBC) Molding Compounds"* Presented to the Composites and Advanced Materials Expo, October 13-16 2014

4. Kia, Hamid G. *Sheet Molding Compounds: Science and Technology, Chapter 4.* Cincinnatti, OH: Hanser/Gardner Publications, 1993

MODE I INTERLAMINAR FRACTURE TOUGHNESS OF BIOLAMINATES COMPOSITES CHARGED WITH REINFORCED PARTICLES

Franco-Urquiza, Edgar A.[1], Renteria, Victoria[2]; Perez-Mora, Ruben[1]; González-García, Pedro[1], Torres-Arellano, Mauricio[1*].

1 CONACYT-CENTA, National Center for Aeronautics Technology
Carretera 200, Parque Aeroespacial Querétaro, Querétaro, MEXICO 76270

2 Center for Engineering and Industrial Development (CIDESI), Av. Pie de la Cuesta 702, Querétaro, MEXICO 76125.

ABSTRACT

Composite biolaminates are a renewable resource, low-cost, with good mechanical properties and are totally biodegradable. Recently, the use of nanoparticles as reinforcement, has allowed the increment of the mechanical and thermal properties these materials. In this work the mechanical properties of biolaminates based on Agave fibers (Henequen) embedded in a bioepoxy resin with 45% natural component and with ZnO nanoparticles at different percentages by weight are evaluated. The mechanical properties show an increase in stiffness and strength with the presence of ZnO nanoparticles. The fracture behavior in mode I registered moderate changes in toughness, related to the type of fibers and the ZnO fraction. The results show the potential application of these biolaminates on commodities, building and automotive industries.

Keywords: Henequen; Bioepoxy; Microparticles; Fracture toughness
Corresponding author: mauricio.torres@cidesi.edu.mx

1. INTRODUCTION

Laminates are composite materials of thermoset matrix reinforced with long fibers. For decades, fiberglass laminates (GFRP) provide the baseline standard for their high specific properties, corrosion resistance and long service life. Lately, disadvantages on the difficulty of finding solutions to relevant issues such as recycling and product end-of-life (EoL) have come up. In the last decade, some research groups have proposed different approaches to combine recycled GFRP materials to obtain comparable properties of pristine composites [1, 2].

An alternative to reduce these serious problems is the development of biolaminates, which are composites with natural fibers and biopolymer resins. Biolaminates are emerging due to the need of environment friendly composites, biodegradability, and renewability.

Among natural fibers, plant fibers are one of the most abundant natural sources of reinforcement. For centuries, cultures have based their technological development on the use of natural fiber applications. In the case of Mexico, Henequen (Agave fourcroydes) has been used in textiles, because of their stiffness and simplicity to knit. These plants have been widely used since

Mexican ancient civilizations, and here now, their use as a reinforcement for plastics in automotive, navy and even aeronautics have gained attention.

Biopolymer resins are thermoset resins, with more than 30% wt of their compounds, obtained from biodiesel or biofuels. In the last decade, researchers have developed resins from different natural sources, obtaining a recyclable epoxy system suitable for resin infusion [3, 4].

The present work evaluates the mechanical and fracture properties of biolaminates with Henequen fibers as reinforcement and bioepoxy resin as matrix. With the goal to produce a multi-functional material, ZnO nanoparticles have been added, to promote thermal and fracture resistance. This research aims to promote the use of natural fibers as potential composite materials for non-structural components in automotive, construction and commodities industries

2. EXPERIMENTATION

2.1 Materials

Henequen natural fibers were employed for this study with a plain weave configuration. Epoxydized Vegetable Oil (EVO) bioepoxy resin by Sicomin Epoxy Systems®, with a mix bio-based carbon content of 45% in weight, was used as matrix.

ZnO nanoparticles were synthesized by coprecipitation method using zinc acetate (Zn(CH3COO)2 2H2O 99.4% from J. T. Baker®), sodium hydroxide (NaOH > 97.0% from MEYER Reactives®), distilled water and ethanol at 70%. Three weight percentages, 1, 2 and 3% wt, identified as ZnO1, ZnO2 and ZnO3 respectively, were downselected as particle reinforcement for the bioresin.

2.2 Manufacturing of biolaminates

Biolaminates were manufactured by Vacuum Assisted Resin Infusion (VARI). The vacuum pressure is fixed in -20 mmHg. The polymerization reaction took 24 h at 25 °C. Biolaminates have two layers, with a $[0/90]_2$ stacking and 30 cm x 30 cm x 0.25 cm as laminate nominal dimensions. ZnO particles were added to the resin by mechanical mixing and infused at room temperature. The resin inlet and outlet positions were at the edges of the biolaminates respectively. Fig. 1 shows the VARI process for the biolaminates.

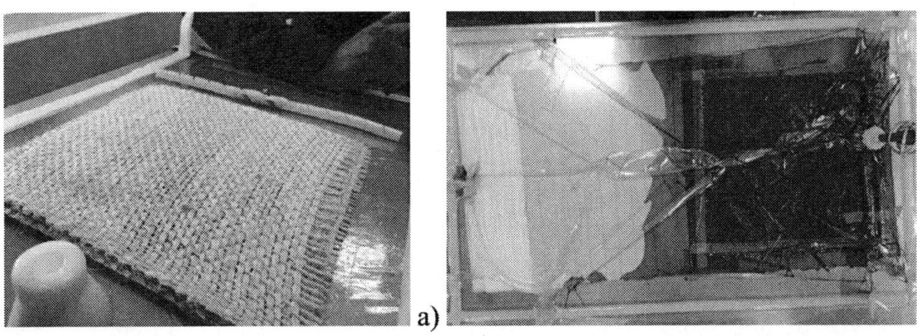

a) b)

Figure 1. a) Henequen plain weave, b) VARI of biolaminates.

2.3 Tensile characterization of biolamiantes

Mechanical tests were carried out in an Instron® 647 universal testing machine with a loading capacity of 100 kN. following ASTM D3039-3039M standard. The nominal dimensions of the tensile coupons were 250 x 25 x 2.5 mm, with a gage length of 150 mm. Testing was performed with a crosshead speed of 0.5"/min (1.27 mm/min). The load displacement-curve was recorded until failure of the coupons (3 repeats of each biolaminate configuration). Videoextensometer BESMAK Manta-AVX® with a 25 mm lens, was employed to record elongation and strain for each test by a non-contact technique.

2.4 Mode I interlaminar fracture toughness of biolaminates

Biolaminates fracture toughness was estimated by Double Cantilever Beam tests following ASTM D5528 standard. All DCB coupons had 100 × 20 × 2.5 mm as nominal dimensions. A 0.02 mm thickness nylon foil was placed the mid-thickness for representing a 25 mm pre-crack. DCB tests were carried out in the universal testing machine cited above at a crosshead speed of 2 mm/min until the fracture of one of the beams or until the crack path detours the middle plane. Coupons were connected to the testing machine by piano hinges. Crack growth was observed with the aid of a 20X magnifying lens and marking the crack extension with a fine ball-pointed highlighter.

The modified beam theory (MBT) method was employed method to calculate the critical energy release rate (G_{IC}). The Mode I fracture toughness, considering a uniform delamination front, is described in Equation 1 as follows:

$$G_{IC} = \frac{P\delta}{2ba} \qquad \text{Eq. (1)}$$

where P is the load, δ is the load point displacement, b is the specimen width, and a is the delamination length. Fig. 2, shows the mechanical setup for Henequen based biolaminates.

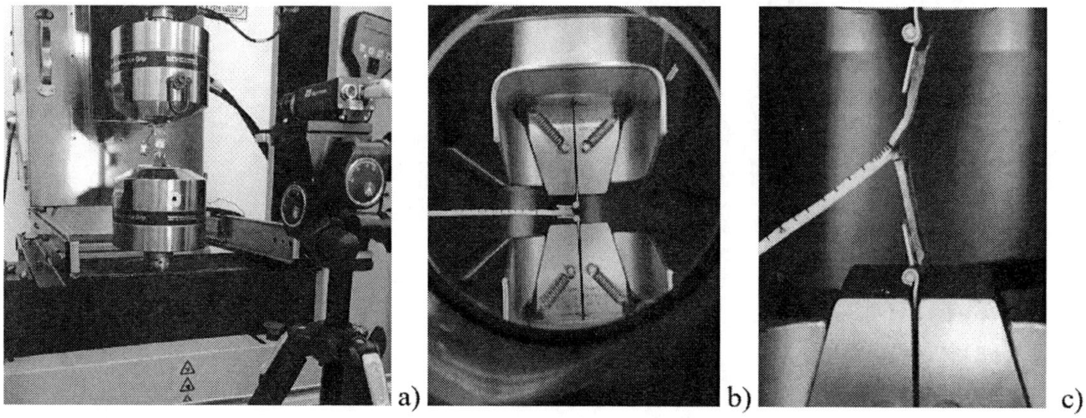

Figure 2. Setup for mechanical characterization, a) tensile test, b) DCB test, c) failure of DCB coupon

3. RESULTS

3.1 Characterization of ZnO particles by SEM

SEM micrograph were taken to examine ZnO nano powder, (see Fig. 3a) where some areas show nanoparticles agglomeration. ZnO particle size is between 50 - 100 nm with hexagonal morphology, typical of wurzite (see Fig. 3b).

Figure 3. Morphology of ZnO particles at a) 35,000 X and b) 100,000 X.

3.2 Biolaminates' tensile properties

Fig. 4 show respectively the average stress-strain curve for Henequen biolaminates with different ZnO weight percentage. For all specimens, two regions were perfectly identified. First region is an elastic linear behavior, from the beginning of the test until reaching 2500 microstrains. Hereafter, the second region shows a non-linear behavior, due to the typical sticking-slipping mechanism of woven laminates. In this second region, a progressive elasto-plastic behavior is exhibited, prior to sudden failure.

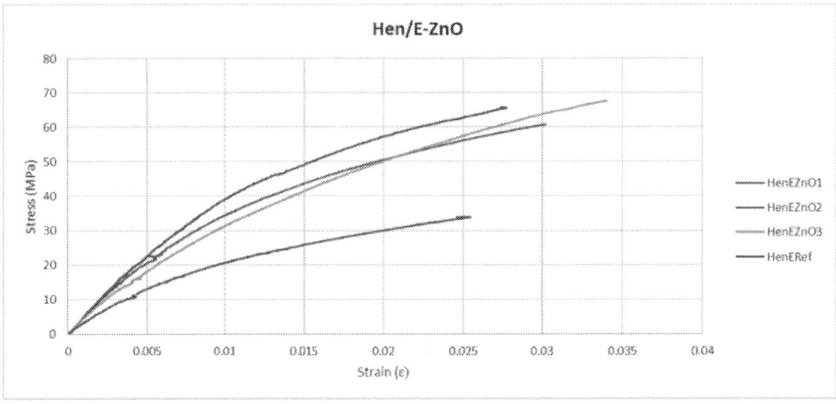

Figure 4. Average tensile stress-strain curves for Henequen biolaminates with different ZnO weight percentages.

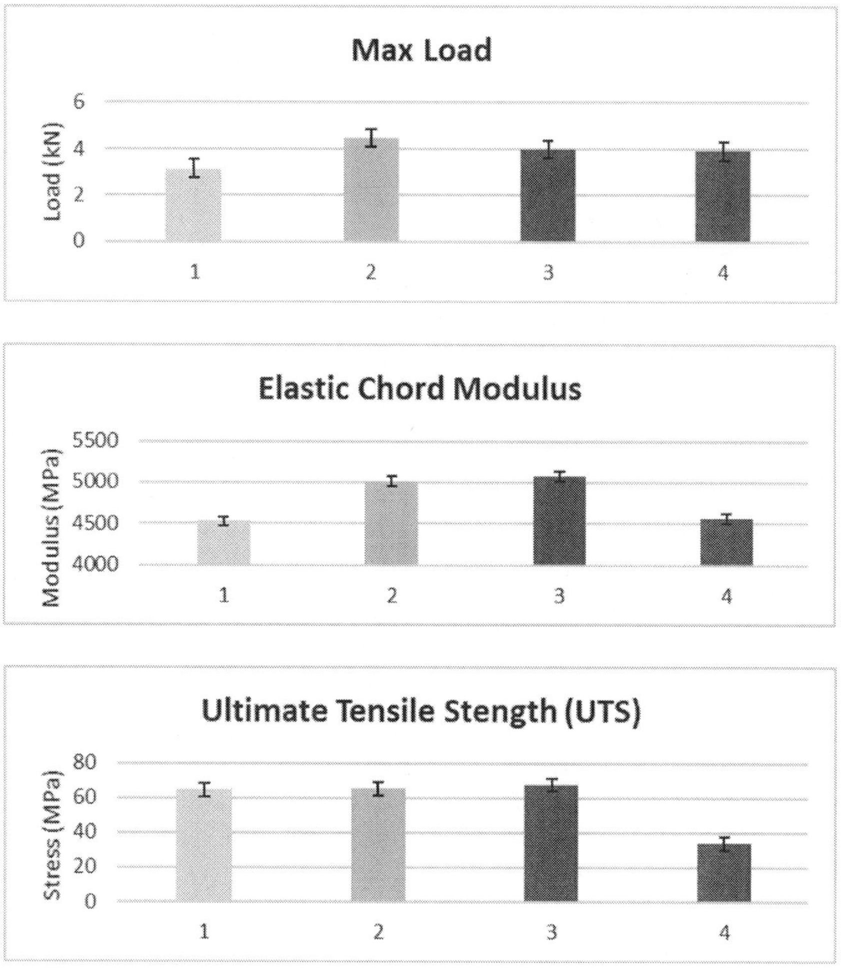

a) b) c)

Figure 5. Mechanical properties of Henequen biolaminates with different ZnO weight percentages. (1, 2 and 3 are for ZnO % wt., 4 is for reference).

All Henequen-EVO-ZnO composites show an increment on maximum load, elastic modulus and ultimate tensile strength, compared with Henequen-epoxy reference (see Fig. 5). At least an increment of 15% is recognizable. As a point to remark is the 95% increment on the UTS for the 3% ZnO wt biolaminates.

It is evident that the addition of ZnO increases the mechanical properties of Henequen, biolaminates. Further test must be run in order to find clear evidence of a proportional dependence between mechanical properties and ZnO weight amounts.

With the observed behavior, we speculate that ZnO increase the stiffness by favoriting the crosslinking between epoxy chains, also, by increasing the in-plane shear strength. For a thermoset polymer, the more numerous the crosslinking are, the higher stiffness is obtained [5, 6]. Therefore, the addition of foreign particles in the epoxy network prohibits microcrack growth and contributes to a diminution of the sticking-slipping failure mechanism.

3.3 Biolaminates' mode I interlaminar fracture toughness

The force-displacement curves for each biolaminate with different ZnO weight percentage are compared. For all coupons, the slope of the force-displacement curve is linear till the maximum force was reached.

Maximum peak load increases as ZnO weight percentage increases (see Fig. 6). The maximum peak load is 31 N, 33 N and 53 N for 1%, 2% and 3% ZnO wt respectively. Henequen composites average peak load is three times lower than carbon-epoxy plain weave composites (85 N) reported by Torres et al. [7]. After the first delamination has grown, the average subsequent loading–unloading steps are well defined. The average subsequent loading–unloading steps are between 10 and 25 N, for the 1% and 2% ZnO wt., and between 35 and 45 N for the 3% ZnO wt. before failure by beam's collapsing.

G_{IC} depends inversely of ZnO weight fraction for short crack's extensions (below 28 mm); however, this behavior changes for larger crack's extensions (above 29 mm). Average G_{IC} is 0.69 ± 0.14 Nmm/mm^2, 0.62 ± 0.07 Nmm/mm^2 and 0.74 ± 0.24 Nmm/mm^2 for 1%, 2% and 3% ZnO wt. respectively (see Fig. 7).

Woven composites have significantly greater scatter on fracture toughness, compared to unidirectional composites. This behavior is associated with the interaction of crack with interlaminar resin pockets, fiber weaviness, fabric humps and matrix-fibre decohesion, as delamination grows. Nonetheless, the GI average values are in the same order of magnitude, regardless the amount of ZnO nanoparticles.

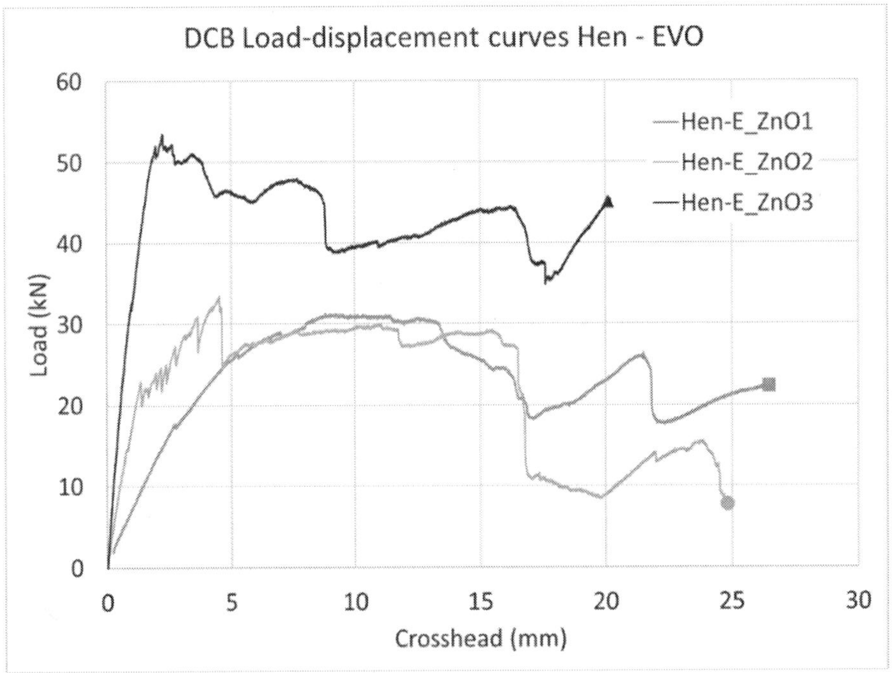

Figure 6. Average load-displacement curves using DCB for Henequen biolaminates with different ZnO weight percentages.

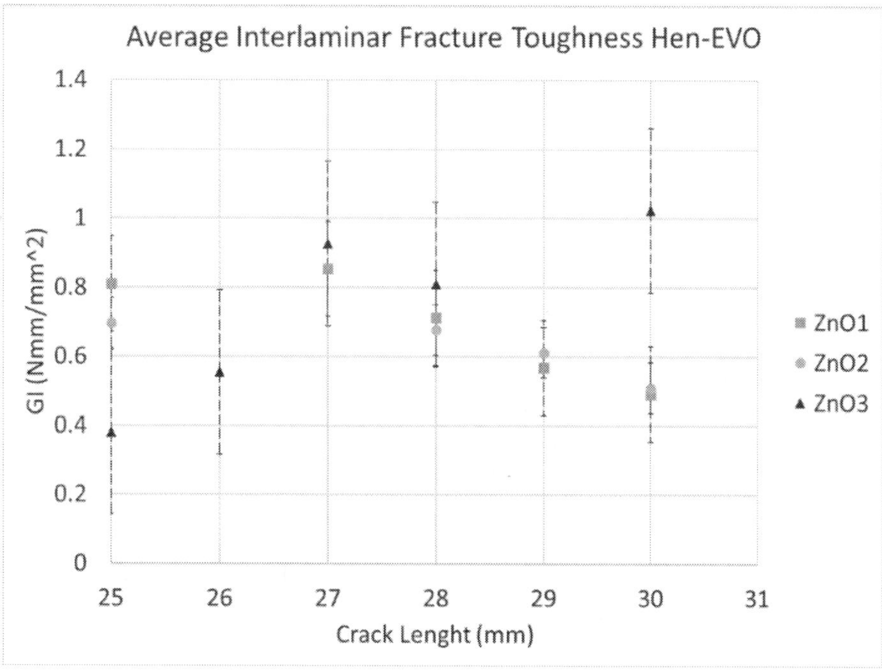

Figure 7. Average Mode I interlaminar fracture toughness (G_{IC}) for Henequen biolaminates with different ZnO weight percentages.

4. CONCLUSIONS

Biolaminates composed of Henequen natural fiber embedded in a bioepoxy resin and with ZnO nanoparticles at different percentages by weight are evaluated.

ZnO nanoparticles were obtained by coprecipitation method using zinc acetate, with a hexagonal morphology and a particle size between 50 and 100 nm.

All Henequen-EVO-ZnO composites show an increment of at least 15% on maximum load and elastic modulus, increment in ultimate tensile strength could reach 80%.

Fracture toughness of henequen biolaminates show that maximum peak load increases as ZnO amount raises. After the first delamination has grown, subsequent loading–unloading steps are well defined, and a larger crack extension is observed.

With these results, biolaminates with Agave fibers, bioepoxy resin and ZnO nanoparticles show promising features to be applied in automotive and construction industries.

Acknowledgements

This research was funded by the Secretary of Public Education and CONACYT through the Basic Science Fund (SEP-CONACYT Ciencia Básica), grant number 257458.

V. Renteria thanks CONACYT for the scholarship to pursuit her master's degree.

Edgar A. Franco-Urquiza, Mauricio Torres, Pedro González-García and Ruben Perez convey their special appreciation to the "CONACYT Researchers Program (Cátedras CONACYT)".

Manufacturing procedure was inspected by M. Sc. Víctor Gómez. Mechanical tests were performed by Eng. Ricardo Lozada.

5. REFERENCES

1. Aono Y., Murae, S., Kubo T. Static Mechanical Properties of GFRP Laminates with Waste GFRP Interleaf. 2011. Procedia Engineering **10**, 2080–2085.
2. Meira-Castro, A.C., Carvalho, J.P., Ribeiro, M.C.S., Meixedo, J.P., Silva F.J.G., Fiúza, A., Dinis, M.L. 2014. An Integrated Recycling Approach for GFRP Pultrusion Wastes: Recycling and Reuse Assessment into New Composite Materials Using Fuzzy Boolean Nets. Journal of Cleaner Production **66**, 420–430.
3. Kuo, P.Y., Saina, M., Yan, N. Synthesis, and characterization of an extractive-based bio-epoxy resin from beetle infested Pinus contorta bark. 2014. Green Chemistry **7**, doi.org/10.1039/C4GC00459K.
4. Cicala, G., Pergolizzi, E., Piscopo, F., Carbone, D., Recca, G. Hybrid composites manufactured by resin infusion with a fully recyclable bioepoxy resin. 2018. Composites Part B: Engineering **132**, 69-76.
5. Gu, J., Yang, X., Lv, Z., Li, N., Liang, C., Zhang, Q. Functionalized graphite nanoplatelets/epoxy resin nanocomposites with high thermal conductivity. 2016. International Journal of Heat and Mass Transfer **92**, 15-22.
6. Mittal, G., Rhee, K.Y., Miskovic-Stankovic, V., Hui, D. Reinforcements in multiscale polymer composites: processing, properties, and applications. 2018. Composites Part B: Engineering **138**, 122-139.
7. Torres, M., Tellez R.A., Hernández, H., Camps, T. Mode I interlaminar fracture toughness of carbon-epoxy coupons with embedded ceramic sensors. 2018. Adv Polym Technol 37, 2294–2302.

THIN-PLY: EXPLORATION AND MANUFACTURING WITH AUTOMATED FIBER PLACEMENT

Ramy Harik[1,2], Andrew Lovejoy[1], Catherine Yokan[3], Dawn Jegley[1]

[1]NASA Langley Research Center
Structural Mechanics and Concepts Branch
8 West Taylor Street
Hampton, VA 23681

[2]University of South Carolina
McNAIR Center, Department of Mechanical Engineering
1000 Catawba Street
Columbia, SC 29201

[3]Universität Bremen
Bibliothekstraße 1 Bremen, Bremen Area,
Bremen, 28359, Germany

ABSTRACT

Highly repeatable and nearly defect-free fabrication of composite parts is critical to the success and widespread acceptance of composite materials. Through optimization using thin-ply materials, composite parts can be manufactured to be lighter and tailored more specifically to anticipated design loads than with standard prepreg materials alone. However, defects arising from the thin-ply manufacturing process are not always similar to defects found with standard tows. These new defects warrant evaluation. At NASA Langley Research Center, the manufacturing process parameters associated with automated fiber placement (AFP), a slit tape-based composite manufacturing process, were optimized for the use of a thin-ply prepreg carbon-epoxy material. Carbon-epoxy tows with areal weights of 30 g/m^2 and 70 g/m^2 were used in these manufacturing trials. The AFP process parameters of heater output, compaction force, tow feed rate, and tow tension were adjusted and optimized for successful manufacturing. This article documents an exploration of thin-ply fabrication on both flat and complex-shaped surfaces. Ultimately, aerospace-quality laminates were made from the 70-g/m^2 material, but imperfections in the 30-g/m^2 material itself and the fact that the AFP machine was not designed for such a thin material meant that more research and trials are required to obtain flight-quality 30-g/m^2 laminates.

1. INTRODUCTION

As the demand for lighter, faster, and more fuel-efficient aircraft increases, manufacturers are seeking state-of-the-art materials and processes to cost-efficiently produce composite parts. The high specific strength and stiffness of carbon fiber composite materials make them an ideal substitute for many metal components in aeronautical and space applications. A basic composite material comprises a matrix, such as an epoxy, in combination with a fiber reinforcement. The resulting inherent directional dependency of the material allows engineers to tailor their designs with greater specificity to the loads, and thereby reduce overall weight [1] [2].

As is the case across almost all industries, composites manufacturers are moving to faster manufacturing methods, which frequently use automated processes as a more cost-effective option compared to traditional hand layup. One cost- and time-effective approach is automated fiber placement (AFP), which uses a robotic or gantry system to additively manufacture a part using slit composite prepreg tapes, also known as "tows". The process of laying tows and consolidating "plies", or layers of thermoset prepreg, with various fiber orientation angles is performed by the AFP system before the part is cured in an autoclave. As a result, fabrication is faster than with traditional hand layup, and the overall quality and reproducibility of the part is improved. The tows typically used in AFP have widths ranging from 3.175 mm to 12.7 mm (0.125 in. to 0.5 in.), which can reduce material waste and match the part boundary more closely than with many other automated processes such as automated tape laying.

At the NASA Langley Research Center (LaRC), AFP is conducted with the Integrated Structural Assembly of Advanced Composites (ISAAC) robotic system. Through research at LaRC with ISAAC, the boundaries of AFP capabilities are constantly being tested, including in the areas of new material forms and the steering of tows for variable stiffness designs. One new material form currently being studied is thin-ply prepreg material, which has less than half the areal weight of standard-ply material. By manufacturing composite parts with thinner plies, the performance could be improved and the weight reduced compared to parts built with only standard thickness plies. The use of thinner plies for selected structures widens the design space and provides opportunities for structural tailoring to specific design loads beyond what can be achieved with standard thickness materials alone.

In addition to the production of lighter parts and expansion of possible laminate designs, the use of thin-ply layers can also provide mechanical advantages such as a delay in the onset of damage and higher ultimate strength values. However, thin-ply fabrication has thus far been largely limited to hand layup, and there has been no literature encountered in the literature review on implementing thin-ply material in an AFP process. New manufacturing-related AFP thin-ply research activities at the LaRC are documented in this paper.

2. LITERATURE REVIEW

With the advent of composite tow spreading technologies, thin-ply material is becoming increasingly commercially available, and has already been implemented in high-performance and weight-sensitive applications such as solar planes and racing catamarans [3]. In comparison to standard prepreg material, thin-ply has a significantly lower fiber areal weight ranging from 15 g/m^2 to 70 g/m^2 compared to the typical 140 g/m^2 to 170 g/m^2. This lower fiber areal weight leads to opportunities in both design and mechanical properties.

Thinner plies permit the design and manufacturing of lighter composite parts and permit the inclusion of a greater number of fiber orientation angles within a given thickness. Composite engineers can propose more optimal laminates by using thinner plies compared to standard-thickness plies. Additionally, hybrid laminates can be created that utilize standard plies along the principal stress directions and thin plies to suppress delamination. Thin plies can also lead to smoother, more gradual laminate thickness changes.

Using thinner plies can increase some laminate mechanical properties compared to using standard-thickness plies. This thin-ply effect can be achieved by varying the laminate laying sequence while keeping the stacking sequence, fiber orientation angles, and fiber volume fraction constant. Laminates can be created with either ply-block stacking, where each fiber orientation angle is repeated "n" times as in $[+45_n/90_n/-45_n/0_n]_s$, or sublaminate stacking, where the sublaminate ply sequence is repeated n times as in $[+45/90/-45/0]_{ns}$. Amacher et. al., was one of the first research groups to observe this thin-ply effect by using a single fiber-resin system, but using a wide range of ply thicknesses, including standard plies, ply-block stacking, and sublaminate stacking [4]. Tsai et al., demonstrated a 10% increase in ultimate strength in unnotched tension of quasi-isotropic thin-ply laminates compared to those with thicker plies [5]. Lovejoy and Scotti [6] showed that combining thin and standard plies in a hybrid laminate can yield weight-savings for aircraft structure while having a reduced impact on manufacturing time.

The failure in composites typically begins with transverse microcracking before leading to catastrophic failure through delamination and possible fiber breakage. Delamination frequently occurs due to interlaminar stresses concentrated near the free edge of the laminate. Kim and Soni showed that the onset of delamination damage increases as the number of plies increases but is dependent on the fiber orientation grouping of plies [7]. The exact cause of the improved mechanical qualities is unknown, but may be attributed to a statistically greater number of defects in thicker plies in comparison to thin-plies that causes them to probabilistically fail at a lower stress, as well as the constraining effect of surrounding plies on intralaminar crack propagation [8].

Most of published thin-ply research at this point has focused on the mechanical properties of thin-ply material in comparison to standard-ply material based on hand layup. The use of AFP for standard thickness composite construction, however, has been well documented with widespread usage, *e.g.*, the Boeing 787 Dreamliner. Similar to the course of metal-working industrialization, AFP processes are based on multi-axis computer numerically controlled (CNC) machines with feeding mechanisms to build up a part layer-by-layer in an additive, rather than subtractive, manufacturing process [9]. Through controlled tow tension, compaction, heat, and tow feed rate, AFP places prepreg tows on a tool, or reverse mold of the intended part, and directly consolidates them with a compaction roller. A variety of resins can be used with AFP including most commonly used thermosets, which are then cured in an autoclave, and thermoplastics.

Although AFP is widely used, especially in the aircraft manufacturing industry, the process is not yet completely automated. The characteristics of the completed composite part are highly dependent upon tool geometry, environmental conditions, and in-coming material quality. Any modifications or fluctuations in the material or process can cause defects in the part to occur. These defects must subsequently be identified and repaired by hand. Such repairs can be quite time-consuming. Without repairs, defects can degrade the mechanical properties. The most common types of defects include gaps/overlaps, puckering, tow wandering, angle deviation, and wrinkles [10]. The presence of gaps decreases mechanical strength through the creation of resin-rich areas, while overlaps tend to locally increase the stiffness of the part [11], [12], [2]. Defects must be kept to a minimum in order to successfully manufacture consistent, high-quality composite parts. The ability to fabricate carbon-epoxy composites using thin-ply material in AFP would significantly expand its potential applications by providing a faster, more cost-effective, and reproducible method of manufacturing thin-ply composite parts.

3. AUTOMATED FIBER PLACEMENT WITH THIN-PLY

At LaRC, AFP research is conducted within the ISAAC facility. The ISAAC robot can be paired with a variety of end effectors, including an AFP head, is shown in Figure 1. Two additional degrees of freedom are provided by a 12.6-m long linear rail and a vertical rotator for the manufacturing of complex geometries [13].

Up to 16 6.35-mm (0.25-in) wide prepreg tows can be laid simultaneously by the AFP head. Following tape backing removal, prepreg tapes are redirected from spools located on the head into a feeding mechanism. The tows are fed onto the surface and cut to the appropriate length while being compacted by a 101.6-mm (4-in.) wide polytetrafluoroethylene-coated roller. Several tows placed at once comprise a course, which can be placed both uni- and bi-directionally. Multiple courses in one layer make up a ply, and several plies constitute the fully formed laminate. All aspects of the robotic system, as well as a 3.7-m by 1.8-m coupon table and additional end effectors, are located within the work cell. The entire system is located in a ISO 7 clean room to maintain control over environmental conditions and airborne particulates. In addition to an AFP end effector, ISAAC also has laser projection, non-destructive inspection (NDI), ultra-sonic cutting, and through-the-thickness stitching capabilities.

Figure 1. ISAAC robotic system.

The process of building a composite part with ISAAC requires a significant amount of project planning and is conducted in several phases as illustrated in Figure 2. A structural engineer initially designs the laminate and defines appropriate fiber orientation angles. A three-dimensional (3D) computer-aided design model is developed to simulate the tool upon which the tows will be laid. The layup design and numerical control (NC) code for ISAAC are created using the programming and simulation programs that can identify potential joint failures or collisions prior to running a program on the robot [14]. The NC code is then imported into ISAAC, and a dry run is conducted without placement of composite material. Upon successful completion of the dry run, the build is carried out. Throughout the build, the process may be halted to allow for inspection and defect repair as necessary. The fully-placed laminate is then vacuum-bagged and cured in an autoclave.

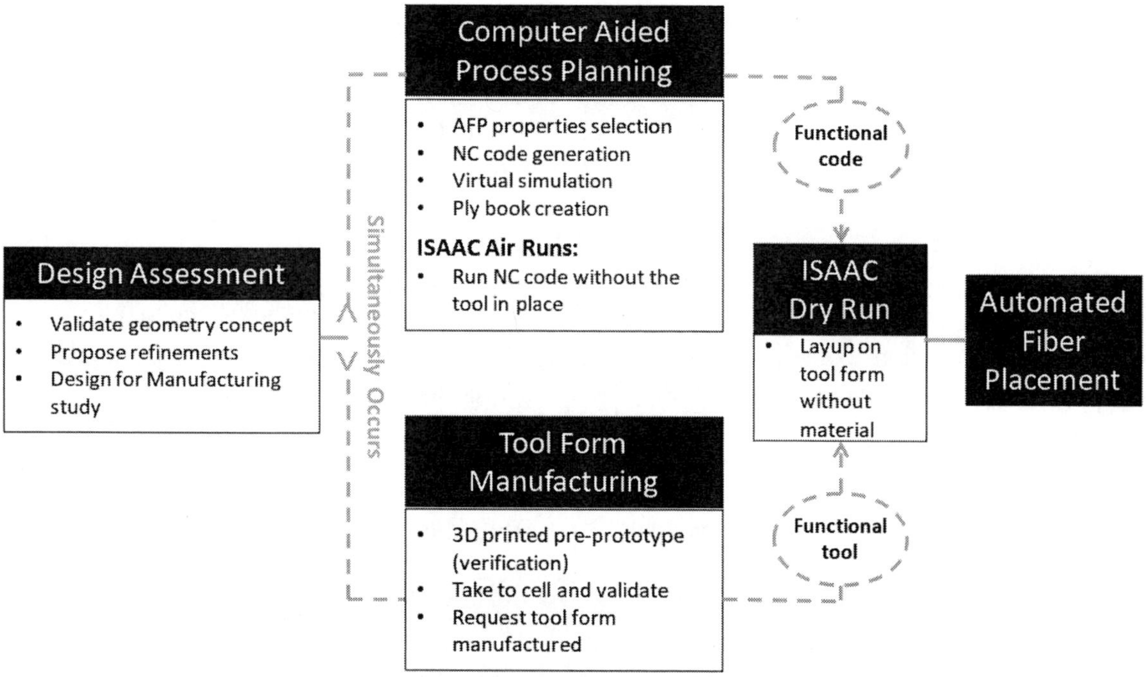

Figure 2. ISAAC manufacturing flowchart [15].

Although the programming controls the majority of the settings, and these settings are established prior to placing material, the operator has real-time control over some manufacturing parameters including tow tension, compaction force, tow feed rate, and heater output during the course of the build. The choice of the values for these parameters are highly dependent on the material system being used and the AFP head orientation relative to the tool geometry. The ideal value of these parameters can vary from one part to another. Therefore, exploration trials must be carried out prior to the manufacturing of a composite part in order to predetermine the manufacturing settings without compromising time or material during the fabrication of the final part.

Identification of appropriate values for process parameters is determined through a qualitative analysis of defects at various fiber orientation angle placements. Ideally, the placement should entail a minimal number of defects while the part is fabricated with a high feed rate. Certain types of defects can be correlated or attributed to a specific process parameter, which can be consequently adjusted. In the thin-ply trials discussed herein, the manufacturing process parameters were determined in order to successfully place plies with areal weights of 30 g/m^2 and 70 g/m^2 material on flat and complex tool surfaces. Parameters were systematically varied individually between trials, which typically consisted of several courses within a ply. Once a ply was laid, trials progressed to full ply placement, and eventually to full laminate manufacturing.

4. THIN-PLY LAYUP OF A FLAT PANEL WITH 70 G/M^2

The goal of these studies was to understand the robot configuration and process parameters necessary to lay plies and manufacture a part with the minimum number of defects requiring manual repair. Thin-ply AFP trials began using the Toray 3900-2/T800S fiber-resin system with tow widths of 6.35 mm (0.25 in.) and an areal weight of 70 g/m^2. Before starting 70 g/m^2 trials, the coupon table surface was cleaned and covered with a Mylar sheet that extended at least 101.6 mm (4 in.) around the perimeter of the two 355.6 mm (14 in.) by 355.6 mm (14 in.) panels intended for both 70 g/m^2 and 30 g/m^2 trials. The edges of the sheet were secured with flash tape, and vacuum was applied to keep the surface flat against the table. Preliminary exploratory trials were conducted. These initial trials were followed by manufacturing trials.

4.1 Exploratory Trials

The initial design of experiments included variations in fiber orientation angles, temperature (heater output), pressure (compaction force), tow tension, and tow feed rate. Parameters were adjusted in succession to facilitate observation of effects on part quality. After each trial, the process was paused, and the ply was qualitatively observed for defects. The fully-laid ply was then removed prior to the next trial. Feed rate is a percentage of the maximum allowable speed, which in this case is 25.4 m/min (1000 inches/min (ipm)). Heater output is proportional to the feed rate. The inverse relationship between heater output and feed rate causes heater output to decrease as feed rate increases. Ten trials were conducted by each placing three courses as shown in Figure 3.

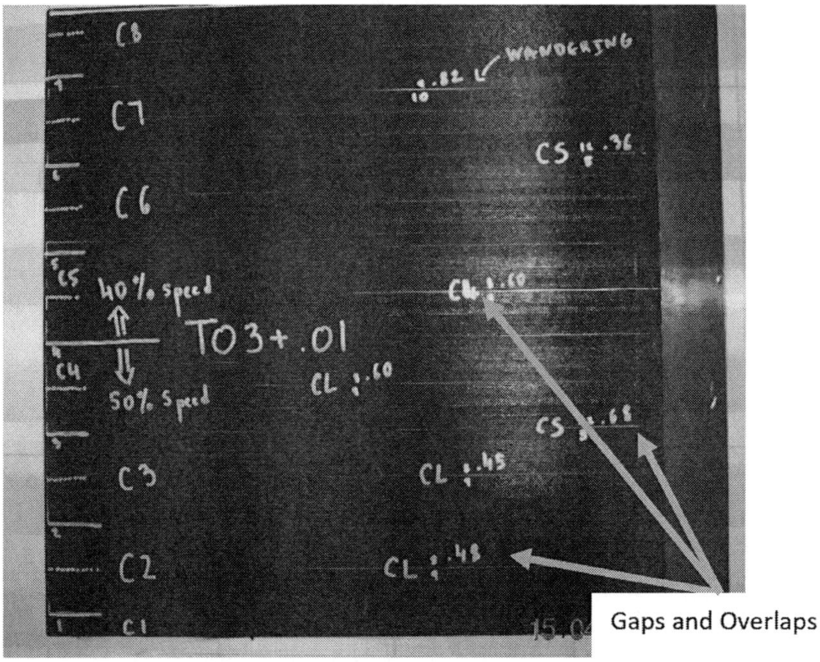

Figure 3. Effect of feed rate on ply quality at 40% feed rate (upper half of panel) and 50% feed rate (lower half of panel).

Heater output varied from 150% to 250%, feed rate from 30% to 100%, and compaction force from 222.41 N to 444.82N (50 lb to 100 lb) while tow tension remained constant at 4.45 N (1 lb). The parameter variations are shown in Table 1. Trials began with lower values of heater output and feed rate and increased as trials progressed, until courses began showing greater number of defects. When heat or feed rate became too high, adhesion to the underlying layer and wrapping of tows around the roller as a result of matrix melting became a common observation. As the AFP process with thin-ply is eventually optimized for manufacturing settings, the ideal process parameters will include the highest feed rate with minimum number of manual repairs necessary.

Table 1. Exploratory trials process parameters.

Trial*	Fiber Orientation Angle [°]	Heater Output [%]	Compaction Force [N (lb)]	Feed Rate [%]	Tension [N (lb)]
1.1	0	150	222.41 (50)	30	4.45 (1)
1.2	0	150	222.41 (50)	30	4.45 (1)
1.3	0	150	222.41 (50)	30	4.45 (1)
2.1	0	250	222.41 (50)	70	4.45 (1)
3.1	0	200	222.41 (50)	50	4.45 (1)
4.1	0	200	444.82 (100)	50	4.45 (1)
2.2	0	250	222.41 (50)	100	4.45 (1)
2.3	0	250	222.41 (50)	70	4.45 (1)
3.2 (C1-4)	0	200	222.41 (50)	50	4.45 (1)
3.2 (C5-8)	0	200	222.41 (50)	40	4.45 (1)

*C stands for course

The effect of feed rate on ply quality, while keeping all other parameters constant, can be observed in Figure 3 which corresponds to Trial 3.2 in Table 1. The upper half of the ply (courses 5-8) was manufactured with a feed rate of 40% and shows fewer tow gaps and less overall tow wandering compared to the lower half of the ply (courses 1-4) that was manufactured with a 50% feed rate. Recurring course gaps are due to material defects. After these initial trials were completed, the exploration progressed from trials based upon sets of three courses to full plies, where the selected variation of process parameters was dependent on the results of the initial trial. One ply was not removed prior to laying the next ply. Visible defects included gaps and overlaps angle deviation, and boundary errors. An initial ply was placed at 45°, as listed in Table 2. For the second ply at a 0° ply angle, the feed rate was kept the same for courses 1-3 at 40% feed rate, increased to 60% for courses 4-6, and then further increased to 70% for courses 7-8 (Figure 4). Some tow wandering was visible at the start of the ply as a result of defects from the underlying ply, as seen in Figure 4.

Table 2. Process parameters during manufacturing trials.

Trial	Fiber Orientation Angle [°]	Courses	Heater Output [%]	Compaction [N (lb)]	Feed Rate [%]	Tension [N (lb)]
1	+45	1-8	200	222.41 (50)	40	4.45 (1)
2.1	0	1-3	200	222.41 (50)	40	4.45 (1)
2.2	0	4-6	150	222.41 (50)	60	4.45 (1)
2.2	0	7-8	150	222.41 (50)	70	4.45 (1)

Based on these results, process parameters were identified to be used for future manufacturing trials with 70 g/m^2 material for this material system. These parameters were 200% heater output, 222.41 N (50 lb) of compaction force, and maximum feed rate values of 15.24 m/min (600 ipm) at all fiber orientation angles.

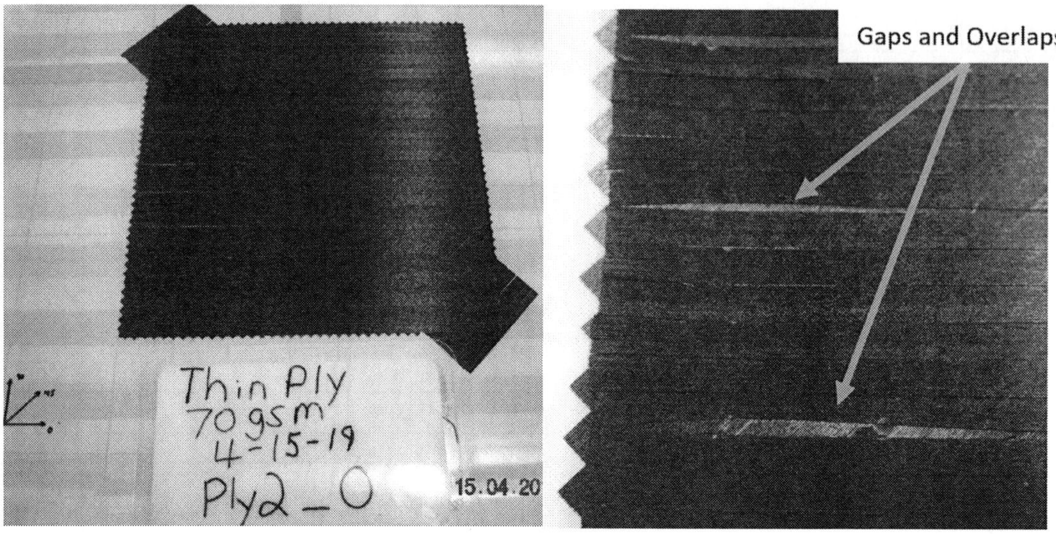

Figure 4. (a) Ply 2 with fiber angle 0°. (b) Tow wandering in ply start.

4.2 Manufacturing Trials

After completing the preliminary exploratory trials, a 35.56 cm by 35.56 cm (14 in. by 14 in.) composite panel with ply stacking sequence $[+45/0/-45/90]_{4S}$ and 8 tows per course was manufactured. Plies were not removed between runs (which make one trial). Courses were programmed at a constant angle with an approach angle of 5° and a unidirectional head movement. The approach angle is lifted from the surface and towards the tool placement. The process parameters for the initial plies were those identified as beneficial by the exploratory trials described in Section 4.1 including heater output of 200% and 222.41 N (50 lb) of compaction force. In the manufacturing trial, the feed rate varied from 7.62 m/min to 15.24 m/min (300 ipm to 600 ipm) to examine the influence of feed rate on tow wandering.

Plies 1-16 of the laminate required significant defect repair and were often subject to tow wandering and gaps/overlaps (Figure 5). Within the plies, tows were manually removed or relocated as necessary to the correct position by hand. Some tows experienced splitting, which may have been caused by the propensity of the material to separate in a resin-rich area (material defect), especially when the thin material was simultaneously subjected to compression by the roller and axial tension. Plies at ±45° required lower feed rates than 0° and 90° plies to obtain minimum-defect courses.

At the midplane of the laminate (between two 90° plies), flash tape was placed along the edges of the panel to facilitate removal of defective tows without damaging the underlying 90° layer. An observation was made that the ply placed over the tape was defect-free. This characteristic was attributed to the use of the flash tape, therefore, flash tape was then applied to just one side of the 45° ply, allowing for a direct comparison of the process with and without the tape, as seen in Figure 6. Since the presence of flash tape significantly reduced tow wandering near the beginning of the course, flash tape was applied to the edge of the preceding ply for all subsequent plies.

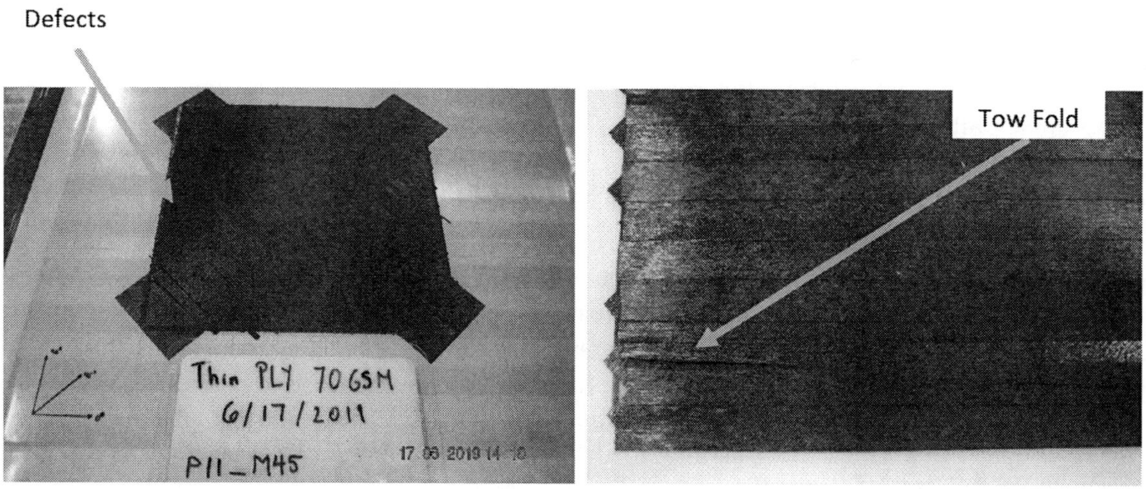

Figure 5. (a) Ply 11 (-45°) with tow wandering and angle deviation defects.

(b) Tow fold, tow wandering, and boundary placement defects in ply 2 (0°).

Figure 6. (a) Defect-free ply with application of flash tape.

(b) Ply with (left) and without (right) application of flash tape.

5. THIN-PLY LAYUP OF A FLAT PANEL WITH 30 G/M^2

The potential of 30 g/m^2 thin-ply material to be used in manufacturing was explored in subsequent trials. The material used in these trials was 6.35-mm (0.25-in.) wide prepreg tows from Hexcel with an IM7/8552 fiber-resin system. It is important to highlight that the material is not optimized for AFP, but was the less tacky hand-layup form of the material system. For 30 g/m^2 thin-ply material, the proposed process parameters to be optimized included variation in temperature (heater output), compaction force, tow tension, and tow feed rate, according to Table 2. Like the 70 g/m^2 study, a 35.56 cm by 35.56 cm (14 in. by14 in.) panel was built. A $[45/0/-45/90]_{4s}$ ply stacking sequence was again used, but 4 or 8 tows per course were placed instead of 16 tows. Like with the 70 g/m^2 material, courses were programmed according to the rosette rule with an approach angle of 5° and a unidirectional head movement. Unlike for the 70 g/m^2 material, a tackifier or double-sided tape was often applied prior to ply placement to increase adhesion of the beginning of a course to the substrate.

5.1 AFP Head Modification

A first backing tape layer had been applied to each 30 g/m^2 spool prior to slitting to support the tows during slitting, followed by the second layer of backing tape used to prevent tow self-adhesion within the spools. Therefore, in order to utilize 30 g/m^2 tows with the existing AFP head, some modification of the head was necessary to remove both layers of backing tape. A secondary take-up spool was designed and prototyped to be used in conjunction with the standard AFP setup. The secondary take-up spool was used to remove the first layer of backing tape that was removed from the tows closer to the feed shoot in order to stabilize the tows and prevent curling. The secondary take-up spools installed on the ISAAC AFP head is shown in Figure 7.

Figure 7: Secondary take-up spools on the ISAAC AFP head

Unfortunately, a noticeable inconsistent variance in tow width on the spools of the 30 g/m^2 material was identified, as shown in Figure 8. This variation led to defects in ply layups, particularly gaps between tows. As this is a material quality issue that may have been caused by a combination of the original material roll quality in combination with the slitting and winding process, this issue must be addressed by the manufacturer and slitter, and therefore cannot be optimized through control of AFP parameters.

Because thinner tows are more sensitive to temperature and mechanical stress during ply layup, lower amounts of heat and compaction force were applied in early trials to reduce the risk to tow jams. However, contrary to expectations, a low heater output and compaction force seemed conducive to causing tows to be pulled in and break within the feed guide chutes and active feed rollers, as seen in Figure 9. Small variations in hardware dimensions among passive guide rollers seemed to also facilitate more frequent jamming of tows in some locations than in others.

Figure 8. (a) Material width of 6.04 mm (0.2378 in) / narrow. (b) Material width of 6.35 mm (0.2500 in) standard.

Figure 9. Tow jams in feed guide rollers.

5.2 Manufacturing Trials

In subsequent trials, compaction, heater output, and tow tension were increased in an effort to better support the tows as they moved through the guide chutes, and the feed rate was varied per course. As shown in Figure 10, the courses with 20% feed rate had the highest quality of layup. However, gaps and tow folds appeared for all feed rates.

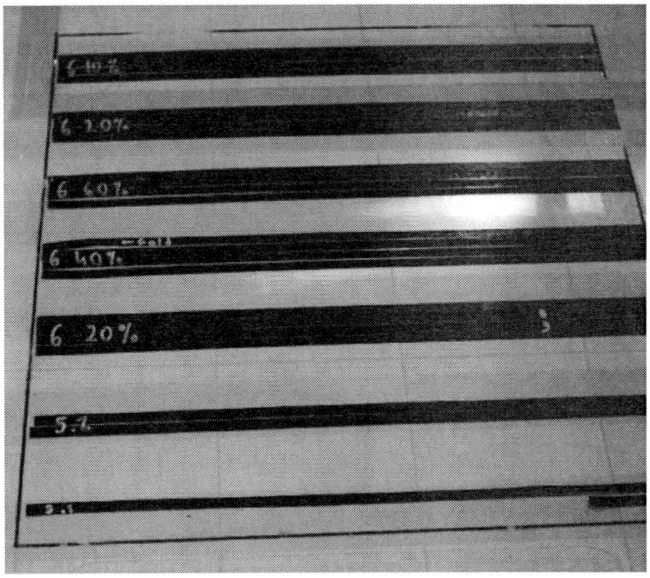

Figure 10. Trial 6, 30 g/m^2 with various feed rates between 10% and 60%.

Because the fiber orientation angle, as well the presence of underlying plies, can have an effect on the number and type of defects, several trials were conducted on overlapping plies at 0°, 90° and +45°. The square area outlined in Figure 11 was relatively free of visible defects, indicating the need for a "lead-in" area during an actual build.

Figure 11. Overlapping trials at various fiber orientation angles.

The process parameters determined to be appropriate for 30 g/m^2 during this study were 200% heater output, 444.82 N (100 lb) of compaction force, 4.45 N (1 lb) of tow tension, and a relatively low feed rate of 5.08 m/min (200 ipm). Using the results from the prior exploration, two overlapping full plies with fiber orientation angles of 0° and 90° were successfully manufactured, as seen in Figure 12. The plies were also relatively defect-free, though some gaps and overlaps between the tows were still present, which could have resulted from the material defects mentioned previously.

Figure 12. Successful placement of 0° and 90° plies.

6. THIN-PLY LAYUP OF A COMPLEX SURFACE WITH 70 G/M^2

After successfully manufacturing a flat panel with 70 g/m^2 Toray 3900-2/T800S fiber-resin material system, an evaluation of placing this same material on a complex-shaped tool surface was conducted as shown in Figure 13. This complex shape contained concave regions that posed a particular challenge to placement. This complex-shaped tool was originally used for placement of standard thickness plies in the study described in [15].

6.1 Exploratory Trials

Four plies of 70 g/m^2 material were sequentially laid up on the complex tool with orientations of 0°, 30°, -60°, and 90°. The processing parameters used in this trial ranged from 200% to 250% heater output, 222.41 N to 444.82 N (50 lb to 100 lb) of compaction force, and a feed rate from 5.08 m/min to 10.16 m/min (200 ipm to 400 ipm). The double convexity of the tool presented an additional challenge, particularly in obtaining sufficient compaction of the plies. Defects were identified between plies, marked, and photographed for documentation in order to examine the effects of defects on subsequent plies. Because this trial was focused on exploration, defects were not repaired.

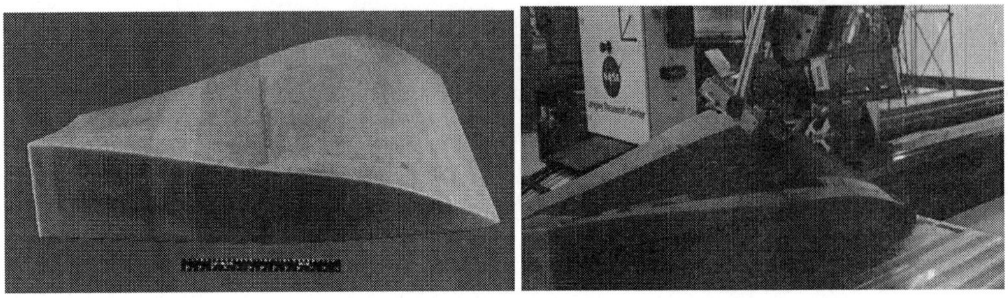

Figure 13. (a) Tool with complex curvature. (b) Layup with ISAAC on complex tool.

6.2 Discussion

A variety of defects were observed in the laminate including puckering, tow wandering, folds, split tows, and boundary errors. The splitting of tows seemed to be caused by resin-rich areas within the tows and was particularly noticeable at the onset of the ply and in the concave area of the tool, which may have been exacerbated by roller compaction or head angle deviations. Puckering was also observed where plies overlaid silver permanent pen markings on previous plies.

The need for a smaller roller (e.g. a roller designed for eight tows and only 5.08-cm (2-in.) wide became evident due to consistent "picking up" of tows by the 10.16-cm (4-in.) wide roller from previously laid courses (Figure 14). Additionally, a build-up of excess resin, as shown in Figure 14, caused several tows to jam, which was resolved by more frequent cleaning of the AFP head. Some tows displayed boundary error defects, showing that the excess resin buildup caused additional dulling of the cutting blades. Both observations indicated a need for more frequent cleaning of the AFP head when working with this thin-ply material system, mostly due to a combination of the fact that this resin system is not AFP optimized as well as that it is thin material.

Figure 14. (a) Roller wrap due to high feed rate or heat. (b) Resin buildup on redirect rollers.

The final processing parameters in order to make fully formed and relatively defect-free plies using the 70 g/m^2 Toray 3900-2/T800S fiber-resin system on a complex-shaped tool were determined to be 200% heater output, 222.41 N (50 lb) of compaction force, and 10.16 m/min (400 ipm). With minimal AFP hardware modification, the ISAAC AFP system could be employed to manufacture this complex geometry with 70 g/m^2 thin-ply material.

7. RESULTS AND DISCUSSION

The results of these trials indicate that both 30 g/m^2 and 70 g/m^2 thin-ply materials have potential for application in relatively defect-free manufacturing once the appropriate process parameters have been identified. During all trials described herein, parameters including tow tension, compaction force, heater output, and tow feed rate were systematically adjusted until the placed plies showed minimal defects. The recommended processing parameter values are shown in Table 3.

Table 3. Final process parameters for all builds.

Material System	Toray 3900-2/T800S			Hexply IM7/8552-1
Areal Weight [g/m^2]	70			30
Build Type	Flat panel	Flat panel	Complex	Flat panel
Ply Orientation [°]	0, ±45, 90	+/-45	all	all
Heater Output [%]	200	200	200	200
Compaction [N (lb)]	444.82 (100)	222.41 (50)	222.41 (50)	444.82 (100)
Feed Rate [m/min (ipm)]	5.08 (200)	10.16 (400)	10.16 (400)	5.08 (200
Tow Tension [N (lb)]	4.45 (1)	4.45 (1)	4.45 (1)	4.45 (1)

A method of eliminating tow wandering was identified for 70 g/m^2 thin-ply material, that allowed the production of a 32-ply panel with minimal defect repair. Upon application of the flash tape along the edge of the ply start, all tow wandering and almost all other defects were eliminated in the ply. This result could be caused by a decrease in friction and roughness within the tow/substrate system as tows were applied in each new ply. The epoxy/epoxy interface of new tows sliding past the edges of previously laid tows exhibited a higher coefficient of friction in comparison to a tow on the smooth polyester flash tape. In AFP, because the tows are cantilevered from the guide tray along the scoop at the beginning of a course placement and before reaching the nip point, they have some freedom to move or change orientation slightly. The increased friction and roughness causes a tow to stick slightly before the nip point, which can introduce an angle deviation of the tow with respect to the nip point that is corrected when movement of the head causes the tows to align. With the flash tape present, the tows are less likely to stick to the underlying surface before the nip point, and therefore, are more likely to achieve the correct orientation through the feed guide.

Orientation sequences that allow for greater contact with underlying tow edges, such placing a 45° ply over a 90° ply, could also increase apparent roughness of the underlying substrate and promote angle deviation. This deviation would cause the 45° ply to have greater tow wandering, as has been previously seen.

The extreme thinness of 30 g/m^2 material introduced a greater sensitivity to the process parameters used in the ISAAC facility and was less compatible with dimensions of AFP head hardware, such as guide feed depth. Similar process parameters were implemented using 30 g/m^2 material but with lower tow feed rates relative to 70 g/m^2 material. Hardware modification to the AFP head was required due to the presence of an extra backing tape provided with the prepreg material. Overlapping plies at various fiber orientation angles were also successfully manufactured with 30 g/m^2, though further adjustment of hardware and optimization of material quality is required before commencing with full-scale manufacturing or complex geometries.

8. CONCLUDING REMARKS

The use of 30 g/m^2 and 70 g/m^2 thin-ply materials was examined using the AFP robotic system at LaRC. The results of these trials indicate that both these material forms have potential for application in relatively defect-free manufacturing once the appropriate process parameters have been identified. During all trials described herein, parameters including tow tension, compaction force, heater output, and tow feed rate were systematically adjusted until the placed plies showed minimal defects. The recommended processing parameter values were determined for Toray 3900-2/T800S and Hexcel IM7/8552 materials through placement of +45°, 90°, -45°, and 0° plies. The use of thin-ply materials is extremely promising as a weight-saving and cost-efficient method to produce novel and high-performance composite parts with minimal time and labor investments.

9. ACKNOWLEDGMENTS

The authors would like to thank Jake Tury, Chris Cusson, Ray Grenoble, Daniel Drake, Benjamin Beltran, and Andrew Anderson for their valuable contributions.

10. REFERENCES

[1] M. Albazzan, R. Harik, B. Tatting, Z. Gurdal, A. Blom-Schieber, M. Rassain and S. Wanthal, "Optimization of Cylinders with Holes under Bending using Nonconventional Laminates," in *AIAA SciTech 2018*, Kissimmee, Florida, 2018. doi: 10.2514/6.2018-1377

[2] M. Albazzan, R. Harik, B. Tatting and Z. Gurdal, "Efficient Design Optimization of Nonconventional Laminated Composites using Lamination Parameters: A State of the Art," *Composite Structures,* vol. 209, pp. 362-374, 2018. doi: 10.1016/j.compstruct.2018.10.095

[3] R. Amacher, W. Smith, J. Botsis, C. Dransfeld and J. Cugnoni, "New Design Opportunities using Thin-Ply Composites," *JEC Composites Magazine,* vol. 52, no. 96, pp. 33-35, 2015.

[4] R. Amacher, J. Cugnoni, J. Botsis, L. Sorensen, W. Smith and C. Dransfeld, "Thin ply Composites: Experimental Characterization and Modeling of Size-Effects," *Composites Science and Technology,* vol. 101, pp. 121-132, 2014. doi: 10.1016/j.compscitech.2014.06.027

[5] S. Sihn, R. Y. Kim, K. Kawabe and S. W. Tsai, "Experimental Studies of Thin-Ply Laminated Composites," *Composites Science and Technology,* vol. 67, no. 6, pp. 996-1008, 2007. doi: 10.1016/j.compscitech.2006.06.008

[6] A. Lovejoy and S. Scotti, "Potential Weight Benefits of IM7/8552 Hybrid Thin-ply Composites for Aircraft Structures," in *AIAA Scitech 2019 Forum*, San Diego, CA, 2019. doi: 10.2514/6.2019-0774

[7] R. Y. Kim and S. R. Soni, "Experimental and Analytical Studies On the Onset of Delamination in Laminated Composites," *Journal of Composite Materials,* 1984. doi: 10.1177/002199838401800106

[8] P. P. Camanho, C. G. Dávila, S. T. Pinho, L. Iannucci and P. Robinson, "Prediction of In Situ Strengths and Matrix Cracking in Composites under Transverse Tension and In-Plane Shear," *Composites Part A: Applied Science and Manufacturing,* vol. 37, no. 2, pp. 165-176, 2006. doi: 10.1016/j.compositesa.2005.04.023

[9] J. Frketic, T. Dickens and S. Ramakrishnan, *Automated Manufacturing And Processing Of Fiber-Reinforced Polymer (FRP) Composites: An Additive Review of Contemporary and Modern Techniques for Advanced Materials Manufacturing,* vol. 14, Elsevier B.V., 2017, pp. 69-86. doi: 10.1016/j.addma.2017.01.003

[10] R. Harik, C. Saidy, S. J. Williams, Z. Gurdal and B. Grimsley, "Automated Fiber Placement Defect Identity Cards: Cause, Anticipation, Existence, Significance, and Progression," in *International SAMPE Technical Conference*, 2018.

[11] K. Fayazbakhsh, M. Arian Nik, D. Pasini and L. Lessard, "Defect Layer Method To Capture Effect of Gaps and Overlaps in Variable Stiffness Laminates Made by Automated Fiber Placement," *Composite Structures,* 2013. doi: 10.1016/j.compstruct.2012.10.031

[12] M. Lan, D. Cartié, P. Davies and C. Baley, "Influence of Embedded Gap and Overlap Fiber Placement Defects on The Microstructure and Shear And Compression Properties of Carbon-Epoxy Laminates," *Composites Part A: Applied Science and Manufacturing,* 2016. doi: 10.1016/j.compositesa.2015.12.007

[13] "NASA ISAAC Fact Sheet, FS-2016-12-273-LaRC," 2016.

[14] "CGTech," [Online]. Available: http://www.cgtech.com. [Accessed 17 01 2020].

[15] R. Harik, J. Halbritter, D. Jegley, R. Grenoble and B. Mason, "Automated Fiber Placement of Composite Wind Tunnel Blades: Process Planning and Manufacturing," in *SAMPE 2019 Conference & Exhibition*, Charlotte, North Carolina, May 20-23, 2019. doi: 10.33599/nasampe/s.19.1538

3D MODELING AND PRINTING OF AUTOMATED FIBER PLACEMENT DEFECTS

Alex Brasington[1], Trevor Schachner[1], Ramy Harik[1]

[1]University of South Carolina
McNAIR Center, Department of Mechanical Engineering
1000 Catawba Street
Columbia, SC 29201

ABSTRACT

Automated Fiber Placement (AFP) is a composite manufacturing technique used to fabricate lightweight air and space vehicles. AFP uses tows, or strips of composites, laid side by side to build plies and laminates. A major consequence of this technique is the defects induced via the AFP process. Knowledge and detection of defects requires some expertise on the size, shape, and significance of the considerable number of possible defects. Detection then becomes increasingly difficult due to the substrate and incoming material both being a dark black color. The inability to detect defects leads to a part with poor quality. This report aims to provide some education on these defects using 3D modeling and 3D printing to visualize each defect. To ease visualization, the model can be printed in three different colors as follows: First, the tool surface is printed in one color, then the tows are printed in a second color. The contrast of these two colors can then be used to visualize defects such as boundary coverage or tow drops. Defects associated with single tows such as gaps, overlaps, and twist can be modeled and printed using a third color. This method creates 3D models with easily identifiable defects that can be used to educate or train AFP personnel.

Keywords: 3D modeling, 3D printing, Automated Fiber Placement, Defects
Corresponding author: Ramy Harik

1. INTRODUCTION

The applications for composite materials are continuously growing as research in design and manufacturing advances. With these advancements comes a need to produce the composite structures. Automated fiber placement (AFP) is a composite manufacturing technique often used to produce large aerospace structures. In this process strips of composite material, referred to as tows, are laid onto a tool surface at various fiber angles using a robotic/gantry platform and a fiber placement head. A heating mechanism attached to the head is used to ensure proper tackiness of the substrate. Directly after the heat is applied to the substrate, a compaction roller runs over the tows to adhere the tows to the substrate. Successive passes create plies, and the combination of the plies produces the overall structure.

Even with the constant improvements of AFP machines and the AFP process, it is far from perfected. One of the main side effects of this process is unavoidable defects that occur due to tool geometry, fiber steering, and material imperfections [1]. Even with recent developments in automated inspection technologies [2] [3], a majority of ply inspection is done by a human. AFP defects such as wrinkling caused by steering [4] can lead to up to a 36% decrease in overall strength of a structure [5]. Two of the most common defects present are gaps and overlaps due to the imperfect placement of tows [6]. Gaps tend to leave a resin rich area where a decrease in strength at the site can occur while overlaps behave oppositely by creating fiber rich areas where the local strength increases [7] [8]. Due to the large effect of defects to the structure, a thorough understand of each defect and its significance is crucial. Defects can be broken into 4 main categories: (1) positioning defects, (2) bonding defects, (3) tow defects, (4) foreign bodies [9]. A complete list of the defects and their category is presented in Table 1, and a complete description of each can be found in [1]. The identification of the defects requires a great deal of manufacturing experience and expertise. Noticing the defects becomes especially difficult due to the black-on-black nature of the composite substrate and current ply. This project aims to ease the difficulty of defect identification by using 3D models of each defect to learn the characteristics of each. All modeling was accomplished using Dassault Systemes' CATIA V5. These models will allow training of personnel on detecting defects before having any AFP experience, while also further spreading AFP knowledge.

Table 1: List of defects and their associated category

Defect	Category
Gap/overlap	1
Twist	1
Missing tow	1
Boundary coverage	1
Angle deviation	1
Wandering tow	1
Position error	1
Fold	2
Pucker	2
Wrinkle	2
Bridging	2
Loose tow	2
Splice	3
Foreign object detection	4

2. MODELLING

Modelling of the defects was done in multiple steps. First the tool was created for the tows to be placed onto. After the tools was created, a set of 5 tows was modeled with each one being a separate body. The tows can be modeled with or without defects to demonstrate good and bad quality layups. Each of the defects in modelled in its own file and then combined into a single model for printing. The process is detailed in the sections below.

2.1 Tool Surface Modeling

To begin modelling, a tool was created with a length and width of 100 mm and 75 mm respectively and was termed "Base Model" (Figure 1). The name of each model is embossed on the side of the tool to quickly know which model is in hand. A tool refers to the surface that the AFP head will place the tows onto. The geometry of the tool surface was created to demonstrate what would typically be seen on a complex tool. Complex tools have curvatures of varying values at each point along the tool. Such a geometry makes fiber placement difficult due to the defects it induces. The tool surface is then projected straight down to create the base of the tool where it is mounted to a flat surface or mandrel.

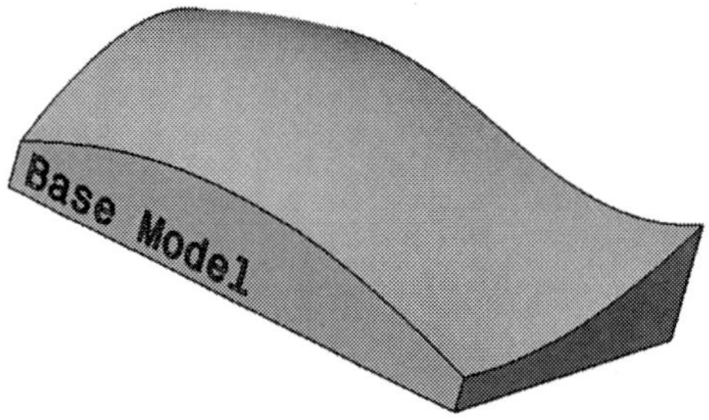

Figure 1: 3D model of the tool surface

Apart from the defect modeling, tool surface modeling can be used to teach personnel about the effects of geometry such as concavity or possible collision areas. Concave portions of a tool are prone to bridging defects, and severe curvatures are probable collision areas. Understanding the limits on tool geometry is the first step in producing high quality AFP structures.

2.2 Tow Modeling

As previously mentioned, tows are the strips of composite material that is placed onto the tool surfaced using the AFP machine and attached head. The tows are modeled following the geometry of the tool surface modeled in the last step. Five large tows are used instead of smaller ones to ease visualization of defect and non-defect tows. In a real manufacturing environment, between 8 and 32 tows are usually deposited at once. The tows "blend" together and it is challenging to decipher the appearance of minor defects. Learning what defects to look for and how to spot them on the larger models makes for an easier transition into locating defects in a real manufacturing environment.

2.2.1 Tows Without Defects

Defect free tows are created with equally sized small spaces between them to identify one tow from another, and these gaps are not to be interpreted as defects. To create each tow straight lines are projected onto the tool surface using reference points on the edge of the tool to define the beginning and end. Each tow is a closed curve, and therefore can be extruded to create the thickness of the tow. The tows can be created in the same file as the base, or they can be modeled in a separate file and combined with the base in a final model (Figure 2). The distinction between these two methods will become clear in Section 3 when the printing of the models is discussed.

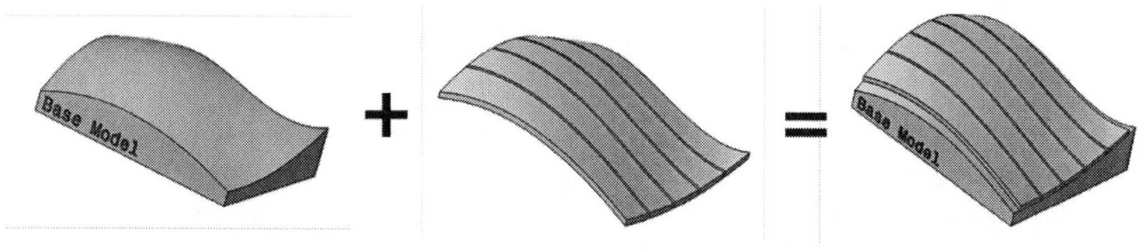

Figure 2: Explanation of combining the base model with the non-defect tows

2.2.2 Tows with Defects

Modelling tows with defects is a similar process to modelling tows without defects. The first step is deciding on which tow, or tows, are going to have the specific defect. The defect is then modeled using the geometry of the previously created tool surface and surrounding tows. This process is demonstrated in Figure 3 where a gap/overlap is modeled. In this model the center tow is modeled to have a gap on one side and a resulting overlap onto the tow on the opposing side. Combining the defect and the other tows with the base model creates the complete gap/overlap model. The benefit of creating the tool and each tow separately will be clear when the 3D printing methods are discussed (Section 3). Using the same method, any desired defect can be modeled.

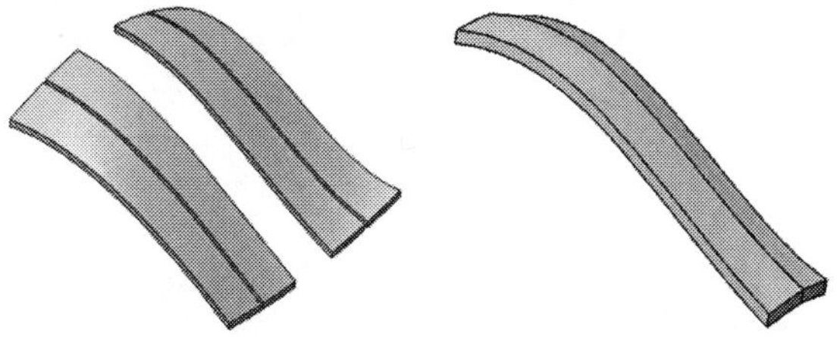

(a) (b)

Figure 3: Example of modeling tows with a gap/overlap

3. 3D PRINTING

3D printing of these educational models creates a hands-on experience for personnel. Printing can be accomplished with any available printer and in two methods: (1) single print and (2) multiple prints. The single print method combines the tool and the tows into a single model for printing. Although this method is faster and easier, it results in the entire model being one color. Using method 2 the model can be printed in multiple colors with the base, non-defect tows, and defected tows each being a separate color. Printing with multiple colors allows for the defects to be highlighted for easy detection. The method chosen in solely based on the desired appearance.

3.1 Single Print Method

In this method, the tows and tool will be combined into a single STL file for printing. Depending on how the model was created, this may require creation of an assembly before exporting the model. Once exported the STL model is imported into the desired slicer software that will create the individual layers and G-code for the printer to follow. The authors used the Ultimaker Cura slicer and have provided an example of two sliced models in Figure 4. Specific printer configuration properties such as nozzle diameter, layer height, infill, etc. are not required. The chosen properties should be based on the knowledge of the printer's performance. If there is not a lot of experience with the printer, the default options will provide a quality print.

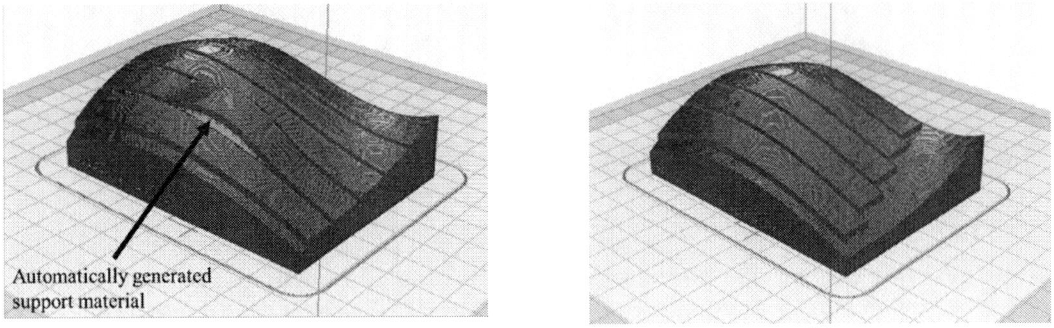

(a) (b)

Figure 4: Sliced models of a (a) loose tow and (b) boundary coverage defect

A major advantage of this method is that no support material is required, except for the defects that are not in contact with the tool surface (Figure 4.a). Support structures are not a part of the model and are generally created through a slicer software. These are generally used in places where the model is not supported by any underlying material, therefore the printer has nothing to deposit the material onto. With the ability to use one continuous print with minimal support material, this method will be faster and more efficient. A depiction of the final printed model using the single print method is shown in Figure 5.

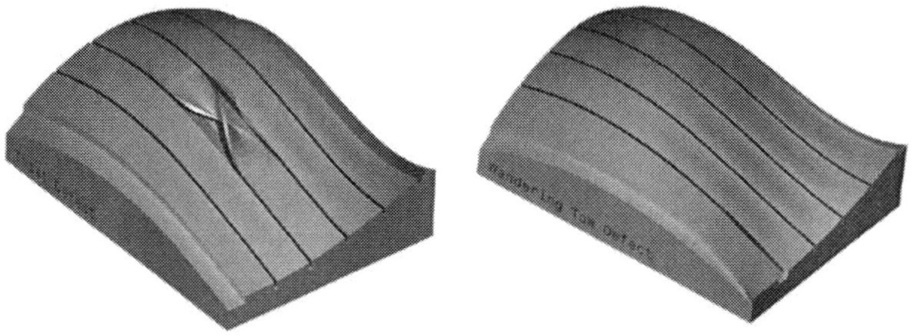

(a) (b)

Figure 5: Example of a (a) twist and (b) angle deviation using the single print method

3.2 Multiple Prints Method

Observing the angle deviation defect in Figure 5.b, it is apparent that some defects are hard to notice in a model printed using a single color. To create a more visually appealing printed model, multiple prints, or a dual nozzle 3D printer, can be used with various colors to highlight the defect. However, this method will be more time and labor intensive due to the amount of prints and assembly that is required. To begin the base with the desired defect name embossed on the side should be printed. Once the base is printed, each tow associated with that defect should be printed separately. Depending on the tows being printed and the placement on the printer bed, support material will be required. If the printer's capabilities allow for multi-color printing it may be possible to print all the tows in the two desired colors at one time.

After the base and each tow are printed, they can be assembled to create the final model. The assembly is accomplished by using an adhesive to attach the tows to the tool. Each tow should fit precisely in place due to the matching curvature of the tool and the tow. Once assemble, the model should present a series of tows with a highlighted defected tow as it was modeled. Figure 6 shows a twist and angle deviation defect using this method. Comparing Figure 6.b with Figure 5.b demonstrates the benefit of simpler defect detection using multiple colors.

(a) (b)

Figure 6: Example of a (a) twist and (b) angle deviation using the multiple prints method

3.3 Completed Defect Models

Following the modeling, printing, and assembly procedures outlined above, each defect presented in Table 1 was modeled and printed using 3 different colors. Gray was used for the tool, black was used for normal tows, and red was used to highlight defected tows. Each model has been presented below in Figure 6 for reference.

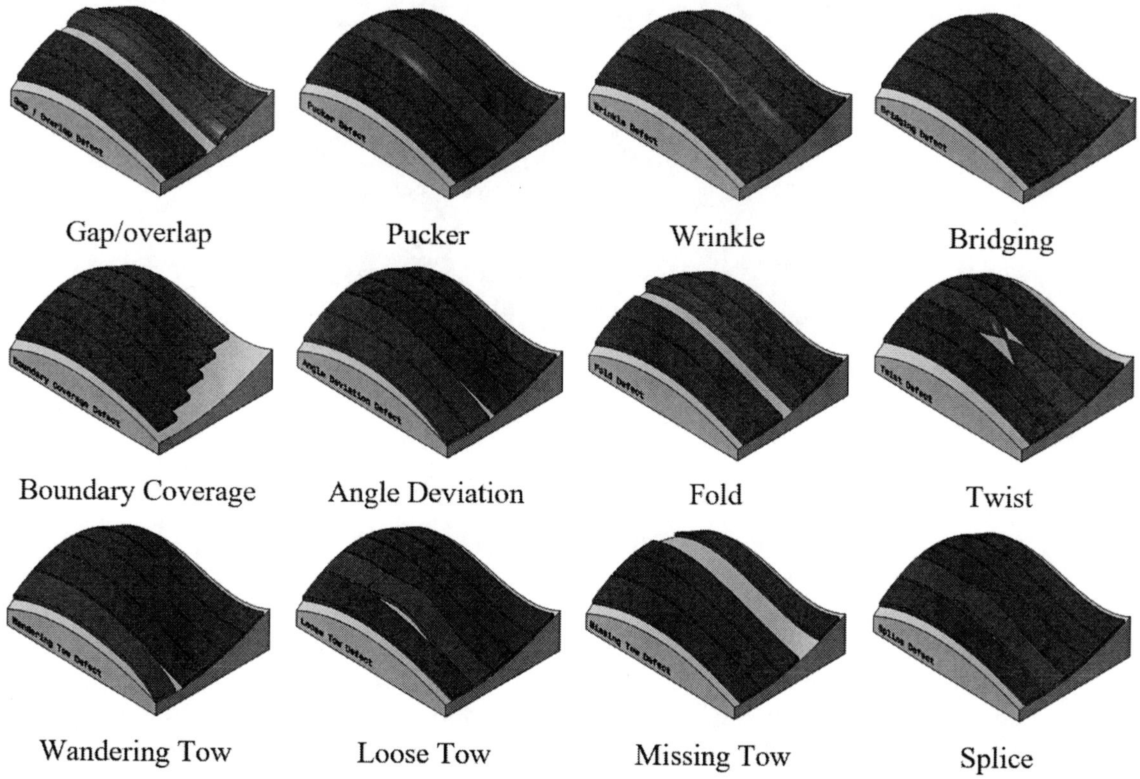

Position Error FOD

Figure 7: 3D models of each defect

4. MODEL DISSEMINATION

The main goal of the creation of these models is to share them with the public. This will allow for anyone to view and print the models seen above in Figure 7. With open access to these models, AFP knowledge can be spread to anyone wanting to learn. For reference, some models previously printed by the authors are shown in Figure 8. The models were printed using the multiple prints method described in Section 3.2 above. Visiting the following link will provide downloadable files of each defect model that can be printed on most 3D printers. Link: https://drive.google.com/file/d/1_mlr6cPgVzUw2Q60watgZ-FQHDa1nVqs/view?usp=sharing

Figure 8: Examples of some 3D printed defect models

5. CONCLUSIONS

The presented methods for 3D modeling and printing of AFP defects can play an integral role in spreading the knowledge associated with these defects. This knowledge is typically gained through experience with inspecting numerous AFP manufactured plies. Exposing personnel to these models introduces the types of AFP defects seen during manufacturing along with their geometry without having any prior experience. Identification of defects is just a small step in a complete understanding of the AFP process. Combining the education gained from modeling and printing the defects with their actual effects and significance will bring a broad knowledge of the process of inspecting and reworking tows.

6. REFERENCES

1. Harik, R., Saidy, C., Williams, S. J., Gurdal, Z., & Grimsley, B. *Automated fiber placement defect identity cards: cause, anticipation, existence, significance, and progression*. In SAMPE Conference & Exhibition, 2018.
2. Sacco, C., Radwan, A. B., Beatty, T., & Harik, R. M*achine learning based AFP inspection: a tool for characterization and integration*. In SAMPE Conference & Exhibition, 2019. DOI: 10.33599/nasampe/s.19.1594
3. Sacco, C., A. B. Radwan, R. Harik, and M. Van Tooren. *Automated fiber placement defects: automated inspection and characterization*. In SAMPE Conference & Exhibition, 2018.
4. Wehbe, R., Tatting, B., Rajan, S., Harik, R., Sutton, M., & Gürdal, Z. (2020). *Geometrical modeling of tow wrinkles in automated fiber placement*. Composite Structures, 112394, 2020. DOI: 10.1016/j.compstruct.2020.112394
5. O'Hare Adams, D., & Hyer, M. W. *Effects of layer waviness on the compression strength of thermoplastic composite laminates*. Journal of Reinforced Plastics and Composites, 12(4) (1993): 414-429. DOI: 10.1016/0142-1123(94)90450-2
6. Blom, A. W., Lopes, C. S., Kromwijk, P. J., Gurdal, Z., & Camanho, P. P. *A theoretical model to study the influence of tow-drop areas on the stiffness and strength of variable-stiffness laminates*. Journal of composite materials, 43(5), 403-425, 2009. DOI: 10.1177/0021998308097675
7. Fayazbakhsh, K., Nik, M. A., Pasini, D., & Lessard, L. *Defect layer method to capture effect of gaps and overlaps in variable stiffness laminates made by automated fiber placement*. Composite Structures, 97, 245-251, 2013. DOI: 10.1016/j.compstruct.2012.10.031
8. Lan, M., Cartié, D., Davies, P., & Baley, C. *Influence of embedded gap and overlap fiber placement defects on the microstructure and shear and compression properties of carbon–epoxy laminates*. Composites Part A: Applied Science and Manufacturing, 82: 198- 207, 2016. DOI: 10.1016/j.compositesa.2015.12.007
9. Denkena, B., Schmidt, C., Völtzer, K., & Hocke, T. *Thermographic online monitoring system for automated fiber placement processes*. Composites Part B: Engineering, 97, 239-243, 2016. DOI: 10.1016/j.compositesb.2016.04.076

STITCHED COMPOSITES WITH THREE-DIMENSIONAL STITCH PATHS

Andrew E. Lovejoy, Dawn C. Jegley
NASA Langley Research Center
Hampton, VA/USA

ABSTRACT

Stitched composites have been shown to exhibit damage tolerance and to reduce weight compared to traditional layered composites through unitization of the structure by elimination of fasteners. Stitching capabilities have been incorporated into the Integrated Structural Assembly of Advanced Composites (ISAAC) system at NASA Langley Research Center with the introduction of two stitching heads. Stitching path control was initially implemented as straight lines in space, as was done for previous stitching development. However, more complex stitched structures such as a wind tunnel blade or around cutouts within a fuselage or wing skin, require that the stitching paths be implemented as three-dimensional (3D) stitching paths in space. Unfortunately, control programming output by an existing preprocessor program cannot stitch these curved paths due to problems that arise in stitch formation and the introduction of side forces on the needles using the conventional programming approach whereby the head is simultaneously controlled through translations and rotations. This lack of capability is most significant for the single-sided stitching head, where two needles are in the preform at the same time for the majority of the stitching process. A means to program 3D stitching paths in space was developed whereby the translation and rotation of each stitch were decoupled, thereby eliminating the problems associated with current control programming approach. Using this newly developed stitching path definition and control programming, complex stitching paths have successfully been stitched at the ISAAC facility. The ability to stitch general 3D stitching paths in space enables the use of stitching on more complex parts.

Keywords: Stitched, Composite, Preform
Corresponding author: Andrew E. Lovejoy (Andrew.E.Lovejoy@nasa.gov)

1. INTRODUCTION

Composite materials are now commonly used in high-performance, light-weight aerospace structures. Typical composite laminates exhibit very good in-plane properties, but their performance can be limited by their low through-the-thickness properties, such as interlaminar shear and interlaminar tensile strengths. These reduced through-the-thickness properties make such laminates susceptible to the formation of delaminations, which are often caused by impacts that occur during part assembly or service operations. As a result, laminated composite parts typically undergo costly nondestructive inspections and evaluations prior to insertion into service and during service. These inspections are necessary to evaluate the delamination sizes and locations to determine if a part needs to be repaired, scrapped, or replaced. Therefore, the need to improve the through-the-thickness properties of composite laminates has led to the development of various methodologies to increase these properties.

Through-the-thickness reinforcement methods have been developed in order to increase the resistance of laminated composites to delamination and other forms of impact damage. These methods are grouped into two main categories, namely matrix augmentation or architecture augmentation [Ref. 1]. Matrix augmentation, or toughening, includes rubber toughening, nanoparticle toughening, and carbon nanotube treatments. Architectural augmentation includes stitching, tufting, z-pinning, three-dimensional (3D) weaving/braiding, and needling. Stitching typically consists of one or two threads or yarns that penetrate through the thickness and connect the through-the-thickness reinforcement across the surfaces of the laminate. Stitches are usually through the entire thickness of the laminate, but blind stitching that penetrates through most of the laminate thickness and only connects the through-the-thickness reinforcement across a single surface is possible. Tufting is similar to stitching except that it uses a single thread or yarn that may or may not connect the through-the-thickness reinforcement across one of the laminate surfaces. Z-pinning consists of inserting precured pins through the thickness of the laminate prior to curing. 3D architecture such as 3D weaving and braiding provide through-the-thickness reinforcement incorporated into the composite architecture, and, depending on the architecture can remove some or most of the unidirectionality associated with typical composite constructions. Lastly, needling is a process whereby barbed needles are pressed through a preform, thereby pushing fibers from one layer down through the layer below, and in the process can cause significant damage to the preform fibers. Stitching was chosen by NASA Langley Research Center (LaRC) as a means to address through-the-thickness strength limitations and was extensively studied.

To achieve the goal of reducing the cost of air travel, NASA has pursued the development of technologies needed for future low-cost, light-weight composite structures for commercial transport aircraft. Through-the-thickness reinforcement was selected for study because the use of traditional layered composite material systems requires fasteners to suppress delaminations and to join structural elements, ultimately leading to fastener pull-through problems or heavy pad-ups in the fastener regions. In contrast, through-the-thickness reinforcement can replace these fasteners throughout each integral panel. This unitization eliminates fasteners and their associated holes, which significantly simplifies the assembly process, reduces part-count, and reduces crack initiation and curtails crack growth throughout the life of the aircraft. The through-the-thickness method chosen was to combine stitching with toughened epoxy in a stitched/resin-infused approach. Stitched/resin-infusion has been extensively researched at LaRC since the 1990s, especially through the Advanced Composite Technology (ACT) [Refs. 2-5] and Environmentally Responsible Aviation (ERA) [Refs. 6-16] projects, as a means to improve the performance, reduce the weight, and reduce the fabrication cost of aircraft structures. Placing stitches at discontinuities, such as along flange edges, has been shown to suppress delamination and turn cracks, which increases the design space and leads to lighter designs [Ref. 16]. Additionally, the most recent infusion and cure processes for stitched panels require high temperatures, but only vacuum pressure, which eliminates the need for an autoclave. The use of stitched dry fabric with resin infusion and oven cure leads to substantial cost savings and eliminates the out-time concerns associated with traditional prepreg. Through these studies, stitched composite structures have been shown to exhibit damage tolerance and to reduce weight through unitization of the structure by elimination of fasteners.

As a consequence of this effort, a stitched graphite-epoxy material system was identified with the potential for reducing the weight and cost of commercial transport aircraft wing structure. By stitching through the thickness of dry carbon fabric and then infusing with epoxy resin, the labor associated with wing cover panel fabrication and assembly can be significantly reduced. By stitching through the thickness of pre-stacked skin and then stitching together stringers, intercostals and spar caps with the skin, the need for mechanical fasteners is almost eliminated. This manufacturing approach reduces part count, and therefore, cost of the structure. At the same time, newer and more novel aircraft configurations have been proposed that can lead to more complex, 3D stitching paths.

Due to efficiencies and benefits of stitched composites, other composite structure applications could also benefit from stitching, such as launch vehicle dry structure, stiffened conformal pressure vessels and wind tunnel blades. Wind tunnel blades for the National Transonic Facility (NTF) [Ref. 17] are of particular interest at LaRC. A wind tunnel blade from the NTF that was part of a study to use automated fiber placement (AFP) to replace the hand-laid NTF blades [Refs. 18-20] is shown in Figure 1. Stitched wind tunnel blades offer a manufacturing option, however, their highly curved surfaces necessitate the ability to stitch curves in three dimensions over continuous paths. The use of 3D stitching paths could facilitate the fabrication of the interior blade spar (Figure 1a) or the complete blade assembly (Figure 1b) as a single unitized structure with improved performance (blade profile to improve aerodynamics and improved damage tolerance, for example). The need for 3D stitching path capability for wind tunnel blades, aircraft structures, spacecraft structures, and other composite structures with highly curved surfaces was the motivation for formal development of the process to define and program these nonlinear stitching paths at LaRC. The remainder of this paper presents an overview of the stitching capabilities at LaRC, the development of the 3D stitching path definitions required to stitch the complex paths, and conclusions related to the stitching path development and definitions.

a) Interior blade and spar

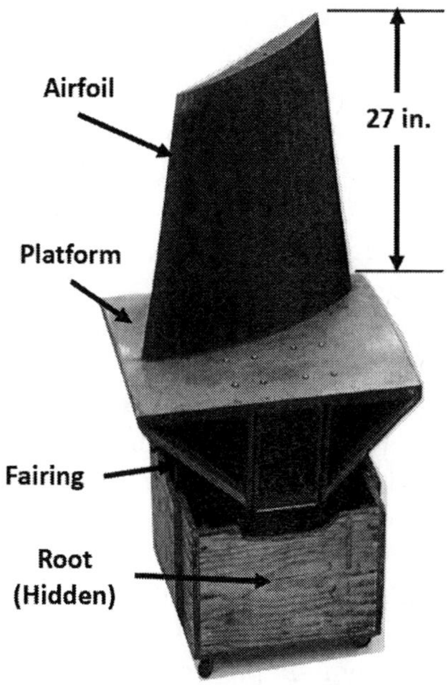

b) Blade assembly

Figure 1: NTF wind tunnel blade.

2. ISAAC FACILITY

Stitching of composite material structures at LaRC is performed at the Integrated Structural Assembly of Advanced Composites (ISAAC) facility. The ISAAC facility houses a robotic system designed to facilitate the development of new technologies related to aerospace composites manufacturing [21]. The robotic system is a six-axis industrial robot capable of accommodating a variety of end effectors. A 41-ft-long linear rail and a vertical rotator provide two additional degrees of freedom. The workcell and operator station are enclosed within an International Organization for Standardization (ISO) Level 7 clean room that provides positive control over airborne particulates and enables consistent temperature and humidity to be maintained. a photograph of the ISAAC robotic system is shown in Figure 2 with the original head capable of performing AFP using up to 16 spools of one quarter-inch wide tows of material. Since the initial installation of the ISAAC facility, several additional heads have been incorporated onto the system to increase functionality.

Because of the success of previous work conducted on stitched composites under LaRC programs, two stitching heads were added to the capabilities of the ISAAC facility to continue development of stitching technology and application to composite structures. Incorporating the stitching heads required coordination between multiple vendors. The stitching end effectors, the machines that perform the stitching process, were manufactured by PFAFF Industriesysteme und Maschinen GmbH Branch Office KSL (KSL)[ϕ] (Lorsch, Germany) through DAP America[ϕ] (Norcross, GA).

[ϕ] The mentioning of a company is not an endorsement by the National Aeronautics and Space Administration (NASA).

The two end effectors comprise a double-sided end effector and a single-sided end effector. The double sided end effector uses one needle and one thread to create a chain stitch to join layers of fabric. The double-sided end effector, termed the RS566 (Figure 3), was primarily an off-the-shell design and requires access to both sides of the preform by the end effector. The single sided end effector uses two needles and one thread to create a modified chain stitch to join layers of fabric that can include more complicated structural arrangements. This single-sided end effector, termed the RS535 (Figure 4), was modified based on past experience with its use during the ERA project and only requires access to one side of the preform by the end effector. Development of the RS535 modifications was facilitated by input from the stitching expert at Boeing that provided stitching development and support throughout the ERA project. LaRC engineers worked with KSL, the needle manufacturer and the Boeing stitching expert, who provided information based upon the experience developed during the ERA project, to develop improvements to the stitching end effector and the catcher needle. The different chain stitch patterns produced by the two stitching heads is shown Figure 5.

The stitching end effectors were incorporated into the ISAAC facility by Electroimpact[ϕ] (EI) (Mukilteo, WA) by joining the stitching end effectors with head plates to create the two stitching heads. In addition, EI provided robot control programming for the ISAAC system. To support that programming, CGTech[ϕ] (Irvine, CA) provided preprocessor capability for the stitching heads in their VERICUT software [22] to enable engineers and technicians to input stitching path geometry to create the G-code, the numerical control code used in computer-controlled manufacturing, for the control system required to perform the stitching operations using the ISAAC robot. Initial implementation of stitching paths in VERICUT was generally limited to straight-line paths, as were used throughout recent stitching development in the ERA project. While VERICUT has the capability to define paths in space, the ability to operate the stitching heads along such paths was not present, and necessitated the development for the logic needed to stitch general paths in three dimensions. Development of these three-dimensional (3D) paths and the associated G-code programming requirements is the focus of this paper, and is presented in the next section.

[ϕ] The mentioning of a company is not an endorsement by the National Aeronautics and Space Administration (NASA).

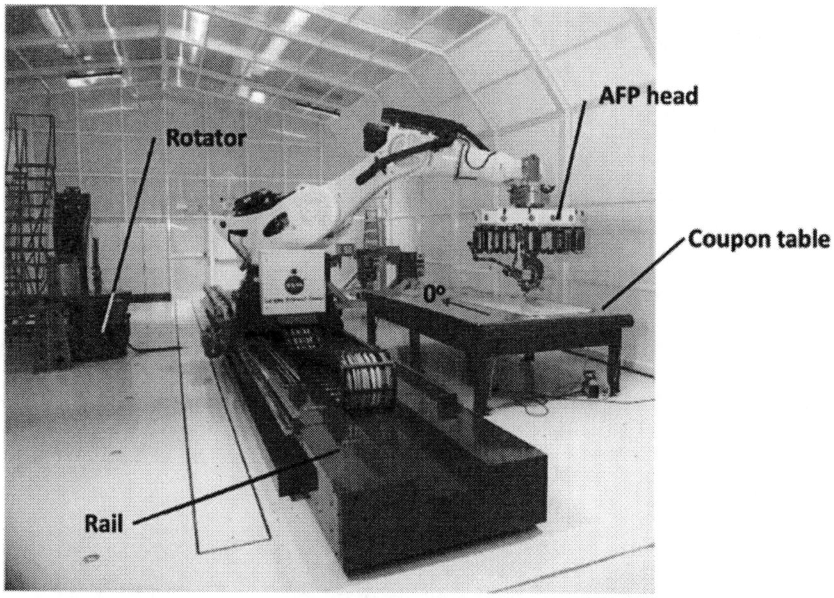

Figure 2: ISAAC robotic system and ISO 7 clean room.

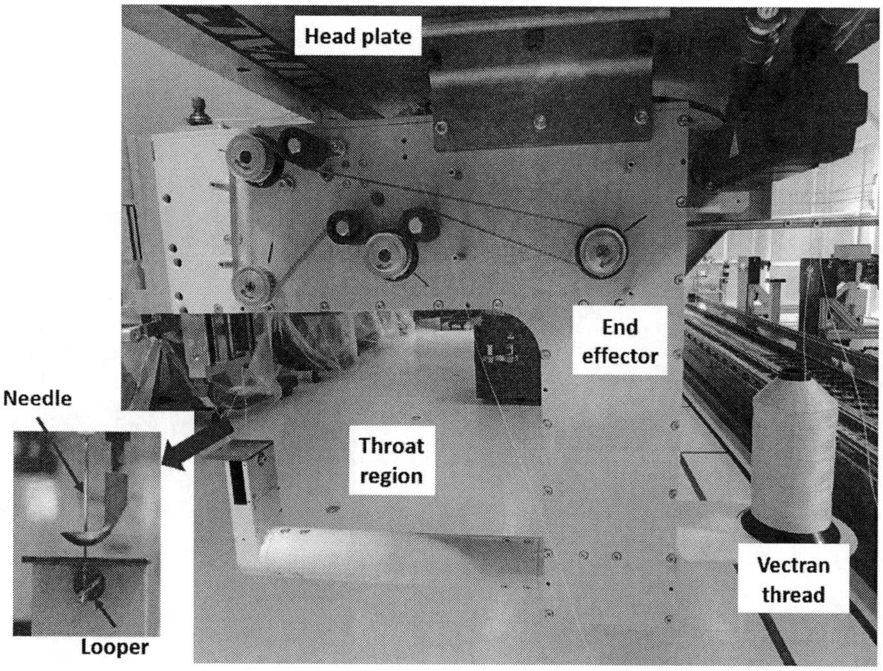

Figure 3. RS566 (double-sided) stitching head.

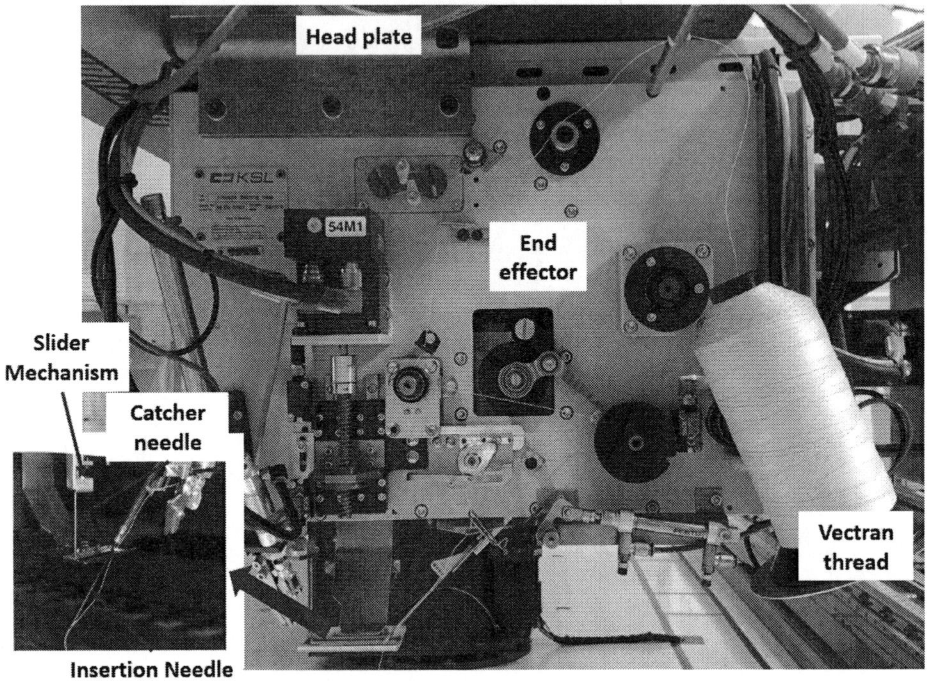

Figure 4. RS535 (single-sided) stitching head.

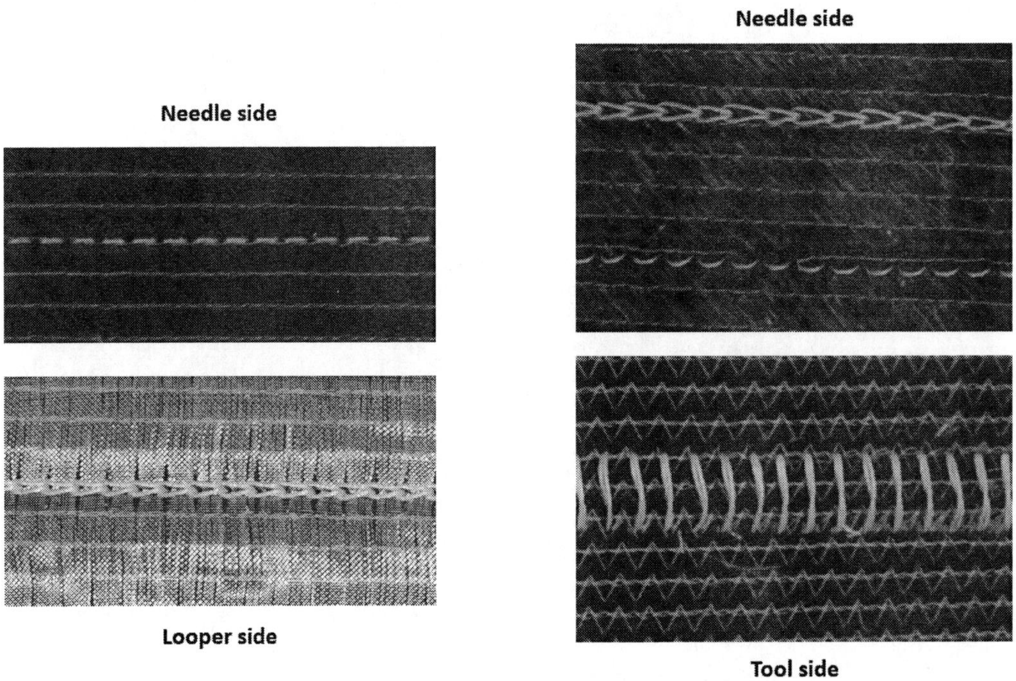

a) RS566 (double-sided)

b) RS535 (single-sided)

Figure 5. Stitching patterns for the stitching heads.

3. STITCHING PATH DEVELOPMENT

When stitching operations are performed on ISAAC, a stitching head is placed on the robot, replacing the AFP head shown in Figure 2. Stitching is performed with the aid of a support structure for the fabric preform for double-sided stitching or a "stitch tool" for single-sided stitching. Either of these is typically attached to the coupon table. The robot is programmed to follow stitch paths that are consistent with support structure or grooves in the stitch tool that allow the needles to pass below the carbon fabric without touching the stitch tool itself. The geometry of the support structure or stitch tool is generally consistent with the desired shape of the final part. A further discussion of stitch tools is presented in Ref. 17.

As mentioned in the previous section, control for stitching paths during the manufacturing and installation process for the two stitching heads using ISAAC was limited to straight-line paths. This capability was consistent with the stitching that Boeing conducted during ERA, where paths consisted of a collection of short, straight-line stitching paths. Since future stitching activities are expected to require longer, 3D curved stitching paths in space, further stitching path development has been conducted to develop these 3D curved stitching paths. This section describes the stitching path development that has been accomplished. First, the description of the initial stitching path definition is presented to provide a baseline for stitching path definitions and programming. Next, the development and definition of the complex 3D stitching path in space, which is the primary focus of this paper, is presented. Lastly, some examples of 3D stitching paths requiring the new path definition are presented.

3.1 Initial Stitching Path Definition

The control for the stitching heads on the ISAAC system had to be developed from scratch since the LaRC stitching heads are placed on a different base robot compared to the robots that Boeing used in the ERA project, and they use different controllers and software. Development of the stitching control laws required definition of the "tool point" that is defined as the point associated with a head that controls the motion of the head.

The tool point is shown in Figures 6 and 7 for the double-sided and single-sided stitching heads, respectively. The control programming provides the ISAAC robotic system the information about the tool point and the required motion that guides the stitching head while traversing the stitching path. The tool point for the double-sided RS566 stitching head has been defined as the point where the needle pierces the top surface of the end effector table as shown in Figure 6. The preform is placed between the presser foot and the end effector table.

The tool point for the single-sided RS535 stitching head has been defined as the point on stitching tool surface where the catcher needle penetrates tool surface, as shown in Figure 7. The preform is placed between the stitching tool surface and the presser foot surface. During the stitching process, the end effector table supports the preform for double-sided stitching, while the stitching tool supports the preform for single-sided stitching. Note that the needle side shown in Figure 5b is the presser foot surface identified in Figure 7 and shown in Figure 8. Preforms with straight stitching paths have successfully been stitched using the ISAAC system with both the RS566 and RS535 heads.

A sketch of the presser foot surface stitch pattern for the RS535 head is shown in Figure 8, where the circles on the stitch pattern indicate the needle penetration points. In the figure the green lines represent thread placed by the insertion needed and the red lines represent thread placed by the catcher needle. The programmed stitch path is shown by the black line. As indicated in the figure, the lengths of the stitch patterns produced by the catcher and insertion needles are equal to the defined stitch length. Straight stitching paths have been used on curved preforms, but their use constrains the design space due to potential stitching problems.

An example of where the straight-line stitching path was successfully used for a curved preform is presented next. A close-up of a stringer/frame bay end on a cylindrical curved panel fabricated by Boeing (Ref. 10) using single-sided stitching is shown in Figure 9. The inner mold line (IML) or tool side is shown. The bay was bounded by the stringers on two sides parallel to the cylinder axis and the frames on the other two sides that were in the circumferential direction. Therefore, the frame webs were curved within their planes, and the flanges of the frames matched the curvature of the skin. The panel had an outer mold line (OML) radius of 90 inches with a stringer spacing of 7.8 inches and frame spacing of 24 inches. The stitch lines connected stringer and frame flanges to the skin within the bay end. Because the stich line on the frame flange does not extend to the stringer centerlines, the approximate length of the frame flange stitch line was about 5.4 inches. The frame flange path was stitched as a straight line in space. The straight line path led to the IML stitch width, the distance of the length of the stitching where the needles penetrated the IML, to be longer at the center of the path (indicated by orange line in Figure 9) than at the end of the path (indicated by red line in Figure 9). Despite this very small chord length, the difference in stitch width at the center and end of the stitch path was about 6%, and is noticeable in the figure where the orange and red stitch width lines are brought close together for comparison, with the blow up inset emphasizing that length difference. As the radius of a preform reduces or the arc length of the stitching path increases, this discrepancy in stitch path width increases and leads to difficulties with stitch formation. Therefore, stitching preforms along a small radius of curvature along a circumferential cylindrical path requires that the stitch path be defined as a 3D curve in space to enable proper stitch formation. The same difficulties arise for any curved stitching path in a plane in which the catcher needle lies throughout the entire stitching path, which is equivalent of the stitching head rotating about the x-axis of a coordinate system affixed to the stitching head at the tool point as shown in Figure 7. Therefore, a stitching path that contains rotation of the head about the x-axis of the local tool point coordinate system (shown in Figure 7) can create a stitch with an insufficient loop for the catcher needle to hook. The severity of the mal-formation is increased as the length of the stitching path increases and as the radius of the stitching path decreases. Additional, and more severe, stitching difficulties arise when the stitching path includes rotations of the stitching head about the local tool point coordinate system z-axis (see Figure 7). Therefore, improvement of the stitching path development and definition are required for generally curved stitching paths, which is presented in the next section.

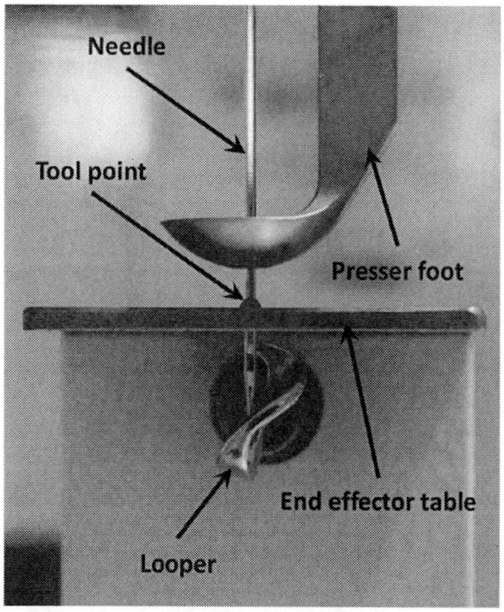

Figure 6. Tool point for the RS566 stitching head.

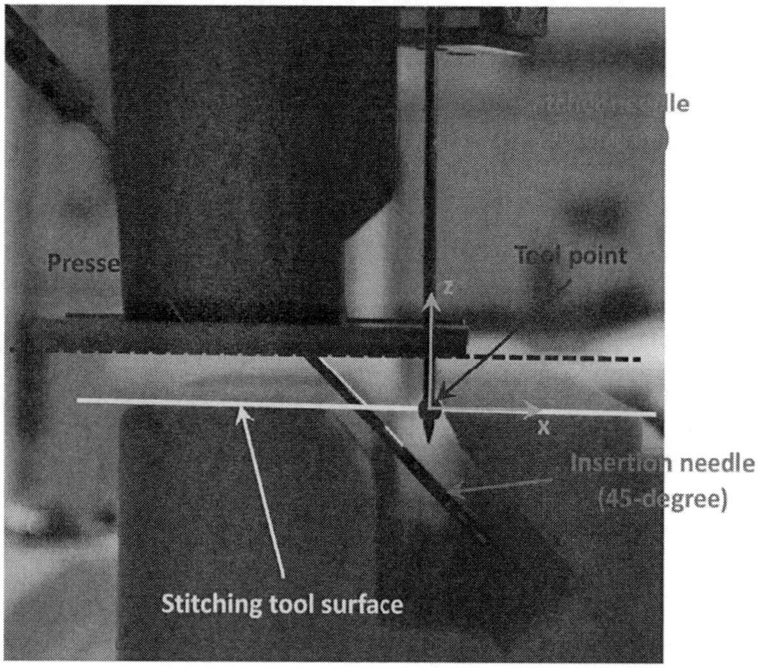

Figure 7. Tool point for the RS535 stitching head.

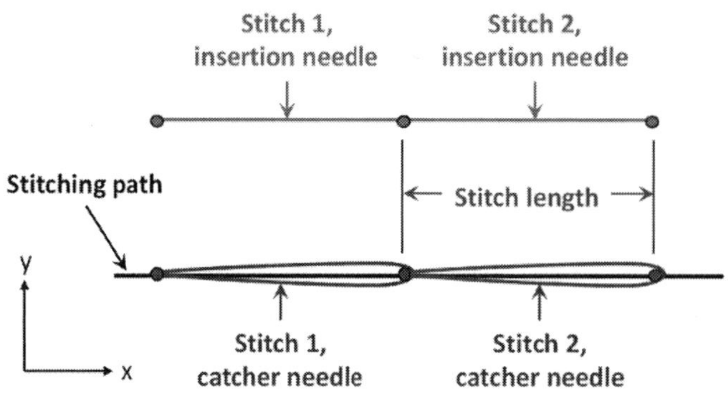

Figure 8. RS535 stitch pattern on the presser foot surface of the preform.

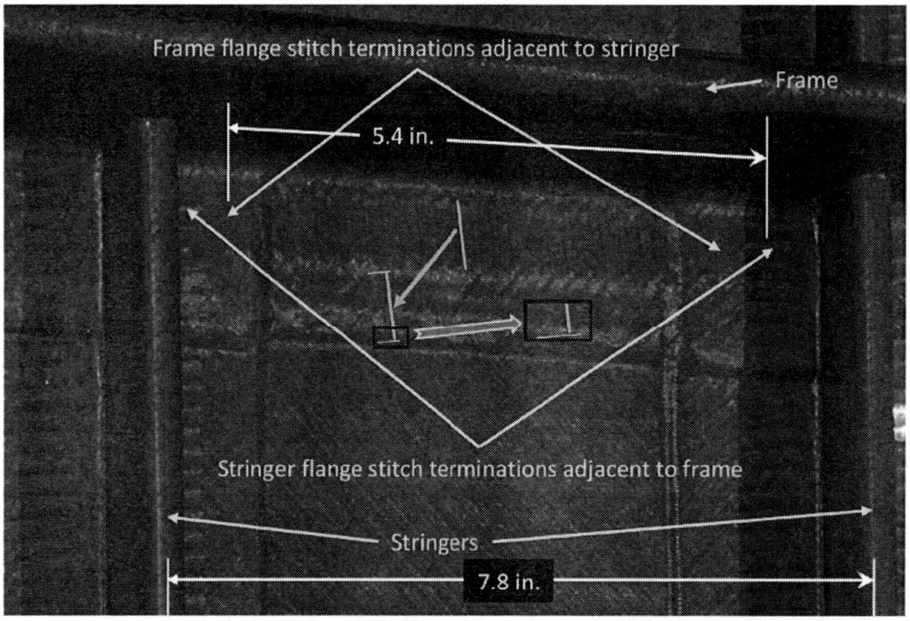

Figure 9. Stitch path terminations for frames and stringers in curved panel test article.

3.2 Path Definition Process

Developing 3D stitching paths consists of two basic steps, namely, 1) developing the stitching control logic, and 2) producing the required control programming. These two items are intimately related to each other, but the development of the paths is presented herein by first describing the logic by which the stitching is controlled, followed by the description of how that logic informs and dictates the control programming. Additionally, because the double-sided RS566 stitching head has only a single needle, it is less affected by the path curvature, and can be considered to be a subset of the single-sided RS535 stitching head as it applies to developing and defining the stitching path. Therefore, the remainder of this section focusses mainly on the RS535 stitching head, and where appropriate, how the RS566 relates to the RS535.

3.2.1 Effects of including path curvature (head rotations)

An important consideration in understanding why the stitching path control is greatly affected by the curvature of the stitching path is to first understand the way that the stitching head works. Specifically, the motion of the needles during the stitching process are the key factor. As originally programmed, the stitching head traverses the stitching path continuously during the stitching process. However, the needles penetrate the material during the stitching process, and to avoid side forces on the needles, a needle walk mechanism moves the needles within the presser foot during the stitching process as shown in Figure 10. In the figure, the yellow line indicates the position of the two needles that can be seen to move within the presser foot, as indicated by the green arrows, in the opposite direction of the presser foot/head motion indicated by the red arrows. The amount of walk required depends on the length of the stitch, and for a typical 5-mm stitch length the needle walk is between 3.5 mm and 4.0 mm. For straight stitching paths, the walk ensures that the needles are not subjected to side forces during the stitching process. However, for a curved stitching path in space, side forces develop on the needles that increase with increasing rotations.

While following a curved stitching path, the stitching head rotates about the tool point around the three axes of the local tool point coordinate system shown in Figure 7. First, consider a stitching path where the only curvature is in the yz-plane, that is, a rotation about the x-axis. A rotation about the local tool point coordinate system x-axis has been discussed previously regarding a cylindrical panel. During this rotation, the catcher and insertion needles have the same translation and rotation throughout the stitching cycle. Therefore, the stitching pattern is the same as a straight seam shown in Figure 8 where both needles create a stitch pattern equal to the stitch length. Very little side force is exerted on the needles during this motion. However, stitch formation is affected because the distance between the catcher and insertion needles shortens on the stitching surface at the ends of the stitching path for a convex surface, and lengthens at the ends of the stitch path for a concave surface when a straight path is used. By introducing curvature to the stitching path by including x-axis rotation, the stitch formation problem exhibited by a straight path on the curved surface is eliminated regardless of the stitch path length or radius of curvature because the distance between the needle penetration points on the stitching surface remains the same.

Second, consider a stitching path that only rotates the stitching head about the local tool point coordinate system y-axis. Rotation about the y-axis tries to rotate the needles within the preform, but introduces small side loads on the needles. However, without including the rotation about the y-axis where the surface normal of the stitching surface does rotate, problems with stitch formation can occur. Stitch formation problems occur because the distance between the needle penetration points on the stitching surface can change dramatically for large rotations as shown in Figure 11. In the case where a rotation about the y-axis is not included in a stitch path, the stitch formation can have problems due to the thread loop being too small for the case shown in part b) of the figure, or too large if the rotation is in the opposite direction.

Third, consider a rotation about the local tool point coordinate system z-axis that creates a curved path in the xy-plane. For the RS535 stitching head we define a convex path as having the insertion needle on the convex side of the curve as shown in Figure 12a, and a concave path as having the insertion needle on the concave side of the curve as shown in Figure 12b. This stitching path motion results in a stitch pattern for the insertion needle that is much longer or much shorter than that of the catcher needle for convex and concave paths, respectively.

A difference in length is shown in Figure 13 for the two cases. The difference in stitch pattern length for the two needles results in a significant side force on the insertion needle during the stitching process. As previously noted, the needles are in the preform for a significant portion of the stitch formation process. The walk mechanism of the stitching head moves both needles the same translational distance, and therefore cannot alleviate the side force on the insertion needle for this curved path. This inability is because the tool point is on the catcher needle, so the walk mechanism controls the translation of the catcher needle the proper distance for the stitch length and keeps side forces off the catcher needle.

The rotation of the stitching head about the local tool point coordinate system z-axis imposes the most severe problems on the stitching process for the RS535 head. A means of controlling the stitching path, given the restrictions of the stitching head mechanisms, is necessary to alleviate the side forces on the insertion needle for a path containing rotation of the head about the local tool point coordinate system z-axis. Contrary to this, for the double-sided RS566 stitching head, a rotation about the local tool point coordinate system z-axis has no significant effect because such a motion simply results in a drilling motion of the needle and introduces no side forces or formation problems.

In summary, rotations about the local tool point coordinate system axes affect the two stitching heads if a path is curved in 3D space. A rotation about the local tool point coordinate system x-axis or y-axis imparts similar forces on the RS566 needle and the RS535 catcher needle, and the RS535 insertion needle also is subjected to side forces. In addition, the stitch formation for the RS535 can be affected if a straight path without rotation included is used on a curved surface. If these rotation-induced shortening or lengthening of the distance between the catcher and insertion needle penetration points on the stitching surface leads to improper loop formation for the catcher needle to hook the thread.

A rotation about the local tool point coordinate system z-axis does not impart a force on the RS566 needle and the RS535 catcher needle, however, the RS535 insertion needle can be subjected to significant side forces due to the increased or decreased translation resulting from the rotation. Stitching a series of straight segments for a curved path within the tool point coordinate system xy-plane is not practical, since this series of straight segments would not be a smooth stitched path. Special programming is necessary to stitch a curved path with rotation about the z-axis and for general paths that require rotations about all three tool point coordinate system axes. The method for developing the curved stitch paths in space is presented next.

a) At needle insertion b) During stitch c) At needle retraction

Figure 10. Needle walk evidenced by yellow line between needles traversing within presser foot (green arrow) as head moves along material (red arrow).

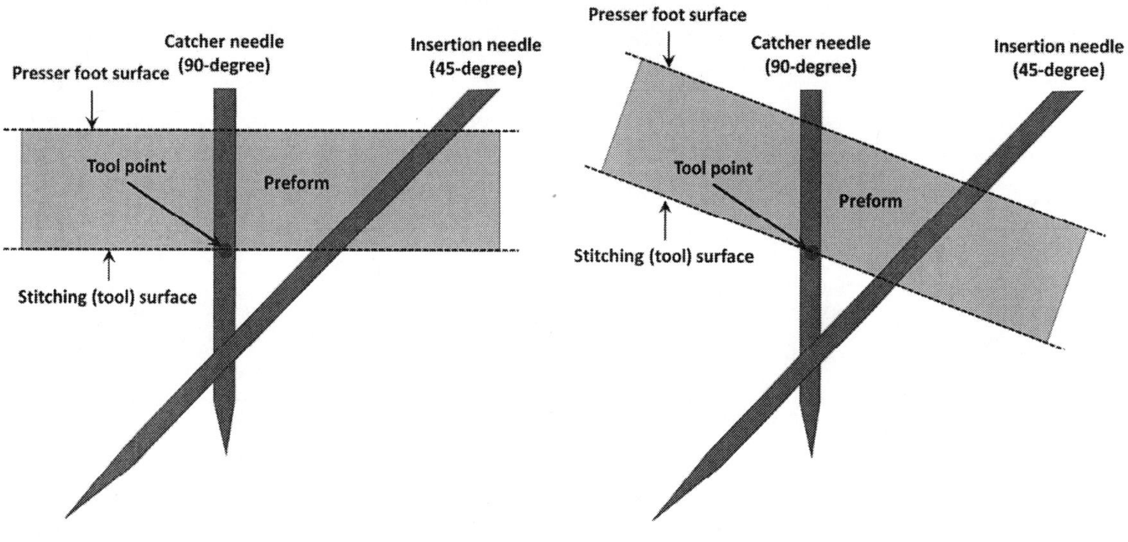

a) Starting orientation of path b) Without head rotation along path

Figure 11. Effect of not including a rotation about the tool point coordinate system y-axis.

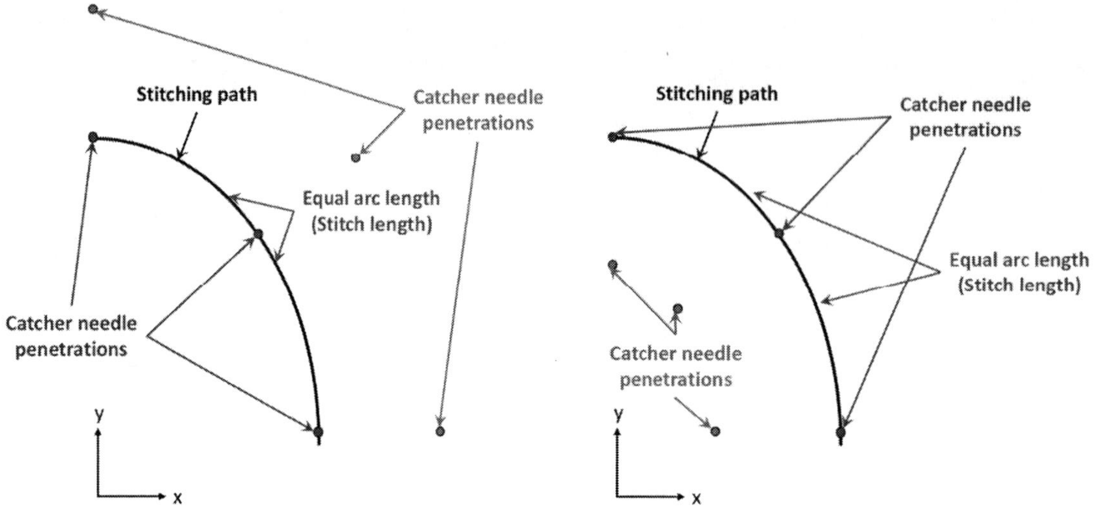

a) Convex path b) Concave path

Figure 12. Definition of convex and concave stitching paths for RS535 stitching head with head rotation about tool point coordinate system z-axis.

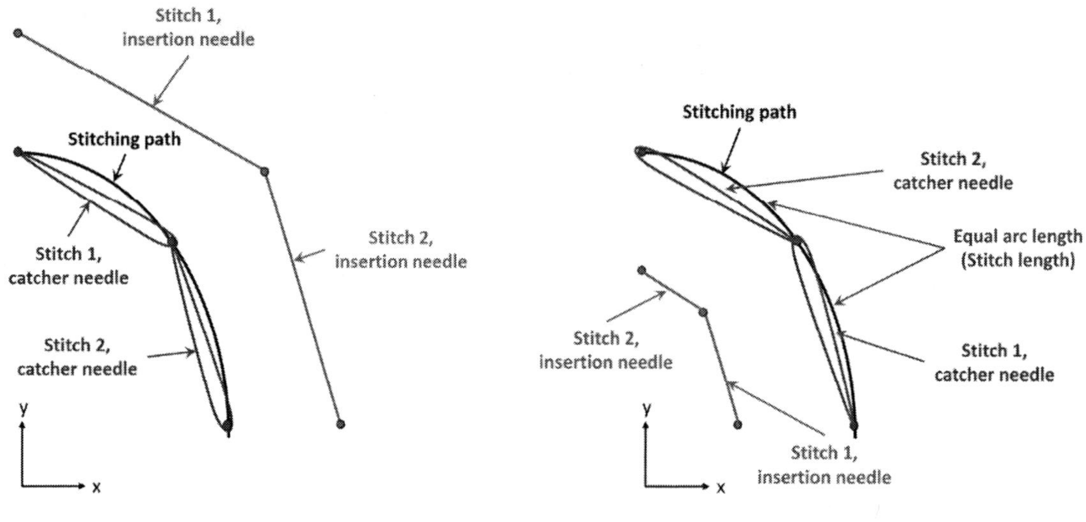

a) Convex path b) Concave path

Figure 13. Stitch patterns for RS535 stitching head with head rotation about tool point coordinate system z-axis.

3.2.2 Curved stitching path programming

The control programming, or G-code, provides information by which the robotic system carries out the stitching process. In general, the preprocessor provides location information along a curve for the tool point to follow and the points output by the preprocessor are more than sufficient to determine a unique path for the tool point to follow. The total distance travelled along the stitching path and a parameter that is tied to the rotation of the drive motor for the stitching head are included with the position points. The drive motor operates cams that complete the stitching process such as needle movement, needle walk, thread tension, and numerous other operations during a full revolution. The drive motor is synchronized with the path motion such that the drive motor completes a single rotation for each stitch length portion of the stitching path traversed by the tool point. As programmed for straight stitching paths, the number and spacing of points to define the curve is not important, and the control system matches the distance along the stitching path with the positions along the path at which the drive motor is to complete revolutions. While this system works for the straight stitching paths, this system does not work for curved stitching paths in 3D space.

In addition to point locations, path length and stitching drive motor information, head rotations in the global coordinate system are specified for curved paths in space. For non-stitching heads, such as the AFP and ultrasonic cutting knife, the preprocessor output controls the head motion satisfactorily because they are only dependent on putting the tool point at a certain location. To control the total motion of the ISAAC heads, the control system interpolates between the programmed points, with the head controlling the tool point to traverse the specified path while rotating the head at the same time. The output currently provided by the preprocessor does not work for stitching curved paths because of the problems associated with rotating the stitching head while the needles are in the preform. Therefore, a method to prevent the problems associated with stitching while applying continuous translations and rotations simultaneously along the stitching path was developed.

Two main considerations informed the choice of how to produce a stitching path definition that permits the proper stitching process along the path. First, the stitch path is not dependent upon the complete path curve, but rather only upon the starting and ending points of each stitch, that is, the locations where the needles penetrate the preform. This requirement is contrary to other heads like the AFP head and ultrasonic knife that must traverse the entire path while being in contact with the part. Second, since the needles cannot be in the preform while the head is rotating, rotations must occur in the parts of the stitching process when the needles are not in the preform. To accomplish this feat, the translation and rotation for each stitch were decoupled. In terms of programming, this decoupling resulted in only the end points of each stitch being defined, and the rotations being applied at the end of each stitch after the translation has been completed. The process for determining the points and rotations is provided next.

The process for developing the stitching path definition and G-code development is depicted in Figure 14. The process begins with input of the stitching surface, stitching path curve (curve the tool point will follow) and the nominal stitch length ($l_{s,n}$). It is desirable for a stitching path to have an integer number of stitches, so a stitch length for the path is calculated for the integer number of stitches that provides a stitch length closest to the nominal stitch length (l_s). As the length of the stitching path increases, the difference between the path-calculated stitch length and the nominal stitch length decreases, and in general the difference between nominal and calculated stitch lengths is negligible. Points along the stitching path are extracted that are separated by equal arc lengths that are defined to be equal to l_s. This definition guarantees that each stitch has the same length along the path, but results in the loop produced by the catcher needle to be equal to (for straight segment) or shorter than (curved segment) the stitch length since the catcher needle is coincident with the tool point and the end points represent a chord of the curved path. The length of the insertion needle stitch pattern depends on whether the curve is convex or concave as discussed previously.

Each stitch progresses from its starting point to the ending point, which is the starting point for the next stitch (with the exception of the last stitch that ends at the stitching path end point). Using the set of calculated stitch points, a direction vector (V_i) for each stitch is calculated. This direction vector is coincident with the local tool point coordinate system y-axis and represents the direction that the stitching head will traverse during the stitch cycle for each stitch. In addition to the orientation vector for each stitch, the tool point coordinate system z-axis and x-axis orientations are calculated for each stitch at the stitch starting point, with the z-axis being coincident with the surface normal at the stitch starting point. The local tool point coordinate system axis vectors are defined in the global coordinate system used by the control system to control the head motion during the stitching process. Given the stitching path starting point and the local tool point coordinate system for the first stitch, the translational and rotational positions for the head at the start of the stitching process for the stitch path are calculated.

Programming begins by writing out the code for the stitching path starting point calculated translational and rotational positions. Also, written out on each line of code are the cumulative path length and the drive axis distance that are both set to zero for the starting point. The drive axis distance is defined as a distance that identifies when full rotations of the drive motor occur, when provided a stitch length. That is, the drive motor to completes a full rotation, and therefore operates the stitching mechanism through an entire stitch formation process, when the drive axis distance increments by the stitch length. Programming continues for each stitch by defining the code for the end point of the stitch (path ending point for the last stitch) and consists of two lines of code for each stitch. Prior to writing out the two lines of code for a stitch, the Euler angles required for rotating the local tool point coordinate system at the starting point of the current stitch to be coincident with the local tool point coordinate system at the starting point of the next stitch are calculated. These Euler angles represent the total rotation of the head that occurs from the start of the stitch to the end of the stitch. Next, the first line of code for a stitch consists of the global translational position of the stitch end point replacing the values in the previous line of code. The rotational position values on the first line of stitch code are identical to the previous line of code, indicating that only translation of the head occurs with no rotation. The drive axis and the stitching path distances on the first line of stitch code are incremented by the stitch length, so that for the nth stitch, the drive axis and stitching path cumulative distances are defined as $n \times l_s$. The second line of code is written out with the translational position values, the drive axis distance and the stitching path distance equal to those on the first line of stitch code. The rotational positions of the second line of stitch code are calculated by adjusting the rotational values on the first line by the calculated Euler angles for the stitch. The second line of code is purely a rotation of the head after the stitch translation occurs in order to line up the head for stitching of the next stitch. Therefore, the two lines of code represent the uncoupling of the stitch translation that occurs in the first line of stitch code and the total rotation of the stitch that occurs in the second line of stitch code. Generation of these two lines of code for each stitch progresses until the final stitch lines of code are generated.

The process described provides the coding necessary to decouple the translation and rotation for each stitch to ensure that the problems associated with rotations of the head while the stitching needles are in the preform do not occur. During the course of stitching trials, it has been seen that if the curvature is shallow (large radius), the problems do not manifest to a degree to raise concern. Thus, these shallow curved stitching paths can be stitched without difficulty using the control programming produced by the existing preprocessor, and decoupling of translations and rotations for each stitch are not required. Avoiding decoupling the translations and rotations is advantageous because executing the single line of code versus the two lines of code is faster. However, the described process enables the stitching when the curvature does require decoupling of the translation and rotation. Determining when to use the decoupled approach is part of ongoing stitching trials and is highly dependent on the rotation about the local tool point coordinate system z-axis for the stitch.

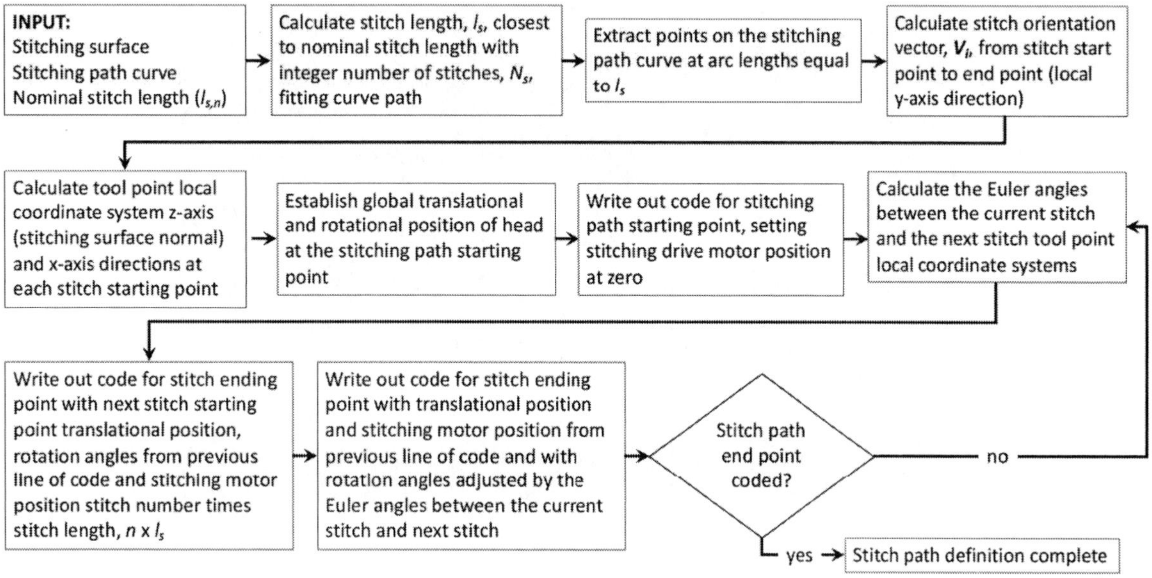

Figure 14. Stitching path development process flow chart.

3.3 Example Stitched Paths

Several stitching paths are presented to demonstrate when decoupling of the translation and rotation are necessary to produce 3D stitched paths. a closed planar circular path with an 8-inch radius is shown in Figure 15. To stitch this path, it was necessary to use the presented decoupling method, and preliminary information indicates that using the decoupled approach is preferable when the radius of curvature is less than 24 inches. A closed path is shown in Figure 16 that consists of straight edges that are connected using elliptical arcs with 1-inch radii where the ellipse segments meet the short ends of the shape. This type of path is representative of an access cutout in a wing lower skin. The stitching can be used to attach reinforcement to the cutout and to provide damage tolerance around the cutout. To stitch this path, the straight segments were not decoupled, but the elliptical segments were decoupled using the presented method. Lastly, a curve on a preform representative of the shape of the NTF wind tunnel blade is shown in Figure 17. For this demonstrator, the two main path segments did not require decoupling of the translation and rotation, but the rotation that occurred between the two main segments was decoupled from the two paths. Other, more severe paths on the same demonstrator article do require decoupling of the stitch translations and rotations.

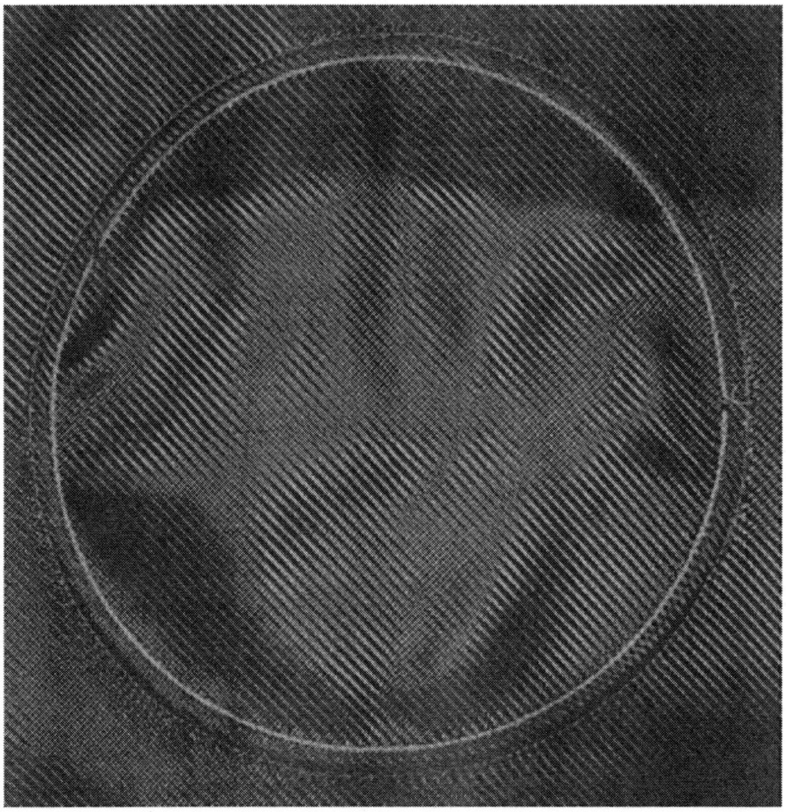

Figure 15. Closed, planar, circular path.

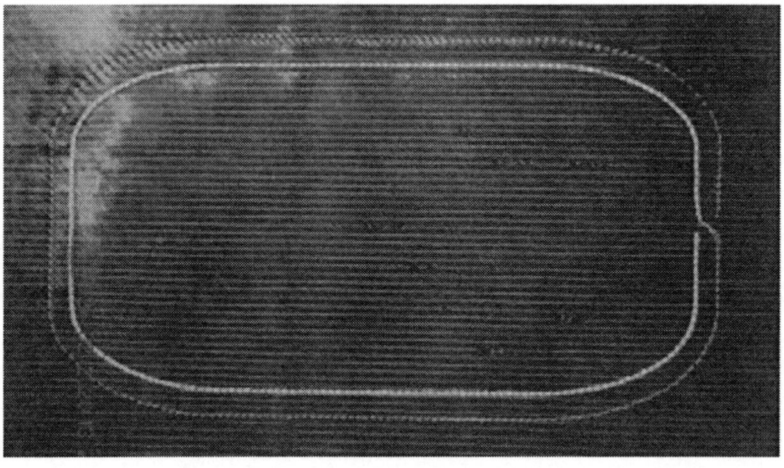

Figure 16. Closed path with straight and elliptical segments.

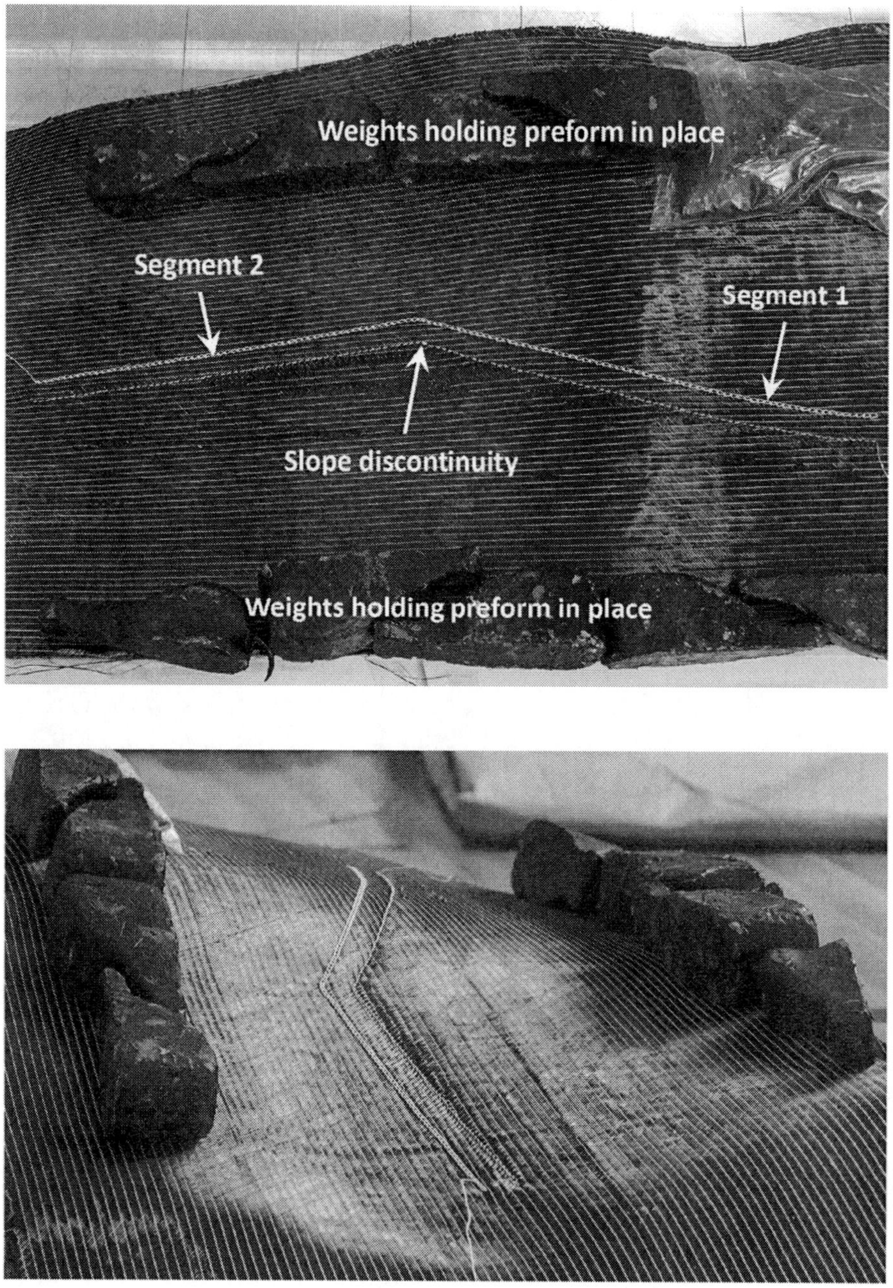

Figure 17. Two views of doubly-curved path demonstrator article representative of the NTF wind tunnel blade.

4. CONCLUSIONS

Stitching capabilities have been incorporated into the ISAAC system at NASA LaRC through the introduction double-sided and a single-sided stitching heads. Initial stitching path control was implemented as straight lines in space as was done for previous stitching development. However, more complex stitched structures such as a wind tunnel blade or cutouts within a fuselage or wing skin require that the stitching paths be implemented as 3D stitching paths in space. Unfortunately, current control programming output by a preprocessor where translations and rotations are applied to the head motion simultaneously cannot stitch these curved paths due to problems that arise in stitch formation and the introduction of side forces on the needles. This inability to stitch is most pronounced for the single-sided stitching head, where two needles are in the preform at the same time for the majority of the stitching process. A means to program 3D stitching paths in space was developed whereby the translation and rotation of each stitch were decoupled, thereby eliminating the problems associated with current control programming approaches. Using this newly developed stitching path definition and control programming, complex stitching paths have successfully been stitched at the ISAAC facility. In order to mitigate the time penalty that is associated with decoupling the stitching path motion, research is ongoing to develop rules to determine when this translation and rotation decoupling method should be applied.

5. ACKNOWLEDGEMENTS

The authors wish to thank Mr. Patrick J. Thrash of The Boeing Company for his valuable contributions to the development of the stitching heads and their installation at the LaRC ISAAC facility.

6. REFERENCES

1. A. P. Mouritz, "Tensile Fatigue Properties of 3D Composites with Through-Thickness Reinforcement," Composites Science and Technology, Vol. 68, 2008, pp. 2503-2510.
2. M. B. Dow and H. B. Dexter, "Development of Stitched, Braided and Woven Composite Structures in the ACT Program and at Langley Research Center (1985 to 1997)," NASA TP-97-206234.
3. D. C. Jegley, H. G. Bush, and A. E. Lovejoy, "Structural Response and Failure of a Full-Scale Stitched Graphite–Epoxy Wing," Journal of Aircraft, Vol. 40, No. 6, November–December 2003, pp. 1192-1199.
4. M. Rouse, D. C. Jegley, D. M. McGowan, H. G. Bush, and W. A. Waters, "Utilization of the Building-Block-Approach in Structural Mechanics Research," 46th AIAA/ASME/ASCE/AHS/ASC Structures, Structural Dynamics & Materials Conference, 18 - 21 April 2005, Austin, Texas. (Available as AIAA Paper 2005-1874.)
5. M. Karal, "AST Composite Wing Program – Executive Summary," NASA CR-2001-210650.
6. K. Gould, A. E. Lovejoy, D. C. Jegley, A. L. Neal, K. A. Linton, A. C. Bergan, and J. G. Bakuckas, Jr., "Nonlinear Analysis and Experimental Behavior of a Curved Unitized Stitched Panel" Journal of Aircraft, Vol. 52, No. 2, March–April 2015, pp. 628-637.
7. D. C. Jegley, "Structural Efficiency and Behavior of Pristine and Notched Stitched Structure," presented at the annual SAMPE meeting October 17-20, 2011.
8. N. Yovanof and D. C. Jegley, "Compressive Behavior of Frame-Stiffened Composite Panels," presented at the 52nd AIAA Structures Dynamics and Materials Conference, April, 2011.
9. D. C. Jegley, "Behavior of Frame-Stiffened Composite Panels with Damage," presented at the 54th AIAA Structures, Structural Dynamics, and Materials Conference, April, 2013.

10. A. Allen and A. Przekop. 2012. "Vibroacoustic Characterization of a New Hybrid Wing-Body Fuselage Concept," presented at the INTER-NOISE 2012 Conference, August, 2012.

11. A. E. Lovejoy, M. Rouse, K. Linton, and V. Li, "Pressure Testing of a Minimum Gauge PRSEUS Panel," presented at the 52nd AIAA Structures Dynamics and Materials Conference, April, 2011.

12. N. Yovanof, A. E. Lovejoy, J. Baraja, and K. Gould, "Design Analysis and Testing of a PRSEUS Pressure Cube to Investigate Assembly Joints," Aircraft Airworthiness and Sustainment Conference, April, 2012.

13. A. Przekop, "Repair Concepts as Design Constraints of a Stiffened Composite PRSEUS Panel," presented at the 53rd AIAA/ASME/ASCE/AHS/ASC Structures, Structural Dynamics and Materials Conference, May, 2012.

14. A. Przekop, D. C. Jegley, M. Rouse, and A. E. Lovejoy, "Analysis and Testing of a Metallic Repair Applicable to Pressurized Composite Aircraft Structure," Presented at the Society for the Advancement of Material and Process Engineering (SAMPE)Technical Conference, June, 2014.

15. D. C. Jegley, "Experimental Behavior of Fatigued Single Stiffener PRSEUS Specimens," NASA-TM-2009-215955.

16. P. J. Thrash, "Manufacturing of a Stitched Resin Infused Fuselage Test Article," Presented at the Society for the Advancement of Material and Process Engineering (SAMPE) Fall Technical Conference, Orlando FL, October, 2014.

17. R. A. Wahls, "The National Transonic Facility: A Research Retrospective," AIAA Paper 2001-0754, 39th AIAA Aerospace Sciences Meeting and Exhibit," January 8-11, 2001, Reno, Nevada.

18. R. M. Grenoble, R. Harik, D. M. Munden, J. Halbritter, D. C. Jegley, and B. H. Mason, "Assessment of Automated Fiber Placement for the Fabrication of Composite Wind Tunnel Blades," Proceedings of Society for the Advancement of Material and Process Engineering (SAMPE) 2019 Conference and Exhibition, Charlotte, North Carolina, May 20-23, 2019.

19. R. Harik, J. Halbritter, D. C. Jegley, R. W. Grenoble, and B. H. Mason, "Automated Fiber Placement of Composite Wind Tunnel Blades: Process Planning and Manufacturing," Proceedings of Society for the Advancement of Material and Process Engineering (SAMPE) 2019 Conference and Exhibition, Charlotte, North Carolina, May 20-23, 2019.

20. B. H. Mason, D. M. Munden, D. C. Jegley, R. W. Grenoble, and R. Harik, "Design and Analysis of a Tool for Automated Fiber Placement of Composite Wind Tunnel Blades," American Society of Composites (ASC) 34th Technical Conference, Atlanta, GA, September 23-25, 2019.

21. NASA ISAAC Fact Sheet, FS-2016-12-273-LaRC, 2016.

22. https://www.cgtech.com/ Accessed 12/17/2018.

NONDESTRUCTIVE EVALUATION (NDE) DATA CORRELATION FOR MANUFACTURING IMPROVEMENT

Roger W. Engelbart
Boeing Research & Technology
St. Louis, Missouri, USA

ABSTRACT

Nondestructive evaluation (NDE) is often considered to be strictly a quality assurance tool, used exclusively as a mechanism for acceptance or rejection of parts. NDE processes may be viewed as having a negative cost impact to any program, due to rejections that may result in repairs, scrap, and additional labor and documentation. However, NDE can also be a valuable engineering tool for process improvement if the data are utilized for correlation to manufacturing parameters and events. This can result in a positive cost advantage to the program by avoiding future rejections.

Under the RApid High-Performance Molding (RAPM) for Small Parts program, a variety of different composite part configurations were manufactured utilizing fabrication processes that were geared toward automotive applications. Ultrasonic and radiographic NDE were performed on these parts to understand the general output of the chosen manufacturing processes. This paper will show how correlation of the NDE data to destructive analysis led to refinement of manufacturing parameters and, for the resin infusion challenge part, showed the path to implementation for an existing aerospace program.

Keywords: Composites, nondestructive evaluation, press molding, resin infusion, resin transfer molding, thermoset
Corresponding author: Gail Hahn

1. INTRODUCTION

The overall goal of the RAPM program is to enable affordability of small composite in trade studies against machined aluminum by achieving manufacturing flexibility to rate and part mix, while retaining aerospace structural performance [1]. For NDE operations, it would be desirable to employ technologies that are reasonably priced, that can be incorporated in line with a manufacturing operation, and that can still deliver aerospace quality in the inspection results. Since aerospace quality typically implies 100% inspection coverage and very conservative acceptance criteria, the requirements of low cost, "in line," and aerospace capable tend to be mutually exclusive. In view of what is needed to prove the capability of an NDE method and qualify it to the application, the technology options become limited.

2. RESIN INFUSION – DEEP DRAW PAN

The pan challenge part, RI-RAPM-004, was inspired by a production component on a current rotorcraft program and is shown in Figure 1 [2, 3]. Three resin infusion preform sections can replace 54 individual prepreg pieces used for the conventional hand lay-up and autoclave process.

The resin transfer molding (RTM) time was two hours. This was important in evaluating the potential impact of the NDE operation on the cycle time of a fabrication cell.

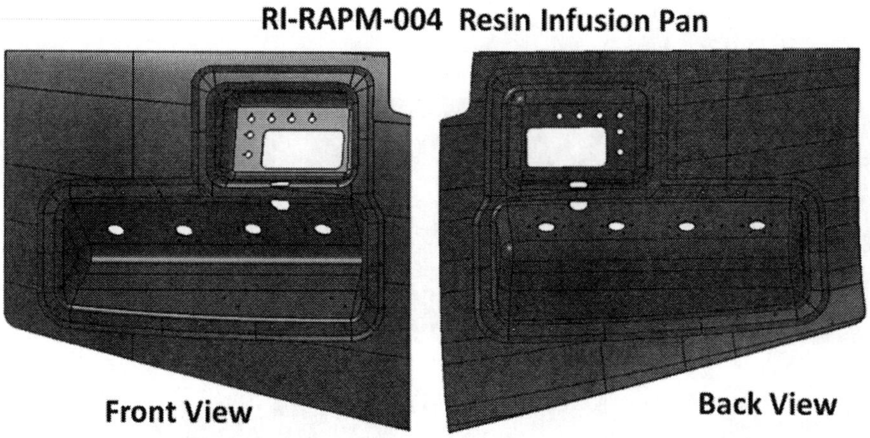

Figure 1. RI-RAPM-004 Challenge Part – Resin Infusion Deep Draw pan

2.1 Nondestructive Evaluation

Ultrasonic inspection of the pan using the Boeing Automated Ultrasonic Scanning System (AUSS) is shown in Figure 2. Post-manufacturing ultrasonic nondestructive evaluation of composite

Figure 2. Automated ultrasonic scanning system (AUSS) used for C-scans of challenge parts (RI-RAPM-004)

structures has traditionally been performed in dedicated laboratories or inspection areas, on such large stationary squirter or immersion systems. The advantage to such systems is the large scan envelope which allows positioning of large complex structures for scanning. The drawback is that parts must be taken out of the manufacturing flow for the time necessary to transport them to the inspection area, complete the inspection, and return them to the next station. Significant savings could be realized if the same quality of inspection could be obtained by using portable equipment in the manufacturing cell. Figure 3 shows various positions of the AUSS water nozzle with respect

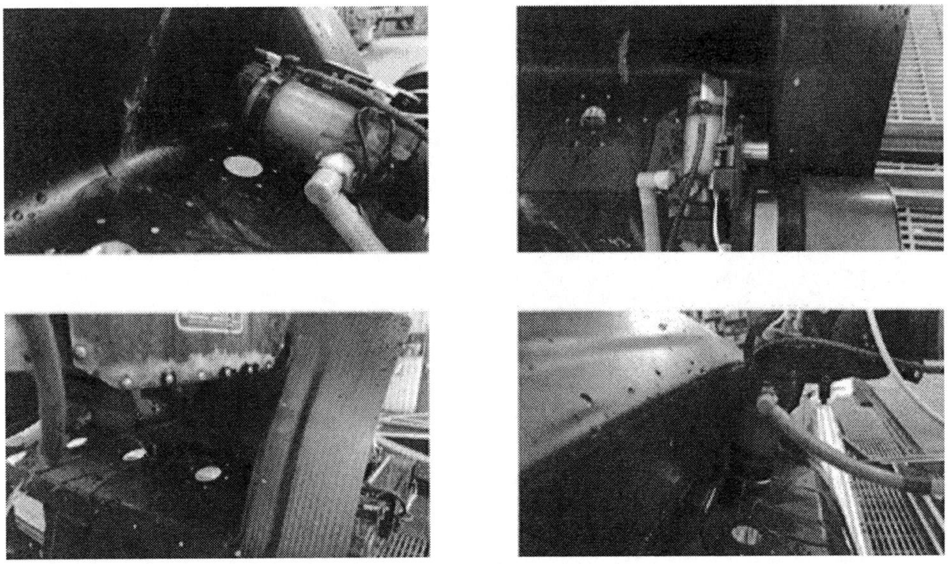

Figure 3. AUSS Nozzle Access to Pan Features

to the complex features of the pan. As seen in Figure 4, the quality of the scan data is very good,

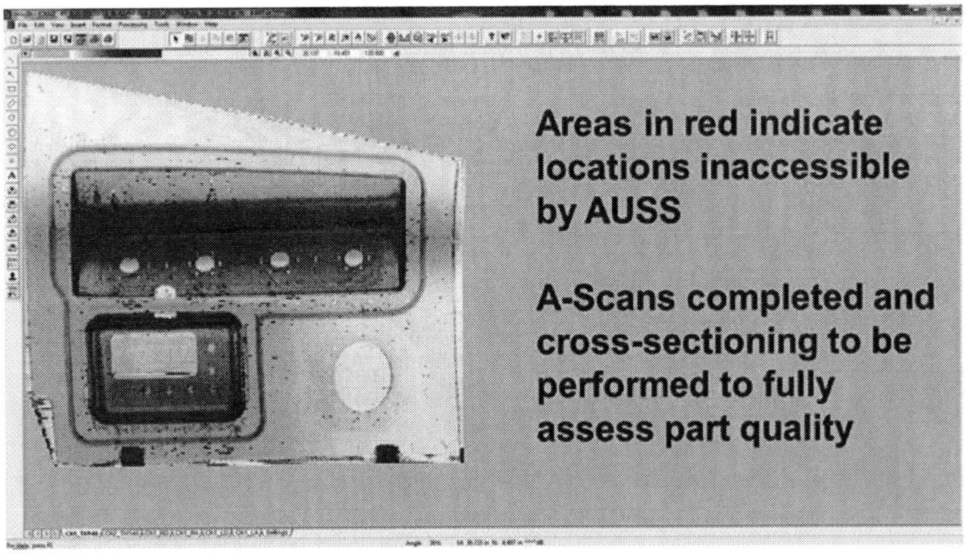

Figure 4. Typical AUSS C-Scan of RI-RAPM-004 parts

but not all locations were accessible, which necessitated additional coverage by portable A-scan. The need to use a series of sub-scans in order to cover the complex geometries contributed to the overall inspection cycle time.

2.2 Destructive Correlation

A series of cut sections was taken to obtain photomicrographs for assessment of laminate equality. The map is shown in Figure 5; nine locations, designated as A through I. These were used to

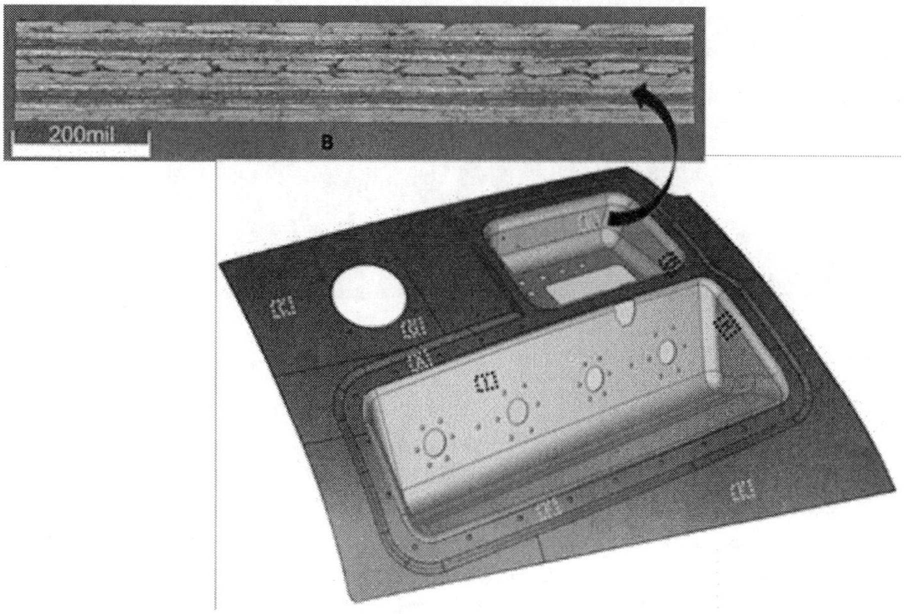

Figure 5. Photomicrograph Locations for Laminate Quality

evaluate the laminate for porosity content and resin pockets. The figure also shows the photomicrograph from location B. The laminate quality was very good and very consistent across all the locations. Several of the pans exhibited ultrasonic A-scan signal responses from anomalies on the walls or in the radii of the deep draw features. An example of one of these is shown in Figure 6, with the cross section. The ultrasonic signal from the anomaly was characteristic of a delamination, and this was confirmed by the photomicrograph. This was considered to have been induced by out-of-plane stresses in the radii generated thermally or mechanically during part removal from the tool. Since there were temperature limitations on the tool that was used for this effort, it was believed that either a modified cure cycle with full cure on the tool or the use of a resin system with improved interlaminar toughness would eliminate this type of flaw from the fabrication of this part.

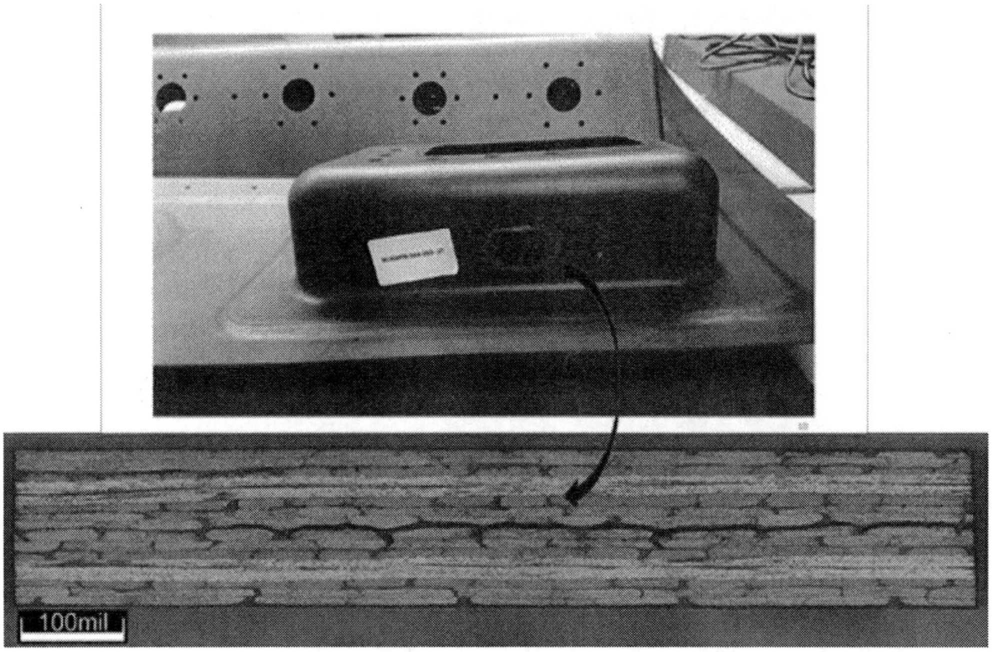

Figure 6. Verification of Delamination Indication

2.3 Evaluation for Cell-Based Inspection

The production counterpart which inspired the RI-RAPM-004 challenge part has a requirement for 100% coverage by ultrasonic inspection to specific criteria for porosity, and the production approach is to inspect using portable A-scan. Because of the need for separate AUSS sub-scans to cover portions of the deep draw features, the inspection cycle time for the pan was approximately two hours. This is equivalent to the cycle time currently experienced in production. With a two-hour RTM time, the inspector would finish one part just as the next one comes out of the mold. Any rejection or re-evaluation would dramatically slow or even stop the flow of the manufacturing cell. With the stated criteria, the only way to fit inspection within the manufacturing cycle time is to have a primary inspector form parts, and one or more secondary inspectors to handle re-evaluations and non-conformance documentation.

2.4 Path to Implementation

Implementation of RI-RAPM-004 technology into a current program would of course involve consideration of allowables, material costs, and production verification to establish process parameters and manufacturing repeatability. This would also be an opportunity to re-evaluate the inspection coverage and criteria in view of the part function, loading, and current cost of inspection. Reducing the coverage from 100% to just the deep draw features would ensure that the inspection time would support the expected manufacturing cycle. The focus would be on such anomalies as delaminations or embedded foreign material, for which there may be a risk. To accomplish this, the porosity criteria must be removed. Multiple cross-sections for laminate quality are taken during the production verification activity, using a cut plan similar to the one shown in Figure 5. These would be used over a number of parts to establish consistency of low porosity and establish a basis for reconsidering the requirements.

3. THERMOSET PRESS FORMING – C-CHANNEL

3.1 Nondestructive Evaluation

Figure 7 shows the TS-RAPM-007 challenge part, a thick wall C channel fabricated by Solvay.

Figure 7. TS-RAPM-007 C-Channel

Parts were manufactured by press forming using two different material systems.

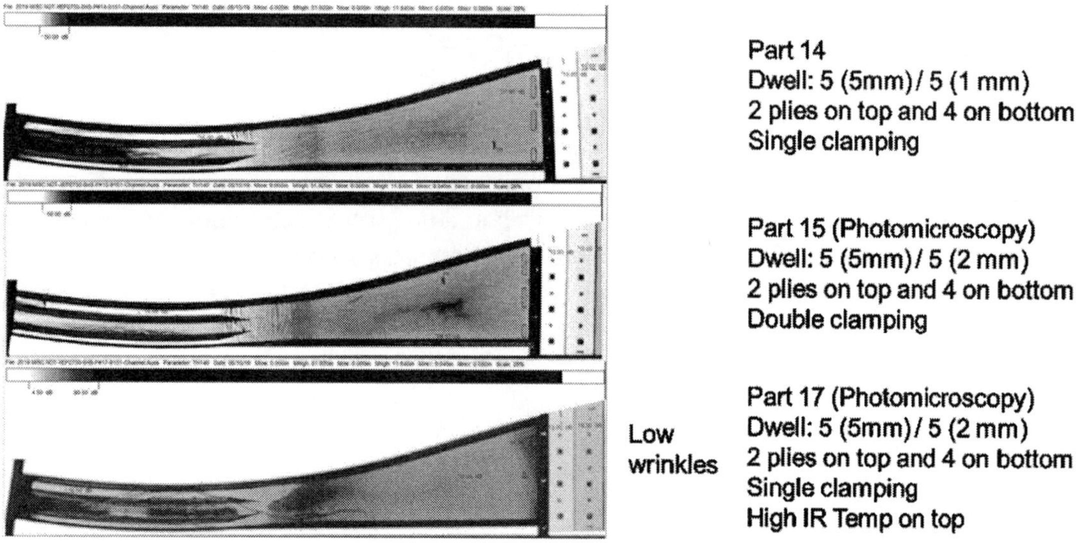

Figure 8. Ultrasonic C-Scans of TS-RAPM-007 Channel Webs with Process Parameters

All parts were ultrasonically inspected by Boeing, but this was limited to the channel web, in order to hold down inspection time and maintain the required development schedule. Inspection of the flanges and radii is addressed in Section 3.3. Solvay maintained a matrix of the C-scans with the corresponding key processing parameters, as shown in Figure 8. This provided a convenient comparison for visualizing what parameter changes produced the best or worst results.

3.2 Destructive Correlation

The Solvay matrix also indicated which parts were to be used for photomicrographs. The cut plan is shown in Figure 9 for Part 17; for each channel, the sections were taken from the same locations

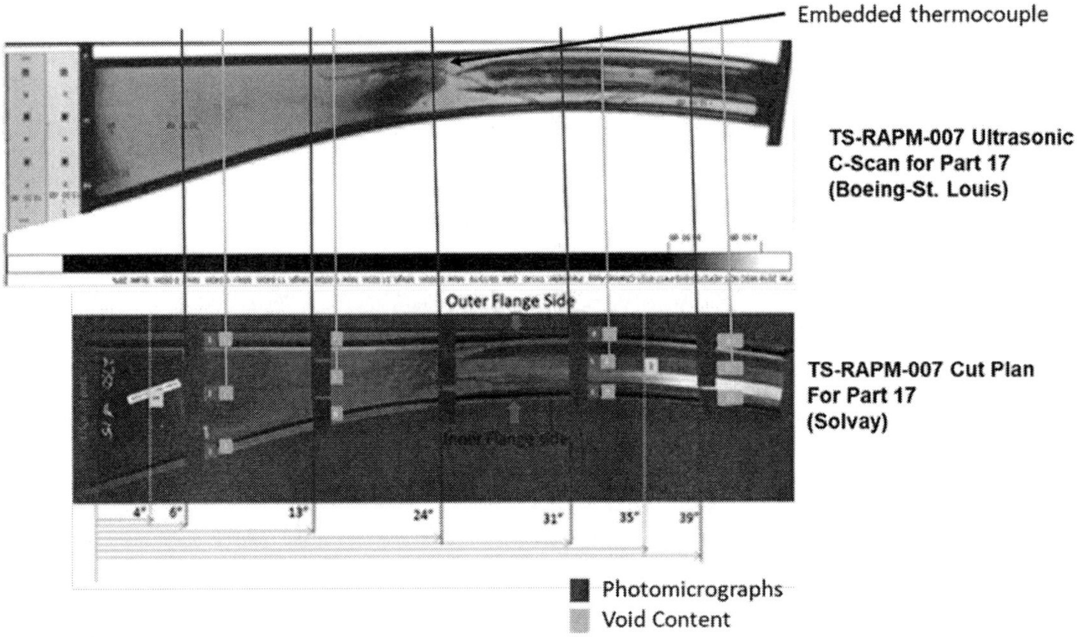

Figure 9. Cut Plan for TS-RAPM-007 Photomicrographs and Void Content

along the part length. At each location, three photomicrograph sections and three void content specimens were taken – one from the center of the web and one from each flange, encompassing the radius. A system of identification numbers for the sections was used to ensure commonality of position across all parts. Void content was determined by acid digestion; sample photomicrographs are shown in Figure 10.

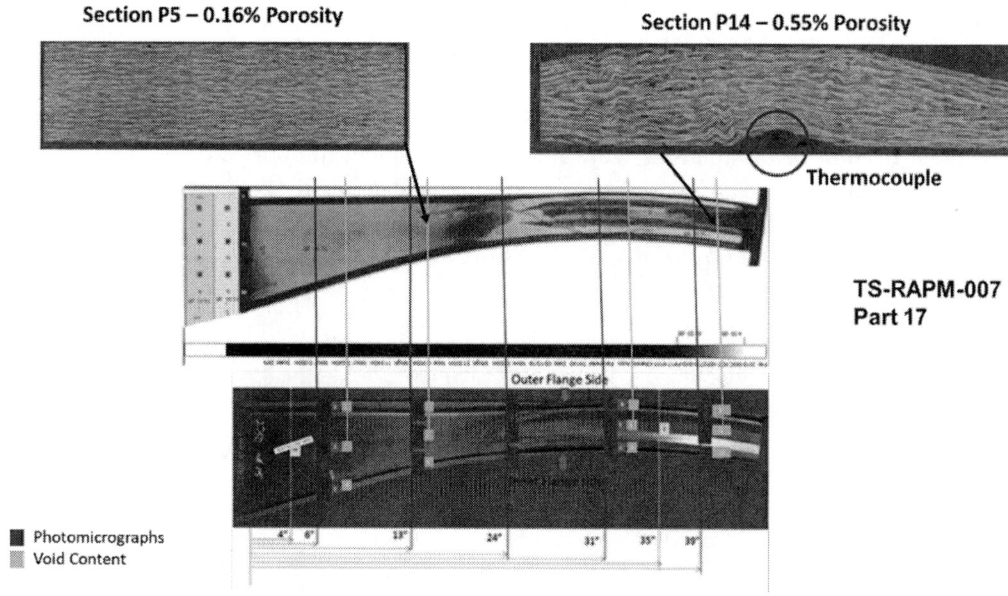

Figure 10. Example Photomicrographs for TS-RAPM-007 Part 17 by Location

3.3 Evaluation for Cell-Based Inspection

The RAPM cost analysis sought to develop an understanding of the cost drivers for manufacturing operations and of the improvements that might be gained from varying levels of automation and capital investment. For NDE, the key drivers are geometric complexity, requirements for 100% inspection, and acceptance criteria. For composite structures, the complex geometry is the radius – it is considered to be a critical feature, it carries tight criteria, it requires complete inspection coverage, and it may present access challenges in terms of tightness and curvature. There are gains associated with increasing the capability and sophistication of the ultrasonic system, but much of depends on the quality of the tooling: positional accuracy, ease of set-up, and ability to hold

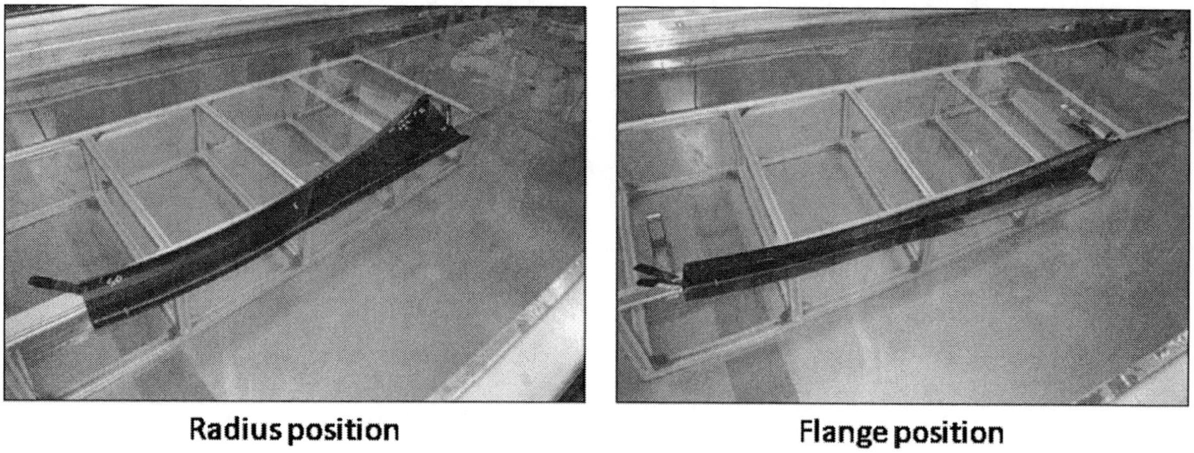

Figure 11. Immersion Ultrasonic Set-Up for TS-RAPM-007 Part 31

position without introducing error through slippage.

In order to understand the full inspection cycle time for TS-RAPM-007, all surfaces of Part 31 were ultrasonically inspected. Figure 11 shows the part positioned both for inspection of the flanges and the radii. Five position changes were required to inspect the part – web, each flange, and each radius. All surfaces were inspected using a flat linear array, a transducer that features numerous "elements", each of which functions like a miniature transducer. This enables coverage of more of the part surface with each pass. Figures 12 and 13 show the individual scan segments for the web,

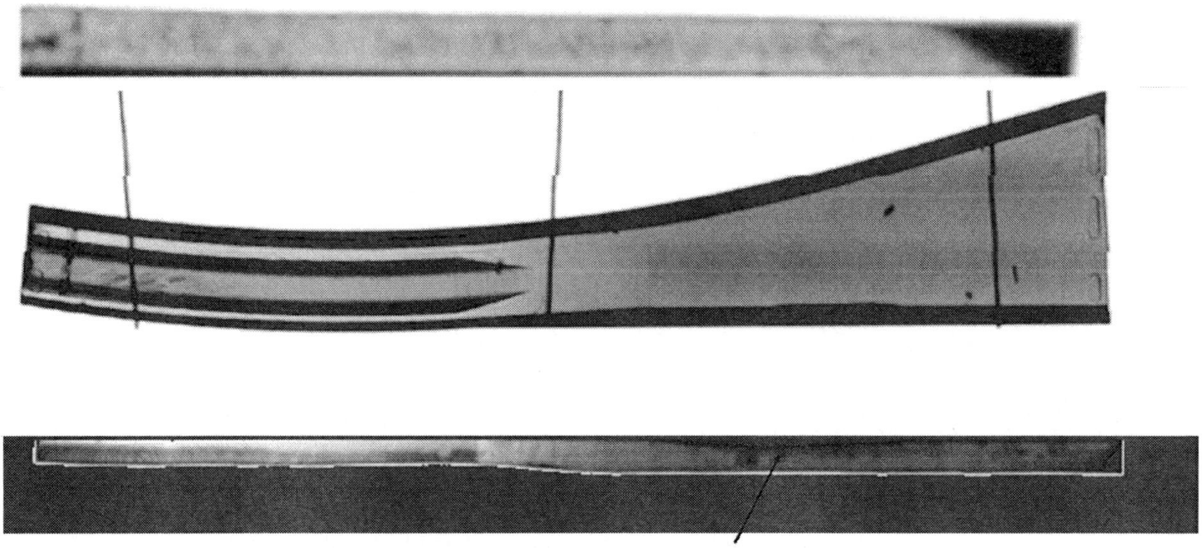

Figure 12. Flange Data for TS-RAPM-007 Part 31

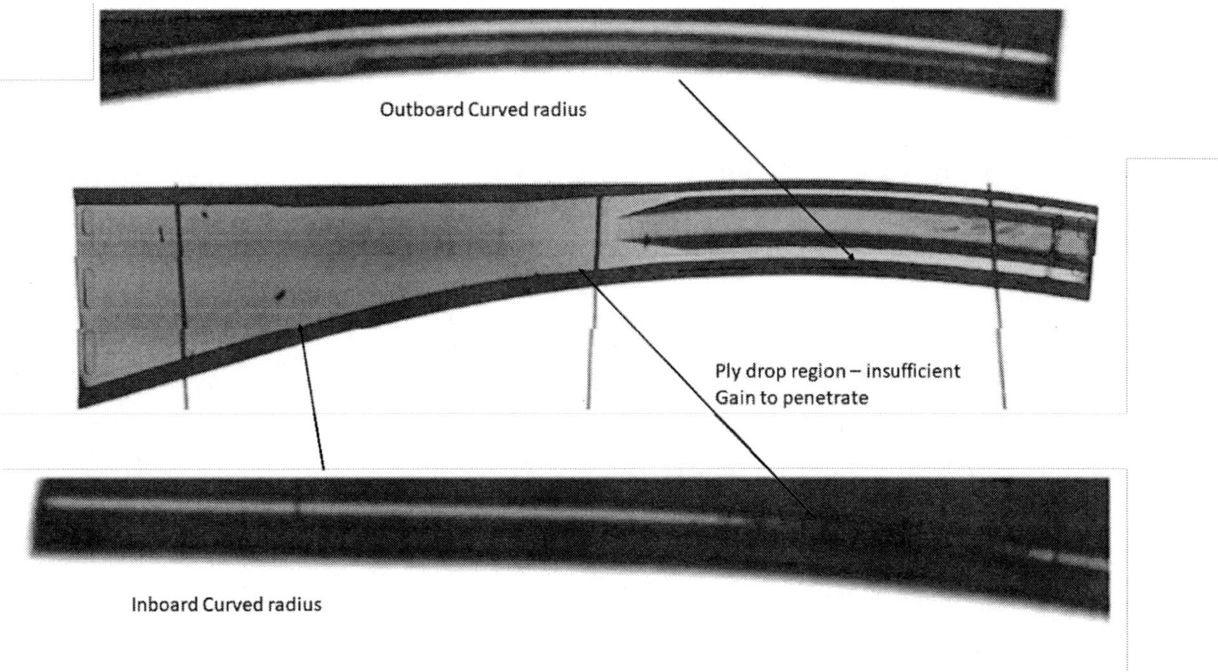

Figure 13. Radius Data for TS-RAPM-007 Part 31

flanges, and radii. With manual set up changes, the part required over two hours to fully inspect. This part is not a good candidate for in line cell-based inspection. The number of set up changes, and the resulting cycle time would immediately stop the line. Considering the likely requirement for 100% coverage, one approach would be the use of a single system and dedicated tooling for each required position. The part would move from one station to the next and would be inserted into fixed tooling for that position, with no changes needed from the operator. While this would improve flow, the capital investment would be prohibitive.

7. CONCLUSIONS

The NDE support of the manufacturing tasks under the RAPM program was essential in revealing the condition of fabricated articles through high quality data and for showing the progress in the development of manufacturing processes. The NDE data was correlated to internal condition and to specific anomalies, in order to guide future interpretations and understanding of the processes. The NDE approaches for the various parts were evaluated for their suitability to be integrated "in line" into the flow of a manufacturing cell. NDE cycle time was shown to be shown to be very dependent on acceptance criteria and inspection coverage requirements; this also drives the suitability for cell integration.

8. ACKNOWLEDGEMENTS

This effort was jointly accomplished by a Boeing-led team and the United States Government (Defense Advanced Research Projects Agency, DARPA) under the guidance of the United States Army Research Office, ARO. The author would like to acknowledge the guidance and support of Dr. Jan Vandenbrande of DARPA/DSO and Dr. David Stepp of Army Research Office

(ARO). The information in this paper was approved for public release by 20-164281-ETT and DISTAR 32948. Research was sponsored by the Defense Advanced Research Projects Agency (DARPA) and was accomplished under Cooperative Agreement Number W911NF-16-2-0087. The views and conclusions contained in this document are those of the authors and should not be interpreted as representing the official policies, either expressed or implied, of the Defense Advanced Research Projects Agency or the U.S. Government. The U.S. Government is authorized to reproduce and distribute reprints for Government purposes notwithstanding any copyright notation herein.

9. REFERENCES

[1] Hahn, G.L., Tsotsis, T.K., "Rapid High-Performance Molding (RAPM) of Small Parts," SAMPE 2019 Conference Proceedings. Charlotte, NC, May 20-23, 2019.

[2] Tsotsis, T.K., Cespedes-Gonzalez, G., Wiener, M., Cohen, L., Calamito, D., Costantino, S., Klunker, F. "Fabrication of a Complex Part with Deep-Draw Sections by Resin Transfer Molding," SAMPE 2019 Conference Proceedings. Charlotte, NC, May 20-23, 2019.

[3] Tsotsis, T.K., Cespedes-Gonzalez, G., Wiener, M., Frank, M., Cohen, L., Calamito, D., Hahn, G.L. "Preform Developments for Resin Infusion for Rapid High-Performance Molding (RAPM) for Small Parts," CAMX 2019 Conference Proceedings, Anaheim, CA, 23-26 September 2019.

DEVELOPMENT OF A THERMOPLASTIC CONTINUOUS FIBER REINFORCED ACCESS PANEL

Peter Levesque[1], Justin Schell[2], Jonathan Sourkes[1], Claire Steggall-Murphy[1], Nicholas Turcotte[1], Gail Hahn[2]
[1]TxV Aerospace Composites, Bristol, Rhode Island, USA
[2]The Boeing Company, Saint Louis, Missouri, USA

ABSTRACT

TxV Aerospace Composites, LLC (TxV) developed and manufactured a continuous fiber reinforced access panel, with joggle features and injection overmolded plastic, in collaboration with Boeing Research & Technology (BR&T) in support of the RApid high-Performance Molding (RAPM) for Small Parts, a program under the Defense Advanced Research Projects Agency (DARPA) Tailorable Feedstock and Forming (TFF) initiative.

Overmolding composites integrates the mechanical performance of continuously reinforced thermoplastic composites with the design flexibility of injection-molding to create complex geometries and features. The hybrid molding process offers continuous manufacturing with the potential for short cycle times and mechanical performance equivalent to or better than traditional aerospace metals.

TxV's hybrid overmolding process is made possible by VICTREX AE™ 250, a low-melt temperature polyaryletherketone (LM-PAEK). The polymer matrix offers substantially similar fatigue performance, chemical and corrosion resistance as polymers in the polyaryletherketone family but with a melt temperature of about 40°C less than VICTREX™ PEEK. This temperature delta is responsible for a strong bond during the overmolding process. These materials combined with TxV's highly automated manufacturing facility, designed for high-rate production, enables the manufacture of composites parts at potentially lower cost than typical aerospace metals.

This paper outlines the development and manufacturing of an overmolded thermoplastic composite access panel demonstrator part following a design-for-manufacture approach covering: automated tape laying, consolidation, forming, over-molding, finishing and inspection.

Corresponding author: Gail Hahn

1. INTRODUCTION

It is no secret that the upcoming demands on the aerospace supply chain are challenging. With predictions that the world will need to about double aircraft in approximately 20 years [1], it is necessary to develop the capability to make aerospace-grade parts more quickly than traditional processes. Within the last few years the aerospace industry has started to adopt more wide-spread use of thermoplastic matrix composite materials to replace metal and thermoset incumbent materials. Thermoplastic composite materials can be used to fabricate very high-quality parts without the use of an autoclave or other lengthy manufacturing processes. Aerospace Original Equipment Manufacturers (OEM) and their Tiers are participating in programs geared toward out-of-autoclave (OOA) solutions using thermoplastic composites to build large-scale technology demonstrators and prove that thermoplastic composites have a place in aircraft beyond brackets and small parts. Programs like these demonstrate dedication from the very top of the supply chain to innovating with thermoplastic composites.

Beyond large aircraft, the needs of the advanced air mobility (AAM) industry are poised to grow at a rapid pace. Electric powered vehicles with a vertical take-off and landing capabilities (eVTOLs) are extremely weight-sensitive, opening the door for heavy use of composite technologies in the fabrication of the vehicles. The fast cycle times of thermoplastic composites will be desirable to keep up with full-scale production rates once these vehicles start to see more widespread adoption. A NASA funded Booz Allen Hamilton study estimates the UAM market size to have an annual value of USD500 Billion [2]. This market potential has driven industry to explorer new material and processing solutions including hybrid overmolding.

1.1 Hybrid Overmolding Introduction

The hybrid overmolding process leverages the strength of continuous fiber composites and the functionality and flexibility of injection molding. This process produces parts that have complex geometries that cannot otherwise be molded with continuous fiber. [3] The success of the process comes from the use of two kinds of thermoplastics that are in the polyaryletherketone (PAEK) family. The matrix of the continuous fiber composite prepreg, VICTREX AE™ 250, has a melting temperature of 305 °C (581 °F). The melting temperature of the VICTREX™ PEEK injection molding plastic has a melting temperature of 343 °C (649 °F). The 40 °C (68 °F) difference in melting temperature allows for the surface of the continuous fiber to melt enough to form a fusion bond with the injection molded portion of the part. This technology can eliminate the need for mechanical fasteners, adhesive, or welding to create a highly complex, bonded part.

1.2 Motivation for Work

This work was supported as part of the RApid high-Performance Molding (RAPM) [4], a program under the Defense Advanced Research Projects Agency (DARPA) Tailorable Feedstock and Forming (TFF) initiative [5].

2. EXPERIMENTATION

2.1 Part Design

The access panel design is meant to simulate a generic opening in a large, unsupported panel. This design incorporates a six-sided central opening in the 6.86 mm (0.270 in) thick, 50 ply base laminate. Approximately 25 mm (1 in) outside of the central hex, the geometry steps up sharply creating a secondary 6 mm (0.240 in) high hex simulating a locating feature. The entire 66.1 cm x 66.1 cm (26 in x 26 in) panel has single curvature of a 254 cm (100 in) radius. See Figure 1 for isometric drawing.

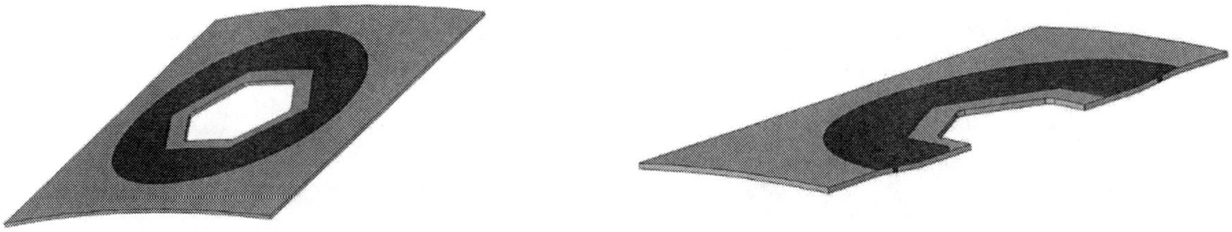

Figure 1: 3-dimension drawing of access door panel.

2.2 Manufacturing Process

The access panels were made with a highly automated four step process: prepreg tape lay-up, consolidation, stamp forming, and injection molding. This multi-step approach combined with automation keeps manual touch time to a minimum and promotes rapid manufacturing when compared to other available manufacturing technologies. Each of these stages will be described.

2.2.1 Tailored Blank Fabrication

Thirty-five tailored blanks were fabricated using an automated tape lay-up (ATL) machine. TxV uses a Dieffenbacher FiberForge 4.0 machine capable of tape lay-down rates as high as 150 kg/hr. The prepreg tape used combined VICTREX AE™ 250 low melt PAEK with Hexcel AS4 12K carbon fiber. It had a fiber areal weight of 143 grams/m2 and a resin content of 34%. The tape width was 50 mm (1.95 inches). The blanks consisted of 50 plies laid up in a quasi-isotropic orientation. The 78.7 cm x 78.7 cm (31 x 31 in) tailored blank can be seen in Figure 2. A hole was left in the middle where an eventual hexagon shaped cut-out would be machined. At 50 plies, this hole represented a significant amount of material savings. This demonstrates the flexibility of the ATL process, which allows for both ply drops and pad ups creating the potential to design with variable wall thicknesses. The FiberForge 4.0 employs a vacuum table to keep initial courses of material in place until they can be tacked with an ultrasonic welder to the layer above them. It then keeps the layers in place as the rest of the tailored blank is fabricated. The full 360° rotation of the table allows for tape to be placed accurately in any direction required by the application.

Figure 2: Tailored blank fabricated using ATL.

Ply gap measurement was performed using a digital microscope with calibrated measurement software. Ply gaps were measured on plies 1, 25, 26, and 49 on a sample size of 5 panels. Ten measurements per ply were taken in the places indicated in Figure 3. A summary of the ply gap measurement results can be found in Table 1 below. Note that the allowable ply gap measurement for this project was 0.762 mm (0.03 in).

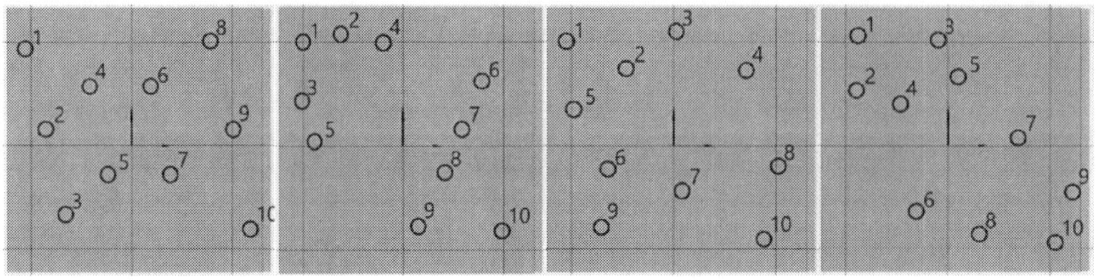

Figure 3: Ply gap measurement locations for (from left to right) ply 1, 25, 26, and 49.

Table 1: Ply gap measurement summary.

Panel ID	Mean Ply Gap [mm] (in)	Standard Deviation [mm] (in)
00077_0-1	0.00093 (0.0236)	0.00048 (0.0123)
00077_0-2	0.00106 (0.0268)	0.00054 (0.0138)
00077_0-3	0.00105 (0.0266)	0.00046 (0.0118)
00077_0-4	0.00097 (0.0247)	0.00054 (0.0136)
00077_0-5	0.00094 (0.0240)	0.00044 (0.0112)

2.2.2 Consolidation

After the ATL process the blanks were then dried in an oven set to 135-149 ºC (275-300 ºF) for at least 15 hours prior to panel consolidation. TxV uses a 2-stage compression molding cell shown in Figure 4. The first stage is an 80-ton press heated with electric heating elements in each platen. The second stage is a 500-ton press heated with circulating hot oil. This press configuration is commonly referred to as a shuttle press. The cycle times achieved in a shuttle press are a fraction of a typical static press configuration with a traditional temperature ramp up, hold, and cool down cycle. The 50 ply laminate construction was 2x to 5x the thickness of the typical composite lay-up. If this were to be a thermoset laminate the press recipe would need to account for off gassing, thermal runaway and/or exothermic event. Utilizing a thermoplastic polymer allows for rapid consolidation without these concerns. The quality of these laminates was assessed in part using phased array ultrasonic scanning. Of particular note, was the consistent quality of the laminate as the scan approached the perimeter of the center hole where tape placement was utilized effectively to reduce material consumption. Figure 5 shows a consolidated panel.

Figure 4: Two-stage consolidation cell at TxV Aerospace Composites.

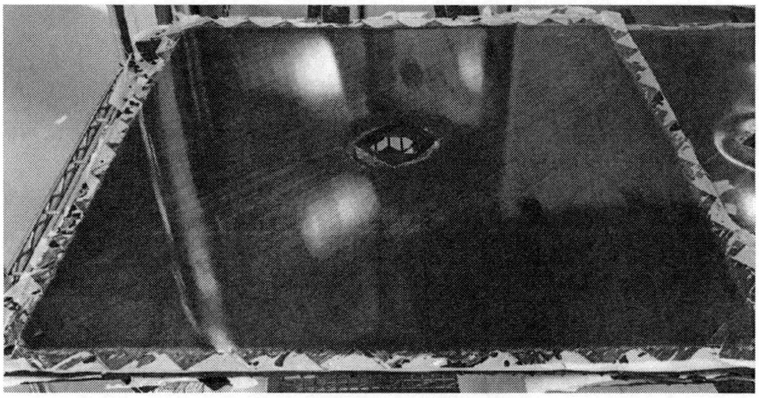

Figure 5: Consolidated panel.

After consolidation the panels were trimmed using a waterjet machine. In this stage the hexagonal cut-out was defined in the center of the panel. Tabs were also cut in order to facilitate alignment in the shuttle for the stamp forming process.

2.2.3 Stamp Forming

2.2.3.1 Tooling Design and Fabrication

The stamp forming tool was designed to fit into TxV's Master Unit Die (MUD) base. TxV ran several iterations of forming analysis to predict deformation and spring-back due to the forming process. Given the comparatively large size and thickness of this part compared to all other known hybrid overmolded parts, the software provider determined that the prediction tool was not well suited to reliably predict the required modifications of tool geometry to form a part to nominal geometry. As a result, the tool was cut to the exact geometry of the formed part with considerations made for the expansion of the tool steel at elevated temperatures. The tool design allows for transport and long-term storage that protects the forming surface through alignment features and binding straps. The tool was designed to receive heat from internal electric heating cartridges as well as from the oil heated MUD base.

The stamp forming tool, shown in Figure 6, was made from P20 hardened tooling steel. Each forming tool half contained 6 electric heating elements. The total stamp forming tool weight, outside of the MUD, was about 770 kg (1700 lbs.).

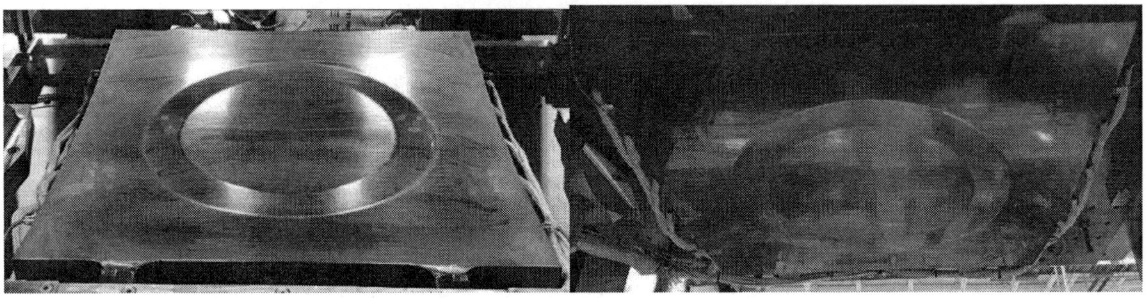

Figure 6: Stamp forming tool inside forming press. Bottom half on left, top half on right.

2.2.3.2 Process

The stamp forming process has two parts: infrared (IR) oven heating and then forming in a heated hydraulic press. The cold laminate is first secured onto a shuttle frame before it is placed in the IR oven by a robot. Once the laminate is above the melt temperature of the matrix it is then moved swiftly from the oven into the forming press by the robot. TxV has 4 IR ovens and 2 robots allowing for high rate production. Figure 7 shows the detail of the stamp formed laminate.

Figure 7: Detail of stamp formed laminate.

A second waterjet trimming process after stamp forming refined the edges on all sides and again on the hexagonal feature in the center of the part.

2.2.4 Injection Molding

2.2.4.1 Tooling Design, Modeling, and Fabrication

In the design of the injection mold tool TxV used a predictive software package to simulate filling of the mold cavity. This analysis guided the location of sprues as well as initial molding settings.

Moldex3D software was used to predict filling, packing, and cooling of the overmolded plastic during injection mold. Images shown below in Figure 8 show the melt front temperatures at time steps between 1.541, 2.280, and 2.991 s during fill. Temperatures range from 370 to 402 °C.

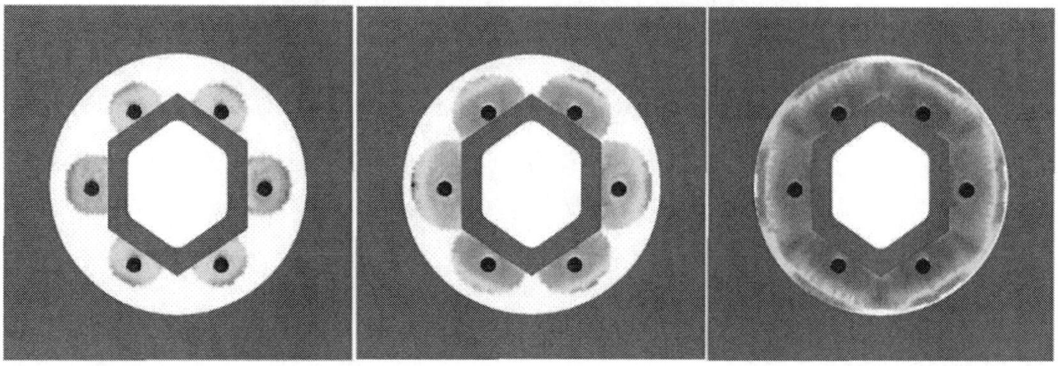

Figure 8: MoldEx 3D cavity fill prediction at (left to right): 1.541, 2.28, and 2.991 seconds.

A six drop, hot runner design was chosen as the most effective way to fill the cavity with plastic at a temperature and pressure that would promote bonding between the plastic and underlying laminate. This tool utilized A-side ejection. The ejectors are depicted in the left side of Figure 9 below as the rectangular shaped inserts supported by three springs. The tool was equipped with hot oil lines throughout. The B-side had a hexagon shaped male alignment feature best shown on the right side of Figure 10 where laminate inserts are hung prior to tool closing.

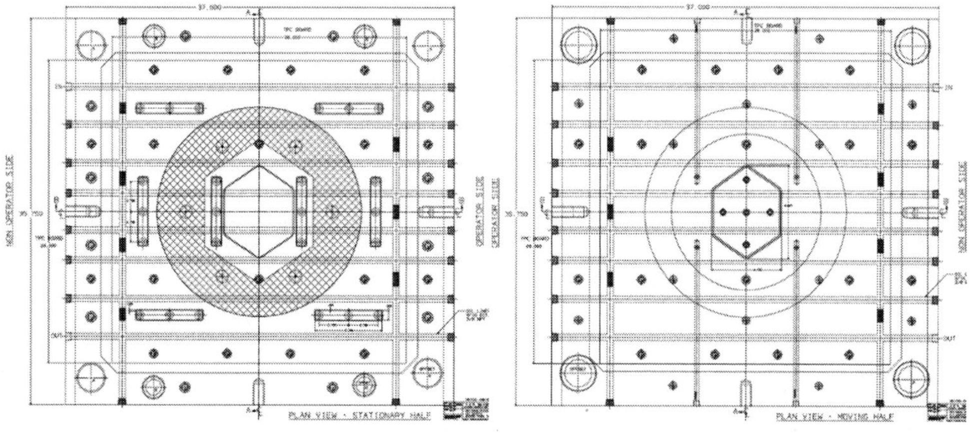

Figure 9: Drawing of A-side (left) and B-side (right) of injection molding tool.

The injection molding tool was machined from P20 hardened tool steel. A Polyshot hot runner system feeds the 6 drops in the A-side. The tool weighed approximately 3629 kg (8,000 lbs).

Figure 10: A-side of injection molding tool.

2.2.4.2 Process

The injection molding process was done using a 750-ton Engel injection molding machine. The overmold was formed using VICTREX™ PEEK Polymer 150CA30 injection molding compound pellets. This product is a PEEK-based material containing 30% short carbon fiber by weight. The thermoplastic composites stamp formed inserts were pre-heated in an oven adjacent to the injection molding press in order to allow the injection molding compound to fill out the entire mold before freezing.

The injection molding process was optimized over the course of three days. Injection molding parameters were determined by performing a series of test shots onto a sheet of machined metal and placed inside the tool. This practice allowed TxV to dial-in the injection molding settings and get on cycle while minimizing startup scrap. A few samples from the optimization process can be seen below in

Figure 11.

Figure 11: Test shots from the injection molding optimization process.

Each of the first two days were used to explore the machine capabilities as they related to the size of the part and the dynamics of the mold. The goal of the first day was to establish a crossover without any packing pressure. During the second day, the machined sheet metal was placed on top of a dummy formed insert using polyimide tape and room temperature vulcanizing rubber adhesive. Adding in the laminate backing allowed for the mold to shut off on the hexagonal feature in the center of the part.

On day three, with some further adjustments to temperature settings, after a series of test shots onto metal, TxV created a large series of fully molded, no-flash parts. A key to the success of the final day of molding was the use of the metal sheet over a dummy composite insert, allowing the team to confirm that the machine was set up for consistent injection molded shots without material waste. Once the consistent shot was confirmed three shots in a row, the first heated insert was placed in the tool and the machine was on cycle. Injection molding allows for rapid manufacturing, with a door to door cycle time of only 166 seconds and a mere 22 seconds of that being the unloading and reloading of the press.

The injection molding process settings can be seen below in Table 2. The surface temperature of each laminate was recorded with a surface temperature probe prior to being inserted into the injection molding machine (see section 3.1, Table 3 for the significance).

Table 2: Injection molding process settings and laminate temperature.

Part #	**Cycle #**	**Metering Stroke [in]**	**Cross Over [in]**	**Pack Time [s]**	**Pack Pressure [psi]**	**Coolin g Time [s]**	**Clamp Force [tons]**	**Surface Temp [F]**
7	1	6.5	2.9	25	600	100	730	430
26	2	6.5	2.9	25	600	100	730	470
21	3	6.5	2.9	25	600	100	730	414
10	4	6.5	2.9	25	600	100	730	465
14	5	6.5	2.9	25	600	100	730	367
5	6	6.5	2.9	25	600	100	730	470
22	7	6.5	2.9	25	600	100	730	354
23	8	6.5	2.9	25	600	100	730	353
29	9	6.5	2.9	25	600	100	730	455
9	10	6.5	2.9	25	600	100	730	322

2.2.5 Final Part Machining

Given that the access panel was a technology demonstrator, there were intermediate machining steps that were required to support a limited production run. These steps would likely get designed out or automated prior to commercialization. Having stated this, the overmolded parts were trimmed one final time along the sides and inside the hexagonal cut-out with the waterjet machine. Then the injection molding sprues were removed by hand. Figure 12 is a finished part.

Figure 12: Finished part ready for delivery to BR&T.

3. RESULTS AND OBSERVATIONS

3.1 Non-destructive Testing (NDT)

3.1.1 Ultra-sonic Scanning

An in-house ultra-sonic scanner was used to detect anomalies within the part after consolidation and stamp forming. An example of part of a scan at the panel level is shown in Figure 13. On the left, the scan shows 0 indications with strong back wall echo. On the right, an example of part of scan with an indication.

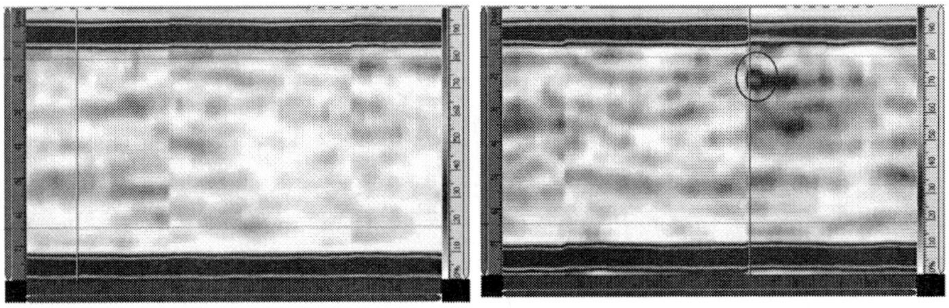

Figure 13: Sample of S scan on consolidated panel. Showing 0 indications on the left and an indication inside the red circle on the right.

3.1.2 Surface Topography Measurements

Additionally, seventeen molded parts were measured with a Hexagon Metrology, Global 9128 coordinate measurement machine (CMM). Surface profile measurements were taken at 48 points shown in Figure 14. The CMM data shows the panel flattened during injection molding resulting in a larger overall bend radius. Further, the straight edges bent such that the panel corners are higher than the edge middle. In Figure 14 the red leader lines indicate measurements higher than nominal while orange leaders are lower than nominal. Green leader lines are dimensions within tolerance. This behavior was directionally in line with software predictions, though much lower in magnitude. The scale of predicted deformations was such that TxV determined the accuracy of deformation prediction was not high enough to alter the construction of the tooling. These data will be fed back to the software provider to refine their approach to modeling thick/large parts.

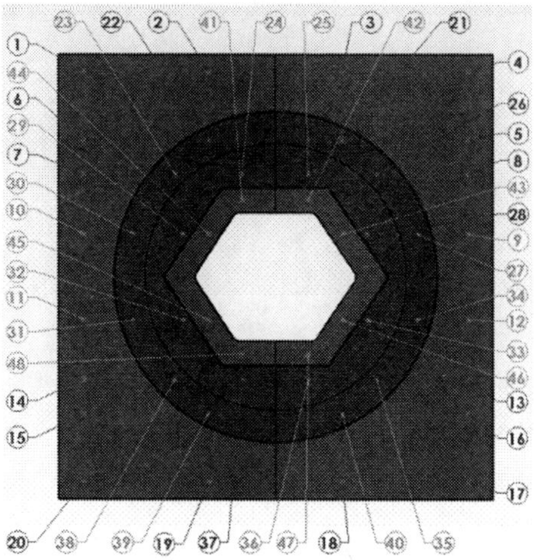

Figure 14 CMM surface profile measurement points.

For overmolded part inspection, a 5 MHz single crystal with delay line probe used to find indications. Due to the geometry of the injection mold and area of contact, only the area of the injection molded plastic was inspected. Additionally, the laminate had been inspected twice previously- once after consolidation and once after stamp forming. The injection molding tool did not contact the laminate outside of the overmolding area during the injection molding. Figure 15 shows a sample of the A-scan signal at the outermost edge of the overmolded plastic. In Figure 16 the left-hand arrow indicates the signal from the bond surface between the overmold and laminate.

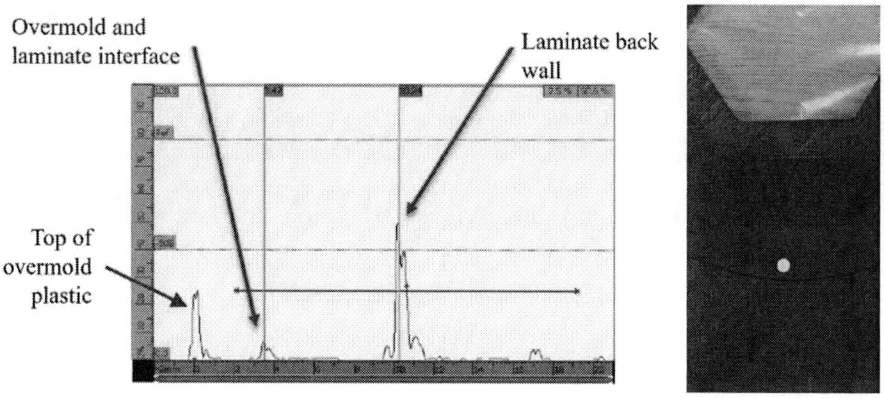

Figure 15: A-Scan sample signal at the outermost edge of the overmolded plastic on the left. The yellow spot on the right-hand image is representative of where the signal on the left-hand image were collected.

In general, bond integrity appears strong throughout the overmolded surface. A representative scan of these areas is shown in Figure 16. Scans of the overmolded areas revealed two patterns of bond.

The first pattern, shown on the left in Figure 17, resembled a "starburst" where bond strength radiates from each sprue and blends into areas of adjacent sprues. The second pattern referred to as "perimeter" style is shown on the right in Figure 17. This pattern has a more distinct line approximately 30 mm (1.25 in) in from the outermost edge of the overmold. Finally, the combination of both styles is represented in Figure 18.

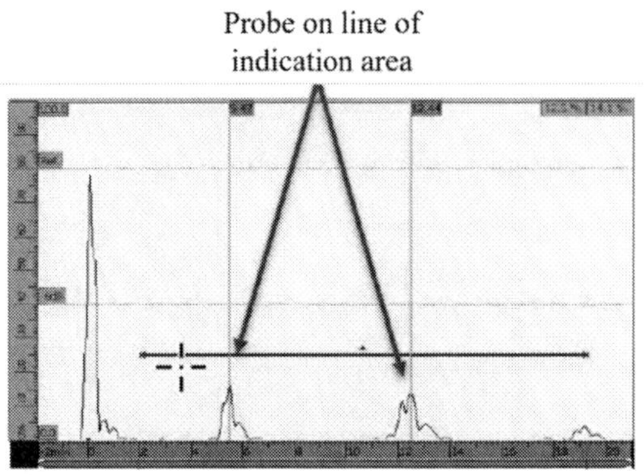

Figure 16: Arrow on left indicated bond surface between overmold and laminate.

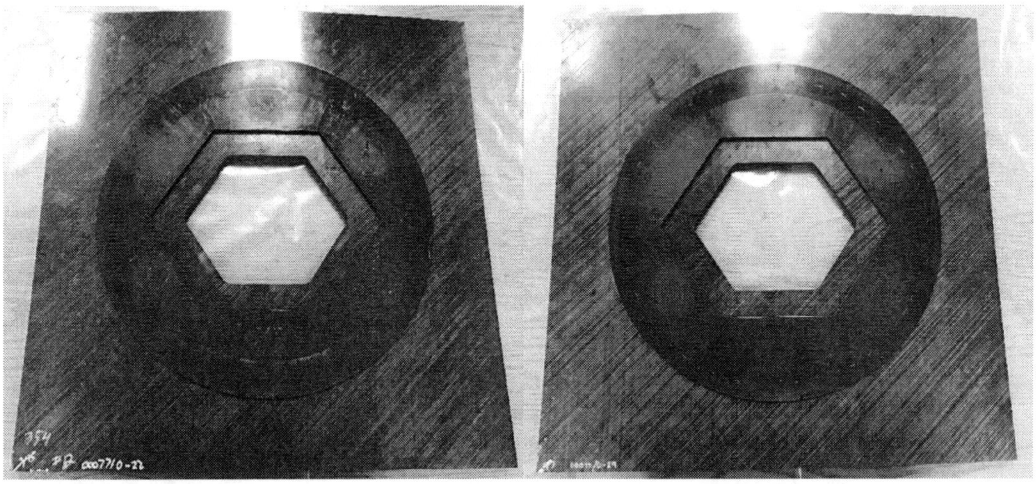

Figure 17: Starburst (left) and perimeter (right) indications marked with orange dotted lines.

Figure 18: Combination of both starburst and perimeter styles of indication.

As per the manufacturing plan, the injection molding process operated at steady state and within established processing windows. During the molding process a laminate surface temperature was measured using a surface temperature probe and recorded. After NDT was performed on the molded parts the two groups of indication styles were noted. Upon reviewing the data captured during molding, there appears to be a correlation between the pattern style and surface temperature of the insert. The laminates were pre-heated using two different ovens. The surface temperature for laminates heated in one oven was range of 157-186 ºC (314-367 ºF) while the other oven heated laminates to a range of 212-250 ºC (414-482 ºF). Both of which yielded laminate surface temperatures within the established processing window. The laminate surface data along with indication style can be found in Table 3.

Table 3: Indication style and laminate surface temperature.

Laminate Surface Temperature [F]	**Indication Style**
322	Starburst
353	Starburst
354	Starburst
367	Starburst/Perimeter Combination
414	Starburst
430	Perimeter
465	Perimeter
470	Perimeter
470	Perimeter

3.2 Project Summary and Conclusions

TxV along with Boeing Research & Technology (BR&T) successfully developed and manufactured one of the thickest and largest known hybrid overmolded part in the form of an access panel in support of the RAPM project. See Figure 19 for a relative comparison of hybrid overmolded parts in the public domain.

Figure 19: Comparison of access door panel to other hybrid over molded parts (left). TxV Aerospace Composites employee holding a finished access panel (right).

The level of automation and rapid manufacturing processes employed offer the potential for both high rates of production and the ability to scale-up and scale-down a program in a repeatable and consistent manner.

- ATL and shuttle press consolidation utilizing rapid manufacturing cycles may be a valid production method for thick laminates.
- Shuttle press consolidation of variable thickness laminates is achievable in a rapid cycle with consistent and low void content.
- Predictive tools can be utilized to make design improvement to tooling with thicker / larger parts, as TxV has done successfully with smaller parts.
- Higher processing temperatures can be utilized to lower viscosity of both the polymer contained on the surface of the laminate and the injection molding grade of polymer; and together this seems to improve the fusion bond.
- Controlling the cooling of the part post-injection manages the residual stresses in the part and may reduce deformation.
- In this case, the hybrid overmolded access panel achieved up to 45% weight savings compared to a part machined from aluminum in an alternative construction.

4. FUTURE WORK

There is opportunity for future work relating to this project. First, the injection molding process can be further refined in order to achieve the most uniform interface between the overmolded plastic and composites insert. A design of experiments can be employed to systematically evaluate the effect of changing the injection molding process settings on a large part. Deeper understanding of the laminate surface temperature phenomenon that was discovered post-NDT of the molded parts needs to be achieved. Investigation techniques to be used with further refinement of the injection molding process may include x-ray, CT imaging and a microscopy study of the bond-line in relation to the sprue locations. Additionally, a mechanical testing regimen with specialized fixturing can be developed in order to evaluate bond strength with different injection parameters.

The forming and injection molding tooling was cut to the exact part drawings, taking into account expansion of tooling at processing temperatures but not laminate spring-back. As a result, some areas of the part were out of tolerance. Perhaps this is due to spring-back after stamp forming and then heating and clamping of the center of the laminate during injection molding. Modifications can be made to the tooling to account for these issues with the use of advanced predictive software.

5. ACKNOWLEDGEMENTS

This effort was jointly accomplished by a Boeing-led team and the United States Government (Defense Advanced Research Projects Agency, DARPA) under the guidance of the United States Army Research Office, ARO. The authors would like to acknowledge the guidance and support of Dr. Jan Vandenbrande of DARPA/DSO and Dr. David Stepp of Army Research Office (ARO). The information in this paper was approved for public release by 20-164281-ETT and DISTAR 32948. Research was sponsored by the Defense Advanced Research Projects Agency (DARPA) and was accomplished under Cooperative Agreement Number W911NF-16-2-0087. The views and conclusions contained in this document are those of the authors and should not be interpreted as representing the official policies, either expressed or implied, of the Defense Advanced Research Projects Agency or the U.S. Government. The U.S. Government is authorized to reproduce and distribute reprints for Government purposes notwithstanding any copyright notation herein.

6. REFERENCES

1. Reid, David. "Airbus ups 20-year new jet forecast with mega-cities tipped to drive growth." *CNBC*, September 18, 2019. Accessed March 27, 2020. <tinyurl.com/y6oqvyyc>
2. "Urban Air Mobility (UAM) Market Study Final Report." NASA, November 21, 2018.
3. Ferfecki, Frank, Smithson, Tyler, Bjerregaard, Chris. "Developing a Hybrid PAEK Overmolded Thermoplastic Composite Bracket." *CAMX Conference Proceedings*, Anaheim, CA, September 23-26, 2019.
4. Hahn, G.L., Tsotsis, T.K., "Rapid High-Performance Molding (RAPM) of Small Parts," SAMPE 2019 Conference Proceedings. Charlotte, NC, May 20-23, 2019.
5. Vandenbrande, J. "Tailorable Feedstock and Forming (TFF)" https://www.darpa.mil/program/tailorable-feedstock-and-forming

PULTRUDABLE BIO-BASED EPOXY RESINS

Terry S. McQuarrie, Martin Sutton, Yaroslav Germanov

EcoPoxy, Inc.

Morris, Manitoba, Canada R0G 1K0

ABSTRACT

Numerous studies have reported on the properties of replacing a portion of a Bis A epoxy resin with bio-based components. An extensive review of these studies included epoxidized soybean oil resins (ESO), blends with alternatives to ESO such as epoxidized glycidyl soyate (EGS), and allyl soyate (EAS) blended with standard petroleum-based epoxies along with alternative bio-based hardeners and accelerators. These studies supply excellent data with regards to the casting and some laminate properties of these alternative composite materials. The experimental work done in this product development program consisted of preparing ESO and epoxidized linseed oil (ELO) resins, selecting the best curing agent to maximize properties, running dynamic mechanical analysis (DMA), liquid resin and neat casting properties, ASTM D6866 Biobased Carbon Content and scanning electron microscopy (SEM) to evaluate resin/glass interface bonding. Most importantly, pultrusion processing data was developed for these ESO and ELO bio-based epoxy resin systems. This development of both ESO and ELO based epoxy resin systems supplies the pultrusion industry with all the necessary technical data that includes the full package of resin gel and cure data, casting properties, pultrusion line design as well as in-line curing parameters before and after the heated die and includes cured composite properties. The result is pultrudable ESO and ELO bio-based epoxy resin systems with zero styrene emissions, bio-based contents of 33% and full pultrusion processing parameters.

Keywords: Pultrusion, bio-based epoxy resins
Corresponding author: Terry S. McQuarrie, Email: terry@ecopoxy.com, phone: (408) 668-9266

1. INTRODUCTION

A review of the current state of bio-based resin development has yielded many and varied formulations. However, there is limited data on these bio-based epoxies, the hardeners and accelerators required to convert these formulations into composites by the pultrusion process. This article focuses on the development and processing of both soybean oil and linseed oil bio-based epoxy resins for the pultrusion process and includes, formulation, resin characteristics, composite properties, pultrusion line design and most importantly the processing details required in order to maximize line speed and quality.

1.1 Background

As background for this development, an extensive review of the literature on the development of bio-based polymers as replacements for petroleum-based counterparts was carried out. This review included unsaturated polyester resins, epoxidized soybean oil resins (ESO), blends with alternatives to ESO such as epoxidized glycidyl soyate, (EGS) and epoxidized allyl soyate (EAS) blended with standard petroleum-based epoxies along with alternative bio-based hardeners and accelerators.

A short summary of these alternatives included the following:

Unsaturated Polyester Resins: Unsaturated polyester resin manufacturers offer bio-based products that exhibit essentially equivalent properties to their petroleum-based counterparts. For pultrusion, examples include Envirez [1] resins which range up to 22% bio-based content with a recycled content up to 44% and the Envirolite [2] resins with a 25% bio-content.

Epoxidized Soybean Oil (ESO) Resins: The studies of ESO resins in composites have resulted in mechanical properties, primarily tensile strength and stiffness, that have been too low and thus limited in the marketplace [3].

Alternatives to ESO Resins: Polymer blends replacing a portion of the DGEBA resin with alternatives to ESO include Epoxidize Allyl Soyate (EAS) and Epoxidized Methyl Soyate (EMS) resulting in improved damping properties [4].

Alternative Hardeners: Research has been done on polymers replacing 10-30% of the DGEBA with (ESO) and curing with sebacic acid [5]. In addition, by using maleic anhydride (MA) as a hardener the maximum crosslink density, heat distortion and Vicat softening temperature were achieved [6].

Epoxy Accelerators: Improvements have been reported in the glass transition temperature (Tg) and storage modulus (E') of an ESO/methylhexahydrophthalic anhydride (MHHPA)/2-ethyl-4 methylimidazole (EMI) resin by increasing the EMI [7].

1.2 Scope

The objectives of this work are to develop ESO and ELO bio-based epoxy resins designed for the pultrusion process and to supply the line design and processing parameters in order to maximize the properties of these bio-based epoxy resins by this manufacturing process.

1.3 Audience

The target audience for this paper includes raw material suppliers, pultruders, pultrusion machine manufacturers and composite design engineers.

2. EXPERIMENTATION

2.1 Materials

The materials used in this study consisted of the control DGEBA /MHHPA system, blends of EGS/ESO/ELO all replacing a portion of the DGEBA and cured with MHHPA and accelerated with imidazole. The mold release was Axel 1888LS and the filler was ASP 900.

2.2 Resin Properties

Liquid resin properties were determined for the control DGEBA/MHHPA system, Biopoxy 30S, 35S and 30L, as well as the 150^{0}C gel, cure and peak exotherm using 3 different accelerator systems. In addition, liquid resin samples of Biopoxy 30S, 35S and 30L were submitted to Beta Analytic Testing Laboratory for ASTM D6866-18 Method B (AMS) Biobased Carbon content.

Neat casting properties were determined for the Biopoxy 30S, 35S, 30L, and compared to those of the control DGEBPA/MHHPA, rigid and flexible polyesters and a vinyl ester resin. The product designations refer to the estimated bio-based content, i.e., 30, 35, etc., and the S and L refer to either the soybean oil or linseed oil-based resin systems.

2.3 Dynamic Mechanical Analysis

Dynamic mechanical analysis was used to screen the various formulations for Loss Modulus and Tan Delta using a TA 850 DMA test instrument run at 5^{0}C/minute heating rate as per ASTM D7028.

2.4 Izod Impact Strength

Izod impact strength was measured for the 30S and 30L as per ASTM D256.

2.5 Scanning Electron Microscopy (SEM)

A SEM was run on the Biopoxy 30S to evaluate the fracture surface.

2.6 Processing

A pultrusion run was performed at the Martin Pultrusion Group facility in North Randall, Ohio. The glass used consisted of 55 ends of HYBON 2025-113 single end roving and the resin systems were the Biopoxy 30S and 30L

3. RESULTS

3.1 Materials and Resin Properties

After a full review of all options for this development program, the initial system evaluated was based on the work done by Dr. Thomas P. Schuman and Rongpeng Wang of the Missouri University of Science and Technology Department of Chemistry. In this work, glycidyl esters of epoxidized fatty acids derived from soybean oil (EGS) and epoxidized linseed oil (EGL) were synthesized and compared to a standard bisphenol A epoxy (DGEBA [8].

Liquid properties were determined for the Biopoxy 30S, 35S, 30L and the Bis A control resin (with anhydride) and compared to petroleum based rigid and flexible polyesters and a vinyl ester resin as seen in Table 1A.

Table 1A: Biopoxy bio-based resin test results

Typical Properties of Unreinforced Castings					
	Tensile Strength, (Mpa)	Tensile Modulus, (Gpa)	Elongation, %	DMA Tg/ HDT^{0}C	% Bio
Control*	72.2	2.87	3.38	Tg 136	0*
Biopoxy 35S	53.9	2.32	3.62	Tg 127	38
Biopoxy 30S	33.2	1.99	1.87	Tg 145	33
Biopoxy 30L	52.9	2.59	2.47	Tg 154	33
Rigid UPR**	52.4	3.79	1.7	HDT 113	0
Flex UPR***	73.1	3.19	4.9	HDT 80	0
Vinyl Ester****	79.9	3.17	5.2	HDT 104	0

Liquid Resin Properties				
	Viscosity, cps	180^{0}F gel/cure/peak^{0}C	310F gel/cure/peak,^{0}C	VOCs
Control*	690	-----	1.8/2.4/214	0*
BioPoxy 35S	355	-----	2.0/3.6/228	0*
BioPoxy 30S	340	-----	1.6/2.4/222-225	0*
BioPoxy 30L	510	-----	1.8/2.7/223	0*
Rigid UPR**	2300-2700	3.0-4.0/0.6-1.6/237	-----	34
Flex UPR***	1000-1200	4.0-4.5/1.0-2.0/223	-----	35
Vinyl Ester****	500	3.0-4.0	-----	44

*Epoxy + anhydride

**DION 31020-03, January 2006

***DION 31031, January 2006

**** DION 9100, July 2012

This Table shows that the liquid viscosity of both the ESO and ELO modified systems are lower than the polyester and vinyl ester systems. This allows for a higher filler content. In addition, there no VOC's in the ESO and ELO modified systems while the polyester and vinyl ester resins contain 34-44% styrene. Table 1B shows the 150^{0}C gel and cure times and peak exotherm with 3 different accelerator systems used with the 30S and 30L

Table 1B. 150⁰ C gel, cure and peak exotherm with 3 different accelerators

Accelerator: AC-PI

	Biopoxy 30s (1)	Biopoxy 30s (2)	Biopoxy 30l (1)	Biopoxy 30l (2)
Gel time seconds	96.5	96.0	89.0	90.0
length to peak temperature seconds	154.0	146.0	143.5	143.0
peak temperature degrees C.	225.0	222.0	231.6	232.1

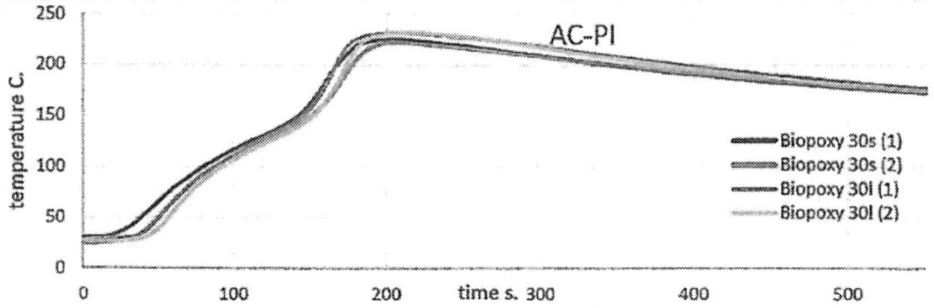

Accelerator: BV Cat 7

	Biopoxy 30s (1)	Biopoxy 30s (2)	Biopoxy 30s (3)	Biopoxy 30s (4)
Gel time seconds	96.5	96.0	89.0	90.0
length to peak temperature seconds	154.0	146.0	143.5	143.0
peak temperature degrees C.	225.0	222.0	231.6	232.1

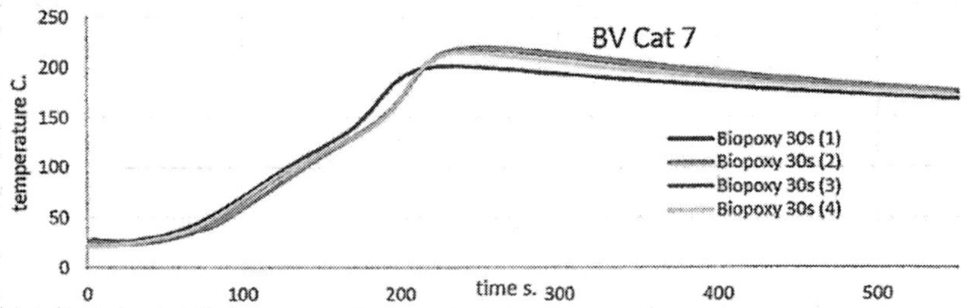

Accelerator: Pre-accelerated MHHPA

	Biopoxy 30s (1)	Biopoxy 30s (2)
Gel time seconds	82	88
length to peak temperature seconds	150.5	162.5
peak temperature degrees C.	227.8	225.6

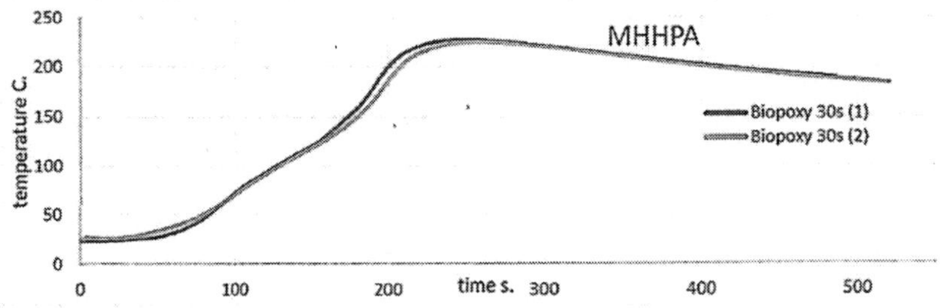

Neat castings of the Biopoxy 30S and 30L were submitted to Beta Analytic Testing Laboratory for ASTM D6866-18 Method B (AMS) Biobased Carbon content and the results of both systems were 33% as a fraction of total organic content [9]. A comparison of the neat castings as seen in Table 1A indicates that the Biopoxy 35S and the Biopoxy 30L systems match up very well with the rigid polyester and have better tensile elongation and improved heat resistant properties.

3.2 Dynamic Mechanical Analysis

Dynamic mechanical analysis was used to screen the various formulations for Loss Modulus and Tan Delta using a TA 850 test instrument. The test results of this screening study are seen in Table 2. This table indicates that the EGS, 30S, and 35S modified systems exhibit lower Tan Delta and Loss Modulus than the control system and the ELO modified systems.

Table 2. DMA on cast resin samples

Sample	Storage modulus, °C	Tan delta peak, °C	Loss modulus max, °C
Control*		160.8-169.9	153.3-159.9
EGS**		139.0-139.5	127.5-128.2
35S***		141.0-142.4	126.5-127.1
35L***		168.5-168.6	154.0-154.1
30S***		140.7	128.9
30L***		162	147

*Control = Bis A epoxy + MHHPA
**EGS = Epoxidized glycidyl soyate
***30S/L and 35S/L designations refer to the estimated bio-based content.

All samples tested with TA 850 DMA test instrument run at 5°C/minute heating rate as per ASTM D7028

3.3 Izod Impact Strength and Scanning Electron Microscopy

Work done by Dr. Hiroki Miyagawa of Michigan State University demonstrated phase out of the ESO modified system vs. the ELO system as seen in Figures 1A-B.

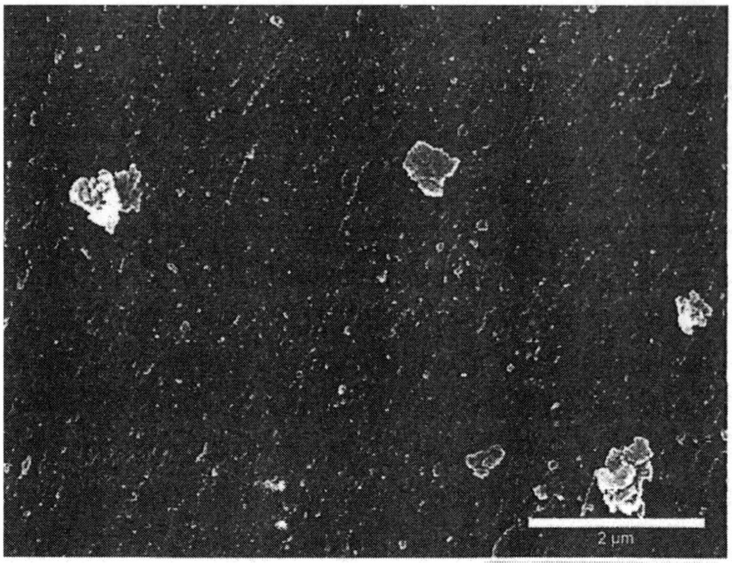

Figure IA

Impact failed surface of anhydride cured biobased neat casting containing 50% weight percent ELO at high magnification

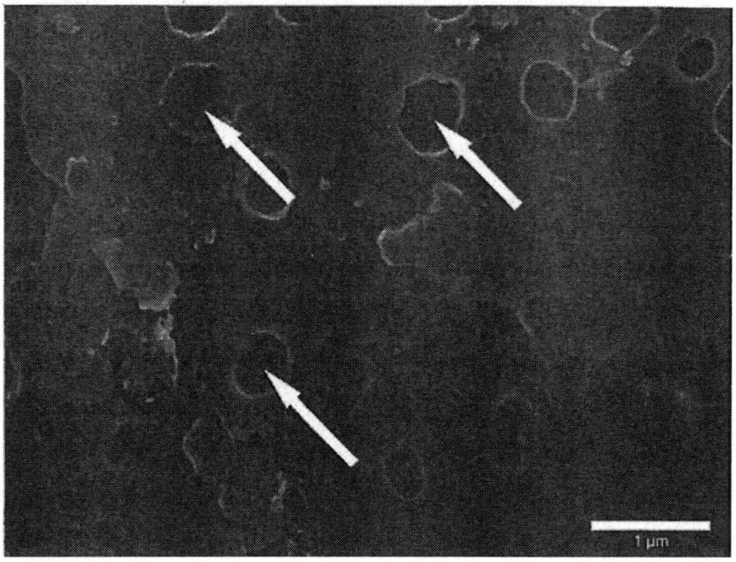

Figure 1B

Impact failed surface of anhydride cured biobased neat casting containing 30% weight percent ESO at high magnification

This resulted in higher impact strength for the ESO modified system vs. the ELO modified system while the ELO system maintained better heat resistant properties as demonstrated by the higher glass transition temperature [10]. Similar results were obtained on the ESO modified system, Biopoxy 30S and the ELO modified system, Biopoxy 30L The impact strength of the 30S was determined to be 2.1 kJ/m^2 vs. 1.4kJ/m^2 for the 30L system, or 1.5 times better impact strength for the ESO based system as per the CIC Engineering Report No. 19-276-004 [11].

3.4 Processing

As stated earlier, a pultrusion line run was performed at the Martin Pultrusion Group facility in North Randall, Ohio. A drawing of the pultrusion line is seen in Figure 2.

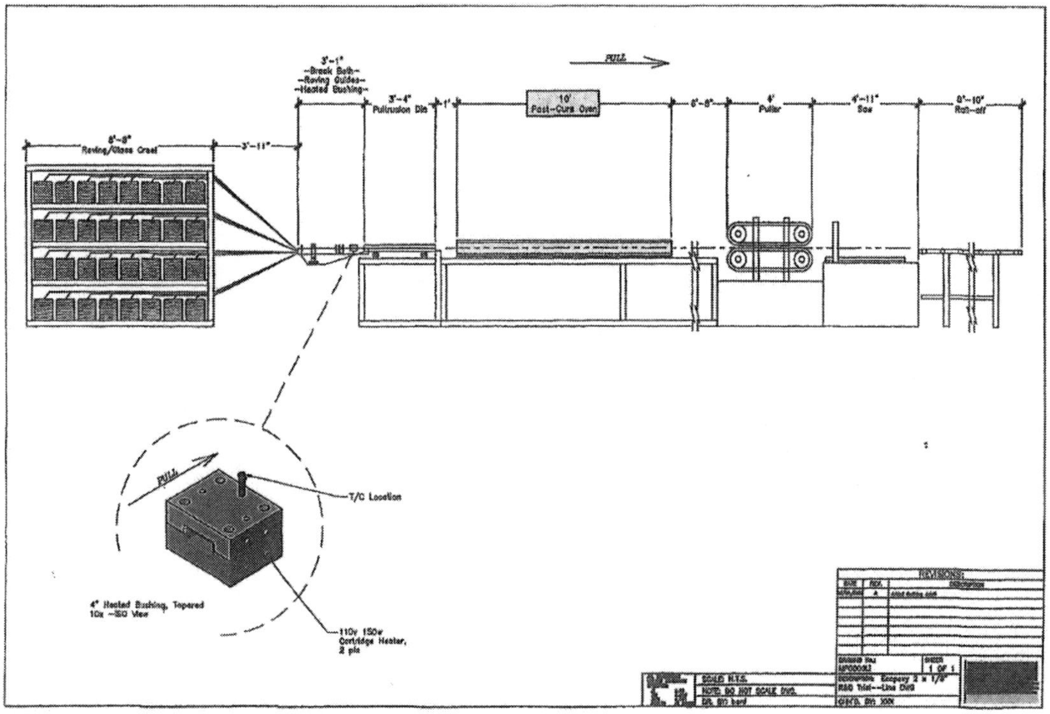

Figure 2. Pultrusion Line

The modifications to a standard pultrusion line were the addition of a heated bushing prior to the die entrance and a post cure oven after the exit of the die. The die used was 2.0 inches wide by 0.125 inch thick. The glass used consisted of 55 ends of HYBON 2025-113 single end roving. The glass was heated to remove any excess moisture.

In order to maximize line speed and cure a Differential Scanning Calorimeter (DSC) test was performed on the fully catalyzed resin mix in order to determine the onset temperature. As seen in Figure 3 the onset temperature for the resin mix was 74⁰C (166⁰F).

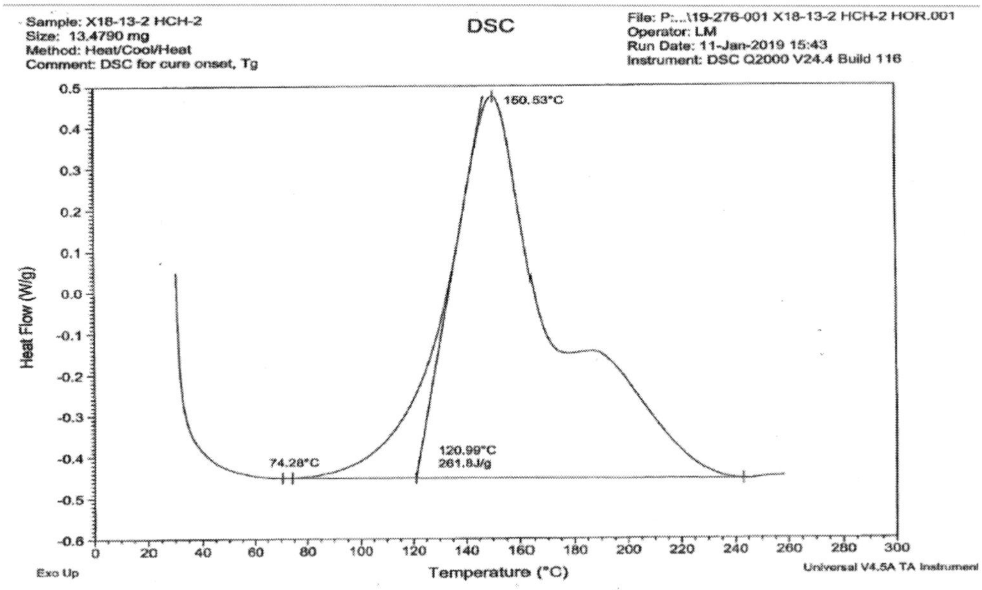

Figure 3. DSC on fully catalyzed resin mix

This is important since this tells us the temperature that we must not exceed in order to prevent premature gelation. Ideally, we want to heat the resin mix to approximately 20⁰F below this temperature in order to maximize glass wet out and shorten the time to initial gelation of the resin system i.e., 146⁰F. The resin drip off temperature was measured between the heated bushing and the die entrance. The resin mix drip off was not returned to the wet out tank bath in order to prevent gelation in the wet out bath. Once the resin mix has returned to ambient it can be added to the wet out tank. The full processing conditions are seen in Table 3.

Table 3. Processing conditions on 2.0 inch x 0.125 inch flat strip

Processing Conditions	
Pull Speed, inches/minute:	12-30
Entrance die heat, ^{0}F	334
Exit die heat, ^{0}F:	360
Bath temperature, ^{0}F:	86-96
Bushing heater set point, ^{0}F:	185
Bushing drip off temperature, ^{0}F:	148-149
Die entrance drip off temperature, ^{0}F:	145-150
Post cure oven temperature, ^{0}F:	352

3.5 Laminate Properties

A summary of the laminate properties is seen in Table 4.

Table 4. Biobased laminate properties

Test	Property	Biopoxy 30S	Biopoxy 30L
ASTM D256	Notched Impact Resistance(J/m)	816	868
ASTM D3171	Glass Content (wt %)	80.5	81.7
ASTM D7028	Glass Transition Temperature by DMA (^{0}C)	129	145

Reference: Composites Innovation Center Report N0. 19-276-004-Summary -27 May 2019

Both samples have glass contents of 80% by weight or better and the DMA Tg for the Biopoxy 30S is 129^{0}C vs. 145^{0}C for the Biopoxy 30L

In addition, Figure 4 shows the excellent resin/glass bonding at the glass/resin interface.

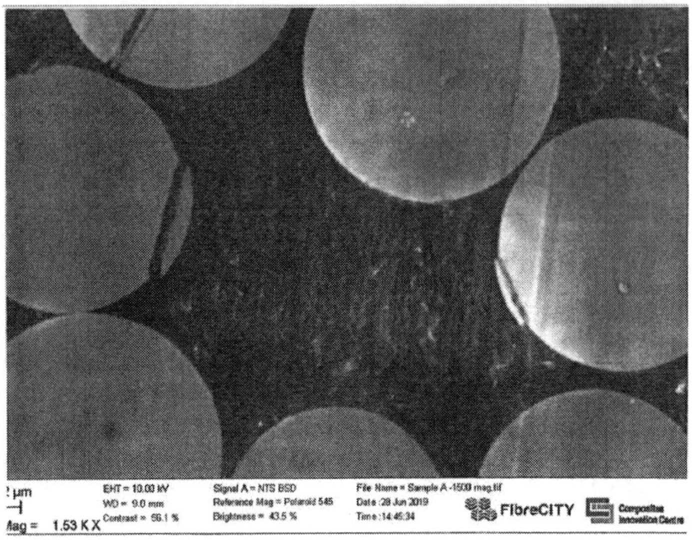

Figure 4. Resin/glass interface

4. CONCLUSIONS

This paper has described the development of both ESO and ELO based epoxy resins that contain zero volatile organic compounds (VOC) and greater than 30% biobased content. Pultrusion processing parameters were presented that include the basic resin gel and cure properties, the recommended pultrusion line design, processing conditions and laminate properties. By utilizing biobased components in these epoxy resins, the result is resins that offer equivalent or better physical and thermal properties along with the environmental advantages of no VOCs and a lower carbon footprint.

5. REFERENCES

1. "Envirez Product Line and Properties", The first Step to Sustainable Composites.

2. "ENVIROLITE" 31325-00 Soy-Oil Based Molding Resin Product Bulletin", February 2010.

3. Phuong Tran, Daniel Graiver, Ramani Narayan, "Biocomposites Synthesized from Chemically Modified Soy Oil and Biofibers",Journal of Applied Polymer Science, Volume 102, 69-75 (2006).

4. S. Ahamed, A. Garg, S. Sundararaman, K. Chandrashekhara, V. Flanagan and S. Kapila, "Dynamic Mechanical Characterization of Soy Based Epoxy Resin System", Proceedings of the SAMPE Conference, Long Beach, CA, May 16-20, 2004.

5. Ghoudsieh Mashouf Roudsari, Amar K. Mohanty and Manjusri Misra, "Study of the Curing Kinetics of Epoxy Resins with Biobased Hardener and Epoxidized Soybean Oil", ACS Sustainable Chemistry & Engineering 2014, 2, 2111-2116.

6. J.M. Espana, L. Sanchez-Nacher, T. Boronat, V. Fombuena and R. Balart, "Properties of Biobased Epoxy Resins from Epoxidized Soybean Oil (ESBO) Cured with Maleic Anhydride (MA)", Journal of the American Oil Chemists' Society, Volume 89, Issue 11, pp 2067-2075, November 2012.

7. S. G. Tan and W.S. Chow, "Curing Characteristics and Thermal Properties of Epoxidized Soybean Oil Based Thermosetting Resin", Journal of the American Oil Chemists' Society, Volume 88, Issue 7, pp 915-923, July 2011.

8. R. Wang and T.P. Schuman, "Vegetable oil-derived monomers and polymer blends: A comparative study with review", eXPRESS Polymer Letters, Vol 7, No. 3, March (2013) 272-292.

9. Beta Analytic Testing Laboratory, Laboratory Numbers Beta-54726 and Beta-547266, dated December 20, 2019.

10.Hiroaki Miyagawa, Manjusri Misra and Lawrence T. Drzai, "Fracture Toughness and Impact Strength of Anhydride-Cured Biobased Epoxy", Polymer Engineering and Science, 2005 Society of Plastics Engineers.

11. CIC Engineering, Report No. 19-276-004-Summary- 6 Mar 2020.

CREATION OF A DIGITAL TWIN FOR AUTOMATED FIBER PLACEMENT

Max Kirkpatrick[1,2], Alex Brasington[1], Andrew D. Anderson[1], Ramy Harik[1]

[1]University of South Carolina
McNAIR Center, Department of Mechanical Engineering
1000 Catawba Street
Columbia, SC 29201

[2]Siemens Digital Industries Software
5101 Westinghouse Blvd
Charlotte, NC 28273

ABSTRACT

Automated Fiber Placement (AFP) is a manufacturing process used to produce large, lightweight composite structures at high rates and increasing reliability. Currently, the process is black box where paths and process parameters are sent to the machine, and it is trusted to maintain the correct values. There are limited methods of monitoring real time process parameter values for current and future analyzing. Even when the exact same parameters are used on the same part multiple times, the resulting manufacturing quality can differ largely. This report will discuss the creation of a digital twin with the capability to simulate and monitor the entire AFP process. A digital twin will provide the ability to examine the process in real time, in addition to after the process is completed. The data collected will then be compared to expected data to examine any differences. Analysis of each point during manufacturing will lead to a deeper understanding of AFP.

Keywords: Digital Twin, Project Lifecycle Management, Industry 4.0, Automated Fiber Placement
Corresponding author: Ramy Harik; harik@cec.sc.edu

1. INTRODUCTION

1.1 State of the Art

Rapid developments in technological capabilities in recent years have enabled the marriage between virtual models of complex physical systems and their real-world counterparts. Digital Twin (DT) is a term that embodies this integration of data between the virtual and physical realms. While not at full scale, portions of DT technologies are currently utilized in various industries to enhance existing product lifecycle management (PLM) tools. PLM tools create a system that allows companies and organizations to monitor the progression of a product, beginning from ideation, to manufacture, and ultimately finishing at the product's end of life. In the composites industry today, the activity of PLM is more difficult due to the inherent complexity of a PLM system and the disconnect of information from the phases of PLM. There have been some successes at developing portions of a DT for the composites industry PLM, however no attempts at creating a DT for the entire PLM cycle were found in the available literature [1] [2] [3] [4] [5]. This gap can be understood through the context of the definition of the term Digital Twin. A general definition of a Digital Twin is given as: "*An integrated, multiphysics, multiscale, probabilistic simulation of an as-built system that uses the best available physical models, sensor updates, product history, etc., to mirror the life of its corresponding twin* [6]." Based on this definition, the foundational components of a DT consist of three things: (1) a physical model, (2) a virtual model, and (3) the data connecting them [7]. The linking between these three features demonstrates that a DT is not a precocious all-in-one model, but a cutting-edge interconnection of data between the virtual and physical domains. The linking of the lifecycle data into a comprehensive virtual system enables accurate, data backed predictions of the physical product by the DT to enhance significant components of the PLM, such as product design optimization and development of manufacturing systems [8] [9].

1.2 Goals of an AFP Digital Twin

The target of this research is to create a complete DT of the entire AFP manufacturing process. This includes, but is not limited to, post processing of initial design, monitoring and inspecting data from the design, design process planning, manufacturing, and post-production phases. The software selected for development of this DT model was Tecnomatix Process Simulate (PS) by SIEMENS. Process Simulate was chosen as the tool for building the DT system due to its robustness and accuracy as a simulation environment, as well as a number of reporting tools that simplify data acquisition and analysis. In addition, the Process Simulate Application Programming Interface (API) was useful in allowing the researchers to integrate the simulation environment with other tools, such as path planning and post processing routines. Process Simulate provides many tools required to build a DT model. Among these are 3D simulation and editing, collision detection, joint monitoring, and data communication. These abilities allow for programming and simulation of the AFP process, all within a single software suite, on any CNC controlled AFP machine. Additionally, all pertinent process parameters that can be streamed from the AFP machine CNC controller will be directly input real-time into Process Simulate, creating a real-time digital model of the manufacturing process. Furthermore, data from external partner software programs such as process planning [10] and inspection [11] will be integrated into Process Simulate. This data will be used to produce a further enhanced DT from the design phase to the end of manufacturing. The data can then flow freely between all phases of the DT to improve the overall product lifecycle. The expected data flow is presented below in Figure 1.

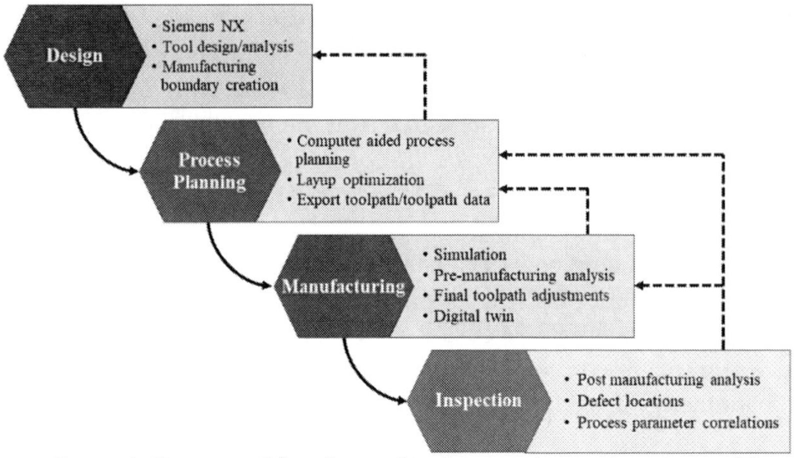

Figure 1: Diagram of data flow within the virtual manufacturing model

1.2.1 Virtual Commissioning

An additional goal of this project is to allow for "Virtual Commissioning" of the entire AFP system. Virtual Commissioning (VC) as described in [12] refers to the partial digitization of the commissioning process for traditional manufacturing systems. More specifically, VC refers to the integration of both kinematic and logical components of a cell in one unified simulation, such that automation code on both robots and programmable logic controllers (PLCs) can be validated against the virtual model before implementation on the physical system. This VC approach can drastically reduce the amount of time required to launch a manufacturing system, in addition to time reduction for making changes to an existing system.

AFP systems are rarely a single machine that performs exactly as programmed, and as such VC is applicable to these systems. An AFP manufacturing cell may contain both a layup system and a robotic inspection system, or such a system may contain multiple robots performing layup simultaneously. Other components like additional actuators, axes, sensors, and inspection devices can also be integrated. The introduction of VC to this type of system allows engineers to test the integration of these various components of the cell using a virtual model before they are implemented. This high level of fidelity in the simulation model is a target of the DT system.

2. EXPERIMENTATION

2.1 Building a Digital Twin of AFP

In order to create an accurate Digital Twin of the AFP manufacturing process, existing CAD models and software were referenced and imported into the Process Simulate model. The University of South Carolina McNAIR Center is an Advanced Composites research facility equipped with a Lynx Horizontal AFP machine by Ingersoll Machine Tools (IMT). For this reason, the McNAIR Center's AFP manufacturing cell was used as the target model for the DT. The first step to create the Digital Twin of the IMT AFP system was to import a CAD model and construct a kinematic tree of the AFP machine. This model, shown in Figure 2, was based on CAD geometry developed by IMT. Elements of the assembly were associated with "links" that describe groups of components that move together as a single unit in context of the machine's kinematic tree, as shown in Figure 2. These links were then connected with "joints" that represent how the links can move with respect to each other. This movement is specified using an axis and can be either prismatic or rotational. The exact joint location, limit, velocity, and acceleration parameters were set to IMT specifications. These specifications were included in a separate file called "motionparameters.e." This file is associated with the DT model and instructs Process Simulate how to emulate the motion of the machine. Frames were defined at the tool center point (TCP) of the roller and the origin of the machine as specified by IMT. A frame is a location including both the cartesian and the rotational component of the location. The TCP frame allows location of the roller on the machine relative to the origin frame of the machine.

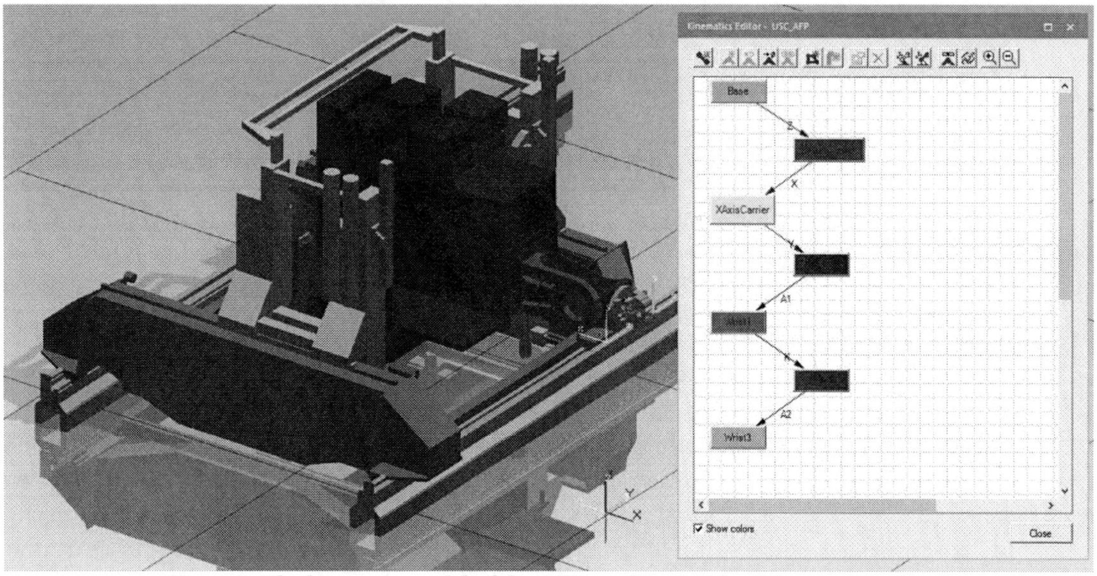

Figure 2: The kinematic model of the AFP machine inside of Process Simulate.

The machine's mandrel and tool were then added to the DT environment, using the same procedure as described above. The tool was located on the mandrel based on the mounting locations used on the actual machine. The mandrel was then added as an external axis to the AFP system. The resulting system is shown in Figure 3.

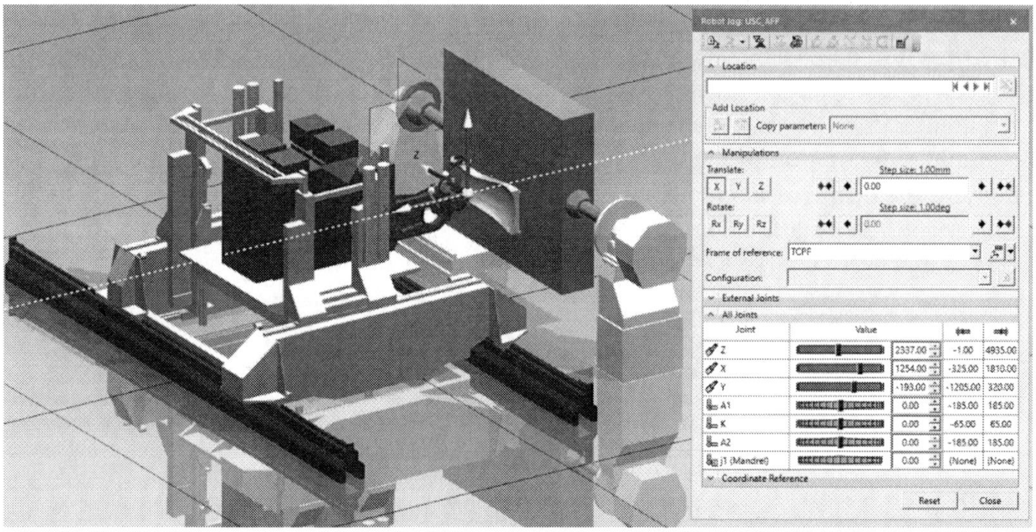

Figure 3: The finalized kinematic model for the AFP system and mandrel. The right shows the joints of the machine and their limits, as well as the external axis

To further improve the accuracy of the DT environment, additional components in the cell were modeled. The rail mounted robotic inspection system and fencing that are present in the physical cell were added to the simulation model. The CAD and kinematic models for these components were available directly from an existing library of resources distributed with Process Simulate. These components were loaded into the cell and placed in their correct locations based on the physical cell. Adding these additional components allows the engineers to assess their layup strategy in the context of the entire cell, considering possible collisions with the surrounding equipment, or to evaluate integration with the inspection system. The completed cell layout can be seen in Figure 4.

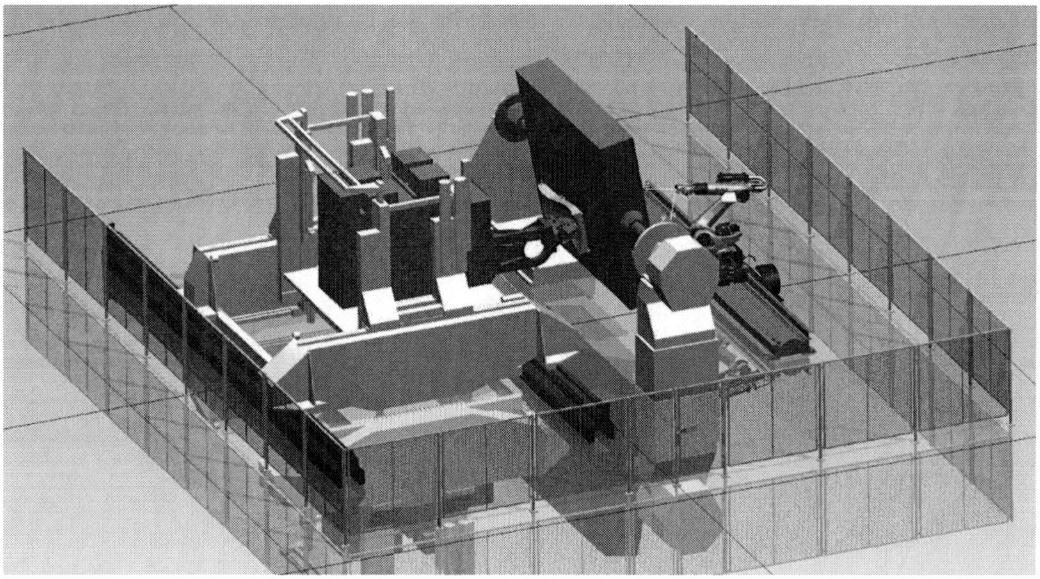

Figure 4: The completed Digital Twin cell layout

1.1.1 2.1.1 Kinematic Simulation

After the DT layout was complete, basic kinematic simulation could be performed. The IMT Lynx has a total of 7 axes, designated as X, Y, Z (prismatic), A1, A2 (rotary), K (wrist), and C1 (mandrel rotary). The inverse kinematic solver implemented within Process Simulate was used to solve axis positions for the AFP machine, given a TCP location. This functionality was tested using the "Robot Jog" and "Jump to Location" tools to verify Process Simulate would properly solve for joint values based on cartesian input (Figure 5).

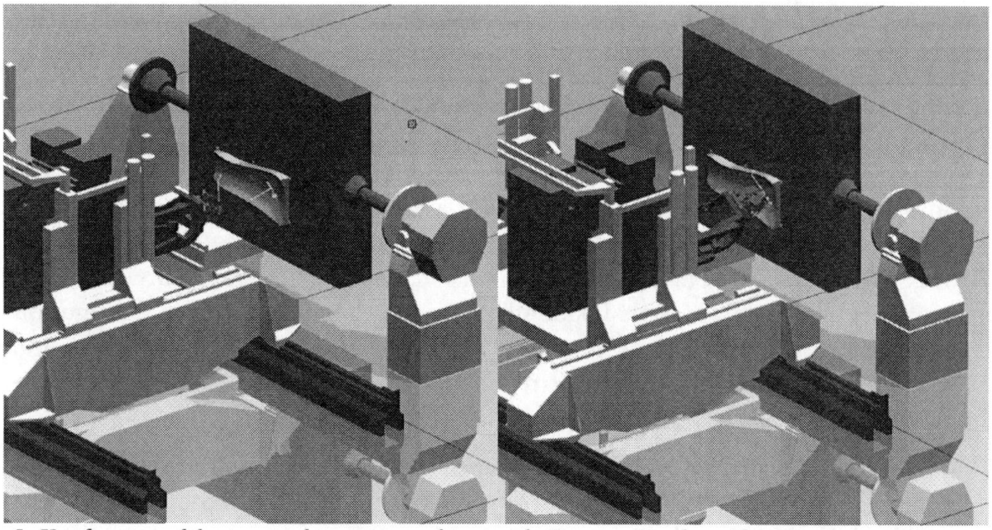

Figure 5: Verification of the inverse kinematics solver inside Process Simulate. The images depict "jumping" the AFP machine to the target location

Due to the inherent nature of the kinematic chain of the IMT AFP machine, there are possible ambiguities in this inverse kinematic solver. This is shown in Figure 6, which depicts two possible solutions for the same target location.

Figure 6: Two different kinematic solutions for the same target location

When programming this machine, engineers need a way to specify the kinematic solutions they wish to use for the layup. In order to allow this specification, the "**motionparameters.e**" file associated with the AFP model was altered to include syntax that differentiates the solutions based on the joint values of A1 and A2 (the primary and secondary rotary wrist joints). This ensured a unique classification of kinematic solutions and allows the user to specify the best solution for their needs. Process Simulate uses this information to present these solutions to the user as options during path planning.

1.1.2 2.1.2 Path Planning

Two workflows were proposed to generate the required course paths for simulation and eventually programming purposes. The first methodology uses an external path planning software developed by IMT called iCPS to generate course center line definitions as a series of coordinates defined relative to the tool surface. These definitions were exported from iCPS in their native format. The Process Simulate .NET API was then used to build a tool inside of Process Simulate to load this series of points as a continuous robotic operation that could be executed by the kinematic model inside Process Simulate. The paths were then verified inside of Process Simulate, external axis positions were set using built in tools inside of Process Simulate, and robot configurations were set to avoid any potentially anomalous kinematic solutions. The paths could then be simulated and checked for any potential layup problems, such as collisions, rapid axis accelerations, discontinuities, etc.

A very similar process was followed for our other proposed workflow. In this case, the design of the toolpaths was performed entirely inside of Process Simulate by using the Continuous Process Generator (CPG). The CPG tool inside of Process Simulate allowed us to create a continuous coverage pattern over the surface of the curved tool with even spacing of courses at a defined angle. These courses were then analyzed in much the same method as described above, using the built-in tools inside Process Simulate to set external axes and configs, and then testing using kinematic simulation and collision detection. This process was more useful for rapid iteration and testing, and likely would not be employed for actual composite path programming. The final proposed process would involve path planning performed in an external tool and then loading the course centerline definitions into Process Simulate as described above.

3. RESULTS

The current state of the virtual AFP cell can be seen in Figure 4. Using this model, we can monitor aspects of the layup such as collisions, joint values, layup time, and velocity profile which can then be used to identify potential manufacturing problems. From the images presented in Figure 7, it is shown that a collision set has been defined between the AFP machine and the mandrel with the attached tool. A collision was detected through layup simulation, and Process Simulate displays this collision by highlighting the 2 colliding parts and providing the penetration distance (1.33 mm). Once a collision is detected, the user can directly edit the path in question or select a different kinematic solution method.

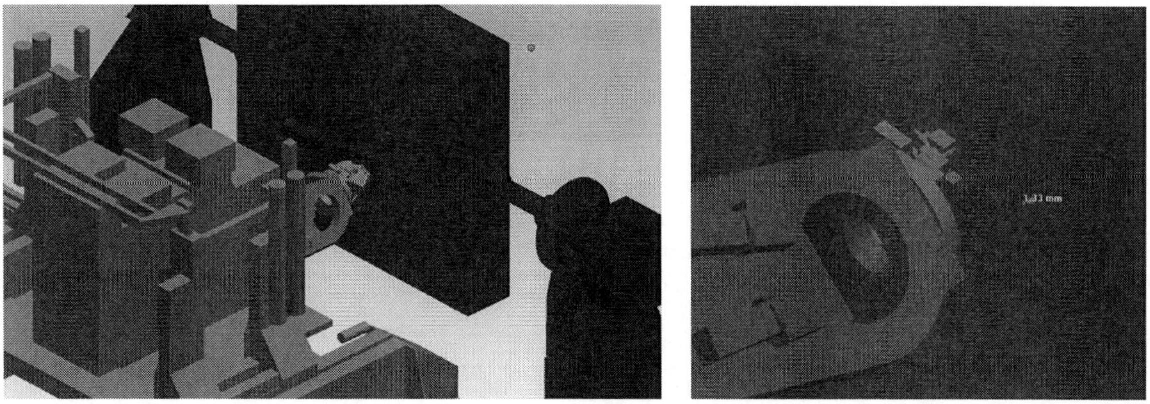

(a) (b)

Figure 7: (a) Emphasized collision set and (b) an identified collision with penetration distance

Similarly, Process Simulate can monitor joint positions and velocities and display graphical analyses before manufacturing. Figure 8.a presents data that shows instantaneous large axis movements during a course. These movements are typically caused by the machine passing through a singularity or a kinematically restricted point. Through adjustment of the kinematic solver technique, the axis movements presented in Figure 8.b can be achieved. These changes are usually made prior to pushing the program to the machine because an instantaneous movement such as the one presented can damage the machine.

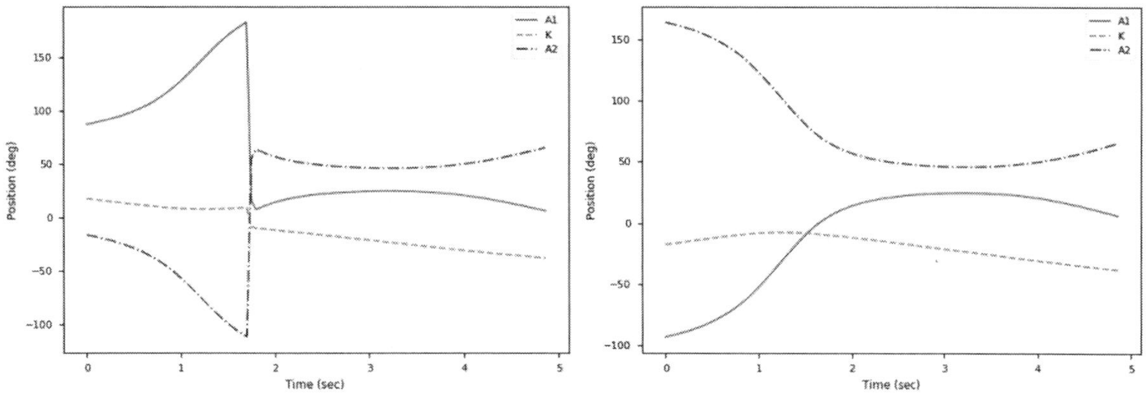

(a) (b)

Figure 8: Graphs showing the radial axis position over time for a course (a) with and (b) without a large position jump

Another potential problem during manufacturing is a rapid axis movement (Figure 9) which can cause defects if not resolved. Such an issue is normally noticed through careful observation of a simulation. Again, using joint analysis within Process Simulate, the user can immediately notice a rapid change in position and velocity and adjust accordingly. These tools allow for a deeper knowledge of the machine's activity throughout a manufacturing process and can prevent defects. Correlation of this data with actual manufacturing data assists in better programming of the AFP courses.

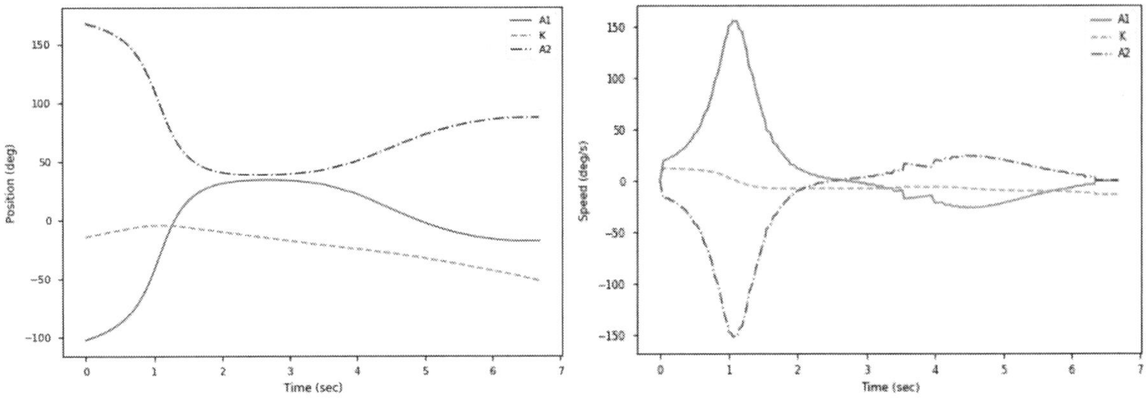

(a) (b)

Figure 9: Graphs showing the (a) position and (b) speed of a course with a rapid axis movement

4. CONCLUSIONS

4.1 Conclusion

Process Simulate is an extremely powerful and robust software suite which allowed the initial steps in the creation of a Digital Twin for the McNAIR IMT Lynx AFP machine. Within this paper, it is clear that a full Digital Twin should not only emulate actual machine performance and characteristics via rapid data transfer but be capable of monitoring and solving issues that occur when commissioning a manufacturing cell. With accurate models for both machine geometry and kinematics, it was shown the Digital Twin can emulate any item within a 3D environment. Tools within the Digital Twin give the user the ability to make rapid developments and changes in the function and simulation of the machine, thus reducing down time and increasing efficiency of all aspects of PLM. This optimization of the PLM via integration of the Digital Twin is a necessity of manufacturing within Industry 4.0.

4.2 Future Work

In this paper only a small portion of the capabilities of Process Simulate were utilized. Future work on the Digital Twin will lead to enhanced cell integration, including modeling of the robotic inspection system in the same manner as the AFP machine. With the inclusion of manufacturing and inspection simulation in the Digital Twin, the entire cell can be commissioned as one, unified environment, allowing further optimization of machine use. Movements of the inspection system will be able to be coordinated with the AFP machine. This is a very important aspect of the Digital Twin, as it will allow for virtual retrofitting of standalone inspection systems into in-process inspection.

Additionally, the Digital Twin allows for integration of software from external partners, with one example being machine tool path optimization. This software will be implemented for optimized path validation against the original model, and lead to a drastic reduction in unnecessary machine travel. Further augmenting this ability will be a proprietary machine learning algorithm, which will input data analytics streamed from Process Simulate into a neural network, allowing the system to make predictive models for machine movement optimization. With the implementation of neural network machine learning coupled with tool path optimization, the Digital Twin may potentially lead to the capability of semi-automated process planning for Automated Fiber Placement.

5. REFERENCES

1. Burov, A. E., & Burova, O. G. *Development of digital twin for composite pressure vessel.* In Journal of Physics: Conference Series (Vol. 1441, No. 1, p. 012133). IOP Publishing. 2020. DOI: 10.1088/1742-6596/1441/1/012133

2. Seon, G., Nikishkov, Y., Makeev, A., & Ferguson, L. *Towards a digital twin for mitigating void formation during debulking of autoclave composite parts.* Engineering Fracture Mechanics, 225, 106792. 2020. DOI: 10.1016/j.engfracmech.2019.106792

3. Polini, W., & Corrado, A. *Digital twin of composite assembly manufacturing process.* International Journal of Production Research, 1-15. 2020. DOI: 10.1080/00207543.2020.1714091

4. Hughes, D., Keir, S., & Meggs, C. *Digital twin methodology for compression moulded thermoplastic composite optimisation.* Flow Processes in Composite Materials (FPCM), 2018.

5. Harik, R., Rivest, L., Bernard, A., Eynard, B., & Bouras, A. (Eds.). *Product Lifecycle Management for Digital Transformation of Industries.* In 13th IFIP WG 5.1 International Conference, PLM 2016, Columbia, SC, USA, July 11-13, 2016, Revised Selected Papers (Vol. 492). Springer, 2017. DOI: 10.1007/978-3-319-54660-5

6. Glaessgen, E., & Stargel, D. *The digital twin paradigm for future NASA and US Air Force vehicles.* In 53rd AIAA/ASME/ASCE/AHS/ASC structures, structural dynamics and materials conference 20th AIAA/ASME/AHS adaptive structures conference 14th AIAA (p. 1818), 2012. DOI: 10.2514/6.2012-1818

7. Tao, F., Cheng, J., Qi, Q., Zhang, M., Zhang, H., & Sui, F. *Digital twin-driven product design, manufacturing and service with big data.* The International Journal of Advanced Manufacturing Technology, 94(9-12), 3563-3576. 2018. DOI: 10.1007/s00170-017-0233-1

8. Rosen, R., Von Wichert, G., Lo, G., & Bettenhausen, K. D. (2015). *About the importance of autonomy and digital twins for the future of manufacturing.* IFAC-PapersOnLine, *48*(3), 567-572. 2015 DOI: 10.1016/j.ifacol.2015.06.141

9. Schleich, B., Anwer, N., Mathieu, L., & Wartzack, S. *Shaping the digital twin for design and production engineering.* CIRP Annals, 66(1), 141-144. 2017. DOI: 10.1016/j.cirp.2017.04.040

10. Halbritter, Joshua & Harik, Ramy & Saidy, Clint & Noevere, August & Grimsley, Brian. *Automation of AFP Process Planning Functions: Importance and Ranking.* In SAMPE Conference & Exhibition, 2019. DOI: 10.33599/nasampe/s.19.1592.

11. Sacco, C. *Machine Learning Methods for Rapid Inspection of Automated Fiber Placement Manufactured Composite Structures.* (Master's thesis). Retrieved from https://scholarcommons.sc.edu/etd/5475, 2019.

12. Lee, C., Park, S. Survey of the virtual commissioning of manufacturing systems. *Journal of Computational Design and Engineering 1(3)*, 213-222, 2014. DOI: 10.7315/JCDE.2014.021.

LOW- COST MANUFACTURING FOR CARBON FIBER BULK MOLDING COMPOUND

Connor W. Lowry, Dean C. Chan, Raymond Choi
The Cooper Union for the Advancement of Science and Art
New York, NY/USA

ABSTRACT

Our research explores the development of a low-cost alternative to carbon fiber bulk molding compound (CFBMC). The motivation is to replace aluminum components used in a vehicle for Formula SAE®, a collegiate automotive design competition. Our goal is to reduce the overall weight of the competition vehicle with the constraint of retaining comparable strength and stiffness to existing aluminum components. We adapted existing manufacturing processes to utilize cheaper, commercially available products such as low-cost carbon fiber tow and high-strength epoxy instead of expensive prepreg systems. Stiffness, strength, and weight targets were set based upon commercially available CFBMC products.

Test specimens were produced and tested according to ASTM standards D3039, D6641, D5379, D7264, D256, and D671. These properties were used to inform the design of components and the expected part-to-part deviation due to manufacturing differences.

A suspension upright was designed as a proof of concept for future development. This component was selected because of its weight-savings potential, complexity, and its wide breadth of loading cases. From a monetary perspective, per-part costs are expected to be only 20-40% of commercial CFBMC using this alternative method. The CFBMC upright is 20% lighter and ten times stiffer than a comparable aluminum upright.

1. INTRODUCTION

1.1 Project Motivation

The information contained within this paper was compiled for the purpose of proposing a new method of carbon fiber compression molding for an academic setting. Using an affordable process and resources available at the school, specifications similar to commercially available products have been achieved. The paper helps readers in academia, specifically at the undergraduate or graduate level, to gain a better understanding of the process and its benefits. CFBMC is generally a cost-prohibitive process, and thus reducing the barrier of entry will benefit the next generation of composites engineers. Contained within are a suggested process to creating CFBMC components, materials testing performed to validate expected properties, as well as a possible application for this process.

1.2 The Carbon Fiber Compression Molding Process

1.2.1 Past Work

Bulk molding compounds (BMCs) are glass-fiber reinforced thermoset polymer materials prepared to be ready to mold upon purchase for compression molding applications. BMCs offer more flexibility in creating parts with complex geometries. However, the application of glass-fiber is inferior to carbon fiber in terms of strength to weight, so the use of CFBMC over BMC would be more applicable for the development of low weight racecar components.

CFBMC is a relatively new manufacturing process for composites that came along to offer faster and cheaper methods to take advantage of the material properties of carbon fiber. Research into CFBMC started in 2007 as a collaboration between Automobili Lamborghini Advanced Composite Structures Laboratory, Innovative Composite Engineering, Quantum Composites, Callaway and Automobili Lamborghini Advanced Composites Research Center where they called the material Forged Carbon®. In 2010, Lamborghini showed off their Sesto Elemento which included many Forged Carbon® parts such as the control arms. In 2011, Callaway released the Callaway Diablo Octane Driver [1] [2]. Since then, their production method for Forged Carbon® has not changed much publicly, though Quantum Composites have released data on testing and variability of the product, as well as how it compares to various other composite manufacturing techniques [3].

1.2.2 Material Selection

For our project, carbon fiber and epoxy were required to be reasonably priced and reliably available to the team could consistently buy material from so that the exact same material could be used in the future. It was decided that T800 12K tow carbon fiber would be bought from Soller Composites, LLC. T800 boasts a tensile modulus of 295 GPa and an ultimate tensile strength of 5860 MPa [4]. As a composite these properties will be significantly less impressive compared to aluminum but should still provide adequate stiffness and strength.

Commercial CFBMC is typically sold as a thermoset prepreg. This means that the carbon fiber comes saturated with resin that will only cure after being heated to a certain temperature above ambient. The epoxy typically used in prepreg is not widely available on the market. Therefore, Fibreglast's System 2000 Epoxy Resin with 2020 Hardener was chosen because of its high strength and ability to be heated to 75°C which significantly reduces the epoxy curing time. [5]

1.2.3 Manufacturing Process

Manufacturing CFBMC requires high heat and pressure. The current procedure developed by Quantum Composites on their data sheet for CFBMC is to heat the part to 55-70°F and compress it under 14 MPa of pressure [6] [7]. This pressure magnitude allows excess epoxy to be squeezed out from the mold while retaining the correct carbon to epoxy ratio (by weight). In addition, high pressure minimizes the air gaps in between the fibers that would lead to imperfections in the finished part. To facilitate this, a two-part aluminum mold was made. One side has a cavity and the other has a matching follower. Proper draft angles, ejector pins, and flashing rings are included in the mold to help with part removal and resin overflow [8] [9]. The tow carbon fiber is cut into 1.5-2 inch strands and mixed with epoxy at a ratio of 1:1 by weight. This mixture is then placed into the cavity of the aluminum mold and the aluminum follower is placed on top.

Heat is applied using cartridge mold heaters and pressure is applied using a universal testing machine (UTM), specifically the Instron 5500R, as a press. After the part has cured, it is left to cool, pressure is released, and the part is taken out using the ejector pins. Light sanding cleans up any excess flashing, and a coating of polyurethane protects against water ingress.

1.2.4 Material Properties

Research into CFBMC by Toray Advanced Composites has shown that the material is able to reach a tensile modulus of 131.0 GPa, a tensile strength of 289.6 MPa, and a density of 1.5 g/cc [10]. Due to its higher stiffness, it is an ideal material for manufacturing the suspension upright. In addition, its high strength to weight ratio means there is weight to be saved by switching to CFBMC.

2. EXPERIMENTATION

2.1 Validation

A series of tests were conducted to validate this process in tension, bending, and compression and ensure it maintains similar material properties. These tests were conducted under ASTM standards to ensure a repeatable testing process.

2.2 Testing Standards

The following ASTM tests were performed in a Universal Testing Machine (UTM), specifically the Instron 5500R.

2.2.1 Tension (ASTM D3039)

This standard requires the use of tensile jaws in a UTM (Figure 1), in which a standard sample (17.78 cm x 2.54 cm x 0.635 cm) is pulled until yield. Additionally, steel plates are bonded to the composite sample at each end to prevent slipping. Finally, an extensometer is used to measure strain, and a load cell to measure loading. [11]

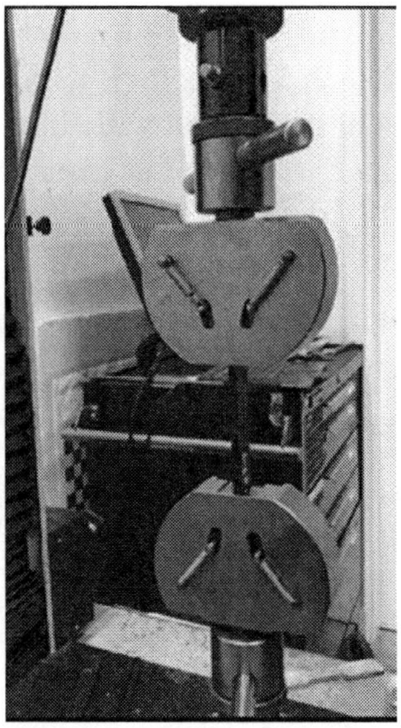

Figure 1: Tensile Test Setup

2.2.2 Combined Loading Compression (ASTM D6641)

Combined Loading Compression (CLC) uses a similar sized sample (15.24 cm x 2.54 cm x 0.635 cm) placed between a set of supports (Figure 2). These supports are then pressed between compression platens. The supports enable a small section to be under unsupported compression, eliminating risk of bending stress [12].

Figure 2: CLC Test Setup
*Note: Sample is not to size

2.2.3 Four Point Bending (ASTM D7264)

Four Point Bending places a sample between two stationary supports and loads it equally between two rolling noses (Figure 3). This creates a constant stress region in between the noses, giving a more repeatable result and more closely replicating real world conditions than a three-point bending setup. A three-point setup creates a high stress region at the center of the sample, which does not model real world loading and can overly highlight imperfections in composite manufacturing. This test requires a high length to thickness ratio, which drove the decision to create a second follower for this test [13].

Figure 3: Four Point Bending Test Setup

2.3 Mold Design

2.3.1 Mold Geometry

Test specimens needed to be produced quickly and repeatably to fulfill test requirements. A standard sample size was used to minimize cost and mold machining time. This standardization enabled the production of samples for all four testing regimes using only three major components. These components include one mold cavity, and two followers (Figure 4). These two followers modify the overall thickness of the sample, allowing for a much thinner sample required by the four-point bending standard.

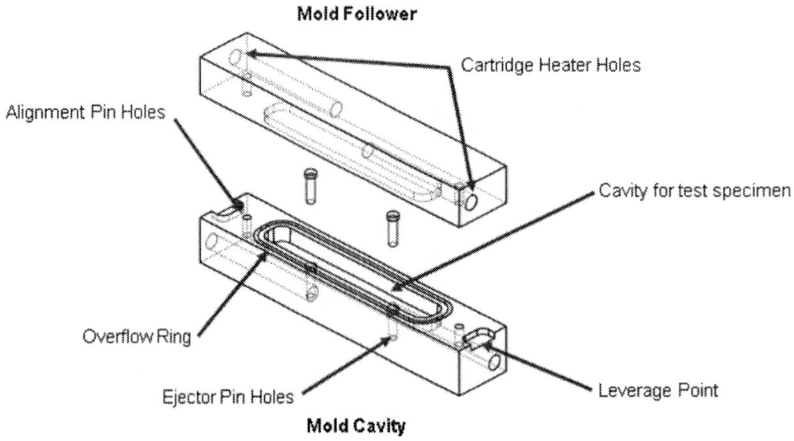

Figure 4: Mold Diagram

The common specimen cavity models the sample geometry. The walls are drafted by 1.5° to promote ejection and mold separation. Other features such as ejector pins and alignment pins are similar to other injection or compression mold designs. Holes for cartridge heaters allow for preheating of mold to speed up this process from many hours at room temperature to 20-25 minutes.

2.3.2 Heater Design

As part of this mold, a temperature controller (Figure 5) was designed to maintain the mold temperature at 60°C. Temperature accuracy is not necessarily important, but precision throughout the mold is desired. Thus, long cartridge heaters were selected to aid in the distribution of heat energy throughout the mold. To control these, a set of On-Off controllers were selected to actuate the heaters. On-Off control was found to be adequate, but a PID system will help prevent overshoot, enabling a quicker heating time overall. To combat overshoot, the setpoint was slowly ramped, enabling the mold to heat up slowly and evenly.

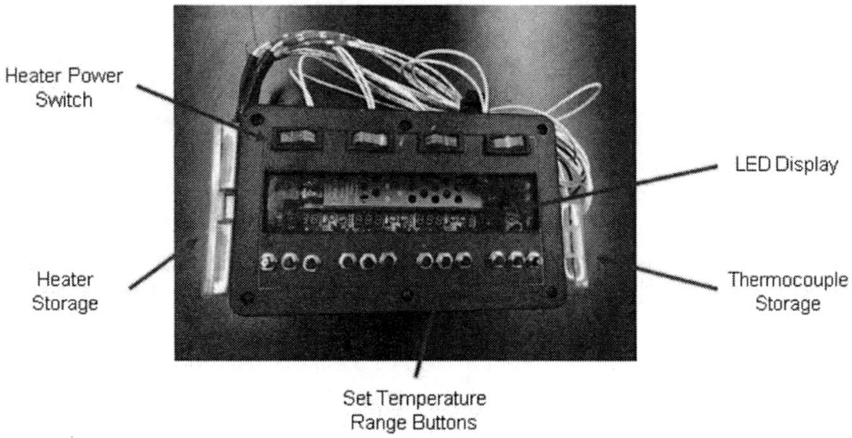

Figure 5: Heater Overview

3. RESULTS AND DISCUSSION

3.1 Tension Testing

Data was collected for three samples using a universal testing machine that pulled each sample until failure at a displacement rate of 2 mm/min. A clip-on extensometer was used and was set at an initial gauge length of approximately 50 mm. Given this initial extensometer gauge length (L_0), the change in extensometer length (dL), the cross-sectional area of each sample (A) and the load at which the sample fails, it is possible to calculate the Tensile Modulus (E) using Eq. 1.

$$E = \frac{F \cdot L_0}{A \cdot dL} \qquad [1]$$

Figure 6 shows the compiled tensile stress-strain curves of the three samples. It should be noted that sample 006T was cured without reaching full compression, resulting in a greater epoxy to carbon ratio than desired and pockets of air in the sample. As a result, its performance is notably worse than samples 005T and 007T. 006T was included as an example of a worst-case scenario which would typically be avoided if full compression is achieved.

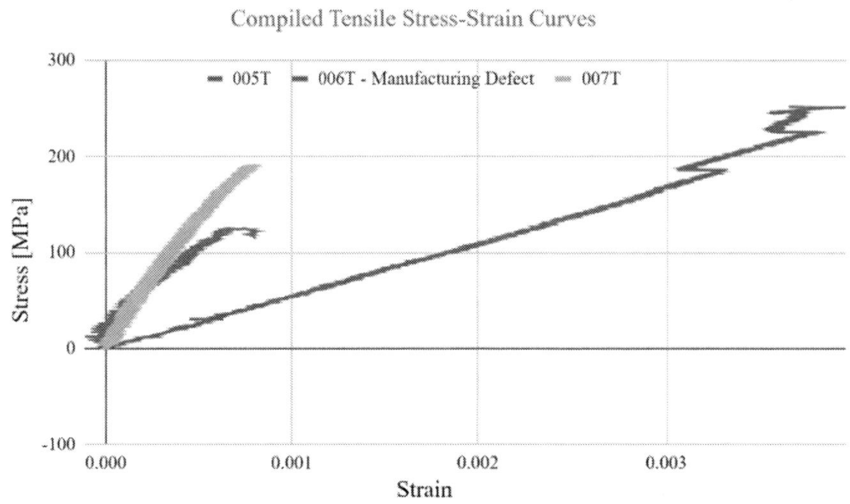

Figure 6: Tensile Stress-Strain Data for In-House CFBMC

Table 1 provides the mechanical properties calculated for each of the CFBMC samples and how they compare to AA7075. Comparing the two materials shows that the average CFBMC sample was able to take 39% of the load an identical AA7075 sample would be able to take. Conversely, the CFBMC sample achieved 230% the stiffness of AA7075 before failing.

Table 1: Comparison of Tensile Results

	Sample	In-House CFBMC		Aluminum 7075 [14]
		Individual	Average	
Yield Stress [MPa]	005T	151	196	503
	006T	252		
	007T	185		
Modulus [GPa]	005T	192	164	71.7
	006T	69		
	007T	231		

3.2 Compression Testing

Three samples were tested on the universal testing machine by applying end-loading and shear loading. The CLC fixture (Figure 2) applies said load to the sample to determine the material properties of CFBMC in compression. During the test, a combined load was applied to an unsupported region of 16mm. The compression modulus was calculated from Eq. 1. Figure 7 shows the compiled compressive stress-strain curves of the three samples. Sample 010C was purposefully run well beyond its failure point to see how CFBMC would perform under failure. It was determined that after the initial yield point, the sample would have a minor fracture before continuing to take more load, with the pattern repeating multiple times. This is suspected to be the result of the random orientation of the carbon fibers in the sample where one group of strands would fail before another set of strands take up the load again. This behavior mimics that of plastic region where a component will not immediately break apart if taken beyond its yield point, making it a natural safety feature for load bearing parts.

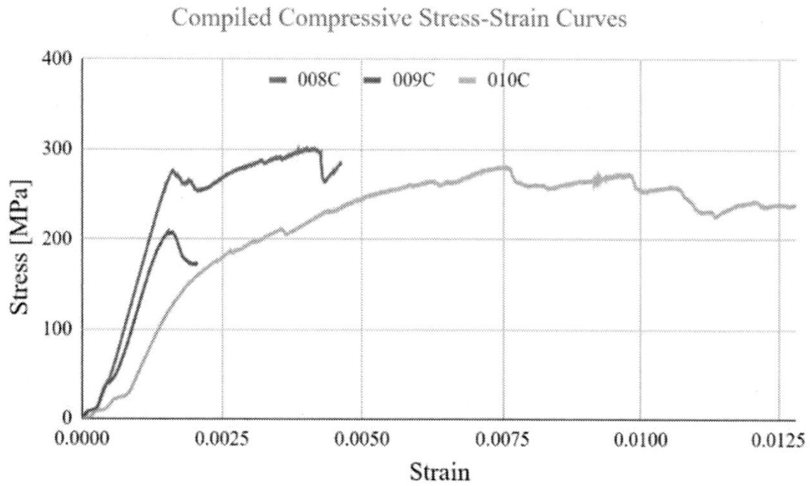

Figure 7: Compression Stress-Strain curves for In-House CFBMC

Table 2 provides the mechanical properties calculated for each sample of CFBMC and how they compare to AA7075. The average CFBMC sample was able to take up to 55% of the yield stress of an AA7075 sample before failing. Based on the compressive modulus calculations, CFBMC was determined to be 200% stiffer than AA7075.

Table 2: Comparison of Compressive Results

	Sample	In-House CFBMC		Aluminum 7075 [14]
		Individual	Average	
Yield Stress [MPa]	008C	276	278	503
	009C	285		
	010C	274		
Modulus [GPa]	008C	170	144	71.7
	009C	186		
	010C	77		

3.3 4-Point Bending Testing

Three samples were tested in 4-point bending to determine the flexural modulus of CFBMC. For a load span of one half the support span, the strain of the sample (r) can be calculated using Eq. 2.

$$r = \frac{D \cdot d}{0.23L^2} \quad [2]$$

where D is midspan deflection, L is the support span, and d is the depth of the beam. To calculate the stress of the outer fiber through the load span (S), Eq. 3 is used.

$$S = \frac{3P \cdot L}{4b \cdot d^2} \quad [3]$$

where P is the load at a given point on the load-deflection curve and b is the width of the beam. Once stress and strain have been calculated, Eq. 1 can be used to determine the flexural modulus of each sample.

It should be noted that of the three samples, 011B was damaged during the manufacturing process and was expected to have a lower flexural modulus compared to 012B and 013B as seen in Figure 8.

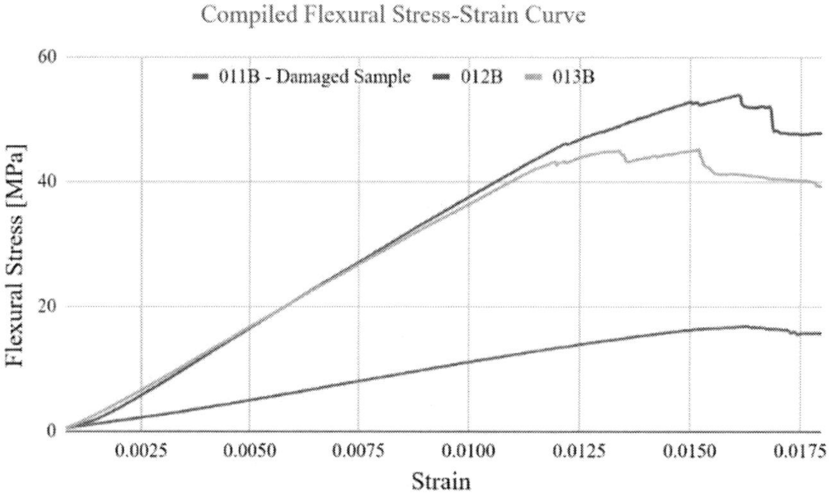

Figure 8: Flexural Stress-Strain curves for In-House CFBMC

Table 3 shows the calculated mechanical properties of CFBMC compared to AA7075. The average CFBMC sample yields at 7% the load an equivalent AA7075 sample could take and has 4.3% the stiffness. The flexural performance of CFBMC is significantly worse than AA7075 in both the amount of load it can take and the stiffness, even disregarding the damaged sample. The samples used for flexural testing are half the thickness of the samples used in the previous two tests to follow ASTM D5379. This was necessary in order to fit the test sample into the flexural fixture. However, it had been noted by Toray that ideal testing should use a sample that is the same thickness as the actual part, which suggests that thinner samples perform significantly worse than a thicker sample with the same span to depth ratio. [6] The results of Table 2 support this theory.

Table 3: Comparison of Flexural Results

	Sample	In-House CFBMC		Aluminum 7075 [14]
		Individual	Average	
Yield Stress [MPa]	011B	16.4	35.3	503
	012B	46.2		
	013B	43.2		
Modulus [GPa]	011B	1.1	3.1	71.7
	012B	4.4		
	013B	3.9		

3.4 Strength to Weight

Although the yield strength of CFBMC cannot exceed the performance of AA7075, a better metric for its potential application is specific strength. Specific strength is defined as yield strength divided by the density of a material and provides a better understanding of how strong a material is relative to how much mass it has. Figure 9 shows the comparison of specific strengths between CFBMC, aluminum 6061 (AA6061), and AA7075. CFBMC exceeds the specific strength of AA6061 by over 88% and shows comparable performance to AA7075. At the same time, since its stiffness exceeds that of AA7075 by over 100%, it becomes advantageous to use in situations where the goal is to reduce the weight of a component while trying to minimize its deflection.

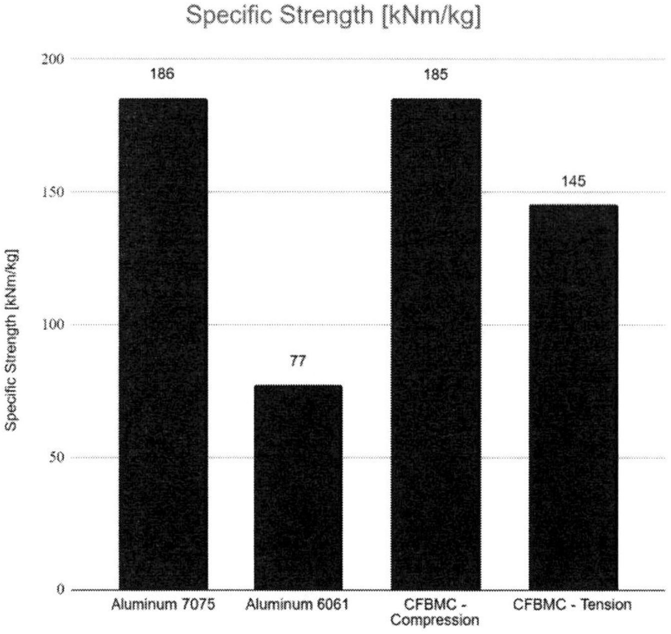

Figure 9: Specific strength comparison of CFBMC and aluminum

3.5 Variability

In each test, three samples were manufactured and tested, giving a total of nine samples overall. Of course, this does not give a good picture of the overall variability of these samples. Three samples per test was selected to enable the development of the manufacturing method while providing enough samples to conduct a baseline test in each loading case. Additional testing was planned, but testing was cut short due to current events. Additional samples must be conducted to attain a better grasp of variability. This can likely be performed in a later study alongside manufacture of the applied component, discussed below.

4. APPLICATION

4.1 Motivation

CFBMC's specific strength to AA7075 show that it can be used as a potential material for load bearing parts. Based on this result, it was decided to design a test component to determine the cost and weight saving potential when compared to aluminum. This was done in conjunction with The Cooper Union's Formula SAE Team, which offered their suspension upright as a test case. The upright is the bracket which holds the wheel to the rest of the car, and as a result it is a critical load bearing component of the vehicle. Traditionally, it is manufactured out of AA7075. Additionally, it is a part which is required to have a very low deflection, making the stiffness properties of CFBMC ideal for this use case.

4.2 Design Considerations

A major drawback to CFBMC is that it is considerably softer than AA7075, making it prone to wear when used in dynamic situations. Thus, care must be taken to ensure protection against part-on-part contact as seen in the design in Figure 10. To combat this issue, sleeves and inserts were used between bearings and bolted connections, providing a much more resistant surface. These inserts were then bonded in to ensure a rigid connection between the carbon and insert. Where concentricity is important, such as the central wheel bearing on this component, a bored CFBMC surface and glass beads specifically designed for maintaining concentricity in composite bonds ensure a tight tolerance. The design for the CFBMC upright (Figure 10) is similar to the aluminum upright (Figure 11). The main difference between the designs is that the CFBMC upright's truss sections are about 50% thicker than the aluminum one.

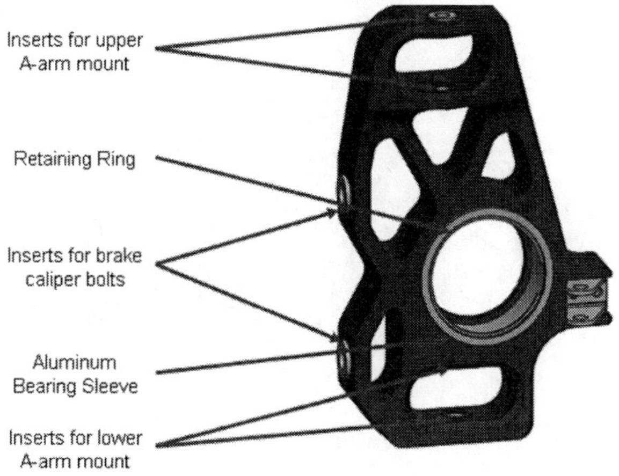

Figure 10: CFBMC Left Upright Design with Inserts and Sleeves

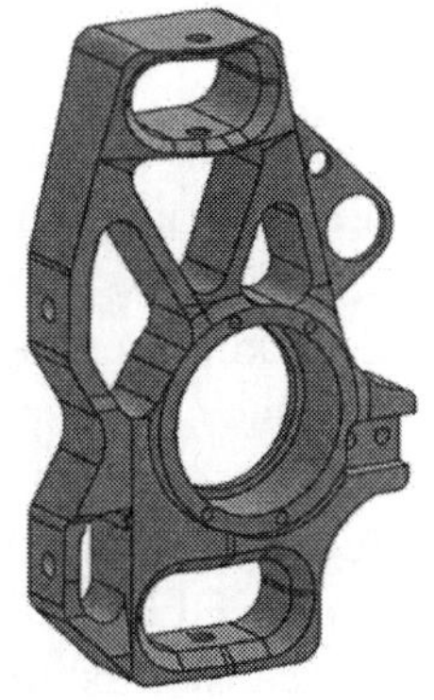

Figure 11: Aluminum Left Upright Design

4.3 Analysis

Once the general geometry of the component was completed, finite element analysis was conducted to refine the design to maintain strength and stiffness while optimizing for weight. These simulations assume random fiber orientation, and thus an approximately isotropic material. Based on initial simulation results, the thicknesses of trusses inside the upright were adjusted to ensure the component was designed to an acceptable factor of safety. The optimized design is an order of magnitude stiffer than its aluminum counterpart, and 23% lighter.

A static structural simulation was run to show the effect that the upright stiffness has on hub compliance. Figure 12 shows that a CFBMC upright reduces hub compliance by approximately half. The aluminum upright significantly contributes to the hub compliance while the CFBMC upright does not.

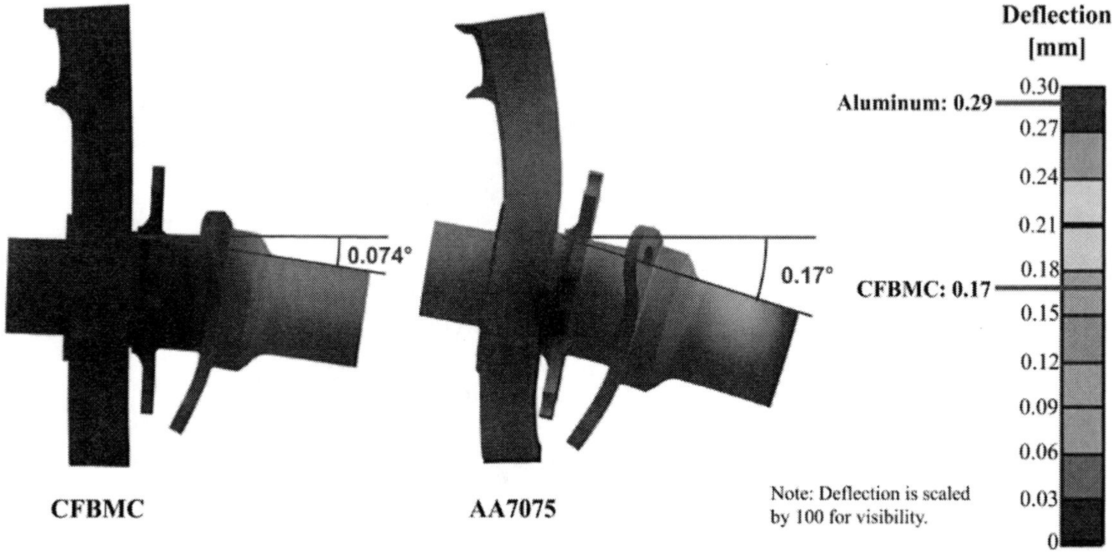

Figure 12: Effect of upright stiffness on hub compliance

4.4 Cost Comparison

Of equal concern for both the client and the research was the expected expenses between the in-house CFBMC, commercially available product, and aluminum. Table 4 shows a cost analysis on the cost to the client to make two uprights in each respective material. Ideally, the client would want to have two uprights made to have a backup prepared for their competition. In-house CFBMC is considerably less expensive than the commercial product and is not significantly more expensive to produce than aluminum. Machining and labor costs are not considered in this cost analysis because the client does not incur any costs here that would inhibit their ability to use in-house CFBMC. Commercially available CFBMC is over 7 times as expensive as our in-house version for carbon and epoxy costs because it is highly regulated for aerospace use, which creates high overhead costs driving up the cost of low-volume orders. This is, of course, not a true cost comparison, but an expected cost to be seen by an academic department wishing to use this process using students and faculty to handle labor. It is for this reason that things like labor and machining time are not considered.

Table 4: Cost Analysis

	7075 Aluminum	In-House CFBMC	Commercial CFBMC
Aluminum Cost (recurring per upright) [15]	$29.77 (Upright Stock)	$4.30 (Insert and Sleeve Stock)	$4.30 (Insert and Sleeve Stock)
Tooling Cost (non-recurring per upright) [16]	$0	$47.40	$47.40
Carbon/Epoxy Costs	$0	$8.64	$62.84
Machining Cost	There is no cost to the client as machining is provided by the school.		
Labor Cost	Because this process is for students in an academic setting, labor costs are not meaningful in this cost comparison.		
Total Cost (For 2 uprights)	$29.77	$36.64	$90.84

5. CONCLUSIONS

Creating samples for testing the ASTM standards validated the manufacturing process for CFBMC. Testing of CFBMC made from this alternative process provided reliable values for the material properties. Lastly, the design and simulation of a load bearing CFBMC upright show that it is a viable material to use in the design of components. While not matching the material properties of commercially available CFBMC, when certain design considerations are taken into account to accommodate its lower yield strength, this alternative process can create parts that are lighter and stiffer than aluminum parts while retaining a similar strength with heavily reduced costs. The upright design shows 30% reduction in cost from commercial product, putting it close in cost to an aluminum component. The design for the CFBMC upright is a success for the team as well as the material. This lower-cost alternative broadens the user-base considerably, especially for academic institutions and student design teams that wish to use this material for components where weight and stiffness are paramount.

6. REFERENCES

[1] Joseph, Noah. "Driving Force: Lamborghini and Callaway Golf collaborate on new forms of carbon fiber". Autoblog. Accessed May 6, 2020

[2] Feraboli, Paolo; Gasco, Federico; Wade, Bonnie; Maier, Steve; Kwan, Roger; Masini, Attilio; DeOto, Luciano; Reggiani, Maurizio. Lamborghini "Forged Composite" *Technology for the Suspension Arms of the Sesto Elemento.* Montreal: American Society for Composites, 2011. Accessed September 30, 2019. http://www.lambolab.org/wp-content/uploads/03research/pub/05chop/2011-ASC-montreal-forged-suspens-ICE.pdf

[3] Kaczmarczyk, Matt, and Langschwager, Tim. "Carbon Fiber SMC Technology for Lightweight Structures." Quantum Composites. Accessed October 3, 2019. http://www.temp.speautomotive.com/SPEA_CD/SPEA2013/pdf/CF/CF3.pdf.

[4] "T800S Intermediate Modulus Carbon Fiber." Toray Composite Materials America, Inc. Accessed May 6, 2020. https://www.toraycma.com/page.php?id=661

[5] "System 2000 Laminating Epoxy Resin." Fibre Glast Developments Corporation. Accessed May 6, 2020. https://s3.amazonaws.com/cdn.fibreglast.com/downloads/00343.pdf

[6] De Oto, Luciano. "Carbon Fibre Innovation for High Volumes: The Forged Composite." Quantum Composites. Accessed October 2, 2019. https://www.quantumcomposites.com/pdf/papers/IICS-JEC-presentation-2011.pdf.

[7] Quantum Composites, *AMC® 8590, 126-76-8 Engineered Structural Composite® (ESC®) Molding Compound.* Bay City: The Composites Group, 2013. Accessed September 22, 2019.https://www.quantumcomposites.com/pdf/datasheets/amc/Quantum-AMC-8590%20126-76-8.pdf

[8] Quantum Composites. *AMC® 8595, 126-76-75 Engineered Structural Composite® (ESC®) Molding Compound.* Bay City: The Composites Group, 2012. Accessed September 22, 2019. https://www.quantumcomposites.com/pdf/datasheets/amc/Quantum-AMC-8595%20126-76-75.pdf

[9] "Thermoset Compression Mold Design Tips." Plenco, March 16, 2015. https://www.plenco.com/plenco_processing_guide/Sect 13 Compression Mold Design Tips.pdf.

[10] "Toray MS-1A." Toray Advanced Composites. Accessed May 6, 2020. https://www.toraytac.com/media/d5298565-a969-44d9-93a4-772c0f4e4eda/xnbf9w/TAC/Documents/Data_sheets/Bulk%20molding%20compounds/Bulk%20molding%20compounds/MS-1A_Epoxy_PDS.pdf

[11] "ASTM D3039 / D3039M-17, Standard Test Method for Tensile Properties of Polymer Matrix Composite Materials", ASTM International, West Conshohocken, PA, 2017, www.astm.org DOI: 10.1520/D3039_D3039M-17

[12] "ASTM D6641 / D6641M-16e1, Standard Test Method for Compressive Properties of Polymer Matrix Composite Materials Using a Combined Loading Compression (CLC) Test Fixture", ASTM International, West Conshohocken, PA, 2016, www.astm.org DOI: 10.1520/D6641_D6641M-16E0

[13] "ASTM D7264 / D7264M-15, Standard Test Method for Flexural Properties of Polymer Matrix Composite Materials", ASTM International, West Conshohocken, PA, 2015, www.astm.org DOI: 10.1520/D7264_D7264M-15

[14] "Aluminum 7075-T6; 7075-T651." *MatWeb.* 2020. Accessed July 21, 2020. http://www.matweb.com/search/DataSheet.aspx?MatGUID=4f19a42be94546b686bbf43f79c51b7d

[15] "7075 Aluminum Plate." *Midwest Steel Supply.* 2020. Rogers, MN. Accessed July 21, 2020. https://www.midweststeelsupply.com/store/7075aluminumplate

[16] "6061 Aluminum Plate." *Midwest Steel Supply.* 2020. Rogers, MN. Accessed July 21, 2020. https://www.midweststeelsupply.com/store/6061aluminumplate

DEVELOPMENT OF NOVEL TOOLING MATERIALS FOR NEXT GENERATION AIR MOBILITY COMPONENTS WITH INTRICATE STRUCTURE

Hiromichi T. Fujii*, Naoki Sakaguchi, Haruyasu Ohno, Kotaro Ona
Shinhokoku Steel Corporation
Kawagoe, JAPAN

ABSTRACT

Thermal expansion characteristics between room and molding temperatures were analyzed in low thermal expansion alloys developed as novel tooling materials for small CFRP components with intricate structure. Low thermal expansion characteristics of the developed alloys could be maintained up to 400 °C. The developed alloys prepared by casting were processed by cryogenic treatment and annealing to improve the mechanical properties. The mechanical properties were examined by tensile tests of round-bar specimens. Yield stress of the processed alloys showed more than twice of as-cast alloys. The drastic improvement of the mechanical properties was found to be attributed to the grain refinement and the high density dislocations through microstructural characterization. Cryogenic treatment and annealing process was also found to possibly reduce variation of mechanical properties. These discoveries suggested that the developed alloys could improve the tooling lifetime and reduce the manufacturing cost in CFRP molding for next generation air mobility.

* Corresponding author: H.T. Fujii, fujii@shst.co.jp

1. INTRODUCTION

Tooling technology for precise molding of small carbon fiber reinforced plastic (CFRP) components with intricate structure can be one of the key technologies for next generation air mobility. Low thermal expansion alloys such as Fe-36mass%Ni (Invar®: a registered trademark of ArcelorMittal - Stainless & Nickel Alloys, formerly known as Imphy Alloys (US Trademark #63970)) [1, 2] and Fe-29%Ni-17%Co (Kovar®: a registered trademark of Carpenter Technology Corporation) [3] alloys are known as the tooling materials for CFRP molding [4, 5]. The reason why the low thermal expansion alloys are applied as CFRP tooling can be illustrated from the viewpoint of dimensional change between room and molding temperatures. Figure 1 shows relative thermal expansion curves of Al alloy 6061, tool steel, invar alloy and commercial CFRP to a dimension at room temperature. Focusing attention on the dimension of these materials at 125 °C, that is a typical temperature for autoclave molding of CFRP, there are large differences between the Al alloy or the steel and the CFRP. In contrast, the difference between the invar alloy and the CFRP is much smaller. This indicates that shape accuracy of the resultant components molded by the invar alloy tooling is much higher than those molded by the other materials.

The CFRP components for conventional airplanes have a large size and a simple shape with a small curvature, so that the invar alloy molds produced by sheet metal working process have been

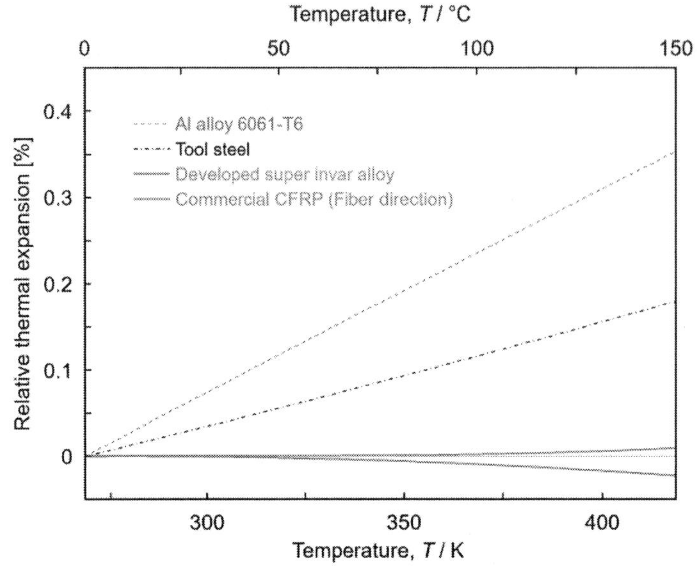

Figure 1 Comparison of relative thermal expansion curves to a dimension at room temperature between Al alloy 6061, tool steel, invar alloy and commercial CFRP. A curve of commercial CFRP was measured along the fiber direction.

used for molding. On the other hand, the CFRP components for next generation air mobility have a small size and intricate shape with large curvature. In addition, mass production technologies such as press molding will be required with increasing air mobility users in the future. Novel tooling materials that support mass production of intricate molding are therefore also required. Near-net shape forming by casting is one of the solutions for precise molding of small CFRP components with intricate structure. Yield strength and Young's modulus of the cast invar alloys, however, will not be enough high to use as tooling materials for mass production. This is because accumulation of the micro-deformation of tooling materials at large curvature portion during mass production significantly jeopardizes the shape accuracy of the CFRP components. The deterioration of mechanical properties in the cast alloys is caused by the formation of coarse columnar grain microstructure during solidification. In general, plastic working process is known to be able to improve the mechanical properties through grain refinement [6, 7]. Nevertheless, the plastic working cannot be applied to the intricate structured components because it destroys the intricate structure by plastic deformation. The metallurgical approach was therefore utilized to improve the mechanical properties of the cast invar alloys in this study.

The application of reverse martensitic transformation is known to be one of strengthening techniques for steels without plastic deformation [8-11]. Figure 2 shows schematic illustration of strengthening mechanisms by the application of reverse martensitic transformation. Martensite phase with body centered cubic (bcc) or body centered tetragonal (bct) structures transforms into austenite phase with face centered cubic (fcc) structure in the reverse martensitic transformation. When the austenite phase transformed from martensite phase was appropriately annealed, fine and equiaxed grain microstructure was reported to be formed through recrystallization [9, 12, 13]. Utilizing these phenomena, the mechanical properties of the low thermal expansion cast alloys are expected to be improved through grain refinement caused by recrystallization. Furthermore,

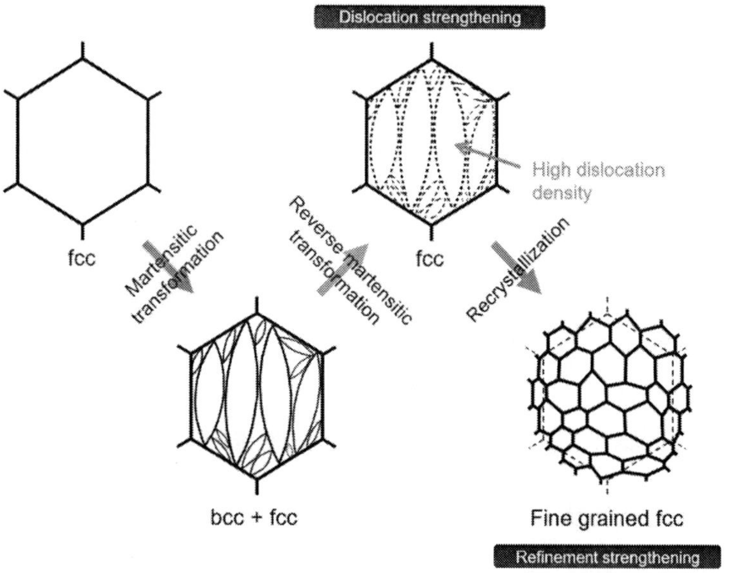

Figure 2 Schematic illustration of dislocation strengthening and grain refinement strengthening caused by reverse martensitic transformation and recrystallization in austenite steels.

martensite phase consists of a lot of regions with high dislocation density because martensitic transformation occurs with microscopic shear deformation. Increase in the dislocation density is known to improve yield stress of metals [14]. That is, the mechanical properties of the cast alloys is considered to be improved also by the formation of residual regions with high dislocation density by appropriate annealing.

The aim of the current work is to reveal the effects of reverse martensitic transformation and annealing on the mechanical properties in the low thermal expansion cast alloys. Future prospects of the developed alloys as tooling materials of CFRP components for next generation air mobility is also discussed on the basis of understandings on the thermal expansion characteristics and the mechanical properties.

2. EXPERIMENTATION

2.1 Sample preparation

This study used the $Fe_{65}Ni_{35}$ (invar [1, 2]), $Fe_{64}Ni_{31}Co_{5}$ (super invar [15, 16]) and $Fe_{55}Ni_{28}Co_{17}$ (kovar [3]) based alloys that can be used at different molding temperatures. In general, low thermal expansion characteristics disappear above approximately 200 °C in the invar and the super invar alloys, and 400 °C in the kovar alloy. Co and Ni elements were therefore adjusted in 0.1% accuracy rate in order to expand the temperature range of low thermal expansion in each alloy to the desired CFRP molding temperature. Target molding temperature of each tooling material was determined to be 200 °C, 300 °C and 400 °C on the basis of temperature in press molding. Table 1 summarizes the alloys prepared for experiments in this study. As the reference, the average coefficient of thermal expansion (CTE) between 20 °C (room temperature) and molding temperature in conventional invar, super invar and kovar alloys are also shown in table 1. The symbol of each alloy was determined on the basis of type of alloy and aimed molding temperature.

Table 1 The alloys used in this study. For reference, average CTE between room and molding temperatures in conventional invar, super invar and kovar alloys are also shown.

Symbol	Chemical composition	Type of alloys	Target molding temperature [°C]	Average CTE in conventional alloys [10^{-6} / °C]
I-200	$Fe_{65.1+x}Ni_{34.9-x}$	Invar	200	2.3 (R.T. ~ 200 °C)
SI-300	$Fe_{64.2+x+y}Ni_{31.0-x}Co_{4.8-y}$	Super invar	300	4.1 (R.T. ~ 300 °C)
K-400	$Fe_{55.3+x+y}Ni_{28.2-x}Co_{16.5-y}$	Kovar	400	5.0 (R.T. ~ 400 °C)

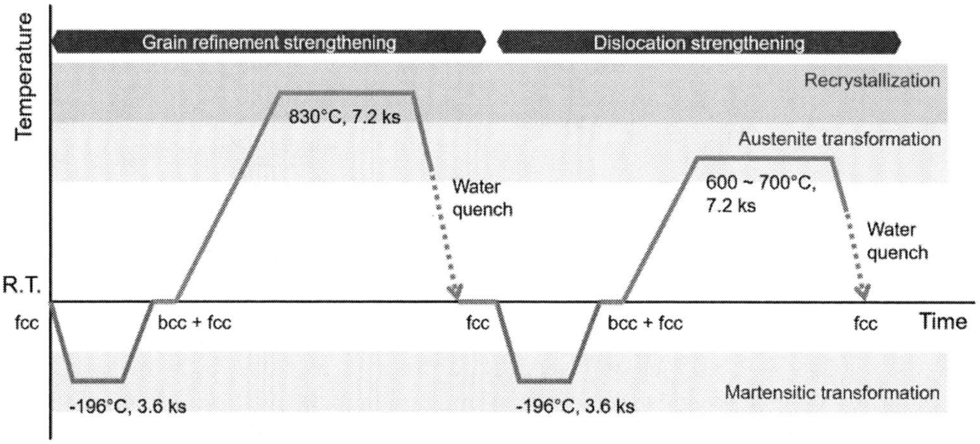

Figure 3 Schematic illustration of dislocation strengthening and grain refinement strengthening in the developed low thermal expansion alloys through cryogenic treatment and annealing.

All alloys were prepared by casting in air from molten metal at 1500 °C. Shape of the casting blocks followed ISO 4990:2015. The specimens cut from the center of the obtained cast blocks were cooled in a liquid nitrogen at -196 °C for 3.6 ks to allow for the formation of the martensite phase. The specimens with martensite phase were annealed at 830 °C for 7.2 ks to induce the reverse martensitic transformation and recrystallization. Thereafter, same cryogenic treatment and annealing at 600 °C or 700 °C for 7.2 ks were conducted to cause dislocation strengthening. These annealing temperatures were above austenite transformation temperature and below recrystallization temperature, which were determined by preliminary measurements of thermal expansion curves. The above thermal treatment process for microstructural control is schematically shown in Fig. 3.

2.2 Microstructural characterization

The specimens were mechanically polished with waterproof papers of #200-#1500 and buff polished with alumina particles of 3 and 1 μm. The surface microstructures were observed by optical microscopy in the polished specimens etched by chemical solution ($Cu_2SO_4 \cdot 5H_2O$: HCl : H_2O = 10 g : 50 ml : 50 ml). The average grain size of austenite phase in each specimen was estimated the quadrature method using optical micrographs. An X-ray diffractometer (XRD) was used to identify the crystallographic phases and to clarify the crystallographic orientation

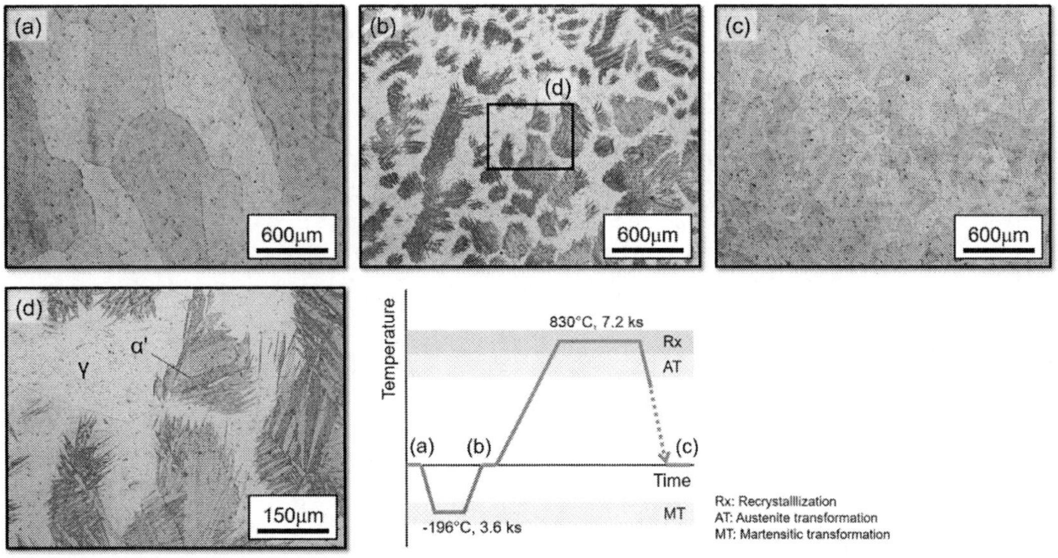

Figure 4 Optical micrographs observed in (a) as-cast, (b) cryogenic treated and (c) cryogenic treated and recrystallized specimens obtained from the I-200 ($x = 1.7$) cast block. (d) shows a magnified micrograph of (b). For reference, each stage of thermal treatment process is also shown schematically.

distribution in as-cast and microstructural controlled specimens.

2.3 Evaluation of thermal and mechanical properties

The specimens for the measurement of thermal expansion curves were cut from the center of the microstructural controlled specimens into a rod shape with $\phi 5 \times 20$ mm^3 in dimension. They were annealed at 315 °C for 7.2 ks to eliminate the residual stress produced by cutting process. The micro-elongations of the rod specimens were precisely measured by the highly vacuum tight pushrod dilatometer. Relative thermal expansion curves to a dimension at room temperature were calculated from the measured micro-elongation. The measurements were carried out at a heating rate of 0.083 °C/s (5.0 °C/min) in the temperature range between room and target molding temperatures.

Mechanical properties of the as-cast and the microstructural controlled specimens with round-bar shapes were measured by tensile tests at each target molding temperature. These tests were performed according to ISO 6892-2:2018. 0.2% yield strength and Young's modulus were estimated from the stress-strain curve in each specimen.

3. RESULTS AND DISCUSSION

3.1 Control of microstructure by the application of reverse martensitic transformation

Microstructural observations of the specimens at each stage of thermal treatment process shown in Fig. 3 were conducted to review the feasibility of reverse martensitic transformation for improvement of the mechanical properties. Figure 4 shows microstructures of (a) as-cast, (b) cryogenic treated and (c) cryogenic treated and recrystallized specimens obtained from the I-200 ($x = 1.7$) cast block. In addition, the magnified micrograph of (b) is also shown by (d). Typical

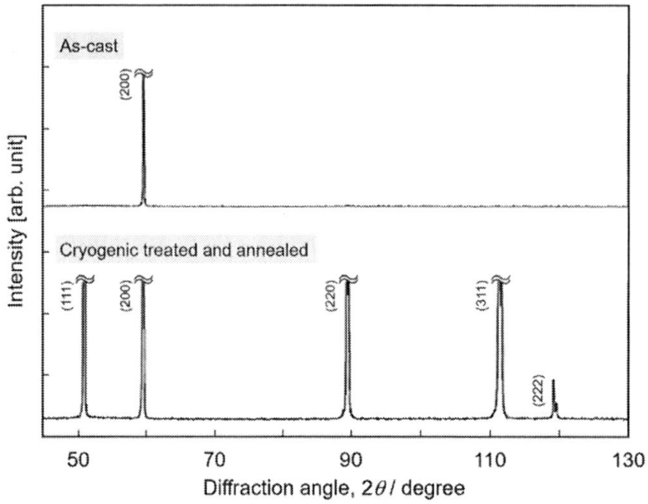

Figure 5 XRD profiles obtained from the as-cast, and cryogenic treated and recrystallized SI-300 specimens in the diffraction angle range between 45° and 130°.

cast microstructure of coarse columnar grain microstructure formed by solidification process was observed in the as-cast specimen (Fig. 4 (a)). When cryogenic treatment was applied to the as-cast specimens by liquid nitrogen, formation of the martensite phase was induced by rapid cooling (Dark regions in Fig. 4 (b)). In general, martensitic transformation temperature M_s of typical invar alloy $Fe_{65}Ni_{35}$ is known to be less than -196 °C of liquid nitrogen temperature. Whereas, M_s of typical super invar and kovar alloys are known to be higher than -196 °C. It is therefore difficult to cause martensitic transformation in typical invar alloy by liquid nitrogen. In this study, Ni content was adjusted at 0.1% accuracy rate to increase the martensitic transformation temperature higher than -196 °C. Looked at Fig. 4 (d) showing the magnified micrograph of (b), a lot of lenticular martensites were observed in the austenite matrix. As growth of lenticular martensite phase causes microscopic shear deformation in many regions around itself [17, 18], dislocation density significantly increases in the residual austenite phase. When the cryogenic treated specimen was annealed above recrystallization temperature, microstructure of fine equiaxed grains was formed as shown in Fig. 4 (c). This would be attributed to increase in the nucleation rate of recrystallized grains in the regions with high dislocation density. Average grain size was estimated to be 1660 μm and 101 μm in the as-cast and recrystallized specimens, respectively. This drastic decrease in average grain size would significantly improve the yield strength in thermally treated specimens according to Hall-Petch relationship [6, 7]. Moreover, cryogenic treatment and annealing above austenite transformation temperature and below recrystallization temperature in the grain refined specimens would cause further improvement of mechanical properties by formation of high dislocation density regions mentioned above. The effects of these thermal treatments on the mechanical properties will be described in section 3.3.

Crystal orientation distributions were analyzed by XRD to aid in understanding the microstructural differences between as-cast and thermally treated SI-300 specimens. Figure 5 shows XRD profiles obtained from them. Typical peaks obtained from fcc crystal structure were observed in both specimens. (200) crystallographic orientations of coarse columnar grains were sharply oriented to the casting direction in the as-cast alloy. This is a typical texture caused in cast alloys and

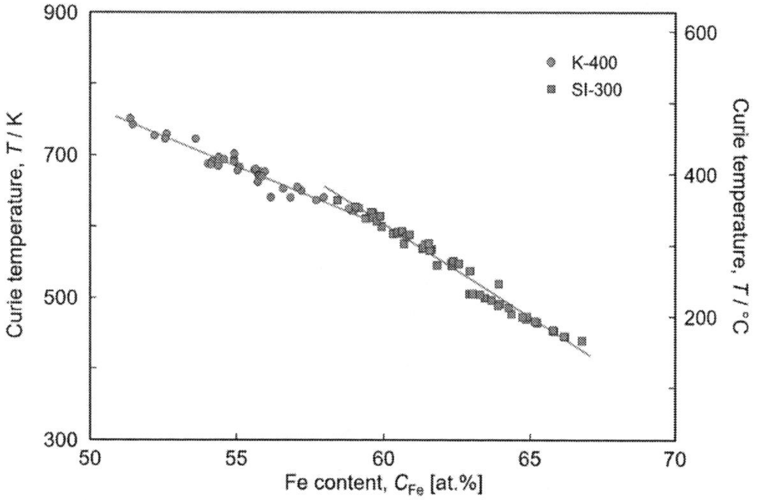

Figure 6 Relationship between Curie temperature and Fe contents in K-400 and SI-300. Curie temperatures were estimated from inflection points of thermal expansion curves.

known to produce large scatter and anisotropy of the mechanical properties [19]. In contrast, typical diffraction patterns of randomly oriented grains microstructure were observed in the thermally treated specimens. It can be concluded that cryogenic treatment and annealing can produce fine equiaxed grain microstructure with random texture in the low thermal expansion cast alloys. In other words, low thermal expansion cast alloys with excellent mechanical properties can be produced by the application of reverse martensitic transformation. These results indicate that novel low thermal expansion cast alloys developed by above thermal treatment can be one of key tooling technologies in CFRP molding for the next generation air mobility.

3.2 Thermal expansion characteristics between room and molding temperatures

Low thermal expansion characteristics of the invar, the super invar and the kovar type alloys have been reported to occur in ferromagnetic fcc Fe based phase [20, 21]. That is, those alloys can be used as low thermal expansion alloys only below Curie temperature. Control of Curie temperature is therefore essential to use the invar type alloys as tooling materials for CFRP molding at desired temperature. Figure 6 shows relationship between Curie temperatures and Fe contents in K-400 and SI-300. They were estimated from inflection points of thermal expansion curves. Linear relationship between Curie temperature and Fe content was found to exist in each alloy. This means that Curie temperature can be empirically predicted from Fe content in super invar type and kovar type alloys. On the basis of Fig. 6, novel low thermal expansion alloys I-200, SI-300 and K-400 were developed as tooling materials for CFRP molding at 200 °C, 300 °C and 400 °C, respectively. These temperatures were determined from the temperatures assumed to be used in future press forming of carbon fiber reinforced thermo plastics (CFRTP) components [22, 23]. Figure 7 shows relative thermal expansion curves to a dimension at room temperature in I-200, SI-300, K-400 and a typical invar alloy $Fe_{65}Ni_{35}$. As the references, assumed tooling temperature ranges in the autoclave molding and press molding are shown in the figure. Low thermal expansion in each alloy was maintained from room temperature to each desired temperature. Table 2 shows the CTE between room temperature and each desired temperature estimated from

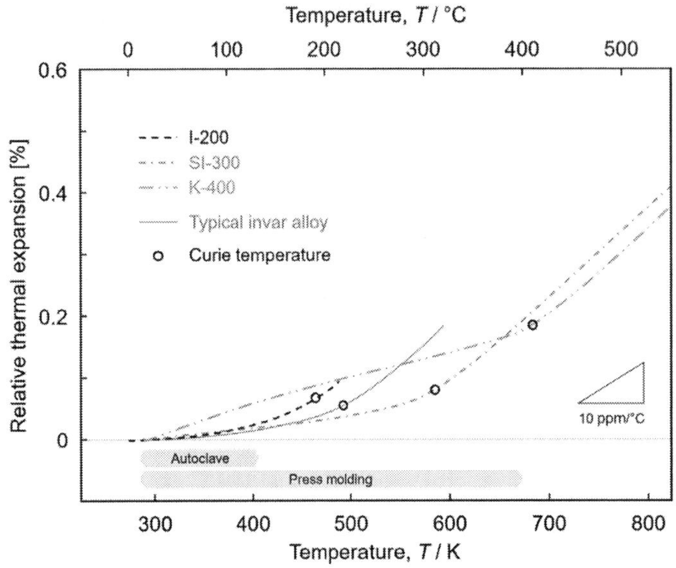

Figure 7 Relative thermal expansion curves to a dimension at 20 °C (room temperature) obtained from I-200, SI-300, K-400 and a typical invar alloy $Fe_{65}Ni_{35}$. For reference, assumed tooling temperature ranges in the autoclave molding and the press molding are also shown.

Table 2 Average CTEs between room and each desired temperatures in I-200, SI-300 and K-400. For comparison, those in typical invar, super invar and kovar alloys are also shown.

Material	Chemical composition	Average CTE between room and desired temperatures [10^{-6} / °C]		
		R.T. ~ 200 °C	R.T. ~ 300 °C	R.T. ~ 400 °C
I-200	$\mathbf{Fe_{65.1+x}Ni_{34.9-x}}$	**4.0**	-	-
Invar	$Fe_{65}Ni_{35}$	2.4	5.5	-
SI-300	$\mathbf{Fe_{64.2+x+y}Ni_{31.0-x}Co_{4.8-y}}$	**2.1**	**2.9**	**5.3**
Super invar	$Fe_{64}Ni_{31}Co_5$	1.7	4.1	6.9
K-400	$\mathbf{Fe_{55.3+x+y}Ni_{28.2-x}Co_{16.5-y}}$	**5.0**	**4.7**	**4.4**
Kovar	$Fe_{55}Ni_{28}Co_{17}$	5.2	4.9	5.4

the thermal expansion curve in each alloy with higher M_s than -196 °C. For comparison, those in typical invar, super invar and kovar alloys are also shown in the table. The average CTE between room temperature and 200 °C in I-200 with higher M_s than -196 °C was found to be apparently larger than that in a typical invar alloy. This was attributed that Ni content was decreased from the optimum composition for low thermal expansion in order to increase M_s. In contrast, all average CTEs in SI-300 and K-400 were able to decrease by control of chemical composition in comparison with typical super invar and kovar alloys.

The effects of cryogenic treatment and annealing on thermal expansion characteristics were examined by comparison of the as-cast and the thermally treated specimens. Figure 8 shows

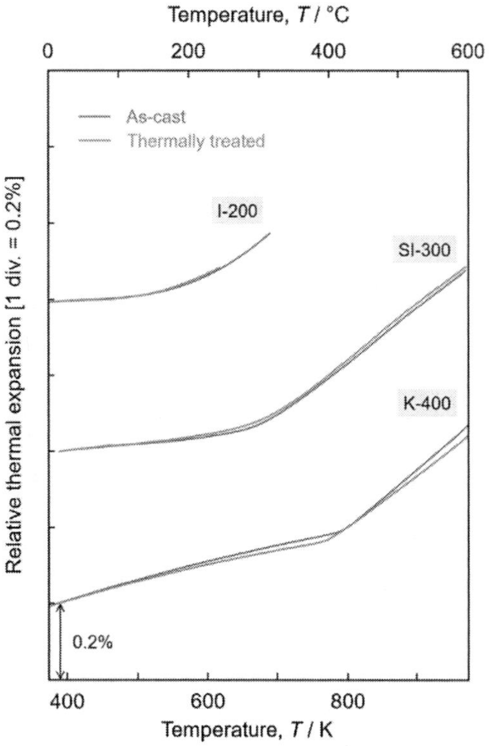

Figure 8 Comparison of relative thermal expansion curves to the dimensions at 20 °C (room temperature) between the as-cast and the thermally treated I-200, SI-300 and K-400 specimens.

relative thermal expansion curves obtained from the as-cast and the thermally treated I-200, SI-300 and K-400 specimens to the dimensions at room temperature. According to a previous study [24], a microstructural change by heat treatment was reported to affect its thermal expansion characteristics. However, there were no significant differences of the curves between the as-cast and the thermally treated specimens in every type of alloy. This indicates that thermal treatments in this study is as effective technique to obtain the fine equiaxed grains microstructure with high dislocation density without deterioration of low thermal expansion characteristics.

3.3 Improvement of mechanical properties and future prospects

Tensile tests at 200 °C, 300 °C and 400 °C were conducted in the as-cast and the thermally treated I-200, SI-300 and K-400 specimens, respectively, in order to clarify the effects of reverse martensitic transformation on the mechanical properties. Figure 9 shows stress-strain curves obtained from their specimens. Maximum stress (tensile strength) was found to increase by thermal treatment in every specimen. As a result, 0.2% yield strength in every specimen also increased by thermal treatment as shown in Fig. 9 (b). These results would be attributed to grain refinement and increase in dislocation density. That is, high density of grain boundaries and dislocations increased the yield strength by acting as barriers of dislocation motion that is fundamental process of plastic deformation. The retardation of dislocation motion would also increase Young's modulus of the specimens. 0.2% yield strength and Young's modulus were therefore quantitatively estimated from the stress-strain curves. Figure 10 shows those values of

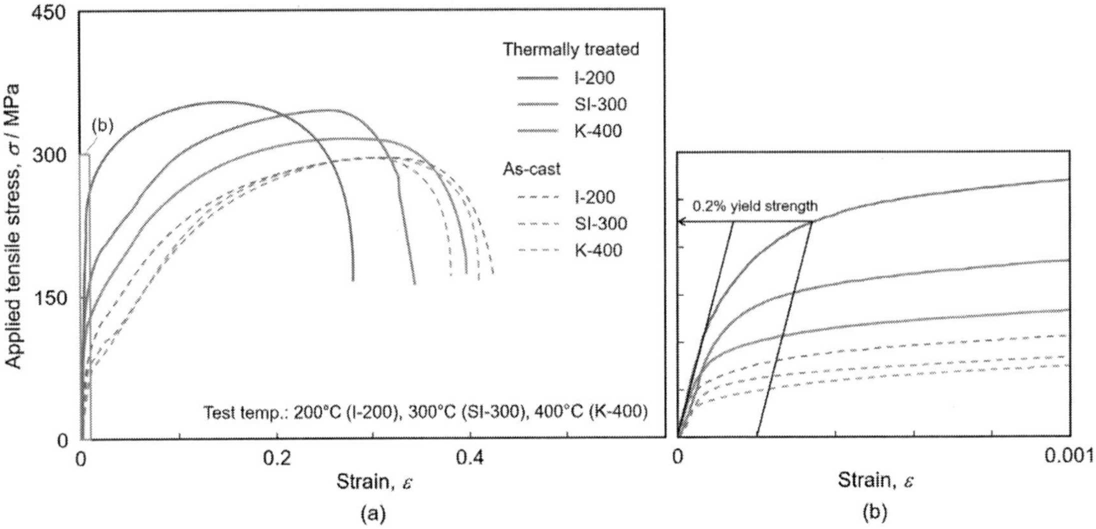

Figure 9 (a) Stress-strain curves obtained from the as-cast and the thermally treated I-200, SI-300 and K-400 specimens at 200 °C, 300 °C and 400 °C, respectively. (b) A magnified view of the region in low strain range indicated by orange square in (a).

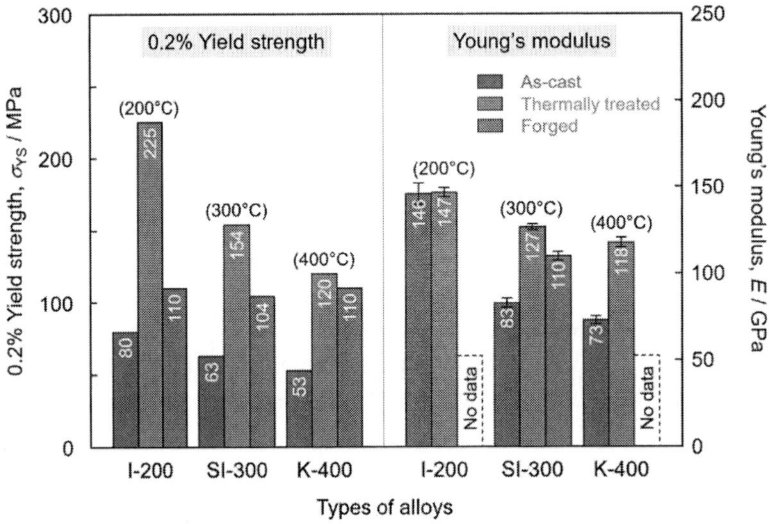

Figure 10 0.2% yield strength and Young's modulus of the as-cast and the thermally treated I-200, SI-300 and K-400 specimens estimated from the stress-strain curves shown in Fig. 9. For comparison, those of the forged specimens were also shown.

the as-cast and the thermally treated I-200, SI-300 and K-400 specimens. For comparison, those of the forged specimens were also shown by light blue bar. The data of the forged I-200 and K-400 refers to literature values [25] of typical invar and kovar alloys, respectively. That of the forged SI-300 were measured values under same conditions as the other specimens in this study. 0.2% yield strength of every specimen was improved more than 2.2 times by thermal treatment.

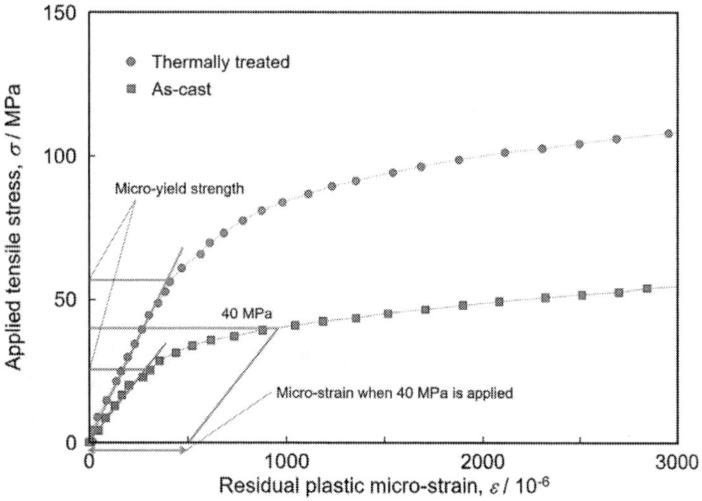

Figure 11 Stress-strain curves in low strain range obtained from the as-cast and the thermally treated K-400 specimens at 400 °C. Micro-yield strength estimated from the threshold stress at which micro-strain starts to occur are shown. Micro-strain that occurs in the as-cast specimen when 40 MPa stress is applied is also shown as an example.

It is remarkable that those of I-200 and SI-300 specimens were more than twice and 1.5 times higher than those of the forged specimens, respectively. As the difference of average grain size between the thermally treated and the forged specimens was considered to be small, improvement of the yield strength would be mainly attributed to increase in the dislocation density. Thermal treatment also improved Young's moduli of SI-300 and K-400 specimens. Although Young's modulus of the I-200 specimen was not improved, the anisotropy and scatter of Young's modulus would be drastically decreased by fine equiaxed grain microstructure. Coarse columnar grain microstructure with sharp (200) texture in the as-cast I-200 specimen shown by Figs. 4 and 5 is well-known to shows large anisotropy of Young's modulus [19, 26]. The anisotropy can deteriorate the shape accuracy at an intricate structured portion of a CFRP component. As a result, the quality and reliability of next generation air mobility are degraded, which may make it difficult to obtain acceptance in society.

The technical advantages of the thermally treated specimens as tooling materials for small CFRP components with intricate structure can be explained from the viewpoint of plastic micro-strain that occurs in the tooling materials during molding. Figure 11 shows the stress-strain curves of the as-cast and the thermally treated K-400 specimens at 400 °C in low strain range less than 3.0 $\times$ 10^{-3}. In general, plastic deformation of metals occurs when the applied stress is more than yield strength of 0.2% offset. However, micro-strain should be caused even by the applied stress less than 0.2% yield strength as shown in Fig. 11. The threshold stress at which micro-strain occurs (Micro-yield strength) were approximately 25 MPa and 63 MPa in the as-cast specimen and the thermally treated specimen, respectively. The local applied stress can reach several tens MPa by stress concentration at high curvature portion in press molding of a small CFRP component. If the local applied stress is assumed to be 40 MPa, micro-strain of 5.0 $\times$ 10^{-4} occurs in the as-cast specimen. In contrast, no micro-strain occurs in the thermally treated specimen. Accumulation of micro-strain of the tooling materials during cyclic CFRP molding must deteriorate shape accuracy of resultant components. In addition, reduction of tooling lifetime by accumulation of micro-strain will be also key concern in mass production. These results suggest that the developed low thermal expansion cast alloys by the application of reverse martensitic transformation can be a powerful tool for precise molding of small CFRP components with intricate structure.

4. CONCLUSIONS

In order to develop novel tooling materials for precise molding of small CFRP components with intricate structure, characterizations of microstructure, thermal expansion and mechanical behavior were conducted in the low thermal expansion cast alloys. The effects of reverse martensitic transformation on thermal expansion characteristics and mechanical properties were clarified. Consequently, this technique was found to be possibly one of key technologies for manufacturing high quality and reliability next generation air mobility components. The most significant results obtained from these experiments are as follows.

1. Cryogenic treatment and annealing can cause grain refinement and form high dislocation density region through reverse martensitic transformation in cast low thermal expansion alloys.

2. Low thermal expansion characteristics are maintained even after microstructural control by cryogenic treatment and annealing.

3. 0.2% yield strength of the low thermal expansion cast alloys at desired CFRP molding temperature can increase more than 2.2 times by the application of reverse martensitic transformation. Anisotropy and scatter of Young's modulus in the thermally treated ones can also be improved.

5. REFERENCES

[1] C.E. Guillaume, "Transformations magnétiques et dilatations", *CR Hebd. Seances Acad. Sci.*, **125** (1897), 235-238. (in French)

[2] Guillaume, C.E., "Invar and its applications", *Nature*, **71** (1904), 134-139.
DOI: https://doi.org/10.1038/071134a0

[3] H. Scott, "Recent developments in metals sealing into glass", *J. Frankl. Inst.*, **220** (1935), 733-753.
DOI: https://doi.org/10.1016/S0016-0032(35)91141-3

[4] C. Giolli, M. Turbil, G. Rizzi, M. Rosso and A. Scrivani, "Wear resistance improvement of small dimension invar massive molds for CFRP components", *J. Therm. Spray Technol.*, **18** (2009), 652-664.
DOI: https://doi.org/10.1007/s11666-009-9397-z

[5] Sean Henson, "Cure Tooling Considerations and Materials Selection Guide", *CAMX 2019, Anaheim, CA*, Presentation file: ES19-0181, (2019).

[6] E.O. Hall, "The Deformation and Ageing of Mild Steel: III Discussion of Results", *Proc. Phys. Soc. Lond.*, **64** (1951), 747-753.

[7] N.J. Petch, "The Cleavage Strength of Polycrystals", *J. Iron Steel Inst.*, **174** (1953), 25-28.

[8] G. Saul, J.A. Roberson and A.M. Adair, "The effects of thermal treatment on the austenitic grain size and mechanical properties of 18 pct Ni maraging steels ", *Met. Trans.*, **1** (1970), 383-387.
DOI: https://doi.org/10.1007/BF02811546

[9] T. Maki, H. Morimoto and I. Tamura, "Recrystallization of reversed austenite and subsequent martensitic transformation in 18%Ni maraging steel", *Trans. Iron Steel Inst. Jap.*, **20** (1980), 700-706.
DOI: https://doi.org/10.2355/isijinternational1966.20.700

[10] K. Tomimura, S. Takaki, S. Tanimoto and Y. Tokunaga, "Optimal chemical composition in Fe-Cr-Ni alloys for ultra grain refining by reversion from deformation induced martensite", *ISIJ Intern.*, **31** (1991), 721-727.
DOI: https://doi.org/10.2355/isijinternational.31.721

[11] A. Di Schino and J.M. Kenny, "Grain refinement strengthening of a micro-crystalline high nitrogen austenitic stainless steel", *Mater. Lett.*, **57** (2003), 1830-1834.
DOI: https://doi.org/10.1016/S0167-577X(02)01076-5

[12] V.D. Sadovskii, G.N. Bogacheva and V.M. Umova, "Recrystallization of maraging alloys", *Metal Sci. Heat Treat.*, **11** (1969), 258-259.
DOI: https://doi.org/10.1007/BF00653187

[13] E. Sato, G. Itoh, Y. Takayama, K. Kitazono, K. Morita, T. Itoi and J. Kobayashi, "Formation of Ultrafine-Grained Structures in 304L and 316L Stainless Steels by Recrystallization and Reverse Phase Transformation", *Mater. Sci. Forum*, **838-839** (2016), 410-415.
DOI: https://doi.org/10.4028/www.scientific.net/MSF.838-839.410

[14] J.E. Bailey and P.B. Hirsch, "The dislocation distribution, flow stress, and stored energy in cold-worked polycrystalline silver", *Philos. Mag.*, **5** (1960), 485.
DOI: https://doi.org/10.1080/14786436008238300

[15] H. Masumoto, "On the thermal expansion of the alloys of iron, nickel and cobalt and the cause of the small expansibility of alloys of the invar type", *Sci. Rep. Tohoku Imp. Univ.*, **20** (1931), 101.

[16] Z. Rao, D. Ponge, F. Körmann, Y. Ikeda, O. Schneeweiss, M. Friák, J. Neugebauer, D. Raabe and Z. Li, "Invar effects in FeNiCo medium entropy alloys: From an Invar treasure map to alloy design", *Intermetallics*, **111** (2019), 106520.
DOI: https://doi.org/10.1016/j.intermet.2019.106520

[17] A. Shibata, S. Morito, T. Furuhara and T. Maki, "Local orientation change inside lenticular martensite plate in Fe-33Ni alloy", *Scripta Mater.*, **53** (2005), 597-602.
DOI: https://doi.org/10.1016/j.scriptamat.2005.04.023

[18] A. Shibata, S. Morito, T. Furuhara and T. Maki, "Substructures of lenticular martensites with different martensite start temperatures in ferrous alloys", *Acta Mater.*, **57** (2009), 483-492.
DOI: https://doi.org/10.1016/j.actamat.2008.09.030

[19] N. Tsuji, H. Takebayashi, T. Takiguchi, K. Tsuzaki and T. Maki, "Effect of initial orientation on the cold-rolling behavior of solidified columnar crystals in an Fe-36%Ni austenitic alloy", *Acta Metall. Mater.*, **43** (1995), 743-754.
DOI: https://doi.org/10.1016/0956-7151(94)00264-I

[20] Weiss, R.J., "The origin of the 'Invar' effect", *Proc. Phys. Soc.*, **82** (1963), 281-288.
DOI: https://doi.org/10.1088/0370-1328/82/2/314

[21] Schilfgaarde, M., Abrikosov, I.A. and Johansson, B., "Origin of the Invar effect in iron-nickel alloys", *Nature*, **400** (1999), 46-49.
DOI: https://doi.org/10.1038/21848

[22] M.D. Wakeman, C.D. Rudd, T.A. Cain, R. Brooks and A.C. Long, "Compression moulding of glass and polypropylene composites for optimized macro- and micro-mechanical properties, 4: Technology demonstrator – a door cassette structure", *Compos. Sci. Tech.*, **60** (2000), 1901-1918.
DOI: https://doi.org/10.1016/S0266-3538(00)00066-X

[23] A. Levy and P. Hubert, "Interstrand void content evolution in compression moulding of randomly oriented strands (ROS) of thermoplastic composites", *Compos. Part A: Appl. Sci. Manuf.*, **70** (2015), 121-131.
DOI: https://doi.org/10.1016/j.compositesa.2014.11.017

[24] M. Hatate, T. Shiota, H. Sumitomo and K. Nakamura, "Effects of heat treatment on the thermal expansion characteristics and mechanical properties in low expansion austenitic cast iron", *Intern. J. Cast Metals Res.*, **11** (1999), 545-549.
DOI: https://doi.org/10.1080/13640461.1999.11819331

[25] Special Metals webpage, "High performance alloys literature"
URL: https://www.specialmetals.com/tech-center/alloys.html

[26] A. Kagawa, T. Okamoto and H. Masumoto, "Young's modulus and thermal expansion of pure iron-cementite alloy casting", *Acta Metall.*, **35** (1987), 797-803.
DOI: https://doi.org/10.1016/0001-6160(87)90157-X

HIGH CONDUCTIVE MULTI FILAMENT FIBER MATERIAL MADE OF INORGANIC FIBERS COATED WITH ALUMINIUM

Alexander Lüking[1], Richard Haas[1], Robert Brüll[1], Francesco Biscaglia[2], Danilo Marini[2], Francesco Marra[2, 3], Marco Valente[2], Thomas Gries[1]

[1]RWTH Aachen University, Aachen, North Rhine-Westphalia/Germany

[2]Department of Chemical Engineering, Material and Environment, University of Sapienza, Rome, Italy

[3]National Interuniversity Consortium of Materials Science and Technology (INSTM), Florence, Italy

ABSTRACT

Conductive textiles are used in applications such as battery shielding in electric vehicles, EM-shielding from 5G emission or the mounting of a solar panel. Electrical textiles can be used to build a Faraday cage to protect high sensitive electronic devices or behave as a wire or sensor. However, conductive textiles are a bottle neck in achieving the goals of the electricmobility. [9] Electric vehicles are to be produced in large quantities and offered at affordable prices. However, the current state of research does not allow such a low-cost production. Currently used metal fibers have a diameter of 2 – 40 µm. The fibers are usually produced by the drawing process, the Taylor process, by melt spinning or extraction. Due to the slow and manual production processes, metal fibers with a diameter of 50 µm cost about 100 €/kg. To realize frequency shielding from 3 – 300 GHz new shielding materials are required. The decisive factor here is that simple EMI shielding by reflection is not sufficient, but for a shielding effect (SE) above 50 dB, the absorption of EM radiation supported by multiple internal reflections is of primary importance. [3, 9]

The aim of the presented work is to develop a low-cost conductive fiber material for electromagnetic shielding. The approach of this technology is the metallic coating of inorganic fibers in the spinning process. The conductivity is achieved by a thin aluminum sheath of only 2 - 4 µm. A flexible inorganic fiber is used as the core, which enables the formation of fabrics like wovens or nonwovens. These fabrics can be draped into complex geometries. The multifilament yarn production is running at a speed of up to 1000 m/min to achieve high productivity. The aluminum coating is happening in the spinning process and covers each single filament. The resulting yarns electrical resistance is 100 Ωm and its temperature resistants is up to 500 °C. Fiber and fabric design can be adapted to the requirement of different applications. [5]

Keywords: Conductive fiber, bi-component fiber, Aluminum, Aluminum coating process, Faraday cage, Inorganic filaments, glass fiber Reinforcement fibers, Matrix, Coating module, electrical resistance

INTRODUCTION

First 2G to 3G. Then the change from 3G to 4G. Each new generation of mobile phones conjured up previously unimagined possibilities on the displays of our mobile devices with a tremendous surge in speed. LTE was introduced as the fourth generation. LTE was planned as a quality name for high performance internet and it worked. Any connection speed under LTE is perceived by people as a sense of loss - lower speed leads to impatience and an emotional and technological disadvantage. [6]

Switching from LTE to 5G, a new level of digitization will be achieved in the next years. High speed streaming to generate virtual realities in high definition. Reality and virtual reality start blurring. These new features are nice to have but the major impact will follow from the guarantee of wide bandwidth and short latency. Application with real time data traffic requirements can start operating. For example, the possibilities for self-driving cars or tele-surgeries gets unlocked because by overcoming the delay of data traffic. These technologies fulfil safety requirements and will change our life. [4,6]

The significantly improved bandwidths and transmission speeds are based on frequency ranges of 3 - 300 GHz. However, this also results in new and increased requirements for the shielding of components in the corresponding electronic components to prevent electromagnetic interference (EMI). As a result of electromagnetic radiation, not only is the performance of the respective devices is reduced, but the radiation is also harmful to our health, depending on its intensity. At 5G and a frequency of 3.5 GHz, about 3.5 billion electromagnetic waves/vibrations are hitting and radiating through the human body per second. The body absorbs a large part of this energy. Thermal and athermal effects on the human body are not fully researched. The effect on high sensitive technical devices like surgery equipment, measurement systems or batteries was proven by several studies. EMI protective barriers are used to protect against unwanted emissions. These shielding materials have to be affordable to get access to every days applications like electrical cars or electronic consumer devices.

State of the art and deficit

The EMI shielding materials currently available on the market are mainly focused on the X-band (4G). New shielding materials are required for the broadband and high frequency range of 3 - 300 GHz. The decisive factor here is that simple EMI shielding by reflection is not sufficient. For a shielding effect (SE) > 50 dB, the absorption of EM radiation supported by multiple internal reflections is of primary importance. Therefore, pure metal sheathings are not the material of choice. Affordable, thin, light and drapeable films or textiles which have a low susceptibility to corrosion and also allow heat to be dissipated by thermal convection are needed. One research topic in that field are composites with conductive fillers made of carbon nanotubes and graphene. Apart from the fact that carbon composites are still a technical challenge, it is difficult for them to achieve SE > 40 dB even with high filler contents (up to 60 %). In addition, the high percolation

threshold also makes these products difficult to process and reduces their mechanical strength. These material are still under development and have not penetrated the market yet. The currently price estimation is round about 100 €/kg. [3,4]
Textiles made of metal fibers of a diameter of 2 - 40 µm are highly conductive and suitable for 5G shielding purpose. The fibers are usually produced by the drawing process, the Taylor process, by melt spinning or extraction. Due to the slow and manual production processes, metal fibers with a diameter of 50 µm are commercial available for 100 €/kg. aluminum filaments with the same diameter cost > 500 €/kg. Conductive polymer fibers using fillers as well or carbon fibers are although used for electromagnetic shielding applications. Carbon fibers their specific electrical resistance of 1.6 - 10.5 Ωm does not met SE > 20 dB. Beside conductivity of round about 100 Ωm, a fiber diameter suitable for the frequency range typically between 15 µm – 25 µm and a bending stiffness for fabric production similar to uncoated glass fibers is necessary. [6]

EXPERIMENTATION

In the following chapter the material choice is described followed by an explanation of the testing procedure.

Selection of a coating metal and a reinforcing core fiber material

A bi-component fiber made of an inorganic core coated with a metal shell is developed by this work. The metal used is a high purity aluminum. The reason for choosing such an alloy is due to high electrical conductivity that is suitable for EMF shielding. Another reason is because of no magnesium content in the alloy, as it leads to major oxide formation which hinders the coating process. However, the process will be transferred to other metals like copper in the future. The metal melting temperature has to be less than 800 °C in order to avoid a thermal damage of the inorganic fibers in additional the process has to be feasible and economically viable.

This coating module was realized for the high purity aluminum alloy with following fluid parameters as listed in the table 1 below.

Properties	Value
Density (kg/m^3)	2700 (solid),2375(liquid)
Melt Viscosity (Pa.s)	0.00126
Surface Tension (N/m)	0.85
Melting Temperature (° C)	660
Specific Heat (J/Kg.K)	971
Latent heat of Fusion (KJ/Kg)	396
Elastic Modulus (GPa)	70-80

Tensile Strength (MPa)	110
Yield Strength (MPa)	105
Elongation (%)	12
Shear Strength (MPa)	69

Table 1: - Properties of aluminum alloy

As inorganic fiber material the choice is glass fiber. Worldwide, approximately six million tons of glass fibers are produced each year. The reason for that is the good ratio between performance and price. The low price of 1 €/kg for glass fiber roving's is about 1/10 th of the purchase price of carbon fiber roving's. Glass fibers are used as reinforcing fibers in more than 90 % of the fiber reinforced plastics. [SG16] Depending on the chemical composition, the modulus of elasticity is 72 to 77 GPa and the tensile strength 3400 to 3700 MPa [4,5] The mechanical properties of the fiber component are almost exclusively in the longitudinal direction of the fiber.

Glass fibers are produced in a spinning process, glass is melted in a platinum-rhodium bushing. The bottom of the bushing is equipped with up to 8000 tips with a diameter of 1 to 2 mm. At the beginning of production, the hydrostatic pressure of the glass melt forms a drop under the tip. The glass filament follows the gob and forms a concave melt cone at the tip outlet. The winder pulls on the filament, resulting in continuous fibers with a nominal diameter of 5 to 25 μm. The glass filaments cool down in the ambient air or by using cooling fins and water mist. After cooling, the filaments are coated with a sizing. Afterwards the filaments are bundled and wound on a winder with up to 4.000 m/min. The application of the sizing serves to increase the cohesion between the glass fibers and to improve further processing. In addition, brittle fractures caused by friction between the fibers are avoided. [4]

Sisercam® pellets were used to obtain the glass filaments. The fiber properties are related to the glass composition. In the following table 2 the oxides making up the glass is presented. The glass is called E-Glass and is chosen because of the low price. Chemical resistance (C-Glass) or extended mechanical properties (S-Glass) are not required for this application.

Oxides	% weight
SiO_2	55
Al_2O_3	15.5
B_2O_3	8
CaO	17
MgO	3.5

K_2O, Na2O	<1

Table 2: Composition of glass used in the experimental study [4].

Spinning plant

The first step of the experiment is the production of glass at the Institut für Textiltechnik of RWTH Aachen University. A single filament spinning plant of the ELSKLO LVTS1 type is used. The plant is set up on two floors, on the upper floor, it is equipped with an electrical heating system. The bushing made of 90 % Platin and 10 % Rhodium where the raw glass marbles are fed into the bushing. The tip on the bottom of the bushing has a diameter of 2 mm. The distance between the bushing and the winder is about 2 m. That distance ensures cooling of the aluminum on the filament before reaching the winding. Figure 1 shows the spinning plant and a used single filament bushing.

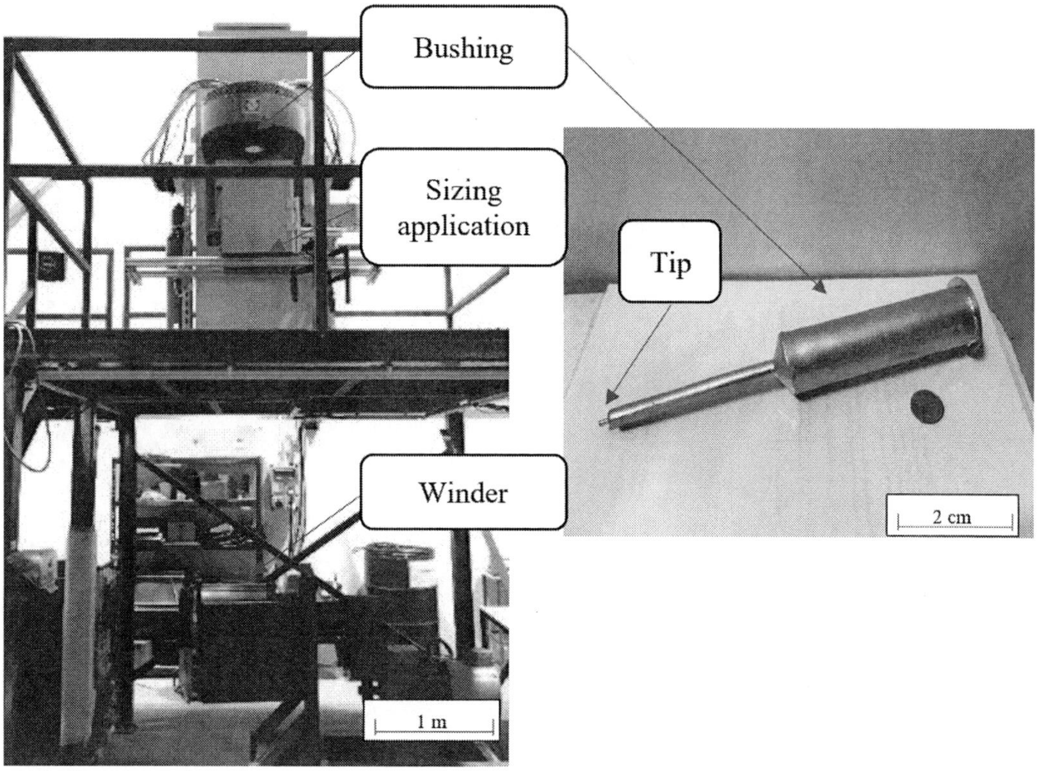

Figure 1: Spinning plant (left) and the Pt/Rh bushing (90/10) (right) used in the experimental study.

Experimental Setup of the aluminum coating module

The single filament spinning plant is extended by an aluminum coating unit. That unit consists of an aluminum bath placed in an oven, with a roller partially immersed to transfer the aluminum melt from bath to the fiber. The schematic sketch is shown in figure 2 below. Solid aluminum blocks are heated through radiation. Once the melting point of 660 °C is reached, the aluminum is ready for coating. aluminum bath is heated to a temperature of 750 °C to ensure a liquid coating material. A partially immersed roller on top of the bath is now rotated, by the wetting between the roller and the melt, the melt is transported to the fiber.

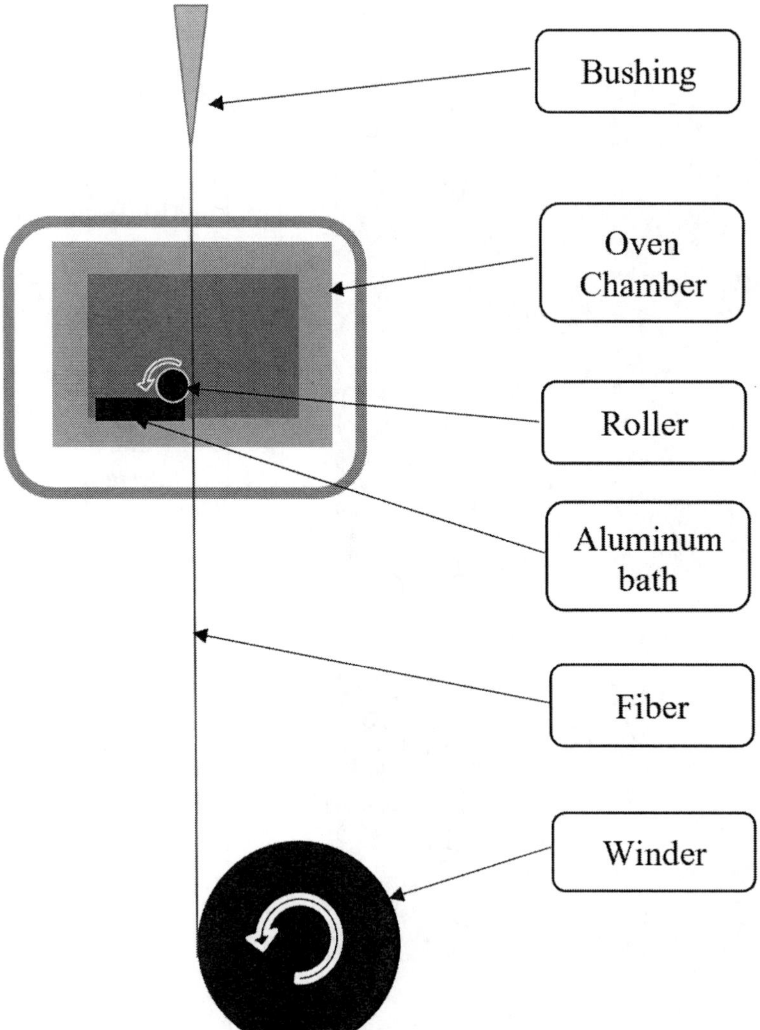

Fig:2 Schematic drawing of the coating process

Modification of a high temperature oven

Certain modifications on a high temperature oven are done to realize the aluminum coating process. Figure 3 and 4 show the equipment used. A hole is drilled on the top and the bottom of the oven, from where the fibers enters and exits inside and outside the oven. Inside the oven chamber fiber gets coated by the partially immersed roller. A rotary unit is installed besides the oven to induce variable rotation speeds to the roller. That roller gets wetted by the aluminum in the bath and coats the passing fiber. The roller speed was set to 30 m/min because the aluminum film of the roller is homogeneous at that speed, although the aluminum is not splashing off because centrifugal force and the required amount of aluminum is available to achieve fiber volume fraction of 50 %. A temperature control unit is added to control the temperature inside the oven.

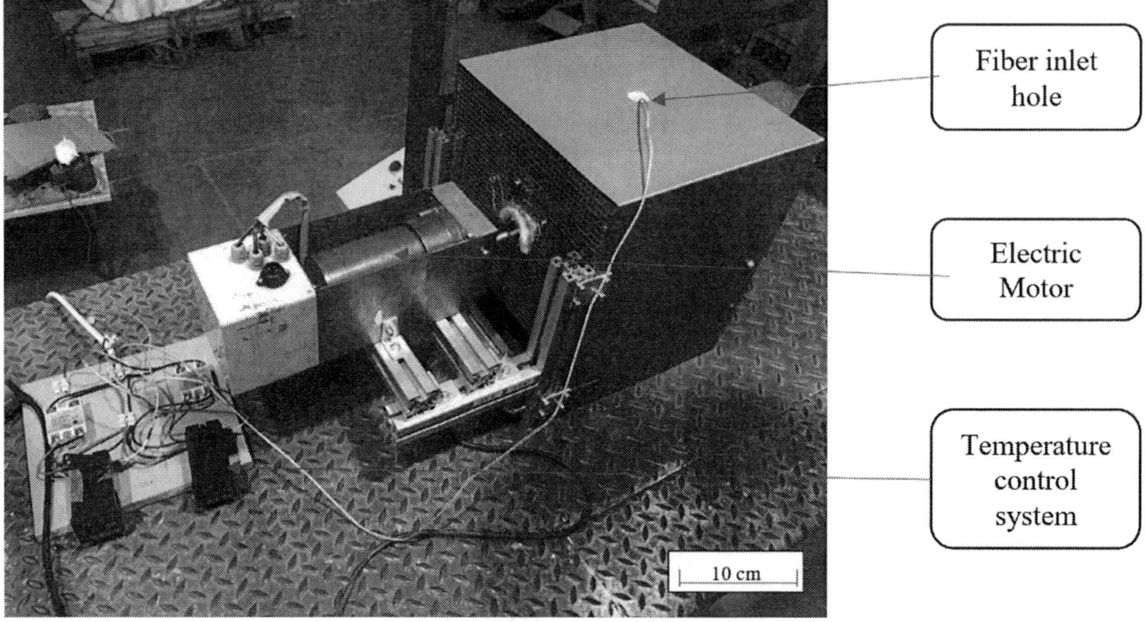

Fig: - Modified high temperature oven setup for coating process

Fig: - Inside view of the coating unit

Mechanism of Coating

Molten aluminum is characterized by high surface tension of 880 mN/m and low flow viscosity of $3 * 10^{-3}$ Pas at the coating temperature of 750 °C. The roller coating module has been realized for these two properties. The main goal is to achieve inertial forces of fiber higher than the surface tension forces of aluminum melt. Since the velocity of the fiber is very high in the order of 1000 m/min, the coating happens in the inertial regime. Inertial forces of fiber cause a liquid film entrainment at the fiber surface which leads to a uniform coating thickness in the range of few microns. The film coating thickness is a factor of fiber speed. However this stands true only if the arbitrary velocity is higher than the critical velocity ($v = 464$ m/min), where the inertial forces overcome the surface tension forces. To facilitate easy transfer of molten aluminum to the fiber, the roller is coated with a ceramic which reduces the surface energy on the roller. The vertical distance between the coating module and the winder are calculated in such a way that, the fibers are completely cooled before being wound, otherwise there is a welding effect between the fibers. Also, the tangential velocity of the roller determines the volume of melt that is picked up for coating. The trials were conducted at a fiber speed of 1000 m/min to achieve a core fiber diameter of 12 μm, to overcome the critical velocity and to achieve high productivity. The coated fiber material on the winder is presented in figure 5.

Fig 5: Aluminum coated glass fiber on the winder.

The performance of the coated fibers is presented in the following chapter.

RESULTS

In this chapter the new fiber material is analyzed. The aluminum coated fibers are developed to create a new generation of shielding material e.g. in 5G applications. In these applications it is required to have a full coating of each single filament to achieve the required multiple internal reflections. In additional a low electrical resistance is required to build a Faraday cage. A thick layer of aluminum oxide on the filament surface has to be voided to achieve high conductivity. Microscope images of the fiber cross section are conducted to confirm the full coating of the individual filaments and to check for an aluminum oxide layer on the filaments surface. Although the amount of aluminum on the filaments can be calculated by doing measurements of the cross sections pictures. In additional electrical conductivity tests are performed on filament bundles to compare the results with the theoretical conductivity calculated by the amount of aluminum that was detected in the cross section. Mechanical performance of the yarn is not presented in that paper. The performance is majorly driven by the core fiber material and meets the requirements for the created textiles.

Trials were conducted for a fiber speed of 1.000 m/min with the above mentioned roller coating setup. The expectations were to achieve complete coating of aluminum on the fibers with low electrical resistance along the length. Since molten aluminum being a high surface tension non-wetting fluid, it is challenging to coat on a low surface energy glass surface. The key here is to solidify aluminum before it forms a droplet on the surface of glass. Low coating thickness in the order of microns helps us achieve this goal.

Microscopy – cross section analysis

A bundle of coated fibers is stacked together, embedded into Epoxy resin, cut and polished to analyze the cross section of the filaments and tested in an optical microscope. Figure 6 shows nearly complete coating of aluminum on the fibers. The dark grey circles represent the glass fiber with a 14 µm diameter in average. The white circles around represent the aluminum coat. Figure 7 shows a top view along the coated filament. A homogeneous film along the fiber can be observed. Even though theoretically there is little or no wetting of aluminum melt on glass, due to their difference in surface energies, there is a good interface bonding. This is due to the fact of coating in inertial regime, which means that coating happens irrespective of surface characteristics. Another factor is, that the solidification time is so low that aluminum melt will not have enough time to constrict and form a droplet on the surface of the fiber. The average volume fraction of aluminum is 40 %. There are several factors that can lead to the incomplete coating of the fibers, one being the oxides on the melt arresting the melt flow on to the fibers. Second factor might be due to its low melt viscosity, the wind blowing onto fibers in one direction can cause the melt to deposit on another side. Third being the fact that aluminum is not completely solidified when it reached the winder, leading it to build up coating on one side. However, this is less likely, as no welding effect can be documented.

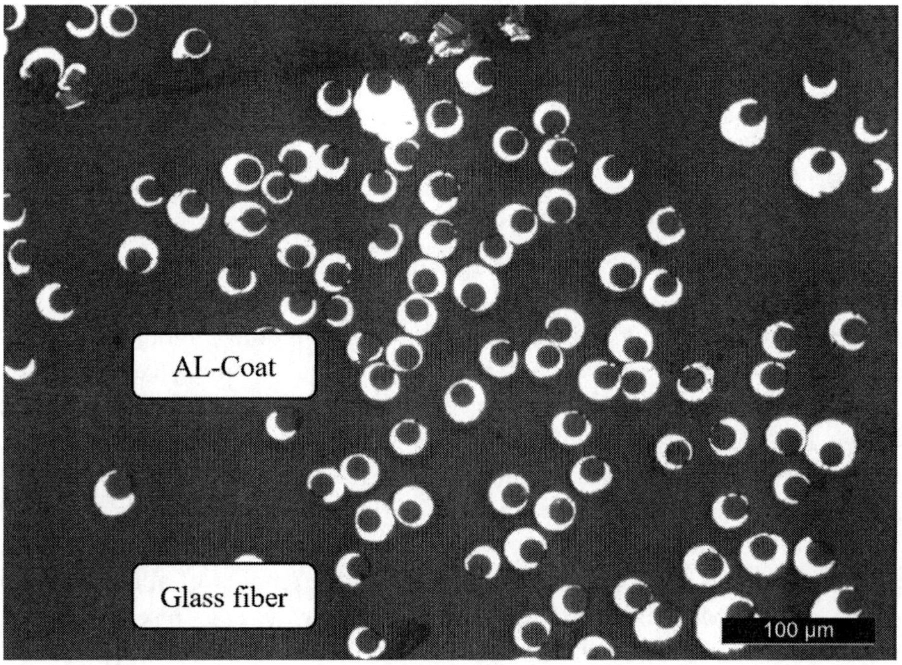

Fig 6: - Cross Section of aluminum coated glass fiber roving

Fig 7: - Top view on of aluminum coated glass fiber roving

Electrical Resistance test

The resistance in the aluminum coated fiber is directly proportional to sample length and inversely proportional to the cross sectional area of the aluminum coating what is known by physical laws [8]. A pure aluminum rod serves as benchmark material because it represents an oxide free wire. The aim is to achieve conductivity as similar as possible to a pure aluminum rod. The theoretical resistance of a rod with a cross section area of 0.23 mm^2 is 0.00175 Ω. Such an area is corresponding to the test sample area test in this work.

Fiber bundles of a 150 mm sample length, were stacked together for determining the electrical resistance along the length. A digital Multimeter is used to determine the resistances in the fibers like shown in figure 8.

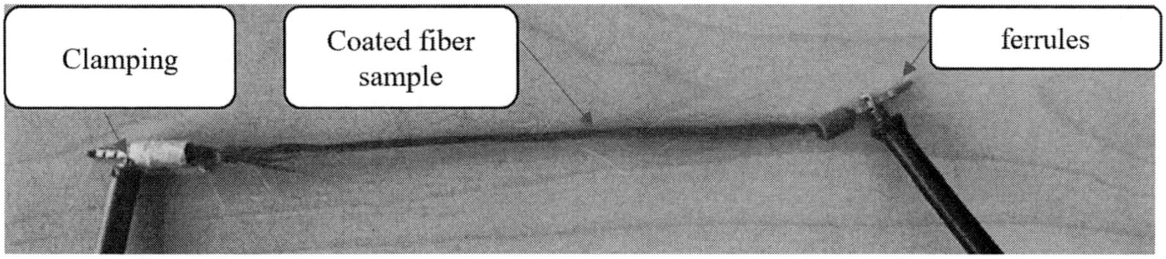

Fig 8: - Measurement of the electrical resistance

The first measurements results are close to the theoretical calculated values but the number of tested samples was not enough to present statistically robust results today.

CONCLUSIONS

High conductive textiles are one possible solution to create a drapeable EMF shield. In this paper a new approach is presented to produce a high conductive fiber material in a highly productive process. The new fiber is a bi-component fiber made of a glass fiber core coated with aluminum on a single filament level. That kiss-roller process is implemented into the glass fiber spinning process and runs at a velocity of 1.000 m/min what creates the high productivity and the low price of the new material. The two main quality criteria to achieve EMF shielding properties are having a closed coating of molten aluminum on fibers to achieve the required multiple internal reflections and low electrical resistance along the length near to pure aluminum wires. The microscopy results show that there is a completely closed coat on the majority of the fibers, expect a few, where there were some discontinuities in the temperature profile in the process. The electrical resistance along the fibers is similar to pure aluminum wire, which ensures maximum electrical conductivity which is vital for EMF Shielding. The new affordable high conductive fiber is the first step to create a new shielding technology. In the following two steps are required to introduce the new material into market. First of all, the presented lab scale coating module has to be scaled to industrial level. The number of simultaneous coated filaments have to be increased and the coating module has to be implemented into industrial scale glass fiber spinning lines. In additional the new fiber material has to be processed into a fabric. Woven and nonwoven fabric production is currently under development.

REFERENCES

[1] F. T. Wallenberger, J. B. MacChesney, R. Naslain, H. D. Ackler, ''Advanced inorganic fibers: processes-structure-properties-applications'' Springer, cap. 4, 2000

[2] D. K. Koli, G. Agnihotri and R. Purohit, ''Advanced aluminium matrix composites: the critical need of automotive and aerospace engineering fields'' Materials Today: Proceedings 2, pp. 3032 – 3041, 2015

[3] M. Haghshenas ''Metal–Matrix Composites'', Elsevier, 2016

[4] Y. Yue, Q. Zheng, ''Fiber spinnability of glass melts'' International Journal of Applied Glass Science, 2017

[5] L. Yang, J. L. Thomason, "The thermal behaviour of glass fibre investigated by thermomechanical analysis", vol. 48, pp. 5768–5775, 2013

[6] K. Neumann, F. Queder, "Mobilfunknetze: Erfolgsvoraussetzungen für 5G", Econstore, 2019

[7] T. Mac, S. Houis, T. Gries "Metallfasern", Technische Textilien, 2004

[8] D. Altenpohl, "Aluminium und Aluminiumlegierungen", Springer Verlag, 1965

[9] C. Schludi, J. Joos "Leichte und sichere Batteriegehäuse aus Faserverbundwerkstoffen", Lightweight Design 12, 44 – 47 Springer Verlag, 2019

Cure Characterization of a Fast Curing Epoxy System For Resin Transfer Molding

Prashanth Badrinarayanan
Olin Epoxy, 2301 N. Brazosport Blvd.
Freeport, TX 77548

ABSTRACT

Resin Transfer Molding (RTM) is a high throughput method to produce composite structures. In the high pressure RTM process, resins are mixed, rapidly injected, and cured in a mold cavity at a high pressure (>100 bar) in typically less than 10 minutes. The ability to achieve high fiber volume fractions, combined with the short cycle times through RTM, has resulted in extensive applications ranging from automotive to high-volume industrial composites. Epoxy resins offer an excellent balance of processing ease for composite applications along with good thermal and mechanical properties following cure; however, fundamental to the effectiveness of epoxy resins for RTM applications is the development of chemistries that enable the fabricators to achieve fast curing of the parts in a desired processing window. Epoxy resins for RTM should exhibit an ideal thermorheological profile for effective infusion during the injection step along with fast reactivity during the molding step. Hence, it is crucial to gain an understanding of the cure kinetics for such fast reacting resins to predict and optimize curing conditions for improved RTM process speed. In this work, we explored the ability of a model-free, isoconversion approach to predict the cure progress for a fast-curing epoxy resin system developed for RTM. The isoconversion approach was found to provide an excellent description of the dynamic DSC measurements. The predictions of cure progress from the isoconversion calculations were found to be in excellent agreement with isothermal DSC measurements for a wide range of cure progress from 66% to full cure. Since fabricators and formulators for processes such as RTM have a significant interest in achieving the highest levels of cure at the shortest cure schedule, this approach could offer a quick and effective screening method to identify the process conditions that would lead to high levels of cure in an accurate manner.

1. INTRODUCTION

Resin Transfer Molding (RTM) is a composite fabrication method, which enables high throughput production of parts for a variety of industrial applications. The primary steps in a typical RTM process are the production of a fiber preform, resin injection, curing, and demolding. RTM is widely preferred for the fabrication of complex composite parts through the use of appropriately designed molds and preforms. Various adaptations of the RTM process include vacuum assisted resin transfer molding (VARTM) and high-pressure resin transfer molding (HPRTM). VARTM involves vacuum-assisted resin injection for the production of thicker laminates such as those used in aerospace applications. Epoxy resins are used in RTM applications due to their ease of processability and excellent thermomechanical properties following cure. Epoxy resins can be formulated with a variety of processing and performance characteristics depending on the choice of the resin and curing agent components. However, for a given formulation it is critical to ensure complete cure of the resin to achieve the desired levels of fiber to resin translation, which in turn would dictate the physical and mechanical performance of composite parts. This is particularly essential in processes, such as RTM, where the emphasis on rapid cure cycles and high throughput may lead to the risk of incompletely cured parts, especially if the cure cycles are not optimized prior to production. At the other end of the spectrum, since resins for RTM can cure very quickly, this can lead to a rapid increase in resin viscosity (Figure 1), which can create issues in processes such as HP-RTM if injection temperatures rates are not selected with appropriate knowledge of the resin cure kinetics.

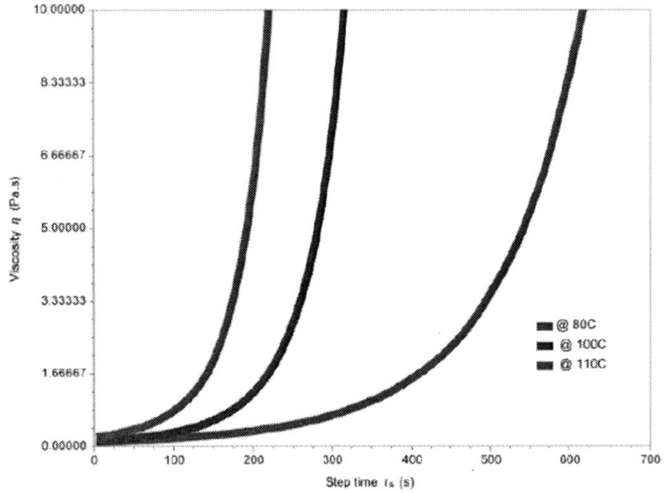

Figure 1: Viscosity profiles as a function of temperature for fast-reacting epoxy systems

Hence, a framework that can enable prediction of epoxy cure progress in fast-curing resin systems such as those used in RTM can enable the selection of appropriate processing conditions for achieving injection, wet-out, and cure completion.

Differential Scanning Calorimetry (DSC) has been a widely adopted technique for exploring the cure kinetics of thermosetting resins. The technique monitors the exothermic heat flow as a function of time and temperature during the cure process. The enthalpy obtained from dynamic DSC measurements performed at a constant heating rate may be used to determine the conversion (α), which is defined as:

$$\alpha(t) = \frac{H(t)}{H_R} \quad (1)$$

H(t) refers to the enthalpy at time t and H_R refers to the total enthalpy or heat of the reaction. α(t) varies from 0 in uncured state to 1 after completion of cure [1]. The kinetics of DSC cure measurements have been described by multi-parameter phenomenological reaction models [2, 3, 4]. However, such models involve multiple fitting parameters and assume a constant activation energy throughout the cure process, which may not be descriptive of some complex cure reactions [5]. More detailed assessments of the strengths and limitations of various models may be found elsewhere in the literature [6,7]. In contrast, model-free isoconversion analysis is another approach to describe DSC cure kinetics, without the constraint of a constant activation energy. The approach involves the determination of the temperature dependent activation energy, (E_α) corresponding to each conversion level. These values of activation energy may be used to predict the extent of conversion or the degree of cure at various other thermal conditions. More details on the isoconversion approach can be found elsewhere [8, 9].

In this work, we will employ DSC measurements to characterize the cure behavior of a fast curing epoxy-amine system specifically developed for RTM. Unlike DSC measurements of typical epoxy resins, the resins used for RTM can pose some challenges due to the tendency of these resins to start curing within a short while at room temperature. To ensure that the resin cure did not advance prior to the DSC measurement, we developed a sample preparation protocol to freeze the resin immediately after mixing and until the commencement of the DSC measurements. We will also employ the isoconversion approach to determine the kinetic parameters of the cure process for the novel toughened system, which will subsequently be used to generate predictions on the degree of cure at potential process conditions.

2. EXPERIMENTATION

2.1 Materials

The epoxy system used was LITESTONE™ 3100 resin with OXH 1201 amine hardener.

2.2 Characterization

Cure Kinetics Measurement: The DSC measurements for the cure kinetics of the resin were conducted using a Q200 DSC from TA Instruments (New Castle, DE). Samples were prepared for the DSC measurements by homogenizing the resin and hardener components in a FlackTeck mixer at 2500 rpm for one minute. The homogenized sample was transferred to a freezer at -40 °C. The semisolid, frozen sample was frozen into a hermetic DSC pan and placed within the designated sample chamber to begin testing. DSC measurements were obtained with multiple heating rates from 0.5 to 10 °C/min. Isothermal DSC measurements were conducted by heating the sample to the desired cure temperature at 50 °C/min, followed by a brief isothermal hold (2-5 minutes), and then the sample was cooled to room temperature at 10 °C/min. The effect of this cure schedule was examined by ramping the sample immediately from 25 °C to 275 °C at 10 °C/min.

Modeling: The isoconversion analysis of the DSC data was conducted using a commercial software package that implements an advanced differential isoconversion approach for analysis. The method has been applied to model and predict the kinetics of a variety of chemical reactions and processes [10,11].

3. RESULTS

DSC measurements obtained with multiple heating rates from 0.5 to 10 °C/min are shown in Figure 2. A typical method of comparing cure kinetics through DSC measurements is through the onset and peak temperatures of the cure exotherm. The onset temperature (To) is determined as the point of intersection of tangents drawn from points just before and after the initial rise of the cure exotherm. The peak temperature (Tp) represents the temperature where the cure exotherm attains its maximum magnitude. The values of To and Tp shift to higher temperatures with increasing heating rate, as shown in Figure 3.

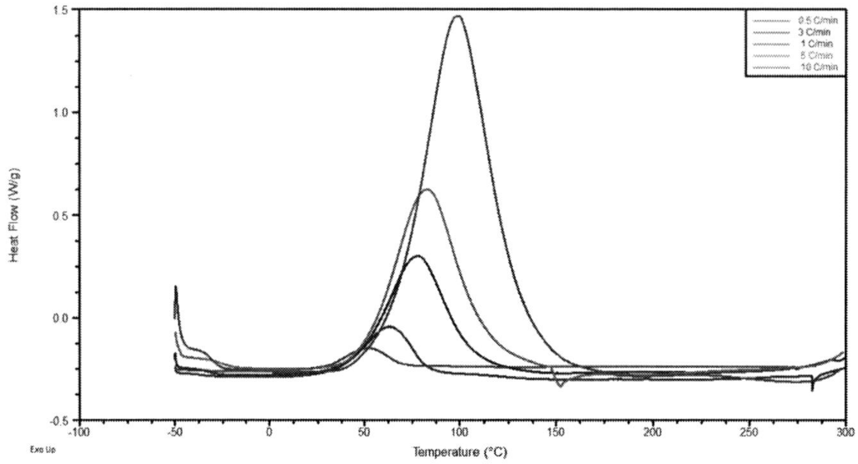

Figure 2: DSC heat flow curves at various rates

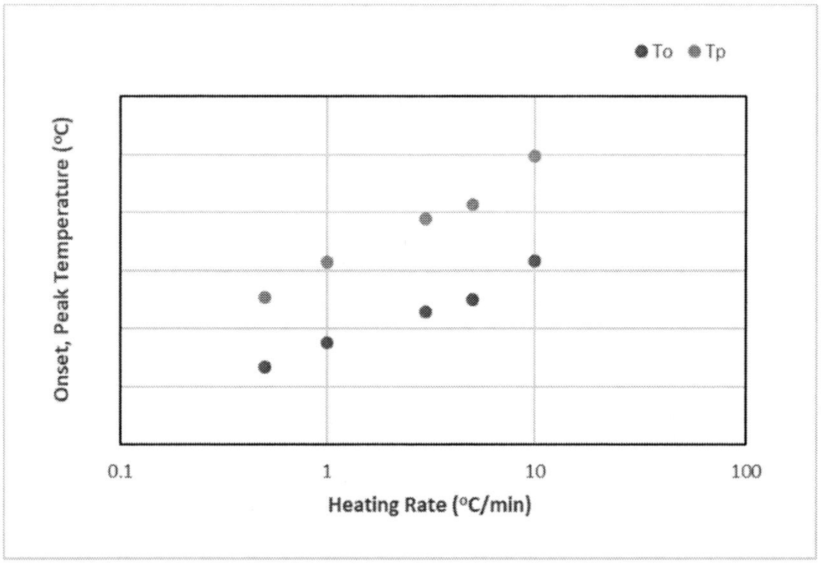

Figure 3: Comparison of onset and peak temperatures of DSC exotherms at various rates from 0.5 to 10 °C/minutes

The integration of the differential heat flow curves in Figure 1 yields the reaction progress or conversion as a function of temperature. Prior to the onset of the cure, the conversion is zero and it increases to 0.5 close to the exotherm peak temperature (Tp), and finally attains a value of 1 at the end of the exotherm. The isoconversion approach employs the heating rate dependence of the conversion to determine the apparent activation energy, without assuming any model function. Typically, the temperature dependence of the conversion may be described, as shown in Equation 2 below:

$$\beta \frac{d\alpha}{dT} = A(\alpha)\exp\left(-\frac{E(\alpha)}{RT}\right) f(\alpha) \tag{2}$$

In the above equation, β represents the heating rate, E represents the activation energy, and A represents the pre-exponential factor. Incorporating β to transform the temperature dependence to time dependence and applying logarithm on both sides yields Equation 3, as shown below:

$$\ln\frac{d\alpha}{dt} = \ln(A(\alpha)) - \frac{E(\alpha)}{RT(t)} + \ln(f(\alpha)) \tag{3}$$

In traditional model-based analyses, the function f(α) would assume a form such as n^{th} order, n^{th} order with autocatalysis, etc. However, in the isoconversion approach, proposed by Friedman [12], the plot of the logarithm of the conversion rate (dα/dt) vs 1/T would yield –E/R as the slope and ln (A(α).f (α)) as the intercept. More detailed discussion on the calculation procedure may be found elsewhere [11, 12].

The isoconversion lines for the dynamic DSC data at multiple heating rates are shown in Figure 4. The cure process is divided into several thousand incremental steps from a conversion of 0.01 to 0.99 to determine the temperature dependence of conversion for each conversion, as shown in Figure 4. The isoconversion lines corresponding to only a few selected values of conversion are shown in the Figure for the sake of visual clarity. Unlike traditional nth order models, isconversion method does not assume a constant activation energy throughout the entire cure process and the isoconversion lines may be used to determine the activation energy corresponding to each conversion level. The relationship between activation energy and conversion or reaction progress is shown in Figure 5 The activation energy increases sharply at low levels of conversion ($\alpha<0.1$) and continues to decrease steadily with increasing conversion until $\alpha = 0.8$ after which, the activation energy once again increases sharply, prior to the eventual near full cure. The changes in activation energy indicate that there are at least three different steps involved in the cure reaction of this epoxy system. Following the initial increase in activation energy, the cure kinetics is autocatalytic from 10% until around 80% conversion. The subsequent increase in activation energy beyond 80% conversion could be due to the restricted chain mobility of the gelled, glassy state or vitrified network. This creates a higher activation threshold for the cure to proceed further.

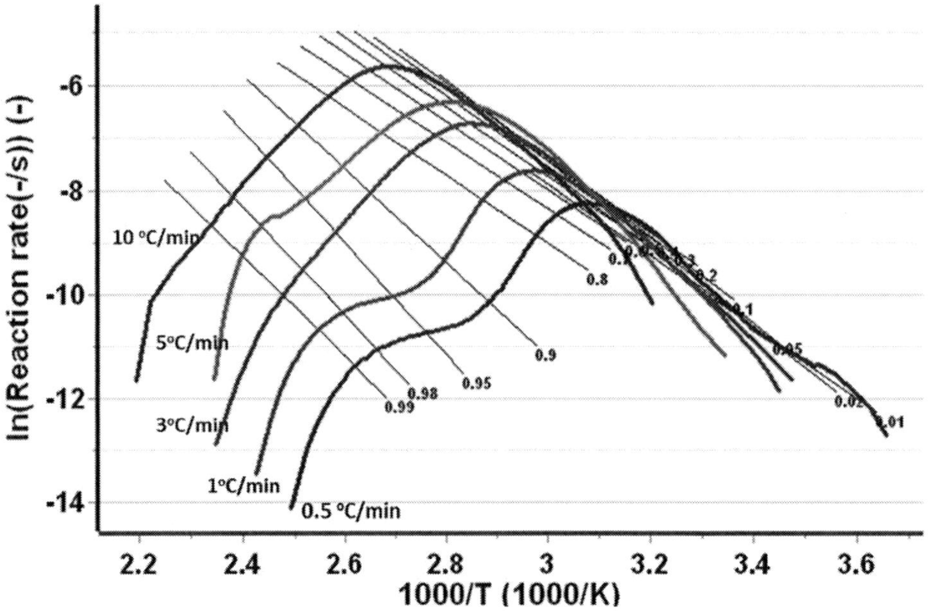

Figure 4: Friedman plot demonstrating determination of isoconversion lines. The colored lines represent the natural logarithm of reaction rate at each heating rate used and the black lines are the isocoversion lines, which capture the temperature dependence of the reaction rate at each heating rate.

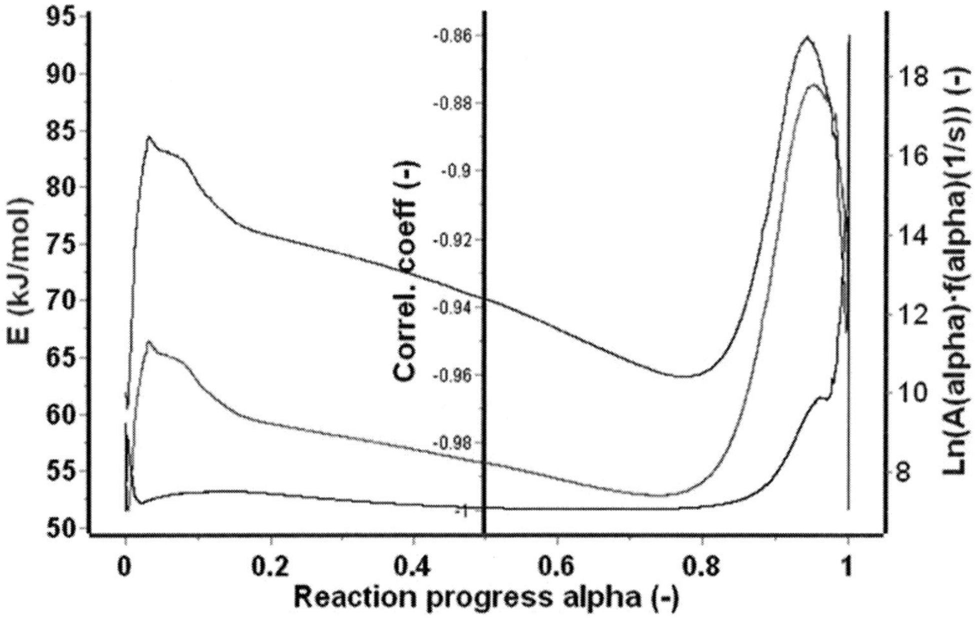

Figure 5: Values of kinetic parameters as a function of conversion.

The fits to the actual experimental using the kinetic parameters obtained from the isoconversion analysis are shown in Figure 7. In the isoconversion approach, the f(α) function is further optimized during the fit to provide the best fit to the data, rather than assuming any specific model function. The overall R value (0.99) confirms the excellent fit to the experimental data. Based on the knowledge of the fitting parameters that provide an excellent description of the experimental data, the cure profile at any other thermal profile may be predicted. The predicted progress in cure for the fast reacting epoxy system for a hypothetical thermal history such as those utilized for a HP-RTM process is shown in Figure 8, involving a quick resin injection (30s) followed by a brief isothermal hold (6 minutes).

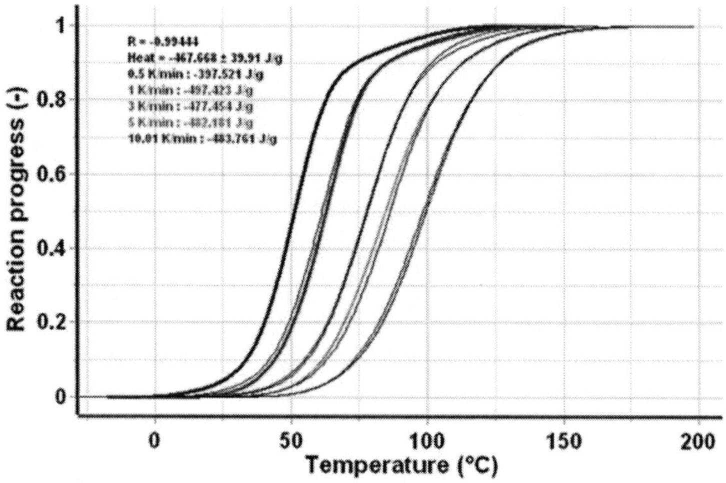

Figure 7: Model fits to experimental data. The model calculations are shown as the solid black line.

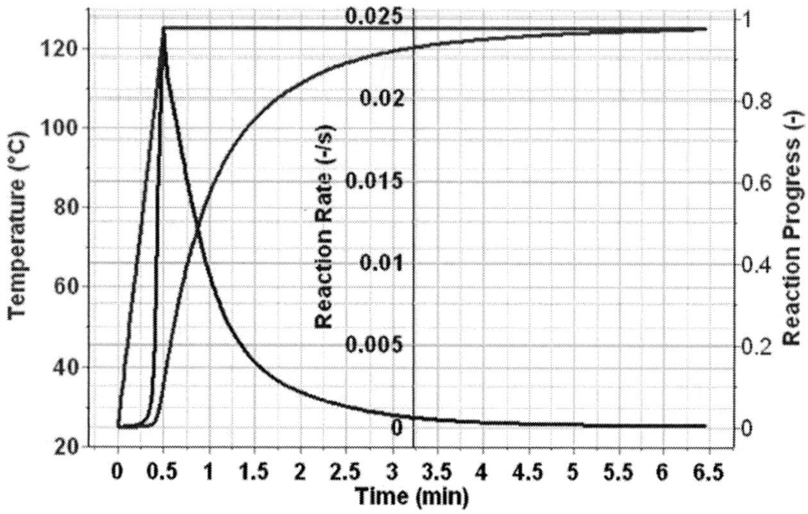

Figure 8: Model fits to experimental data. The model calculations are shown as the solid black line.

DSC measurements were conducted with various cure schedules in order to compare and validate the degree of cure predicted by the model. These measurements consisted of ramping the temperature to a cure temperature of interest, where it was held briefly, and then cooled down to room temperature. The subsequent heating scan, immedietely following this schedule. would reveal a residual exotherm corresponding to the reaction of the uncured monomer. The residual exotherm from DSC measurements may be used to estimate a degree of cure, as shown in Equation 4 below:

$$\alpha = \frac{\Delta Ht - \Delta Hr}{\Delta Ht} \tag{4}$$

Where ΔHt refers to the exothermic heat of a completely uncured resin (and ΔHr refers to the residual exotherm of a partly cured sample. The comparison between experimentally determined degree of cure from the residual exotherm and prediction from the model for the degree of cure following the cool down to room temperature at the end of the isothermal hold is compared in Table 1 below.The same ramp rates (heating/cooling), isothermal temperatures, and hold times were used in the model calculations and the DSC experiments. The isothermal temperatures represented typical temperatures used in processes such as RTM and HP-RTM, where the goal is to achieve high levels of conversion within the shortest period of time. In general, excellent agreement is observed between the cure progress predicted by the isoconversion calculations and the actual values of conversion from the isothermal DSC measurements, as shown in Table 1. A potential source for the some of the slight deviations (8% or lower) could be the errors associated with the integration of the exotherms particularly for the isothermal measurements where the onset of the exotherm may overlap with the glass transition temperature. The modelling approach explored in this work demonstates the potential to identify the degree of cure with changes in cure temperature or time, which would faciliate optimization of process conditions for RTM prior to actual part production. For example, apart from ensuring high levels of conversion during the cure process, the results from the calculations could also be used to predict the potential for the resin to cure during the resin injection step. For a hypothetical process, where the mixed resin system is maintained at 25 °C for 10 minutes followed by a ramp to injection temperature (40 °C) and held there for three (3) minutes during the injection, the calculations suggest that the resin cure advances by less than 5%, which is indicative of good stability prior the cure process.

Table 1: Comparison of predicted and experimental conversion of various cure schedules

Cure Schedule	Degree of Cure (%) (model predicted)	Degree of Cure (%) (DSC- Residual Exotherm)
90 °C for 2 minutes	66.0	62.1
90 °C for 5 minutes	82.0	89.1
100 °C for 2 minutes	81.1	84.4
100 °C for 5 minutes	89.3	90.8
125 °C for 5 minutes	98.0	98.6

The potential to combine the thermal analysis measurements with viscosity results from rheological studies to further improve the prediction of cure progress of composite laminates during RTM will be explored in future work.

4. SUMMARY

In this work, the cure kinetics of a fast curing epoxy system designed for the Resin Transfer Molding process was examined. An isoconversion approach was found to provide an excellent description of the dynamic DSC measurements of the fast curing system and was used to determine the kinetic parameters of the cure. The kinetic evaluation revealed that the reaction mechanism proceeds in three steps. The activation energy increased initally at low values of conversion in the first step. It decreased until 80% conversion indicative of autocatalytic kinetics in the second step, before increasing further due to the onset of diffusion controlled reaction mechanisms in the final step. The predictions of cure progress from the isoconversion calculations were found to be in excellent agreement with isothermal DSC measurements for a wide range of cure progress from 66% to full cure. The predictions were also used to determine the potential level of cure advancement for a hypothetical mixing and injection step. This approach offers a quick and effective way for formulators and fabricators to determine the degree of cure at various process conditions for fast-reacting epoxy systems. Future work will explore an approach to combine rheology measurements, along with DSC measurements, to develop models with further improvements in precisions for predictions at the composite scale

ACKNOWLEDGEMENT

The authors would like to acknowledge analytical and sample preparation support provided by Tyler Runk (Olin Epoxy, R&D).

REFERENCES

1. Menczel J.D. and Prime R. *Thermal Analysis of Polymers: Fundamentals and Applications*, Chapter 1, pg 130, 2009.
2. Yousefi A., Lafleur P. J., and Gauvin R. *Polym Compos* 18,157, 1997.
3. Hale A. *Handbook of Thermal Analysis and Calorimetry*, vol. 3, ed. By Cheng SZD. Elsevier, pp. 295–353, 2002.
4. Bilyeu B., Brostow W., and Menard K.P. *J. Mat. Edu.* 23, 189, 2001.
5. Sbirrazuoli N., and Vyazovkin S. *Thermochim. Acta* 388, 289, 2002.
6. Vyazovkin S., and Sbirrazzuoli N. *Macromol. Chem. Phys.* 200, 2294, 1999.
7. Sbirrazzuoli N., Vyazovkin S., Mititelu A., Sladic C., Vincent L. *Macromol. Chem. Phys.* 204, 1815, 2003.
8. Vyazovkin, S., Sbirrazzuoli N. *Macromol. Rapid Comm.* 21, 85, 2000.
9. Vyazovkin, S., Sbirrazzuoli N. *Macromol. Chem. Phys.* 200, 2294, 1999.
10. Roduit B., Xia L., Folly P., Berger B., Mathieu J., Sarbach A., Andres H., Ramin M., Vogelsanger B., Spitzer D., Moulard H., and Dilhan D. *J. Therm. An. Calorim.* 93, 143, 2008.
11. Roduit B., Hartmann M., Folly P., Sarbach A., Brodard P., and Baltensperger R. *Thermochim. Acta* 621, 6, 2015.
12. Friedman H.L. J. Polym. Sci, Part C, Polymer Symposium 6PC, 183, 1964.

USING DIGITAL TECHNOLOGY IN COMPOSITES FABRICATION TO CREATE A COMPREHENSIVE AS-BUILT DIGITAL TWIN

Scott Blake
Aligned Vision
27 Industrial Avenue
Chelmsford, Massachusetts 01824 USA

ABSTRACT

As composites manufacturers advance the state of the art by applying automation, laser templating, sensors and other data-dependent and data-generating systems to their fabrication processes, an important opportunity is developing to create a comprehensive as-built digital twin. A digital twin is a virtual representation of all critical aspects and attributes of a component or process. The digital twin of product is already a mature technology, consisting of CAD and other digital design data. Likewise, the digital twin of process is well-developed, consisting of manufacturing engineering data, electronic work instructions, digital process control settings, and more. Of course, for any digital twin to be more than a label applied to raw data, the data must be aggregated and contextualized to produce a unified, searchable digital representation. To a significant extent, the digital twins of product and process are achieving this status. Still emerging is the as-built digital twin, which consists of all available information about each individual finished component – materials used, processing details, personnel involved in its production, in-situ and post-processing inspection data, non-conformances and corrective actions, and more. In the composites industry, the as-built digital twin requires two advancements, each of which is currently being developed. First, inspection technology must be applied not only to commonly inspected attributes (e.g., edge location, fiber orientation, absence of FOD) across full plies, but also to quality-critical attributes that were economically and/or technological infeasible to inspect in the past (e.g. anomalies in raw material). Second, the data generated by these inspections, along with process data, must be used to populate a digital twin for each component as it is built. The as-built digital twin will also include quality events and corrective actions taken during manufacture. This paper will present the current status of developments leading to comprehensive as-built digital twin technology and will discuss the implications of the as-built digital twin for closed-loop manufacturing and continuous improvement of product design and manufacturing processes.

Keywords: Digital Twin, As-Built Digital Twin, Automatic Inspection, Laser Projection, FOD, Composites Fabrication, Raw Material Inspection
Corresponding author: Scott Blake

1. INTRODUCTION

Coined in 2002 [1], the term "digital twin" has grown in use and significance as manufacturing digitalization technologies (i.e. Industry 4.0, the digital factory, or smart manufacturing) have evolved and spread. A digital twin is a virtual representation of all critical aspects and attributes of a component or process. The term may be applied to product design, systems [2] or process engineering, and finished products or components. In this last application area, the "as-built digital twin" uniquely extends the virtual realm to the full lifecycle of each individual component or process. A digital twin of product consists of CAD and other design data and is applied to all the components a manufacturer will build in accordance with that design. Likewise, a digital twin of process virtually represents the set of procedures and activities that will be used to fabricate multiple components. An as-built digital twin, on the other hand, virtually represents what actually occurred during the fabrication of one completed component: the material lot and out time for each ply, the specific operator(s) involved in each fabrication step, measured fiber orientations and edge locations from each inspection, each non-conformance event (such as the presence of foreign objects and debris (FOD)) and corrective action taken, actual vacuum achieved and duration of each debulk cycle, temperature and pressure of cure cycles, the finished component's actual dimensions and weight, and more.

The technologies and procedures that enable manufacturers to generate as-built digital twins have lagged behind those used to create other digital twins (perhaps even more so in manufacturing with composite materials than with materials for which automated and digitalized processes are more mature); but they are now commercially available and being implemented by a growing number of composites fabricators. They include automated fabrication equipment, laser templating, electronic work instructions and process control, and automatic inspection.

Commercialized within the past decade, automatic inspection technology is especially critical to the development of a comprehensive as-built digital twin. While automation, laser templating and process control technologies generate process data that inform the as-built digital twin of equipment settings and processing parameters for each individual component, automatic inspection provides information about the actual product's or component's attributes. To enable automatic inspection to contribute a full data set for a truly comprehensive as-built digital twin, two technology advancements are underway. First, artificial intelligence (AI) is accelerating application of automatic inspection to quality-critical attributes that were economically or technologically infeasible to inspect in the past (e.g. anomalies or FOD in raw material). Second, raw data from inspections is being aggregated and contextualized to create intelligent, searchable virtual representations of each component, as built.

1.1 Automated Fabrication, Laser Projection, Electronic Process Control and the As-Built Digital Twin

Automated fabrication, laser projection and electronic process control systems are benefactors of the digital twins of product and process: they receive data from these digital twins that determine equipment settings and process parameters. In automated fiber placement (AFP), for example, these digital twins provide tow characteristics, ply and course data, gap and overlap settings, fiber steering data and so on. Automated equipment, laser projection systems and electronic process control also generate data that can be aggregated in the as-built digital twin. An electronic process control system, for example – even one associated with hand layup of composite component – may automatically populate a layup report that includes material lot, out time, debulk information, operator and inspector buyoffs, and more. In the development of as-built digital twins, data included in such layup reports would also be included in the digital twin. Likewise, the as-built digital twin may incorporate sensor data from an AFP or other automated fabrication system.

1.2 Automatic Inspection and the As-Built Digital Twin

Primarily two kinds of automatic inspection systems are being applied currently to the production of composite products and components. One is laser profilometry. A laser profilometer may be mounted to the deposition head of an AFP, automated tape laying (ATL) system, composite 3D printer or other automated fabrication system. From this position, it collects data in parallel with material lay down. Profilometry relies on distance measurement by parallax: by projecting a laser line and measuring the distance to points along that laser line, the profilometer creates a profile of the surface that is analyzed to identify and measure features, such as overlaps and gaps between tow lanes.

Because the profilometer travels with the deposition head, laser profilometry is an effective means of verifying correct relative position of laid down material – relative to other material, courses or plies in the layup. It may also identify non-conformance events (e.g. out-of-tolerance gap width) in real time. Once a non-conformance is found by the profilometer, the next step is to have an operator investigate and correct the non-conformance. This can be accomplished efficiently by employing a laser projector. Using the same location data as the profilometer, the laser projector pinpoints the area of interest so that the operator knows where to investigate. Since the AFP must stop while the operator is in the work cell, pinpointing saves cycle time by swiftly guiding the operator. The combination of profilometry and laser projection affords the fabricator an opportunity to take corrective action and prevent non-conformances from becoming part of the finished laminate.

The other kind of automatic inspection system available to composites fabricators is a large-field automatic inspection system that incorporates machine vision and laser projection in one unit. This system is capable of inspecting in detail any area on the surface of a layup, independent of any deposition head. As a result, it is able to measure absolute position and orientation of a ply within the work cell, and to measure it at any time before it is covered by another ply (whereas the profilometer inspects tows only immediately after the deposition head places them). The system (Figure 1) is designed to inspect detailed regions of interest anywhere on a large complex surface (typically 5m by 5m). Each unit of the system projects its laser and captures images within a 60° (±30°) angle. Meeting Boeing D6-55902 requirements, the unit achieves an accuracy of 0.38 mm (0.015 in.) in a standard 3m by 3m (10 ft by 10 ft) field at a distance of 3m (10 ft) from the unit.

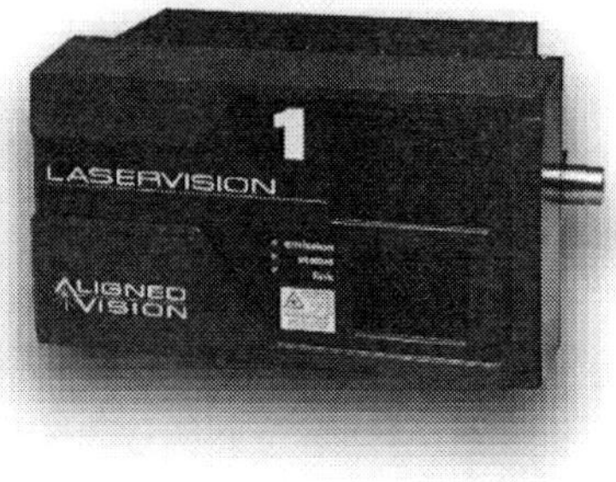

Figure 1. A large-field automatic inspection system employs a laser projector with a built-in machine vision system. Model-directed software captures and analyzes detailed images of an in-process composite layup.

To perform automatic inspection, the vision component of the system directly accesses and applies manufacturing data to aim a high-resolution, high-magnification camera, which captures "calibrated images" under data control. "Calibrated imaging" refers to the images being captured along with a photogrammetric transform that defines the relationship between the camera and the feature being imaged. This enables measurements in the image that correspond accurately to features on the surface of the part. That is, the transform enables each pixel in the 2D image to be dimensioned relative to the 3D surface and in the coordinate system of the part. The analyzed images support a "gauging" function, in which the surface containing the features being imaged is assumed to be correct and the location of features on that surface are assessed for translations and rotations.

Because the large-field automatic inspection system is completely independent of the AFP head, the AFP can continue to add material even when the inspection system detects a non-conformance or FOD. Operators may remotely conduct a detailed examination of inspection images (Figure 2) while layup continues. If entry into the work cell is warranted, the system's laser projector pinpoints the area of interest, minimizing AFP downtime while near real-time corrective measures are taken.

Figure 2. Analysis enabled by transform. Out-of-tolerance tow placement identified.

2. COMPREHENSIVE ATTRIBUTE INSPECTION

Automatic inspection technology currently measures tow end and edge positions, which define ply boundaries in AFP layup, as well as ply boundaries created through other layup methods (manual, automation-assisted, Automated Tape Placement, etc.). As with many new technologies, the first attributes to which automatic inspection has been applied are those commonly inspected by the previous state of the art, in this case, manual inspection. In addition to gap and overlap widths, attributes to which laser profilometers have been applied include pits, bulges and distortions in filament-wound cylinders [3]; and automated dry material placement [4]. Large-field automatic inspection has been used to verify (or flag) material type (unidirectional versus woven plies, for example), fiber orientation (Figure 3), and FOD. To detect FOD, the automatic inspection system's analysis algorithm identifies any abrupt changes in optical surface characteristics where none is expected. Such changes prompt the system to flag the location, and the laser projector draws a boundary around the FOD on the surface (Figure 4).

```
visionuni version 20180105

All read in: 0.547000
 Calcs done: 1.687000
angle expected -89.727594
angle found -89.283803
angle diff 0.443792  tol 5.000000
REFERENCE_ANGLE -89.727594
IMAGE_RESULT?ORIENT+ANGLE+-89.283803+PASS
```

Figure 3. Raw (left) and analyzed (right) image from large-field automatic inspection system verifying that fiber orientation is within tolerance. Red lines in the analyzed image represent tolerance band for fiber orientation; green line shows actual orientation. Measurements are quantified in data read-out (below analyzed image).

Figure 4. Foreign objects and debris (FOD) detected by automatic inspection are pinpointed on the surface by laser outline.

2.1 Accelerating Application of Automatic Inspection to New Attributes

The identification of non-conformances and corrective actions taken are critical aspects of a comprehensive as-built digital twin, informing OEMs and end users of each quality event associated with each component. Advancements in the identification of FOD events illustrate one of the two key technological developments contributing to a comprehensive as-built digital twin. Application of automatic inspection to each specific kind of FOD in each specific fabrication is unique. For example, detecting peel ply in critical bonding applications requires the development of algorithms specific to the environment in which bonding will take place, the kinds of composite surfaces to be bonded, and the type of peel ply being used. In the past, each algorithm has been painstakingly hand-engineered, but AI is able to eliminate this bottleneck.

With the high number of composites fabrication variables, plus additional factors with which automatic inspection technology must contend (like lighting within the work cell), developing an analysis algorithm for a particular application via hand-engineering typically takes several weeks of dedicated effort. On the other hand, AI-enabled development of a similar analysis algorithm has been shown in recent experimentation to require only one to two days [5].

AI-enabled algorithm development uses the type of AI known as machine learning, which is the ability of a computer to use data to improve performance of a particular task without being explicitly programmed to do so [6]. One helpful characteristic of machine learning with automatic inspection is that it can be conducted away from the manufacturing floor, which means that manufacturers may be able to transition to automatic inspection quickly and with little to no disruption of shop floor activities. Manufacturers employing laser templating, for example, could replace a standard laser projector with a laser projection/machine vision unit and continue to conduct laser templating operations while machine learning uses harvested images to develop the specific application. The only additional requirement would be the deliberate creation of flaws to provide a full data set for the machine learning system.

To create an algorithm that is ready to detect all potential flaws, the fabricator would create artifacts containing those flaws, as well as correctly fabricated regions. Then in an environment similar to the manufacturing floor, the machine vision system would capture images that would be used to train the machine learning system and produce the application-specific algorithms.

2.2 Attributes to Include in an As-Built Digital Twin

The commercial availability of automatic inspection systems, combined with accelerated algorithm development through machine learning, opens up the potential to include in an as-built digital twin a full range of attributes critical to part performance. Many of these attributes are extremely difficult, if not impossible, to detect through human visual inspection.

Prior to the commercial availability of automatic inspection for composites, the only alternative for measuring some of these attributes has been metrology, which is infeasible for most applications because it is so labor- and time-intensive. (Additionally, some metrology systems provide a costly and unnecessary level of precision.) To measure some of these attributes with metrology, a "point cloud" comprised of hundreds of thousands or millions of points has to be captured and converted into a surface. The modeled surface must then be aligned with a nominal surface and the two compared. In contrast, the large-field automatic inspection system uses much less data-intensive expectation-driven analysis: employing the laser projector, it generates an acceptable pattern and its tolerances directly from the CAD model. The system's machine vision captures an as-built laser pattern from the actual surface being inspected. Automatic image analysis then compares the CAD-generated and as-built patterns, producing a go/no-go metric.

One significant data set that automatic inspection could add to the as-built digital twin is raw material attributes. Despite the impact of raw material quality on critical component attributes, common practice today is to inspect raw material manually or not at all. Manual inspection here faces the same difficulties as in the fabrication cell: human limitations such as distractedness, fatigue and verification bias decrease inspection effectiveness. Moreover, manual inspection is very time consuming. In contrast, an as-built digital twin fed by automatic inspection as well as material tracking data could incorporate raw material attributes in an economical and technically practicable way.

Beyond raw material data, the as-built digital twin may also be enhanced by inspection of additional fabrication attributes such as wrinkles, bridging and shear. Inspection of these attributes has been assessed at the Advanced Composites Centre for Innovation and Science (ACCIS), University of Bristol, U.K. [7]. Wrinkles occur when material does not conform completely to the tool surface. To detect wrinkles, the automatic inspection system's analysis algorithm identifies discontinuities in a laser reference pattern that would not be expected when viewing an unwrinkled surface. The approximate size of a detected wrinkle is determined through the defined geometric relationships among the projection source and pattern, the camera, and each surface (Figure 5). The wrinkle detection mechanism is similar to the profilometer's distance measurement by parallax: the system generates a parallax by projecting the laser pattern from one location to the area of interest (anywhere within the large envelope that the system inspects) and capturing the image from another location.

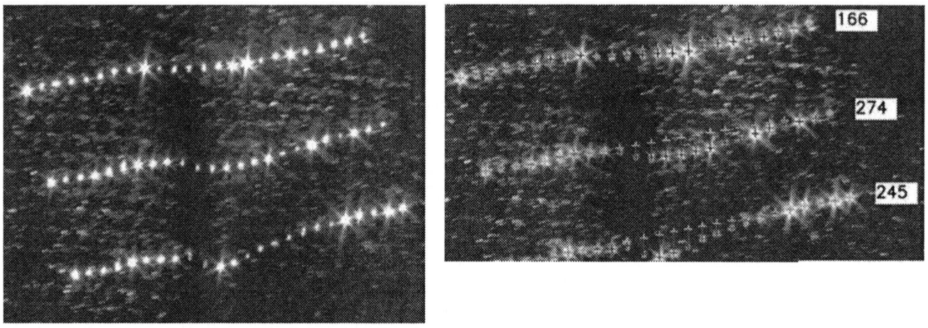

Figure 5. Automatic detection of 0.8mm wrinkle in robotic layup system

Bridging occurs when material does not conform fully to the tool surface in a concave region. With secondary bridging, material that initially conforms is pulled out of conformity as material is draped or creased elsewhere. Using the same mechanism as with wrinkle detection, the automated inspection system may be programmed not only to detect bridging but also to automatically re-inspect areas that are susceptible to secondary bridging (Figure 6).

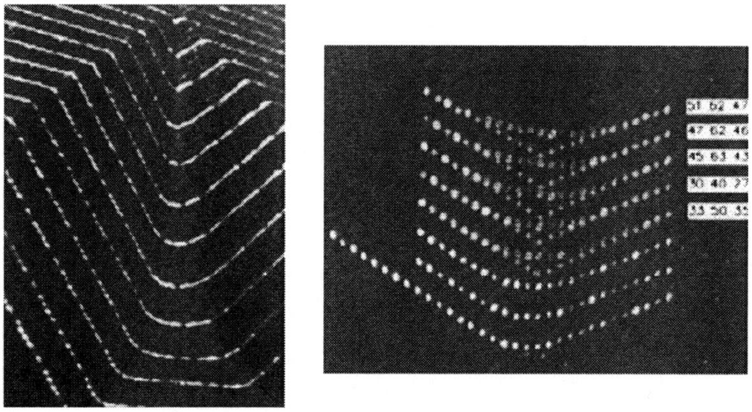

Figure 6. Difference between expected and actual location of projected reference points indicates proportion of bridge displacement.

Shear – the distortion of the material's fiber orientation as it is deformed in three dimensions – often occurs by design in composite fabrication. Excessive shear, however, may weaken the finished product. To detect and measure shear, the automatic inspection system illuminates the surface to maximize the visibility of each tow. Analysis algorithms track the appearance of each tow in images captured by the system's aimed camera and calculate the angular change in direction along the length of the tow. Maximum positive and negative angular changes are automatically calculated (Figure 7).

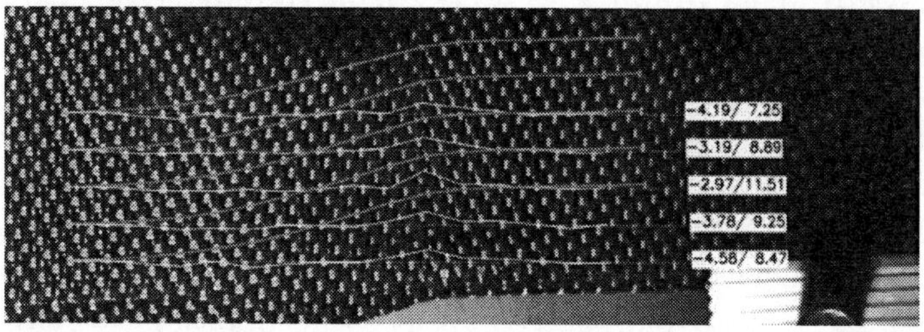

Figure 7. Automatic discovery of fiber paths in material sheared in press

Automatic inspection technologies may also generate post-processing data for the as-built digital twin. This may include machining, assembly and other manufacturing processes. The large-field automatic inspection system described here has the capability to inspect for any visible feature's presence or absence, as well as its location on a surface. For example, it may be used to verify not only the position of fasteners in an assembly, but also the type of fasteners being used (Figure 8).

Figure 8. Automatic inspection detects fastener position and type. The captured image (left, from 10 ft away) is high resolution and shows excellent detail under zoom (center and right).

3. CREATING THE AS-BUILT DIGITAL TWIN

In a paper on its automated inspection system used in large-scale automated fiber placement (AFP) cells, Electroimpact reports that inspection data can approach 1 TB (terabyte) per part for large complex composite parts [8]. How does a fabricator tap this large "data lake" to generate the as-built digital twin? This is the second key advancement occurring in the development of a comprehensive digital twin, and the approach depends on the level of digitalization the fabricator has implemented. The more integrated the automatic inspection system is with manufacturing operations management (MOM) digital systems, the more visible, accessible and useful the as-built digital twin becomes.

Initially, a fabricator might implement automatic inspection as a "digital point solution" that imports design and control data from the MOM system. Automatic inspection data, along with data from a resident electronic process control system, is aggregated with automatic inspection system's own software. Aggregation and contextualization software was first developed for a predecessor technology to the large-field automatic inspection system described here. Developed for the U.S. Air Force and employed by Sikorsky in rotor blade fabrication, that system automatically archived data verifying critical fiber orientation characteristics. Used for more than a decade, the system has enabled immediate access to inspection data for any rotor blade, because all the data has been automatically organized so as to be traceable by work order or traveler number. An example of contextualization is that the fiber orientation data for a specific blade is cross-referenced to the specific inspection device used and that device's certification and operational data.

Software development has continued, and in the current system, inspection and process data are both captured and organized. One specific feature of the software is its handling of imaging data: while the automatic inspection system captures detailed images of small regions within the surface of a large component, it has the capacity to combine these images and produce a single image of a complete ply surface (Figure 9).

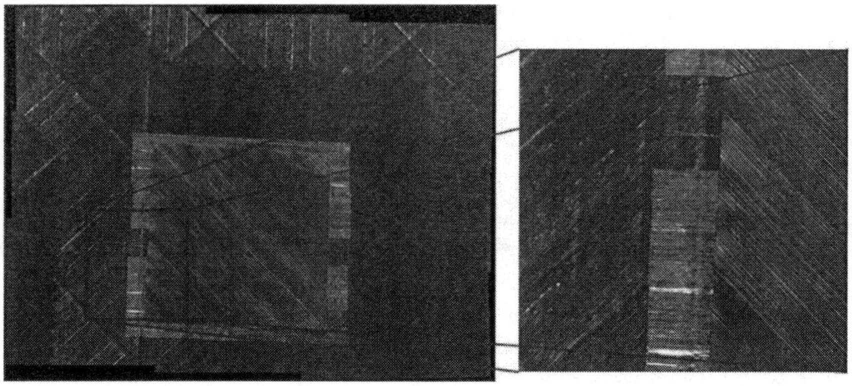

Figure 9. Nine captured images combined into one (left).
Algorithm in the automatic inspection system aligns images seamlessly.

As a stand-alone digital point solution, the system processes generated images and data, creating an as-built digital twin that is exported back to the MOM system (Figure 10). Within the MOM system, this as-built digital twin could be combined with additional data (e.g. material supply chain information, machining or assembly data) to generate a more complete as-built digital twin.

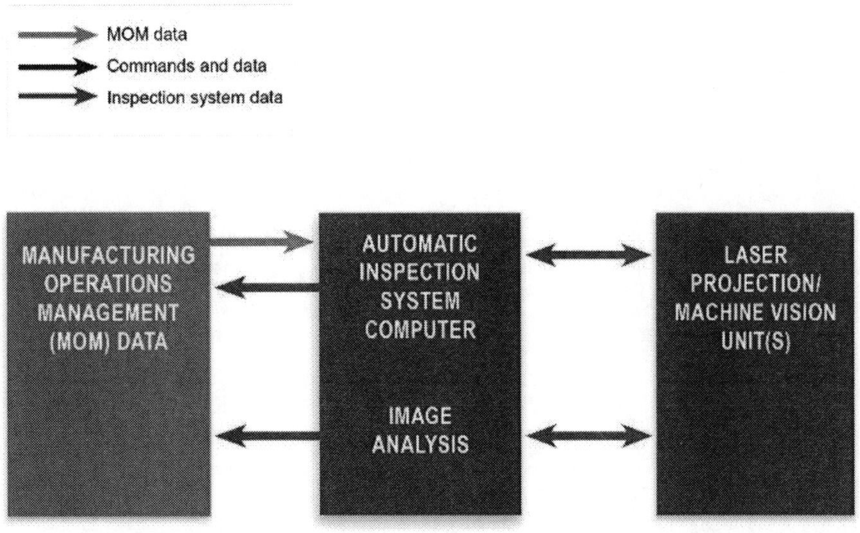

Figure 10. As a digital point solution, an automatic inspection system operates as a stand-alone, importing design and control data from the manufacturing operations management (MOM) system and exporting as-built data back to the MOM system.

Fabricators that integrate automatic inspection into their MOM solution are able to more immediately aggregate as-built data from various sources, including the automatic inspection system. Automatic inspection can be integrated into a MOM system (Figure 11) using a software development kit (SDK) or other integration process. On an integrated MOM platform, automatic inspection becomes a critical function within the quality ecosystem, maximizing a fabricator's ability to respond to real-time manufacturing quality events. The integrated system creates a comprehensive as-built digital twin because it is able to aggregate and contextualize data from all relevant sources: material lot and out-time, automated equipment setting and sensor readings, laser templating settings, automatic inspection images and data, and post-processing component features and attributes.

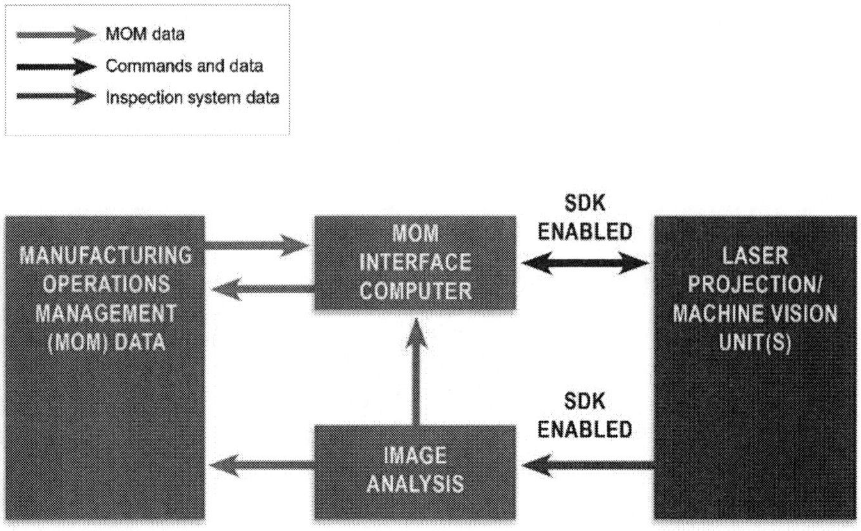

Figure 11. Integrated into a manufacturing operations management (MOM) system, automatic inspection becomes one of the critical functions within a quality ecosystem.

4. THE AS-BUILT DIGITAL TWIN AND CLOSED-LOOP MANUFACTURING

Closed-loop manufacturing is the second primary opportunity to apply AI to composites fabrication (the first being algorithm development through machine learning). The form of AI used in closed-loop manufacturing is "deep learning." While machine learning usually is applied to a specific process or task, deep learning typically uses larger data sets and more complex logic networks [6] to improve more broad-ranging aspects of a process or product. A deep learning system will tap all comprehensive as-built digital twins from a production run and perform analysis that also involves the digital twins of product and process. The root-cause, predictive and prescriptive analysis performed in deep learning generates manufacturing insights that may be used to improve product design and/or process engineering. For example, deep learning may identify an error-prone steering sequence in a component's AFP fabrication, the solution for which may involve a redesign of fiber orientation in that portion of the component, a modification of component tooling, a change in the AFP's lay-down rate or tow tension, or some combination of these changes.

An example of closed-loop process improvement (though a deep learning tool has not yet been applied to this particular circumstance) is the use of laser projector alignment data to detect fabrication trends. A laser projector establishes itself within a common coordinate system with fabrication tooling through scans of reference targets on the tool. Of course, the physical position of each target is limited in accuracy, as is the scanning function of the laser projection system. Each time a laser projector scans targets, it generates data that can be compared to nominal values for target locations. By analyzing this data, a fabricator can detect trends regarding how well the positions of multiple targets agree with the nominal data and the positional data of other targets. If a trend toward non-conformity is detected, root-cause analysis can be used to determine whether the tool surface or shape has been compromised (e.g. by tool drops) or whether the laser projector requires recalibration and recertification. By collecting this generated data and using it in analysis, a fabricator may improve the geometric (shape) quality of a component production run.

Applying deep learning to a set of comprehensive digital twins and feeding insights back into design and manufacturing processes, a fabricator gains ongoing opportunities for improvements in design, simulation, modeling, product quality and process efficiency and efficacy.

5. CONCLUSIONS

The comprehensive as-built digital twin is an essential piece of Industry 4.0 for composites fabricators. The technology needed to generate an as-built digital twin for each composite component – automatic inspection technology, in particular – has been commercialized only recently. Ongoing developments are enabling accelerated data generation, aggregation, contextualization, and analysis. As a result, as-built digital twins of composite components are becoming more comprehensive, thus giving fabricators the opportunity to gain greater manufacturing insights and to employ those insights in continuous product and process improvement.

6. REFERENCES

1. M. Grieves, "Digital Twin: Manufacturing Excellence through Virtual Factory Replication." White Paper. Melbourne, Fla.: Michael Grieves, LLC, 2014.
2. A. Madni, C. Madni and S Lucero, "Leveraging Digital Twin Technology in Model-Based Systems Engineering." *Systems* 7(1), 2019. https://doi.org/10.3390/systems7010007.
3. R. Saulsberry, "Development of a Composite Pressure Vessel Laser Profilometry and Eddy Current Scanning System to Meet Manufacturing and Analytical Needs." NASA Engineering & Safety Center news release, 30 November 2016.
4. S. Black, "Improving Composites Processing with Automated Inspection, Part II." *CompositesWorld* 4:6 (June 2018).
5. S. Blake, "Recent Advances in Artificial Intelligence Applications to Composites Fabrication," presented at CAMX, 26 September 2019.
6. "Artificial Intelligence (AI) vs. Machine Learning vs Deep Learning." 25 Jan 2019: https://skymind.ai/wiki/ai-vs-machine-learning-vs-deep-learning.
7. K. Potter, C. Ward, D. Crowley, M. Elkington, S Blake, and D. Whittington, "Real Time Gauging of Wrinkle, Shear and Bridge Formation in an Automated Composite Manufacturing Process," presented at the Composites and Advanced Materials Expo, September 27-29, 2016

8. J. Cemenska, T. Rudberg, M. Henscheid, A. Lauletta, et al., “AFP Automated Inspection System Performance and Expectations,” SAE Technical Paper 2017-01-2150, 2017, DOI:10.4271/2017-01-2150.

PERFORMANCE EVALUATION OF 2.5D NEEDLE-PUNCHED CFRP/PHENOLICS COMPOSITES

Esmerehildo Trevino[2] Jitendra Tate[1,2,*], and Camila Belduque[2]

Texas State University
601 University Drive, San Marcos, TX, 78666
[1]Materials Science Engineering and Commercialization Program, Texas State University
[2]Ingram School of Engineering, Texas State University
JT31@txstate.edu; 512-245-1826

ABSTRACT

With the increase of composite components being introduced into the aerospace industry, the adoption of new manufacturing processes that improve the performance of the created parts when compared to traditional methods can be highly advantageous for a company. In the case of 2D composites, where reinforcement fibers are oriented along two axes, the result is a composite with anisotropic properties, which are typically weaker in the out-of-plane direction. Under high stress conditions, composite parts can delaminate resulting in a drastic loss of the load bearing capabilities of the part. One of the solutions to improve the thru-thickness mechanical properties is to add non-woven fiber layers to a dry composite preform stack and process it through a needle-punching operation. Needle-punching consists of rows of evenly spaced, barbed needles which repeatedly penetrate through the preform. The needles re-orient the loose web fibers through the thickness of the composite laminates, providing the z-axis or out-of-plane fiber reinforcement. In this project, Carbon felt and plain-woven carbon fabric preform is needle-punched and infused with a Phenolics resin, suitable for the Vacuum Assisted Resin Transfer Molding (VARTM) technique. The design and manufacturing parameters of the needle-punching process such as the number of non-woven web layers added to the preform, alternate layup configurations, and varying levels of needle-punch density are analyzed. To evaluate the in-plane and out-of-plane mechanical properties of the 2D and 2.5D needle-punched composites, the following mechanical testing will be performed per ASTM standards: tensile test, flexure, interlaminar shear and impact testing. Additionally, scanning electron microscope (SEM) analysis is conducted on the specimens.

1. INTRODUCTION

Composite materials are being widely applied in aerospace, automotive, sports, medical and marine industries because of their high strength to weight ratio, ease of design and processing, and high performance. A composite material is comprised by a matrix and a reinforcement placed in one or more specific directions, which gives the final part an anisotropic mechanical behavior. Composites are traditionally built in a layer-by-layer manufacturing method in which reinforcement fabrics are laid on top of each other until the desired thickness is achieved, and infused with a resin to create a 2D composite laminate. The fabric plies consist of continuous fibers that lie in a 2D plane and are oriented in the direction where the highest strength is required. When high stresses or impact forces are applied in the out-of-plane direction to the composite, it can result in mechanical failure characterized by a separation of the fabric plies and often occurs between the resin rich interlayer that exists between the stacked fabric plies. This mechanism is referred to as delamination, once it occurs there is a significant loss in the composite materials strength and toughness.

Different manufacturing methods have been developed to produce composite parts whit enhanced through-thickness (TT) mechanical properties, interlaminar shear strength, and impact toughness. These improvements increase the composites ability to absorb the energy imparted by an impact. Some examples of 2.5D/3D manufacturing methods are 3D weaving, stitching, z-pinning, and needle-punching. Techniques that will be concisely explained in the following paragraphs.

3D weaving components can be obtained through different methods such as 2D weaving, 3D weaving and nobbing. However, a true 3D weaving process is recognized when three orthogonal yarns are interlaced by shedding the multi-layer warp-*z* across the fabric thickness and fabric width (*x* and *y* wefts, row and column wise, respectively) as shown in figure 3 obtained from Behera et al., 2008 [1]. 3D weaving yields parts designed to fulfill high structural requirements, which makes it a processing method highly implemented in the automotive, construction, and marine industries, etc. It is not suitable to produce large quantities of parts because it is a low speed process [2]. Additionally, the in-plane mechanical properties of 3D weaving parts are decreased because of fiber crimp and resin rich regions (RRR) caused by weave designs and weave defects [3].

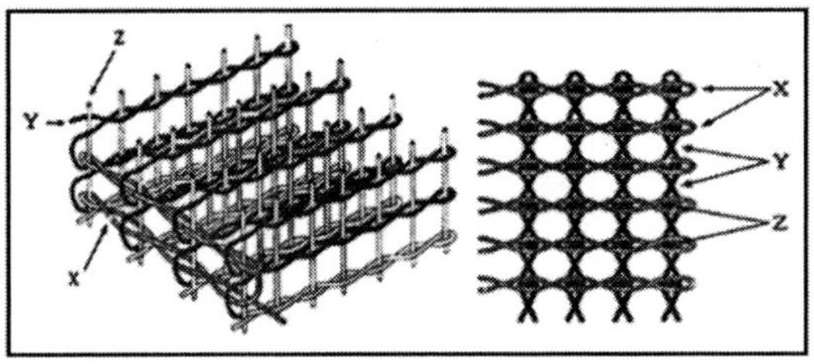

Figure 3. 3D weaving setting.

Regarding the stitching technique, it consists on sewing 1% to 5% of a high tensile strength yarn into a composite. Needles with different sewing angles can be employed, it is a cheap process that reduces the use of mechanical fasteners, highly implemented to build wing structures [4]. Figure 4 obtained from Aktas et al., 2009 displays on the left side the stitching process with multiple needles sewing Kevlar yarn into a woven carbon skin. The structure created is shown on the right side. It has proven to be an efficient process during Mode-I loading, fracture toughness and impact resistance. However, as with 3D weaving, stitching causes crimp and fabric and RRR distortion. Several experiments have concluded that this process diminishes in-plane stiffness and strength and decreases fatigue life in composites due to damages induced during the fabric stitching process. [5]

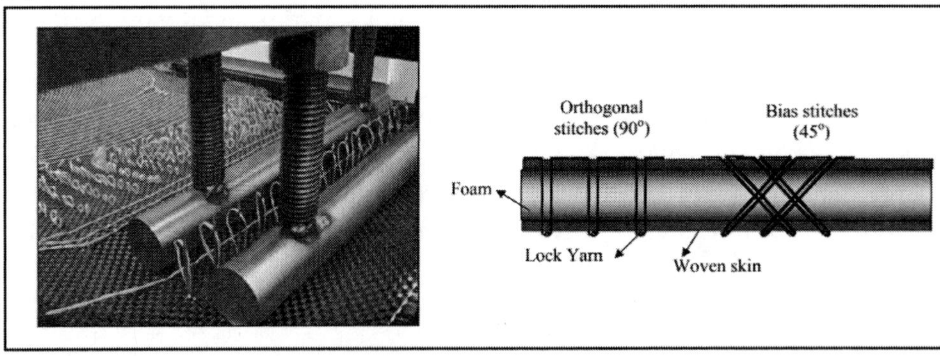

Figure 4. Stitching process machine and fiber setting.

Z-pin process consists of stiff pins inserted into an uncured prepreg and is highly useful to improve pull-off failures and Mode-I and Mode-II fracture toughness. As with Stitching, Z-pinning eliminates the use of mechanical fasteners. It is a time-consuming process because it is more efficient with a large quantity of small pins. This increases the surface contact area with the out-of-plane reinforcements, effect not reachable with big pins or stitches. Z-pinning requires different tools according to the complexity of the geometry being processed, which limits the size and design level of the components that undergo this method. Additionally, to modify the pin orientation is a difficult process [6]. Figure 5 illustrates the Ultrasonically Assisted Z-Fiber® insertion (UAZe insertion) explained by Partridge and Cartié in [7] which consists of a preform that holds the z-pins in position, which will be inserted by an ultrasonic horn at 20kHz under speed and controlled pressure into the laminate. The pin parts above the surface of the final laminate are removed by the cutter at the end of the process.

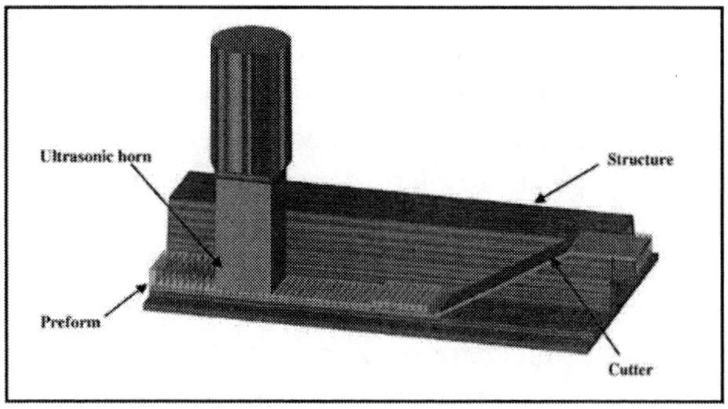

Figure 5. Ultrasonically Assisted Z-Fiber® insertion.

Needle punching is a process where several reinforcing fabric layers are joined by fibers getting displaced through the thickness of the laminate by a needle board consisting of several rows of barbed needles. This machine can be observed in figure 6.

Figure 6. Feltcraft Industrial Needle-Punching Machine.

The needle board repeatedly punches into the multiple layered fabric plies covering the entire surface area, creating a dense preform. The barbed needles transfer the fibers which become entangled and are held in place by frictional forces. Figure 7, obtained by Chen et al., 2016 represents the post needle-punched laminate, the short-cut fibers or long-cut fibers of the web cloth layers are transplanted through the thickness of the laminate until a homogenous structure is created [8]. The z-axis fibers create loop formations which creates the mechanical bond between the multiple layers. Among its advantages, it is possible to say that needle-punching is an attractive manufacturing process for creating 2.5D composite preforms at a relatively low price and can maintain high production rates.

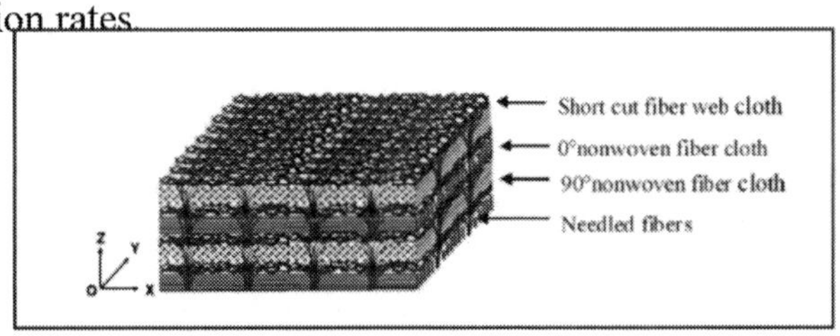

Figure 7. Laminate setting after being processed through needle punching.

The mechanical properties of 2.5/3D composites have been extensively investigated. Kim et al. studied the mechanical properties of carbon fiber fabrics reinforced with woven fabrics and needle-punched carbon fiber felts consisting of loose oxidized polyacrylonitrile (PAN) fibers. The loose fibers are displaced during needle-punching process to create the z-axis fibers for the mechanical joining of the adjacent plies. Three different wave patters were analyzed, plain weave, 2/2 twill weave and 8-harness satin weave. The effects of needle-density (ND) on the in-plane and out-of-plane mechanical properties such as in-plane tensile strength, in-plane compressive strength, interlaminar shear strength, interlaminar fracture toughness and compressive strength after impact were analyzed. It was concluded that the felt/plain-woven-fabrics composites and a high ND improves the damage tolerance of the matrix and does not reduce the in-plane compressive strength significatively. Additionally, the felts/carbon composites compressive strength increases as ND increases independently of the woven pattern [9]. Kang and Lee compared the mechanical and impact resistant properties of woven E-glass to needle punch non-woven E-glass fiber composites with a major focus on the ND. Here, the needle punched composites had a lower energy absorption capability as the needle density increased with a smaller area of damage. The Mode-I fracture toughness and compressive strength after impact of the needle punch composite was much greater than the composite constructed of woven fabric due to the TT fibers [10]. Kang et al., have shown that the needle punch density is an important variable in the needle-punching process and can change the thermal and mechanical properties in needle-punched carbon/phenolic composites. Increase in needle punching density was shown to improve the flexural strength, interlaminar shear strength, tensile strength and even increased the resistance to thermal erosion. The equation bellow can be used to calculate the needle punching density (F) for a needle punching operation where S is the stroke frequency of the needle board (strokes/minute), N is the number of needles per square area of the needle board (needles/cm^2), W is the width of the needle board machine (cm), and V is the material feed rate (cm/min) [11].

$$F = \frac{SNW}{V}$$

In this study, needle-punching is implemented to improve the in-plane and out-of-plane mechanical properties of rayon-based-carbon/Phenolics composites, which are manufactured via Vacuum Assisted Resin Transfer Molding (VARTM). The mechanical properties regarding tension, flexure, short-beam shear and impact behavior of the laminates are evaluated following ASTM standards. Additionally, the failure modes and morphology of the specimens are analyzed with scanning electron microscopy (SEM)

2. MATERIAL SYSTEM

The material system implemented in this study is comprised of 8-harness satin rayon based carbon fiber fabric, which is obtained from Cytec Engineered Materials Inc. Its thickness is 0.48mm with an 8-harness satin weave style and one non-woven felt, the areal weight of the carbon fiber is 261g/m^2. A Durite SC1008 liquid Phenolic resin from Hexion, with a viscosity of 180 – 300 cps, a pH of 7.9 - 8.5, and an operational temperature of up to 500°C is used to impregnate the reinforcement and is suitable for VARTM because of its low viscosity.

3. MANUFACTURING PROCESS

Wet hand layup was used to manufacture the single lap shear composite test coupons with a short room temp cure resin. An overlap region of 3in, 2in, and 1in were made in that order to evaluate the failure mode and to minimize the overlap area as to where the failure would occur prior to tensile failure in the coupons. The wet hand layup specimens did not have complete wet-out after cure so we then went with an open mold process to completely wet-out the specimens as shown in Figure 8.

Figure 8. Open Mold SLS Coupons

Four 12" x 12" (305 x 305 mm) panels are processed thru the needle-punching operation to create a dry preform, each will have a needle punch time of 0, 2, 4 and 6 minutes, respectively. These values are reached using the Feltcraft industrial table-top needle felting machine (Figure 6). 13 woven layers and 1 loose fiber layer are stacked together and passed through the motor driven needle-punch process.

To manufacture the panels, the VARTM technique is followed. A release agent is applied to the surface where the preform is placed to prevent the part from sticking to the mold once the composite part hardens and is cured. Subsequently, the carbon preform is laid into a mold, which has two openings, one with a tube through which the Phenolics resin is infused into the fabric, and a second one placed on the opposite end in which a vacuum source is placed. Additionally, this mold is carefully sealed around to prevent trapped air into the laminate. The change in pressure from the vacuum source to the resin inlet source causes the resin to flow along the length and width of the preform, filling the void space between the fabric layers. A resin distribution media is placed on top of the preform and peel ply to allow the resin to quickly flow through the upper layers of the preform before slowing down and penetrating through the thickness. The VARTM setting can be observed in figure 9.

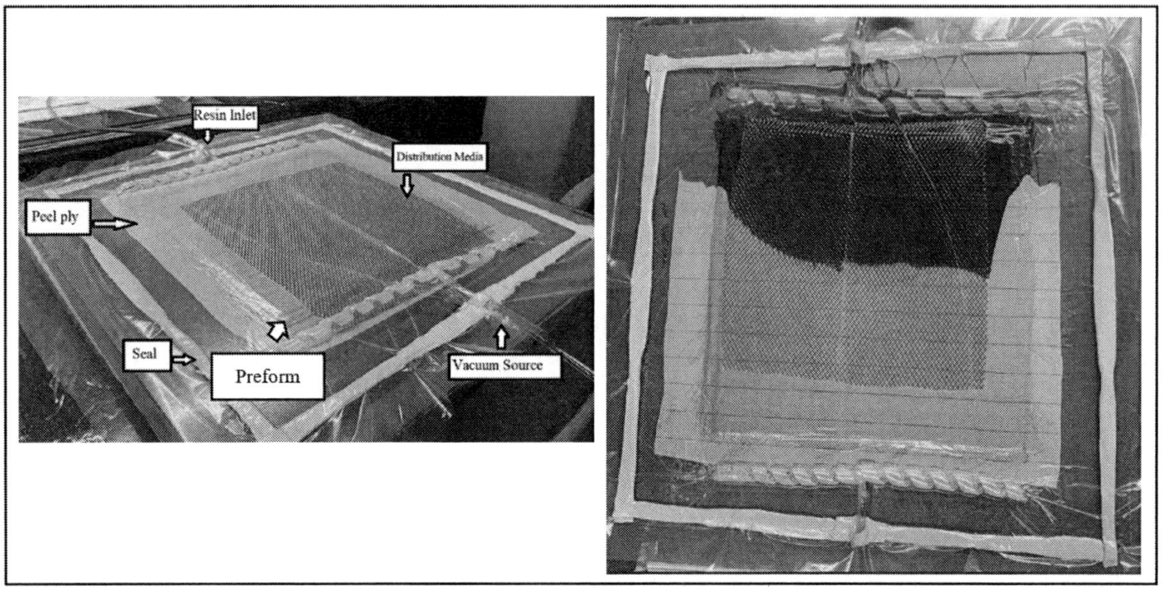

Figure 9. VARTM configuration and phenolics resin flowing.

After a complete impregnation of the preform, the panel is taken to the Genesis compression press at 350°F for 2h at a pressure of 2tons. After the two hour cycle the mold is open and the panels are cured at room temperatue.

All test specimens are extracted from the panels using a CNC water-jet machine which produces minimal heat generated in the samples. Subsequently, a visual inspection for edge delamination under magnification, and measurements for dimensional accuracy using calipers is implemented.

4. EXPERIMENTATION

Experimentation was initially performed with polyester felt fabrics to understand the needle punching machine parameters such as needle punch speed, material feed rates and needle punch density with regards to time. Two separate plies were joined in an overlap region that was subsequently needle-punched for various times as illustrated in Figure 10. This was completed for dry fabric testing.

Later experiments were performed to create Single Lap Shear composite test specimens by wet hand layup and followed by open mold infusion process. The overlap area having been needle-punched to evaluate the impact of the single lap shear strength. The specimens were tested using Mechanical Test System (MTS).

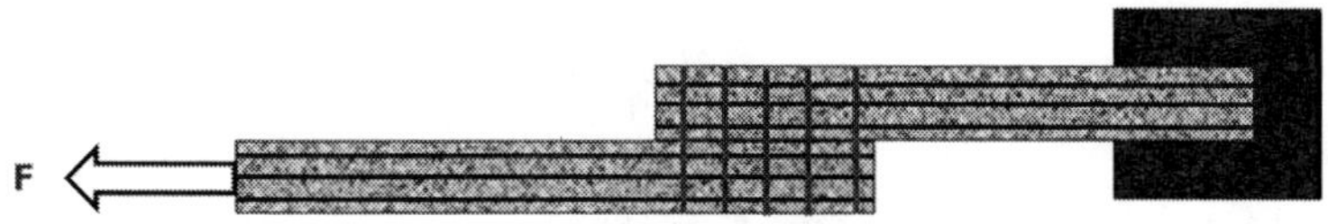

Figure 10. Lap Shear Composite Coupons

Four mechanical properties are evaluated following ASTM standards, tension, flexure, short-beam shear and impact. Additionally, Ultrasonic testing and Scanning Electron Microscopy analysis are implemented.

The composites tension behavior will be evaluated following ASTM standard D3039, for which five samples of 10.00 x 1.00 x 't' in (254 x 25.4 x t mm) are obtained. The flexure analysis requires five samples of 5.5 x 0.5 x 't' in (140 x 13 x 't' mm), according to ASTM D7264. The short-beam shear test explained in the ASTM D2344 requires five samples of 6t x 2t x 't' or below. These properties are evaluated using an MTS 810 Servo Hydraulic System displayed in figure 11.

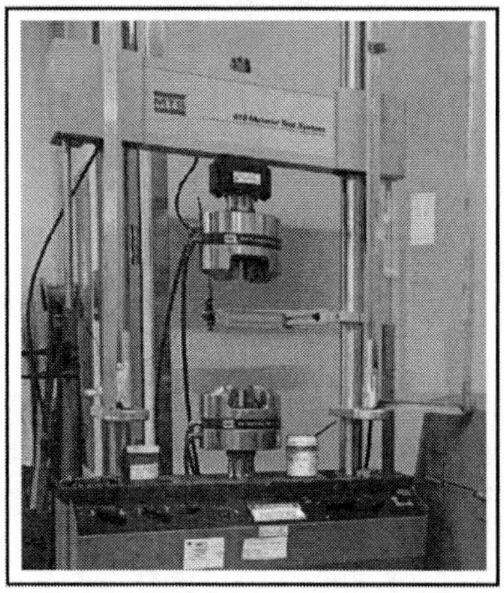

Figure 11. MTS 810 Servo Hydraulic System to evaluate tension, flexure and short-beam shear test.

The impact resistance is evaluated on five samples of 0.5 x 2.5 x 0.16 in according to ASTM D256 using an Instron CEAST 9340 Drop Impact Test System shown in figure 12. The tower can deliver 0.30 to 405 joules of energy which produces high strain rates in the composite and can cause a failure that differs from static forces. The energy level will be such that internal delamination occurs in the panels without the impactor fully penetrating the composite specimen. The delamination will be constrained mostly to the impact surface and the internal structure of the laminate.

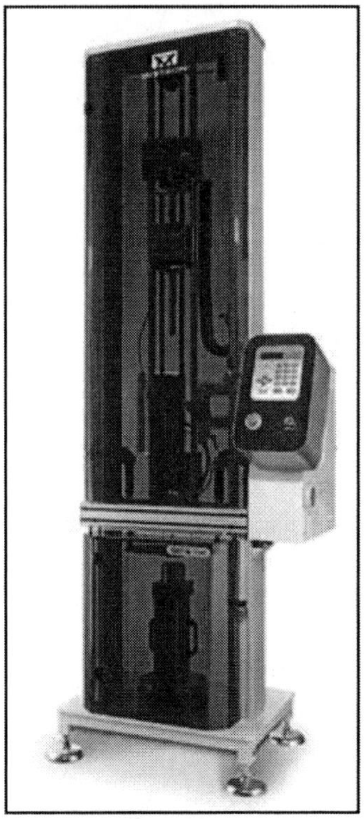

Figure 12. CEAST 9340 Drop Impact Test System to evaluate impact performance of Carbon/Phenolics composites.

5. RESULTS AND DISCUSSION

From the dry fabric needle-punching of polyester felt there was a near linear improvement in the max force that was required to completely separate the two joined needle-punched layers from each other. Graph 1 illustrates the improvement of the dry fabric interlayer pull strength with regards to the needle punch time. This demonstrates that the higher the needle-punch density, the increase in reinforcing fibers that are being displaced and re-oriented thru the thickness of the laminate or the Z-axis.

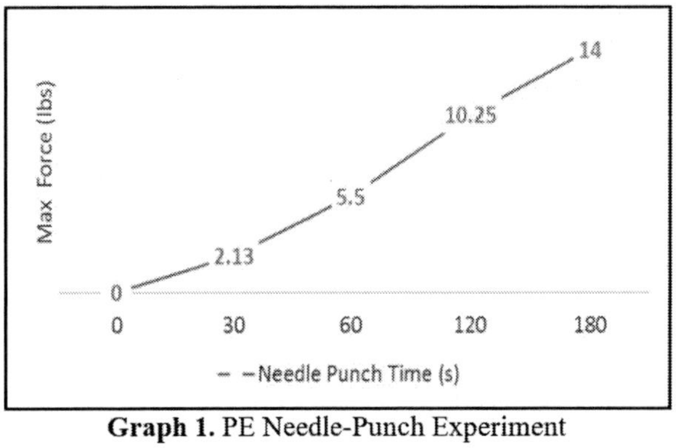

Graph 1. PE Needle-Punch Experiment

Figure 13 shows the post-test results of a dry lap shear needle-punched test sample. It is evident that there are fibers joining the individual layers and once the preform is joined to the matrix in a composite sample these reinforcing z fibers will remain in place and improve the interlaminar properties of the composite.

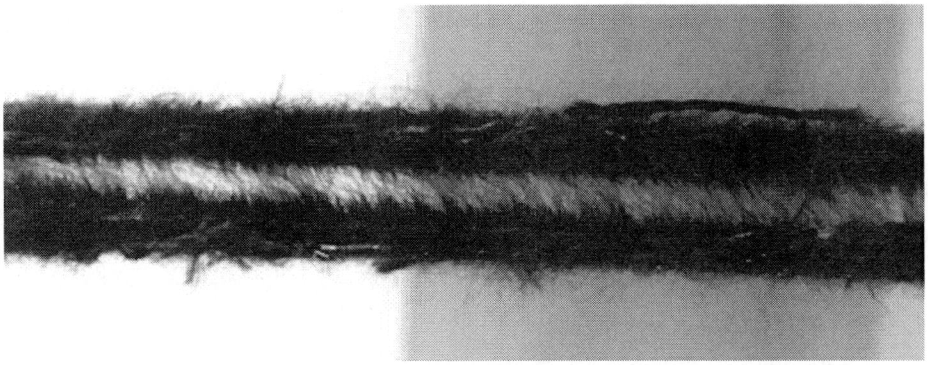

Figure 13. SLS post-test of Needle-punched carbon fabric

Failure modes for the hand layup and open mold infused single lap shear (SLS) composite coupon tests were not desirable. The failure mode occurring outside of the test region and the overlap could no longer be reduced do to needle-punching process constraints. The failures occurred at the neck outside of the over-lap area in tensile failure mode versus the desired shear failure mode. Figure 10 shows the lap shear configuration with the overlap of the two plies being the needle-punched region.

The load over displacement curves are shown in Graph 2 for the SLS coupons where three samples of each group were tested. The loads seen were representative of a tensile failure. The same results were witnessed for both the wet hand layup and open mold infusion coupons.

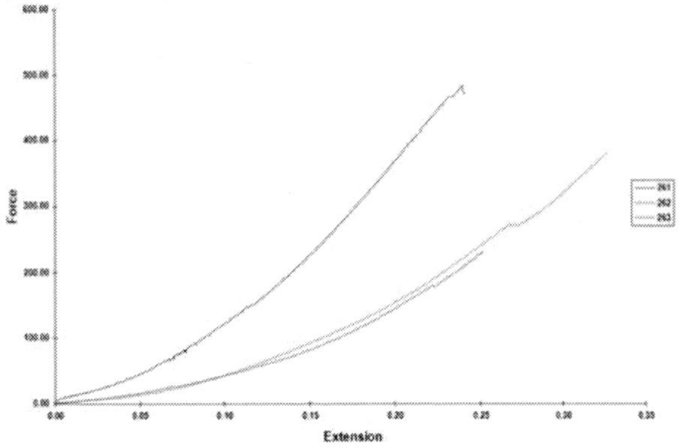

Graph 2. Load vs Displacement Curve of SLS Coupons

The next phase of experimentation is to produce full sized needle-punched panels using VARTM and heated compression press processing. Table 1 shows the plan to continue the research in 2.5D needle-punched composites with regards to mechanical and impact testing.

Phase 2 - Tension; Flexure; ILSS
Labs Re-open
Control Panel Manufacturing
Needle Punch Panel 1 Manufacturing (gauge 1; density 2min)
Needle Punch Panel 2 Manufacturing (gauge 1; density 4min)
Needle Punch Panel 3 Manufacturing (gauge 1; density 4 min)
Change needle-size on needle-board to gauge 2
Needle Punch Panel 4 Manufacturing
Needle Punch Panel 5 Manufacturing
Needle Punch Panel 6 Manufacturing
Waterjet Cut all 7 panels
Mechanical Testing

6. CONCLUSIONS

The feltcraft needle-punching process is an effective method of reinforcing a laminate in the z axis by displacing fibers from the parent fabric layers with the use of barbed needles. These z axis fibers will assist in preventing severe delamination in composites upon impact. It has been shown by Chen et al. 2016 that fiber breakage or damage is inherent in the needle-punching process for the tows that run in the x & y planes. If the needle-density is too high, the in-plane properties will begin to degrade [8]. There must be a balance of needle-density that produced the desired thru-thickness improvements while maintaining required in-plane mechanical properties. A design of experiment will yield the best combination of parameters for the needle-punching manufacturing process.

ACKNOWLEDGMENTS

The authors are grateful to Mr. David Gerald of Interplastic Corporation of Hexion, Inc., and Mr. Luis Arroys of Owens Corning, for their kind support in this research. The authors highly appreciate the help of Dr Ray Cook, Jason Wagner, and Ruben Villarreal for waterjet cutting of test samples. The authors would also like to thank the ARSC at Texas State University for the training and support for TEM and SEM Imaging. The authors are thankful to members of Advanced Composites Laboratory for their support and help at various stages of testing.

7. REFERENCES

[1] Behera, B. K., & Mishra, R. "3-Dimensional weaving". *Indian Journal of Fibre and Textile Research, 33*, 274-287. 2008.

[2] Gokarneshan, N., & Alagirusamy, R. "Weaving of 3D fabrics: A critical appreciation of the developments". *Textile Progress, 41*(1), 1-58. 2009; doi:10.1080/00405160902804239

[3] Ansar, M., Xinwei, W., & Chouwei, Z. "Modeling strategies of 3D woven composites: A review". *Composite Structures, 93*(8), 1947-1963. 2011; doi:https://doi.org/10.1016/j.compstruct.2011.03.010

[4] Aktas, A., Potluri, P., & Porat, I. "Multi-needle stitched composites for improved damage tolerance". *ICCM International Conferences on Composite Materials*. 2009.

[5] Saboktakin, A. "3D textile preforms and composites for aircraft strcutures: A review". *International Journal of Aviation, Aeronautics, and Aerospace, 6*. 2019; doi:10.15394/ijaaa.2019.1299

[6] Tsotsis, T. K. (2012). 8 - Considerations of failure mechanisms in polymer matrix composites in the design of aerospace structures. In P. Robinson, E. Greenhalgh, & S. Pinho (Eds.), *Failure Mechanisms in Polymer Matrix Composites* (pp. 227-278): Woodhead Publishing.

[7] Partridge, I. K., & Cartié, D. D. R. "Delamination resistant laminates by Z-Fiber® pinning: Part I manufacture and fracture performance". *Composites Part A: Applied Science and Manufacturing, 36*(1), 55-64. 2005; doi:https://doi.org/10.1016/j.compositesa.2004.06.029

[8] Chen, X., Chen, L., Zhang, C., Song, L., & Zhang, D. "Three-dimensional needle-punching for composites – A review". *Composites Part A: Applied Science and Manufacturing, 85*, 12-30. 2016; doi:https://doi.org/10.1016/j.compositesa.2016.03.004

[9] Kim, J., Shioya, M., Kobayashi, H., Kaneko, J., & Kido, M. "Mechanical properties of woven laminates and felt composites using carbon fibers. Part 1: in-plane properties". *Composites Science and Technology, 64*(13), 2221-2229. 2004; doi:https://doi.org/10.1016/j.compscitech.2004.03.012

[10] Tae Jin, K., & Sung Ho, L. E. E. "Mechanical properties of non-woven glass fibre composites". *Polymers & polymer composites, 5*(1), 29-39. 1997.

[11] Kang, T. J., Jung, K. H., Park, J. K., Youn, J. R., & Lee, S. G. "Effect of Punching Density on the Mechanical and Thermal Properties of Needle-punched Nonwoven

Carbon/Phenolic Composites". *Polymers and Polymer Composites, 10*(7), 521-530. 2002; doi:10.1177/096739110201000704

Effects of Thermal Cure Protocols on Gelation Behavior and Cure-Induced Shrinkage of Benzoxazine Networks

Bernardo Barea López and Dr. Jeffrey S. Wiggins*
The School of Polymers and High Performance Materials, The University of Southern Mississippi, 118 College Dr. #5050, Hattiesburg, MS 39406, USA.

1. ABSTRACT

Volumetric shrinkage of glassy amorphous polymer matrix networks during cure leads to internal strain in carbon fiber composite parts which resultantly translate to warpage in the final parts. Polybenzoxazine thermomechanical properties such as high glass transition temperatures, superior UV resistance, low coefficient of thermal expansion (CTE), and low shrinkage upon cure compared to more traditional aerospace thermosetting polymers. Recent advancements in continuous prepolymer reactor science provide new opportunities to prepare benzoxazine monomers with high degree of purity without the necessity of performing tedious work-up protocols. However, the polymerization conditions of this promising class of thermosets are still poorly understood with regard to cure-induced shrinkage. This research seeks to establish the relationship between cure shrinkage and cure protocol of polybenzoxazine networks. Herein, three different isothermal curing procedures were selected to achieve different degrees of conversion for the same network. The impact of the curing protocols on the frequency-independent gelation times of polybenzoxazines is determined. Futhermore, the effect of the three different cure protocols have on the chemical shrinkage is investigated through a novel technique which allows to track in real time the volumetric change of the polymer network. This research will ultimately contribute towards the understanding of the relationship between network formation of thermoset materials and the creation of internal strain in the manufacturing process of composite materials.

2. INTRODUCTION

Benzoxazine networks have earned the attention of the aerospace science and engineering community due to their enhanced combination of desirable properties as compared to epoxies which include reductions in fire, flame and smoke, increase in T_g, and enhanced UV stability with comparable mechanical properties and molecular versatility.[1–5] One additional feature that has made benzoxazine particularly interesting as an aerospace composite matrix material is their near-zero shrinkage upon polymerization.[5,6] The challenge associated with benzoxazines is monomers of interest for aerospace applications are intractable crystalline solids, so manufacturing processability has limited their ability to be fully evaluated in structure. Recent advancements in continuous reactor engineering provide a new approach for preparing 100% benzoxazine monomers and prepolymers in the appropriate physical state to prepare high quality aerospace composites. Compared to a synthesis performed in a batch reactor, the main advantages for the synthesis using a continuous reactor presented in this work are the possibility to homogeneously react a broad range of viscosities and molecular weights simultaneously. Twin screw extrusion is remarkably versatile due to the precision feeding, mixing, controlled shear energies, excellent heat transfer, insensitivity to viscosity changes and devolatization. This new approach "*opens the door*" to resolve the unique attributes benzoxazine networks potentially offer the aerospace composites industry.[7–9] This research prepares 100% benzoxazine networks with chemo-rheological properties to quantify matrix cure shrinkage effects upon structural mechanics of aerospace quality fiber reinforced composites.

Volumetric expansion and contractions in composite matrix polymers that occur from chemical shrinkage during cure, and thermal cycling in application becoming important phenomenon in strain states inherent to aerospace composites. Strains result at the interface from differences in thermally controlled volumetric variances between carbon fiber and organic matrix polymers. Chemical shrinkage of amorphous glassy polymer network matrices arise during the polymerization of the network. During crosslinking, small molecules and oligomers undergo step-growth polymerization reactions to create a highly crosslinked structure of infinite molecular weight. The polymerization and resulting secondary interactions between molecules leads to densification and shrinkage throughout cure.[10] Polymerization of glassy polymer networks are conducted at elevated temperatures with glass transition temperature, crosslink density and network density increasing through vitrification and conversion. Often, the polymer is cured at a temperature well beyond the ultimate glass transition temperature, and in the rubbery state. Increases and decreases in temperature throughout cure lead to a dynamic and complex volumetric environment of the matrix leading to volumetric expansion in the material, known as the coefficient of thermal expansion (CTE).[11] Thermoset networks are reported with volumetric shrinkage between 3-15%, including aerospace epoxies in the 2-7% during cure, while BOX is reported to exhibit near-zero shrinkage upon cure.[5] Outcomes from this research will contribute to further understand the volumetric shrinkage during polymerization of benzoxazine networks using different cure protocols.

3. EXPERIMENTATION

a. Materials

2,2-Bis(4-hydroxyphenyl)propane (BPA, TCI America, MW = 228.29 g/mol, purity > 99.0%), aniline (Sigma Aldrich, MW = 93.13 g/mol, purity > 99.5 %) and paraformaldehyde powder (Sigma Aldrich, MW = 30.03 (as monomer), purity 95%) were used as received.

b. BA-a Synthesis

A patented continuous high-shear reactor (CSHR) method was used for the synthesis of the benzoxazine monomer studied in this research.[7,8] This method consisted of a Prism 16 mm co-rotating intermeshing twin-screw reactor (L/D=25) and enclosed by a walk-in fume hood. This reactor features 5 different zones which can be independently heated and cooled. The screw configuration was designed to promote the conversion of reactants to monomers by balancing high-shear mixing and residence time. The efficacy of transporting the feed through the reactor was qualitatively determined by the lack of agglomerated feed in the final product as well as the absence of build-up or clogs on the screws or inside the barrel of the reactor. An automatic feeding method was designed which was comprised of a liquid feeder and a solid feeder. The liquid feeder was used to pump a solution of BPA and aniline (1:2 mole ratio respectively) and the solid feed charged paraformaldehyde at a 4:1 mole ratio with respect to the BPA added. All the reagents were fed into zone 1 of the twin-screw reactor. Figure 1 shows a synthesis scheme.

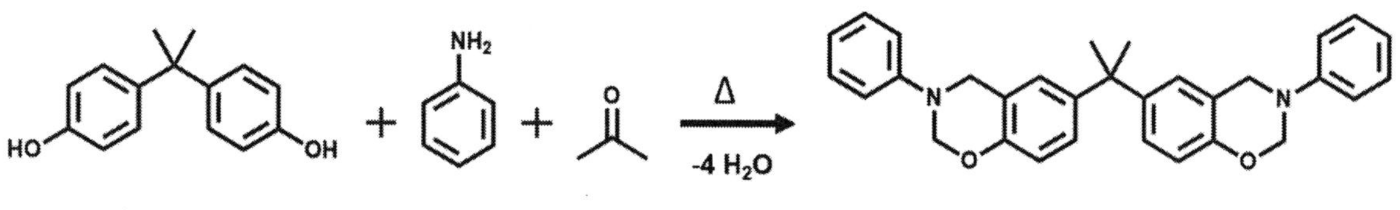

Figure 1. BA-a monomer synthesis scheme.

c. Thermogravimetric Analysis (TGA)

Thermogravimetric Analysis (TGA) was performed on a TA Instruments Q50 TGA. The samples were heated isothermally at 160 °C, 170 °C and 180 °C for 4 hours in air and nitrogen atmospheres.

d. Rheological and Volumetric Shrinkage Analyses

All the gelation and volumetric shrinkage measurements were obtained using a TA Instruments ARES-G2 rheometer equipped with a Forced Convection Oven (FCO). Disposable 25 mm parallel aluminum plates were used for every run performed with this instrument. The rheological and volumetric shrinkage studies for our samples were obtained under isothermal conditions at 160 °C, 170 °C and 180 °C.

Fourier Transform Mechanical Spectroscopy (FTMS)[12] was the method chosen to calculate the gel point of the BOX networks following the Winter et al. criterion.[13] During FTMS experiments, a compound strain waveform was applied to the sample which allowed for the simultaneously collection of G' and G" data for multiple frequencies. The gel point of each sample was then estimated at the time which tan δ is independent of frequency. Furthermore, the critical relaxation exponent (n_c) at the point of gelation of the benzoxazine networks was also calculated (Eq.1) and analyzed.

$$n_c = \frac{2\delta}{\pi} \; ; \; \delta = \tan^{-1}\frac{G''}{G'} \quad \text{(Eq.1)}$$

Once the gelation times for the different isothermal conditions were obtained, it was possible to proceed with the volumetric shrinkage experiment. This experiment had three different stages: Gelation, Cure and Cool Down. During the Gelation stage the conditions used to obtain the gelation times were replicated. At the point where the network achieved the point of gelation previously calculated, the experiment moved to the Cure stage. The network at this point was going through a liquid-solid transition. Therefore, the strain applied by the rheometer at this stage was reduced. Furthermore, a normal constant force of 0 N between the aluminum plates was set and the gap between the parallel was tracked. This gap was plotted in function of time to get a real-time linear dimensional shrinkage upon cure. After 3 hours (Gelation and Cure stages combined) of experiment under isothermal conditions, the Cool Down stage began. During this stage, the temperature was brought down from the cure temperature to 30 °C by purging air at room temperature inside the FCO. The gap between both plates at room temperature was not affected by the thermal expansion occurring at the high curing temperatures, exhibiting the true gelation shrinkage.[14] Finally, in order to transform the linear gap separation into a volumetric change the equation derived by M. Haider et al. was used (Eq.2). ε_υ is the volumetric change, h_0 is the initial gap and h is the gap over time.[15]

$$\epsilon_v = \left[1 + \frac{1}{3}\left(\frac{h-h_0}{h_0}\right)\right]^3 - 1 \quad \text{(Eq. 2)}$$

4. RESULTS

The BA-a obtained through the continuous reactive synthesis was degassed for 1 hour under high vacuum and no further purification processes were performed previous to any of the characterization analysis presented in this work. TGA tests were needed to establish how thermally stable BA-a during the selected cure protocol. Figure 2 shows the weight loss during three different isothermal curing protocols under nitrogen and air conditions. It was noticeable a higher weight loss when air was present during the cure which agrees with past research done on this compound.[1] Furthermore, in both cases the majority of the weight loss occurred within the first 30 minutes of the cure. Since the resultant product from the reactor was not purified, it is believed that this initial loss in weight at the beginning of the cure is related to the degradation of initial reagents from the synthesis and the evaporation of humidity. Most importantly, it was observed that after the initial drop of weight, the samples stay thermally stable during the rest of the curing time.

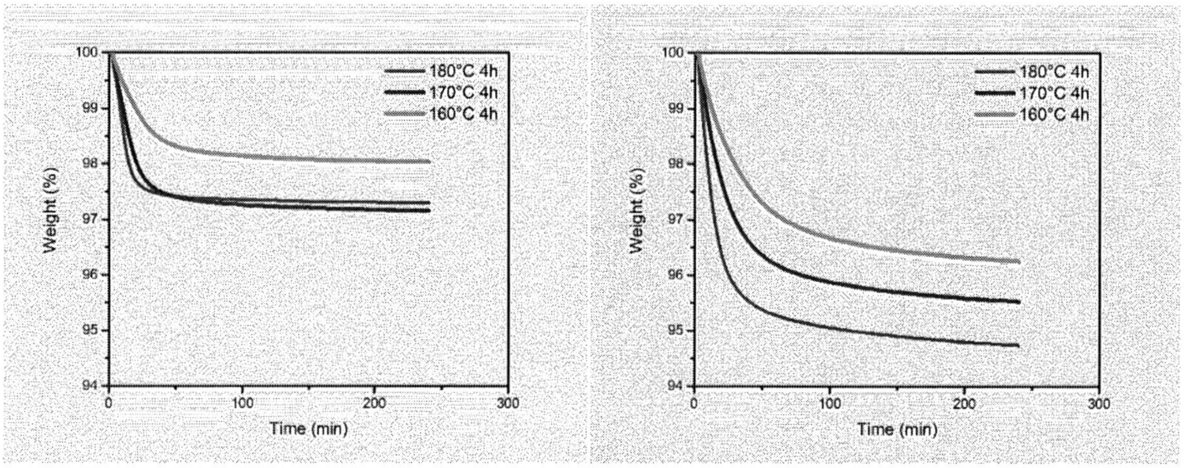

Figure 2. Weight loss during the isothermal cure through TGA in Nitrogen (left) and air (right)

Once it was proved the thermal stability of BA-a, a rheological study was performed to measure the gelation time at different isothermal polymerization temperatures. Firstly, to validate the get points of our compound, every rheological test had to operate within the linear viscoelastic regime (LVR). A small amplitude oscillatory shear (SAOS) was ran to obtain the LVR of the BA-a network studied. Nevertheless, a fundamental issue arrived to determine the LVR of thermoset networks at the point of gelation. The LVR of the BA-a monomer extended past 100% strain, while the LVR of the cured BA-a network is much lower. For the purpose of this research, the gelation studies of BA-a were performed at a 25% strain since it yielded highly defined crossover points regardless the criteria used to obtain them.

Two criteria to define the gel point of BA-a were obtained and compared. The three graphs on Figure 3 show the crossover point between the storage modulus (G') and the loss modulus (G") of two different frequencies at three different isothermal curing temperatures. Historically, this point has been consider the point of gelation of a given polymer network. However, the crossover point obtained from this criterion is frequency dependent due to stress relaxation during the data acquisition time.[13] The results obtained in this research confirmed the findings done by Winters[13] and they showed the gelation times of BA-a at different temperatures was frequency independent as shown in Figure 4. The frequency dependency on the gelation time seemed to be more attenuated the longer the sample took to reach gelation. For instance, during the 160 °C isothermal cure, the crossover time discrepancy between the two frequencies is 251 seconds. Yet, this time discrepancy appears to be marginal and within the data time acquisition error of the instrument during the 180 °C isothermal cure.

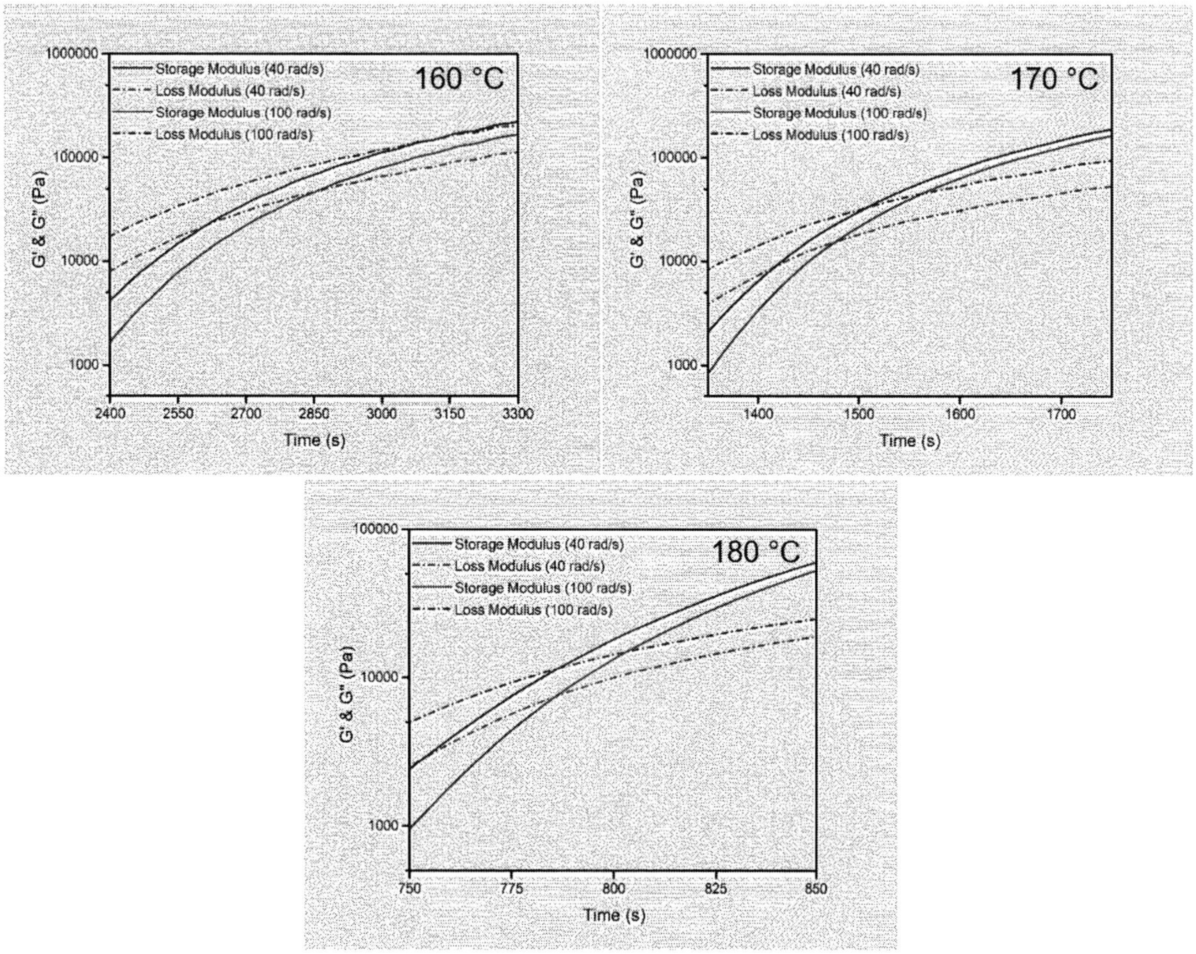

Figure 3. G' and G" crossover graphs at two different frequencies (40 rad/s and 100 rad/s) at 25% strain under three different isothermal cures.

To address this gel time discrepancy especially present at lower curing temperatures, Winter *et al.* proposed that the point of true gel or critical gelation point is found at the tan δ crossover at different frequencies.[13] It is not the first time the critical gel point has been studied for BA-a. Ishida *et al.* tried to calculated previously.[16] However, they reported that BA-a did not display a critical gel behavior in the multi-frequency experiments. In addition, this work done by Ishida in 2001 required the repetition of lengthy SAOS experiments at a variety of frequencies. Fortunately, Fourier transform mechanical spectroscopy (FTMS) can bypass the necessity of running multiple SAOS to obtain the tan δ crossover point of the sample. FTMS applies compound strain waves resultant from the addition of numerous sinusoidal strain waves at different frequencies. Upon Fourier transformation of the resulting stress response into individual stress values for each frequency component, standard G'/G" data at multiple frequencies can be obtained for each oscillation cycle.[12] Figure 4 shows defined tan δ crossover points for every isothermal cure temperature when the frequencies applied were 40 rad/s, 60 rad/s, 80 rad/s and 100 rad/s. It was observed that frequencies outside the vicinity studied in the present work did not exhibit a clear crossover point. In addition, Figure 4 proves that BA-a displayed a critical gel behavior when FTMS was applied as opposed to what it was believed in the past.[16] The critical gel points for these three cases was shown in Table 1.

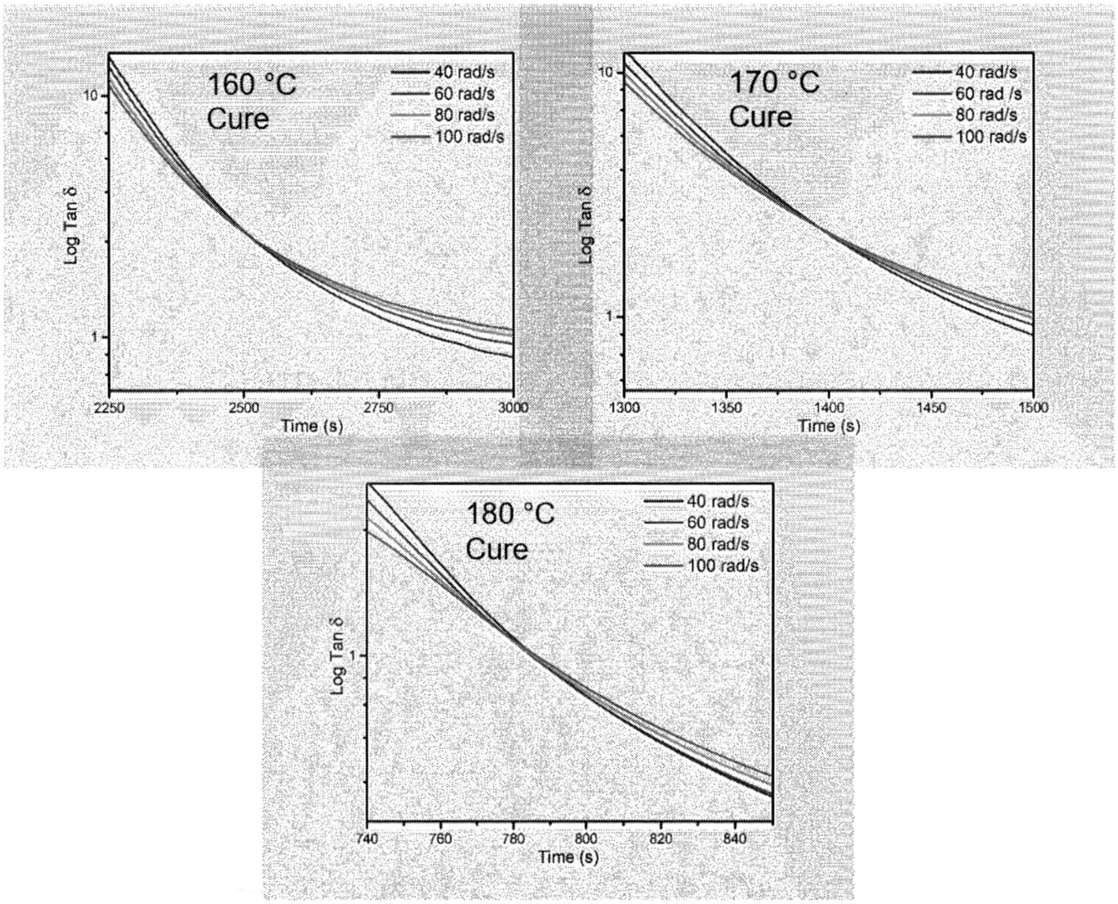

Figure 4. Tan δ crossover graphs at 4 different frequencies (40 rad/s, 60 rad/s, 80 rad/s and 100 rad/s) at 25 % strain under three different isothermal cures.

FTMS was also used to calculate the critical relaxation exponent (n_c) of the different BA-a networks that underwent different curing protocols. This parameter was calculated using Eq.1, hence n_c is proved to be frequency independent. Additionally, n_c has also been shown to be temperature independent for systems where the network growth is invariant to temperature profile.[17] However, BA-a networks are thermally polymerized so n_c will inherently dependent to the cure temperature since this dictates the network growth of the sample. The value range for n_c goes from 0 to 1. A n_c value close to 0 is representative of a stiff network at the point of gelation which implies a low degree of conversion at the point of gelation. On the other hand, a n_c value close to 1 is representative of a soft network or a network with low degree of conversion at this point. In the present work, we compared the values obtained for n_c using three different cure profiles which were summarized in Table 1. The n_c for the 160 °C and the 170 °C cure were very similar, 0.773 and 0.738 respectively. The similarity between these two values might suggest an also similar degree of conversion of the network in both cases. However, the n_c for the isothermal cure at 180 °C was 0.514. Interestingly, this value was expected since only when n_c is 0.5 the G' and G" crossover becomes frequency independent. Figure 3 shows that the G' and G" crossovers at two different frequencies during this cure protocol occur within the error vicinity of the instrument data time collection.

Table 1. Summary of the rheological parameters.

Cure Temperature	**Tan δ Crossover Time**	**G' & G" crossover time (40 rad/s)**	**G' & G" crossover time (100 rad/s)**	n_c
160 °C	2505 s	2853 s	3104 s	0.773
170 °C	1396 s	1477 s	1512 s	0.738
180 °C	783 s	787 s	787 s	0.514

The gel studies did not only provide insightful information regarding the network morphology and kinetics of BA-a systems. It was also used to complement the technique used in the present work to calculate the volumetric variances of polymer networks. This technique firstly developed by M. Haider *et al.* consisted of calculating the volumetric variances of a polymer network upon cure by tracking the gap between parallel rheometer plates where the sample was placed.[15] The conversion from a linear change to a volumetric change is calculated by using Eq. 2 located in the experimental part of this manuscript. In this technique, the volumetric starts to be recorded after the gel point of the sample. Therefore, the gel studies data previously gathered will be fundamental to determine with precision the exact point where the volumetric shrinkage data collection should begin. Figure 5 shows the volumetric changes of BA-a networks after gelation during three different cure protocols. The point of gelation of these samples was determined as the tan δ crossover of these samples shown in Table 2. To our knowledge, the use of FTMS as a complement of this novel technique to measure the volumetric changes of a polymer network upon curing has never been reported.

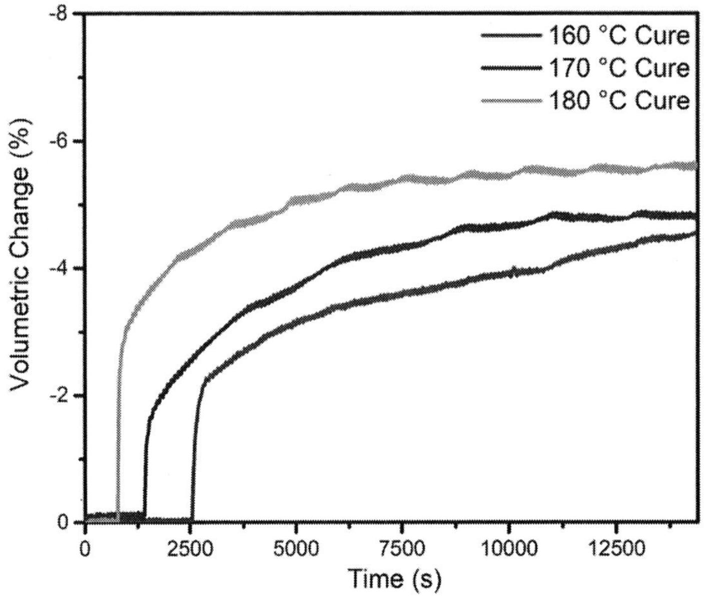

Figure 5. Volumetric change of BA-a networks after gelation during three different cure protocols

Table 2. Data summary of volumetric change experiment.

Cure Temperature	Volumetric Shrinkage after 4 hours
160 °C	4.57 %
170 °C	4.83 %
180 °C	5.59 %

Figure 5 reveals that BA-a networks experience a considerable reduction in volume after gelation. The gel point of a network is considered to be the point when the network virtually achieves infinite molecular weight. In other words, most of the monomer unit are covalently linked. At this point, the network is formed but it is far away from being fully converted. Since all the monomeric units are already linked, the successive bond formation occurs at a faster rate due to the proximity between the reactive units. Hence, this increase in reactivity after gelation explains why the volumetric shrinkage rate is greater right after the sample achieves the gel point. Additionally, the graph shows that the curing temperature has a clear effect on the shrinkage of the BA-a polymer network. The cure temperature will determine the degree of conversion of thermally activated polybenzoxazine networks.[18] Covalent bonds among the repeating units of the thermoset network are created as the degree of conversion goes up. Since BA-a monomers are solely thermally reactive, a higher cure temperature for these systems will imply a higher degree of network conversion. Therefore, it is expected to observe a higher volumetric shrinkage as the cure temperature of the BA-a network increases.

5. CONCLUSIONS AND FUTURE WORK

This work aimed to demonstrate the importance of the cure protocol in the network formation of BA-a systems. The recent findings in TSE reactions made possible the synthesis of large amounts of high purity benzoxazine monomers. Three different isothermal cure schedules were selected for this research: 160 °C, 170 °C and 180 °C for four hours. The thermal stability of BA-a systems undergoing these curing schedules were demonstrated by thermogravimetric analyses. Futhermore, the gel times of BA-a systems at different isothermal curing temperatures were calculated by FTMS using the crossover of tan δ at different frequencies as the point of critical gel. This approach was proved to solve gel time discrepancies that more traditional gel point definitions exhibited. Additionally, FTMS revealed that the relaxation exponent at gel changed when the systems were cured at different isothermal temperatures. Finally, the volumetric shrinkage of BA-a networks was real-time tracked from the point of gelation. It was observed that volumetric shrinkage was not a linear process and it was temperature dependent. Future studies will involve the use of Differential Scanning Calirometry (DSC) and Real-Time Fourier-Transform Infrared (RT-FTIR) to track the degree of cure and how the BA-a network evolves during polymerization.

6. REFERENCES

1. Ran, Q.; Gu, Y.; Ishida, H. *Advanced and Emerging Polybenzoxazine Science and Technology*. Ch. 12. **2017**.
2. Yee Low, H.; Ishida, H. *Polymer (Guildf)*. **1999**, *40* (15), 4365–4376.
3. Wang, Y. X.; Ishida, H. *J. Appl. Polym. Sci.* **2002**.
4. Ohashi, S.; Ishida, H. *Advanced and Emerging Polybenzoxazine Science and Technology*. Ch. 1. **2017**.
5. Ishida, H.; Allen, D. J. *J. Polym. Sci. Part B Polym. Phys.* **1996**, *34* (6), 1019–1030.
6. Ishida, H.; Low, H. Y. *Macromolecules* **1997**, *30* (4), 1099–1106.
7. Frazee, A. S.; Wiggins, J. S. U.S. Patent 10106511, October 23. **2018**.
8. Frazee, A. S.; Wiggins, J. S. U.S. Patent 10106653, October 23. **2018**.
9. Frazee, A. Ph.D. Dissertation, The University of Southern Mississippi. **2017**.
10. Billotte, C.; Bernard, F. M.; Ruiz, E. *Eur. Polym. J.* **2013**, *49* (11), 3548–3560.
11. Knowles, K. R. Ph.D. Dissertation, The University of Southern Mississippi. **2017**.
12. Holly, E. E.; Venkataraman, S. K.; Chambon, F.; Henning Winter, H. *J. Nonnewton. Fluid Mech.* **1988**, *27* (1), 17–26.

13. Winter, H, H. *Polym. Eng. Sci.* **1987**, *27* (22).

14. Shah, D. U.; Schubel, P. J. *Polym. Test.* **2010**, *29* (6), 629–639.

15. Haider, M.; Hubert, P.; Lessard, L. *Compos. Part A Appl. Sci. Manuf.* **2007**, *38* (3), 994–1009.

16. Ishida, H.; Allen, D. J. *J. Appl. Polym. Sci.* **2001**, *79* (3), 406–417.

17. Izuka, A.; Winter, H. H.; Hashimoto, T. *Macromolecules* **1994**, *27* (23), 6883–6888.

18. Jungman, M. J.; Cobb, J. S.; Lawler, D. J.; Sholar, J. D.; Johnson, C. B.; Patton, D. L. *Macromol. Symp.* **2013**, *329* (1), 133–141.

THE USE OF POST CONSOLIDATION TO IMPROVE THE MECHANICAL PROPERTIES OF LIGHT WEIGHT REINFORCED THERMOPLASTIC COMPOSITES

Hongyu Chen, Ruomiao Wang, Mark O. Mason
Hanwha Azdel
2000 Enterprise Drive
Forest, VA 24551, USA

ABSTRACT

Light weight reinforced thermoplastic (LWRT) composites are widely used in the automotive industry because of the advantage of light weight resulting in high fuel efficiency. This paper discusses the use of post consolidation to improve the mechanical performances of the fiber composite material developed at Hanwha Azdel. This high performance composite material is comprised of a porous core with chopped glass fiber and thermoplastic resin, and surface finish layers on each side. A post process was developed to improve the mechanical properties by further consolidation of the porous core. Mechanical performance was investigated by both flexural and tensile tests, and the surface morphology was examined by scanning electronic microscope. Both mechanical and morphology investigations indicate that, especially for the lighter materials, this post process significantly improved the consolidation level of the porous core. Therefore, this promising post consolidation method is expected to further reduce weight without sacrificing the mechanical strength of the material.

Corresponding author: Hongyu Chen. Electronic mail: hchen@azdel.com

1. INTRODUCTION

It is an ever-lasting trend to pursue lightweight materials in the automotive industry to increase fuel efficiency for vehicles [1-2]. There has been a fast growth in the use of reinforced thermoplastic composites for automotive parts, such as seat backs, door panels, trunk trim, interior parts, and headliners [3]. Such fiber reinforced thermoplastic composites mainly consist of polymer resin matrix and fiber reinforcement. The thermoplastic matrix can be polyolefin, Polyethylene terephthalate (PET), polyurethane, nylon, and other engineering polymers. The reinforcing fibers are glass, natural fibers from plants, carbon, and synthetic polymeric fibers.

The primary topic of this paper is glass fiber reinforced polypropylene (PP) composites manufactured by wet-laid process for thermo-forming use. They are excellent candidates as headliner materials due to their light weight, good formability, and low coefficient of thermal expansion [4]. To produce the LWRT sheet through wet-laid process, an aqueous suspension of well-dispersed PP resin in powder form and chopped glass fiber is transferred onto a web-forming belt. After the removal of water, heating the materials to above the resin melting point, lamination and consolidation, a LWRT sheet is manufactured as the final product by Hanwha Azdel. A headliner, for example, can be produced by thermo-forming the produced LWRT sheet as shown in Figure 1. During the thermo-forming process, after the resin melts, the residual stress among fibers is released leading to the expansion of the LWRT sheet which is a phenomenon called "loft". The "lofted" sheet is then mold to the targeted thickness that is normally thicker than the as-produced thickness of the LWRT sheet. As a result, the final part (headliner) normally has a very porous structure in which resin serves the function of adhesive to hold the chopped fiber together. The intrinsic high porosity structure of the LWRT products provides opportunities for improvement of mechanical properties. One promising approach investigated in this work is the use of post/secondary consolidation that can notably enhance the "wet-out" of resin on the surface of glass fibers in the final parts. As a result the mechanical properties, especially for the lighter weight materials, have been largely improved. This promising post consolidation method is expected to further reduce the weight of wet-laid process related LWRT materials without sacrificing the mechanical performances.

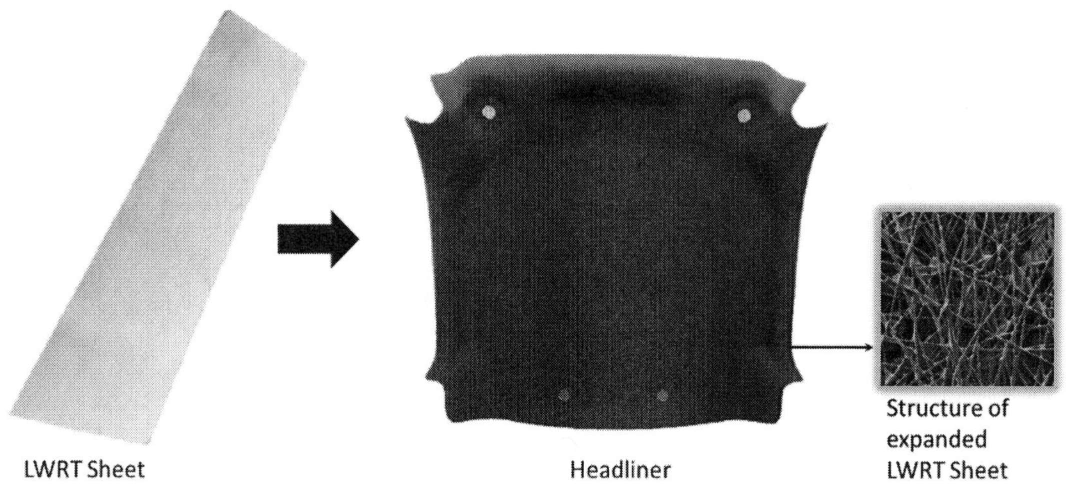

Figure 1. Headliner made of LWRT sheet.

2. EXPERIMENTATION

2.1 LWRT Sheet Manufacturing Process

The LWRT core was manufactured by using the wet-laid process. Polypropylene powder, chopped glass fiber and other additives were dispersed in water. The aqueous slurry was transferred onto a web forming section. The resulting web was drained and then heated to be above the melting point of the polymer resin. According to the end-use application, the LWRT core was then laminated with surface materials (non-woven scrim on top and polymer film on bottom) on both sides as shown in Figure 2. Finally, the material was consolidated to produce a flat LWRT composite sheet. Materials with various basis weight (areal densities) can be produced by adjusting the manufacturing parameters. In this study, three samples with different basis weight (gsm, grams per square meter) were produced as shown in Table 1.

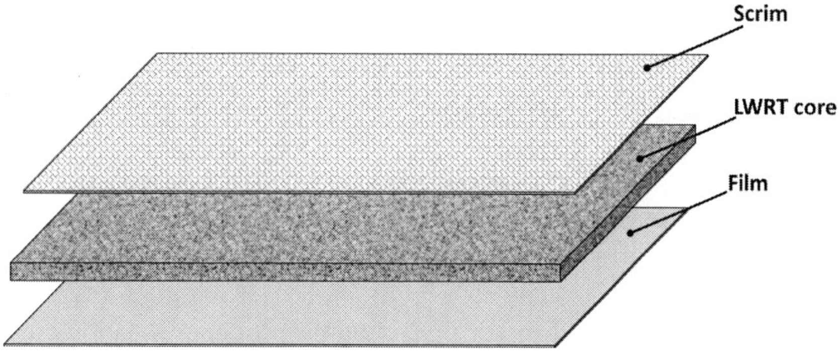

Figure 2. LWRT sheet construction.

Table 1. Basis weight design for the three LWRT sheet samples

Sample	Core's basis weight (gsm)	Skin on top	Skin on bottom
A	450	20 gsm scrim	98 gsm film
B	1000	20 gsm scrim	70 gsm film
C	1200	20 gsm scrim	70 gsm film

2.2 Post consolidation and molding

The general processing flow diagram for post consolidating the as-produced LWRT composite board is outlined in Figure 3. Briefly, the as-produced LWRT composite board is pressed in the hot press with specified pressure and dwell time. In this study the hot press is heated up to 195 °C and a pressure of 3.8 bar is applied to the board. Three dwell times were investigated: 5, 10, and 20 seconds. After heating in the hot press, the board is transferred to the cold press to cool down, where the same pressure and dwell time are used as in the hot press. The use of the cold press is to maintain the thickness of the hot pressed board and prevent the occurrence of lofting while cooling. After consolidation, the board is lofted and molded to the targeted thickness through the thermoforming process as shown in Figure 4. For purposes of comparison, a control for each sample was directly lofted and molded to the targeted thickness without post consolidation. The post consolidation settings and targeted molding thicknesses are listed in Table 2.

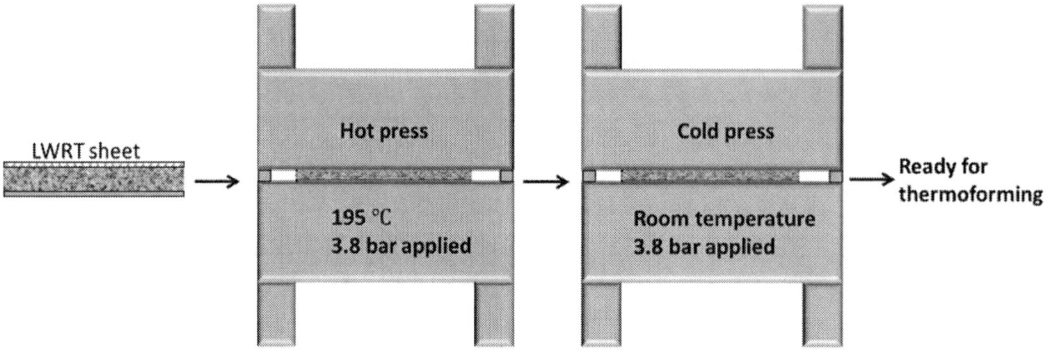

Figure 3. Schematic diagram explaining the post consolidation process.

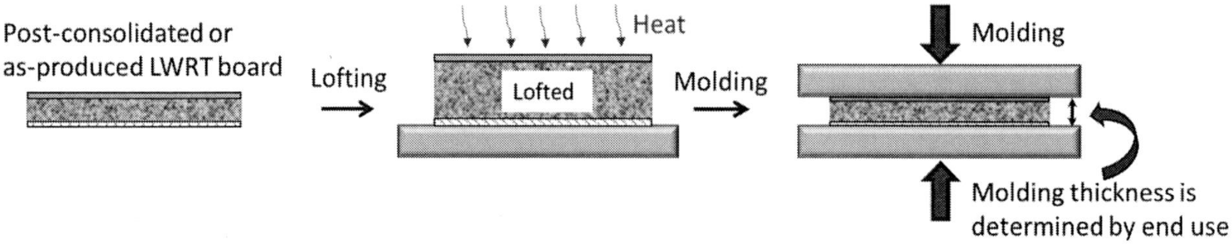

Figure 4. Schematic diagram explaining the lofting and molding process.

Table 2. Post consolidation settings and targeted molding thickness

Sample	Dwell time in hot/cold press (seconds)	Pressure of hot/cold press (bar)	Targeted molding thickness (mm)
A	5, 10, and 20	3.8	4.5
Control of A	N/A	N/A	4.5
B	5, 10, and 20	3.8	2.75
Control of B	N/A	N/A	2.75
C	5, 10, and 20	3.8	3
Control of C	N/A	N/A	3

2.3 Scanning electron microscope (SEM)

The surface morphologies on the scrim side of the molded LWRT boards with and without post consolidation were investigated by scanning electron microscope. A small rectangular specimen was cut from the molded panel, coated with a thin layer of gold on the surface of interest (scrim side), and the surface morphology was examined under vacuum.

2.4 Mechanical properties

Mechanical tests were conducted to evaluate the effects of the post consolidation process. Specimens for tests were cut out from the molded LWRT boards. The flexural properties of all the specimens were evaluated according to ASTM D790. Tensile tests were carried out as well, according to ASTM D5034.

3. RESULTS

3.1 Surface morphology (SEM)

For LWRT composite, mechanical performance is highly dependent on the "wet-out" of resin on the surface of glass fibers. Therefore, an extra post consolidation process was introduced to improve the adhesion between glass fiber and PP resin. Surface morphology from SEM micrographs is a direct indicator of the degree of "wet-out" in the LWRT composite. Comparisons between molded samples without post consolidation and those with post consolidation using dwell times of 20 s, are made in Figure 5 to Figure 7 by checking the surface morphology of the scrim sides. It turns out that, for all the molded samples (450, 1000 and 1200 core gsm), the extra post consolidation process significantly changed the surface morphology of the scrim side. Without post consolidation, the scrim sides are highly porous, while the porosity is decreased significantly by the post consolidation process. It is an indication that the extra pressure and heating during post consolidation made the resin spread out better among the fibers in the core.

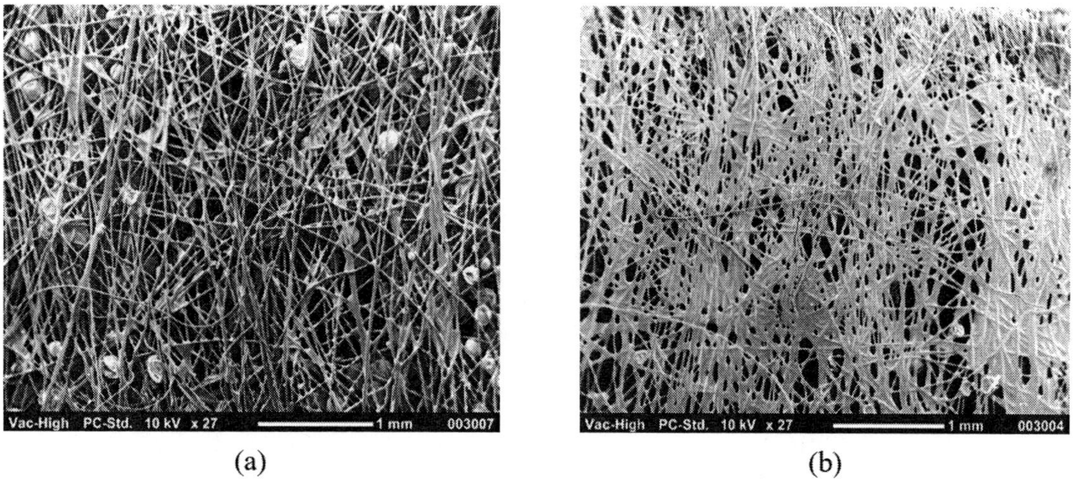

Figure 5. Surface morphology (SEM micrographs) on the scrim side of the molded LWRT board, (a) without post consolidation (control), and (b) with post consolidation, for sample A: 450 core gsm materials.

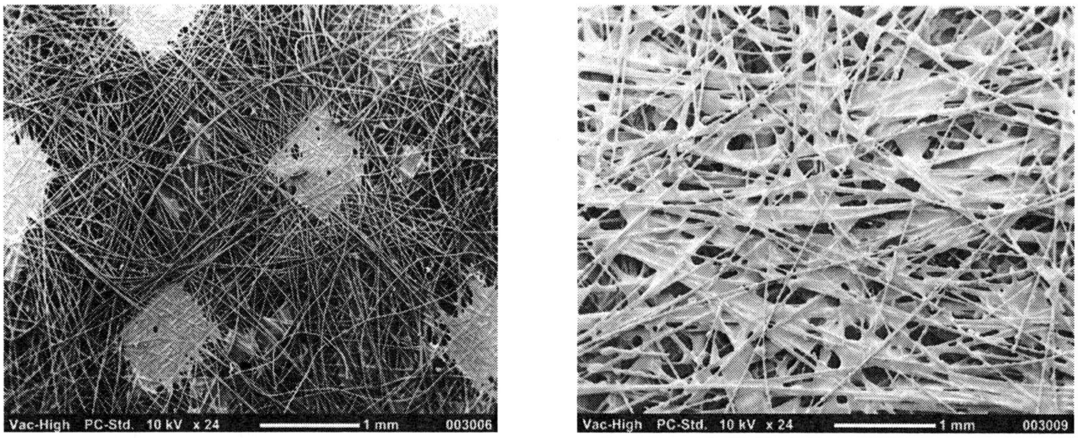

(a) (b)

Figure 6. Surface morphology (SEM micrographs) on the scrim side of the molded LWRT board, (a) without post consolidation (control), and (b) with post consolidation, for sample B: 1000 core gsm materials.

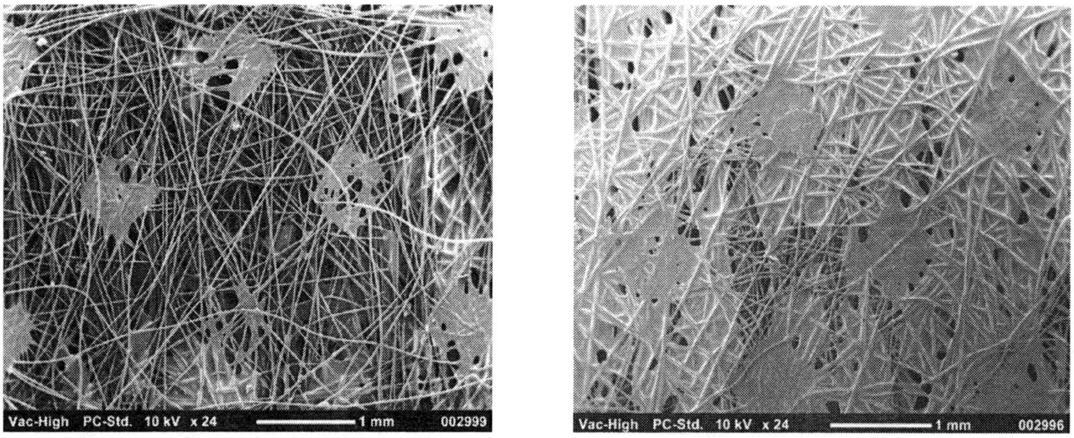

(a) (b)

Figure 7. Surface morphology (SEM micrographs) on the scrim side of the molded LWRT board, (a) without post consolidation (control), and (b) with post consolidation, for sample C: 1200 core gsm materials.

3.2 Flexural properties

Flexural properties are important for the handling of the LWRT sheets during the thermoforming process, in which the sheets are molded into headliners and attached to the fabric A surface. If the sheets are not stiff or strong enough, wrinkles might appear on the A surface due to bending deformation during handling, and even catastrophic failure of the sheets is possible. The flexural properties of all the specimens were evaluated and discussed here. The flexural peak load and stiffness of the control and the post consolidated specimens for 450, 1000 and 1200 core gsm samples are given in Figure 8, Figure 9, and Figure 10, respectively. All the samples show significantly better results in machine direction (MD) than those in the cross-machine direction (CD). The aligning of fibers happens primarily in the head box during the web-forming part of the wet-laid process, and it favors the machine direction remarkably. The flow in the head-box is a mix of both shear and extension. There is strong shearing flow close to the walls and extensional flow toward the machine direction within the entire domain. As a result, the fibers are strongly aligned toward the flow direction leading to better mechanical performance in MD.

For the lightest 450 core gsm materials, the post consolidation remarkably improved both of the peak load and stiffness for all three dwell times. The longer the dwell time, the better the flexural performance. The use of the shortest time (5 s) improved the flexural properties from ~ 39% to 60%. With the longest time (20 s), the least improvement of 65% was seen in CD for the stiffness, and the most improvement of 89% was achieved for peak load in MD. For the 1000 core gsm materials, dwell time seems to have much less influence on the mechanical properties. According to t-test results, both peak load and stiffness of the post consolidated specimens still get improved by 14% (stiffness in CD) to 44% (peak load in CD) compared to the control. For the heaviest 1200 core gsm materials, improvements were only realized for the peak loads according to the t-test, and there is no enhancement for stiffness by the use of post consolidation. It is worth increasing the hot press temperature in future to see whether the stiffness can get improved as well for the heavier materials.

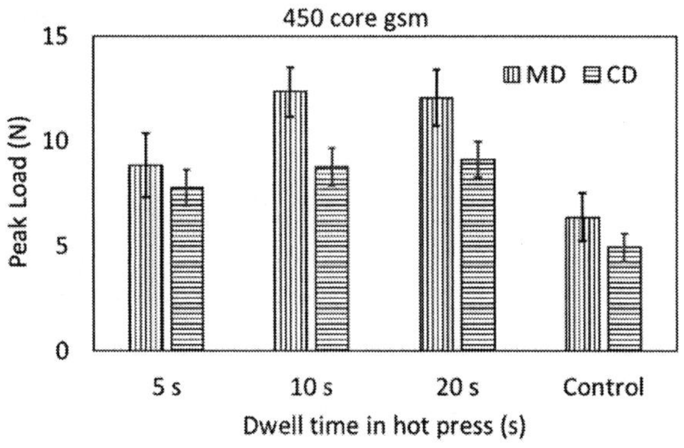

(a)

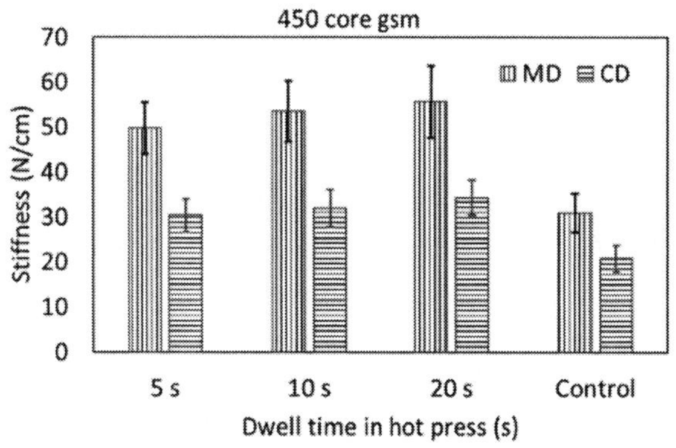

(b)

Figure 8. Flexural property comparisons between the control (without post consolidation), and the use of 5 s, 10 s, and 20 s dwell time for post consolidated samples, of (a) peak load, and (b) stiffness, for sample A: 450 core gsm materials.

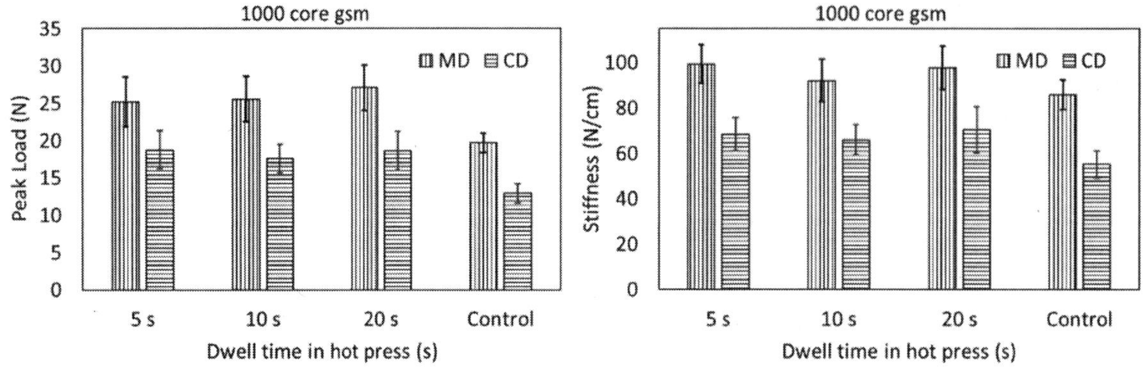

(a) (b)

Figure 9. Flexural property comparisons between the control (without post consolidation), and the use of 5 s, 10 s, and 20 s dwell time for post consolidated samples, of (a) peak load, and (b) stiffness, for sample B: 1000 core gsm materials.

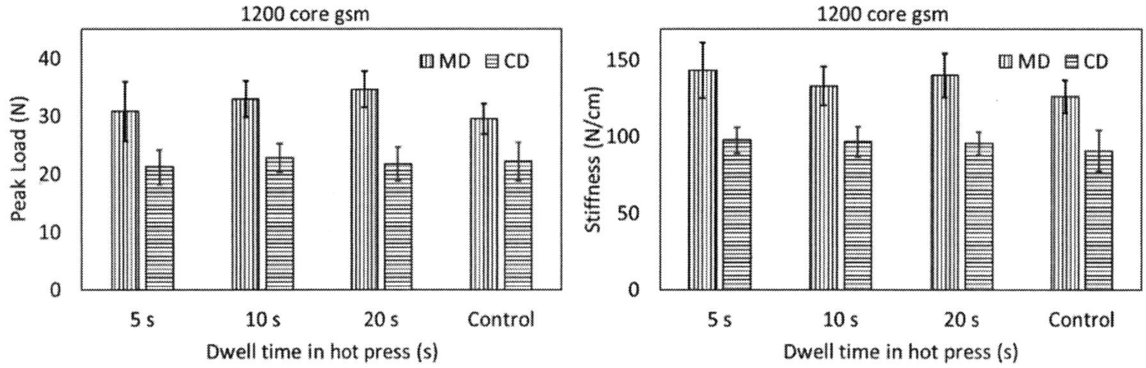

(a) (b)

Figure 10. Flexural property comparisons between the control (without post consolidation), and the use of 5 s, 10 s, and 20 s dwell time for post consolidated samples, of (a) peak load, and (b) stiffness, for sample C: 1200 core gsm materials.

3.3 Tensile properties

Tensile properties were also investigated in this study, because they are also highly dependent on the bonding between resin and fibers [5]. Figure 11, Figure 12, and Figure 13 compare the tensile properties of strength and modulus between post consolidated and control specimens of 450, 1000 and 1200 core gsm samples. The results are very similar to those of the flexural properties. The dwell time only affects the properties of the lightest 450 core gsm materials. As the basis weight increases, the improvement due to post consolidation decreases. For the 1200 core gsm materials, tensile strength in MD with 20 s dwell time and the properties in CD with 20 s are improved slightly, while there is no enhancement for the rest. It seems that with hot press temperature of 195 °C, the post consolidation process is only suitable for lighter materials than 1200 core gsm. Use of a higher temperature will be studied in our future work.

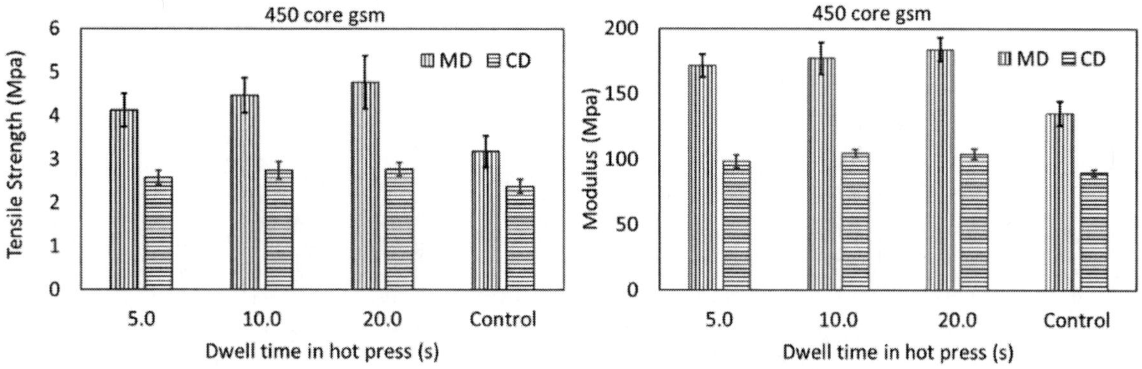

(a) (b)

Figure 11. Tensile property comparisons between the control (without post consolidation), and the use of 5 s, 10 s, and 20 s dwell time for post consolidated samples, of (a) peak load, and (b) stiffness, for sample A: 450 core gsm materials.

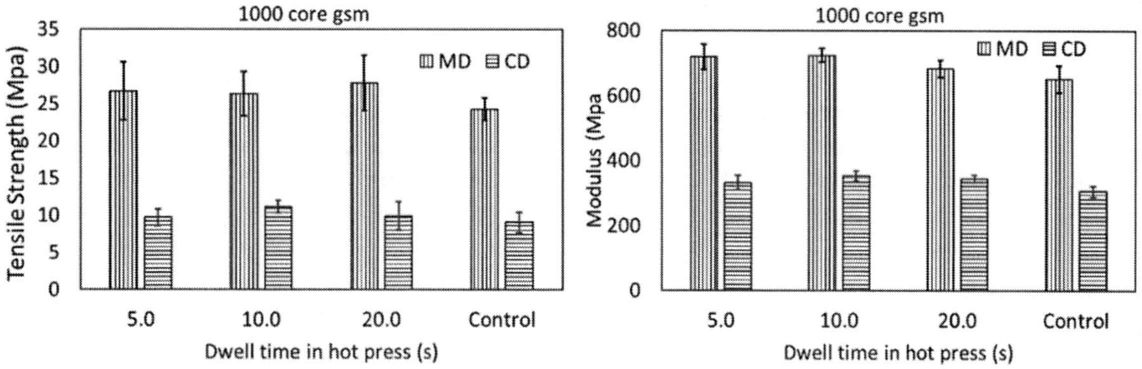

(a) (b)

Figure 12. Tensile property comparisons between the control (without post consolidation), and the use of 5 s, 10 s, and 20 s dwell time for post consolidated samples, of (a) peak load, and (b) stiffness, for sample B: 1000 core gsm materials.

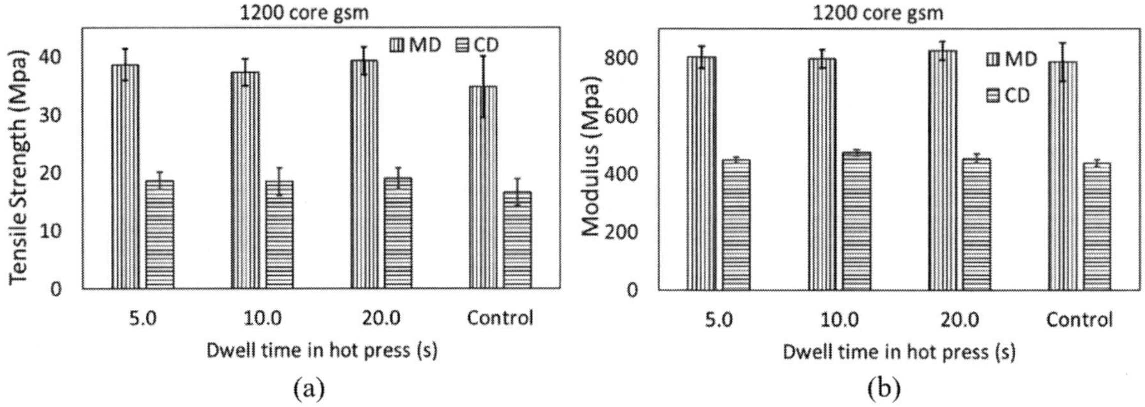

Figure 13. Tensile property comparisons between the control (without post consolidation), and the use of 5 s, 10 s, and 20 s dwell time for post consolidated samples, of (a) peak load, and (b) stiffness, for sample C: 1200 core gsm materials.

4. CONCLUSIONS

A post process of secondary consolidation has been developed by compressing the as-produced LWRT sheet in both hot and cold presses. This work demonstrated the feasibility of this process with practical processing conditions to improve mechanical properties for the LWRT composites with light basis weight. This is because the post process significantly improved the consolidation level of the porous core. Therefore, this promising post consolidation method is expected to further reduce weight without sacrificing the mechanical strength of the material.

5. REFERENCES

1. Swift TK, Moore M, Sanchez E, *Plastics and Polymer Composites in Light Vehicles*. Google Scholar, 2015.
2. R. Stewart, *Lightweighting the Automotive Market*. Reinforced Plastics, 2009.
3. T. Ishikawa, K. Amaoka, et al., *Overview of Automotive Structural Composites Technology Developments in Japan*. Composites Science and Technology, 2018
4. Y. Araki, T. Suzuki and S. Hanatani, *Composite Material for Automotive Headliners—Expandable Stampable Sheet with Light Weight and High Stiffness*. JFE Tech Report, 2004.
5. M. Taya and T. Mura, *On stiffness and Strength of An Aligned Short-Fiber Reinforced Composite Containing Fiber-End Cracks under Uniaxial Applied Stress*. Journal of Applied Mechanics, 1981.

Resin Infusion of a Hollow Airfoil Using an In-situ Polymerizing Low Viscosity Thermoplastic Matrix

Samuel D. Strassler, Patrick A. Rodriguez, Mark E. Bourgeois, Donald W. Radford
Composite Materials, Manufacturing and Structures Laboratory
Colorado State University
Fort Collins, CO 80523-1374

ABSTRACT

Thermoplastic polymer resins that can be in-situ polymerized have been demonstrated, on a number of occasions, as low viscosity alternatives for composite matrices, enabling the manufacture of composites using resin infusion molding (RIM). Elium from Arkema has been commercialized which, when processed, results in a methacrylate thermoplastic. This in-situ polymerized thermoplastic promises the advantages of a thermoplastic matrix in terms of toughness and recyclability with the ability to be used in production of relatively large composite articles in reduced amounts of time and with no requirement for added heat during the initial phase of processing the composite. The Elium resin has a viscosity of 100 cP which has the potential to reduce processing time and consumable expenditure; however, this low viscosity can introduce challenges when following common practices developed for thermoset infusion resins which are typically in the 250 – 300 cP range.

This effort, focusing on process optimization for a low viscosity, resin infused in-situ polymerized thermoplastic resin, considers various styles of expendables and process parameters. The demonstration article is a small hollow airfoil geometry of high length to chord aspect ratio that is representative of a small vertical axis wind turbine blade. The manufacturing process under evaluation uses a RIM technique to produce hollow blades, with molded exterior surfaces. Processed composite blade sections, infused with variations in process parameters and consumable arrangement, were evaluated.

The results describe how flow rate and flow path manipulation can be used to reduce the number and size of defects observed on the wind turbine blade. More specifically, this study shows that variations in process parameters and consumable arrangement can be used to limit the defects in a resin infused carbon fiber reinforced thermoplastic wind turbine blade section using a resin with a low viscosity.

1. INTRODUCTION

Resin infusion of a carbon fiber reinforced thermoplastic wind turbine blade section is presented, and evaluated, specifically with respect to manufacturing defects. The work described provides a detailed description of a manufacturing process capable of producing hollow airfoils, with a high length to chord aspect ratio using a low viscosity resin, and the effects of process parameters and expendable materials on the resulting quality.

1.1 Resin Infusion Molding - Overview

The resin infusion molding (RIM) process encompasses a variety of different manufacturing methods and has been proven to be cost-effective for producing composite structures. In RIM, a dry fiber preform is placed between a rigid mold and a flexible polymer film that serves as a vacuum bag. Once the preform is placed and resin inlets are incorporated, a vacuum is drawn on the system that compacts the preform and generates a pressure gradient between the resin inlet ports and the vacuum outlet. This pressure gradient draws the resin through the compacted preform, wetting the reinforcement with resin.

RIM is performed because of a number of unique aspects, including eased reinforcement placement which is not limited by pot life/shelf life of a wet resin system, low VOC's and low void content. It can compete with autoclave processing in terms of void content and can achieve relatively high fiber fractions under appropriate conditions. The RIM process is a contact molding operation that has been shown to be cost effective for producing large structures due to the relatively low-cost tooling associated with it.

1.2 Resin Infusion Molding – Wind Energy

Thermoset resins have been almost exclusively used as the matrix materials in industrial RIM processes, where epoxies, vinylesters, and polyester resins with viscosities ranging from 250 – 300 cP are most common[1]. The wind energy industry uses the RIM process to manufacture Gigawatt scale wind turbine blades which can reach lengths in excess of 100 meters[2]. Traditionally, wind turbine blades have been manufactured as two separate skins and a central spar, which were chemically bonded together. Siemens Gamesa has recently developed the "One-Shot" process, capable of manufacturing 60-meter blades in a single infusion process[3]. This process promises significant reductions in weight and assembly time as compared to traditional methods. However, demands for a more cost-effective recycling process have resulted in research on a comparable process using a thermoplastic matrix. Recent advances have resulted in the development of low viscosity thermoplastic resin monomer systems which are in-situ polymerized, making them a viable alternative to thermoset matrix materials in the RIM process. While some early examples of in-situ polymerization required significant heating to drive the reaction, the thermoplastic resin under investigation allows for polymerization at room temperature, eliminating the need for the heated tooling that can be common, and costly, in thermoset RIM [4], [5]. Difficulties in generating uniform wetout of the fiber reinforcement using conventional infusion practices have been described in conjunction with resins with viscosities as low as 100 cP. While a variety of solutions have been evaluated, developments in processing approaches and in expendables have been very effective in enabling the use of lower viscosity matrix resins. The work presented demonstrates that the number and size of defects observed using a low viscosity thermoplastic matrix can be reduced through flow rate and flow path manipulation.

2. EXPERIMENTATION

The following section presents experimental efforts directed at evaluating the relationships between infused composite material quality and a combination of process parameters and expendables utilized. The hollow composite airfoil sections evaluated in this study were processed inside a clamshell mold set to generate a fully molded exterior finish. The molded airfoil is relatively small, with a 15.24 cm chord and a 2.3 cm thickness, making the implementation of internal expendables and vacuum bagging a challenge. The evaluation encompassed in this work includes preliminary and advanced infusion trials, representing the processing of both 81 cm and 162 cm long airfoils respectively. The manufacturing process developed in this study provides the foundation for further efforts related to the production of a full size VAWT prototype.

2.1 Materials

Various styles of expenditures and reinforcements make up the experimental variables shown in the initial consumable arrangement study and remain a critical factor for all tests performed.

2.1.1 Reinforcement & matrix

The matrix material used in this study, Elium 150, is an in-situ polymerized thermoplastic resin that has been developed specifically for the RIM process. This resin system relies on a low viscosity monomer that, when combined with the LUPEROX AFR40 initiator, results in a polymerization reaction producing an amorphous thermoplastic with a glass transition temperature of 119 °C.

The seven distinct iterations of the manufacturing process presented in this work demonstrate how variations in consumable arrangement and process parameters can be used to reduce manufacturing related defects. Table 1 provides information on the four reinforcement materials used in this work.

Table 1: Reinforcement Materials

Reinforcement	Manufacturer	Product Code	Weight (gsm)	Iteration #
Braided Biax-Glass sleeve 12.7 cm Dia. [±45]	A&P Technology	Y29L500R	531	1
Non-crimp Triax-glass fabric [0, ±45]	VectorPly	E-TLX 2400	885	2-5
Braided Biax-carbon sleeve 10.2 cm Dia. [±45]	Highlands	C400-15	610	6-7
Non-crimp UD-carbon fabric [0]	VectorPly	C-LA 0912	352	6-7

2.1.2 Consumables

Consumable materials of interest for this study include any disposable products that influence critical process parameters such as flow rate, flow path, vacuum integrity, and demolding feasibility. While the majority of consumables had some influence on the quality of the final product, figures 1 and 2 provide images of the consumables that were considered experimental variables throughout the process development.

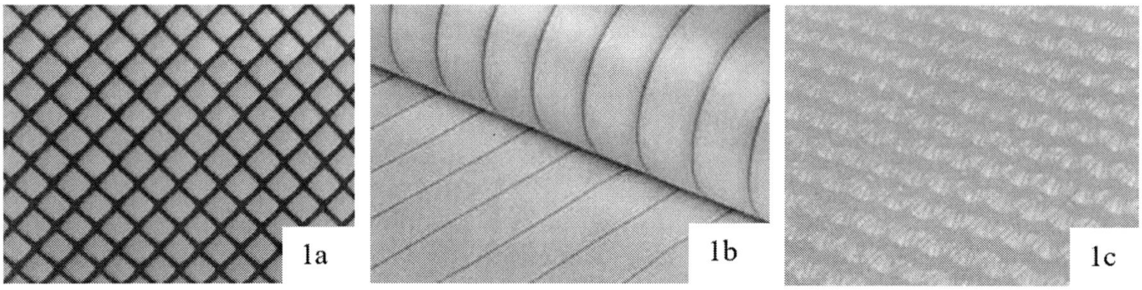

Figure 1: Flow media and peel ply consumables*[6]–[8]*

Figure 1 shows the various styles of flow media and peel ply that are used in the iterations of the manufacturing process. The primary purpose of the flow media is to provide a region of relatively high permiability to allow for the resin to flow and wet out the reinforcement. The peel ply is placed between the flow media and the part, preventing adhesion between the two and thus, allowing for the removal of consumables following polymerization. The ResinFlow 60 and Econostitch G shown in figures 1A and 1B respectivly are commonly used in conjunction for processing composite laminates. The Compoflex RF3, shown in figure 1C, is a combination of peel ply (white fabric) and flow media (yellow lines) with a lower permeability and release force then the combination of ResinFlow 60 and Econostitch G.

Figure 2: Inlet and outlet materials*[9], [10]*

The various styles of inlet/outlet materials that were included in the investigation are shown in Figure 2. The 1.3 cm OD spiral tubing, shown in figure 2A, is commonly used to both rapidly supply resin to the flow media and to serve as a vacuum outlet for the RIM process. The jacketed Enkafusion, shown in figure 2B, is a lower profile resin inlet material that incorporates directional flow media (black material), promoting rapid transport of the resin along its length. The DAHLPAC MC79, shown in figure 2C, is an alternative outlet material that uses fabric (blue material) that is permeable to air, while preventing the flow of resin. The DAHLPAC allows for an increase in control in the amount of resin present in the final product as it prevents resin from leaving the system while providing a vacuum path throughout the polymerization process.

2.2 Preliminary Infusion Trials

In order to perform preliminary manufacturing trials, an 81 cm long master was machined from medium density fiberboard (MDF). The surface was sealed to enable the production of molds. Two fiberglass reinforced epoxy mold halves were produced, which, because the airfoil is symmetric, results in a complete mold set.

Using the 81 cm mold set, initial experiments were performed to determine areas of principal difficulty and, in general, the overall practicality of the proposed manufacturing approach. In this approach flow media, resin inlets, and a vacuum outlet were positioned on a tubular plastic vacuum bag that is placed over a simple flat plastic removable tool serving as a manufacturing aid. The reinforcement is wrapped around the flow media, bagging and flat plastic tool and placed in the mold. A larger tubular vacuum bag is pulled over the mold and sealed to the inner bag, resulting in an envelope bag that consolidates the fibers to the inside mold surface and holds the mold halves together. Once the bagging is complete the flat manufacturing aid is removed. The material sequence is shown schematically in Figure 3.

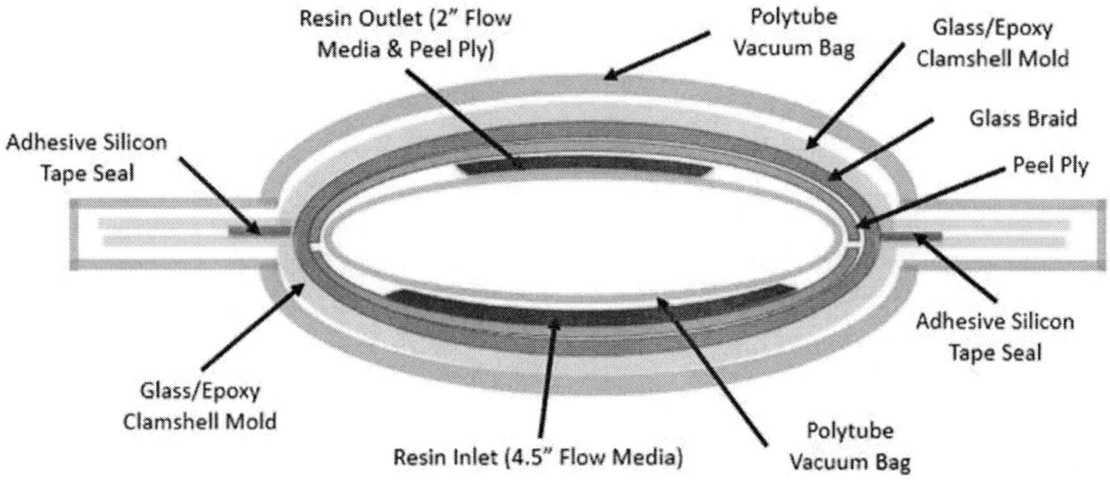

Figure 3: Schematic of preliminary materials stack-up in the assembled mold halves

The preliminary process experiments encompass the manufacture of the 81 cm long glass fiber reinforced airfoils. These preliminary infusions are separated into three unique iterations of the manufacturing process that represent major improvements. The variables investigated in the preliminary process development can be separated into three categories including reinforcement, consumable materials, and process parameters. The preliminary process development focused on consumable materials, demonstrating the effects of various types of flow media, peel ply, and inlet/outlet materials. Additionally, various geometries of the flow media were tested to investigate effects on the resin flow path and flow rate. The only process parameter varied in the 81 cm long airfoils was the drop test criterion, which was made more stringent to increase confidence in the vacuum integrity of the bagged system. Table 2 provides a summary of the variations in consumables and process parameters between the iterations of the preliminary manufacturing process.

Table 2: Experimental Procedure for the 81 cm long glass fiber reinforced blades

	Iteration #1	Iteration #2	Iteration #3
Reinforcement	Braided Biax-Glass sleeve 12.7 cm Dia.	Non-crimp Triax-glass fabric	Non-crimp Triax-glass fabric
Flow media	Resinflow 60	Resinflow 60	Compoflex RF3
Flow media width	Full circumference	11.4 cm top 15.2 cm bottom	10.2 cm top 10.2 cm bottom
Inlet material	Spiral tube	Spiral tube	Enkafusion
Outlet material	Spiral tube	DahlPac	DahlPac
Peel Ply	Econostitch	Econostitch	Compoflex RF3
Drop Test Criteria	3.4 kPa/10 min	1.7 kPa /10 min	0.34 kPa /10 min

The preliminary infusion trials were concluded when the defect concentration was substantially reduced in the 81 cm long glass fiber reinforced. The consumable arrangement determined in the preliminary infusion trials was then utilized for all advanced infusion trials.

2.3 Advanced Infusion Trials

The advanced infusion trials evaluated the effects of reinforcement types and process parameters on the quality of 162 cm long airfoils. In order to perform the advanced infusion trials, two additional 81 cm long molds were produced and joined, end-to-end, to the previous mold set, resulting in a 162 cm long mold set.

The advanced infusion trials included four additional iterations detailing the development of the manufacturing process for 162 cm long fiber reinforced airfoils. The airfoils produced in the advanced trials demonstrate the effects of various reinforcements ranging from a stitched triaxial glass fabric to a Braided Biax-carbon sleeve. The advanced infusion trials were used to evaluate the quality of airfoils infused with variations in process parameters, including vacuum pressure, debulk time, and exotherm control procedures as detailed in Table 3.

Table 3: Process Parameter Variables

	Iteration #4	Iteration #5	Iteration #6	Iteration #7
Reinforcement	Non-crimp Triax-glass fabric	Non-crimp Triax-glass fabric	Braided Biax-carbon sleeve + UD carbon fabric	Braided Biax-carbon sleeve + UD carbon fabric
Vacuum Pressure	84.7 kPa	84.7 kPa	84.7 kPa	77.9 kPa
Debulk Time	0 min	30 min	60 min	60 min
Exotherm Control	No	Yes	Yes	Yes

3. RESULTS AND DISCUSSION

Notable trends in the size, location, and quantity of defects are presented and discussed for each iteration. The variations in composite quality and repeatability between iterations are evaluated, and the effects of the experimental variables on the resin flow rate and flow path are compared to relevant analogues from literature.

3.1 Preliminary Infusion Trials – 81 cm mold set

Many different expendable materials and process techniques were attempted before a final process was defined that resulted in successful and repeatability part infusion, without using excessive quantities of the Elium® resin. The initial manufacturing process, shown in Figure *4*, was developed based on common lab practices for manufacturing flat composite laminates. The initial preform configuration and bagged system, shown in Figures 4a to 4d, used spiral tubing as the inlet and outlet material to infuse the airfoil from the leading edge to trailing edge with the flow media and peel ply in contact with the entire inner surface of the part. A 12.7 cm diameter braided fiberglass biaxial sleeving with an epoxy sizing was chosen as the reinforcing material for the proof of concept testing. Figure *4*e shows preliminary infusion trial #4 and is representative of the defects commonly observed in the first iteration of each of the manufacturing process iterations.

The relatively small dimensions of the hollow airfoil cross section (15.24 cm chord and 2.3 cm thickness) proved to be a dry reinforcement positioning challenge, both in the preparation for resin infusion molding, as well as during removal of the infusion media after processing. The initial consumable arrangement was susceptible to unintended pleating and shifting of the consumables, leading to both locally resin rich and resin poor regions throughout the airfoils. Additionally, with the inlet and outlet located at the leading and trailing edge insufficient space remained inside the airfoil to remove the consumables and negatively impacted reinforcement consolidation. The resin used in this investigation does not fully polymerize in the presence of oxygen, and the flat sheets of vacuum bag used to make the envelope bags were found to be susceptible to leaks during the infusion.

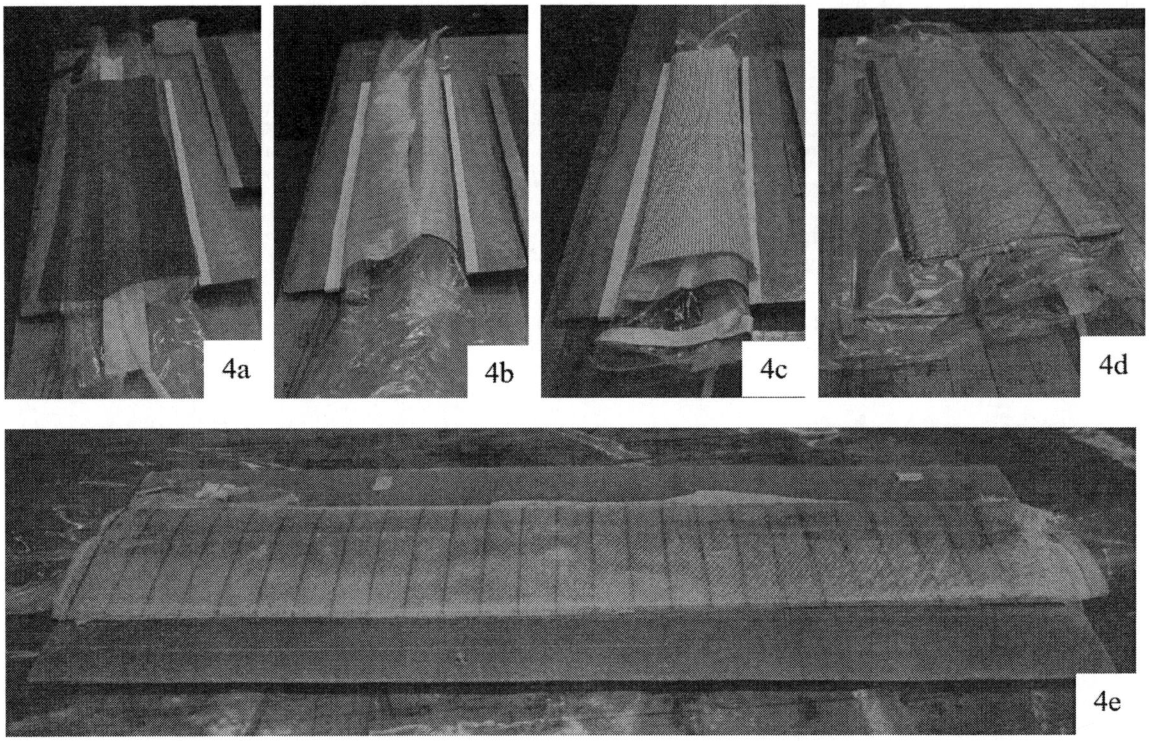

Figure 4: Manufacturing process iteration #1

The airfoil section shown in Figure 4e had a higher degree of polymerization than earlier infusions; however, this airfoil still showed regions of incomplete polymerization. Additionally, the airfoils produced during the first iteration consistently included dry spots, particularly along the trailing edge and by the outlet. The defects observed in the first iteration were similar to defects presented in the literature when the resin is allowed to flow too fast, resulting in micro voids located within the tows of the reinforcement[11]–[13]. Thus, to address the difficulties noted in removing the expendables, and the dry spots, a second group of manufacturing trials focused on facilitating the removal of consumables, as well as reducing the resin flow rate.

The second manufacturing iteration, shown in Figure 5a to 5d, demonstrates effects of changes in reinforcement, consumable materials, and preform set-up. In this second iteration the reinforcement was a non-crimp triaxial stitched glass fabric that more closely resembled the final desired stacking sequence. The resin manufacturer recommended an F0E sizing which was unavailable in a braided fiberglass biaxial sleeving, resulting in a shift from braided to stitched fabric reinforcements. The location of the inlet and outlet were changed to the bottom and the top of the airfoil, respectively, nominally at the point of maximum thickness to maximize the space available for the removal of consumables. In addition, the width of the flow media was reduced, creating a break zone at both the leading and trailing edges to slow the flow of resin. The break zones, while important for any infusion, are critical to improving the consistency of the wet-out when using a low viscosity resin. The placement of the break zones was chosen to prevent the flow media from becoming physically trapped in the leading or trailing edge. The DAHLPAC outlet was chosen as it allows improved control of the resin remaining in the composite and has a lower profile than the spiral tubing, further facilitating the removal of consumables, post polymerization. Vacuum bagging specific lay flat tubing was procured to reduce areas susceptible to leaks, and the acceptable drop test criterion was changed from 3.4 kPa/10 min to 1.7 kPa/10 min to address the oxygen sensitivity of the resin.

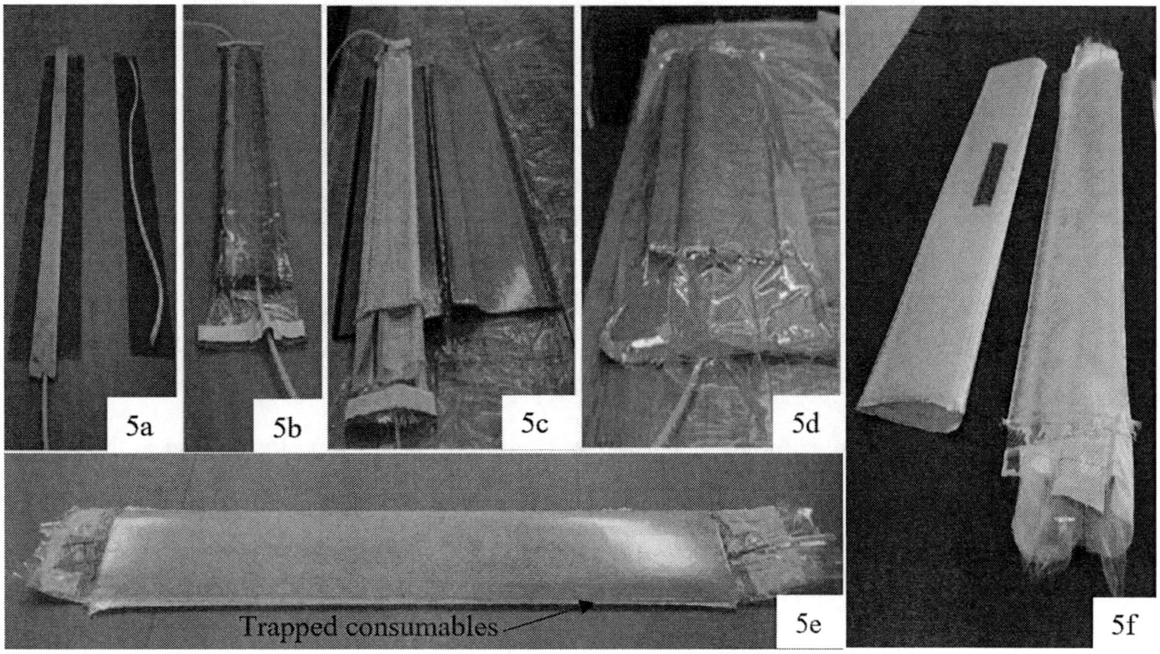

Figure 5: Manufacturing Iteration #2

Figure 5e demonstrates a significant improvement in composite quality as the top and bottom of the airfoil were fully wet-out and the surface finish of the part matched the mold. While this part incorporated break zones at the leading and trailing edge, flow media and peel ply became physically trapped in the trailing edge near the outlet. Additionally, this process did not show acceptable repeatability, as subsequent infusions, shown in 5f, had significant defects at the leading and trailing edges as well as on the bottom of the airfoil.

The final iteration in the preliminary process development, shown in figure 6, was found to successfully produce repeatable 81 cm long glass fiber reinforced airfoils with few observed defects. The preform configuration shown in figures 6a to 6c, as well as the removed consumables shown on the right of Figure 6d show the Compoflex RF3 that replaced both the flow media and peel ply used in previous iterations. This combination product, specifically designed for RIM, simplified pre-infusion preparations and improved control of the placement of consumables. Further, in this iteration, the inlet material was changed to Enkafusion, enabling a rapid flow rate down the length of the blade but with a lower profile than the spiral tubing, enhancing the ease of removal of the consumables. Minor adjustments to the width of the flow media, and subsequently the width of the break zones, located at the leading and trailing edges allowed for an increased repeatability in the wet-out of the reinforcement. The drop test criterion was again changed, this time from 1.7 kPa/10 min to 0.34 kPa/10 min, as some areas of incomplete polymerization were still observed.

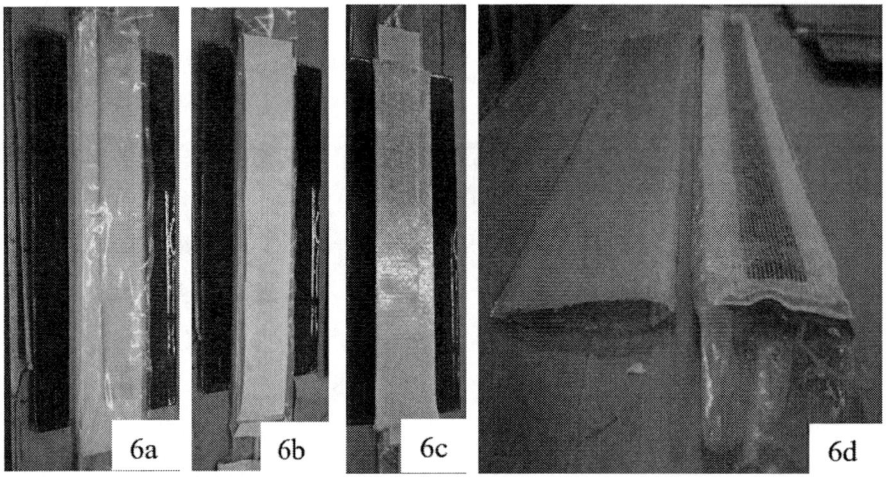

Figure 6: Manufacturing process iteration #3

The common defects located along the leading and trailing edge were not observed in the airfoils produced using manufacturing process iteration #3. The combination of Enkafusion and Compoflex RF3 allowed for the resin to flow rapidly down the length of the airfoil, yet significantly reduced the resin flow rate radially from the inlet to the outlet, increasing the time it took for resin to be observed at the outlet, from nominally 5 minutes to approximately 9 minutes. Additionally, the Compoflex RF3 had a lower release force than the Econostich peel ply, easing the removal of consumables, post polymerization. The resulting quality and uniformity of wet-out was acceptable and the process was shown to be repeatable.

3.2 Advanced Infusion Trials – 162 cm mold set

The advanced infusion trials built upon the results of the preliminary trials to develop a manufacturing process capable of producing 162 cm carbon fiber reinforced thermoplastic airfoils of reproducible quality. The consumable configuration used in the final manufacturing process for 81 cm long airfoils was maintained throughout the 162 cm airfoil manufacturing investigation. The development of the manufacturing process in the advanced infusion trials focused on varying process parameters including, vacuum pressure, debulk time, and exotherm control procedures. The updated materials stack-up in the 162 cm molds is schematically shown in Figure 7.

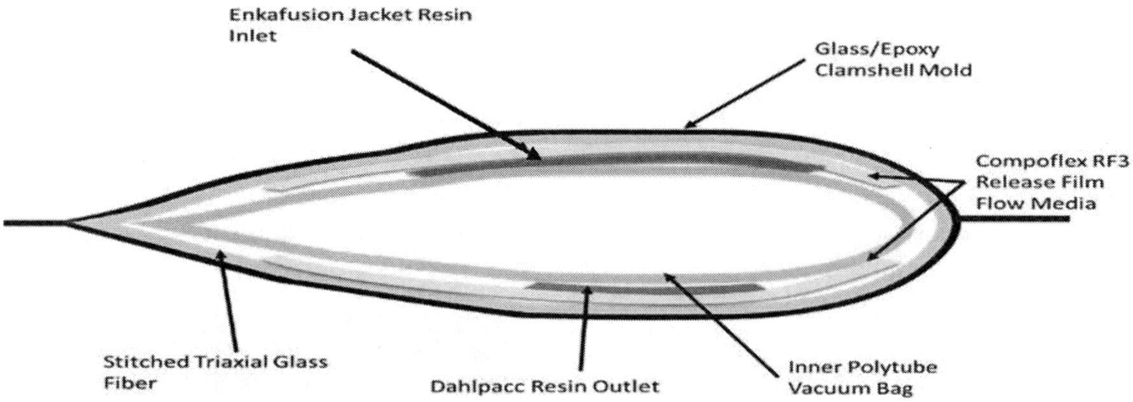

Figure 7: Materials stack-up in the assembled mold halves used in the advanced infusion trials.

The preform configuration for manufacturing iterations #4 and #5 are demonstrated in Figure 8a and 9b, and airfoil cross sections with, and without the consumables, are shown in Figure 9c. The initial attempt to manufacture a 162 cm long airfoil demonstrated the robustness of the manufacturing process developed in preliminary trials, resulting in composite parts with a relatively uniform wet-out and a good surface finish.

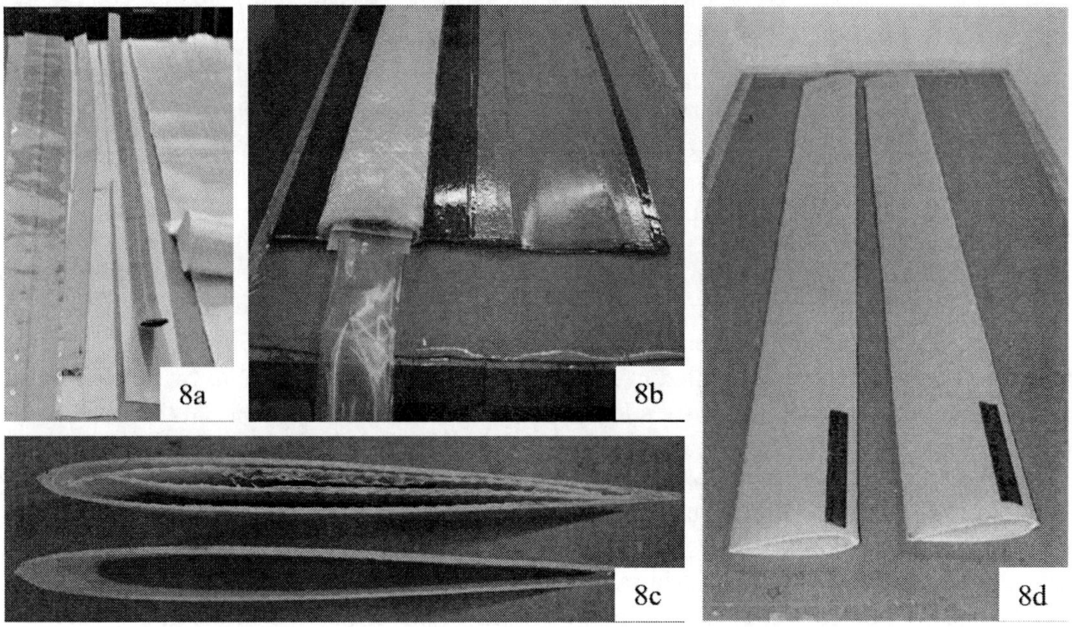

Figure 8: Manufacturing Process Iterations #4 and #5

Figure 8d shows two 162 cm long airfoils processed using iteration #4 that resulted in a uniform wet-out. However, the increase in volume of the resin in the Enkafusion resulted in peak exotherm temperatures exceeding the boiling point of the resin in these longer blades. An exothermic reaction strong enough to compromise the inner polyethylene lay flat tubing was observed along the top of the airfoil. The initial mitigation technique was the procurement of Dahlar® lay flat tubing, which is designed to maintain vacuum integrity at temperatures up to 140 °C.

Thermocouples placed along the top, bottom, leading edge, and trailing edge, as well as a FLIR thermal imaging camera, recorded temperatures in excess of 100 °C during the polymerization process of the 162 cm airfoils, while a maximum exotherm temperature of only 77 °C was observed in the 81 cm airfoils. The temperature profile along the length of a 162 cm airfoil is shown in Figures 9a and 9b indicating the higher temperatures measured along the top of the airfoil near the resin inlet. Additionally, the temperature near the resin inlet was high enough to cause the resin to boil and generate localized regions of defects and an observed high void content, as indicated in 9c. These results demonstrated the need for exotherm control measures.

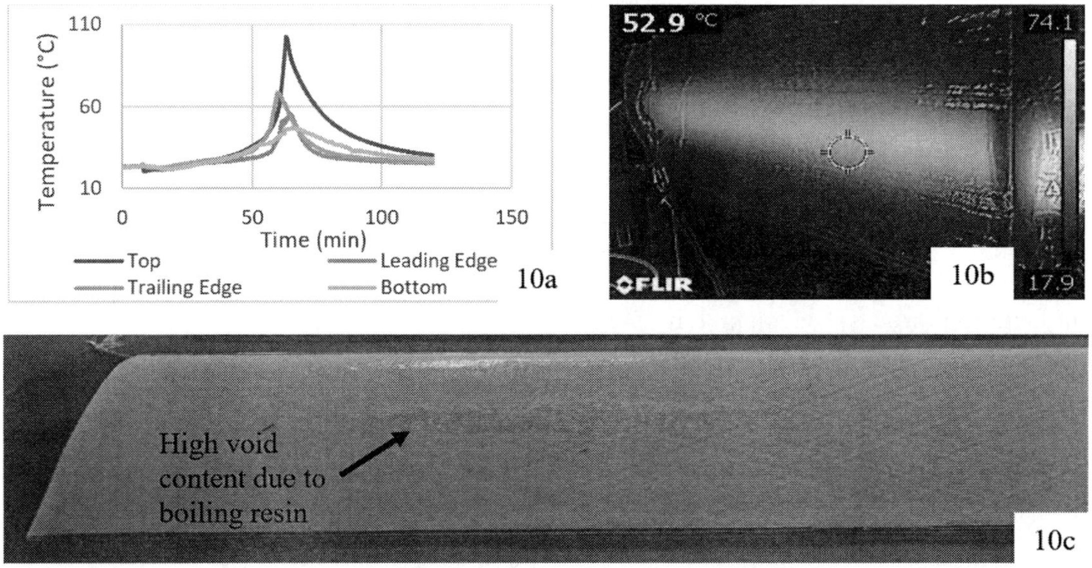

Figure 9: Effects of temperature of polymerization in a 162 cm long airfoil

To overcome the effects of this exotherm, various cooling methods were considered, including forced convection and a reduction in the initiator concentration in the resin; however, these methods failed to prevent the measured exotherm from exceeding 100 °C. Ultimately, the temperature was adequately controlled in manufacturing process iteration #5 by placing ice along the top of the outside of the mold once a temperature of 50 °C was observed at the inlet.

Concerns related to the possibility of small amounts of oxygen remaining in the reinforcement and expendables, affecting the resin during the polymerization reaction, were also identified. To remove this possible source of oxygen within the vacuum bag, and thus reduce the likelihood of incomplete polymerization, manufacturing iteration 5 introduced a 30-minute debulk into the process to ensure the removal of any air trapped in the preform, prior to infusion

Figure 10a through 10d show the sequence of steps used in preparation of the carbon fiber preform and a resulting 162 cm carbon fiber reinforced Elium® 150 hollow airfoil. The braided carbon biaxial sleeving, with a general purpose (GP) sizing, used in manufacturing iterations 6 and 7 simplified the manufacturing process. The braid could be rolled back over itself capturing the three plies of UD carbon fabric, creating a more manageable preform. This resulted in the final desired stacking sequence [±45,0_3,±45] along the top and bottom of the airfoil, and [±45, ±45] at the leading edge and trailing edge. This preform configuration was consistent with that planned for the final demonstration blade set. Manufacturing iteration 6 included a 60-minute debulk time rather than the 30 minutes previously used to further reduce the potential for incomplete polymerization.

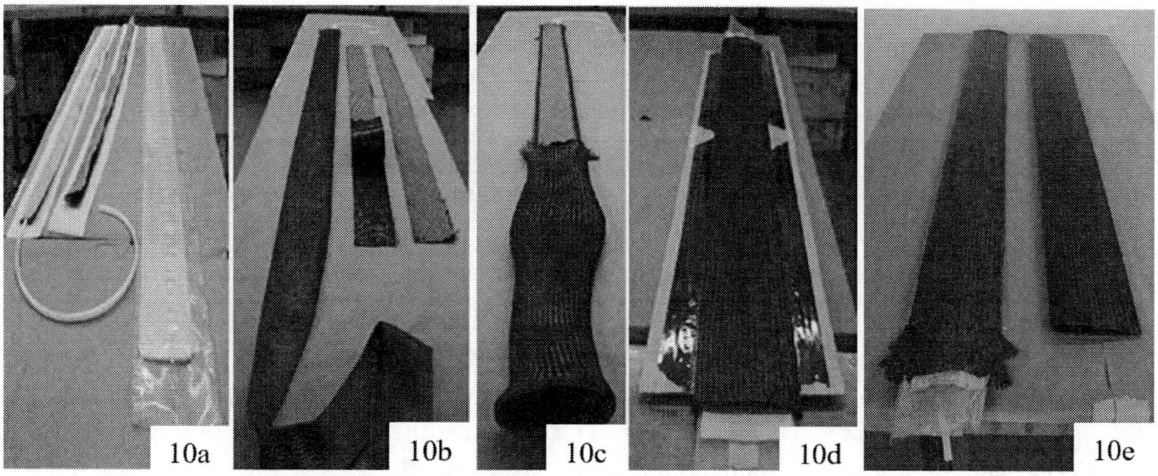

Figure 10: Manufacturing Process Iterations #6 and #7

The 162 cm long airfoils, shown in figure 10e, were processed using the parameters introduced during manufacturing iteration #6 and demonstrated comparable quality to the glass fiber reinforced airfoils processed in the earlier trials. The carbon fiber airfoils demonstrated uniform wet out and a good surface finish along the majority of the blade length; however, small localized regions with voids were observed in the break zone near the leading edge. Dry regions located between the tows of the reinforcement in the carbon blades, shown in Figure 11, were similar to defects observed in literature which were a sign of a slow resin flow rate [11]–[13]

Figure 11: Defects observed in airfoils processed using manufacturing iteration #6

Manufacturing process iteration #7 focused on increasing the flow rate within the break zone, without increasing the flow rate along the top and bottom of the airfoil to alleviate the void formation in the break zones noted in iteration #6. The vacuum pressure was decreased from 84.8 kPa to 77.9 kPa and reducing consolidation pressure and resulting in the removal of defects, consistent with an increase in the permeability. The decrease in vacuum pressure had a relatively small impact to the permeability of regions containing flow media as compared to the break zones, resulting in a more consistent flow rate throughout the airfoil. Figure *11* shows a section of an airfoil processed using manufacturing iteration #7, immediately after demolding, demonstrating a uniform wet-out and a good overall surface finish. Several high-quality airfoils were produced using manufacturing process iteration #7 with a repeatable reduction in the number and size of defects.

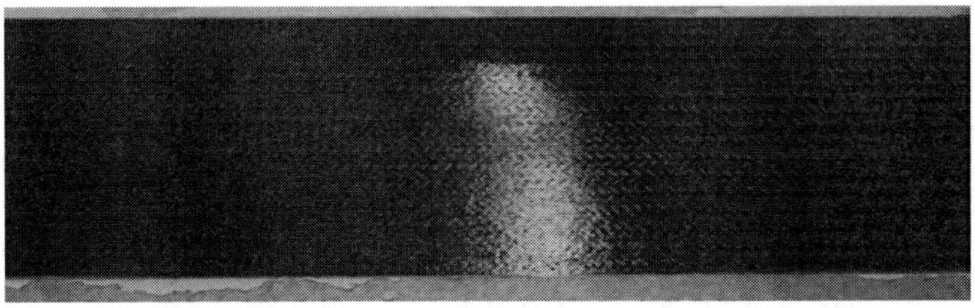

Figure 12: 162 cm long carbon fiber airfoil processed using manufacturing iteration #7

4. FUTURE WORK

The work presented here represents a preliminary study in a project to manufacture a prototype of a vertical axis wind turbine. This work formed the basis for the development of a manufacturing process targeted for use in the production of longer, 350 cm airfoils. The full-scale prototype will combine concurrent studies on the mechanical response of fiber reinforced Elium, post-molding deformation, and fusion bonding of required end fittings.

5. CONCLUSIONS

A Resin Infusion Molding process to produce a hollow carbon fiber airfoil using an in-situ polymerizing thermoplastic matrix has been developed. The evolution of the manufacturing process, detailed in this work, shows how variations in consumables, consumable arrangement and process parameters can relate to a reduction in defects present in the composite product.

This investigation resulted in a robust manufacturing process that generated repeatable hollow composite braided airfoils, with high length to cord ratio, having minimal defects. It was determined that a form of active cooling was required to remove the heat of exotherm, which increased with airfoil length, from the airfoils during polymerization. Additionally, techniques to achieve a controlled, relatively uniform flow rate, of the resin, within the reinforcement preform, were determined to be critical in reducing defects in the infused composite airfoils and in generating a high quality molded external surface.

In summary, this work increases our understanding of how variations in consumables, consumable arrangement and process parameters effect manufacturing quality of an infused composite structure. Ultimately, this investigation provides insight into the manufacture of fiber reinforced thermoplastic structures manufactured using a low viscosity in-situ polymerizing resin in an infusion molding process.

6. ACKNOWLEDGMENT

The authors would like to acknowledge the support of the Institute for Advanced Composites Manufacturing and Innovation (IACMI), which provided funding for the project, Task 4.5, through US Department of Energy Award DE-EE006926, the funding from the Colorado Office of

Economic Development and International Trade (OEDIT), the expertise of the industry lead, Steelhead Composites and Arkema for support with the Elium resin used in this effort.

7. REFERENCES

[1] L. Mishnaevsky, K. Branner, H. N. Petersen, J. Beauson, M. McGugan, and B. F. Sørensen, "Materials for Wind Turbine Blades: An Overview," *Materials*, vol. 10, no. 11, p. 1285, Nov. 2017, doi: 10.3390/ma10111285.

[2] D. T. Griffith and T. D. Ashwill, "The Sandia 100-meter All-glass Baseline Wind Turbine Blade: SNL100-00," p. 67. (2011)

[3] M. Damiano, A. Russo, A. Sellitto, E. Vecchio, T. Stellato, and A. Riccio, "Design of a composite wind turbine blade manufactured with the ONE SHOT BLADE® technology," *Materials Today: Proceedings*, p. S2214785320304673, Feb. 2020, doi: 10.1016/j.matpr.2020.01.366.

[4] R. E. Murray *et al.*, "Manufacturing and Flexural Characterization of Infusion-Reacted Thermoplastic Wind Turbine Blade Subcomponents," *Applied Composite Materials*, Jan. 2019, doi: 10.1007/s10443-019-9760-2.

[5] K. van Rijswijk and H. E. N. Bersee, "Reactive processing of textile fiber-reinforced thermoplastic composites – An overview," *Composites Part A: Applied Science and Manufacturing*, vol. 38, no. 3, pp. 666–681, Mar. 2007, doi: 10.1016/j.compositesa.2006.05.007.

[6] "Red Resin Infusion Mesh | Resin Flow Media | Resinflow 60 | Per Yd | ACP Composites." https://store.acpsales.com/products/7787/resin-infusion-mesh-red (accessed Jun. 11, 2020).

[7] "Fibre Glast Econostitch Peel Ply - 60 Inches Wide - 1 Yard Package - Visible Before and After Cure - vehicleshopcenter." https://vehicleshopcenter.com/fibre-glast-econostitch-peel-ply-60-inches-wide-1-yard-package-visible-before-and-after-cure (accessed Jun. 11, 2020).

[8] "COMPOFLEX SB RF 3 - Non-Coated Peel Ply, Release Film & Flow Media in One - 60." https://www.rockwestcomposites.com/3045-d-group (accessed Jun. 11, 2020).

[9] "SPIRAL-WRAP-5/8"O.D.-100'." https://estore.airtechintl.com/spiral-wrap-5-8-o-d-100.html (accessed Jun. 11, 2020).

[10] "EnkaFusion Filter Jacket," *Fibre Glast.* https://www.fibreglast.com/product/EnkaFusion_Filter_Jacket_1400 (accessed Jun. 11, 2020).

[11] N. Patel and L. J. Lee, "Effects of fiber mat architecture on void formation and removal in liquid composite molding," *Polymer Composites*, vol. 16, no. 5, pp. 386–399, 1995, doi: 10.1002/pc.750160507.

[12] N. Patel, V. Rohatgi, and L. J. Lee, "Micro scale flow behavior and void formation mechanism during impregnation through a unidirectional stitched fiberglass mat," *Polymer Engineering & Science*, vol. 35, no. 10, pp. 837–851, 1995, doi: 10.1002/pen.760351006.

[13] J. S. Leclerc and E. Ruiz, "Porosity reduction using optimized flow velocity in Resin Transfer Molding," *Composites Part A: Applied Science and Manufacturing*, vol. 39, no. 12, pp. 1859–1868, Dec. 2008, doi: 10.1016/j.compositesa.2008.09.008.

FILAMENT WINDING AND COMPRESSION MOLDING PROCESS FOR HIGH PERFORMANCE COMPOSITE STRUCTURES

Dan B. Hannula
Advanced Composites, Inc.
Salt Lake City, Utah 84119

ABSTRACT

A mandrel-less filament winding and compression molding method was developed to manufacture high performance composite structures. Goals of the process were to create high fiber volume and low void content solid composite structures with control of fiber orientation from 0 to 90 degrees to the winding axis utilizing a highly automated and repeatable process. To generate the material, a filament winding machine was modified to allow the material to be wound between a driven headstock and a driven tailstock. During filament winding, material that was part of the manufactured end use component would also serve as the support material and tooling for fabrication. After filament winding, the produced material was compression molded to enhance material properties. For demonstration of this technology a high-performance ball (also known as a frac ball) used as a seal/valve for oilfield applications was manufactured and tested with this material. The use of this material to construct the ball would overcome several of the issues with current composite frac balls in the market related to fiber orientation or short strand fiber reinforcement. The frac ball was developed to withstand a service temperature of 177 °C (350 °F) and pressure of 69 MPa (10 ksi) in any orientation. Testing of the ball demonstrated excellent performance at the required temperature and pressure, independent of the specific tested orientation of the ball. This mandrel-less filament winding and compression molding technique successfully produced high performance composite material in a scalable and efficient production process.

1. INTRODUCTION

Filament winding is a mature advanced composite manufacturing technology that can produce high performance parts with a highly automated and efficient process. In traditional filament winding, fibers are wrapped around a rotating tool with a CNC machine that places the continuous fibers in a controlled pattern. With the inherent flexibility of filament winding, a designer can choose all the individual components of the composite being produced- fiber and matrix materials can be independently selected for the best combination of properties. Additionally, as the machine is wrapping individual filaments around the tool (or mandrel), full control of fiber orientation can be achieved.

Filament winding has proven to be a successful technology for reliable production of composite structures in multiple industry sectors for decades. The filament winding process works well for hollow or tubular structures but has limitations for the generation of solid structures and more complex shapes. Also, at higher volume production a large quantity of winding tools may be needed to keep up with the high speed of the winding process.

This paper investigates the development of a novel winding and compression molding process to manufacture high performance composite structures without a winding mandrel. The intended audience should have some familiarity with the filament winding process, but it is not intended that the reader is an expert. The process as described in this paper will provide a high-level overview.

In this paper the process concept will be detailed followed by a specific example- a frac ball that was fabricated at Advanced Composites, Inc. located in Salt Lake City, Utah. This process was utilized to produce a solid cylindrical composite structure with continuous fiber to be used as billet material to construct the high performance frac ball.

1.1 Process Development

Process development began with the need to rapidly produce a composite billet material that was a solid cylindrical shape. In the application that was being developed, it was desired to have a high degree of continuous fiber, no shear planes that went across the entire structure, and nearing quasi-isotropic material properties. Additionally, the process had to be scalable and efficient for a high-volume production process.

Commercially available composite products in a cylindrical geometry currently exist as pultruded rod, stacked laminate material, or a combination of technologies such as a filament wound tube with a bonded-in core. The available products did not meet all the outlined requirements- pultruded rod has fiber primarily unidirectional along the axis of the geometry, stacked laminate would have a shear plane across the laminate plies, and a combination of current products would not be desired for efficient production.

Filament winding can largely address the issues described above but as the composite material needs to be wound around a tool there would be a hollow cavity in the inner diameter of the part. If the initial winding tool could be constructed from the same material as the subsequent windings, the tool for winding the part could be integral to the part itself. A standard filament winding machine utilizing a resin bath (wet winding) is illustrated in Figure 1.

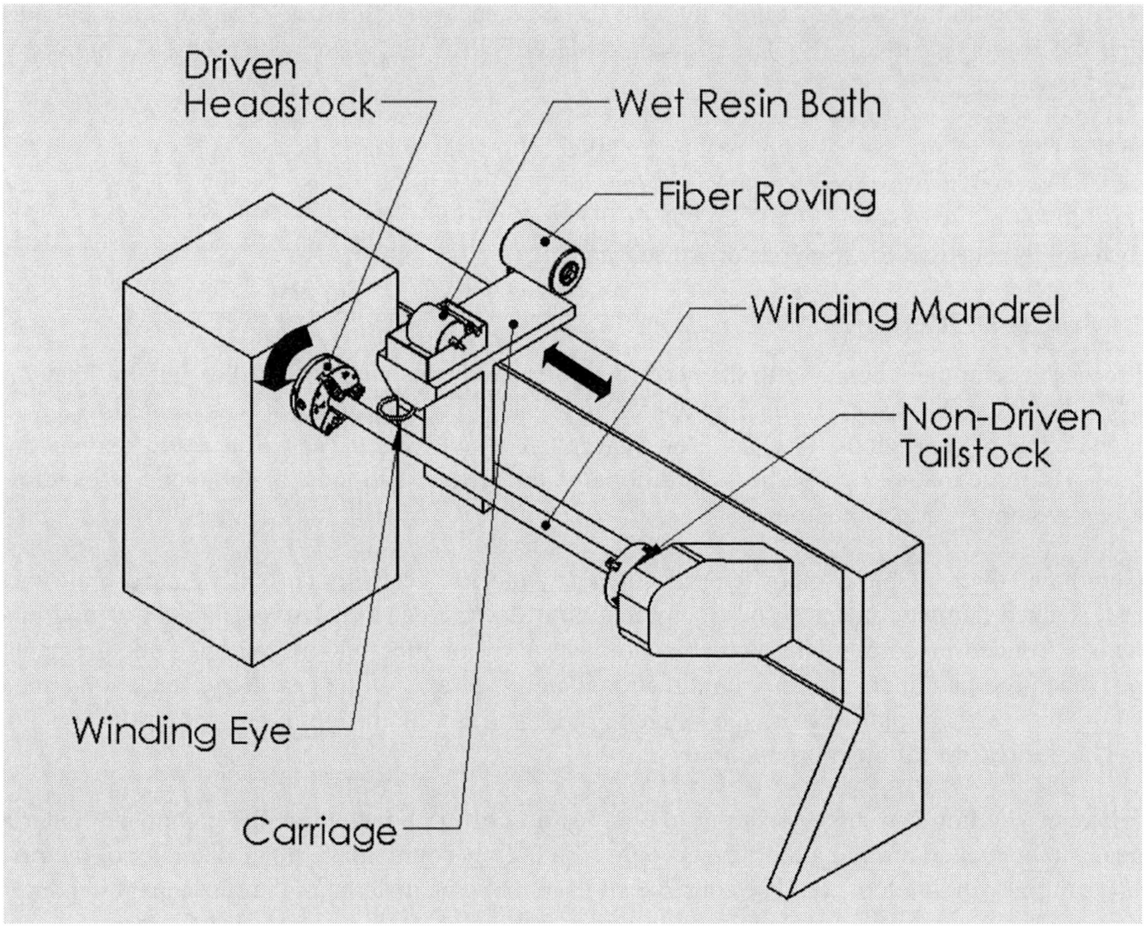

Figure 1. Standard wet filament winding machine with a rotating mandrel and a carriage that translates axially to manufacture the desired shape. After winding over the tool and curing, the tool is extracted, and a hollow inner diameter remains.

To generate a solid composite structure utilizing a typical filament winding machine a couple of challenges needed to be overcome. First, the support structure would need to be created without support tools. Second, the winding would not have a tool to transmit the torque needed to overcome the tension of the fiber wound onto the part. Third, there would be no rigid structure to support the material during the cure. To address these issues a new filament winding machine was designed (patent pending) in addition to new tools generated to support the composite during cure.

1.2 Equipment Setup and Process

1.2.1 Dual Drive Winding Machine and Tooling

To address issues related to the lack of support tooling for the winding process, the winding machine was designed to drive both the headstock and the tailstock. To create the integral fiber structure to wind the part over, custom pin-rings were created that would catch the fiber that was being wound as it came to the end of the axial travel on the winding machine as illustrated in

Figure 2. Small amounts of fiber were strung between the pin-rings under tension and layers were generated in the subsequent winding patterns/layers.

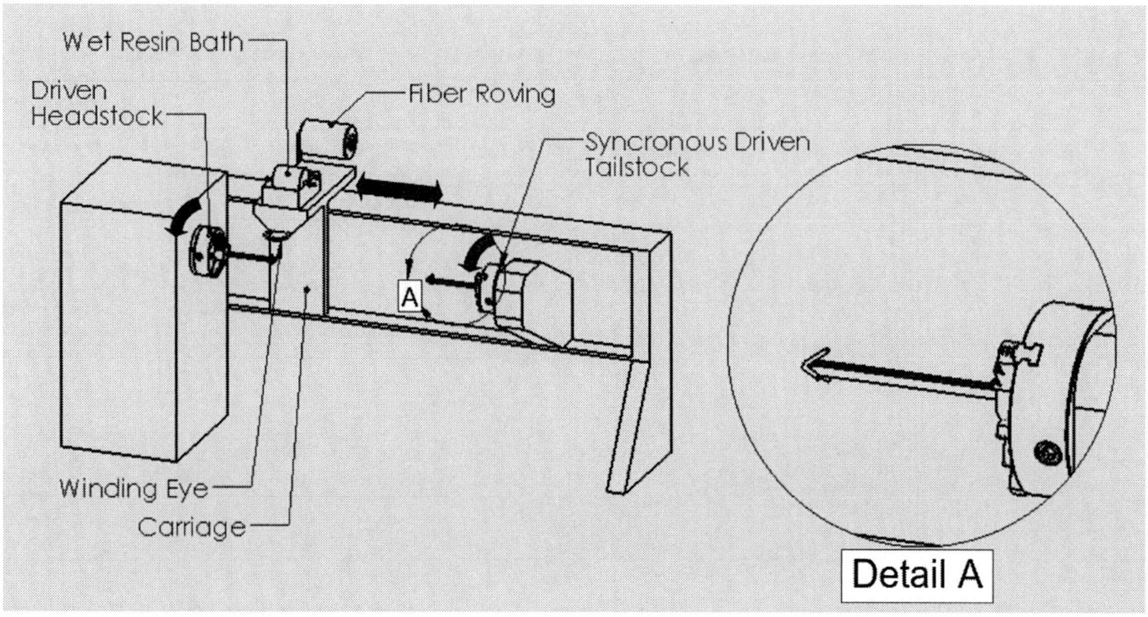

Figure 1. Dual drive wind machine setup and pin ring winding tools (Detail A) winding tools. After winding the component is removed from the machine and cured, there is no tool to remove after curing.

1.2.2 Mandrel-less Winding Process

For the example shown in the following process, a solid composite cylinder is detailed. The cylinder demonstrates the winding of a laminate with the following winding pattern:

Table 1. Winding Layer Table

Layer	Winding Angle
1	0°
2	88°
3	± 45°
4	± 15°
5	± 60°
6	-88°

The mandrel-less winding process begins as two custom pin-rings are installed in the driven headstock and tailstock illustrated in Figure 3.

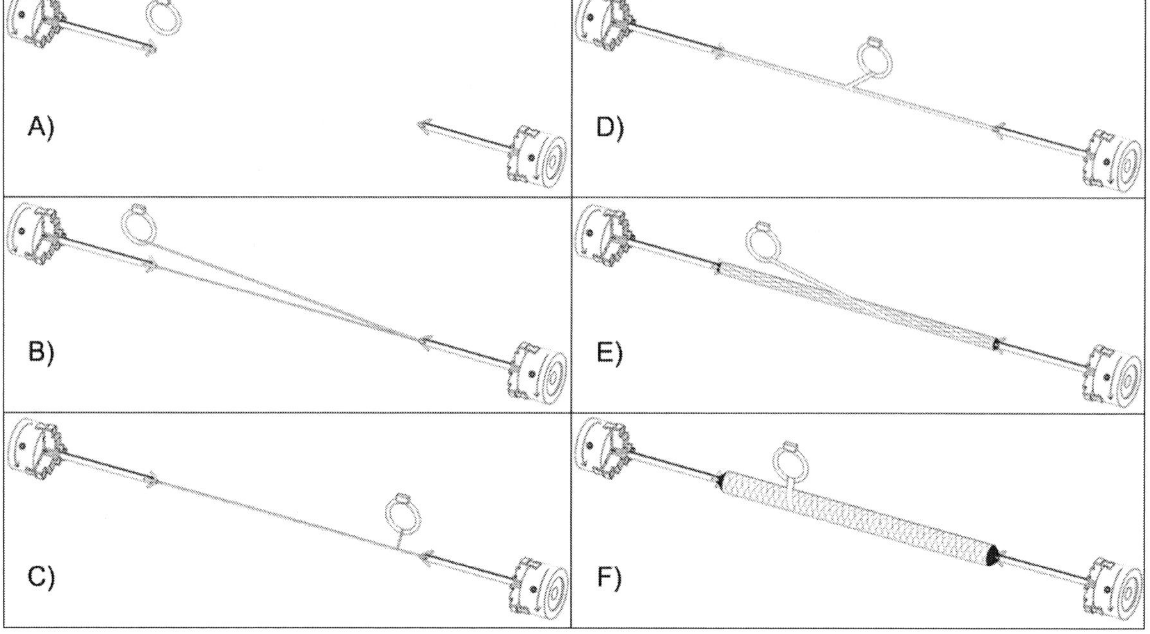

Figure 3. Mandrel-less winding process for solid cylindrical shape. A) Initial fibers are wound between hooks. B) Generation of support material- 0° pattern, layer 1. C) 88° pattern, layer 2. D) ± 45° pattern, layer 3. E) ± 15° pattern, layer 4. F) ± 60° pattern, layer 5 (layer 6 will be the same as layer 2 but reverse direction).

After the winding process is completed, the composite is removed from the machine and either allowed to cure at ambient conditions or is placed into an oven or autoclave for a higher temperature cure. The wound component may be cured in a simple supportive tray or placed into a mold or similar structure to control the shape during the cure.

1.2.3 Compression Molds and Molding Process

In the process (patent pending) discussed in this paper, the mandrel-less filament winding technique is described to create cylindrical material that was placed into a compression mold and oven cured. For the cylindrical material that was wound in this example, straight tubular compression molds were constructed out of a suitable metallic material. As seen in Figure 4, the material is removed from the winding machine and then gets placed into the inner diameter of the metal mold. A top plunger assembly and a bottom cap are installed in the ends of the mold. The larger top plunger has a small through bore that remains open to atmosphere while the plunger is installed. This plunger compacts the composite and allows excess resin and entrapped air to escape. After initial compression, resin is added through the bore of the larger top plunger and the smaller diameter plunger is installed. The process creates a pressure 5-10 times what is achievable in an autoclave. The high pressure inside the mold greatly reduces voids, increases fiber volume, and contributes to more of a 3-Dimensional composite structure. The compressed mold is secured in position and placed into an oven for curing.

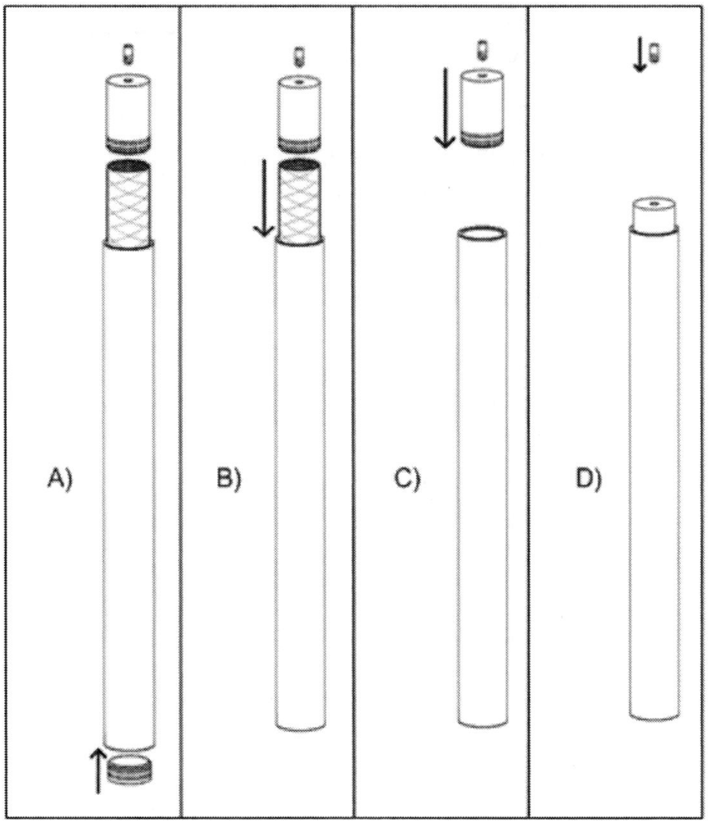

Figure 2. Compression molding the wound material. A) Bottom cap is installed in the steel mold and wound material is installed from the top. B) The wound material is fully installed in the tool to rest on the top of the bottom cap. C) The top plunger assembly is installed, and force is applied to compress the material. D) Additional epoxy may be added, and a smaller plunger used to apply additional pressure to the epoxy.

2. APPLICATION- HIGH PERFORMANCE FRAC BALL

Frac balls are used in oil field work to seal off fluids from the top of a frac plug during hydraulic fracturing operations. The ball is placed on top of a frac plug that acts like a one-way check valve/plug that is installed in a wellbore casing and holds back the hydraulic pressure of the fluid above it. When in use this ball needs to withstand fluid pressure of 69 MPa (10 ksi) and operating temperatures of 177 °C (350 °F) without failure or permanent deformation that would cause leakage. Composite is preferred over metallic as this ball and the plug itself will need to be drilled out in later operations and this material significantly reduces drilling time.

Advanced Composites, Inc. has tested other composite balls and materials for suitability in this application. Balls that have been constructed with a stacked ply laminate or pultruded rod can fail when loaded in the wrong orientation as illustrated in Figure 5. Non-reinforced high temperature plastics do not have enough strength, neither does short strand random-orientation reinforcement composite.

Figure 5. Failures of stacked laminate composite balls.

To overcome the issues identified through testing, Advanced Composites, Inc. (ACI) developed a billet material to be machined to the final product. With the development of the mandrel-less filament winding process, a cylindrical billet could be produced that would have continuous fiber and control of fiber orientation. Subsequent compression molding would yield a part that would have high fiber volume and low void content mitigating leak paths and failure points. Finally, filament winding would yield a scalable and efficient process for high volume production.

2.1 Frac Ball Requirements

The frac ball that was to be produced with this new process (patent pending) would have the following specifications (see Figure 6):

- Construction consisting of E-Glass/Epoxy
- Withstand a pressure of 69 MPa (10 ksi) and operating temperatures of 177 °C (350 °F). without failure or permanent deformation
- Diameter of 57.15 mm +/- 0.05 mm (2.250" +/-0.002")
- Sphericity of 0.025 mm (0.001")
- Surface finish of 1.25 μm Ra (50 μin Ra)

Figure 3. As manufactured frac ball meeting performance requirements.

2.2 Production of Frac Balls

After the cylindrical billet material was wound and cured, balls were machined to rough shape - a CNC lathe and diamond grinding tools performed rough machining. The ball was then ground to final diameter and roundness.

2.3 Testing

2.3.1 Test Overview

Five completed frac balls were produced and a test procedure provided to a 3rd party lab that specializes in oil field tool testing. The test was designed to establish satisfactory performance of frac ball sealing at 69 MPa pressure (10 ksi) and 177 °C (350 °F). For the test, the ball was placed on a ball seat within a pressure chamber designed by ACI. The test fixture consisting of the pressure chamber, ball, high pressure lines, and oil (working fluid) would be submerged in a 177 °C (350 °F) test well (hot oil bath). When the internal temperature in the test fixture was up to test temperature 171 °C-182 °C (340 °F-360 °F), pressure would be applied to the top of the ball until the pressure target of 69 MPa was reached. Upon reaching target temperature and pressure, a 30-minute dwell time would be maintained. If the 5 balls withstood the maintained temperature and pressure without damage the test would be deemed successful.

2.3.2 Test Apparatus and Procedure

The test fixture consisted of a top piece with a port for pressure to be applied from above the ball, and a port for a thermocouple probe to read the internal temperature at the time of the test. The bottom piece of the test fixture had a port in the bottom for a bleed valve to relieve pressure and contain fluids from being introduced into the test well. The bottom piece also has a ball seat and an o-ring groove for sealing the bolted-together assembly. Oil was the test medium used to apply internal pressure inside the test fixture and apply a uniform temperature around the submerged fixture in the test well. High-pressure lines were coupled to the test fixture to apply internal pressure from the test apparatus control panel. See Figure 7 for test assembly. Figure 8 shows the test fixture being assembled and hoisted in the test well.

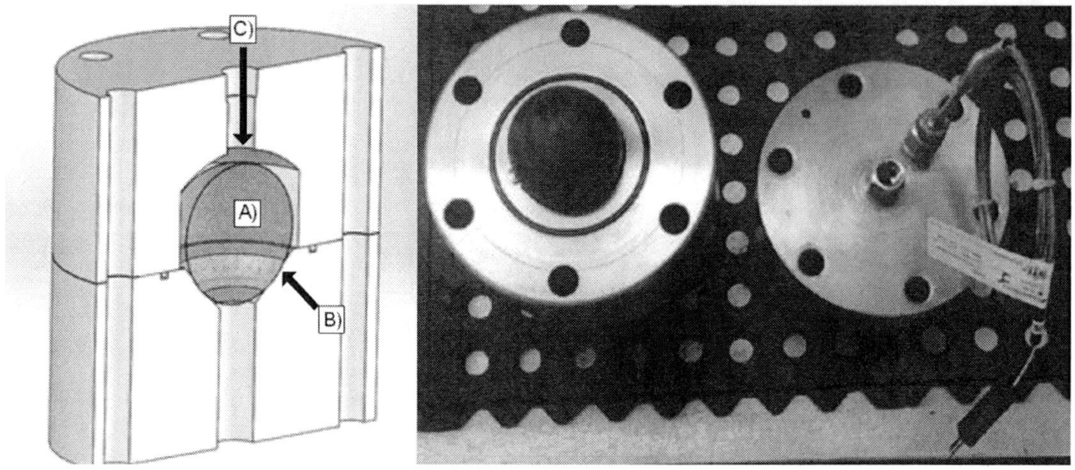

Figure 4. Left: Cross sectional view of pressure test fixture detailing A) ball location, B) ball seat, and C) pressure port. Right: actual test fixture in assembly process.

Figure 5. Assembled test fixture on bench and hoisted into test well (hot oil bath).

2.3.3 Test Procedure with Test Results:

The test well had stabilized at 177 °C (±5 °C) with hot oil surrounding the test fixture. Additionally, the temperature as measured inside the test fixture stabilized at 177 °C (±5 °C) prior to the application of pressure to the top of the ball.

The chart in Figure 9 below shows the temperature on the Frac Ball 1 during the pressure cycles at 177 °C (±5 °C).

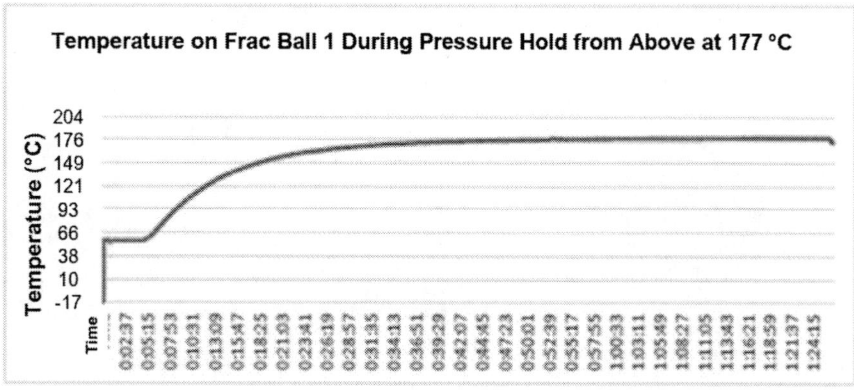

Figure 6. Temperature for Frac Ball 1 test.

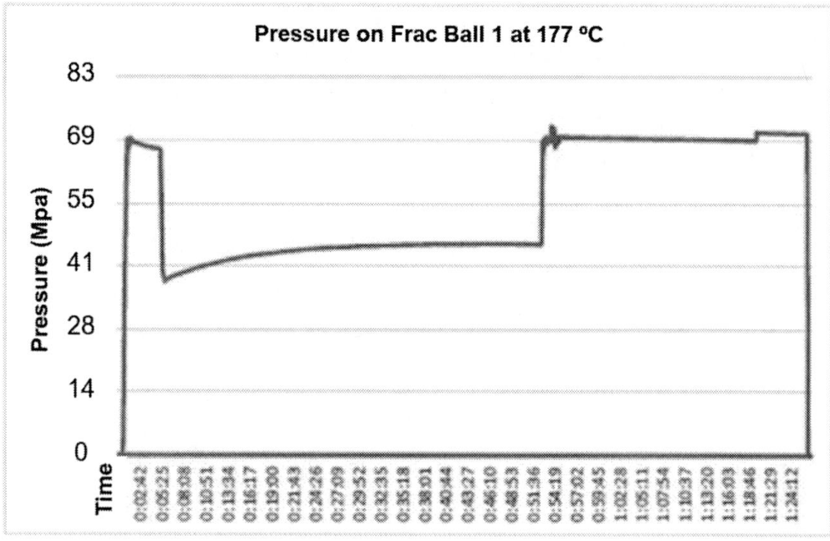

Figure 7. Pressure from Frac Ball 1 test.

Figure 10 shows the pressure sequence applied to Frac Ball 1 during the pressure cycle up to the target pressure of 69 MPa. Hydraulic pressure would be applied to the top of the ball beginning at 69 MPa to check that the ball would "seat" and no leaks had occurred. Pressure was then reduced to 41 MPa to wait for the internal temperature to reach 177 °C. Once the target temperature was reached, pressure was increased to 69 MPa and held for 30 minutes. The ball successfully held the target pressure for 30 minutes and all pressure was removed. The fixture was then cooled so it could be disassembled and handled after the test by quenching it in a basin of cool oil. Once cooled to an internal temperature of 71 °C, it was taken apart and the ball removed for inspection (see Ball 1 in Figure 11-after test).

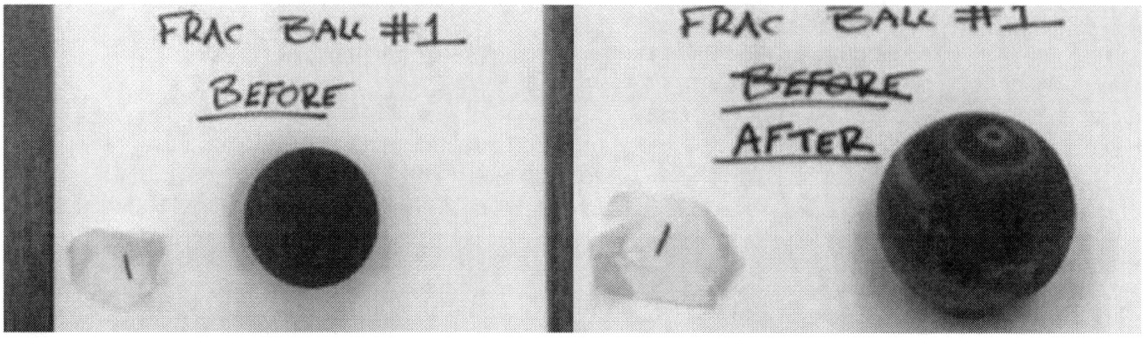

Figure 8. Frac Ball 1 test- before and after. Ball has some cosmetic marking but no permanent deformation or damage. Test passed.

2.3.4 Frac Ball Test Conclusions

ACI Frac Ball 1 successfully completed the pressure hold at 69 MPa and at 177 °C and did not exhibit permanent deformation or damage. After the successful hold at the target temperature and pressure, all pressure was removed and the test was brought to completion. The test fixture was removed from the bath and Frac Ball 1 was inspected for damage. Additional tests were conducted on Frac Balls 2-5 with all balls passing the test. In the subsequent tests, balls were oriented in different positions to test any worst-case orientation on the ballseat.

3. FUTURE WORK

There are many other opportunities to explore with both tool-less filament winding and combined processes with compression molding or other techniques. This process could be used to generate high performance composite forms that are used in subsequent molding or forming stages of which one example is illustrated in Figure 12.

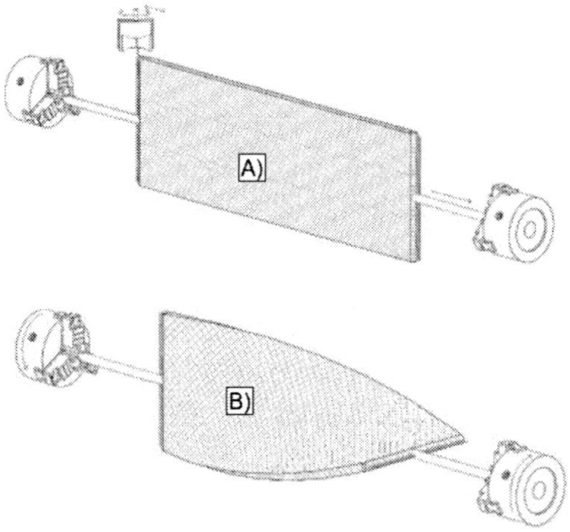

Figure 9. Potential opportunities with controlled orientation preform- A) flat panel B) shaped fiber preform.

4. CONCLUSIONS

The process developed in this paper for mandrel-less filament winding has been proven to be a viable production process. Through the development of a new patent pending filament winding machine design and additional mold tooling, filament winding was able to produce solid composite structures with highly controlled fiber orientation and low void content. This paper has discussed the initial process concept and machine design, all the way to the qualification of a commercially available product made with this process.

Comparison of Hand-Layup and Automated Tape Placement Processing of Thin Ply IM7/8552 Composites

Uday Kiran Balaga[1], Verena Gargitter[1,2], Dirk Heider[1,3], John Tierney[1,3], Suresh Advani[1,2], Shridhar Yarlagadda[1,3]
1 Center for Composite Materials, University of Delaware, 101 Academy St, Newark, DE 19716
2 Dept. of Mechanical Engineering, University of Delaware, Spencer Lab, Newark, DE 19711
3 Composites Automation LLC, 9 Adelaide Court, Newark, DE 19702

ABSTRACT

Composite structures offer higher lightweight performance when compared to traditional metals yet lack sufficient tolerance for damage onset. The focus of this effort is on thin-ply thermoset prepreg materials manufactured by Hexcel, which have shown higher resistance to damage initiation under applied load when compared to standard ply-thickness materials. To manufacture large structures currently Automated Tape placement (ATP) process is used. However, composites manufactured using thin-ply materials require placement of many more layers to achieve the same net volume of coverage. Gaps and overlaps between adjacent prepreg layers are unavoidable with tape placement and can result in degradation of mechanical properties due to these changes in the microstructure. Placement of thin-ply tape will also cause microstructural variability due to the low handling stiffness of this material. Modifications to existing automated tape placement equipment were made to address the steering of the thin-ply tapes which resulted in improved placement accuracy. Unidirectional and quasi-isotropic panels were fabricated from 12′′ wide unidirectional prepreg sheet with traditional hand layup method and compared with panels fabricated with ATP process using a ¼′′ unidirectional prepreg tape from the same 12′′ prepreg stock under varying processing conditions. All panels were cured in the autoclave with the recommended cycle. This paper compares the effects of processing parameters on the mechanical properties and the microstructure of the processed panels. This was achieved by characterizing the microstructure using C-scans, confocal microscopy and conducting mechanical tests.

Keywords: Automated Tape Placement, Mechanical Characterization, Microstructure Evaluation, Gaps and Overlaps
Corresponding author: Uday Kiran Balaga[1]

1. INTRODUCTION

Composite materials have been used for many years in the aerospace industry to reduce aircraft weight and to optimize performance. The traditional hand layup techniques produced high variability, defects and required higher skill levels to produce complex shapes. In order to address this, new processes such as ATP (Automated Tape Placement) have been developed, which allow for less variability of reinforcement lay-up and close control of process parameters [1].

ATP is a manufacturing process in which a tape usually ¼" to 6" wide rather than a fibre tow is placed onto the substrate or the tool. Subsequent layers are placed and consolidated layer by layer to fabricate a panel or a complex structure. This process can otherwise be also referred to as 'layer by layer additive manufacturing' [2]. This process of manufacturing is generally used to produce larger structures which are difficult to lay-up manually. The resin can be a thermoset or thermoplastic. Unlike the Automated fibre placement (AFP) process, by ATP, it is possible to process not only rectangular ply patches but also customized shapes including curved parts or any irregular forms. By controlling the shape, it is possible to obtain continuous laminates with less defects in the structure. When the layup process of the prepreg sheets onto the tool is done manually, the manufacturing method is called hand layup. The limitation of the traditional layup process is that the speed and the quality of the process itself highly depends on the skill level of the worker. Moreover, the process is hardly repeatable, which can be improved by automation [3].

The ATP process is very similar to the hand layup method with the difference that the process of laying in ATP is done automatically by an automated robotic head with a gradient in temperature and force. This manufacturing method is highly beneficial over the hand layup because of its excellent material layup rates. Prepreg tapes are placed onto substrate or on previously placed layers to build the desired shape of the part and then the part is cured either in an autoclave or an oven. Curing in an autoclave is advantageous because it could prevent the formation of internal voids since higher pressures can be used for consolidation. The tape widths that are generally used in ATP are usually 75, 150, and 300 mm and 250 μm thick [3]. However, we used a thin and narrow tape which is approximately 6.35 mm wide and 70 μm thick. To handle this thin ply tape, several hardware developments were added to the existing robotic design [4][5]. Generally, ATP process is dependent of three main parameters which are pressure, layup speed and the temperature. When the material is exposed to heat, it becomes tacky and can be consolidated under the roller pressure. However, as the material thickness is reduced it becomes more flexible and requires proper steering control. Therefore, hardware changes were done to the existing robotic head by adding different components such as re-feeder and cutter which have the capability to handle thin prepreg tapes unlike the earlier designs which were suitable for wider tapes. The most common defects during ATP process are the gaps and overlaps that occur between the adjacent layers [1][6][7][8][9][10]. As our tape is much narrower and thinner, it will require a larger number of repetitions to achieve the required dimensions which could possibly increase the probability of more gaps and overlaps which could have an adverse effect on the quality of the manufactured composite part. Another issue that needed to be addressed was the tape path. As the tape refeeds after cutting, there is no specific path for the tape to take its position under the compaction roller. It required re-adjusting the distance to align itself after consolidation which is known as ``fibre straightening''[7][8]. We programmed our robot to account for this correction. The modified ATP system is shown in Fig.1 with its individual components that were modified.

The goal of the present work is to develop an Automated Tape Placement (ATP) system for the processing of thin prepreg composite laminates, which exhibits equivalent or superior properties compared to the traditional hand layup laminates manufactured from the same material. Study on different processing defects such as gaps and overlaps of different geometries, have been examined in the laminates with different stacking sequences, as well as their effect on mechanical properties and microstructure. Initially, laminates were processed with hand layup in different stacking sequences to compare with the laminates processed from ATP with the same stacking sequence with the presence of gaps and overlaps. C-scan inspections were used to check the quality of all the laminates for any internal voids and surface porosity. Then confocal microscopy was used to examine the microstructure of the laminates. Finally, mechanical tests were conducted to compare and contrast the extent of defect on the mechanical properties of the composite laminate processed from ATP and hand layup methods.

a. Material stocker, b. Vacuum suction for backing paper, c. Idlers for tape accuracy and alignment, d. Refeed arm for tape input, e. Hot plate, f. Consolidation roller

Figure 1: Modified Automated Tape Placement system with modified system shown within the boxes that are magnified on the right side

2. EXPERIMENTATION

2.1 Material

The material used is carbon-reinforced prepregs (8552/IM7/RC41/AW70) manufactured by Hexcel Composites and has been characterized by NASA Langley and Glenn research centres [11]. The nominal baseline fibre volume fraction is 54 %. For the hand layup method, the prepreg is used in its original form as unidirectional sheets (12′′ wide). For the ATP process, the prepreg was slit from the 12′′ unidirectional sheet as a tape of width ¼′′ (6.35 mm). The unidirectional sheet and the tapes come from the same batch of material. The thickness of the material is approximately 70 μm as it has an areal weight of 70 gsm.

2.2 Manufacturing of Composite Panels

The robotic system used for manufacturing of panels is from ABB and the head was designed at the Centre for Composite Materials at the University of Delaware and is capable of placing a tape that has a width of ¼′′ (6.35 mm) in each iteration until the required dimensions are realized. A hot gun and a hot plate, placed beneath the substrate, locally heats each layer for adhesion between the subsequent plies. The pressure input to the components like compaction roller, cutter and the backing paper suction is from the same source. The roller has pneumatic actuators present on either side which suppresses the higher forces for nominal consolidation. So, compaction force of 414 kN/m^2 (60 psi) which is also efficient for the other components was applied during the layup for consolidation of tapes. The layup speed was 200 mm/s. An aluminium plate was used as a substrate with double sided tapes used on the edges to facilitate adhesion of the first layer. The stacking sequences used for various panels are $[0]_3$ (three layers of unidirectional), $[0]_{16}$, $[45/0/-45/90]_s$ (s - symmetry in mid plane) and $[-45/90/45/0]_s$ which were programmed using an offline platform from ABB [12]. All the layup sequences are programmed for a brick face pattern to attenuate defects in every individual layer. The layup pattern is shown in Fig. 2. The manufacturing process of thin prepreg tapes using ATP is shown in Fig. 2. The manufactured panel was then cut into required dimensions discarding the edges. The panels were prepared from the same material used in the hand layup method in which 12′′ wide panel was manufactured as well for comparison with the ATP panels.

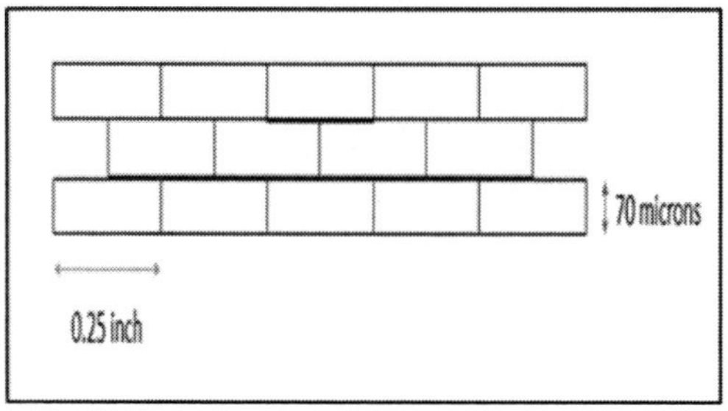

Figure 2: Example of a Quasi-isotropic laminate with the ATP process (left); Brick face layup pattern (right).

2.3 Processing Cycle

The autoclave cure cycle recommended by the supplier (temperature ramp to 110 ^{0}C in 60 min followed by a slower ramp to 180 ^{0}C for 120 min with a maximum autoclave pressure of 690 kN/m^2 (100 psi)) was used to manufacture the panels after the prepregs were placed in the desired sequence either by hand layup or by ATP process. However, in an attempt to eliminate any internal voids created by the layup process; the autoclave pressure was increased to 2068 kN/m^2 (300 psi). Higher pressure may also assist in decreasing the width of any defects present within the laminates [13]. A pressure transducer and thermocouple were placed in the autoclave to record the temperature and pressure during the entire curing and consolidation process. A steel caul plate was placed on top of the laminate positioned inside the vacuum bag to distribute the pressure evenly across the caul plate which will result in smooth surface finish and uniform thickness [1]. The schematic representation of the bagging and autoclave is shown in Fig. 3

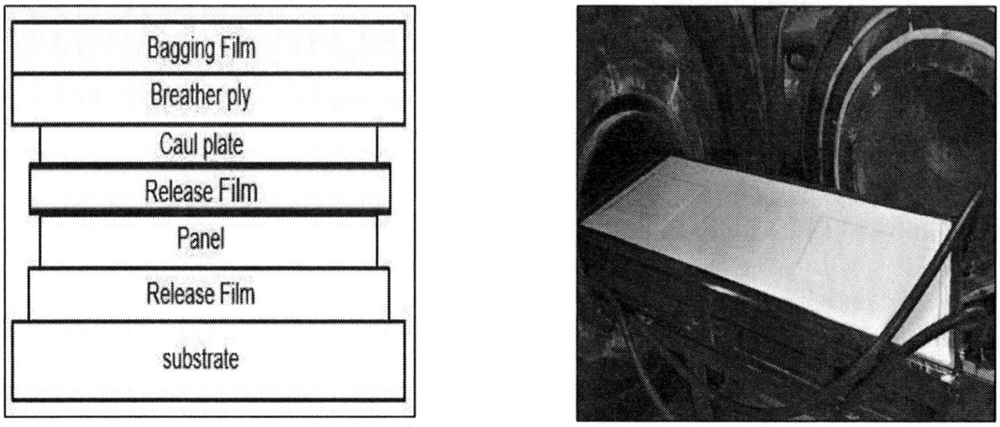

Figure 3: Schematic diagram of bagging procedure (left); Autoclave for curing (right)

2.4 Ultrasonic C-Scan Inspection

For this study, ultrasonic C-Scan was chosen for analysis of the laminate quality. The ultrasonic C-scan equipment used for the inspection is equipped with a transducer and a receiver. For these inspections, the laminates were placed in between them and immersed in a tank containing water. The control is performed by a focused transducer, which scans in the X and Y directions that allows the entire surface of the part to be covered. This helps determine the defects in the laminate such as gaps, overlaps, internal voids, and porosity.

2.5 Confocal Microscopy and Image Analysis

A VK-X250 3D Laser Scanning Confocal Microscope was used to examine the defects within and in between the plies on the microscopic scale. The samples in the mid-section were cut and polished down to 0.05 μm grit to obtain flat, clear surfaces before investigation. Fibre volume fraction and void content were determined using image analysis of SCM images with ImageJ software [14] and Smartree software [15]. Stitching of several images was necessary to re-create all the regions which would shed light on the shape and size of the defect areas.

2.6 Mechanical Property Characterization Experiments

Tensile tests were conducted according to the ASTM D3039 [16]. The test coupons were cut from the manufactured panel as per the test standards of ASTM. These tests were carried on an Instron test machine, with load cell varying from 30 kN to 250 kN depending on the type of test, at a loading rate of 2 mm/min. In order to reduce stress concentrations in tensile samples that may arise by the clamping system of the testing machine and to avoid premature failure, end tabs of 2 mm thickness were bonded to each end of the specimen using an epoxy adhesive [1]. To obtain the strain information from the tensile samples, a strain gage was bonded in the centre of each sample which was connected to a strain acquisition system for data collection during the tests. For each configuration, an average of five to six samples were tested.

3. RESULTS & DISCUSSION

3.1 Comparison of Hand Layup with ATP for Unidirectional laminate

The laminates (panels) with stacking sequence $[0]_3$ and $[0]_{16}$ was prepared with both hand layup and ATP methods. These uncured laminates were then transferred to an autoclave to cure with the manufacturer's recommended cycle. The laminates were weighed before and after cure to calculate the resin loss during its presence in the autoclave. The fibre volume fraction was calculated by accounting for the resin loss. The average thickness of each laminate was measured from the cured panels then subjected to C-scan for quality inspection. The values obtained were attenuated into smaller decibel values to determine the defects. No signs of defects from the images were noticed in both cases. The average thickness and fibre volume fraction are listed in Table 1. The images obtained from C-scan are shown in Fig. 4.

Table 1: Average thickness and fiber volume fraction of unidirectional panels

	Thickness (in.)	**Fibre Volume Fraction (%)**
$[0]_3$ ATP	0.0085 ± 0.0002	53.3
$[0]_3$ Hand layup	0.0087 ± 0.0005	52.8
$[0]_{16}$ ATP	0.0419 ± 0.0025	53.8
$[0]_{16}$ Hand layup	0.0427 ± 0.0030	53.7

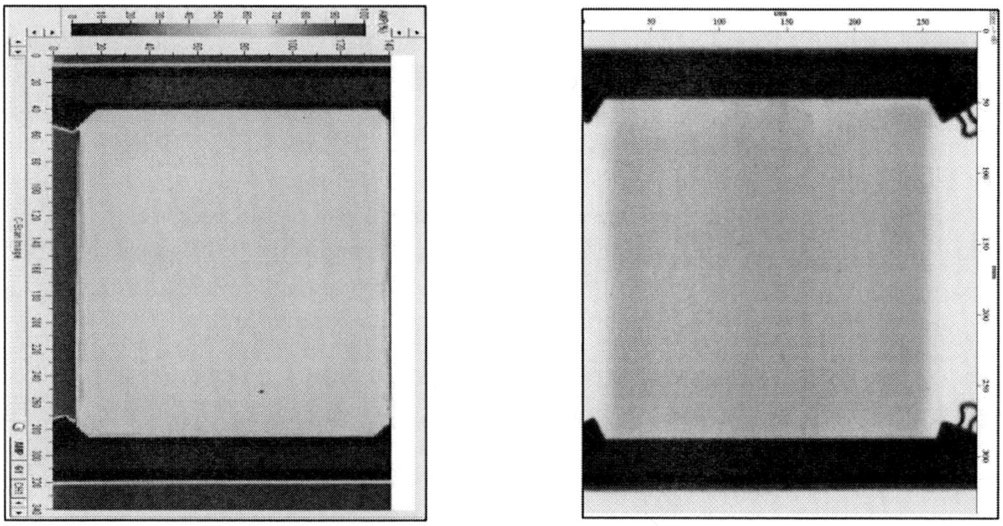

a) b)

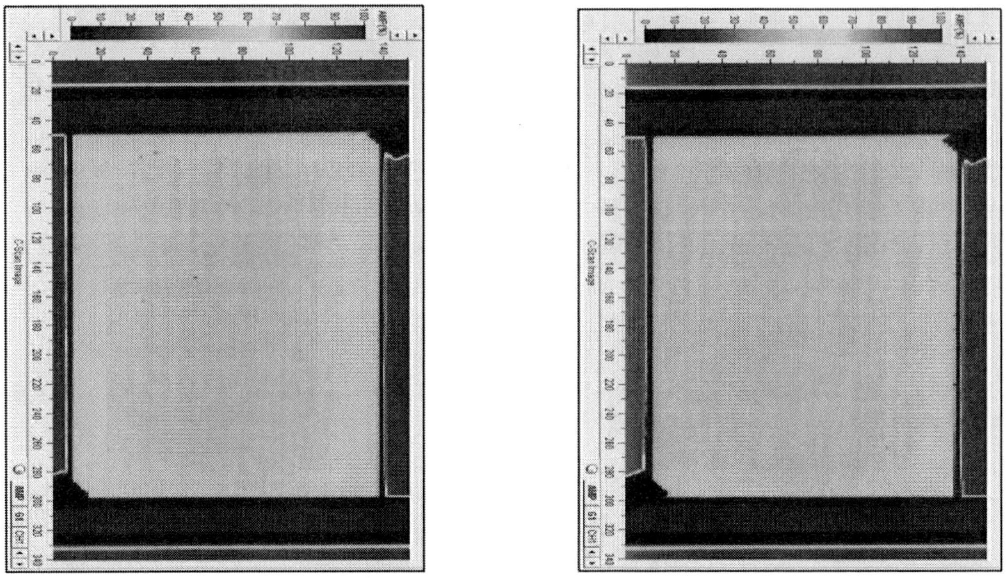

c) d)

Figure 4: C-Scan images: a) $[0]_3$ ATP, b) $[0]_3$ Hand layup, c) $[0]_{16}$ ATP and d) $[0]_{16}$ Hand layup

A small sample was cut from the centre of the panel and potted for microscopy. The microscopic images were very consistent for both laminates without any signs of internal voids, but the ATP sample did exhibit some gaps and overlaps in its structure which were filled with the fibres from adjacent layers. This in turn caused in-plane waviness which may lead to damage initiation [8]. From the results of mechanical tests, fluctuation in strain values are seen, which is caused by the in-plane waviness. This could result in non-uniform fibre volume fraction which may affect the mechanical properties. The void content was found to be negligible as the pressure used in the autoclave was 2068 kN/m^2 (300 psi) which could have prevented voids from expanding or initiating during the cure process. The microscopy images are shown in Fig. 5.

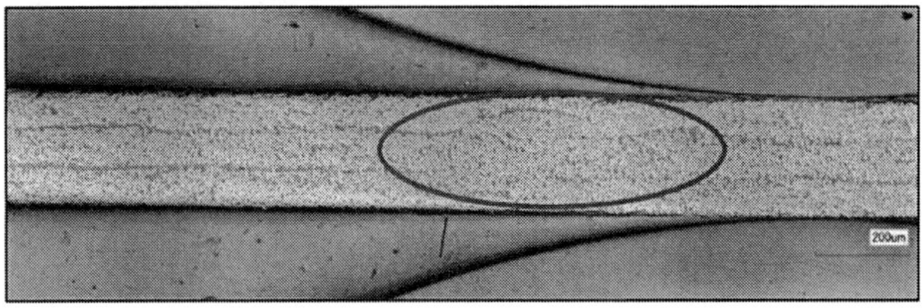

a)

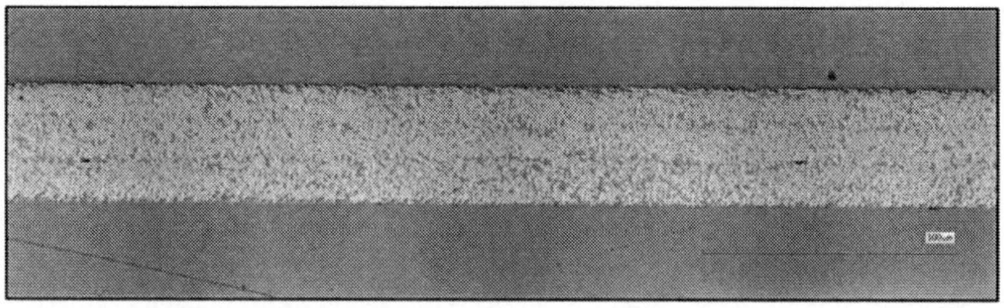

b)

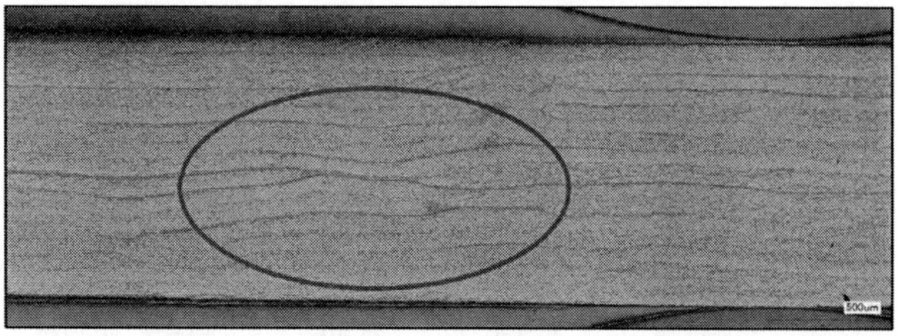

c)

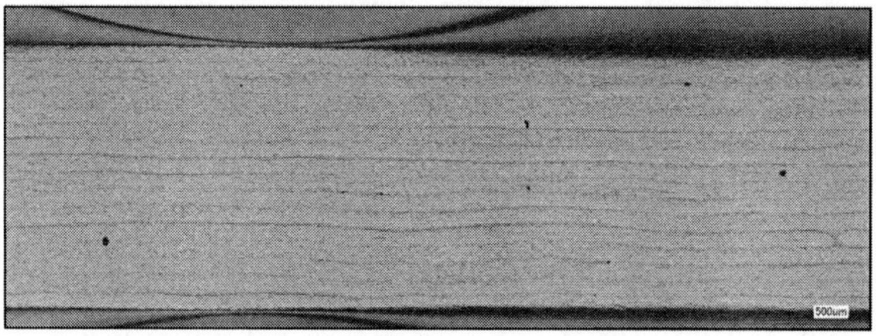

d)

Figure 5: Microscopy images: a) $[0]_3$ ATP– Overlapping of fibres adjusted due to autoclave pressure (region in red circle), b) $[0]_3$ Hand layup, c) $[0]_{16}$ ATP – in plane waviness caused due to gaps (region in red circle) and d) $[0]_{16}$ Hand layup

The average strength and modulus are approximately the same for both layup methods of $[0]_{16}$ but modulus in panel $[0]_3$ is higher in ATP which is probably due to the overlapping of fibres which can be seen in the microscopy could increase the fibre volume fraction in turn increasing the modulus of the panel. Even the strain values were fluctuating in case of ATP for both layups since the defects cause undulation as discussed earlier along with resin rich areas which is possibly initiating the microcrack [17][18]. The variability coefficient of strength, modulus, and strain to failure in ATP are 3.9 %, 2.8 % and 5.45 % respectively, whereas for hand layup samples they are 3.35 %, 3.4 % and 3.08 % respectively. The variability coefficient of strength, modulus, and strain to failure for the sample with 16 layers with ATP are 5.6 %, 3.8 % and 5.06 % respectively whereas for hand layup samples they are 2 %, 3.9 % and 3.13 % respectively. These results clearly indicate that slightly higher degree of variability is encountered with the ATP manufacturing method. The failure mode of the samples was catastrophic for both cases which is approximately similar to the results of the tests conducted by Andrew E. Lovejoy [11]. The typical failure in the sample can be seen in Fig. 6 The mechanical properties of the panels are presented in Fig. 7.

Figure 6: Failed Unidirectional sample from the ATP laminate

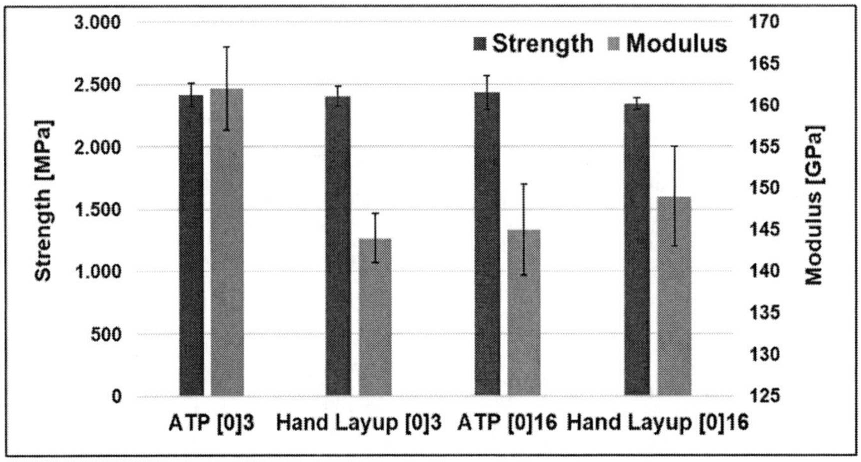

Figure 7: Strength and Modulus in Unidirectional samples with three and sixteen plies manufactured by ATP and Hand Layup process

3.2 Comparison of Hand Layup with ATP for Quasi - isotropic laminate

Laminates with $[45/0/-45/90]_s$, $[-45/90/45/0]_s$ stacking sequences were manufactured by both ATP and hand layup techniques. From the C-scan analysis, no internal voids or surface porosity was observed in panels made by both the hand layup and the ATP process. The measured average thickness and the fibre volume fraction of the panels are shown in Table 2. The images obtained from C scan are shown in the Fig. 8.

Table 2: Average thickness and fiber volume fraction of Quasi isotropic panels

	Thickness (in.)	Fibre Volume Fraction (%)
$[45/0/-45/90]_s$ ATP	0.0219 ± 0.0010	56.4
$[45/0/-45/90]_s$ Hand layup	0.0211 ± 0.0020	56.4
$[-45/90/45/0]_s$ ATP	0.0226 ± 0.0002	53.9
$[-45/90/45/0]_s$ Hand layup	0.0240 ± 0.0003	51.6

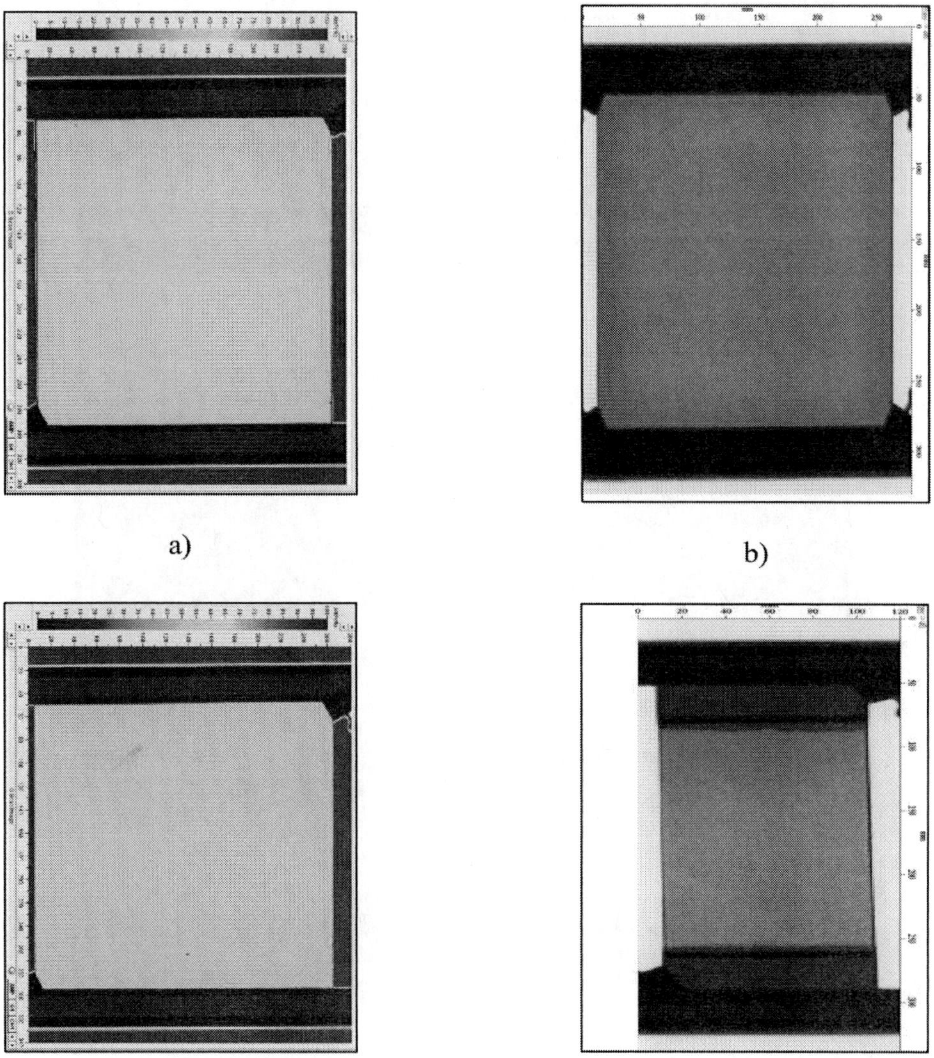

a) b)

c) d)

Figure 2: C-scan images: a) $[45/0/-45/90]_s$ ATP, b) $[45/0/-45/90]_s$ Hand layup, c) $[-45/90/45/0]_s$ ATP and d) $[-45/90/45/0]_s$ Hand layup

The microstructure is consistent with little gaps and overlaps which could be seen in the microscopic images (Fig. 9). The average fibre volume fraction was the average of values measured at different locations along the 0^0 direction. There is no sign of any internal voids in the stack, but the 0^0 plies seem to fill the gaps with the fibres from adjacent tapes due to the application of the high autoclave pressure.

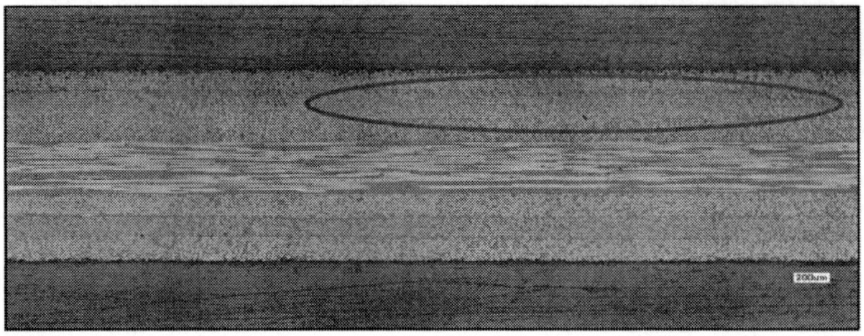

a)

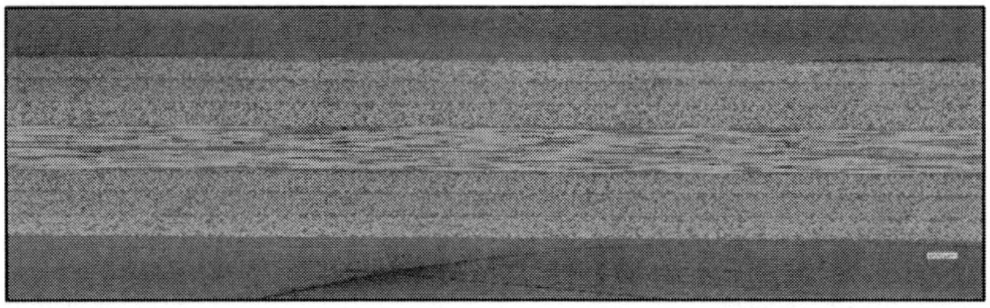

b)

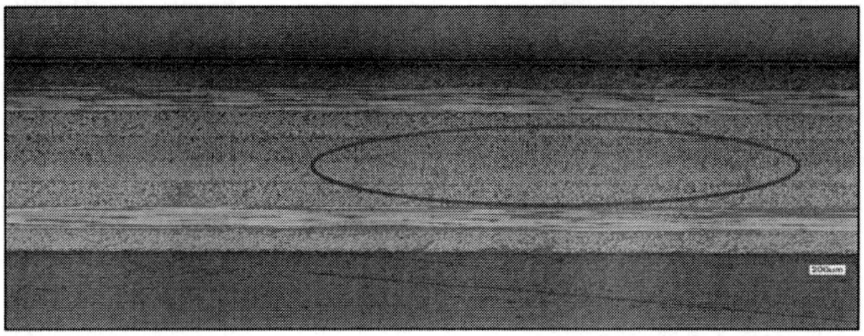

c)

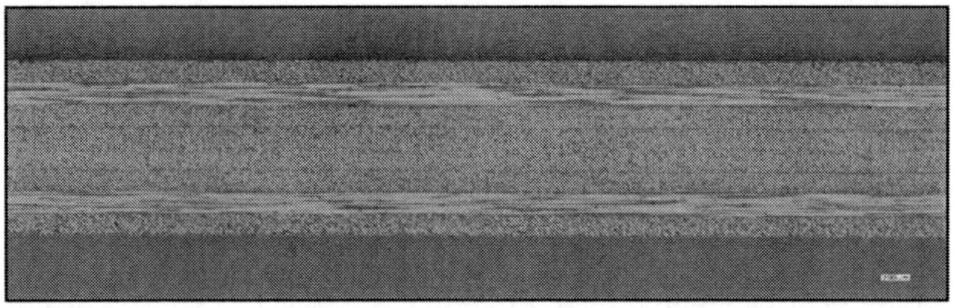

d)

Figure 9: Microscopy images: (a) $[45/0/-45/90]_s$ ATP – Fibres adjusting the defects (region in red circle), (b) $[45/0/-45/90]_s$ Hand layup, (c) $[-45/90/45/0]_s$ ATP – defect filling in the mid plane (region in red circle), (d) $[-45/90/45/0]_s$ Hand layup

The average strength and modulus depend on the stacking sequence of the laminate. The stacking sequence which has 0^0 in the midplane will always have better strength and modulus compared to the stack with 90^0 in the midplane. Since the laminate configuration is symmetric, the areal weight of the material doubles in the mid plane making the centre ply equal to 140 gsm. Therefore, for the layup that has two 90^0 plies in the midplane, the microcracks initiate earlier than the other layup pattern which has two layers of 0^0 in the midplane. The results from the ATP and hand layup samples with 0^0 in the midplane show very small coefficients of variation as compared to the samples in which the 90^0 ply is in the middle. Our results are consistent with the quasi layup for thin prepregs. The failure mode of the samples is catastrophic multimode for the samples having 0^0 in midplane and multiple failure for the samples having 90^0 in the mid plane. Typical failure in the specimen are shown in Fig. 10 whereas property values along with their variations are presented in Fig. 11.

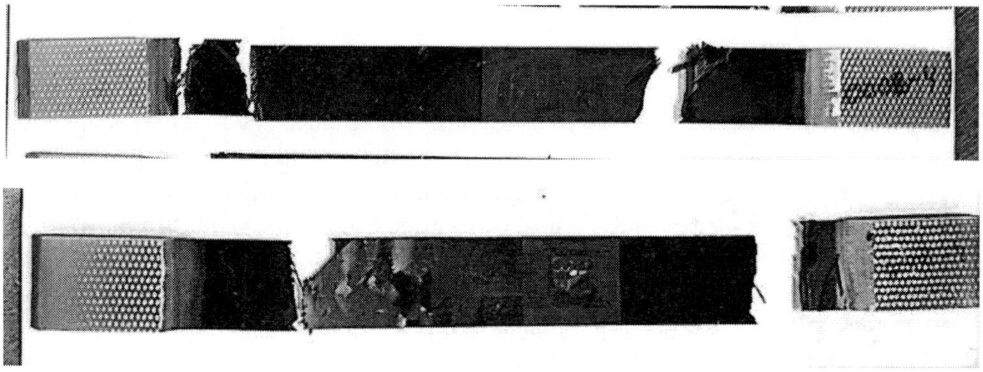

Figure 10: Failed Quasi isotropic samples, $[-45/90/45/0]_s$ – top, $[45/0/-45/90]_s$ - bottom

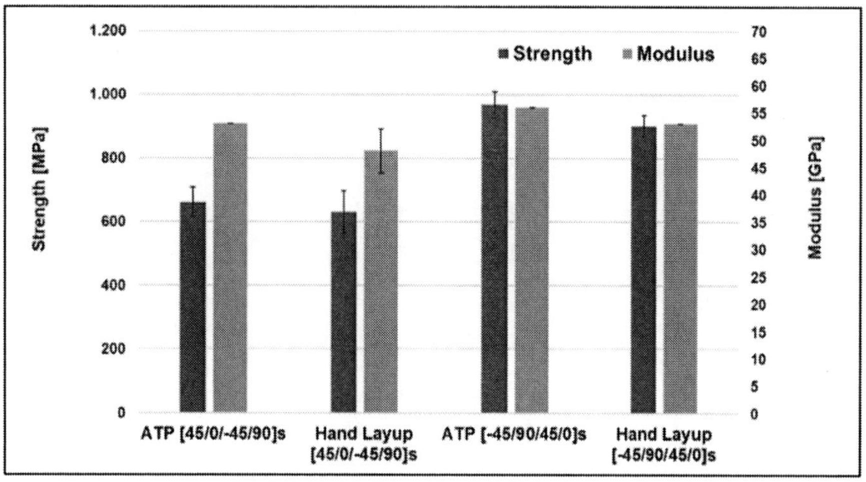

Figure 11: Strength and Modulus in Quasi isotropic samples with 8 layers

4. SUMMARY & CONCLUSIONS

In the present research study, the comparison of mechanical properties and microstructure has been carried out for two different manufacturing methods hand layup and Automated Tape Placement (ATP) for thin prepreg material. Also, the extent of defect formation in ATP processing was analysed. Initially, panels were prepared by traditional layup method for baseline properties comparison with the ATP panels. Different layup sequences such as unidirectional layup and Quasi isotropic layup were used for mechanical tests and microstructure analysis.

In the first comparison of unidirectional panels, both methods have shown no significant variation in their properties, however the microstructure of ATP panels varies from hand layup since they present processing defects such as gaps and overlaps. The influence of these defects is higher in the unidirectional layup which is evident from the microstructure and the mechanical properties. The coefficient of variability seems to be higher in ATP because of these processing defects. Later the same process was carried out for Quasi isotropic layup with two different stacking sequences $[45/0/-45/90]_s$, and $[-45/90/45/0]_s$. For the first sequence, there are two 90^0 plies in the middle whereas in the second sequence there are two 0^0 in the mid plane. The mechanical properties with the first sequence as expected show lower properties due to doubling of the thickness in the midplane. However there is no significant difference and the variability is less than the reference hand layup panel and in some cases the panel processed from ATP exhibit better properties than hand layup from which we can draw the conclusion that thin prepreg processing through ATP could be viable for Quasi Isotropic layup. The microstructure quality looks good despite the presence of defects in the ATP during the placement process.

The present study has shown that presence of random processing defects of ATP with the thin prepreg materials does not have significant effect on the mechanical properties compared to traditional layup methods when considering thick parts in a cross ply pattern but it shows significant variability for panels in unidirectional pattern due to the microstructure variation caused by the defects.

5. ACKNOWLEDGEMENT

The authors would like to thank the National Aeronautics and Space Administration (NASA USA) for their funding support. The authors are also pleased to express their grateful acknowledgements to Nicholas Shevchenko for the ultrasonic c-scan analysis and Lake Edward for technical support in autoclave.

6. REFERENCES

[1] M. Lan, D. Cartié, P. Davies, and C. Baley, “Microstructure and tensile properties of carbon-epoxy laminates produced by automated fibre placement: Influence of a caul plate on the effects of gap and overlap embedded defects,” *Compos. Part A Appl. Sci. Manuf.*, vol. 78, pp. 124–134, 2015, doi: 10.1016/j.compositesa.2015.07.023.

[2] M. H. Nguyen, A. A. Vijayachandran, P. Davidson, D. Call, D. Lee, and A. M. Waas, “Effect of automated fiber placement (AFP) manufacturing signature on mechanical performance,” *AIAA Scitech 2019 Forum*, no. January, pp. 1–43, 2019, doi: 10.2514/6.2019-0516.

[3] H. B. T.-S. and V. of T. W. C. S. Abramovich, Ed., “7 - Stability of composite shell–type

structures," Woodhead Publishing, 2017, pp. 253–428.
[4] H. B. Olsen, "Automated composite tape lay-up using robotic devices," *Proc. - IEEE Int. Conf. Robot. Autom.*, vol. 3, pp. 291–297, 1993, doi: 10.1109/robot.1993.292190.
[5] J. Tierney, "Material quality development during the automated tow placement process," Jan. 2003.
[6] K. Croft, L. Lessard, D. Pasini, M. Hojjati, J. Chen, and A. Yousefpour, "Experimental study of the effect of automated fiber placement induced defects on performance of composite laminates," *Compos. Part A Appl. Sci. Manuf.*, vol. 42, no. 5, pp. 484–491, 2011, doi: 10.1016/j.compositesa.2011.01.007.
[7] F. Heinecke and C. Willberg, "Manufacturing-Induced Imperfections in Composite Parts Manufactured via Automated Fiber Placement," *J. Compos. Sci.*, vol. 3, no. 2, p. 56, 2019, doi: 10.3390/jcs3020056.
[8] E. Oromiehie, B. G. Prusty, P. Compston, and G. Rajan, "Automated fibre placement based composite structures: Review on the defects, impacts and inspections techniques," *Compos. Struct.*, vol. 224, no. May, p. 110987, 2019, doi: 10.1016/j.compstruct.2019.110987.
[9] A. J. Sawicki and P. J. Minguet, "Effect of intraply overlaps and gaps upon the compression strength of composite laminates," *Collect. Tech. Pap. - AIAA/ASME/ASCE/AHS/ASC Struct. Struct. Dyn. Mater. Conf.*, vol. 1, pp. 744–754, 1998, doi: 10.2514/6.1998-1786.
[10] M. Lan, D. Cartié, P. Davies, and C. Baley, "Influence of embedded gap and overlap fiber placement defects on the microstructure and shear and compression properties of carbon-epoxy laminates," *Compos. Part A Appl. Sci. Manuf.*, vol. 82, pp. 198–207, 2016, doi: 10.1016/j.compositesa.2015.12.007.
[11] A. E. Lovejoy, S. J. Scotti, S. G. Miller, P. Heimann, and S. S. Miller, "Characterization of IM7/8552 thin-ply and hybrid thin-ply composites," *AIAA Scitech 2019 Forum*, no. January, 2019, doi: 10.2514/6.2019-0773.
[12] ABB, "Operating manual RobotStudio," p. 588, 2018, doi: 10.1016/j.msec.2014.05.077.
[13] M. Abouhamzeh, D. Nardi, R. Leonard, and J. Sinke, "Effect of prepreg gaps and overlaps on mechanical properties of fibre metal laminates," *Compos. Part A Appl. Sci. Manuf.*, vol. 114, no. April, pp. 258–268, 2018, doi: 10.1016/j.compositesa.2018.08.028.
[14] "ImageJ User Guide User Guide ImageJ."
[15] I. T. Management, "SMARTREE USER."
[16] ASTM, "Astm D3039/D3039M," *Annu. B. ASTM Stand.*, pp. 1–13, 2014, doi: 10.1520/D3039.
[17] M. A. Azmah Hanim, D. Brabazon, and M. S. J. Hashmi, "8 - Cracks, microcracks, and fracture toughness of polymer composites: Formation, testing method, nondestructive detection, and modifications," in *Woodhead Publishing Series in Composites Science and Engineering*, M. Jawaid, M. Thariq, and N. B. T.-F. A. in B. Saba Fibre-Reinforced Composites and Hybrid Composites, Eds. Woodhead Publishing, 2019, pp. 157–180.
[18] H. Ghayoor, C. C. Marsden, S. V. Hoa, and A. R. Melro, "Numerical analysis of resin-rich areas and their effects on failure initiation of composites," *Compos. Part A Appl. Sci. Manuf.*, vol. 117, no. November 2018, pp. 125–133, 2019, doi: 10.1016/j.compositesa.2018.11.016.

DEVELOPMENT OF A LOW-COST SILICONE MOLD TOOL FOR INJECTION MOLDING PLASTIC PARTS

Irfan Tahir, John Rapinac, Abdulaziz Abutunis, Venkatagireesh Menta

Department of Mechanical and Industrial Engineering,

University of Minnesota Duluth, Duluth MN 55812

ABSTRACT

The design and development of injection molding (IM) tools is a very expensive and time-consuming process which makes it uneconomic to incorporate traditional IM into the prototyping stage of a product. Currently, the most widely researched method used for rapid prototyping of IM tools is 3D printing with engineered plastics. This project investigates an alternative to 3D printed IM tools by investigating a cost-effective mold tool made of silicone. Design of Experiment (DOE) is used to measure the main and interaction effects of design parameters (e.g., durometer hardness, geometry, and design complexity) on the performance of the silicone mold. It was found that a durometer of Shore A Hardness 40 is the most optimal value for a silicone mold tool. Using 3D printed inserts and a short runner improved the mold performance. Comparison of mechanical properties of the silicone mold test coupons with those produced using a metallic mold tool revealed that there was a 7.3% decrease in Ultimate Tensile Strength, better than those previously reported for some 3D printed mold tools. Results show that the silicone mold tool is a promising alternative to 3D printed mold tools for low-volume injection molding.

1. INTRODUCTION

Injection molding (IM) is a widely used polymer processing technique for fabricating plastic parts of different shapes and profiles. The main working principle of the process involves injecting molten plastic material into a mold tool where it cools and hardens to take the shape of the mold. Due to the ability to mass-produce at high speeds, high accuracy for complicated shapes, and lower mass-production costs, the IM process has become one of the most critical processing methods in the polymer industry [1]. In addition to this, plastic parts produced via injection molding offer excellent repeatability [2]. Some of the products manufactured using injection molding are bottles, toys, automotive components, storage containers, and medical device components. One of the major limitations of the IM process is the high initial start-up cost during the design and development stage of a product. Any design improvements or modifications of the part at later stages is quite difficult and expensive. Overall, the design of the mold tool is time-consuming and expensive during product development. Thus, injection molding is typically avoided in making prototypes [3].

Rapid prototyping (RP) is a well-known process that has come to light with the advancements in three-dimensional (3D) printing and lower cost associated with Computer Numerical Control (CNC) machining. The additively manufactured molds that are created using RP are built with shorter lead time and at a fraction of the cost when compared to traditional tooling. However, mold tools created using RP have challenges and part design limitations that must be addressed. The material that is chosen to replace the traditional metallic tooling must be able to withstand the appropriate pressures and temperatures needed to inject the thermoplastic material. Likewise, the mold must be rigid and maintain its shape under the clamping pressure of the injection molding machine, so the final product comes out as intended. Moreover, the thermal conductivity of the rapidly prototyped mold tool is less than that of the metallic mold as RP mold tools are made of polymer. The lower cooling rate (lower thermal conductivity) of thermoplastics can alter the microstructure and change the mechanical properties of the injected part. Another issue with RP is the limitation of injection materials for alternative methods such as 3D Printing, in contrast to IM where there is a very large range of plastic and biodegradable plastic materials easily available.

As 3D printing is the current standard for rapid prototyping of mold tools, it is worth highlighting some of the work that has been done in this area. Bartlett investigated the effects on the mechanical properties of PP plastic parts produced using a desktop 3D printed mold tool with Digital ABS material [4]. The results showed that the Ultimate Tensile Strength (UTS) decreased by 11%, Young's modulus of the parts increased by 9.4%, and strain at break, or ductility of the parts, decreased significantly by 87% when compared to parts made using a metallic mold. Annealing the samples before tensile testing helped increase ductility but it was not as good as steel mold parts. Simpson et al. also used Digital ABS with PolyJet 3D printing and investigated the effects of different 3D printing materials and the injection material on the mold tool [5]. The authors reported that the strain at break of Digital ABS was consistent with that of the steel mold, but stiffness was reduced by 28.5%.

Another technique for rapid prototyping injection mold tools is to use 3D printed inserts, also known as steel/plastic hybrid molds. Dawson et al. investigated composite molds made out of atactic polystyrene and compared the properties of the plastic sample with a traditional steel mold [6]. Results showed that parts produced using a composite mold had 17% lower UTS, similar Young's modulus, 19% higher flexural strength, and 20% lower ultimate elongation than parts produced using a steel mold. Mischkot et al. used the design of experiments to study a disk-shaped insert built using PolyJet 3D printing technology [7]. The material that was injected into the mold was Low-Density Polyethylene (LDPE). The authors reported that a cooling time of 50 seconds was necessary to enable the injection of the parts without any major deformations or discoloration. The inserts lasted between 25 and 116 production runs with a cycle time of 300 seconds. The authors concluded that a 3D printed insert is a viable option for medium-sized molds (80 x 60 x 20 mm^3).

Starting with the most basic question of whether or not it is possible to inject plastic material into a silicone mold tool, this project investigates the development of a mold tool made out of silicone for simple and cost-effective prototyping of injection molded parts. These mold tools have the potential of enabling researchers and materials scientists to add variability to their designs. The manufacturing cost of a silicone mold tool can be as much as 10 times cheaper than the manufacturing of a metallic mold tool. The low cost associated with silicone also attracts customers who are not experts in designing a mold tool as the process is relatively fast and simple and uses commercially available materials. An additional advantage of silicone is the wide operating range for high-temperature injection materials. For biopolymers such as lignin, alginate, PLA, and other "sticky" materials, silicone mold tools can prove to be beneficial as materials tend not to stick to it and there is no need for a mold release agent. This is especially useful for medical device components used in vivo studies.

2. EXPERIMENTATION

2.1 Materials

The silicone materials used in this project were sourced from the Smooth-On company from their location in Pennsylvania, United States. In total, Mold Star^{TM} silicone of Shore A Durometer Hardness 30 and Mold Max^{TM} silicone of Shore A Durometer Hardness of 40 and 60 respectively were used. Polypropylene (PP) in pellet form was obtained with a donation from the University of Minnesota Duluth's Materials Science Laboratory. The casting mold to manufacture the silicon mold was 3D printed using Hatchbox PLA filament with a diameter of 1.75mm.

2.2 Silicone Mold Tool Process

An outer cast (Figure 1 (a)) for the silicone mold tool was 3D printed with a Creality Ender 3 using PLA and an infill of 20%. The liquid silicone components were poured into a measuring cup then mixed and vacuum degassed for 180 seconds. The mixture was then poured in the 3D printed cast cavity at a single point and left to cure for 24 hours. The same process was followed for each silicone mold tool. Figures 1 (b)-(d) illustrate these steps.

(a) (b) (c) (d)

Figure 1: Making the silicone mold tool: (a) 3D printed mold tool cast (b) Pouring the liquid silicone into the cast (c) Making sure the liquid is level (d) Final shape of the mold after curing for 24 hours

2.3 Injection Molding

A Morgan Press G-125T injection molding machine was used for producing ASTM D638 Type IV standard tensile test coupons. Injection molding parameters for each run were changed according to the DOE studies. A metallic outer frame was used to encompass the silicone mold as shown in Figure 2.

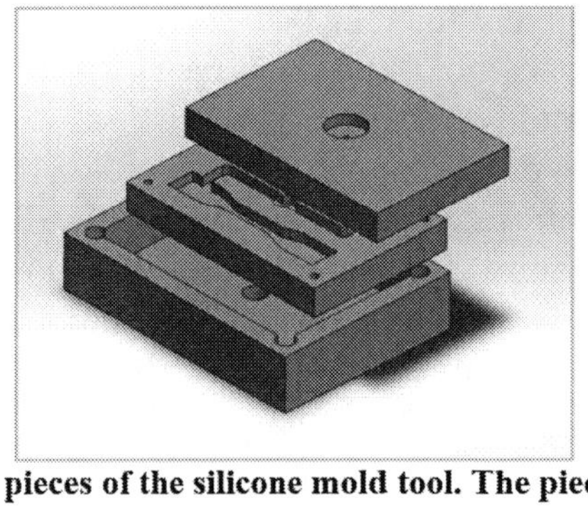

Figure 2: Schematic showing the pieces of the silicone mold tool. The pieces in gray are made out of metal

2.4 Mechanical Tests

Tensile tests were carried out as per the ASTM D638 standard using the Applied Testing Systems (ATS) Universal Testing Machine. All the tests were performed at a cross head speed of 5 mm/min using a 500 lbs. load cell. Jaw grippers without extensometers were used for holding the tensile samples.

2.5 Design of Experiments

The purpose of performing the DOE was two-fold: (a) to determine the proper silicone durometer and (b) to determine the right operating parameters for a successful injection run.

2.5.1 DOE #1 Durometer vs. Mold Height vs. Corner Radius

The first DOE was conducted to determine the right durometer hardness of the silicone mold tool. Eight variables kept the constant in the first DOE (Table 1).

Table 1: Constant parameters for DOE #1

Parameter	Value
Nozzle Temperature (F)	440
Barrel Temperature (F)	420
Plate Temperature (F)	250
Hold Time (min)	4
Injection Pressure (psi)	1500
Injection Time (sec)	5
Clamping Pressure	8000
Material	Polypropylene

A full factorial DOE study with $2^3 = 8$ injection runs was conducted with three factors and two levels each (Table 2). The factors were durometer hardness, mold height, and outside corner radius (Figure 3). The reason for selecting the height of the mold tool as a factor was to investigate whether having more material between the mold cavity and the bottom of the mold will have any effect on the rigidity of the mold. The outside corner radius was investigated to see if increasing the radius of the curve, where the melted plastic flows from the runner to the tensile specimen, will help the plastic flow with ease.

A quantitative response, fill percentage, was used to measure the performance of each injection run. The fill percentage was calculated by taking a photo of the injected part and calculating its surface area using the software ImageJ. By measuring a real distance on the sample, a relationship was established in ImageJ of known distance vs. the same distance in pixels. Depending on the type of injection (underfill or overfill), this surface area was compared to the original 100% fill percentage. In addition to this, other responses such as mold deterioration, warpage, sink marks, flash, and discoloration were also taken into account.

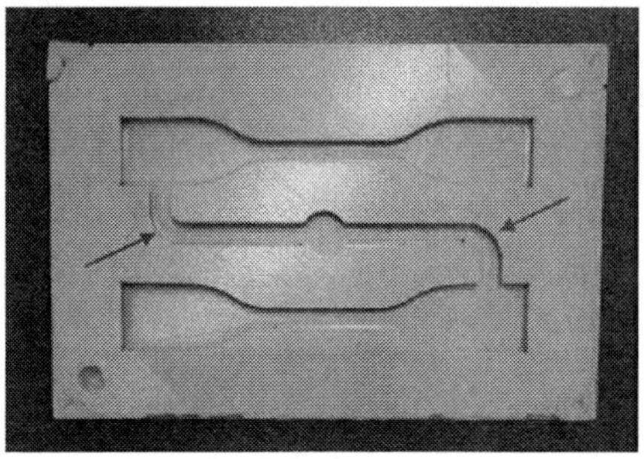

Figure 3: A 40 durometer mold with a mold height of 16 mm and outside corner radius of 10 mm. The arrows show the location of the outside corner radius.

Table 2: Factors and levels for DOE #1

Factors		**Durometer**	**Mold Height (mm)**	**Outside Corner Radius (mm)**
Levels	1	30	16	5
	2	40	26	10

2.5.2 DOE #2 Injection Pressure vs. Time vs. Clamping Pressure

The second DOE was conducted to determine the right injection molding parameters for a successful injection run. Eight variables kept the constant in this DOE (Table 3).

Table 3: Constant parameters for DOE #2

Parameter	**Value**
Nozzle Temperature (F)	440
Barrel Temperature (F)	420
Plate Temperature (F)	250
Hold Time (min)	4
Mold Height (mm)	16
Outside Corner Radius (mm)	10
Durometer	40
Material	Polypropylene

A full factorial DOE study with $2^3 = 8$ injection runs was conducted with three factors and two levels each (Table 4). The first chosen DOE factor was injection pressure. In the authors' previous work [8], injection pressure was found to have the greatest effect on the fill percentage. In this previous study [8], the same injection molding machine was used to inject high-density polyethylene (HDPE) material into a metallic mold. The second chosen DOE factor was injection time. If injection time is too short or too long, there will be an injection failure caused by the short shot or flash, respectively. The third factor was clamping pressure. It was hypothesized that besides the long injection time, the failure due flash was caused by the imperfect/weak contact between the two mold surfaces. If there are any gaps or weak contact, the melted plastic will leak from the runner as it is injected.

Table 4: Factors and levels for DOE #1

Factors		Injection Pressure (psi)	Injection Time (s)	Clamping Pressure (psi)
Levels	1	1500	5	8000
	2	3000	10	10000

2.5 Other Mold Designs: Center Injection and 3D Printed Inserts

To add rigidity to the area around the runner, a 3D printed material was inserted into the mold as shown in Figure 5. The materials used for 3D printed inserts were PLA and Nylon. Additionally, a short runner geometry with a center injection was also carried out (Figure 6).

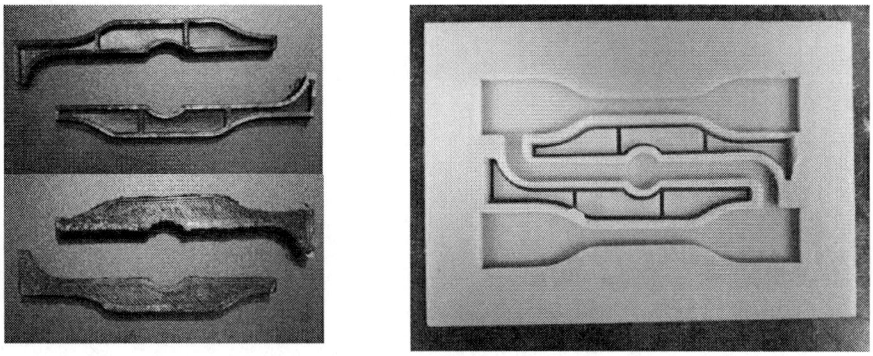

Figure 5: Updated mold tool design to incorporate the 3D printed inserts

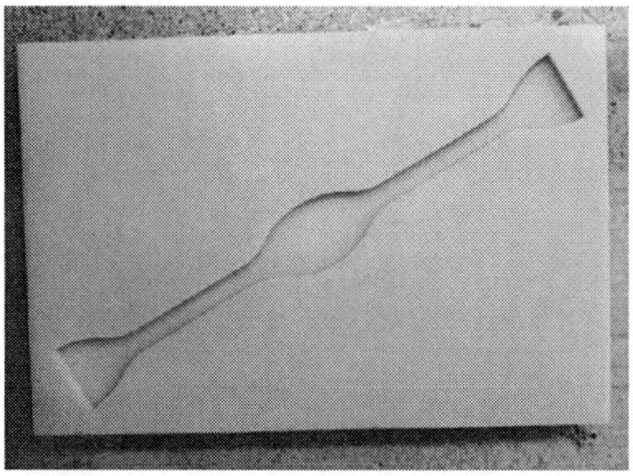

Figure 6: Mold tool design simplifying the geometry by eliminating the runner and injecting in the center

3. RESULTS AND DISCUSSIONS

3.1 Design of Experiments

3.1.1 Analysis of Response for DOE #1

After running the experiments and calculating the surface area of each test sample, the fill percentage was obtained using ImageJ and input into Minitab. The Pareto chart of the first DOE study is shown in Figure 7. The dotted line represents the standardized effect value. Any effect with bars reaching beyond the standardized effect line is considered a significant effect. In this Pareto chart, it can be seen that the durometer significantly affects the response (the fill percentage). The outside corner radius and mold height did not have as large of an effect and in the case of mold height, the effect is almost negligible as seen in the Pareto chart.

An ideal sample will have a fill percentage of 100%. The main effects plots shown in Figure 8 can aid in determining which specific levels will result in a 100% fill percentage. Each slope has a maximum and minimum value represented by blue dots. The vertical axis of the main effects plot shows us the mean fill percentage of the DOE study. To determine which level should be used in future DOE study, the factor level which is closest to a 100% mean fill percentage should be used. In this case, those factors and levels are durometer 40, mold height 16mm, and outside corner radius of 10mm.

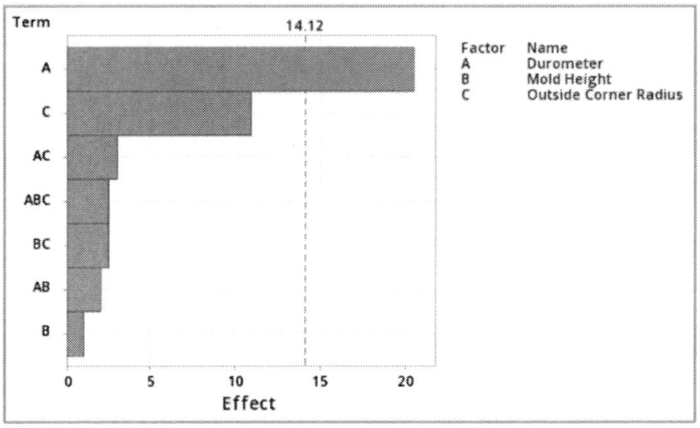

Figure 7: Pareto chart for DOE #1

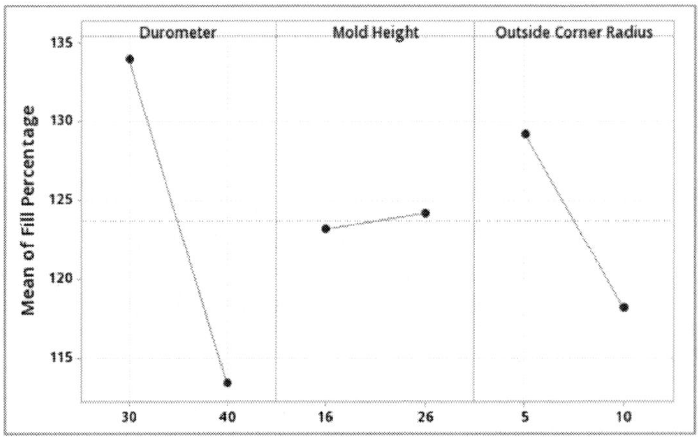

Figure 8: Main effects plot for DOE #1

3.1.2 Mold Deterioration and Sample Quality

In addition to exhibiting a high fill percentage and consequently failure by flash, the 30 durometer mold tool also showed deterioration at the four corners (Figure 9 (a)) more than the 40 durometer mold tool (Figure 9 (b)) after 8 runs. This can be attributed to the fact that the mold tools had an outer metal frame which limits the silicone mold tool's deformation. In contrast to this, the durometer 40 mold showed a slight deterioration in the corner, but it was much less than the one made using durometer 30. This shows that the durometer 30 silicone is more prone to deformations which in part led to a failure by flash.

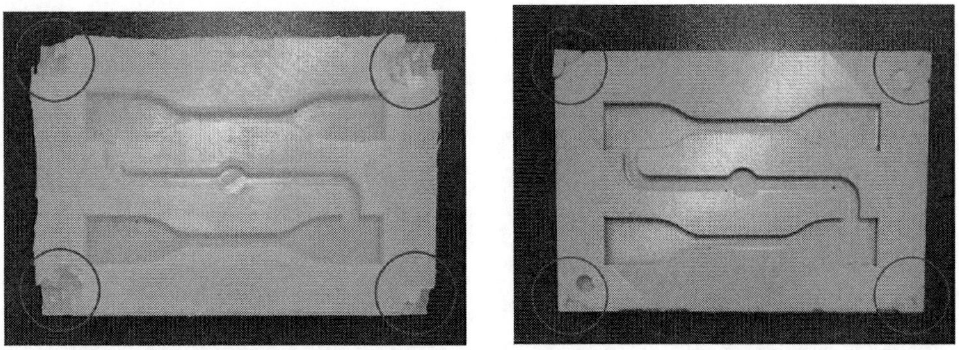

(a) (b)

Figure 9: Comparison of mold deterioration for (a) durometer 30 and (b) durometer 40 mold tool

The samples which were overfilled did not show any shrinkage or dimensional instability as shown in Figure 10a. There was also no warpage of the samples observed. This can be explained by the fact that after the material has finished injecting, it is not subjected to high pressure in the gate location. So, even if the mold tool is deformed while injecting, it returns to its original flat shape once the injection is finished. For samples that were underfilled, there was a slight shrinkage that was observed in the necking area of the sample as shown in Figure 10 (b). This can be attributed to the material not completely filling the cavity of the mold tool.

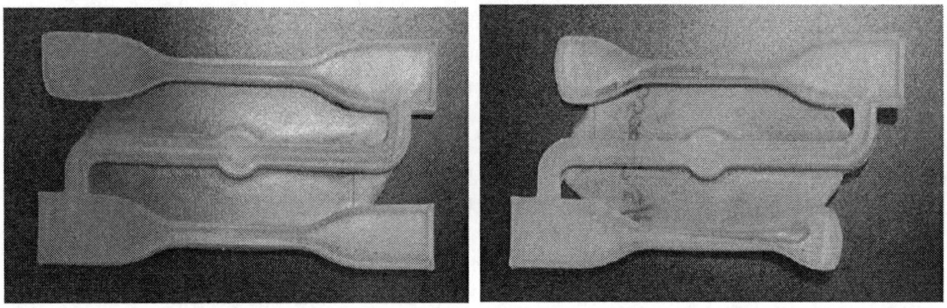

(a) (b)

Figure 10: Photos of the sample from (a) Run #1 and (b) Run #7

3.1.3 Effect of Increasing Durometer Hardness

Durometer was found to have the most significant effect on the mold tool response. The main effects plot (Figure 8) showed that to reach an ideal fill percentage of 100%, the durometer should increase. Hence, along with 30 and 40, durometer hardness 60 was used to conduct further experiments. The reason for not using durometer 50 was because that durometer number is not commercially available from the manufacturers of the silicone material.

The minimum possible injection molding pressure of 1500 psi an injection time of 5 seconds was used to avoid brittle failure of the mold. However, on the first injection run, a brittle fracture propagated through the runner of the mold (Figure 11). The experiment was repeated two more times. For all three of these experiments, a brittle failure to the mold caused the melted plastic to seep through the mold as soon as it was injected. Thus, it was concluded that a durometer of 60 is too high for a silicone mold and it was eliminated from any further DOE studies. From the results of DOE #1, it was concluded that a durometer 40 silicone is more suitable to use in the experiments than a durometer 30 silicone as the latter is more prone to deformation and deterioration. In addition, a mold height of 16 mm was chosen as it uses less silicone material and an outside corner radius of 10 mm was chosen as samples obtained from that type of mold were the closest to the ideal case.

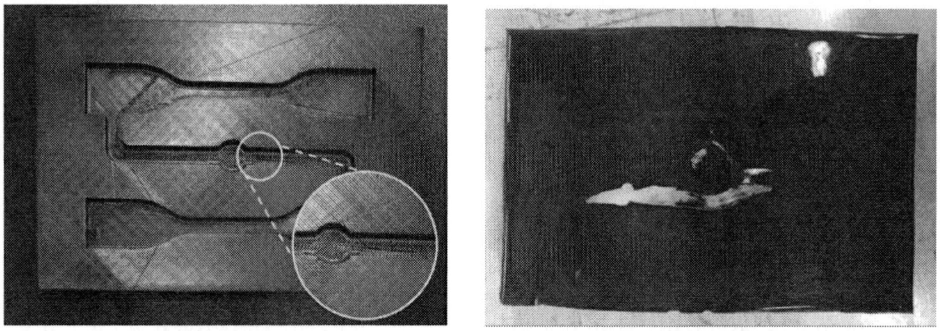

(a) (b)

Figure 11: (a) Durometer 60 mold tool, where the circle shows the location of the crack. (b) Bottom of the Durometer 60 mold tool showing the leaked PP material

3.1.4 Analysis of Response for DOE #2

During DOE #2, For an injection pressure of 3000 psi, all the samples failed by flash. When the injection pressure was half of that, there were short shots (or underfill) but the fill percentage was much closer to 100%. This made it clear that an injection pressure of 3000 psi was too high for the silicone mold tool. Additionally, an injection time of 10 seconds when combined with an injection pressure of 1500 psi and clamping pressure of 10000 psi gave the closest fill percentage (108%) to an ideal sample.

The Pareto chart in Figure 12 shows that two of the employed factors, injection pressure and injection time, significantly influence the fill percentage. A third significant effect came from the interaction between injection pressure and clamping pressure. The main effects plots in Figure 13 also show how injection pressure and injection time had the greatest effect on the fill percentage. Changing the clamping pressure from 8000 psi to 10000 psi did not result in a large change in the fill percentage. This was contrary to past experiments on metal mold in which increasing the clamping pressure reduced the flash. This meant that the upper limit that was selected for the clamping pressure was not high enough.

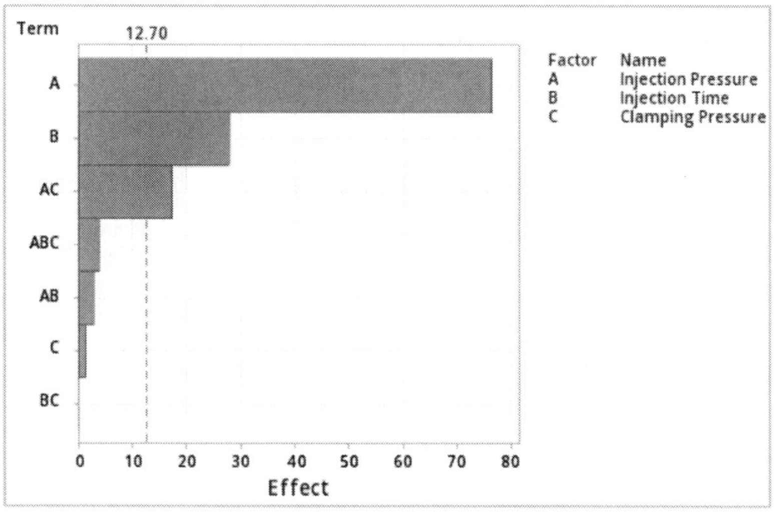

Figure 12: Pareto chart for DOE #2

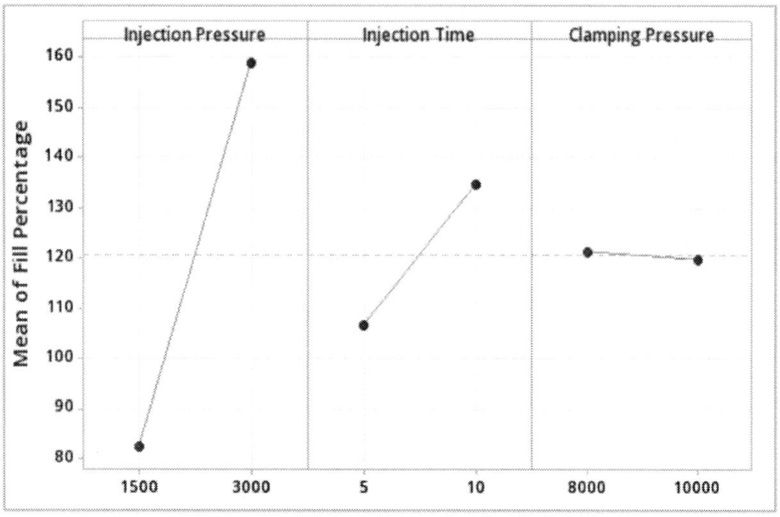

Figure 13: Main effects plot for DOE #2

The bottom left quadrant of Figure 14 shows the interaction of injection pressure with clamping pressure in relation to fill percentage. It can be observed that an injection pressure of 3000 psi with a clamping pressure of 8000 psi resulted in a higher fill percentage than with a clamping pressure of 10000 psi. When the injection pressure was 1500 psi, the opposite was observed. Also, the higher clamping pressure (10000 psi) with the injection pressure of 1500 psi resulted in a fill percentage closer to 100%, as seen in the y-axis of Figure 14. As injection pressure was increased and clamping pressure was lowered, the molten plastic flowed outside the cavity with a higher rate and resulted in higher failure by flash. This can be observed from the steeper slope of 8000 psi clamping pressure compared to the 10000 psi. Even when there was an underfill, the melted plastic "leaked" from the runner when applied pressures were not appropriate. This can be seen in Figure 15 in which as the melted plastic flowed through the mold cavity, it simultaneously flashed from the runner.

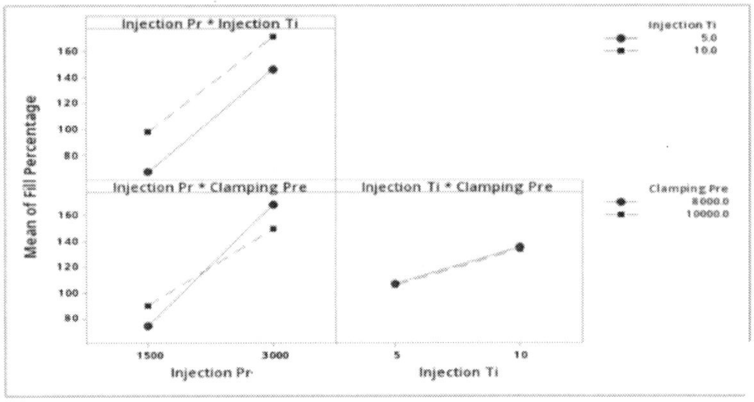

Figure 14: Interaction effects plot for DOE #2

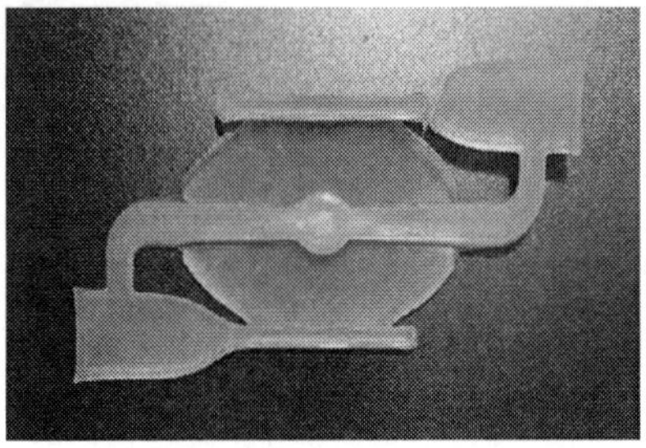

Figure 15: Photo of Run # 2 sample from DOE #2 showing the leakage of material from runner

3.1.5 Effect of Increasing Clamping Pressure

Based on the interaction plot results (Figure 14) of DOE # 2, additional experiments were conducted to measure the response of the mold tool to the variation in the clamping pressure.

Results showed that increasing the clamping pressure to 14000 psi with an increased injection time did result in the reduction of flash as seen in Figure 16. However, it also led to dimensional instabilities in the shape of the sample. The runner area of the mold tool expanded due to the high clamping pressure. As the problem of leakage from the runner was not solved, a change in mold geometry was implemented.

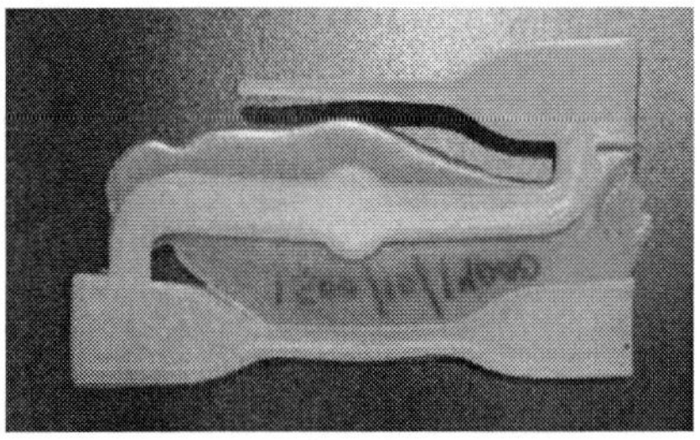

Figure 16: Photos of the samples Run#1 after increasing the clamping pressure

3.2 Effects of 3D Printed Inserts and Short Runner Length

3.2.1 3D Printed Inserts

The PLA insert was melted into the PP plastic as it has a melting point (150°C) lower than the nozzle temperature (220°C) as shown in Figure 17 (a). The Nylon insert was able to sustain the high nozzle temperature without melting as Nylon melting point is 240°C. Results showed that the insert was able to control failure by the flash from the runner and this kind of mold tool gave the best samples in terms of quality so far (Figure 17b). The brown coloring of the sample was due to oxidation. This occurred when the sample was taken out of the mold earlier than the specified hold time. Mischkot et al. injected LDPE into a disk-shaped steel mold and found 50 seconds to be a sufficient hold time to avoid any loss in the original color [7]. However, in this mold tool, whenever the hold time was below 240 seconds, a slight brown coloration of the samples was observed.

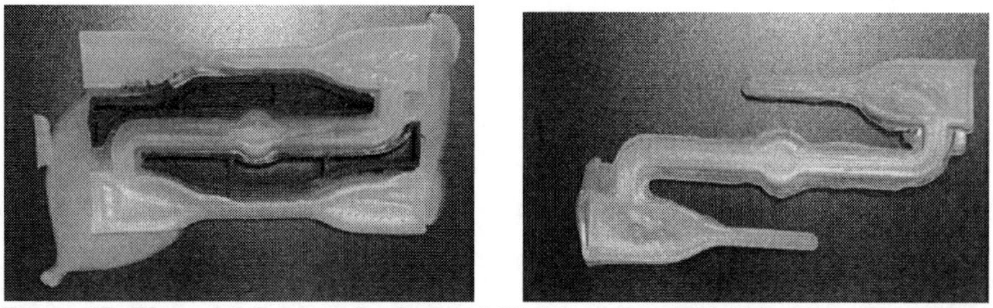

Figure 17: Photos of the samples from (a) PLA insert and (b) Nylon insert mold tool

Using 3D printed nylon inserts solved the issue of flash but there was a failure by short shot as the plastic failed to fill the mold cavity completely. To investigate this issue, a new DOE study on injection molding parameters should be conducted. As seen in the previous DOE studies, different injection molding parameters resulted in different fill percentages. With the flash controlled, increasing the injection pressure or injection time may result in a 100% or more fill percentage. This DOE could not be conducted due to the closure of university labs due to the worldwide COVID-19 pandemic in March 2020.

3.2.2 Short Runner

A new approach was implemented in parallel with the 3D printed inserts. In the revised mold geometry, the runner was eliminated while the neck length of the tensile specimen was kept the same (see Figure 6). For the first injection run, a small amount of flash was observed at the center. The leakage of the molten plastic from the center was attributed to the imperfect contact between the top metallic plate and the bottom silicon mold. When the material was injected into the center of the mold with high pressure, the silicone mold tried to expand. However, the outer metallic frame prevented it from doing so, which resulted in a slight curve in the mold. This curve, in turn, created gaps between the silicone mold and the top metallic plate and hence led to the leakage of the melted plastic. Taking these factors into account, an injection run was carried out with a borderless straight metallic plate so that the mold is free to expand and does not curve.

Using an injection pressure of 1500 psi, injection time of 5 sec, and clamping pressure of 10000 psi, an ideal sample with no flash was obtained as shown in Figure 18. This injection run was repeated 3 more times, and each time an ideal sample was obtained. The samples were then post-processed and cut into half so that they can be tested using a tensile testing machine.

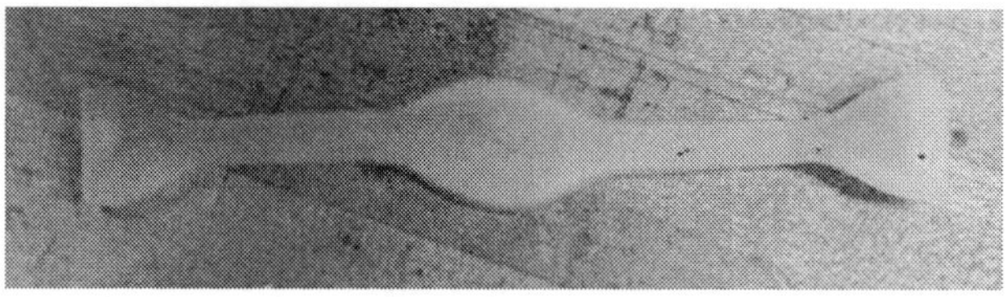

Figure 18: Sample obtained with center injection without outer frame without any failure by flash

3.3 Mechanical Properties

Firstly, to check if the tensile tests were done correctly, the UTS values of the PP samples obtained from the metal mold were compared with those present in online databases. In total, three sources were used as references. The comparisons showed that there was a 2.4%, 3.7%, and 8.0% error in UTS respectively for each of the three sources [9-11]. Overall, the tensile tests of metal mold samples showed a better consistency than the silicone mold samples. This can be quantitatively assessed by calculating the standard deviation of UTS of each sample set, where the metal mold and silicone mold samples showed a standard deviation of 0.527 and 0.912 respectively (Table 5).

Table 5: Mechanical properties comparison with standard deviations

Property	Metal Mold	Silicone Mold	% Decrease
UTS (MPa)	32.197 ± 0.527	29.854 ± 0.912	7.3 %
Young's Modulus (MPa)	360.255 ± 14.41	142.511 ± 31.75	60.4 %
Yield Strength (MPa)	18.194 ± 0.242	11.935 ± 1.158	34.4 %

Comparing the average UTS, it was observed that there was a 7.3% decrease for the silicone mold tool samples (Figure 19). This decrease in UTS is less than the ones reported in the literature, where the decrease in UTS was as high as 11% for 3D printed mold [4] and 21% for composite mold [12]. On average, there was a 60.4% decrease in Young's modulus and a 34.4% decrease in the yield strength when going from metal to silicone mold tool.

For PP tensile samples produced using a Fused Deposition Modeling (FDM) 3D printed mold tool made out of ABS Digital, Bartlett reported a 9.4% increase in Young's modulus compared to parts produced using a steel mold [4]. León-Cabezas et al. reported a 45.98% decrease in Young's modulus for tensile specimens produced using a Polyjet 3D printed mold tool when compared to parts produced using a metal mold tool [13]. In another study of polystyrene composite molds, Young's modulus remained unchanged [12]. The difference in silicone mold tool and 3D printed material's thermal conductivity can be attributed to such a difference.

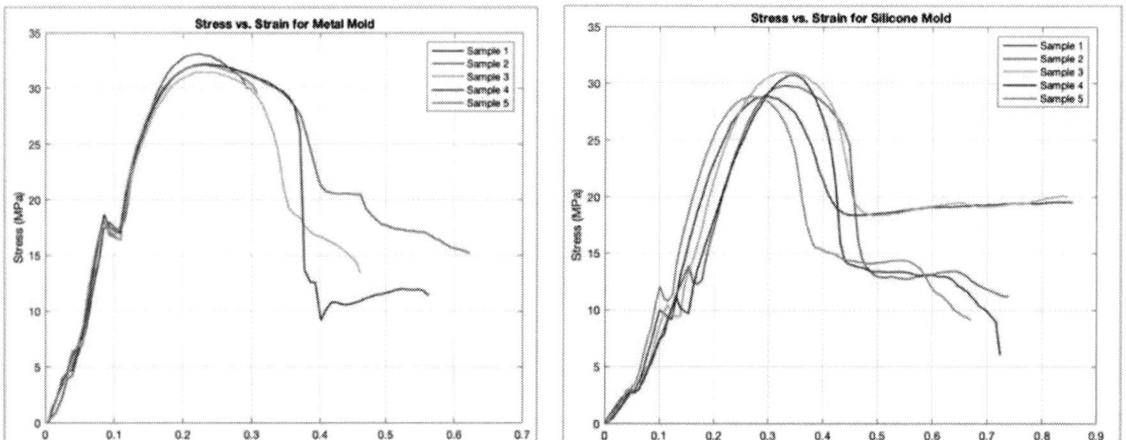

Figure 19: Stress vs. Strain plots produced by(a) metal mold and (b) silicone mold

4. CONCLUSIONS

It has been shown that it is possible to develop a mold tool made out of a flexible material like silicone to inject Polypropylene. Results show that not only silicone can withstand the high temperatures and pressures, but also able to resist mold deterioration at certain durometer hardness. It was found that the plastic parts produced from silicone mold did not show many types of common failures found in injection molded samples like sink marks, warpage, or blisters.

There are limitations to using a silicone mold tool for injection molded parts. Mainly, these are related to the geometry of the mold cavity. It was found that even with the right type of durometer, the length of the runner of the mold tool can have a significant effect on whether or not the part will fail due to flash. Another type of failure due to a short shot was also observed when the injection time was too low. The most ideal free defect sample was obtained using a combination of a short runner, low injection pressure, short time, high clamping pressure, and the absence of any outer frame.

The use of silicone mold tools is promising especially for material scientists and manufacturers experimenting with plastic parts in the early stages of their research. In its solid form, silicone is an excellent choice when dealing with materials that tend to stick to other surfaces. Examples of these include medical-grade biomaterials such as PLA, PGLA, lignin, alginate, and bio-based PET. Another advantage of using a silicone mold tool to produce simple medical device components is that the mold surface does not require any type of release agent to be sprayed prior to injection. This eliminates the risk of any possible contamination in the medical components. Compared to most types of 3D printed molds, the silicone mold is also able to withstand higher temperatures. It has the potential to inject resins with high melting temperatures such as PEEK, PPA, and PEI. Also, commercially available materials for the silicone mold tool are 3 times less expensive than those used in 3D printed molds. The decrease in UTS was also relatively better for silicone mold tool parts when compared to 3D printed mold tool parts.

For future work, it is recommended that a DOE study is conducted with 3D printed inserts as that proved to be a viable solution to failure through flash without compromising the mold geometry. Flexural, Izod impact, and hardness tests of the injected plastic should be conducted. Repetitions of all the tensile tests should be carried out to verify the results with larger sample size. DSC tests should be conducted to measure the crystallinity of the injected parts. Using the silicone mold tool, researchers can continue using injection molding for their prototypes instead of reverting to other techniques such as casting, 3D printing, or compression molding. For design projects where the final part will be made using injection molding, the silicone mold tool provides consistency during the design process. In conclusion, this project can be expanded further by conducting tests and experimenting with more complex geometries so that the gap between preliminary research and production of a part can be filled.

5. REFERENCES

[1] M. P. Groover, "Shaping Processes for Plastics," in Fundamentals of Modern Manufacturing: Materials Processes, and Systems, 5th ed., New York: John Wiley & Sons, 2012, pp. 316-326.

[2] J. C. Viana, N. M. Alves, and J. F. Mano, "Morphology and Mechanical Properties of Injection Molded Poly (Ethylene Terephthalate)," Polymer Engineering & Science, vol. 44, no. 12, pp. 2174-2184, 2004.

[3] T. Rogers, "Everything You Need To Know About Injection Molding," in Creative Mechanisms Blog, ed. Creative Mechanisms, 2015. Available: http://www.creativemechanisms.com/blog/everything-you-need-to-know-about-injection-molding

[4] L. P. Bartlett, "A Preliminary Study of Using Plastic Molds in Injection Molding," The Ohio State University, 2017.

[5] P. Simpson, A. D. Zakula, J. Nelson, J. K. Dworshak, E. M. Johnson, and C. A. Ulven, "Injection Molding with an Additive Manufactured Tool," Polymer Engineering & Science, vol. 59, no. 9, pp. 1911-1918, 2019.

[6] E. Dawson and J. Muzzy, "The Effect of Rapid Prototype Tooling on Final Product Properties," SPE ANTEC Technical Paper, no. 45, pp. 456-460, 1999.

[7] M. Mischkot et al., " Injection Moulding Pilot Production: Performance Assessment of Tooling Process Chains Based on Tool Inserts Made from Brass and a 3D Printed Photopolymer," in Annual Technical Conference 2017 of the Society of Plastic Engineers, pp. 1898-1902, 2017.

[8] Mahmood S, Tahir I, Menta V. Thermo-Mechanical Characterization of HDPE-Tobacco Lignin Blends. The Composites and Advanced Materials Expo; California 2019.

[9] "Polypropylene (PP)." MakeItFrom.com, 8 Sept. 2018. Available: https://www.makeitfrom.com/material-properties/Polypropylene-PP-Copolymer

[10] Polypropylene (PP) Mechanical Properties at 23 °C." Polymer, Database. Available: https://polymerdatabase.com/Commercial%20Polymers/PP.html

[11] "Typical Properties of Polypropylene." Tap Plastics, Available: https://www.tapplastics.com/uploads/pdf/Polypropylene_Data.pdf.

[12] M. Damle, " Effect of Fiber Orientation on the Mechanical Properties of an Injection Molded Part and a Stereolithography Insert Molded Part," in Annual Technical Conference (ANTEC 1998) of the Society of Plastic Engineers, Atlanta, USA, pp. 584 - 588, 1998.

[13] M. León-Cabezas, A. Martínez-García, and F. Varela-Gandía, "Innovative Advances in Additive Manufactured Moulds for Short Plastic Injection Series," Procedia Manufacturing, vol. 13, pp. 732-737, 2017.

THE CONVERGENCE OF COMPOSITES AND TOPOLOGY OPTIMIZATION, USHERING IN THE NEXT ERA OF AIRCRAFT LIGHTWEIGHT STRUCTURES

Riley Reese, Sana Elyas, Yang Shen, Luis G. Bahamonde Jácome
Arris Composites
Berkeley, CA/USA

ABSTRACT

Although advanced composite material outperforms metal on material data sheets, actual composite structures often fail to provide a significant improvement. In part, this is due to the application of design approaches that were originally meant for metallic constructions. As a result, advanced composite structures end up having a redundant layup, with a quasi-isotropic stacking sequence that eliminates anisotropy, instead of leveraging it, so called black aluminum. Today's approach to take better advantage of continuous carbon fiber's mechanical properties, fibers are aligned based on the anticipated loading conditions. This can be achieved using hand layup or automated tape layup (ATL) / automated fiber placement (AFP) techniques. Though this provides a significant improvement over the "black aluminum" approach, it still falls short of realizing the full potential of continuous fiber anisotropy. Since carbon fibers perform best in tension, the part itself should be redesigned to take advantage of this effect. Though this exercise may seem intuitive for simple parts, in the aerospace industry these coupled design activities easily become non-intuitive due to the complex loading conditions the aircraft structures are subjected to.

Arris Composites has developed a new process, additive molding^{TM}, capable of manufacturing complex geometries, using continuous fiber. This paper presents optimizing topology and fiber orientation for an aerospace bracket, having complex 3D load cases. These optimized structures are shown to outperform current composite structures as well as structures machined and 3D printed from metal, making them ideal for next generation aerospace brackets and joining structures.

Corresponding author: Riley Reese <riley@arriscomposites.com>

1. INTRODUCTION

The A350XWB and B787 programs are the first large-scale commercial programs to develop composite-intensive aircraft with a composition of about 80% composite materials by volume [1]. Reports on the costly investment of these clean-sheet designs [2] suggest that next-generation aircraft structures will likely be based upon derivatives of these models. Yet light weighting challenges still abound within these programs and the upcoming ones. Particularly, for brackets whose function is to transfer load from one mechanical interface to another. These aerospace applications still lack a complete advanced composites solution. While commercial support of simulation tools is currently emerging for metallic additive manufacturing processes, advanced composites manufacturing processes still lack such support. For instance, most leading vendors of finite element analysis packages have recently added topology optimization features for isotropic materials, yet topology optimization for fiber-reinforced composite structures is still not supported.

CAMX Virtual Conference Proceedings. September 21-24, 2020.

On the academic front, several research papers have already focused on topology optimization of continuous fiber-reinforced composites [3-4] and the academic community appears to have settled on a few benchmark problems, like the three-point flex problem, so-called the Messerschmitt-Bölkow-Blohm (MBB) beam. However, these problems are simplified for purposes of benchmarking. Thus, the challenge of applying topology optimization and finite element analysis techniques to aerospace composite structures remains.

This paper aims to explore the potential benefits of applying a simultaneous topology and fiber orientation optimization toolset to 3D aerospace composite structures. Although care to experimentally verifiable performance measures has been given, the exploration is numerical in nature. The remainder of this paper is structured as follows. Section 2 presents the methodology used to apply this toolset during the design process. The results of applying the methodology of Section 2 on two case studies are presented and discussed in Section 3. Finally, Section 4 overviews the benefits and limitations of using a toolset based on topology optimization and finite element analysis techniques.

2. METHODOLOGY

Computer-driven optimization requires a parametric model of the product under design. These parameters can represent geometry like the thickness of a plate under sizing optimization, or material properties like free material optimization [5], where the stiffness coefficients are under design. Moreover, these parameters can be lumped like the layer orientation of uni-directional (UD) composite laminates or spatially distributed. For instance, topology optimization, as formulated by Bendsoe and Sigmund [6], parameterizes the shape of a structural component by assigning a fictitious density to all the points of the design space and labeling them as being part of the component or not. Thus, topology optimization is a spatially distributed parameterization of geometry. Subsection 2.1 presents the design parameterization used by Arris Composites toolset. The probability of significantly improving the performance of the design increases with more parameters. However, the computational cost and non-convexity increases as well. Non-convexity is simply defined in this paper as the number of local optima that arise and where the optimizer might get stuck.

Once the design is parameterized, the design variables are to be optimized based on key performance criteria. These criteria are mathematically formulated in terms of objective functions and constraints in what Subsection 2.2 calls an optimization problem statement.

2.1 Design parameterization

To formulate a topology optimization problem, one must define a volumetric space where the simulation is allowed to add or remove material, and onto which load and boundary conditions are applied. This design space is then discretized into finite elements to make the problem amenable to finite element analysis. Moreover, these finite elements are used to spatially discretize the so-called density field into variables that describe the presence of material in a given element. A density value of 0 denotes a void (i.e., material is removed), while the value of 1 denotes the presence of material in that finite element. Each finite element is also parameterized with a vector, u that describes the orientation of the fiber at the centroid of such finite element [3-4].

The parameterization of the stiffness matrix as a function of the design variables x and u is accomplished by using a stiffness matrix, computed as:

$$C = E_L(x)[T_\sigma(u)][\hat{C}][T_\epsilon(u)] \qquad [1]$$

where, E_L is the Young modulus along the direction of the fiber,
$[T_\sigma]$ and $[T_\epsilon]$ are the stress and strain coordinate transformation matrices, and
$\hat{C}$ is a normalized transversely isotropic stiffness matrix given by

$$C^{-1} = \begin{bmatrix} 1 & -v_{LT} & -v_{LT} & & & \\ -v_{LT} & \frac{E_L}{E_T} & -\frac{E_L v_{TT}}{E_T} & & \ddots & \\ -v_{LT} & -\frac{E_L v_{TT}}{E_T} & \frac{E_L}{E_T} & & & \\ & & & \frac{2(1+v_{TT})E_L}{E_T} & 0 & 0 \\ & \ddots & & 0 & \frac{E_L}{G_{LT}} & 0 \\ & & & 0 & 0 & \frac{E_L}{G_{LT}} \end{bmatrix} \qquad [2]$$

Where $E_L, E_T, v_{LT}, v_{TT}, G_{LT}$ are the engineering constants of a transversely isotropic material, and their subscripts L and T denote the fiber direction and the plane of isotropy perpendicular to it, respectively.

2.2 Optimization problem statement

The challenge of light-weighting a load bearing structure made with continuous carbon fiber composite can be formulated as a multi-objective minimization problem

$$\begin{aligned} &\min_{x,u}(U(x), r(u)) \\ &\textit{subject} \quad \frac{V}{V_0} \leq \eta \\ &\qquad\qquad 0 \leq x \leq 1 \end{aligned} \qquad [3]$$

Where $\boldsymbol{U}$ denotes the strain energy and measures global stiffness
$\boldsymbol{r}$ denotes a vector of failure indices, one per each finite element, and measures local strength.

The light-weighting criteria is formulated as a constraint to achieve a target volume fraction, η. Finally, the densities are bounded to values between 0 and 1 [7].

A traditional approach to solving this problem is to sequentially design the topology using a proxy isotropic material and thereafter optimize the fiber orientation for the previously optimized shape. This approach, called hereafter sequential design, decouples each design activity and provides a flexible toolbox to design from functional requirements or with legacy structural shapes. However, it does not account for the anisotropy of the reinforcement during the shape definition stage, resulting in not leveraging the full design latitude of design for functional requirements.

Taking advantage of the anisotropy of the reinforcement requires solving the topology and fiber orientation simultaneously. Moreover, the solution to this optimization problem shall be implemented in such a way that both a sequential or simultaneous approach is available to the user. The solution to this problem must use computer resources efficiently to scale up to many parameters. Finally, it also requires a manufacturing process capable of aligning the fibers along the complex shapes that may result thereof.

3. RESULTS

This section presents the results of applying the Arris Composites toolset to a couple case studies. Section 3.1 studies the design of a three-point bending beam, so called MBB beam, with the purpose of verifying the Arris Composites toolset against state-of-the-art topology optimization results. Section 3.2, on the other hand, serves as a demonstrator of the light weighting benefits that can be obtained on aerospace bracket design. Finally, Section 3.3 discusses the experimental results from the comparative testing, done by Northrop Grumman, between a titanium baseline T-bracket and an Additive Molded Aligned T800/PEEK composite bracket.

3.1 MBB Beam

The classic MBB problem is a simply supported 2D beam, subjected to a vertical point load at mid-length. As shown by Andreassen et al. [7], the optimal shape after performing an isotropic topology optimization on the MBB beam is a Warren-type truss, where the bar arrangement shows equilateral triangle patterns. Figure 1 verifies that the same topology is obtained when applying the herein proposed methodology with an isotropic material. However, Figure 2 shows that when using a simultaneous approach, the optimized shape converges to a topology reminiscent of a Bowstring-type truss. A Bowstring truss connects the top curved chord with the bottom horizontal chord with diagonal struts.

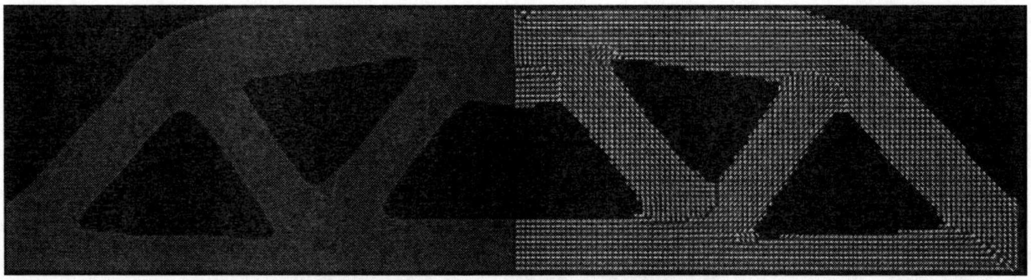

Figure 1. Optimized shape of an isotropically optimized MBB beam. The left side shows a solid view of the shape, while the right side plots the a-posteriori aligned fiber orientations.

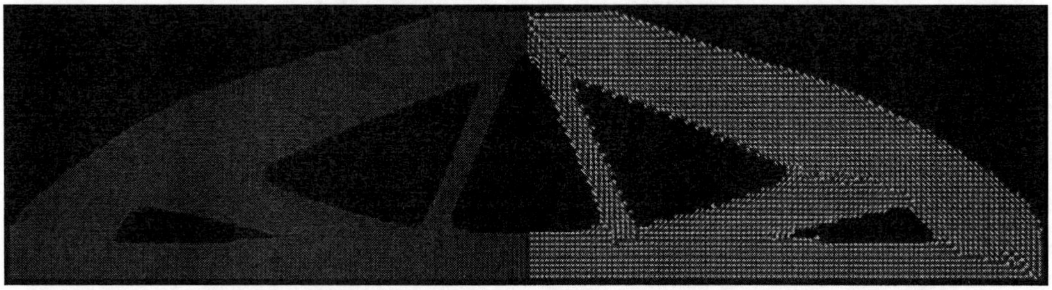

Figure 2. Optimized shape of a simultaneously optimized MBB beam. The left side shows a solid view of the shape, while the right side plots the aligned fiber orientations.

Figure 3 shows the convergence history of both optimizations by plotting the variation of the objective function, normalized with respect to the initial value, as the number of iterations increases. Compared to the isotropic optimization run, the simultaneous optimization converges to a smaller value of the objective function, suggesting improved stiffness for the same weight reduction.

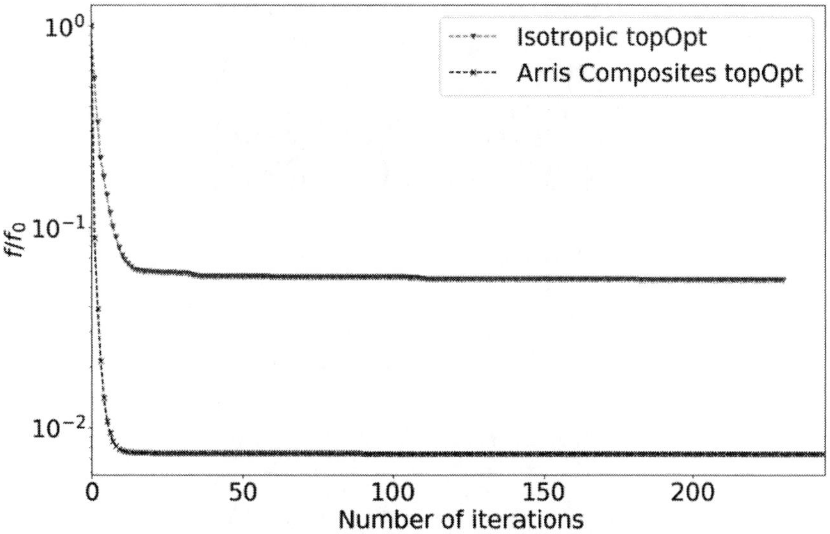

Figure 3. Convergence history of the normalized objective function for both MBB optimizations.

3.2 Aerospace bracket

The aerospace bracket under study is shown in Figure 4. The bracket has 5 mechanical interfaces. The top bolt hole is connected to a tie rod which applies a horizontal force (contained in the midplane and parallel to the plane of the bottom face), while the remaining four bolt holes are mounted onto an infinitely rigid plate. The finite element model applies the horizontal force using rigid body constraints on the inner face of the bolt hole, and fixes all degrees of freedom of the remaining four bottom holes.

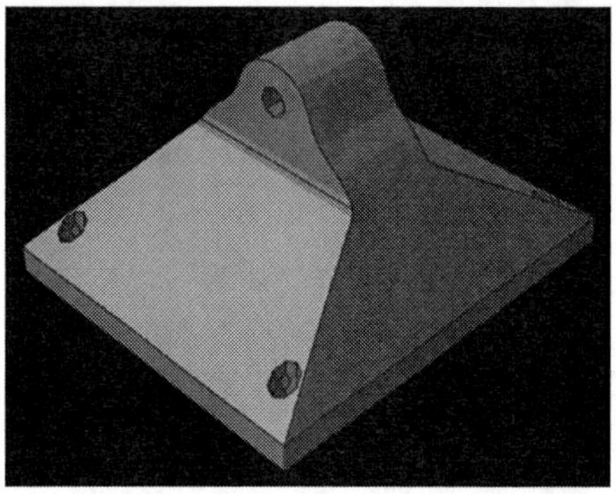

Figure 4. Design space of an aerospace bracket.

Figure 5 shows the resulting shape after running an isotropic topology optimization for 100 iterations with a light weighting target of 70% weight reduction. The isotropic material was the common aluminum alloy 6061-T6. On the other hand, the optimized shape obtained after running 100 iterations a simultaneous optimization, with the same light weighting target, is shown in Figure 6. The thermoplastic T800/PEEK material was used for the simultaneous optimization. Note Figure 6 also shows the fiber orientation vector plot.

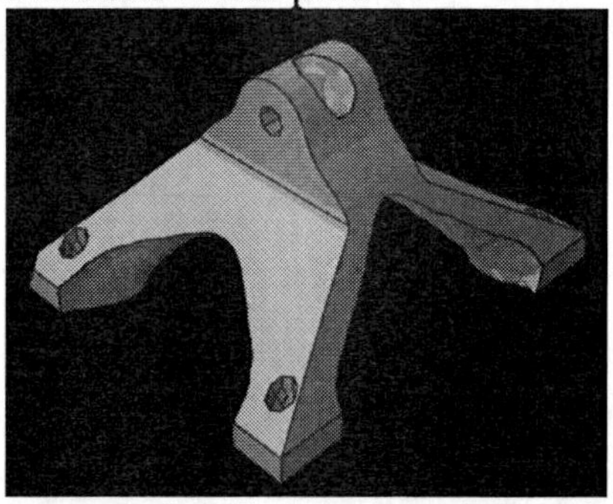

Figure 5. Resulting aerospace bracket design after an isotropic optimization.

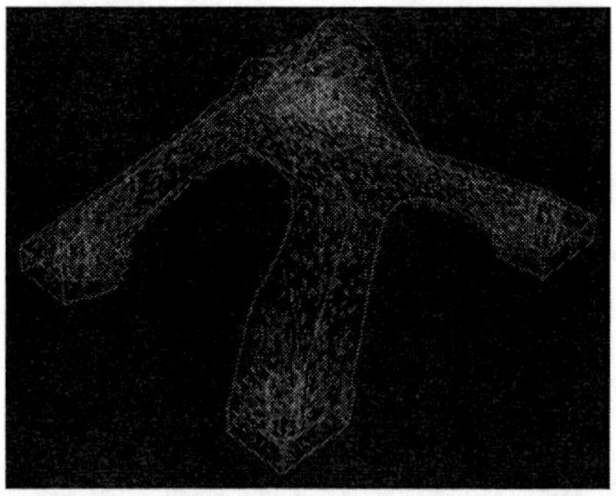

Figure 6. Resulting aerospace bracket design after a simultaneous optimization.

Both shapes connect the fixed interfaces with the top loaded bolt hole using uniaxial members. However, while the isotropically optimized bracket exhibits a webbing between the two lateral legs, the simultaneous optimization prefers to shift the anisotropic material along the axis of each leg.

Figure 7 shows the convergence history of both optimizations by plotting the variation of the objective function, normalized with respect to the initial value, as the number of iterations increase. Compared to the isotropic optimization run, the simultaneous optimization converges to a smaller value of the objective function, suggesting improved stiffness for the same weight reduction.

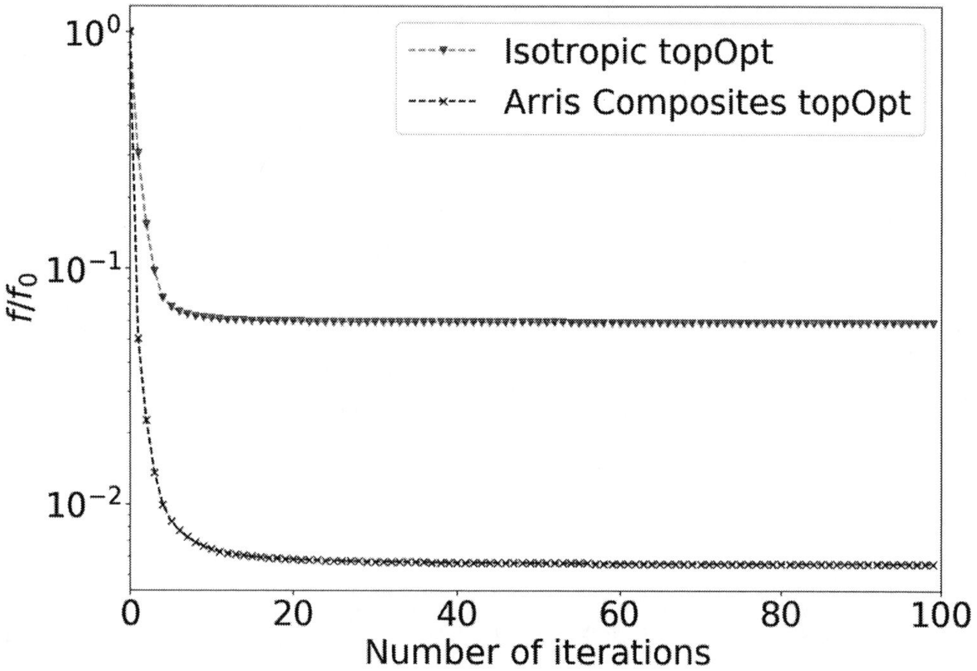

Figure 7. Convergence history of the normalized objective function for both bracket optimizations.

To more accurately compare these designs in terms of stiffness, a post-optimality analysis was performed and the displacement of the centerline of the top bolt hole was used as the measurement of stiffness. This measurement is more amenable to correlation with experimental data. Figure 8 shows the tip displacement and weight, normalized with respect to the displacement and weight of the baseline design of Figure 4, for four configurations:

1. the isotropically optimized design shows a reduction in stiffness of 35.67% when the weight is reduced by 70.00%;

2. if the same isotropically optimized (QI) shape where to be made with a non-continuous carbon fiber composite that has randomly oriented chopped fibers, and thus can be modeled

as quasi-isotropic in 3D, the weight reduces by 81.52% but the stiffness also suffers a large reduction of 72.37%. Thus, having a very light yet very flexible bracket.

3. Aligning fibers along the uni-directional members of the isotropically optimized shape shows a significant numerical increase of stiffness retention. The stiffness is now reduced by only 19.25%. Since the material has not been changed, but optimally placed, the weight reduction is roughly the same as 81.52%.

4. However, using the simultaneous optimization approach the design weight is also reduced by 81.52% but the stiffness with 96.67% retention remains practically the same as the original aluminum baseline. And compared to a sequential approach of case 3 is 16.10% stiffer.

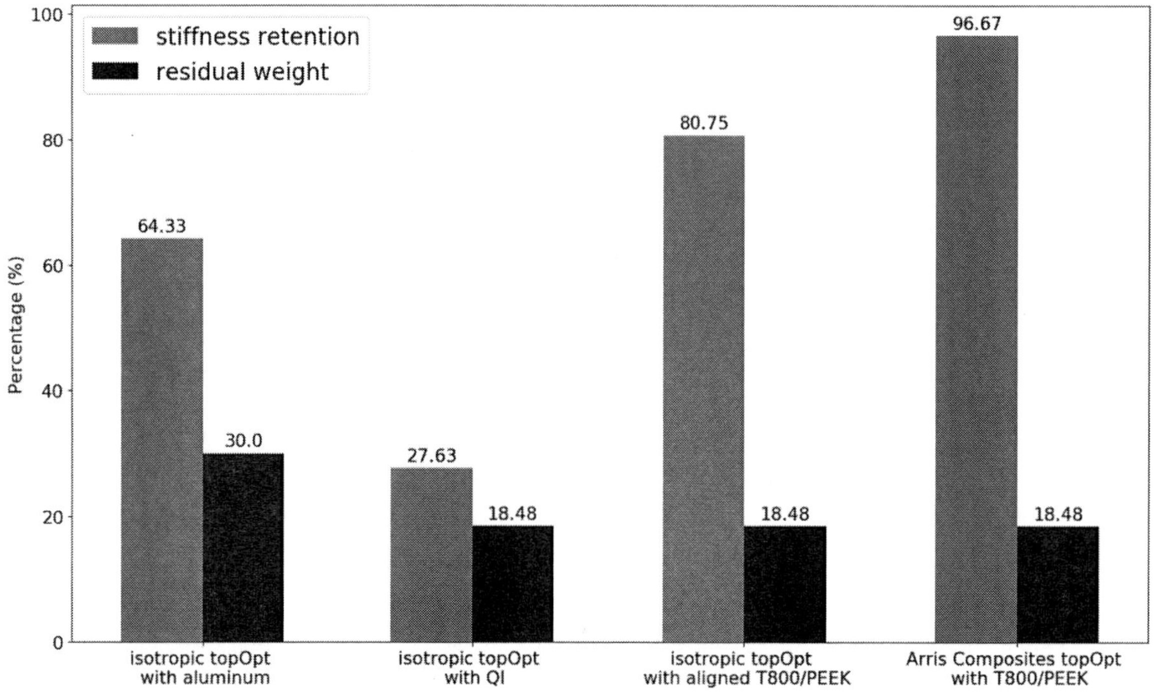

Figure 8. Stiffness and weight reduction of 3 light weighted design configurations as compared to the baseline design of Figure 4.

3.3 Comparative Testing by Northrop Grumman of Titanium Baseline T- Bracket to Additive Molded Aligned T800/PEEK Composite Bracket

Northrop Grumman [8] performed a comparative mechanical destructive testing of its typical baseline titanium T-Plate bracket to the additive molded aligned T800 Carbon Fiber/PEEK bracket design supplied by Arris. From the Figure 9 comparative Force vs. Deflection curves, the additive molded 4-prong T800/PEEK bracket failed at ~93% of the force at which the titanium bracket failed but revealed similar stiffness as shown by the similar slopes of their respective Force vs. Deflection curves. More important, the composite bracket weighed only 16 grams in comparison to the 56 grams for the optimized titanium bracket; an approximate 71% weight reduction. The results are summarized in Figure 10.

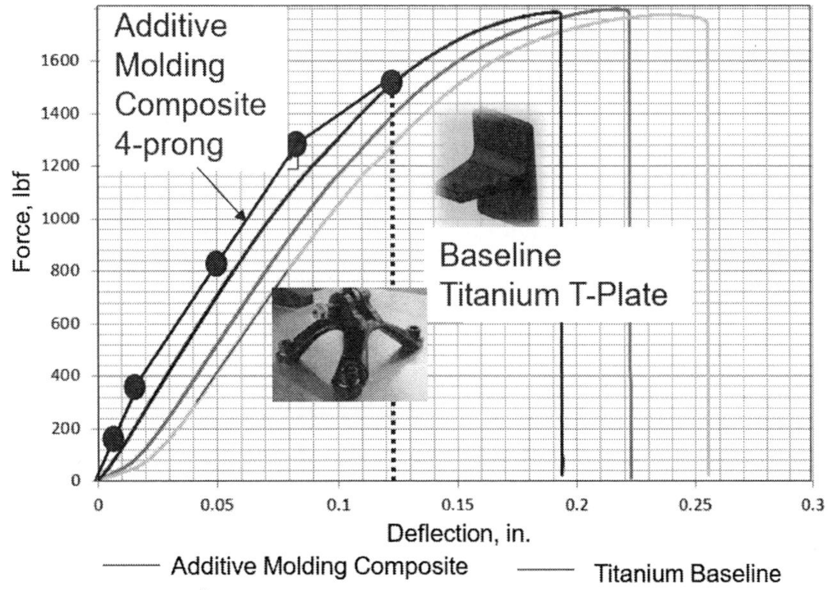

Figure 9. Comparative Force vs. Deflection Curves between Baseline Titanium T-Plate Bracket under bending and the Additive Molded Composite 4-Prong T800/PEEK Arris Bracket under tension.

		Weight	Comments
Baseline 3D Printed Titanium T-Bracket		76 gms	Part that is Not Optimized for Weight Savings
Optimized 3D Printed Titanium 6-Prong Bracket		56 gms	30% Lighter Weight vs. Baseline 3D Printed Titanium
Additive Molded Continuous Carbon Fiber Composite 4-Prong Bracket	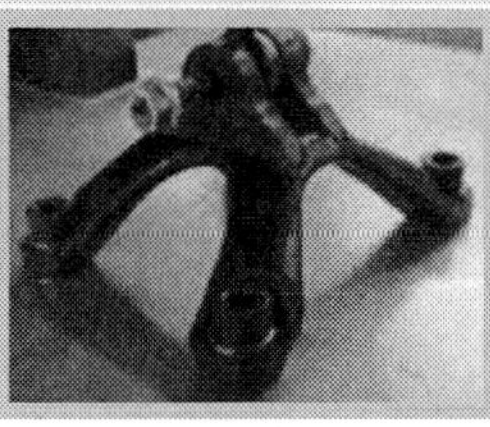	16 gms	71% Lighter Compared to Optimized 3D Printed Titanium

Figure 10. Comparative Weight Between Titanium and Additive Molded Composite Brackets

4. CONCLUSIONS

A design parameterization based on discretizing the design space with finite elements was proposed in Section 2.1. This design parameterization was applied on 2D MBB and 3D aerospace bracket case studies, in Section 3, by using the Arris Composites toolset.

Numerical results suggest that using a simultaneous optimization approach not only provides a T800/PEEK bracket design that is lighter than an isotropically optimized titanium, but also can be just as stiff as the original overweight titanium baseline. Experimental results further confirmed these numerically predicted benefits.

5. ACKNOWLEDGEMENT

The authors would like to acknowledge Hunter Lee, Structural Engineer and Dr. Ed Silverman, NG Fellow of the Northrop Grumman Space Systems Division, for their contribution in testing and comparing the results of the 4-prong additive molded composite brackets to their previously assessment of the baseline titanium 6-prong bracket.

6. REFERENCES

[1] Giurgiutiu, Victor. Introduction. *Structural Health Monitoring of Aerospace Composites.* 2016

[2] Gates, Dominic. "Boeing celebrates 787 delivery as program's costs top $32 billion". *The Seattle Times.* Published September 25, 2011. Accessed on January 21 2020: https://www.seattletimes.com/business/boeing-celebrates-787-delivery-as-programs-costs-top-32-billion/.

[3] Safonov, Alexander, A, "3D topology optimization of continuous fiber-reinforced structures via natural evolution method", *Composite Structures: Elsevier* (2019): 289-297.

[4] Hoglund, R. and Smith, D.E., "Continuous fiber angle topology optimization for polymer fused filament fabrication", Solid Freeform Fabrication Symposium. Austin, TX, 2016.

[5] Kočvara M, Stingl M and Zowe J. "Free material optimization: recent progress†". *Optimization.* 2008;57(1):79-100.

[6] Bendsøe MP, Sigmund O. *Topology Optimization.* Berlin, Heidelberg: Springer Berlin Heidelberg; 2004.

[7] Andreassen E., Clausen A., Schevenels M., Lazarov BS, Sigmund O. "Efficient topology optimization in MATLAB using 88 lines of code", *Struct Multidisc Optim.* 2011;43(1):1-16.

[8] Lee, Hunter and Silverman, Ed, Northrop Grumman Space Systems Division; 2019.

EFFECT OF NANOCALCITE-MODIFIED EPOXY RESINS ON BURST PRESSURE AND DESIGN OF COMPOSITE OVERWRAPPED PRESSURE VESSELS FOR HYDROGEN STORAGE

James Nelson, Douglas Goetz, Jay Lomeda, Nicholas Larson, Wendy Thompson, Paul Sedgwick, Dave Sewell
3M Company
3M Center, Saint Paul, MN 55144

Daryl Thompson, Jake Walker
HyPerComp Engineering Incorporated
156 S. 800 W. Suite H, Brigham City, UT 84302

William King, PE
in-Finite Solutions, Inc.
1102 N. Arlington Dr., Eagle, ID

ABSTRACT

A study was undertaken to investigate the effect of nanocalcite matrix modification on Composite Overwrapped Pressure Vessels (COPVs) for hydrogen gas storage. Thick-walled type III COPVs were prepared via filament winding using a conventional unfilled epoxy matrix as a control material and an epoxy matrix resin modified with surface-treated nanocalcite particles at 35% by weight (3M Developmental Resin AMD 931). The COPVs were evaluated for burst pressure. The results show that the modification of the filament winding matrix resin with nanocalcite significantly improved burst pressure relative to the unmodified control resin. Additionally, impregnated strand tensile strength testing was performed on Toray T700S carbon fiber to investigate the effect of matrix resin on realized fiber strength. Data from impregnated tow specimens made using 0 and 35 weight% nanocalcite in the epoxy show that nanocalcite modification shifts the delivered fiber strength (DFS) distribution. The higher performance demonstrated for a given design suggests that use of nanocalcite-modified matrix resin could enable improved margins of safety, qualification at increased burst pressure, or redesign to reduce COPV weight and cost. To examine the implications of the measured increase in hydroburst pressure on article design, a design optimization study based on finite element analysis (FEA) was performed for a nylon-liner 52L COPV design using the increased DFS values. The optimized tank having the same external envelope had overall weight reduction of 12% and increased gas-carrying volume of 12 % due to reduction in composite wall-thickness enabling increasing the internal volume. The impact of increased hydrogen tank capacity on fuel cell electric vehicle range will be discussed as well as additional design options that are enabled.

1. INTRODUCTION

The subject of this study is the effect of nanocalcite-modified epoxy resins on the burst performance of Composite Overwrapped Pressure Vessels (COPVs). The focus is tanks used for hydrogen storage for Fuel Cell Electric Vehicles (FCEVs) and the implications for their design. First, the context of this application is discussed. A brief review of the characteristics of this type of nanoparticle-modified matrix resin composite follows, and the rationale and plan for the study are given.

1.1 Hydrogen Storage for Fuel Cell Electric Vehicles

Fuel Cell Electric Vehicles currently rely on compressed hydrogen storage employing COPVs with polymeric liners (Type IV). In order to meet range and functional requirements, high pressure storage is required, typically at 700 bar (10,000 psi) for light duty vehicles. The hydrogen storage system is a major cost component for FCEVs. Of this, over half is the cost of carbon fiber for the vessel composite overwrap [1]. Therefore, reduction of carbon fiber cost and efficient use of carbon fiber are vital cost-reduction goals for impacting FCEV cost. This work addresses the efficiency of carbon fiber utilization.

The amount of carbon fiber required for a COPV design depends on several factors, such as the nominal service pressure, fiber properties (e.g., tensile strength), and translation of fiber strength in the composite. Translation refers to the in-situ fiber strength that can be realized. It is typically expressed as a percentage of the fiber strength reported by fiber manufacturers. The manufacturers determine fiber lot strength using impregnated tow testing. Translation is a key design driver for overwrap design. Generally, COPV manufacturers have found that translation decreases as the composite overwrap wall thickness increases (as is required by higher operating pressure or larger tank diameter). Translation is determined experimentally through burst testing of COPVs and is then used to guide COPV design. The main goal of the current work is to increase translation of fiber strength through matrix modification in order to optimize the fiber/matrix composite *as a system* to achieve reductions in COPV cost and provide options for cost/performance optimization.

1.2 Inorganic Nanoparticle Modifications of Matrix Resins

3M has developed thermosetting composite matrix resins with high loadings of low aspect-ratio inorganic nanoparticles [e.g., 2-5]. The nanoparticles are surface-modified to stabilize the particles in the liquid resin and control particle-particle interactions, resulting in minimal impact on viscosity. Particle size is small enough to avoid filtering of the particles when the resin is infiltrated into reinforcing fibers during composite processing. Therefore, the resins can be used in conventional composite manufacturing processes such as filament winding. Typically, particles are used at a volume fraction of 15-25% in the cured resin. (Note that high aspect-ratio carbon nanotubes, graphene, or clay particles cannot be incorporated at such high loadings and still maintain composite processability.) This level of stiff inorganic particle incorporation leads to significant increases in cured resin modulus. Notably, modulus can be increased well beyond what can be achieved without inorganic particles while maintaining or increasing fracture resistance. Coefficient of thermal expansion, cure shrinkage, and cure exotherm are reduced. These property improvements for the modified matrix resins have been shown to translate into significant mechanical property improvements in composite material properties and products [2-5].

1.2.1 Composite Overwrapped Pressure Vessels

Previous studies considered the effect of matrix modification with nanoparticles on COPV performance for thin-walled, aluminum liner (Type III) vessels. Both nanosilica [2] and nanocalcite [3] modifications were shown to increase burst pressure, burst pressure after impact damage, and fatigue performance relative to unmodified control resin. Matrix modification with nanosilica was shown to shift the impregnated tow tensile strength so that the increase in mean strength was similar to the increase in burst pressure [3]. Limited data on COPVs using nanocalcite-modified matrix resin showed greater increase in burst pressure for a thicker composite tank design. For a tank design having an average control burst pressure of 43.9 MPa, the increase was 4.4% and 9.7 % for two lots of fiber. For a thicker-wall tank design having a control burst pressure of 92.4 MPa, the increase was 18%. In the context of hydrogen storage for FCEVs, much thicker, higher-pressure designs are of interest. This study addresses this type of design.

1.2.2 Micromechanics of Tensile Failure

In a cylindrical COPV, burst failure pressure depends on the in-situ strength of fibers in hoop wraps of the cylindrical section, if dome and cylinder-to-dome transition regions are adequately designed. Typically, the interior hoop wraps are at the highest stress, and they control the burst pressure of a COPV. In the cylindrical section the through-thickness stress distribution is uniform along the length and around the circumference. Therefore, the inner hoop wraps represent a large volume of material under nominally uniform stress. Failure is driven by the statistical strength distribution of fibers. As individual fibers fail, the load near the fiber break is transferred to the neighboring fibers.

Since the earliest models of unidirectional composite strength by Rosen [6] and colleagues in the 1960s, the role of matrix modulus as a factor in load transfer between fibers near an individual fiber break has been recognized. Higher matrix modulus tends to increase the observed strength of a composite. While conventional thermosetting resins have modulus values in a narrow range, nanoparticle modification increases modulus enough to access this aspect of the micromechanics of tensile failure. Because of the dependence of burst strength on hoop layers and the large volume of material at essentially uniform stress in those layers, pressure vessels are unique structures for taking advantage of the tensile performance of composites with nanoparticle-modified matrix resins.

1.3 Rationale and Plan of Study

The goal of this study is to examine the significance of matrix modification for the performance and design of hydrogen storage tanks for FCEVs. A key feature of the study is the need to determine fiber strength translation for thick-wall COPVs in order to work through the design implications for the target application. First, the effect of nanoparticle modification on neat resin properties was determined. The primary intent is to demonstrate differences in mechanical properties relative to conventional resins and characteristics relevant to processability. Next, COPV burst testing was conducted using subscale thick-composite Type III vessels designed to burst at an appropriate pressure to represent hydrogen storage. From these tests the delivered fiber strength is determined. Impregnated tow tensile strength testing was done to examine the effect of matrix resin modification on the fiber strength distribution. Finally, the measured delivered fiber strength was used to conduct a design study to explore the impact of increased delivered fiber strength on the design and characteristics of a representative Type IV hydrogen tank.

2. EXPERIMENTATION

2.1 Materials and Resin Sample Preparation and Testing

The carbon fiber used in this study is Toray T700S C-24000-50C. This is a high-performing fiber widely used for pressure vessel applications.

The unmodified epoxy used as a control material was an amine-cured Epon 826 bisphenol-A epoxy (Hexion). The nanocalcite-modified epoxy formulation AMD 931 is a 3M developmental resin. AMD 988-2 is a 3M developmental, aromatic amine curative system.

Resin/curative samples were generated using a DAC 600 SpeedMixer (Flacktek) at 2350 rpm for 45 seconds to produce well-dispersed blends. These blends were degassed under vacuum for 3-5 minutes prior to being poured into appropriate molds for neat resin tensile testing, dynamic mechanical analysis (DMA), and determination of hardness, coefficient of thermal expansion (CTE), density, and fracture toughness. The samples were cured in a forced air oven for 1 hour at 75 °C, 2 hours at 125 °C, and 2 hours at 150 °C. Samples were cured with a 1.3 overcharge of amine to epoxide functionality. The cured resin specimen preparation and testing methods used to obtain neat resin properties have been described previously [2-5].

2.2 COPV Design, Fabrication and Testing

To allow testing the effect of matrix resin modification on COPV burst strength using an application-relevant COPV format, a wrap pattern for a subscale 7.5-liter Type III test vessel was developed that would prevent rupture at greater than 155 MPa (22,500 psi). The intent was to design a subscale vessel having a composite wall thickness that would be similar to a 69 MPa (10,000 psi) pressure vessel with a factor of safety of 2.25. An initial design was evaluated and refined using burst test results. The observed mode of failure was sidewall burst, per the design intention. This failure mode is desirable because it stresses the hoop fibers to their maximum tensile capability and allows for the best extraction of data regarding the actual delivered fiber strength in the COPV application. The final wrap pattern had 37 wraps, with 21 hoop plies and the remainder being helical and knuckle wraps with transition patterns. The necks of the vessel were also wrapped to prevent excessive expansion during pressurization. The composite wall thickness of the vessels for this study is much greater than for previous studies of nanoparticle-modified resins which had burst pressures about 30% of those of the present study [2, 3].

All vessels were fabricated using two tows from the same lot of fiber. The manufacturer's tow strength determined by impregnated tow tensile testing was 5081 MPa (737 ksi). A nylon peel ply was applied after fiber winding to assist in composite consolidation and excess resin removal.

Curing was done in an oven using a two-step cure schedule: ramp to 83 °C and hold for 3 hours, then ramp to 126 °C and hold for 8 hours. The length of the final curing step was longer than the minimum time necessary for cure but was chosen to ensure complete curing of the resins.

Thermocouples installed on the vessels were used to verify the desired part temperatures were achieved. This is particularly important for large ovens, thick walls, or for larger vessels.

Hydroburst testing was performed by HyPerComp. Pressure was increased continuously to failure and recorded with time. Two vessels had to be pressurized twice to achieve failure due to seal failures in the initial pressurization. Finite Element Analysis (FEA) was used to calculate the tensile stress in the hoop fibers at peak pressure to determine the demonstrated fiber strength. This value was divided by the manufacturer's reported fiber strength to obtain the % translation. The % translation was used in the subsequent design study.

2.3 Impregnated Tow Tensile Testing

Impregnated tow tensile testing was performed to determine how the matrix resin affects the fiber strength distribution. The intention was to determine if impregnated tow testing showed changes in composite strength that correlated to COPV burst results. In this test, a carbon fiber tow is impregnated with liquid resin and then the resin is cured with the tow under tension to form a composite. A set of specimens are loaded in tension to failure to produce statistical information from the strength distribution. Because the COPVs for this study were designed to fail by sidewall burst, failure is governed by hoop wrap strength. Tow tensile strength corresponds to hoop strength. The probabilistic nature of COPV failure and size effects mean that there is not necessarily a direct correlation between tow tensile mean strength and delivered fiber strength in COPVs. However, it is reasonable to expect that positive changes in burst pressure will be reflected in positive changes in tow tensile strength. Tow tensile testing was performed for the control resin and developmental resin AMD 931. The fiber was T700S 24k with the same sizing as was used for the COPV burst testing, but using a different lot of fiber. The manufacturer's impregnated tow strength for the lot is 5160 MPa (748 ksi).

The general guidelines of ASTM D4018, "Standard Test Methods for Properties of Continuous Filament Carbon and Graphite Fiber Tows" [7] were followed. Fiber volume fraction for all specimens was controlled by pulling the impregnated tow through a die with a hole diameter such that fiber volume fraction was uniformly 52 to 53%. Fiber volume fraction was calculated for each specimen by weighing cured specimens and using densities of the matrix resins and fiber along with the mass per unit length of the fiber. The tow impregnation for the two resins was done sequentially and the specimens were cured side by side in the same oven. The cure state of the tow specimens was checked by determining the glass transition temperature of the matrix resins by differential scanning calorimetry. These values matched the glass transition temperature of neat resin retained from the impregnation bath that was cured in small aluminum pans in the oven alongside the tow tensile specimens. Tabs were adhesively bonded with epoxy to each specimen to form a gage length of 150 mm. A load frame equipped with hydraulic grips was used to load the specimens to failure at a displacement rate of 5 mm/min. Fiber strength was calculated using a rule of mixtures using the modulus of the matrix and fiber, assuming linearity up to failure. This calculation accounts for the differences in matrix modulus. The usual ASTM D4018 strength calculation neglects load carried by the matrix, and calculates strength based only on fiber cross-sectional area. Sets of 22 and 33 specimens were tested for the Control and AMD 931 resins, respectively.

3. RESULTS

The following sections present the results of neat resin characterization, COPV burst testing, and impregnated carbon fiber tow tensile testing.

3.1 Neat Resin Characterization

Tables 1-2 summarize neat resin data for this study.

Table 1. Properties of Neat Resins

Resin	Tensile Properties				Fracture Toughness (MPa√m)	Tg by DMA onset (°C)	Viscosity at 30 °C (Pa)
	Temp. (°C)	Modulus (GPa)	Peak Stress (MPa)	Strain at Break (%)			
Epon 826/ AMD 988-2	-40	3.3	70	3.2	0.74	130	3.9
	25	2.5	74	5.6			
	85	2.1	55	5.7			
AMD 931/ AMD 988-2	-40	5.4	78	2.0	1.21	120	6.3
	25	4.2	81	5.3			
	85	3.2	50	7.2			

Table 2. Properties of Neat Resins (continued).

Resin	Barcol Hardness (HB)	CTE (μm/m/°C)		Cure Exotherm (J/g)	Cure Exotherm Reduction (%)	Cure Shrinkage (%)	Density (g/cc)
		-25-25 °C	25-75 °C				
Epon 826/ AMD 988-2	28.8	58.0	69.7	318	-	0.8	1.161
AMD 931/ AMD 988-2	55.1	45.7	52.4	201	34	0.5	1.403

The neat resin properties of commercial nanocalcite matrix resins and the effect of nanoparticle concentration were reported previously [4, 5]. In summary, the inclusion of nanocalcite leads to resin systems with:

- High modulus
- High hardness
- Increased fracture resistance
- Low Coefficient of Thermal Expansion (CTE)
- Reduced cure shrinkage
- Reduced cure exotherm.

Enhanced modulus relative to the control resin is retained over the temperature range of interest.

3.2 COPV Burst Testing

Results of COPV burst testing are shown in Table 3. For these vessels, the fiber lot, winding pattern, and all aspects of fabrication were kept the same except for the matrix resin used.

The effect of matrix resin is clearly seen on the burst pressure. The average burst pressure is 28% higher for the AMD 931 vessels relative to the control vessels.

The delivered fiber strength calculated using finite element analysis from the burst pressures is shown; the manufacturer's impregnated tow tensile strength for this lot of fiber was 5081 MPa (737 ksi). Increases in average translation are slightly greater than increases in burst pressure, 32% higher for the AMD 931 vessels relative to the control vessels. This greater increase reflects the effect of load sharing with the liner and non-hoop plies.

The translation of fiber strength to in-situ strength highlights a common result for high-pressure thick-walled COPVs. The translation of strength is reduced relative to usual translation seen for thinner-wall, lower-pressure vessels. The increased translation of fiber strength for nanocalcite-modified matrix resins is essentially a reduction in the knock-down of translation suffered by the control. One implication of this observation is that comparison of resins should be done in a winding configuration that reflects the actual use scenario of the vessel. Previous studies using nanoparticle-modified resins for thin-walled type III COPVs have shown increases in burst and translation of fiber strength, but of smaller magnitude [2, 3].

Table 3. Results of COPV Burst Testing (Toray T700S Fiber)

Resin	Burst Pressure (MPa)	Increase in Avg. Burst Pressure (%)	Delivered Fiber Strength (MPa)	Translation (%)	Avg. Translation (%)
Control	149.0	--	3680	72.4	73.5
Control	153.2		3794	74.7	
AMD 931	193.7	28	4945	97.3	96.7
AMD 931	192.4*		4906	96.5	
AMD 931	192.1		4897	96.4	

*Pressurized twice; highest pressure achieved during first pressurization

Figure 1 shows the appearance of representative vessels after burst testing. The failure appearance for the vessels made using nanocalcite-modified resin has some differences from the control vessels. Figure 1a shows a control vessel after burst. The crack initiation occurred in the sidewall and propagated to the domes with bifurcation. The composite failure along the sidewall is along a well-defined line, and the composite plies have failed mostly along well-defined lines. In Figure 1b the composite of the AMD 931 vessel has a less well-defined fracture locus, which was characteristic of the three AMD 931 vessels tested. In previous COPV burst tests the post-burst appearance of nanocalcite-modified matrix resin COPVs was also seen to be characteristically different than the control [3].

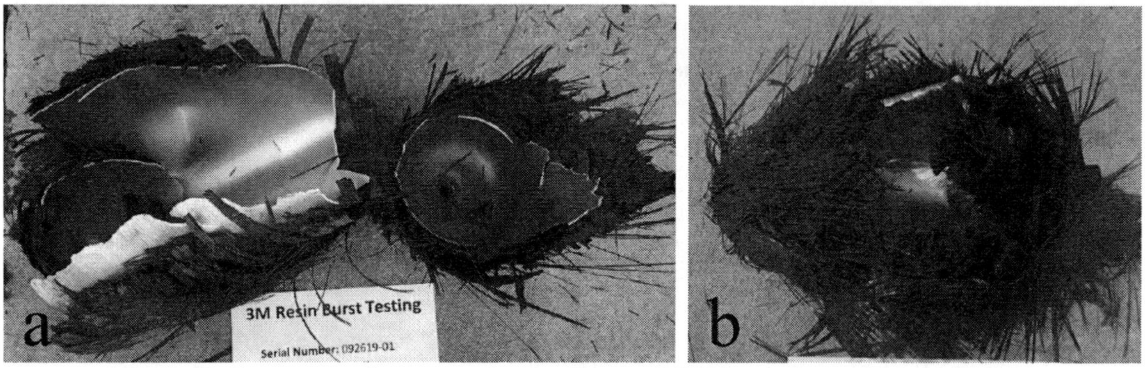

Figure 1. Representative appearance of burst COPVs: a) Control b) AMD 931

3.3 Impregnated Tow Tensile Testing

Figure 2 shows the results of impregnated tow tensile testing. In this probability plot the strength value for each specimen is displayed versus the percentage of values in the sample that are less than or equal to it, along a fitted distribution line. The y-axis is scaled so that the fitted normal distribution forms a straight line. 95% confidence intervals for the fitted distribution are shown for each fit line. Each of the strength data sets are seen to follow a normal distribution very well. A Weibull distribution has often been used in the literature to fit tow tensile data (e.g., [8]). For this data, a normal distribution gives a better fit. The probability distribution is seen to be shifted to higher strength for the specimens made using AMD 931.

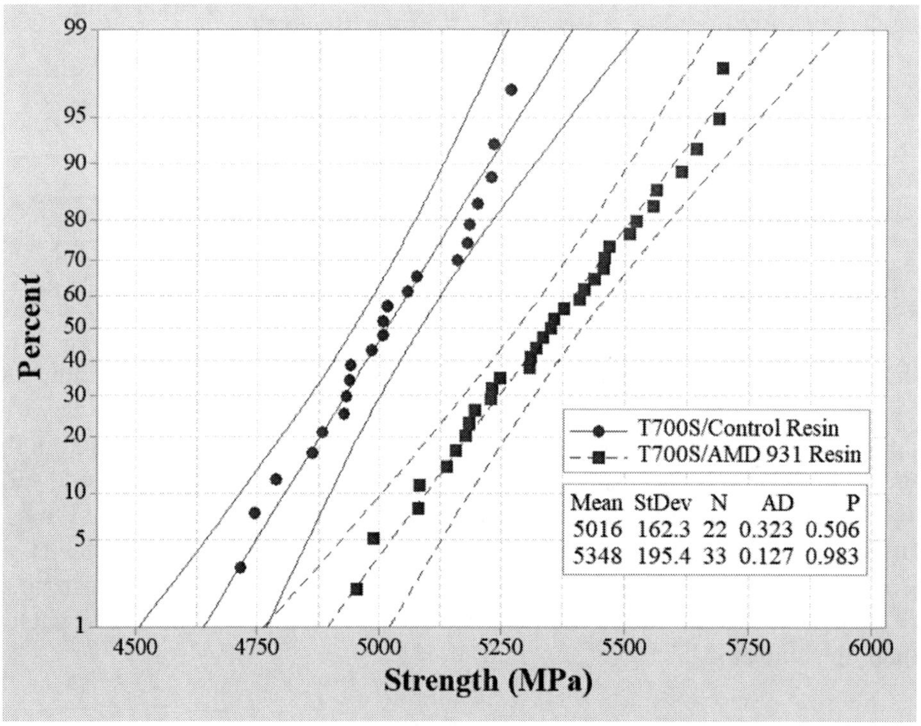

Figure 2. Probability of Failure for Impregnated Tow Tensile Testing. Normal distribution.

Figure 3 shows the mean values of 5016 for the control and 5348 MPa for AMD 931 plotted with 95% confidence intervals. The manufacturer's lot average was 5160 MPa. For the purpose of this study, the two primary features of the results are that the implementation of tow testing reflected the manufacturer's practice, and matrix modification produced a positive shift in the strength distribution. Previous tow tensile testing of nanosilica-modified matrix resins suggested the importance of the shift in the lower end of the distribution for correlation with COPV burst pressure [2, 3].

The shifted probability of failure gives a qualitative correlation with the increased burst pressure observed for nanocalcite-modified matrix resins. But if the impregnated tow tensile strength is representative of the tow strength in the pressure vessels, the observed increase in burst pressure is not a simple expression of the in-situ strength, but must also depend on additional factors because the magnitude of the increases are different. The idea that additional factors affect the translation of fiber strength is supported by the observation that translation varies with vessel thickness. Two broad categories of potential factors are processing effects such as fiber tension variations through the thickness and statistical scaling such as the so-called size effect of material strength.

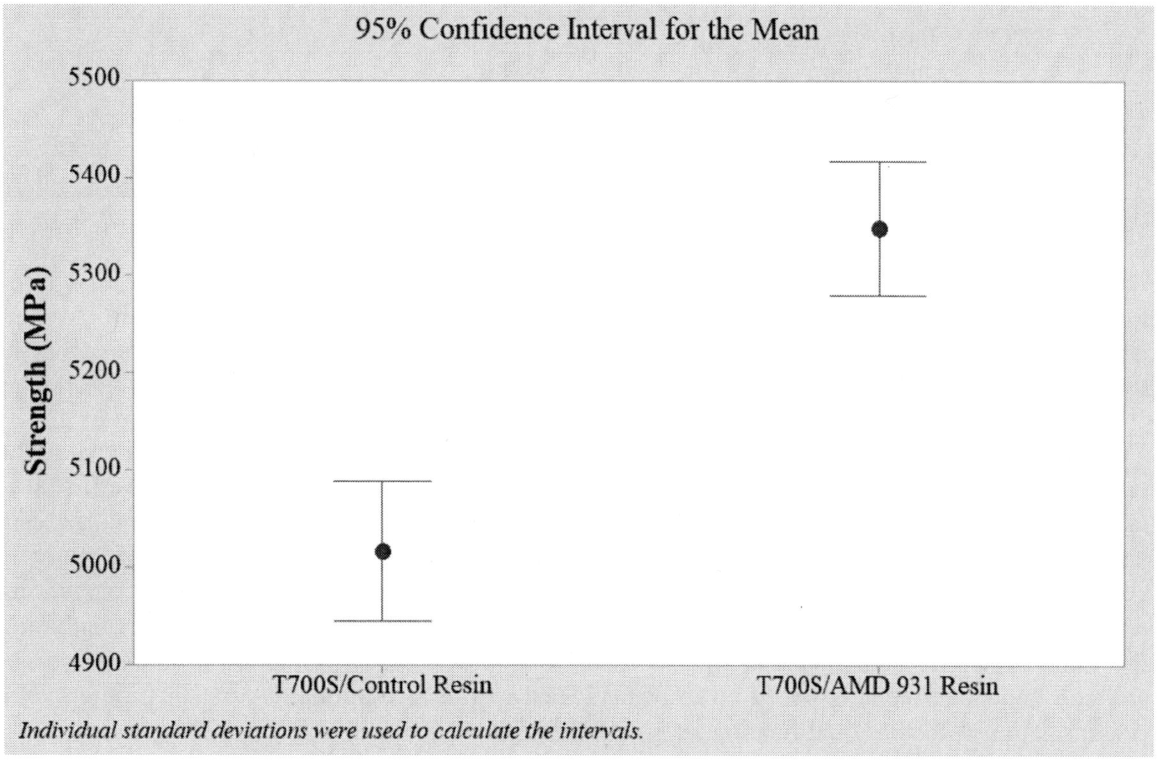

Figure 3. Confidence interval plot for tow tensile mean strength

4. DESIGN STUDY

Having established the magnitude of the effect of matrix resin on delivered fiber strength in an appropriate subscale COPV, a design study was performed to indicate the potential design and performance implications for a baseline hydrogen fuel tank for a FCEV.

4.1 Design Problem, Constraints, and Goals

A baseline design specification was assembled for a COPV representing a current commercial tank. The maximum design operating pressure is 69 MPa (10,000 psi). The minimum required burst pressure at 2.25 times the operating pressure is 155 MPa (22,500 psi). For this design study, the average targeted burst pressure was chosen to be 190 MPa (27,500 psi). This value accounts for material and manufacturing variations. Table 4 lists the physical values of the baseline design. For the new design, the external dimensions (diameter and length) were kept constant. The desired outcome for this choice of constraints is a reduction of composite overwrap weight which corresponds to a reduction in carbon fiber and cost, as well as an increase in interior volume which corresponds to greater fuel capacity and therefore vehicle range. The tank redesigned based on higher in-situ fiber strength would be a "drop-in" replacement for the baseline tank that offered cost reduction along with the benefits of weight reduction and increased range.

Table 4. Specifications of Baseline Design and Results of Revised Design

Design Feature (Pressure = 0 for volume, diam., length)	**Baseline Design**	**Revised Design**	**% Change**
Vented Internal Volume (liters)	52	58.3	+12
Outer Diameter (mm)	362	362	--
Length (mm)	889	889	--
Mass (kg)	37	32.4	-12
(Operating Pressure x Volume)/Mass (MPa x L/kg)	96.9	124.1	+28

Alternative design constraints could have been chosen. For example, for a new vehicle design the internal volume could be fixed to satisfy a given vehicle range. Then design with higher-performance composite would reduce cost, weight, and reduce the exterior volume required for hydrogen storage, impacting available space in the vehicle.

The design study demonstrates important steps in an approach to evaluating and using the enhanced fiber performance accessible with nanocalcite matrix resin modification. The first step is determination of delivered fiber strength for the candidate fiber using the incumbent matrix resin and nanocalcite resin in an application-appropriate COPV format. Second, a design study to generate optimized designs for both material sets should be performed, potentially considering alternative design criteria. The design study should inform price/performance analysis including the value of tank enhancements as well as cost. Next, experimental validation of the lead design and subsequent refinement based on the COPV test results. Additional testing including cyclic pressurization and other qualification tests would follow.

4.2 Design Optimization

The vessel preliminary design employed traditional netting analysis. Subsequently the preliminary design was refined using nonlinear 2D axisymmetric finite element analysis (FEA) that was able to capture the stresses in the critical dome and transition areas. A proprietary dome contour was employed. Lamina wind angles and thicknesses were generated using an in-house code. Detailed analysis considered not only critical fiber stresses calculated from element stresses, but also stresses in the metal boss. The factors of safety for the operating pressure, proof pressure (1.5x the operating pressure), and targeted burst pressure were all considered in arriving at the final design.

The orthotropic composite material properties were based on T700S carbon fiber assuming 65% fiber volume fraction that is typically achieved in practice. The material properties accounted for possible matrix stiffness degradation but used the demonstrated fiber strength of the T700S/AMD 931 composite of 4916 MPa. Therefore, the design results are impacted only by the observed increase in in-situ fiber strength and not the increased transverse and shear stiffness of composite plies. The increase of off-axis stiffness has been shown to be significant for filament-wound structures [4]. Increased load sharing of non-hoop plies is believed to be a factor in increased fatigue life observed for Type III COPVs [3]. The liner material was polyamide. The bosses were 17-4PH H1075 stainless steel.

Figure 4 shows a representative 2D model and fiber stress contours used in the design analysis. Such stress profiles were used to refine the winding pattern to achieve most efficient use of the fiber subject to the constraints in overall length and diameter.

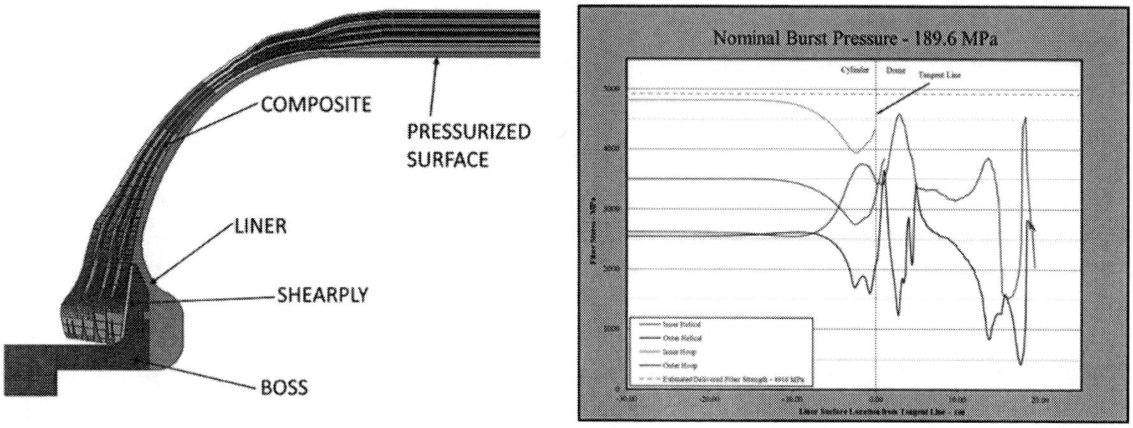

Figure 4. Axisymmetric FEA model and representative stress contours near dome transition

4.3 Optimized Design

The final design used a liner having a nominal sidewall thickness of 2.54 mm and minimum sidewall thickness of 1.78 mm with a nominal mass of 2.68 kg . The boss mass was 1.13 kg. The composite overwrap had a total sidewall thickness of 18.6 mm with a total mass of 26.78 kg.

Table 5 shows the maximum hoop fiber stress and margin of safety for the pressurization states considered.

Table 5. Maximum Sidewall Fiber Stress at Key Pressurization States

Pressure State	Hoop Fiber Stress (MPa)	Margin of Safety (Factor of Safety -1)
Operating Pressure (69 MPa)	1662	1.96
Proof Pressure (103.5 MPa)	2527	0.95
Nominal Design Burst Pressure (190 MPa)	4812	0.02

The final internal volume and mass of the revised design are shown in Table 4 and are compared to the baseline design specification. The resulting design had unpressurized internal volume 12% greater than the baseline and an overall decreased mass 12% lower than the baseline. A useful figure of merit for comparing designs is the (working pressure x volume)/mass. The design achieved a 28% improvement over the baseline design.

5. SUMMARY AND SIGNIFICANCE

This study examined the effect of matrix modification using nanocalcite on in-situ fiber strength demonstrated in a thick-wall COPV and demonstrated the opportunity to optimize vessel design.

5.1 Summary of Results

The specific matrix modification implemented in this work was accomplished using a high loading (35 weight %) of surface-modified low aspect ratio nanocalcite particles. Neat resin characterization showed that the stiffness of the resin was greatly increased; modulus was 68% higher than the unmodified control resin at room temperature. The modified resin had enhanced modulus over the range of temperatures of interest for FCEV systems (-40 to 85 °C). Other positive changes in matrix resin properties were also reported.

Testing of subscale thick-walled COPVs provided a direct comparison of burst performance for vessels made using a conventional resin and the nanoparticle-modified matrix resin. For otherwise identical vessels, the average burst pressure increased 28%. The translation of fiber strength in the two sets of vessels was 73.5 % and 96.7% for the control and nanocalcite vessels, respectively. The increased translation was found to be a decreased knock-down for the thick composite overwrap. This illustrates the importance of using an application-appropriate composite construction for evaluating the effect of matrix modification on in-situ fiber strength and burst pressure.

Impregnated tow tensile testing was used to determine the fiber strength distribution for the resins used in the COPV burst tests. Nanocalcite matrix modification was seen to shift the strength distribution to higher values. The magnitude of the shift was smaller than the increase in burst pressure produced by matrix modification. This implies that scaling and perhaps other factors associated with thick-walled composite vessel design and fabrication are important for expression of the shift in tensile failure stress in COPVs.

A rigorous design study for a baseline design representing a Type IV hydrogen storage tank for a Fuel Cell Electric Vehicle illustrated the implications for vessel design optimization. For fixed external dimensions, the use of the measured fiber delivered strength for the modified matrix resin system allowed a reduction in vessel mass of 12% while increasing the storage volume 12%. The resulting improvement in the figure of merit (working pressure x volume)/mass was 28% over the baseline.

This study illustrated a series of steps for using the enhanced performance made possible by matrix modification. As a preliminary step formulation of appropriately surface-functionalized matrix resin is required such that the mechanical properties of the matrix are significantly enhanced while preserving processability. Next, determination of in-situ delivered fiber strength is made by burst testing COPVs with characteristics appropriate for the target application. For the FCEV target application chosen for this work, the composite wall thickness appropriate for 700 bar hydrogen storage is a key characteristic. Location of the failure initiation in the vessel sidewall is important to allow the greatest confidence in the fiber translation that is determined. The next step is design optimization using detailed stress analysis using chosen constraints and design goals. A final step not illustrated here is burst testing of the final design for validation, testing for additional certification requirements, and refinement of the design as needed.

5.2 Significance for Hydrogen Storage for Fuel Cell Electric Vehicles

Numerous strategies for cost reduction of hydrogen storage have been examined [9]. The most important objective for reducing hydrogen storage cost for FCEVs is the reduction in the total cost of the carbon fiber used in the hydrogen tanks. Efficient use of carbon fiber depends on COPV design as well as manufacturing practices, and a primary mechanical driver of design optimization is fiber strength translation. The main goal of the current work is to increase translation of fiber strength through matrix modification in order to optimize the fiber/matrix composite *as a system* to achieve reductions in COPV cost and provide options for cost/performance optimization.

Cost/performance optimization goals will depend on the overall vehicle performance and price goals. In the design study of this work the outer dimensions of a baseline design were taken as constraints. This is appropriate if the desire is to design a drop-in replacement for a hydrogen tank in an existing vehicle design. The cost of the tank is reduced by reducing the mass of the composite overwrap while providing advantages of reduced overall vehicle weight and increased hydrogen storage capacity. Increased capacity allows extended vehicle range. Because hydrogen fueling stations are currently limited, extended range addresses the so-called "range anxiety" that is a barrier to consumer adoption of FCEVs and provides an easily quantifiable, highly-valued performance metric for the consumer. An alternative design target is maximum cost reduction. A design constraint is maintaining a minimum fuel capacity to achieve a target range. For this goal the thinning of the composite overwrap will reduce the cost through reduced fiber usage, with the additional value of reduced space requirements for tanks and overall vehicle weight. Figure 5 illustrates these two design scenarios, each reducing the cost of hydrogen storage, but with different additional advantages.

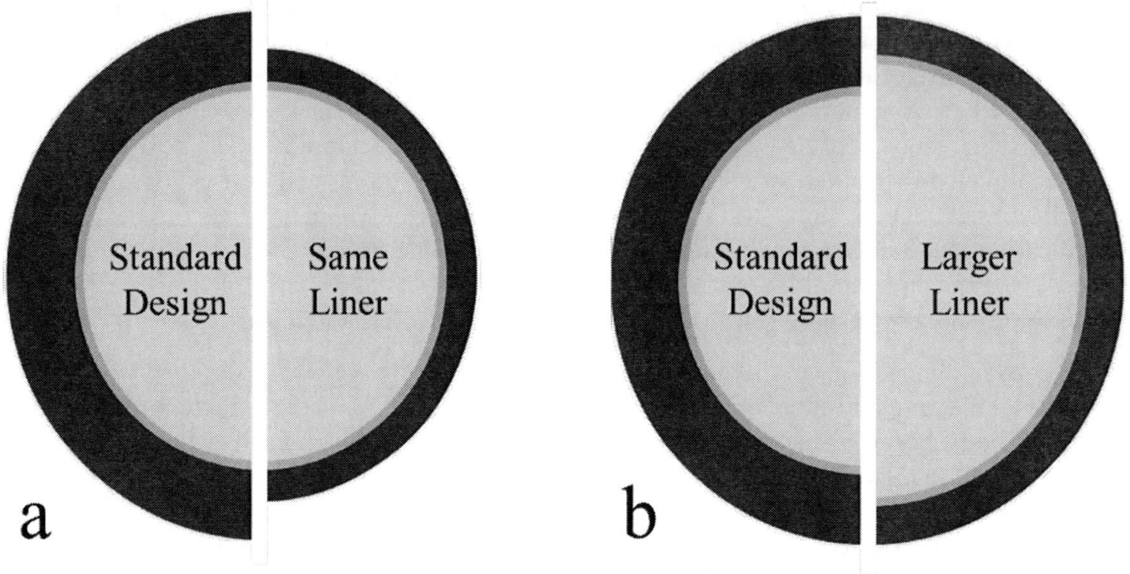

Figure 5. Schematic cross-sections of pressure vessels showing possible design optimization results: a) Same liner, reducing cost and weight. b) Larger liner, reducing cost, weight and increasing vehicle range

6. REFERENCES

1. Gangloff, J., Ordaz, G., Randolph, K, Adams, J., and Stetson, N., "Carbon Fiber Composite Material Cost Challenges for Compressed Hydrogen Storage On-Board Fuel Cell Electric Vehicles". *Proceedings of the International SAMPE Tech. Conf.* Seattle, WA, May 22-25, 2017. Society for the Advancement of Materials and Process Engineering.

2. Thunhorst, K., et al. "Nanosilica Concentration Effect on Epoxy Resins and Filament-Wound Composite Overwrapped Pressure Vessels." *Proceedings of the International SAMPE Tech. Conf.* Long Beach, CA, May 23-26, 2011. Society for the Advancement of Materials and Process Engineering.

3. Goenner, E., Thunhorst, K., Goetz, D., Nelson, J. "The Effect of Nanocalcite on the Mechanical Performance of Filament-Wound Composite Overwrapped Pressure Vessels." CAMX Conference Proceedings. Anaheim, CA, Dec. 12-14, 2017. CAMX – The Composites and Advanced Materials Expo.

4. Sharma, A., et al. "Improved Performance of Filament-Wound Composite Drive Shafts with Next Generation Inorganic Nanoparticle-Filled Epoxy Resins." CAMX Conference Proceedings. Anaheim, CA, Sept. 26-29, 2016. CAMX – The Composites and Advanced Materials Expo.

5. Nelson, J., et al. "Nanocalcite-Modified Thermosetting Resins for Fiber-Reinforced Composites: Properties and Applications." *Proceedings of the International SAMPE Tech. Conf.* Seattle, WA, May 22-25, 2017. Society for the Advancement of Materials and Process Engineering.

6. Rosen, B.W., "Tensile Failure of Fibrous Composites." *AIAA Journal* 2 (1964):1985-91.
7. ASTM D4018-17, 2017, "Standard Test Methods for Properties of Continuous Filament Carbon and Graphite Fiber Tows" ASTM International, West Conshohocken, PA, 2017, DOI: 10.1520/D4018-17
8. Cohen, D., and Lowe, K., "The Influence of Epoxy Matrix Properties on Delivered Fiber Strength in Filament Wound Composite Pressure Vessels." *J of Reinforced Plastics and Composites* 10 (1991): 112-131.
9. Johnson, K., Veenstra, M., et al., "Advancements and Opportunities for On-Board 700 Bar Compressed Hydrogen Tanks in the Progression Towards the Commercialization of Fuel Cell Vehicles," *SAE Int. J. Alt. Power.* 6 (2):2017, DOI:10.4271/2017-01-1183.

Carbon Fiber- Fiberglass SMC Hybrids in Relationship to Resin Chemistry Systems

James Bono
Polynt Composites LLC
Kansas City, MO

ABSTRACT

With the onset of light weighting opportunities across multiple markets, the demand for advanced composites is growing. Reduction in component weights have forced material engineers to evaluate high modulus materials- reducing the thickness of parts thus reducing their weights.

Carbon fibers are widely used for the exception al high strength and stiffness to weight ratios. This has resulted in carbon fiber composites being used in aerospace and wind energy markets where the material cost is not as critical. The problem with carbon fiber in higher volume applications has been its unit cost does not break through the cost to performance barrier. One of these higher volume markets is the automotive sector where light weighting opportunities are in high demand for electric and hybrid car designs.

Fiberglass composites, while having similar component costs to metal and aluminum substrates, does not have the same weight reduction opportunities. Usually, standard density fiberglass components must be designed with higher wall thicknesses based on their modulus in comparison to carbon fiber materials. Carbon fiber composites- specifically sheet molding compounds (SMC)- can reduce the laminate wall thickness based on their higher stiffness. Hybrid combinations of fiberglass and carbon fibers used in SMC applications would result in minimal increase in the cost of the components and increased modulus to aid in the weight reduction.

This paper will evaluate carbon fiber: fiberglass SMC hybrids and compare their mechanical properties relationships to both density and generic cost structures. In addition, three alternate resin systems- including vinylester and urethane modified resin chemistries- will be compared to determine their effect on the mechanical properties for the carbon fiber: fiberglass SMC hybrids.

Keywords: Carbon Fiber SMC, Hybrid fiberglass carbon fiber, Urethane modified resin

Corresponding author: James Bono

1. INTRODUCTION

Sheet molding compounds are used extensively for high production volume applications- including automotive and aerospace opportunities. In these markets, there exists a need to take advantage of light weighting opportunities. Sheet molding compounds (SMC) are fiber-reinforced thermosetting products- usually used in a compression molded high temperature application. While they offer some light weighting, standard density SMC can offer more by replacing higher density fiberglass reinforcements with lower density carbon fibers. Glass fibers are the major reinforcements used as reinforcements for SMC manufacturing. Carbon fiber has been used to a limited extent but has failed to be accepted in broad applications. However, carbon fibers produce composites with lighter weight and higher mechanical properties than those obtained using conventional glass fibers.

The properties of CFSMC composites can be dependent on more than just resin chemistries. Factors such as fiber sizing for adhesion of the resin systems and bundle and tex dimensions significantly affect the performance of the fiber when mechanical properties of the composite are studied. Fiberglass, on the contrary has developed processing for optimizing its performance in SMC applications- fabrication of many bundles into assembled roving's- split when cut through the SMC chopper assembly as well as fine tuned sizing chemistries to improve adhesion of a given resin system onto the glass filament.

In this paper, we will study the effect of incorporating carbon fiber in different ratios of the overall reinforcement content – thus evaluation carbon fiber fiberglass SMC hybrids. The intent of this paper is to give an initial evaluation of a 15K carbon fiber tow and optimized its use with fiberglass. Key fitness for use characteristics were identified as these materials were being developed including:

1. Mechanical property improvements- flexural and tensile properties.
2. Study the effect on specific gravity to improve weight reduction in composite parts.
3. Determine how shrinkage is affected with the incorporation of carbon fiber into SMC formulations
4. Maintain cost effectiveness in high performance composites market while still offering improved light weighting opportunities

2. EXPERIMENTATION

2.1 Materials

The resin types discussed in this paper includes:

DION® ITP 31638-00 + 31100: urethane modified hybrid resin system,

POLYLITE® 31600 + AM9033: Bisphenol A thickenable vinyl ester, and
ACPOL® 042-1035 + 31100: urethane modified hybrid resin system

All the above resins were manufactured at Polynt-Reichhold facilities, and are proprietary of Polynt-Reichhold based, Carpentersville, IL. Catalysts used to cure the resins are benzoyl peroxide (BPO) (Arkema), and 1,1-Di(tert-butylperoxy)-3,3,5-trimethylcyclohexane (NOROX 500-90-AL3) from United Initiators. BYK-P 9085 as an internal mold release was obtained from BYK. AM-9033 as a maturating agent was provided by Chromaflo for the Polylite 31600-00 resin system. Carbon fiber tow Pyrofil TR 50S15L was obtained from Mitsubishi Chemical Carbon Fiber and Composites. SMC fiberglass roving OCV 905A Advantex was obtained from Owens Corning OCV Reinforcements.

Clear cast resin specimen preparation: These lab panels were unreinforced and fabricated by curing the resin between glass plates. BPO was used as the free radical initiator to cure the resin at 54.4 °C for 16 hours in a forced hot air lab oven. Panels then were processed through a post cure cycle of 2 hours at 82.2 °C followed by 2 hours at 121.1 °C. Test coupons were then cut from the panels for mechanical testing.

Compounding for SMC and molding procedure: The SMC was prepared on a Finn and Fram SMC laboratory machine. The A and B-side pastes for each formula were blended on a Bowers high shear mixing system. The formulations were targeted to achieve the same volume fraction reinforcement content – either fiberglass or carbon fiber or combinations of both. The targeted volume fraction was between 39.1 and 39.4%. These SMC formulations also contained no fillers- i.e. calcium carbonate, aluminum trihydrate, etc. When paste viscosity reached to anywhere between 8 - 25 MMCP (MMCP refers to million centipoise), panels from SMC were molded on a 100-ton hydraulic press in a 14" X 14" plaque tool for samples. The molding was conducted at 300°F for 3 minutes with 1000 psi molding pressure. Molding charges were adjusted to yield panels of 3 mm nominal thickness.

2.2 Characterization

2.2.1 Mechanical Properties

An Instron Series 4206 mechanical testing system was used to determine tensile, flexural, and impact properties of the materials. Tensile testing was performed in accordance to ASTM D638. Flexural tests were conducted in accordance to ASTM D790. Specific Gravity tests were conducted in accordance to ASTM D792. At least eight specimens of each sample are tested.

2.2.2 Viscosity test

Viscosity tests of the liquid resins were performed on a LV viscometer, and viscosity tests of the SMC pastes were performed on a Brookfield DV-III Ultra Programmable Rheometer.

2.2.3 Mold Shrinkage test

Panel shrinkage was determined on flat laboratory molded panel specimens using an ASTM D6289 test method.

2.2.4 Glass transition temperatures – Dynamic Mechanical Analysis (DMA)

Glass transition temperatures were determined by dynamic mechanical analysis. The samples were measured with the TA Instruments Solids Analyzer (RSA-III) along with the three point bending test fixtures. The oven was purged with nitrogen during the analyses. To attempt to determine the Tg, the sample was scanned in dynamic temperature ramp mode, from 30 °C to 250 °C at a heating rate of 3 °C/minute and using a dynamic oscillation frequency of one hertz.

2.2.5 Generic Cost Model Structure

Standard calculations of cost were determined based on marketing information for raw material selling prices. For purposes of this paper, all similar raw materials were held constant (resins, and minor raw materials). Fiberglass pricing was set to $1.00 USD per pound and carbon fiber was set to $10 USD per pound. Costing was reported as a percent increase by either weight or volume against the all fiberglass reinforced SMC material as a control.

3. RESULTS

3.1 Liquid, thermal, physical, and mechanical properties of resins

Table 1: Comparisons of resins in terms of liquid, thermal, physical, and mechanical properties.

		POLYLITE ® 31600-00	**DION ® ITP 31638-00**	**ACPOL ® 042-1035**
		Bisphenol A Vinylester Thickenable	Urethane Modified Hybrid Resin	Urethane Modified Hybrid Resin
Tensile strength	psi	9,610	12,000	13,460
Tensile modulus	psi	542,000	493,000	605,100
Tensile elongation	%	2.31	5.8	2.88
Flexural strength	psi	20,082	20,000	17,800
Flexural modulus	psi	534,600	537,000	717,400
Heat deflection temperature	^{o}C	130.4	82	110
DMA Tg	^{o}C	161	119	136

3.1.1 Physical properties

Clear cast unreinforced mechanical properties for the resins used in the study to prepare the SMC samples is given above. This information is used as a point of reference.

3.1.2 Mechanical Properties

Samples for tensile, and HDT tests are cut from clear castings prepared with mixtures of resin with 1% BPO. Results in table 1 showed that the resins have a wide range of mechanical properties and may have a good balance of stiffness and toughness.

3.1.3 Thermal Properties

Glass transition temperature (Tg) of clear cast samples was measured by dynamic mechanical analysis (DMA) using a TA Instruments Solids Analyzer (RSA-III). As seen from Table 1, all resins have a Tg above 110 °C, which indicates that these resins and the resulting CFSMC may have excellent thermal properties.

3.2 Mechanical and thermal properties of Fiberglass; Fiberglass: Carbon Fiber Hybrids; and Carbon Fiber SMC

3.2.1 *Effect of percent carbon fiber added on mechanical properties of Fiberglass: Carbon Fiber SMC Hybrids with Various Resin Chemistries*

The properties of CFSMC composites depend on many factors, including the characteristics of the resin matrix and fibers, the size and volume of the fibers, and the nature of the interfacial adhesion between the fibers and resin matrix.

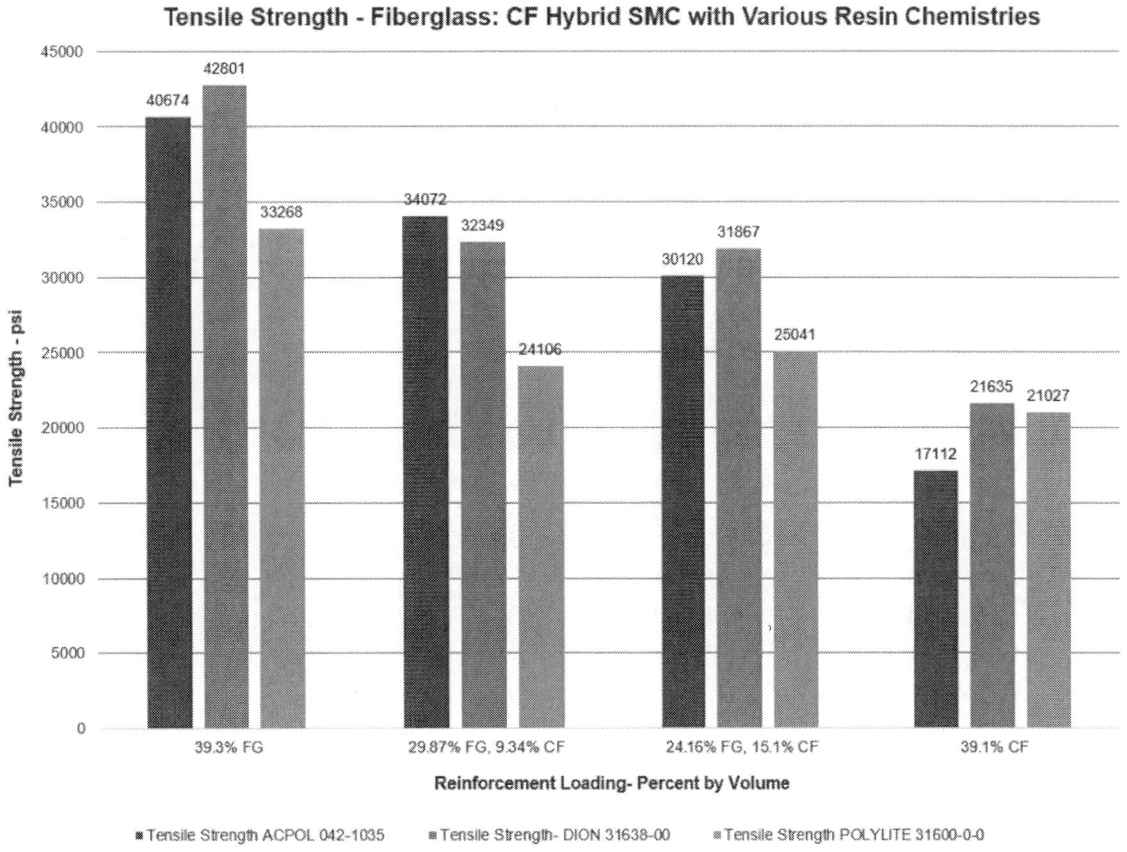

Figure 1. Tensile strength as a function of reinforcement types, for FG:CF SMC Hybrids – Various Resin Systems

Figure 1 shows the effect on tensile strength of the addition of a 15K tow carbon fiber as a reinforcement keeping the overall fiber reinforcement percent by volume fraction the same. In this case, for all three resin systems tested, tensile strength is reduced. It is hypothesized that the carbon fiber is reducing the tensile strength based on the size of the tow used in comparison to the dispersion of the SMC roving used. In traditional SMC formulations using assembled fiberglass rovings, the roving can "split" into the individual assembled bundles upon being cut in the SMC chopper. By using a 15K carbon fiber tow, the carbon fiber fraction of the composite was not efficiently split or filamentized upon being chopped and was distributed amongst the final laminate in large chopped bundles. It is theorized that a smaller size tex (3K tow) would yield high tensile strength values as 3K tow would disperse more evenly throughout the laminate. Tensile strengths are reduced with increasing levels of the 15K carbon fiber tow from 16.5%-49.5% of their original value using all fiberglass as the reinforcement in the SMC formulations.

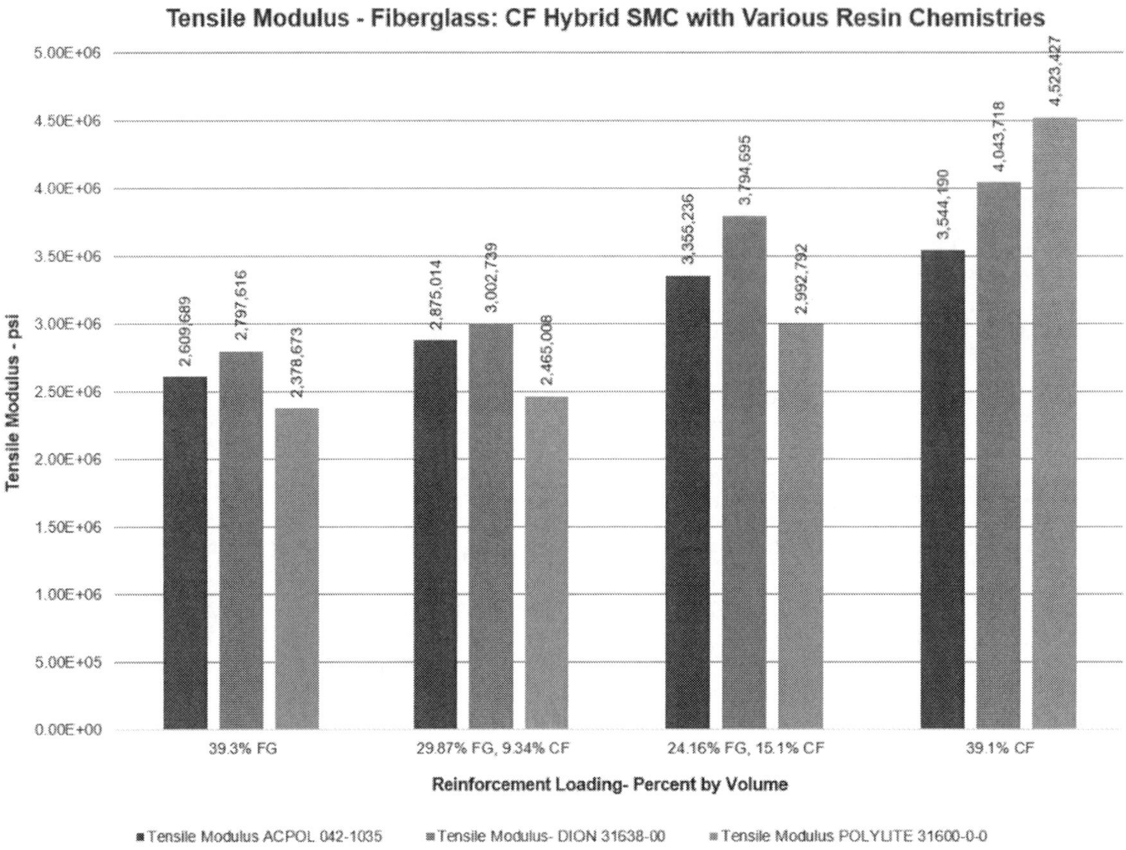

Figure 2. Tensile modulus as a function of reinforcement types, for FG:CF SMC Hybrids – Various Resin Systems

The tensile modulus results show results that increase in proportion to the amount of the 15K carbon fiber tow used to replace the fiberglass roving. For tensile modulus, as the carbon fiber increases in the volume of reinforcement it replaces, the tensile modulus dramatically increases by 10.2- 90.2% in stiffness values. This indicates that the stiffness can be affected by the addition of carbon fiber. This demonstrates that the stiffness of the laminate can be increased which in turn should allow for finite element analysis models to design thinner nominal composite wall sections and maintain the same part performance. This results in a reduction in the overall mass of the composite part.

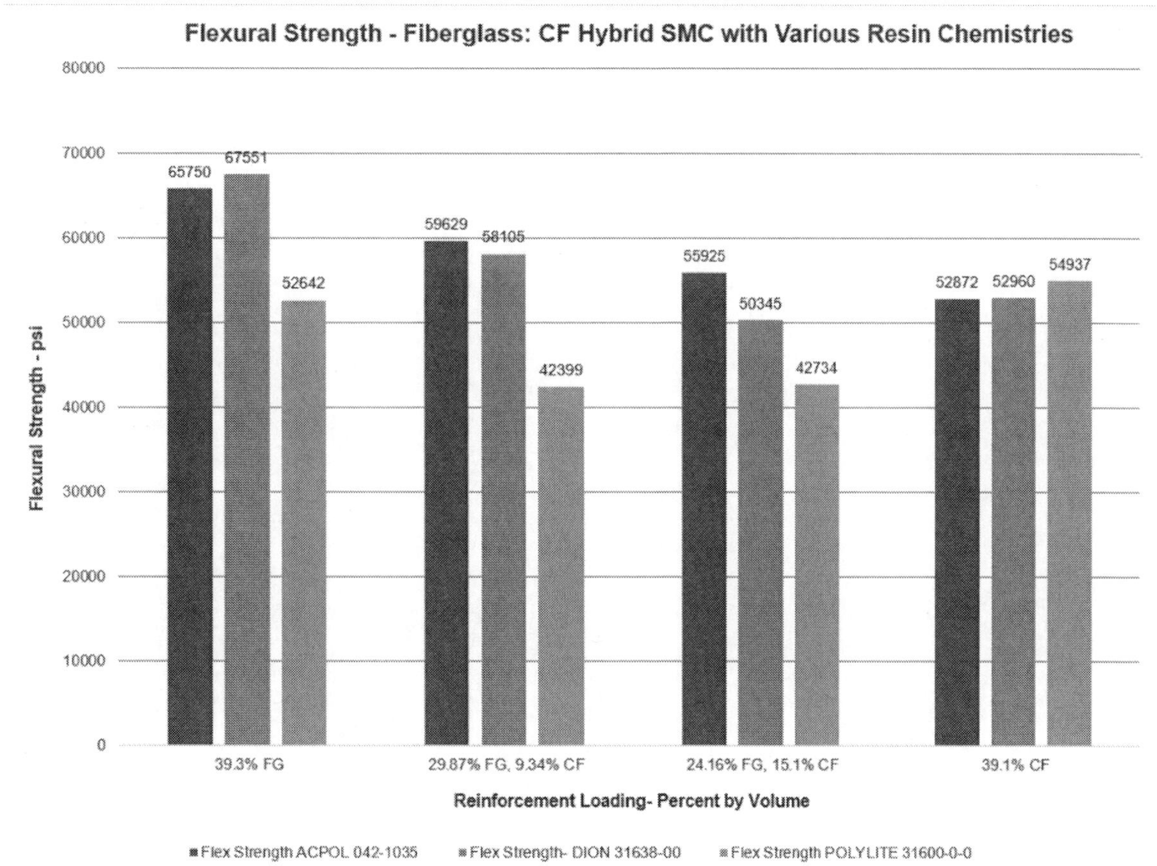

Figure 3. Flexural strength as a function of reinforcement types, for FG:CF SMC Hybrids – Various Resin Systems

Similar to the tensile strength, the flexural strength is adversely affected by the addition of a 15K carbon tow although the effect is not as significant for the measured flexural strength of the laminates. Flexural strengths are reduced by 9.3% - 25.5% as the level of the reinforcement used in these SMC formulations has carbon fiber sustituted for the typical fiberglass roving.

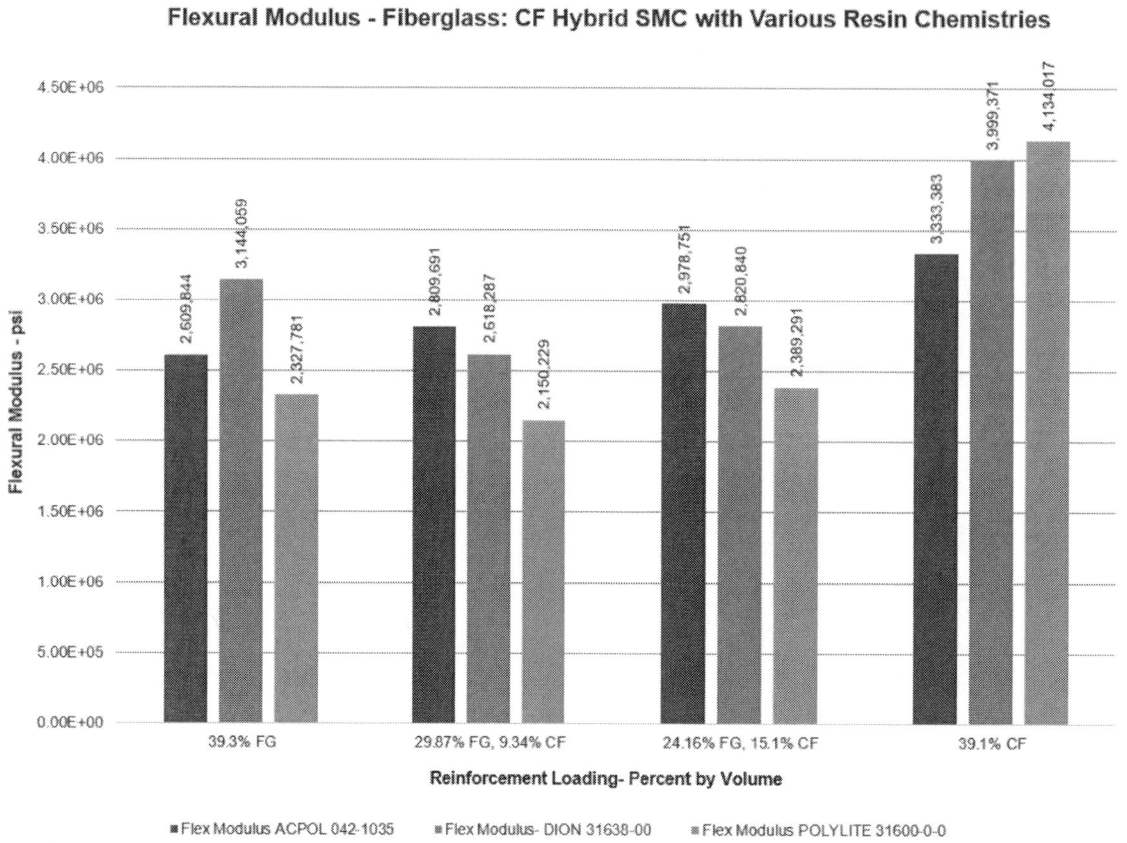

Figure 4. Flexural modulus as a function of reinforcement types, for FG:CF SMC Hybrids – Various Resin Systems

Similar to the tensile modulus, the flexural modulus increases in proportion to the addition of a 15K carbon tow for two of the resin systems- ACPOL 042-1035 and POLYLITE 31600-00. Flexural moduli are increased by 9.3% - 25.5% as the level of the reinforcement used in these SMC formulations has carbon fiber sustituted for the typical fiberglass roving. The modulus increases from 7.7% to 77.6% for these two resin systems. For DION 31638-00, the flexural modulus decreases slightly with the initial lower loadings of carbon fiber reinforcement, then increase when all of the reinforcement is replaced by carbon fiber.

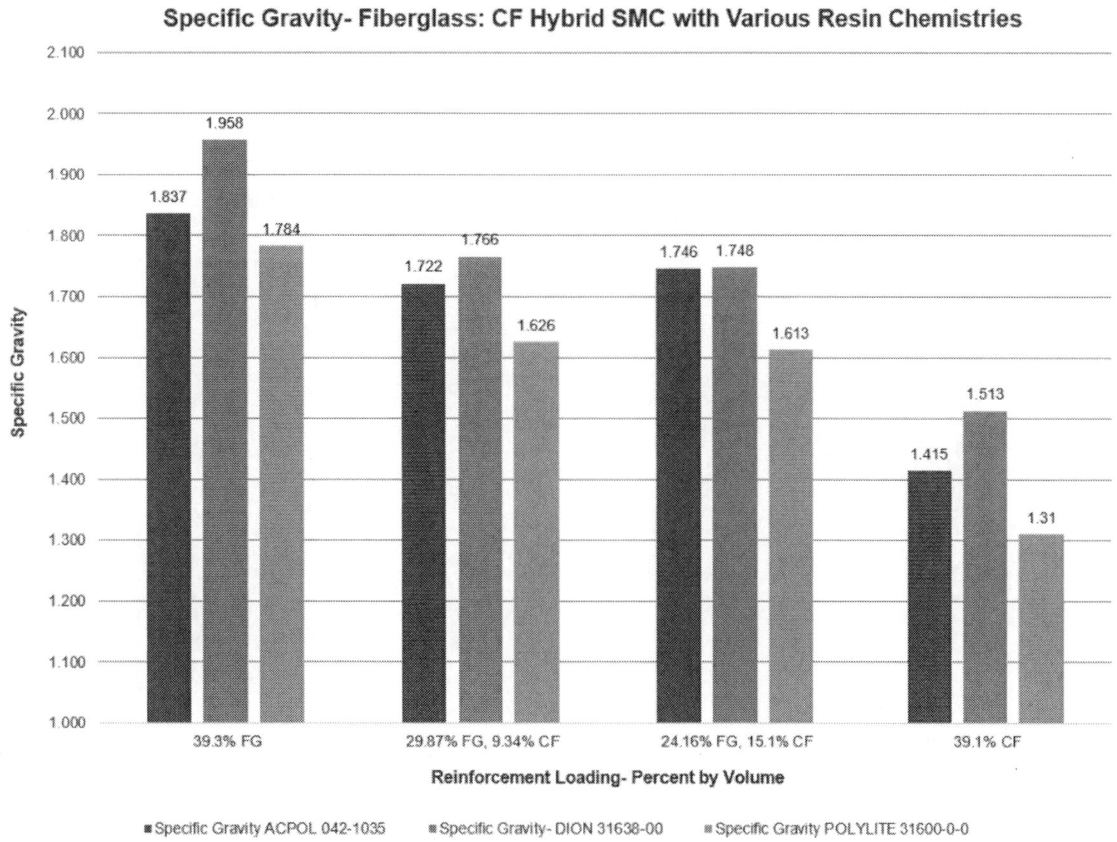

Figure 5. Measured Specific Gravity as a function of reinforcement types, for FG:CF SMC Hybrids – Various Resin Systems

Specific gravity is a function of the density of the components which make up the SMC formulation- from resins, thickeners, minor constituents, and reinforcements. Since carbon fiber has a lower density than that of fiberglass, it is expected to see a reduction in the specific gravity as the level of carbon fiber is increased in the SMC formulations. The measured specific gravities validate that as show in figure 5. In this figure, specific gravity can be seen to be reduced as the level of carbon fiber increases. With 9.34% CF replacing fiberglass in the overall 39.3% volume fraction of reinforcement used, the specific gravity reduces from 6.3% to 9.5%. Increasing the carbon fiber content to a level of 15.1%, the measured specific gravity is still reduced from 5.0%-10.7%. Once all the fiberglass is reduced, the specific gravity is reduced by 22.7% to 26.7%.

3.2.1 *Effect of percent carbon fiber added on Shrinkage from Mold Dimensions of Fiberglass: Carbon Fiber SMC Hybrids with Various Resin Chemistries*

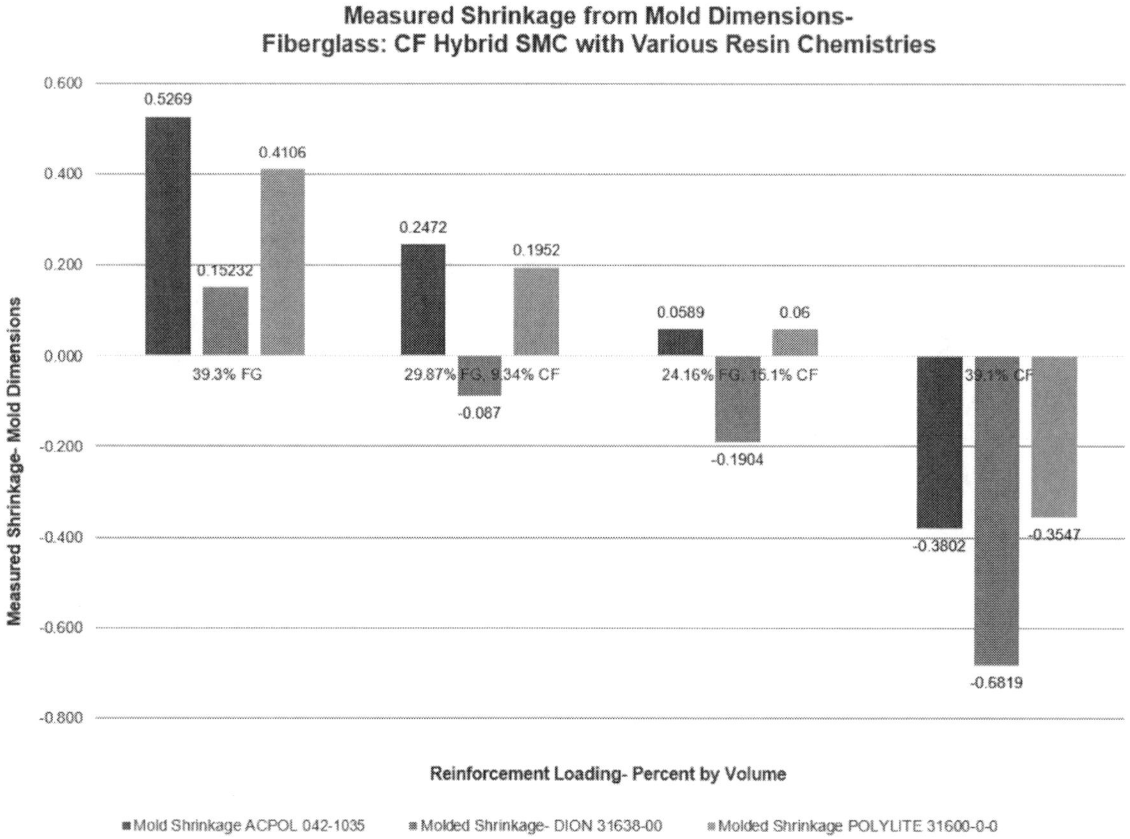

Figure 6. Measured Shrinkage from Mold Dimensions as a function of reinforcement types, for FG:CF SMC Hybrids

Figure 6 shows the effect of the addition of carbon finer as it relates to the measured shrinkage from mold dimensions. In this case, the measured mold dimensions is given in mils/ inch of shrink observed of a 14.0050 inch flat panel lab tool. A negative connotation indicates expansion of the thermoset system rather than shrinkage. As can be seen in the graph, the shrinkage of the system is affected proportionally to the amount of carbon fiber replacement for the fiberglass reinforcement. Each resin system has unique shrinkage associated with their chemistries. Carbon fiber has a negative coefficient of thermal expansion which means composites constructed with carbon fiber reinforcements don't expand or shrink as much as fiberglass when the temperature changes. This effect is having a definite influence on the measured shrinkage of the composite. This is a desirable quality for applications that have to operate in a wide range of temperatures. In each resin system case, the end result of the shrinkage is reduced with the addition of carbon fiber to the SMC formulations.

3.3 Generic Cost Model Structure of Fiberglass; Fiberglass: Carbon Fiber Hybrids; and Carbon Fiber SMC

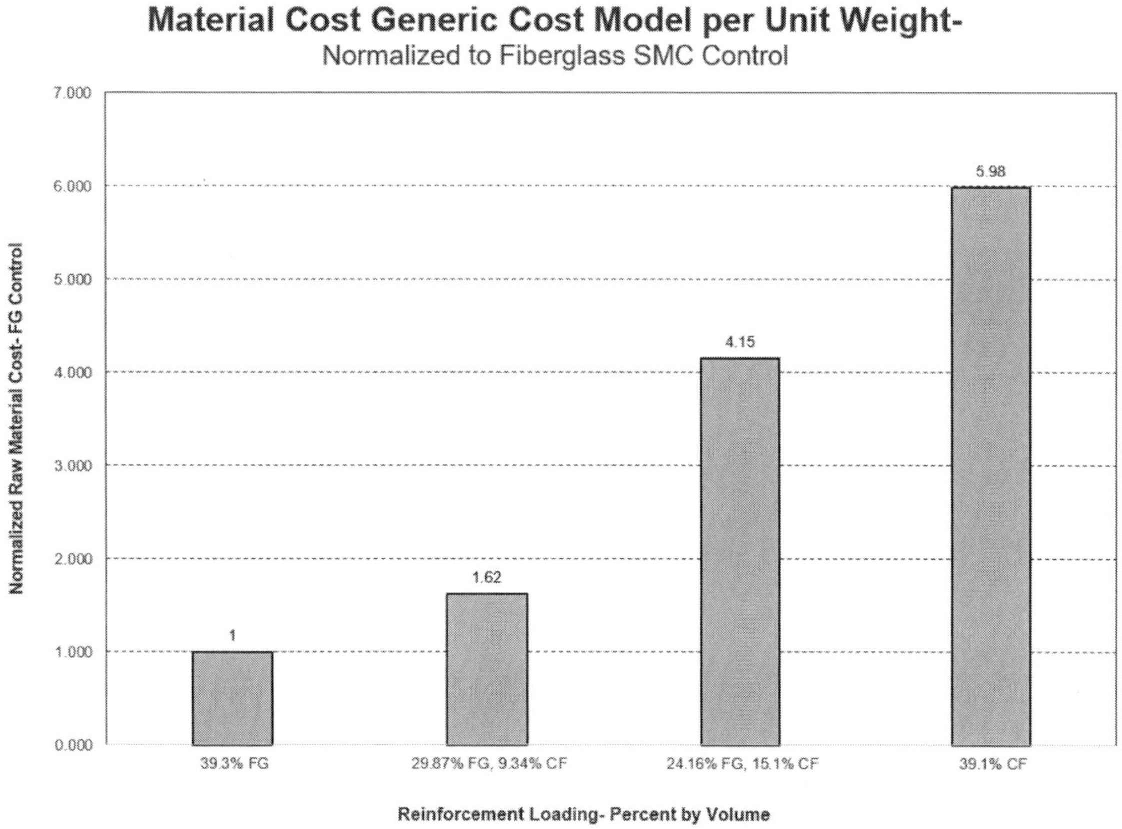

Figure 7. Normalized Raw Material Cost Model per unit weight as a function of reinforcement types, for FG:CF SMC Hybrids

Using marketing information for raw material costs and setting the fiberglass raw material cost to $1 USD per pound and the carbon fiber reinforcement pricing at $10 USD per pound, it is possible to calculate a general cost model for the replacement of fiberglass with carbon fiber in the SMC formulations studied here. Since the resin and other liquid components were kept constant across multiple formulations- the cost model looked at only changes to the reinforcement. All raw material cost models for the formulations were normalized around the all fiberglass SMC control. As expected, as the carbon fiber content is increased the raw material cost of the SMC increases as well.

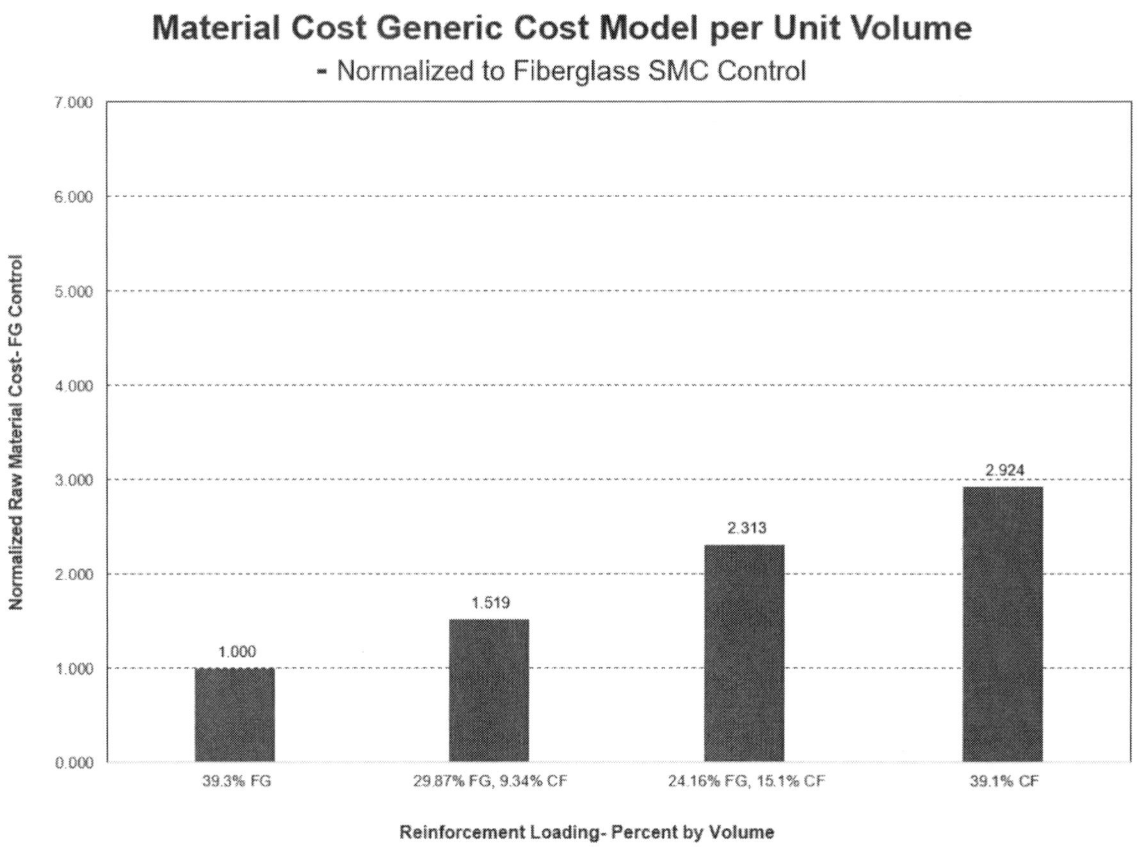

Figure 8. Normalized Raw Material Cost Model per unit volume as a function of reinforcement types, for FG:CF SMC Hybrids

Figure 7 shows the same cost model but in this case, the cost model is developed taken into account the specific gravity of the material which converts the chart to a cost model per unit volume of composite. In this case, the cost is affected to a lower degree and is more representative to the actual cost factors for a laminate part.

4. CONCLUSIONS

A series of fiberglass: carbon fiber hybrid combinations have been successfully developed for SMC applications. The physical properties, mechanical properties, and cost modeling of those resins were studied, and the results showed that these resins have excellent properties which vary due to resin chemistry. The modulus of each resin system was shown to increase with the increasing level of carbon fiber tow incorporated into the fiberglass; carbon fiber SMC hybrids. This increase in modulus values will subsequently increase the stiffness of the composite components made with such SMC formulations. The density of each system was reduced as the carbon fiber level in the SMC formulation was increased. The shrinkage attributes of each resin and their relationship with the addition of carbon fiber as a reinforcement with combinations of fiberglass were studies and shown that the measured shrinkage of the SMC formulations independent of the resin system were reduced proportionally as the carbon fiber content in the formulation was increased. In addition, generic cost models were calculated to show the resultant effect of the addition of a 15K tow carbon fiber incorporated into SMC formulations. The cost models were calculated as a function of unit weight and unit volume- taking into account the changes in density for each composite. The presented approach shows the opportunity for improved mechanical properties and reductions in component weights which represents a commercially viable and competitive approach to light weighting opportunities that can be used which will enhance the competitiveness to the SMC industry.

5. REFERENCES

1. Rodriguez, Ferdinand. Principles of Polymer Systems. New York, NY: Hemisphere Publishing Corporation, 1989.
2. "Plastics and Polymer Composites Technology Roadmap for Automotive Markets." March 2014. Copyright American Chemical Council 2014.
3. ASTM Standard D638-10. "Standard Test Method for Tensile Properties of Plastics." ASTM International, West Conshohocken, PA, 2010, DOI: 10.1520/D0638-10, www.astm.org .
4. ASTM Standard D3039-14. "Standard Test Method for Tensile Properties of Polymer Matrix Composite Materials." ASTM International, West Conshohocken, PA, 2010, DOI: 10.1520/D3039-D3039M-14, www.astm.org .
5. ASTM Standard D790-10. "Standard Test Methods for Flexural Properties of Unreinforced and Reinforced Plastics and Electrical Insulating Materials." ASTM International, West Conshohocken, PA, 2010, DOI: 10.1520/D0790-10, www.astm.org .
6. ASTM Standard D648-07. "Standard Test Method for Deflection Temperature of Plastics Under Flexural Load in the Edgewise Position." ASTM International, West Conshohocken, PA, 2007, DOI: 10.1520/D0648-07, www.astm.org .

EFFECTS OF MATERIAL CHARACTERISTICS AND EQUIPMENT CONFIGURATION ON PROFILOMETRY SCANNING RESULTS FOR ERROR MITIGATION

Jacob O. Ondeck, Wout De Backer, Michel J. L. van Tooren
McNAIR Research Center, University of South Carolina
1000 Catawba Street Columbia
SC 29201

ABSTRACT

The Automated Fiber Placement (AFP) process is a composite manufacturing process for constructing layered parts by placing tapes of material on a tool using a compaction roller and heat to make the material tacky. Manufacturing artefacts, such as unintentional tow twists, gaps, overlaps and even missing tows during the layup process may occur during layup. These defects deviate the manufactured structure from the as-designed structure, and have been proven to introduce stress concentration sources, which can ultimately undermine the performance of a structure. To detect and avert these defects during manufacturing, a profilometry driven topology analysis system can be used to scan the placed tows, check for layup defects, and record a history of the part. However, for certain materials and environmental conditions certain profilometers do not currently return reliable readings of the material topology, resulting in noisy or missing topology data. An experimental investigation into the feasibility of improving scan results of specific thermoset composite materials is summarized by investigating settings on commercially available profilometry scanners. Additionally, the impacts of material characteristics including surface quality are explored. Presented are the challenges, analysis, and potential solutions discovered to improve scanning results.

1. INTRODUCTION

1.1 Problem Statement

AFP manufacturing artefacts can arise due to manufacturing parameters such as part geometry and material characteristics, manifesting as the misplacement or unintended modifications to the asdesigned layup geometry [1]. To reduce the workload of operator s during the manufacturing process, Ingersoll Machine Tools (IMT) developed an automated composite structure inspection system (ACSIS) that scans the placed tows of each ply, checks for layup defects, and records a history of the part—employing digital profilometry scanners to measure the surface of the structure [2–4]. During operationalizing of the ACSIS system, IMT observed that the scanning of certain materials produced noisy scan image results. In some cases, materials were found to be completely unreadable (resulting in a void in the scan data). This scanning inspection system, due to the various material manufacturers and types used in composite manufacturing and the known impacts of layup defects on structure performance [1,5–7], is required to provide consistent results across a broad selection. For this reason, the source of the errors needs to be determined, as well as a set of possible solutions implemented to mitigate the impact of these errors during manufacturing. The focus of this paper is to present an investigation into the reason for which certain materials produce noisy image results when inspected by IMT's ACSIS, as well as to propose methods for

reducing the occurrence of these profilometry scanning errors. The investigations presented in this paper were made utilizing Keyence LJ-V7060, which was found to be equivalent to the equipment used in the ACSIS system with similar ideal scan distance, and resulting scan resolution [2].

1.2 Scan Noise & Material Selection

It is important to note that throughout this document, two types of noise signatures are referenced. It was observed that scanning noise manifests as both "Voids" and "Spikes". Examples of each of these can be seen in Figure 1 and Figure 2. Within the data matrix exported by the profilometers, each variant of scanning error displays itself as either a sudden decrease or increase in value compared to the local average or neighboring values. Furthermore, voids appear as sudden decreases in value, trending towards negative infinity, and spikes appear as sudden increases in value, trending towards positive infinity.

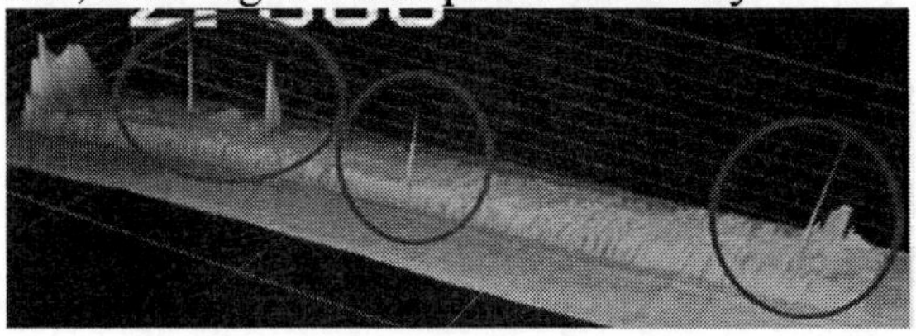

Figure 1. Sample scan with Spike Noise

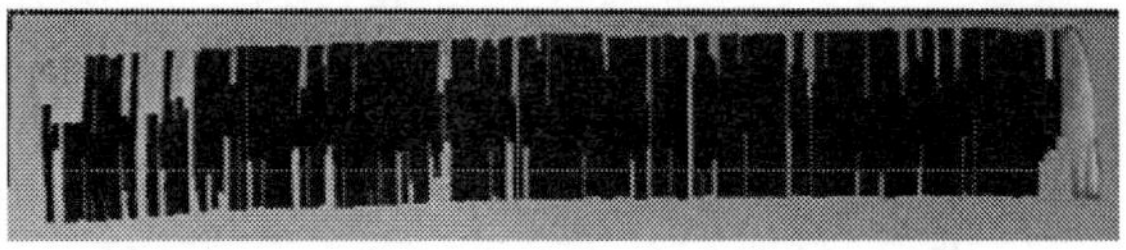

Figure 2. Sample Scan with Prominent Void Noise

The materials of interest to be studied in this presented research were selected based on preliminary scan data. Three materials were selected, each displaying varying levels of scan noise; i.e. "Very Noisy", "Noisy", and "Good". The noise of the three materials selected were initially done so qualitatively and were done so prior to the development of a method to quantify materials scanning error. Initial scans of the materials selected can be seen in Figure 3.

Figure 3. Initial Scans of Materials Selected

1.3 Noise Replication & Theory

To understand the cause of this noise, an attempt was made to replicate these results, but by scanning an object other than a thermoset tow. It was hypothesized that a mirrored surface may replicate the same "black hole" results that was observed when scanning noisy material. The results of this successful investigation can be seen in Figure 4, where each the replicated noise and the metal edges of the mirror are highlighted with red arrows.

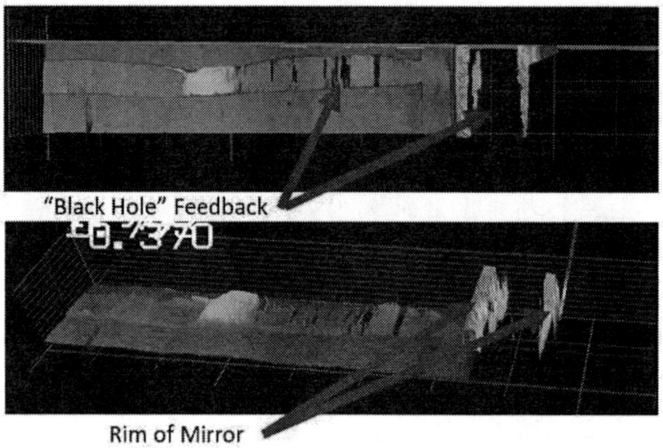

Figure 4. Noise Replication with Use of Mirror

The cause of this effect is a result of the scanning tool's data measurement method. From analysis of the scanner's manual and data sheet [2], the scanner's measurement relies on the target object's ability to scatter light. All light emitted from the scanner, once it reaches the target object's surface, is either reflected, refracted, or absorbed (Figure 5). Information or light reflected, the focus of this report, experiences either specular reflection of diffuse reflection. The mirror, and similarly the "noisy" reflective materials, have high specular reflectivity rather than diffuse reflectivity. Therefore, these targets do not scatter, i.e. diffuse, light well. It is this reflection which causes scanning error.

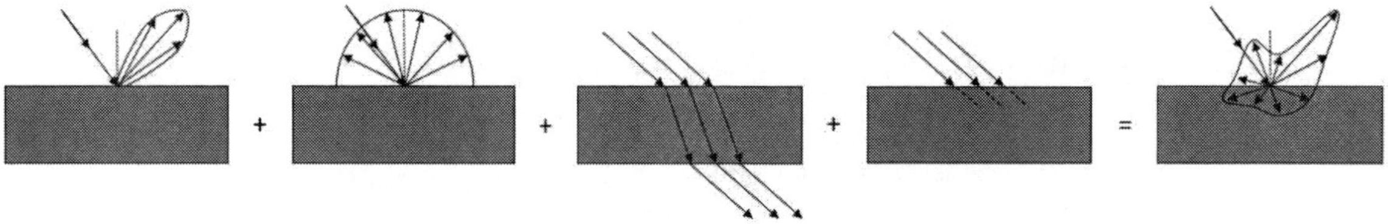

Figure 5. The Summation of (from left) Specular Reflection, Diffuse Reflection, Refraction, and Absorption

1.4 Variables to be Investigated

To explore the effects of the mirror-like material surface on profilometry scanning, several related variables were investigated. Pitch angle, or the incidence angle at which the profilometry scanner is sending and receiving information, was investigated to determine whether varying this parameter could positively influence the results of profilometry scanning. Based on the scanner geometry, it was hypothesized that noise could be reduced by introducing an ideal pitch angle when scanning objects with highly specular reflective surfaces. Similarly, the intensity of the emitted laser can be varied. As light is reflected off the target surface and disbursed, its intensity is diminished. This impacts the signal reaching the scanner's receiver. Furthermore, intensity is investigated at varying angles to determine if a compound effect is present at some ideal combination of settings.

An impactful variable of significant interest is the exposure time of the material prior to scanning. Thermoset materials can cure at room temperature, and their chemical and physical properties change as they are exposed over time. There was an interest to evaluate how scan results of the composites can vary over time, as the thermoset tows were stored. Material handling, storage, and out time considerations are an integral aspect of AFP manufacturing, and these practices impacting the feasibility of layup error detection would be significant. As curing directly effects the material surface, it was of experimental interest as to quantify the impact this curing has on error detection.

2. EXPERIMENTATION

2.1 Setup

The profilometer was mounted to a gantry system provided a high level of repeatability and consistency between scans (Figure 6). To investigate the effects of scanning at varying angles of attack, a modular scanner mount was added to this system, allowing for mounting of the profilometer and controllable pitch angles. Additionally, guidance lasers were used to ensure as little alignment variation as possible between scans. These guidance lasers were not operational during scans. A sample mount system was also designed which allowed for height adjustability as well as easily swapping out samples without disrupting the experimental setup. These samples were made by laying-up thermoset tows onto aluminum stages. Two geometries were laid-up by hand; single tow samples and samples which simulated a common layup error (Figure 7).

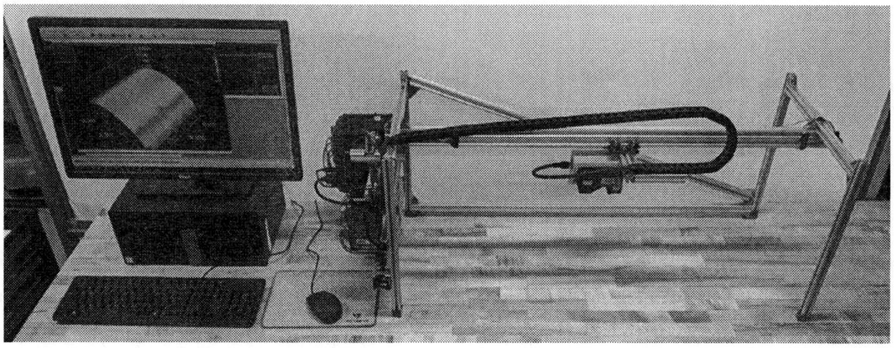

Figure 6. Scanning Gantry System

The layup of each stage was done in an identical manner. Special care was taken in order to maintain a clean environment. Each aluminum stage was first cleaned with acetone and then airdried completely. Thermoset materials were removed from a storage freezer (held at a set temperature of 0° Fahrenheit) for the duration of the setup yet were not left in room temperature conditions for longer than twenty minutes at a time. Tows were cut to length using clean stainlesssteel shears and were then placed on the aluminum stage. During this step, care was taken to eliminate contact between the surface of the tows and any object other than the aluminum stage. Once placed, the tows were compacted by using rollers. A similar method was followed for each material, however material C had initial difficulty adhering. To compensate for this difficulty, a heat gun was used to warm the plate prior to tow placement and compaction. Once the two stages were completed for each material, both were scanned for day one results. Following this initial scan, one stage was placed in the freezer and the other in ambient conditions. A picture of a layup (Material A) can be seen in Figure 7.

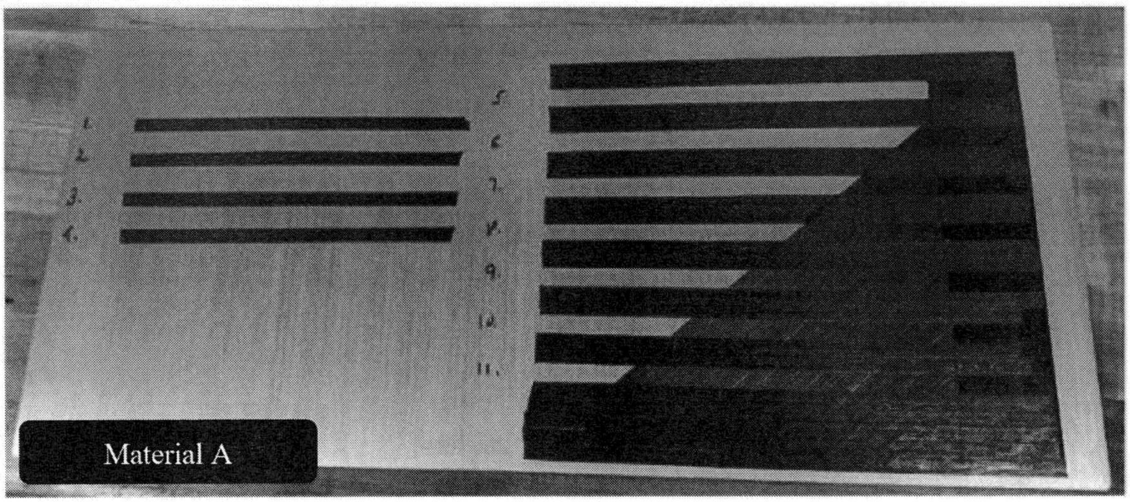

Figure 7. Exposure Test Material A Layup

2.2 Data Analysis Script

To quantify noise data from material scans, a python processing script was developed. This script accepts raw scan data (in the form of CSV input files) and returns both meaningful information about each void and spike noise represented within the scan and a generated grayscale image of the scan. These CSV files are the default output format of Keyence's controller software and store topographical data collected by the scan. This script was used throughout the research presented to create fair comparisons between possible solution approaches and error mitigation methods.

Additionally, this script was utilized in establishing the inherent standard deviation in scanning noise caused by varying the experimental setup or the variation in noise levels of experimentally identical, consecutive scans. This was done by comparing noise levels between scans in which the target object was reset between trials versus scans in which the target was unmoved between trials, allowing for a better understanding of what constitutes a significant change in noise level (section 2.3).

2.3 Baseline Variability Analysis

Before analyzing any scan results, it is important to understand the effect of variations of the stage location on the results of the analysis script. This influence occurs during swapping out of samples on the stage or moving the stage to focus on another target area for scanning. An investigation into standard noise deviation was performed. For this, scans were performed for each material at 0° pitch under two cases. Firstly, with no stage movement between each scan, and secondly, while resetting the target between each scan. The goal was to determine what constitutes a significant change in noise levels, through observing base levels of fluctuation between identically setup scans. The results of this short experiment can be seen plotted as percent noise in Figure 8 through Figure 11.

No Stage Movement	Stage Movement

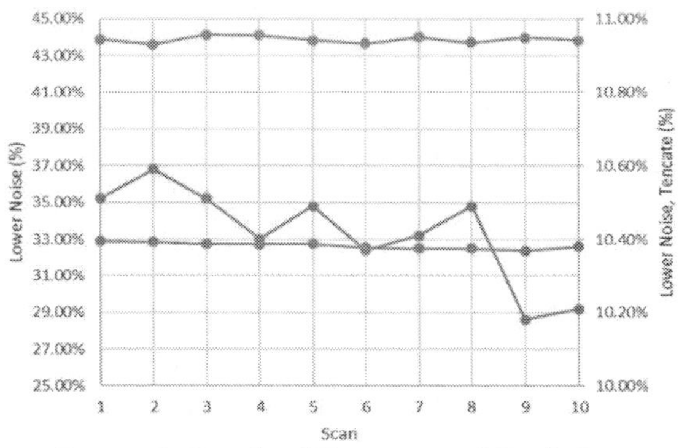

Figure 8. Standard Deviation of Void Noise across Identical Scans

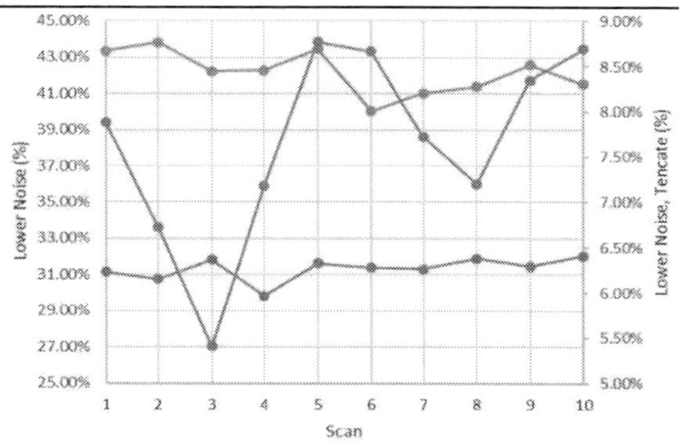

Figure 9. Standard Deviation of Void Noise across Identical Scans

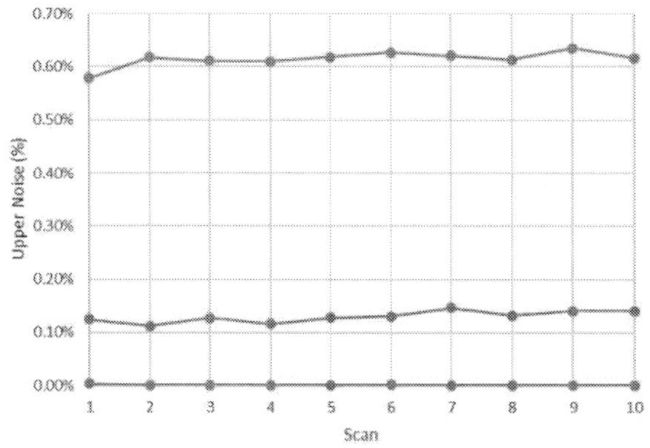

Figure 10. Standard Deviation of Spike Noise across Identical Scans

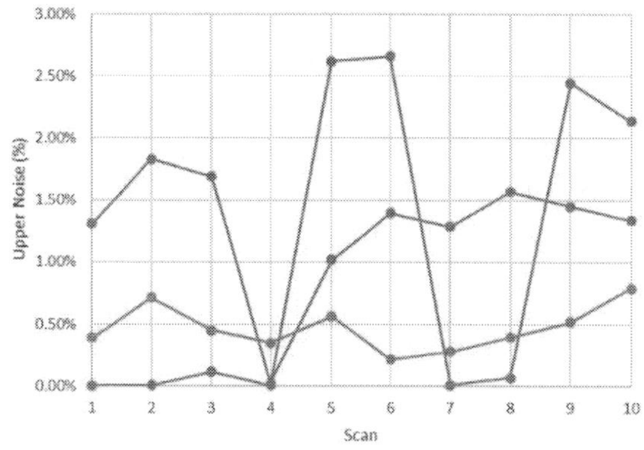

Figure 11. Standard Deviation of Spike Noise across Identical Scans

Graph Legend:

—●— Material A

—●— Material B

—●— Material C

Results from the stage movement experiment display minimal deviation between samples in the case of no base movement between scans. This is true for all material types and for each void and spike noise types. Specifically, for the case of no base movement, void noise standard deviation was all within 0.2%. Spike noise was even lower, with standard deviations between scans all within 0.015% for all materials (Table 1).

Table 1. Standard Deviation of Noise per Material, Base Movement, and Noise Type

	No Base Movement		With Base Movement	
Material	**Void Noise SD**	**Spike Noise SD**	**Void Noise SD**	**Spike Noise SD**
A	0.17%	0.01%	0.65%	0.50%
B	0.19%	0.01%	1.20%	0.18%
C	0.13%	0.00%	1.07%	1.26%

In the practical case—with stage movement between scans to replicate a typical scan setup—a marginal increase in standard noise can be seen. For void noise, the standard deviation between scans was calculated to be equal or less than 1.2% for all material types, where spike noise standard deviation was similarly less than 1.3% across all materials. The detailed conclusions can be seen in Table 1. These results can inform the conclusions drawn from other experiments, and display what this report considers a base-level expected acceptable standard for noise. Any results with noise variance within these thresholds can be regarded as insignificant.

3. RESULTS

3.1 Effect of Pitch Angle

The pitch test was done by scanning simulated layup error samples of each material of interest at incremental angles between -30° and 30°. The results of these scans were processed for each spike and void noise. While the spike noise all remained representative of below 2.5% of the total scan area and was deemed insignificant, the plotted results of void noise can be seen in Figure 12.

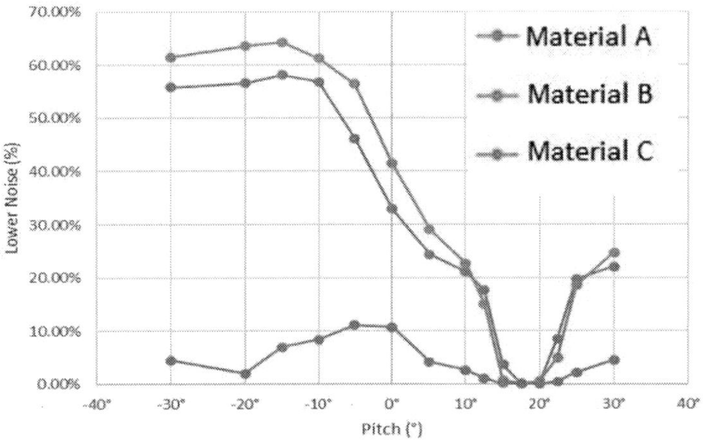

Figure 12. Void Noise by Percent of Total Data as a Function of Pitch Angles

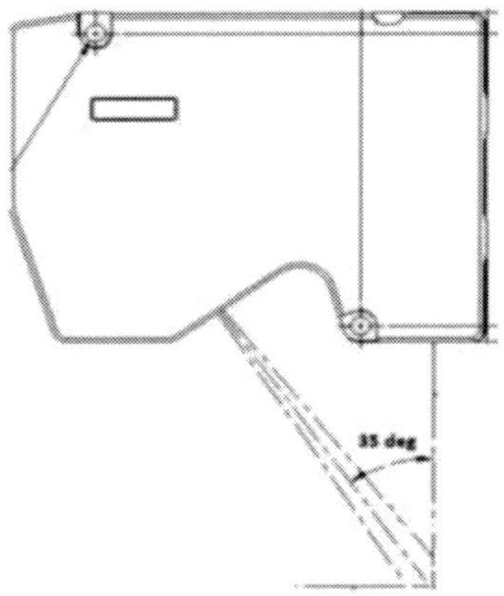

Figure 13. Profilometry Scanner Geometry

Results of the pitch tests clearly display a benefit of scanning at the bisecting angle of 17.5° (Figure 13). For all materials and samples studied, the void noise results fall to their lowest calculated values when scanning at this pitch (Figure 12). The change in calculated noise presence is significant when compared to the baseline anticipated noise deviation determined to be 1.3% in section 2.3, Table 1. As shown in Figure 12, there exists an error reduction of over 40% for the noisiest material scanned. The conclusion drawn is that a pitch benefits profilometry scanning error, and when possible, introducing a pitch angle of the bisecting angle, in this case 17.5°, is the preferred angle to resolve scanning noise when scanning highly specular reflective materials.

3.2 Effect of Environmental Exposure

A selection of noise results of samples plotted over the life of the exposure test are presented. The noise is presented as a percentage of total scan data within a sample for simulated layup samples, and a simple count of occurrences for single tow samples. The goal is to observe the noise quality as a function of time for sets of similar scanning and storage conditions for each material of interest.

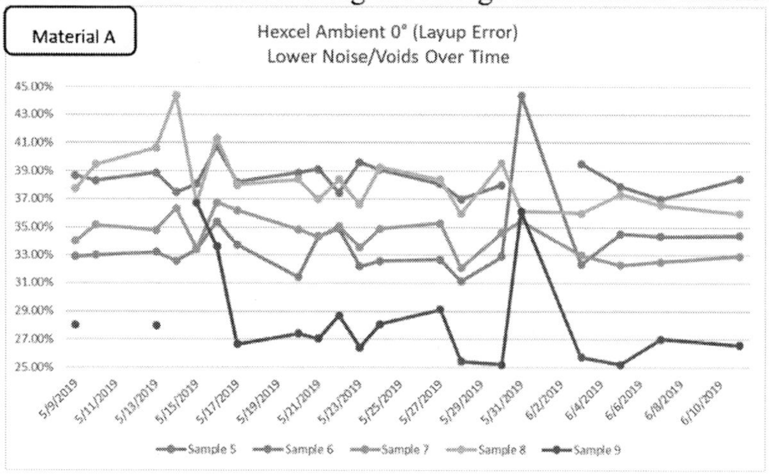

Figure 14. Void Noise for Material A
Ambient Simulated Layup Error Samples Scanned at 0°

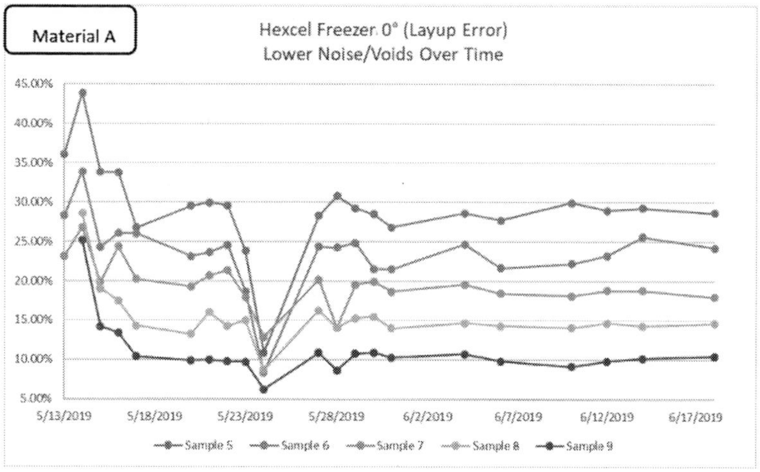

Figure 15. Void Noise for Material A
Freezer Simulated Layup Error Samples Scanned at 0°

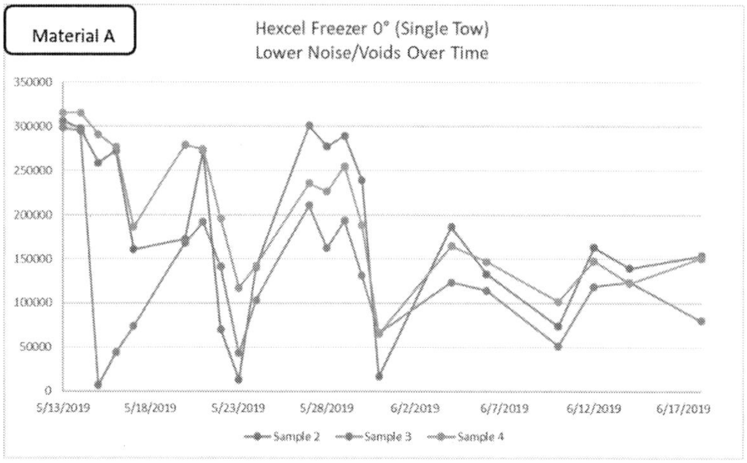

Figure 16. Void Noise for Material A
Freezer Single Tow Samples Scanned at 0°

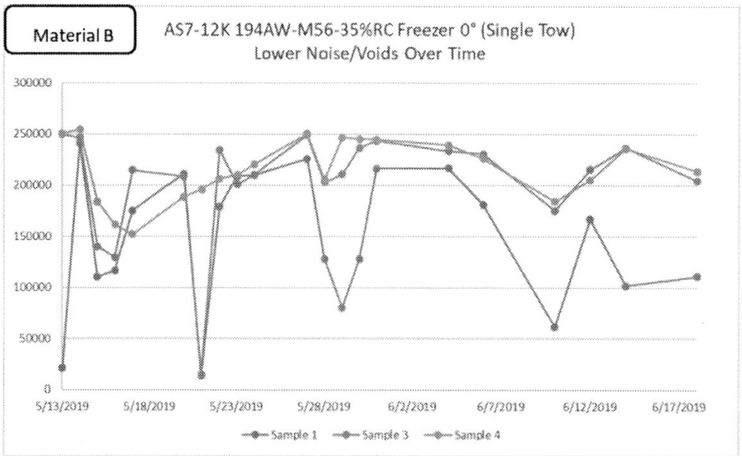

Figure 17. Void Noise for Material B
Freezer Single Tow Samples Scanned at 0°

When daily scan data is plotted based on samples' shared material type, storage condition, scanning pitch angle, and noise type, there is an obvious consistency in results. For many plots there exist days on which all samples either dramatically spike or drop in noise levels. Plots displaying such occurrences can be seen in Figure 14, Figure 15, Figure 16, and Figure 17. These plots show dramatic changes in noise values which align for most of or all samples within the set. However, these occurrences do not necessarily display an overall trend. In fact, the expected result of noise levels dropping due to the samples' ambient curing over the life of the experiment was, in most cases, not observed.

It is concluded from this that ambient air effects have an impact of noise levels. Comparing between plots shows that specific dates (or small ranges of dates) appear to have increased noise across most samples. This can be attributed to factors outside of the setup's control, such as ambient humidity temperature. This effect seems to override any one variable's effect, including storage condition. This indicates that the impact of ambient air effects seems to effect material rapidly, as samples were only removed from storage for the duration of the scans. These ambient effects could have also been responsible for the potential presence of condensation on the surface of freezer samples, impacting scanning errors.

3.3 Effect of Intensity Range

Experiments were performed to determine the effect of the laser's Intensity Range by scanning layup error samples of each material at varying levels while also observing the combined effect of pitch angle. Scans were first performed at 0° at settings between 80 and 99 in increments of 1, then at a pitch of 17.5° between settings of 11 and 99 in increments of 11, as shown in Figure 18 and Figure 19.

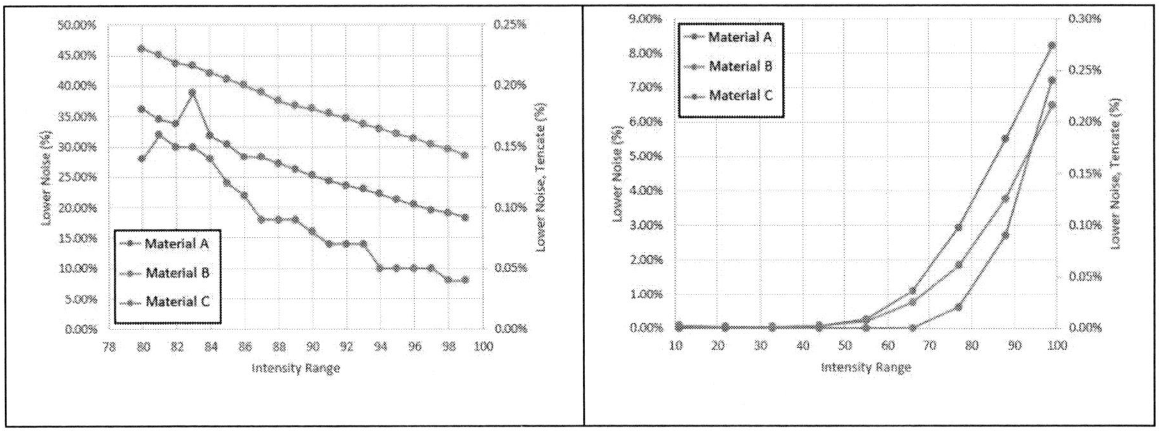

Figure 18. Void Noise as a Function of Intensity Range (Left: 0° Pitch, Right: 17.5° Pitch)

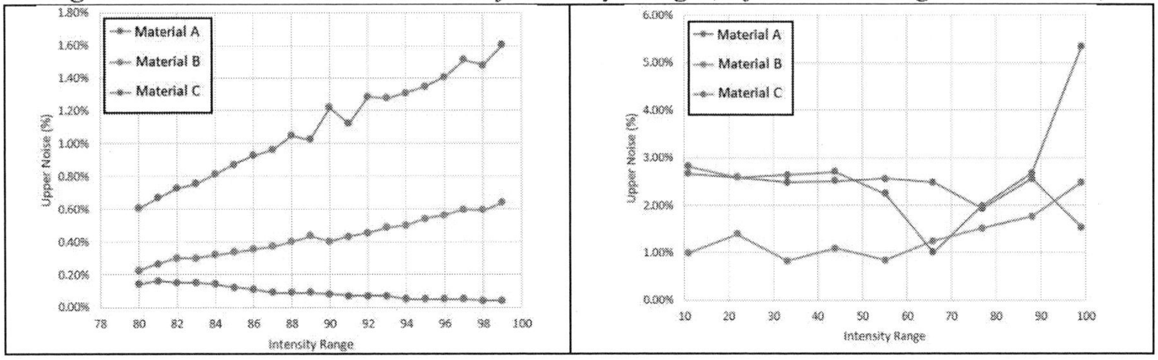

Figure 19. Spike Noise as a Function of Intensity Range (Left: 0° Pitch, Right: 17.5° Pitch)

Modifying the Intensity Range setting had varied results depending on pitch angle. Additionally, void noise and spike noise had opposite results in each trial. While scanning with a pitch of 0°, increasing the intensity range decreased void noise significantly, while increasing spike noise marginally. Conversely, while scanning with a pitch of 17.5°, increasing intensity range increased void noise yet seemed to have no consistent effect on spike noise. The results are summarized in Table 2. Additionally, in Figure 20 a series of images generated from this experiment are shown, detailing the degradation in scanning quality while increasing the Intensity Range for a sample scanned at a pitch angle of 17.5°.

Table 2. Noise Level Effects of Increasing Intensity Range per Pitch Angle

Pitch Angle	Effect on Void Noise	Effect on Spike Noise
0°	Decreased	Increased
17.5°	Increased	Inconsistent

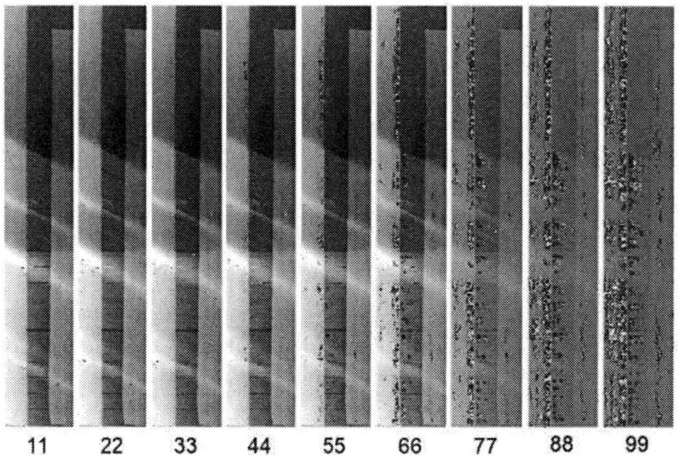

Figure 20. Images Generated at Various Intensity Ranges
Material B – Scanned at 17.5°

3.4 Scanner Setting Analysis

An investigation into the effects of the software's scanner settings on resultant noise while scanning at 0° was conducted for each material through an iterative trial and error process. Various combinations of settings were changed throughout the experiment, until visual results of the scan data improved (as generated by the Keyence software). This was performed until some "ideal settings" were realized. The default settings are presented below in Figure 21.

- **Trigger Settings.**
  Trigger Mode : Encoder
  Input Mode : 2- phase 2 times
  Sample Frequency : 1 kHz
  Pitch : 0.075.
  Pitch between triggers : Set at 0.075.
  Batch Points : 9383.
- **Common Measure Settings.**
  Operation Mode : High Speed.
  TRG : 100 µs.
  Encoder : 500 ns.
  Control Terminal : 250 µs.
- **Imaging Settings.**
  Common head Binning : On.
  CMOS Sensitivity : High Dynamics range 3
  Exposure time : 960 µs.
  Control light setting : Auto.
  Peak Detec. Sensitivity : 5(High).
  Intrpol pts inval data : 75 points.
  Peak selection : Standard (Max Peak)
  Imaging Mode : Multi Emission.
  Emission Times : (2 times).

Figure 21. Keyence Profilometer LJ-V7060 Default Settings

The experiment into scanner settings yielded consistent results across all material types investigated. For each material, in both simulated layup error and single tow samples, the default settings yielded noisy results. Although the iterative investigation was run for each material in the study, it was observed that these "ideal settings" converged to a common set, i.e. for each material, investigated while scanning at 0°, ideal setting was determined to be the same. All changed settings in this new ideal set were categorized as image settings, relating to sensitivity, exposure time, and peak selection. All settings not changed were held at values informed by the scanning software's default settings. An example of how a scan can be improved by varying the settings can be seen in Figure 22.

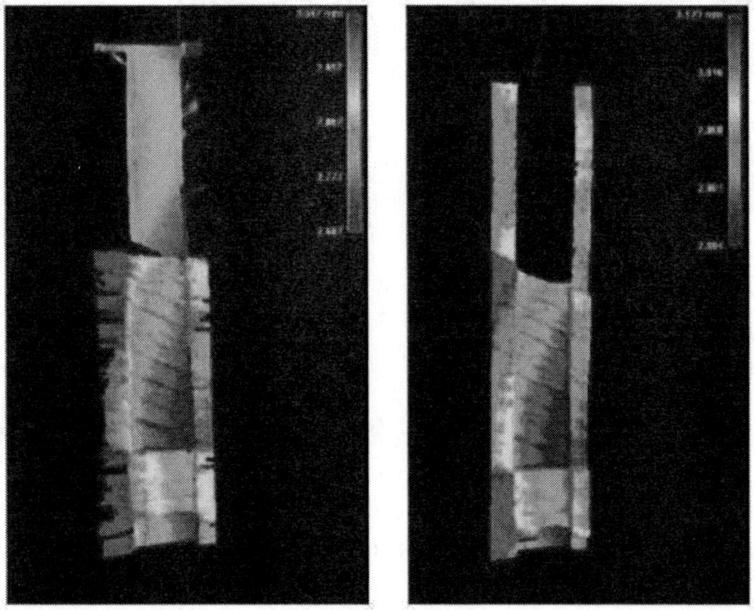

Figure 22. Results of Ideal Settings (Left: Default Settings, Right: Ideal Settings) Material A

3.5 Combined Variable Analysis

An experiment to determine the combined effects of variables was performed, this includes both scanner settings and pitch angle. In this approach, the experimentally determined "ideal settings" were tested while scanning at a pitch angle of 17.5°. The results of each study (one scanning while pitched at 0° and the other at 17.5°) could then be compared to determine if pitching the scanner and implementing the ideal settings has a combined effect. Additionally, multiple geometries of each material type were investigated. Scans were performed for each single tow samples as well as simulated layup error samples. For each material in this comparison, eight groups were analyzed, shown in Table 3.

Table 3. Combinations of Variables to be Analyzed

1. Layup Error – Default Settings 0°
2. Layup Error – Default Settings 17.5°
3. Layup Error – Ideal Settings 0°
4. Layup Error – Ideal Settings 17.5°
5. Single Tow – Default Settings 0°
6. Single Tow – Default Settings 17.5°
7. Single Tow – Ideal Settings 0°
8. Single Tow – Ideal Settings 17.5°

When samples were scanned with default settings at 0°, void noise dominated the scan, due to the reflectivity of the material. In comparison, when scans were performed with default settings and a pitch of 17.5° was introduced, the void noise was mitigated significantly, however in many cases not entirely. Additionally, spike noise was increased in many cases (typically this increase in spike noise due to an introduction of pitch angle can be observed in simulated layup error samples).

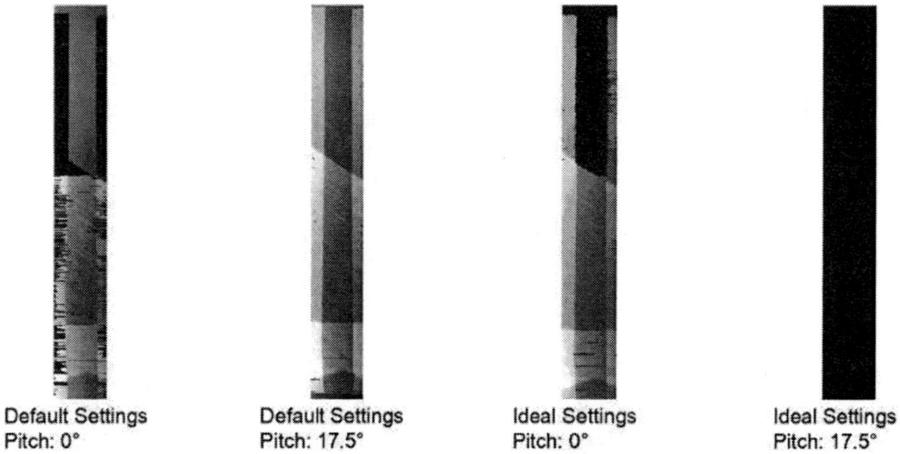

Figure 23. Script Created Images Default/Ideal Settings and 0°/17.5° Pitch – Material A

Scanning at the ideal settings had a significant improvement over default settings when scanning at 0°. Visually, all noise on the target object's surface appears to be completely mitigated for all material types (both void noise and spike noise). Results are presented in Figure 23, comparing each default and ideal settings investigated at each 0° and a pitch angle of 17.5°. However, introducing the pitch angle while scanning with "ideal" settings does not have a positive cumulative effect. In fact, utilizing the defined "ideal" settings in conjunction with a pitch of 17.5° results in unreadable scan data. This is consistent for all materials and for each single tow and simulated layup error samples.

Additionally, scanning with ideal settings at 0°, while resulting in material legibility, introduces a unique effect on scan data. When analyzing these results, it was observed that the brushed 5052H32 Aluminum backing plate does not register in the scan data. The regions that otherwise would manifest as relatively consistent values—effectively measuring the flatness of the backing plate—instead, appear similarly to pure void noise, recorded as values trending towards the lowest reportable value by the profilometer. This anomaly could not be resolved by the developed analysis script. For this reason, only a visual analysis is presented in this report.

The final observation on this topic is that utilizing the ideal settings while scanning at 0° are only a marginal improvement to scanning with default settings at 17.5°. Additionally, the previously mentioned issue of the backing plate introducing simulated noise suggests that this method of noise reduction may introduce secondary erroneous noise sources which cannot be predicted at this time. Scanning at 17.5° while utilizing the described default settings is preferred to scanning at 0° with ideal settings. However, since the two are comparable, the use of them shall likely be situationally determined (part geometry, collision prevention, feasibility, etc.).

4. CONCLUSIONS

Performance of profilometry scanning is predominately dependent on light diffuse reflection. As such, mitigation techniques improve a target's ability to reflect light into the receiver of the profilometry scanner. Equipment configuration, such as scanner orientation or settings, and material characteristics impact the success of laser topography scanning, controlled by material selection and the ambient conditions to which they are exposed. The results from the initial selection of thermoset composite materials highlight the variations in profilometry noise. The three materials selected performed differently under identical equipment configurations, highlighting that material characteristics influenced performance. Experiments proved that least noisy material yielded the least scanning error and was the most challenging to layup by hand (it was the least tacky).

Environmental conditions impact the physical and chemical properties of a thermoset composite which affects performance in profilometry scanning, mostly affected by tackiness. This is in line with common practices in AFP manufacturing in which material out time is restricted prior to laying up due to its impact of the material's ability to adhere to tooling. Scanning error was proven to be correlated to ambient environmental conditions.

Pitch angle has a significant effect of profilometry scanning error. The bisecting angle between the laser emitter and the receiver is the pitch angle which gives the best scanning results. The reduction in calculated error far surpasses the anticipated noise variance between scans of identical setup.

The effect of laser intensity range resulted in varied effects dependent on whether a pitch was introduced during scanning. The conclusions are drawn with regards to void noise, as it was consistently determined to be the dominating scanning error. When scans are performed with a pitch angle of 0°, intensity and profilometry scanning error has an inverse relationship. When scans are performed with a bisecting pitch angle present, intensity and scanning data error have a direct relationship.

The intensity of the laser signal and other combinations of scanner settings were iteratively tested and a user-defined set of "ideal settings" was determined. These "ideal" scanner settings are proven to only benefit without the presence of pitch, i.e. a pitch angle of 0°, which performed similar to ideal settings at a pitch of 17.5° with regard to scanning the surface of the thermoset tow sample. However, utilizing the "ideal settings" at 0° results in the backing plate returning meaningless data. Further investigations are needed to determine the cause(s) of this effect, and to determine how the removal of the tooling from meaningful scan data would impact the performance of the ACSIS system's ability to automatically detect or categorize layup defects. Lastly, when scanning with these "ideal settings" at a pitch angle of 17.5, the data collected was meaningless and resulted in pure signal error. Combining the ideal settings with the ideal pitch angle ultimately has a negative effect of the presence of profilometry scanning error.

5. ACKNOWLEDGEMENTS

The authors would like to thank Ingersoll Machine Tools Inc., in particular Kristopher Czaja and Piergiorgio Assandri and the South Carolina SmartState Center for Multifunctional Materials and Structures for their funding and support.

6. REFERENCES

[1] Croft, K., Lessard, L., Pasini, D., Hojjati, M., Chen, J., and Yousefpour, A., 2011, "Experimental Study of the Effect of Automated Fiber Placement Induced Defects on Performance of Composite Laminates," Compos. Part A Appl. Sci. Manuf., **42**(5), pp. 484–491.

[2] Keyence Corporation, 2013, "High-Speed 2D/3D Laser Scanner LJ-V7000 Series."

[3] Lukaszewicz, D. H. J. A., Ward, C., and Potter, K. D., 2012, "The Engineering Aspects of Automated Prepreg Layup: History, Present and Future," Compos. Part B Eng., **43**(3), pp. 997–1009.

[4] Cemenska, J., Rudberg, T., and Henscheid, M., 2015, "Automated In-Process Inspection System for AFP Machines," SAE Int. J. Aerosp., **8**(2).

[5] Fayazbakhsh, K., Arian Nik, M., Pasini, D., and Lessard, L., 2013, "Defect Layer Method to Capture Effect of Gaps and Overlaps in Variable Stiffness Laminates Made by Automated Fiber Placement," Compos. Struct., **97**, pp. 245–251.

[6] Mukhopadhyay, S., Jones, M. I., and Hallett, S. R., 2015, "Tensile Failure of Laminates Containing an Embedded Wrinkle; Numerical and Experimental Study," Compos. Part A Appl. Sci. Manuf., **77**, pp. 219–228.

[7] Mukhopadhyay, S., Jones, M. I., and Hallett, S. R., 2015, "Compressive Failure of Laminates Containing an Embedded Wrinkle; Experimental and Numerical Study," Compos. Part A Appl. Sci. Manuf., **73**, pp. 132–142.

A Minimally Intrusive Impact Detection System for Aircraft Moveable using Random Forest

Li Ai1a, Vafa Soltangharaei2a, Wout de Backer3b, Paul Ziehl4a, Michel van Tooren5b

[a] Department of Civil and Environmental Engineering, University of South Carolina, Columbia, SC, USA
[b] Department of Mechanical Engineering, University of South Carolina, Columbia, SC, USA

ABSTRACT

Impact events are of interest during the service life of commercial and military aircraft and reliably assessing the location of impact damage is important for aircraft maintenance. Traditional impact detection for aircraft is generally confined to visual inspections between the flights intermittently followed by more comprehensive and detailed inspections. The purpose of the study described herein is to investigate a path forward for the automated detection and localization of impact events. Machine learning is investigated as a means to improve upon traditional inspection methods. A flight amenable impact monitoring system was developed and utilized to detect and localize impact events on a thermoplastic aircraft elevator in a laboratory setting at the McNAIR Aerospace Center. Steel spheres were dropped from a controlled height on the elevator skin to simulate damaging events that may occur during flight. To keep weight, power, and cabling to a minimum a single sensor was attached to the spar of the elevator. A source localization approach based on random forest is proposed. Several features were extracted, and feature importance was ranked using random forest. The selected features were gathered as a dataset to train and test the performance of the proposed source localization approach. Results demonstrate the efficacy and potential of the random forest-based approach for localization of impact event monitoring for the application of a thermoplastic aircraft elevator.

Keywords: Acoustic emission, Source localization, Random forest
Corresponding author: Paul Ziehl

1. INTRODUCTION

Aircraft structures are exposed to impact damage during their service life caused by debris and hail. The primary design concern in composite structures is the resistance of layered surfaces against accidental damage such as impact. Therefore, the localization and damage quantification of damage impacts should be studied and considered to guarantee flight safety and prevent severe structural failure. Since the aircraft structural components are large scale, the visual inspection and monitoring of them are challenging and subject to human errors. A real time health monitoring system can be applied to automatically recognize and localize the impact damage instead of manually inspection.

Acoustic emission (AE) is a promising nondestructive structural health monitoring technology which can be applied on the impact monitoring of aircraft. This method is sensitive and has continuous monitoring capabilities [1-3]. It has been widely used for the detection and evaluation of composite damage [for example, 4-10]. Ono et al. [4] evaluated the damage of carbon fiber reinforced polymer (CFRP) plates under dynamic load using AE signals, Impact tests were conducted on four types of CFRP plates with a steel sphere. AE sensors were attached on both surfaces of the specimens. The results indicated that several failure modes were detected under dynamic loading in comparison to quasi-static loading. Matrix crack propagation was the dominated failure mode, while other failure modes such as delamination could not be discerned clearly. Marec et al [5] identified the damage mechanisms in polymer-based composite materials by using AE monitoring. A three-point bending test was performed on a fiber-matrix composite specimen, while two AE sensors were applied for signals collection. Unsupervised pattern recognition analysis and principal component analysis (PCA) were utilized for classification of the AE signals. The classification results performed a good correlation to the damage mechanisms of the composite specimen being monitored. Liu et al. [6] studied on the failure mechanisms and damage evolution of carbon fiber/epoxy composite laminates. The acoustic emission data during tensile test was mapping with the results that watched and analyzed by scanning electron microscope (SEM). The study demonstrated that failure modes including the splitting matrix cracking, fiber/matrix interface debonding, fiber pull-out and breakage and delamination could be represented by creating true mapping based on the data recorded by AE monitoring. Whitlow et al. [7] proposed an approach for associating in situ AE detection with final failure in continuous fiber reinforced ceramic matrix composites (CMCs). Digital image correlation (DIC) was used for obtaining surface strain measurements. The results shown there was good agreement between the two techniques. Saidane et al. [8] investigated the evaluation of damage mechanisms during tensile tests in hybrid flax-glass fibres reinforced epoxy composites. AE monitoring was conducted for providing signals during the tests. Comparing with the results observed by SEM, a conclusion was made: the cumulative of AE energy could indicate the overall failure of composite. Dia et al. [9] analyzed the characterization of damages in a hybrid laminate aluminum /glass during quasi-static and fatigue tests. Principal Components Analysis (PCA), k-means unsupervised clustering analysis, classification and regression Trees (CART) were used for damage identification. Results shown it was possible to identify damage in fibre metal laminates (FML) during both quasi-static and fatigue test. Khamedi et al. [10] identified failure mechanisms of unidirectional carbon/epoxy composites by studying the wavelet packet transform of AE signal processing. AE events were recorded during tensile test. The collected AE data was converted to wavelet for comparing with SEM observations. The results of this study pointed out that the wavelet transformed from AE waveform could link to the damage mechanisms of unidirectional carbon/epoxy composites.

However, there is limited access to attach AE sensor on the aircraft. One of the main challenges to accomplish the AE-based real time monitoring is developing a method to localize the impacts with a maximum level of certainty and using a single sensor. To solve this problem. Random forest is investigated as a mean to achieve the localization of impact events

Random forest is an ensemble algorithm. By combining multiple decision trees, the final result can be voted or averaged to make the overall model results with high accuracy and generalization performance. Random forest has an advantage that the impotence of input variables can be ranked. A features selection can be conducted based on this. Random forest has been successfully applied in the fields of faults fault diagnosis based on vibration and AE signals. Cerrada et al. [11] built a robust system for the multi-class fault diagnosis in spur gears using genetic algorithm and random forest. An acceptable diagnose accuracy was obtained based on the real vibration signals. Patel et al. [12] presented random forest classifier as an approach for classification of bearing fault and features selection. The most important features of vibration signals were selected and assigned to the random forest model. Results indicated the random forest is turn out to be a suitable approach for fault diagnosis of any rotating machine. Shevchik et al. [13] utilized AE and random forest as tool to investigate the prediction of the scuffing failure in lubricated mechanical components. A good performance was observed by using wavelet-derived features as input data. Those previous research shows using random forest might be an approach to select appropriate features and improve the passive health monitoring system.

In this study, a minimally intrusive impact detection system is explored through acoustic emission monitoring and random forest model. A single AE sensor was employed to detect and collect AE data during impact. Random forest model was utilized for source localization. An impact experiment was conducted on a thermoplastic aircraft elevator in the laboratory environment to verity the efficiency of the proposed impact detection system. The results show that the impact monitoring system using AE and random forest is reliable and has high accuracy in locating the impact area.

2. METHODOLOGY

2.1 Acoustic Emission monitoring

AE is a physical phenomenon that stress waves are generated by the rapid release of elastic energy when the object is under the condition of external force or deformation. AE signals can reflect some properties of the object. By attaching AE sensors to the surface of the object, the AE signal can be detected. The technique of collecting, analyzing and using AE signals to diagnose the status of the object is called AE nondestructive monitoring. By processing the AE signal, it can be transformed into different AE features. Commonly used AE features such as "Amplitude", "Counts", "Energy", "Rise time" and "Duration" are shown in Figure 1. In this study, A single AE sensor was employed to detect and collect AE data during impact. 15 features were extracted from the signals and formed an AE dataset. The source localization approach by using these AE features from one sensor is introduced in Section 2.2.

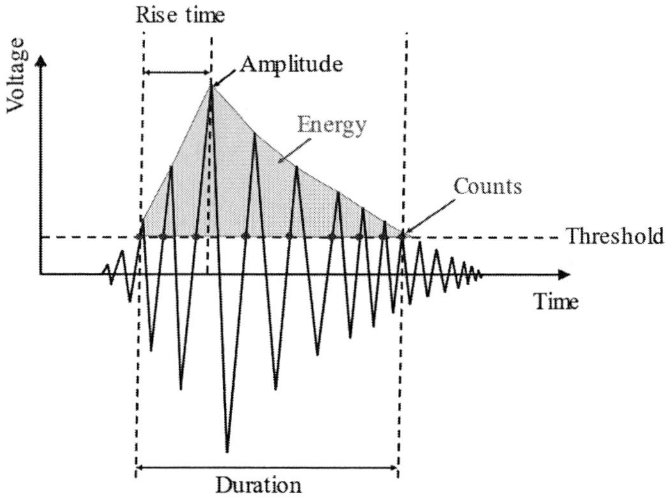

Figure 1. AE waveform and typical features

2.2 Random forest model

A random forest classification model contains 100 decision trees was utilized in this study as the approach for source localization. The AE dataset contain 15 features. They are "Amplitude", "Count", "Rise time", "Duration", "Average frequency", "Root mean square", "Average signal level", "Energy", "Absolute energy", "Peak frequency", "Reverberation frequency", "Initial frequency", "Signal strength", "Frequency centroid" and "Counts to peak". Those features were utilized for features selection. A ranking of the importance of features is given by random forest model. The features with the majority importance are gathered as input dataset to train and test the impact source localization random forest. The reason for conducting the feature selection is that: impacts occur frequently during the flight and the AE system collects a large amount of AE signals while the data storage space is limited. By reducing the dimension, the required storage can be saved for more AE collecting. The input dataset after feature selection is utilized to build 100 subsets by using bootstrapping method. Each subset is assigned to its own decision tree. These decision trees inside the random forest work independently and generate their own localization results. The final result (Zone number) is given by voting. Figure 2 shows the random forest used in this article. The output of the random forest model is the zone number of the corresponding AE event. The determination of zones is presented in Section 4.1

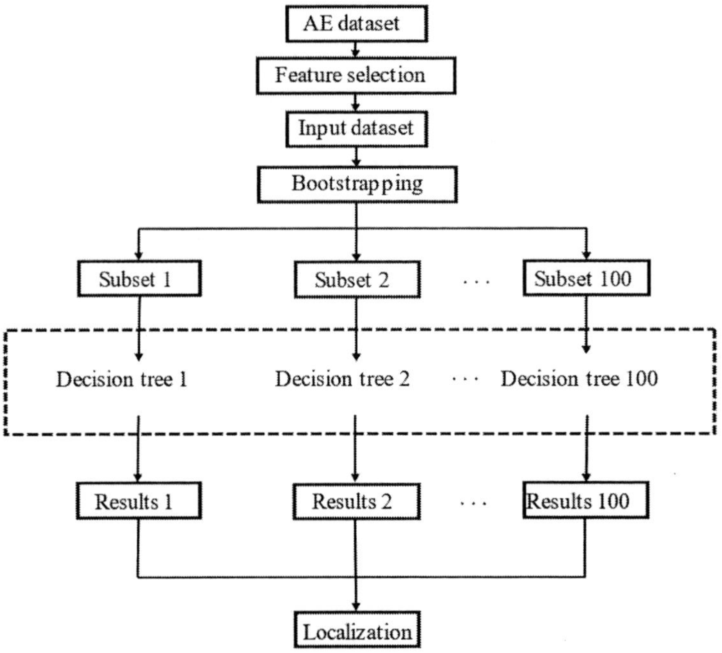

Figure 2. Configuration of random forest model

3. EXPERIMENTATION

A steel ball impact test was conducted on a real sized aircraft elevator (Figure 5) to verify the effectiveness of the source localization method using random forest. The elevator was mounted on a steel frame, made of 5-inch channels, with an overall length of 240 inches and a height of 24 inches. The hinge brackets on the elevator spar were connected to hinge points located on the frame. A turnbuckle was used to apply bending in the elevator to simulate the flexure of the horizontal tail during flight.

To collect the dataset for random forest. An impact tests were conducted using steel balls of 1/2 inch in diameter. The drop height of the steel ball is kept constant at 2 feet (Figure 3). A tube was used as a guide to control each impact's location and height.

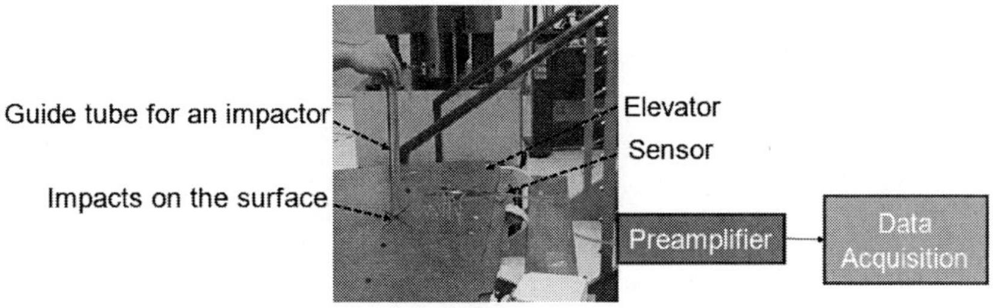

Figure 3. Steel ball dropping test

Three locations were marked on each rib. The impact locations are shown in Figure 4, marked as red points. Each location was impacted 60 times by the steel ball. An AE sensor was attached on the front spar of the elevator as shown in Figure 4.

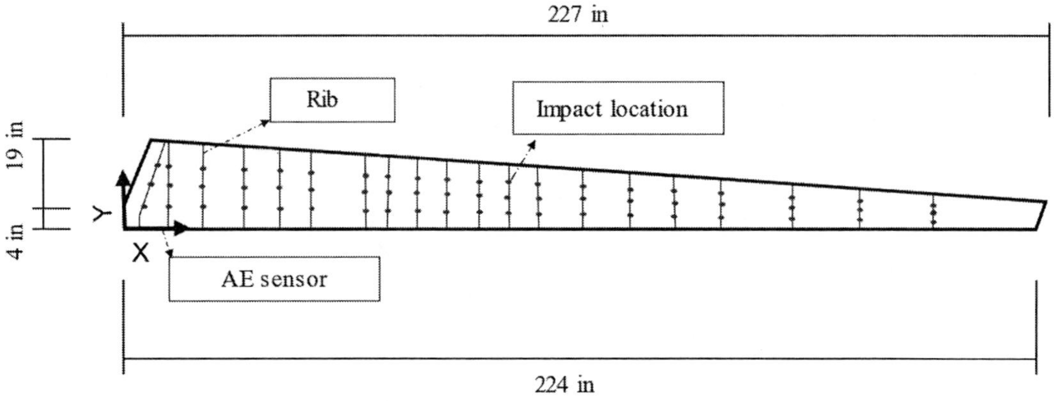

Figure 4. Impact and sensor locations

4. RESULTS

4.1 Determination of zones

As mentioned before, this study is focusing on zonal inspection. To determine the appropriate division of the zones, unsupervised pattern recognition method (k-means) was utilized for clustering the AE data. To reduce the dimension of features before sending for the unsupervised pattern recognition, a Principal component analysis (PCA) was conducted to the data. 2 principal components (PC) were utilized to present the 15 features. The data was classified into three clusters and presented in the PC space in Figure 5. This clustering was used to define the zones for the random forest model as shown in Figure 6. From left to right, the elevator is divided by zone 1, zone 2 and zone 3.

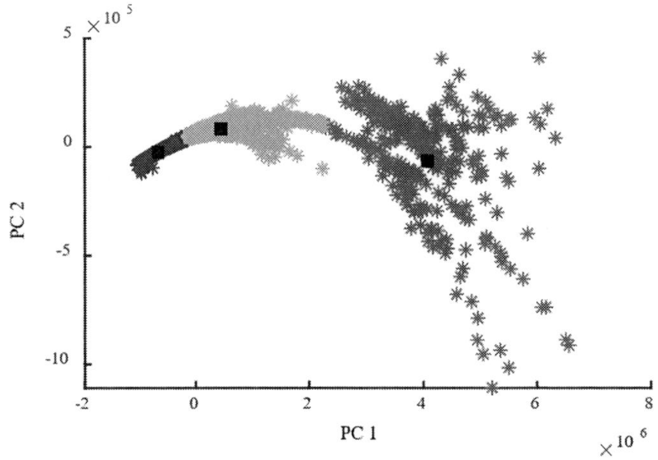

Figure 5. Clustered data in PC space

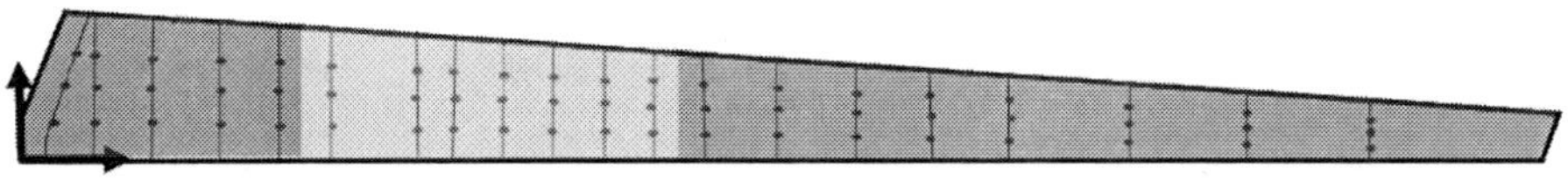

Figure 6. Zone division

4.2 Source localization

Dataset with 15 features was assigned to the random forest model as input. The ratios for training and testing dataset are 66.7% and 33.3%. The overall accuracy of impact localization is 98.33%, as shown in the confusion matrix (Figure 7). The localized accuracy of zone 1 to 3 is 98.67%, 98.33% and 98.20%. This result is compared with the case when using input dataset after feature selection. The details are shown in Section 4.3.

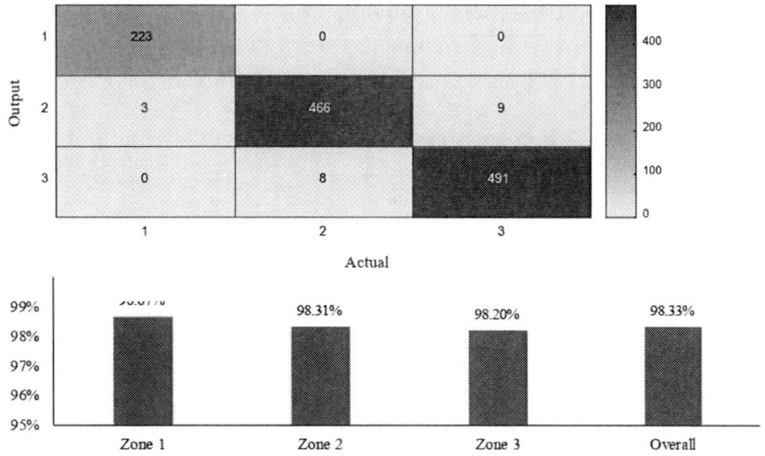

Figure 7. Accuracy of impact source localization

4.3 Features selection

By using the random forest model, this study analyzes the importance of the AE signal features. The names of these 15 features and their corresponding percentages of importance are shown in Figure 8. By observing Figure 8, it can be noticed that the features "count", "amplitude", "duration", "signal strength" and "energy" account for a large proportion of the overall importance of the feature, their cumulative importance reaches 65% out of the overall importance. Those features have a major impact on the source localization results. The importance of the rest features is relatively low. Deleting them will not have a significant impact on localization performance.

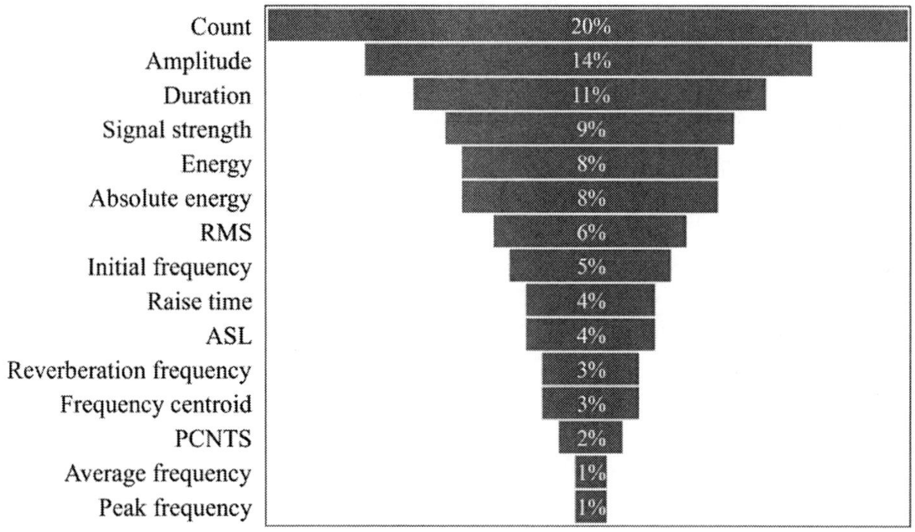

Figure 8. Ranking of the features

An input dataset made of the top 5 features was assigned to the random forest model. The localized accuracy of is 97.75%. The localization accuracy and required input storage were compared with the case when using AE data before feature selection (15 features). The detail is shown in Table 3.

It can be noticed that reducing the input dimension from 15 to 5 lead to a slightly decreasing in the accuracy of impact localization while the required storage for input decrease from 299,008 byte to 90,112 bytes. Almost 68.86% of the storage can be saved for more AE signals collection.

Table.1 Comparison of the accuracy by different input

Input	Storage (byte)	Accuracy
15 features	299,008	98.33%
5 features	90,112	97.75%

5. CONCLUSIONS

In this study, impact experiment was conducted on a real-scale thermoplastic elevator specimen. Acoustic emission monitoring was utilized to capture signals during impact. Several AE features were extracted from the acquired AE signals. Random forest model was assigned to select appropriate features and give the source localization results based on the selected features

Pertinent conclusions are:

1. By employing AE monitoring and using random forest as a source localization approach. A good performance on the impact detection and localization can be observed when a single AE sensor is used.

2. The localization accuracy decreases slightly after deleting the features with relative low importance. Meanwhile the required data storage is significantly reduced. During the flight, the AE acquisition can keep several necessary features and delete the others to save the storage space for long term monitoring.

Further work could be an investigation of the influence of the steel ball's size during the impact testing. Other advanced machine learning technique like deep learning could be investigated for the impact detection system in the future.

References:

[1] Ono K. *Acoustic emission in materials research-a review.* Journal of acoustic emission. 2011 Jan 1;29.

[2] Peters K, Wee J. *Acoustic Modes in Optical Fibers for Lamb Wave and Acoustic Emission Detection.* Journal of Acoustic Emission. 2019 Jan 1;36.

[3] Ono K. *Review on structural health evaluation with acoustic emission.* Applied Sciences. 2018 Jun;8(6):958.

[4] Ono K, Mizutani Y, Takemoto M. *Analysis of acoustic emission from impact and fracture of CFRP laminates.* Journal of Acoustic Emission. 2007 Jan 1;25(1):179-86.

[5] Marec A, Thomas JH, El Guerjouma R. *Damage characterization of polymer-based composite materials: Multivariable analysis and wavelet transform for clustering acoustic emission data.* Mechanical systems and signal processing. 2008 Aug 1;22(6):1441-64.

[6] Liu P.F, Chu J.K, Liu Y.L, Zheng J.Y. *A study on the failure mechanisms of carbon fiber/epoxy composite laminates using acoustic emission.* Materials & Design. 2012 May 1;37:228-35.
[7] Whitlow T, Jones E, Przybyla C. *In-situ damage monitoring of a SiC/SiC ceramic matrix composite using acoustic emission and digital image correlation.* Composite Structures. 2016 Dec 15;158:245-51.
[8] Saidane EH, Scida D, Assarar M, Ayad R. *Damage mechanisms assessment of hybrid flax-glass fibre composites using acoustic emission.* Composite Structures. 2017 Aug 15;174:1-1.
[9] Dia A, Dieng L, Gaillet L, Gning PB. *Damage detection of a hybrid composite laminate aluminum/glass under quasi-static and fatigue loadings by acoustic emission technique.* Heliyon. 2019 Mar 1;5(3):e01414.
[10] Khamedi R, Abdi S, Ghorbani A, Ghiami A, Erden S. *Damage characterization of carbon/epoxy composites using acoustic emission signals wavelet analysis.* Composite Interfaces. 2020 Jan 2;27(1):111-24.
[11] Cerrada M, Zurita G, Cabrera D, Sánchez RV, Artés M, Li C. *Fault diagnosis in spur gears based on genetic algorithm and random forest.* Mechanical Systems and Signal Processing. 2016 Mar 1;70:87-103.

[12] Patel RK, Giri VK. *Feature selection and classification of mechanical fault of an induction motor using random forest classifier.* Perspectives in Science. 2016 Sep 1;8:334-7.

[13] Shevchik SA, Saeidi F, Meylan B, Wasmer K. *Prediction of failure in lubricated surfaces using acoustic time–frequency features and random forest algorithm.* IEEE Transactions on Industrial Informatics. 2016 Dec 2;13(4):1541-53.

CHARACTERIZATION OF ANISOTROPIC THERMAL CONDUCTIVITY FOR BIG AREA ADDITIVE MANUFACTURING WITH POLYMERS

Artem A. Trofimov[1], Hsin Wang[1], Ahmed Arabi Hassen[1], Halil Tekinalp[2], Vlastimil Kunc[2], Seokpum Kim[2], Soydan Ozcan[2]

[1] Materials Science & Technology Division, Oak Ridge National Laboratory, Oak Ridge, TN

[2] Energy and Transportation Science Division, Oak Ridge National Laboratory, Oak Ridge, TN

ABSTRACT

Additive manufacturing with polymers has been used mainly for prototyping. A recent development of Big Area Additive Manufacturing (BAAM) at Oak Ridge National Laboratory has opened its applications in the mold and die industry. A numerical simulation and prediction for a mold heating performance requires accurate anisotropic thermal properties of the printed material, which are challenging to obtain and often requires the use of multiple techniques. The transient plane source (TPS) technique has been widely used due to its ability to measure the thermal properties of an extensive range of materials (solids, liquids, powder). Despite the capability to characterize thermal conductivity *(k)* of isotropic and anisotropic materials, the measurements of latter materials are limited to the cases, where the samples have the same thermal conductivity *(k)* along *x*- and *y*-axis that form the radial plane. In this work, the method for a characterization of *k* in all three dimensions is developed, and the application of TPS is extended to the determination of thermal properties along the *x*-, *y*-, and *z*-axis individually. The materials are represented by additively manufactured polymers including polylactic acid (PLA) and styrene maleic anhydride (SMA). The developed method consists of (1) a determination of the heat capacity of the polymers by means of TPS in combination with the model developed in this work for the data analysis procedure, (2) a machining three types of cylindrical samples from the same material, with the height corresponding either to *x*-, *y*-, or *z*-direction of printing, and (3) a determination of axial thermal conductivity employing anisotropic model and using previously determined heat capacity

Corresponding authors: A. Trofimov[1], trofimovaa@ornl.gov; H. Wang[1], wangh2@ornl.gov

1. INTRODUCTION

Additive manufacturing (AM) has become a significant part of the industry due to the rapid production of various parts [1, 2] and the accurate control of deposition [3]. The recent development of large-scale printing systems has opened new applications of additive manufacturing in the mold and die industry. For example, using Big Area Additive Manufacturing (BAAM), Oak Ridge National Laboratory (ORNL) has printed a trim tool for Boeing 777X, a hand layup tool for Shelby Cobra hood, a tool for vacuum-assisted resin transfer molding (VARTM), and a mold for a 13-meter long windmill blade [4].

Important part of AM application in the aforementioned industry is a numerical simulation and prediction under service loading for molding or tooling. As seen in Ref. [5], ORNL has developed a computational capability to predict the thermally expanded geometry of an additively manufactured in-autoclave mold and to machine the surface with the updated geometry. The accuracy of this approach highly depends on the measurement of the thermal properties. Obtaining these properties is often challenging since extruded specimen can have anisotropic thermal properties, *e.g.*, due to printing direction or reinforcement with the fibers. As a result, multiple techniques and instruments are frequently needed to measure such properties as thermal conductivity in three different directions (along x-, y-, and z-axis). One choice is to perform the measurements by means of laser flash technique [6] and differential scanning calorimetry (DSC) [7], where the former provides a thermal diffusivity of the material along the single axis parallel to the laser beam, and the latter allows to obtain the specific heat capacity. In such manner, the thermal conductivity is obtained by combining the heat capacity, the directional thermal diffusivity, and the separately measured density.

Alternatively, Transient Plane Source (TPS) technique, commonly referred as the Hot Disk method [8-11], has been one of the popular tools in recent years. TPS allows for the evaluation of thermal properties at a fast rate and using the samples as small as a few millimeters. The method has been used for a wide variety of materials including bulk solids and thin films [9, 12-14], liquids [15, 16] and powders [17]. Additionally, TPS can test both isotropic and anisotropic samples [18, 19]; however, latter requires prior knowledge of material's heat capacity as well as the same thermal conductivity in the radial direction.

In this work, the heat capacity measurements using TPS have been modified to provide more accurate results for the materials with low thermal conductivity. The TPS method has been extended to the evaluation of thermal conductivity along three individual axes on the specimen of additively manufactured polylactic acid (PLA) and styrene maleic anhydride (SMA).

2. EXPERIMENTATION

2.1 Principle of TPS technique: thermal conductivity

Thermal conductivity measurements using TPS commonly consist of the following steps: (1) the sensor, represented by a double spiral made from nickel and sealed between two thin Kapton layers, is sandwiched between two pieces of the same sample to be tested, (2) using an input power defined by the user, a constant current is passed through a sensor, and (3) simultaneously, the change of the sensor's temperature is recorded as a function of time. An experimental setup is illustrated in Figure 1.

The sensor is a resistor/thermometer and acts both as a heater and as a temperature sensor. If the tested samples are isotropic then the temperature increase of the sensor is related to the thermal properties of the surrounding material as [20]:

$$\Delta T(\tau) = P_0(\pi^{3/2}rk)^{-1}D(\tau) \quad (1)$$

where k is the thermal conductivity of the tested material, P_0 is the input power, $\Delta T(\tau)$ is the time-dependent temperature increase of the TPS sensor, $\tau = (t/\theta)^{1/2}$, t is the time from the beginning of the transient measurement, $\theta = r^2/\alpha$ is the characteristic time, r is the radius of the sensor, α is the thermal diffusivity of the tested material, and $D(\tau)$ is a dimensionless time function that takes into account the conducting pattern of the disk-shaped sensor consisting of n number of concentric rings of nickel; n depends on the used sensor, and $D(\tau)$ is defined elsewhere [8, 20].

Using θ as an adjustable parameter, experimental data can be linearly fitted to Eq. (1) when $\Delta T(\tau) \sim D(\tau)$, which allows to obtain the thermal diffusivity α of the sample and the value of the slope, $P_0(\pi^{3/2}rk)^{-1}$, which is used to extract a thermal conductivity k.

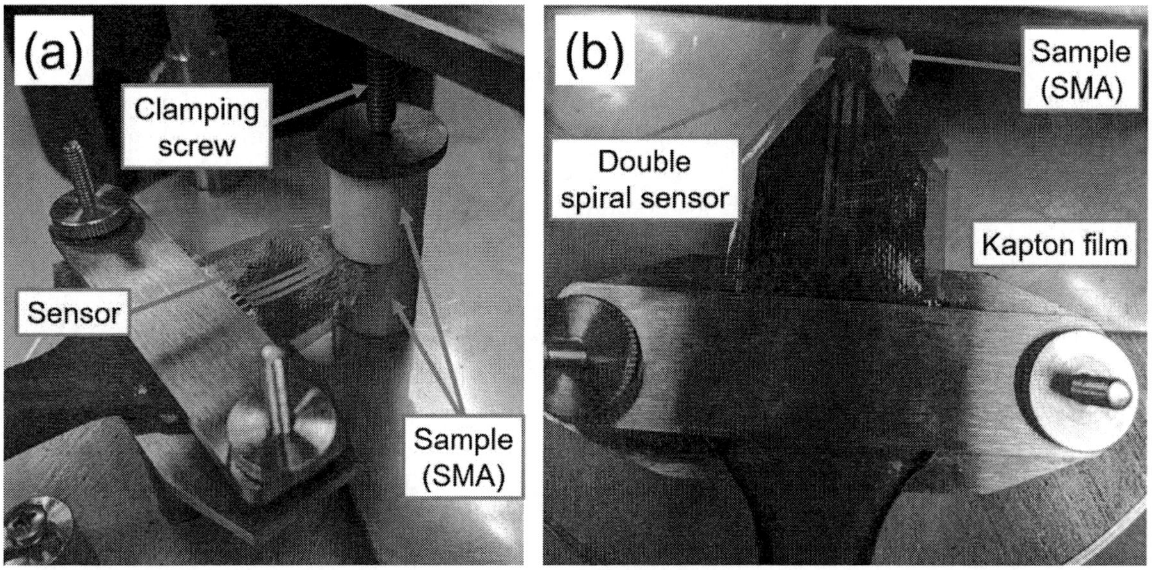

Figure 1. (a) TPS set-up depicting the sensor positioned between two reference samples forming a sample-sensor-sample arrangement. Brass free weight is used to ensure a good contact between the sensor and the samples. (b) Double spiral sensor sealed by Kapton.

In the case of anisotropic material, where the thermal properties are the same along the two axes forming a radial plane (*e.g.*, x- and y-axis) but different from those along the third axis (*e.g.*, z-axis), the performed experiments are the same, but the fitting equation is

$$\Delta T(\tau_x) = P_0[\pi^{3/2}r(k_x k_z)^{1/2}]^{-1}D(\tau_x)\,, \quad (2)$$

where $k_x = k_y \neq k_z$, $\tau_x = (t/\theta_x)^{1/2}$, and $\theta_x = r^2/\alpha_x$. Accordingly, θ_x is used as a fitting parameter to calculate thermal diffusivity α_x along the x-axis. Next, the previously determined density, ρ, and the heat capacity, C_p, are used to calculate $k_x = \alpha_x \rho C_p$, and the axial thermal conductivity, k_z, is obtained from the determined k_x and the slope of the line described in Eq. (2) [18, 19].

The situation of anisotropic material with $k_x \neq k_y \neq k_z$ is subject of this investigation and, therefore, is discussed in the Results section.

2.2 Principle of TPS technique: heat capacity

In the specific heat capacity experiments, a TPS sensor is attached to the bottom side of the high conductivity sample holder with a diameter of 20 mm and a height of 5 mm. Evaluation of the heat capacity consists of at least two measurements: reference measurement in order to acquire transient heating profile of the sample holder, and the sample+reference measurement (sample holder with the tested material placed inside). The method is based on the assumption of constant temperature gradients inside the holder-sample assembly, neglecting the heat losses to the surrounding. Comparative analysis of two transient profiles (reference versus sample+reference) allows evaluation of sample's heat capacity [21]:

$$mC_p = P_2/\delta_2 - P_1/\delta_1 \,, \tag{3}$$

where P_1 and P_2 are the input power for the reference and sample+reference measurements, respectively, and $\delta = \mathrm{d}T/\mathrm{d}t$ is the time derivative of the temperature within the analyzed segment of the transient curve.

2.3 Materials and Methods

Neat polylactic acid (PLA) and styrene maleic anhydride (SMA) were used as the test materials. A part with a hexagon shape was printed for each material. The samples were prepared from the hexagons by cutting and machining. The samples for the thermal conductivity experiments had a cylindrical shape with the dimensions of 20 mm in diameter and 20 mm in height, where the height corresponded either to *x*-, or *y*-, or *z*-directions of the printing. The samples for the heat capacity measurement had a disk shape with the dimensions of 20 mm in diameter and 4 mm in height. Three groups of samples were prepared to repeat the same test three times for a statistical analysis. For each group and each material, six samples were prepared for the conductivity measurement (two samples for each direction) and one sample was prepared for heat capacity measurement (= 7 samples for each test group). Since we are testing three times for two materials, the total number of samples prepared are 42 samples (=7×3×2). Figure 2(a-b) shows the shapes and the dimensions of the samples, and Figure 2(c-d) shows the actual appearance of the samples.

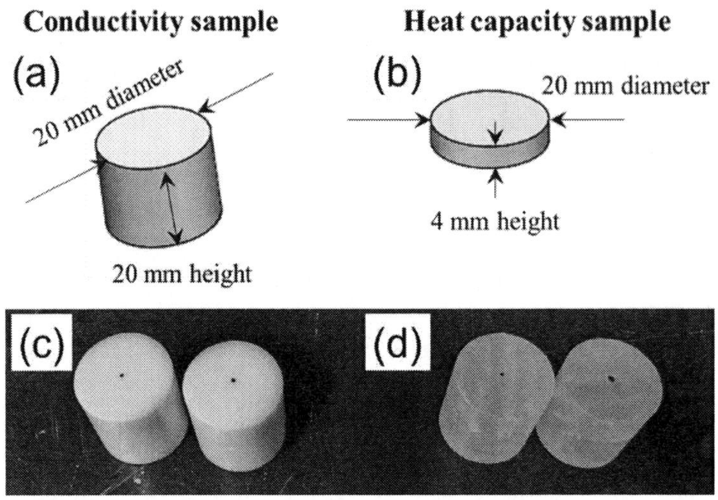

Figure 2: Sample shape and dimensions for (a) thermal conductivity measurement and (b) heat capacity measurement. (c) Styrene maleic anhydride (SMA) and (d) polylactic acid (PLA).

The heat capacity and the thermal conductivity of the investigated materials were measured by a hot disk thermal constant analyzer (Hot Disk Inc., Sweden). A TPS 3500 Hot Disk system with a bridge circuit, a digital voltmeter, and a data analysis module was used.

The heat capacity experiments were sensitive to the change of the surrounding temperature and, in order to maintain constant initial temperature of the holder-sample assembly, it was placed inside the temperature platform (see Figure 3). The input power was 160-410 mW, and the duration of the measurements was 40-160 sec. Two hours of waiting time was allowed between the measurements in order to establish reproducible isothermal conditions at the start of the measurements. The analysis procedure of the data to extract the heat capacity was modified as part of this work, and, therefore, is discussed in Results section. MACOR glass ceramic with the known heat capacity was used as a reference in order to establish the procedure.

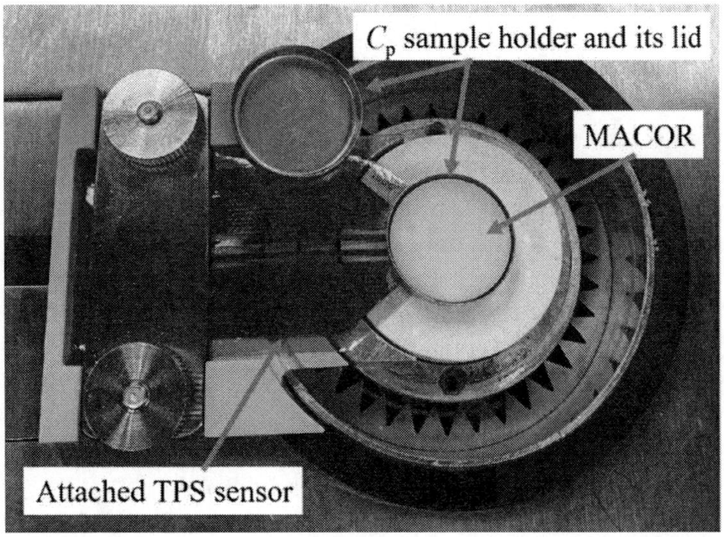

Figure 3. Heat capacity holder-sample (MACOR) assembly with the sensor attached to the bottom of the holder and placed inside the temperature platform. During experiments, the lid is closed, and the assembly is covered with thermal isolation to provide constant initial temperature

The thermal conductivity measurements were performed using anisotropic module and the Hot Disk sensors with the radius of 3.189 mm. The input power was 30-35 mW, and the duration of the measurement was 40-80 sec. Thirty minutes of waiting time was allowed between the measurements in order to establish the reproducible isothermal conditions. The measurements were repeated three times for each set of parameters. The experiments were performed at room temperature and ambient conditions under stainless steel cover to avoid temperature fluctuations from drafts of air to the sample.

3. RESULTS

3.1 Specific heat capacity

Comparative analysis of two transient heating profiles (sample holder alone and holder-sample assembly) allows evaluation of sample's heat capacity. Analysis procedure requires to establish constant temperature gradients inside the holder and sample assembly during the experiment and analyze corresponding portions of the experimental curves. During the experiment, 200 datapoints are recorded regardless the measurement duration, and the standard procedure recommends analyzing last 100 points, as shown in Figure 4a, assuming that the sample has a uniform temperature during this time window. Nevertheless, this assumption is often invalid, especially for low conductive materials, due to slow heat propagation and increasing role of heat loss for long measuring time. As a result, the window of last 100 points does not always correspond to the portion of the transient profile with the constant temperature gradients and can yield erroneous or even negative heat capacity values.

Adl-Zarrabi *et al* [22] suggested to analyze the data in a variety of windows, where the data point intervals are defined as [A; 2A], [A+2; 2(A+2)], *etc.*, where A is the first selected datapoint (for example, points [10;20], [12;24], …, [98-196], [100-200]). The results should give a curve with a maximum value corresponding to the specific heat of the analyzed material (Figure 4b). The presence of aforementioned heat capacity curve can be assigned to the fact that during earlier time windows the calculated heat capacity increases because the sample has not reached a uniform temperature, while at later times the heat loss starts to dominate the measurement and the heat capacity decreases. While, the procedure might give the correct value in some cases, it might also overestimate the heat capacity as shown in Figure 4b on the example of reference MACOR sample.

Therefore, in addition to the aforementioned improvement, the conditions of the analysis need to be modified even further. Besides the analysis of the variety of windows, it should be equal time windows (*e.g.*, every five seconds), and one needs to monitor the temperature increase of both the sample holder, ΔT_{ref}, and the holder-sample assembly, $\Delta T_{\mathrm{sample}}$, within the time window of analysis. The portion of the transient profiles, where both assemblies have similar temperature increase (*i.e.*, $\Delta T_{\mathrm{ref}} - \Delta T_{\mathrm{sample}} \rightarrow 0$) corresponds not only to the time window where their temperature gradients are constant, but also to the window where the gradients are the same, providing accurate heat capacity evaluation (Figure 4c).

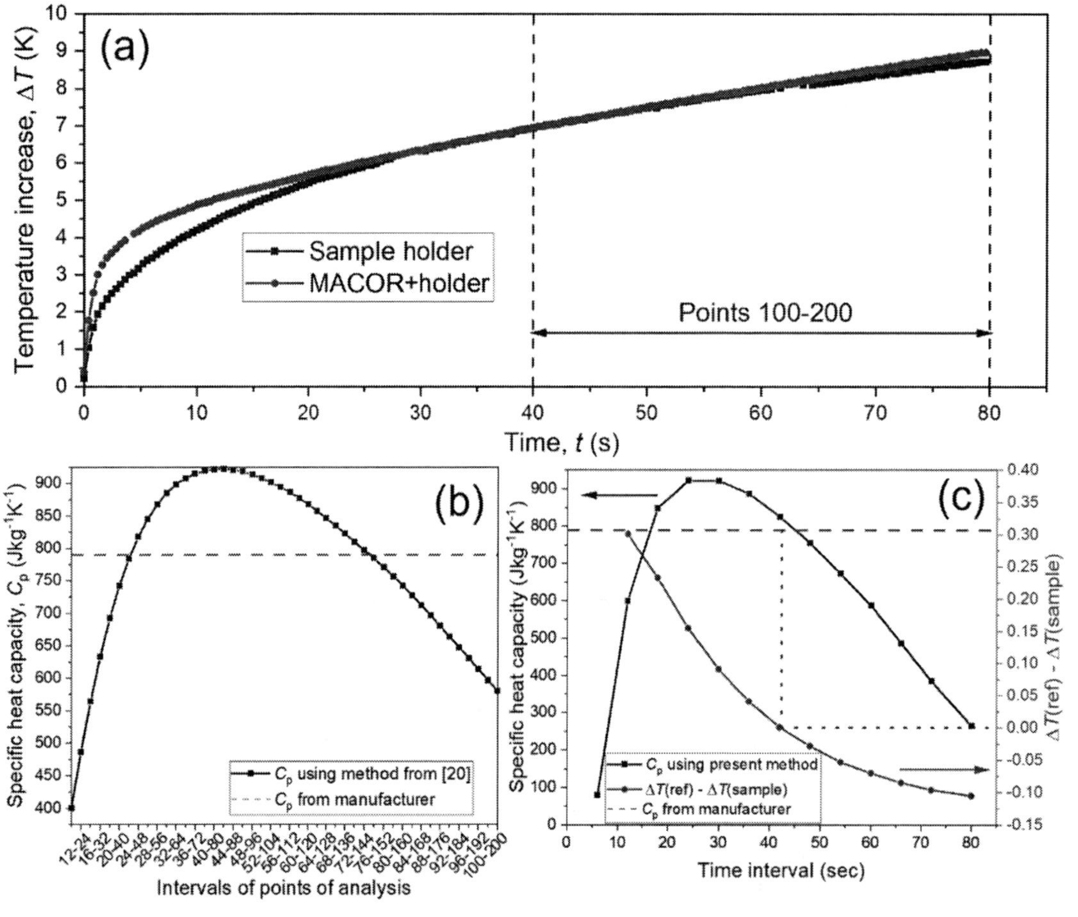

Figure 4. (a) Transient heating profiles of the sample holder and holder-sample (MACOR) assembly during the heat capacity measurements for 80 s using TPS. (b) Specific heat capacity curve obtained using the method from [22]. (c) Specific heat capacity curve obtained using the

method presented in this work, and the corresponding difference between the temperature increase of the sample holder, ΔT_{ref}, and the holder-sample assembly, ΔT_{sample}. Time intervals in (c) were 5 sec each and the x-axis values correspond to the end of the given interval.

Using proposed analysis, specific heat capacity of MACOR reference and two additively manufactured polymers (PLA and SMA) were evaluated as shown in Figure 5a. Results for the MACOR showed a good agreement with the reported value by manufacturing company. The specific heat capacity of PLA was within the range of literature values; however, it should be noted that the reports on C_p of PLA have a significant variation from 1300 J/kgK to 1800 J/kgK [23-25]. Specific heat capacity of SMA was not available in the literature to the best of authors' knowledge.

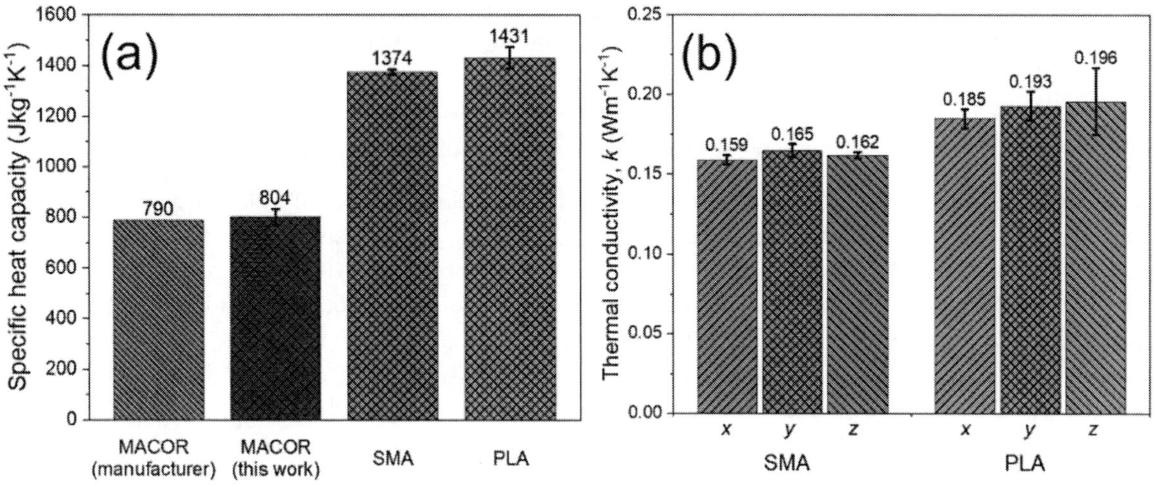

Figure 5. (a) Specific heat capacity of MACOR, SMA, and PLA obtained using the analysis proposed in this work. (b) Thermal conductivity of SMA and PLA for x-, y-, and z-axis.

3.2 Thermal conductivity

Initially, the attempts were made to obtain a thermal conductivity in three distinct directions (x-, y-, and z-axis) using one-dimensional measurements with TPS. Three types of samples of the same material were cut in the shape of cylindrical rod having the direction of the sample's length along x-, y-, or z-axis and the diameter of the samples closely matching the diameter of the used TPS sensor. Similar diameters of the sensor and the samples are necessary to establish a one-dimensional heat propagation in the direction of the rod length, and undisturbed air around the samples is typically adequate to minimize heat flow and heat loss in the radial direction. However, it was found that low thermal conductivity of the investigated polymers required longer than recommended measuring time preventing performance of the measurements without the influence of heat loss.

Consequently, the strategy was changed to the use of the TPS anisotropic model, which normally allows the evaluation of two types of thermal conductivity: radial and axial. In the present case, the thermal conductivity was assumed to be different in all three directions, which would result in the erroneous values of radial conductivity, but it would still allow the evaluation of axial thermal conductivity, as long as the heat propagation was limited to within the sample. Therefore, three types of samples having axial direction cut along *x*-, *y*-, or *z*-axis were still required. Further steps of the method are similar to the standard TPS method described in Experimentation section, where the heat capacity was measured directly with TPS (Figure 5a). The transient profile was fitted to the anisotropic model in order to calculate the thermal properties of the materials, and only values of axial thermal conductivity, shown in Figure 5b, were used as a final result.

The thermal conductivity of SMA, averaged over three axes, was $k_{SMA} = 0.163 \pm 0.004$ W/mK, similar to the reported value of 0.170 W/mK [26, 27], and the thermal conductivity of PLA averaged over three axes was $k_{PLA} = 0.191 \pm 0.014$ W/mK, which was within the reported range of 0.130-0.210 W/mK [23, 26, 28]. Additionally, the thermal conductivities of each polymer were close between three different axes (Figure 5b), indicating that the printing direction did not have a strong influence on the thermal conductivity.

4. SUMMARY AND CONCLUSIONS

This work presented a modified data analysis of the Transient Plane Source (TPS) heat capacity experiments in order to accurately evaluate the heat capacity of the tested materials and established a procedure to obtain the thermal conductivity in three different directions using anisotropic model of the TPS technique. In order to accurately evaluate the C_p value, the standard heat capacity experiment should follow by the analysis of the variety of equal time windows with simultaneous monitoring of the temperature increase of both the sample holder, ΔT_{ref}, and the holder-sample assembly, ΔT_{sample}, within the time window of analysis. It was shown on the example of a commercial MACOR glass-ceramic that the time window where $[\Delta T_{ref} - \Delta T_{sample} \rightarrow 0]$ should yield an accurate heat capacity value. The heat capacity of additively manufacture polylactic acid (PLA) and styrene maleic anhydride (SMA) was also evaluated and used for the thermal conductivity measurements. The developed method to extract the thermal conductivity in three different directions included (1) a determination of the heat capacity as described above, (2) a machining of three types of cylindrical samples from the same material, with the height corresponding either to *x*-, or *y*-, or *z*-direction of printing, and (3) a determination of the axial thermal conductivity using anisotropic model. The results were in an agreement with the literature and showed that the printing direction did not have a strong influence on the thermal conductivity of SMA and PLA.

Acknowledgment

This research is sponsored by the US Department of Energy (DOE), Office of Energy Efficiency and Renewable Energy, Advanced Manufacturing Office, under contract DE-AC05-00OR22725 with UT-Battelle LLC. For large-scale additive manufacturing, the printing equipment was provided by Cincinnati Incorporated, a manufacturer of metal and additive manufacturing equipment, headquartered in Harrison, Ohio (www.e-ci.com). The printing material for large-scale additive manufacturing was provided by Techmer PM, a material design and manufacture company headquartered in Clinton, TN.

References

[1] S.H. Ahn, M. Montero, D. Odell, S. Roundy, K. Wright Paul, Anisotropic material properties of fused deposition modeling ABS, Rapid Prototyping Journal, 8 (2002) 248-257.

[2] V. Kunc, A.A. Hassen, J. Lindahl, S. Kim, B. Post, L. Love, Large scale additively manufactured tooling for composites, 15th Japan International SAMPE Symposium and Exhibition, Tokyo, Japan, 2017.

[3] J. Scott, N. Gupta, C. Wember, S. Newsom, T. Wohlers, T. Caffrey, Additive Manufacturing: Status and Opportunities, Washington DC, 2012, pp. 1-29.

[4] B. Post, P. Chesser, A. Roschli, L. Love, K. Gaul, Large-Scale Additive Manufacturing for Low Cost Small-Scale Wind Turbine Manufacturing, Oak Ridge National Laboratory, Oak Ridge, TN, 2018.

[5] A. Hassen, A. Lambert, J. Lindahl, D. Hoskins, C. Duty, S. Simunovic, C. Chin, V. Oancea, L. Love, V. Kunc, S. Kim, Simulation Assisted Design for an Additively Manufactured Autoclave Tool Accounting for an Anisotropic Expansion, in: The Composites and Advanced Materials Expo (CAMX 2019), Anaheim, CA, 2019.

[6] ASTM E1461-07, Standard Test Method for Thermal Diffusivity by the Flash Method, ASTM International, West Conshohocken, PA, 2007.

[7] ASTM E1269-11, Standard Test Method for Determining Specific Heat Capacity by Differential Scanning Calorimetry, ASTM International, West Conshohocken, PA, 2018.

[8] S.E. Gustafsson, Transient plane source techniques for thermal conductivity and thermal diffusivity measurements of solid materials, Review of Scientific Instruments, 62 (1991) 797-804.

[9] M. Gustavsson, E. Karawacki, S.E. Gustafsson, Thermal conductivity, thermal diffusivity, and specific heat of thin samples from transient measurements with hot disk sensors, Review of Scientific Instruments, 65 (1994) 3856-3859.

[10] T. Log, S.E. Gustafsson, Transient plane source (TPS) technique for measuring thermal transport properties of building materials, Fire and Materials, 19 (1995) 43-49.

[11] V. Bohac, M.K. Gustavsson, L. Kubicar, S.E. Gustafsson, Parameter estimations for measurements of thermal transport properties with the hot disk thermal constants analyzer, Review of Scientific Instruments, 71 (2000) 2452-2455.

[12] H. Zhang, M.-J. Li, W.-Z. Fang, D. Dan, Z.-Y. Li, W.-Q. Tao, A numerical study on the theoretical accuracy of film thermal conductivity using transient plane source method, Applied Thermal Engineering, 72 (2014) 62-69.

[13] M. Ahadi, M. Andisheh-Tadbir, M. Tam, M. Bahrami, An improved transient plane source method for measuring thermal conductivity of thin films: Deconvoluting thermal contact resistance, International Journal of Heat and Mass Transfer, 96 (2016) 371-380.

[14] S.A. Al-Ajlan, Measurements of thermal properties of insulation materials by using transient plane source technique, Applied Thermal Engineering, 26 (2006) 2184-2191.

[15] R.J. Warzoha, A.S. Fleischer, Determining the thermal conductivity of liquids using the transient hot disk method. Part I: Establishing transient thermal-fluid constraints, International Journal of Heat and Mass Transfer, 71 (2014) 779-789.

[16] R.J. Warzoha, A.S. Fleischer, Determining the thermal conductivity of liquids using the transient hot disk method. Part II: Establishing an accurate and repeatable experimental methodology, International Journal of Heat and Mass Transfer, 71 (2014) 790-807.

[17] S. Flueckiger, T. Voskuilen, Y. Zheng, T.e. Pourpoint, Advanced Transient Plane Source Method for the Measurement of Thermal Properties of High Pressure Metal Hydrides, in ASME

2009 Heat Transfer Summer Conference collocated with the InterPACK09 and 3rd Energy Sustainability Conferences, Vol. 1: Heat Transfer in Energy Systems; Thermophysical Properties; Heat Transfer Equipment; Heat Transfer in Electronic Equipment, 2009, pp. 215-221.
[18] M. Gustavsson, S.E. Gustafsson, On the use of Transient Plane Source sensors for studying materials with direction dependent properties in: R. Dinwiddie, R. Mannello (eds.) 26th International Thermal Conductivity Conference/14th International Thermal Expansion Symposium, Vol. 26, Cambridge, MA, 2005, pp. 367-377.
[19] ISO 22007-2:2015, Plastics — Determination of thermal conductivity and thermal diffusivity — Part 2: Transient plane heat source (hot disc) method, International Organization for Standardization, Switzerland, 2015.
[20] Y. He, Rapid thermal conductivity measurement with a hot disk sensor: Part 1. Theoretical considerations, Thermochimica Acta, 436 (2005) 122-129.
[21] M. Gustavsson, N.S. Saxena, E. Karawacki, S.E. Gustafsson, Specific heat measurements with the Hot Disk thermal constants analyser, in: K.E. Wilkes, R.B. Dinwiddie, R.S. Graves (eds.) Thermal conductivity 23, Interantional thermal conductivity conference, Nashville, TN, 1995, pp. 56-65.
[22] A. Berge, B. Adl-Zarrabi, C.-E. Hagentoft, Determination of specific heat capacity by transient plane source, Frontiers of Architectural Research, 2 (2013) 476-482.
[23] S. Farah, D.G. Anderson, R. Langer, Physical and mechanical properties of PLA, and their functions in widespread applications — A comprehensive review, Advanced Drug Delivery Reviews, 107 (2016) 367-392.
[24] Technical data sheet available at <https://polymerdatabase.com/polymers/>.
[25] M. Pyda, R.C. Bopp, B. Wunderlich, Heat capacity of poly(lactic acid), The Journal of Chemical Thermodynamics, 36 (2004) 731-742.
[26] Technical data sheet available at <https://omnexus.specialchem.com/polymer-properties/properties/thermal-insulation#PS-X>.
[27] A. Sarı, A. Biçer, C. Alkan, Poly(styrene-co-maleic anhydride)-graft-fatty acids as novel solid–solid PCMs for thermal energy storage, Polymer Engineering & Science, 59 (2019) E337-E347.
[28] R. Guo, Z. Ren, H. Bi, M. Xu, L. Cai, Electrical and Thermal Conductivity of Polylactic Acid (PLA)-Based Biocomposites by Incorporation of Nano-Graphite Fabricated with Fused Deposition Modeling, Polymers, 11 (2019).

THERMOPLASTIC UNI-TAPE QUALITY INSPECTION

Joseph P. Heil[1], Andi C. Meyer[1], Waruna P. Senevirante[2], Brandon L. Saathoff[2]
[1]Spirit AeroSystems, Inc
[2]National Institute for Aviation Research- Wichita State University

Wichita, Kansas USA

ABSTRACT

Recently, high performance thermoplastic resins from the Poly Ether Ketone family have been integrated into carbon fiber unidirectional prepreg (uni-tape) formats for use in aerospace manufacturing. However, the quality level of thermoplastic uni-tape is much more influential on processing and final part quality compared to thermoset uni-tape. The primary goal of this research is to document thermoplastic uni-tape quality from different suppliers and determine the impact these characteristics have on the ability to fabricate high-quality high-performance structural parts per a defined process. Six material systems derived from five suppliers and three polymer systems are characterized. Parts will be fabricated using oven consolidation, autoclave consolidation, and stamp forming. Differences in void content and crystallinity are expected as a result of the varying pressure and thermal history. Results from this work will support future processing methods such as automated fiber placement (AFP) in-situ consolidation.

1. INTRODUCTION

Thermoplastic carbon fiber composites (thermoplastics) are attracting interest in the aerospace industry as the next big advancement in composite materials. Thermoplastics offer short cycle times, opportunities for fastener-less joining, as well as increased toughness and environmental resistance compared to thermosets. New material systems using PolyEtherKetoneKetone (PEKK) or PolyArylEtherKetone (PAEK) as the polymer matrix have been introduced that melt at a lower temperature but have similar mechanical properties to PolyEtherEtherKetone (PEEK). These polymer matrices are supplied in unidirectional pre-pregs, referred to as uni-tape, much the same way as thermoset materials are. Uni-tapes are well suited for automated manufacturing and enable highly tailorable composite designs. However, much is still to be learned about their manufacture affects quality and consequently how quality affects processability and the mechanical performance. A variety of industrial and academic publications have shown that not all uni-tapes are created equal [1] [2]. Similarly, the level of variability and defects in these uni-tapes are not well established. Given the recent introduction of these materials the assumption is their quality is not as high as thermoset uni-tapes. An overview of the properties of thermoplastic uni-tapes is given below with an explanation of why or how the property is relevant. Thickness, fiber and resin distribution, and void content are specific measures of quality while viscosity and degree of crystallinity also have significant impacts on the manufacturability of thermoplastics.

1.1 Thickness

Uni-tape needs to have a consistent thickness for design and manufacturing purposes. Having thick and thin spots can create fiber waviness in the composite cross section and lead to resin pockets or voids. To take advantage of thermoplastics needing minutes instead of hours to form, using autoclaves to press out any inconsistencies in material thickness is undesirable.

1.2 Fiber and Resin Distribution

Both the overall fiber content and how well the fiber is dispersed throughout the resin matrix are important. Due to the high melt viscosity of thermoplastics, the impregnation techniques developed for thermosets need to be modified when using thermoplastic resins. The pre-pregging process should be carefully controlled to ensure there are not resin pockets or clumps of fibers where resin has not infiltrated. The combination of how evenly fibers and resin are distributed and where any resin rich areas are in the uni-tape seems to have an effect on how well the material can be fed through machinery or how entrapped air is removed during the consolidation process [3].

1.3 Void Content

Voids reduce the mechanical properties of composites; an acceptable level of porosity for thermoplastics has yet to be set. If voids are in the form of entrapped air, then when the material is heated the void could grow considerably causing deconsolidation of the composite. Depending on the fabrication process void volume is reduced by compressing and or removing voids. Comparing the void content of as received material versus consolidated laminates will give insight into the ability of each fabrications process to reduce voids for a given material.

1.4 Crystallinity

The high performance engineering thermoplastics discussed in this paper are semi-crystalline. The crystallinity of these polymers influences their mechanical strength and in part provides the superior environmental and chemical resistance associated with thermoplastics. While high crystallinity is desirable for a part, high crystallinity slows down the interfacial welding of polymer chains that occurs during consolidation [4]. Balancing these competing interests will be an ongoing challenge for material suppliers. Crystallinity is a complex phenomenon as it is path dependent; not only is there the degree of crystallinity to consider but also the shape and size of the crystals. Depending on the application, targets are set for the level of crystallinity and/or mechanical properties. Therefore, the process and the polymer used must be selected in conjunction such that the thermal cycles selected for manufacturing rate must be within the temperature range and cooling rates that the polymer can accommodate and still achieve sufficient crystallinity.

2. EXPERIMENTATION

All materials were received in uni-tape format and are described in Table 1, below. The materials came from multiple suppliers and covers PEKK, PAEK, and Polyphenylene Sulfide (PPS) resin systems. Thickness variation, resin and fiber distribution, and void content were selected as attributes describing quality.

Table 1. Material Overview

Product Name	Tg (°C)	Tm (°C)	Consolidation Temperature (°C)
PEKK-1	160	337	371- 400
PEKK-2	160	337	377
PEKK-3	160	337	371 – 385
PAEK-1	147	305	320 – 382
PAEK-2	147	581	300 - 400
PPS-1	90	280	330

2.1 Optical Microscopy Image Analysis

In order to investigate the thickness variation, void content, and resin-fiber distribution of the uni-tape, 9 samples were removed from each roll of material. This consisted of 3 locations along the width (left, center, right) and 3 locations along the length of the roll (Figure 1). In most cases, the materials were acquired with 305 mm (12") wide widths, with the exception of one material (PAEK-2). Samples were removed with a width of 38 mm (1.5") and potted with the fibers oriented perpendicular to the surface (Figure 2). These various locations were used to ensure the quality was assessed at multiple locations, and a single area of the material was not used to draw concrete conclusions. After potting and polishing, thickness measurements were taken at every 0.1" from the left edge using Keyence VKX-1000 laser confocal microscope. Image analysis procedures using Fuji Image J were used to measure fiber and void content and thickness variation while the overall resin and fiber distribution was qualitatively assessed using the images.

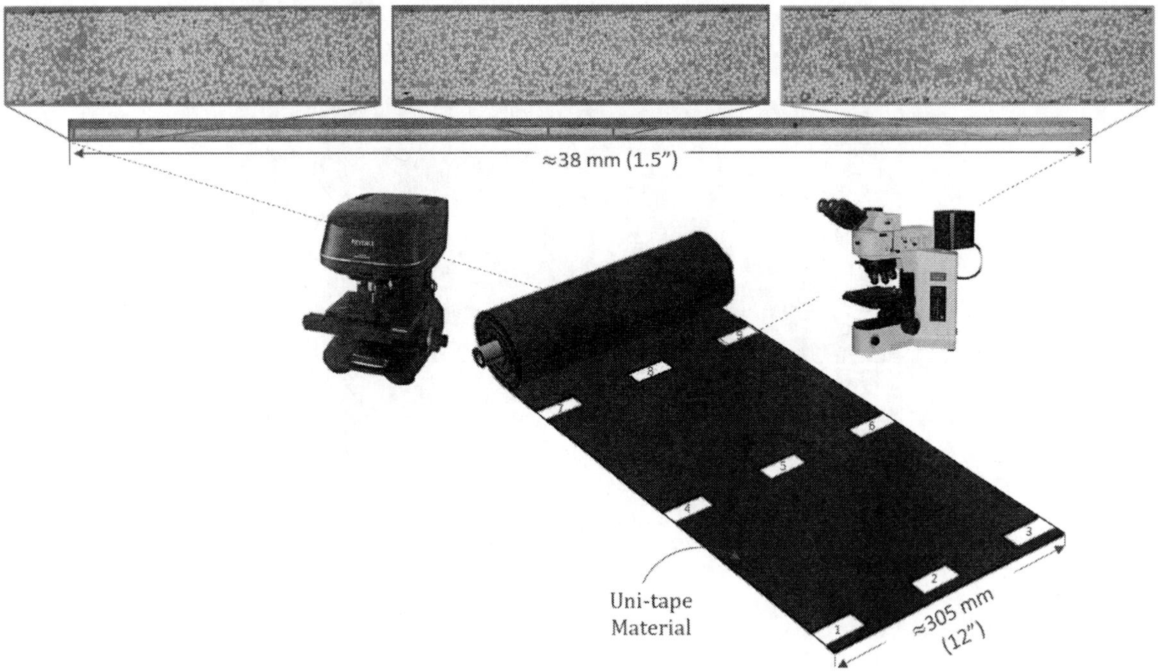

Figure 1. Photomicrograph sampling scheme

Double-sided tape on potting mold base (top view)

Epoxy-separated strips centered on mold (side view)

Resin mix poured into mold (top view)

Figure 2. Potting Procedure

3. RESULTS

All six materials in their uni-tape form underwent a receiving type inspection for quality covering thickness variation, resin and fiber distribution, and void content. Several examples of non-uniformities witnessed in the materials are shown in Figure 3. In order to highlight fiber pattern variations in the photomicrograph, fibers were highlighted in red for visual interpretation.

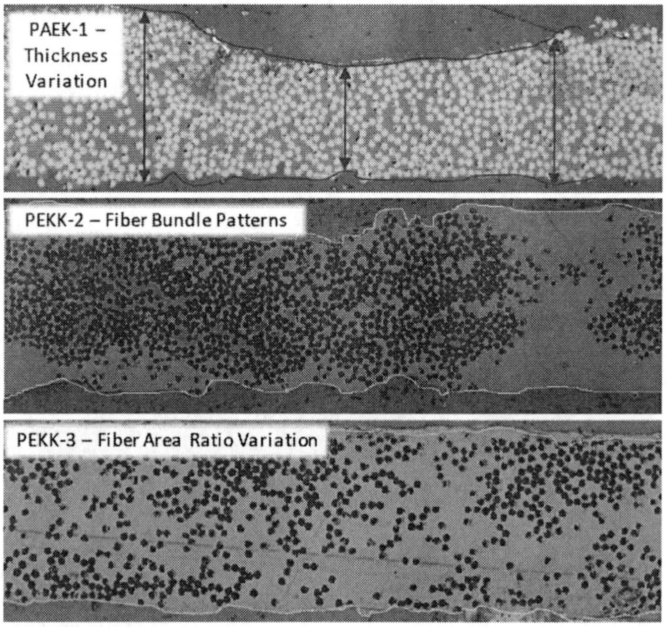

Figure 3. Representative non-uniformities witnessed in microscopic inspections

3.1 Thickness Uniformity

Material thicknesses measured along the width of the 305 mm roll width in 38 mm segments (1.5") from the left, middle, and right of the roll are shown in Figure 4. PAEK-2 had the most uniform thickness while the most significant variation was noticed in PEKK-2 material. No material had a disproportionate amount of thickness variation in the length or width direction. Based on the intimate contact model proposed by Lee and Springer [5], larger thickness variations (with all other factors held constant) increase the time required to form an intimate bond across a ply interface. PAEK-1, a new material, showed a similar level of thickness variation as more mature products such as PPS-1 and PEKK-1. PEKK-3 was marginally thinner than its specification of 0.14 mm (0.0055") consolidated ply thickness and has a slightly higher level of thickness variation than PAEK-1/PPS-1/PEKK-1 but less than PEKK-2. No significant variation was witnessed between the roll left, middle, or right locations. Similarly, similar variation levels were witnessed along the roll length locations.

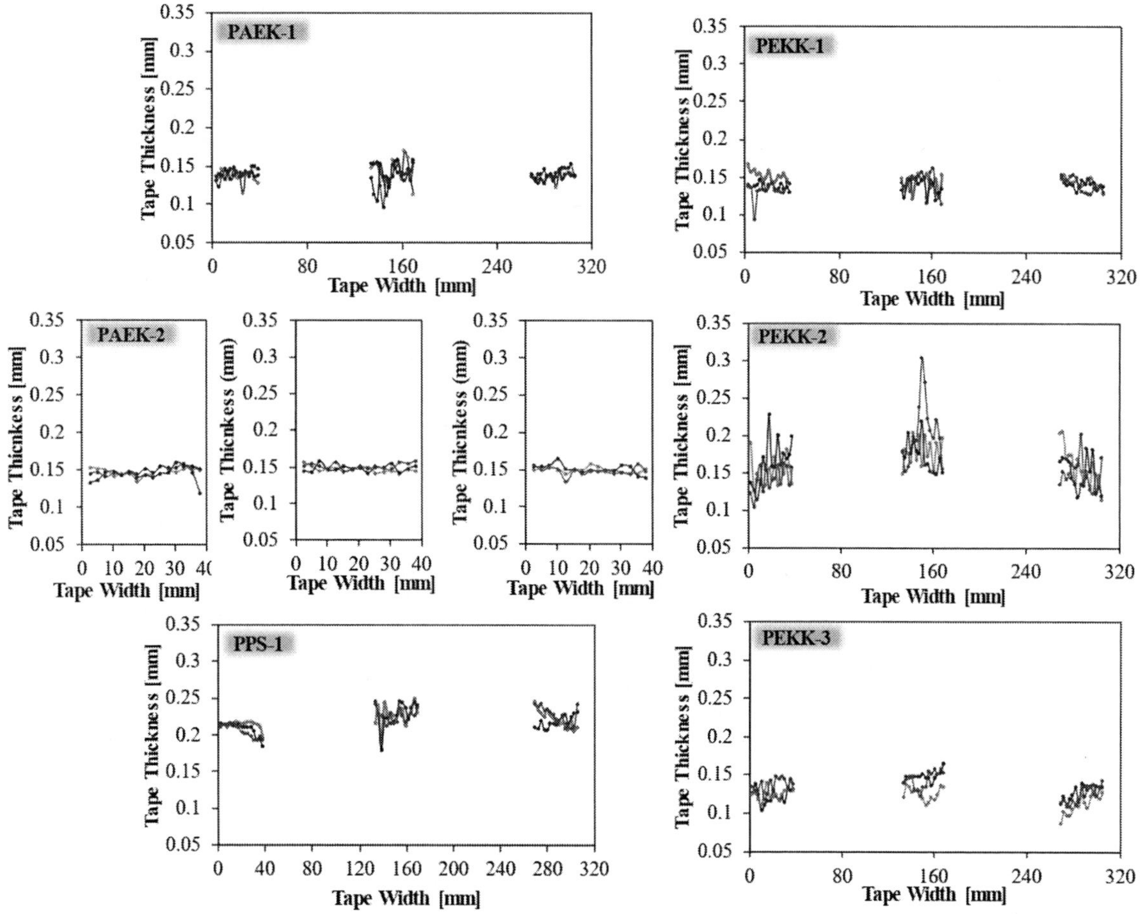

Figure 4. Thickness Uniformity Summary

3.2 Resin and Fiber Distribution

Resin and fiber content for each material is shown as bar graphs in Figure 7. Photomicrographs shown in Figure 8 are used to examine the distribution of resin and fiber. Visually analyzing the photomicrographs revealed PEKK-2 has distinct clustering of the fibers. The top and bottom exhibited a thin resin rich layer; the resin richness of the top side of PEKK-2, is well known to organizations with a heavy thermoplastic focus. Distinct waviness in the top and bottom surfaces was visible which correlated to the high degree of thickness variation. As a consolidated laminate some fiber gaps/resin rich areas were still seen; although the clustering of fibers relaxed to some degree. Resin rich areas were seen in between the fiber bundles as shown in Figure 5.

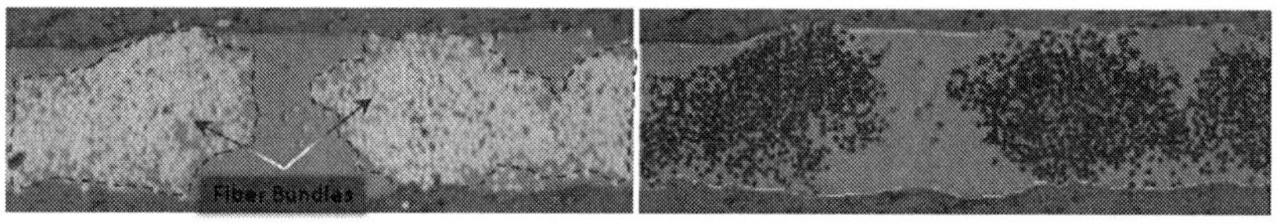

Figure 5. Fiber Bundle Pattern

The method followed in this study to understand the resin to fiber ratio uses image analysis of photomicrographs to compute an area ratio instead of an acid digestion method, which incorporates volumetric measurements. Photomicrographs taken of selected regions of the uni-tape roll are analyzed using software to detect fibers in a selected Region of Interest (ROI). The ROI contains two significant peaks that relate to resin and fiber within that region (as shown on the right in Figure 6). Once the lower and upper thresholds are set to capture the higher gray scale value peak representing fibers, the areal fiber percentage is calculated as the ratio of pixel counts for the fiber region to the total number of pixels. This process heavily depends on the quality of photomicrographs obtained using the optical microscope. Results can differ from manufacturer datasheet values for two reasons; the method followed to capture ratios or the quality of the selected tape region for example if it contains fiber bundles or resin rich areas.

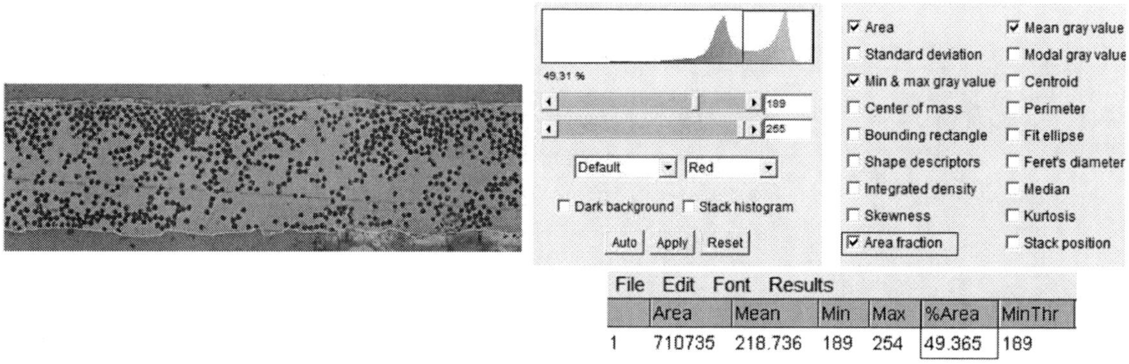

Figure 6. Representative ImageJ greyscale threshold analysis

The resin/fiber area percentages are compared in Figure 7 at three locations along the length of each roll. Figure 7 shows resin having a higher area percentage than fiber, which may be counterintuitive given aerospace prepreg materials have fiber weight percents around 60%. The area percent scaled by the density of the selected fiber or resin region yields fiber weight percentages within supplier specifications.

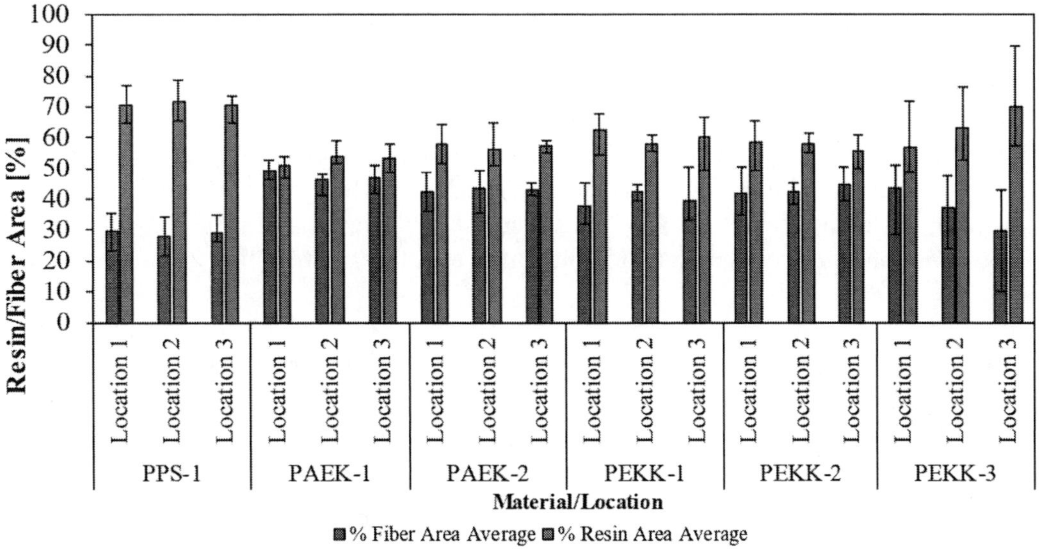

Figure 7. Resin and Fiber Content from Image Analysis

Significant resin and fiber content variation was noticed in PEKK-3. Although PEKK-2 showed a significant amount of fiber bundling, the average fiber-area percentages were more uniform in comparison to PEKK-3. Figure 8 shows the representative images used for analyzing the resin-fiber distributions for each material system.

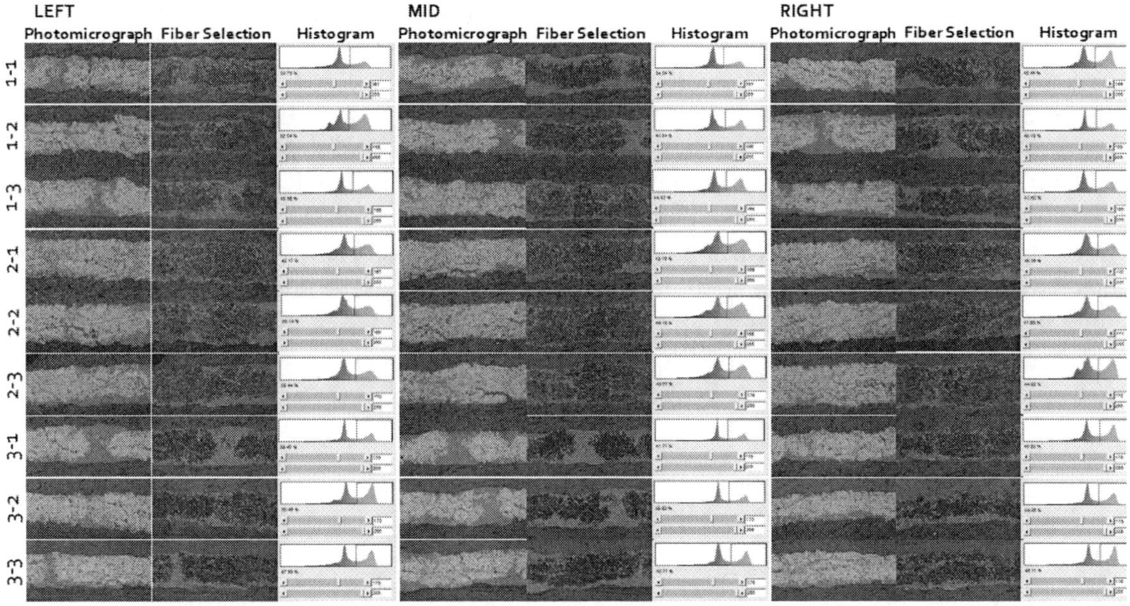

Figure 8. PEKK-2 Representative Resin and Fiber Distribution

PPS-1 was very regular with smooth edges and no noticeable voids or fiber clumping. Fibers in PAEK-1 are not as well distributed as in PPS-1. However, the distribution is consistent throughout and very few voids could be identified. PEKK-1 had a similar fiber distribution as PAEK-1. Fibers appeared more closely packed at top and bottom than in the center of the thickness. Compared to PAEK-1 the top and bottom edges were smoother.

PEKK-3 registered the most variation in the appearance of the cross section from location to location across and along the uni-tape. The central thickness region of PEKK-3 appeared to have lower fiber content. While some surface waviness was seen, the fiber and resin were well distributed in the surface layer. PAEK-2 was incredibly uniform having an array like appearance. The top and bottom surfaces demonstrated the same resin/fiber distribution as the central thickness region.

In addition to the photomicrograph inspections, x-ray computed tomography (XRCT) inspections using a Zeiss Xradia 520 Versa was performed to evaluate the quality of the prepreg tapes. A scan resolution of approximately 0.4-micron was achieved. Results from the XRCT inspections showed each material contained internal voids that were randomly situated along the width and length of each material, which contributed to a significant amount of variation in analyzing the voids with photomicrographs alone.

4. CONCLUSIONS

The quality of thermoplastic uni-tape, while variable, is certainly not poor. Photomicrographs serve as an invaluable tool to describe the quality of the material. Features such as a resin rich surface or an incredibly uniform fiber distribution visually distinguish one material from another, and may provide further insight into understanding the processing behavior. OoA research has suggested irregularities in the uniformity of uni-tape may improve permeability for void evacuation during oven consolidation [6]. Even though two or more uni-tapes may share similar fibers and the same polymer matrix, the quality of these tapes will not be the same. Use of image analysis for measuring voids in uni-tape material showed very low void content and these measurements are being evaluated along with X-ray CT and acid digestion measurements. Two materials to watch as they undergo further evaluation will be PEKK-2 and PAEK-2. These 2 materials are on the opposite ends of the uniformity spectrum; with PEKK-2 having waviness and fiber clumping while PAEK-2 has a very regular appearance. As this work continues the most quality critical elements will be identified and how attributes of the uni-tape make it more or less suitable for a given consolidation process.

5. REFERENCES

[1] D. Guillon, Y. Le Friant and L. Poitevin, "Measuring thermoplastic prepreg tape quality for part process control," *Composites World,* 14 May 2019.

[2] T. K. Slange, "Rapid Manufacturing of Tailored Thermoplastic Composites by Automated Lay-up and Stamp Forming: A Study on the Consolidation Mechanisms," University of Twente, 2019.

[3] T. Kok, "On the Consolidation Quality in Laser Assisted Fiber Placement," Enschede, The Netherlands, 2018.

[4] Y. Boiko, G. Guerin, V. Marikhin and R. Prud'homme, "Healing of interfaces of amorphous and semi-crystalline poly(ethylene terephthalate) in the vicinity of the glass transion temperature," *Polymer,* vol. 42, pp. 8695-8702, 2001.

[5] W. Lee and G. Springer, "A Model of the manufacturing process of thermoplastic matrix composites," *Journal of Composite Materials,* vol. 21, no. 11, pp. 1017-1055, 1987.

[6] J. Swamy, "VBO Consolidation of AFP Strucutres," TPRC Spring Technical Advisory Board Meeting, 2020.

COMPARISON OF ACCELERATED UV LAB TESTS FOR UV EFFECTS ON FIBERGLASS CROSSARMS

Jiayi Jenny Zhu and Michael S. Schoenoff
GEOTEK, LLC
Stewartville, MN USA

ABSTRACT

Fiberglass crossarms are built with fiber-reinforced polymer (FRP) composite materials that are mounted on utility poles or transmission towers to support overhead power lines. The service life of a fiberglass crossarm is influenced by outdoor weathering, which includes ultraviolet (UV) radiation, temperature, and water from rain and condensing humidity (dew), etc. In this paper, we compared two standard accelerated UV lab tests based on the commonly used Cycle 1 specified in both ASTM G154 and ASTM G155 to evaluate the effects of UV radiation on samples of fiberglass crossarms. Upon completion of 10,000 hr of accelerated UV lab test and along with inspections of the test specimens at intervals of every 1,000 hr, we were able to monitor the onset of fiber blooming and the progressions of both loss in gloss and fade in color. We also evaluated changes in sample weight and surface roughness of each test specimen at the end of the accelerated UV lab tests. We found that the data collected from the 10,000 hr of accelerated UV lab tests were comparable to the results we published previously from a different study of FRP crossarm samples which was based on 12 months of accelerated outdoor weathering test per ASTM G90. These results demonstrated that the progression of UV effects on FRP composite materials can be monitored using standardized accelerated UV lab tests and that they are comparable to the UV effects under sunlight on fiberglass crossarms.

Keywords: Accelerated UV Test, UV Stability of FRP Composites, Fiberglass Crossarms
Corresponding author: Jiayi Jenny Zhu

1. INTRODUCTION

Over the past 30 years, fiberglass crossarms have been used to replace traditional wood crossarms due to the superior properties of fiber-reinforced polymer (FRP) composite materials such as lighter weight, higher strength, consistent properties, and longer service life. While in service, FRP fiberglass crossarms are required to support electrical conductors under prolonged mechanical loading in an outdoor environment and in a variety of climates. It is critical to understand and to assess the effects of outdoor weathering on the service life of fiberglass crossarms.

It is well known that all polymers undergo UV-induced degradations under outdoor weathering [1-4] which can lead to chemical and physical changes including molecular chain scission, oxidation, reduction in gloss retention, color change, and fiber blooming. Fiber blooming is especially problematic for fiberglass crossarms because the exposed fibers on the surface of fiberglass crossarms present service and handling issues for utility line workers who work on or near the fiber bloomed crossarms. Furthermore, the increased surface roughness with exposed fibers on the surface of crossarms can trap electrically conductive contaminants which, in the presence of the electric potential found in overhead electrical conductors, can lead to arc tracking and potentially pole top fires.

In general, outdoor weathering is mainly influenced by UV radiation from sunlight, water from rain and condensing humidity (dew), and temperature, as well as wind and pollutants, among others. Effects of water and temperature on FRP composite materials have been well studied [5-8]. Effects of UV radiation on polymers have been well documented [1-4], but effects of UV-induced polymer degradation on FRP composite materials and their impacts to the service life of fiberglass crossarms have not been well studied. A study published by the same authors [9-10] investigated effects of accelerated outdoor weathering on FRP composite materials by completing 12 months of accelerated outdoor weathering testing based on Cycle 1 specified in ASTM G90 [11]. It is noted that due to the seasonal changes, the duration of the study based on ASTM G90 is limited to the intervals of 12 months in order to obtain comparable results. Since the outdoor weathering is unpredictable, it is difficult to reproduce the same outdoor weathering conditions for the continuation of the study.

In order to conduct repeatable and reproducible lab tests to evaluate the effects of UV radiation on FRP composite materials, this paper compared two standard accelerated UV lab tests based on the commonly used Cycle 1 in both ASTM G154 [12] and ASTM G155 [13], both of which specify the use of UVA-340 lamps to simulate UV radiation from sunlight during the cycling tests inside the test chambers. The study revealed resemblances and correlations between UV radiation from sunlight and through these standard accelerated UV lab tests. It also estimated equivalents between the durations of total test times through these standard accelerated UV lab tests and the equivalent sun years established for Arizona and Florida. These findings are very valuable for the understanding and comparison of UV effects on the service life of fiberglass crossarms.

The study presented in this paper also demonstrated the progression of UV effects on samples of commercially available FRP crossarms by completing 10,000 hr of accelerated UV lab tests based on Cycle 1 of both ASTM G154 and ASTM G155 along with inspection for the onset of fiber blooming and measurement of gloss and color at 1,000 hr intervals of the total test time. It further compared changes due to the loss of degraded polymers by recording sample weights and surface roughness on non-exposed and UV exposed samples, and found that they were comparable to those published previously by the same authors [9-10] from a different study of FRP crossarm samples which was based on 12 months of accelerated outdoor weathering test per ASTM G90.

2. EXPERIMENTATION

2.1 Test Specimens of Fiberglass Crossarms

Four commercially available fiberglass crossarms were selected for the accelerated UV lab tests. They were produced by four different manufacturers and were labeled as A, B, C, and D. Note that A and B had no outer coating on the outer surface of the crossarms while C and D had a UV protective coating on the outer surface of the crossarms.

Test specimens for the accelerated UV lab tests were cut from one side of the fiberglass crossarms. The cut edges of each test specimen were sealed with a water resistance coating to minimize the effects of moisture inside the UV test chambers on the test specimens through the cut edges. The test specimens were then placed inside the UV test chambers with the original outer surface of crossarms exposed directly to the UV radiation.

2.2 Accelerated UV Lab Tests

Two test standards were selected for the accelerated UV lab tests.

One test standard was based on the commonly used Cycle-1 specified in ASTM G154 [12]. It will be referred to simply as ASTM G154 in the rest of this paper. Each test cycle is 12 hr of test time with 8 hr of UV exposure under UVA-340 fluorescent lamps at 0.89 $W/m^2/nm$ irradiance and 60°C of black panel temperature followed by 4 hr of condensation at 50°C of black panel temperature without UV exposure. The accelerated UV lab test based on ASTM G154 was conducted using a QUV fluorescent test chamber (Model QUV/se, by Q-Lab, Cleveland, OH) for total test times of 5,000 hr and 10,000 hr, respectively. Test specimens were inspected for fiber blooming and measured for gloss and color at intervals of every 1,000 hr of the test time. Sample weights and images of water drop test were recorded on test specimens with 0 hr, 5,000 hr, and 10,000 hr of the accelerated UV lab test.

Another test procedure was based on the commonly used Cycle-1 specified in ASTM G155 [13]. It will be referred to simply as ASTM G155 in the rest of this paper. Each test cycle is 2 hr of test time with UV exposure under UVA-340 xenon arc lamp with a daylight filter at 0.35 $W/m^2/nm$ irradiance. During each cycle, 102 minutes were maintained at 63°C of black panel temperature and 18 minutes were cooled with water spray without temperature control. The accelerated UV lab test based on ASTM G155 was conducted using a Q-Sun Xenon Arc Test Chamber (Model Xe-1-S, by Q-Lab, Cleveland, OH) for a total of 10,000 hr. Test specimens were inspected for fiber blooming and measured for color and gloss at intervals of every 1,000 hr of the test time. Sample weights and images of water drop test were recorded on test specimens with 0 hr and 10,000 hr of the accelerated UV lab test.

2.3 Fiber Blooming Inspection

Each test specimen was visually inspected for fiber blooming at intervals of every 1,000 hr of the test time. The onset for fiber blooming was defined as when traces of exposed fibers were first observed on the surface of a test specimen.

2.4 Sample Weight Measurement

Sample weights for test specimens were measured using an analytical balance before the start of the accelerated UV test and at the end of the 5,000 or 10,000 hr of the total test time. Sample weight for each test specimen was recorded after gently rinsing the test surface with DI water using a wash bottle and letting the test specimen air-dried for at least 48 hr. Weight change of a test specimen was calculated by equation (1):

$$\Delta Wt\ (\%) = 100 \times (Wt_{final} - Wt_{initial}) / Wt_{initial} \quad (1)$$

Where Wt_{final} and $Wt_{initial}$ were measured weights of each test specimen after and before the accelerated UV lab tests, respectively.

2.5 Gloss Measurement

Changes in gloss on test specimens were monitored by using a glossmeter (Model ETB-0833, by M & I Instruments Inc., Mississauga, ON Canada) measured at the 60 degree geometry based on ASTM D523 [14]. Each test specimen was measured at the start of the accelerated UV test and at an interval of every 1,000 hr of the test time. Gloss retention of a test specimen was calculated by equation (2):

$$GR (\%) = 100 \times Gloss_{final} / Gloss_{initial} \quad (2)$$

Where $Gloss_{final}$ and $Gloss_{initial}$ were measured values of 60 degree gloss of each test specimen after and before the accelerated UV lab tests, respectively.

2.6 Color Measurements

Changes in color on test specimens were monitored using a Precise Color Reader (Model WR-10, by ShenZhen Wave Optoelectronics Technology Co., Ltd., China) based on ASTM D2244 [15]. Each test specimen was measured at the start of the accelerated UV test and at an interval of every 1,000 hr of the test time. CIELAB (L*, a*, b*) color scale was used for the color measurements and total color change of a test specimen was calculated by equation (3):

$$\Delta E^* = [(L^* - L_o^*)^2 + (a^* - a_o^*)^2 + (b^* - b_o^*)^2]^{1/2} \quad (3)$$

Where L*, a*, and b* were measured on each test specimen after the accelerated UV lab tests; L_o^*, a_o^*, and b_o^* were measured on each test specimen before the accelerated UV lab tests.

2.7 Water Drop Test

Changes in surface roughness on test specimens were analyzed by comparing digital images of water drops, which were taken at intervals of 0, 1, and 5 min. after the water drops were placed onto the surfaces of test specimens. Each UV-exposed test specimen was tested side-by-side with a non-exposed control specimen of the same fiberglass crossarm sample. To enhance the contrast of the water drops in digital images, a colored water solution was prepared for the water drop test by adding one drop of food die (Red Food Color, McCormick) into 50 mL of DI water. Water drops were placed onto the surfaces of test specimens by using a 1 mL disposable pipette.

3. RESULTS AND DISCUSSIONS

3.1 UV Exposure under Sunlight and through Accelerated UV Lab Tests

Sunlight refers to the full spectra of solar irradiance, which is in the wavelengths between about 295 nm and 4050 nm [16]. It is commonly divided into UV radiation (295 nm - 400 nm), visible light (400 - 760 um), and infra radiation (760 – 4050 nm). Since solar energy is electromagnetic energy which is made up of photons and is inversely proportional to wavelength, UV radiation in the short wavelength region is accompanied with small high energy photons that can be easily absorbed by polymer molecules and causes polymers to undergo UV-induced degradations which can lead to chemical and physical changes of the polymers, resulting in shortened service life of the FRP composite materials. Therefore, UV radiation among total solar radiation is the primary cause of UV-induced polymer degradations under sunlight. UV effects on samples of fiberglass crossarms and their service lives will be further discussed in section 3.2.

UV exposure under sunlight refers to the dosage of UV radiant energy emitted by sunlight to a given area of an object in a given time. Both solar radiation and UV radiation from sunlight can be measured and recorded in the unit of joules per square meter (J/m^2) or million joules per square meter (MJ/m^2). The recorded values of these radiation dosages can vary according to the geographical locations and altitudes of the measuring sites as well as seasonal and climate changes. Based on many years of recorded data by Q-Lab located in Arizona and in Florida, it was reported that the average values of annual UV radiations in the wavelengths of 295 – 385 nm remained relatively unchanged for these two locations and that these values can be referenced as one equivalent sun year based on 333 MJ/m^2 for Arizona or 280 MJ/m^2 for Florida [17].

Since UV radiation from sunlight is unpredictable at any given time or location, a Benchmark Solar Spectra which is defined as a standard UV spectral irradiance distribution has been recommended to be used as a guide for the comparison of UV exposure between natural sunlight and artificial lights from the UV lamps used in the accelerated UV lab tests [18]. Figure 1 shows spectra irradiance distributions for a) sunlight; b) a fluorescent lamp UVA-340 used in a QUV test chamber per ASTM G154; and c) a xenon arc lamp UVA-340 with a daylight filter used in a Q-Sun tester per ASTM G155 [19]

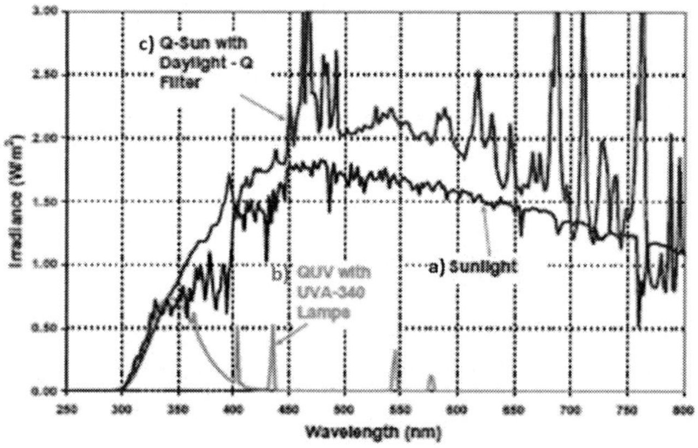

Figure-1. Spectral irradiance distributions of a) sunlight, b) fluorescent lamp UVA-340 used in a QUV test chamber per ASTM G154, and c) xenon arc lamp UVA-340 with a daylight filter used in a Q-Sun tester per ASTM G155 [19]

Compared to the unpredictable UV radiation from sunlight, UV radiation from the UV lamps used in accelerated lab tests is not affected by changes in climate or altitude because it is the accumulation of total UV irradiance over the distinctive spectral irradiance distributions of the selected UV lamps. A UV test chamber with selected UV lamps can be placed in any lab and can be run continuously at a specified UV irradiance under controlled temperatures and/or moistures inside the UV test chamber. Therefore, UV exposure received by the test specimens inside a UV test chamber under specified test standards through accelerated UV lab tests is considerably repeatable and reproducible.

Total UV radiation (UVR) emitted by the UV lamps between wavelength 1 and 2 onto the test specimens through an accelerated UV lab test can be calculated by equation (4):

$$UVR_{1\sim2} = 0.0036 \times E_{1\sim2} \times T_{uv} \quad (4)$$

Where $UVR_{1\sim2}$ is in the unit of MJ/m^2, $E_{1\sim2}$ is the total UV irradiance of the UV lamps used in an accelerated UV lab test between wavelength 1 and 2 in the unit of W/m^2, and T_{uv} is the duration of total UV exposure time through an accelerated UV lab test in the unit of hr. Since the average annual UV radiation from sunlight for the same location does not change substantially from one year to another, the calculated total UVR from an accelerated UV lab test can be compared to the average annual UV radiation from sunlight or to a referenced equivalent sun year which is established for a specific location.

Calculated values of total UVR received by the test specimens and estimated corresponding equivalent sun years in Arizona or in Florida through accelerated UV lab tests based on ASTM G154 and G155 were listed in Table 1. Note that the listed values of irradiance at 340 nm for the UV lamps were specified by cycle 1 of the two test standards while the listed values of irradiance between 300 – 400 nm for the UV lamps were provided by Q-Lab that supplied both the UV test chambers and the UV lamps for the accelerated UV lab tests used for this study. Average values of annual UV radiation for the estimation of corresponding equivalent sun years were based on the data recorded by Q-Lab at 333 MJ/m^2 for Arizona and 280 MJ/m^2 for Florida, respectively. For the purpose of comparison, Table 1 showed that one year of UV exposure under sunlight in Arizona or in Florida can be considered as comparable to 2,800 hr or 2,400 hr of accelerated UV lab tests based on ASTM G154, respectively. It can also be considered as comparable to 2,400 hr or 2,100 hr of accelerated UV lab tests based on ASTM G155, respectively.

Table 1 Calculated values of total UVR between wavelength 300 – 400 nm after accelerated UV lab tests based on ASTM G154 and ASTM G155 and estimated corresponding equivalent sun years in Arizona or in Florida.

Test Standard	UV Source	Cycles	Irradiance at 340 nm	Irradiance at 300-400 nm	Total Test Time, hr	UV Exp. Time, hr	Cal. UVR, MJ/m^2	Eq Sun Yr in Az	Eq Sun Yr in Fl
ASTM G154	Fluorescent Lamp UVA-340	Cycle-1	0.89 $W/m^2/nm$	50.6 W/m^2	**2,400**	1600	291	0.9	**1.0**
					2,800	1867	340	**1.0**	1.2
					10,000	6667	1214	**3.6**	**4.3**
ASTM G155	Xenon Arc with Daylight Filter UVA-340	Cycle-1	0.35 W/m2/nm	38.5 W/m^2	**2,100**	2100	291	0.9	**1.0**
					2,400	2400	333	**1.0**	1.2
					10,000	10000	1386	**4.2**	**5.0**

It is noticed that the calculated values of total UVR listed in Table 1 for 10,000 hr of testing based on ASTM G154 and ASTM G155 were slightly different. This is due to the different test conditions specified for each cycle of the two test standards – i.e., different UV irradiance and spectra irradiance distributions for the UV lamps used in the two test standards and different duration of UV exposure for each cycle of the two test standards.

It must be noted that the calculated values listed in Table 1 were based on total UV radiation discussed above. As stated earlier in this paper, UV effects from natural sunlight can be influenced by a number of unpredictable factors which makes it difficult to be evaluated with reproducible results. By referencing a well-recognized Benchmark Solar Spectrum and by following well-defined test standards, correlations between UV radiation from sunlight and through specified accelerated UV lab tests can be established. Therefore, UV effects on fiberglass crossarms, which are placed into service in direct sunlight, can be evaluated by comparing UV effects on test specimens through standard accelerated UV lab tests with considerably repeatable and reproducible results.

3.2 Progressive Changes on Samples after Accelerated UV Lab Tests

UV effects on samples of four commercially available fiberglass crossarms were evaluated by exposing prepared test specimens to accelerated UV lab tests based on ASTM G154 or ASTM G155. After a few thousand hours of testing, Samples A and B which had no outer coating showed fiber blooming on UV exposed surfaces while Samples C and D which had a protective coating showed no fiber blooming upon completion of 10,000 hr of testing. Weight changes for all the tested specimens were not significant due to the small portion of degraded polymers on the UV exposed surface comparing to the overall weight for each test specimen. All samples showed varied degree of reduction in gloss retention and fade in color. Table-2 lists the recorded onset time period when fiber blooming (FB) was first observed, percent of weight change after 10,000 hr of testing, percent of 60 degree gloss retention, and total color changes measured at intervals of every 1,000 hr up to 10,000 hr of testing. Note that the results of 60 degree gloss retention and total color change after 4k, 6k and 8k hr of testing were not listed in Table-2 due to the relatively insignificant changes in recorded values and the limited column space for the table.

Table-2 Results of onset for fiber blooming (FB), percent of weight change, 60 degree gloss retention, and total color change for Samples A, B, C, and D after accelerated UV lab tests based on ASTM G154 and ASTM G155.

Sample IDs	Test Std	FB Onset	ΔWt, %		60 Degree Gloss Retention, %							Total Color Change, ΔE						
			5k hr	**10k hr**	1k hr	2k hr	3k hr	**5k hr**	7k hr	9k hr	**10k hr**	1k hr	2k hr	3k hr	**5k hr**	7k hr	9k hr	**10k hr**
A	G154	4k-5k hr	**-0.4%**	**-0.7%**	9%	5%	4%	**4%**	5%	5%	**6%**	8	7	8	**8**	7	7	**9**
	G155	5k hr	N/A	-1.2%	6%	2%	3%	**4%**	6%	13%	**17%**	2	2	2	**2**	2	2	**4**
B	G154	5k-6k hr	**-0.3%**	**-0.5%**	104%	11%	3%	**3%**	4%	6%	**5%**	5	7	8	**7**	7	7	**8**
	G155	9k hr	N/A	-0.6%	94%	55%	6%	**4%**	5%	6%	**5%**	3	4	5	**5**	6	7	**8**
C	G154	None	**-0.1%**	**-0.2%**	109%	94%	57%	**31%**	21%	22%	**23%**	2	2	2	**1**	3	6	**7**
	G155	None	N/A	-0.6%	98%	75%	49%	**32%**	28%	24%	**23%**	0.2	0.7	0.8	**0.7**	2	5	**6**
D	G154	None	**-0.2%**	**-0.2%**	94%	77%	35%	**20%**	14%	13%	**13%**	2	2	2	**1**	1	3	**3**
	G155	None	N/A	-0.3%	88%	71%	47%	**11%**	9%	8%	**7%**	0.3	0.4	0.5	**0.2**	2	3	**4**

Figure 2 shows plots of progressive changes for 60 degree gloss retention (as shown on the left) and total color changes (as shown on the right) for the test specimens of the four samples vs total test hours of accelerated UV lab tests per ASTM G154 and ASTM G155, respectively. It appeared that Samples A and B, which had no outer coating, showed more rapid decrease in their average values of 60 degree gloss retention and more significant increase in the average values of total color changes than Samples C and D which had a protective coating, indicating UV-induced changes in gloss and color on samples without an outer coating were more advanced than those with a protective coating.

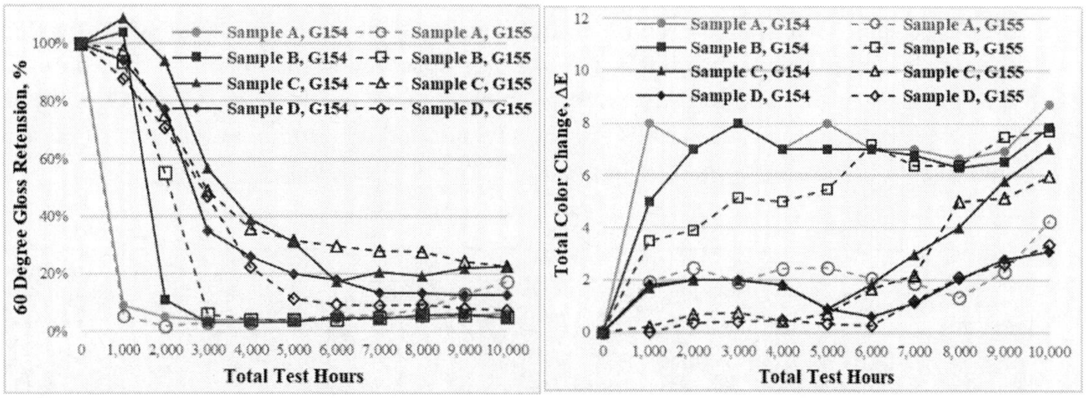

Figure-2. Plots of 60 degree gloss retention and total color change for Samples A, B, C, and D vs total test hours of accelerated UV lab tests per ASTM G154 and ASTM G155, respectively.

It was noticed that for the same sample tested by the two test standards, the average values of 60 degree gloss retention showed no significant differences while the average values of total color changes appeared to be advancing differently. This was likely due to the different responses by the test specimens to the different test conditions between these two test standards as discussed in section 3.1.

It should be noted that the onset of fiber blooming on the surface of FRP crossarms indicates the outer layer of resin rich polymers have been degraded noticeably, resulting in a rough surface with exposed glass fibers and increased surface porosity due to the loss of degraded polymers. The exposed glass fibers present service challenges for utility workers who work on or near the fiber bloomed crossarms while the increased surface roughness or porosity can lead to the accumulation of contaminates and cause arc tracking or flashover which can shorten the service life of the FRP crossarms.

To compare changes in surface roughness on the test specimens due to the loss of degraded polymers, Figure 3 and 4 showed images of water drops on Samples A, B, C, and D after accelerated UV lab tests based on ASTM G154 and ASTM G155, respectively. Each UV-exposed test specimen was tested side-by-side with a non-exposed control specimen of the same sample and the images were taken at intervals of 0, 1, and 5 min. after the water drops were placed onto the surfaces of test specimens. Apparently, water drops placed on all the non-exposed control samples showed well-defined circular edges and remained unchanged in size and shape since the surface of a fresh fiberglass crossarm was smooth and hydrophobic. After UV exposure through accelerated UV lab tests, water drops on Sample A looked much larger due to the water spreading and/or submerging into the substrate immediately after being placed onto the surface. The changes of water drops on UV exposed Sample A became more significant when the test time increased from 5,000 hr to 10,000 hr based on ATM G154, indicating a significant increase in surface roughness due to the massive loss of degraded polymers. Water drops on UV exposed Sample B were similar to those observed on UV exposed Sample A, except that water spreading and/or submerging into the substrate around the edges of water drops appeared to be milder than those on UV exposed Sample A, indicating a moderate increase in surface roughness due to the loss of less amount of degraded polymers. On UV exposed Samples C and D, water drops showed slight changes in the size and shape with little or no signs of water spreading around the edges of water drops, indicating no significant increase in surface roughness. It appeared that the increase in surface roughness observed in the images of water drops were similar to the trends of progressive changes observed in the onset for fiber blooming, reduction in gloss retention, and increase in total color changes for the four UV-exposed samples – that is, UV-induced changes in Samples A and B which had no outer coating progressed more rapidly than those in Samples C and D which had a protective coating.

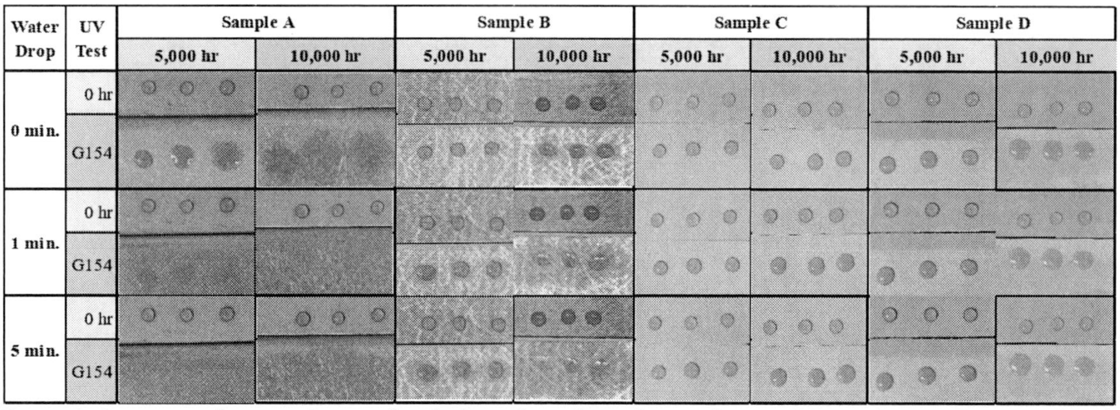

Figure 3. Images of water drops after being placed on the surfaces of Samples A, B, C, and D for 0 min., 1 min., and 5 min., respectively. Each sample was tested side-by-side with a non-exposed control specimen and a UV-exposed test specimen after 5,000 or 10,000 hr of accelerated UV lab tests based on ASTM G154.

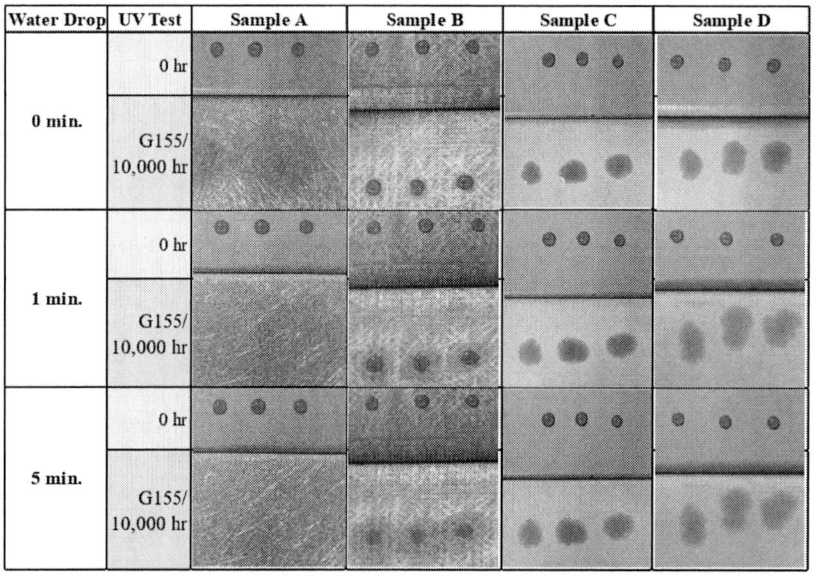

Figure 4. Images of water drops after being placed on the surfaces of Samples A, B, C, and D for 0 min., 1 min., and 5 min., respectively. Each sample was tested side-by-side with a non-exposed control specimen and a UV-exposed test specimen after 10,000 hr of accelerated UV lab tests based on ASTM G155.

It worth noting that the surface changes discussed in this paper on samples after 10,000 hr of accelerated UV lab tests based on ASTM G154 or ASTM G155 were comparable to the study published by the same authors on similar samples of fiberglass crossarms after 12 months of accelerated UV outdoor test based on ASTM G90 [9-10]. It was further noticed that based on equation (4), the 12 months of accelerated UV outdoor test per ASTM G90 which were evaluated by the same authors with a reported total UVR value at 1,363 MJ/m^2 [9-10] can be considered as equivalent to approximately 11,200 hr or 9,800 hr of accelerated UV lab test based on ASTM G154 or ASTM G155, respectively.

4. CONCLUSIONS

Under methodical considerations, UV radiation through accelerated UV lab tests based on Cycle 1 in both ASTM G154 and ASTM G155 can be correlated to the UV radiation from sunlight for an established location. This makes it possible to conduct repeatable and reproducible accelerated UV lab tests in the environmentally controlled test chambers for the evaluation of UV effects on FRP composite materials. The duration of test times based on ASTM 154 and ASTM G155 are readily adjustable and can be correlated to the equivalent sun years for the duration of UV exposure under natural sunlight in Arizona or Florida. Therefore, the progressive changes observed on samples of FRP composite materials after these standard accelerated UV lab tests can provide valuable information for the understanding and comparison of the service life of fiberglass crossarms in the field.

The results presented in this paper demonstrate that the progressive changes due to UV-induced polymer degradations on samples of fiberglass crossarms can be evaluated by conducting standard accelerated UV lab tests based on ASTM G154 or ASTM G155 along with visual inspecting for fiber blooming and measuring for gloss, color, and surface roughness at regular intervals during the testing. These results were in good agreement with the findings published earlier by the same authors on similar samples of fiberglass crossarms which were evaluated by completing 12 months of accelerated outdoor weathering test based on Cycle 1 in ASTM G90. It was apparent that upon exposure to UV radiation, fiberglass crossarms without an outer coating showed more advanced progressions for the onset of fiber blooming, reduction in gloss retention, total color change, and increase in surface roughness than those with a protective coating.

5. REFERENCES

[1] A.L. Andrady, et al *Effects of Increased Solar Ultraviolet Radiation on Materials*, J. Photochemistry and Photobiology B: Biology 46 (1998), 96 - 103

[2] J.F. Rabek, *Polymer Photodegradation*, London: Chapman and Hall, 1995

[3] G. Scott, *Polymer Degradation and Stabilization*, London: Elsevier Applied Science, 1990

[4] J. Wypych, *Weathering Handbook*, Toronto, Canada: Chemtec Publishing, 1990

[5] J. Yao and G. Ziegmann, *Water Absorption Behavior and Its Influence on Properties of GRP Pipe*, J. of Composite Materials, Vol. 41, No. 8/2007, DOI: 10.1177/0021998306067265

[6] E. Faguaga, et al, *Effect of water absorption on the dynamic mechanical properties of composites used for windmill blades*, Mater. Des 2012, 36, 609 – 616.

[7] S.A. Grammatikos, et al, *Thermal cycling effects on the durability of a pultruded GFRP material for off-shore civil engineering structures*, Composite Structures, 153 (2016) 297 – 310.

[8] M.A. Sawpan, et al, *Glass transitions of hygrothermal aged pultruded glass fibre reinforced polymer rebar by dynamic mechanical thermal analysis*, Mater Des 2012, 42, 272 - 278.

[9] J.J. Zhu, et al, *Outdoor Weathering Stability of Fiber Reinforced Polymers for Fiberglass Crossarms*, CAMX Conference Proceedings, Anaheim, CA, Sept. 26 -29, 2016. CAMX – The Composites and Advanced Materials Expo.

[10] J.J. Zhu and M.S. Schoenoff, *Effects of Natural Sunlight on Fiberglass Reinforced Polymers for Crossarms*, 2018 IEEE REPC, Memphis, TN, May 6 – 8, 2018, 101 - 105 DOI 10.1109/REPC.2018.00023

[11] ASTM G90-17, *Standard Practice for Performing Accelerated Outdoor Weathering of Materials Using Concentrated Natural Sunlight*, ASTM International, West Conshohocken, PA, 2017, www.astm.org

[12] ASTM 154-16, *Standard Practice for Operating Fluorescent Ultraviolet (UV) Lamp Apparatus for Exposure of Nonmetallic Materials*, ASTM International, West Conshohocken, PA, 2016, www.astm.org

[13] ASTM G155-13, *Standard Practice for Operating Xenon Arc Light Apparatus for Exposure of Non-Metallic Materials, ASTM International*, West Conshohocken, PA, 2013, www.astm.org

[14] ASTM D523 -14 (2018), *Standard Test Method for Specular Gloss, ASTM International*, West Conshohocken, PA, 2018, www.astm.org

[15] ASTM D2244-16, *Standard Practice for Calculation of Color Tolerances and Color Differences from Instrumentally Measured Color Coordinates*, ASTM International, West Conshohocken, PA, 2016, www.astm.org

[16] ASTM G113-16, *Standard Terminology Relating to Natural and Artificial Weathering Tests of Nonmetallic Materials*, ASTM International, West Conshohocken, PA, 2016, www.astm.org

[17] Technical Bulletin LL-9031.3 © 2011, *Q-TRAC Natural Sunlight Concentrator*, Q-Lab Corporation, www/q-lab.com, accessed on Feb. 3, 2016. http://www.q-lab.com/documents/public/002697f8-4405-4421-9055-b86f7430e6a2.pdf

[18] ASTM G177-03 (2012), *Standard Tables for Reference Solar Ultraviolet Spectral Distributions: Hemispherical on 37° Tilted Surface*, ASTM International, West Conshohocken, PA, 2012, www.astm.org

[19] Technical Bulletin LU-8009.4 ©2017 Q-Lab Corporation, QUV & Q-SUN, *A Comparison of Two Effective Approaches to Accelerated Weathering & Light Stability Testing*, https://www.q-lab.com/documents/public/e9cd0ed7-86fa-4e1f-bed5-af90d81e53e3.pdf

TEST SAMPLE PREPARATION OF PULTRUDED UNIDIRECTIONAL CARBON PARTS – ISSUES IN MACHINING PRACTICE AND ALTERNATIVES

Leigh Nolen, James Orr, Christiaan Mauldin, Dr. Mac Puckett, and Dr. Lei Zhao
PolyOne Advanced Composites

3850 Pinson Valley Parkway

Birmingham, AL 35217

ABSTRACT

The advantages of carbon fiber composites are well known – lightweight, corrosion resistant, high strength and high modulus. The economic advantages of producing composites by a continuous pultrusion process are also well understood. But testing these unidirectional carbon composite parts can be problematic. Most composite testing methods call out the use of thin flat rectangular strips or in some cases cylindrical sections. Unfortunately, shapes produced by pultrusion are not limited to cross-sectional areas or geometries that are easily or directly tested. Unlike isotropic metals or quasi-isotropic prepreg lay-ups, the unidirectional parts typically made in pultrusion are very anisotropic, not perfectly symmetrical in the Z-axis, and have a propensity to fail in shear during a variety of testing.

It is often the case that full scale sample testing is required to meet consumer or regulatory compliance. Some end market specifications require testing that can be challenging or nearly impossible to perform based on composite structural shape or testing limitations. In order to complete testing, samples are manipulated through traditional machining methods to create a sample that can be tested. One such test is tensile strength as measured by ISO 527-5 (ASTM D3039). For carbon composites made of a standard modulus carbon, this test is particularly difficult to perform if the part tensile strength exceeds the shear strength of the tabbing adhesive and shear area combination. Traditional practice would dictate that the part thickness be ground down to achieve a breakable cross-sectional area for this carbon composite. However, grinding unidirectional pultruded composites can terminate fibers producing a false material failure point.

This paper will highlight the challenges of using the tensile test to qualify carbon unidirectional composites and discuss the merits of producing and testing a thinner representative sample in place of a ground sample for qualifications.

1. INTRODUCTION

1.1 The Market

An end use application that values the nature of pultruded parts is infrastructure support, such as concrete surface and near-surface-mount (NSM) reinforcement and rebar. With long spanning bridges and parking decks made of high compression strength concrete, reinforcement made of carbon fiber reinforced epoxy provides an ideal companion for load bearing members, particularly in concrete repair applications. With the loads of the structure already calculated, adding high weight and corrodible metallic materials is not an option. Low weight, high strength carbon is added to reinforce the concrete structure without increasing the weight. However, conflict arises during the material qualification of unidirectional carbon composites for use in a market that is more accustomed to the qualification of traditional materials.

1.2 Material Selection and Process Overview

The high strength-to-weight ratio of carbon fiber composite parts provide an ideal property offering for use in many strength bearing applications. Possessing properties that rival those of steel at a fraction of the weight, carbon fiber is utilized when weight is at a premium and strength cannot be sacrificed. Carbon fiber reinforced composites can be found in applications from aircraft structures, vehicle components and bodies, and infrastructure supports. For many applications, epoxy resin is an ideal choice to pair with carbon fiber for strength bearing members. It boasts a high shear strength, corrosion resistance, and can be easily bonded to other members during onsite installation using well established practices. Pultrusion is a well-known process for the manufacture of continuous fiber reinforced products. Continuous fibers are pulled through liquid resin that are then formed and consolidated in a heated die. The heat from the die activates the thermoset resin, creating a fully formed continuous fiber reinforced shape with the constant cross sectional dimensions of the die. The continuous nature of pultrusion affords the process two benefits over other composite processes. Since the process does not stop (continuous), it is very economical to manufacture large quantities of parts at relatively high speeds and low conversion costs compared with other batch wise composite processes. In addition, the nature of the process allows for production of parts with very long length that are not limited to the length of a mold or autoclave, providing continuous length composites that are not feasible with other manufacturing techniques.

However, the process has some limitations. Pultrusion part shape is controlled by the die shape, meaning the parts are limited to the shape of the die in the pull direction. All pultruded profiles have a constant cross section, and therefore cannot have tapers or curves prevalent in many modern designs and components. This makes pultruded parts similar in many ways to metallic structural members that are extruded or roll formed. Pultrusion also limits the input fiber orientations when compared to other FRP processes. When a prepreg layup is constructed, fibers can be placed and balanced in any direction, producing a quasi-isotropic part. Pultrusion requires that a high percentage of the fiber content be placed in the direction of pull. The high shear forces from both the liquid resin consolidation and the drag on the die surface dictate that the part be strong enough in the pull direction to withstand the manufacturing process and be pulled through the die. With these limitations, this does not mean that pultrusion is impractical, it simply means that parts that are pultruded ideally have high strength in the longitudinal direction and are long, constant cross section parts, which describes many strength bearing members and stiffness support members.

1.3 Testing Hurdles

When a traditional, well known material such as steel is being qualified for use in construction, engineers have many well-defined practices. Alloys of steel are well defined and controlled by industry standards, meaning that engineers can have high confidence in the material based on the individual material designation. For example, 1018 steel from one supplier will test very similarly to 1018 steel from a competitor. The isotropic nature of metals also makes qualification relatively straight forward. In order to gain an understanding of the material properties, it is not necessary to test the entire specimen. A small sample can either be purchased with the same material designation or a sample can be easily machined out of a larger part, with little concern that this manipulation will affect the material property. This reality is not the case for pultruded composites. If mature markets of steel are characterized by tight industry standards, by comparison the relatively immature composite market is characterized by trade secrets and variability from both base materials and processes. A one-half inch carbon rod with epoxy resin from one manufacturer could test very differently than the same part from a different manufacturer.

This fact requires then that individual materials must be stringently vetted and qualified before use. However, the anisotropic nature of carbon fiber reinforced material makes testing to a specification difficult. To begin, testing of composite tensile strength (ASTM D3039) (Reference 1) is accomplished by adhesively bonding a tabbing material at the ends of the test specimen corresponding to the grip interface. This reduces damage to the part caused by the large forces at the gripper interface. Material such as G10 is glued to the ends of the part on both sides of the specimen to protect the surface during testing to reduce the likelihood of a grip failure or local crushing of the part, both of which would cause a test failure and register a low tensile strength.

This tabbing process potentially introduces a different mode of test failure. The shear strength of the glue must exceed that of the tensile strength of the specimen, or the tabs will fail before the part. The two equations below display this relationship. Equation 1 describes the strength of the bond surface, doubled for a double bond layer in tabs, while Equation 2 characterizes the tensile strength of the sample in the gauge region.

Equation 1:

$$\delta_i = \frac{F_i}{l \times w}$$

Where

δ_i is the shear stress in the bond, MPa (psi)

F_i is the applied load, N (lbf)

l is the overlap length of the bond, mm (in)

and w is the total width of the bond, in this case both sides of the sample, mm (in)

(Reference 2)

Equation 2:

$$F^{tu} = \frac{P^{max}}{w \times t}$$

Where

F^{tu} is the ultimate tensile strength, MPa (psi)

P^{max} is the maximum load before failure, N (lbf)

w is the sample width, mm (in)

and t is the sample thickness, mm (in)

(Reference 1)

The failure characterized by Equation 2 is not possible for carbon fiber test samples with too large a cross-sectional area. If the part is too thick relative to the available surface area for gluing, a proper tensile failure cannot be completed on the full cross section. When this failure is expected, an attempt is made to reduce the part thickness. While reducing the width of the part would reduce the cross-sectional area being tested, it also dramatically reduces the surface area for adhesive bonding of tabs. Traditional machining methods (grinding, milling) are used to prepare the sample and reduce the thickness. The problem arises when such a significant amount of material is removed that the strength bearing fibers are terminated, damaging the integrity of the sample and drastically reducing the tensile strength. This reality will be examined and an outline for selecting an appropriate sampling method will be proposed.

2. EXPERIMENTATION

2.1 Sample Size and Specifications

The part requirements dictated that in order to meet the specified code, the part must have the following:

Tensile Modulus: 145 GPa (21.0 msi)

Tensile Strength: 2480 MPa (360 ksi)

Approximate-Cross sectional area: 80.7 mm^2 (0.125 in^2)

Surface Prep: Polyester Peel Ply both sides, at least ½ inch total coverage

PolyOne Advanced Composites had a die that was 13.6 mm x 6.38 mm that the customer accepted to meet their needs and timeline. Parts were produced with confidence that they would meet these specifications in end use. However, the customer required testing to validate the parts would perform for their infrastructure application. Tensile modulus testing was conducted on the full section using ISO 527-5 (Reference 3) with the following results:

Table 1:

Full Sample Tensile Modulus Test Results

Sample	Width (mm)	Thickness (mm)	Modulus (GPa)
1	13.4	5.97	159
2	13.4	5.97	156
3	13.4	5.97	158
4	13.4	5.97	158
5	13.4	5.97	156
Mean	13.4	5.97	157
Std	0.00	0.00	1.2

While tensile modulus can be easily tested full section, tensile strength cannot.

2.2 Sample Preparation of Tensile Specimen

In order to validate the assumed tensile strength of the part, the sample had to be ground down to a breakable thickness and tabbed. The estimated tensile strength of the part was 2760 MPa. Since the width of the part is 13.6 mm and the length of the tabs prescribed by the test is 50 mm, the tabbed area of both sides of the part is 1360 mm^2. Using 3M AF 163-2 for the bond lines, the theoretical yield stress of the glue line is 36.2 MPa (Reference 4), giving a maximum load the double bonded tabs can withstand, calculated using Equation 1, of approximately 50 kN. From this maximum force, the required thickness can be calculated using Equation 2 of the samples to be a maximum of 1.33 mm. The parts were ground down below this level, to approximately 1.27mm (.050 in) thickness to ensure proper tensile failure below the force of tab failure. A Sharp SG-820-2AN surface grinder with a grinding stone of 46 grit and coolant was used to grind the parts down. Since a large amount of material removal was required, the parts were ground from both sides, removing 1.25 mm on side A, flipping the sample, removing 1.25mm on Side B, flipping the sample, removing 1 mm on Side A, flipping the sample, removing 1mm on Side B, flipping the sample, and then a final grind of approximately .25 mm on each side to reach the desired thickness. The parts were glued down to the grinder table at each step to reduce the chance of part release.

2.3 Surrogate Parts

To test an unground sample, the exact same carbon fiber and epoxy formulation was pultruded using a die with a cavity dimension of 50 mm x 1.35 mm. At this thickness, the samples can be slit to a tensile specimen width and then tabbed without the need for additional grinding, avoiding the potential for damage to the fibers.

3. RESULTS

3.1 Results of Ground Parts

The parts were tested following ISO 527-5, and the following results were obtained.

Table 2:

Tensile test 1, ground samples

Sample	Width (mm)	Thickness (mm)	Tensile Strength (MPa)	Modulus (GPa)
1	13.6	1.35	2214	156
2	13.6	1.35	2364	161
3	13.6	1.35	2017	162
4	13.6	1.35	2234	136
5	13.6	1.35	2306	153
Mean	13.6	1.35	2227	153
Std	0	0	118	9.4

With a mean of 2227 MPa, the parts were falling short of the specified tensile property. The parts passed tensile modulus and fiber volume expectations, but failed tensile strength. The parts did not show signs of under cured resin or tabbing failure, and the samples had a full brooming fiber failure, shown in Figures 1 and 2 below.

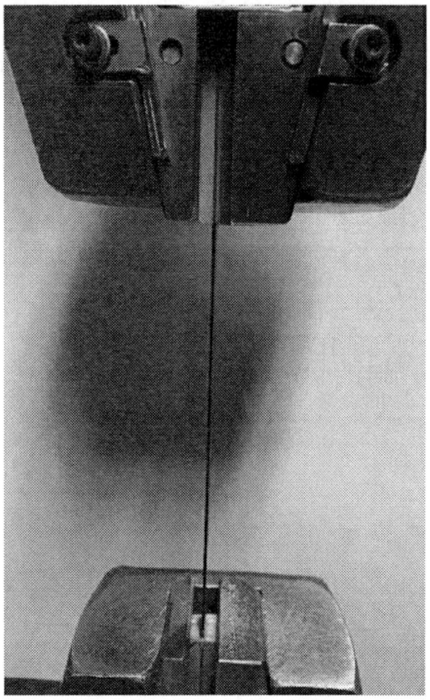

Figure 1: Test Setup

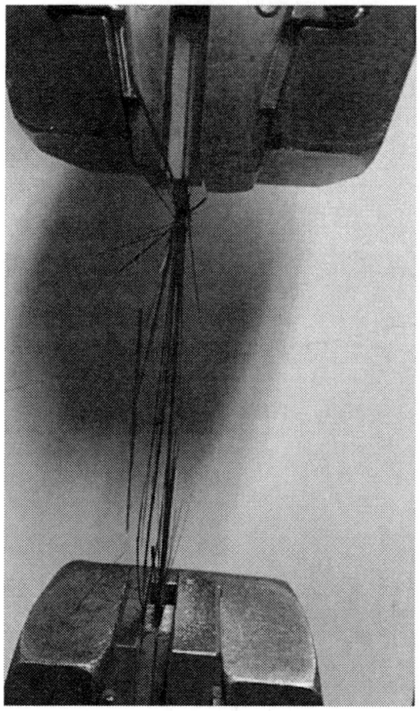

Figure 2: Failed Sample

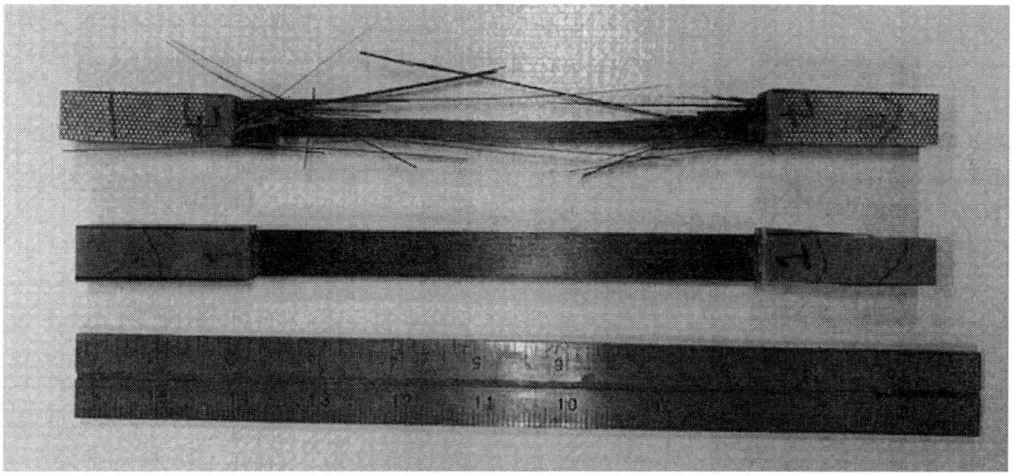

Figure 3: One un-failed and one failed sample

The failure of the sample shows signs of a true tensile failure, indicating the parts strength was much lower than anticipated. To investigate this, the glass transition temperature of the samples was checked to confirm complete cure, shown below in Figure 4.

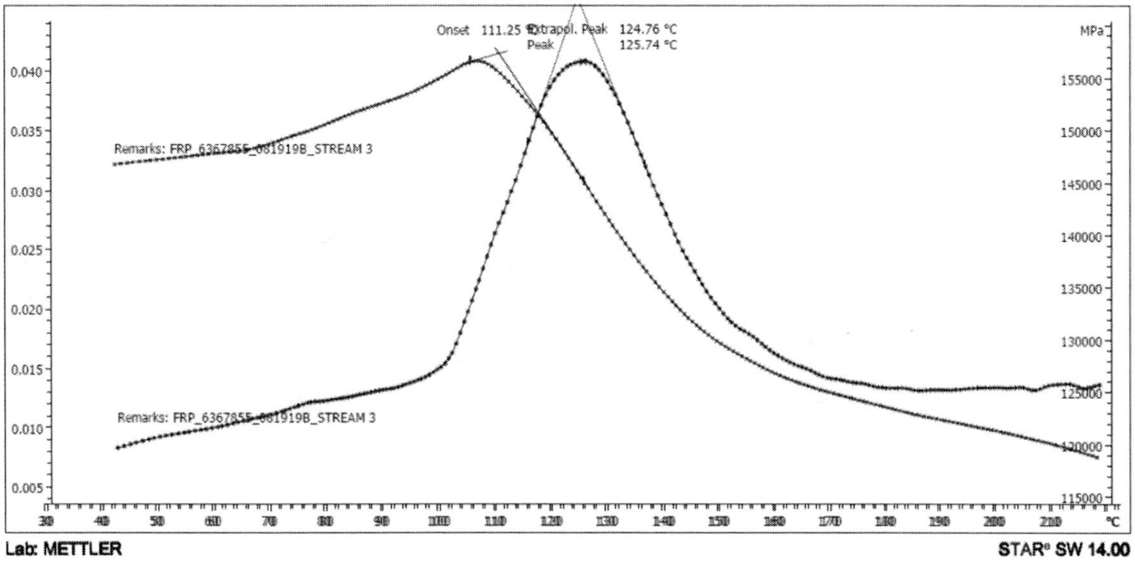

Figure 4: DMA Curve for Samples with Low Tensile Strength

The samples showed a full cure, with an onset Tg of 111 °C (tan δ = 125 °C) in line with many other parts using the same resin formula. In the absence of a failed cure, the samples were prepared again, with extra care taken to grind specimens as carefully as possible.

Table 3:

Tensile test 2, ground samples

Sample	Width (mm)	Thickness (mm)	Tensile Strength (MPa)	Modulus (GPa)
1	13.6	1.12	2082	168
2	13.6	1.12	2248	147
3	13.6	1.12	2303	161
4	13.6	1.12	2234	154
Mean	13.6	1.12	2217	157
Std	0	0	81.8	7.7

After this second set of tests confirmed the first test set, the samples were investigated to learn why these samples had visibly good tensile failures but low strength numbers. The culprit was found by looking at the side of the sample, shown in Figure 5 below.

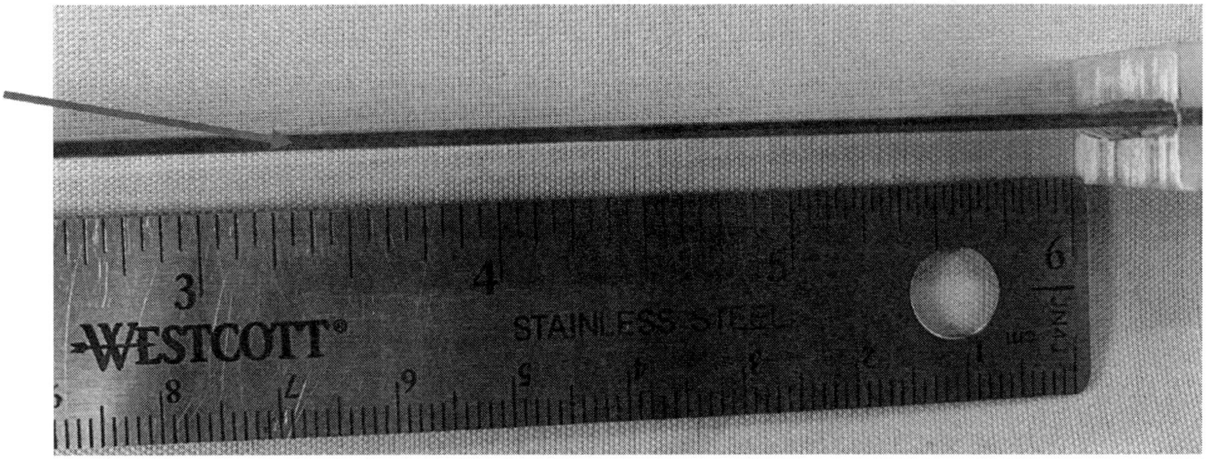

Figure 5: Side view of ground sample

The white line seen (marked with a blue arrow) in Figures 5 and 6 is a result of the parting line of the die. Due to the design of the split cavity pultrusion die, this line is a controlled, straight line at the center of the part, shown below in an unground sample.

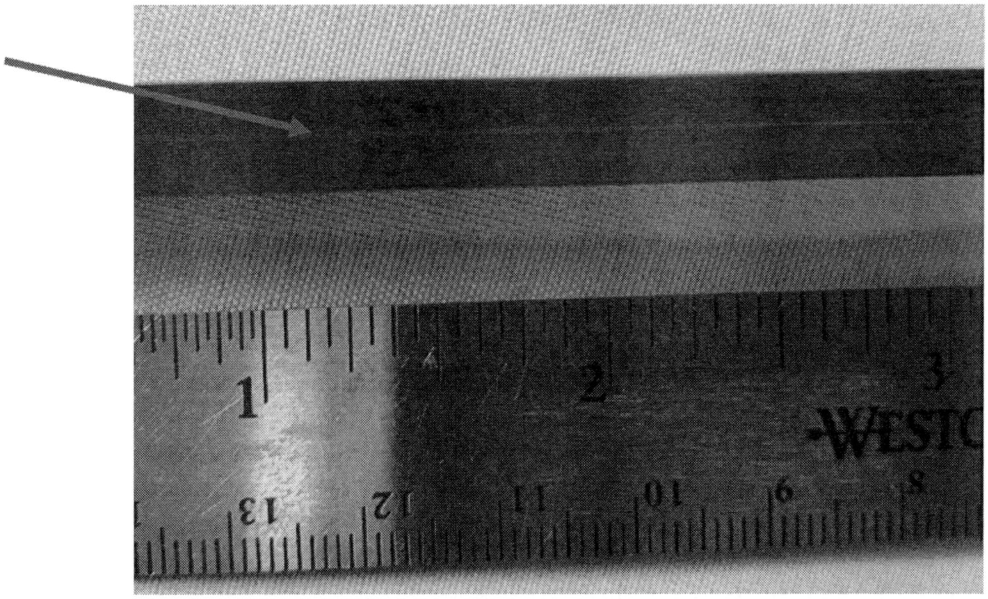

Figure 6: Side view of un-ground sample

However, after grinding, the parting line was no longer straight or in the center of the part. This means that many of the strength fibers of the part had been terminated, as illustrated below in Figures 7 and 8.

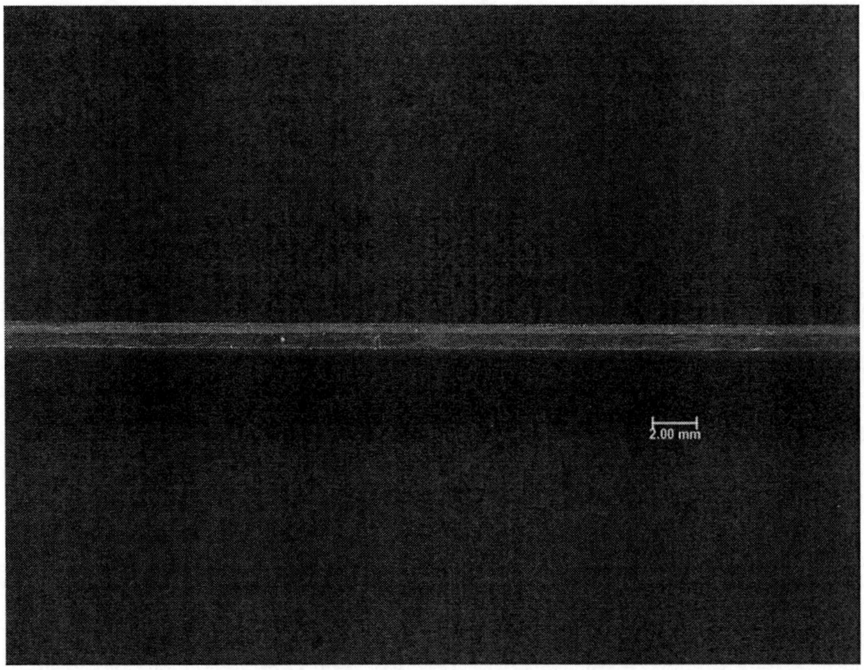

Figure 7: Close up of sloping parting line

Figure 8: Dramatization of sloping parting line after grinding

After repeated attempts to grind the part flat and even, the parting line did not remain equidistant from both ground sides of the part and was closer to one edge, predominately in the center of the part.

3.2 Results of Surrogate Parts

In order to properly represent the tensile strength of the parts, the decision was made to manufacture test coupons of the same carbon volume fraction using a die with a thickness of 1.37 mm. This die would allow for parts to be made using the same carbon content, the same resin composition, and the same cure profile to be tested in tension without the requirement to grind down the parts, thus ensuring the fibers would be unterminated and straight relative to the test direction. The parts were produced, the volume fraction and cure profile were confirmed, and the tensile test shown below.

Table 4:

Tensile Results of samples from Surrogate Die

Sample	Width (mm)	Thickness (mm)	Tensile Strength (Mpa)	Modulus (Gpa)
1	12.2	1.37	2979	172
2	12.1	1.37	2930	156
3	11.9	1.37	2882	163
4	11.6	1.37	2841	174
5	12.0	1.37	2820	152
Mean	12.0	1.37	2890	163
Std	0.22	0.00	58.1	8.8

The fiber volume testing confirmed that the surrogate samples were representative of the full-size product, shown below in Table 5. The fiber volume was slightly lower than the initial samples (< 0.5%), but given that the strength was still passing the specified failure load, the parts were shipped.

Table 5:

Comparative Fiber Volume Fraction

Full Sample	Surrogate Sample
65.20%	64.80%

Testing the undamaged carbon fiber allowed the fibers to carry more load than the ground and terminated fibers of the original sample set. If the original parts had been significantly wider or a significantly higher strength glue could be obtained to perform a test on a thicker sample, the original samples would have shown very similar tensile strength values. However, there was too much damage to the parts during the sample preparation to properly qualify the part strength, without using the surrogate samples.

4. CONCLUSIONS

This test data is all consistent with damage to the load bearing fibers due to imperfect machining of unidirectional fibers. If the fibers are not held exactly parallel to each other either during the pultrusion or the machining, then machining of the finished part to alter the thickness, will inevitably lead to fiber termination. The strength of the part will continue to decrease as the number of fibers terminated increases. The danger of using machining practices to modify test samples is the ease with which these parts can simulate tensile failure that are actually delamination failure of terminated fibers. Untrained testers could easily misinterpret this as a true tensile failure and either disqualify the part or require more material be used when less is needed, driving up costs and undervaluing the benefits of carbon fiber.

In the absence of industry standards surrounding carbon fiber part manufacture, and in light of the damage that is easy to cause with sample preparation of anisotropic carbon parts, it is recommended that the industry adopt standard practice for sampling parts produced on a thinner die. In metal testing and prepreg testing, this practice is already fairly standard. Smaller samples can be bought of the same type of metal and tested. Thinner plates of prepreg can be made, forgoing the need for grinding. Producing thinner samples of pultruded parts should follow the same conditions, limited to the following: The fiber volume remains the same, the input materials are identical, and the resin glass transition temperature confirms the matrix cure profile matches that of the full-size sample.

5. REFERENCES

1. ASTM D3039/D3039M-17, "*Standard Test Method for Tensile Properties of Polymer Matrix Composite Materials*"
2. ASTM D5656-10, "*Standard Test Method for Thick-Adherend Metal Lap-Shear Joints for Determination of the Stress-Strain Behavior of Adhesives in Shear by Tension Loading*"
3. ISO 527-5:2009, "*Plastics — Determination of tensile properties — Part 5: Test conditions for unidirectional fibre-reinforced plastic composites*"
4. *3M Scotch Weld™ Structural Adhesive Film,* AF 162-2, 2009. [online]. Available: https://www.3m.com/3M/en_US/company-us/all-3m-products/~/3M-Scotch-Weld-Structural-Adhesive-Film-AF-163-2/?N=5002385+3292667667&rt=rud

FATIGUE LIFE ASSESSMENT OF GFRP FOR LONGITUDINAL LEAF SPRINGS

Fabian Becker[1)], Christian Hopmann[1)], Francesco Italiano[2)], Alberto Girelli[2)]
1) Institute for Plastics Processing at RWTH Aachen University
Aachen, Germany

2) Ford Research and Innovation Center
Aachen, Germany

ABSTRACT

Due to the high specific properties, fiber reinforced plastics are ideal candidates for the substitution of steel parts in the automotive sector enabling high weight savings and increasing the efficiency of automobiles. One application are composite leaf springs made of glass fiber reinforced plastic (GFRP). Because of the simple bending load case, the fiber direction can be aligned with the loading direction. For longitudinal leaf springs, the axle is clamped in the center to the leaf spring component leading to a different stress state compared to pure bending.
The aim of the research presented (in a cooperation between Institute for Plastics Processing and Ford Research and Innovation Center) is to develop a testing method on specimen level that is able to determine the fatigue life including the complex stress state in the clamping section. For this purpose, three point bending experiments on unidirectional specimens made of GFRP epoxy pre-preg material are performed including a miniaturized clamping system. This results in a more realistic representation of the leaf spring stress state in the fatigue experiments. The applied force in the clamping section is measured during experiments and reveals the creeping behavior of the composite material. Fatigue testing also shows a significant change in damage evolution compared with the pure bending stress state resulting in shorter fatigue lives. With the results of this research, higher safety in the design of this component and less testing time is achieved.

Keywords: Composite leaf spring, GFRP, fatigue, longitudinal, transverse loading, damage

Corresponding author: Fabian Becker, fabian.becker@ikv.rwth-aachen.de; +49 241 80 27136

1. INTRODUCTION

The trend towards e-mobility and higher efficiency of combustion motor vehicles requires lightweight structures and materials. One possibility for mass savings up to 60 % compared to steel is the composite leaf spring [1]. A simple geometry, better damping properties and the abscence of corrosion issues compared to steel allows for a competitive design of the composite leaf spring. Glass fibers are frequently used due to their higher strain-to-failure and lower costs compared to carbon fibers. As material, epoxy resin as well as polyurethan resins are in use. Transverse and longitudinal leaf springs are already used in series production of automobiles [2].

Longitudinal leaf springs allow for a simple and robust design of a vehicle suspension system. The GFRP leaf spring in the research presented replaces a steel part. For an economic replacement, no or little modifications are made to the connecting parts. The axle is connected via a clamping group to the leafspring. The leafsprings' ends are respectively connected to the chassis frame through a y-axis rotational-free rubber joint on the one hand side and through a connecting link on the other hand side (shackle). The shackle enables free-rotation about the y-axis and compensation of the leaf springs' change in support span during bending (Figure 1). The load case of bending leads to a well defined stress distribution in the leaf spring, allowing a design with glass fiber reinforced plastics (GFRP), where the fibers are oriented in main stress direction.

Material testing for leaf springs is done taking into account the boundary conditions close to the real load case in three point bending experiments under pure bending stress state, neglecting the complex load case in the clamping section. Fatigue testing and modelling for multi axial stress states is frequently done on thin walled laminates for different combinations of stresses acting in the lamina plane [3-6], but not for thick walled laminates. Static failure models exist for complex stress states with combined shear stress, transverse stress and fiber parallel stress for inter fiber failure and fiber parallel compressive failure [7-8]. An experimental approach performed by *Bastian* aims to reproduce the stress state of GFRP leaf spring materials via one sided clamped cantilever beam [9]. However, the approach does not resemble completely the symmetry in the clamping section of a longitudinal leaf spring. Basic strengths measured with unidirectional material as input for laminate failure criteria may underestimate the strength under bending load cases [14].

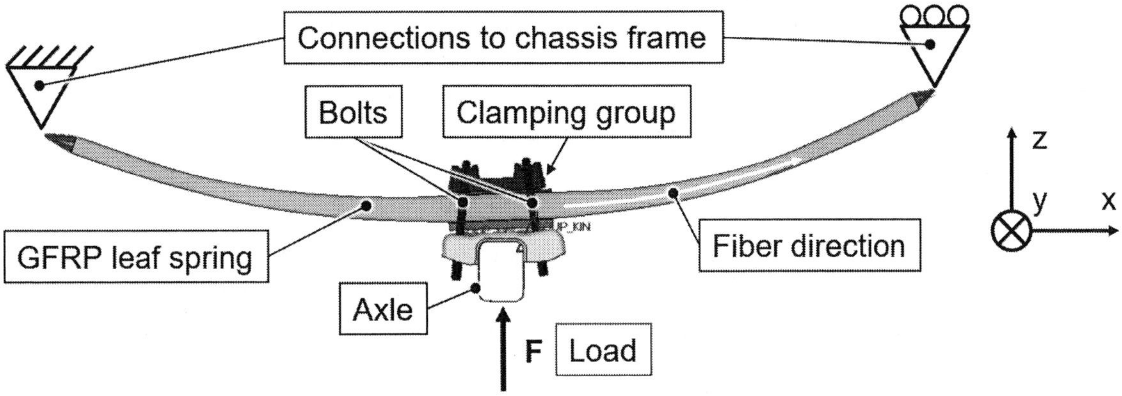

Figure 1: GFRP longitudinal leaf spring with clamping group and load case, adapted figure from [10]

In the following, an experimental approach will be presented that aims to reproduce the complex stress state, which occurs in the clamping region of a leaf spring. The clamping section is derived from the real component (Figure 1).

2. EXPERIMENTATION

2.1 Materials

The material used in this research is a unidirectional E-glass fiber reinforced epoxy, which is manufactured by pre-preg laying in a hot molding press. The tool has three molds to produce unidirectional bars. After demolding, each of the bar is sawn into three specimens, which are further prepared manually by deburring the edges on the compression side. After this process step, the specimens are cured. The final specimen geometry has a non-rectangular cross section based on the work on *Feldten* [11]. The fillet radius on the tension side of the specimen is 5 mm and causes a shift of the neutral axis during bending to guarantee a gradual damaging behavior on the tension side during fatigue testing without damages caused by stress inhomogeneity near the edges (Figure 2).The specimens have a thickness of h = 5.57 mm +/-0.097 mm and a width of w = 20.11 mm +/- 0.037 mm. Their length was not measured individually, but larger than 170 mm, since the stress state depends on the support span in the three point bending. For the scope of the research presented, the coordinate system is defined as "1" in fiber direction, "2" in the lamina plane and "3" in thickness direction.

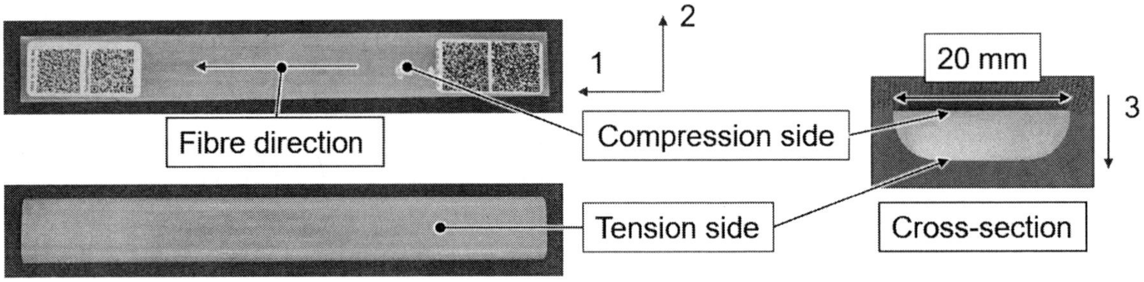

Figure 2: Used specimen: Top left: Compression side; bottom left: Tension side; Right: Cross section

2.2 Clamping system to introduce transverse stress

To introduce transverse stresses during the three point bending fatigue test, a clamping system is developed. The clamping system consists of an upper and a lower steel plate, which are identical in the geometric dimensions (Figure 3). The upper half is connected with an adapter to the testing rig. Between the two steel plates, aluminum bodies are placed, covered with a 30 % glass fiber reinforced polyamide layer. At the location of specimen contact, the PA layer has a thickness of 1 mm. The PA layer aims to reduce stress peaks that may lead to a premature failure due to localized fiber breaks. The aluminum support allows the application of thermoelements via drilled holes at two different locations, with one location being in the center of the support and one location close to the specimen. The use of the support of aluminum between steel plate and PA layer was implemented after we published first results of the first version of the clamping [12].

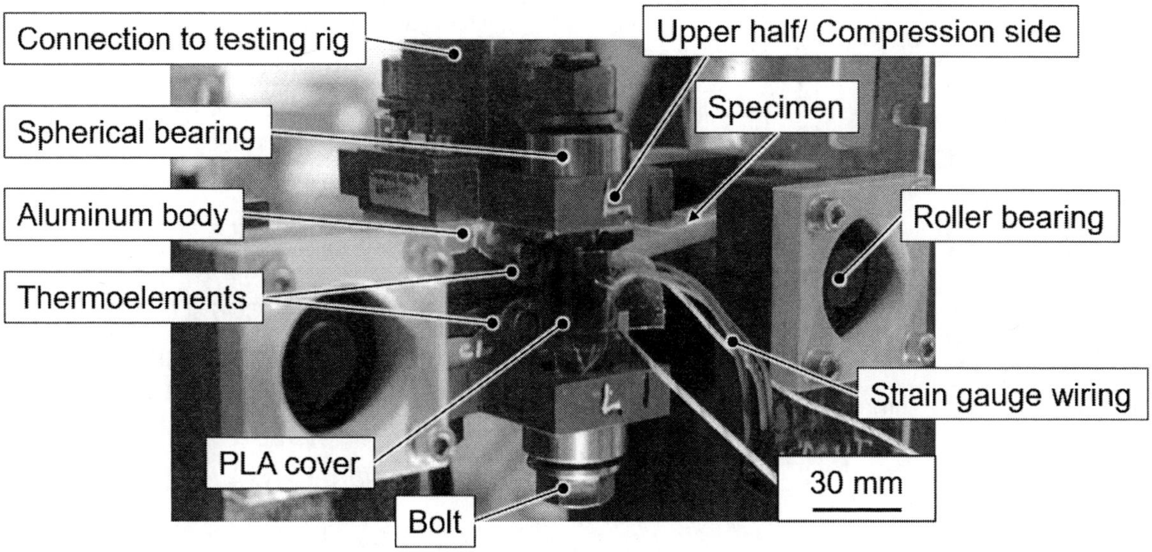

Figure 3: Clamping system to introduce transverse stress

The upper and the lower half of the clamping are connected by two M12 bolts (one for each side) made of Ti–6Al–4V alloy for lower mass and necessary durability of the bolts. A lower Young's modulus enhances the sensitivity of the measured strain gauge values. The endings of the bolts are supported by spherical bearings, which avoid any bending moment acting on the bolts guaranteeing a pure tensile load application. The load on the bolts is achieved by tightening the two nuts at the end of the bolts. The bolt loads are measured with applied strain gauges on the bolt shafts in Wheatstone bridge configuration. To protect the strain gauges, they are covered with 3D-printed parts of PLA plastic filled with epoxy resin.

2.2.1 Calibration of bolt load cells

The force signal was calibrated in the same configuration as in the clamping group using a 150 kN universal tensile testing machine from Zwick, Ulm, Germany (Figure 4, left). The signal was processed with a VLink 200 amplifier from Lord MicroStrain, Williston, USA. To distinguish the two signals, the bolt loads are denoted with "green" and "black" in the following, according to the color of the plastic cover of the strain gauges.

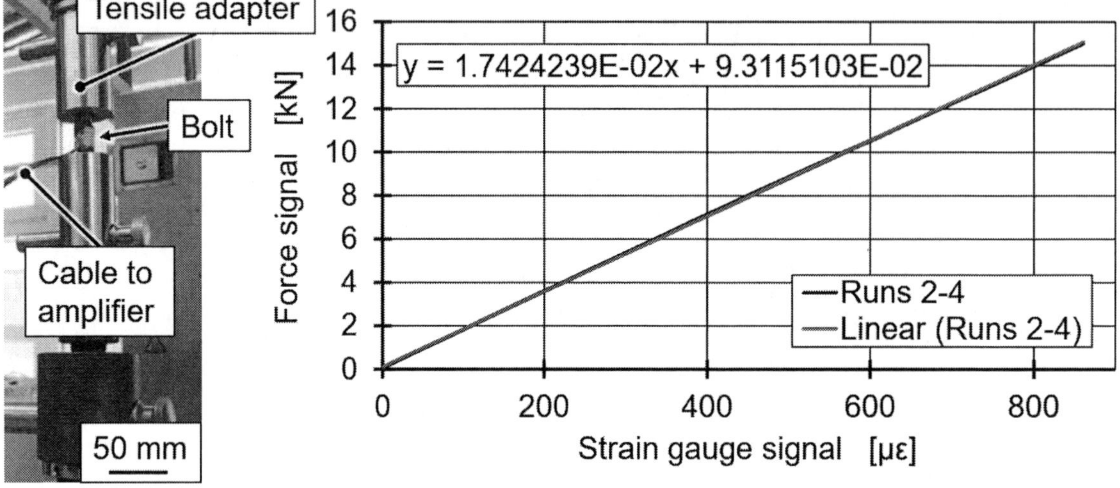

Figure 4: Left: Testing setup for bolt load calibration; Right: Calibration fit polynomial

The bolts were successively loaded with 10 kN in a first step and then up to 15 kN with 5 repetitions. The last 4 repetitions were used for the correlation of force signal to strain gauge signal to avoid slipping or nonlinearities occurring in the first repetition. Since the behavior of the bolts is linear-elastic, a simple linear function was fit for subsequent calculations (Figure 4, right). The entire system was designed for a minimum strength of 30 kN. From the determined transmission factors between strain gauge signal and force signal, the corresponding force could be determined directly from the strain gauge signals. It is not necessary to determine the modulus of elasticity of the bolt shafts with this method. After calibration of the bolt load cells, the clamping system was pre-loaded with 15 kN bolt loads on each side to allow creeping of the PA layer and the formation of a flat section in the center of the clamping. Pre-loading was performed with the application of a steel specimen with the same edge fillet as the specimens.

2.3 Finite Element Simulation to determine the stress state

To determine the stress state caused by the transverse bolt load in the clamping group, a quarter model of the experimental setup was implemented with the software Abaqus, Simulia, Providence, USA, including symmetry conditions. The material parameters were selected as follows for its components:

Material	Young's modulus [GPa]	Poisson number [-]
Steel	210	0.2
Aluminum	70	0.35
Ti–6Al–4V	116	0.34
PA-GF30	8.5	0.3

Table 1: Material parameters used in the Finite Element simulation

The material parameters and the geometry of the specimens are adjusted according to the experimental results for the bending modulus. All other elastic constants of the GFRP material are calculated with the rules of mixtures presented by *Bleier* [13] and under the assumption of an isotropic E-glass fiber and epoxy resin with the following material parameters:

Material	Young's modulus [GPa]	Poisson ratio [-]
E-glass fiber	73	0.2
Epoxy resin	3.3	0.3

Table 2: Material parameters of fiber and matrix to determine the elastic constants

Spherical bearings at the end of the bolts were modelled by couplings at the ends of the bolts that only transfer tensile loading.

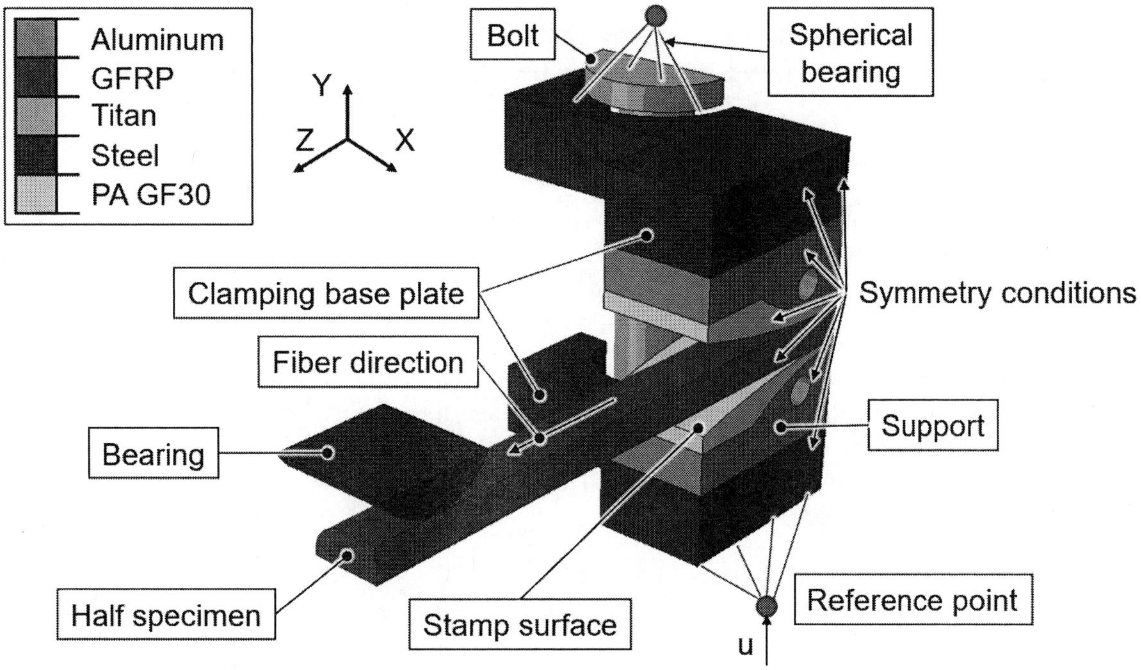

Figure 5: Quarter model of clamping system in FE-analysis

The Finite element model was loaded subsequently. In a first step, the simulation bolt loads were adjusted to the half of the bolt load measured in the experiments due to the half representation of the bolts in the model. In a second step, the bolts were fixed at their current length and the bending of the specimen was achieved via the displacement of a reference point. The reference node was coupled via continuum distribution constraints to one steel plate on the compression side of the specimen. A maximum bending displacement of 32 mm was set. From the finite element model, the maximum strain and stress on the tension side were found via a Python script and were used to define the relation of force and stress as well as the relation between displacement and strain:

$$\sigma = f(F) \qquad \text{Eq. 1}$$

$$\varepsilon = f(u) \quad \text{Eq. 2}$$

From the definition of the function, the necessary machine displacement of the testing rig during the fatigue tests could be calculated. Figure 6 shows the resulting function for Eq. 2, normalized to the strain of first failure events from the unclamped static loadcase under pure bending.

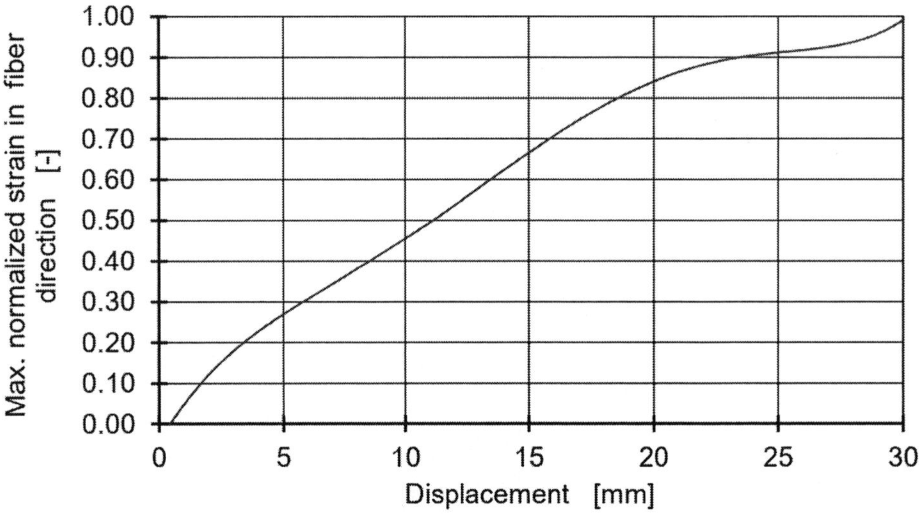

Figure 6: Master-curve for calculation of test rig displacement during the fatigue test to obtain the desired strain in fiber direction

2.4 Static testing

The static testing is divided into three different testing series with bolt loads of 5 kN, 7.5 kN and 10 kN on each bolt. Each series was conducted with six specimens. During testing, the failure behavior of the specimens was recorded with a camera system EOS 700 D, Canon, Tokyo/ Japan. The camera system was moving together with the transverse of the testing rig, observing the damaging region via a mirror (Figure 7).

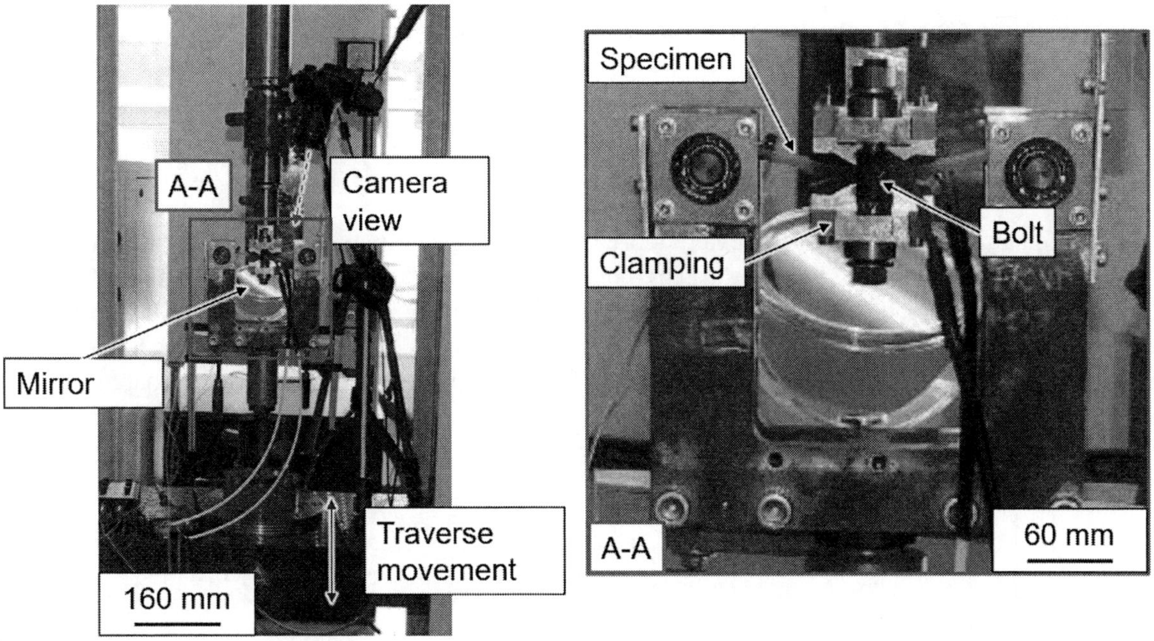

Figure 7: Experimental setup of static testing

All tests were run displacement controlled until a bending displacement of 32 mm. Static testing was performed under ambient conditions. Prior to the testing, each specimen was bent individually to displacements of 2 mm and 4 mm to calculate the bending modulus from both resulting bending force values. The specimen was then placed into the clamping system and the torque was applied until the desired bolt loads. Each individual bolt load was recorded for each specimen.

2.5 Fatigue testing series

The fatigue tests were performed in four different testing series. In the reference series, the open clamping was selected to cause a pure bending stress state. The tests were performed displacement controlled with $R = 0.1$ and a frequency of 10 Hz. As failure criterion a residual stiffness criterion based on the dynamic stiffness measured during the test was chosen. The resulting S-N curve serves as a reference line.

In the following testing series, specimens were loaded with bolt forces of 5 kN, 7.5 kN and 10 kN each and then tested in a fatigue test under displacement controlled 3-point bending with $R = 0.1$ until the residual stiffness criterion was reached. To account for the stiffness of the clamping group, the Master-curve from the FE-analysis was used to calculate the necessary machine displacement for reaching the desired strain in fiber direction.

The fatigue tests of the bolt load series were carried out at a frequency of 2 Hz to reduce local heating of the specimens by friction. The test was automatically interrupted at constant intervals and images were taken to document the state of damage of the tension side. The bolt loads were recorded with a frequency of 32 Hz. Additional temperature data was recorded with type K thermoelements in the upper half and the lower half of the clamping, with one thermoelement close to the specimen and one inside the clamping group. One was placed at the servohydraulic loading pin. All correlations of measured data were performed using the software Matlab, The MathWorks, Natick, USA.

3. RESULTS

3.1 Static testing

Since the static testing series was evaluated with the FE-analysis, the output of the reaction force in the model should agree with the measured bending force of the experiment until the first damage in the experiment occurs. Figure 8 shows the correlation between the FE-analysis and the measured values for a specimen clamped with 10 kN bolt load. Since a good agreement is reached by this approach, the stress-strain curves creation from FE-data appears as reasonable approach.

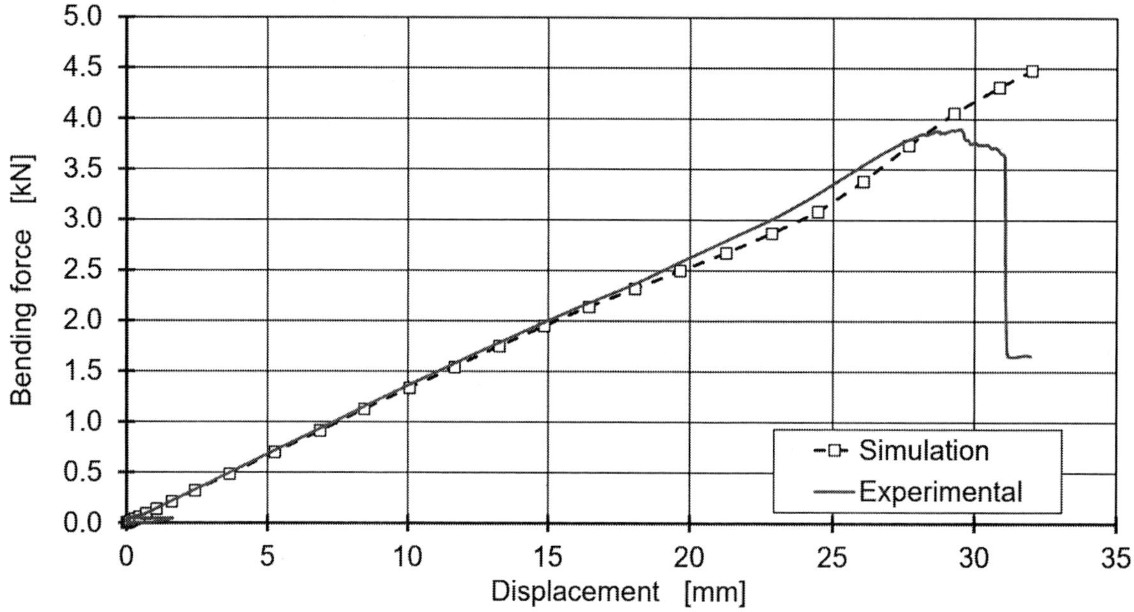

Figure 8: Experimental bending force compared to the reaction force in the reference point of the FE-analysis for a specimen with 10 kN bolt load

In the stress-strain curve, the sudden drop in stiffness is visible as the specimens fails on the compression side. Stress-strain curve as calculated by Eq. 1 and Eq. 2 is displayed in Figure 9.

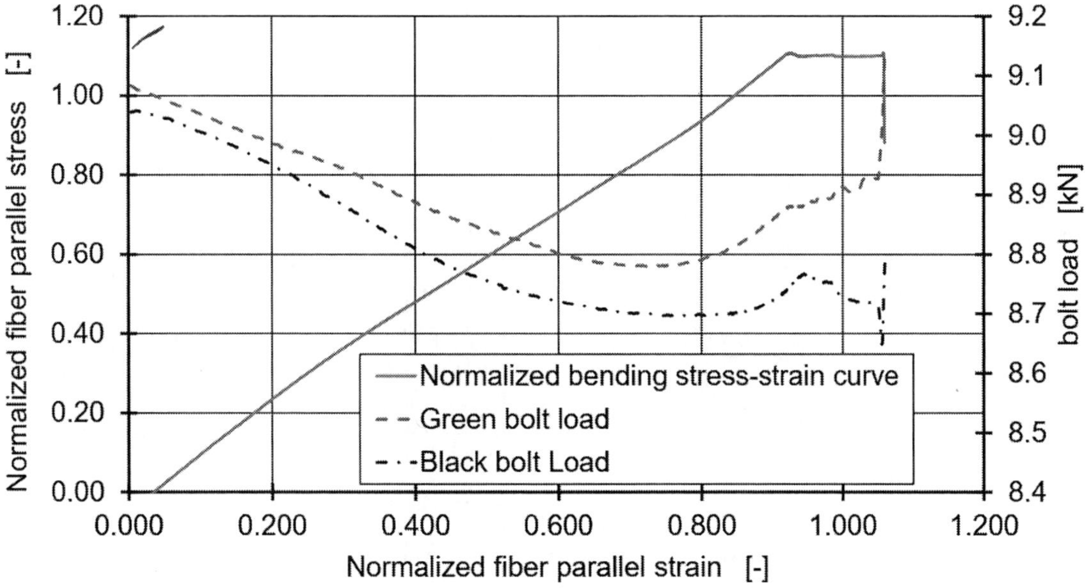

Figure 9: Stress-strain response of the specimen calculated with Eq. 1 and Eq. 2 normalized to static values

The bolt loads during the static test for the specimen shown in Figure 9 on the right y-axis are lower than the initially adjusted bot loads of 10 kN due to creeping of the specimen prior to the bending test. During increasing displacement, the bolt loads show a decrease which may be assigned to transverse contraction effects of the material. After contact of the specimen with the compression side of the clamping group, the bolt load increases because the specimen is constraint in separating the compression and tension side of the clamping.

All specimens failed due to compressive failure after contact of the clamping surface on the compression side. Prior to the failure on the compression side, some of the specimens showed visible damage on the tension side. However, due to the transverse clamping, a shear stress failure was expected. Since the shear failure did not occur in any of the specimens, but the overall failure was constraint by the geometry of the clamping, the conclusion can be drawn that under static load cases the clamping does not alter the damaging behavior on the tension side (Figure 10). The calculated strain-to-failure, however, is in the same range as the strain-to-failure on the tension side in the unclamped case. This indicates that the strain-to-failure on the compression side may be higher under three point bending than measured in unidirectional specimens since micro buckling is absent and true fiber failure occurs. This agrees with the findings of *Preller et al.* [14].

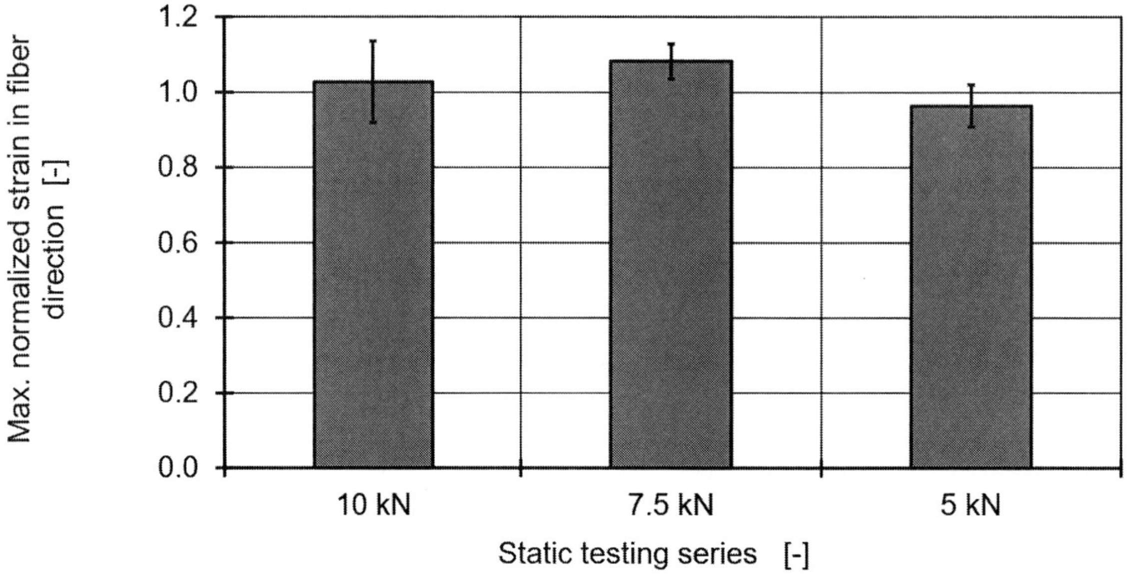

Figure 10: Strain-to-failure in the clamping testing series, normalized to the unclamped strain-to-failure

3.2 Fatigue testing

During fatigue testing, the dynamic stiffness of the testing rig, the bolt loads and the temperature data of five positions at the experimental setup as well as the ambient temperature are recorded. Together with the automatically taken images, data correlation allows for the determination of the damaging and failure behavior of the specimens.

Figure 11 shows the results for an initially with 5 kN bolt loads clamped specimen. In the top diagram, the loss in dynamic stiffness during the experiment is plotted over the normalized cycles until reaching the residual stiffness failure criterion. Since the test was run without an automatic stopping criterion on the testing rig, the test duration exceeds the value of one on the x-axis. The numerical value of the normalized dynamic stiffness is related to the right y-axis on the right side of the diagram. The dynamic stiffness curve shows a plateau in the beginning of the fatigue test, where no or very little loss of dynamic stiffness is observed. This is followed by a steep loss of dynamic stiffness (0.7 times of the life of the specimen). The steep loss is again followed by a gradual loss in dynamic stiffness until the next steep loss in dynamic stiffness can be observed (1.1 time the life of the specimen), again followed by a gradual loss in dynamic stiffness. For the displayed specimen, this behavior repeats again after 1.2 times of the life of the specimen.

The bolt loads showed an oscillating behavior during each cycle as expected in the fatigue tests. However, the R-ratio of the bolt loads for all specimens was higher than 0.95 being close to a quasi-static preloading. For the display of the bolt loads, the bolt load signal was averaged over every single cycle. This averaged bolt loads are shown in Figure 11 in the top diagram with values on the left y-axis. Although the bolts were tightened to 5 kN prior to the fatigue test, both bolt loads show a loss at the beginning of the experiment. This behavior does not correlate with the stiffness curve and is interpreted as relaxation of the matrix material since the specimen is loaded in transverse direction by the clamping system. Relaxation effects decrease during the fatigue test as expected for this time dependent property. The first steep loss in stiffness is not visible on the bolt loads. However, the second and third loss of dynamic stiffness is also visible in the bolt load data indicating a significant amount of damage of the specimen.

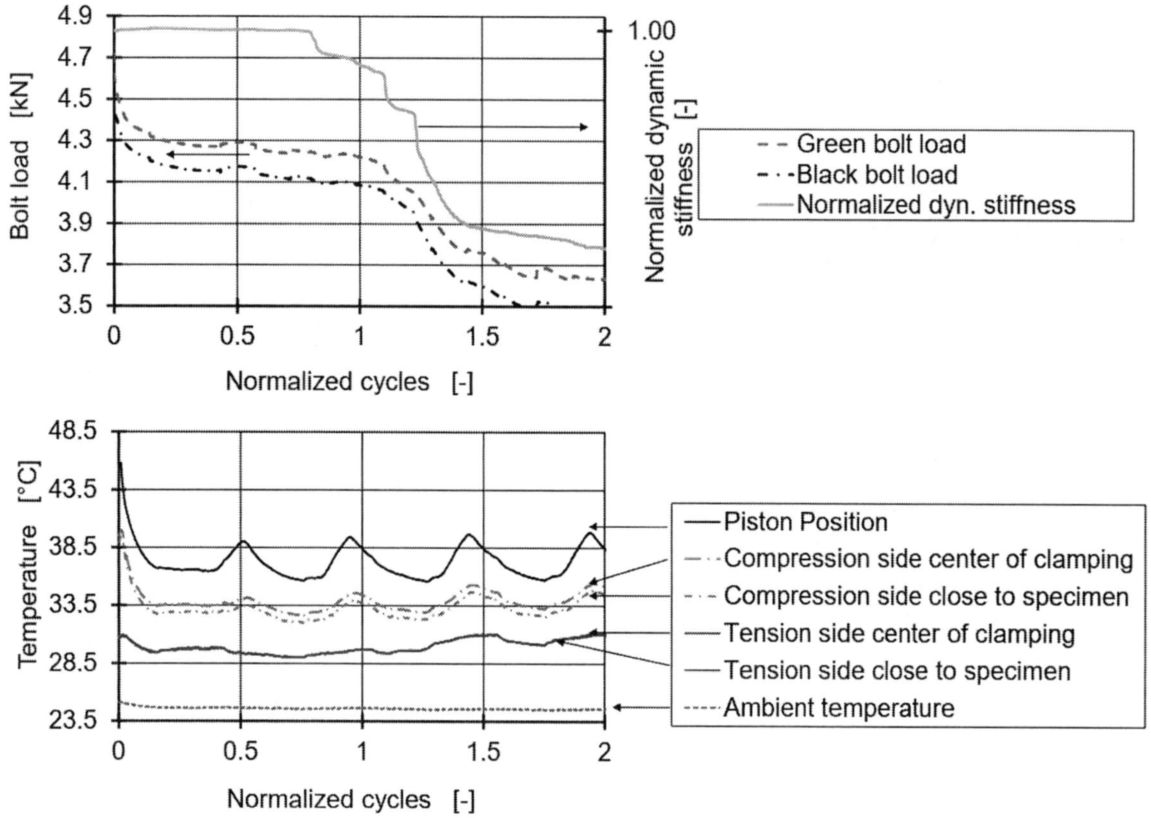

Figure 11: Fatigue testing results of a specimen loaded with 5 kN bolt load; Top: Loss in dynamic stiffness and bolt load during fatigue testing; Bottom: Temperature measurement at different position of the clamping

The long-term oscillation of temperature is visible on the loading pin position of the thermoelement (Figure 11, bottom). Since the slopes of the temperatures on the compression side of the clamping follow the slope from the loading pin position but with a lower absolute value, it is concluded that heat conduction takes place. A variable oil temperature during the fatigue testing may lead to this behavior causing a subsequent thermal expansion of the compression side of the clamping system. The thermoelements on the tension side of the clamping do not reveal a similar oscillation but show higher values compared to the ambient temperature, indicating that heat transfer over the bolts may be present with a high thermal inertia. The temperature difference between the compression side and the tension side of the clamping system correlates with slight temporary increases in the bolt load signal. It is concluded that a different thermal expansion between compression side and tension side of the clamping system led to an increasing bolt load, but not the damaging behavior of the specimen itself.

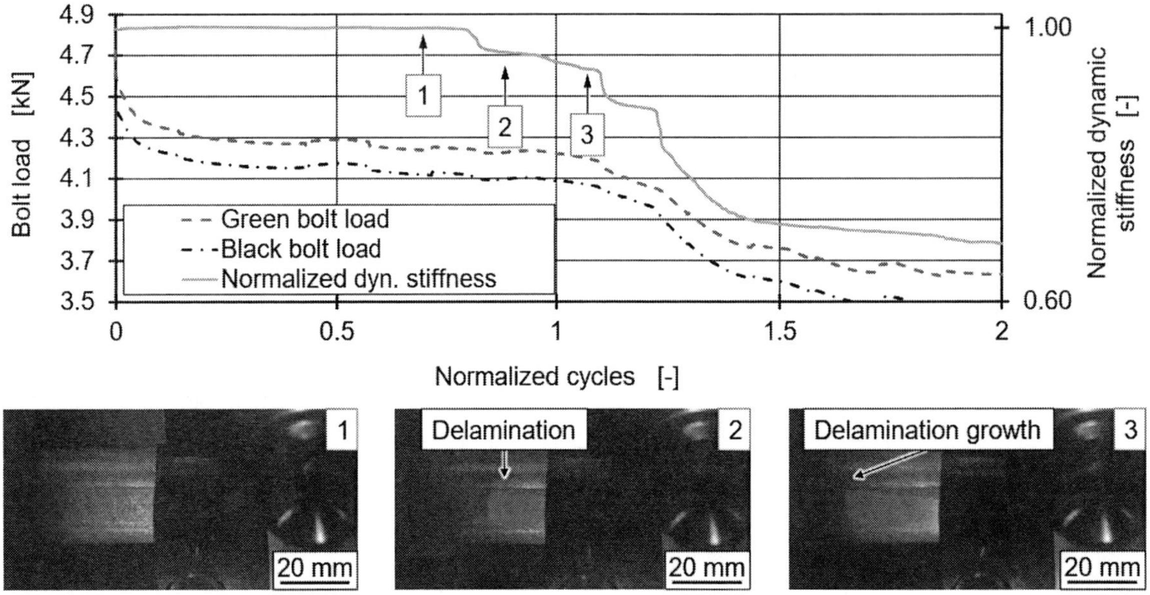

Figure 12: Delamination initiation and growth during the fatigue experiment

From the images, the steeper loss in stiffness can be attributed to the beginning of a delamination on the tension side of the specimen. Three images are shown that were taken before and after the first loss in stiffness (Figure 12). Compared to the pristine state in the plateau of the stiffness curve, where no delamination is present (Figure 12, 1) a delamination originating from the edge of the clamping system can be seen in the following image (Figure 12, 2). Compared to the length of the delamination in Figure 12, 3, the velocity of the delamination growth decreases, leading to the subsequent slower loss in dynamic stiffness.

The second and third steep loss in dynamic stiffness can be attributed to a second delamination in another layer of the specimen. After the fatigue experiment, the tension layers of the specimens remained intact, although they were found separated from the neighboring layers. This leads to the final conclusion that a shear induced delamination growth was present. The figures are representative for all specimens that were fatigue tested with clamping system.

The resulting fatigue lives judged solely by the residual stiffness criterion and related to the maximum fiber parallel bending strain are shown in Figure 13. The double logarithmic diagram includes all three clamped testing series as well as the reference testing series without clamped specimens. The maximum fiber parallel bending strain is normalized to the strain of the first damage event from static tests. The x-axis is normalized to the maximum number of cycles of testing. The maximum numbers of cycles as normalization value was taken from one specimen on a low strain level that did not reach the residual stiffness criterion (marked with arrow in Figure 13). All specimens are assigned to their initial clamping force prior to the start of the fatigue test.

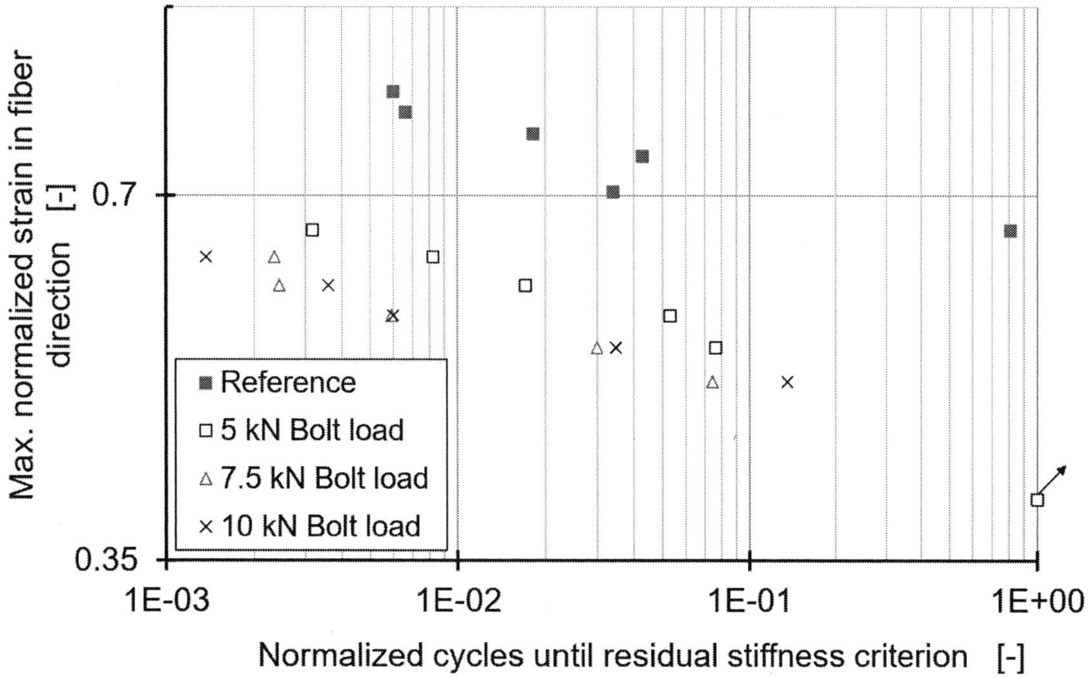

Figure 13: Reference S-N data (without clamping) and S-N data of 5 kN, 7.5 kN and 10 kN bolt load testing series

Based on the altered damaging behavior, all specimen series reveal a significantly shorter life compared to the unclamped case. This also has been reported for a different testing setup also by *Bastian* [9], although the damage mechanism was different. For the specimens of the 5 kN bolt loads, the S-N data shows the tendency that delaminations are initiated later during the fatigue test compared to the 7.5 kN and 10 kN testing series. The testing series of 7.5 kN and 10 kN are in the same range. An increasing bolt load therefore does lead to a nonlinear shift of the S-N data. The cause for this behavior is yet undetermined.

4. CONCLUSIONS

A clamping system has been designed to introduce additional transverse stresses on the middle section of a glass fiber reinforced plastics material under three point bending loading. The desired clamping load is introduced through two surfaces by tightening two bolts, which are positioned on the left and right hand side of the specimen. The bolt loads are measured via strain gauges. The whole system aims to reproduce the clamping situation in a longitudinal leaf spring.

During the static testing series, the final failure was found on the compression side, which was induced by specimen contact to the surface of the clamping group. However, the calculated strain-to-failure agrees with the static strain-to-failure in tensile direction of the GFRP material without clamping. All three levels of bolt load do not alter the failure behavior in the static load case.

Under three point bending fatigue loading, a different damaging and failure behavior was introduced by the clamping group. After a plateau visible in the loss of stiffness curve, where micro damages are accumulated, a steep loss in stiffness is observed for all specimens. From images during the fatigue tests, the drop in stiffness could be assigned to the initiation and growth of a delamination. Since the fiber layer does not fail in this case, shear induced delamination are the main reason for the loss in stiffness. Due to testing with different bolt loads, a nonlinear relation between bolt load and fatigue lives is seen in the S-N data.

The results show that a life estimation of a longitudinal leaf spring based on three point bending experiments on specimen level without considering the influence of a clamping will overestimate the fatigue life leading to a non-conservative fatigue design.

5. ACKNOWLEDGEMENTS

The depicted research has been funded by the Ford Motor Company as part of the "RWTH Ford Alliance" contract. We would like to extend our thanks to the Ford Motor Company.

6. REFERENCES

[1] F. D. Meatto and E. D. Pilpel, *Durability Comparison of Fiberglass Monoleaf Hybrid and Multileaf Steel Springs*: Steering and Suspension Technology Symposium. Detroit, Michigan, USA, 1999

[2] K. Wood, *Composite leaf springs: Saving weight in production.* Composites Technology 19 1, S. 34-39, 2014

[3] M. Gude, W. Hufenbach, I. Koch, and R. Protz, *Fatigue failure criteria and degradation rules for composites under multiaxial loadings.* Mechanics of Composite Materials 42 5, S. 443-450, 2006

[4] Z. Fawaz and F. Ellyin, *Fatigue Failure Model for Fibre-Reinforced Materials under General Loading Conditions.* Journal of Composite Materials 28 15, S. 1432-1451, 1994

[5] X. Diao, L. B. Lessard and M. M. Shokrieh, *Statistical model for multiaxial fatigue behavior of unidirectional plies.* Composites Science and Technology 59 13, S. 2025-2035, 1993

[6] J. Petermann and A. Plumtree, *A unified fatigue failure criterion for unidirectional laminates.* Composites Part A 32, S. 107-118, 2001

[7] S. T. Pinho, C. G. Dávila, P. P. Camanho, Iannucci and P. Robinson, *Failure models and criteria for FRP under in-plane or three-dimensional stress states including shear non-linearity.* NASA Langley Research Center, Report TM-2005-213530, 2004

[8] A. Puck and H. Schürmann, *Failure Analysis of FRP Laminates by Means of Physically Based Phenomenological Models.* Composites Science and Technology 58 7, S. 1045-1067, 1998

[9] P. Bastian, Klemm-Krafteinleitungen für biegebelastete Faserverbund-Balken. Technische Universität Darmstadt, PhD thesis, 2001

[10] A. Preisler, F. Wolf-Monheim, A. Dafnis, K.-U. Schröder, P. van der Jagt, W. David and P. Zandbergen, *Faseroptische Schadensüberwachung von GFK-Blattfedern mittels struktureller Schadensindikatoren*. Proceedings of 4SMARTS Conference. Darmstadt, 2019

[11] D. Feldten, *Beitrag zur Steigerung der Werkstoffausnutzung von UD-GFK bei Biegeschwellbeanspruchung*. Technische Universität Darmstadt, PhD thesis, 2014

[12] F. Becker, Ch. Hopmann, F. Italiano and A. Girelli, *Fatigue testing of GFRP materials for the application in automotive leaf springs:* Proceedings of the 8th International Conference on Fatigue Design. Senlis, Frankreich, 2019

[13] A. Bleier, *Prüfverfahren zur Ermittlung exakter Werkstoffkennwerte einer unidirektionalen Schicht unter besonderer Berücksichtigung physikalischer Nichtlinearitäten*. Technische Universität Darmstadt, PhD thesis, 2011

[14] E. Schmachtenberg, W. Michaeli and F. Preller, *Ermittlung der faserparallelen Druckfestigkeit von CFK durch 4-Punkt-Biegung*. Zeitschrift Kunststofftechnik 3 5, S. 1-15, 2007

IN-SITU COMPOSITE ROTOR DAMAGE DETECTION

Dan Kominsky, Brian Rife, Elias D. Bearinger, Nur Aida Abdul Rahim
Luna Innovations Inc.
Blacksburg, VA

ABSTRACT

A Rapid In-Place Composite Rotor Damage Detection (RIPCoRDD) System was developed for determining and tracking the structural health of composite rotorcraft blades. This fulfills the need for accurate and reliable assessment of the condition of composite parts which may have been damaged through impacts or fatigue, especially for cases where damage is not visible from the surface. The approach utilizes high-definition fiber optic strain sensors (HD-FOS). Sensor lengths up to 50 m can be embedded during fabrication and provides spatially dense strain measurements (sub-millimeter) from within the composite structure. In this work, HD-FOS is used to build a system of spatially distributed strain rosettes over flat and curved surfaces. The software and algorithmic work developed for interpreting the RIPCoRDD system output is described. Finally, the system is characterized through a series of impact tests performed on panels and beams. This data is used as a training data set to distinguish impact based on distance to the fiber sensor as well as energy content. Blind testing of the impact damage detection algorithms is then shown to achieve a success rate of at least 51 out of 54 trials. This approach can be applied to a variety of composite structures as a means of assessing if they have sustained any damage during service.
Corresponding author: Nur Aida Abdul Rahim
DOI: (will be filled by SAMPE)

1. INTRODUCTION

Damage detection and assessment in composite rotorcraft blades is critical to ensuring their safe operation. The size and location of damage must be accurately identified in order to determine the type and size of repair that needs to be performed. Visual inspection is typically the first and most rapid inspection performed after a suspected impact. However, visual inspection requires expertise from an inspector, is dependent on the distance and amount of light available during inspection, does not provide means of determining the true severity of damage, and furthermore damage may not even be visible from the exterior. Another method is ultrasonic inspection, which is a time-consuming manual inspection method that requires an experienced operator, and where the quality of the results is highly structure-dependent [1]. Digital Image Correlation (DIC) can also be used, but is affected by vibration and speckle noise in the inspection environment, with the largest limitation relating to the need for line-of-sight to the test article, negating the ability to measure strains on, under, and inside a composite structure simultaneously [2].

High definition fiber optic sensing (HD-FOS) uses inexpensive telecommunications grade optical fiber as the sensor, and results in a truly spatially continuous measurement every 0.65 mm along the length of the fiber sensor (maximum length of 50 m per channel). This results in sensors that are lightweight, flexible, and minimally invasive, making them ideal for embedding in composite materials during manufacturing and can be used as an in-situ structural health monitoring (SHM) tool. HD-FOS has been used extensively over the past few years for composite damage detection.

Fiber optic sensors have been embedded in flat composite panels to detect delamination events, impact damage, as well as other failure modes during testing [3-5]. These sensors have been embedded into composite an overwrapped pressure vessel (COPV) that was then subjected to various impact damage scenarios [6]. Fiber optic sensors have also been used to characterize the adhesive joint between two dissimilar materials including single and double lap shear joints, doubler plate joints with artificial delamination, and composite patch repairs of a steel I-beam [7].

This work extends the application space into the ability to resolve directionality of damage-induced strain field changes and implements an automated damage detection and validation scheme.

2. EXPERIMENTATION

2.1 High Definition Fiber Optic Sensing (HD-FOS)

The key technology that has enabled this research is high-definition fiber optic sensing (HD-FOS). HD-FOS uses optical frequency domain reflectometry (OFDR) to turn commercially available optical fiber into a distributed sensor capable of measuring strain or temperature every 0.65 mm along its length. The methodology behind this process has been previously described [8-10]. In brief, the slight variations in refractive index of a single mode fiber, created during the fiber manufacturing process, results in a random but stable back-reflection fingerprint through a process known as Rayleigh backscattering. External stimuli such as strain or temperature causes a distortion in this backscattering pattern of the fiber in a manner proportional to the stimulus being applied. Luna's Optical Distributed Sensor Interrogator (ODiSI) system takes advantage of the backscatter phenomenon to measure strain and temperature. The Rayleigh backscatter amplitude along the length of the entire sensor is divided into segments called gages. Using an inverse Fourier transform, the data inside this gage-length is converted to the spectral domain. By running a cross-correlation between the measurement data and a baseline taken at a state of known temperature or strain, an optical frequency shift in the spectral content of the gage is calculated. This frequency shift corresponds to the distortion of the backscatter signature of the fiber and can be related to either strain or temperature by a set of coefficients.

For this work, initial testing revealed some anomalies in measurement data. The curing process creates a significant level of stress-induced birefringence. Local birefringence in the sensor fiber causes the apparent strain to exhibit different values depending on whether the incident light is polarized in the same direction or perpendicular to the direction of the transverse stress. Because the polarization state of the light at any given point along the sensor is uncontrolled and the local transverse stress may add or subtract to the spectral shift due to axial strain or temperature depending on the orientation of the incident polarization state, local transverse stress contributes uncertainty to the axial strain or temperature measurement. In effect, every measurement that is made without accounting for this polarization state-dependent uncertainty is the superposition of two independent solutions. To address this, an integrated polarization switch was used in the ODiSI, which allows for the injection of orthogonal polarization states into the sensor fiber. The use of the switch results in the ability to compensate for moderate levels of birefringence in the fiber, collapsing the two solutions into a single value.

2.2 Distributed Strain Rosette

One of the key questions addressed in this effort was how to measure strain in an arbitrary part without knowing in advance what the orientation of the strain will be in any given location. The optical fiber of the RIPCoRDD system is a locally single axis sensor that is only sensitive to the component of strain aligned in the fiber direction at that location. A similar situation exists for conventional single axis foil strain gages. Vector strain is obtained by creating a sensor rosette, which is a grouping of multiple sensors aligned in different orientations to discriminate the two orthogonal strain components and the shear strain (or orientation of the principal strain components). In order to fully characterize the strain profile of the test articles built here, an equivalent distributed fiber strain rosette was developed. The pattern that was selected (equivalent to the 3rd arrangement shown in Figure 1) requires the greatest amount of fiber per unit area but provides significant advantages with respect to coverage and uniformity of measurement on the various axes.

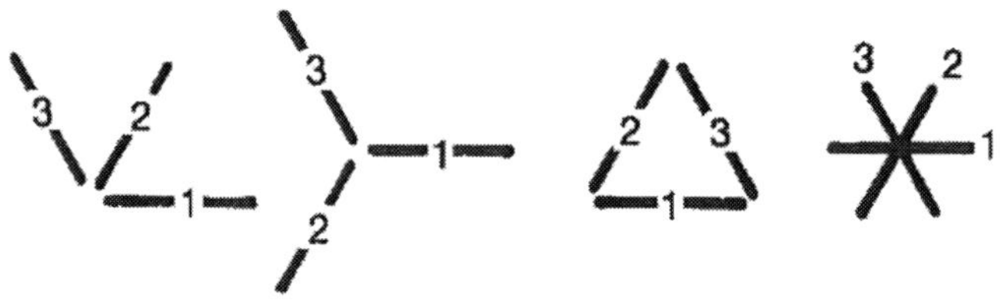

Figure 1: Different arrangements of point strain gage rosettes.

The incoming data is a single stream of strains along the length of the sensing fiber. First that data is split into its portions based on the orientation of the fiber. Each of those three data sets is spatially interpolated to generate full field strain maps for each of those three axes independently, $\varepsilon_1, \varepsilon_2, \varepsilon_3$. From those values, the normal strains and orientation can be determined using the relationships shown in equations 1 and 2 below.

$$\varepsilon_{P,Q} = \frac{\varepsilon_1 + \varepsilon_3}{2} \pm \frac{1}{\sqrt{2}} \sqrt{(\varepsilon_1 + \varepsilon_2)^2 + (\varepsilon_2 - \varepsilon_3)^2} \quad [1]$$

$$\theta = \frac{1}{2} tan^{-1}\left(\frac{\varepsilon_1 - 2\varepsilon_2 + \varepsilon_3}{\varepsilon_1 - \varepsilon_3}\right) \quad [2]$$

The distributed rosette technology was incorporated into all the test articles constructed for this project. The methods used to reproduce repeatable rosette patterns and accurately map the fiber route across the test article was previously described [11].

2.3 Design and Fabrication

The test articles were designed to facilitate testing related to measuring the response to impact energy, distance from impact to sensing fiber, impact localization, and system repeatability.

2.3.1 Coupon

Dog bone coupons with multiple embedded fibers were fabricated, to quantify the robustness of carbon fiber composite structures with embedded optical fiber. Figure 3 illustrates a panel which was fabricated to obtain this data. A single composite panel was fabricated with a series of regions with varying numbers of optical fibers embedded within them – 0, 1, 2, and 3 fibers in a repeating fashion. Five regions were present with each of the fiber counts, in every case with the fibers aligned with the axis of the coupon. Once the panel was complete, it was sent to Metallurgical Engineering Service, Inc. where the individual dog-bone coupons were water jet cut from the panel (per ASTM standard D638) and tested with a technique in accordance with ASTM standard D3039. The results of this testing were the modulus and breaking strength of the coupons as a function of the embedded fiber count.

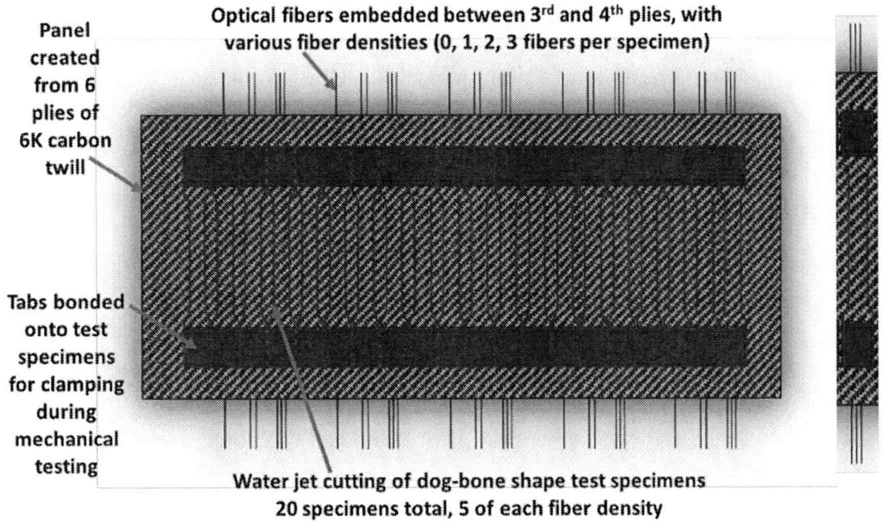

Figure 2: Strength coupon panel design. Left: Overall panel design. Right: Panel subset showing the dog-bone profile (in red) of a single test coupon.

2.3.2 Panel

The panel design includes two fiber sensors. One sensor is laid out with the distributed rosette pattern. This was used to establish the RIPCoRDD system's localization capabilities, as well as determine how well the full field strain mapping can be accomplished. A second sensor is laid out linear to the panel length where eight passes of optical fiber are routed through the panel at different depths, to establish the penetration effects of impact damage. Although the "stack-up" portion of Figure 4 (left) shows only five layers, the panels actually have either nine layers for those made with braided carbon fiber, or 12 layers for those using a 2 x 2 woven carbon fiber.

2.3.3 Beam

Figure 4 (right) shows the design of the impact beams (rectangular tubes). In the beams, the optical fiber is located at a depth of two or three carbon fiber plies below the outer surface (again determined by the type of prepreg used for each construction. Due to radius of curvature constraints, the straight longitudinal runs of fiber are confined to the top and bottom surfaces of the beam (3 longitudinal passes each), and the right and left surfaces have a single sawtooth pass each.

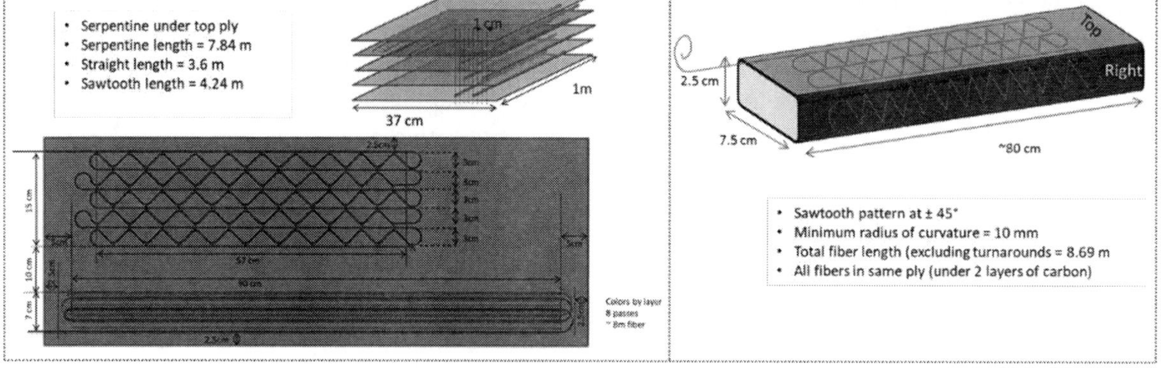

Figure 3: Test article design. Left: Panel. Right: Beam.

2.4 Impact Test Rig

An impact test rig was designed to adhere to the ASTM D7146 Standard: Test Method for Measuring the Damage Resistance of a Fiber Reinforced Polymer Matrix Composite to a Drop-Weight Impact Event. The rig consists of a weight with a fixed mass that can be released from various heights based on the amount of energy being imparted to the test article. The base of the impact test rig was designed to secure the three different geometries of composite test articles under the point of impact. Efforts were made to balance the requirements of the ASTM D7136 Standard (which only specifies test methods for panel geometry parts) with the different test article geometries to ensure the testing is performed accurately and in a repeatable manner. Shown in Figure 6 is a 19 mm thick aluminum base with a 100 mm x 150 mm cut through that can be configured using a variety of clamps to position the panel and beam under the impactor.

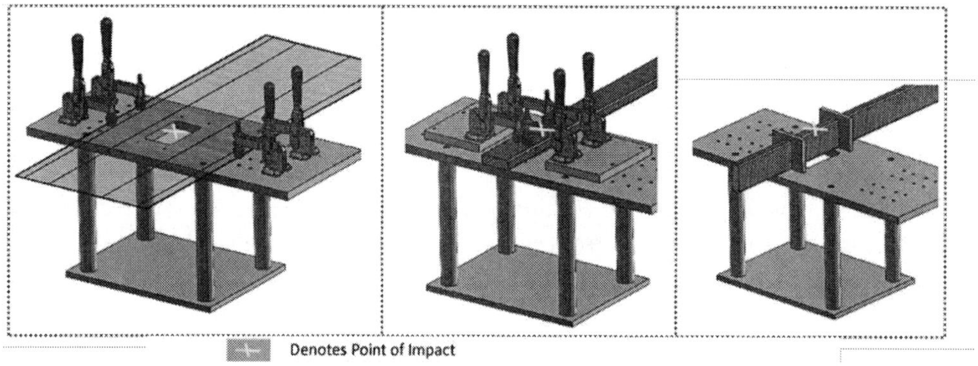

Figure 4: Different configurations of the impact test base to accommodate the test article geometries.

The thickness of the base and diameter and length of the support columns are dictated by the standard and are 13 mm thick for the base, 19 mm thick for the impact plate, and 38 mm in diameter and 300 mm long for the supports. The clamps are required to have a holding capacity of 1100 N with a neoprene rubber foot. Clamps were selected that can be adjusted with respect to the location of the grip to provide flexibility in the position of the placement of the test article. These clamps have a maximum holding capacity of 3300 N when the grip is at the point closest to the handle and 1155 N when fully extended. Several sets of holes have been placed in the impact base to accommodate different positions for the clamps. This provides the most amount of movement to enable all edges of the plate to be struck.

An impact weight was also designed to meet the ASTM standard. The diameter of the impact weight is 31mm and has a spherical tip 16mm in diameter. The spherical tip is a tooling ball which inserts into the main steel weight and is welded in place. This tip has been tempered to have a Rockwell C60 hardness. Provisions were also made in the impact weight to have threaded holes on the side (to attach a handle) and top (to add more weight, if needed). The assembly has a total mass of 2094 grams (4.61 pounds).

2.5 Impact Testing Sequence

Three basic categories of impact testing were conducted. Two types of training sets were used to ascertain the sensitivity limits of the RIPCoRDD system. In the first, all impacts were conducted from the nominal height specified in the ASTM standard, with the impacts occurring at varying distances from the sensing fibers. In the second, the impacts were conducted at a uniform distance from the fibers, but the drop height (and hence the impact energy) was varied with a maximum value of that specified in the standard and a minimum value of 10% of the standard drop height. A total of 65 training set impact configurations were carried out across the test articles as listed in Table 1.

Table 1: Impact testing matrix.

Test Article Type	Type of Test	Number of Trials within Test
Panel P02	Constant height, variable distance	16
Panel P04	Variable height, constant distance	18
Panel P03	Blind testing (variable height, variable distance)	35
Beam B01	Constant height, variable distance	15
Beam B04	Variable height, constant distance	16
Beam B05	Blind testing (variable height, variable distance)	19

The third and final version of the drop testing were the "blind" tests which were used to characterize the accuracy of the system. In these tests, an engineer collected a series of measurement data, some of which contained data from damaging impacts and some of which contained data from non-damaging "placebo" impacts. This collection of files was then provided, with no explanation, to the principal investigator who ran it through the automated signal processing algorithms. Once the predictions were made of which files contained impacts, along with the confidence of each prediction, the ground truth information was compared to ascertain the degree to which RIPCoRDD is successful with no prior knowledge on the part of the user.

2.6 Automated Data Analysis

An automated signal processing approach was developed and implemented to provide a reliable detection of damage. A summary of this process is shown in Figure 7. Strain measurements every 0.65 mm along the entire sensor length can be either streamed live or evaluated after the fact. At least two scans are needed: one preceding and one succeeding the damage event. A mild low pass spatial filter is applied to the data to reduce the sensitivity to extremely local signals. If several scans are available for each period, temporal filtering can be applied for further noise reduction.

The strain difference between preceding and succeeding scans is calculated to determine the change in strain by location. A qualification metric is then calculated by breaking the strain differential profile into spatial subsections and calculating the standard deviation of each subsection. This is reduced to the median standard deviation. A peak (or valley for back-side impact) detection routine was run with the following key parameters: the minimum allowable spacing between peaks is 3.2 cm and the minimum threshold that a peak must reach is the lesser of twice the overall standard deviation of the data or 5 microstrain.

Peak down-selection is carried out to remove false positives, for example peaks generated by the underlying weave substructure. In this work, any peak with a prominence of less than 1% of the maximum peak prominence was discarded. Of any remaining peaks, only the 15 strongest were retained. At this point, an assortment of peaks has been extracted, some of which may represent the true changes in the strain state, and others of which are spurious. The peaks are then qualified based on the magnitude of each peak as measured in median standard deviations. The candidate peaks were then categorized as follows: 1. Any peak that is more than 100 median standard deviations high is considered to be valid with 100% confidence; 2. Any peak of magnitude less than 5 median standard deviations is considered to be noise; 3. Peaks with magnitudes between 5 and 100 are assigned a confidence percentage based on the relationship shown in equation 3.

$$confidence = log_{10}(Peak\ height) - 1 \qquad [3]$$

This combination of denoising and processing allows the system to be robust to a variety of signals which could otherwise confound damage detection. The use of the peak finding and assessment based on prominence eliminates large scale variations which may arise from overall component loading or thermal variations. Additionally, for some specific cases, it may be desirable to subtract any very slowly varying spatial signal to account for static loading behavior of the overall part.

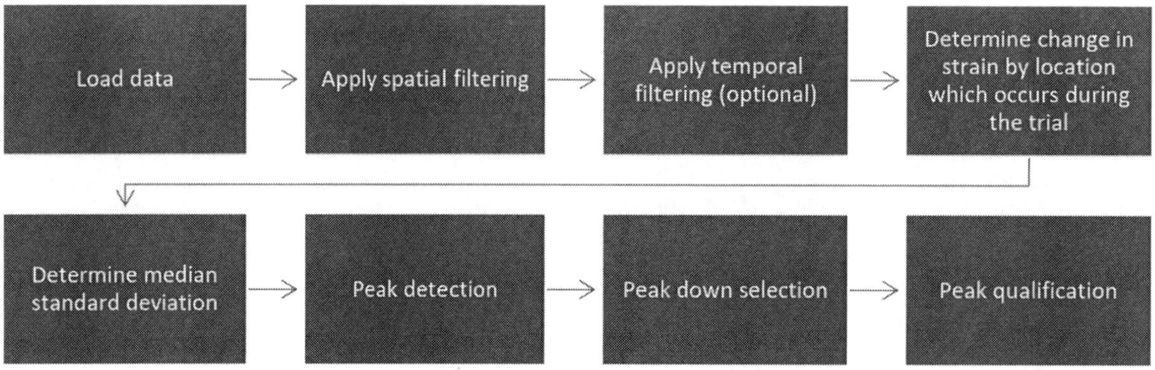

Figure 5: Signal processing flow for automated data analysis.

3. RESULTS

3.1 Coupon Strength Testing

One of the first tests run during this project was the validation of the robustness of carbon fiber composite structures with embedded optical fiber. A question frequently asked by the composites industry is whether embedding sensing fiber in a structure degrades its mechanical strength. Published articles indicate that there is "minimal" effect in this regard [12]. Test results validate this conclusion. Figure 8 (left) shows the ultimate strength of the coupons at breaking grouped by the number of fibers present within the coupon. It is noted that there was a single outlier in the group with three fibers, but otherwise the clustering is quite tight, and the embedded optical fibers have negligible impact on the strength of the coupons. It is interesting to note that the coupons with the greatest ultimate strength were those which included a single optical fiber. Figure 8 (right) shows the corresponding maximum elongation of the coupons prior to breaking, which again demonstrates that the optical fiber resulted in minimal change to the composite's mechanical attributes.

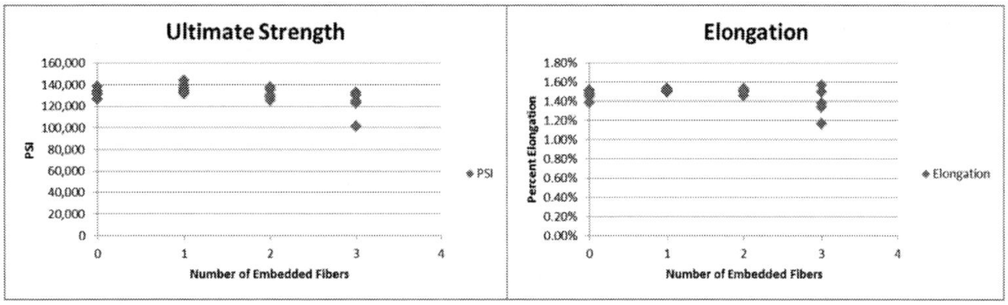

Figure 6: Coupon strength testing results. Left: Ultimate breaking strength of test coupons as a function of the number of embedded sensing fibers. Right: Maximum elongation prior to breaking as a function of the number of sensing fibers.

3.2 Impact Testing: Training Test

Two different impact profiles were carried out for both the panels and beam, one being a constant height (variable distance) training set, and the other being a variable height (constant distance) training set. In the constant height training set, all impacts were conducted from the nominal height specified in the ASTM standard (94 cm), with the impacts occurring at varying distances from the sensing fibers. In the variable height training set, the impacts were conducted at a uniform distance from the fibers, but the drop height (and hence the impact energy) was varied with a maximum value of that specified in the standard and a minimum value of 10% of the standard drop height.

3.2.1 Panel: Constant Height Training Set

The strain change for all impact events, along the entire sensor length (encompassing both linear and rosette sections), is shown in Figure 9. From this data set, the automated analysis system extracted the features that are most likely to be meaningful data and labeled them with the location in the fiber (circles along the plot). For visualization purposes, each spatial strain profile is offset by 100 microstrain to create the vertically stacked plot. Take impact event labeled 'C' as an example. This is one where the drop event had an energy of 10.31 J and impacted the bottom of the plate, at x- and y- coordinates of 40 cm and 8 cm respectively. The automated analysis system identified 8 locations along the fiber as having meaningful change due to the impact event. However, it is not immediately obvious how to interpret this linear strain plot.

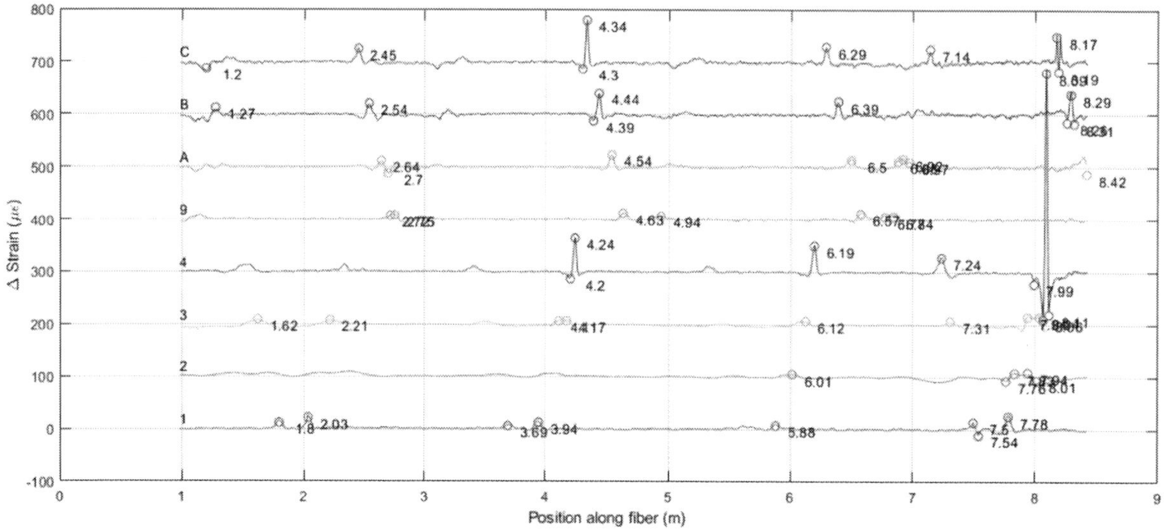

Figure 7: Raw impact data from linear fiber segment of the constant height training test panel. Each colored line represents the entire change in strain profile for a given impact. For visualization purposes, each has had a constant offset of 100 microstrain applied to create a stacked plot.

To assist in result interpretation, the data from Figure 9 is converted into a true-coordinate representation as shown in Figure 10 where the linear spatial profile is mapped back onto the actual path of the fiber. The blue line shows the sensor route within the panel. The cyan stars indicate the locations of the actual impacts, and the circles on top of the blue lines are the automatically detected impacts. Circle size corresponds to the magnitude of the detected damage.

As expected, the strain response is higher for impacts that occur closer to the fiber. However, even impacts 1 and 9, which are furthest away from the fiber (4 cm away), show a consistently detectable response to the impacts.

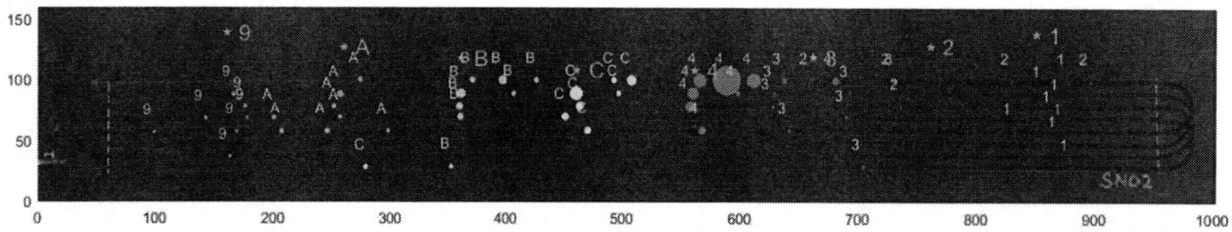

Figure 8: Impact map for linear fiber segment of the constant height training test panel. The cyan stars are the actual impact locations as measured by hand on the physical test article. The colored circles are the locations where the automated processing software generated a report of an impact related change, with the color of the circle corresponding to the impact in question, and the size of the circle indicating the magnitude of the detected damage.

A similar data set is available for impacts in the serpentine portion of the panel. Figure 11 shows the impact map for these segments. Once again, there is a very strong correlation between the locations at which impacts are detected and the actual locations where the impact occurred.

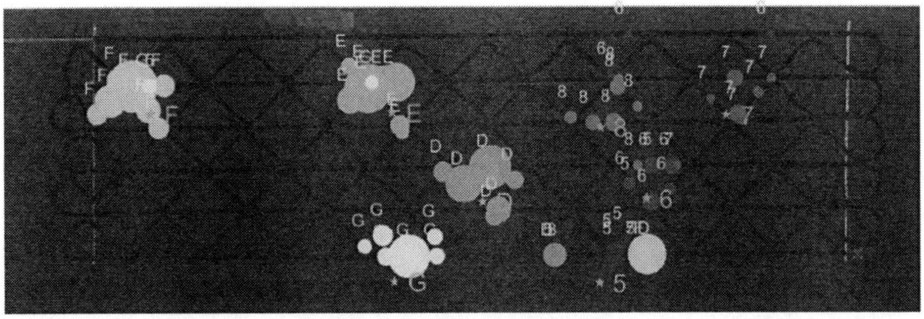

Figure 9: Impact map for serpentine fiber segment of the constant height training test panel.

3.2.2 Panel: Variable Height Training Set

A similar analysis was carried out for the training panel impacted at a uniform distance from the fibers, but with varying drop energies. A plot of spatial strain profile across 8 impact events is shown in Figure 12, and this data is presented as an impact map in Figure 13. The impact map again shows clustering of automatically detected impact events about the actual impact locations.

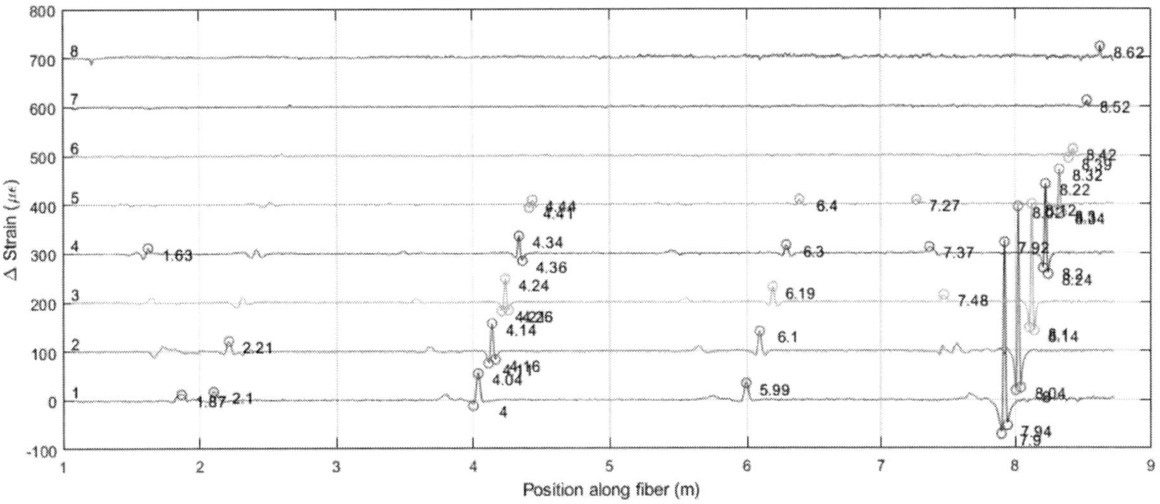

Figure 10: Raw impact data from linear fiber segment of the variable height training test panel.

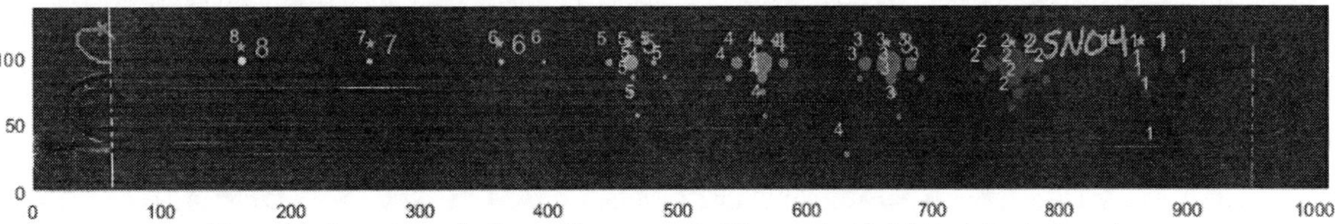

Figure 11: Impact map for linear fiber segment of the constant height training test panel.

3.2.3 Beam: Constant Height Training Set

A pair of beams were also used as training sets. The first of those beams was impacted with constant height drops for all locations. Unlike the panels, there is no region of the beams that is "outside" the fiber pattern, so this trial set does not provide the ability to discern the drop-off of sensor response with distance from the fiber. An overall summary of the impact map is shown in

Figure 14, where the beam has been "unwrapped" to show all faces in a single image. One consequence of this unwrapped display is that impacts which occur close to the corner between the top and left faces (corresponding to the top and bottom of the figure) will show some response at the opposite edge of the figure. For example, there is a damage site attributed to impact 5 at the very bottom right-hand portion of the image, although the cyan star showing the actual impact 5 location is near the top of the image. However, in actuality those two locations are quite close to each other.

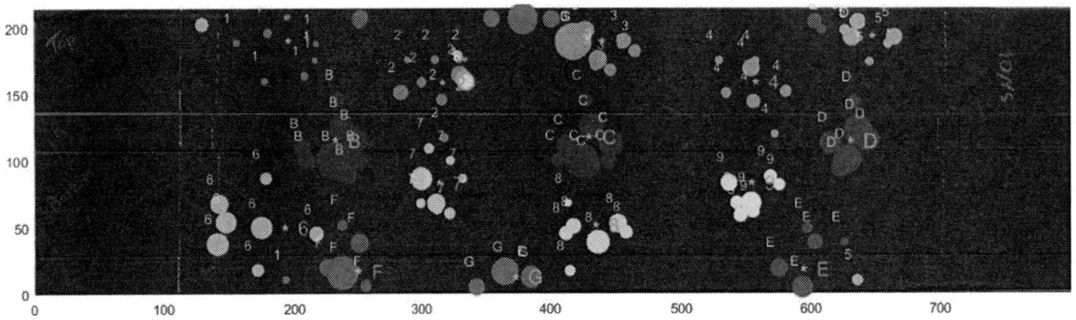

Figure 12: Unwrapped impact map for constant height training test beam.

3.2.4 Beam: Variable Height Training Set

A second beam was used to produce a training set with varying heights (energies) of impact. In this case the impacts ranged from a drop height of 80 cm down to 10 cm. The impact map for this test article is seen in Figure 15. The RIPCoRDD system correctly identified all trials except for trial 8 which corresponds to the 10 cm drop height. In the case of impact 7 (from 20 cm) the system detected two points which had sufficient change in strain to count as a response, and overall reported a 67% confidence that the impact had occurred. It is particularly interesting to note, however that only the drop from 80 cm resulted in visible or measurable damage to the surface. Thus, while the RIPCoRDD system achieved a detection success rate of 94%, direct visual inspection only reported damage in 6.25% of the trials.

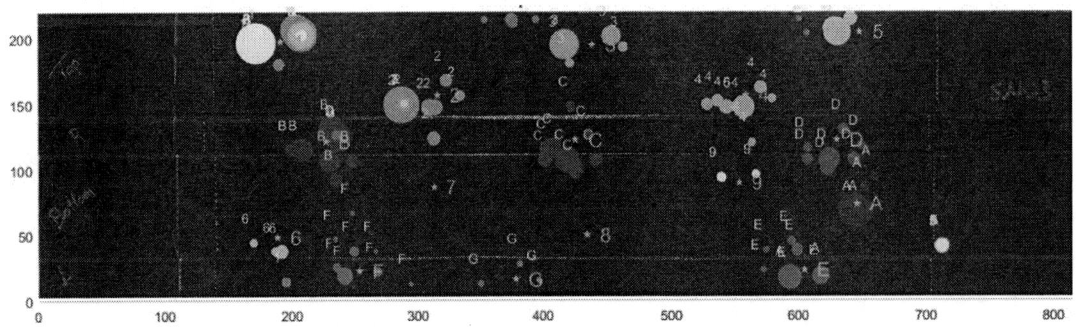

Figure 13: Unwrapped impact map for variable height training test beam.

3.2.5 Training Set Analysis

The constant height training sets were carried out to determine the main parameters relating to peak qualification. The three main parameters of interest are: 1. The variations in response based on distance from the fiber at constant impact energy; 2. The variations in response based on impact surface (top vs bottom); 3. The uniformity of physical damage signifiers (dent presence, depth, and width) given a uniform impact profile. Impacts 1-4 were carried out on the top surface of the panel whereas impacts 9-C were carried out on the bottom.

Figure 16A shows the relationship between the distance from the impact site to the closest fiber and the confidence which the automated system provides in its prediction of the presence or absence of an impact event. As the impacts get farther from the closest fiber, there is a drop in the confidence of the assessment. In other words, less damage is detected. It is also noted that the confidence is significantly higher for the four impacts which occurred on the top face of the panel (1-4) than those on the back (9-C). This is ascribed to the fact that the closest fiber to the point of impact is closer to the back face of the panel than to the front face of the panel.

The conclusion that can be drawn from this data is that for this test article geometry and material set, fiber sensors need to be at most 4 cm away from impact events in order to reliably detect the maximum strength impact events. If the detection requirement is adjusted to include lower impact energy levels, a denser sensor layout would be required. It is worth noting that the sensitivity as a function of distance is also coupled to the underlying geometry of the composite part. For example, as a composite part is made thicker and thicker, then the strain of impact damage would be expected to propagate a shorter distance (as there is more volume to support the stresses.)

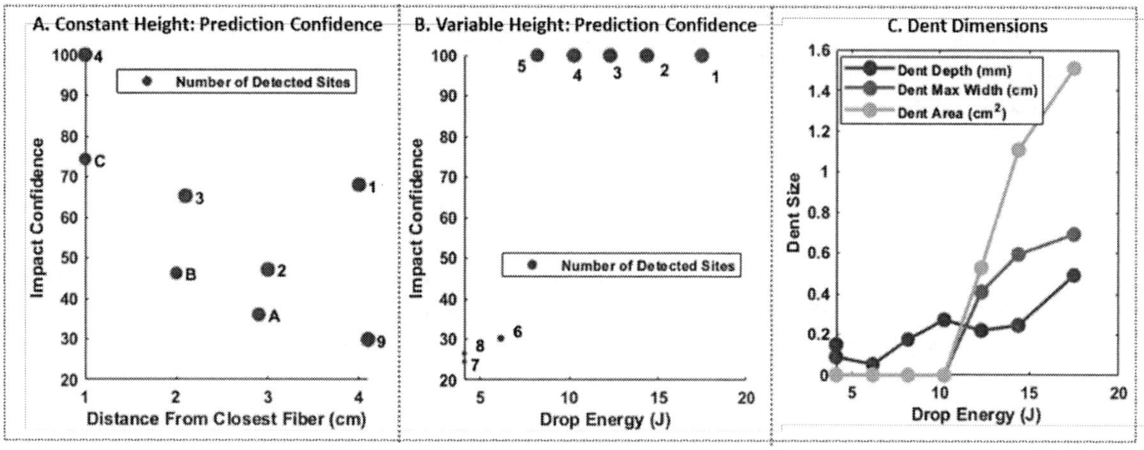

Figure 14: Results from training set analysis. A. Confidence of prediction of impacts on linear panel (constant height training test) as a function of distance from impact to closest fiber. B. Confidence of prediction of impacts on linear panel (variable height training test) as a function of drop energy. C. Relationship between drop energy and dent dimensions.

Figure 16B is a plot of the confidence of detecting hits as a function of the drop energy. There is a very strong dependency of the confidence on the drop energy with total confidence being achieved at drop energies of 8.2 Joules or above.

Figure 16C shows the relationship between the drop energy which was applied to each impact and the size of the visible damage that was detected as a result. There appears to be a slight correlation between the energy of the impacts and the depth of the dent which forms in the surface. There is, however, and extremely clear correlation between the impact energy and the width of the dent that is measured. For energies of less than about 10 J, there is no detectable width to the impact mark.

For energies above the threshold there is a nearly linear relationship between the area of the dent and the drop energy (or a square root relationship with the dent's maximum width).

Figure 17 shows the relevance of the distributed rosette pattern in terms of the ability to understand the total local strain field that results from an impact. This figure plots the change in strain state in the serpentine panel caused by impact "D". It shows that there is a residual strain pattern which is largely circularly symmetric about the point of impact.

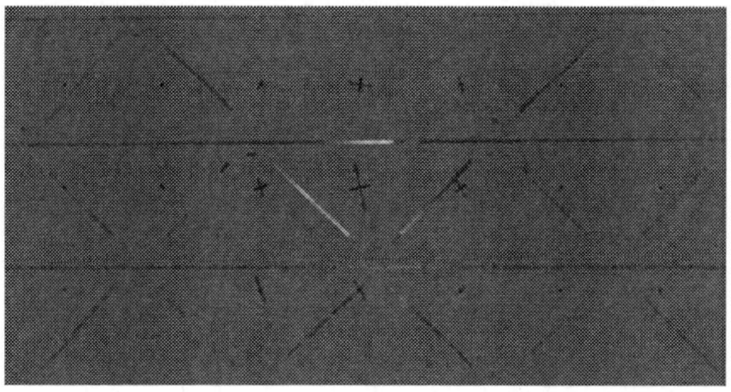

Figure 15: Visualization showing the full residual strain state in the panel after impact "D".

3.3 Impact Testing: Blind Test

The capstone test for the RIPCoRDD system was a blind validation test where the impact magnitudes and locations were designed and performed by one engineer while a different engineer ran the data through the same automated detection algorithms that were developed using the training set data. The blind test was conducted within the parameters shown in Table 2. Results from the automated data analysis on the blind tests were excellent, as shown in Table 3, with an overall success rate of > 94%. An impact map of the serpentine section of the blind panel test (Figure 18) similarly represents the excellent alignment between the detected impacts and the actual impact locations. The rates of false positives and false negatives are very low, at 5% and 9% respectively. These results indicate that the algorithms developed through the training test data are very well suited for automatically detecting and qualifying perturbations in the strain measurements for impact-induced damage events.

Table 2: Blind test parameters.

Distance to Closest Fiber (cm)	Drop Height Range (cm)	Drop Energy Range (J)
0 – 1	20 – 50	4.1 – 10.3
1 – 2	50 – 94	10.3 – 19.3
2 – 3	94	19.3

Table 3: Blind test results.

Test Article	No. of Trials	No. of Hits	No. of Placebos	No. of False Positives	No. of False Negatives	Accuracy (%)
Panel: Linear	15	12	3	1	2	80
Panel: Serpentine	20	15	5	0	0	100
Beam	19	16	3	0	0	100
Total	54	43	11	1	2	94

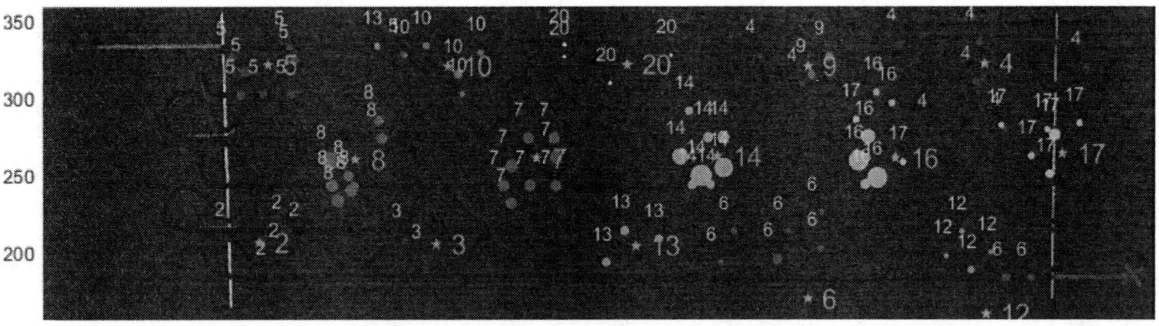

Figure 16: Impact map for blind panel test.

4. CONCLUSIONS

The Rapid In-Place Composite Rotor Damage Detection (RIPCoRDD) System described here was embedded within carbon composite panels and used to detect impact damage imparted at various energy levels and spatial distributions over the panel surface. Using a training data set where impact damage was imparted at known energy levels and locations, on both panels and beams, an automatic detection algorithm was developed for detecting and qualifying spatial strain features that corresponded to damage. A subsequent set of impact events were evaluated blind by the detection algorithm and resulted in a > 94% success rate.

This approach can be applied to a variety of composite structures as a means of assessing the structure's health and whether they have sustained any damage during service. One such example is a helicopter rotor blade (Figure 19). This implementation was previously described by Bearinger et. al. [11]. A similar detection and analysis scheme can be developed to account for the more complex curvatures that exist on real-world parts. It would be interesting to evaluate which parameter would bring most benefit in terms of improving the event detection accuracy for a more complex structure – a denser fiber sensor layout pattern, finer spatial sampling resolution, or better strain magnitude measurement accuracy.

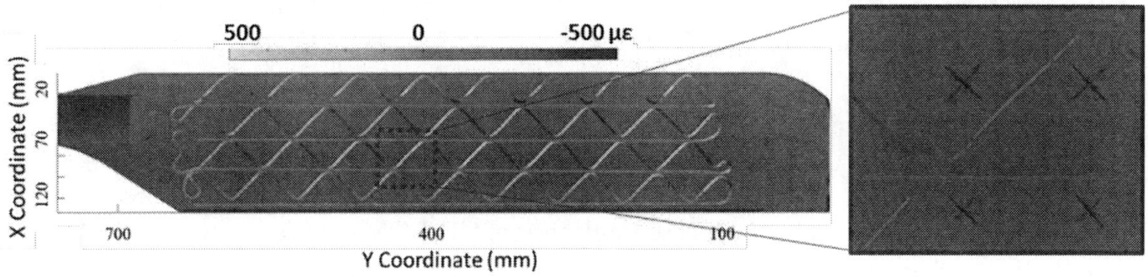

Figure 17: RIPCoRDD system implemented on the suction side of a composite rotor blade. Inset arrows map magnitude and orientation of distributed principal strains across the blade under torsion (fixed at the base (left) while lifting the trailing edge).

5. ACKNOWLEDGEMENTS

The authors would like to express their appreciation of NASA having funded portions of this work through a Phase II SBIR, contract number NNX17CC21C.

6. REFERENCES

1. Shull, P. J. Nondestructive Evaluation: Theory, Techniques, and Applications. CRC Press; 1st edition, 2002

2. Zhu, Y-K., Tian, G-Y., Lu, R-S., and Zhang, H. "A review of optical NDT technologies." *Sensors* 11 (2011): 7773-7798

3. Alaziz, R., and Sullivan, R. W. "Investigation of Damage in Composite Structures Under Vibration Using Hilbert-Huang Transform." *Proc. of the American Society of Composites*. Oct. 2019

4. Batte, L. K., Sullivan, R. W., Ranatunga, V. and Brown. K. "Impact Response in Polymer Composites from Embedded Optical Fibers." *J. of Comp. Materials* 2018, 52(25) 3415-3427

5. Kominsky, D., Abdul Rahim, N. A., Davis, M. A., Garg, N., Castellucci, M. A., Bryson, R. L., Beaty, N. B. "Extracting Information From Damaged Carbon Fiber Composites Using High Definition Fiber Optic Sensing (HD-FOS)." *Proc. of CAMX* 2017. Orlando Dec 12-14, 2017

6. Saeter, E., Lasn, K., Nony, F., and Echtermeyer, A. T. "Embedded Optical Fibres for Monitoring Pressurization and Impact of Filament Wound Cylinders." *Composite Structures* 210 (2019) 608-817

7. Grave, J. H. L. and Echtermeyer, A. T. "Strain fields in adhesively bonded patch repairs of damaged Metallic beams." *Polymer Testing* 48 (2015) 50-58.

8. Froggatt, Mark, and Moore, Jason. "High resolution strain measurement in optical fiber with Rayleigh scatter." *Appl. Opt.* 37 (1998): 1735-1740.

9. Kreger, S., Gifford, D. K., Froggatt, M. E., Soller, B. J., and Wolfe, M. S. "High resolution distributed strain or temperature measurements in single- and multi-mode fiber using swept-wavelength interferometry." *Optical Fiber Sensors, OSA Technical Digest.* Cancun, Mexico, October 23, 2006. Optical Society of America, 2006. Paper ThE42. CD-ROM.

10. Kreger, S. T., Abdul Rahim, N. A., Garg, N., Klute, S. M., Metrey, D. R., Beaty, N., Jeans, J. W., and Gamber, R. "Optical Frequency Domain Reflectometry: Principles and Applications in Fiber Optic Sensing." *Proc. SPIE 9852, Fiber Optic Sensors and Applications XIII.* May 12, 2016.

11. Bearinger, E.D., Kominsky, D., Bryson, R. L., and Abdul Rahim, N. A. "A Fully Distributed Strain Rosette Using High Definition Fiber Optic Sensing." *SAMPE Technical Conference Proceedings: Design, Analysis, and Testing.* Charlotte, NC, May 20-23, 2019. Society for the Advancement of Material and Process Engineering. CD-ROM—15 pp.

12. Meadows, L., Sullivan, R. W., and Vehorn, K. "Distributed Optical Sensing in Composite Laminate Adhesive Bonds." *AIAA SciTech.* San Diego, CA, January 4-8, 2016. 57th AIAA/ASCE/AHS/ASC Structures, Structural Dynamics, and Materials Conference.